Electrical Change Order Costs with RSMeans data

Antonio D'Aulerio, Senior Editor

GORDIAN®

2020
32nd annual edition

Chief Data Officer
Noam Reininger

Vice President, Data
Tim Duggan

Principal Engineer
Bob Mewis (1, 4)

Contributing Editors
Brian Adams (21, 22)
Paul Cowan
Christopher Babbitt
Sam Babbitt
Stephen Bell
Michelle Curran
Antonio D'Aulerio (26, 27, 28, 48)

Matthew Doheny (8, 9, 10)
John Gomes, CCT (13, 41)
Derrick Hale, PE (2, 31, 32, 33, 34, 35, 44, 46)
Barry Hutchinson
Joseph Kelble (14, 23, 25)
Scott Keller (3, 5)
Charles Kibbee
Gerard Lafond, PE
Thomas Lane (6, 7)
Thomas Lyons
Jake MacDonald (11, 12)
John Melin, P.E.
Elisa Mello
Matthew Sorrentino

Kevin Souza
David Yazbek

Production Manager
Debbie Panarelli

Production
Jonathan Forgit
Sharon Larsen
Sheryl Rose
Janice Thalin

Data Quality Manager
Joseph Ingargiola

Innovation
Ray Diwakar
Kedar Gaikwad
Srini Narla

Cover Design
Blaire Collins

Data Analytics
David Byars
Ellen D'amico
Thomas Hauger
Cameron Jagoe
Matthew Kelliher-Gibson
Renee Rudicil

Numbers in italics are the divisional responsibilities for each editor. Please contact the designated editor directly with any questions.

RSMeans data from Gordian
Construction Publishers & Consultants
1099 Hingham Street, Suite 201
Rockland, MA 02370
United States of America
1.800.448.8182
RSMeans.com

Copyright 2019 by The Gordian Group Inc.
All rights reserved.
Cover photo © iStock.com/yenwen

Printed in the United States of America
ISSN 1540-630X
ISBN 978-1-950656-05-9

0230 $339.99 per copy (in United States)
Price is subject to change without prior notice.

Related Data and Services

Our engineers recommend the following products and services to complement *Electrical Change Order Costs with RSMeans data:*

Annual Cost Data Books
2020 Electrical Costs with RSMeans data
2020 Building Construction Costs with RSMeans data
2020 Mechanical Costs with RSMeans data
2020 Plumbing Costs with RSMeans data

Reference Books
Electrical Estimating Methods
Mechanical Estimating Methods
Designing & Building with the IBC
Estimating Building Costs
RSMeans Estimating Handbook
Green Building: Project Planning & Estimating
How to Estimate with RSMeans data
Plan Reading & Material Takeoff
Project Scheduling and Management for Construction

Seminars and In-House Training
Unit Price Estimating
Training for our online estimating solution
Practical Project Management for Construction Professionals
Scheduling with MSProject for Construction Professionals
Mechanical & Electrical Estimating

RSMeans data
For access to the latest cost data, an intuitive search, and an easy-to-use estimate builder, take advantage of the time savings available from our online application.

To learn more visit: **RSMeans.com/online**

Enterprise Solutions
Building owners, facility managers, building product manufacturers, and attorneys across the public and private sectors engage with RSMeans data Enterprise to solve unique challenges where trusted construction cost data is critical.

To learn more visit: **RSMeans.com/Enterprise**

Custom Built Data Sets
Building and Space Models: Quickly plan construction costs across multiple locations based on geography, project size, building system component, product options, and other variables for precise budgeting and cost control.

Predictive Analytics: Accurately plan future builds with custom graphical interactive dashboards, negotiate future costs of tenant build-outs, and identify and compare national account pricing.

Consulting
Building Product Manufacturing Analytics: Validate your claims and assist with new product launches.

Third-Party Legal Resources: Used in cases of construction cost or estimate disputes, construction product failure vs. installation failure, eminent domain, class action construction product liability, and more.

API
For resellers or internal application integration, RSMeans data is offered via API. Deliver Unit, Assembly, and Square Foot Model data within your interface. To learn more about how you can provide your customers with the latest in localized construction cost data visit:
RSMeans.com/API

Table of Contents

Foreword

The Value of RSMeans data from Gordian

Since 1942, RSMeans data has been the industry-standard materials, labor, and equipment cost information database for contractors, facility owners and managers, architects, engineers, and anyone else that requires the latest localized construction cost information. More than 75 years later, the objective remains the same: to provide facility and construction professionals with the most current and comprehensive construction cost database possible.

With the constant influx of new construction methods and materials, in addition to ever-changing labor and material costs, last year's cost data is not reliable for today's designs, estimates, or budgets. Gordian's cost engineers apply real-world construction experience to identify and quantify new building products and methodologies, adjust productivity rates, and adjust costs to local market conditions across the nation. This adds up to more than 22,000 hours in cost research annually. This unparalleled construction cost expertise is why so many facility and construction professionals rely on RSMeans data year over year.

About Gordian

Gordian originated in the spirit of innovation and a strong commitment to helping clients reach and exceed their construction goals. In 1982, Gordian's chairman and founder, Harry H. Mellon, created Job Order Contracting while serving as chief engineer at the Supreme Headquarters Allied Powers Europe. Job Order Contracting is a unique indefinite delivery/indefinite quantity (IDIQ) process, which enables facility owners to complete a substantial number of repair, maintenance, and construction projects with a single, competitively awarded contract. Realizing facility and infrastructure owners across various industries could greatly benefit from the time and cost saving advantages of this innovative construction procurement solution, he established Gordian in 1990.

Continuing the commitment to providing the most relevant and accurate facility and construction data, software, and expertise in the industry, Gordian enhanced the fortitude of its data with the acquisition of RSMeans in 2014. And in an effort to expand its facility management capabilities, Gordian acquired Sightlines, the leading provider of facilities benchmarking data and analysis, in 2015.

Our Offerings

Gordian is the leader in facility and construction cost data, software, and expertise for all phases of the building life cycle. From planning to design, procurement, construction, and operations, Gordian's solutions help clients maximize efficiency, optimize cost savings, and increase building quality with its highly specialized data engineers, software, and unique proprietary data sets.

Our Commitment

At Gordian, we do more than talk about the quality of our data and the usefulness of its application. We stand behind all of our RSMeans data—from historical cost indexes to construction materials and techniques—to craft current costs and predict future trends. If you have any questions about our products or services, please call us toll-free at 800.448.8182 or visit our website at gordian.com.

MasterFormat® 2016/ MasterFormat® 2018 Comparison Table

This table compares the 2016 edition of the Construction Specifications Institute's MasterFormat® to the expanded 2018 edition. For your convenience, all revised 2016 numbers and titles are listed along with the corresponding 2018 numbers and titles.

CSI 2016 MF ID	CSI 2016 MF Description	CSI 2018 MF ID	CSI 2018 MF Description
015632	Temporary Security	015733	Temporary Security
019308	Facility Maintenance Equipment	019308	Facilities Maintenance, Equipment
024200	Removal and Salvage of Construction Materials	024200	Removal and Diversion of Construction Materials
040130	Unit Masonry Cleaning	04012052	Cleaning Masonry
068010	Composite Decking	067300	Composite Decking
072127	Reflective Insulation	072153	Reflective Insulation
072610	Above-Grade Vapor Retarders	072613	Above-Grade Vapor Retarders
074473	Metal Faced Panels	074433	Metal Faced Panels
075430	Ketone Ethylene Ester Roofing	075416	Ketone Ethylene Ester Roofing
077280	Vents	077280	Vent Options
087125	Weatherstripping	087125	Door Weatherstripping
087530	Weatherstripping	087530	Window Weatherstripping
096223	Bamboo Flooring	096436	Bamboo Flooring
099103	Paint Restoration	090190	Maintenance of Painting and Coating
102833	Laundry Accessories	102823	Laundry Accessories
117610	Operating Room Equipment	117610	Equipment for Operating Rooms
117710	Radiology Equipment	117710	Equipment for Radiology
122310	Wood Interior Shutters	122313	Wood Interior Shutters
123580	Commercial Kitchen Casework	123539	Commercial Kitchen Casework
124636	Desk Accessories	124113	Desk Accessories
141210	Dumbwaiters	141000	Dumbwaiters
211113	Facility Water Distribution Piping	211113	Facility Fire Suppression Piping
233715	Louvers	233715	Air Outlets and Inlets, HVAC Louvers
260580	Wiring Connections	260583	Wiring Connections
270110	Operation and Maintenance of Communications Systems	270110	Operation and Maintenance of Communication Systems
272123	Data Communications Switches and Hubs	272129	Data Communications Switches and Hubs
283149	Carbon-Monoxide Detection Sensors	284611.21	Carbon-Monoxide Detection Sensors
284621	Fire Alarm	284620	Fire Alarm
316233	Drilled Micropiles	316333	Drilled Micropiles
323420	Fabricated Pedestrian Bridges	323413	Fabricated Pedestrian Bridges
337543	Shunt Reactors	337253	Shunt Reactors
350100	Operation and Maint. of Waterway & Marine Construction	350100	Operation and Maintenance of Waterway and Marine Construction

How the Cost Data Is Built: An Overview

Unit Prices*

All cost data have been divided into 50 divisions according to the MasterFormat® system of classification and numbering.

Assemblies*

The cost data in this section have been organized in an "Assemblies" format. These assemblies are the functional elements of a building and are arranged according to the 7 elements of the UNIFORMAT II classification system. For a complete explanation of a typical "Assembly", see "RSMeans data: Assemblies—How They Work."

Residential Models*

Model buildings for four classes of construction—economy, average, custom, and luxury—are developed and shown with complete costs per square foot.

Commercial/Industrial/Institutional Models*

This section contains complete costs for 77 typical model buildings expressed as costs per square foot.

Green Commercial/Industrial/Institutional Models*

This section contains complete costs for 25 green model buildings expressed as costs per square foot.

References*

This section includes information on Equipment Rental Costs, Crew Listings, Historical Cost Indexes, City Cost Indexes, Location Factors, Reference Tables, and Change Orders, as well as a listing of abbreviations.

- **Equipment Rental Costs:** Included are the average costs to rent and operate hundreds of pieces of construction equipment.
- **Crew Listings:** This section lists all the crews referenced in the cost data. A crew is composed of more than one trade classification and/or the addition of power equipment to any trade classification. Power equipment is included in the cost of the crew. Costs are shown both with bare labor rates and with the installing contractor's overhead and profit added. For each, the total crew cost per eight-hour day and the composite cost per labor-hour are listed.

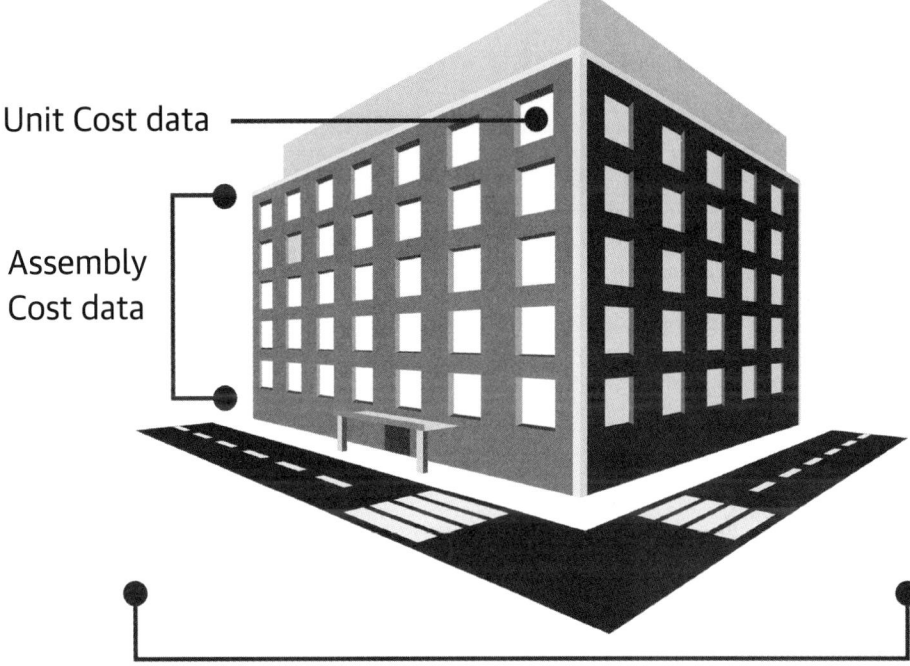

Unit Cost data

Assembly Cost data

Square Foot Models

- **Historical Cost Indexes:** These indexes provide you with data to adjust construction costs over time.
- **City Cost Indexes:** All costs in this data set are U.S. national averages. Costs vary by region. You can adjust for this by CSI Division to over 730 cities in 900+ 3-digit zip codes throughout the U.S. and Canada by using this data.
- **Location Factors:** You can adjust total project costs to over 730 cities in 900+ 3-digit zip codes throughout the U.S. and Canada by using the weighted number, which applies across all divisions.
- **Reference Tables:** At the beginning of selected major classifications in the Unit Prices are reference numbers indicators. These numbers refer you to related information in the Reference Section. In this section, you'll find reference tables, explanations, and estimating information that support how we develop the unit price data, technical data, and estimating procedures.
- **Change Orders:** This section includes information on the factors that influence the pricing of change orders.

- **Abbreviations:** A listing of abbreviations used throughout this information, along with the terms they represent, is included.

Index (printed versions only)

A comprehensive listing of all terms and subjects will help you quickly find what you need when you are not sure where it occurs in MasterFormat®.

Conclusion

This information is designed to be as comprehensive and easy to use as possible.

The Construction Specifications Institute (CSI) and Construction Specifications Canada (CSC) have produced the 2018 edition of MasterFormat®, a system of titles and numbers used extensively to organize construction information.

All unit prices in the RSMeans cost data are now arranged in the 50-division MasterFormat® 2018 system.

---

*   Not all information is available in all data sets

Note: The material prices in RSMeans cost data are "contractor's prices." They are the prices that contractors can expect to pay at the lumberyards, suppliers'/distributors' warehouses, etc. Small orders of specialty items would be higher than the costs shown, while very large orders, such as truckload lots, would be less. The variation would depend on the size, timing, and negotiating power of the contractor. The labor costs are primarily for new construction or major renovation rather than repairs or minor alterations. With reasonable exercise of judgment, the figures can be used for any building work.

Estimating with RSMeans data: Unit Prices

Following these steps will allow you to complete an accurate estimate using RSMeans data: Unit Prices.

1. Scope Out the Project

- Think through the project and identify the CSI divisions needed in your estimate.
- Identify the individual work tasks that will need to be covered in your estimate.
- The Unit Price data have been divided into 50 divisions according to CSI MasterFormat® 2018.
- In printed versions, the Unit Price Section Table of Contents on page 1 may also be helpful when scoping out your project.
- Experienced estimators find it helpful to begin with Division 2 and continue through completion. Division 1 can be estimated after the full project scope is known.

2. Quantify

- Determine the number of units required for each work task that you identified.
- Experienced estimators include an allowance for waste in their quantities. (Waste is not included in our Unit Price line items unless otherwise stated.)

3. Price the Quantities

- Use the search tools available to locate individual Unit Price line items for your estimate.
- Reference Numbers indicated within a Unit Price section refer to additional information that you may find useful.
- The crew indicates who is performing the work for that task. Crew codes are expanded in the Crew Listings in the Reference Section to include all trades and equipment that comprise the crew.
- The Daily Output is the amount of work the crew is expected to complete in one day.
- The Labor-Hours value is the amount of time it will take for the crew to install one unit of work.
- The abbreviated Unit designation indicates the unit of measure upon which the crew, productivity, and prices are based.
- Bare Costs are shown for materials, labor, and equipment needed to complete the Unit Price line item. Bare costs do not include waste, project overhead, payroll insurance, payroll taxes, main office overhead, or profit.
- The Total Incl O&P cost is the billing rate or invoice amount of the installing contractor or subcontractor who performs the work for the Unit Price line item.

4. Multiply

- Multiply the total number of units needed for your project by the Total Incl O&P cost for each Unit Price line item.
- Be careful that your take off unit of measure matches the unit of measure in the Unit column.
- The price you calculate is an estimate for a completed item of work.
- Keep scoping individual tasks, determining the number of units required for those tasks, matching each task with individual Unit Price line items, and multiplying quantities by Total Incl O&P costs.
- An estimate completed in this manner is priced as if a subcontractor, or set of subcontractors, is performing the work. The estimate does not yet include Project Overhead or Estimate Summary components such as general contractor markups on subcontracted work, general contractor office overhead and profit, contingency, and location factors.

5. Project Overhead

- Include project overhead items from Division 1-General Requirements.
- These items are needed to make the job run. They are typically, but not always, provided by the general contractor. Items include, but are not limited to, field personnel, insurance, performance bond, permits, testing, temporary utilities, field office and storage facilities, temporary scaffolding and platforms, equipment mobilization and demobilization, temporary roads and sidewalks, winter protection, temporary barricades and fencing, temporary security, temporary signs, field engineering and layout, final cleaning, and commissioning.
- Each item should be quantified and matched to individual Unit Price line items in Division 1, then priced and added to your estimate.
- An alternate method of estimating project overhead costs is to apply a percentage of the total project cost—usually 5% to 15% with an average of 10% (see General Conditions).
- Include other project related expenses in your estimate such as:
  - Rented equipment not itemized in the Crew Listings
  - Rubbish handling throughout the project (see section 02 41 19.19)

6. Estimate Summary

- Include sales tax as required by laws of your state or county.
- Include the general contractor's markup on self-performed work, usually 5% to 15% with an average of 10%.
- Include the general contractor's markup on subcontracted work, usually 5% to 15% with an average of 10%.
- Include the general contractor's main office overhead and profit:
  - RSMeans data provides general guidelines on the general contractor's main office overhead (see section 01 31 13.60 and Reference Number R013113-50).
  - Markups will depend on the size of the general contractor's operations, projected annual revenue, the level of risk, and the level of competition in the local area and for this project in particular.
- Include a contingency, usually 3% to 5%, if appropriate.
- Adjust your estimate to the project's location by using the City Cost Indexes or the Location Factors in the Reference Section:
  - Look at the rules in "How to Use the City Cost Indexes" to see how to apply the Indexes for your location.
  - When the proper Index or Factor has been identified for the project's location, convert it to a multiplier by dividing it by 100, then multiply that multiplier by your estimated total cost. The original estimated total cost will now be adjusted up or down from the national average to a total that is appropriate for your location.

Editors' Note:

We urge you to spend time reading and understanding the supporting material. An accurate estimate requires experience, knowledge, and careful calculation. The more you know about how we at RSMeans developed the data, the more accurate your estimate will be. In addition, it is important to take into consideration the reference material such as Equipment Listings, Crew Listings, City Cost Indexes, Location Factors, and Reference Tables.

How to Use the Cost Data: The Details

What's Behind the Numbers? The Development of Cost Data

RSMeans data engineers continually monitor developments in the construction industry in order to ensure reliable, thorough, and up-to-date cost information. While overall construction costs may vary relative to general economic conditions, price fluctuations within the industry are dependent upon many factors. Individual price variations may, in fact, be opposite to overall economic trends. Therefore, costs are constantly tracked and complete updates are performed yearly. Also, new items are frequently added in response to changes in materials and methods.

Costs in U.S. Dollars

All costs represent U.S. national averages and are given in U.S. dollars. The City Cost Index (CCI) with RSMeans data can be used to adjust costs to a particular location. The CCI for Canada can be used to adjust U.S. national averages to local costs in Canadian dollars. No exchange rate conversion is necessary because it has already been factored in.

G The processes or products identified by the green symbol in our publications have been determined to be environmentally responsible and/or resource-efficient solely by RSMeans data engineering staff. The inclusion of the green symbol does not represent compliance with any specific industry association or standard.

Material Costs

RSMeans data engineers contact manufacturers, dealers, distributors, and contractors all across the U.S. and Canada to determine national average material costs. If you have access to current material costs for your specific location, you may wish to make adjustments to reflect differences from the national average. Included within material costs are fasteners for a normal installation. RSMeans data engineers use manufacturers' recommendations, written specifications, and/or standard construction practices for the sizing and spacing of fasteners. Adjustments to material costs may be required for your specific application or location. The manufacturer's warranty is assumed. Extended warranties are not included in the material costs. **Material costs do not include sales tax.**

Labor Costs

Labor costs are based upon a mathematical average of trade-specific wages in 30 major U.S. cities. The type of wage (union, open shop, or residential) is identified on the inside back cover of printed publications or selected by the estimator when using the electronic products. Markups for the wages can also be found on the inside back cover of printed publications and/or under the labor references found in the electronic products.

- If wage rates in your area vary from those used, or if rate increases are expected within a given year, labor costs should be adjusted accordingly.

Labor costs reflect productivity based on actual working conditions. In addition to actual installation, these figures include time spent during a normal weekday on tasks, such as material receiving and handling, mobilization at the site, site movement, breaks, and cleanup.

Productivity data is developed over an extended period so as not to be influenced by abnormal variations and reflects a typical average.

Equipment Costs

Equipment costs include not only rental but also operating costs for equipment under normal use. The operating costs include parts and labor for routine servicing, such as the repair and replacement of pumps, filters, and worn lines. Normal operating expendables, such as fuel, lubricants, tires, and electricity (where applicable), are also included. Extraordinary operating expendables with highly variable wear patterns, such as diamond bits and blades, are excluded. These costs are included under materials. Equipment rental rates are obtained from industry sources throughout North America—contractors, suppliers, dealers, manufacturers, and distributors.

Rental rates can also be treated as reimbursement costs for contractor-owned equipment. Owned equipment costs include depreciation, loan payments, interest, taxes, insurance, storage, and major repairs.

Equipment costs do not include operators' wages.

Equipment Cost/Day—The cost of equipment required for each crew is included in the Crew Listings in the Reference Section (small tools that are considered essential everyday tools are not listed out separately). The Crew Listings itemize specialized tools and heavy equipment along with labor trades. The daily cost of itemized equipment included in a crew is based on dividing the weekly bare rental rate by 5 (number of working days per week), then adding the hourly operating cost times 8 (the number of hours per day). This Equipment Cost/Day is shown in the last column of the Equipment Rental Costs in the Reference Section.

Mobilization, Demobilization—The cost to move construction equipment from an equipment yard or rental company to the job site and back again is not included in equipment costs. Mobilization (to the site) and demobilization (from the site) costs can be found in the Unit Price Section. If a piece of equipment is already at the job site, it is not appropriate to utilize mobilization or demobilization costs again in an estimate.

Overhead and Profit

Total Cost including O&P for the installing contractor is shown in the last column of the Unit Price and/or Assemblies. This figure is the sum of the bare material cost plus 10% for profit, the bare labor cost plus total overhead and profit, and the bare equipment cost plus 10% for profit. Details for the calculation of overhead and profit on labor are shown on the inside back cover of the printed product and in the Reference Section of the electronic product.

General Conditions

Cost data in this data set are presented in two ways: Bare Costs and Total Cost including O&P (Overhead and Profit). General Conditions, or General Requirements, of the contract should also be added to the Total Cost including O&P when applicable. Costs for General Conditions are listed in Division 1 of the Unit Price Section and in the Reference Section.

General Conditions for the installing contractor may range from 0% to 10% of the Total Cost including O&P. For the general or prime contractor, costs for General Conditions may range from 5% to 15% of the Total Cost including O&P, with a figure of 10% as the most typical allowance. If applicable, the Assemblies and Models sections use costs that include the installing contractor's overhead and profit (O&P).

Factors Affecting Costs

Costs can vary depending upon a number of variables. Here's a listing of some factors that affect costs and points to consider.

Quality—The prices for materials and the workmanship upon which productivity is based represent sound construction work. They are also in line with industry standard and manufacturer specifications and are frequently used by federal, state, and local governments.

Overtime—We have made no allowance for overtime. If you anticipate premium time or work beyond normal working hours, be sure to make an appropriate adjustment to your labor costs.

Productivity—The productivity, daily output, and labor-hour figures for each line item are based on an eight-hour work day in daylight hours in moderate temperatures and up to a 14' working height unless otherwise indicated. For work that extends beyond normal work hours or is performed under adverse conditions, productivity may decrease.

Size of Project—The size, scope of work, and type of construction project will have a significant impact on cost. Economies of scale can reduce costs for large projects. Unit costs can often run higher for small projects.

Location—Material prices are for metropolitan areas. However, in dense urban areas, traffic and site storage limitations may increase costs. Beyond a 20-mile radius of metropolitan areas, extra trucking or transportation charges may also increase the material costs slightly. On the other hand, lower wage rates may be in effect. Be sure to consider both of these factors when preparing an estimate, particularly if the job site is located in a central city or remote rural location. In addition, highly specialized subcontract items may require travel and per-diem expenses for mechanics.

Other Factors—

- season of year
- contractor management
- weather conditions
- local union restrictions
- building code requirements
- availability of:
  - adequate energy
  - skilled labor
  - building materials
- owner's special requirements/restrictions
- safety requirements
- environmental considerations
- access

Unpredictable Factors—General business conditions influence "in-place" costs of all items. Substitute materials and construction methods may have to be employed. These may affect the installed cost and/or life cycle costs. Such factors may be difficult to evaluate and cannot necessarily be predicted on the basis of the job's location in a particular section of the country. Thus, where these factors apply, you may find significant but unavoidable cost variations for which you will have to apply a measure of judgment to your estimate.

Rounding of Costs

In printed publications only, all unit prices in excess of $5.00 have been rounded to make them easier to use and still maintain adequate precision of the results.

How Subcontracted Items Affect Costs

A considerable portion of all large construction jobs is usually subcontracted. In fact, the percentage done by subcontractors is constantly increasing and may run over 90%. Since the workers employed by these companies do nothing else but install their particular products, they soon become experts in that line. As a result, installation by these firms is accomplished so efficiently that the total in-place cost, even with the general contractor's overhead and profit, is no more, and often less, than if the principal contractor had handled the installation. Companies that deal with construction specialties are anxious to have their products perform well and, consequently, the installation will be the best possible.

Contingencies

The allowance for contingencies generally provides for unforeseen construction difficulties. On alterations or repair jobs, 20% is not too much. If drawings are final and only field contingencies are being considered, 2% or 3% is probably sufficient and often nothing needs to be added. Contractually, changes in plans will be covered by extras. The contractor should consider inflationary price trends and possible material shortages during the course of the job. These escalation factors are dependent upon both economic conditions and the anticipated time between the estimate and actual construction. If drawings are not complete or approved, or a budget cost is wanted, it is wise to add 5% to 10%. Contingencies, then, are a matter of judgment.

Important Estimating Considerations

The productivity, or daily output, of each craftsman or crew assumes a well-managed job where tradesmen with the proper tools and equipment, along with the appropriate construction materials, are present. Included are daily set-up and cleanup time, break time, and plan layout time. Unless otherwise indicated, time for material movement on site (for items that can be transported by hand) of up to 200' into the building and to the first or second floor is also included. If material has to be transported by other means, over greater distances, or to higher floors, an additional allowance should be considered by the estimator.

While horizontal movement is typically a sole function of distances, vertical transport introduces other variables that can significantly impact productivity. In an occupied building, the use of elevators (assuming access, size, and required protective measures are acceptable) must be understood at the time of the estimate. For new construction, hoist wait and cycle times can easily be 15 minutes and may result in scheduled access extending beyond the normal work day. Finally, all vertical transport will impose strict weight limits likely to preclude the use of any motorized material handling.

The productivity, or daily output, also assumes installation that meets manufacturer/designer/ standard specifications. A time allowance for quality control checks, minor adjustments, and any task required to ensure proper function or operation is also included. For items that require connections to services, time is included for positioning, leveling, securing the unit, and making all the necessary connections (and start up where applicable) to ensure a complete installation. Estimating of the services themselves (electrical, plumbing, water, steam, hydraulics, dust collection, etc.) is separate.

In some cases, the estimator must consider the use of a crane and an appropriate crew for the installation of large or heavy items. For those situations where a crane is not included in the assigned crew and as part of the line item cost, then equipment rental costs, mobilization and demobilization costs, and operator and support personnel costs must be considered.

Labor-Hours

The labor-hours expressed in this publication are derived by dividing the total daily labor-hours for the crew by the daily output. Based on average installation time and the assumptions listed above, the labor-hours include: direct labor, indirect labor, and nonproductive time. A typical day for a craftsman might include but is not limited to:

- Direct Work
  - Measuring and layout
  - Preparing materials
  - Actual installation
  - Quality assurance/quality control
- Indirect Work
  - Reading plans or specifications
  - Preparing space
  - Receiving materials
  - Material movement
  - Giving or receiving instruction
  - Miscellaneous
- Non-Work
  - Chatting
  - Personal issues
  - Breaks
  - Interruptions (i.e., sickness, weather, material or equipment shortages, etc.)

If any of the items for a typical day do not apply to the particular work or project situation, the estimator should make any necessary adjustments.

Final Checklist

Estimating can be a straightforward process provided you remember the basics. Here's a checklist of some of the steps you should remember to complete before finalizing your estimate.

Did you remember to:

- factor in the City Cost Index for your locale?
- take into consideration which items have been marked up and by how much?
- mark up the entire estimate sufficiently for your purposes?
- read the background information on techniques and technical matters that could impact your project time span and cost?
- include all components of your project in the final estimate?
- double check your figures for accuracy?
- call RSMeans data engineers if you have any questions about your estimate or the data you've used? Remember, Gordian stands behind all of our products, including our extensive RSMeans data solutions. If you have any questions about your estimate, about the costs you've used from our data, or even about the technical aspects of the job that may affect your estimate, feel free to call the Gordian RSMeans editors at 1.800.448.8182.

Unit Price Section

Table of Contents

RSMeans data: Unit Prices— How They Work

All RSMeans data: Unit Prices are organized in the same way.
NOTE: data used in this example is from *Electrical Costs with RSMeans data*.

03 30 Cast-In-Place Concrete

03 30 53 – Miscellaneous Cast-In-Place Concrete

03 30 53.40 Concrete In Place	Crew	Daily Output	Labor-Hours	Unit	Material	2020 Bare Costs Labor	Equipment	Total	Total Incl O&P
0010 CONCRETE IN PLACE									
0020 Including forms (4 uses), Grade 60 rebar, concrete (Portland cement									
0050 Type I), placement and finishing unless otherwise indicated									
3540 Equipment pad (3000 psi), 3' x 3' x 6" thick	C-14H	45	1.067	Ea.	50.50	55	.60	106.10	139
3550 4' x 4' x 6" thick		30	1.600		78	82.50	.90	161.40	210
3560 5' x 5' x 8" thick		18	2.667		138	138	1.49	277.49	360
3570 6' x 6' x 8" thick		14	3.429		190	177	1.92	368.92	475
3580 8' x 8' x 10" thick		8	6		395	310	3.36	708.36	905
3590 10' x 10' x 12" thick		5	9.600		695	495	5.40	1,195.40	1,500
3800 Footings (3000 psi), spread under 1 C.Y.	C-14C	28	4	C.Y.	203	201	.96	404.96	525
3813 Install new concrete (3000 psi) light pole base, 24" diam. x 8'	C-1	2.66	12.030		325	605		930	1,275
3825 1 C.Y. to 5 C.Y.	C-14C	43	2.605		240	131	.63	371.63	465
3850 Over 5 C.Y.	"	75	1.493		220	75	.36	295.36	355
3900 Footings, strip (3000 psi), 18" x 9", unreinforced	C-14L	40	2.400		154	118	.67	272.67	350

It is important to understand the structure of RSMeans data: Unit Prices so that you can find information easily and use it correctly.

❶ Line Numbers

Line Numbers consist of 12 characters, which identify a unique location in the database for each task. The first 6 or 8 digits conform to the Construction Specifications Institute MasterFormat® 2018. The remainder of the digits are a further breakdown in order to arrange items in understandable groups of similar tasks. Line numbers are consistent across all of our publications, so a line number in any of our products will always refer to the same item of work.

❷ Descriptions

Descriptions are shown in a hierarchical structure to make them readable. In order to read a complete description, read up through the indents to the top of the section. Include everything that is above and to the left that is not contradicted by information below. For instance, the complete description for line 03 30 53.40 3550 is "Concrete in place, including forms (4 uses), Grade 60 rebar, concrete (Portland cement Type 1), placement and finishing unless otherwise indicated; Equipment pad (3000 psi), 4' × 4' × 6" thick."

❸ RSMeans data

When using **RSMeans data**, it is important to read through an entire section to ensure that you use the data that most closely matches your work. Note that sometimes there is additional information shown in the section that may improve your price. There are frequently lines that further describe, add to, or adjust data for specific situations.

❹ Reference Information

Gordian's RSMeans engineers have created **reference** information to assist you in your estimate. **If** there is information that applies to a section, it will be indicated at the start of the section. The Reference Section is located in the back of the data set.

❺ Crews

Crews include labor and/or equipment necessary to accomplish each task. In this case, Crew C-14H is used. Gordian's RSMeans staff selects a crew to represent the workers and equipment that are typically used for that task. In this case, Crew C-14H consists of one carpenter foreman (outside), two carpenters, one rodman, one laborer, one cement finisher, and one gas engine vibrator. Details of all crews can be found in the Reference Section.

Crews - Standard

Crew No.	Bare Costs		Incl. Subs O & P		Cost Per Labor-Hour	
Crew C-14H	Hr.	Daily	Hr.	Daily	Bare Costs	Incl. O&P
1 Carpenter Foreman (outside)	$55.15	$441.20	$82.85	$662.80	$51.65	$77.36
2 Carpenters	53.15	850.40	79.85	1277.60		
1 Rodman (reinf.)	56.40	451.20	84.90	679.20		
1 Laborer	42.10	336.80	63.25	506.00		
1 Cement Finisher	49.95	399.60	73.45	587.60		
1 Gas Engine Vibrator		26.85		29.54	.56	.62
48 L.H., Daily Totals		$2506.05		$3742.74	$52.21	$77.97

6 Daily Output

The **Daily Output** is the amount of work that the crew can do in a normal 8-hour workday, including mobilization, layout, movement of materials, and cleanup. In this case, crew C-14H can install thirty 4' × 4' × 6" thick concrete pads in a day. Daily output is variable and based on many factors, including the size of the job, location, and environmental conditions. RSMeans data represents work done in daylight (or adequate lighting) and temperate conditions.

7 Labor-Hours

The figure in the **Labor-Hours** column is the amount of labor required to perform one unit of work—in this case the amount of labor required to construct one 4' × 4' equipment pad. This figure is calculated by dividing the number of hours of labor in the crew by the daily output (48 labor-hours divided by 30 pads = 1.6 hours of labor per pad). Multiply 1.6 times 60 to see the value in minutes: 60 × 1.6 = 96 minutes. Note: the labor-hour figure is not dependent on the crew size. A change in crew size will result in a corresponding change in daily output, but the labor-hours per unit of work will not change.

8 Unit of Measure

All RSMeans data: Unit Prices include the typical **Unit of Measure** used for estimating that item. For concrete-in-place the typical unit is cubic yards (C.Y.) or each (Ea.). For installing broadloom carpet it is square yard and for gypsum board it is square foot. The estimator needs to take special care that the unit in the data matches the unit in the take-off. Unit conversions may be found in the Reference Section.

9 Bare Costs

Bare Costs are the costs of materials, labor, and equipment that the installing contractor pays. They represent the cost, in U.S. dollars, for one unit of work. They do not include any markups for profit or labor burden.

10 Bare Total

The **Total column** represents the total bare cost for the installing contractor in U.S. dollars. In this case, the sum of $78 for material + $82.50 for labor + $.90 for equipment is $161.40.

11 Total Incl O&P

The **Total Incl O&P column** is the total cost, including overhead and profit, that the installing contractor will charge the customer. This represents the cost of materials plus 10% profit, the cost of labor plus labor burden and 10% profit, and the cost of equipment plus 10% profit. It does not include the general contractor's overhead and profit. Note: See the inside back cover of the printed product or the Reference Section of the electronic product for details on how the labor burden is calculated.

National Average

*The RSMeans data in our print publications represent a "national average" cost. This data should be modified to the project location using the **City Cost Indexes** or **Location Factors** tables found in the Reference Section. Use the Location Factors to adjust estimate totals if the project covers multiple trades. Use the City Cost Indexes (CCI) for single trade projects or projects where a more detailed analysis is required. All figures in the two tables are derived from the same research. The last row of data in the CCI—the weighted average—is the same as the numbers reported for each location in the location factor table.*

RSMeans data: Unit Prices— How They Work (Continued)

Project Name: Pre-Engineered Steel Building				Architect: As Shown				
Location:	**Anywhere, USA**						**01/01/20**	**STD**
Line Number	**Description**	**Qty**	**Unit**	**Material**	**Labor**	**Equipment**	**SubContract**	**Estimate Total**
03 30 53.40 3940	Strip footing, 12" x 24", reinforced	15	C.Y.	$2,565.00	$1,770.00	$8.40	$0.00	
03 30 53.40 3950	Strip footing, 12" x 36", reinforced	34	C.Y.	$5,610.00	$3,196.00	$15.30	$0.00	
03 11 13.65 3000	Concrete slab edge forms	500	L.F.	$165.00	$1,345.00	$0.00	$0.00	
03 22 11.10 0200	Welded wire fabric reinforcing	150	C.S.F.	$2,872.50	$4,350.00	$0.00	$0.00	
03 31 13.35 0300	Ready mix concrete, 4000 psi for slab on grade	278	C.Y.	$35,306.00	$0.00	$0.00	$0.00	
03 31 13.70 4300	Place, strike off & consolidate concrete slab	278	C.Y.	$0.00	$5,309.80	$136.22	$0.00	
03 35 13.30 0250	Machine float & trowel concrete slab	15,000	S.F.	$0.00	$9,900.00	$750.00	$0.00	
03 15 16.20 0140	Cut control joints in concrete slab	950	L.F.	$47.50	$418.00	$57.00	$0.00	
03 39 23.13 0300	Sprayed concrete curing membrane	150	C.S.F.	$1,867.50	$1,065.00	$0.00	$0.00	
Division 03	**Subtotal**			**$48,433.50**	**$27,353.80**	**$966.92**	**$0.00**	**$76,754.22**
08 36 13.10 2650	Manual 10' x 10' steel sectional overhead door	8	Ea.	$11,000.00	$3,760.00	$0.00	$0.00	
08 36 13.10 2860	Insulation and steel back panel for OH door	800	S.F.	$4,000.00	$0.00	$0.00	$0.00	
Division 08	**Subtotal**			**$15,000.00**	**$3,760.00**	**$0.00**	**$0.00**	**$18,760.00**
13 34 19.50 1100	Pre-Engineered Steel Building, 100' x 150' x 24'	15,000	SF Flr.	$0.00	$0.00	$0.00	$367,500.00	
13 34 19.50 6050	Framing for PESB door opening, 3' x 7'	4	Opng.	$0.00	$0.00	$0.00	$2,320.00	
13 34 19.50 6100	Framing for PESB door opening, 10' x 10'	8	Opng.	$0.00	$0.00	$0.00	$9,600.00	
13 34 19.50 6200	Framing for PESB window opening, 4' x 3'	6	Opng.	$0.00	$0.00	$0.00	$3,450.00	
13 34 19.50 5750	PESB door, 3' x 7', single leaf	4	Opng.	$2,920.00	$736.00	$0.00	$0.00	
13 34 19.50 7750	PESB sliding window, 4' x 3' with screen	6	Opng.	$2,940.00	$630.00	$67.80	$0.00	
13 34 19.50 6550	PESB gutter, eave type, 26 ga., painted	300	L.F.	$2,415.00	$864.00	$0.00	$0.00	
13 34 19.50 8650	PESB roof vent, 12" wide x 10' long	15	Ea.	$570.00	$3,465.00	$0.00	$0.00	
13 34 19.50 6900	PESB insulation, vinyl faced, 4" thick	27,400	S.F.	$11,782.00	$10,138.00	$0.00	$0.00	
Division 13	**Subtotal**			**$20,627.00**	**$15,833.00**	**$67.80**	**$382,870.00**	**$419,397.80**
	Subtotal			$84,060.50	$46,946.80	$1,034.72	$382,870.00	$514,912.02
Division 01	**General Requirements @ 7%**			5,884.24	3,286.28	72.43	26,800.90	
	Estimate Subtotal			$89,944.74	$50,233.08	$1,107.15	$409,670.90	$514,912.02
	Sales Tax @ 5%			4,497.24		55.36	10,241.77	
	Subtotal A			94,441.97	50,233.08	1,162.51	419,912.67	
	GC O & P			9,444.20	25,769.57	116.25	41,991.27	
	Subtotal B			103,886.17	76,002.64	1,278.76	461,903.94	$643,071.51
	Contingency @ 5%							32,153.58
	Subtotal C							$675,225.09
	Bond @ $12/1000 +10% O&P							8,912.97
	Subtotal D							$684,138.06
	Location Adjustment Factor				115.50			106,041.40
	Grand Total							**$790,179.46**

This estimate is based on an interactive spreadsheet. You are free to download it and adjust it to your methodology.
A copy of this spreadsheet is available at **RSMeans.com/2020books.**

4

Sample Estimate

This sample demonstrates the elements of an estimate, including a tally of the RSMeans data lines and a summary of the markups on a contractor's work to arrive at a total cost to the owner. The Location Factor with RSMeans data is added at the bottom of the estimate to adjust the cost of the work to a specific location.

---

① Work Performed

The body of the estimate shows the RSMeans data selected, including the line number, a brief description of each item, its take-off unit and quantity, and the bare costs of materials, labor, and equipment. This estimate also includes a column titled "SubContract." This data is taken from the column "Total Incl O&P" and represents the total that a subcontractor would charge a general contractor for the work, including the sub's markup for overhead and profit.

② Division 1, General Requirements

This is the first division numerically but the last division estimated. Division 1 includes project-wide needs provided by the general contractor. These requirements vary by project but may include temporary facilities and utilities, security, testing, project cleanup, etc. For small projects a percentage can be used—typically between 5% and 15% of project cost. For large projects the costs may be itemized and priced individually.

③ Sales Tax

If the work is subject to state or local sales taxes, the amount must be added to the estimate. Sales tax may be added to material costs, equipment costs, and subcontracted work. In this case, sales tax was added in all three categories. It was assumed that approximately half the subcontracted work would be material cost, so the tax was applied to 50% of the subcontract total.

④ GC O&P

This entry represents the general contractor's markup on material, labor, equipment, and subcontractor costs. Our standard markup on materials, equipment, and subcontracted work is 10%. In this estimate, the markup on the labor performed by the GC's workers uses "Skilled Workers Average" shown in Column F on the table "Installing Contractor's Overhead & Profit," which can be found on the inside back cover of the printed product or in the Reference Section of the electronic product.

⑤ Contingency

A factor for contingency may be added to any estimate to represent the cost of unknowns that may occur between the time that the estimate is performed and the time the project is constructed. The amount of the allowance will depend on the stage of design at which the estimate is done and the contractor's assessment of the risk involved. Refer to section 01 21 16.50 for contingency allowances.

⑥ Bonds

Bond costs should be added to the estimate. The figures here represent a typical performance bond, ensuring the owner that if the general contractor does not complete the obligations in the construction contract the bonding company will pay the cost for completion of the work.

⑦ Location Adjustment

Published prices are based on national average costs. If necessary, adjust the total cost of the project using a location factor from the "Location Factor" table or the "City Cost Index" table. Use location factors if the work is general, covering multiple trades. If the work is by a single trade (e.g., masonry) use the more specific data found in the "City Cost Indexes."

Pre-Installation Costs

Pre-Installation
Change Order Costs

This section contains cost data for electrical change order work that occurs before the actual installation (new construction) has begun but after the contract documents have been completed, and, possibly, after the crews have been formed and material purchases made. For change order work that occurs after the original installation is substantially complete, see the "Post-Installation Change Orders" section.

These costs are categorized according to the CSI MasterFormat by division, subdivision, and line number. This numbering system is explained in detail in "How RSMeans Unit Price Data Works." The meaning of each column (Crew, Daily Output, Labor-hours, Unit, Material, Labor, Equipment, Total, and Total IncludingO&P) is also discussed at the beginning of this data set in "How RSMeans Unit Price Data Works."

The unit prices in this Pre-Installation Change Orders section are for change order additions to the base contract.

In most cases, change orders resulting in a deduction or a credit reflect only the bare costs. The contractor retains the anticipated overhead and profit based on the original bid.

Estimating Tips
01 20 00 Price and Payment Procedures

- Allowances that should be added to estimates to cover contingencies and job conditions that are not included in the national average material and labor costs are shown in Section 01 21.

- When estimating historic preservation projects (depending on the condition of the existing structure and the owner's requirements), a 15–20% contingency or allowance is recommended, regardless of the stage of the drawings.

01 30 00 Administrative Requirements

- Before determining a final cost estimate, it is good practice to review all the items listed in Subdivisions 01 31 and 01 32 to make final adjustments for items that may need customizing to specific job conditions.

- Requirements for initial and periodic submittals can represent a significant cost to the General Requirements of a job. Thoroughly check the submittal specifications when estimating a project to determine any costs that should be included.

01 40 00 Quality Requirements

- All projects will require some degree of quality control. This cost is not included in the unit cost of construction listed in each division. Depending upon the terms of the contract, the various costs of inspection and testing can be the responsibility of either the owner or the contractor. Be sure to include the required costs in your estimate.

01 50 00 Temporary Facilities and Controls

- Barricades, access roads, safety nets, scaffolding, security, and many more requirements for the execution of a safe project are elements of direct cost. These costs can easily be overlooked when preparing an estimate. When looking through the major classifications of this subdivision, determine which items apply to each division in your estimate.

- Construction equipment rental costs can be found in the Reference Section in Section 01 54 33. Operators' wages are not included in equipment rental costs.

- Equipment mobilization and demobilization costs are not included in equipment rental costs and must be considered separately.

- The cost of small tools provided by the installing contractor for his workers is covered in the "Overhead" column on the "Installing Contractor's Overhead and Profit" table that lists labor trades, base rates, and markups. Therefore, it is included in the "Total Incl. O&P" cost of any unit price line item.

01 70 00 Execution and Closeout Requirements

- When preparing an estimate, thoroughly read the specifications to determine the requirements for Contract Closeout. Final cleaning, record documentation, operation and maintenance data, warranties and bonds, and spare parts and maintenance materials can all be elements of cost for the completion of a contract. Do not overlook these in your estimate.

Reference Numbers

Reference numbers are shown at the beginning of some major classifications. These numbers refer to related items in the Reference Section. The reference information may be an estimating procedure, an alternate pricing method, or technical information.

Note: Not all subdivisions listed here necessarily appear. ■

Same Data. Simplified.

Enjoy the convenience and efficiency of accessing your costs anywhere:

- **Skip the multiplier** by setting your location
- **Quickly search,** edit, favorite and share costs
- **Stay on top of price changes** with automatic updates

Discover more at rsmeans.com/online

01 21 Allowances

01 21 53 – Factors Allowance

01 21 53.60 Security Factors		Crew	Daily Output	Labor-Hours	Unit	Material	2020 Bare Costs Labor	Equipment	Total	Total Incl O&P
0010	**SECURITY FACTORS**									
0100	Additional costs due to security requirements									
0110	Daily search of personnel, supplies, equipment and vehicles									
0120	Physical search, inventory and doc of assets, at entry				Costs		30%			
0130	At entry and exit						50%			
0140	Physical search, at entry						6.25%			
0150	At entry and exit						12.50%			
0160	Electronic scan search, at entry						2%			
0170	At entry and exit						4%			
0180	Visual inspection only, at entry						.25%			
0190	At entry and exit						.50%			
0200	ID card or display sticker only, at entry						.12%			
0210	At entry and exit				▼		.25%			
0220	Day 1 as described below, then visual only for up to 5 day job duration									
0230	Physical search, inventory and doc of assets, at entry				Costs		5%			
0240	At entry and exit						10%			
0250	Physical search, at entry						1.25%			
0260	At entry and exit						2.50%			
0270	Electronic scan search, at entry						.42%			
0280	At entry and exit				▼		.83%			
0290	Day 1 as described below, then visual only for 6-10 day job duration									
0300	Physical search, inventory and doc of assets, at entry				Costs		2.50%			
0310	At entry and exit						5%			
0320	Physical search, at entry						.63%			
0330	At entry and exit						1.25%			
0340	Electronic scan search, at entry						.21%			
0350	At entry and exit				▼		.42%			
0360	Day 1 as described below, then visual only for 11-20 day job duration									
0370	Physical search, inventory and doc of assets, at entry				Costs		1.25%			
0380	At entry and exit						2.50%			
0390	Physical search, at entry						.31%			
0400	At entry and exit						.63%			
0410	Electronic scan search, at entry						.10%			
0420	At entry and exit				▼		.21%			
0430	Beyond 20 days, costs are negligible									
0440	Escort required to be with tradesperson during work effort				Costs		6.25%			

01 51 Temporary Utilities

01 51 13 – Temporary Electricity

01 51 13.50 Temporary Power Equip (Pro-Rated Per Job)		Crew	Daily Output	Labor-Hours	Unit	Material	2020 Bare Costs Labor	Equipment	Total	Total Incl O&P
0010	**TEMPORARY POWER EQUIP (PRO-RATED PER JOB)**									
0020	Service, overhead feed, 3 use									
0030	100 Amp	1 Elec	1.15	6.960	Ea.	430	425		855	1,100
0040	200 Amp		.92	8.700		470	535		1,005	1,300
0050	400 Amp		.69	11.590		830	710		1,540	1,975
0060	600 Amp	▼	.46	17.390	▼	1,200	1,075		2,275	2,925
0100	Underground feed, 3 use									
0110	100 Amp	1 Elec	1.84	4.350	Ea.	445	267		712	885
0120	200 Amp		1.06	7.560		520	465		985	1,275
0130	400 Amp		.92	8.700		875	535		1,410	1,750
0140	600 Amp	▼	.69	11.590	▼	1,075	710		1,785	2,225

Pre-Installation Change Order Costs

01 51 13.50 Temporary Power Equip (Pro-Rated Per Job)	Crew	Daily Output	Labor-Hours	Unit	Material	2020 Bare Costs Labor	Equipment	Total	Total Incl O&P
0150	800 Amp	1 Elec	.46	17.390	Ea.	1,525	1,075		2,600	3,275
0160	1000 Amp		.32	24.840		1,700	1,525		3,225	4,125
0170	1200 Amp		.23	34.780		1,875	2,125		4,000	5,225
0180	2000 Amp	↓	.18	43.480	↓	2,250	2,675		4,925	6,450
0200	Transformers, 3 use									
0210	30 kVA	1 Elec	.92	8.700	Ea.	1,925	535		2,460	2,925
0220	45 kVA		.69	11.590		2,450	710		3,160	3,750
0230	75 kVA		.46	17.390		3,725	1,075		4,800	5,700
0240	112.5 kVA	↓	.37	21.740	↓	4,100	1,325		5,425	6,475
0250	Feeder, PVC, CU wire in trench									
0260	60 Amp	1 Elec	88.32	.091	L.F.	2.73	5.55		8.28	11.25
0270	100 Amp		78.20	.102		4.96	6.30		11.26	14.80
0280	200 Amp		54.28	.147		12	9.05		21.05	26.50
0290	400 Amp	↓	38.64	.207	↓	32	12.70		44.70	54
0300	Feeder, PVC, aluminum wire in trench									
0310	60 Amp	1 Elec	88.32	.091	L.F.	2.67	5.55		8.22	11.20
0320	100 Amp		78.20	.102		3.19	6.30		9.49	12.85
0330	200 Amp		54.28	.147		8.65	9.05		17.70	23
0340	400 Amp	↓	38.64	.207	↓	15.75	12.70		28.45	36
0350	Feeder, EMT, CU wire									
0360	60 Amp	1 Elec	82.80	.097	L.F.	3.10	5.95		9.05	12.25
0370	100 Amp		73.60	.109		5.75	6.65		12.40	16.25
0380	200 Amp		55.20	.145		12.85	8.90		21.75	27.50
0390	400 Amp	↓	32.20	.248	↓	32.50	15.25		47.75	58
0400	Feeder, EMT, aluminum wire									
0410	60 Amp	1 Elec	82.80	.097	L.F.	4.59	5.95		10.54	13.90
0420	100 Amp		73.60	.109		5.05	6.65		11.70	15.50
0430	200 Amp		55.20	.145		9.25	8.90		18.15	23.50
0440	400 Amp	↓	32.20	.248	↓	17.30	15.25		32.55	41.50
0500	Equipment, 3 use									
0510	Spider box, 50 Amp	1 Elec	7.36	1.090	Ea.	785	67		852	965
0520	Lighting cord, 100'		7.36	1.090		126	67		193	239
0530	Light stanchion	↓	7.36	1.090	↓	72	67		139	179
0540	Temporary cords, 100', 3 use									
0550	Feeder cord, 50 Amp	1 Elec	14.72	.543	Ea.	237	33.50		270.50	310
0560	Feeder cord, 100 Amp		11.04	.725		1,200	44.50		1,244.50	1,400
0570	Tap cord, 50 Amp		11.04	.725		630	44.50		674.50	760
0580	Tap cord, 100 Amp	↓	5.52	1.450	↓	1,100	89		1,189	1,325
0590	Temporary cords, 50', 3 use									
0600	Feeder cord, 50 Amp	1 Elec	14.72	.543	Ea.	305	33.50		338.50	385
0610	Feeder cord, 100 Amp		11.04	.725		625	44.50		669.50	755
0620	Tap cord, 50 Amp		11.04	.725		305	44.50		349.50	400
0630	Tap cord, 100 Amp	↓	5.52	1.450	↓	615	89		704	810
0700	Connections									
0710	Compressor or pump									
0720	30 Amp	1 Elec	6.44	1.240	Ea.	17	76		93	132
0730	60 Amp		4.88	1.640		30	101		131	184
0740	100 Amp	↓	3.68	2.170	↓	80	133		213	286
0750	Tower crane									
0760	60 Amp	1 Elec	4.14	1.930	Ea.	30	118		148	210
0770	100 Amp	"	2.76	2.900	"	80	178		258	355
0780	Manlift									
0790	Single	1 Elec	2.76	2.900	Ea.	34.50	178		212.50	305

01 51 13.50 Temporary Power Equip (Pro-Rated Per Job)	Crew	Daily Output	Labor-Hours	Unit	Material	2020 Bare Costs Labor	Equipment	Total	Total Incl O&P
0800 Double	1 Elec	1.84	4.350	Ea.	76	267		343	480
0810 Welder with disconnect									
0820 50 Amp	1 Elec	4.60	1.740	Ea.	205	107		312	385
0830 100 Amp		3.50	2.290		395	140		535	640
0840 200 Amp		2.30	3.480		735	214		949	1,125
0850 400 Amp		.92	8.700		1,575	535		2,110	2,525
0860 Office trailer									
0870 60 Amp	1 Elec	4.14	1.930	Ea.	55	118		173	237
0880 100 Amp		2.76	2.900		83.50	178		261.50	355
0890 200 Amp		1.84	4.350		292	267		559	715

 Pre-Installation Change Order Costs

Estimating Tips
General

- The items in this division are usually priced per square foot or each. Most of these items are purchased by the owner and installed by the contractor. Do not assume the items in Division 12 will be purchased and installed by the contractor. Check the specifications for responsibilities and include receiving, storage, installation, and mechanical and electrical hookups in the appropriate divisions.

- Some items in this division require some type of support system that is not usually furnished with the item. Examples of these systems include blocking for the attachment of casework and heavy drapery rods. The required blocking must be added to the estimate in the appropriate division.

Reference Numbers

Reference numbers are shown at the beginning of some major classifications. These numbers refer to related items in the Reference Section. The reference information may be an estimating procedure, an alternate pricing method, or technical information.

Same Data. Simplified.

Enjoy the convenience and efficiency of accessing your costs anywhere:

- **Skip the multiplier** by setting your location
- **Quickly search,** edit, favorite and share costs
- **Stay on top of price changes** with automatic updates

Discover more at rsmeans.com/online

12 46 Furnishing Accessories

12 46 19 - Clocks

12 46 19.50 Wall Clocks	Crew	Daily Output	Labor-Hours	Unit	Material	2020 Bare Costs Labor	Equipment	Total	Total Incl O&P
0010 **WALL CLOCKS**									
0080 12" diameter, single face	1 Elec	7.36	1.090	Ea.	109	67		176	220
0100 Double face	"	5.70	1.400	"	161	86		247	305

Pre-Installation Change Order Costs

Estimating Tips
General

- The items and systems in this division are usually estimated, purchased, supplied, and installed as a unit by one or more subcontractors. The estimator must ensure that all parties are operating from the same set of specifications and assumptions, and that all necessary items are estimated and will be provided. Many times the complex items and systems are covered, but the more common ones, such as excavation or a crane, are overlooked for the very reason that everyone assumes nobody could miss them. The estimator should be the central focus and be able to ensure that all systems are complete.

- It is important to consider factors such as site conditions, weather, shape and size of building, as well as labor availability as they may impact the overall cost of erecting special structures and systems included in this division.

- Another area where problems can develop in this division is at the interface between systems.

The estimator must ensure, for instance, that anchor bolts, nuts, and washers are estimated and included for the air-supported structures and pre-engineered buildings to be bolted to their foundations. Utility supply is a common area where essential items or pieces of equipment can be missed or overlooked because each subcontractor may feel it is another's responsibility. The estimator should also be aware of certain items which may be supplied as part of a package but installed by others, and ensure that the installing contractor's estimate includes the cost of installation. Conversely, the estimator must also ensure that items are not costed by two different subcontractors, resulting in an inflated overall estimate.

13 30 00 Special Structures

- The foundations and floor slab, as well as rough mechanical and electrical, should be estimated, as this work is required for the assembly and erection of the structure. Generally, as noted in the data set, the pre-engineered building comes as a shell. Pricing is based on the size and structural design parameters stated in the reference section. Additional features, such as windows and doors with their related structural framing, must also be included by the estimator. Here again, the estimator must have a clear understanding of the scope of each portion of the work and all the necessary interfaces.

Reference Numbers

Reference numbers are shown at the beginning of some major classifications. These numbers refer to related items in the Reference Section. The reference information may be an estimating procedure, an alternate pricing method, or technical information.

Note: Not all subdivisions listed here necessarily appear. ■

Same Data. Simplified.

Enjoy the convenience and efficiency of accessing your costs anywhere:

- **Skip the multiplier** by setting your location
- **Quickly search,** edit, favorite and share costs
- **Stay on top of price changes** with automatic updates

Discover more at rsmeans.com/online

13 47 13.16 Cathodic Prot. for Underground Storage Tanks	Crew	Daily Output	Labor-Hours	Unit	Material	2020 Bare Costs Labor	Equipment	Total	Total Incl O&P
0010 **CATHODIC PROTECTION FOR UNDERGROUND STORAGE TANKS**									
1000 Anodes, magnesium type, 9 #	R-15	17.02	2.820	Ea.	38.50	169	16.40	223.90	315
1010 17 #		11.96	4.010		81.50	241	23.50	346	475
1020 32 #		9.20	5.220		123	315	30.50	468.50	635
1030 48 #		6.62	7.250		160	435	42	637	875
1100 Graphite type w/epoxy cap, 3" x 60" (32 #)	R-22	7.73	4.820		148	271		419	570
1110 4" x 80" (68 #)		5.52	6.750		246	380		626	835
1120 6" x 72" (80 #)		4.78	7.790		1,525	435		1,960	2,325
1130 6" x 36" (45 #)		8.83	4.220		770	237		1,007	1,200
2000 Rectifiers, silicon type, air cooled, 28 V/10 A	R-19	3.22	6.210		2,325	380		2,705	3,125
2010 20 V/20 A		3.22	6.210		2,325	380		2,705	3,150
2100 Oil immersed, 28 V/10 A		2.76	7.250		2,875	445		3,320	3,850
2110 20 V/20 A		2.76	7.250		3,250	445		3,695	4,250
3000 Anode backfill, coke breeze	R-22	3542	.011	Lb.	.30	.59		.89	1.21
4000 Cable, HMWPE, No. 8		2.21	16.880	M.L.F.	605	950		1,555	2,075
4010 No. 6		2.21	16.880		820	950		1,770	2,300
4020 No. 4		2.21	16.880		1,225	950		2,175	2,750
4030 No. 2		2.21	16.880		1,900	950		2,850	3,500
4040 No. 1		2.02	18.420		2,575	1,025		3,600	4,400
4050 No. 1/0		2.02	18.420		3,350	1,025		4,375	5,225
4060 No. 2/0		2.02	18.420		5,250	1,025		6,275	7,325
4070 No. 4/0		1.84	20.260		7,225	1,150		8,375	9,650
5000 Test station, 7 terminal box, flush curb type w/lockable cover	R-19	11.04	1.810	Ea.	79.50	111		190.50	254
5010 Reference cell, 2" diam. PVC conduit, cplg., plug, set flush	"	4.42	4.530	"	158	278		436	590

Estimating Tips
22 10 00 Plumbing Piping and Pumps

This subdivision is primarily basic pipe and related materials. The pipe may be used by any of the mechanical disciplines, i.e., plumbing, fire protection, heating, and air conditioning.

Note: CPVC plastic piping approved for fire protection is located in 21 11 13.

- The labor adjustment factors listed in Subdivision 22 01 02.20 apply throughout Divisions 21, 22, and 23. CAUTION: the correct percentage may vary for the same items. For example, the percentage add for the basic pipe installation should be based on the maximum height that the installer must install for that particular section. If the pipe is to be located 14' above the floor but it is suspended on threaded rod from beams, the bottom flange of which is 18' high (4' rods), then the height is actually 18' and the add is 20%. The pipe cover, however, does not have to go above the 14' and so the add should be 10%.

- Most pipe is priced first as straight pipe with a joint (coupling, weld, etc.) every 10' and a hanger usually every 10'. There are exceptions with hanger spacing such as for cast iron pipe (5')

and plastic pipe (3 per 10'). Following each type of pipe there are several lines listing sizes and the amount to be subtracted to delete couplings and hangers. This is for pipe that is to be buried or supported together on trapeze hangers. The reason that the couplings are deleted is that these runs are usually long, and frequently longer lengths of pipe are used. By deleting the couplings, the estimator is expected to look up and add back the correct reduced number of couplings.

- When preparing an estimate, it may be necessary to approximate the fittings. Fittings usually run between 25% and 50% of the cost of the pipe. The lower percentage is for simpler runs, and the higher number is for complex areas, such as mechanical rooms.

- For historic restoration projects, the systems must be as invisible as possible, and pathways must be sought for pipes, conduit, and ductwork. While installations in accessible spaces (such as basements and attics) are relatively straightforward to estimate, labor costs may be more difficult to determine when delivery systems must be concealed.

22 40 00 Plumbing Fixtures

- Plumbing fixture costs usually require two lines: the fixture itself and its "rough-in, supply, and waste."

- In the Assemblies Section (Plumbing D2010) for the desired fixture, the System Components Group at the center of the page shows the fixture on the first line. The rest of the list (fittings, pipe, tubing, etc.) will total up to what we refer to in the Unit Price section as "Rough-in, supply, waste, and vent." Note that for most fixtures we allow a nominal 5' of tubing to reach from the fixture to a main or riser.

- Remember that gas- and oil-fired units need venting.

Reference Numbers

Reference numbers are shown at the beginning of some major classifications. These numbers refer to related items in the Reference Section. The reference information may be an estimating procedure, an alternate pricing method, or technical information.

Note: Not all subdivisions listed here necessarily appear. ■

Same Data. Simplified.

Enjoy the convenience and efficiency of accessing your costs anywhere:

- **Skip the multiplier** by setting your location
- **Quickly search,** edit, favorite and share costs
- **Stay on top of price changes** with automatic updates

Discover more at rsmeans.com/online

---

22 05 29.10 Hangers & Supp. for Plumb'g/HVAC Pipe/Equip.	Crew	Daily Output	Labor-Hours	Unit	Material	2020 Bare Costs Labor	Equipment	Total	Total Incl O&P
0010 **HANGERS AND SUPPORTS FOR PLUMB'G/HVAC PIPE/EQUIP.**									
0011 TYPE numbers per MSS-SP58									
2650 Rods, carbon steel									
2660 Continuous thread									
2670 1/4" thread size	1 Plum	132.48	.060	L.F.	2.43	3.89		6.32	8.45
2680 3/8" thread size		132.48	.060		2.59	3.89		6.48	8.65
2690 1/2" thread size		132.48	.060		4.07	3.89		7.96	10.30
2700 5/8" thread size		132.48	.060		5.75	3.89		9.64	12.15
2710 3/4" thread size		132.48	.060		10.60	3.89		14.49	17.50
2720 7/8" thread size		132.48	.060		13.35	3.89		17.24	20.50
2721 1" thread size	Q-1	147.20	.109		22.50	6.30		28.80	34.50
2722 1-1/8" thread size	"	110.40	.145		24.50	8.40		32.90	39.50
2730 For galvanized, add					40%				
8800 Wire cable support system									
8810 Cable with hook terminal and locking device									
8830 2 mm (.079") diam. cable (100 lb. cap.)									
8840 1 m (3.3') length, with hook	1 Shee	88.32	.091	Ea.	4.68	5.65		10.33	13.70
8850 2 m (6.6') length, with hook		77.28	.104		5.45	6.45		11.90	15.80
8860 3 m (9.9') length, with hook		66.24	.121		5.35	7.50		12.85	17.30
8900 3 mm (.118") diam. cable (200 lb. cap.)									
8910 1 m (3.3') length, with hook	1 Shee	88.32	.091	Ea.	11.80	5.65		17.45	21.50
8920 2 m (6.6') length, with hook		77.28	.104		13.05	6.45		19.50	24
8930 3 m (9.9') length, with hook		66.24	.121		14.20	7.50		21.70	27
9000 Cable system accessories									
9010 Anchor bolt, 3/8", with nut	1 Shee	128.80	.062	Ea.	1.36	3.87		5.23	7.35
9020 Air duct corner protector		147.20	.054		1.81	3.39		5.20	7.15
9030 Air duct support attachment		128.80	.062		2.18	3.87		6.05	8.25
9040 Flange clip, hammer-on style									
9044 For flange thickness 3/32"-9/64", 160 lb. cap.	1 Shee	165.60	.048	Ea.	.53	3.01		3.54	5.15
9048 For flange thickness 1/8"-1/4", 200 lb. cap.		147.20	.054		.38	3.39		3.77	5.55
9052 For flange thickness 5/16"-1/2", 200 lb. cap.		138	.058		.78	3.61		4.39	6.35
9056 For flange thickness 9/16"-3/4", 200 lb. cap.		128.80	.062		1.07	3.87		4.94	7

Estimating Tips

The labor adjustment factors listed in Subdivision 22 01 02.20 also apply to Division 23.

23 10 00 Facility Fuel Systems

- The prices in this subdivision for above- and below-ground storage tanks do not include foundations or hold-down slabs, unless noted. The estimator should refer to Divisions 3 and 31 for foundation system pricing. In addition to the foundations, required tank accessories, such as tank gauges, leak detection devices, and additional manholes and piping, must be added to the tank prices.

23 50 00 Central Heating Equipment

- When estimating the cost of an HVAC system, check to see who is responsible for providing and installing the temperature control system. It is possible to overlook controls, assuming that they would be included in the electrical estimate.

- When looking up a boiler, be careful on specified capacity. Some manufacturers rate their products on output while others use input.

- Include HVAC insulation for pipe, boiler, and duct (wrap and liner).

- Be careful when looking up mechanical items to get the correct pressure rating and connection type (thread, weld, flange).

23 70 00 Central HVAC Equipment

- Combination heating and cooling units are sized by the air conditioning requirements. (See Reference No. R236000-20 for the preliminary sizing guide.)

- A ton of air conditioning is nominally 400 CFM.

- Rectangular duct is taken off by the linear foot for each size, but its cost is usually estimated by the pound. Remember that SMACNA standards now base duct on internal pressure.

- Prefabricated duct is estimated and purchased like pipe: straight sections and fittings.

- Note that cranes or other lifting equipment are not included on any lines in Division 23. For example, if a crane is required to lift a heavy piece of pipe into place high above a gym floor, or to put a rooftop unit on the roof of a four-story building, etc., it must be added. Due to the potential for extreme variation—from nothing additional required to a major crane or helicopter—we feel that including a nominal amount for "lifting contingency" would be useless and detract from the accuracy of the estimate. When using equipment rental cost data from RSMeans, do not forget to include the cost of the operator(s).

Reference Numbers

Reference numbers are shown at the beginning of some major classifications. These numbers refer to related items in the Reference Section. The reference information may be an estimating procedure, an alternate pricing method, or technical information.

Note: Not all subdivisions listed here necessarily appear. ■

Same Data. Simplified.

Enjoy the convenience and efficiency of accessing your costs anywhere:

- **Skip the multiplier** by setting your location
- **Quickly search,** edit, favorite and share costs
- **Stay on top of price changes** with automatic updates

Discover more at rsmeans.com/online

---

23 83 Radiant Heating Units

23 83 33 – Electric Radiant Heaters

23 83 33.10 Electric Heating	Crew	Daily Output	Labor-Hours	Unit	Material	2020 Bare Costs Labor	Equipment	Total	Total Incl O&P
0010 **ELECTRIC HEATING**, not incl. conduit or feed wiring									
1100 Rule of thumb: Baseboard units, including control	1 Elec	4.05	1.980	kW	116	121		237	310
1300 Baseboard heaters, 2' long, 350 watt		7.36	1.090	Ea.	29	67		96	131
1400 3' long, 750 watt		7.36	1.090		35	67		102	138
1600 4' long, 1,000 watt		6.16	1.300		40.50	80		120.50	164
1800 5' long, 935 watt		5.24	1.530		44	94		138	189
2000 6' long, 1,500 watt		4.60	1.740		57	107		164	222
2200 7' long, 1,310 watt		4.05	1.980		61	121		182	249
2400 8' long, 2,000 watt		3.68	2.170		70	133		203	275
2600 9' long, 1,680 watt		3.31	2.420		86.50	148		234.50	315
2800 10' long, 1,875 watt		3.04	2.640		162	162		324	420
2950 Wall heaters with fan, 120 to 277 volt									
3160 Recessed, residential, 750 watt	1 Elec	5.52	1.450	Ea.	94.50	89		183.50	236
3170 1,000 watt		5.52	1.450		94.50	89		183.50	236
3180 1,250 watt		4.60	1.740		107	107		214	277
3190 1,500 watt		3.68	2.170		107	133		240	315
3210 2,000 watt		3.68	2.170		99	133		232	305
3230 2,500 watt		3.22	2.480		236	152		388	485
3240 3,000 watt		2.76	2.900		365	178		543	670
3250 4,000 watt		2.48	3.220		370	198		568	700
3260 Commercial, 750 watt		5.52	1.450		211	89		300	365
3270 1,000 watt		5.52	1.450		211	89		300	365
3280 1,250 watt		4.60	1.740		211	107		318	390
3290 1,500 watt		3.68	2.170		195	133		328	410
3300 2,000 watt		3.68	2.170		211	133		344	430
3310 2,500 watt		3.22	2.480		211	152		363	460
3320 3,000 watt		2.76	2.900		425	178		603	735
3330 4,000 watt		2.48	3.220		500	198		698	845
3600 Thermostats, integral		14.72	.543		29	33.50		62.50	81.50
3800 Line voltage, 1 pole		7.36	1.090		16.30	67		83.30	117
3810 2 pole		7.36	1.090		23.50	67		90.50	126
4000 Heat trace system, 400 degree									
4020 115 V, 2.5 watts/L.F.	1 Elec	487.60	.016	L.F.	11	1.01		12.01	13.60
4030 5 watts/L.F.		487.60	.016		11	1.01		12.01	13.60
4050 10 watts/L.F.		487.60	.016		9.20	1.01		10.21	11.60
4060 208 V, 5 watts/L.F.		487.60	.016		11	1.01		12.01	13.60
4080 480 V, 8 watts/L.F.		487.60	.016		11	1.01		12.01	13.60
4200 Heater raceway									
4260 Heat transfer cement									
4280 1 gallon				Ea.	64.50			64.50	71
4300 5 gallon				"	252			252	277
4320 Cable tie									
4340 3/4" pipe size	1 Elec	432.40	.019	Ea.	.02	1.14		1.16	1.71
4360 1" pipe size		408.48	.020		.02	1.20		1.22	1.81
4380 1-1/4" pipe size		368	.022		.02	1.33		1.35	2.01
4400 1-1/2" pipe size		326.60	.024		.02	1.50		1.52	2.26
4420 2" pipe size		294.40	.027		.05	1.67		1.72	2.54
4440 3" pipe size		147.20	.054		.05	3.33		3.38	5
4460 4" pipe size		92	.087		.06	5.35		5.41	8
4480 Thermostat NEMA 3R, 22 amp, 0-150 degree, 10' cap.		7.36	1.090		193	67		260	315
4500 Thermostat NEMA 4X, 25 amp, 40 degree, 5-1/2' cap.		6.44	1.240		248	76		324	385
4520 Thermostat NEMA 4X, 22 amp, 25-325 degree, 10' cap.		6.44	1.240		660	76		736	840
4540 Thermostat NEMA 4X, 22 amp, 15-140 degree		5.52	1.450		560	89		649	745

For customer support on your Electrical Change Order Costs with RSMeans data, call 800.448.8182. **Pre-Installation Change Order Costs**

23 83 Radiant Heating Units

23 83 33 – Electric Radiant Heaters

23 83 33.10 Electric Heating	Crew	Daily Output	Labor-Hours	Unit	Material	2020 Bare Costs Labor	Equipment	Total	Total Incl O&P
4580	Thermostat NEMA 4, 7, 9, 22 amp, 25-325 degree, 10' cap.	1 Elec	3.31	2.420	Ea.	810	148		958	1,100
4600	Thermostat NEMA 4, 7, 9, 22 amp, 15-140 degree		2.76	2.900		815	178		993	1,150
4720	Fiberglass application tape, 36 yard roll		10.12	.791		75	48.50		123.50	155
5000	Radiant heating ceiling panels, 2' x 4', 500 watt		14.72	.543		405	33.50		438.50	495
5050	750 watt		14.72	.543		430	33.50		463.50	525
5200	For recessed plaster frame, add		29.44	.272		138	16.65		154.65	177
5300	Infrared quartz heaters, 120 volts, 1,000 watt		6.16	1.300		350	80		430	505
5350	1,500 watt		4.60	1.740		335	107		442	530
5400	240 volts, 1,500 watt		4.60	1.740		405	107		512	605
5450	2,000 watt		3.68	2.170		335	133		468	570
5500	3,000 watt		2.76	2.900		330	178		508	630
5550	4,000 watt		2.39	3.340		380	205		585	725
5570	Modulating control	▼	.74	10.870	▼	141	665		806	1,150
5600	Unit heaters, heavy duty, with fan & mounting bracket									
5650	Single phase, 208-240-277 volt, 3 kW	1 Elec	5.52	1.450	Ea.	705	89		794	905
5750	5 kW		5.06	1.580		800	97		897	1,025
5800	7 kW		4.60	1.740		1,375	107		1,482	1,650
5850	10 kW		3.68	2.170		1,100	133		1,233	1,400
5950	15 kW		3.50	2.290		1,850	140		1,990	2,225
6000	480 volt, 3 kW		5.52	1.450		470	89		559	645
6020	4 kW		5.34	1.500		700	92		792	905
6040	5 kW		5.06	1.580		710	97		807	925
6060	7 kW		4.60	1.740		1,150	107		1,257	1,425
6080	10 kW		3.68	2.170		1,100	133		1,233	1,425
6100	13 kW		3.50	2.290		1,950	140		2,090	2,350
6120	15 kW		3.40	2.350		2,150	144		2,294	2,600
6140	20 kW		3.22	2.480		3,075	152		3,227	3,625
6300	3 phase, 208-240 volt, 5 kW		5.06	1.580		575	97		672	780
6320	7 kW		4.60	1.740		940	107		1,047	1,175
6340	10 kW		3.68	2.170		805	133		938	1,075
6360	15 kW		3.40	2.350		1,600	144		1,744	1,975
6380	20 kW		3.22	2.480		2,500	152		2,652	2,975
6400	25 kW		3.04	2.640		3,725	162		3,887	4,350
6500	480 volt, 5 kW		5.06	1.580		650	97		747	860
6520	7 kW		4.60	1.740		930	107		1,037	1,175
6540	10 kW		3.68	2.170		860	133		993	1,150
6560	13 kW		3.50	2.290		1,650	140		1,790	2,025
6580	15 kW		3.40	2.350		1,475	144		1,619	1,850
6600	20 kW		3.22	2.480		2,325	152		2,477	2,775
6620	25 kW		3.04	2.640		3,450	162		3,612	4,050
6630	30 kW		2.76	2.900		3,025	178		3,203	3,600
6640	40 kW		1.84	4.350		4,050	267		4,317	4,850
6650	50 kW	▼	1.47	5.430	▼	6,675	335		7,010	7,850
6800	Vertical discharge heaters, with fan									
6820	Single phase, 208-240-277 volt, 10 kW	1 Elec	3.68	2.170	Ea.	860	133		993	1,150
6840	15 kW		3.40	2.350		1,650	144		1,794	2,025
6900	3 phase, 208-240 volt, 10 kW		3.68	2.170		1,050	133		1,183	1,375
6920	15 kW		3.40	2.350		1,600	144		1,744	2,000
6940	20 kW		3.22	2.480		2,925	152		3,077	3,450
6960	25 kW		3.04	2.640		3,175	162		3,337	3,750
6980	30 kW		2.76	2.900		4,350	178		4,528	5,050
7000	40 kW		1.84	4.350		4,325	267		4,592	5,150
7020	50 kW		1.47	5.430		6,950	335		7,285	8,150

23 83 Radiant Heating Units

23 83 33 – Electric Radiant Heaters

23 83 33.10 Electric Heating		Crew	Daily Output	Labor-Hours	Unit	Material	2020 Bare Costs Labor	Equipment	Total	Total Incl O&P
7100	480 volt, 10 kW	1 Elec	3.68	2.170	Ea.	1,200	133		1,333	1,525
7120	15 kW		3.40	2.350		1,950	144		2,094	2,375
7140	20 kW		3.22	2.480		2,650	152		2,802	3,150
7160	25 kW		3.04	2.640		3,450	162		3,612	4,050
7180	30 kW		2.76	2.900		3,800	178		3,978	4,450
7200	40 kW		1.84	4.350		3,925	267		4,192	4,725
7410	Sill height convector heaters, 5" high x 2' long, 500 watt		6.16	1.300		365	80		445	525
7420	3' long, 750 watt		5.98	1.340		435	82		517	595
7430	4' long, 1,000 watt		5.70	1.400		500	86		586	680
7440	5' long, 1,250 watt		5.06	1.580		570	97		667	770
7450	6' long, 1,500 watt		4.42	1.810		645	111		756	875
7460	8' long, 2,000 watt		3.31	2.420		875	148		1,023	1,175
7470	10' long, 2,500 watt		2.76	2.900		1,100	178		1,278	1,475
7900	Cabinet convector heaters, 240 volt, three phase,									
7920	3' long, 2,000 watt	1 Elec	4.88	1.640	Ea.	2,300	101		2,401	2,675
7940	3,000 watt		4.88	1.640		2,400	101		2,501	2,800
7960	4,000 watt		4.88	1.640		2,475	101		2,576	2,850
7980	6,000 watt		4.23	1.890		2,675	116		2,791	3,125
8000	8,000 watt		4.23	1.890		2,750	116		2,866	3,200
8020	4' long, 4,000 watt		4.23	1.890		2,500	116		2,616	2,925
8040	6,000 watt		3.68	2.170		2,600	133		2,733	3,050
8060	8,000 watt		3.68	2.170		2,675	133		2,808	3,125
8080	10,000 watt		3.68	2.170		2,700	133		2,833	3,175
8100	Available also in 208 or 277 volt									
8200	Cabinet unit heaters, 120 to 277 volt, 1 pole,									
8220	wall mounted, 2 kW	1 Elec	4.23	1.890	Ea.	2,075	116		2,191	2,475
8230	3 kW		4.23	1.890		2,075	116		2,191	2,450
8240	4 kW		4.05	1.980		2,150	121		2,271	2,550
8250	5 kW		4.05	1.980		2,275	121		2,396	2,675
8260	6 kW		3.86	2.070		2,250	127		2,377	2,675
8270	8 kW		3.68	2.170		2,300	133		2,433	2,725
8280	10 kW		3.50	2.290		2,350	140		2,490	2,775
8290	12 kW		3.22	2.480		2,575	152		2,727	3,075
8300	13.5 kW		2.67	3		3,650	184		3,834	4,300
8310	16 kW		2.48	3.220		3,025	198		3,223	3,650
8320	20 kW		2.12	3.780		3,825	232		4,057	4,550
8330	24 kW		1.75	4.580		4,000	281		4,281	4,850
8350	Recessed, 2 kW		4.05	1.980		2,275	121		2,396	2,675
8370	3 kW		4.05	1.980		2,325	121		2,446	2,725
8380	4 kW		3.86	2.070		2,200	127		2,327	2,625
8390	5 kW		3.86	2.070		2,300	127		2,427	2,725
8400	6 kW		3.68	2.170		2,350	133		2,483	2,775
8410	8 kW		3.50	2.290		2,525	140		2,665	3,000
8420	10 kW		3.22	2.480		3,150	152		3,302	3,700
8430	12 kW		2.67	3		3,175	184		3,359	3,775
8440	13.5 kW		2.48	3.220		3,050	198		3,248	3,650
8450	16 kW		2.12	3.780		3,250	232		3,482	3,925
8460	20 kW		1.75	4.580		3,775	281		4,056	4,600
8470	24 kW		1.47	5.430		3,725	335		4,060	4,600
8490	Ceiling mounted, 2 kW		2.94	2.720		2,100	167		2,267	2,550
8510	3 kW		2.94	2.720		2,200	167		2,367	2,675
8520	4 kW		2.76	2.900		2,250	178		2,428	2,750
8530	5 kW		2.76	2.900		2,325	178		2,503	2,825

For customer support on your Electrical Change Order Costs with RSMeans data, call 800.448.8182.

Pre-Installation Change Order Costs

23 83 33 – Electric Radiant Heaters

23 83 33.10 Electric Heating		Crew	Daily Output	Labor-Hours	Unit	Material	2020 Bare Costs Labor	Equipment	Total	Total Incl O&P
8540	6 kW	1 Elec	2.58	3.110	Ea.	2,325	191		2,516	2,850
8550	8 kW		2.21	3.620		2,450	222		2,672	3,025
8560	10 kW		2.02	3.950		3,050	242		3,292	3,700
8570	12 kW		1.84	4.350		2,625	267		2,892	3,300
8580	13.5 kW		1.38	5.800		2,550	355		2,905	3,325
8590	16 kW		1.20	6.690		2,675	410		3,085	3,525
8600	20 kW		.83	9.660		3,700	595		4,295	4,950
8610	24 kW	↓	.55	14.490	↓	3,725	890		4,615	5,425
8630	208 to 480 V, 3 pole									
8650	Wall mounted, 2 kW	1 Elec	4.23	1.890	Ea.	2,025	116		2,141	2,425
8670	3 kW		4.23	1.890		2,375	116		2,491	2,800
8680	4 kW		4.05	1.980		2,425	121		2,546	2,850
8690	5 kW		4.05	1.980		2,500	121		2,621	2,925
8700	6 kW		3.86	2.070		2,525	127		2,652	2,975
8710	8 kW		3.68	2.170		2,625	133		2,758	3,075
8720	10 kW		3.50	2.290		2,775	140		2,915	3,250
8730	12 kW		3.22	2.480		2,825	152		2,977	3,325
8740	13.5 kW		2.67	3		2,825	184		3,009	3,375
8750	16 kW		2.48	3.220		2,925	198		3,123	3,525
8760	20 kW		2.12	3.780		4,475	232		4,707	5,275
8770	24 kW		1.75	4.580		4,200	281		4,481	5,025
8790	Recessed, 2 kW		4.05	1.980		2,250	121		2,371	2,650
8810	3 kW		4.05	1.980		2,300	121		2,421	2,725
8820	4 kW		3.86	2.070		2,400	127		2,527	2,850
8830	5 kW		3.86	2.070		2,425	127		2,552	2,875
8840	6 kW		3.68	2.170		2,350	133		2,483	2,775
8850	8 kW		3.50	2.290		2,425	140		2,565	2,875
8860	10 kW		3.22	2.480		2,450	152		2,602	2,925
8870	12 kW		2.67	3		2,750	184		2,934	3,275
8880	13.5 kW		2.48	3.220		2,375	198		2,573	2,925
8890	16 kW		2.12	3.780		5,175	232		5,407	6,050
8900	20 kW		1.75	4.580		4,300	281		4,581	5,150
8920	24 kW		1.47	5.430		4,350	335		4,685	5,300
8940	Ceiling mount, 2 kW		2.94	2.720		2,325	167		2,492	2,800
8950	3 kW		2.94	2.720		2,150	167		2,317	2,600
8960	4 kW		2.76	2.900		2,475	178		2,653	3,000
8970	5 kW		2.76	2.900		2,175	178		2,353	2,650
8980	6 kW		2.58	3.110		2,475	191		2,666	3,000
8990	8 kW		2.21	3.620		2,550	222		2,772	3,150
9000	10 kW		2.02	3.950		2,350	242		2,592	2,950
9020	13.5 kW		1.38	5.800		3,150	355		3,505	4,000
9030	16 kW		1.20	6.690		4,300	410		4,710	5,350
9040	20 kW		.83	9.660		4,300	595		4,895	5,600
9060	24 kW	↓	.55	14.490	↓	5,325	890		6,215	7,175

Division Notes

		CREW	DAILY OUTPUT	LABOR-HOURS	UNIT	BARE COSTS				TOTAL INCL O&P
						MAT.	LABOR	EQUIP.	TOTAL	

Estimating Tips

- When estimating material costs for special systems, it is always prudent to obtain manufacturers' quotations for equipment prices and special installation requirements that may affect the total cost.

- For cost modifications for elevated tray installation, add the percentages to labor according to the height of the installation and only to the quantities exceeding the different height levels, not to the total tray quantities. Refer to subdivision 26 01 02.20 for labor adjustment factors.

- Do not overlook the costs for equipment used in the installation.

Reference Numbers

Reference numbers are shown at the beginning of some major classifications. These numbers refer to related items in the Reference Section. The reference information may be an estimating procedure, an alternate pricing method, or technical information.

Same Data. Simplified.

Enjoy the convenience and efficiency of accessing your costs anywhere:

- **Skip the multiplier** by setting your location
- **Quickly search,** edit, favorite and share costs
- **Stay on top of price changes** with automatic updates

Discover more at rsmeans.com/online

Note: Trade Service, in part, has been used as a reference source for some of the material prices used in Division 25.

25 05 Common Work Results for Integrated Automation

25 05 28 – Pathways for Integrated Automation

25 05 28.39 Surface Raceways for Integrated Automation	Crew	Daily Output	Labor-Hours	Unit	Material	2020 Bare Costs Labor	Equipment	Total	Total Incl O&P
0010 **SURFACE RACEWAYS FOR INTEGRATED AUTOMATION**									
1000 Din rail, G type, 15 mm, steel, galvanized, 3', for termination blocks	1 Elec	44.16	.181	Ea.	12.45	11.10		23.55	30
1010 32 mm, steel, galvanized, 3', for termination blocks		44.16	.181		16.20	11.10		27.30	34.50
1020 35 mm, steel, galvanized, 3', for termination blocks		44.16	.181		14.55	11.10		25.65	32.50
1030 Din rail, rail mounted, 35 mm, aluminum, 3', for termination blocks		44.16	.181		20.50	11.10		31.60	39
1040 35 mm, aluminum, 6', for termination blocks		44.16	.181		25.50	11.10		36.60	45
2000 Terminal block, Din mounted, 35 mm, medium screw, connector end bracket		44.16	.181		2.19	11.10		13.29	18.95
2010 35 mm, large screw, connector end bracket		44.16	.181		3.43	11.10		14.53	20.50
2020 Large screw, connector end bracket		44.16	.181		4.88	11.10		15.98	22
3000 Terminal block, feed through, 35 mm, Din mounted, 600V, 20A		44.16	.181		1	11.10		12.10	17.65
3010 35 mm, Din mounted, 600V, 30A		44.16	.181		1.29	11.10		12.39	17.95
3020 Terminal block, fuse holder, LED indicator, Din mounted, 110V, 6A		44.16	.181		9.50	11.10		20.60	27
3030 Din mounted, 220V, 6A		44.16	.181		7.80	11.10		18.90	25
3040 Din mounted, 300V, 6A		44.16	.181		7.65	11.10		18.75	25

25 36 Integrated Automation Instrumentation & Terminal Devices

25 36 19 – Integrated Automation Current Sensors

25 36 19.10 Integrated Automation Current Sensors

	Crew	Daily Output	Labor-Hours	Unit	Material	Labor	Equipment	Total	Total Incl O&P
0010 **INTEGRATED AUTOMATION CURRENT SENSORS**									
1000 Current limiter inrush, 100 to 480 VAC, 10A, power supply	1 Elec	44.16	.181	Ea.	86.50	11.10		97.60	112

Pre-Installation Change Order Costs

Estimating Tips

26 05 00 Common Work Results for Electrical

- Conduit should be taken off in three main categories—power distribution, branch power, and branch lighting—so the estimator can concentrate on systems and components, therefore making it easier to ensure all items have been accounted for.

- For cost modifications for elevated conduit installation, add the percentages to labor according to the height of installation and only to the quantities exceeding the different height levels, not to the total conduit quantities. Refer to subdivision 26 01 02.20 for labor adjustment factors.

- Remember that aluminum wiring of equal ampacity is larger in diameter than copper and may require larger conduit.

- If more than three wires at a time are being pulled, deduct percentages from the labor hours of that grouping of wires.

- When taking off grounding systems, identify separately the type and size of wire, and list each unique type of ground connection.

- The estimator should take the weights of materials into consideration when completing a takeoff. Topics to consider include: How will the materials be supported? What methods of support are available? How high will the support structure have to reach? Will the final support structure be able to withstand the total burden? Is the support material included or separate from the fixture, equipment, and material specified?

- Do not overlook the costs for equipment used in the installation. If scaffolding or highlifts are available in the field, contractors may use them in lieu of the proposed ladders and rolling staging.

26 20 00 Low-Voltage Electrical Transmission

- Supports and concrete pads may be shown on drawings for the larger equipment, or the support system may be only a piece of plywood for the back of a panelboard. In either case, they must be included in the costs.

26 40 00 Electrical and Cathodic Protection

- When taking off cathodic protection systems, identify the type and size of cable, and list each unique type of anode connection.

26 50 00 Lighting

- Fixtures should be taken off room by room using the fixture schedule, specifications, and the ceiling plan. For large concentrations of lighting fixtures in the same area, deduct the percentages from labor hours.

Reference Numbers

Reference numbers are shown at the beginning of some major classifications. These numbers refer to related items in the Reference Section. The reference information may be an estimating procedure, an alternate pricing method, or technical information.

Note: Not all subdivisions listed here necessarily appear. ∎

Same Data. Simplified.

Enjoy the convenience and efficiency of accessing your costs anywhere:

- **Skip the multiplier** by setting your location
- **Quickly search,** edit, favorite and share costs
- **Stay on top of price changes** with automatic updates

Discover more at rsmeans.com/online

---

26 01 02.20 Labor Adjustment Factors	Crew	Daily Output	Labor-Hours	Unit	Material	2020 Bare Costs Labor	Equipment	Total	Total Incl O&P
0010 **LABOR ADJUSTMENT FACTORS** (For Div. 26, 27 and 28)									
0100 Subtract from labor for Economy of Scale for Wire									
0110 4-5 wires						25%			
0120 6-10 wires						30%			
0130 11-15 wires						35%			
0140 over 15 wires						40%			
0150 Labor adjustment factors (For Div. 26, 27, 28 and 48)									
0200 Labor factors: The below are reasonable suggestions, but									
0210 each project must be evaluated for its own peculiarities, and									
0220 the adjustments be increased or decreased depending on the									
0230 severity of the special conditions.									
1000 Add to labor for elevated installation (above floor level)									
1010 10' to 14.5' high						10%			
1020 15' to 19.5' high						20%			
1030 20' to 24.5' high						25%			
1040 25' to 29.5' high						35%			
1050 30' to 34.5' high						40%			
1060 35' to 39.5' high						50%			
1070 40' and higher						55%			
2000 Add to labor for crawl space									
2010 3' high						40%			
2020 4' high						30%			
3000 Add to labor for multi-story building									
3100 For new construction (No elevator available)									
3110 Add for floors 3 thru 10						5%			
3120 Add for floors 11 thru 15						10%			
3130 Add for floors 16 thru 20						15%			
3140 Add for floors 21 thru 30						20%			
3150 Add for floors 31 and up						30%			
3200 For existing structure (Elevator available)									
3210 Add for work on floor 3 and above						2%			
4000 Add to labor for working in existing occupied buildings									
4010 Hospital						35%			
4020 Office building						25%			
4030 School						20%			
4040 Factory or warehouse						15%			
4050 Multi-dwelling						15%			
5000 Add to labor, miscellaneous									
5010 Cramped shaft						35%			
5020 Congested area						15%			
5030 Excessive heat or cold						30%			
6000 Labor factors: the above are reasonable suggestions, but									
6100 each project should be evaluated for its own peculiarities									
6200 Other factors to be considered are:									
6210 Movement of material and equipment through finished areas						10%			
6220 Equipment room min security direct access w/authorization						15%			
6230 Attic space						25%			
6240 No service road						25%			
6250 Poor unloading/storage area, no hydraulic lifts or jacks						20%			
6260 Congested site area/heavy traffic						20%			
7000 Correctional facilities (no compounding division 1 adjustment factors)									
7010 Minimum security w/facilities escort						30%			
7020 Medium security w/facilities and correctional officer escort						40%			

For customer support on your Electrical Change Order Costs with RSMeans data, call 800.448.8182.

Pre-Installation Change Order Costs

26 01 Operation and Maintenance of Electrical Systems

26 01 02 – Labor Adjustment

26 01 02.20 Labor Adjustment Factors	Crew	Daily Output	Labor-Hours	Unit	Material	2020 Bare Costs Labor	Equipment	Total	Total Incl O&P
7030	Max security w/facilities & correctional officer escort (no inmate contact)						50%			

26 05 Common Work Results for Electrical

26 05 13 – Medium-Voltage Cables

26 05 13.10 Cable Terminations

		Crew	Daily Output	Labor-Hours	Unit	Material	2020 Bare Costs Labor	Equipment	Total	Total Incl O&P
0010	**CABLE TERMINATIONS**, 5 kV to 35 kV									
0100	Indoor, insulation diameter range 0.64" to 1.08"									
0300	Padmount, 5 kV	1 Elec	7.36	1.090	Ea.	87.50	67		154.50	196
0400	10 kV		5.89	1.360		109	83.50		192.50	243
0500	15 kV		5.52	1.450		142	89		231	288
0600	25 kV		5.15	1.550		210	95		305	375
0700	Insulation diameter range 1.05" to 1.8"									
0800	Padmount, 5 kV	1 Elec	7.36	1.090	Ea.	36.50	67		103.50	140
0900	10 kV		5.52	1.450		163	89		252	310
1000	15 kV		5.15	1.550		169	95		264	330
1100	25 kV		4.88	1.640		242	101		343	415
1200	Insulation diameter range 1.53" to 2.32"									
1300	Padmount, 5 kV	1 Elec	6.81	1.180	Ea.	152	72.50		224.50	275
1400	10 kV		5.15	1.550		175	95		270	335
1500	15 kV		4.88	1.640		194	101		295	365
1600	25 kV		4.60	1.740		315	107		422	510
1700	Outdoor systems, #4 stranded to 1/0 stranded									
1800	5 kV	1 Elec	6.81	1.180	Ea.	108	72.50		180.50	226
1900	15 kV		4.88	1.640		120	101		221	282
2000	25 kV		4.60	1.740		178	107		285	355
2100	35 kV		4.42	1.810		325	111		436	525
2200	#1 solid to 4/0 stranded, 5 kV		6.35	1.260		146	77.50		223.50	275
2300	15 kV		4.60	1.740		184	107		291	360
2400	25 kV		4.42	1.810		254	111		365	445
2500	35 kV		4.23	1.890		355	116		471	565
2600	2/0 solid to 350 kcmil stranded, 5 kV		5.89	1.360		193	83.50		276.50	335
2700	15 kV		4.42	1.810		210	111		321	395
2800	25 kV		4.23	1.890		294	116		410	500
2900	35 kV		4.05	1.980		405	121		526	625
3000	400 kcmil compact to 750 kcmil stranded, 5 kV		5.52	1.450		175	89		264	325
3100	15 kV		4.23	1.890		231	116		347	425
3200	25 kV		4.05	1.980		335	121		456	550
3300	35 kV		3.86	2.070		340	127		467	560
3400	1,000 kcmil, 5 kV		5.15	1.550		250	95		345	415
3500	15 kV		4.05	1.980		390	121		511	605
3600	25 kV		3.86	2.070		340	127		467	560
3700	35 kV		3.68	2.170		470	133		603	720

26 05 13.16 Medium-Voltage, Single Cable

		Crew	Daily Output	Labor-Hours	Unit	Material	2020 Bare Costs Labor	Equipment	Total	Total Incl O&P
0010	**MEDIUM-VOLTAGE, SINGLE CABLE** Splicing & terminations not included									
0040	Copper, XLP shielding, 5 kV, #6	2 Elec	4.05	3.950	C.L.F.	205	242		447	585
0050	#4		4.05	3.950		266	242		508	655
0100	#2		3.68	4.350		305	267		572	730
0200	#1		3.68	4.350		405	267		672	845
0400	1/0		3.50	4.580		400	281		681	860
0600	2/0		3.31	4.830		490	296		786	980
0800	4/0		2.94	5.430		655	335		990	1,225

26 05 13 – Medium-Voltage Cables

26 05 13.16 Medium-Voltage, Single Cable	Crew	Daily Output	Labor-Hours	Unit	Material	2020 Bare Costs Labor	Equipment	Total	Total Incl O&P
1000	250 kcmil	3 Elec	4.14	5.800	C.L.F.	785	355		1,140	1,400
1200	350 kcmil		3.59	6.690		1,025	410		1,435	1,725
1400	500 kcmil		3.31	7.250		1,300	445		1,745	2,100
1600	15 kV, ungrounded neutral, #1	2 Elec	3.68	4.350		440	267		707	880
1800	1/0		3.50	4.580		525	281		806	1,000
2000	2/0		3.31	4.830		600	296		896	1,100
2200	4/0		2.94	5.430		795	335		1,130	1,375
2400	250 kcmil	3 Elec	4.14	5.800		880	355		1,235	1,500
2600	350 kcmil		3.59	6.690		1,100	410		1,510	1,825
2800	500 kcmil		3.31	7.250		1,250	445		1,695	2,025
3000	25 kV, grounded neutral, 1/0	2 Elec	3.31	4.830		730	296		1,026	1,250
3200	2/0		3.13	5.120		800	315		1,115	1,350
3400	4/0		2.76	5.800		1,000	355		1,355	1,625
3600	250 kcmil	3 Elec	3.86	6.210		1,250	380		1,630	1,950
3800	350 kcmil		3.31	7.250		1,475	445		1,920	2,275
3900	500 kcmil		3.04	7.910		1,725	485		2,210	2,625
4000	35 kV, grounded neutral, 1/0	2 Elec	3.13	5.120		715	315		1,030	1,250
4200	2/0		2.94	5.430		905	335		1,240	1,500
4400	4/0		2.58	6.210		1,150	380		1,530	1,825
4600	250 kcmil	3 Elec	3.59	6.690		1,325	410		1,735	2,075
4800	350 kcmil		3.04	7.910		1,600	485		2,085	2,500
5000	500 kcmil		2.76	8.700		1,900	535		2,435	2,875
5050	Aluminum, XLP shielding, 5 kV, #2	2 Elec	4.60	3.480		291	214		505	640
5070	#1		4.05	3.950		300	242		542	690
5090	1/0		3.68	4.350		350	267		617	780
5100	2/0		3.50	4.580		390	281		671	850
5150	4/0		3.31	4.830		460	296		756	950
5200	250 kcmil	3 Elec	4.42	5.430		560	335		895	1,100
5220	350 kcmil		4.14	5.800		660	355		1,015	1,250
5240	500 kcmil		3.59	6.690		845	410		1,255	1,550
5260	750 kcmil		3.31	7.250		1,125	445		1,570	1,875
5300	15 kV aluminum, XLP, #1	2 Elec	4.05	3.950		375	242		617	770
5320	1/0		3.68	4.350		385	267		652	820
5340	2/0		3.50	4.580		415	281		696	880
5360	4/0		3.31	4.830		510	296		806	1,000
5380	250 kcmil	3 Elec	4.42	5.430		610	335		945	1,175
5400	350 kcmil		4.14	5.800		685	355		1,040	1,275
5420	500 kcmil		3.59	6.690		945	410		1,355	1,650
5440	750 kcmil		3.31	7.250		1,250	445		1,695	2,025

26 05 19 – Low-Voltage Electrical Power Conductors and Cables

26 05 19.13 Undercarpet Electrical Power Cables

		Crew	Daily Output	Labor-Hours	Unit	Material	Labor	Equipment	Total	Total Incl O&P
0010	**UNDERCARPET ELECTRICAL POWER CABLES**									
0020	Power System									
0100	Cable flat, 3 conductor, #12, w/attached bottom shield	1 Elec	903.44	.009	L.F.	5.90	.54		6.44	7.25
0200	Shield, top, steel		1626.56	.005	"	5.30	.30		5.60	6.30
0250	Splice, 3 conductor		44.16	.181	Ea.	16.10	11.10		27.20	34.50
0300	Top shield		88.32	.091		1.49	5.55		7.04	9.90
0350	Tap		36.80	.217		20.50	13.35		33.85	42.50
0400	Insulating patch, splice, tap & end		44.16	.181		51.50	11.10		62.60	73.50
0450	Fold		211.60	.038			2.32		2.32	3.45
0500	Top shield, tap & fold		88.32	.091		1.49	5.55		7.04	9.90
0700	Transition, block assembly		70.84	.113		75.50	6.95		82.45	93.50

26 05 19.13 Undercarpet Electrical Power Cables	Crew	Daily Output	Labor-Hours	Unit	Material	2020 Bare Costs Labor	Equipment	Total	Total Incl O&P
0750	Receptacle frame & base	1 Elec	29.44	.272	Ea.	41.50	16.65		58.15	70.50
0800	Cover receptacle		110.40	.072		3.49	4.45		7.94	10.45
0850	Cover blank		147.20	.054		4.11	3.33		7.44	9.50
0860	Receptacle, direct connected, single		23	.348		88.50	21.50		110	130
0870	Dual		14.72	.543		145	33.50		178.50	210
0880	Combination high & low, tension		19.32	.414		107	25.50		132.50	156
0900	Box, floor with cover		18.40	.435		89.50	26.50		116	138
0920	Floor service w/barrier		3.68	2.170		254	133		387	475
1000	Wall, surface, with cover		18.40	.435		59	26.50		85.50	105
1100	Wall, flush, with cover		18.40	.435	▼	41.50	26.50		68	85
1450	Cable flat, 5 conductor #12, w/attached bottom shield		736	.011	L.F.	9.65	.67		10.32	11.65
1550	Shield, top, steel		1626.56	.005	"	9.60	.30		9.90	11
1600	Splice, 5 conductor		44.16	.181	Ea.	26	11.10		37.10	45
1650	Top shield		88.32	.091		1.49	5.55		7.04	9.90
1700	Tap		44.16	.181		34.50	11.10		45.60	54.50
1750	Insulating patch, splice tap & end		76.36	.105		60.50	6.45		66.95	76
1800	Transition, block assembly		70.84	.113		54	6.95		60.95	70
1850	Box, wall, flush with cover		18.40	.435	▼	55	26.50		81.50	100
1900	Cable flat, 4 conductor #12		858.36	.009	L.F.	6.80	.57		7.37	8.30
1950	3 conductor #10		903.44	.009		6.55	.54		7.09	8
1960	4 conductor #10		858.36	.009		8.60	.57		9.17	10.30
1970	5 conductor #10	▼	813.28	.010	▼	10.45	.60		11.05	12.40
2500	Telephone System									
2510	Transition fitting wall box, surface	1 Elec	22.08	.362	Ea.	66.50	22		88.50	106
2520	Flush		22.08	.362		66.50	22		88.50	106
2530	Flush, for PC board		22.08	.362		66.50	22		88.50	106
2540	Floor service box	▼	3.68	2.170		262	133		395	485
2550	Cover, surface					24			24	26.50
2560	Flush					24			24	26.50
2570	Flush for PC board					24			24	26.50
2700	Floor fitting w/duplex jack & cover	1 Elec	19.32	.414		60.50	25.50		86	105
2720	Low profile		48.76	.164		19.95	10.05		30	37
2740	Miniature w/duplex jack		48.76	.164		37.50	10.05		47.55	56.50
2760	25 pair kit		19.32	.414		56	25.50		81.50	99.50
2780	Low profile		48.76	.164		20.50	10.05		30.55	37.50
2800	Call director kit for 5 cable		17.48	.458		104	28		132	157
2820	4 pair kit		17.48	.458		106	28		134	159
2840	3 pair kit		17.48	.458		132	28		160	187
2860	Comb. 25 pair & 3 conductor power		19.32	.414		104	25.50		129.50	153
2880	5 conductor power		19.32	.414		102	25.50		127.50	150
2900	PC board, 8 per 3 pair		148.12	.054		77.50	3.31		80.81	90
2920	6 per 4 pair		148.12	.054		77.50	3.31		80.81	90
2940	3 pair adapter		148.12	.054		70.50	3.31		73.81	82.50
2950	Plug		70.84	.113		3.53	6.95		10.48	14.20
2960	Couplers		295.32	.027	▼	9.95	1.66		11.61	13.40
3000	Bottom shield for 25 pair cable		4066.40	.002	L.F.	.95	.12		1.07	1.22
3020	4 pair		4066.40	.002		.45	.12		.57	.68
3040	Top shield for 25 pair cable		4066.40	.002		.95	.12		1.07	1.22
3100	Cable assembly, double-end, 50', 25 pair		10.86	.737	Ea.	273	45		318	370
3110	3 pair		21.71	.368		86.50	22.50		109	129
3120	4 pair		21.71	.368	▼	104	22.50		126.50	149
3140	Bulk 3 pair		1355.16	.006	L.F.	1.47	.36		1.83	2.16
3160	4 pair		1355.16	.006	"	1.83	.36		2.19	2.55

26 05 19.13 Undercarpet Electrical Power Cables

		Crew	Daily Output	Labor-Hours	Unit	Material	2020 Bare Costs Labor	2020 Bare Costs Equipment	Total	Total Incl O&P
3500	Data System									
3520	Cable 25 conductor w/connection 40', 75 ohm	1 Elec	13.34	.600	Ea.	94	37		131	159
3530	Single lead		20.24	.395		257	24.50		281.50	320
3540	Dual lead		20.24	.395		273	24.50		297.50	335
3560	Shields same for 25 conductor as 25 pair telephone									
3570	Single & dual, none required									
3590	BNC coax connectors, plug	1 Elec	36.80	.217	Ea.	16.80	13.35		30.15	38.50
3600	TNC coax connectors, plug	"	36.80	.217	"	16.90	13.35		30.25	38.50
3700	Cable-bulk									
3710	Single lead	1 Elec	1355.16	.006	L.F.	4.07	.36		4.43	5
3720	Dual lead	"	1355.16	.006	"	5.90	.36		6.26	7.05
3730	Hand tool crimp				Ea.	570			570	625
3740	Hand tool notch				"	22.50			22.50	25
3750	Boxes & floor fitting same as telephone									
3790	Data cable notching, 90°	1 Elec	89.24	.090	Ea.		5.50		5.50	8.20
3800	180°		55.20	.145			8.90		8.90	13.25
8100	Drill floor		147.20	.054		4.86	3.33		8.19	10.30
8200	Marking floor		1472	.005	L.F.		.33		.33	.50
8300	Tape, hold down		5888	.001	"	.20	.08		.28	.34
8350	Tape primer, 500' per can		88.32	.091	Ea.	54	5.55		59.55	68
8400	Tool, splicing				"	184			184	202

26 05 19.20 Armored Cable

		Crew	Daily Output	Labor-Hours	Unit	Material	2020 Bare Costs Labor	2020 Bare Costs Equipment	Total	Total Incl O&P
0010	**ARMORED CABLE**									
0050	600 volt, copper (BX), #14, 2 conductor, solid	1 Elec	2.21	3.620	C.L.F.	49	222		271	385
0100	3 conductor, solid		2.02	3.950		82	242		324	450
0120	4 conductor, solid		1.84	4.350		116	267		383	525
0150	#12, 2 conductor, solid		2.12	3.780		51.50	232		283.50	400
0200	3 conductor, solid		1.84	4.350		86	267		353	490
0220	4 conductor, solid		1.66	4.830		119	296		415	570
0250	#10, 2 conductor, solid		1.84	4.350		102	267		369	505
0300	3 conductor, solid		1.47	5.430		136	335		471	645
0320	4 conductor, solid		1.29	6.210		210	380		590	795
0340	#8, 2 conductor, stranded		1.38	5.800		320	355		675	885
0350	3 conductor, stranded		1.20	6.690		320	410		730	965
0370	4 conductor, stranded		1.01	7.910		460	485		945	1,225
0380	#6, 2 conductor, stranded		1.20	6.690		315	410		725	955
0400	3 conductor with PVC jacket, in cable tray, #6		2.85	2.810		665	172		837	985
0450	#4	2 Elec	4.97	3.220		815	198		1,013	1,200
0500	#2		4.23	3.780		1,000	232		1,232	1,450
0550	#1		3.68	4.350		1,275	267		1,542	1,800
0600	1/0		3.31	4.830		1,375	296		1,671	1,950
0650	2/0		3.13	5.120		1,400	315		1,715	2,025
0700	3/0		2.94	5.430		2,175	335		2,510	2,900
0750	4/0		2.76	5.800		2,625	355		2,980	3,425
0800	250 kcmil	3 Elec	3.31	7.250		3,300	445		3,745	4,275
0850	350 kcmil		3.04	7.910		4,100	485		4,585	5,250
0900	500 kcmil		2.76	8.700		5,675	535		6,210	7,050
0910	4 conductor with PVC jacket, in cable tray, #6	1 Elec	2.48	3.220		1,725	198		1,923	2,200
0920	#4	2 Elec	4.23	3.780		1,875	232		2,107	2,425
0930	#2		3.68	4.350		1,500	267		1,767	2,050
0940	#1		3.31	4.830		1,900	296		2,196	2,550
0950	1/0		3.13	5.120		1,775	315		2,090	2,425

26 05 19 – Low-Voltage Electrical Power Conductors and Cables

26 05 19.20 Armored Cable		Crew	Daily Output	Labor-Hours	Unit	Material	2020 Bare Costs Labor	Equipment	Total	Total Incl O&P
0960	2/0	2 Elec	2.94	5.430	C.L.F.	2,175	335		2,510	2,900
0970	3/0		2.76	5.800		2,875	355		3,230	3,700
0980	4/0		2.21	7.250		2,625	445		3,070	3,525
0990	250 kcmil	3 Elec	3.04	7.910		4,000	485		4,485	5,125
1000	350 kcmil		2.76	8.700		5,375	535		5,910	6,700
1010	500 kcmil		2.48	9.660		7,450	595		8,045	9,075
1050	5 kV, copper, 3 conductor with PVC jacket,									
1060	non-shielded, in cable tray, #4	2 Elec	349.60	.046	L.F.	9.20	2.81		12.01	14.30
1100	#2		331.20	.048		11.95	2.96		14.91	17.55
1200	#1		276	.058		15.20	3.56		18.76	22
1400	1/0		266.80	.060		17.55	3.68		21.23	25
1600	2/0		239.20	.067		20.50	4.10		24.60	28.50
2000	4/0		220.80	.072		27	4.45		31.45	36
2100	250 kcmil	3 Elec	303.60	.079		37	4.85		41.85	48
2150	350 kcmil		289.80	.083		45.50	5.10		50.60	57.50
2200	500 kcmil		248.40	.097		67.50	5.95		73.45	83
2400	15 kV, copper, 3 conductor with PVC jacket galv., steel armored									
2500	grounded neutral, in cable tray, #2	2 Elec	276	.058	L.F.	20	3.56		23.56	27.50
2600	#1		257.60	.062		20.50	3.81		24.31	28
2800	1/0		239.20	.067		24	4.10		28.10	32.50
2900	2/0		202.40	.079		31	4.85		35.85	41.50
3000	4/0		174.80	.092		35.50	5.60		41.10	47.50
3100	250 kcmil	3 Elec	248.40	.097		39.50	5.95		45.45	52.50
3150	350 kcmil		220.80	.109		46.50	6.65		53.15	61
3200	500 kcmil		193.20	.124		62.50	7.60		70.10	80
3400	15 kV, copper, 3 conductor with PVC jacket,									
3450	ungrounded neutral, in cable tray, #2	2 Elec	239.20	.067	L.F.	21	4.10		25.10	29
3500	#1		211.60	.076		23	4.64		27.64	32.50
3600	1/0		184	.087		26.50	5.35		31.85	37
3700	2/0		174.80	.092		32.50	5.60		38.10	44
3800	4/0		147.20	.109		39	6.65		45.65	53
4000	250 kcmil	3 Elec	193.20	.124		45.50	7.60		53.10	61.50
4050	350 kcmil		179.40	.134		60	8.20		68.20	78
4100	500 kcmil		165.60	.145		73	8.90		81.90	93.50
4200	600 volt, aluminum, 3 conductor in cable tray with PVC jacket									
4300	#2	2 Elec	496.80	.032	L.F.	4.96	1.98		6.94	8.40
4400	#1		423.20	.038		5.90	2.32		8.22	9.90
4500	1/0		368	.043		7.35	2.67		10.02	12
4600	2/0		331.20	.048		7.40	2.96		10.36	12.55
4700	3/0		312.80	.051		8.70	3.14		11.84	14.20
4800	4/0		294.40	.054		10.55	3.33		13.88	16.55
4900	250 kcmil	3 Elec	414	.058		12.65	3.56		16.21	19.20
5000	350 kcmil		331.20	.072		15.10	4.45		19.55	23
5200	500 kcmil		303.60	.079		18.75	4.85		23.60	27.50
5300	750 kcmil		262.20	.092		24	5.60		29.60	35
5400	600 volt, aluminum, 4 conductor in cable tray with PVC jacket									
5410	#2	2 Elec	478.40	.033	L.F.	5.90	2.05		7.95	9.50
5430	#1		404.80	.040		7.20	2.43		9.63	11.50
5450	1/0		349.60	.046		8.40	2.81		11.21	13.45
5470	2/0		312.80	.051		8.60	3.14		11.74	14.10
5480	3/0		294.40	.054		10.10	3.33		13.43	16.05
5500	4/0		276	.058		11.85	3.56		15.41	18.30
5520	250 kcmil	3 Elec	386.40	.062		12.70	3.81		16.51	19.65

26 05 19.20 Armored Cable	Crew	Daily Output	Labor-Hours	Unit	Material	2020 Bare Costs Labor	Equipment	Total	Total Incl O&P
5540	350 kcmil	3 Elec	303.60	.079	L.F.	16.40	4.85		21.25	25.50
5560	500 kcmil		276	.087		20.50	5.35		25.85	30.50
5580	750 kcmil		248.40	.097		28.50	5.95		34.45	40
5600	5 kV, aluminum, unshielded in cable tray, #2 with PVC jacket	2 Elec	349.60	.046		7.40	2.81		10.21	12.35
5700	#1 with PVC jacket		331.20	.048		8.30	2.96		11.26	13.55
5800	1/0 with PVC jacket		276	.058		8.50	3.56		12.06	14.65
6000	2/0 with PVC jacket		266.80	.060		8.70	3.68		12.38	15.05
6200	3/0 with PVC jacket		239.20	.067		10.45	4.10		14.55	17.60
6300	4/0 with PVC jacket		220.80	.072		12.25	4.45		16.70	20
6400	250 kcmil with PVC jacket	3 Elec	303.60	.079		13.45	4.85		18.30	22
6500	350 kcmil with PVC jacket		289.80	.083		15.75	5.10		20.85	25
6600	500 kcmil with PVC jacket		276	.087		18.65	5.35		24	28.50
6800	750 kcmil with PVC jacket		248.40	.097		23	5.95		28.95	34.50
6900	15 kV, aluminum, shielded-grounded, #2 with PVC jacket	2 Elec	294.40	.054		16.90	3.33		20.23	23.50
7000	#1 with PVC jacket		276	.058		17.40	3.56		20.96	24.50
7200	1/0 with PVC jacket		257.60	.062		18.75	3.81		22.56	26
7300	2/0 with PVC jacket		239.20	.067		19	4.10		23.10	27
7400	3/0 with PVC jacket		220.80	.072		21.50	4.45		25.95	30
7500	4/0 with PVC jacket		202.40	.079		22	4.85		26.85	31
7600	250 kcmil with PVC jacket	3 Elec	276	.087		24	5.35		29.35	34.50
7700	350 kcmil with PVC jacket		248.40	.097		28	5.95		33.95	40
7800	500 kcmil with PVC jacket		220.80	.109		34	6.65		40.65	47
8000	750 kcmil with PVC jacket		187.68	.128		40	7.85		47.85	55.50
8200	15 kV, aluminum, shielded-ungrounded, #1 with PVC jacket	2 Elec	230	.070		21	4.27		25.27	30
8300	1/0 with PVC jacket		211.60	.076		22	4.64		26.64	31
8400	2/0 with PVC jacket		193.20	.083		24	5.10		29.10	34
8500	3/0 with PVC jacket		184	.087		24.50	5.35		29.85	35
8600	4/0 with PVC jacket		174.80	.092		26.50	5.60		32.10	37.50
8700	250 kcmil with PVC jacket	3 Elec	248.40	.097		28.50	5.95		34.45	40.50
8800	350 kcmil with PVC jacket		220.80	.109		32.50	6.65		39.15	46
8900	500 kcmil with PVC jacket		193.20	.124		39.50	7.60		47.10	54.50
8950	750 kcmil with PVC jacket		160.08	.150		49	9.20		58.20	67.50
9010	600 volt, copper (MC) steel clad, #14, 2 wire	R-1A	7.51	2.130	C.L.F.	52	118		170	232
9020	3 wire		6.63	2.410		96.50	133		229.50	305
9030	4 wire		6.15	2.600		250	144		394	490
9040	#12, 2 wire		6.63	2.410		71.50	133		204.50	277
9050	3 wire		6.15	2.600		84.50	144		228.50	305
9060	4 wire		5.56	2.880		145	159		304	395
9070	#10, 2 wire		5.56	2.880		129	159		288	380
9080	3 wire		5.20	3.080		165	170		335	435
9090	4 wire		4.81	3.330		260	184		444	560
9100	#8, 2 wire, stranded		3.62	4.410		201	244		445	585
9110	3 wire, stranded		3.13	5.120		231	283		514	675
9120	4 wire, stranded		2.58	6.210		380	345		725	925
9130	#6, 2 wire, stranded		3.17	5.040		279	278		557	720
9200	600 volt, copper (MC) aluminum clad, #14, 2 wire		7.51	2.130		51	118		169	231
9210	3 wire		6.63	2.410		93	133		226	300
9220	4 wire		6.15	2.600		115	144		259	340
9230	#12, 2 wire		6.63	2.410		52	133		185	256
9240	3 wire		6.15	2.600		90	144		234	315
9250	4 wire		5.56	2.880		185	159		344	440
9260	#10, 2 wire		5.56	2.880		111	159		270	360
9270	3 wire		5.20	3.080		143	170		313	410

26 05 Common Work Results for Electrical

26 05 19 – Low-Voltage Electrical Power Conductors and Cables

26 05 19.20 Armored Cable	Crew	Daily Output	Labor-Hours	Unit	Material	2020 Bare Costs Labor	Equipment	Total	Total Incl O&P
9280	4 wire	R-1A	4.81	3.330	C.L.F.	271	184		455	570
9600	Alum (MC) aluminum clad, #6, 3 conductor w/#6 grnd		2.53	6.320		292	350		642	840
9610	4 conductor w/#6 grnd		2.11	7.590		261	420		681	910
9620	#4, 3 conductor w/#6 grnd	R-1B	2.88	8.330		249	445		694	935
9630	4 conductor w/#6 grnd		2.46	9.770		269	520		789	1,075
9640	#2, 3 conductor w/#4 grnd		2.38	10.070		445	535		980	1,275
9650	4 conductor w/#4 grnd		2.03	11.800		500	630		1,130	1,475
9660	#1, 3 conductor w/#4 grnd		1.98	12.130		370	645		1,015	1,375
9670	4 conductor w/#4 grnd		1.70	14.100		680	750		1,430	1,875
9680	1/0, 3 conductor w/#4 grnd		1.99	12.080		420	640		1,060	1,425
9690	4 conductor w/#4 grnd		1.59	15.080		510	800		1,310	1,775
9700	2/0, 3 conductor w/#4 grnd		1.76	13.660		690	725		1,415	1,825
9710	4 conductor w/#4 grnd		1.41	17.050		490	905		1,395	1,900
9720	3/0, 3 conductor w/#4 grnd		1.61	14.910		885	795		1,680	2,150
9730	4 conductor w/#4 grnd		1.29	18.630		610	990		1,600	2,150
9740	4/0, 3 conductor w/#2 grnd		1.42	16.940		990	900		1,890	2,450
9750	4 conductor w/#2 grnd		1.13	21.210		430	1,125		1,555	2,150
9760	250 kcmil, 3 conductor w/#1 grnd	R-1C	1.75	18.310		1,125	1,000	58	2,183	2,800
9770	4 conductor w/#1 grnd		1.40	22.880		1,525	1,275	73	2,873	3,625
9780	350 kcmil, 3 conductor w/1/0 grnd		1.62	19.760		1,500	1,100	63	2,663	3,350
9790	4 conductor w/1/0 grnd		1.30	24.670		1,150	1,375	78.50	2,603.50	3,375
9800	500 kcmil, 3 conductor w/#1 grnd		1.51	21.210		1,900	1,175	67.50	3,142.50	3,925
9810	4 conductor w/2/0 grnd		1.21	26.550		1,575	1,475	84.50	3,134.50	4,000
9840	750 kcmil, 3 conductor w/1/0 grnd		1.21	26.550		3,025	1,475	84.50	4,584.50	5,600
9850	4 conductor w/3/0 grnd		.96	33.440		3,125	1,850	106	5,081	6,325

26 05 19.25 Cable Connectors

		Crew	Daily Output	Labor-Hours	Unit	Material	2020 Bare Costs Labor	Equipment	Total	Total Incl O&P
0010	**CABLE CONNECTORS**									
0100	600 volt, nonmetallic, #14-2 wire	1 Elec	147.20	.054	Ea.	2.14	3.33		5.47	7.30
0200	#14-3 wire to #12-2 wire		122.36	.065		2.14	4.01		6.15	8.30
0300	#12-3 wire to #10-2 wire		104.88	.076		2.14	4.68		6.82	9.30
0400	#10-3 wire to #14-4 and #12-4 wire		92	.087		2.14	5.35		7.49	10.30
0500	#8-3 wire to #10-4 wire		73.60	.109		4.05	6.65		10.70	14.40
0600	#6-3 wire		36.80	.217		5.20	13.35		18.55	25.50
0800	SER, 3 #8 insulated + 1 #8 ground		29.44	.272		3.94	16.65		20.59	29.50
0900	3 #6 + 1 #6 ground		22.08	.362		4.92	22		26.92	38.50
1000	3 #4 + 1 #6 ground		20.24	.395		6.15	24.50		30.65	43
1100	3 #2 + 1 #4 ground		18.40	.435		13.20	26.50		39.70	54
1200	3 1/0 + 1 #2 ground		16.56	.483		26.50	29.50		56	73
1400	3 2/0 + 1 #1 ground		14.72	.543		37	33.50		70.50	90
1600	3 4/0 + 1 #2/0 ground		12.88	.621		36.50	38		74.50	97
1800	600 volt, armored, #14-2 wire		73.60	.109		1.58	6.65		8.23	11.70
2200	#14-4, #12-3 and #10-2 wire		36.80	.217		.99	13.35		14.34	21
2400	#12-4, #10-3 and #8-2 wire		29.44	.272		4.35	16.65		21	30
2600	#8-3 and #10-4 wire		23.92	.334		5.15	20.50		25.65	36
2650	#8-4 wire		20.24	.395		7.65	24.50		32.15	44.50
2700	PVC jacket connector, #6-3 wire, #6-4 wire		14.72	.543		7.50	33.50		41	57.50
2800	#4-3 wire, #4-4 wire		14.72	.543		7.50	33.50		41	57.50
2900	#2-3 wire		11.04	.725		7.50	44.50		52	74
3000	#1-3 wire, #2-4 wire		11.04	.725		16.95	44.50		61.45	84.50
3200	1/0-3 wire		10.12	.791		16.95	48.50		65.45	90.50
3400	2/0-3 wire, 1/0-4 wire		9.20	.870		16.95	53.50		70.45	98
3500	3/0-3 wire, 2/0-4 wire		8.28	.966		22	59.50		81.50	113

26 05 19.25 Cable Connectors

		Crew	Daily Output	Labor-Hours	Unit	Material	2020 Bare Costs Labor	Equipment	Total	Total Incl O&P
3600	4/0-3 wire, 3/0-4 wire	1 Elec	6.44	1.240	Ea.	22	76		98	138
3800	250 kcmil-3 wire, 4/0-4 wire		5.52	1.450		43	89		132	179
4000	350 kcmil-3 wire, 250 kcmil-4 wire		4.60	1.740		43	107		150	206
4100	350 kcmil-4 wire		3.68	2.170		248	133		381	470
4200	500 kcmil-3 wire		3.68	2.170		248	133		381	470
4250	500 kcmil-4 wire, 750 kcmil-3 wire		3.22	2.480		274	152		426	525
4300	750 kcmil-4 wire		2.76	2.900		274	178		452	565
4400	5 kV, armored, #4		7.36	1.090		83.50	67		150.50	192
4600	#2		7.36	1.090		83.50	67		150.50	192
4800	#1		7.36	1.090		109	67		176	220
5000	1/0		5.89	1.360		135	83.50		218.50	272
5200	2/0		4.88	1.640		135	101		236	298
5500	4/0		3.68	2.170		181	133		314	400
5600	250 kcmil		3.31	2.420		181	148		329	420
5650	350 kcmil		2.94	2.720		214	167		381	485
5700	500 kcmil		2.30	3.480		279	214		493	625
5720	750 kcmil		2.02	3.950		335	242		577	730
5750	1,000 kcmil		1.84	4.350		400	267		667	835
5800	15 kV, armored, #1		3.68	2.170		174	133		307	390
5900	1/0		3.68	2.170		217	133		350	435
6000	3/0		3.31	2.420		320	148		468	570
6100	4/0		3.13	2.560		350	157		507	620
6200	250 kcmil		2.94	2.720		440	167		607	730
6300	350 kcmil		2.48	3.220		395	198		593	725
6400	500 kcmil		1.84	4.350		565	267		832	1,025

26 05 19.30 Cable Splicing

		Crew	Daily Output	Labor-Hours	Unit	Material	2020 Bare Costs Labor	Equipment	Total	Total Incl O&P
0010	**CABLE SPLICING** URD or similar, ideal conditions									
0100	#6 stranded to #1 stranded, 5 kV	1 Elec	3.68	2.170	Ea.	166	133		299	380
0120	15 kV		3.31	2.420		224	148		372	470
0140	25 kV		2.94	2.720		272	167		439	545
0200	#1 stranded to 4/0 stranded, 5 kV		3.31	2.420		171	148		319	410
0210	15 kV		2.94	2.720		245	167		412	515
0220	25 kV		2.58	3.110		272	191		463	585
0300	4/0 stranded to 500 kcmil stranded, 5 kV		3.04	2.640		171	162		333	430
0310	15 kV		2.67	3		405	184		589	720
0320	25 kV		2.30	3.480		465	214		679	830
0400	500 kcmil, 5 kV		2.94	2.720		228	167		395	500
0410	15 kV		2.58	3.110		405	191		596	730
0420	25 kV		2.12	3.780		465	232		697	855
0500	600 kcmil, 5 kV		2.67	3		228	184		412	525
0510	15 kV		2.21	3.620		430	222		652	800
0520	25 kV		1.84	4.350		495	267		762	940
0600	750 kcmil, 5 kV		2.39	3.340		228	205		433	555
0610	15 kV		2.02	3.950		430	242		672	830
0620	25 kV		1.75	4.580		495	281		776	965
0700	1,000 kcmil, 5 kV		2.12	3.780		228	232		460	595
0710	15 kV		1.75	4.580		465	281		746	935
0720	25 kV		1.47	5.430		475	335		810	1,025

26 05 19.35 Cable Terminations

		Crew	Daily Output	Labor-Hours	Unit	Material	2020 Bare Costs Labor	Equipment	Total	Total Incl O&P
0010	**CABLE TERMINATIONS**									
0015	Wire connectors, screw type, #22 to #14	1 Elec	239.20	.033	Ea.	.08	2.05		2.13	3.14
0020	#18 to #12		220.80	.036		.09	2.22		2.31	3.41

26 05 19 – Low-Voltage Electrical Power Conductors and Cables

26 05 19.35 Cable Terminations		Crew	Daily Output	Labor-Hours	Unit	Material	2020 Bare Costs Labor	Equipment	Total	Total Incl O&P
0035	#16 to #10	1 Elec	211.60	.038	Ea.	.34	2.32		2.66	3.82
0040	#14 to #8		193.20	.041		.43	2.54		2.97	4.25
0045	#12 to #6		165.60	.048		.61	2.96		3.57	5.10
0050	Terminal lugs, solderless, #16 to #10		46	.174		.48	10.65		11.13	16.45
0100	#8 to #4		27.60	.290		.96	17.80		18.76	27.50
0150	#2 to #1		20.24	.395		1.03	24.50		25.53	37
0200	1/0 to 2/0		14.72	.543		2.05	33.50		35.55	52
0250	3/0		11.04	.725		6.05	44.50		50.55	72.50
0300	4/0		10.12	.791		5.45	48.50		53.95	78
0350	250 kcmil		8.28	.966		4.07	59.50		63.57	93
0400	350 kcmil		6.44	1.240		5.55	76		81.55	119
0450	500 kcmil		5.52	1.450		10.20	89		99.20	143
0500	600 kcmil		5.34	1.500		11.90	92		103.90	150
0550	750 kcmil		4.78	1.670		13.20	102		115.20	168
0600	Split bolt connectors, tapped, #6		14.72	.543		4.73	33.50		38.23	54.50
0650	#4		12.88	.621		4.43	38		42.43	61.50
0700	#2		11.04	.725		8.20	44.50		52.70	75
0750	#1		10.12	.791		9.20	48.50		57.70	82
0800	1/0		9.20	.870		9.20	53.50		62.70	89.50
0850	2/0		8.28	.966		16	59.50		75.50	106
0900	3/0		6.62	1.210		29.50	74		103.50	144
1000	4/0		5.89	1.360		36	83.50		119.50	164
1100	250 kcmil		5.24	1.530		36	94		130	180
1200	300 kcmil		4.88	1.640		44	101		145	198
1400	350 kcmil		4.23	1.890		44	116		160	221
1500	500 kcmil	↓	3.68	2.170	↓	101	133		234	310
1600	Crimp 1 hole lugs, copper or aluminum, 600 volt									
1620	#14	1 Elec	55.20	.145	Ea.	.77	8.90		9.67	14.10
1630	#12		46	.174		1.30	10.65		11.95	17.35
1640	#10		41.40	.193		1.30	11.85		13.15	19.10
1780	#8		33.12	.242		1.96	14.80		16.76	24
1800	#6		27.60	.290		2.28	17.80		20.08	29
2000	#4		24.84	.322		5.25	19.75		25	35.50
2200	#2		22.08	.362		3.61	22		25.61	37
2400	#1		18.40	.435		6.50	26.50		33	46.50
2500	1/0		16.10	.497		7.85	30.50		38.35	54
2600	2/0		13.80	.580		10.55	35.50		46.05	64.50
2800	3/0		11.04	.725		10.30	44.50		54.80	77.50
3000	4/0		10.12	.791		8.35	48.50		56.85	81
3200	250 kcmil		8.28	.966		9.35	59.50		68.85	99
3400	300 kcmil		7.36	1.090		14.40	67		81.40	115
3500	350 kcmil		6.44	1.240		15.20	76		91.20	130
3600	400 kcmil		5.98	1.340		20	82		102	144
3800	500 kcmil		5.52	1.450		26.50	89		115.50	161
4000	600 kcmil		5.34	1.500		35	92		127	176
4200	700 kcmil		5.06	1.580		42.50	97		139.50	191
4400	750 kcmil	↓	4.78	1.670	↓	42.50	102		144.50	200
4500	Crimp 2-way connectors, copper or alum., 600 volt									
4510	#14	1 Elec	55.20	.145	Ea.	3.29	8.90		12.19	16.85
4520	#12		46	.174		3.50	10.65		14.15	19.75
4530	#10		41.40	.193		3.69	11.85		15.54	21.50
4540	#8		24.84	.322		3.73	19.75		23.48	33.50
4600	#6	↓	23	.348	↓	5.30	21.50		26.80	38

Pre-Installation Change Order Costs   For customer support on your Electrical Change Order Costs with RSMeans data, call 800.448.8182.

26 05 19.35 Cable Terminations

26 05 19.35 Cable Terminations		Crew	Daily Output	Labor-Hours	Unit	Material	2020 Bare Costs Labor	Equipment	Total	Total Incl O&P
4800	#4	1 Elec	21.16	.378	Ea.	5.15	23		28.15	40
5000	#2		18.40	.435		7.35	26.50		33.85	47.50
5200	#1		14.72	.543		7.40	33.50		40.90	57.50
5400	1/0		11.96	.669		7.40	41		48.40	69
5420	2/0		11.04	.725		10.20	44.50		54.70	77.50
5440	3/0		10.12	.791		9.60	48.50		58.10	82.50
5460	4/0		9.20	.870		10	53.50		63.50	90.50
5480	250 kcmil		8.28	.966		22	59.50		81.50	113
5500	300 kcmil		7.82	1.020		21	62.50		83.50	116
5520	350 kcmil		7.36	1.090		19.95	67		86.95	122
5540	400 kcmil		6.72	1.190		26	73		99	138
5560	500 kcmil		5.70	1.400		31	86		117	162
5580	600 kcmil		5.06	1.580		62	97		159	212
5600	700 kcmil		4.14	1.930		65	118		183	248
5620	750 kcmil		3.68	2.170		72	133		205	277
7000	Compression equipment adapter, aluminum wire, #6		27.60	.290		9.80	17.80		27.60	37.50
7020	#4		24.84	.322		9.95	19.75		29.70	40.50
7040	#2		22.08	.362		9.20	22		31.20	43
7060	#1		18.40	.435		10.50	26.50		37	51
7080	1/0		16.56	.483		17.45	29.50		46.95	63
7100	2/0		13.80	.580		18.35	35.50		53.85	73
7140	4/0		10.12	.791		21.50	48.50		70	95.50
7160	250 kcmil		8.28	.966		23	59.50		82.50	114
7180	300 kcmil		7.36	1.090		24	67		91	126
7200	350 kcmil		6.44	1.240		33	76		109	150
7220	400 kcmil		5.98	1.340		39	82		121	165
7240	500 kcmil		5.52	1.450		39.50	89		128.50	176
7260	600 kcmil		5.34	1.500		45	92		137	187
7280	750 kcmil		4.78	1.670		54.50	102		156.50	213
8000	Compression tool, hand					2,000			2,000	2,200
8100	Hydraulic					2,500			2,500	2,750
8500	Hydraulic dies					465			465	510

26 05 19.50 Mineral Insulated Cable

26 05 19.50 Mineral Insulated Cable		Crew	Daily Output	Labor-Hours	Unit	Material	2020 Bare Costs Labor	Equipment	Total	Total Incl O&P
0010	**MINERAL INSULATED CABLE** 600 volt									
0100	1 conductor, #12	1 Elec	1.47	5.430	C.L.F.	425	335		760	960
0200	#10		1.47	5.430		645	335		980	1,200
0400	#8		1.38	5.800		805	355		1,160	1,425
0500	#6		1.29	6.210		855	380		1,235	1,500
0600	#4	2 Elec	2.21	7.250		1,275	445		1,720	2,050
0800	#2		2.02	7.910		1,725	485		2,210	2,625
0900	#1		1.93	8.280		2,000	510		2,510	2,950
1000	1/0		1.84	8.700		2,325	535		2,860	3,350
1100	2/0		1.75	9.150		2,925	560		3,485	4,050
1200	3/0		1.66	9.660		3,525	595		4,120	4,750
1400	4/0		1.47	10.870		4,075	665		4,740	5,475
1410	250 kcmil	3 Elec	2.21	10.870		4,575	665		5,240	6,050
1420	350 kcmil		1.79	13.380		5,275	820		6,095	7,025
1430	500 kcmil		1.79	13.380		6,650	820		7,470	8,525
1500	2 conductor, #12	1 Elec	1.29	6.210		1,125	380		1,505	1,800
1600	#10		1.10	7.250		1,375	445		1,820	2,150
1800	#8		1.01	7.910		1,575	485		2,060	2,450
2000	#6		.97	8.280		2,150	510		2,660	3,125

26 05 19 – Low-Voltage Electrical Power Conductors and Cables

26 05 19.50 Mineral Insulated Cable		Crew	Daily Output	Labor-Hours	Unit	Material	2020 Bare Costs Labor	Equipment	Total	Total Incl O&P
2100	#4	2 Elec	1.84	8.700	C.L.F.	3,050	535		3,585	4,175
2200	3 conductor, #12	1 Elec	1.10	7.250		1,300	445		1,745	2,075
2400	#10		1.01	7.910		1,475	485		1,960	2,350
2600	#8		.97	8.280		1,975	510		2,485	2,925
2800	#6		.92	8.700		2,600	535		3,135	3,650
3000	#4	2 Elec	1.66	9.660		3,425	595		4,020	4,625
3100	4 conductor, #12	1 Elec	1.10	7.250		1,375	445		1,820	2,150
3200	#10		1.01	7.910		1,650	485		2,135	2,525
3400	#8		.92	8.700		2,300	535		2,835	3,325
3600	#6		.83	9.660		3,125	595		3,720	4,325
3620	7 conductor, #12		1.01	7.910		1,875	485		2,360	2,800
3640	#10		.92	8.700		2,575	535		3,110	3,650
3800	Terminations, 600 volt, 1 conductor, #12		7.36	1.090	Ea.	22	67		89	124
4000	#10		6.99	1.140		22	70		92	128
4100	#8		6.72	1.190		22	73		95	133
4200	#6		6.16	1.300		22	80		102	143
4400	#4		5.70	1.400		22	86		108	152
4600	#2		5.24	1.530		33	94		127	177
4800	#1		4.88	1.640		33	101		134	187
5000	1/0		4.60	1.740		33	107		140	196
5100	2/0		4.32	1.850		33	114		147	206
5200	3/0		3.96	2.020		33	124		157	222
5400	4/0		3.68	2.170		73.50	133		206.50	279
5410	250 kcmil		3.68	2.170		73.50	133		206.50	279
5420	350 kcmil		3.68	2.170		126	133		259	335
5430	500 kcmil		3.68	2.170		126	133		259	335
5500	2 conductor, #12		6.16	1.300		22	80		102	143
5600	#10		5.89	1.360		33	83.50		116.50	161
5800	#8		5.70	1.400		33	86		119	165
6000	#6		5.24	1.530		33	94		127	177
6200	#4		4.88	1.640		73.50	101		174.50	231
6400	3 conductor, #12		5.24	1.530		33	94		127	177
6500	#10		5.06	1.580		33	97		130	181
6600	#8		4.78	1.670		33	102		135	190
6800	#6		4.42	1.810		33	111		144	202
7200	#4		4.23	1.890		73.50	116		189.50	254
7400	4 conductor, #12		4.23	1.890		37	116		153	214
7500	#10		4.05	1.980		37	121		158	222
7600	#8		3.86	2.070		37	127		164	230
8400	#6		3.68	2.170		77	133		210	283
8500	7 conductor, #12		3.22	2.480		36.50	152		188.50	268
8600	#10		2.76	2.900		79.50	178		257.50	355
8800	Crimping tool, plier type					92.50			92.50	102
9000	Stripping tool					350			350	385
9200	Hand vise					104			104	114

26 05 19.55 Non-Metallic Sheathed Cable

		Crew	Daily Output	Labor-Hours	Unit	Material	2020 Bare Costs Labor	Equipment	Total	Total Incl O&P
0010	**NON-METALLIC SHEATHED CABLE** 600 volt									
0100	Copper with ground wire (Romex)									
0150	#14, 2 conductor	1 Elec	2.48	3.220	C.L.F.	20	198		218	315
0200	3 conductor		2.21	3.620		29	222		251	360
0220	4 conductor		2.02	3.950		46.50	242		288.50	410
0250	#12, 2 conductor		2.30	3.480		29	214		243	350

26 05 19.55 Non-Metallic Sheathed Cable		Crew	Daily Output	Labor-Hours	Unit	Material	2020 Bare Costs Labor	Equipment	Total	Total Incl O&P
0300	3 conductor	1 Elec	2.02	3.950	C.L.F.	45	242		287	410
0320	4 conductor		1.84	4.350		71.50	267		338.50	475
0350	#10, 2 conductor		2.02	3.950		49.50	242		291.50	415
0400	3 conductor		1.66	4.830		73	296		369	520
0420	4 conductor		1.47	5.430		115	335		450	620
0430	#8, 2 conductor		1.47	5.430		96.50	335		431.50	600
0450	3 conductor		1.38	5.800		128	355		483	670
0500	#6, 3 conductor		1.29	6.210		197	380		577	780
0520	#4, 3 conductor	2 Elec	2.21	7.250		440	445		885	1,150
0540	#2, 3 conductor	"	2.02	7.910		1,000	485		1,485	1,825
0550	SE type SER aluminum cable, 3 RHW and									
0600	1 bare neutral, 3 #8 & 1 #8	1 Elec	1.47	5.430	C.L.F.	69	335		404	570
0650	3 #6 & 1 #6	"	1.29	6.210		78	380		458	650
0700	3 #4 & 1 #6	2 Elec	2.21	7.250		86.50	445		531.50	755
0750	3 #2 & 1 #4		2.02	7.910		158	485		643	900
0800	3 #1/0 & 1 #2		1.84	8.700		184	535		719	995
0850	3 #2/0 & 1 #1		1.66	9.660		219	595		814	1,125
0900	3 #4/0 & 1 #2/0		1.47	10.870		320	665		985	1,350
1000	URD - triplex underground distribution cable, alum. 2 #4 + #4 neutral		2.58	6.210		92.50	380		472.50	665
1010	2 #2 + #4 neutral		2.44	6.560		117	400		517	730
1020	2 #2 + #2 neutral		2.35	6.820		116	420		536	755
1030	2 1/0 + #2 neutral		2.21	7.250		143	445		588	820
1040	2 1/0 + 1/0 neutral		2.12	7.560		162	465		627	870
1050	2 2/0 + #1 neutral		2.02	7.910		161	485		646	900
1060	2 2/0 + 2/0 neutral		1.93	8.280		169	510		679	940
1070	2 3/0 + 1/0 neutral		1.79	8.920		214	545		759	1,050
1080	2 3/0 + 3/0 neutral		1.79	8.920		218	545		763	1,050
1090	2 4/0 + 2/0 neutral		1.70	9.400		207	575		782	1,100
1100	2 4/0 + 4/0 neutral		1.70	9.400		248	575		823	1,125
1450	UF underground feeder cable, copper with ground, #14, 2 conductor	1 Elec	3.68	2.170		39.50	133		172.50	242
1500	#12, 2 conductor		3.22	2.480		37	152		189	268
1550	#10, 2 conductor		2.76	2.900		58.50	178		236.50	330
1600	#14, 3 conductor		3.22	2.480		40.50	152		192.50	272
1650	#12, 3 conductor		2.76	2.900		54.50	178		232.50	325
1700	#10, 3 conductor		2.30	3.480		88.50	214		302.50	415
1710	#8, 3 conductor		1.84	4.350		153	267		420	565
1720	#6, 3 conductor		1.66	4.830		235	296		531	700
2400	SEU service entrance cable, copper 2 conductors, #8 + #8 neutral		1.38	5.800		109	355		464	650
2600	#6 + #8 neutral		1.20	6.690		152	410		562	780
2800	#6 + #6 neutral		1.20	6.690		171	410		581	800
3000	#4 + #6 neutral	2 Elec	2.02	7.910		244	485		729	995
3200	#4 + #4 neutral		2.02	7.910		257	485		742	1,000
3400	#3 + #5 neutral		1.93	8.280		350	510		860	1,150
3600	#3 + #3 neutral		1.93	8.280		335	510		845	1,125
3800	#2 + #4 neutral		1.84	8.700		395	535		930	1,225
4000	#1 + #1 neutral		1.75	9.150		415	560		975	1,300
4200	1/0 + 1/0 neutral		1.66	9.660		935	595		1,530	1,900
4400	2/0 + 2/0 neutral		1.56	10.230		895	630		1,525	1,925
4600	3/0 + 3/0 neutral		1.47	10.870		995	665		1,660	2,100
4620	4/0 + 4/0 neutral		1.33	11.990		1,950	735		2,685	3,250
4800	Aluminum 2 conductors, #8 + #8 neutral	1 Elec	1.47	5.430		61.50	335		396.50	565
5000	#6 + #6 neutral	"	1.29	6.210		62.50	380		442.50	635
5100	#4 + #6 neutral	2 Elec	2.30	6.960		77	425		502	720

26 05 19 – Low-Voltage Electrical Power Conductors and Cables

26 05 19.55 Non-Metallic Sheathed Cable	Crew	Daily Output	Labor-Hours	Unit	Material	2020 Bare Costs Labor	Equipment	Total	Total Incl O&P
5200	#4 + #4 neutral	2 Elec	2.21	7.250	C.L.F.	89	445		534	760
5300	#2 + #4 neutral		2.12	7.560		116	465		581	815
5400	#2 + #2 neutral		2.02	7.910		119	485		604	855
5450	1/0 + #2 neutral		1.93	8.280		177	510		687	950
5500	1/0 + 1/0 neutral		1.84	8.700		187	535		722	1,000
5550	2/0 + #1 neutral		1.75	9.150		197	560		757	1,050
5600	2/0 + 2/0 neutral		1.66	9.660		198	595		793	1,100
5800	3/0 + 1/0 neutral		1.56	10.230		267	630		897	1,225
6000	3/0 + 3/0 neutral		1.56	10.230		218	630		848	1,175
6200	4/0 + 2/0 neutral		1.47	10.870		218	665		883	1,225
6400	4/0 + 4/0 neutral	▼	1.47	10.870	▼	242	665		907	1,250
6500	Service entrance cap for copper SEU									
6600	100 amp	1 Elec	11.04	.725	Ea.	8.65	44.50		53.15	75.50
6700	150 amp		9.20	.870		12.90	53.50		66.40	93.50
6800	200 amp	▼	7.36	1.090	▼	19.60	67		86.60	121

26 05 19.70 Portable Cord

		Crew	Daily Output	Labor-Hours	Unit	Material	Labor	Equipment	Total	Total Incl O&P
0010	**PORTABLE CORD** 600 volt									
0100	Type SO, #18, 2 conductor	1 Elec	901.60	.009	L.F.	.39	.54		.93	1.23
0110	3 conductor		901.60	.009		.37	.54		.91	1.22
0120	#16, 2 conductor		772.80	.010		.46	.64		1.10	1.46
0130	3 conductor		772.80	.010		4.75	.64		5.39	6.20
0140	4 conductor		772.80	.010		.58	.64		1.22	1.59
0240	#14, 2 conductor		772.80	.010		.58	.64		1.22	1.59
0250	3 conductor		772.80	.010		.62	.64		1.26	1.63
0260	4 conductor		772.80	.010		1.22	.64		1.86	2.29
0280	#12, 2 conductor		772.80	.010		.80	.64		1.44	1.83
0290	3 conductor		772.80	.010		.99	.64		1.63	2.04
0320	#10, 2 conductor		703.80	.011		1.09	.70		1.79	2.24
0330	3 conductor		703.80	.011		1.61	.70		2.31	2.81
0340	4 conductor		703.80	.011		1.75	.70		2.45	2.96
0360	#8, 2 conductor		510.60	.016		1.88	.96		2.84	3.50
0370	3 conductor		496.80	.016		2.27	.99		3.26	3.96
0380	4 conductor		483	.017		2.37	1.02		3.39	4.11
0400	#6, 2 conductor		483	.017		2.86	1.02		3.88	4.66
0410	3 conductor		450.80	.018		3.39	1.09		4.48	5.35
0420	4 conductor	▼	381.80	.021		3.58	1.29		4.87	5.85
0440	#4, 2 conductor	2 Elec	763.60	.021		4.20	1.29		5.49	6.55
0450	3 conductor		644	.025		4.40	1.52		5.92	7.10
0460	4 conductor		607.20	.026		5.85	1.62		7.47	8.85
0480	#2, 2 conductor		414	.039		5.50	2.37		7.87	9.60
0490	3 conductor		322	.050		7.90	3.05		10.95	13.25
0500	4 conductor	▼	257.60	.062	▼	10.10	3.81		13.91	16.75
2000	See 26 27 26.20 for Wiring Devices Elements									

26 05 19.75 Modular Flexible Wiring System

		Crew	Daily Output	Labor-Hours	Unit	Material	Labor	Equipment	Total	Total Incl O&P
0010	**MODULAR FLEXIBLE WIRING SYSTEM**									
0020	Commercial system grid ceiling									
0100	Conversion Module	1 Elec	29.44	.272	Ea.	9	16.65		25.65	35
0120	Fixture cable, for fixture to fixture, 3 conductor, 15' long		36.80	.217		37	13.35		50.35	60.50
0150	Extender cable, 3 conductor, 15' long		44.16	.181		36	11.10		47.10	56.50
0200	Switch drop, 1 level, 9' long		29.44	.272		39	16.65		55.65	68
0220	2 level, 9' long		29.44	.272		38	16.65		54.65	67
0250	Power tee, 9' long	▼	29.44	.272	▼	31.50	16.65		48.15	59.50

26 05 19.75 Modular Flexible Wiring System		Crew	Daily Output	Labor-Hours	Unit	Material	2020 Bare Costs Labor	Equipment	Total	Total Incl O&P
1020	Industrial system open ceiling									
1100	Converter, interface between hardwiring and modular wiring	1 Elec	29.44	.272	Ea.	22	16.65		38.65	49
1120	Fixture cable, for fixture to fixture, 3 conductor, 21' long		29.44	.272		126	16.65		142.65	163
1125	3 conductor, 25' long		22.08	.362		162	22		184	211
1130	3 conductor, 31' long		17.48	.458		143	28		171	200
1150	Fixture cord drop, 10' long	↓	36.80	.217	↓	39	13.35		52.35	62.50

26 05 19.90 Wire

		Crew	Daily Output	Labor-Hours	Unit	Material	2020 Bare Costs Labor	Equipment	Total	Total Incl O&P
0010	**WIRE**, normal installation conditions in wireway, conduit, cable tray									
0020	600 volt, copper type THW, solid, #14	1 Elec	11.96	.669	C.L.F.	7.50	41		48.50	69.50
0030	#12		10.12	.791		11.85	48.50		60.35	85
0040	#10		9.20	.870		19.15	53.50		72.65	101
0050	Stranded, #14		11.96	.669		9.10	41		50.10	71
0100	#12		10.12	.791		16.50	48.50		65	90
0120	#10		9.20	.870		26	53.50		79.50	108
0140	#8		7.36	1.090		41.50	67		108.50	145
0160	#6	↓	5.98	1.340		69	82		151	198
0180	#4	2 Elec	9.75	1.640		105	101		206	266
0200	#3		9.20	1.740		141	107		248	315
0220	#2		8.28	1.930		163	118		281	355
0240	#1		7.36	2.170		208	133		341	425
0260	1/0		6.07	2.640		271	162		433	540
0280	2/0		5.34	3		470	184		654	795
0300	3/0		4.60	3.480		335	214		549	685
0350	4/0	↓	4.05	3.950		540	242		782	955
0400	250 kcmil	3 Elec	5.52	4.350		585	267		852	1,050
0420	300 kcmil		5.24	4.580		825	281		1,106	1,325
0450	350 kcmil		4.97	4.830		800	296		1,096	1,325
0480	400 kcmil		4.69	5.120		1,450	315		1,765	2,050
0490	500 kcmil		4.42	5.430		1,150	335		1,485	1,775
0500	600 kcmil		3.59	6.690		1,525	410		1,935	2,275
0510	750 kcmil		3.04	7.910		1,375	485		1,860	2,250
0520	1,000 kcmil	↓	2.48	9.660		2,250	595		2,845	3,325
0540	600 volt, aluminum type THHN, stranded, #6	1 Elec	7.36	1.090		24.50	67		91.50	127
0560	#4	2 Elec	11.96	1.340		30	82		112	155
0580	#2		9.75	1.640		58	101		159	214
0600	#1		8.28	1.930		68.50	118		186.50	252
0620	1/0		7.36	2.170		96.50	133		229.50	305
0640	2/0		6.62	2.420		98.50	148		246.50	330
0680	3/0		6.07	2.640		123	162		285	375
0700	4/0	↓	5.70	2.810		157	172		329	430
0720	250 kcmil	3 Elec	8	3		165	184		349	455
0740	300 kcmil		7.45	3.220		228	198		426	545
0760	350 kcmil		6.90	3.480		231	214		445	575
0780	400 kcmil		6.35	3.780		262	232		494	635
0800	500 kcmil		5.52	4.350		296	267		563	720
0850	600 kcmil		5.24	4.580		375	281		656	835
0880	700 kcmil		4.69	5.120		460	315		775	980
0900	750 kcmil		4.42	5.430		450	335		785	990
0910	1,000 kcmil	↓	3.48	6.900		800	425		1,225	1,500
0920	600 volt, copper type THWN-THHN, solid, #14	1 Elec	11.96	.669		8	41		49	70
0940	#12		10.12	.791		12.05	48.50		60.55	85.50
0960	#10	↓	9.20	.870	↓	18.70	53.50		72.20	100

Pre-Installation Change Order Costs

26 05 19 – Low-Voltage Electrical Power Conductors and Cables

26 05 19.90 Wire		Crew	Daily Output	Labor-Hours	Unit	Material	2020 Bare Costs Labor	Equipment	Total	Total Incl O&P
1000	Stranded, #14	1 Elec	11.96	.669	C.L.F.	9.25	41		50.25	71
1200	#12		10.12	.791		12.75	48.50		61.25	86
1250	#10		9.20	.870		24.50	53.50		78	107
1300	#8		7.36	1.090		32.50	67		99.50	136
1350	#6		5.98	1.340		53	82		135	181
1400	#4	2 Elec	9.75	1.640		84	101		185	242
1450	#3		9.20	1.740		105	107		212	275
1500	#2		8.28	1.930		130	118		248	320
1550	#1		7.36	2.170		166	133		299	380
1600	1/0		6.07	2.640		240	162		402	505
1650	2/0		5.34	3		249	184		433	550
1700	3/0		4.60	3.480		310	214		524	660
2000	4/0		4.05	3.950		395	242		637	790
2200	250 kcmil	3 Elec	5.52	4.350		460	267		727	900
2400	300 kcmil		5.24	4.580		620	281		901	1,100
2600	350 kcmil		4.97	4.830		645	296		941	1,150
2700	400 kcmil		4.69	5.120		725	315		1,040	1,275
2800	500 kcmil		4.42	5.430		930	335		1,265	1,525
2802	600 kcmil		3.59	6.690		935	410		1,345	1,625
2804	750 kcmil		3.04	7.910		1,825	485		2,310	2,750
2805	1,000 kcmil		1.78	13.450		2,950	825		3,775	4,475
2900	600 volt, copper type XHHW, solid, #14	1 Elec	11.96	.669	C.L.F.	13.25	41		54.25	75.50
2920	#12		10.12	.791		21	48.50		69.50	95
2940	#10		9.20	.870		32.50	53.50		86	115
3000	Stranded, #14		11.96	.669		11.65	41		52.65	74
3020	#12		10.12	.791		16.45	48.50		64.95	90
3040	#10		9.20	.870		24.50	53.50		78	107
3060	#8		7.36	1.090		40.50	67		107.50	144
3080	#6		5.98	1.340		65.50	82		147.50	194
3100	#4	2 Elec	9.75	1.640		92.50	101		193.50	252
3120	#2		8.28	1.930		144	118		262	335
3140	#1		7.36	2.170		190	133		323	410
3160	1/0		6.07	2.640		236	162		398	500
3180	2/0		5.34	3		295	184		479	600
3200	3/0		4.60	3.480		370	214		584	725
3220	4/0		4.05	3.950		460	242		702	865
3240	250 kcmil	3 Elec	5.52	4.350		540	267		807	990
3260	300 kcmil		5.24	4.580		640	281		921	1,125
3280	350 kcmil		4.97	4.830		700	296		996	1,200
3300	400 kcmil		4.69	5.120		860	315		1,175	1,425
3320	500 kcmil		4.42	5.430		1,000	335		1,335	1,600
3340	600 kcmil		3.59	6.690		1,250	410		1,660	1,975
3360	750 kcmil		3.04	7.910		2,200	485		2,685	3,150
3380	1,000 kcmil		2.21	10.870		2,925	665		3,590	4,225
5020	600 volt, aluminum type XHHW, stranded, #6	1 Elec	7.36	1.090		31	67		98	134
5040	#4	2 Elec	11.96	1.340		38.50	82		120.50	165
5060	#2		9.75	1.640		48	101		149	203
5080	#1		8.28	1.930		76.50	118		194.50	261
5100	1/0		7.36	2.170		91.50	133		224.50	298
5120	2/0		6.62	2.420		112	148		260	345
5140	3/0		6.07	2.640		134	162		296	390
5160	4/0		5.70	2.810		121	172		293	390
5180	250 kcmil	3 Elec	8	3		188	184		372	480

26 05 19 – Low-Voltage Electrical Power Conductors and Cables

26 05 19.90 Wire		Crew	Daily Output	Labor-Hours	Unit	Material	2020 Bare Costs Labor	Equipment	Total	Total Incl O&P
5200	300 kcmil	3 Elec	7.45	3.220	C.L.F.	250	198		448	570
5220	350 kcmil		6.90	3.480		265	214		479	610
5240	400 kcmil		6.35	3.780		297	232		529	670
5260	500 kcmil		5.52	4.350		325	267		592	750
5280	600 kcmil		5.24	4.580		420	281		701	880
5300	700 kcmil		4.97	4.830		475	296		771	960
5320	750 kcmil		4.69	5.120		450	315		765	970
5340	1,000 kcmil		3.31	7.250		710	445		1,155	1,450
5390	600 volt, copper type XLPE-USE (RHW), solid, #14	1 Elec	11.04	.725		14.25	44.50		58.75	81.50
5400	#12		10.12	.791		21.50	48.50		70	96
5420	#10		9.20	.870		28	53.50		81.50	111
5440	Stranded, #14		11.96	.669		18.05	41		59.05	81
5460	#12		10.12	.791		21.50	48.50		70	96
5480	#10		9.20	.870		33.50	53.50		87	117
5500	#8		7.36	1.090		40.50	67		107.50	144
5520	#6		5.98	1.340		68	82		150	197
5540	#4	2 Elec	9.75	1.640		104	101		205	264
5560	#2		8.28	1.930		168	118		286	360
5580	#1		7.36	2.170		226	133		359	445
5600	1/0		6.07	2.640		283	162		445	550
5620	2/0		5.34	3		370	184		554	685
5640	3/0		4.60	3.480		465	214		679	830
5660	4/0		4.05	3.950		505	242		747	915
5680	250 kcmil	3 Elec	5.52	4.350		610	267		877	1,075
5700	300 kcmil		5.24	4.580		700	281		981	1,200
5720	350 kcmil		4.97	4.830		785	296		1,081	1,300
5740	400 kcmil		4.69	5.120		920	315		1,235	1,475
5760	500 kcmil		4.42	5.430		1,000	335		1,335	1,600
5780	600 kcmil		3.59	6.690		1,500	410		1,910	2,250
5800	750 kcmil		3.04	7.910		2,325	485		2,810	3,275
5820	1,000 kcmil		2.48	9.660		3,125	595		3,720	4,325
5840	600 volt, aluminum type XLPE-USE (RHW), stranded, #6	1 Elec	7.36	1.090		38.50	67		105.50	142
5860	#4	2 Elec	11.96	1.340		45	82		127	172
5880	#2		9.75	1.640		63.50	101		164.50	220
5900	#1		8.28	1.930		90	118		208	275
5920	1/0		7.36	2.170		106	133		239	315
5940	2/0		6.62	2.420		124	148		272	355
5960	3/0		6.07	2.640		146	162		308	400
5980	4/0		5.70	2.810		162	172		334	435
6000	250 kcmil	3 Elec	8	3		220	184		404	515
6020	300 kcmil		7.45	3.220		285	198		483	610
6040	350 kcmil		6.90	3.480		289	214		503	640
6060	400 kcmil		6.35	3.780		350	232		582	730
6080	500 kcmil		5.52	4.350		390	267		657	825
6100	600 kcmil		5.24	4.580		500	281		781	970
6110	700 kcmil		4.97	4.830		545	296		841	1,050
6120	750 kcmil		4.69	5.120		585	315		900	1,125

26 05 23 – Control-Voltage Electrical Power Cables

26 05 23.10 Control Cable

0010	**CONTROL CABLE**									
0020	600 volt, copper, #14 THWN wire with PVC jacket, 2 wires	1 Elec	8.28	.966	C.L.F.	53.50	59.50		113	148
0030	3 wires		7.36	1.090		63	67		130	169

26 05 23 – Control-Voltage Electrical Power Cables

26 05 23.10 Control Cable	Crew	Daily Output	Labor-Hours	Unit	Material	2020 Bare Costs Labor	Equipment	Total	Total Incl O&P
0100	4 wires	1 Elec	6.44	1.240	C.L.F.	65	76		141	185
0150	5 wires		5.98	1.340		89	82		171	220
0200	6 wires		5.52	1.450		120	89		209	264
0300	8 wires		4.88	1.640		167	101		268	335
0400	10 wires		4.42	1.810		199	111		310	385
0500	12 wires		3.96	2.020		256	124		380	465
0600	14 wires		3.50	2.290		271	140		411	505
0700	16 wires		3.22	2.480		310	152		462	565
0800	18 wires		3.04	2.640		360	162		522	635
0810	19 wires		2.85	2.810		335	172		507	625
0900	20 wires		2.76	2.900		355	178		533	655
1000	22 wires		2.58	3.110		430	191		621	755

26 05 23.20 Special Wires and Fittings

		Crew	Daily Output	Labor-Hours	Unit	Material	Labor	Equipment	Total	Total Incl O&P
0010	**SPECIAL WIRES & FITTINGS**									
0100	Fixture TFFN 600 volt 90°C stranded, #18	1 Elec	11.96	.669	C.L.F.	11.25	41		52.25	73.50
0150	#16		11.96	.669		14.80	41		55.80	77.50
0500	Thermostat, jacket non-plenum, twisted, #18-2 conductor		7.36	1.090		12.20	67		79.20	113
0550	#18-3 conductor		6.44	1.240		15.20	76		91.20	130
0600	#18-4 conductor		5.98	1.340		24	82		106	149
0650	#18-5 conductor		5.52	1.450		29.50	89		118.50	165
0700	#18-6 conductor		5.06	1.580		33.50	97		130.50	181
0750	#18-7 conductor		4.60	1.740		43.50	107		150.50	207
0800	#18-8 conductor		4.42	1.810		42.50	111		153.50	212
2460	Tray cable, type TC, copper, #16-2 conductor		8.65	.925		37	57		94	126
2464	#16-3 conductor		7.73	1.040		61.50	64		125.50	163
2468	#16-4 conductor		6.72	1.190		65.50	73		138.50	181
2472	#16-5 conductor		6.16	1.300		51	80		131	175
2476	#16-7 conductor		5.06	1.580		70	97		167	221
2480	#16-9 conductor		4.60	1.740		87.50	107		194.50	256
2484	#16-12 conductor	2 Elec	8.10	1.980		152	121		273	350
2488	#16-15 conductor		6.62	2.420		152	148		300	390
2492	#16-19 conductor		6.07	2.640		188	162		350	450
2496	#16-25 conductor		5.89	2.720		265	167		432	540
2500	#14-2 conductor	1 Elec	8.28	.966		42.50	59.50		102	136
2520	#14-3 conductor		7.36	1.090		41	67		108	145
2540	#14-4 conductor		6.44	1.240		58.50	76		134.50	178
2560	#14-5 conductor		5.98	1.340		89	82		171	220
2564	#14-7 conductor		4.88	1.640		110	101		211	271
2568	#14-9 conductor	2 Elec	8.83	1.810		127	111		238	305
2572	#14-12 conductor		7.91	2.020		177	124		301	380
2576	#14-15 conductor		6.62	2.420		209	148		357	450
2578	#14-19 conductor		5.70	2.810		263	172		435	545
2582	#14-25 conductor		4.88	3.280		350	201		551	685
2590	#12-2 conductor	1 Elec	7.73	1.040		62	64		126	164
2592	#12-3 conductor		6.99	1.140		67.50	70		137.50	178
2594	#12-4 conductor		6.07	1.320		126	81		207	260
2596	#12-5 conductor		5.70	1.400		137	86		223	279
2598	#12-7 conductor	2 Elec	9.57	1.670		144	102		246	310
2602	#12-9 conductor		8.10	1.980		260	121		381	465
2604	#12-12 conductor		7.18	2.230		264	137		401	495
2606	#12-15 conductor		6.26	2.560		315	157		472	580
2608	#12-19 conductor		5.52	2.900		385	178		563	685

26 05 23 – Control-Voltage Electrical Power Cables

26 05 23.20 Special Wires and Fittings	Crew	Daily Output	Labor-Hours	Unit	Material	2020 Bare Costs Labor	Equipment	Total	Total Incl O&P
2610	#12-25 conductor	3 Elec	7.08	3.390	C.L.F.	520	208		728	885
2618	#10-2 conductor	1 Elec	7.36	1.090		101	67		168	211
2622	#10-3 conductor	"	6.72	1.190		101	73		174	221
2624	#10-4 conductor	2 Elec	11.78	1.360		192	83.50		275.50	335
2626	#10-5 conductor		10.86	1.470		163	90		253	315
2628	#10-7 conductor		8.65	1.850		235	114		349	430
2630	#10-9 conductor		7.73	2.070		298	127		425	520
2632	#10-12 conductor		6.62	2.420		330	148		478	585
2640	300 V, copper braided shield, PVC jacket									
2650	2 conductor #18 stranded	1 Elec	6.44	1.240	C.L.F.	36	76		112	153
2660	3 conductor #18 stranded		5.52	1.450		53.50	89		142.50	191
2670	4 conductor #18 stranded		5.52	1.450		194	89		283	345
3000	Strain relief grip for cable									
3050	Cord, top, #12-3	1 Elec	36.80	.217	Ea.	12.85	13.35		26.20	34
3060	#12-4		36.80	.217		12.85	13.35		26.20	34
3070	#12-5		35.88	.223		15.20	13.70		28.90	37
3100	#10-3		35.88	.223		15.20	13.70		28.90	37
3110	#10-4		34.96	.229		15.20	14.05		29.25	37.50
3120	#10-5		34.96	.229		18.35	14.05		32.40	41
3200	Bottom, #12-3		36.80	.217		30.50	13.35		43.85	53.50
3210	#12-4		36.80	.217		30.50	13.35		43.85	53.50
3220	#12-5		35.88	.223		32	13.70		45.70	55.50
3230	#10-3		35.88	.223		32	13.70		45.70	55.50
3300	#10-4		34.96	.229		32	14.05		46.05	56
3310	#10-5		34.96	.229		40	14.05		54.05	65
3400	Cable ties, standard, 4" length		174.80	.046		.18	2.81		2.99	4.38
3410	7" length		147.20	.054		.22	3.33		3.55	5.20
3420	14.5" length		82.80	.097		.49	5.95		6.44	9.40
3430	Heavy, 14.5" length		73.60	.109		.71	6.65		7.36	10.75
3500	Cable gland, nylon, 1/4" NPT thread		55.20	.145		2.11	8.90		11.01	15.55
3510	Cable gland, nylon, 3/8" NPT thread		55.20	.145		2.43	8.90		11.33	15.90
3520	Cable gland, nylon, 1/2" NPT thread		55.20	.145		3.04	8.90		11.94	16.60
3530	Cable gland, nylon, 3/4" NPT thread		55.20	.145		4.59	8.90		13.49	18.30
3540	Cable gland, nylon, 1" NPT thread		55.20	.145		7.25	8.90		16.15	21.50
3550	Cable gland, nylon, 1-1/4" NPT thread		55.20	.145		8.55	8.90		17.45	22.50
3970	U-cable guards									
3980	.75" x 5' U-cable guard, 14 gauge	1 Elec	22.08	.362	Ea.	11.30	22		33.30	45.50
3990	.75" x 8' U-cable guard, 14 gauge		22.08	.362		21	22		43	56
4000	1-1/8" x 5' U-cable guard		22.08	.362		16.40	22		38.40	51
4010	1-1/8" x 8' U-cable guard		22.08	.362		19.30	22		41.30	54
4020	2-3/16" x 5' U-cable guard		22.08	.362		33.50	22		55.50	69.50
4030	2-3/16" x 8' U-cable guard		22.08	.362		32.50	22		54.50	69
4040	3-3/16" x 5' U-cable guard		22.08	.362		37	22		59	74
4050	3-3/16" x 8' U-cable guard		22.08	.362		48	22		70	86
4060	4" x 8' U-cable guard		22.08	.362		42	22		64	79
4070	2" x 9' flanged-cable guard		18.40	.435		98.50	26.50		125	148
4080	3" x 9' flanged-cable guard		18.40	.435		125	26.50		151.50	178
4090	3.5" x 9' flanged-cable guard		18.40	.435		151	26.50		177.50	206
5000	2" x 5' flanged-cable guard, extension		18.40	.435		66.50	26.50		93	113
5010	3" x 5' flanged-cable guard, extension		18.40	.435		91	26.50		117.50	140
5020	.05" x 8' U-cable guard, black plastic		23.92	.334		5.55	20.50		26.05	36.50
5030	.75" x 8' U-cable guard, black plastic		23.92	.334		7.50	20.50		28	39
5040	1" x 8' U-cable guard, black plastic		23.92	.334		11.95	20.50		32.45	43.50

For customer support on your Electrical Change Order Costs with RSMeans data, call 800.448.8182.

Pre-Installation Change Order Costs

26 05 26.80 Grounding		Crew	Daily Output	Labor-Hours	Unit	Material	2020 Bare Costs Labor	Equipment	Total	Total Incl O&P
0010	**GROUNDING**									
0030	Rod, copper clad, 8' long, 1/2" diameter	1 Elec	5.06	1.580	Ea.	26	97		123	173
0040	5/8" diameter		5.06	1.580		26	97		123	173
0050	3/4" diameter		4.88	1.640		41	101		142	195
0080	10' long, 1/2" diameter		4.42	1.810		28	111		139	196
0090	5/8" diameter		4.23	1.890		25.50	116		141.50	202
0100	3/4" diameter		4.05	1.980		61	121		182	249
0130	15' long, 3/4" diameter		3.68	2.170		63.50	133		196.50	268
0150	Coupling, bronze, 1/2" diameter					5.80			5.80	6.40
0160	5/8" diameter					6.25			6.25	6.90
0170	3/4" diameter					16.95			16.95	18.65
0190	Drive studs, 1/2" diameter					14.65			14.65	16.15
0210	5/8" diameter					23.50			23.50	25.50
0220	3/4" diameter					26.50			26.50	29
0230	Clamp, bronze, 1/2" diameter	1 Elec	29.44	.272		6.05	16.65		22.70	31.50
0240	5/8" diameter		29.44	.272		5.35	16.65		22	31
0250	3/4" diameter		29.44	.272		6.50	16.65		23.15	32
0260	Wire ground bare armored, #8-1 conductor		1.84	4.350	C.L.F.	85.50	267		352.50	490
0270	#6-1 conductor		1.66	4.830		96.50	296		392.50	545
0280	#4-1 conductor		1.47	5.430		168	335		503	680
0320	Bare copper wire, #14 solid		12.88	.621		8.15	38		46.15	65.50
0330	#12		11.96	.669		14.80	41		55.80	77.50
0340	#10		11.04	.725		20.50	44.50		65	88.50
0350	#8		10.12	.791		29	48.50		77.50	104
0360	#6		9.20	.870		45	53.50		98.50	130
0370	#4		7.36	1.090		113	67		180	224
0380	#2		4.60	1.740		146	107		253	320
0390	Bare copper wire, stranded, #8		10.12	.791		46	48.50		94.50	123
0400	#6		9.20	.870		48.50	53.50		102	133
0450	#4	2 Elec	14.72	1.090		98.50	67		165.50	208
0600	#2		9.20	1.740		102	107		209	271
0650	#1		8.28	1.930		149	118		267	340
0700	1/0		7.36	2.170		180	133		313	395
0750	2/0		6.62	2.420		291	148		439	540
0800	3/0		6.07	2.640		355	162		517	630
1000	4/0		5.24	3.050		485	187		672	815
1200	250 kcmil	3 Elec	6.62	3.620		505	222		727	885
1210	300 kcmil		6.07	3.950		645	242		887	1,075
1220	350 kcmil		5.52	4.350		630	267		897	1,100
1230	400 kcmil		5.24	4.580		695	281		976	1,175
1240	500 kcmil		4.69	5.120		905	315		1,220	1,475
1260	750 kcmil		3.31	7.250		1,500	445		1,945	2,300
1270	1,000 kcmil		2.76	8.700		1,800	535		2,335	2,775
1360	Bare aluminum, stranded, #6	1 Elec	8.28	.966		16.75	59.50		76.25	107
1370	#4	2 Elec	14.72	1.090		36.50	67		103.50	140
1380	#2		11.96	1.340		52.50	82		134.50	180
1390	#1		9.75	1.640		45.50	101		146.50	201
1400	1/0		8.28	1.930		58	118		176	240
1410	2/0		7.36	2.170		89.50	133		222.50	297
1420	3/0		6.62	2.420		99	148		247	330
1430	4/0		6.07	2.640		126	162		288	380
1440	250 kcmil	3 Elec	8.56	2.810		132	172		304	400

26 05 26 – Grounding and Bonding for Electrical Systems

26 05 26.80 Grounding		Crew	Daily Output	Labor-Hours	Unit	Material	2020 Bare Costs Labor	Equipment	Total	Total Incl O&P
1450	300 kcmil	3 Elec	8	3	C.L.F.	139	184		323	425
1460	400 kcmil		6.90	3.480		183	214		397	520
1470	500 kcmil		6.35	3.780		222	232		454	590
1480	600 kcmil		5.52	4.350		264	267		531	685
1490	700 kcmil		5.24	4.580		284	281		565	735
1500	750 kcmil		4.69	5.120		315	315		630	815
1510	1,000 kcmil		4.42	5.430		400	335		735	935
1800	Water pipe ground clamps, heavy duty									
2000	Bronze, 1/2" to 1" diameter	1 Elec	7.36	1.090	Ea.	33	67		100	136
2100	1-1/4" to 2" diameter		7.36	1.090		35.50	67		102.50	139
2200	2-1/2" to 3" diameter		5.52	1.450		69.50	89		158.50	209
2400	Grounding, exothermic welding									
2500	Exothermic welding reusable mold, cable to cable, parallel, vertical	1 Elec	7.36	1.090	Ea.	390	67		457	525
2505	Splice single		7.36	1.090		335	67		402	470
2510	Splice single to double		7.36	1.090		390	67		457	530
2520	Cable to cable, termination, Tee		7.36	1.090		380	67		447	520
2530	Cable to rod, termination, Tee		7.36	1.090		330	67		397	460
2535	Cable to rebar over, termination, Tee		7.36	1.090		111	67		178	222
2540	Cable to rod, termination, 90 Deg		7.36	1.090		380	67		447	520
2550	Cable to verticle flat steel		7.36	1.090		385	67		452	525
2555	Cable to 45 Deg		7.36	1.090		390	67		457	530
2730	Exothermic weld, 4/0 wire to 1" ground rod		6.44	1.240		10.90	76		86.90	125
2740	4/0 wire to building steel		6.44	1.240		10.90	76		86.90	125
2750	4/0 wire to motor frame		6.44	1.240		10.90	76		86.90	125
2760	4/0 wire to 4/0 wire		6.44	1.240		10.90	76		86.90	125
2770	4/0 wire to #4 wire		6.44	1.240		10.90	76		86.90	125
2780	4/0 wire to #8 wire		6.44	1.240		10.90	76		86.90	125
2790	Mold, reusable, for above					162			162	179
2800	Brazed connections, #6 wire	1 Elec	11.04	.725		18.55	44.50		63.05	86.50
3000	#2 wire		9.20	.870		25	53.50		78.50	107
3100	3/0 wire		7.36	1.090		37	67		104	140
3200	4/0 wire		6.44	1.240		42	76		118	160
3400	250 kcmil wire		4.60	1.740		49	107		156	213
3600	500 kcmil wire		3.68	2.170		60.50	133		193.50	265
3700	Insulated ground wire, copper #14		11.96	.669	C.L.F.	8.90	41		49.90	71
3710	#12		10.12	.791		12.20	48.50		60.70	85.50
3720	#10		9.20	.870		23.50	53.50		77	105
3730	#8		7.36	1.090		31.50	67		98.50	134
3740	#6		5.98	1.340		51	82		133	178
3750	#4	2 Elec	9.75	1.640		80.50	101		181.50	239
3770	#2		8.28	1.930		125	118		243	315
3780	#1		7.36	2.170		159	133		292	375
3790	1/0		6.07	2.640		230	162		392	495
3800	2/0		5.34	3		239	184		423	535
3810	3/0		4.60	3.480		297	214		511	645
3820	4/0		4.05	3.950		375	242		617	775
3830	250 kcmil	3 Elec	5.52	4.350		440	267		707	875
3840	300 kcmil		5.24	4.580		590	281		871	1,075
3850	350 kcmil		4.97	4.830		620	296		916	1,125
3860	400 kcmil		4.69	5.120		695	315		1,010	1,225
3870	500 kcmil		4.42	5.430		890	335		1,225	1,475
3880	600 kcmil		3.59	6.690		1,475	410		1,885	2,225
3890	750 kcmil		3.04	7.910		1,325	485		1,810	2,175

26 05 Common Work Results for Electrical

26 05 26 – Grounding and Bonding for Electrical Systems

26 05 26.80 Grounding		Crew	Daily Output	Labor-Hours	Unit	Material	2020 Bare Costs Labor	Equipment	Total	Total Incl O&P
3900	1,000 kcmil	3 Elec	2.48	9.660	C.L.F.	2,150	595		2,745	3,225
3960	Insulated ground wire, aluminum, #6	1 Elec	7.36	1.090		23.50	67		90.50	125
3970	#4	2 Elec	11.96	1.340		28.50	82		110.50	154
3980	#2		9.75	1.640		55.50	101		156.50	211
3990	#1		8.28	1.930		65.50	118		183.50	249
4000	1/0		7.36	2.170		92.50	133		225.50	299
4010	2/0		6.62	2.420		94.50	148		242.50	325
4020	3/0		6.07	2.640		118	162		280	370
4030	4/0		5.70	2.810		151	172		323	425
4040	250 kcmil	3 Elec	8	3		158	184		342	450
4050	300 kcmil		7.45	3.220		219	198		417	535
4060	350 kcmil		6.90	3.480		221	214		435	565
4070	400 kcmil		6.35	3.780		251	232		483	620
4080	500 kcmil		5.52	4.350		283	267		550	705
4090	600 kcmil		5.24	4.580		360	281		641	815
4100	700 kcmil		4.69	5.120		440	315		755	955
4110	750 kcmil		4.42	5.430		435	335		770	970
5000	Copper electrolytic ground rod system									
5010	Includes augering hole, mixing bentonite clay,									
5020	Installing rod, and terminating ground wire									
5100	Straight vertical type, 2" diam.									
5120	8.5' long, clamp connection	1 Elec	2.46	3.260	Ea.	720	200		920	1,100
5130	With exothermic weld connection		1.79	4.460		730	274		1,004	1,200
5140	10' long		2.16	3.700		710	227		937	1,125
5150	With exothermic weld connection		1.64	4.890		890	300		1,190	1,425
5160	12' long		1.99	4.030		1,025	247		1,272	1,500
5170	With exothermic weld connection		1.54	5.210		1,125	320		1,445	1,700
5180	20' long		1.60	5		1,400	305		1,705	1,975
5190	With exothermic weld connection		1.29	6.210		1,375	380		1,755	2,100
5195	40' long with exothermic weld connection	2 Elec	1.84	8.700		3,050	535		3,585	4,175
5200	L-shaped, 2" diam.									
5220	4' vert. x 10' horiz., clamp connection	1 Elec	4.90	1.630	Ea.	1,325	100		1,425	1,625
5230	With exothermic weld connection	"	2.83	2.820	"	1,325	173		1,498	1,700
5300	Protective box at grade level, with breather slots									
5320	Round 12" long, fiberlyte	1 Elec	29.44	.272	Ea.	70	16.65		86.65	102
5330	Concrete	"	14.72	.543		112	33.50		145.50	173
5400	Bentonite clay, 50# bag, 1 per 10' of rod					56			56	61.50
5500	Equipotential earthing bar	1 Elec	1.84	4.350		138	267		405	545
7000	Exothermic welding kit, multi vertical	"	7.36	1.090		720	67		787	890

26 05 29 – Hangers and Supports for Electrical Systems

26 05 29.20 Hangers		Crew	Daily Output	Labor-Hours	Unit	Material	2020 Bare Costs Labor	Equipment	Total	Total Incl O&P
0010	**HANGERS**									
0015	See section 22 05 29.10 for additional items									
0030	Conduit supports									
0050	Strap w/2 holes, rigid steel conduit									
0100	1/2" diameter	1 Elec	432.40	.019	Ea.	.11	1.14		1.25	1.81
0150	3/4" diameter		404.80	.020		.13	1.21		1.34	1.95
0200	1" diameter		368	.022		.18	1.33		1.51	2.18
0300	1-1/4" diameter		326.60	.024		.29	1.50		1.79	2.56
0350	1-1/2" diameter		294.40	.027		.38	1.67		2.05	2.89
0400	2" diameter		244.72	.033		.50	2.01		2.51	3.54
0500	2-1/2" diameter		147.20	.054		.78	3.33		4.11	5.80

26 05 29.20 Hangers		Crew	Daily Output	Labor-Hours	Unit	Material	2020 Bare Costs Labor	Equipment	Total	Total Incl O&P
0550	3" diameter	1 Elec	122.36	.065	Ea.	1.04	4.01		5.05	7.10
0600	3-1/2" diameter		92	.087		2.73	5.35		8.08	10.95
0650	4" diameter		73.60	.109		1.05	6.65		7.70	11.10
0700	EMT, 1/2" diameter		432.40	.019		.20	1.14		1.34	1.91
0800	3/4" diameter		404.80	.020		.19	1.21		1.40	2.02
0850	1" diameter		368	.022		.36	1.33		1.69	2.39
0900	1-1/4" diameter		326.60	.024		.74	1.50		2.24	3.05
0950	1-1/2" diameter		294.40	.027		.99	1.67		2.66	3.57
1000	2" diameter		244.72	.033		1	2.01		3.01	4.09
1100	2-1/2" diameter		147.20	.054		1.49	3.33		4.82	6.60
1150	3" diameter		122.36	.065		1.85	4.01		5.86	8
1200	3-1/2" diameter		92	.087		4.11	5.35		9.46	12.45
1250	4" diameter		73.60	.109		2.96	6.65		9.61	13.20
1400	Hanger, with bolt, 1/2" diameter		184	.043		.44	2.67		3.11	4.45
1450	3/4" diameter		174.80	.046		.46	2.81		3.27	4.69
1500	1" diameter		161.92	.049		.49	3.03		3.52	5.05
1550	1-1/4" diameter		147.20	.054		.61	3.33		3.94	5.65
1600	1-1/2" diameter		128.80	.062		.81	3.81		4.62	6.55
1650	2" diameter		119.60	.067		1.41	4.10		5.51	7.65
1700	2-1/2" diameter		92	.087		1.81	5.35		7.16	9.95
1750	3" diameter		58.88	.136		2.38	8.35		10.73	15
1800	3-1/2" diameter		46	.174		2.34	10.65		12.99	18.45
1850	4" diameter		36.80	.217		3.60	13.35		16.95	24
1900	Riser clamps, conduit, 1/2" diameter		36.80	.217		8.95	13.35		22.30	29.50
1950	3/4" diameter		33.12	.242		8.55	14.80		23.35	31.50
2000	1" diameter		27.60	.290		7.75	17.80		25.55	35
2100	1-1/4" diameter		24.84	.322		11.75	19.75		31.50	42.50
2150	1-1/2" diameter		24.84	.322		10.95	19.75		30.70	41.50
2200	2" diameter		18.40	.435		11.30	26.50		37.80	52
2250	2-1/2" diameter		18.40	.435		12.30	26.50		38.80	53
2300	3" diameter		16.56	.483		13.75	29.50		43.25	59
2350	3-1/2" diameter		16.56	.483		12.75	29.50		42.25	58
2400	4" diameter		12.88	.621	▽	16.55	38		54.55	74.50
2500	Threaded rod, painted, 1/4" diameter		239.20	.033	L.F.	1.96	2.05		4.01	5.20
2600	3/8" diameter		184	.043		7.40	2.67		10.07	12.05
2700	1/2" diameter		128.80	.062		8.45	3.81		12.26	14.90
2800	5/8" diameter		92	.087		10.65	5.35		16	19.70
2900	3/4" diameter	▽	55.20	.145	▽	7.80	8.90		16.70	22
2940	Couplings painted, 1/4" diameter				C	390			390	425
2960	3/8" diameter					570			570	625
2970	1/2" diameter					485			485	530
2980	5/8" diameter					955			955	1,050
2990	3/4" diameter					1,200			1,200	1,325
3000	Nuts, galvanized, 1/4" diameter					13.05			13.05	14.35
3050	3/8" diameter					21			21	23
3100	1/2" diameter					58			58	64
3150	5/8" diameter					222			222	244
3200	3/4" diameter					157			157	172
3250	Washers, galvanized, 1/4" diameter					17.60			17.60	19.35
3300	3/8" diameter					20			20	22
3350	1/2" diameter					51.50			51.50	56.50
3400	5/8" diameter					102			102	112
3450	3/4" diameter					178			178	196

For customer support on your Electrical Change Order Costs with RSMeans data, call 800.448.8182.

Pre-Installation Change Order Costs

26 05 29 – Hangers and Supports for Electrical Systems

26 05 29.20 Hangers	Crew	Daily Output	Labor-Hours	Unit	Material	2020 Bare Costs Labor	Equipment	Total	Total Incl O&P
3500	Lock washers, galvanized, 1/4" diameter				C	11.40			11.40	12.55
3550	3/8" diameter					19.50			19.50	21.50
3600	1/2" diameter					29			29	32
3650	5/8" diameter					51.50			51.50	56.50
3700	3/4" diameter					101			101	111
3710	304 stainless steel, 1' long, 12GA, 1-5/8" x 1-5/8"	1 Elec	46	.174	Ea.	22.50	10.65		33.15	41
3715	304 stainless steel, 1'6" long, 12GA, 1-5/8" x 1-5/8"		30.66	.261		30.50	16		46.50	57.50
3720	304 stainless steel, 2' long, 12GA, 1-5/8" x 1-5/8"		23	.348		47	21.50		68.50	84
3725	304 stainless steel, 3' long, 12GA, 1-5/8" x 1-5/8"		15.33	.522		60.50	32		92.50	115
3730	304 stainless steel, 4' long, 12GA, 1-5/8" x 1-5/8"		11.50	.696		75	42.50		117.50	146
3735	304 stainless steel, 5' long, 12GA, 1-5/8" x 1-5/8"		9.20	.870		113	53.50		166.50	205
3740	304 stainless steel, 10' long, 12GA, 1-5/8" x 1-5/8"		4.60	1.740		171	107		278	350
3745	Aluminum, 1' length, 12GA, 1-5/8" x 1-5/8"		46	.174		6.85	10.65		17.50	23.50
3750	Aluminum, 1'6" length, 12GA, 1-5/8" x 1-5/8"		30.67	.261		10.15	16		26.15	35
3755	Aluminum, 2' length, 12GA, 1-5/8" x 1-5/8"		23	.348		12.80	21.50		34.30	46
3760	Aluminum, 3' length, 12GA, 1-5/8" x 1-5/8"		15.33	.522		19.95	32		51.95	69.50
3765	Aluminum, 4' length, 12GA, 1-5/8" x 1-5/8"		11.50	.696		27	42.50		69.50	93
3770	Aluminum, 5' length, 12GA, 1-5/8" x 1-5/8"		9.20	.870		31.50	53.50		85	114
3775	Aluminum, 10' length, 12GA, 1-5/8" x 1-5/8"		4.60	1.740		57.50	107		164.50	223
3800	Channels, steel, 3/4" x 1-1/2", 14 ga.		73.60	.109	L.F.	3.73	6.65		10.38	14.05
3900	1-1/2" x 1-1/2", 12 ga.		64.40	.124		4.60	7.60		12.20	16.40
4000	1-7/8" x 1-1/2"		55.20	.145		26.50	8.90		35.40	42.50
4100	3" x 1-1/2"		46	.174		25	10.65		35.65	43.50
4200	Spring nuts, long, 1/4"		110.40	.072	Ea.	.81	4.45		5.26	7.50
4250	3/8"		92	.087		.94	5.35		6.29	9
4300	1/2"		73.60	.109		1.04	6.65		7.69	11.10
4350	Spring nuts, short, 1/4"		110.40	.072		1.70	4.45		6.15	8.45
4400	3/8"		92	.087		2.13	5.35		7.48	10.30
4450	1/2"		73.60	.109		2.26	6.65		8.91	12.45
4500	Closure strip		184	.043	L.F.	4.58	2.67		7.25	9
4550	End cap		55.20	.145	Ea.	1.51	8.90		10.41	14.90
4600	End connector 3/4" conduit		36.80	.217		8.95	13.35		22.30	29.50
4650	Junction box, 1 channel		14.72	.543		46.50	33.50		80	101
4700	2 channel		12.88	.621		64.50	38		102.50	127
4750	3 channel		11.04	.725		81.50	44.50		126	156
4800	4 channel		9.20	.870		107	53.50		160.50	198
4850	Splice plate		36.80	.217		11.75	13.35		25.10	33
4900	Continuous concrete insert, 1-1/2" deep, 1' long		14.72	.543		25	33.50		58.50	77
4950	2' long		12.88	.621		30	38		68	89.50
5000	3' long		11.04	.725		30	44.50		74.50	99
5050	4' long		9.20	.870		43	53.50		96.50	127
5100	6' long		7.36	1.090		64.50	67		131.50	171
5150	3/4" deep, 1' long		14.72	.543		25.50	33.50		59	77.50
5200	2' long		12.88	.621		25.50	38		63.50	84.50
5250	3' long		11.04	.725		27	44.50		71.50	95.50
5300	4' long		9.20	.870		43.50	53.50		97	128
5350	6' long		7.36	1.090		65.50	67		132.50	172
5400	90° angle fitting 2-1/8" x 2-1/8"		55.20	.145		4.73	8.90		13.63	18.45
5450	Supports, suspension rod type, small		55.20	.145		77	8.90		85.90	98
5500	Large		36.80	.217		82	13.35		95.35	110
5550	Beam clamp, small		55.20	.145		35.50	8.90		44.40	52.50
5600	Large		36.80	.217		32	13.35		45.35	55
5650	U-support, small		55.20	.145		6.60	8.90		15.50	20.50

26 05 29 – Hangers and Supports for Electrical Systems

26 05 29.20 Hangers	Crew	Daily Output	Labor-Hours	Unit	Material	2020 Bare Costs Labor	Equipment	Total	Total Incl O&P
5700	Large	1 Elec	36.80	.217	Ea.	13.55	13.35		26.90	35
5750	Concrete insert, cast, for up to 1/2" threaded rod		14.72	.543		14.75	33.50		48.25	65.50
5800	Beam clamp, 1/4" clamp, for 1/4" threaded drop rod		29.44	.272		3.40	16.65		20.05	28.50
5900	3/8" clamp, for 3/8" threaded drop rod		29.44	.272		3.98	16.65		20.63	29.50
6000	Strap, rigid conduit, 1/2" diameter		496.80	.016		1.50	.99		2.49	3.12
6050	3/4" diameter		404.80	.020		1.56	1.21		2.77	3.53
6100	1" diameter		386.40	.021		1.86	1.27		3.13	3.94
6150	1-1/4" diameter		368	.022		2.19	1.33		3.52	4.40
6200	1-1/2" diameter		368	.022		2.53	1.33		3.86	4.77
6250	2" diameter		245.64	.033		2.89	2		4.89	6.15
6300	2-1/2" diameter		245.64	.033		2.78	2		4.78	6.05
6350	3" diameter		147.20	.054		3.58	3.33		6.91	8.90
6400	3-1/2" diameter		122.36	.065		3.95	4.01		7.96	10.30
6450	4" diameter		92	.087		4.63	5.35		9.98	13.05
6500	5" diameter		73.60	.109		11.35	6.65		18	22.50
6550	6" diameter		55.20	.145		38.50	8.90		47.40	56
6600	EMT, 1/2" diameter		496.80	.016		1.26	.99		2.25	2.86
6650	3/4" diameter		404.80	.020		1.33	1.21		2.54	3.27
6700	1" diameter		386.40	.021		1.39	1.27		2.66	3.42
6750	1-1/4" diameter		368	.022		1.76	1.33		3.09	3.93
6800	1-1/2" diameter		368	.022		2.25	1.33		3.58	4.47
6850	2" diameter		245.64	.033		2.15	2		4.15	5.35
6900	2-1/2" diameter		245.64	.033		2.31	2		4.31	5.50
6950	3" diameter		147.20	.054		2.93	3.33		6.26	8.20
6970	3-1/2" diameter		122.36	.065		3.90	4.01		7.91	10.25
6990	4" diameter		92	.087		4.36	5.35		9.71	12.75
7000	Clip, 1 hole for rigid conduit, 1/2" diameter		460	.017		.08	1.07		1.15	1.67
7050	3/4" diameter		432.40	.019		.93	1.14		2.07	2.71
7100	1" diameter		404.80	.020		2.26	1.21		3.47	4.30
7150	1-1/4" diameter		368	.022		2.08	1.33		3.41	4.27
7200	1-1/2" diameter		326.60	.024		2.58	1.50		4.08	5.05
7250	2" diameter		294.40	.027		1.99	1.67		3.66	4.67
7300	2-1/2" diameter		244.72	.033		6	2.01		8.01	9.60
7350	3" diameter		147.20	.054		5.10	3.33		8.43	10.55
7400	3-1/2" diameter		122.36	.065		7.55	4.01		11.56	14.25
7450	4" diameter		92	.087		15.80	5.35		21.15	25.50
7500	5" diameter		73.60	.109		121	6.65		127.65	144
7550	6" diameter		55.20	.145		266	8.90		274.90	305
7600	Hammer on purlin clip, 1/4" hole, 1/16" flange		55.20	.145		.76	8.90		9.66	14.10
7610	1/4" hole threaded,1/4" flange, rod hanger		55.20	.145		1.70	8.90		10.60	15.10
7620	3/8" hole threaded,1/4" flange, rod hanger		55.20	.145		1.78	8.90		10.68	15.20
7630	Z purlin clip, 1/4" & 3/8" flange threaded rod hanger		55.20	.145		1.36	8.90		10.26	14.75
7640	1/2" rigid, 3/4" conduit, 1" flange		55.20	.145		1.79	8.90		10.69	15.20
7650	1" conduit, 1" flange		55.20	.145		2.03	8.90		10.93	15.50
7660	1-1/4" conduit, 1" flange		55.20	.145		2.39	8.90		11.29	15.90
7670	Push-in type conduit clip 3/4" EMT, with 9/32" hole		55.20	.145		.58	8.90		9.48	13.90
7680	3/4" EMT, 1/2" rigid		55.20	.145		1.56	8.90		10.46	14.95
7690	Push-in type tandem conduit clip 3/4" - 1" EMT, 1/2" - 3/4" rigid		55.20	.145		1.66	8.90		10.56	15.10
7820	Conduit hangers, with bolt & 12" rod, 1/2" diameter		138	.058		7.80	3.56		11.36	13.90
7830	3/4" diameter		133.40	.060		7.85	3.68		11.53	14.10
7840	1" diameter		124.20	.064		7.85	3.95		11.80	14.55
7850	1-1/4" diameter		110.40	.072		8	4.45		12.45	15.40
7860	1-1/2" diameter		101.20	.079		8.20	4.85		13.05	16.20

For customer support on your Electrical Change Order Costs with RSMeans data, call 800.448.8182.   **Pre-Installation Change Order Costs**

26 05 29.20 Hangers		Crew	Daily Output	Labor-Hours	Unit	Material	2020 Bare Costs Labor	Equipment	Total	Total Incl O&P
7870	2" diameter	1 Elec	92	.087	Ea.	9.85	5.35		15.20	18.75
7880	2-1/2" diameter		73.60	.109		10.25	6.65		16.90	21
7890	3" diameter		55.20	.145		10.80	8.90		19.70	25
7900	3-1/2" diameter		41.40	.193		10.75	11.85		22.60	29.50
7910	4" diameter		32.20	.248		12.05	15.25		27.30	36
7920	5" diameter		27.60	.290		14.25	17.80		32.05	42
7930	6" diameter		23	.348		15.70	21.50		37.20	49.50
7950	Jay clamp, 1/2" diameter		29.44	.272		9.40	16.65		26.05	35.50
7960	3/4" diameter		29.44	.272		12.35	16.65		29	38.50
7970	1" diameter		29.44	.272		13.75	16.65		30.40	40
7980	1-1/4" diameter		27.60	.290		13.65	17.80		31.45	41.50
7990	1-1/2" diameter		27.60	.290		15.25	17.80		33.05	43.50
8000	2" diameter		27.60	.290		22.50	17.80		40.30	51.50
8010	2-1/2" diameter		25.76	.311		28	19.05		47.05	59.50
8020	3" diameter		25.76	.311		28	19.05		47.05	59.50
8030	3-1/2" diameter		23	.348		78.50	21.50		100	119
8040	4" diameter		23	.348		78.50	21.50		100	119
8050	5" diameter		18.40	.435		141	26.50		167.50	195
8060	6" diameter		14.72	.543		269	33.50		302.50	345
8070	Channels, 3/4" x 1-1/2" w/12" rods for 1/2" to 1" conduit		27.60	.290		12.95	17.80		30.75	41
8080	1-1/2" x 1-1/2" w/12" rods for 1-1/4" to 2" conduit		25.76	.311		13.15	19.05		32.20	43
8090	1-1/2" x 1-1/2" w/12" rods for 2-1/2" to 4" conduit		23.92	.334		16.40	20.50		36.90	48.50
8100	1-1/2" x 1-7/8" w/12" rods for 5" to 6" conduit		22.08	.362		61	22		83	100
8110	Beam clamp, conduit, plastic coated steel, 1/2" diam.		27.60	.290		43	17.80		60.80	74
8120	3/4" diameter		27.60	.290		42.50	17.80		60.30	73.50
8130	1" diameter		27.60	.290		50	17.80		67.80	81.50
8140	1-1/4" diameter		25.76	.311		62	19.05		81.05	96.50
8150	1-1/2" diameter		25.76	.311		73.50	19.05		92.55	110
8160	2" diameter		25.76	.311		73.50	19.05		92.55	109
8170	2-1/2" diameter		23.92	.334		101	20.50		121.50	142
8180	3" diameter		23.92	.334		112	20.50		132.50	155
8190	3-1/2" diameter		21.16	.378		84.50	23		107.50	127
8200	4" diameter		21.16	.378		116	23		139	163
8210	5" diameter		16.56	.483		250	29.50		279.50	320
8220	Channels, plastic coated									
8250	3/4" x 1-1/2", w/12" rods for 1/2" to 1" conduit	1 Elec	25.76	.311	Ea.	43.50	19.05		62.55	76.50
8260	1-1/2" x 1-1/2", w/12" rods for 1-1/4" to 2" conduit		23.92	.334		54.50	20.50		75	90.50
8270	1-1/2" x 1-1/2", w/12" rods for 2-1/2" to 3-1/2" conduit		22.08	.362		59	22		81	98
8280	1-1/2" x 1-7/8", w/12" rods for 4" to 5" conduit		20.24	.395		98	24.50		122.50	144
8290	1-1/2" x 1-7/8", w/12" rods for 6" conduit		18.40	.435		108	26.50		134.50	158
8320	Conduit hangers, plastic coated steel, with bolt & 12" rod, 1/2" diam.		128.80	.062		28	3.81		31.81	36
8330	3/4" diameter		124.20	.064		28.50	3.95		32.45	37.50
8340	1" diameter		115	.070		29.50	4.27		33.77	39
8350	1-1/4" diameter		101.20	.079		31	4.85		35.85	41
8360	1-1/2" diameter		92	.087		34.50	5.35		39.85	46
8370	2" diameter		82.80	.097		38.50	5.95		44.45	51
8380	2-1/2" diameter		64.40	.124		45.50	7.60		53.10	61.50
8390	3" diameter		46	.174		55	10.65		65.65	76.50
8400	3-1/2" diameter		32.20	.248		50	15.25		65.25	78
8410	4" diameter		23	.348		81.50	21.50		103	122
8420	5" diameter		18.40	.435		89.50	26.50		116	138
9000	Parallel type, conduit beam clamp, 1/2"		29.44	.272		7.75	16.65		24.40	33.50
9010	3/4"		29.44	.272		8.95	16.65		25.60	35

26 05 29.20 Hangers

		Crew	Daily Output	Labor-Hours	Unit	Material	Labor	Equipment	Total	Total Incl O&P
9020	1"	1 Elec	29.44	.272	Ea.	9.05	16.65		25.70	35
9030	1-1/4"		27.60	.290		12.90	17.80		30.70	40.50
9040	1-1/2"		27.60	.290		15.90	17.80		33.70	44
9050	2"		27.60	.290		18.95	17.80		36.75	47.50
9060	2-1/2"		25.76	.311		24	19.05		43.05	55
9070	3"		25.76	.311		29	19.05		48.05	60.50
9090	4"		23	.348		19.05	21.50		40.55	53
9110	Right angle, conduit beam clamp, 1/2"		29.44	.272		3.15	16.65		19.80	28.50
9120	3/4"		29.44	.272		3.39	16.65		20.04	28.50
9130	1"		29.44	.272		3.05	16.65		19.70	28.50
9140	1-1/4"		27.60	.290		4.66	17.80		22.46	31.50
9150	1-1/2"		27.60	.290		4.76	17.80		22.56	32
9160	2"		27.60	.290		6.95	17.80		24.75	34
9170	2-1/2"		25.76	.311		8.55	19.05		27.60	38
9180	3"		25.76	.311		9.35	19.05		28.40	39
9190	3-1/2"		23	.348		12.95	21.50		34.45	46.50
9200	4"		23	.348		14.40	21.50		35.90	48
9230	Adjustable, conduit hanger, 1/2"		29.44	.272		9.45	16.65		26.10	35.50
9240	3/4"		29.44	.272		7	16.65		23.65	32.50
9250	1"		29.44	.272		9.10	16.65		25.75	35
9260	1-1/4"		27.60	.290		9.90	17.80		27.70	37.50
9270	1-1/2"		27.60	.290		11.80	17.80		29.60	39.50
9280	2"		27.60	.290		4.91	17.80		22.71	32
9290	2-1/2"		25.76	.311		28	19.05		47.05	59.50
9300	3"		25.76	.311		33	19.05		52.05	65
9310	3-1/2"		23	.348		33	21.50		54.50	68
9320	4"		23	.348		39	21.50		60.50	75
9330	5"		18.40	.435		47	26.50		73.50	91.50
9340	6"		14.72	.543		59	33.50		92.50	115
9350	Combination conduit hanger, 3/8"		29.44	.272		18.45	16.65		35.10	45.50
9360	Adjustable flange, 3/8"		29.44	.272		21.50	16.65		38.15	49

26 05 29.30 Fittings and Channel Support

		Crew	Daily Output	Labor-Hours	Unit	Material	Labor	Equipment	Total	Total Incl O&P
0010	**FITTINGS & CHANNEL SUPPORT**									
0020	Rooftop channel support									
0200	2-7/8" L x 1-5/8" W, 12 ga. pre galv., dbl. base	1 Elec	42.32	.189	Ea.	4.78	11.60		16.38	22.50
0210	2-7/8" L x 1-5/8" W, 12 ga. hot dip galv., dbl. base		42.32	.189		2.78	11.60		14.38	20.50
0220	2-7/16" L x 1-5/8" W, 12 ga. pre galv., sngl base		44.16	.181		2.29	11.10		13.39	19.05
0230	2-7/16" L x 1-5/8" W, 12 ga. hot dip galv., sngl base		44.16	.181		8.15	11.10		19.25	25.50
0240	13/16" L x 1-5/8" W, 14 ga. pre galv., sngl base		51.52	.155		1.95	9.55		11.50	16.35
0250	13/16" L x 1-5/8" W, 14 ga. hot dip galv., sngl base		51.52	.155		7.95	9.55		17.50	23
0260	1-5/8" L x 1-5/8" W, 12 ga. pre galv., sngl base		49.68	.161		2.11	9.90		12.01	17
0270	1-5/8" L x 1-5/8" W, 12 ga. hot dip galv., sngl base		49.68	.161		2.24	9.90		12.14	17.15
0280	1-5/8" x 28", 12 ga. pre galv., dbl. H block base		38.64	.207		4.84	12.70		17.54	24
0290	1-5/8" x 36", 12 ga. pre galv., dbl. H block base		36.80	.217		5.10	13.35		18.45	25.50
0300	1-5/8" x 42", 12 ga. pre galv., dbl. H block base		31.28	.256		5.75	15.70		21.45	30
0310	1-5/8" x 50", 12 ga. pre galv., dbl. H block base		34.96	.229		6.15	14.05		20.20	28
0320	1-5/8" x 60", 12 ga. pre galv., dbl. H block base		33.12	.242		6.75	14.80		21.55	29.50
0330	13/16" x 8" H, threaded rod pre galv., H block base		32.20	.248		3.24	15.25		18.49	26
0340	13/16" x 12" H, threaded rod pre galv., H block base		31.83	.251		4.05	15.40		19.45	27.50
0350	1-5/8" x 16" H, threaded rod dbl. H block base		31.28	.256		6.05	15.70		21.75	30
0360	1-5/8" x 12" H, pre galv., dbl. H block base		31.83	.251		11.60	15.40		27	36
0370	1-5/8" x 24"H, pre galv., dbl. H block base		31.28	.256		12.25	15.70		27.95	37

26 05 29 – Hangers and Supports for Electrical Systems

26 05 29.30 Fittings and Channel Support	Crew	Daily Output	Labor-Hours	Unit	Material	2020 Bare Costs Labor	Equipment	Total	Total Incl O&P
0380 3-1/4" x 24" H, pre galv., dbl. H block base	1 Elec	29.44	.272	Ea.	28	16.65		44.65	55.50
0390 3-1/4" x 36" H, pre galv., dbl. H block base	↓	27.60	.290	↓	30.50	17.80		48.30	60

26 05 33 – Raceway and Boxes for Electrical Systems

26 05 33.13 Conduit

	Crew	Daily Output	Labor-Hours	Unit	Material	2020 Bare Costs Labor	Equipment	Total	Total Incl O&P
0010 **CONDUIT** To 10' high, includes 2 terminations, 2 elbows,									
0020 11 beam clamps, and 11 couplings per 100 L.F.									
0300 Aluminum, 1/2" diameter	1 Elec	92	.087	L.F.	2.18	5.35		7.53	10.35
0500 3/4" diameter		82.80	.097		3.22	5.95		9.17	12.40
0700 1" diameter		73.60	.109		4.29	6.65		10.94	14.65
1000 1-1/4" diameter		64.40	.124		5.25	7.60		12.85	17.10
1030 1-1/2" diameter		59.80	.134		6.40	8.20		14.60	19.20
1050 2" diameter		55.20	.145		8.70	8.90		17.60	23
1070 2-1/2" diameter	↓	46	.174		11	10.65		21.65	28
1100 3" diameter	2 Elec	82.80	.193		15.40	11.85		27.25	34.50
1130 3-1/2" diameter		73.60	.217		19.40	13.35		32.75	41.50
1140 4" diameter		64.40	.248		23	15.25		38.25	48
1150 5" diameter		46	.348		43.50	21.50		65	79.50
1160 6" diameter	↓	36.80	.435	↓	76.50	26.50		103	124
1161 Field bends, 45° to 90°, 1/2" diameter	1 Elec	48.76	.164	Ea.		10.05		10.05	15
1162 3/4" diameter		43.24	.185			11.35		11.35	16.90
1163 1" diameter		40.48	.198			12.10		12.10	18.05
1164 1-1/4" diameter		21.16	.378			23		23	34.50
1165 1-1/2" diameter		19.32	.414			25.50		25.50	38
1166 2" diameter		14.72	.543			33.50		33.50	49.50
1170 Elbows, 1/2" diameter		36.80	.217		5.85	13.35		19.20	26.50
1200 3/4" diameter		29.44	.272		7.65	16.65		24.30	33.50
1230 1" diameter		25.76	.311		10.85	19.05		29.90	40.50
1250 1-1/4" diameter		22.08	.362		23	22		45	58
1270 1-1/2" diameter		18.40	.435		30	26.50		56.50	73
1300 2" diameter		14.72	.543		33.50	33.50		67	86
1330 2-1/2" diameter		11.04	.725		64	44.50		108.50	137
1350 3" diameter		7.36	1.090		104	67		171	215
1370 3-1/2" diameter		5.52	1.450		154	89		243	300
1400 4" diameter		4.60	1.740		177	107		284	355
1410 5" diameter		3.68	2.170		535	133		668	785
1420 6" diameter		2.30	3.480		735	214		949	1,125
1430 Couplings, 1/2" diameter		294.40	.027		2.14	1.67		3.81	4.83
1450 3/4" diameter		176.64	.045		3.24	2.78		6.02	7.70
1470 1" diameter		147.20	.054		4.20	3.33		7.53	9.60
1500 1-1/4" diameter		110.40	.072		4.50	4.45		8.95	11.55
1530 1-1/2" diameter		88.32	.091		5.90	5.55		11.45	14.75
1550 2" diameter		73.60	.109		8.50	6.65		15.15	19.30
1570 2-1/2" diameter		63.48	.126		14.50	7.75		22.25	27.50
1600 3" diameter		58.88	.136		24.50	8.35		32.85	39.50
1630 3-1/2" diameter		51.52	.155		27.50	9.55		37.05	44.50
1650 4" diameter		48.76	.164		41	10.05		51.05	60
1670 5" diameter		44.16	.181		101	11.10		112.10	128
1690 6" diameter	↓	42.32	.189	↓	236	11.60		247.60	277
1691 See note on line 26 05 33.13 9995									
1750 Rigid galvanized steel, 1/2" diameter	1 Elec	82.80	.097	L.F.	3.32	5.95		9.27	12.50
1770 3/4" diameter		73.60	.109		6.15	6.65		12.80	16.75
1800 1" diameter	↓	59.80	.134	↓	9.25	8.20		17.45	22.50

26 05 33.13 Conduit		Crew	Daily Output	Labor-Hours	Unit	Material	2020 Bare Costs Labor	Equipment	Total	Total Incl O&P
1830	1-1/4" diameter	1 Elec	55.20	.145	L.F.	6.40	8.90		15.30	20.50
1850	1-1/2" diameter		50.60	.158		11.20	9.70		20.90	27
1870	2" diameter		41.40	.193		11.75	11.85		23.60	30.50
1900	2-1/2" diameter		32.20	.248		17.10	15.25		32.35	41.50
1930	3" diameter	2 Elec	46	.348		19	21.50		40.50	53
1950	3-1/2" diameter		40.48	.395		24.50	24.50		49	62.50
1970	4" diameter		36.80	.435		29.50	26.50		56	72
1980	5" diameter		27.60	.580		47	35.50		82.50	105
1990	6" diameter		18.40	.870		72.50	53.50		126	160
1991	Field bends, 45° to 90°, 1/2" diameter	1 Elec	40.48	.198	Ea.		12.10		12.10	18.05
1992	3/4" diameter		36.80	.217			13.35		13.35	19.85
1993	1" diameter		33.12	.242			14.80		14.80	22
1994	1-1/4" diameter		17.48	.458			28		28	42
1995	1-1/2" diameter		16.56	.483			29.50		29.50	44
1996	2" diameter		11.96	.669			41		41	61
2000	Elbows, 1/2" diameter		29.44	.272		4.64	16.65		21.29	30
2030	3/4" diameter		25.76	.311		5.35	19.05		24.40	34.50
2050	1" diameter		22.08	.362		7.05	22		29.05	41
2070	1-1/4" diameter		16.56	.483		13.60	29.50		43.10	59
2100	1-1/2" diameter		14.72	.543		23	33.50		56.50	74.50
2130	2" diameter		11.04	.725		33	44.50		77.50	103
2150	2-1/2" diameter		7.36	1.090		36	67		103	139
2170	3" diameter		5.52	1.450		47	89		136	184
2200	3-1/2" diameter		3.86	2.070		78	127		205	275
2220	4" diameter		3.68	2.170		88.50	133		221.50	295
2230	5" diameter		3.22	2.480		269	152		421	525
2240	6" diameter		1.84	4.350		375	267		642	810
2250	Couplings, 1/2" diameter		245.64	.033		1.25	2		3.25	4.35
2270	3/4" diameter		147.20	.054		1.84	3.33		5.17	7
2300	1" diameter		122.36	.065		3.20	4.01		7.21	9.45
2330	1-1/4" diameter		92	.087		3.10	5.35		8.45	11.35
2350	1-1/2" diameter		73.60	.109		6.55	6.65		13.20	17.15
2370	2" diameter		61.64	.130		8.55	7.95		16.50	21.50
2400	2-1/2" diameter		52.44	.153		12.60	9.35		21.95	28
2430	3" diameter		48.76	.164		16.25	10.05		26.30	33
2450	3-1/2" diameter		43.24	.185		21	11.35		32.35	40
2470	4" diameter		40.48	.198		35	12.10		47.10	56.50
2480	5" diameter		36.80	.217		40.50	13.35		53.85	65
2490	6" diameter		34.96	.229		67	14.05		81.05	95
2491	See note on line 26 05 33.13 9995									
2500	Steel, intermediate conduit (IMC), 1/2" diameter	1 Elec	92	.087	L.F.	2.29	5.35		7.64	10.45
2530	3/4" diameter		82.80	.097		3.17	5.95		9.12	12.35
2550	1" diameter		64.40	.124		4.28	7.60		11.88	16.05
2570	1-1/4" diameter		59.80	.134		4.74	8.20		12.94	17.40
2600	1-1/2" diameter		55.20	.145		6.90	8.90		15.80	21
2630	2" diameter		46	.174		8.20	10.65		18.85	25
2650	2-1/2" diameter		36.80	.217		10.85	13.35		24.20	32
2670	3" diameter	2 Elec	55.20	.290		16.25	17.80		34.05	44.50
2700	3-1/2" diameter		49.68	.322		22.50	19.75		42.25	54.50
2730	4" diameter		46	.348		14.45	21.50		35.95	48
2731	Field bends, 45° to 90°, 1/2" diameter	1 Elec	40.48	.198	Ea.		12.10		12.10	18.05
2732	3/4" diameter		36.80	.217			13.35		13.35	19.85
2733	1" diameter		33.12	.242			14.80		14.80	22

26 05 33 – Raceway and Boxes for Electrical Systems

26 05 33.13 Conduit		Crew	Daily Output	Labor-Hours	Unit	Material	2020 Bare Costs Labor	Equipment	Total	Total Incl O&P
2734	1-1/4" diameter	1 Elec	17.48	.458	Ea.		28		28	42
2735	1-1/2" diameter		16.56	.483			29.50		29.50	44
2736	2" diameter		11.96	.669			41		41	61
2750	Elbows, 1/2" diameter		29.44	.272		6.55	16.65		23.20	32
2770	3/4" diameter		25.76	.311		14.10	19.05		33.15	44
2800	1" diameter		22.08	.362		15.95	22		37.95	50.50
2830	1-1/4" diameter		16.56	.483		21.50	29.50		51	67.50
2850	1-1/2" diameter		14.72	.543		46	33.50		79.50	100
2870	2" diameter		11.04	.725		72.50	44.50		117	146
2900	2-1/2" diameter		7.36	1.090		48	67		115	153
2930	3" diameter		5.52	1.450		66.50	89		155.50	206
2950	3-1/2" diameter		3.86	2.070		236	127		363	450
2970	4" diameter		3.68	2.170		88	133		221	295
3000	Couplings, 1/2" diameter		269.56	.030		1.25	1.82		3.07	4.08
3030	3/4" diameter		161.92	.049		1.84	3.03		4.87	6.55
3050	1" diameter		135.24	.059		3.20	3.63		6.83	8.90
3070	1-1/4" diameter		101.20	.079		3.10	4.85		7.95	10.60
3100	1-1/2" diameter		80.96	.099		6.55	6.05		12.60	16.25
3130	2" diameter		67.16	.119		8.55	7.30		15.85	20.50
3150	2-1/2" diameter		57.96	.138		12.60	8.45		21.05	26.50
3170	3" diameter		54.28	.147		16.25	9.05		25.30	31.50
3200	3-1/2" diameter		47.84	.167		21	10.25		31.25	38.50
3230	4" diameter		45.08	.177		35	10.90		45.90	54.50
3231	See note on line 26 05 33.13 9995									
4100	Rigid steel, plastic coated, 40 mil thick									
4130	1/2" diameter	1 Elec	73.60	.109	L.F.	11.65	6.65		18.30	23
4150	3/4" diameter		64.40	.124		12	7.60		19.60	24.50
4170	1" diameter		50.60	.158		15.05	9.70		24.75	31
4200	1-1/4" diameter		46	.174		21	10.65		31.65	39
4230	1-1/2" diameter		41.40	.193		23	11.85		34.85	42.50
4250	2" diameter		32.20	.248		27.50	15.25		42.75	52.50
4270	2-1/2" diameter		23	.348		45.50	21.50		67	82
4300	3" diameter	2 Elec	40.48	.395		50	24.50		74.50	91
4330	3-1/2" diameter		36.80	.435		60	26.50		86.50	106
4350	4" diameter		33.12	.483		74.50	29.50		104	126
4370	5" diameter		27.60	.580		145	35.50		180.50	213
4400	Elbows, 1/2" diameter	1 Elec	25.76	.311	Ea.	24	19.05		43.05	55
4430	3/4" diameter		22.08	.362		21.50	22		43.50	57
4450	1" diameter		16.56	.483		25.50	29.50		55	72
4470	1-1/4" diameter		14.72	.543		37	33.50		70.50	90.50
4500	1-1/2" diameter		11.04	.725		38.50	44.50		83	109
4530	2" diameter		7.36	1.090		50	67		117	155
4550	2-1/2" diameter		5.52	1.450		103	89		192	245
4570	3" diameter		3.86	2.070		161	127		288	365
4600	3-1/2" diameter		3.68	2.170		253	133		386	475
4630	4" diameter		3.50	2.290		235	140		375	465
4650	5" diameter		3.22	2.480		595	152		747	880
4680	Couplings, 1/2" diameter		195.96	.041		6.95	2.50		9.45	11.40
4700	3/4" diameter		117.76	.068		6.80	4.17		10.97	13.65
4730	1" diameter		98.44	.081		8.25	4.99		13.24	16.45
4750	1-1/4" diameter		73.60	.109		12.15	6.65		18.80	23.50
4770	1-1/2" diameter		58.88	.136		12.10	8.35		20.45	25.50
4800	2" diameter		48.76	.164		17.20	10.05		27.25	34

26 05 33.13 Conduit		Crew	Daily Output	Labor-Hours	Unit	Material	2020 Bare Costs Labor	Equipment	Total	Total Incl O&P
4830	2-1/2" diameter	1 Elec	42.32	.189	Ea.	49	11.60		60.60	71.50
4850	3" diameter		39.56	.202		46.50	12.40		58.90	69.50
4870	3-1/2" diameter		34.96	.229		69.50	14.05		83.55	97.50
4900	4" diameter		33.12	.242		69	14.80		83.80	98
4950	5" diameter	↓	29.44	.272	↓	256	16.65		272.65	305
4951	See note on line 26 05 33.13 9995									
5000	Electric metallic tubing (EMT), 1/2" diameter	1 Elec	156.40	.051	L.F.	.91	3.14		4.05	5.65
5020	3/4" diameter		119.60	.067		1.29	4.10		5.39	7.50
5040	1" diameter		105.80	.076		2.16	4.64		6.80	9.30
5060	1-1/4" diameter		92	.087		3.52	5.35		8.87	11.80
5080	1-1/2" diameter		82.80	.097		4.16	5.95		10.11	13.40
5100	2" diameter		73.60	.109		5.25	6.65		11.90	15.70
5120	2-1/2" diameter	↓	55.20	.145		5.80	8.90		14.70	19.65
5140	3" diameter	2 Elec	92	.174		7.40	10.65		18.05	24
5160	3-1/2" diameter		82.80	.193		9.25	11.85		21.10	28
5180	4" diameter	↓	73.60	.217	↓	16.25	13.35		29.60	38
5200	Field bends, 45° to 90°, 1/2" diameter	1 Elec	81.88	.098	Ea.		6		6	8.90
5220	3/4" diameter		73.60	.109			6.65		6.65	9.95
5240	1" diameter		67.16	.119			7.30		7.30	10.90
5260	1-1/4" diameter		34.96	.229			14.05		14.05	21
5280	1-1/2" diameter		33.12	.242			14.80		14.80	22
5300	2" diameter		23.92	.334			20.50		20.50	30.50
5320	Offsets, 1/2" diameter		59.80	.134			8.20		8.20	12.20
5340	3/4" diameter		57.04	.140			8.60		8.60	12.80
5360	1" diameter		48.76	.164			10.05		10.05	15
5380	1-1/4" diameter		27.60	.290			17.80		17.80	26.50
5400	1-1/2" diameter		25.76	.311			19.05		19.05	28.50
5420	2" diameter		18.40	.435			26.50		26.50	39.50
5700	Elbows, 1" diameter		36.80	.217		3.91	13.35		17.26	24
5720	1-1/4" diameter		29.44	.272		5.25	16.65		21.90	31
5740	1-1/2" diameter		22.08	.362		6.10	22		28.10	39.50
5760	2" diameter		18.40	.435		8.75	26.50		35.25	49
5780	2-1/2" diameter		11.04	.725		21.50	44.50		66	89.50
5800	3" diameter		8.28	.966		29.50	59.50		89	121
5820	3-1/2" diameter		6.44	1.240		40.50	76		116.50	158
5840	4" diameter		5.52	1.450		47.50	89		136.50	184
6200	Couplings, set screw, steel, 1/2" diameter		216.20	.037		1.93	2.27		4.20	5.50
6220	3/4" diameter		172.96	.046		2.74	2.84		5.58	7.25
6240	1" diameter		144.44	.055		4.18	3.40		7.58	9.65
6260	1-1/4" diameter		108.56	.074		8.95	4.52		13.47	16.60
6280	1-1/2" diameter		86.48	.093		1.96	5.70		7.66	10.60
6300	2" diameter		71.76	.111		17.40	6.85		24.25	29.50
6320	2-1/2" diameter		61.64	.130		7.65	7.95		15.60	20.50
6340	3" diameter		54.28	.147		7.20	9.05		16.25	21.50
6360	3-1/2" diameter		43.24	.185		10.05	11.35		21.40	28
6380	4" diameter	↓	39.56	.202	↓	70	12.40		82.40	95.50
6381	See note on line 26 05 33.13 9995									
6500	Box connectors, set screw, steel, 1/2" diameter	1 Elec	110.40	.072	Ea.	.43	4.45		4.88	7.10
6520	3/4" diameter		101.20	.079		.71	4.85		5.56	8
6540	1" diameter		82.80	.097		1.33	5.95		7.28	10.30
6560	1-1/4" diameter		64.40	.124		2.38	7.60		9.98	13.95
6580	1-1/2" diameter		55.20	.145		3.96	8.90		12.86	17.60
6600	2" diameter	↓	46	.174	↓	4.72	10.65		15.37	21

Pre-Installation Change Order Costs

26 05 33.13 Conduit		Crew	Daily Output	Labor-Hours	Unit	Material	2020 Bare Costs Labor	Equipment	Total	Total Incl O&P
6620	2-1/2" diameter	1 Elec	33.12	.242	Ea.	9.50	14.80		24.30	32.50
6640	3" diameter		24.84	.322		41.50	19.75		61.25	75
6680	3-1/2" diameter		19.32	.414		12.85	25.50		38.35	52
6700	4" diameter		14.72	.543		15.35	33.50		48.85	66.50
6740	Insulated box connectors, set screw, steel, 1/2" diameter		110.40	.072		.82	4.45		5.27	7.50
6760	3/4" diameter		101.20	.079		1.81	4.85		6.66	9.20
6780	1" diameter		82.80	.097		3.48	5.95		9.43	12.70
6800	1-1/4" diameter		64.40	.124		4.14	7.60		11.74	15.90
6820	1-1/2" diameter		55.20	.145		5.95	8.90		14.85	19.80
6840	2" diameter		46	.174		7.95	10.65		18.60	24.50
6860	2-1/2" diameter		33.12	.242		36	14.80		50.80	61.50
6880	3" diameter		24.84	.322		30.50	19.75		50.25	63
6900	3-1/2" diameter		19.32	.414		58	25.50		83.50	102
6920	4" diameter		14.72	.543		43	33.50		76.50	96.50
7000	EMT to conduit adapters, 1/2" diameter (compression)		64.40	.124		4.19	7.60		11.79	15.95
7020	3/4" diameter		55.20	.145		6.55	8.90		15.45	20.50
7040	1" diameter		46	.174		10.35	10.65		21	27.50
7060	1-1/4" diameter		36.80	.217		20	13.35		33.35	42.50
7080	1-1/2" diameter		27.60	.290		24	17.80		41.80	53
7100	2" diameter		23	.348		37.50	21.50		59	73.50
7200	EMT to Greenfield adapters, 1/2" to 3/8" diameter (compression)		82.80	.097		2.59	5.95		8.54	11.70
7220	1/2" diameter		82.80	.097		5.85	5.95		11.80	15.25
7240	3/4" diameter		73.60	.109		7.10	6.65		13.75	17.80
7260	1" diameter		64.40	.124		18.65	7.60		26.25	32
7270	1-1/4" diameter		55.20	.145		23.50	8.90		32.40	39.50
7280	1-1/2" diameter		46	.174		26.50	10.65		37.15	45
7290	2" diameter		36.80	.217		39.50	13.35		52.85	63.50
7400	EMT, LB, LR or LL fittings with covers, 1/2" diameter, set screw		22.08	.362		5.80	22		27.80	39.50
7420	3/4" diameter		18.40	.435		8.10	26.50		34.60	48.50
7440	1" diameter		14.72	.543		12.05	33.50		45.55	63
7450	1-1/4" diameter		11.96	.669		17.95	41		58.95	81
7460	1-1/2" diameter		10.12	.791		24	48.50		72.50	98.50
7470	2" diameter		8.28	.966		39.50	59.50		99	132
7600	EMT, "T" fittings with covers, 1/2" diameter, set screw		14.72	.543		6.75	33.50		40.25	57
7620	3/4" diameter		13.80	.580		8.70	35.50		44.20	62.50
7640	1" diameter		11.04	.725		11	44.50		55.50	78
7650	1-1/4" diameter		10.12	.791		16.35	48.50		64.85	90
7660	1-1/2" diameter		9.20	.870		21	53.50		74.50	103
7670	2" diameter		7.36	1.090		28.50	67		95.50	131
8000	EMT, expansion fittings, no jumper, 1/2" diameter		22.08	.362		102	22		124	145
8020	3/4" diameter		18.40	.435		150	26.50		176.50	205
8040	1" diameter		14.72	.543		161	33.50		194.50	227
8060	1-1/4" diameter		11.96	.669		183	41		224	262
8080	1-1/2" diameter		10.12	.791		242	48.50		290.50	340
8100	2" diameter		8.28	.966		405	59.50		464.50	535
8110	2-1/2" diameter		6.44	1.240		490	76		566	655
8120	3" diameter		5.52	1.450		665	89		754	860
8140	4" diameter		4.60	1.740		945	107		1,052	1,200
8200	Split adapter, 1/2" diameter		101.20	.079		3.92	4.85		8.77	11.50
8210	3/4" diameter		82.80	.097		3.77	5.95		9.72	13
8220	1" diameter		64.40	.124		5.85	7.60		13.45	17.80
8230	1-1/4" diameter		55.20	.145		9.95	8.90		18.85	24
8240	1-1/2" diameter		46	.174		13.90	10.65		24.55	31

26 05 33.13 Conduit		Crew	Daily Output	Labor-Hours	Unit	Material	2020 Bare Costs Labor	Equipment	Total	Total Incl O&P
8250	2" diameter	1 Elec	33.12	.242	Ea.	39	14.80		53.80	65
8300	1 hole clips, 1/2" diameter		460	.017		.50	1.07		1.57	2.14
8320	3/4" diameter		432.40	.019		.23	1.14		1.37	1.94
8340	1" diameter		408.48	.020		.34	1.20		1.54	2.16
8360	1-1/4" diameter		368	.022		.48	1.33		1.81	2.52
8380	1-1/2" diameter		326.60	.024		1.19	1.50		2.69	3.55
8400	2" diameter		294.40	.027		1.75	1.67		3.42	4.41
8420	2-1/2" diameter		244.72	.033		4.12	2.01		6.13	7.50
8440	3" diameter		147.20	.054		4.62	3.33		7.95	10.05
8460	3-1/2" diameter		122.36	.065		8.95	4.01		12.96	15.75
8480	4" diameter		92	.087		8.85	5.35		14.20	17.65
8500	Clamp back spacers, 1/2" diameter		460	.017		.97	1.07		2.04	2.66
8510	3/4" diameter		432.40	.019		1.21	1.14		2.35	3.02
8520	1" diameter		408.48	.020		2.27	1.20		3.47	4.28
8530	1-1/4" diameter		368	.022		4.73	1.33		6.06	7.20
8540	1-1/2" diameter		326.60	.024		9.15	1.50		10.65	12.35
8550	2" diameter		294.40	.027		6.55	1.67		8.22	9.70
8560	2-1/2" diameter		244.72	.033		12.30	2.01		14.31	16.50
8570	3" diameter		147.20	.054		32	3.33		35.33	40.50
8580	3-1/2" diameter		122.36	.065		50.50	4.01		54.51	61.50
8590	4" diameter		92	.087		123	5.35		128.35	143
8600	Offset connectors, 1/2" diameter		36.80	.217		3.34	13.35		16.69	23.50
8610	3/4" diameter		29.44	.272		4.37	16.65		21.02	30
8620	1" diameter		22.08	.362		6.05	22		28.05	39.50
8650	90° pulling elbows, female, 1/2" diameter, with gasket		22.08	.362		9.35	22		31.35	43.50
8660	3/4" diameter		18.40	.435		16.60	26.50		43.10	58
8700	Couplings, compression, 1/2" diameter, steel		71.76	.111		1.57	6.85		8.42	11.95
8710	3/4" diameter		61.64	.130		1.70	7.95		9.65	13.70
8720	1" diameter		54.28	.147		2.77	9.05		11.82	16.50
8730	1-1/4" diameter		43.24	.185		6.10	11.35		17.45	23.50
8740	1-1/2" diameter		34.96	.229		9	14.05		23.05	31
8750	2" diameter		28.52	.281		16.15	17.20		33.35	43.50
8760	2-1/2" diameter		22.08	.362		45.50	22		67.50	83
8770	3" diameter		19.32	.414		33	25.50		58.50	74.50
8780	3-1/2" diameter		17.48	.458		133	28		161	188
8790	4" diameter		14.72	.543		81.50	33.50		115	139
8791	See note on line 26 05 33.13 9995									
8800	Box connectors, compression, 1/2" diam., steel	1 Elec	110.40	.072	Ea.	2.94	4.45		7.39	9.85
8810	3/4" diameter		101.20	.079		4.03	4.85		8.88	11.65
8820	1" diameter		82.80	.097		5.70	5.95		11.65	15.10
8830	1-1/4" diameter		64.40	.124		14.40	7.60		22	27
8840	1-1/2" diameter		55.20	.145		17.60	8.90		26.50	32.50
8850	2" diameter		46	.174		24	10.65		34.65	42.50
8860	2-1/2" diameter		33.12	.242		57.50	14.80		72.30	85
8870	3" diameter		24.84	.322		77.50	19.75		97.25	115
8880	3-1/2" diameter		19.32	.414		121	25.50		146.50	171
8890	4" diameter		14.72	.543		122	33.50		155.50	184
8900	Box connectors, insulated compression, 1/2" diam., steel		110.40	.072		2.14	4.45		6.59	8.95
8910	3/4" diameter		101.20	.079		2.44	4.85		7.29	9.90
8920	1" diameter		82.80	.097		4.51	5.95		10.46	13.80
8930	1-1/4" diameter		64.40	.124		9.40	7.60		17	21.50
8940	1-1/2" diameter		55.20	.145		13.90	8.90		22.80	28.50
8950	2" diameter		46	.174		20.50	10.65		31.15	38.50

26 05 33.13 Conduit		Crew	Daily Output	Labor-Hours	Unit	Material	2020 Bare Costs Labor	2020 Bare Costs Equipment	Total	Total Incl O&P
8960	2-1/2" diameter	1 Elec	33.12	.242	Ea.	79	14.80		93.80	109
8970	3" diameter		24.84	.322		70.50	19.75		90.25	107
8980	3-1/2" diameter		19.32	.414		120	25.50		145.50	170
8990	4" diameter		14.72	.543		106	33.50		139.50	166
9100	PVC, schedule 40, 1/2" diameter		174.80	.046	L.F.	1.04	2.81		3.85	5.35
9110	3/4" diameter		133.40	.060		1.23	3.68		4.91	6.85
9120	1" diameter		115	.070		1.85	4.27		6.12	8.40
9130	1-1/4" diameter		101.20	.079		2.38	4.85		7.23	9.80
9140	1-1/2" diameter		92	.087		2.86	5.35		8.21	11.10
9150	2" diameter		82.80	.097		4.03	5.95		9.98	13.30
9160	2-1/2" diameter		59.80	.134		5.50	8.20		13.70	18.30
9170	3" diameter	2 Elec	101.20	.158		5.80	9.70		15.50	21
9180	3-1/2" diameter		92	.174		7.65	10.65		18.30	24.50
9190	4" diameter		82.80	.193		6.90	11.85		18.75	25
9200	5" diameter		64.40	.248		10	15.25		25.25	33.50
9210	6" diameter		55.20	.290		15	17.80		32.80	43
9220	Elbows, 1/2" diameter	1 Elec	46	.174	Ea.	.70	10.65		11.35	16.65
9225	3/4" diameter		38.64	.207		.72	12.70		13.42	19.70
9230	1" diameter		32.20	.248		1.45	15.25		16.70	24
9235	1-1/4" diameter		25.76	.311		3.26	19.05		22.31	32
9240	1-1/2" diameter		18.40	.435		3.25	26.50		29.75	43
9245	2" diameter	R-1A	33.49	.478		4.25	26.50		30.75	44
9250	2-1/2" diameter		24.56	.651		9.30	36		45.30	64
9255	3" diameter		21.07	.759		14.70	42		56.70	78.50
9260	3-1/2" diameter		20.42	.783		22.50	43.50		66	89
9265	4" diameter		16.74	.956		27.50	53		80.50	109
9270	5" diameter		11.13	1.440		41	79.50		120.50	164
9275	6" diameter		10.21	1.570		42	86.50		128.50	176
9312	Couplings, 1/2" diameter	1 Elec	46	.174		.13	10.65		10.78	16.05
9314	3/4" diameter		38.64	.207		.37	12.70		13.07	19.30
9316	1" diameter		32.20	.248		.30	15.25		15.55	23
9318	1-1/4" diameter		25.76	.311		.42	19.05		19.47	29
9320	1-1/2" diameter		18.40	.435		.65	26.50		27.15	40
9322	2" diameter	R-1A	33.49	.478		1.51	26.50		28.01	41
9324	2-1/2" diameter		24.56	.651		2.16	36		38.16	56
9326	3" diameter		21.07	.759		1.78	42		43.78	64.50
9328	3-1/2" diameter		20.42	.783		2.46	43.50		45.96	67
9330	4" diameter		16.74	.956		3.18	53		56.18	82
9332	5" diameter		11.13	1.440		8.40	79.50		87.90	127
9334	6" diameter		10.21	1.570		12.20	86.50		98.70	142
9335	See note on line 26 05 33.13 9995									
9340	Field bends, 45° & 90°, 1/2" diameter	1 Elec	41.40	.193	Ea.		11.85		11.85	17.65
9350	3/4" diameter		36.80	.217			13.35		13.35	19.85
9360	1" diameter		32.20	.248			15.25		15.25	22.50
9370	1-1/4" diameter		29.44	.272			16.65		16.65	25
9380	1-1/2" diameter		24.84	.322			19.75		19.75	29.50
9390	2" diameter		18.40	.435			26.50		26.50	39.50
9400	2-1/2" diameter		14.72	.543			33.50		33.50	49.50
9410	3" diameter		11.96	.669			41		41	61
9420	3-1/2" diameter		11.04	.725			44.50		44.50	66
9430	4" diameter		9.20	.870			53.50		53.50	79.50
9440	5" diameter		8.28	.966			59.50		59.50	88.50
9450	6" diameter		7.36	1.090			67		67	99.50

26 05 33.13 Conduit		Crew	Daily Output	Labor- Hours	Unit	Material	2020 Bare Costs Labor	Equipment	Total	Total Incl O&P
9460	PVC adapters, 1/2" diameter	1 Elec	46	.174	Ea.	.31	10.65		10.96	16.25
9470	3/4" diameter		38.64	.207		.44	12.70		13.14	19.40
9480	1" diameter		34.96	.229		.47	14.05		14.52	21.50
9490	1-1/4" diameter		32.20	.248		.91	15.25		16.16	23.50
9500	1-1/2" diameter		29.44	.272		.85	16.65		17.50	26
9510	2" diameter		24.84	.322		1.67	19.75		21.42	31.50
9520	2-1/2" diameter		21.16	.378		2.41	23		25.41	37
9530	3" diameter		16.56	.483		3.42	29.50		32.92	48
9540	3-1/2" diameter		11.96	.669		6.25	41		47.25	68
9550	4" diameter		10.12	.791		5.75	48.50		54.25	78.50
9560	5" diameter		7.36	1.090		11.60	67		78.60	112
9570	6" diameter	▼	5.52	1.450	▼	22.50	89		111.50	157
9580	PVC-LB, LR or LL fittings & covers									
9590	1/2" diameter	1 Elec	18.40	.435	Ea.	2.82	26.50		29.32	42.50
9600	3/4" diameter		14.72	.543		5.75	33.50		39.25	56
9610	1" diameter		11.04	.725		5.55	44.50		50.05	72
9620	1-1/4" diameter		8.28	.966		10.40	59.50		69.90	100
9630	1-1/2" diameter		6.44	1.240		12.30	76		88.30	127
9640	2" diameter		5.52	1.450		17.80	89		106.80	152
9650	2-1/2" diameter		5.52	1.450		96.50	89		185.50	238
9660	3" diameter		4.60	1.740		51	107		158	215
9670	3-1/2" diameter		3.68	2.170		68	133		201	273
9680	4" diameter	▼	2.76	2.900	▼	74	178		252	345
9690	PVC-tee fitting & cover									
9700	1/2"	1 Elec	12.88	.621	Ea.	6.35	38		44.35	63.50
9710	3/4"		11.96	.669		6.85	41		47.85	68.50
9720	1"		9.20	.870		6.40	53.50		59.90	86.50
9730	1-1/4"		8.28	.966		13.15	59.50		72.65	103
9740	1-1/2"		7.36	1.090		16.05	67		83.05	117
9750	2"	▼	6.44	1.240		21.50	76		97.50	137
9760	PVC-reducers, 3/4" x 1/2" diameter					1.21			1.21	1.33
9770	1" x 1/2" diameter					3.83			3.83	4.21
9780	1" x 3/4" diameter					3.36			3.36	3.70
9790	1-1/4" x 3/4" diameter					2.88			2.88	3.17
9800	1-1/4" x 1" diameter					5.20			5.20	5.70
9810	1-1/2" x 1-1/4" diameter					4.13			4.13	4.54
9820	2" x 1-1/4" diameter					5.40			5.40	5.95
9830	2-1/2" x 2" diameter					22			22	24
9840	3" x 2" diameter					25.50			25.50	28
9850	4" x 3" diameter					27.50			27.50	30.50
9860	Cement, quart					18.65			18.65	20.50
9870	Gallon					85			85	93.50
9880	Heat bender, to 6" diameter				▼	1,800			1,800	1,975
9900	Add to labor for higher elevated installation									
9905	10' to 14.5' high, add						10%			
9910	15' to 20' high, add						20%			
9920	20' to 25' high, add						25%			
9930	25' to 30' high, add						35%			
9940	30' to 35' high, add						40%			
9950	35' to 40' high, add						50%			
9960	Over 40' high, add						55%			
9995	Do not include labor when adding couplings to a fitting installation	R260533-30								

26 05 33 – Raceway and Boxes for Electrical Systems

26 05 33.14 Conduit		Crew	Daily Output	Labor-Hours	Unit	Material	2020 Bare Costs Labor	Equipment	Total	Total Incl O&P
0010	**CONDUIT** To 10' high, includes 11 couplings per 100'									
0200	Electric metallic tubing, 1/2" diameter	1 Elec	400.20	.020	L.F.	.70	1.23		1.93	2.61
0220	3/4" diameter		232.76	.034		1.05	2.11		3.16	4.29
0240	1" diameter		190.44	.042		1.80	2.58		4.38	5.80
0260	1-1/4" diameter		159.16	.050		3.13	3.08		6.21	8.05
0280	1-1/2" diameter		140.76	.057		3.64	3.49		7.13	9.20
0300	2" diameter		119.60	.067		4.62	4.10		8.72	11.20
0320	2-1/2" diameter		84.64	.095		4.74	5.80		10.54	13.85
0340	3" diameter	2 Elec	136.16	.118		5.50	7.20		12.70	16.80
0360	3-1/2" diameter		123.28	.130		7.70	7.95		15.65	20.50
0380	4" diameter		104.88	.153		14.45	9.35		23.80	30
0500	Steel rigid galvanized, 1/2" diameter	1 Elec	134.32	.060		2.23	3.65		5.88	7.90
0520	3/4" diameter		115	.070		4.74	4.27		9.01	11.55
0540	1" diameter		85.56	.094		7.60	5.75		13.35	16.95
0560	1-1/4" diameter		80.96	.099		4.65	6.05		10.70	14.15
0580	1-1/2" diameter		73.60	.109		9.05	6.65		15.70	19.95
0600	2" diameter		59.80	.134		8.60	8.20		16.80	21.50
0620	2-1/2" diameter		44.16	.181		14.65	11.10		25.75	32.50
0640	3" diameter	2 Elec	58.88	.272		16.15	16.65		32.80	43
0660	3-1/2" diameter		55.20	.290		21	17.80		38.80	49.50
0680	4" diameter		47.84	.334		25.50	20.50		46	58.50
0700	5" diameter		46	.348		31.50	21.50		53	66.50
0720	6" diameter		44.16	.362		48.50	22		70.50	86
1000	Steel intermediate conduit (IMC), 1/2" diameter	1 Elec	142.60	.056		1.14	3.44		4.58	6.35
1010	3/4" diameter		119.60	.067		1.39	4.10		5.49	7.65
1020	1" diameter		92	.087		2.15	5.35		7.50	10.30
1030	1-1/4" diameter		85.56	.094		2.44	5.75		8.19	11.25
1040	1-1/2" diameter		78.20	.102		3.69	6.30		9.99	13.40
1050	2" diameter		64.40	.124		4.21	7.60		11.81	16
1060	2-1/2" diameter		48.76	.164		7.55	10.05		17.60	23.50
1070	3" diameter	2 Elec	73.60	.217		11.60	13.35		24.95	32.50
1080	3-1/2" diameter		64.40	.248		13.90	15.25		29.15	38
1090	4" diameter		55.20	.290		9.30	17.80		27.10	37

26 05 33.15 Conduit Nipples

		Crew	Daily Output	Labor-Hours	Unit	Material	2020 Bare Costs Labor	Equipment	Total	Total Incl O&P
0010	**CONDUIT NIPPLES** With locknuts and bushings									
0100	Aluminum, 1/2" diameter, close	1 Elec	33.12	.242	Ea.	8.45	14.80		23.25	31.50
0120	1-1/2" long		33.12	.242		7.95	14.80		22.75	31
0140	2" long		33.12	.242		9.85	14.80		24.65	33
0160	2-1/2" long		33.12	.242		12.45	14.80		27.25	35.50
0180	3" long		33.12	.242		10.70	14.80		25.50	34
0200	3-1/2" long		33.12	.242		8.60	14.80		23.40	31.50
0220	4" long		33.12	.242		9	14.80		23.80	32
0240	5" long		33.12	.242		11.75	14.80		26.55	35
0260	6" long		33.12	.242		12.40	14.80		27.20	35.50
0280	8" long		33.12	.242		12.20	14.80		27	35.50
0300	10" long		33.12	.242		13.35	14.80		28.15	36.50
0320	12" long		33.12	.242		22	14.80		36.80	46.50
0340	3/4" diameter, close		29.44	.272		11.60	16.65		28.25	38
0360	1-1/2" long		29.44	.272		11.60	16.65		28.25	38
0380	2" long		29.44	.272		11.60	16.65		28.25	38
0400	2-1/2" long		29.44	.272		13.75	16.65		30.40	40
0420	3" long		29.44	.272		13.80	16.65		30.45	40

26 05 33.15 Conduit Nipples		Crew	Daily Output	Labor-Hours	Unit	Material	2020 Bare Costs Labor	Equipment	Total	Total Incl O&P
0440	3-1/2" long	1 Elec	29.44	.272	Ea.	18.90	16.65		35.55	46
0460	4" long		29.44	.272		14.25	16.65		30.90	40.50
0480	5" long		29.44	.272		15.75	16.65		32.40	42.50
0500	6" long		29.44	.272		15.55	16.65		32.20	42
0520	8" long		29.44	.272		15.80	16.65		32.45	42.50
0540	10" long		29.44	.272		17.90	16.65		34.55	44.50
0560	12" long		29.44	.272		24.50	16.65		41.15	52
0580	1" diameter, close		24.84	.322		16.50	19.75		36.25	47.50
0600	2" long		24.84	.322		24.50	19.75		44.25	56.50
0620	2-1/2" long		24.84	.322		17.45	19.75		37.20	48.50
0640	3" long		24.84	.322		20.50	19.75		40.25	52.50
0660	3-1/2" long		24.84	.322		30	19.75		49.75	62.50
0680	4" long		24.84	.322		23.50	19.75		43.25	55
0700	5" long		24.84	.322		23	19.75		42.75	54.50
0720	6" long		24.84	.322		22.50	19.75		42.25	54
0740	8" long		24.84	.322		33.50	19.75		53.25	66.50
0760	10" long		24.84	.322		60	19.75		79.75	96
0780	12" long		24.84	.322		56.50	19.75		76.25	91.50
0800	1-1/4" diameter, close		21.16	.378		25.50	23		48.50	62.50
0820	2" long		21.16	.378		33	23		56	70.50
0840	2-1/2" long		21.16	.378		23.50	23		46.50	60.50
0860	3" long		21.16	.378		29	23		52	66.50
0880	3-1/2" long		21.16	.378		29	23		52	66.50
0900	4" long		21.16	.378		26	23		49	63
0920	5" long		21.16	.378		33	23		56	71
0940	6" long		21.16	.378		37	23		60	75.50
0960	8" long		21.16	.378		57.50	23		80.50	98
0980	10" long		21.16	.378		35.50	23		58.50	73.50
1000	12" long		21.16	.378		47.50	23		70.50	86.50
1020	1-1/2" diameter, close		18.40	.435		33.50	26.50		60	76.50
1040	2" long		18.40	.435		35.50	26.50		62	78.50
1060	2-1/2" long		18.40	.435		36	26.50		62.50	79
1080	3" long		18.40	.435		35.50	26.50		62	78.50
1100	3-1/2" long		18.40	.435		35	26.50		61.50	78
1120	4" long		18.40	.435		35	26.50		61.50	78
1140	5" long		18.40	.435		42.50	26.50		69	86.50
1160	6" long		18.40	.435		44	26.50		70.50	88
1180	8" long		18.40	.435		43.50	26.50		70	87.50
1200	10" long		18.40	.435		51	26.50		77.50	95.50
1220	12" long		18.40	.435		65	26.50		91.50	111
1240	2" diameter, close		16.56	.483		48	29.50		77.50	96.50
1260	2-1/2" long		16.56	.483		46.50	29.50		76	95
1280	3" long		16.56	.483		49	29.50		78.50	98
1300	3-1/2" long		16.56	.483		49.50	29.50		79	98.50
1320	4" long		16.56	.483		51.50	29.50		81	101
1340	5" long		16.56	.483		57	29.50		86.50	107
1360	6" long		16.56	.483		52.50	29.50		82	102
1380	8" long		16.56	.483		76	29.50		105.50	128
1400	10" long		16.56	.483		76	29.50		105.50	128
1420	12" long		16.56	.483		91	29.50		120.50	144
1440	2-1/2" diameter, close		13.80	.580		106	35.50		141.50	169
1460	3" long		13.80	.580		98	35.50		133.50	161
1480	3-1/2" long		13.80	.580		101	35.50		136.50	164

Pre-Installation Change Order Costs

26 05 33 – Raceway and Boxes for Electrical Systems

26 05 33.15 Conduit Nipples	Crew	Daily Output	Labor-Hours	Unit	Material	2020 Bare Costs Labor	Equipment	Total	Total Incl O&P
1500	4" long	1 Elec	13.80	.580	Ea.	112	35.50		147.50	176
1520	5" long		13.80	.580		142	35.50		177.50	210
1540	6" long		13.80	.580		107	35.50		142.50	171
1560	8" long		13.80	.580		112	35.50		147.50	176
1580	10" long		13.80	.580		140	35.50		175.50	207
1600	12" long		13.80	.580		126	35.50		161.50	191
1620	3" diameter, close		11.04	.725		113	44.50		157.50	191
1640	3" long		11.04	.725		119	44.50		163.50	197
1660	3-1/2" long		11.04	.725		143	44.50		187.50	224
1680	4" long		11.04	.725		148	44.50		192.50	229
1700	5" long		11.04	.725		113	44.50		157.50	190
1720	6" long		11.04	.725		124	44.50		168.50	202
1740	8" long		11.04	.725		155	44.50		199.50	236
1760	10" long		11.04	.725		186	44.50		230.50	271
1780	12" long		11.04	.725		179	44.50		223.50	263
1800	3-1/2" diameter, close		10.12	.791		249	48.50		297.50	345
1820	4" long		10.12	.791		283	48.50		331.50	380
1840	5" long		10.12	.791		257	48.50		305.50	355
1860	6" long		10.12	.791		270	48.50		318.50	370
1880	8" long		10.12	.791		251	48.50		299.50	350
1900	10" long		10.12	.791		284	48.50		332.50	380
1920	12" long		10.12	.791		360	48.50		408.50	465
1940	4" diameter, close		8.28	.966		279	59.50		338.50	395
1960	4" long		8.28	.966		300	59.50		359.50	420
1980	5" long		8.28	.966		300	59.50		359.50	420
2000	6" long		8.28	.966		355	59.50		414.50	480
2020	8" long		8.28	.966		345	59.50		404.50	470
2040	10" long		8.28	.966		360	59.50		419.50	485
2060	12" long		8.28	.966		330	59.50		389.50	450
2080	5" diameter, close		6.44	1.240		560	76		636	730
2100	5" long		6.44	1.240		655	76		731	835
2120	6" long		6.44	1.240		660	76		736	840
2140	8" long		6.44	1.240		705	76		781	895
2160	10" long		6.44	1.240		590	76		666	765
2180	12" long		6.44	1.240		610	76		686	790
2200	6" diameter, close		5.52	1.450		850	89		939	1,075
2220	5" long		5.52	1.450		910	89		999	1,125
2240	6" long		5.52	1.450		835	89		924	1,050
2260	8" long		5.52	1.450		990	89		1,079	1,200
2280	10" long		5.52	1.450		1,050	89		1,139	1,275
2300	12" long		5.52	1.450		1,075	89		1,164	1,325
2320	Rigid galvanized steel, 1/2" diameter, close		29.44	.272		3.03	16.65		19.68	28.50
2340	1-1/2" long		29.44	.272		3.24	16.65		19.89	28.50
2360	2" long		29.44	.272		3.34	16.65		19.99	28.50
2380	2-1/2" long		29.44	.272		3.49	16.65		20.14	29
2400	3" long		29.44	.272		3.64	16.65		20.29	29
2420	3-1/2" long		29.44	.272		3.78	16.65		20.43	29
2440	4" long		29.44	.272		3.93	16.65		20.58	29.50
2460	5" long		29.44	.272		4.14	16.65		20.79	29.50
2480	6" long		29.44	.272		4.58	16.65		21.23	30
2500	8" long		29.44	.272		6.45	16.65		23.10	32
2520	10" long		29.44	.272		7.50	16.65		24.15	33.50
2540	12" long		29.44	.272		7.90	16.65		24.55	33.50

26 05 33.15 Conduit Nipples		Crew	Daily Output	Labor-Hours	Unit	Material	2020 Bare Costs Labor	Equipment	Total	Total Incl O&P
2560	3/4" diameter, close	1 Elec	24.84	.322	Ea.	4.39	19.75		24.14	34.50
2580	2" long		24.84	.322		4.64	19.75		24.39	34.50
2600	2-1/2" long		24.84	.322		4.80	19.75		24.55	35
2620	3" long		24.84	.322		4.96	19.75		24.71	35
2640	3-1/2" long		24.84	.322		5.05	19.75		24.80	35
2660	4" long		24.84	.322		5.30	19.75		25.05	35.50
2680	5" long		24.84	.322		5.60	19.75		25.35	35.50
2700	6" long		24.84	.322		6.05	19.75		25.80	36
2720	8" long		24.84	.322		7.95	19.75		27.70	38.50
2740	10" long		24.84	.322		9.55	19.75		29.30	40
2760	12" long		24.84	.322		9.70	19.75		29.45	40
2780	1" diameter, close		21.16	.378		6.80	23		29.80	42
2800	2" long		21.16	.378		6.95	23		29.95	42
2820	2-1/2" long		21.16	.378		7.10	23		30.10	42.50
2840	3" long		21.16	.378		7.40	23		30.40	42.50
2860	3-1/2" long		21.16	.378		7.85	23		30.85	43
2880	4" long		21.16	.378		8	23		31	43.50
2900	5" long		21.16	.378		8.40	23		31.40	44
2920	6" long		21.16	.378		8.75	23		31.75	44
2940	8" long		21.16	.378		11.05	23		34.05	46.50
2960	10" long		21.16	.378		12.85	23		35.85	48.50
2980	12" long		21.16	.378		13.90	23		36.90	50
3000	1-1/4" diameter, close		18.40	.435		9.35	26.50		35.85	50
3020	2" long		18.40	.435		9.60	26.50		36.10	50
3040	3" long		18.40	.435		10.10	26.50		36.60	50.50
3060	3-1/2" long		18.40	.435		10.45	26.50		36.95	51
3080	4" long		18.40	.435		10.75	26.50		37.25	51.50
3100	5" long		18.40	.435		11.40	26.50		37.90	52
3120	6" long		18.40	.435		11.95	26.50		38.45	52.50
3140	8" long		18.40	.435		15.10	26.50		41.60	56
3160	10" long		18.40	.435		17.30	26.50		43.80	58.50
3180	12" long		18.40	.435		18.90	26.50		45.40	60.50
3200	1-1/2" diameter, close		16.56	.483		13.55	29.50		43.05	59
3220	2" long		16.56	.483		13.75	29.50		43.25	59
3240	2-1/2" long		16.56	.483		14.15	29.50		43.65	59.50
3260	3" long		16.56	.483		14.40	29.50		43.90	60
3280	3-1/2" long		16.56	.483		15	29.50		44.50	60.50
3300	4" long		16.56	.483		15.40	29.50		44.90	61
3320	5" long		16.56	.483		16.05	29.50		45.55	61.50
3340	6" long		16.56	.483		17.30	29.50		46.80	63
3360	8" long		16.56	.483		21	29.50		50.50	67
3380	10" long		16.56	.483		23	29.50		52.50	69.50
3400	12" long		16.56	.483		24	29.50		53.50	70.50
3420	2" diameter, close		14.72	.543		17.50	33.50		51	69
3440	2-1/2" long		14.72	.543		18.10	33.50		51.60	69.50
3460	3" long		14.72	.543		18.85	33.50		52.35	70
3480	3-1/2" long		14.72	.543		19.55	33.50		53.05	71
3500	4" long		14.72	.543		20	33.50		53.50	71.50
3520	5" long		14.72	.543		21	33.50		54.50	73
3540	6" long		14.72	.543		22	33.50		55.50	74
3560	8" long		14.72	.543		26	33.50		59.50	78
3580	10" long		14.72	.543		28.50	33.50		62	80.50
3600	12" long		14.72	.543		30.50	33.50		64	83

For customer support on your Electrical Change Order Costs with RSMeans data, call 800.448.8182.
Pre-Installation Change Order Costs

26 05 33 – Raceway and Boxes for Electrical Systems

26 05 33.15 Conduit Nipples	Crew	Daily Output	Labor-Hours	Unit	Material	2020 Bare Costs Labor	Equipment	Total	Total Incl O&P
3620	2-1/2" diameter, close	1 Elec	11.96	.669	Ea.	60.50	41		101.50	128
3640	3" long		11.96	.669		64.50	41		105.50	132
3660	3-1/2" long		11.96	.669		66.50	41		107.50	134
3680	4" long		11.96	.669		67	41		108	135
3700	5" long		11.96	.669		69.50	41		110.50	138
3720	6" long		11.96	.669		71.50	41		112.50	140
3740	8" long		11.96	.669		77	41		118	146
3760	10" long		11.96	.669		80.50	41		121.50	150
3780	12" long		11.96	.669		85.50	41		126.50	155
3800	3" diameter, close		11.04	.725		60.50	44.50		105	133
3820	3" long		11.04	.725		66	44.50		110.50	139
3900	3-1/2" long		11.04	.725		62.50	44.50		107	135
3920	4" long		11.04	.725		64.50	44.50		109	137
3940	5" long		11.04	.725		66.50	44.50		111	140
3960	6" long		11.04	.725		69.50	44.50		114	143
3980	8" long		11.04	.725		75.50	44.50		120	150
4000	10" long		11.04	.725		80.50	44.50		125	155
4020	12" long		11.04	.725		88	44.50		132.50	163
4040	3-1/2" diameter, close		9.20	.870		105	53.50		158.50	195
4060	4" long		9.20	.870		108	53.50		161.50	199
4080	5" long		9.20	.870		111	53.50		164.50	202
4100	6" long		9.20	.870		114	53.50		167.50	206
4120	8" long		9.20	.870		120	53.50		173.50	212
4140	10" long		9.20	.870		126	53.50		179.50	219
4160	12" long		9.20	.870		133	53.50		186.50	227
4180	4" diameter, close		7.36	1.090		167	67		234	284
4200	4" long		7.36	1.090		169	67		236	286
4220	5" long		7.36	1.090		172	67		239	289
4240	6" long		7.36	1.090		174	67		241	291
4260	8" long		7.36	1.090		178	67		245	296
4280	10" long		7.36	1.090		183	67		250	300
4300	12" long		7.36	1.090		188	67		255	305
4320	5" diameter, close		5.52	1.450		240	89		329	395
4340	5" long		5.52	1.450		254	89		343	410
4360	6" long		5.52	1.450		259	89		348	415
4380	8" long		5.52	1.450		261	89		350	420
4400	10" long		5.52	1.450		280	89		369	440
4420	12" long		5.52	1.450		298	89		387	460
4440	6" diameter, close		4.60	1.740		420	107		527	620
4460	5" long		4.60	1.740		440	107		547	645
4480	6" long		4.60	1.740		450	107		557	655
4500	8" long		4.60	1.740		460	107		567	665
4520	10" long		4.60	1.740		480	107		587	685
4540	12" long		4.60	1.740		490	107		597	700
4560	Plastic coated, 40 mil thick, 1/2" diameter, 2" long		29.44	.272		28.50	16.65		45.15	56.50
4580	2-1/2" long		29.44	.272		25	16.65		41.65	52.50
4600	3" long		29.44	.272		25.50	16.65		42.15	53
4680	3-1/2" long		29.44	.272		30	16.65		46.65	58
4700	4" long		29.44	.272		27.50	16.65		44.15	55
4720	5" long		29.44	.272		30.50	16.65		47.15	58.50
4740	6" long		29.44	.272		26	16.65		42.65	53.50
4760	8" long		29.44	.272		36	16.65		52.65	64.50
4780	10" long		29.44	.272		39.50	16.65		56.15	68.50

26 05 33.15 Conduit Nipples		Crew	Daily Output	Labor-Hours	Unit	Material	2020 Bare Costs Labor	Equipment	Total	Total Incl O&P
4800	12" long	1 Elec	29.44	.272	Ea.	44	16.65		60.65	73.50
4820	3/4" diameter, 2" long		23.92	.334		30.50	20.50		51	64
4840	2-1/2" long		23.92	.334		29	20.50		49.50	62.50
4860	3" long		23.92	.334		32.50	20.50		53	66.50
4880	3-1/2" long		23.92	.334		31.50	20.50		52	65
4900	4" long		23.92	.334		34	20.50		54.50	67.50
4920	5" long		23.92	.334		33	20.50		53.50	66.50
4940	6" long		23.92	.334		40.50	20.50		61	75.50
4960	8" long		23.92	.334		36	20.50		56.50	70
4980	10" long		23.92	.334		33	20.50		53.50	67
5000	12" long		23.92	.334		31.50	20.50		52	65
5020	1" diameter, 2" long		20.24	.395		33	24.50		57.50	72.50
5040	2-1/2" long		20.24	.395		32	24.50		56.50	71.50
5060	3" long		20.24	.395		36.50	24.50		61	76.50
5080	3-1/2" long		20.24	.395		37	24.50		61.50	76.50
5100	4" long		20.24	.395		31	24.50		55.50	70
5120	5" long		20.24	.395		29.50	24.50		54	68.50
5140	6" long		20.24	.395		34.50	24.50		59	74
5160	8" long		20.24	.395		40	24.50		64.50	80
5180	10" long		20.24	.395		42.50	24.50		67	83
5200	12" long		20.24	.395		39	24.50		63.50	78.50
5220	1-1/4" diameter, 2" long		16.56	.483		37.50	29.50		67	85
5240	2-1/2" long		16.56	.483		41	29.50		70.50	89
5260	3" long		16.56	.483		42	29.50		71.50	90
5280	3-1/2" long		16.56	.483		45	29.50		74.50	93.50
5300	4" long		16.56	.483		51	29.50		80.50	100
5320	5" long		16.56	.483		46	29.50		75.50	94.50
5340	6" long		16.56	.483		45	29.50		74.50	94
5360	8" long		16.56	.483		48	29.50		77.50	97
5380	10" long		16.56	.483		54	29.50		83.50	104
5400	12" long		16.56	.483		53.50	29.50		83	103
5420	1-1/2" diameter, 2" long		14.72	.543		46	33.50		79.50	100
5440	2-1/2" long		14.72	.543		46	33.50		79.50	101
5460	3" long		14.72	.543		44	33.50		77.50	97.50
5480	3-1/2" long		14.72	.543		48.50	33.50		82	103
5500	4" long		14.72	.543		41.50	33.50		75	95
5520	5" long		14.72	.543		41.50	33.50		75	95
5540	6" long		14.72	.543		49.50	33.50		83	104
5560	8" long		14.72	.543		58	33.50		91.50	114
5580	10" long		14.72	.543		77.50	33.50		111	135
5600	12" long		14.72	.543		70	33.50		103.50	127
5620	2" diameter, 2-1/2" long		12.88	.621		47.50	38		85.50	109
5640	3" long		12.88	.621		51.50	38		89.50	113
5660	3-1/2" long		12.88	.621		59	38		97	121
5680	4" long		12.88	.621		58.50	38		96.50	121
5700	5" long		12.88	.621		63.50	38		101.50	127
5720	6" long		12.88	.621		61	38		99	124
5740	8" long		12.88	.621		73	38		111	137
5760	10" long		12.88	.621		95.50	38		133.50	162
5780	12" long		12.88	.621		93.50	38		131.50	160
5800	2-1/2" diameter, 3-1/2" long		11.04	.725		124	44.50		168.50	203
5820	4" long		11.04	.725		127	44.50		171.50	206
5840	5" long		11.04	.725		152	44.50		196.50	233

26 05 33.15 Conduit Nipples

		Crew	Daily Output	Labor-Hours	Unit	Material	2020 Bare Costs Labor	Equipment	Total	Total Incl O&P
5860	6" long	1 Elec	11.04	.725	Ea.	126	44.50		170.50	205
5880	8" long		11.04	.725		167	44.50		211.50	249
5900	10" long		11.04	.725		182	44.50		226.50	266
5920	12" long		11.04	.725		184	44.50		228.50	268
5940	3" diameter, 3-1/2" long		10.12	.791		141	48.50		189.50	227
5960	4" long		10.12	.791		126	48.50		174.50	211
5980	5" long		10.12	.791		147	48.50		195.50	233
6000	6" long		10.12	.791		161	48.50		209.50	249
6020	8" long		10.12	.791		198	48.50		246.50	290
6040	10" long		10.12	.791		222	48.50		270.50	315
6060	12" long		10.12	.791		218	48.50		266.50	310
6080	3-1/2" diameter, 4" long		8.28	.966		199	59.50		258.50	310
6100	5" long		8.28	.966		201	59.50		260.50	310
6120	6" long		8.28	.966		186	59.50		245.50	293
6140	8" long		8.28	.966		244	59.50		303.50	355
6160	10" long		8.28	.966		284	59.50		343.50	405
6180	12" long		8.28	.966		325	59.50		384.50	450
6200	4" diameter, 4" long		6.90	1.160		260	71		331	390
6220	5" long		6.90	1.160		278	71		349	410
6240	6" long		6.90	1.160		300	71		371	435
6260	8" long		6.90	1.160		320	71		391	460
6280	10" long		6.90	1.160		390	71		461	535
6300	12" long		6.90	1.160		435	71		506	585
6320	5" diameter, 5" long		5.06	1.580		265	97		362	435
6340	6" long		5.06	1.580		325	97		422	505
6360	8" long		5.06	1.580		340	97		437	515
6380	10" long		5.06	1.580		355	97		452	535
6400	12" long		5.06	1.580		390	97		487	575
6420	6" diameter, 5" long		4.14	1.930		515	118		633	745
6440	6" long		4.14	1.930		530	118		648	760
6460	8" long		4.14	1.930		555	118		673	785
6480	10" long		4.14	1.930		580	118		698	810
6500	12" long		4.14	1.930		610	118		728	845

26 05 33.16 Boxes for Electrical Systems

		Crew	Daily Output	Labor-Hours	Unit	Material	2020 Bare Costs Labor	Equipment	Total	Total Incl O&P
0010	**BOXES FOR ELECTRICAL SYSTEMS**									
0020	Pressed steel, octagon, 4"	1 Elec	18.40	.435	Ea.	3.70	26.50		30.20	43.50
0040	For Romex or BX		18.40	.435		7.15	26.50		33.65	47.50
0050	For Romex or BX, with bracket		18.40	.435		10.65	26.50		37.15	51.50
0060	Covers, blank		58.88	.136		1.09	8.35		9.44	13.60
0100	Extension rings		36.80	.217		5.75	13.35		19.10	26
0150	Square, 4"		18.40	.435		8.60	26.50		35.10	49
0160	For Romex or BX		18.40	.435		15.05	26.50		41.55	56
0170	For Romex or BX, with bracket		18.40	.435		11.40	26.50		37.90	52
0200	Extension rings		36.80	.217		6.05	13.35		19.40	26.50
0220	2-1/8" deep, 1" KO		18.40	.435		4.21	26.50		30.71	44
0250	Covers, blank		58.88	.136		.96	8.35		9.31	13.45
0260	Raised device		58.88	.136		1.96	8.35		10.31	14.55
0300	Plaster rings		58.88	.136		2.08	8.35		10.43	14.70
0350	Square, 4-11/16"		18.40	.435		5.50	26.50		32	45.50
0370	2-1/8" deep, 3/4" to 1-1/4" KO		18.40	.435		7.35	26.50		33.85	47.50
0400	Extension rings		36.80	.217		13.25	13.35		26.60	34.50
0450	Covers, blank		48.76	.164		1.41	10.05		11.46	16.55

26 05 33.16 Boxes for Electrical Systems		Crew	Daily Output	Labor-Hours	Unit	Material	2020 Bare Costs Labor	Equipment	Total	Total Incl O&P
0460	Raised device	1 Elec	48.76	.164	Ea.	9.05	10.05		19.10	25
0500	Plaster rings		48.76	.164		7.60	10.05		17.65	23.50
0550	Handy box		24.84	.322		3.15	19.75		22.90	33
0560	Covers, device		58.88	.136		1.26	8.35		9.61	13.80
0600	Extension rings		49.68	.161		4.50	9.90		14.40	19.65
0650	Switchbox		24.84	.322		7	19.75		26.75	37
0660	Romex or BX		24.84	.322		9.80	19.75		29.55	40.50
0670	with bracket		24.84	.322		9	19.75		28.75	39.50
0680	Partition, metal		24.84	.322		4.20	19.75		23.95	34
0700	Masonry, 1 gang, 2-1/2" deep		24.84	.322		10.45	19.75		30.20	41
0710	3-1/2" deep		24.84	.322		10.65	19.75		30.40	41
0750	2 gang, 2-1/2" deep		18.40	.435		22	26.50		48.50	64
0760	3-1/2" deep		18.40	.435		14.70	26.50		41.20	55.50
0800	3 gang, 2-1/2" deep		11.96	.669		23	41		64	86.50
0850	4 gang, 2-1/2" deep		9.20	.870		31.50	53.50		85	114
0860	5 gang, 2-1/2" deep		8.28	.966		46	59.50		105.50	139
0870	6 gang, 2-1/2" deep		7.36	1.090		76.50	67		143.50	184
0880	Masonry thru-the-wall, 1 gang, 4" block		14.72	.543		44.50	33.50		78	98.50
0890	6" block		14.72	.543		58	33.50		91.50	114
0900	8" block		14.72	.543		77.50	33.50		111	135
0920	2 gang, 6" block		14.72	.543		87	33.50		120.50	145
0940	Bar hanger with 3/8" stud, for wood and masonry boxes		48.76	.164		6.65	10.05		16.70	22.50
0950	Concrete, set flush, 4" deep		18.40	.435		10.25	26.50		36.75	51
1000	Plate with 3/8" stud		73.60	.109		10.40	6.65		17.05	21.50
1100	Concrete, floor, 1 gang		4.88	1.640		140	101		241	305
1150	2 gang		3.68	2.170		209	133		342	430
1200	3 gang		2.48	3.220		335	198		533	665
1250	For duplex receptacle, pedestal mounted, add		22.08	.362		146	22		168	194
1270	Flush mounted, add		24.84	.322		43.50	19.75		63.25	77
1300	For telephone, pedestal mounted, add		27.60	.290		141	17.80		158.80	182
1350	Carpet flange, 1 gang		48.76	.164		72	10.05		82.05	94
1400	Cast, 1 gang, FS (2" deep), 1/2" hub		11.04	.725		31.50	44.50		76	101
1410	3/4" hub		11.04	.725		36.50	44.50		81	106
1420	FD (2-11/16" deep), 1/2" hub		11.04	.725		23.50	44.50		68	91.50
1430	3/4" hub		11.04	.725		27.50	44.50		72	96.50
1450	2 gang, FS, 1/2" hub		9.20	.870		62.50	53.50		116	148
1460	3/4" hub		9.20	.870		56.50	53.50		110	142
1470	FD, 1/2" hub		9.20	.870		76	53.50		129.50	164
1480	3/4" hub		9.20	.870		69	53.50		122.50	156
1500	3 gang, FS, 3/4" hub		8.28	.966		101	59.50		160.50	200
1510	Switch cover, 1 gang, FS		58.88	.136		8.05	8.35		16.40	21.50
1520	2 gang		48.76	.164		12.60	10.05		22.65	29
1530	Duplex receptacle cover, 1 gang, FS		58.88	.136		8.70	8.35		17.05	22
1540	2 gang, FS		48.76	.164		13.25	10.05		23.30	29.50
1542	Weatherproof blank cover, 1 gang		58.88	.136		1.41	8.35		9.76	13.95
1544	2 gang		48.76	.164		2.85	10.05		12.90	18.15
1550	Weatherproof switch cover, 1 gang		58.88	.136		7.35	8.35		15.70	20.50
1554	2 gang		48.76	.164		12.20	10.05		22.25	28.50
1600	Weatherproof receptacle cover, 1 gang		58.88	.136		8.95	8.35		17.30	22.50
1604	2 gang		48.76	.164		11.20	10.05		21.25	27.50
1620	Weatherproof receptacle cover, tamper resistant, 1 gang		53.36	.150		17	9.20		26.20	32.50
1624	2 gang		44.16	.181		33	11.10		44.10	53
1750	FSC, 1 gang, 1/2" hub		10.12	.791		34.50	48.50		83	110

For customer support on your Electrical Change Order Costs with RSMeans data, call 800.448.8182.

Pre-Installation Change Order Costs

26 05 33.16 Boxes for Electrical Systems

	26 05 33.16 Boxes for Electrical Systems	Crew	Daily Output	Labor-Hours	Unit	Material	2020 Bare Costs Labor	Equipment	Total	Total Incl O&P
1760	3/4" hub	1 Elec	10.12	.791	Ea.	42	48.50		90.50	119
1770	2 gang, 1/2" hub		8.28	.966		63.50	59.50		123	159
1780	3/4" hub		8.28	.966		69	59.50		128.50	165
1790	FDC, 1 gang, 1/2" hub		10.12	.791		37	48.50		85.50	113
1800	3/4" hub		10.12	.791		47.50	48.50		96	124
1810	2 gang, 1/2" hub		8.28	.966		83	59.50		142.50	180
1820	3/4" hub		8.28	.966		82.50	59.50		142	180
1850	Weatherproof in-use cover, 1 gang		58.88	.136		30.50	8.35		38.85	46
1870	2 gang		48.76	.164		38.50	10.05		48.55	57.50
2000	Poke-thru fitting, fire rated, for 3-3/4" floor		6.26	1.280		212	78.50		290.50	350
2040	For 7" floor		6.26	1.280		209	78.50		287.50	345
2100	Pedestal, 15 amp, duplex receptacle & blank plate		4.83	1.660		190	102		292	360
2120	Duplex receptacle and telephone plate		4.83	1.660		190	102		292	360
2140	Pedestal, 20 amp, duplex recept. & phone plate		4.60	1.740		191	107		298	370
2160	Telephone plate, both sides		4.83	1.660		179	102		281	350
2200	Abandonment plate		29.44	.272		55.50	16.65		72.15	86

26 05 33.17 Outlet Boxes, Plastic

	26 05 33.17 Outlet Boxes, Plastic	Crew	Daily Output	Labor-Hours	Unit	Material	2020 Bare Costs Labor	Equipment	Total	Total Incl O&P
0010	**OUTLET BOXES, PLASTIC**									
0050	4" diameter, round with 2 mounting nails	1 Elec	23	.348	Ea.	3.61	21.50		25.11	36
0100	Bar hanger mounted		23	.348		6.30	21.50		27.80	39
0200	4", square with 2 mounting nails		23	.348		6.95	21.50		28.45	39.50
0300	Plaster ring		58.88	.136		2.63	8.35		10.98	15.30
0400	Switch box with 2 mounting nails, 1 gang		27.60	.290		4.01	17.80		21.81	31
0500	2 gang		23	.348		4.94	21.50		26.44	37.50
0600	3 gang		18.40	.435		6.90	26.50		33.40	47
0700	Old work box		27.60	.290		5.75	17.80		23.55	33
1400	PVC, FSS, 1 gang, 1/2" hub		12.88	.621		16.35	38		54.35	74.50
1410	3/4" hub		12.88	.621		14.25	38		52.25	72
1420	FD, 1 gang for variable terminations		12.88	.621		9.90	38		47.90	67.50
1450	FS, 2 gang for variable terminations		11.04	.725		16.85	44.50		61.35	84.50
1480	Weatherproof blank cover, FS, 1 gang		58.88	.136		4.81	8.35		13.16	17.70
1500	2 gang		48.76	.164		6.40	10.05		16.45	22
1510	Weatherproof switch cover, FS, 1 gang		58.88	.136		10.25	8.35		18.60	23.50
1520	2 gang		48.76	.164		21	10.05		31.05	38
1530	Weatherproof duplex receptacle cover, FS, 1 gang		58.88	.136		16.10	8.35		24.45	30
1540	2 gang		48.76	.164		17	10.05		27.05	33.50
1750	FSC, 1 gang, 1/2" hub		11.96	.669		10.10	41		51.10	72
1760	3/4" hub		11.96	.669		10.60	41		51.60	72.50
1770	FSC, 2 gang, 1/2" hub		10.12	.791		18.80	48.50		67.30	92.50
1780	3/4" hub		10.12	.791		14.55	48.50		63.05	88
1790	FDC, 1 gang, 1/2" hub		11.96	.669		13.10	41		54.10	75.50
1800	3/4" hub		11.96	.669		13.30	41		54.30	75.50
1810	Weatherproof, T box w/3 holes		12.88	.621		13.70	38		51.70	71.50
1820	4" diameter round w/5 holes		12.88	.621		12.85	38		50.85	70.50
1850	In-use cover, 1 gang		58.88	.136		9.05	8.35		17.40	22.50
1870	2 gang		48.76	.164		12.90	10.05		22.95	29

26 05 33.18 Pull Boxes

	26 05 33.18 Pull Boxes	Crew	Daily Output	Labor-Hours	Unit	Material	2020 Bare Costs Labor	Equipment	Total	Total Incl O&P
0010	**PULL BOXES**									
0100	Steel, pull box, NEMA 1, type SC, 6" W x 6" H x 4" D	1 Elec	7.36	1.090	Ea.	12.30	67		79.30	113
0180	8" W x 6" H x 4" D		7.36	1.090		29	67		96	132
0200	8" W x 8" H x 4" D		7.36	1.090		16.15	67		83.15	117
0210	10" W x 10" H x 4" D		6.44	1.240		31.50	76		107.50	148

26 05 33.18 Pull Boxes		Crew	Daily Output	Labor-Hours	Unit	Material	2020 Bare Costs Labor	Equipment	Total	Total Incl O&P
0220	12" W x 12" H x 4" D	1 Elec	5.98	1.340	Ea.	29.50	82		111.50	154
0230	15" W x 15" H x 4" D		4.78	1.670		54.50	102		156.50	213
0240	18" W x 18" H x 4" D		4.05	1.980		49.50	121		170.50	235
0250	6" W x 6" H x 6" D		7.36	1.090		18.35	67		85.35	120
0260	8" W x 8" H x 6" D		6.90	1.160		19	71		90	127
0270	10" W x 10" H x 6" D		5.06	1.580		25.50	97		122.50	172
0300	10" W x 12" H x 6" D		4.88	1.640		38	101		139	192
0310	12" W x 12" H x 6" D		4.78	1.670		31	102		133	187
0320	15" W x 15" H x 6" D		4.23	1.890		49	116		165	227
0330	18" W x 18" H x 6" D		3.86	2.070		58.50	127		185.50	254
0340	24" W x 24" H x 6" D		2.94	2.720		117	167		284	375
0350	12" W x 12" H x 8" D		4.60	1.740		38	107		145	201
0360	15" W x 15" H x 8" D		4.14	1.930		57	118		175	239
0370	18" W x 18" H x 8" D		3.68	2.170		76	133		209	282
0380	24" W x 18" H x 6" D		3.40	2.350		145	144		289	375
0400	16" W x 20" H x 8" D		3.68	2.170		103	133		236	310
0500	20" W x 24" H x 8" D		2.94	2.720		138	167		305	400
0510	24" W x 24" H x 8" D		2.76	2.900		131	178		309	410
0600	24" W x 36" H x 8" D		2.48	3.220		191	198		389	505
0610	30" W x 30" H x 8" D		2.48	3.220		252	198		450	570
0620	36" W x 36" H x 8" D		1.84	4.350		270	267		537	690
0630	24" W x 24" H x 10" D		2.30	3.480		232	214		446	575
0650	Pull box, hinged, NEMA 1, 6" W x 6" H x 4" D		7.36	1.090		21	67		88	123
0660	8" W x 8" H x 4" D		7.36	1.090		29.50	67		96.50	132
0670	10" W x 10" H x 4" D		6.44	1.240		47	76		123	165
0680	12" W x 12" H x 4" D		5.52	1.450		49.50	89		138.50	187
0690	15" W x 15" H x 4" D		4.78	1.670		55	102		157	214
0700	18" W x 18" H x 4" D		4.05	1.980		63	121		184	250
0710	6" W x 6" H x 6" D		7.36	1.090		24.50	67		91.50	127
0720	8" W x 8" H x 6" D		6.90	1.160		33.50	71		104.50	143
0730	10" W x 10" H x 6" D		5.06	1.580		44	97		141	193
0740	12" W x 12" H x 6" D		4.78	1.670		56	102		158	215
0800	12" W x 16" H x 6" D		4.32	1.850		61.50	114		175.50	237
0820	18" W x 18" H x 6" D		3.86	2.070		97	127		224	295
1000	20" W x 20" H x 6" D		3.31	2.420		134	148		282	370
1010	24" W x 24" H x 6" D		2.94	2.720		204	167		371	470
1020	12" W x 12" H x 8" D		4.60	1.740		101	107		208	270
1030	15" W x 15" H x 8" D		4.14	1.930		115	118		233	305
1040	18" W x 18" H x 8" D		3.68	2.170		162	133		295	375
1200	20" W x 20" H x 8" D		2.94	2.720		203	167		370	470
1210	24" W x 24" H x 8" D		2.76	2.900		229	178		407	515
1220	30" W x 30" H x 8" D		2.48	3.220		330	198		528	660
1400	24" W x 36" H x 8" D		2.48	3.220		320	198		518	650
1600	24" W x 42" H x 8" D		1.84	4.350		475	267		742	915
1610	36" W x 36" H x 8" D		1.84	4.350		420	267		687	855
2100	Pull box, NEMA 3R, type SC, raintight & weatherproof									
2150	6" L x 6" W x 6" D	1 Elec	9.20	.870	Ea.	21.50	53.50		75	104
2200	8" L x 6" W x 6" D		7.36	1.090		31	67		98	134
2250	10" L x 6" W x 6" D		6.44	1.240		40	76		116	157
2300	12" L x 12" W x 6" D		4.60	1.740		84	107		191	252
2350	16" L x 16" W x 6" D		4.14	1.930		115	118		233	300
2400	20" L x 20" W x 6" D		3.68	2.170		108	133		241	315
2450	24" L x 18" W x 8" D		2.76	2.900		191	178		369	475

26 05 33 – Raceway and Boxes for Electrical Systems

26 05 33.18 Pull Boxes		Crew	Daily Output	Labor-Hours	Unit	Material	2020 Bare Costs Labor	Equipment	Total	Total Incl O&P
2500	24" L x 24" W x 10" D	1 Elec	2.30	3.480	Ea.	360	214		574	715
2550	30" L x 24" W x 12" D		1.84	4.350		495	267		762	940
2600	36" L x 36" W x 12" D	↓	1.38	5.800	↓	865	355		1,220	1,475
2800	Cast iron, pull boxes for surface mounting									
3000	NEMA 4, watertight & dust tight									
3050	6" L x 6" W x 6" D	1 Elec	3.68	2.170	Ea.	345	133		478	580
3100	8" L x 6" W x 6" D		2.94	2.720		520	167		687	825
3150	10" L x 6" W x 6" D		2.30	3.480		555	214		769	930
3200	12" L x 12" W x 6" D		2.12	3.780		930	232		1,162	1,375
3250	16" L x 16" W x 6" D		1.20	6.690		1,050	410		1,460	1,750
3300	20" L x 20" W x 6" D		.74	10.870		1,200	665		1,865	2,325
3350	24" L x 18" W x 8" D		.64	12.420		3,250	760		4,010	4,700
3400	24" L x 24" W x 10" D		.46	17.390		5,900	1,075		6,975	8,100
3450	30" L x 24" W x 12" D		.37	21.740		8,425	1,325		9,750	11,200
3500	36" L x 36" W x 12" D		.18	43.480		10,300	2,675		12,975	15,400
3510	NEMA 4 clamp cover, 6" L x 6" W x 4" D		3.68	2.170		340	133		473	575
3520	8" L x 6" W x 4" D	↓	3.68	2.170	↓	415	133		548	655
4000	NEMA 7, explosion proof									
4050	6" L x 6" W x 6" D	1 Elec	1.84	4.350	Ea.	935	267		1,202	1,425
4100	8" L x 6" W x 6" D		1.66	4.830		1,325	296		1,621	1,900
4150	10" L x 6" W x 6" D		1.47	5.430		1,650	335		1,985	2,300
4200	12" L x 12" W x 6" D		.92	8.700		3,000	535		3,535	4,100
4250	16" L x 14" W x 6" D		.55	14.490		4,375	890		5,265	6,150
4300	18" L x 18" W x 8" D		.46	17.390		8,100	1,075		9,175	10,500
4350	24" L x 18" W x 8" D		.37	21.740		9,575	1,325		10,900	12,500
4400	24" L x 24" W x 10" D		.28	28.990		14,000	1,775		15,775	18,100
4450	30" L x 24" W x 12" D		.18	43.480		17,900	2,675		20,575	23,700
5000	NEMA 9, dust tight 6" L x 6" W x 6" D		2.94	2.720		515	167		682	820
5050	8" L x 6" W x 6" D		2.48	3.220		615	198		813	970
5100	10" L x 6" W x 6" D		1.84	4.350		815	267		1,082	1,300
5150	12" L x 12" W x 6" D		1.47	5.430		1,575	335		1,910	2,225
5200	16" L x 16" W x 6" D		.92	8.700		2,600	535		3,135	3,650
5250	18" L x 18" W x 8" D		.64	12.420		4,050	760		4,810	5,575
5300	24" L x 18" W x 8" D		.55	14.490		5,750	890		6,640	7,650
5350	24" L x 24" W x 10" D		.37	21.740		7,725	1,325		9,050	10,500
5400	30" L x 24" W x 12" D	↓	.28	28.990	↓	11,600	1,775		13,375	15,500
6000	J.I.C. wiring boxes, NEMA 12, dust tight & drip tight									
6050	6" L x 8" W x 4" D	1 Elec	9.20	.870	Ea.	97	53.50		150.50	186
6100	8" L x 10" W x 4" D		7.36	1.090		161	67		228	277
6150	12" L x 14" W x 6" D		4.88	1.640		122	101		223	284
6200	14" L x 16" W x 6" D		4.32	1.850		158	114		272	345
6250	16" L x 20" W x 6" D		4.05	1.980		410	121		531	635
6300	24" L x 30" W x 6" D		2.94	2.720		525	167		692	825
6350	24" L x 30" W x 8" D		2.67	3		610	184		794	945
6400	24" L x 36" W x 8" D		2.48	3.220		500	198		698	845
6450	24" L x 42" W x 8" D		2.12	3.780		675	232		907	1,100
6500	24" L x 48" W x 8" D	↓	1.84	4.350	↓	740	267		1,007	1,200

26 05 33.23 Wireway

0010	**WIREWAY** to 10' high									
0020	For higher elevations, see Section 26 05 36.40									
0100	NEMA 1, screw cover w/fittings and supports, 2-1/2" x 2-1/2"	1 Elec	41.40	.193	L.F.	13.70	11.85		25.55	33
0200	4" x 4"	"	36.80	.217	↓	14.75	13.35		28.10	36

26 05 33.23 Wireway		Crew	Daily Output	Labor-Hours	Unit	Material	2020 Bare Costs Labor	Equipment	Total	Total Incl O&P
0400	6" x 6"	2 Elec	55.20	.290	L.F.	26	17.80		43.80	55
0600	8" x 8"		36.80	.435		31	26.50		57.50	73.50
0620	10" x 10"		27.60	.580		61.50	35.50		97	121
0640	12" x 12"		18.40	.870		70.50	53.50		124	157
0800	Elbows, 90°, 2-1/2"	1 Elec	22.08	.362	Ea.	35.50	22		57.50	72
1000	4"		18.40	.435		41.50	26.50		68	85.50
1200	6"		16.56	.483		46	29.50		75.50	94.50
1400	8"		14.72	.543		73.50	33.50		107	130
1420	10"		11.04	.725		96.50	44.50		141	172
1440	12"		9.20	.870		136	53.50		189.50	230
1500	Elbows, 45°, 2-1/2"		22.08	.362		37	22		59	73.50
1510	4"		18.40	.435		43.50	26.50		70	87.50
1520	6"		16.56	.483		50	29.50		79.50	99
1530	8"		14.72	.543		73.50	33.50		107	130
1540	10"		11.04	.725		97.50	44.50		142	173
1550	12"		9.20	.870		185	53.50		238.50	284
1600	"T" box, 2-1/2"		16.56	.483		41	29.50		70.50	89
1800	4"		14.72	.543		54	33.50		87.50	109
2000	6"		12.88	.621		60	38		98	123
2200	8"		11.04	.725		102	44.50		146.50	179
2220	10"		9.20	.870		126	53.50		179.50	218
2240	12"		7.36	1.090		179	67		246	297
2300	Cross, 2-1/2"		14.72	.543		45.50	33.50		79	99.50
2310	4"		12.88	.621		55.50	38		93.50	118
2320	6"		11.04	.725		68.50	44.50		113	142
2400	Panel adapter, 2-1/2"		22.08	.362		7.60	22		29.60	41.50
2600	4"		18.40	.435		9.70	26.50		36.20	50
2800	6"		16.56	.483		13.90	29.50		43.40	59.50
3000	8"		14.72	.543		20	33.50		53.50	71.50
3020	10"		12.88	.621		31	38		69	90.50
3040	12"		11.04	.725		45.50	44.50		90	116
3200	Reducer, 4" to 2-1/2"		22.08	.362		13.75	22		35.75	48
3400	6" to 4"		18.40	.435		30.50	26.50		57	73.50
3600	8" to 6"		16.56	.483		37.50	29.50		67	85
3620	10" to 8"		14.72	.543		48	33.50		81.50	103
3640	12" to 10"		12.88	.621		62.50	38		100.50	125
3780	End cap, 2-1/2"		22.08	.362		5.75	22		27.75	39.50
3800	4"		18.40	.435		6.65	26.50		33.15	47
4000	6"		16.56	.483		7.85	29.50		37.35	52.50
4200	8"		14.72	.543		10.05	33.50		43.55	60.50
4220	10"		12.88	.621		17.70	38		55.70	76
4240	12"		11.04	.725		23.50	44.50		68	91.50
4300	U-connector, 2-1/2"		184	.043		4.91	2.67		7.58	9.35
4320	4"		184	.043		5.90	2.67		8.57	10.45
4340	6"		165.60	.048		6.90	2.96		9.86	12
4360	8"		156.40	.051		12.05	3.14		15.19	17.90
4380	10"		138	.058		17.20	3.56		20.76	24
4400	12"		119.60	.067		22	4.10		26.10	30.50
4420	Hanger, 2-1/2"		92	.087		13.50	5.35		18.85	23
4430	4"		92	.087		13.50	5.35		18.85	23
4440	6"		73.60	.109		16.05	6.65		22.70	27.50
4450	8"		59.80	.134		27	8.20		35.20	41.50
4460	10"		46	.174		49	10.65		59.65	70

26 05 33 – Raceway and Boxes for Electrical Systems

26 05 33.23 Wireway		Crew	Daily Output	Labor-Hours	Unit	Material	2020 Bare Costs Labor	Equipment	Total	Total Incl O&P
4470	12"	1 Elec	36.80	.217	Ea.	71.50	13.35		84.85	98.50
4475	NEMA 3R, screw cover w/fittings and supports, 4" x 4"		33.12	.242	L.F.	18.35	14.80		33.15	42
4480	6" x 6"	2 Elec	50.60	.316		20.50	19.40		39.90	51.50
4485	8" x 8"		33.12	.483		36	29.50		65.50	83.50
4490	12" x 12"		16.56	.966		66.50	59.50		126	162
4500	Hinged cover, with fittings and supports, 2-1/2" x 2-1/2"	1 Elec	55.20	.145		22.50	8.90		31.40	38.50
4520	4" x 4"	"	41.40	.193		15.20	11.85		27.05	34.50
4540	6" x 6"	2 Elec	73.60	.217		23	13.35		36.35	45.50
4560	8" x 8"		55.20	.290		40.50	17.80		58.30	71.50
4580	10" x 10"		46	.348		52.50	21.50		74	90
4600	12" x 12"		22.08	.725		93.50	44.50		138	169
4700	Elbows 90°, 2-1/2" x 2-1/2"	1 Elec	29.44	.272	Ea.	51.50	16.65		68.15	81.50
4720	4"		24.84	.322		84	19.75		103.75	122
4730	6"		21.16	.378		65	23		88	106
4740	8"		16.56	.483		78.50	29.50		108	131
4750	10"		12.88	.621		122	38		160	192
4760	12"		11.04	.725		198	44.50		242.50	284
4800	Tee box, hinged cover, 2-1/2" x 2-1/2"		21.16	.378		67.50	23		90.50	109
4810	4"		18.40	.435		97	26.50		123.50	147
4820	6"		16.56	.483		75	29.50		104.50	127
4830	8"		14.72	.543		178	33.50		211.50	246
4840	10"		11.04	.725		181	44.50		225.50	265
4860	12"		9.20	.870		235	53.50		288.50	340
4880	Cross box, hinged cover, 2-1/2" x 2-1/2"		16.56	.483		68	29.50		97.50	119
4900	4"		14.72	.543		96.50	33.50		130	156
4920	6"		11.96	.669		118	41		159	191
4940	8"		10.12	.791		169	48.50		217.50	258
4960	10"		9.20	.870		267	53.50		320.50	375
4980	12"		8.28	.966		259	59.50		318.50	375
5000	NEMA 12, hinged cover, 2-1/2" x 2-1/2"		36.80	.217	L.F.	27	13.35		40.35	49.50
5020	4" x 4"		32.20	.248		30.50	15.25		45.75	56
5040	6" x 6"	2 Elec	55.20	.290		45.50	17.80		63.30	76.50
5060	8" x 8"	"	46	.348		61	21.50		82.50	99
5120	Elbows 90°, flanged, 2-1/2" x 2-1/2"	1 Elec	21.16	.378	Ea.	110	23		133	156
5140	4"		18.40	.435		138	26.50		164.50	192
5160	6"		16.56	.483		165	29.50		194.50	225
5180	8"		13.80	.580		153	35.50		188.50	221
5240	Tee box, flanged, 2-1/2" x 2-1/2"		16.56	.483		143	29.50		172.50	201
5260	4"		14.72	.543		187	33.50		220.50	256
5280	6"		13.80	.580		151	35.50		186.50	219
5300	8"		11.96	.669		212	41		253	294
5360	Cross box, flanged, 2-1/2" x 2-1/2"		13.80	.580		188	35.50		223.50	260
5380	4"		11.96	.669		155	41		196	231
5400	6"		11.04	.725		200	44.50		244.50	286
5420	8"		9.20	.870		250	53.50		303.50	355
5480	Flange gasket, 2-1/2"		147.20	.054		2.83	3.33		6.16	8.05
5500	4"		73.60	.109		3.87	6.65		10.52	14.20
5520	6"		48.76	.164		6.25	10.05		16.30	22
5530	8"		36.80	.217		6.85	13.35		20.20	27.50

26 05 33.25 Conduit Fittings for Rigid Galvanized Steel	Crew	Daily Output	Labor-Hours	Unit	Material	2020 Bare Costs Labor	Equipment	Total	Total Incl O&P
0010 **CONDUIT FITTINGS FOR RIGID GALVANIZED STEEL**									
0050 Standard, locknuts, 1/2" diameter				Ea.	.24			.24	.26
0100 3/4" diameter					.48			.48	.53
0300 1" diameter					.74			.74	.82
0500 1-1/4" diameter					.95			.95	1.04
0700 1-1/2" diameter					1.58			1.58	1.74
1000 2" diameter					2.22			2.22	2.44
1030 2-1/2" diameter					5.45			5.45	6
1050 3" diameter					7.75			7.75	8.55
1070 3-1/2" diameter					13.40			13.40	14.70
1100 4" diameter					18.15			18.15	19.95
1110 5" diameter					37			37	40.50
1120 6" diameter					64.50			64.50	71
1130 Bushings, plastic, 1/2" diameter	1 Elec	36.80	.217		.11	13.35		13.46	19.95
1150 3/4" diameter		29.44	.272		.12	16.65		16.77	25
1170 1" diameter		25.76	.311		.24	19.05		19.29	29
1200 1-1/4" diameter		22.08	.362		.30	22		22.30	33.50
1230 1-1/2" diameter		16.56	.483		.42	29.50		29.92	44.50
1250 2" diameter		13.80	.580		.74	35.50		36.24	54
1270 2-1/2" diameter		11.96	.669		1.38	41		42.38	62.50
1300 3" diameter		11.04	.725		1.61	44.50		46.11	68
1330 3-1/2" diameter		10.12	.791		1.92	48.50		50.42	74
1350 4" diameter		8.28	.966		2.12	59.50		61.62	91
1360 5" diameter		6.44	1.240		14.85	76		90.85	129
1370 6" diameter		4.60	1.740		54.50	107		161.50	219
1390 Steel, 1/2" diameter		36.80	.217		.73	13.35		14.08	20.50
1400 3/4" diameter		29.44	.272		1	16.65		17.65	26
1430 1" diameter		25.76	.311		1.56	19.05		20.61	30
1450 Steel insulated, 1-1/4" diameter		22.08	.362		5.85	22		27.85	39.50
1470 1-1/2" diameter		16.56	.483		7.15	29.50		36.65	52
1500 2" diameter		13.80	.580		8.50	35.50		44	62.50
1530 2-1/2" diameter		11.96	.669		20	41		61	83
1550 3" diameter		11.04	.725		23.50	44.50		68	91.50
1570 3-1/2" diameter		10.12	.791		30.50	48.50		79	106
1600 4" diameter		8.28	.966		38	59.50		97.50	130
1610 5" diameter		6.44	1.240		79.50	76		155.50	200
1620 6" diameter		4.60	1.740		177	107		284	355
1630 Sealing locknuts, 1/2" diameter		36.80	.217		2.08	13.35		15.43	22
1650 3/4" diameter		29.44	.272		1.90	16.65		18.55	27
1670 1" diameter		25.76	.311		5.50	19.05		24.55	34.50
1700 1-1/4" diameter		22.08	.362		6	22		28	39.50
1730 1-1/2" diameter		16.56	.483		6.85	29.50		36.35	51.50
1750 2" diameter		13.80	.580		8.70	35.50		44.20	62.50
1760 Grounding bushing, insulated, 1/2" diameter		29.44	.272		5.90	16.65		22.55	31.50
1770 3/4" diameter		25.76	.311		8.15	19.05		27.20	37.50
1780 1" diameter		18.40	.435		12.05	26.50		38.55	53
1800 1-1/4" diameter		16.56	.483		14.10	29.50		43.60	59.50
1830 1-1/2" diameter		14.72	.543		16.40	33.50		49.90	67.50
1850 2" diameter		11.96	.669		15	41		56	77.50
1870 2-1/2" diameter		11.04	.725		35.50	44.50		80	105
1900 3" diameter		10.12	.791		37	48.50		85.50	113
1930 3-1/2" diameter		8.28	.966		30	59.50		89.50	122

26 05 Common Work Results for Electrical

26 05 33 – Raceway and Boxes for Electrical Systems

26 05 33.25 Conduit Fittings for Rigid Galvanized Steel	Crew	Daily Output	Labor-Hours	Unit	Material	2020 Bare Costs Labor	Equipment	Total	Total Incl O&P
1950	4" diameter	1 Elec	7.36	1.090	Ea.	70	67		137	177
1960	5" diameter		5.52	1.450		91.50	89		180.50	233
1970	6" diameter		3.68	2.170		132	133		265	345
1990	Coupling, with set screw, 1/2" diameter		46	.174		5.15	10.65		15.80	21.50
2000	3/4" diameter		36.80	.217		6.45	13.35		19.80	27
2030	1" diameter		32.20	.248		10.35	15.25		25.60	34
2050	1-1/4" diameter		25.76	.311		14	19.05		33.05	44
2070	1-1/2" diameter		21.16	.378		16.20	23		39.20	52.50
2090	2" diameter		18.40	.435		37.50	26.50		64	80.50
2100	2-1/2" diameter		16.56	.483		84.50	29.50		114	137
2110	3" diameter		13.80	.580		106	35.50		141.50	170
2120	3-1/2" diameter		11.04	.725		183	44.50		227.50	267
2130	4" diameter		9.20	.870		210	53.50		263.50	310
2140	5" diameter		8.28	.966		292	59.50		351.50	410
2150	6" diameter		7.36	1.090		500	67		567	650
2160	Box connector with set screw, plain, 1/2" diameter		64.40	.124		2.66	7.60		10.26	14.30
2170	3/4" diameter		55.20	.145		4.73	8.90		13.63	18.45
2180	1" diameter		46	.174		8.40	10.65		19.05	25
2190	Insulated, 1-1/4" diameter		36.80	.217		12.10	13.35		25.45	33
2200	1-1/2" diameter		27.60	.290		17	17.80		34.80	45
2210	2" diameter		18.40	.435		35.50	26.50		62	78.50
2220	2-1/2" diameter		16.56	.483		104	29.50		133.50	159
2230	3" diameter		13.80	.580		168	35.50		203.50	237
2240	3-1/2" diameter		11.04	.725		207	44.50		251.50	294
2250	4" diameter		9.20	.870		216	53.50		269.50	320
2260	5" diameter		8.28	.966		475	59.50		534.50	610
2270	6" diameter		7.36	1.090		485	67		552	635
2280	LB, LR or LL fittings & covers, 1/2" diameter		14.72	.543		8.80	33.50		42.30	59
2290	3/4" diameter		11.96	.669		10.60	41		51.60	72.50
2300	1" diameter		10.12	.791		15.90	48.50		64.40	89.50
2330	1-1/4" diameter		7.36	1.090		31	67		98	134
2350	1-1/2" diameter		5.52	1.450		30.50	89		119.50	166
2370	2" diameter		4.60	1.740		73	107		180	240
2380	2-1/2" diameter		3.68	2.170		136	133		269	350
2390	3" diameter		3.22	2.480		179	152		331	425
2400	3-1/2" diameter		2.76	2.900		220	178		398	505
2410	4" diameter		2.30	3.480		266	214		480	615
2420	T fittings, with cover, 1/2" diameter		11.04	.725		12.25	44.50		56.75	79.50
2430	3/4" diameter		10.12	.791		13.35	48.50		61.85	86.50
2440	1" diameter		8.28	.966		20.50	59.50		80	111
2450	1-1/4" diameter		5.52	1.450		26.50	89		115.50	161
2470	1-1/2" diameter		4.60	1.740		46.50	107		153.50	210
2500	2" diameter		3.68	2.170		48	133		181	251
2510	2-1/2" diameter		3.22	2.480		111	152		263	350
2520	3" diameter		2.76	2.900		146	178		324	425
2530	3-1/2" diameter		2.30	3.480		270	214		484	615
2540	4" diameter		1.84	4.350		291	267		558	715
2550	Nipples chase, plain, 1/2" diameter		36.80	.217		.46	13.35		13.81	20.50
2560	3/4" diameter		29.44	.272		1.79	16.65		18.44	27
2570	1" diameter		25.76	.311		2.57	19.05		21.62	31.50
2600	Insulated, 1-1/4" diameter		22.08	.362		9.50	22		31.50	43.50
2630	1-1/2" diameter		16.56	.483		14.85	29.50		44.35	60.50
2650	2" diameter		13.80	.580		16.65	35.50		52.15	71.50

26 05 33.25 Conduit Fittings for Rigid Galvanized Steel	Crew	Daily Output	Labor-Hours	Unit	Material	2020 Bare Costs Labor	Equipment	Total	Total Incl O&P
2660	2-1/2" diameter	1 Elec	11.04	.725	Ea.	49	44.50		93.50	120
2670	3" diameter		9.20	.870		47	53.50		100.50	131
2680	3-1/2" diameter		8.28	.966		75	59.50		134.50	171
2690	4" diameter		7.36	1.090		103	67		170	213
2700	5" diameter		6.44	1.240		335	76		411	485
2710	6" diameter		5.52	1.450		530	89		619	710
2720	Nipples offset, plain, 1/2" diameter		36.80	.217		6.90	13.35		20.25	27.50
2730	3/4" diameter		29.44	.272		6.90	16.65		23.55	32.50
2740	1" diameter		22.08	.362		10.45	22		32.45	44.50
2750	Insulated, 1-1/4" diameter		18.40	.435		8.70	26.50		35.20	49
2760	1-1/2" diameter		16.56	.483		10.60	29.50		40.10	55.50
2770	2" diameter		14.72	.543		15	33.50		48.50	66
2780	3" diameter		12.88	.621		99	38		137	166
2850	Coupling, expansion, 1/2" diameter		11.04	.725		59.50	44.50		104	132
2880	3/4" diameter		9.20	.870		70	53.50		123.50	157
2900	1" diameter		7.36	1.090		92	67		159	201
2920	1-1/4" diameter		5.89	1.360		116	83.50		199.50	251
2940	1-1/2" diameter		4.88	1.640		158	101		259	325
2960	2" diameter		4.23	1.890		231	116		347	425
2980	2-1/2" diameter		3.31	2.420		385	148		533	645
3000	3" diameter		2.76	2.900		450	178		628	760
3020	3-1/2" diameter		2.58	3.110		670	191		861	1,025
3040	4" diameter		2.21	3.620		805	222		1,027	1,225
3060	5" diameter		1.84	4.350		1,275	267		1,542	1,800
3080	6" diameter		1.66	4.830		2,175	296		2,471	2,825
3100	Expansion deflection, 1/2" diameter		11.04	.725		315	44.50		359.50	415
3120	3/4" diameter		11.04	.725		355	44.50		399.50	455
3140	1" diameter		9.20	.870		415	53.50		468.50	540
3160	1-1/4" diameter		5.89	1.360		660	83.50		743.50	850
3180	1-1/2" diameter		4.88	1.640		795	101		896	1,025
3200	2" diameter		4.23	1.890		940	116		1,056	1,200
3220	2-1/2" diameter		3.31	2.420		1,050	148		1,198	1,375
3240	3" diameter		2.76	2.900		1,600	178		1,778	2,025
3260	3-1/2" diameter		2.58	3.110		1,575	191		1,766	2,000
3280	4" diameter		2.21	3.620		1,800	222		2,022	2,300
3300	5" diameter		1.84	4.350		3,425	267		3,692	4,175
3320	6" diameter		1.66	4.830		4,925	296		5,221	5,850
3340	Ericson, 1/2" diameter		14.72	.543		4.54	33.50		38.04	54.50
3360	3/4" diameter		12.88	.621		5.50	38		43.50	62.50
3380	1" diameter		10.12	.791		9.20	48.50		57.70	82
3400	1-1/4" diameter		7.36	1.090		17.70	67		84.70	119
3420	1-1/2" diameter		6.44	1.240		23	76		99	138
3440	2" diameter		4.60	1.740		43	107		150	206
3460	2-1/2" diameter		3.68	2.170		77	133		210	283
3480	3" diameter		3.22	2.480		126	152		278	365
3500	3-1/2" diameter		2.76	2.900		165	178		343	445
3520	4" diameter		2.48	3.220		193	198		391	505
3540	5" diameter		2.30	3.480		395	214		609	755
3560	6" diameter		2.12	3.780		505	232		737	900
3580	Split, 1/2" diameter		29.44	.272		5.05	16.65		21.70	30.50
3600	3/4" diameter		24.84	.322		6.20	19.75		25.95	36.50
3620	1" diameter		18.40	.435		13.05	26.50		39.55	54
3640	1-1/4" diameter		14.72	.543		16.90	33.50		50.40	68

For customer support on your Electrical Change Order Costs with RSMeans data, call 800.448.8182.

Pre-Installation Change Order Costs

26 05 33.25 Conduit Fittings for Rigid Galvanized Steel	Crew	Daily Output	Labor-Hours	Unit	Material	2020 Bare Costs Labor	Equipment	Total	Total Incl O&P
3660	1-1/2" diameter	1 Elec	12.88	.621	Ea.	17.90	38		55.90	76
3680	2" diameter		11.04	.725		36.50	44.50		81	107
3700	2-1/2" diameter		9.20	.870		94.50	53.50		148	184
3720	3" diameter		8.28	.966		145	59.50		204.50	249
3740	3-1/2" diameter		7.36	1.090		226	67		293	350
3760	4" diameter		6.44	1.240		220	76		296	355
3780	5" diameter		5.52	1.450		445	89		534	620
3800	6" diameter		4.60	1.740		595	107		702	810
4600	Reducing bushings, 3/4" to 1/2" diameter		49.68	.161		2.94	9.90		12.84	17.95
4620	1" to 3/4" diameter		42.32	.189		3.41	11.60		15.01	21
4640	1-1/4" to 1" diameter		36.80	.217		5.70	13.35		19.05	26
4660	1-1/2" to 1-1/4" diameter		33.12	.242		7.05	14.80		21.85	30
4680	2" to 1-1/2" diameter		29.44	.272		15.60	16.65		32.25	42
4740	2-1/2" to 2" diameter		27.60	.290		19.35	17.80		37.15	48
4760	3" to 2-1/2" diameter		25.76	.311		23.50	19.05		42.55	54.50
4800	Through-wall seal, 1/2" diameter		7.36	1.090		79	67		146	187
4820	3/4" diameter		6.90	1.160		340	71		411	475
4840	1" diameter		5.98	1.340		330	82		412	480
4860	1-1/4" diameter		5.06	1.580		345	97		442	525
4880	1-1/2" diameter		4.60	1.740		495	107		602	705
4900	2" diameter		3.86	2.070		350	127		477	575
4920	2-1/2" diameter		3.22	2.480		630	152		782	915
4940	3" diameter		2.76	2.900		670	178		848	1,000
4960	3-1/2" diameter		2.30	3.480		1,300	214		1,514	1,750
4980	4" diameter		1.84	4.350		1,000	267		1,267	1,500
5000	5" diameter		1.38	5.800		990	355		1,345	1,625
5020	6" diameter	▼	.92	8.700	▼	990	535		1,525	1,900
5100	Cable supports, 2 or more wires									
5120	1-1/2" diameter	1 Elec	7.36	1.090	Ea.	196	67		263	315
5140	2" diameter		5.52	1.450		246	89		335	405
5160	2-1/2" diameter		3.68	2.170		264	133		397	490
5180	3" diameter		3.22	2.480		340	152		492	600
5200	3-1/2" diameter		2.39	3.340		490	205		695	845
5220	4" diameter		1.84	4.350		560	267		827	1,000
5240	5" diameter		1.38	5.800		770	355		1,125	1,375
5260	6" diameter		.92	8.700		1,225	535		1,760	2,150
5280	Service entrance cap, 1/2" diameter		14.72	.543		6.45	33.50		39.95	56.50
5300	3/4" diameter		11.96	.669		7.05	41		48.05	69
5320	1" diameter		9.20	.870		5.75	53.50		59.25	86
5340	1-1/4" diameter		7.36	1.090		5.80	67		72.80	106
5360	1-1/2" diameter		5.98	1.340		11.65	82		93.65	135
5380	2" diameter		5.06	1.580		25.50	97		122.50	172
5400	2-1/2" diameter		3.68	2.170		98.50	133		231.50	305
5420	3" diameter		3.13	2.560		107	157		264	350
5440	3-1/2" diameter		2.76	2.900		169	178		347	450
5460	4" diameter		2.48	3.220		224	198		422	540
5750	90° pull elbows steel, female, 1/2" diameter		14.72	.543		9.50	33.50		43	60
5760	3/4" diameter		11.96	.669		11.95	41		52.95	74
5780	1" diameter		10.12	.791		17.75	48.50		66.25	91.50
5800	1-1/4" diameter		7.36	1.090		28.50	67		95.50	131
5820	1-1/2" diameter		5.52	1.450		39.50	89		128.50	176
5840	2" diameter	▼	4.60	1.740	▼	64	107		171	230
6000	Explosion proof, flexible coupling									

26 05 33.25 Conduit Fittings for Rigid Galvanized Steel	Crew	Daily Output	Labor-Hours	Unit	Material	2020 Bare Costs Labor	Equipment	Total	Total Incl O&P
6010	1/2" diameter, 4" long	1 Elec	11.04	.725	Ea.	161	44.50		205.50	243
6020	6" long		11.04	.725		147	44.50		191.50	227
6050	12" long		11.04	.725		198	44.50		242.50	284
6070	18" long		11.04	.725		255	44.50		299.50	345
6090	24" long		11.04	.725		345	44.50		389.50	445
6110	30" long		11.04	.725		325	44.50		369.50	425
6130	36" long		11.04	.725		365	44.50		409.50	470
6140	3/4" diameter, 4" long		9.20	.870		170	53.50		223.50	267
6150	6" long		9.20	.870		180	53.50		233.50	278
6180	12" long		9.20	.870		234	53.50		287.50	335
6200	18" long		9.20	.870		305	53.50		358.50	415
6220	24" long		9.20	.870		375	53.50		428.50	490
6240	30" long		9.20	.870		375	53.50		428.50	490
6260	36" long		9.20	.870		460	53.50		513.50	585
6270	1" diameter, 6" long		7.36	1.090		340	67		407	475
6300	12" long		7.36	1.090		435	67		502	580
6320	18" long		7.36	1.090		495	67		562	640
6340	24" long		7.36	1.090		740	67		807	910
6360	30" long		7.36	1.090		900	67		967	1,100
6380	36" long		7.36	1.090		805	67		872	985
6390	1-1/4" diameter, 12" long		5.89	1.360		590	83.50		673.50	775
6410	18" long		5.89	1.360		725	83.50		808.50	925
6430	24" long		5.89	1.360		955	83.50		1,038.50	1,175
6450	30" long		5.89	1.360		975	83.50		1,058.50	1,200
6470	36" long		5.89	1.360		1,625	83.50		1,708.50	1,900
6480	1-1/2" diameter, 12" long		4.88	1.640		780	101		881	1,000
6500	18" long		4.88	1.640		940	101		1,041	1,175
6520	24" long		4.88	1.640		1,150	101		1,251	1,400
6540	30" long		4.88	1.640		1,750	101		1,851	2,075
6560	36" long		4.88	1.640		1,450	101		1,551	1,750
6570	2" diameter, 12" long		4.23	1.890		1,050	116		1,166	1,325
6590	18" long		4.23	1.890		1,350	116		1,466	1,650
6610	24" long		4.23	1.890		1,450	116		1,566	1,775
6630	30" long		4.23	1.890		1,725	116		1,841	2,050
6650	36" long		4.23	1.890		1,950	116		2,066	2,325
7000	Close up plug, 1/2" diameter, explosion proof		36.80	.217		2.98	13.35		16.33	23
7010	3/4" diameter		29.44	.272		3.50	16.65		20.15	29
7020	1" diameter		25.76	.311		4.13	19.05		23.18	33
7030	1-1/4" diameter		22.08	.362		4.60	22		26.60	38
7040	1-1/2" diameter		16.56	.483		6.10	29.50		35.60	50.50
7050	2" diameter		13.80	.580		10.35	35.50		45.85	64.50
7060	2-1/2" diameter		11.96	.669		15.85	41		56.85	78.50
7070	3" diameter		11.04	.725		24	44.50		68.50	92.50
7080	3-1/2" diameter		10.12	.791		28	48.50		76.50	103
7090	4" diameter		8.28	.966		48	59.50		107.50	141
7091	Elbow female, 45°, 1/2"		14.72	.543		18.90	33.50		52.40	70.50
7092	3/4"		11.96	.669		24.50	41		65.50	88
7093	1"		10.12	.791		30	48.50		78.50	105
7094	1-1/4"		7.36	1.090		37.50	67		104.50	141
7095	1-1/2"		5.52	1.450		31	89		120	166
7096	2"		4.60	1.740		47	107		154	211
7097	2-1/2"		4.14	1.930		123	118		241	310
7098	3"		3.86	2.070		140	127		267	345

26 05 33.25 Conduit Fittings for Rigid Galvanized Steel	Crew	Daily Output	Labor-Hours	Unit	Material	2020 Bare Costs Labor	Equipment	Total	Total Incl O&P
7099	3-1/2"	1 Elec	3.68	2.170	Ea.	213	133		346	430
7100	4"		3.50	2.290		261	140		401	495
7101	90°, 1/2"		14.72	.543		18.50	33.50		52	70
7102	3/4"		11.96	.669		22	41		63	85
7103	1"		10.12	.791		27	48.50		75.50	102
7104	1-1/4"		7.36	1.090		40.50	67		107.50	144
7105	1-1/2"		5.52	1.450		61	89		150	199
7106	2"		4.60	1.740		115	107		222	285
7107	2-1/2"		4.14	1.930		204	118		322	400
7110	Elbows 90°, long male & female, 1/2" diameter, explosion proof		14.72	.543		27	33.50		60.50	79.50
7120	3/4" diameter		11.96	.669		30	41		71	94
7130	1" diameter		10.12	.791		40.50	48.50		89	117
7140	1-1/4" diameter		7.36	1.090		62.50	67		129.50	169
7150	1-1/2" diameter		5.52	1.450		52.50	89		141.50	190
7160	2" diameter		4.60	1.740		70.50	107		177.50	237
7170	Capped elbow, 1/2" diameter, explosion proof		10.12	.791		27	48.50		75.50	102
7180	3/4" diameter		7.36	1.090		36.50	67		103.50	140
7190	1" diameter		5.52	1.450		39.50	89		128.50	176
7200	1-1/4" diameter		4.60	1.740		86	107		193	254
7210	Pulling elbow, 1/2" diameter, explosion proof		10.12	.791		168	48.50		216.50	257
7220	3/4" diameter		7.36	1.090		231	67		298	355
7230	1" diameter		5.52	1.450		293	89		382	455
7240	1-1/4" diameter		4.60	1.740		380	107		487	580
7250	1-1/2" diameter		4.60	1.740		465	107		572	675
7260	2" diameter		3.68	2.170		545	133		678	800
7270	2-1/2" diameter		3.22	2.480		920	152		1,072	1,225
7280	3" diameter		2.76	2.900		1,425	178		1,603	1,850
7290	3-1/2" diameter		2.30	3.480		1,950	214		2,164	2,475
7300	4" diameter		2.02	3.950		1,925	242		2,167	2,475
7310	LB conduit body, 1/2" diameter		10.12	.791		71.50	48.50		120	151
7320	3/4" diameter		7.36	1.090		94.50	67		161.50	204
7330	T conduit body, 1/2" diameter		8.28	.966		72	59.50		131.50	168
7340	3/4" diameter		5.52	1.450		90.50	89		179.50	232
7350	Explosion proof, round box w/cover, 3 threaded hubs, 1/2" diameter		7.36	1.090		70.50	67		137.50	177
7351	3/4" diameter		7.36	1.090		94.50	67		161.50	204
7352	1" diameter		6.90	1.160		101	71		172	217
7353	1-1/4" diameter		6.44	1.240		106	76		182	229
7354	1-1/2" diameter		6.44	1.240		285	76		361	430
7355	2" diameter		5.52	1.450		284	89		373	440
7356	Round box w/cover & mtng flange, 3 threaded hubs, 1/2" diameter		7.36	1.090		73.50	67		140.50	181
7357	3/4" diameter		7.36	1.090		67	67		134	173
7358	4 threaded hubs, 1" diameter		6.44	1.240		80	76		156	201
7400	Unions, 1/2" diameter		18.40	.435		14.55	26.50		41.05	55.50
7410	3/4" - 1/2" diameter		14.72	.543		23	33.50		56.50	75
7420	3/4" diameter		14.72	.543		24.50	33.50		58	76.50
7430	1" diameter		12.88	.621		45	38		83	106
7440	1-1/4" diameter		11.04	.725		67.50	44.50		112	141
7450	1-1/2" diameter		9.20	.870		85	53.50		138.50	173
7460	2" diameter		7.82	1.020		112	62.50		174.50	216
7480	2-1/2" diameter		7.36	1.090		169	67		236	286
7490	3" diameter		6.44	1.240		248	76		324	385
7500	3-1/2" diameter		5.52	1.450		480	89		569	655
7510	4" diameter		4.60	1.740		440	107		547	645

26 05 33.25 Conduit Fittings for Rigid Galvanized Steel	Crew	Daily Output	Labor-Hours	Unit	Material	2020 Bare Costs Labor	Equipment	Total	Total Incl O&P
7680	Reducer, 3/4" to 1/2"	1 Elec	49.68	.161	Ea.	6.50	9.90		16.40	22
7690	1" to 1/2"		42.32	.189		6.20	11.60		17.80	24
7700	1" to 3/4"		42.32	.189		7.80	11.60		19.40	26
7710	1-1/4" to 3/4"		36.80	.217		10.05	13.35		23.40	31
7720	1-1/4" to 1"		36.80	.217		12	13.35		25.35	33
7730	1-1/2" to 1"		33.12	.242		15.35	14.80		30.15	39
7740	1-1/2" to 1-1/4"		33.12	.242		15.45	14.80		30.25	39
7750	2" to 3/4"		29.44	.272		22	16.65		38.65	49
7760	2" to 1-1/4"		29.44	.272		24.50	16.65		41.15	52
7770	2" to 1-1/2"		29.44	.272		31.50	16.65		48.15	59.50
7780	2-1/2" to 1-1/2"		27.60	.290		27	17.80		44.80	56
7790	3" to 2"		27.60	.290		25.50	17.80		43.30	55
7800	3-1/2" to 2-1/2"		25.76	.311		81.50	19.05		100.55	118
7810	4" to 3"		25.76	.311		84	19.05		103.05	121
7820	Sealing fitting, vertical/horizontal, 1/2" diameter		13.34	.600		23	37		60	80.50
7830	3/4" diameter		12.24	.654		27	40		67	89
7840	1" diameter		10.49	.763		34	47		81	107
7850	1-1/4" diameter		9.20	.870		33.50	53.50		87	117
7860	1-1/2" diameter		8.10	.988		54	60.50		114.50	150
7870	2" diameter		7.36	1.090		66.50	67		133.50	173
7880	2-1/2" diameter		6.16	1.300		98.50	80		178.50	227
7890	3" diameter		5.24	1.530		134	94		228	288
7900	3-1/2" diameter		4.32	1.850		350	114		464	555
7910	4" diameter		3.68	2.170		585	133		718	845
7920	Sealing hubs, 1" by 1-1/2"		11.04	.725		34.50	44.50		79	104
7930	1-1/4" by 2"		9.20	.870		65	53.50		118.50	151
7940	1-1/2" by 2"		8.28	.966		89.50	59.50		149	187
7950	2" by 2-1/2"		7.36	1.090		119	67		186	231
7960	3" by 4"		6.44	1.240		213	76		289	345
7970	4" by 5"		5.52	1.450		455	89		544	630
7980	Drain, 1/2"		29.44	.272		132	16.65		148.65	170
7990	Breather, 1/2"		29.44	.272		142	16.65		158.65	181
7991	Explosion proof sealant compound, hub, fittings, 60 min set time, 1lb. pail		44.16	.181		26.50	11.10		37.60	45.50
7992	5 lb. pail		44.16	.181		75	11.10		86.10	99
7993	2 oz. tube		46	.174		81	10.65		91.65	105
7994	6 oz. tube		46	.174		149	10.65		159.65	180
7995	2.0 oz. cartridge		46	.174		53	10.65		63.65	74.50
7996	6.0 oz. cartridge		46	.174		104	10.65		114.65	131
7997	2.0 oz. box		46	.174		35	10.65		45.65	54
7998	8.0 oz. box		46	.174		107	10.65		117.65	133
7999	1.0 lb. box		46	.174		176	10.65		186.65	210
8000	Plastic coated 40 mil thick									
8010	LB, LR or LL conduit body w/cover, 1/2" diameter	1 Elec	11.96	.669	Ea.	65	41		106	133
8020	3/4" diameter		10.12	.791		84	48.50		132.50	165
8030	1" diameter		7.36	1.090		100	67		167	210
8040	1-1/4" diameter		5.52	1.450		148	89		237	295
8050	1-1/2" diameter		4.60	1.740		155	107		262	330
8060	2" diameter		4.14	1.930		263	118		381	465
8070	2-1/2" diameter		3.68	2.170		520	133		653	770
8080	3" diameter		3.22	2.480		515	152		667	790
8090	3-1/2" diameter		2.76	2.900		935	178		1,113	1,300
8100	4" diameter		2.30	3.480		1,075	214		1,289	1,500
8150	T conduit body with cover, 1/2" diameter		10.12	.791		77	48.50		125.50	157

26 05 33.25 Conduit Fittings for Rigid Galvanized Steel	Crew	Daily Output	Labor-Hours	Unit	Material	2020 Bare Costs Labor	Equipment	Total	Total Incl O&P
8160	3/4" diameter	1 Elec	8.28	.966	Ea.	98	59.50		157.50	197
8170	1" diameter		5.52	1.450		115	89		204	259
8180	1-1/4" diameter		4.60	1.740		172	107		279	350
8190	1-1/2" diameter		4.14	1.930		196	118		314	390
8200	2" diameter		3.68	2.170		251	133		384	475
8210	2-1/2" diameter		3.22	2.480		485	152		637	755
8220	3" diameter		2.76	2.900		660	178		838	990
8230	3-1/2" diameter		2.30	3.480		1,150	214		1,364	1,575
8240	4" diameter		1.84	4.350		1,100	267		1,367	1,625
8300	FS conduit body, 1 gang, 3/4" diameter		10.12	.791		83.50	48.50		132	164
8310	1" diameter		9.20	.870		70.50	53.50		124	157
8350	2 gang, 3/4" diameter		8.28	.966		127	59.50		186.50	229
8360	1" diameter		7.36	1.090		184	67		251	305
8400	Duplex receptacle cover		58.88	.136		62	8.35		70.35	81
8410	Switch cover		58.88	.136		74.50	8.35		82.85	94
8420	Switch, vaportight cover		48.76	.164		227	10.05		237.05	265
8430	Blank, cover		58.88	.136		49	8.35		57.35	66.50
8520	FSC conduit body, 1 gang, 3/4" diameter		9.20	.870		81.50	53.50		135	170
8530	1" diameter		8.28	.966		84.50	59.50		144	181
8550	2 gang, 3/4" diameter		7.36	1.090		142	67		209	256
8560	1" diameter		6.44	1.240		196	76		272	330
8590	Conduit hubs, 1/2" diameter		16.56	.483		48.50	29.50		78	97.50
8600	3/4" diameter		14.72	.543		55	33.50		88.50	110
8610	1" diameter		12.88	.621		71	38		109	135
8620	1-1/4" diameter		11.04	.725		85.50	44.50		130	160
8630	1-1/2" diameter		9.20	.870		94.50	53.50		148	184
8640	2" diameter		8.10	.988		133	60.50		193.50	237
8650	2-1/2" diameter		7.82	1.020		183	62.50		245.50	294
8660	3" diameter		7.36	1.090		285	67		352	415
8670	3-1/2" diameter		6.90	1.160		400	71		471	545
8680	4" diameter		6.44	1.240		390	76		466	545
8690	5" diameter		5.52	1.450		470	89		559	650
8700	Plastic coated 40 mil thick									
8710	Pipe strap, stamped 1 hole, 1/2" diameter	1 Elec	432.40	.019	Ea.	17.40	1.14		18.54	21
8720	3/4" diameter		404.80	.020		15.45	1.21		16.66	18.80
8730	1" diameter		368	.022		16.60	1.33		17.93	20
8740	1-1/4" diameter		326.60	.024		27	1.50		28.50	31.50
8750	1-1/2" diameter		294.40	.027		27.50	1.67		29.17	32.50
8760	2" diameter		244.72	.033		36.50	2.01		38.51	43
8770	2-1/2" diameter		184	.043		38.50	2.67		41.17	46
8780	3" diameter		122.36	.065		52.50	4.01		56.51	63.50
8790	3-1/2" diameter		101.20	.079		120	4.85		124.85	139
8800	4" diameter		82.80	.097		124	5.95		129.95	146
8810	5" diameter		64.40	.124		170	7.60		177.60	198
8840	Clamp back spacers, 3/4" diameter		404.80	.020		24	1.21		25.21	28
8850	1" diameter		368	.022		32.50	1.33		33.83	38
8860	1-1/4" diameter		326.60	.024		31.50	1.50		33	36.50
8870	1-1/2" diameter		294.40	.027		47	1.67		48.67	54
8880	2" diameter		244.72	.033		71.50	2.01		73.51	81.50
8900	3" diameter		122.36	.065		106	4.01		110.01	122
8920	4" diameter		82.80	.097		86	5.95		91.95	103
8950	Touch-up plastic coating, spray, 12 oz.					80			80	88
8960	Sealing fittings, 1/2" diameter	1 Elec	10.12	.791		79.50	48.50		128	160

26 05 33 – Raceway and Boxes for Electrical Systems

	26 05 33.25 Conduit Fittings for Rigid Galvanized Steel	Crew	Daily Output	Labor-Hours	Unit	Material	2020 Bare Costs Labor	Equipment	Total	Total Incl O&P
8970	3/4" diameter	1 Elec	8.28	.966	Ea.	120	59.50		179.50	221
8980	1" diameter		6.90	1.160		91	71		162	206
8990	1-1/4" diameter		5.98	1.340		137	82		219	273
9000	1-1/2" diameter		5.06	1.580		168	97		265	330
9010	2" diameter		4.42	1.810		192	111		303	375
9020	2-1/2" diameter		3.68	2.170		350	133		483	585
9030	3" diameter		3.22	2.480		350	152		502	610
9040	3-1/2" diameter		2.76	2.900		860	178		1,038	1,200
9050	4" diameter		2.30	3.480		1,325	214		1,539	1,800
9060	5" diameter		1.56	5.120		2,425	315		2,740	3,150
9070	Unions, 1/2" diameter		16.56	.483		64	29.50		93.50	115
9080	3/4" diameter		13.80	.580		82	35.50		117.50	144
9090	1" diameter		11.96	.669		85	41		126	155
9100	1-1/4" diameter		10.12	.791		147	48.50		195.50	234
9110	1-1/2" diameter		8.74	.915		222	56		278	330
9120	2" diameter		7.36	1.090		263	67		330	390
9130	2-1/2" diameter		6.90	1.160		345	71		416	485
9140	3" diameter		6.26	1.280		385	78.50		463.50	540
9150	3-1/2" diameter		5.34	1.500		580	92		672	775
9160	4" diameter		4.42	1.810		710	111		821	945
9170	5" diameter		3.68	2.170		1,125	133		1,258	1,425

26 05 33.30 Electrical Nonmetallic Tubing (ENT)

		Crew	Daily Output	Labor-Hours	Unit	Material	2020 Bare Costs Labor	Equipment	Total	Total Incl O&P
0010	**ELECTRICAL NONMETALLIC TUBING (ENT)**									
0050	Flexible, 1/2" diameter	1 Elec	248.40	.032	L.F.	.63	1.98		2.61	3.64
0100	3/4" diameter		211.60	.038		.81	2.32		3.13	4.34
0200	1" diameter		133.40	.060		1.69	3.68		5.37	7.35
0210	1-1/4" diameter		115	.070		1.63	4.27		5.90	8.15
0220	1-1/2" diameter		92	.087		2.83	5.35		8.18	11.05
0230	2" diameter		69	.116		3.39	7.10		10.49	14.35
0300	Connectors, to outlet box, 1/2" diameter		211.60	.038	Ea.	1.17	2.32		3.49	4.74
0310	3/4" diameter		193.20	.041		2.21	2.54		4.75	6.20
0320	1" diameter		184	.043		2.76	2.67		5.43	7
0400	Couplings, to conduit, 1/2" diameter		133.40	.060		1.15	3.68		4.83	6.75
0410	3/4" diameter		119.60	.067		1.33	4.10		5.43	7.55
0420	1" diameter		115	.070		2.52	4.27		6.79	9.10

26 05 33.33 Raceway/Boxes for Utility Substations

		Crew	Daily Output	Labor-Hours	Unit	Material	2020 Bare Costs Labor	Equipment	Total	Total Incl O&P
0010	**RACEWAY/BOXES FOR UTILITY SUBSTATIONS**									
7000	Conduit, conductors, and insulators									
7100	Conduit, metallic	R-11	515.20	.109	Lb.	3.47	6.35	1.24	11.06	14.65
7110	Non-metallic	"	736	.076	"	9.05	4.44	.87	14.36	17.50
7190	See Section 26 05 33									
7200	Wire and cable	R-11	644	.087	Lb.	9.95	5.10	.99	16.04	19.60
7290	See Section 26 05 19									

26 05 33.35 Flexible Metallic Conduit

		Crew	Daily Output	Labor-Hours	Unit	Material	2020 Bare Costs Labor	Equipment	Total	Total Incl O&P
0010	**FLEXIBLE METALLIC CONDUIT**									
0050	Steel, 3/8" diameter	1 Elec	184	.043	L.F.	.44	2.67		3.11	4.45
0100	1/2" diameter		184	.043		.52	2.67		3.19	4.54
0200	3/4" diameter		147.20	.054		.70	3.33		4.03	5.75
0250	1" diameter		92	.087		1.38	5.35		6.73	9.45
0300	1-1/4" diameter		64.40	.124		1.60	7.60		9.20	13.10
0350	1-1/2" diameter		46	.174		2.93	10.65		13.58	19.15
0370	2" diameter		36.80	.217		3.71	13.35		17.06	24

26 05 33.35 Flexible Metallic Conduit		Crew	Daily Output	Labor-Hours	Unit	Material	2020 Bare Costs Labor	Equipment	Total	Total Incl O&P
0380	2-1/2" diameter	1 Elec	27.60	.290	L.F.	4.03	17.80		21.83	31
0390	3" diameter	2 Elec	46	.348		6.90	21.50		28.40	39.50
0400	3-1/2" diameter		36.80	.435		7.75	26.50		34.25	48
0410	4" diameter		27.60	.580		8.90	35.50		44.40	63
0420	Connectors, plain, 3/8" diameter	1 Elec	92	.087	Ea.	2.14	5.35		7.49	10.30
0430	1/2" diameter		73.60	.109		2.32	6.65		8.97	12.50
0440	3/4" diameter		64.40	.124		2.44	7.60		10.04	14.05
0450	1" diameter		46	.174		5.05	10.65		15.70	21.50
0452	1-1/4" diameter		41.40	.193		6.55	11.85		18.40	25
0454	1-1/2" diameter		36.80	.217		9.60	13.35		22.95	30.50
0456	2" diameter		25.76	.311		13.90	19.05		32.95	44
0458	2-1/2" diameter		23	.348		23	21.50		44.50	57.50
0460	3" diameter		18.40	.435		32.50	26.50		59	75.50
0462	3-1/2" diameter		14.72	.543		105	33.50		138.50	166
0464	4" diameter		11.96	.669		116	41		157	189
0490	Insulated, 1" diameter		36.80	.217		9.40	13.35		22.75	30
0500	1-1/4" diameter		36.80	.217		15.15	13.35		28.50	36.50
0550	1-1/2" diameter		29.44	.272		28	16.65		44.65	55.50
0600	2" diameter		21.16	.378		30	23		53	67
0610	2-1/2" diameter		18.40	.435		66	26.50		92.50	112
0620	3" diameter		15.64	.512		134	31.50		165.50	195
0630	3-1/2" diameter		11.96	.669		325	41		366	415
0640	4" diameter		9.20	.870		345	53.50		398.50	460
0650	Connectors 90°, plain, 3/8" diameter		73.60	.109		3.37	6.65		10.02	13.65
0660	1/2" diameter		55.20	.145		6.15	8.90		15.05	20
0700	3/4" diameter		46	.174		8.55	10.65		19.20	25.50
0750	1" diameter		36.80	.217		13.85	13.35		27.20	35
0790	Insulated, 1" diameter		36.80	.217		19.95	13.35		33.30	42
0800	1-1/4" diameter		27.60	.290		27	17.80		44.80	56.50
0850	1-1/2" diameter		21.16	.378		53	23		76	93
0900	2" diameter		16.56	.483		77	29.50		106.50	129
0910	2-1/2" diameter		14.72	.543		194	33.50		227.50	264
0920	3" diameter		12.88	.621		254	38		292	335
0930	3-1/2" diameter		10.12	.791		660	48.50		708.50	795
0940	4" diameter		7.36	1.090		825	67		892	1,000
0960	Couplings, to flexible conduit, 1/2" diameter		46	.174		1.10	10.65		11.75	17.10
0970	3/4" diameter		36.80	.217		1.84	13.35		15.19	22
0980	1" diameter		32.20	.248		2.99	15.25		18.24	26
0990	1-1/4" diameter		25.76	.311		7.25	19.05		26.30	36.50
1000	1-1/2" diameter		21.16	.378		8.60	23		31.60	44
1010	2" diameter		18.40	.435		17.10	26.50		43.60	58.50
1020	2-1/2" diameter		16.56	.483		45	29.50		74.50	94
1030	3" diameter		13.80	.580		114	35.50		149.50	178
1032	Aluminum, 3/8" diameter		193.20	.041	L.F.	.52	2.54		3.06	4.35
1034	1/2" diameter		193.20	.041		.70	2.54		3.24	4.55
1036	3/4" diameter		151.80	.053		1	3.23		4.23	5.90
1038	1" diameter		96.60	.083		1.79	5.10		6.89	9.50
1040	1-1/4" diameter		69	.116		2.08	7.10		9.18	12.90
1042	1-1/2" diameter		48.76	.164		3.64	10.05		13.69	19
1044	2" diameter		38.64	.207		5.25	12.70		17.95	24.50
1046	2-1/2" diameter		29.44	.272		5.05	16.65		21.70	30.50
1048	3" diameter	2 Elec	48.76	.328		8.85	20		28.85	40
1050	3-1/2" diameter		38.64	.414		7.20	25.50		32.70	46

26 05 33.35 Flexible Metallic Conduit

	26 05 33.35 Flexible Metallic Conduit	Crew	Daily Output	Labor-Hours	Unit	Material	2020 Bare Costs Labor	Equipment	Total	Total Incl O&P
1052	4" diameter	2 Elec	29.44	.543	L.F.	10.60	33.50		44.10	61
1070	Sealtite, 3/8" diameter	1 Elec	128.80	.062		.77	3.81		4.58	6.50
1080	1/2" diameter		128.80	.062		1.30	3.81		5.11	7.10
1090	3/4" diameter		92	.087		1.82	5.35		7.17	9.95
1100	1" diameter		64.40	.124		2.68	7.60		10.28	14.30
1200	1-1/4" diameter		46	.174		3.58	10.65		14.23	19.85
1300	1-1/2" diameter		36.80	.217		3.98	13.35		17.33	24
1400	2" diameter		27.60	.290		5.15	17.80		22.95	32
1410	2-1/2" diameter		24.84	.322		9.20	19.75		28.95	39.50
1420	3" diameter	2 Elec	46	.348		11.70	21.50		33.20	45
1440	4" diameter	"	27.60	.580		24	35.50		59.50	79.50
1490	Connectors, plain, 3/8" diameter	1 Elec	64.40	.124	Ea.	2.76	7.60		10.36	14.40
1500	1/2" diameter		64.40	.124		3.62	7.60		11.22	15.35
1700	3/4" diameter		46	.174		5.20	10.65		15.85	21.50
1900	1" diameter		36.80	.217		8.30	13.35		21.65	29
1910	Insulated, 1" diameter		36.80	.217		11.40	13.35		24.75	32.50
2000	1-1/4" diameter		29.44	.272		13.55	16.65		30.20	40
2100	1-1/2" diameter		24.84	.322		25	19.75		44.75	57
2200	2" diameter		18.40	.435		58	26.50		84.50	103
2210	2-1/2" diameter		13.80	.580		124	35.50		159.50	189
2220	3" diameter		11.04	.725		176	44.50		220.50	260
2240	4" diameter		7.36	1.090		232	67		299	355
2290	Connectors, 90°, 3/8" diameter		64.40	.124		5.95	7.60		13.55	17.90
2300	1/2" diameter		64.40	.124		7.95	7.60		15.55	20
2400	3/4" diameter		46	.174		8.65	10.65		19.30	25.50
2600	1" diameter		36.80	.217		16.95	13.35		30.30	38.50
2790	Insulated, 1" diameter		36.80	.217		14.25	13.35		27.60	35.50
2800	1-1/4" diameter		29.44	.272		16.60	16.65		33.25	43.50
3000	1-1/2" diameter		24.84	.322		29	19.75		48.75	61.50
3100	2" diameter		18.40	.435		31	26.50		57.50	73.50
3110	2-1/2" diameter		12.88	.621		147	38		185	219
3120	3" diameter		10.12	.791		415	48.50		463.50	525
3140	4" diameter		6.44	1.240		196	76		272	330
4300	Coupling sealtite to rigid, 1/2" diameter		18.40	.435		4.98	26.50		31.48	45
4500	3/4" diameter		16.56	.483		7	29.50		36.50	51.50
4800	1" diameter		12.88	.621		7.95	38		45.95	65.50
4900	1-1/4" diameter		11.04	.725		19.50	44.50		64	87.50
5000	1-1/2" diameter		10.12	.791		24	48.50		72.50	98.50
5100	2" diameter		9.20	.870		38	53.50		91.50	121
5110	2-1/2" diameter		8.74	.915		177	56		233	279
5120	3" diameter		8.28	.966		180	59.50		239.50	287
5130	3-1/2" diameter		8.28	.966		208	59.50		267.50	320
5140	4" diameter		7.82	1.020		237	62.50		299.50	355

26 05 33.95 Cutting and Drilling

	26 05 33.95 Cutting and Drilling	Crew	Daily Output	Labor-Hours	Unit	Material	2020 Bare Costs Labor	Equipment	Total	Total Incl O&P
0010	**CUTTING AND DRILLING**									
0100	Hole drilling to 10' high, concrete wall									
0110	8" thick, 1/2" pipe size	R-31	11.04	.725	Ea.	.27	44.50	5.65	50.42	72.50
0120	3/4" pipe size		11.04	.725		.27	44.50	5.65	50.42	72.50
0130	1" pipe size		8.74	.915		.35	56	7.15	63.50	91.50
0140	1-1/4" pipe size		8.74	.915		.35	56	7.15	63.50	91.50
0150	1-1/2" pipe size		8.74	.915		.35	56	7.15	63.50	91.50
0160	2" pipe size		4.05	1.980		.53	121	15.45	136.98	199

For customer support on your Electrical Change Order Costs with RSMeans data, call 800.448.8182.　　**Pre-Installation Change Order Costs**

26 05 33 – Raceway and Boxes for Electrical Systems

26 05 33.95 Cutting and Drilling	Crew	Daily Output	Labor-Hours	Unit	Material	2020 Bare Costs Labor	Equipment	Total	Total Incl O&P
0170	2-1/2" pipe size	R-31	4.05	1.980	Ea.	.53	121	15.45	136.98	199
0180	3" pipe size		4.05	1.980		.53	121	15.45	136.98	199
0190	3-1/2" pipe size		3.04	2.640		.60	162	20.50	183.10	264
0200	4" pipe size		3.04	2.640		.60	162	20.50	183.10	264
0500	12" thick, 1/2" pipe size		8.65	.925		.40	57	7.20	64.60	93
0520	3/4" pipe size		8.65	.925		.40	57	7.20	64.60	93
0540	1" pipe size		6.72	1.190		.52	73	9.30	82.82	120
0560	1-1/4" pipe size		6.72	1.190		.52	73	9.30	82.82	120
0570	1-1/2" pipe size		6.72	1.190		.52	73	9.30	82.82	120
0580	2" pipe size		3.31	2.420		.80	148	18.90	167.70	243
0590	2-1/2" pipe size		3.31	2.420		.80	148	18.90	167.70	243
0600	3" pipe size		3.31	2.420		.80	148	18.90	167.70	243
0610	3-1/2" pipe size		2.58	3.110		.90	191	24.50	216.40	310
0630	4" pipe size		2.30	3.480		.90	214	27	241.90	350
0650	16" thick, 1/2" pipe size		6.99	1.140		.54	70	8.90	79.44	114
0670	3/4" pipe size		6.44	1.240		.54	76	9.70	86.24	124
0690	1" pipe size		5.52	1.450		.69	89	11.30	100.99	145
0710	1-1/4" pipe size		5.06	1.580		.69	97	12.35	110.04	158
0730	1-1/2" pipe size		5.06	1.580		.69	97	12.35	110.04	158
0750	2" pipe size		2.76	2.900		1.06	178	22.50	201.56	291
0770	2-1/2" pipe size		2.48	3.220		1.06	198	25	224.06	325
0790	3" pipe size		2.30	3.480		1.06	214	27	242.06	350
0810	3-1/2" pipe size		2.12	3.780		1.20	232	29.50	262.70	380
0830	4" pipe size		1.84	4.350		1.20	267	34	302.20	435
0850	20" thick, 1/2" pipe size		5.89	1.360		.67	83.50	10.60	94.77	136
0870	3/4" pipe size		5.52	1.450		.67	89	11.30	100.97	145
0890	1" pipe size		4.60	1.740		.86	107	13.60	121.46	175
0910	1-1/4" pipe size		4.42	1.810		.86	111	14.15	126.01	182
0930	1-1/2" pipe size		4.23	1.890		.86	116	14.75	131.61	190
0950	2" pipe size		2.48	3.220		1.33	198	25	224.33	325
0970	2-1/2" pipe size		2.21	3.620		1.33	222	28.50	251.83	360
0990	3" pipe size		2.02	3.950		1.33	242	31	274.33	395
1010	3-1/2" pipe size		1.84	4.350		1.50	267	34	302.50	435
1030	4" pipe size		1.56	5.120		1.50	315	40	356.50	515
1050	24" thick, 1/2" pipe size		5.06	1.580		.81	97	12.35	110.16	158
1070	3/4" pipe size		4.69	1.710		.81	105	13.35	119.16	172
1090	1" pipe size		3.96	2.020		1.04	124	15.80	140.84	203
1110	1-1/4" pipe size		3.68	2.170		1.04	133	16.95	150.99	218
1130	1-1/2" pipe size		3.68	2.170		1.04	133	16.95	150.99	218
1150	2" pipe size		2.21	3.620		1.59	222	28.50	252.09	365
1170	2-1/2" pipe size		2.02	3.950		1.59	242	31	274.59	395
1190	3" pipe size		1.84	4.350		1.59	267	34	302.59	435
1210	3-1/2" pipe size		1.66	4.830		1.80	296	37.50	335.30	485
1230	4" pipe size		1.38	5.800		1.80	355	45.50	402.30	580
1500	Brick wall, 8" thick, 1/2" pipe size		16.56	.483		.27	29.50	3.77	33.54	48.50
1520	3/4" pipe size		16.56	.483		.27	29.50	3.77	33.54	48.50
1540	1" pipe size		12.24	.654		.35	40	5.10	45.45	65.50
1560	1-1/4" pipe size		12.24	.654		.35	40	5.10	45.45	65.50
1580	1-1/2" pipe size		12.24	.654		.35	40	5.10	45.45	65.50
1600	2" pipe size		5.24	1.530		.53	94	11.95	106.48	154
1620	2-1/2" pipe size		5.24	1.530		.53	94	11.95	106.48	154
1640	3" pipe size		5.24	1.530		.53	94	11.95	106.48	154
1660	3-1/2" pipe size		4.05	1.980		.60	121	15.45	137.05	199

26 05 33 – Raceway and Boxes for Electrical Systems

26 05 33.95 Cutting and Drilling	Crew	Daily Output	Labor-Hours	Unit	Material	2020 Bare Costs Labor	Equipment	Total	Total Incl O&P
1680	4" pipe size	R-31	3.68	2.170	Ea.	.60	133	16.95	150.55	217
1700	12" thick, 1/2" pipe size		13.34	.600		.40	37	4.68	42.08	60.50
1720	3/4" pipe size		13.34	.600		.40	37	4.68	42.08	60.50
1740	1" pipe size		10.12	.791		.52	48.50	6.15	55.17	79.50
1760	1-1/4" pipe size		10.12	.791		.52	48.50	6.15	55.17	79.50
1780	1-1/2" pipe size		10.12	.791		.52	48.50	6.15	55.17	79.50
1800	2" pipe size		4.60	1.740		.80	107	13.60	121.40	175
1820	2-1/2" pipe size		4.60	1.740		.80	107	13.60	121.40	175
1840	3" pipe size		4.60	1.740		.80	107	13.60	121.40	175
1860	3-1/2" pipe size		3.50	2.290		.90	140	17.90	158.80	230
1880	4" pipe size		3.04	2.640		.90	162	20.50	183.40	264
1900	16" thick, 1/2" pipe size		11.32	.707		.54	43.50	5.50	49.54	71
1920	3/4" pipe size		11.32	.707		.54	43.50	5.50	49.54	71
1940	1" pipe size		8.56	.935		.69	57.50	7.30	65.49	94.50
1960	1-1/4" pipe size		8.56	.935		.69	57.50	7.30	65.49	94.50
1980	1-1/2" pipe size		8.56	.935		.69	57.50	7.30	65.49	94.50
2000	2" pipe size		4.05	1.980		1.06	121	15.45	137.51	199
2010	2-1/2" pipe size		4.05	1.980		1.06	121	15.45	137.51	199
2030	3" pipe size		4.05	1.980		1.06	121	15.45	137.51	199
2050	3-1/2" pipe size		3.04	2.640		1.20	162	20.50	183.70	265
2070	4" pipe size		2.76	2.900		1.20	178	22.50	201.70	291
2090	20" thick, 1/2" pipe size		9.84	.813		.67	50	6.35	57.02	81.50
2110	3/4" pipe size		9.84	.813		.67	50	6.35	57.02	81.50
2130	1" pipe size		7.36	1.090		.86	67	8.50	76.36	110
2150	1-1/4" pipe size		7.36	1.090		.86	67	8.50	76.36	110
2170	1-1/2" pipe size		7.36	1.090		.86	67	8.50	76.36	110
2190	2" pipe size		3.68	2.170		1.33	133	16.95	151.28	218
2210	2-1/2" pipe size		3.68	2.170		1.33	133	16.95	151.28	218
2230	3" pipe size		3.68	2.170		1.33	133	16.95	151.28	218
2250	3-1/2" pipe size		2.76	2.900		1.50	178	22.50	202	292
2270	4" pipe size		2.48	3.220		1.50	198	25	224.50	325
2290	24" thick, 1/2" pipe size		8.65	.925		.81	57	7.20	65.01	93.50
2310	3/4" pipe size		8.65	.925		.81	57	7.20	65.01	93.50
2330	1" pipe size		6.53	1.220		1.04	75	9.55	85.59	123
2350	1-1/4" pipe size		6.53	1.220		1.04	75	9.55	85.59	123
2370	1-1/2" pipe size		6.53	1.220		1.04	75	9.55	85.59	123
2390	2" pipe size		3.31	2.420		1.59	148	18.90	168.49	244
2410	2-1/2" pipe size		3.31	2.420		1.59	148	18.90	168.49	244
2430	3" pipe size		3.31	2.420		1.59	148	18.90	168.49	244
2450	3-1/2" pipe size		2.58	3.110		1.80	191	24.50	217.30	310
2470	4" pipe size	▼	2.30	3.480	▼	1.80	214	27	242.80	350
3000	Knockouts to 8' high, metal boxes & enclosures									
3020	With hole saw, 1/2" pipe size	1 Elec	48.76	.164	Ea.		10.05		10.05	15
3040	3/4" pipe size		43.24	.185			11.35		11.35	16.90
3050	1" pipe size		36.80	.217			13.35		13.35	19.85
3060	1-1/4" pipe size		33.12	.242			14.80		14.80	22
3070	1-1/2" pipe size		29.44	.272			16.65		16.65	25
3080	2" pipe size		24.84	.322			19.75		19.75	29.50
3090	2-1/2" pipe size		18.40	.435			26.50		26.50	39.50
4010	3" pipe size		14.72	.543			33.50		33.50	49.50
4030	3-1/2" pipe size		11.96	.669			41		41	61
4050	4" pipe size	▼	10.12	.791	▼		48.50		48.50	72

For customer support on your Electrical Change Order Costs with RSMeans data, call 800.448.8182.

Pre-Installation Change Order Costs

26 05 36 – Cable Trays for Electrical Systems

26 05 36.10 Cable Tray Ladder Type	Crew	Daily Output	Labor-Hours	Unit	Material	2020 Bare Costs Labor	Equipment	Total	Total Incl O&P
0010 **CABLE TRAY LADDER TYPE** w/ftngs. & supports, 4" dp., to 15' elev.									
0100　　　For higher elevations, see Section 26 05 36.40									
0160　　Galvanized steel tray									
0170　　　　4" rung spacing, 6" wide	2 Elec	90.16	.177	L.F.	14.90	10.90		25.80	32.50
0180　　　　　9" wide		84.64	.189		16.30	11.60		27.90	35
0200　　　　　12" wide		79.12	.202		17.95	12.40		30.35	38
0400　　　　　18" wide		75.44	.212		21	13		34	42.50
0600　　　　　24" wide		71.76	.223		24	13.70		37.70	47
0650　　　　　30" wide		62.56	.256		30.50	15.70		46.20	57
0800　　　　6" rung spacing, 6" wide		92	.174		13.60	10.65		24.25	31
0850　　　　　9" wide		86.48	.185		14.90	11.35		26.25	33.50
0860　　　　　12" wide		80.96	.198		15.65	12.10		27.75	35.50
0870　　　　　18" wide		77.28	.207		17.90	12.70		30.60	38.50
0880　　　　　24" wide		73.60	.217		19.95	13.35		33.30	42
0890　　　　　30" wide		64.40	.248		23.50	15.25		38.75	48.50
0910　　　　9" rung spacing, 6" wide		93.84	.171		12.65	10.45		23.10	29.50
0920　　　　　9" wide		90.16	.177		13	10.90		23.90	30.50
0930　　　　　12" wide		86.48	.185		14.90	11.35		26.25	33.50
0940　　　　　18" wide		82.80	.193		16.85	11.85		28.70	36
0950　　　　　24" wide		79.12	.202		19.15	12.40		31.55	39.50
0960　　　　　30" wide		73.60	.217		21	13.35		34.35	43
0980　　　　12" rung spacing, 6" wide		97.52	.164		12.45	10.05		22.50	28.50
0990　　　　　9" wide		95.68	.167		12.75	10.25		23	29.50
1000　　　　　12" wide		92	.174		13.50	10.65		24.15	31
1010　　　　　18" wide		88.32	.181		14.50	11.10		25.60	32.50
1020　　　　　24" wide		86.48	.185		15.55	11.35		26.90	34
1030　　　　　30" wide		80.96	.198		17.65	12.10		29.75	37.50
1041　　　　18" rung spacing, 6" wide		99.36	.161		12.40	9.90		22.30	28.50
1042　　　　　9" wide		97.52	.164		12.55	10.05		22.60	29
1043　　　　　12" wide		93.84	.171		13.35	10.45		23.80	30.50
1044　　　　　18" wide		90.16	.177		14.05	10.90		24.95	31.50
1045　　　　　24" wide		88.32	.181		14.50	11.10		25.60	32.50
1046　　　　　30" wide		82.80	.193		15.25	11.85		27.10	34.50
1050　　　Elbows horiz. 9" rung spacing, 90°, 12" radius, 6" wide		8.83	1.810	Ea.	57.50	111		168.50	229
1060　　　　　9" wide		7.73	2.070		61	127		188	256
1070　　　　　12" wide		6.99	2.290		66	140		206	282
1080　　　　　18" wide		5.70	2.810		87	172		259	355
1090　　　　　24" wide		4.97	3.220		97.50	198		295.50	400
1100　　　　　30" wide		4.42	3.620		129	222		351	470
1120　　　90°, 24" radius, 6" wide		8.46	1.890		119	116		235	305
1130　　　　　9" wide		7.36	2.170		122	133		255	330
1140　　　　　12" wide		6.62	2.420		125	148		273	360
1150　　　　　18" wide		5.34	3		136	184		320	425
1160　　　　　24" wide		4.60	3.480		146	214		360	480
1170　　　　　30" wide		4.05	3.950		159	242		401	535
1190　　　90°, 36" radius, 6" wide		8.10	1.980		131	121		252	325
1200　　　　　9" wide		6.99	2.290		138	140		278	360
1210　　　　　12" wide		6.26	2.560		145	157		302	395
1220　　　　　18" wide		4.97	3.220		166	198		364	475
1230　　　　　24" wide		4.23	3.780		181	232		413	545
1240　　　　　30" wide		3.68	4.350		218	267		485	635
1260　　　45°, 12" radius, 6" wide		12.14	1.320		44.50	81		125.50	170

26 05 Common Work Results for Electrical

26 05 36 – Cable Trays for Electrical Systems

26 05 36.10 Cable Tray Ladder Type	Crew	Daily Output	Labor-Hours	Unit	Material	2020 Bare Costs Labor	Equipment	Total	Total Incl O&P
1270	9" wide	2 Elec	10.12	1.580	Ea.	46	97		143	195
1280	12" wide		8.83	1.810		48	111		159	218
1290	18" wide		6.99	2.290		50.50	140		190.50	265
1300	24" wide		5.70	2.810		65.50	172		237.50	330
1310	30" wide		4.97	3.220		83.50	198		281.50	385
1330	45°, 24" radius, 6" wide		11.78	1.360		56	83.50		139.50	186
1340	9" wide		9.75	1.640		60	101		161	216
1350	12" wide		8.46	1.890		67	116		183	247
1360	18" wide		6.62	2.420		69	148		217	297
1370	24" wide		5.34	3		83.50	184		267.50	365
1380	30" wide		4.60	3.480		99.50	214		313.50	430
1400	45°, 36" radius, 6" wide		11.41	1.400		80.50	86		166.50	217
1410	9" wide		9.38	1.710		85.50	105		190.50	250
1420	12" wide		8.10	1.980		87	121		208	277
1430	18" wide		6.26	2.560		94	157		251	340
1440	24" wide		4.97	3.220		99.50	198		297.50	405
1450	30" wide	↓	4.23	3.780	↓	129	232		361	485
1470	Elbows horizontal, 4" rung spacing, use 9" rung x 1.50									
1480	6" rung spacing, use 9" rung x 1.20									
1490	12" rung spacing, use 9" rung x 0.93									
1500	Elbows vertical, 9" rung spacing, 90°, 12" radius, 6" wide	2 Elec	8.83	1.810	Ea.	79	111		190	252
1510	9" wide		7.73	2.070		80.50	127		207.50	278
1520	12" wide		6.99	2.290		83.50	140		223.50	300
1530	18" wide		5.70	2.810		87	172		259	355
1540	24" wide		4.97	3.220		90.50	198		288.50	395
1550	30" wide		4.42	3.620		97.50	222		319.50	435
1570	24" radius, 6" wide		8.46	1.890		112	116		228	296
1580	9" wide		7.36	2.170		115	133		248	325
1590	12" wide		6.62	2.420		119	148		267	350
1600	18" wide		5.34	3		125	184		309	410
1610	24" wide		4.60	3.480		132	214		346	465
1620	30" wide		4.05	3.950		138	242		380	510
1640	36" radius, 6" wide		8.10	1.980		150	121		271	345
1650	9" wide		6.99	2.290		155	140		295	380
1660	12" wide		6.26	2.560		157	157		314	405
1670	18" wide		4.97	3.220		171	198		369	480
1680	24" wide		4.23	3.780		173	232		405	535
1690	30" wide	↓	3.68	4.350	↓	195	267		462	610
1710	Elbows vertical, 4" rung spacing, use 9" rung x 1.25									
1720	6" rung spacing, use 9" rung x 1.15									
1730	12" rung spacing, use 9" rung x 0.90									
1740	Tee horizontal, 9" rung spacing, 12" radius, 6" wide	2 Elec	4.60	3.480	Ea.	122	214		336	455
1750	9" wide		4.23	3.780		131	232		363	490
1760	12" wide		4.05	3.950		138	242		380	510
1770	18" wide		3.68	4.350		157	267		424	570
1780	24" wide		3.31	4.830		185	296		481	645
1790	30" wide		3.13	5.120		221	315		536	715
1810	24" radius, 6" wide		4.23	3.780		190	232		422	555
1820	9" wide		3.86	4.140		199	254		453	600
1830	12" wide		3.68	4.350		206	267		473	620
1840	18" wide		3.31	4.830		223	296		519	685
1850	24" wide		2.94	5.430		242	335		577	760
1860	30" wide	↓	2.76	5.800	↓	263	355		618	820

26 05 36 – Cable Trays for Electrical Systems

26 05 36.10 Cable Tray Ladder Type		Crew	Daily Output	Labor-Hours	Unit	Material	2020 Bare Costs Labor	Equipment	Total	Total Incl O&P
1880	36" radius, 6" wide	2 Elec	3.86	4.140	Ea.	284	254		538	690
1890	9" wide		3.50	4.580		296	281		577	745
1900	12" wide		3.31	4.830		315	296		611	785
1910	18" wide		2.94	5.430		340	335		675	865
1920	24" wide		2.58	6.210		380	380		760	985
1930	30" wide		2.39	6.690		415	410		825	1,075
1980	Tee vertical, 9" rung spacing, 12" radius, 6" wide		4.97	3.220		207	198		405	520
1990	9" wide		4.78	3.340		209	205		414	535
2000	12" wide		4.60	3.480		211	214		425	550
2010	18" wide		4.23	3.780		213	232		445	580
2020	24" wide		4.05	3.950		218	242		460	600
2030	30" wide		3.68	4.350		232	267		499	650
2050	24" radius, 6" wide		4.60	3.480		375	214		589	730
2060	9" wide		4.42	3.620		380	222		602	745
2070	12" wide		4.23	3.780		385	232		617	765
2080	18" wide		3.86	4.140		395	254		649	815
2090	24" wide		3.68	4.350		405	267		672	840
2100	30" wide		3.31	4.830		415	296		711	900
2120	36" radius, 6" wide		4.23	3.780		695	232		927	1,100
2130	9" wide		4.05	3.950		710	242		952	1,150
2140	12" wide		3.86	4.140		715	254		969	1,175
2150	18" wide		3.50	4.580		725	281		1,006	1,225
2160	24" wide		3.31	4.830		735	296		1,031	1,250
2170	30" wide	▼	2.94	5.430	▼	760	335		1,095	1,325
2190	Tee, 4" rung spacing, use 9" rung x 1.30									
2200	6" rung spacing, use 9" rung x 1.20									
2210	12" rung spacing, use 9" rung x 0.90									
2220	Cross horizontal, 9" rung spacing, 12" radius, 6" wide	2 Elec	3.68	4.350	Ea.	150	267		417	560
2230	9" wide		3.50	4.580		158	281		439	595
2240	12" wide		3.31	4.830		173	296		469	630
2250	18" wide		3.13	5.120		188	315		503	675
2260	24" wide		2.76	5.800		209	355		564	760
2270	30" wide		2.58	6.210		253	380		633	845
2290	24" radius, 6" wide		3.31	4.830		293	296		589	760
2300	9" wide		3.13	5.120		300	315		615	800
2310	12" wide		2.94	5.430		310	335		645	835
2320	18" wide		2.76	5.800		330	355		685	895
2330	24" wide		2.39	6.690		350	410		760	995
2340	30" wide		2.21	7.250		375	445		820	1,075
2360	36" radius, 6" wide		2.94	5.430		410	335		745	945
2370	9" wide		2.76	5.800		425	355		780	1,000
2380	12" wide		2.58	6.210		445	380		825	1,050
2390	18" wide		2.39	6.690		490	410		900	1,150
2400	24" wide		2.02	7.910		545	485		1,030	1,325
2410	30" wide	▼	1.84	8.700	▼	590	535		1,125	1,450
2430	Cross horizontal, 4" rung spacing, use 9" rung x 1.30									
2440	6" rung spacing, use 9" rung x 1.20									
2450	12" rung spacing, use 9" rung x 0.90									
2460	Reducer, 9" to 6" wide tray	2 Elec	11.96	1.340	Ea.	83.50	82		165.50	214
2470	12" to 9" wide tray		11.04	1.450		85.50	89		174.50	226
2480	18" to 12" wide tray		9.57	1.670		85.50	102		187.50	247
2490	24" to 18" wide tray		8.28	1.930		87	118		205	272
2500	30" to 24" wide tray	▼	7.36	2.170	▼	89	133		222	296

26 05 36.10 Cable Tray Ladder Type	Crew	Daily Output	Labor-Hours	Unit	Material	2020 Bare Costs Labor	Equipment	Total	Total Incl O&P
2510	36" to 30" wide tray	2 Elec	6.44	2.480	Ea.	99.50	152		251.50	335
2511	Reducer, 18" to 6" wide tray		9.57	1.670		89	102		191	251
2512	24" to 12" wide tray		8.28	1.930		90.50	118		208.50	276
2513	30" to 18" wide tray		7.36	2.170		93	133		226	300
2514	30" to 12" wide tray		7.36	2.170		98.50	133		231.50	305
2515	36" to 24" wide tray		6.44	2.480		101	152		253	340
2516	36" to 18" wide tray		6.44	2.480		101	152		253	340
2517	36" to 12" wide tray		6.44	2.480		101	152		253	340
2520	Dropout or end plate, 6" wide		29.44	.543		9.35	33.50		42.85	60
2530	9" wide		25.76	.621		11	38		49	68.50
2540	12" wide		23.92	.669		12.05	41		53.05	74.50
2550	18" wide		20.24	.791		14.10	48.50		62.60	87.50
2560	24" wide		18.40	.870		17.10	53.50		70.60	98.50
2570	30" wide		16.56	.966		18.70	59.50		78.20	109
2590	Tray connector		44.16	.362	▼	16.55	22		38.55	51
3200	Aluminum tray, 4" deep, 6" rung spacing, 6" wide		123.28	.130	L.F.	15.75	7.95		23.70	29
3210	9" wide		117.76	.136		16.75	8.35		25.10	31
3220	12" wide		114.08	.140		17.65	8.60		26.25	32
3230	18" wide		104.88	.153		19.70	9.35		29.05	35.50
3240	24" wide		97.52	.164		23	10.05		33.05	40
3250	30" wide		92	.174		25	10.65		35.65	44
3270	9" rung spacing, 6" wide		128.80	.124		12.80	7.60		20.40	25.50
3280	9" wide		123.28	.130		13.50	7.95		21.45	26.50
3290	12" wide		119.60	.134		13.70	8.20		21.90	27.50
3300	18" wide		112.24	.143		16	8.75		24.75	30.50
3310	24" wide		106.72	.150		18.60	9.20		27.80	34
3320	30" wide		99.36	.161		20.50	9.90		30.40	37
3340	12" rung spacing, 6" wide		134.32	.119		13.65	7.30		20.95	26
3350	9" wide		128.80	.124		13.80	7.60		21.40	26.50
3360	12" wide		123.28	.130		14.15	7.95		22.10	27.50
3370	18" wide		117.76	.136		15.10	8.35		23.45	29
3380	24" wide		114.08	.140		16.75	8.60		25.35	31
3390	30" wide		104.88	.153		17.95	9.35		27.30	33.50
3401	18" rung spacing, 6" wide		138	.116		13.55	7.10		20.65	25.50
3402	9" wide tray		132.48	.121		13.70	7.40		21.10	26
3403	12" wide tray		128.80	.124		14.15	7.60		21.75	27
3404	18" wide tray		123.28	.130		15	7.95		22.95	28.50
3405	24" wide tray		119.60	.134		16.15	8.20		24.35	30
3406	30" wide tray		110.40	.145	▼	16.75	8.90		25.65	31.50
3410	Elbows horiz., 9" rung spacing, 90°, 12" radius, 6" wide		8.83	1.810	Ea.	60.50	111		171.50	232
3420	9" wide		7.73	2.070		62.50	127		189.50	258
3430	12" wide		6.99	2.290		74	140		214	291
3440	18" wide		5.70	2.810		83.50	172		255.50	350
3450	24" wide		4.97	3.220		103	198		301	410
3460	30" wide		4.42	3.620		109	222		331	450
3480	24" radius, 6" wide		8.46	1.890		108	116		224	292
3490	9" wide		7.36	2.170		112	133		245	320
3500	12" wide		6.62	2.420		118	148		266	350
3510	18" wide		5.34	3		128	184		312	415
3520	24" wide		4.60	3.480		141	214		355	475
3530	30" wide		4.05	3.950		153	242		395	530
3550	90°, 36" radius, 6" wide		8.10	1.980		119	121		240	310
3560	9" wide		6.99	2.290		122	140		262	345

Pre-Installation Change Order Costs

26 05 36.10 Cable Tray Ladder Type	Crew	Daily Output	Labor-Hours	Unit	Material	2020 Bare Costs Labor	Equipment	Total	Total Incl O&P
3570	12" wide	2 Elec	6.26	2.560	Ea.	136	157		293	385
3580	18" wide		4.97	3.220		153	198		351	460
3590	24" wide		4.23	3.780		172	232		404	535
3600	30" wide		3.68	4.350		190	267		457	605
3620	45°, 12" radius, 6" wide		12.14	1.320		40	81		121	165
3630	9" wide		10.12	1.580		41.50	97		138.50	190
3640	12" wide		8.83	1.810		43	111		154	213
3650	18" wide		6.99	2.290		51	140		191	265
3660	24" wide		5.70	2.810		63	172		235	325
3670	30" wide		4.97	3.220		72.50	198		270.50	375
3690	45°, 24" radius, 6" wide		11.78	1.360		53.50	83.50		137	183
3700	9" wide		9.75	1.640		60.50	101		161.50	217
3710	12" wide		8.46	1.890		62	116		178	242
3720	18" wide		6.62	2.420		66.50	148		214.50	295
3730	24" wide		5.34	3		77	184		261	360
3740	30" wide		4.60	3.480		92	214		306	420
3760	45°, 36" radius, 6" wide		11.41	1.400		72.50	86		158.50	208
3770	9" wide		9.38	1.710		74	105		179	238
3780	12" wide		8.10	1.980		75.50	121		196.50	264
3790	18" wide		6.26	2.560		83	157		240	325
3800	24" wide		4.97	3.220		102	198		300	405
3810	30" wide	▽	4.23	3.780	▽	108	232		340	465
3830	Elbows horizontal, 4" rung spacing, use 9" rung x 1.50									
3840	6" rung spacing, use 9" rung x 1.20									
3850	12" rung spacing, use 9" rung x 0.93									
3860	Elbows vertical, 9" rung spacing, 90°, 12" radius, 6" wide	2 Elec	8.83	1.810	Ea.	84.50	111		195.50	258
3870	9" wide		7.73	2.070		89	127		216	287
3880	12" wide		6.99	2.290		90.50	140		230.50	310
3890	18" wide		5.70	2.810		93.50	172		265.50	360
3900	24" wide		4.97	3.220		99.50	198		297.50	405
3910	30" wide		4.42	3.620		104	222		326	445
3930	24" radius, 6" wide		8.46	1.890		92	116		208	274
3940	9" wide		7.36	2.170		95	133		228	300
3950	12" wide		6.62	2.420		98	148		246	330
3960	18" wide		5.34	3		104	184		288	390
3970	24" wide		4.60	3.480		110	214		324	440
3980	30" wide		4.05	3.950		116	242		358	485
4000	36" radius, 6" wide		8.10	1.980		142	121		263	340
4010	9" wide		6.99	2.290		145	140		285	370
4020	12" wide		6.26	2.560		154	157		311	405
4030	18" wide		4.97	3.220		160	198		358	470
4040	24" wide		4.23	3.780		166	232		398	530
4050	30" wide	▽	3.68	4.350	▽	179	267		446	590
4070	Elbows vertical, 4" rung spacing, use 9" rung x 1.25									
4080	6" rung spacing, use 9" rung x 1.15									
4090	12" rung spacing, use 9" rung x 0.90									
4100	Tee horizontal, 9" rung spacing, 12" radius, 6" wide	2 Elec	4.60	3.480	Ea.	94	214		308	425
4110	9" wide		4.23	3.780		101	232		333	455
4120	12" wide		4.05	3.950		96	242		338	465
4130	18" wide		3.86	4.140		122	254		376	515
4140	24" wide		3.68	4.350		133	267		400	540
4150	30" wide		3.31	4.830		156	296		452	610
4170	24" radius, 6" wide	▽	4.23	3.780	▽	160	232		392	520

26 05 36 – Cable Trays for Electrical Systems

26 05 36.10 Cable Tray Ladder Type	Crew	Daily Output	Labor-Hours	Unit	Material	2020 Bare Costs Labor	Equipment	Total	Total Incl O&P
4180	9" wide	2 Elec	3.86	4.140	Ea.	169	254		423	565
4190	12" wide		3.68	4.350		173	267		440	585
4200	18" wide		3.50	4.580		190	281		471	630
4210	24" wide		3.31	4.830		207	296		503	670
4220	30" wide		2.94	5.430		225	335		560	745
4240	36" radius, 6" wide		3.86	4.140		227	254		481	630
4250	9" wide		3.50	4.580		233	281		514	675
4260	12" wide		3.31	4.830		255	296		551	720
4270	18" wide		3.13	5.120		268	315		583	765
4280	24" wide		2.94	5.430		296	335		631	820
4290	30" wide		2.58	6.210		375	380		755	975
4310	Tee vertical, 9" rung spacing, 12" radius, 6" wide		4.97	3.220		210	198		408	525
4320	9" wide		4.78	3.340		218	205		423	545
4330	12" wide		4.60	3.480		218	214		432	560
4340	18" wide		4.23	3.780		224	232		456	590
4350	24" wide		4.05	3.950		242	242		484	625
4360	30" wide		3.86	4.140		246	254		500	650
4380	24" radius, 6" wide		4.60	3.480		355	214		569	715
4390	9" wide		4.42	3.620		360	222		582	730
4400	12" wide		4.23	3.780		370	232		602	750
4410	18" wide		3.86	4.140		375	254		629	795
4420	24" wide		3.68	4.350		385	267		652	820
4430	30" wide		3.50	4.580		395	281		676	855
4450	36" radius, 6" wide		4.23	3.780		785	232		1,017	1,200
4460	9" wide		4.05	3.950		800	242		1,042	1,250
4470	12" wide		3.86	4.140		835	254		1,089	1,300
4480	18" wide		3.50	4.580		840	281		1,121	1,350
4490	24" wide		3.31	4.830		860	296		1,156	1,400
4500	30" wide	↓	3.13	5.120	↓	870	315		1,185	1,425
4520	Tees, 4" rung spacing, use 9" rung x 1.30									
4530	6" rung spacing, use 9" rung x 1.20									
4540	12" rung spacing, use 9" rung x 0.90									
4550	Cross horizontal, 9" rung spacing, 12" radius, 6" wide	2 Elec	4.05	3.950	Ea.	130	242		372	505
4560	9" wide		3.86	4.140		145	254		399	540
4570	12" wide		3.68	4.350		139	267		406	550
4580	18" wide		3.31	4.830		178	296		474	635
4590	24" wide		3.13	5.120		184	315		499	670
4600	30" wide		2.76	5.800		216	355		571	770
4620	24" radius, 6" wide		3.68	4.350		250	267		517	670
4630	9" wide		3.50	4.580		259	281		540	705
4640	12" wide		3.31	4.830		267	296		563	735
4650	18" wide		2.94	5.430		280	335		615	805
4660	24" wide		2.76	5.800		305	355		660	865
4670	30" wide		2.39	6.690		325	410		735	970
4690	36" radius, 6" wide		3.31	4.830		296	296		592	765
4700	9" wide		3.13	5.120		310	315		625	810
4710	12" wide		2.94	5.430		325	335		660	855
4720	18" wide		2.58	6.210		350	380		730	950
4730	24" wide	↓	2.39	6.690	↓	430	410		840	1,075
4740	30" wide	↓	2.02	7.910	↓	475	485		960	1,250
4760	Cross horizontal, 4" rung spacing, use 9" rung x 1.30									
4770	6" rung spacing, use 9" rung x 1.20									
4780	12" rung spacing, use 9" rung x 0.90									

Pre-Installation Change Order Costs

26 05 36.10 Cable Tray Ladder Type	Crew	Daily Output	Labor-Hours	Unit	Material	2020 Bare Costs Labor	Equipment	Total	Total Incl O&P
4790	Reducer, 9" to 6" wide tray	2 Elec	14.72	1.090	Ea.	69.50	67		136.50	176
4800	12" to 9" wide tray		12.88	1.240		72.50	76		148.50	193
4810	18" to 12" wide tray		11.41	1.400		72.50	86		158.50	208
4820	24" to 18" wide tray		9.75	1.640		74	101		175	232
4830	30" to 24" wide tray		8.46	1.890		75.50	116		191.50	256
4840	36" to 30" wide tray		7.36	2.170		80	133		213	286
4841	Reducer, 18" to 6" wide tray		11.41	1.400		72.50	86		158.50	208
4842	24" to 12" wide tray		9.75	1.640		74	101		175	232
4843	30" to 18" wide tray		8.46	1.890		75.50	116		191.50	256
4844	30" to 12" wide tray		8.46	1.890		75.50	116		191.50	256
4845	36" to 24" wide tray		7.36	2.170		80	133		213	286
4846	36" to 18" wide tray		7.36	2.170		80	133		213	286
4847	36" to 12" wide tray		7.36	2.170		80	133		213	286
4850	Dropout or end plate, 6" wide		29.44	.543		9.10	33.50		42.60	59.50
4860	9" wide tray		25.76	.621		9.85	38		47.85	67.50
4870	12" wide tray		23.92	.669		10.60	41		51.60	72.50
4880	18" wide tray		20.24	.791		14.30	48.50		62.80	87.50
4890	24" wide tray		18.40	.870		16.75	53.50		70.25	98
4900	30" wide tray		16.56	.966		19.85	59.50		79.35	111
4920	Tray connector		44.16	.362		13.30	22		35.30	47.50
8000	Elbow 36" radius horiz., 60°, 6" wide tray		9.75	1.640		101	101		202	261
8010	9" wide tray		8.28	1.930		110	118		228	297
8020	12" wide tray		7.18	2.230		113	137		250	330
8030	18" wide tray		5.70	2.810		120	172		292	390
8040	24" wide tray		4.60	3.480		141	214		355	475
8050	30" wide tray		4.05	3.950		144	242		386	520
8060	30°, 6" wide tray		12.88	1.240		77	76		153	198
8070	9" wide tray		10.49	1.530		81.50	94		175.50	230
8080	12" wide tray		9.02	1.770		83	109		192	254
8090	18" wide tray		6.81	2.350		87.50	144		231.50	310
8100	24" wide tray		5.34	3		107	184		291	390
8110	30" wide tray		4.42	3.620		110	222		332	450
8120	Adjustable, 6" wide tray		11.41	1.400		102	86		188	241
8130	9" wide tray		9.38	1.710		108	105		213	275
8140	12" wide tray		8.10	1.980		110	121		231	300
8150	18" wide tray		6.26	2.560		120	157		277	365
8160	24" wide tray		4.97	3.220		127	198		325	435
8170	30" wide tray		4.23	3.780		142	232		374	500
8180	Wye 36" radius horiz., 45°, 6" wide tray		4.23	3.780		111	232		343	465
8190	9" wide tray		4.05	3.950		117	242		359	490
8200	12" wide tray		3.86	4.140		122	254		376	515
8210	18" wide tray		3.50	4.580		138	281		419	570
8220	24" wide tray		3.31	4.830		154	296		450	610
8230	30" wide tray		3.13	5.120		172	315		487	660
8240	Elbow 36" radius vert. in/outside, 60°, 6" wide tray		9.75	1.640		111	101		212	272
8250	9" wide tray		8.28	1.930		116	118		234	305
8260	12" wide tray		7.18	2.230		117	137		254	335
8270	18" wide tray		5.70	2.810		122	172		294	390
8280	24" wide tray		4.60	3.480		129	214		343	460
8290	30" wide tray		4.05	3.950		138	242		380	510
8300	45°, 6" wide tray		11.41	1.400		100	86		186	238
8310	9" wide tray		9.38	1.710		106	105		211	273
8320	12" wide tray		8.10	1.980		108	121		229	300

26 05 36.10 Cable Tray Ladder Type	Crew	Daily Output	Labor-Hours	Unit	Material	2020 Bare Costs Labor	Equipment	Total	Total Incl O&P
8330	18" wide tray	2 Elec	6.26	2.560	Ea.	111	157		268	355
8340	24" wide tray		4.97	3.220		117	198		315	425
8350	30" wide tray		4.23	3.780		120	232		352	475
8360	30°, 6" wide tray		12.88	1.240		88	76		164	210
8370	9" wide tray		10.49	1.530		89.50	94		183.50	239
8380	12" wide tray		9.02	1.770		95.50	109		204.50	267
8390	18" wide tray		6.81	2.350		97	144		241	320
8400	24" wide tray		5.34	3		99.50	184		283.50	385
8410	30" wide tray		4.42	3.620		105	222		327	445
8660	Adjustable, 6" wide tray		11.41	1.400		100	86		186	238
8670	9" wide tray		9.38	1.710		106	105		211	273
8680	12" wide tray		8.10	1.980		108	121		229	300
8690	18" wide tray		6.26	2.560		120	157		277	365
8700	24" wide tray		4.97	3.220		127	198		325	435
8710	30" wide tray		4.23	3.780		141	232		373	500
8720	Cross, vertical, 24" radius, 6" wide tray		3.31	4.830		730	296		1,026	1,250
8730	9" wide tray		3.13	5.120		740	315		1,055	1,275
8740	12" wide tray		2.94	5.430		765	335		1,100	1,325
8750	18" wide tray		2.58	6.210		800	380		1,180	1,450
8760	24" wide tray		2.39	6.690		825	410		1,235	1,525
8770	30" wide tray		2.02	7.910		875	485		1,360	1,675
9200	Splice plate	1 Elec	44.16	.181	Pr.	13.95	11.10		25.05	32
9210	Expansion joint		44.16	.181		16.75	11.10		27.85	35
9220	Horizontal hinged		44.16	.181		13.95	11.10		25.05	32
9230	Vertical hinged		44.16	.181		16.75	11.10		27.85	35
9240	Ladder hanger, vertical		25.76	.311	Ea.	4.18	19.05		23.23	33
9250	Ladder to channel connector		22.08	.362		48.50	22		70.50	86.50
9260	Ladder to box connector, 30" wide		17.48	.458		48.50	28		76.50	95.50
9270	24" wide		18.40	.435		48.50	26.50		75	93
9280	18" wide		19.32	.414		47	25.50		72.50	89.50
9290	12" wide		20.24	.395		44	24.50		68.50	84.50
9300	9" wide		21.16	.378		39.50	23		62.50	77.50
9310	6" wide		22.08	.362		31	22		53	67.50
9320	Ladder floor flange		22.08	.362		31.50	22		53.50	67.50
9330	Cable roller for tray, 30" wide		9.20	.870		275	53.50		328.50	385
9340	24" wide		10.12	.791		234	48.50		282.50	330
9350	18" wide		11.04	.725		223	44.50		267.50	310
9360	12" wide		11.96	.669		175	41		216	254
9370	9" wide		12.88	.621		155	38		193	228
9380	6" wide		13.80	.580		126	35.50		161.50	192
9390	Pulley, single wheel		11.04	.725		288	44.50		332.50	380
9400	Triple wheel		9.20	.870		575	53.50		628.50	710
9440	Nylon cable tie, 14" long		73.60	.109		.57	6.65		7.22	10.60
9450	Ladder, hold down clamp		55.20	.145		8.45	8.90		17.35	22.50
9460	Cable clamp		55.20	.145		9	8.90		17.90	23
9470	Wall bracket, 30" wide tray		17.48	.458		51	28		79	98
9480	24" wide tray		18.40	.435		39.50	26.50		66	83
9490	18" wide tray		19.32	.414		35.50	25.50		61	77
9500	12" wide tray		20.24	.395		33	24.50		57.50	72.50
9510	9" wide tray		21.16	.378		29	23		52	66.50
9520	6" wide tray		22.08	.362		28.50	22		50.50	64

26 05 36.20 Cable Tray Solid Bottom	Crew	Daily Output	Labor-Hours	Unit	Material	2020 Bare Costs Labor	Equipment	Total	Total Incl O&P
0010 **CABLE TRAY SOLID BOTTOM** w/ftngs. & supports, 3" deep, to 15' high									
0200 For higher elevations, see Section 26 05 36.40									
0220 Galvanized steel, tray, 6" wide	2 Elec	110.40	.145	L.F.	11.80	8.90		20.70	26.50
0240 12" wide		92	.174		15.15	10.65		25.80	32.50
0260 18" wide		64.40	.248		18.40	15.25		33.65	43
0280 24" wide		55.20	.290		21.50	17.80		39.30	50
0300 30" wide		46	.348		25.50	21.50		47	60.50
0340 Elbow horizontal, 90°, 12" radius, 6" wide		8.83	1.810	Ea.	89.50	111		200.50	264
0360 12" wide		6.26	2.560		100	157		257	345
0370 18" wide		4.97	3.220		115	198		313	420
0380 24" wide		4.05	3.950		140	242		382	515
0390 30" wide		3.50	4.580		167	281		448	605
0420 24" radius, 6" wide		8.46	1.890		128	116		244	315
0440 12" wide		5.89	2.720		148	167		315	410
0450 18" wide		4.60	3.480		172	214		386	510
0460 24" wide		3.68	4.350		200	267		467	615
0470 30" wide		3.13	5.120		234	315		549	730
0500 36" radius, 6" wide		8.10	1.980		186	121		307	385
0520 12" wide		5.52	2.900		210	178		388	495
0530 18" wide		4.23	3.780		257	232		489	630
0540 24" wide		3.31	4.830		278	296		574	745
0550 30" wide		2.76	5.800		335	355		690	900
0580 Elbow vertical, 90°, 12" radius, 6" wide		8.83	1.810		109	111		220	284
0600 12" wide		6.26	2.560		119	157		276	365
0610 18" wide		4.97	3.220		134	198		332	440
0620 24" wide		4.05	3.950		136	242		378	510
0630 30" wide		3.50	4.580		145	281		426	580
0670 24" radius, 6" wide		8.46	1.890		153	116		269	340
0690 12" wide		5.89	2.720		169	167		336	435
0700 18" wide		4.60	3.480		183	214		397	520
0710 24" wide		3.68	4.350		198	267		465	615
0720 30" wide		3.13	5.120		210	315		525	700
0750 36" radius, 6" wide		8.10	1.980		210	121		331	410
0770 12" wide		6.07	2.640		234	162		396	500
0780 18" wide		4.23	3.780		257	232		489	630
0790 24" wide		3.31	4.830		281	296		577	750
0800 30" wide		2.76	5.800		305	355		660	870
0840 Tee horizontal, 12" radius, 6" wide		4.60	3.480		126	214		340	460
0860 12" wide		3.68	4.350		136	267		403	545
0870 18" wide		3.13	5.120		162	315		477	650
0880 24" wide		2.58	6.210		181	380		561	765
0890 30" wide		2.39	6.690		207	410		617	840
0940 24" radius, 6" wide		4.23	3.780		195	232		427	560
0960 12" wide		3.31	4.830		228	296		524	690
0970 18" wide		2.76	5.800		253	355		608	810
0980 24" wide		2.21	7.250		345	445		790	1,025
0990 30" wide		2.02	7.910		370	485		855	1,125
1020 36" radius, 6" wide		3.86	4.140		305	254		559	720
1040 12" wide		2.94	5.430		345	335		680	870
1050 18" wide		2.39	6.690		370	410		780	1,025
1060 24" wide		2.02	7.910		480	485		965	1,250
1070 30" wide		1.84	8.700		520	535		1,055	1,375

26 05 36.20 Cable Tray Solid Bottom		Crew	Daily Output	Labor-Hours	Unit	Material	2020 Bare Costs Labor	Equipment	Total	Total Incl O&P
1100	Tee vertical, 12" radius, 6" wide	2 Elec	4.60	3.480	Ea.	207	214		421	550
1120	12" wide		3.68	4.350		216	267		483	630
1130	18" wide		3.31	4.830		219	296		515	680
1140	24" wide		3.13	5.120		240	315		555	735
1150	30" wide		2.76	5.800		257	355		612	815
1180	24" radius, 6" wide		4.23	3.780		305	232		537	680
1200	12" wide		3.31	4.830		320	296		616	790
1210	18" wide		2.94	5.430		335	335		670	865
1220	24" wide		2.76	5.800		350	355		705	915
1230	30" wide		2.39	6.690		385	410		795	1,025
1260	36" radius, 6" wide		3.86	4.140		475	254		729	905
1280	12" wide		2.94	5.430		485	335		820	1,025
1290	18" wide		2.58	6.210		520	380		900	1,150
1300	24" wide		2.39	6.690		540	410		950	1,200
1310	30" wide		2.02	7.910		630	485		1,115	1,425
1340	Cross horizontal, 12" radius, 6" wide		3.68	4.350		150	267		417	560
1360	12" wide		3.13	5.120		164	315		479	650
1370	18" wide		2.58	6.210		191	380		571	775
1380	24" wide		2.21	7.250		216	445		661	895
1390	30" wide		1.84	8.700		240	535		775	1,050
1420	24" radius, 6" wide		3.31	4.830		274	296		570	740
1440	12" wide		2.76	5.800		305	355		660	870
1450	18" wide		2.21	7.250		345	445		790	1,025
1460	24" wide		1.84	8.700		440	535		975	1,275
1470	30" wide		1.66	9.660		475	595		1,070	1,400
1500	36" radius, 6" wide		2.94	5.430		450	335		785	990
1520	12" wide		2.39	6.690		495	410		905	1,150
1530	18" wide		1.84	8.700		540	535		1,075	1,375
1540	24" wide		1.66	9.660		635	595		1,230	1,575
1550	30" wide		1.47	10.870		690	665		1,355	1,750
1580	Drop out or end plate, 6" wide		29.44	.543		18.30	33.50		51.80	69.50
1600	12" wide		23.92	.669		23	41		64	86
1610	18" wide		20.24	.791		24.50	48.50		73	99
1620	24" wide		18.40	.870		29	53.50		82.50	111
1630	30" wide		16.56	.966		32	59.50		91.50	124
1660	Reducer, 12" to 6" wide		11.04	1.450		88	89		177	229
1680	18" to 12" wide		9.75	1.640		89.50	101		190.50	249
1700	18" to 6" wide		9.75	1.640		89.50	101		190.50	249
1720	24" to 18" wide		8.46	1.890		93	116		209	275
1740	24" to 12" wide		8.46	1.890		93	116		209	275
1760	30" to 24" wide		7.36	2.170		100	133		233	310
1780	30" to 18" wide		7.36	2.170		100	133		233	310
1800	30" to 12" wide		7.36	2.170		100	133		233	310
1820	36" to 30" wide		6.62	2.420		103	148		251	335
1840	36" to 24" wide		6.62	2.420		103	148		251	335
1860	36" to 18" wide		6.62	2.420		103	148		251	335
1880	36" to 12" wide		6.62	2.420		103	148		251	335
2000	Aluminum tray, 6" wide		138	.116	L.F.	10.95	7.10		18.05	22.50
2020	12" wide		119.60	.134		14.40	8.20		22.60	28
2030	18" wide		92	.174		18.05	10.65		28.70	36
2040	24" wide		82.80	.193		23	11.85		34.85	42.50
2050	30" wide		64.40	.248		27	15.25		42.25	52
2080	Elbow horizontal, 90°, 12" radius, 6" wide		8.83	1.810	Ea.	88.50	111		199.50	262

26 05 36.20 Cable Tray Solid Bottom		Crew	Daily Output	Labor-Hours	Unit	Material	2020 Bare Costs Labor	Equipment	Total	Total Incl O&P
2100	12" wide	2 Elec	6.99	2.290	Ea.	99.50	140		239.50	320
2110	18" wide		6.26	2.560		122	157		279	370
2120	24" wide		5.34	3		139	184		323	425
2130	30" wide		4.60	3.480		170	214		384	505
2160	24" radius, 6" wide		8.46	1.890		126	116		242	310
2180	12" wide		6.62	2.420		148	148		296	385
2190	18" wide		5.89	2.720		170	167		337	435
2200	24" wide		4.97	3.220		209	198		407	525
2210	30" wide		4.23	3.780		240	232		472	610
2240	36" radius, 6" wide		8.10	1.980		186	121		307	385
2260	12" wide		6.26	2.560		223	157		380	480
2270	18" wide		5.52	2.900		265	178		443	555
2280	24" wide		4.60	3.480		284	214		498	630
2290	30" wide		3.86	4.140		320	254		574	735
2320	Elbow vertical, 90°, 12" radius, 6" wide		8.83	1.810		105	111		216	280
2340	12" wide		6.99	2.290		109	140		249	330
2350	18" wide		6.26	2.560		123	157		280	370
2360	24" wide		5.34	3		130	184		314	415
2370	30" wide		4.60	3.480		138	214		352	470
2400	24" radius, 6" wide		8.46	1.890		148	116		264	335
2420	12" wide		6.62	2.420		162	148		310	400
2430	18" wide		5.89	2.720		174	167		341	440
2440	24" wide		4.97	3.220		188	198		386	500
2450	30" wide		4.23	3.780		200	232		432	565
2480	36" radius, 6" wide		8.10	1.980		196	121		317	395
2500	12" wide		6.26	2.560		220	157		377	475
2510	18" wide		5.52	2.900		240	178		418	530
2520	24" wide		4.60	3.480		249	214		463	595
2530	30" wide		3.86	4.140		276	254		530	685
2560	Tee horizontal, 12" radius, 6" wide		4.60	3.480		137	214		351	470
2580	12" wide		4.05	3.950		166	242		408	540
2590	18" wide		3.68	4.350		186	267		453	600
2600	24" wide		3.31	4.830		208	296		504	670
2610	30" wide		2.76	5.800		240	355		595	795
2640	24" radius, 6" wide		4.23	3.780		225	232		457	595
2660	12" wide		3.68	4.350		253	267		520	675
2670	18" wide		3.31	4.830		286	296		582	755
2680	24" wide		2.76	5.800		350	355		705	915
2690	30" wide		2.21	7.250		400	445		845	1,100
2720	36" radius, 6" wide		3.86	4.140		365	254		619	780
2740	12" wide		3.31	4.830		400	296		696	880
2750	18" wide		2.94	5.430		460	335		795	1,000
2760	24" wide		2.39	6.690		540	410		950	1,200
2770	30" wide		1.84	8.700		590	535		1,125	1,450
2800	Tee vertical, 12" radius, 6" wide		4.60	3.480		195	214		409	535
2820	12" wide		4.05	3.950		199	242		441	580
2830	18" wide		3.86	4.140		207	254		461	605
2840	24" wide		3.68	4.350		220	267		487	635
2850	30" wide		3.31	4.830		240	296		536	705
2880	24" radius, 6" wide		4.23	3.780		286	232		518	660
2900	12" wide		3.68	4.350		305	267		572	730
2910	18" wide		3.50	4.580		325	281		606	775
2920	24" wide		3.31	4.830		350	296		646	825

26 05 36.20 Cable Tray Solid Bottom

		Crew	Daily Output	Labor-Hours	Unit	Material	2020 Bare Costs Labor	Equipment	Total	Total Incl O&P
2930	30" wide	2 Elec	2.94	5.430	Ea.	385	335		720	920
2960	36" radius, 6" wide		3.86	4.140		450	254		704	875
2980	12" wide		3.13	5.120		470	315		785	985
2990	18" wide		3.13	5.120		495	315		810	1,025
3000	24" wide		2.94	5.430		510	335		845	1,050
3010	30" wide		2.58	6.210		550	380		930	1,175
3040	Cross horizontal, 12" radius, 6" wide		4.05	3.950		174	242		416	550
3060	12" wide		3.68	4.350		196	267		463	610
3070	18" wide		3.13	5.120		223	315		538	715
3080	24" wide		2.58	6.210		253	380		633	845
3090	30" wide		2.39	6.690		286	410		696	925
3120	24" radius, 6" wide		3.68	4.350		305	267		572	730
3140	12" wide		3.31	4.830		350	296		646	825
3150	18" wide		2.76	5.800		380	355		735	950
3160	24" wide		2.21	7.250		460	445		905	1,175
3170	30" wide		2.02	7.910		510	485		995	1,275
3200	36" radius, 6" wide		3.31	4.830		530	296		826	1,025
3220	12" wide		2.94	5.430		560	335		895	1,100
3230	18" wide		2.39	6.690		620	410		1,030	1,300
3240	24" wide		1.84	8.700		675	535		1,210	1,550
3250	30" wide		1.66	9.660		730	595		1,325	1,675
3280	Dropout, or end plate, 6" wide		29.44	.543		17.45	33.50		50.95	68.50
3300	12" wide		23.92	.669		19.75	41		60.75	82.50
3310	18" wide		20.24	.791		23.50	48.50		72	98
3320	24" wide		18.40	.870		26	53.50		79.50	108
3330	30" wide		16.56	.966		30.50	59.50		90	122
3380	Reducer, 12" to 6" wide		12.88	1.240		85.50	76		161.50	207
3400	18" to 12" wide		11.04	1.450		88.50	89		177.50	229
3420	18" to 6" wide		11.04	1.450		88.50	89		177.50	229
3440	24" to 18" wide		9.75	1.640		93.50	101		194.50	253
3460	24" to 12" wide		9.75	1.640		93.50	101		194.50	253
3480	30" to 24" wide		8.46	1.890		102	116		218	285
3500	30" to 18" wide		8.46	1.890		102	116		218	285
3520	30" to 12" wide		8.46	1.890		102	116		218	285
3540	36" to 30" wide		7.36	2.170		107	133		240	315
3560	36" to 24" wide		7.36	2.170		109	133		242	320
3580	36" to 18" wide		7.36	2.170		109	133		242	320
3600	36" to 12" wide		7.36	2.170		111	133		244	320

26 05 36.30 Cable Tray Trough

		Crew	Daily Output	Labor-Hours	Unit	Material	2020 Bare Costs Labor	Equipment	Total	Total Incl O&P
0010	**CABLE TRAY TROUGH** vented, w/ftngs. & supports, 6" deep, to 10' high									
0020	For higher elevations, see Section 26 05 36.40									
0200	Galvanized steel, tray, 6" wide	2 Elec	82.80	.193	L.F.	15.55	11.85		27.40	35
0240	12" wide		73.60	.217		14.20	13.35		27.55	35.50
0260	18" wide		64.40	.248		21.50	15.25		36.75	46
0280	24" wide		55.20	.290		48.50	17.80		66.30	80
0300	30" wide		46	.348		47.50	21.50		69	84.50
0340	Elbow horizontal, 90°, 12" radius, 6" wide		6.99	2.290	Ea.	124	140		264	345
0360	12" wide		5.15	3.110		147	191		338	445
0370	18" wide		4.05	3.950		170	242		412	545
0380	24" wide		3.31	4.830		210	296		506	670
0390	30" wide		2.94	5.430		237	335		572	755
0420	24" radius, 6" wide		6.62	2.420		183	148		331	420

For customer support on your Electrical Change Order Costs with RSMeans data, call 800.448.8182.

Pre-Installation Change Order Costs

26 05 36.30 Cable Tray Trough		Crew	Daily Output	Labor-Hours	Unit	Material	2020 Bare Costs Labor	Equipment	Total	Total Incl O&P
0440	12" wide	2 Elec	4.78	3.340	Ea.	212	205		417	540
0450	18" wide		3.68	4.350		255	267		522	675
0460	24" wide		2.94	5.430		288	335		623	810
0470	30" wide		2.58	6.210		335	380		715	930
0500	36" radius, 6" wide		6.26	2.560		239	157		396	495
0520	12" wide		4.42	3.620		272	222		494	630
0530	18" wide		3.31	4.830		325	296		621	800
0540	24" wide		2.58	6.210		335	380		715	935
0550	30" wide		2.21	7.250		375	445		820	1,075
0580	Elbow vertical, 90°, 12" radius, 6" wide		6.99	2.290		136	140		276	360
0600	12" wide		5.15	3.110		151	191		342	450
0610	18" wide		4.05	3.950		152	242		394	530
0620	24" wide		3.31	4.830		175	296		471	635
0630	30" wide		2.94	5.430		179	335		514	690
0660	24" radius, 6" wide		6.62	2.420		194	148		342	435
0680	12" wide		4.78	3.340		205	205		410	530
0690	18" wide		3.68	4.350		228	267		495	645
0700	24" wide		2.94	5.430		235	335		570	755
0710	30" wide		2.58	6.210		264	380		644	855
0740	36" radius, 6" wide		6.26	2.560		262	157		419	520
0760	12" wide		4.42	3.620		279	222		501	635
0770	18" wide		3.31	4.830		310	296		606	780
0780	24" wide		2.58	6.210		330	380		710	930
0790	30" wide		2.21	7.250		340	445		785	1,025
0820	Tee horizontal, 12" radius, 6" wide		3.68	4.350		158	267		425	570
0840	12" wide		2.94	5.430		177	335		512	690
0850	18" wide		2.58	6.210		200	380		580	785
0860	24" wide		2.21	7.250		232	445		677	915
0870	30" wide		2.02	7.910		264	485		749	1,025
0900	24" radius, 6" wide		3.31	4.830		262	296		558	730
0920	12" wide		2.58	6.210		286	380		666	880
0930	18" wide		2.21	7.250		320	445		765	1,000
0940	24" wide		1.84	8.700		415	535		950	1,250
0950	30" wide		1.66	9.660		450	595		1,045	1,375
0980	36" radius, 6" wide		2.94	5.430		375	335		710	910
1000	12" wide		2.21	7.250		430	445		875	1,125
1010	18" wide		1.84	8.700		470	535		1,005	1,300
1020	24" wide		1.47	10.870		565	665		1,230	1,625
1030	30" wide		1.29	12.420		600	760		1,360	1,775
1060	Tee vertical, 12" radius, 6" wide		3.68	4.350		249	267		516	670
1080	12" wide		2.94	5.430		253	335		588	775
1090	18" wide		2.76	5.800		260	355		615	815
1100	24" wide		2.58	6.210		274	380		654	865
1110	30" wide		2.39	6.690		290	410		700	930
1140	24" radius, 6" wide		3.31	4.830		335	296		631	810
1160	12" wide		2.58	6.210		350	380		730	950
1170	18" wide		2.39	6.690		370	410		780	1,025
1180	24" wide		2.21	7.250		395	445		840	1,100
1190	30" wide		2.02	7.910		415	485		900	1,175
1220	36" radius, 6" wide		2.94	5.430		545	335		880	1,100
1240	12" wide		2.21	7.250		560	445		1,005	1,275
1250	18" wide		2.02	7.910		580	485		1,065	1,350
1260	24" wide		1.84	8.700		610	535		1,145	1,475

26 05 36.30 Cable Tray Trough	Crew	Daily Output	Labor-Hours	Unit	Material	2020 Bare Costs Labor	Equipment	Total	Total Incl O&P
1270	30" wide	2 Elec	1.66	9.660	Ea.	710	595		1,305	1,650
1300	Cross horizontal, 12" radius, 6" wide		2.94	5.430		196	335		531	710
1320	12" wide		2.58	6.210		198	380		578	785
1330	18" wide		2.21	7.250		222	445		667	905
1340	24" wide		1.84	8.700		233	535		768	1,050
1350	30" wide		1.66	9.660		262	595		857	1,175
1380	24" radius, 6" wide		2.58	6.210		310	380		690	905
1400	12" wide		2.21	7.250		325	445		770	1,025
1410	18" wide		1.84	8.700		355	535		890	1,175
1420	24" wide		1.47	10.870		450	665		1,115	1,500
1430	30" wide		1.29	12.420		500	760		1,260	1,675
1460	36" radius, 6" wide		2.21	7.250		530	445		975	1,250
1480	12" wide		1.84	8.700		545	535		1,080	1,400
1490	18" wide		1.47	10.870		560	665		1,225	1,600
1500	24" wide		1.10	14.490		670	890		1,560	2,050
1510	30" wide		.92	17.390		740	1,075		1,815	2,400
1540	Dropout or end plate, 6" wide		23.92	.669		23.50	41		64.50	87
1560	12" wide		20.24	.791		29	48.50		77.50	104
1580	18" wide		18.40	.870		31.50	53.50		85	115
1600	24" wide		16.56	.966		36.50	59.50		96	129
1620	30" wide		14.72	1.090		39	67		106	143
1660	Reducer, 12" to 6" wide		8.65	1.850		95.50	114		209.50	274
1680	18" to 12" wide		7.73	2.070		99	127		226	298
1700	18" to 6" wide		7.73	2.070		99	127		226	298
1720	24" to 18" wide		6.62	2.420		106	148		254	335
1740	24" to 12" wide		6.62	2.420		106	148		254	335
1760	30" to 24" wide		5.89	2.720		110	167		277	370
1780	30" to 18" wide		5.89	2.720		110	167		277	370
1800	30" to 12" wide		5.89	2.720		110	167		277	370
1820	36" to 30" wide		5.34	3		117	184		301	405
1840	36" to 24" wide		5.34	3		117	184		301	405
1860	36" to 18" wide		5.34	3		119	184		303	405
1880	36" to 12" wide		5.34	3		119	184		303	405
2000	Aluminum, tray, vented, 6" wide		110.40	.145	L.F.	16.10	8.90		25	31
2010	9" wide		101.20	.158		18.25	9.70		27.95	34.50
2020	12" wide		92	.174		23.50	10.65		34.15	42
2030	18" wide		82.80	.193		24.50	11.85		36.35	44.50
2040	24" wide		73.60	.217		29	13.35		42.35	51.50
2050	30" wide		64.40	.248		38.50	15.25		53.75	64.50
2080	Elbow horiz., 90°, 12" radius, 6" wide		6.99	2.290	Ea.	102	140		242	320
2090	9" wide		6.44	2.480		110	152		262	350
2100	12" wide		5.70	2.810		117	172		289	385
2110	18" wide		5.15	3.110		137	191		328	435
2120	24" wide		4.23	3.780		158	232		390	520
2130	30" wide		3.68	4.350		195	267		462	610
2160	24" radius, 6" wide		6.62	2.420		147	148		295	385
2180	12" wide		5.34	3		173	184		357	465
2190	18" wide		4.78	3.340		198	205		403	525
2200	24" wide		3.86	4.140		226	254		480	630
2210	30" wide		3.31	4.830		259	296		555	725
2240	36" radius, 6" wide		6.26	2.560		212	157		369	470
2260	12" wide		4.97	3.220		241	198		439	560
2270	18" wide		4.42	3.620		275	222		497	630

26 05 36 – Cable Trays for Electrical Systems

26 05 36.30 Cable Tray Trough	Crew	Daily Output	Labor-Hours	Unit	Material	2020 Bare Costs Labor	Equipment	Total	Total Incl O&P
2280	24" wide	2 Elec	3.50	4.580	Ea.	300	281		581	750
2290	30" wide		3.13	5.120		355	315		670	860
2320	Elbow vertical, 90°, 12" radius, 6" wide		6.99	2.290		123	140		263	345
2330	9" wide		6.44	2.480		132	152		284	370
2340	12" wide		5.70	2.810		133	172		305	405
2350	18" wide		5.15	3.110		137	191		328	435
2360	24" wide		4.23	3.780		151	232		383	510
2370	30" wide		3.68	4.350		158	267		425	570
2400	24" radius, 6" wide		6.62	2.420		167	148		315	405
2420	12" wide		5.34	3		181	184		365	475
2430	18" wide		4.78	3.340		197	205		402	520
2440	24" wide		3.86	4.140		199	254		453	600
2450	30" wide		3.31	4.830		211	296		507	670
2480	36" radius, 6" wide		6.26	2.560		211	157		368	465
2500	12" wide		4.97	3.220		225	198		423	540
2510	18" wide		4.42	3.620		253	222		475	610
2520	24" wide		3.50	4.580		273	281		554	720
2530	30" wide		3.13	5.120		300	315		615	800
2560	Tee horizontal, 12" radius, 6" wide		3.68	4.350		154	267		421	565
2570	9" wide		3.50	4.580		158	281		439	595
2580	12" wide		3.31	4.830		173	296		469	630
2590	18" wide		2.94	5.430		199	335		534	715
2600	24" wide		2.58	6.210		229	380		609	815
2610	30" wide		2.21	7.250		246	445		691	930
2640	24" radius, 6" wide		3.31	4.830		235	296		531	700
2660	12" wide		2.94	5.430		267	335		602	790
2670	18" wide		2.58	6.210		294	380		674	890
2680	24" wide		2.21	7.250		370	445		815	1,075
2690	30" wide		1.84	8.700		395	535		930	1,225
2720	36" radius, 6" wide		2.94	5.430		380	335		715	910
2740	12" wide		2.58	6.210		435	380		815	1,050
2750	18" wide		2.21	7.250		490	445		935	1,200
2760	24" wide		1.84	8.700		575	535		1,110	1,425
2770	30" wide		1.47	10.870		610	665		1,275	1,675
2800	Tee vertical, 12" radius, 6" wide		3.68	4.350		217	267		484	635
2810	9" wide		3.50	4.580		218	281		499	660
2820	12" wide		3.31	4.830		234	296		530	700
2830	18" wide		3.13	5.120		238	315		553	730
2840	24" wide		2.94	5.430		241	335		576	760
2850	30" wide		2.76	5.800		253	355		608	810
2880	24" radius, 6" wide		3.31	4.830		305	296		601	775
2900	12" wide		2.94	5.430		330	335		665	860
2910	18" wide		2.76	5.800		355	355		710	920
2920	24" wide		2.58	6.210		375	380		755	980
2930	30" wide		2.39	6.690		405	410		815	1,050
2960	36" radius, 6" wide		2.94	5.430		465	335		800	1,000
2980	12" wide		2.58	6.210		490	380		870	1,100
2990	18" wide		2.39	6.690		510	410		920	1,175
3000	24" wide		2.21	7.250		545	445		990	1,250
3010	30" wide		2.02	7.910		590	485		1,075	1,375
3040	Cross horizontal, 12" radius, 6" wide		3.31	4.830		192	296		488	650
3050	9" wide		3.13	5.120		207	315		522	695
3060	12" wide		2.94	5.430		212	335		547	730

26 05 36.30 Cable Tray Trough		Crew	Daily Output	Labor-Hours	Unit	Material	2020 Bare Costs Labor	Equipment	Total	Total Incl O&P
3070	18" wide	2 Elec	2.58	6.210	Ea.	222	380		602	810
3080	24" wide		2.21	7.250		248	445		693	930
3090	30" wide		2.02	7.910		315	485		800	1,075
3120	24" radius, 6" wide		2.94	5.430		345	335		680	875
3140	12" wide		2.58	6.210		380	380		760	980
3150	18" wide		2.21	7.250		405	445		850	1,100
3160	24" wide		1.84	8.700		470	535		1,005	1,325
3170	30" wide		1.66	9.660		510	595		1,105	1,450
3200	36" radius, 6" wide		2.58	6.210		600	380		980	1,225
3220	12" wide		2.21	7.250		630	445		1,075	1,350
3230	18" wide		1.84	8.700		685	535		1,220	1,550
3240	24" wide		1.47	10.870		795	665		1,460	1,875
3250	30" wide		1.29	12.420		860	760		1,620	2,075
3280	Dropout, or end plate, 6" wide		23.92	.669		19.65	41		60.65	82.50
3300	12" wide		20.24	.791		23.50	48.50		72	97.50
3310	18" wide		18.40	.870		28	53.50		81.50	111
3320	24" wide		16.56	.966		34	59.50		93.50	126
3330	30" wide		14.72	1.090		36	67		103	139
3370	Reducer, 9" to 6" wide		11.04	1.450		96.50	89		185.50	238
3380	12" to 6" wide		10.49	1.530		101	94		195	251
3390	12" to 9" wide		10.49	1.530		101	94		195	251
3400	18" to 12" wide		8.83	1.810		108	111		219	284
3420	18" to 6" wide		8.83	1.810		108	111		219	284
3430	18" to 9" wide		8.83	1.810		108	111		219	284
3440	24" to 18" wide		7.73	2.070		116	127		243	315
3460	24" to 12" wide		7.73	2.070		116	127		243	315
3470	24" to 9" wide		7.73	2.070		117	127		244	320
3475	24" to 6" wide		7.73	2.070		117	127		244	320
3480	30" to 24" wide		6.62	2.420		119	148		267	350
3500	30" to 18" wide		6.62	2.420		119	148		267	350
3520	30" to 12" wide		6.62	2.420		121	148		269	355
3540	36" to 30" wide		5.89	2.720		123	167		290	385
3560	36" to 24" wide		5.89	2.720		123	167		290	385
3580	36" to 18" wide		5.89	2.720		123	167		290	385
3600	36" to 12" wide		5.89	2.720		123	167		290	385
3610	Elbow horizontal, 60°, 12" radius, 6" wide		7.18	2.230		84	137		221	296
3620	9" wide		6.62	2.420		91.50	148		239.50	320
3630	12" wide		5.89	2.720		101	167		268	360
3640	18" wide		5.34	3		114	184		298	400
3650	24" wide		4.42	3.620		133	222		355	475
3680	Elbow horizontal, 45°, 12" radius, 6" wide		7.36	2.170		75.50	133		208.50	281
3690	9" wide		6.81	2.350		76.50	144		220.50	300
3700	12" wide		6.07	2.640		81	162		243	330
3710	18" wide		5.52	2.900		91	178		269	365
3720	24" wide		4.60	3.480		106	214		320	435
3750	Elbow horizontal, 30°, 12" radius, 6" wide		7.54	2.120		63	130		193	264
3760	9" wide		6.99	2.290		66	140		206	282
3770	12" wide		6.26	2.560		72	157		229	315
3780	18" wide		5.70	2.810		76.50	172		248.50	340
3790	24" wide		4.78	3.340		84	205		289	395
3820	Elbow vertical, 60° in/outside, 12" radius, 6" wide		7.18	2.230		104	137		241	320
3830	9" wide		6.62	2.420		106	148		254	335
3840	12" wide		5.89	2.720		107	167		274	365

Pre-Installation Change Order Costs

26 05 36.30 Cable Tray Trough		Crew	Daily Output	Labor-Hours	Unit	Material	2020 Bare Costs Labor	Equipment	Total	Total Incl O&P
3850	18" wide	2 Elec	5.34	3	Ea.	111	184		295	395
3860	24" wide		4.42	3.620		116	222		338	455
3890	Elbow vertical, 45° in/outside, 12" radius, 6" wide		7.36	2.170		88.50	133		221.50	296
3900	9" wide		6.81	2.350		85.50	144		229.50	310
3910	12" wide		6.07	2.640		91.50	162		253.50	340
3920	18" wide		5.52	2.900		94.50	178		272.50	370
3930	24" wide		4.60	3.480		104	214		318	435
3960	Elbow vertical, 30° in/outside, 12" radius, 6" wide		7.54	2.120		72	130		202	274
3970	9" wide		6.99	2.290		76.50	140		216.50	294
3980	12" wide		6.26	2.560		78	157		235	320
3990	18" wide		5.70	2.810		81	172		253	345
4000	24" wide		4.78	3.340		82.50	205		287.50	395
4250	Reducer, left or right hand, 24" to 18" wide		7.73	2.070		110	127		237	310
4260	24" to 12" wide		7.73	2.070		110	127		237	310
4270	24" to 9" wide		7.73	2.070		110	127		237	310
4280	24" to 6" wide		7.73	2.070		111	127		238	310
4290	18" to 12" wide		8.83	1.810		103	111		214	278
4300	18" to 9" wide		8.83	1.810		103	111		214	278
4310	18" to 6" wide		8.83	1.810		103	111		214	278
4320	12" to 9" wide		10.49	1.530		98.50	94		192.50	248
4330	12" to 6" wide		10.49	1.530		98.50	94		192.50	248
4340	9" to 6" wide		11.04	1.450		95.50	89		184.50	237
4350	Splice plate	1 Elec	44.16	.181		9.10	11.10		20.20	26.50
4360	Splice plate, expansion joint		44.16	.181		9.35	11.10		20.45	27
4370	Splice plate, hinged, horizontal		44.16	.181		7.50	11.10		18.60	25
4380	Vertical		44.16	.181		10.70	11.10		21.80	28.50
4390	Trough, hanger, vertical		25.76	.311		30	19.05		49.05	61.50
4400	Box connector, 24" wide		18.40	.435		36	26.50		62.50	79
4410	18" wide		19.32	.414		34	25.50		59.50	75
4420	12" wide		20.24	.395		32.50	24.50		57	71.50
4430	9" wide		21.16	.378		31	23		54	68.50
4440	6" wide		22.08	.362		29.50	22		51.50	65.50
4450	Floor flange		22.08	.362		31	22		53	67
4460	Hold down clamp		55.20	.145		3.53	8.90		12.43	17.15
4520	Wall bracket, 24" wide tray		18.40	.435		35	26.50		61.50	78
4530	18" wide tray		19.32	.414		34	25.50		59.50	75.50
4540	12" wide tray		20.24	.395		18.25	24.50		42.75	56
4550	9" wide tray		21.16	.378		16.30	23		39.30	52.50
4560	6" wide tray		22.08	.362		14.85	22		36.85	49.50
5000	Cable channel aluminum, vented, 1-1/4" deep, 4" wide, straight		73.60	.109	L.F.	8	6.65		14.65	18.75
5010	Elbow horizontal, 36" radius, 90°		4.60	1.740	Ea.	145	107		252	320
5020	60°		5.06	1.580		111	97		208	266
5030	45°		5.52	1.450		90.50	89		179.50	232
5040	30°		5.98	1.340		77.50	82		159.50	207
5050	Adjustable		5.52	1.450		73.50	89		162.50	213
5060	Elbow vertical, 36" radius, 90°		4.60	1.740		156	107		263	330
5070	60°		5.06	1.580		120	97		217	276
5080	45°		5.52	1.450		98.50	89		187.50	240
5090	30°		5.98	1.340		87.50	82		169.50	218
5100	Adjustable		5.52	1.450		73.50	89		162.50	213
5110	Splice plate, hinged, horizontal		44.16	.181		10.20	11.10		21.30	28
5120	Splice plate, hinged, vertical		44.16	.181		14.45	11.10		25.55	32.50
5130	Hanger, vertical		25.76	.311		15.95	19.05		35	46

26 05 36 – Cable Trays for Electrical Systems

26 05 36.30 Cable Tray Trough	Crew	Daily Output	Labor-Hours	Unit	Material	2020 Bare Costs Labor	Equipment	Total	Total Incl O&P
5140	Single	1 Elec	25.76	.311	Ea.	24.50	19.05		43.55	55
5150	Double		18.40	.435		25	26.50		51.50	67
5160	Channel to box connector		22.08	.362		33	22		55	69.50
5170	Hold down clip		73.60	.109		4.60	6.65		11.25	15
5180	Wall bracket, single		25.76	.311		11.80	19.05		30.85	41.50
5190	Double		18.40	.435		15.15	26.50		41.65	56
5200	Cable roller		14.72	.543		186	33.50		219.50	254
5210	Splice plate		44.16	.181		5.70	11.10		16.80	23

26 05 36.36 Cable Trays for Utility Substations

		Crew	Daily Output	Labor-Hours	Unit	Material	2020 Bare Costs Labor	Equipment	Total	Total Incl O&P
0010	**CABLE TRAYS FOR UTILITY SUBSTATIONS**									
7700	Cable tray	R-11	36.80	1.520	L.F.	19.15	88.50	17.30	124.95	172
7790	See Section 26 05 36									

26 05 36.40 Cable Tray, Covers and Dividers

		Crew	Daily Output	Labor-Hours	Unit	Material	2020 Bare Costs Labor	Equipment	Total	Total Incl O&P
0010	**CABLE TRAY, COVERS AND DIVIDERS** To 10' high									
0011	For higher elevations, see lines 9900 – 9960									
0100	Covers, ventilated galv. steel, straight, 6" wide tray size	2 Elec	478.40	.033	L.F.	8.20	2.05		10.25	12.05
0200	9" wide tray size		423.20	.038		8.70	2.32		11.02	13.05
0300	12" wide tray size		368	.043		14.35	2.67		17.02	19.70
0400	18" wide tray size		276	.058		34	3.56		37.56	43
0500	24" wide tray size		202.40	.079		58.50	4.85		63.35	71.50
0600	30" wide tray size		165.60	.097		26.50	5.95		32.45	38
1000	Elbow horizontal, 90°, 12" radius, 6" wide tray size		138	.116	Ea.	43.50	7.10		50.60	58.50
1020	9" wide tray size		117.76	.136		48	8.35		56.35	65.50
1040	12" wide tray size		99.36	.161		51	9.90		60.90	70.50
1060	18" wide tray size		77.28	.207		70.50	12.70		83.20	96.50
1080	24" wide tray size		60.72	.264		85	16.15		101.15	118
1100	30" wide tray size		55.20	.290		108	17.80		125.80	146
1160	24" radius, 6" wide tray size		125.12	.128		72.50	7.85		80.35	91
1180	9" wide tray size		106.72	.150		74	9.20		83.20	95
1200	12" wide tray size		88.32	.181		82.50	11.10		93.60	107
1220	18" wide tray size		69.92	.229		102	14.05		116.05	133
1240	24" wide tray size		55.20	.290		124	17.80		141.80	163
1260	30" wide tray size		47.84	.334		163	20.50		183.50	210
1320	36" radius, 6" wide tray size		110.40	.145		108	8.90		116.90	132
1340	9" wide tray size		95.68	.167		119	10.25		129.25	146
1360	12" wide tray size		77.28	.207		127	12.70		139.70	159
1380	18" wide tray size		66.24	.242		162	14.80		176.80	200
1400	24" wide tray size		47.84	.334		193	20.50		213.50	243
1420	30" wide tray size		42.32	.378		229	23		252	287
1480	Elbow horizontal, 45°, 12" radius, 6" wide tray size		138	.116		32.50	7.10		39.60	46.50
1500	9" wide tray size		117.76	.136		39	8.35		47.35	55.50
1520	12" wide tray size		99.36	.161		42.50	9.90		52.40	61.50
1540	18" wide tray size		80.96	.198		52	12.10		64.10	75
1560	24" wide tray size		69.92	.229		60.50	14.05		74.55	87.50
1580	30" wide tray size		60.72	.264		71.50	16.15		87.65	103
1640	24" radius, 6" wide tray size		125.12	.128		45.50	7.85		53.35	61.50
1660	9" wide tray size		106.72	.150		51	9.20		60.20	69.50
1680	12" wide tray size		88.32	.181		57.50	11.10		68.60	79.50
1700	18" wide tray size		73.60	.217		66.50	13.35		79.85	93
1720	24" wide tray size		64.40	.248		76	15.25		91.25	106
1740	30" wide tray size		55.20	.290		96.50	17.80		114.30	133
1800	36" radius, 6" wide tray size		110.40	.145		67	8.90		75.90	87

26 05 36.40 Cable Tray, Covers and Dividers		Crew	Daily Output	Labor-Hours	Unit	Material	2020 Bare Costs Labor	Equipment	Total	Total Incl O&P
1820	9" wide tray size	2 Elec	95.68	.167	Ea.	75.50	10.25		85.75	98.50
1840	12" wide tray size		77.28	.207		83	12.70		95.70	110
1860	18" wide tray size		69.92	.229		95.50	14.05		109.55	126
1880	24" wide tray size		57.04	.281		116	17.20		133.20	154
1900	30" wide tray size		47.84	.334		126	20.50		146.50	170
1960	Elbow vertical, 90°, 12" radius, 6" wide tray size		138	.116		38.50	7.10		45.60	53
1980	9" wide tray size		117.76	.136		39	8.35		47.35	55.50
2000	12" wide tray size		99.36	.161		42	9.90		51.90	61
2020	18" wide tray size		80.96	.198		48	12.10		60.10	71
2040	24" wide tray size		62.56	.256		49	15.70		64.70	77.50
2060	30" wide tray size		55.20	.290		54	17.80		71.80	85.50
2120	24" radius, 6" wide tray size		125.12	.128		47	7.85		54.85	63.50
2140	9" wide tray size		106.72	.150		52	9.20		61.20	70.50
2160	12" wide tray size		88.32	.181		54	11.10		65.10	75.50
2180	18" wide tray size		73.60	.217		71.50	13.35		84.85	98.50
2200	24" wide tray size		57.04	.281		80	17.20		97.20	114
2220	30" wide tray size		47.84	.334		88	20.50		108.50	128
2280	36" radius, 6" wide tray size		110.40	.145		53	8.90		61.90	72
2300	9" wide tray size		95.68	.167		68	10.25		78.25	90.50
2320	12" wide tray size		77.28	.207		75.50	12.70		88.20	102
2340	18" wide tray size		69.92	.229		95.50	14.05		109.55	126
2350	24" wide tray size		49.68	.322		106	19.75		125.75	146
2360	30" wide tray size		42.32	.378		127	23		150	174
2400	Tee horizontal, 12" radius, 6" wide tray size		84.64	.189		69.50	11.60		81.10	94
2410	9" wide tray size		73.60	.217		71.50	13.35		84.85	98.50
2420	12" wide tray size		62.56	.256		81	15.70		96.70	113
2430	18" wide tray size		55.20	.290		98	17.80		115.80	135
2440	24" wide tray size		47.84	.334		123	20.50		143.50	167
2460	30" wide tray size		33.12	.483		145	29.50		174.50	203
2500	24" radius, 6" wide tray size		80.96	.198		112	12.10		124.10	141
2510	9" wide tray size		69.92	.229		119	14.05		133.05	151
2520	12" wide tray size		58.88	.272		122	16.65		138.65	159
2530	18" wide tray size		51.52	.311		157	19.05		176.05	201
2540	24" wide tray size		44.16	.362		240	22		262	297
2560	30" wide tray size		29.44	.543		274	33.50		307.50	350
2600	36" radius, 6" wide tray size		77.28	.207		191	12.70		203.70	229
2610	9" wide tray size		66.24	.242		194	14.80		208.80	236
2620	12" wide tray size		55.20	.290		216	17.80		233.80	265
2630	18" wide tray size		47.84	.334		253	20.50		273.50	310
2640	24" wide tray size		40.48	.395		335	24.50		359.50	400
2660	30" wide tray size		25.76	.621		360	38		398	455
2700	Cross horizontal, 12" radius, 6" wide tray size		62.56	.256		104	15.70		119.70	138
2710	9" wide tray size		58.88	.272		110	16.65		126.65	146
2720	12" wide tray size		55.20	.290		122	17.80		139.80	162
2730	18" wide tray size		47.84	.334		145	20.50		165.50	190
2740	24" wide tray size		33.12	.483		171	29.50		200.50	232
2760	30" wide tray size		27.60	.580		194	35.50		229.50	267
2800	24" radius, 6" wide tray size		58.88	.272		187	16.65		203.65	231
2810	9" wide tray size		55.20	.290		203	17.80		220.80	250
2820	12" wide tray size		51.52	.311		219	19.05		238.05	270
2830	18" wide tray size		44.16	.362		262	22		284	320
2840	24" wide tray size		29.44	.543		320	33.50		353.50	400
2860	30" wide tray size		23.92	.669		350	41		391	445

26 05 36.40 Cable Tray, Covers and Dividers	Crew	Daily Output	Labor-Hours	Unit	Material	2020 Bare Costs Labor	Equipment	Total	Total Incl O&P
2900	36" radius, 6" wide tray size	2 Elec	55.20	.290	Ea.	320	17.80		337.80	375
2910	9" wide tray size		51.52	.311		320	19.05		339.05	380
2920	12" wide tray size		47.84	.334		345	20.50		365.50	410
2930	18" wide tray size		40.48	.395		390	24.50		414.50	465
2940	24" wide tray size		25.76	.621		490	38		528	595
2960	30" wide tray size		20.24	.791		525	48.50		573.50	650
3000	Reducer, 9" to 6" wide tray size		117.76	.136		44.50	8.35		52.85	61.50
3010	12" to 6" wide tray size		99.36	.161		46.50	9.90		56.40	65.50
3020	12" to 9" wide tray size		99.36	.161		46.50	9.90		56.40	65.50
3030	18" to 12" wide tray size		80.96	.198		49.50	12.10		61.60	72.50
3050	18" to 6" wide tray size		80.96	.198		49.50	12.10		61.60	72.50
3060	24" to 18" wide tray size		73.60	.217		69.50	13.35		82.85	96
3070	24" to 12" wide tray size		73.60	.217		63.50	13.35		76.85	90
3090	30" to 24" wide tray size		64.40	.248		73.50	15.25		88.75	103
3100	30" to 18" wide tray size		64.40	.248		73.50	15.25		88.75	103
3110	30" to 12" wide tray size		64.40	.248		63.50	15.25		78.75	92.50
3140	36" to 30" wide tray size		58.88	.272		80	16.65		96.65	113
3150	36" to 24" wide tray size		58.88	.272		80	16.65		96.65	113
3160	36" to 18" wide tray size		58.88	.272		80	16.65		96.65	113
3170	36" to 12" wide tray size		58.88	.272		80	16.65		96.65	113
3250	Covers, aluminum, straight, 6" wide tray size		478.40	.033	L.F.	4.79	2.05		6.84	8.30
3270	9" wide tray size		423.20	.038		5.80	2.32		8.12	9.80
3290	12" wide tray size		368	.043		6.90	2.67		9.57	11.55
3310	18" wide tray size		294.40	.054		9.15	3.33		12.48	15
3330	24" wide tray size		239.20	.067		11.40	4.10		15.50	18.65
3350	30" wide tray size		184	.087		12.60	5.35		17.95	22
3400	Elbow horizontal, 90°, 12" radius, 6" wide tray size		138	.116	Ea.	36	7.10		43.10	50
3410	9" wide tray size		117.76	.136		37.50	8.35		45.85	53.50
3420	12" wide tray size		99.36	.161		40.50	9.90		50.40	59
3430	18" wide tray size		80.96	.198		52.50	12.10		64.60	76
3440	24" wide tray size		64.40	.248		66.50	15.25		81.75	95.50
3460	30" wide tray size		58.88	.272		79.50	16.65		96.15	113
3500	24" radius, 6" wide tray size		125.12	.128		46.50	7.85		54.35	62.50
3510	9" wide tray size		106.72	.150		57.50	9.20		66.70	76.50
3520	12" wide tray size		88.32	.181		63	11.10		74.10	86
3530	18" wide tray size		73.60	.217		75	13.35		88.35	102
3540	24" wide tray size		58.88	.272		92.50	16.65		109.15	127
3560	30" wide tray size		51.52	.311		113	19.05		132.05	154
3600	36" radius, 6" wide tray size		110.40	.145		81	8.90		89.90	103
3610	9" wide tray size		95.68	.167		89	10.25		99.25	113
3620	12" wide tray size		77.28	.207		101	12.70		113.70	130
3630	18" wide tray size		69.92	.229		120	14.05		134.05	153
3640	24" wide tray size		51.52	.311		146	19.05		165.05	190
3660	30" wide tray size		46	.348		168	21.50		189.50	216
3700	Elbow horizontal, 45°, 12" radius, 6" wide tray size		138	.116		26.50	7.10		33.60	39.50
3710	9" wide tray size		117.76	.136		27	8.35		35.35	42
3720	12" wide tray size		99.36	.161		30.50	9.90		40.40	48
3730	18" wide tray size		80.96	.198		35	12.10		47.10	56
3740	24" wide tray size		73.60	.217		38.50	13.35		51.85	62.50
3760	30" wide tray size		64.40	.248		48.50	15.25		63.75	76
3800	24" radius, 6" wide tray size		125.12	.128		29	7.85		36.85	43.50
3810	9" wide tray size		106.72	.150		37	9.20		46.20	54
3820	12" wide tray size		88.32	.181		38.50	11.10		49.60	58.50

Pre-Installation Change Order Costs

26 05 36 – Cable Trays for Electrical Systems

26 05 36.40 Cable Tray, Covers and Dividers	Crew	Daily Output	Labor-Hours	Unit	Material	2020 Bare Costs Labor	Equipment	Total	Total Incl O&P
3830	18" wide tray size	2 Elec	73.60	.217	Ea.	46.50	13.35		59.85	71
3840	24" wide tray size		66.24	.242		57.50	14.80		72.30	85
3860	30" wide tray size		58.88	.272		66	16.65		82.65	98
3900	36" radius, 6" wide tray size		110.40	.145		51	8.90		59.90	70
3910	9" wide tray size		95.68	.167		55.50	10.25		65.75	76.50
3920	12" wide tray size		77.28	.207		58.50	12.70		71.20	83.50
3930	18" wide tray size		69.92	.229		70.50	14.05		84.55	98.50
3940	24" wide tray size		58.88	.272		85.50	16.65		102.15	120
3960	30" wide tray size		51.52	.311		97.50	19.05		116.55	136
3970	36" wide tray size		46	.348		111	21.50		132.50	154
4000	Elbow vertical, 90°, 12" radius, 6" wide tray size		138	.116		30.50	7.10		37.60	44
4010	9" wide tray size		117.76	.136		30.50	8.35		38.85	46
4020	12" wide tray size		99.36	.161		33.50	9.90		43.40	51
4030	18" wide tray size		80.96	.198		38.50	12.10		50.60	60.50
4040	24" wide tray size		64.40	.248		40	15.25		55.25	66.50
4060	30" wide tray size		58.88	.272		41	16.65		57.65	70
4070	36" wide tray size		49.68	.322		50	19.75		69.75	84.50
4100	24" radius, 6" wide tray size		125.12	.128		35	7.85		42.85	49.50
4110	9" wide tray size		106.72	.150		38	9.20		47.20	55
4120	12" wide tray size		88.32	.181		41	11.10		52.10	61.50
4130	18" wide tray size		73.60	.217		49.50	13.35		62.85	74.50
4140	24" wide tray size		58.88	.272		54	16.65		70.65	84.50
4160	30" wide tray size		51.52	.311		66	19.05		85.05	102
4170	36" wide tray size		44.16	.362		72	22		94	113
4200	36" radius, 6" wide tray size		110.40	.145		39	8.90		47.90	56.50
4210	9" wide tray size		95.68	.167		48	10.25		58.25	68.50
4220	12" wide tray size		77.28	.207		55.50	12.70		68.20	80
4230	18" wide tray size		69.92	.229		67.50	14.05		81.55	95.50
4240	24" wide tray size		51.52	.311		79.50	19.05		98.55	116
4260	30" wide tray size		46	.348		101	21.50		122.50	143
4270	36" wide tray size		40.48	.395		108	24.50		132.50	155
4300	Tee horizontal, 12" radius, 6" wide tray size		99.36	.161		49.50	9.90		59.40	69
4310	9" wide tray size		80.96	.198		52.50	12.10		64.60	76
4320	12" wide tray size		73.60	.217		58.50	13.35		71.85	84.50
4330	18" wide tray size		62.56	.256		69.50	15.70		85.20	100
4340	24" wide tray size		51.52	.311		88	19.05		107.05	126
4360	30" wide tray size		40.48	.395		103	24.50		127.50	150
4370	36" wide tray size		33.12	.483		125	29.50		154.50	181
4400	24" radius, 6" wide tray size		88.32	.181		80.50	11.10		91.60	105
4410	9" wide tray size		73.60	.217		85.50	13.35		98.85	114
4420	12" wide tray size		66.24	.242		95.50	14.80		110.30	127
4430	18" wide tray size		55.20	.290		112	17.80		129.80	150
4440	24" wide tray size		44.16	.362		178	22		200	229
4460	30" wide tray size		36.80	.435		197	26.50		223.50	256
4470	36" wide tray size		29.44	.543		219	33.50		252.50	291
4500	36" radius, 6" wide tray size		80.96	.198		143	12.10		155.10	175
4510	9" wide tray size		66.24	.242		146	14.80		160.80	183
4520	12" wide tray size		58.88	.272		160	16.65		176.65	201
4530	18" wide tray size		51.52	.311		182	19.05		201.05	229
4540	24" wide tray size		40.48	.395		231	24.50		255.50	290
4560	30" wide tray size		33.12	.483		260	29.50		289.50	330
4570	36" wide tray size		25.76	.621		293	38		331	380
4600	Cross horizontal, 12" radius, 6" wide tray size		73.60	.217		75.50	13.35		88.85	103

26 05 36.40 Cable Tray, Covers and Dividers		Crew	Daily Output	Labor-Hours	Unit	Material	2020 Bare Costs Labor	Equipment	Total	Total Incl O&P
4610	9" wide tray size	2 Elec	66.24	.242	Ea.	80	14.80		94.80	110
4620	12" wide tray size		58.88	.272		89	16.65		105.65	123
4630	18" wide tray size		51.52	.311		108	19.05		127.05	148
4640	24" wide tray size		44.16	.362		125	22		147	170
4660	30" wide tray size		36.80	.435		148	26.50		174.50	203
4670	36" wide tray size		29.44	.543		171	33.50		204.50	238
4700	24" radius, 6" wide tray size		66.24	.242		141	14.80		155.80	178
4710	9" wide tray size		58.88	.272		151	16.65		167.65	192
4720	12" wide tray size		51.52	.311		163	19.05		182.05	208
4730	18" wide tray size		44.16	.362		190	22		212	242
4740	24" wide tray size		36.80	.435		226	26.50		252.50	289
4760	30" wide tray size		29.44	.543		265	33.50		298.50	340
4770	36" wide tray size		22.08	.725		287	44.50		331.50	380
4800	36" radius, 6" wide tray size		58.88	.272		231	16.65		247.65	279
4810	9" wide tray size		51.52	.311		241	19.05		260.05	294
4820	12" wide tray size		46	.348		267	21.50		288.50	325
4830	18" wide tray size		40.48	.395		293	24.50		317.50	360
4840	24" wide tray size		33.12	.483		360	29.50		389.50	445
4860	30" wide tray size		25.76	.621		415	38		453	515
4870	36" wide tray size		20.24	.791		455	48.50		503.50	570
4900	Reducer, 9" to 6" wide tray size		117.76	.136		38	8.35		46.35	54.50
4910	12" to 6" wide tray size		99.36	.161		39.50	9.90		49.40	58
4920	12" to 9" wide tray size		99.36	.161		39.50	9.90		49.40	58
4930	18" to 12" wide tray size		80.96	.198		43.50	12.10		55.60	66
4950	18" to 6" wide tray size		80.96	.198		43.50	12.10		55.60	66
4960	24" to 18" wide tray size		73.60	.217		56	13.35		69.35	81.50
4970	24" to 12" wide tray size		73.60	.217		48	13.35		61.35	73
4990	30" to 24" wide tray size		64.40	.248		58.50	15.25		73.75	86.50
5000	30" to 18" wide tray size		64.40	.248		58.50	15.25		73.75	86.50
5010	30" to 12" wide tray size		64.40	.248		58.50	15.25		73.75	86.50
5040	36" to 30" wide tray size		58.88	.272		64.50	16.65		81.15	96
5050	36" to 24" wide tray size		58.88	.272		64.50	16.65		81.15	96
5060	36" to 18" wide tray size		58.88	.272		64.50	16.65		81.15	96
5070	36" to 12" wide tray size		58.88	.272		64.50	16.65		81.15	96
5710	Tray cover hold down clamp	1 Elec	55.20	.145		10.35	8.90		19.25	24.50
8000	Divider strip, straight, galvanized, 3" deep		184	.043	L.F.	5.55	2.67		8.22	10.05
8020	4" deep		165.60	.048		6.75	2.96		9.71	11.85
8040	6" deep		147.20	.054		8.85	3.33		12.18	14.70
8060	Aluminum, straight, 3" deep		193.20	.041		5.55	2.54		8.09	9.90
8080	4" deep		174.80	.046		6.85	2.81		9.66	11.75
8100	6" deep		156.40	.051		8.75	3.14		11.89	14.30
8110	Divider strip, vertical fitting, 3" deep									
8120	12" radius, galvanized, 30°	1 Elec	25.76	.311	Ea.	26.50	19.05		45.55	58
8140	45°		24.84	.322		33.50	19.75		53.25	66.50
8160	60°		23.92	.334		35.50	20.50		56	69.50
8180	90°		23	.348		44.50	21.50		66	81
8200	Aluminum, 30°		26.68	.300		17.80	18.40		36.20	47
8220	45°		25.76	.311		21	19.05		40.05	51.50
8240	60°		24.84	.322		24.50	19.75		44.25	56.50
8260	90°		23.92	.334		31	20.50		51.50	64.50
8280	24" radius, galvanized, 30°		23	.348		40	21.50		61.50	76
8300	45°		22.08	.362		42.50	22		64.50	80
8320	60°		21.16	.378		54	23		77	93.50

Pre-Installation Change Order Costs

26 05 36.40 Cable Tray, Covers and Dividers	Crew	Daily Output	Labor-Hours	Unit	Material	2020 Bare Costs Labor	Equipment	Total	Total Incl O&P
8340	90°	1 Elec	20.24	.395	Ea.	73	24.50		97.50	117
8360	Aluminum, 30°		23.92	.334		28	20.50		48.50	61.50
8380	45°		23	.348		32.50	21.50		54	67.50
8400	60°		22.08	.362		40.50	22		62.50	77.50
8420	90°		21.16	.378		55	23		78	95
8440	36" radius, galvanized, 30°		20.24	.395		51	24.50		75.50	92
8460	45°		19.32	.414		59.50	25.50		85	103
8480	60°		18.40	.435		69.50	26.50		96	116
8500	90°		17.48	.458		96.50	28		124.50	148
8520	Aluminum, 30°		21.16	.378		42.50	23		65.50	81.50
8540	45°		20.24	.395		55	24.50		79.50	96.50
8560	60°		19.32	.414		71.50	25.50		97	117
8570	90°	▼	18.40	.435	▼	93.50	26.50		120	143
8590	Divider strip, vertical fitting, 4" deep									
8600	12" radius, galvanized, 30°	1 Elec	24.84	.322	Ea.	32	19.75		51.75	64.50
8610	45°		23.92	.334		37	20.50		57.50	71
8620	60°		23	.348		41.50	21.50		63	78
8630	90°		22.08	.362		51	22		73	89
8640	Aluminum, 30°		25.76	.311		24	19.05		43.05	55
8650	45°		24.84	.322		27.50	19.75		47.25	60
8660	60°		23.92	.334		31.50	20.50		52	65.50
8670	90°		23	.348		37.50	21.50		59	73
8680	24" radius, galvanized, 30°		22.08	.362		51	22		73	89
8690	45°		21.16	.378		64	23		87	105
8700	60°		20.24	.395		72.50	24.50		97	116
8710	90°		19.32	.414		96.50	25.50		122	144
8720	Aluminum, 30°		23	.348		37.50	21.50		59	73
8730	45°		22.08	.362		45.50	22		67.50	83
8740	60°		21.16	.378		54	23		77	93.50
8750	90°		20.24	.395		73.50	24.50		98	117
8760	36" radius, galvanized, 30°		21.16	.378		60.50	23		83.50	101
8770	45°		20.24	.395		69	24.50		93.50	112
8780	60°		19.32	.414		87	25.50		112.50	134
8790	90°		18.40	.435		116	26.50		142.50	167
8800	Aluminum, 30°		22.08	.362		58.50	22		80.50	97.50
8810	45°		21.16	.378		75	23		98	117
8820	60°		20.24	.395		91.50	24.50		116	137
8830	90°	▼	19.32	.414	▼	116	25.50		141.50	165
8840	Divider strip, vertical fitting, 6" deep									
8850	12" radius, galvanized, 30°	1 Elec	22.08	.362	Ea.	36.50	22		58.50	73.50
8860	45°		21.16	.378		41	23		64	79.50
8870	60°		20.24	.395		47	24.50		71.50	88
8880	90°		19.32	.414		59	25.50		84.50	103
8890	Aluminum, 30°		23	.348		27.50	21.50		49	62
8900	45°		22.08	.362		32.50	22		54.50	68.50
8910	60°		21.16	.378		34	23		57	72
8920	90°		20.24	.395		40	24.50		64.50	80
8930	24" radius, galvanized, 30°		21.16	.378		51	23		74	90.50
8940	45°		20.24	.395		64	24.50		88.50	107
8950	60°		19.32	.414		73	25.50		98.50	119
8960	90°		18.40	.435		96.50	26.50		123	146
8970	Aluminum, 30°		22.08	.362		38.50	22		60.50	75
8980	45°		21.16	.378		52	23		75	92

26 05 36 – Cable Trays for Electrical Systems

26 05 36.40 Cable Tray, Covers and Dividers	Crew	Daily Output	Labor-Hours	Unit	Material	2020 Bare Costs Labor	Equipment	Total	Total Incl O&P
8990	60°	1 Elec	20.24	.395	Ea.	57.50	24.50		82	99
9000	90°		19.32	.414		78	25.50		103.50	124
9010	36" radius, galvanized, 30°		20.24	.395		59.50	24.50		84	101
9020	45°		19.32	.414		73	25.50		98.50	119
9030	60°		18.40	.435		96.50	26.50		123	146
9040	90°		17.48	.458		127	28		155	181
9050	Aluminum, 30°		21.16	.378		59.50	23		82.50	100
9060	45°		20.24	.395		79.50	24.50		104	124
9070	60°		19.32	.414		94	25.50		119.50	141
9080	90°		18.40	.435		113	26.50		139.50	164
9120	Divider strip, horizontal fitting, galvanized, 3" deep		30.36	.264		36.50	16.15		52.65	64
9130	4" deep		27.60	.290		40.50	17.80		58.30	71.50
9140	6" deep		24.84	.322		52	19.75		71.75	86.50
9150	Aluminum, 3" deep		32.20	.248		27.50	15.25		42.75	52.50
9160	4" deep		29.44	.272		30	16.65		46.65	58
9170	6" deep		26.68	.300		40	18.40		58.40	71.50
9300	Divider strip protector		276	.029	L.F.	2.96	1.78		4.74	5.90
9310	Fastener, ladder tray				Ea.	.52			.52	.57
9320	Trough or solid bottom tray				"	.39			.39	.43
9900	Add to labor for higher elevated installation									
9905	10' to 14.5' high, add						10%			
9910	15' to 20' high, add						20%			
9920	20' to 25' high, add						25%			
9930	25' to 30' high, add						35%			
9940	30' to 35' high, add						40%			
9950	35' to 40' high, add						50%			
9960	Over 40' high, add						55%			

26 05 39 – Underfloor Raceways for Electrical Systems

26 05 39.30 Conduit In Concrete Slab

		Crew	Daily Output	Labor-Hours	Unit	Material	2020 Bare Costs Labor	Equipment	Total	Total Incl O&P
0010	**CONDUIT IN CONCRETE SLAB** Including terminations,									
0020	fittings and supports									
3230	PVC, schedule 40, 1/2" diameter	1 Elec	248.40	.032	L.F.	.66	1.98		2.64	3.67
3250	3/4" diameter		211.60	.038		.70	2.32		3.02	4.22
3270	1" diameter		184	.043		.93	2.67		3.60	5
3300	1-1/4" diameter		156.40	.051		1.26	3.14		4.40	6.05
3330	1-1/2" diameter		128.80	.062		1.53	3.81		5.34	7.35
3350	2" diameter		110.40	.072		1.91	4.45		6.36	8.70
3370	2-1/2" diameter		82.80	.097		3.15	5.95		9.10	12.30
3400	3" diameter	2 Elec	147.20	.109		4.16	6.65		10.81	14.55
3430	3-1/2" diameter		110.40	.145		5.40	8.90		14.30	19.20
3440	4" diameter		92	.174		5.70	10.65		16.35	22
3450	5" diameter		73.60	.217		8.95	13.35		22.30	29.50
3460	6" diameter		55.20	.290		11.55	17.80		29.35	39
3530	Sweeps, 1" diameter, 30" radius	1 Elec	29.44	.272	Ea.	11.50	16.65		28.15	37.50
3550	1-1/4" diameter		22.08	.362		10.30	22		32.30	44.50
3570	1-1/2" diameter		19.32	.414		11.25	25.50		36.75	50.50
3600	2" diameter		16.56	.483		13.50	29.50		43	59
3630	2-1/2" diameter		12.88	.621		96	38		134	162
3650	3" diameter		9.20	.870		81.50	53.50		135	170
3670	3-1/2" diameter		7.36	1.090		62.50	67		129.50	169
3700	4" diameter		6.44	1.240		30	76		106	147
3710	5" diameter		5.52	1.450		66	89		155	205

For customer support on your Electrical Change Order Costs with RSMeans data, call 800.448.8182.   **Pre-Installation Change Order Costs**

26 05 39 – Underfloor Raceways for Electrical Systems

26 05 39.30 Conduit In Concrete Slab	Crew	Daily Output	Labor-Hours	Unit	Material	2020 Bare Costs Labor	Equipment	Total	Total Incl O&P
3730	Couplings, 1/2" diameter				Ea.	.14			.14	.16
3750	3/4" diameter					.18			.18	.20
3770	1" diameter					.28			.28	.30
3800	1-1/4" diameter					.46			.46	.50
3830	1-1/2" diameter					.52			.52	.57
3850	2" diameter					.70			.70	.77
3870	2-1/2" diameter					1.33			1.33	1.47
3900	3" diameter					2.11			2.11	2.32
3930	3-1/2" diameter					2.94			2.94	3.23
3950	4" diameter					3.29			3.29	3.62
3960	5" diameter					6.75			6.75	7.45
3970	6" diameter					10.60			10.60	11.70
4030	End bells, 1" diameter, PVC	1 Elec	55.20	.145		3.22	8.90		12.12	16.80
4050	1-1/4" diameter		48.76	.164		5.45	10.05		15.50	21
4100	1-1/2" diameter		44.16	.181		3.72	11.10		14.82	20.50
4150	2" diameter		31.28	.256		5.25	15.70		20.95	29.50
4170	2-1/2" diameter		24.84	.322		4.86	19.75		24.61	35
4200	3" diameter		18.40	.435		7.60	26.50		34.10	48
4250	3-1/2" diameter		14.72	.543		8.25	33.50		41.75	58.50
4300	4" diameter		12.88	.621		9.30	38		47.30	67
4310	5" diameter		11.04	.725		14.50	44.50		59	82
4320	6" diameter		8.28	.966		14.85	59.50		74.35	105
4350	Rigid galvanized steel, 1/2" diameter		184	.043	L.F.	2.80	2.67		5.47	7.05
4400	3/4" diameter		156.40	.051		5.35	3.14		8.49	10.50
4450	1" diameter		119.60	.067		8.25	4.10		12.35	15.20
4500	1-1/4" diameter		101.20	.079		5.45	4.85		10.30	13.20
4600	1-1/2" diameter		92	.087		10.05	5.35		15.40	19.05
4800	2" diameter		82.80	.097		9.85	5.95		15.80	19.70

26 05 39.40 Conduit In Trench

		Crew	Daily Output	Labor-Hours	Unit	Material	Labor	Equipment	Total	Total Incl O&P
0010	**CONDUIT IN TRENCH** Includes terminations and fittings									
0200	Rigid galvanized steel, 2" diameter	1 Elec	138	.058	L.F.	9.35	3.56		12.91	15.60
0400	2-1/2" diameter	"	92	.087		15.60	5.35		20.95	25
0600	3" diameter	2 Elec	147.20	.109		17.45	6.65		24.10	29
0800	3-1/2" diameter		128.80	.124		23	7.60		30.60	37
1000	4" diameter		92	.174		28	10.65		38.65	46.50
1200	5" diameter		73.60	.217		45	13.35		58.35	69.50
1400	6" diameter		55.20	.290		69	17.80		86.80	103

26 05 43 – Underground Ducts and Raceways for Electrical Systems

26 05 43.10 Trench Duct

		Crew	Daily Output	Labor-Hours	Unit	Material	Labor	Equipment	Total	Total Incl O&P
0010	**TRENCH DUCT** Steel with cover									
0020	Standard adjustable, depths to 4"									
0100	Straight, single compartment, 9" wide	2 Elec	36.80	.435	L.F.	289	26.50		315.50	360
0200	12" wide		29.44	.543		365	33.50		398.50	450
0400	18" wide		23.92	.669		395	41		436	495
0600	24" wide		20.24	.791		415	48.50		463.50	525
0700	27" wide		19.32	.828		215	51		266	315
0800	30" wide		18.40	.870		228	53.50		281.50	330
1000	36" wide		14.72	1.090		261	67		328	385
1020	Two compartment, 9" wide		34.96	.458		132	28		160	187
1030	12" wide		27.60	.580		157	35.50		192.50	226
1040	18" wide		22.08	.725		199	44.50		243.50	285
1050	24" wide		18.40	.870		247	53.50		300.50	350

26 05 43 – Underground Ducts and Raceways for Electrical Systems

26 05 43.10 Trench Duct	Crew	Daily Output	Labor-Hours	Unit	Material	2020 Bare Costs Labor	Equipment	Total	Total Incl O&P
1060	30" wide	2 Elec	16.56	.966	L.F.	315	59.50		374.50	440
1070	36" wide		12.88	1.240		350	76		426	500
1090	Three compartment, 9" wide		33.12	.483		152	29.50		181.50	211
1100	12" wide		25.76	.621		175	38		213	250
1110	18" wide		20.24	.791		213	48.50		261.50	305
1120	24" wide		16.56	.966		267	59.50		326.50	380
1130	30" wide		14.72	1.090		330	67		397	465
1140	36" wide		11.04	1.450		380	89		469	545
1200	Horizontal elbow, 9" wide		4.97	3.220	Ea.	455	198		653	795
1400	12" wide		4.23	3.780		440	232		672	830
1600	18" wide		3.68	4.350		625	267		892	1,075
1800	24" wide		2.94	5.430		895	335		1,230	1,475
1900	27" wide		2.76	5.800		1,075	355		1,430	1,725
2000	30" wide		2.39	6.690		1,175	410		1,585	1,900
2200	36" wide		2.21	7.250		1,575	445		2,020	2,375
2220	Two compartment, 9" wide		3.50	4.580		735	281		1,016	1,225
2230	12" wide		2.76	5.800		810	355		1,165	1,425
2240	18" wide		2.21	7.250		975	445		1,420	1,725
2250	24" wide		1.84	8.700		1,175	535		1,710	2,075
2260	30" wide		1.66	9.660		1,550	595		2,145	2,600
2270	36" wide		1.47	10.870		1,900	665		2,565	3,075
2290	Three compartment, 9" wide		3.31	4.830		725	296		1,021	1,225
2300	12" wide		2.58	6.210		810	380		1,190	1,450
2310	18" wide		2.02	7.910		985	485		1,470	1,800
2320	24" wide		1.66	9.660		1,250	595		1,845	2,250
2330	30" wide		1.47	10.870		1,625	665		2,290	2,775
2350	36" wide		1.29	12.420		1,975	760		2,735	3,300
2400	Vertical elbow, 9" wide		4.97	3.220		179	198		377	490
2600	12" wide		4.23	3.780		172	232		404	535
2800	18" wide		3.68	4.350		197	267		464	610
3000	24" wide		2.94	5.430		245	335		580	765
3100	27" wide		2.76	5.800		256	355		611	810
3200	30" wide		2.39	6.690		271	410		681	910
3400	36" wide		2.21	7.250		297	445		742	985
3600	Cross, 9" wide		3.68	4.350		745	267		1,012	1,200
3800	12" wide		2.94	5.430		785	335		1,120	1,350
4000	18" wide		2.39	6.690		940	410		1,350	1,625
4200	24" wide		2.02	7.910		1,175	485		1,660	2,000
4300	27" wide		2.02	7.910		1,325	485		1,810	2,175
4400	30" wide		1.84	8.700		1,450	535		1,985	2,400
4600	36" wide		1.66	9.660		1,825	595		2,420	2,875
4620	Two compartment, 9" wide		3.50	4.580		730	281		1,011	1,225
4630	12" wide		2.76	5.800		765	355		1,120	1,375
4640	18" wide		2.21	7.250		930	445		1,375	1,675
4650	24" wide		1.84	8.700		1,175	535		1,710	2,100
4660	30" wide		1.66	9.660		1,550	595		2,145	2,600
4670	36" wide		1.47	10.870		1,875	665		2,540	3,075
4690	Three compartment, 9" wide		3.31	4.830		740	296		1,036	1,250
4700	12" wide		2.58	6.210		850	380		1,230	1,500
4710	18" wide		2.02	7.910		1,000	485		1,485	1,825
4720	24" wide		1.66	9.660		1,250	595		1,845	2,250
4730	30" wide		1.47	10.870		1,625	665		2,290	2,800
4740	36" wide		1.29	12.420		2,000	760		2,760	3,325

For customer support on your Electrical Change Order Costs with RSMeans data, call 800.448.8182.

Pre-Installation Change Order Costs

26 05 43.10 Trench Duct		Crew	Daily Output	Labor-Hours	Unit	Material	2020 Bare Costs Labor	Equipment	Total	Total Incl O&P
4800	End closure, 9" wide	2 Elec	13.25	1.210	Ea.	44	74		118	159
5000	12" wide		11.04	1.450		50	89		139	187
5200	18" wide		9.20	1.740		77	107		184	244
5400	24" wide		7.36	2.170		101	133		234	310
5500	27" wide		6.44	2.480		117	152		269	355
5600	30" wide		6.07	2.640		156	162		318	415
5800	36" wide		5.34	3		151	184		335	440
6000	Tees, 9" wide		3.68	4.350		425	267		692	865
6200	12" wide		3.31	4.830		495	296		791	985
6400	18" wide		2.94	5.430		635	335		970	1,200
6600	24" wide		2.76	5.800		910	355		1,265	1,525
6700	27" wide		2.58	6.210		1,000	380		1,380	1,675
6800	30" wide		2.39	6.690		1,175	410		1,585	1,900
7000	36" wide		1.84	8.700		1,550	535		2,085	2,500
7020	Two compartment, 9" wide		3.50	4.580		495	281		776	965
7030	12" wide		3.13	5.120		530	315		845	1,050
7040	18" wide		2.76	5.800		710	355		1,065	1,300
7050	24" wide		2.58	6.210		960	380		1,340	1,625
7060	30" wide		2.21	7.250		1,300	445		1,745	2,075
7070	36" wide		1.75	9.150		1,625	560		2,185	2,625
7090	Three compartment, 9" wide		3.31	4.830		565	296		861	1,050
7100	12" wide		2.94	5.430		595	335		930	1,150
7110	18" wide		2.58	6.210		750	380		1,130	1,400
7120	24" wide		2.39	6.690		1,025	410		1,435	1,750
7130	30" wide		2.02	7.910		1,350	485		1,835	2,225
7140	36" wide		1.66	9.660		1,725	595		2,320	2,775
7200	Riser, and cabinet connector, 9" wide		4.97	3.220		186	198		384	500
7400	12" wide		4.23	3.780		217	232		449	585
7600	18" wide		3.68	4.350		228	267		495	645
7800	24" wide		2.94	5.430		325	335		660	850
7900	27" wide		2.76	5.800		300	355		655	860
8000	30" wide		2.39	6.690		370	410		780	1,025
8200	36" wide		1.84	8.700		430	535		965	1,275
8400	Insert assembly, cell to conduit adapter, 1-1/4"	1 Elec	14.72	.543		74	33.50		107.50	131
8500	Adjustable partition	"	294.40	.027	L.F.	26	1.67		27.67	31.50
8600	Depth of duct over 4", per 1", add					12.10			12.10	13.30
8700	Support post	1 Elec	220.80	.036		27.50	2.22		29.72	34
8800	Cover double tile trim, 2 sides					43			43	47
8900	4 sides					120			120	132
9160	Trench duct 3-1/2" x 4-1/2", add					11.50			11.50	12.65
9170	Trench duct 4" x 5", add					11.50			11.50	12.65
9200	For carpet trim, add					39			39	43
9210	For double carpet trim, add					119			119	131

26 05 43.20 Underfloor Duct

		Crew	Daily Output	Labor-Hours	Unit	Material	2020 Bare Costs Labor	Equipment	Total	Total Incl O&P
0010	**UNDERFLOOR DUCT**									
0100	Duct, 1-3/8" x 3-1/8" blank, standard	2 Elec	147.20	.109	L.F.	12.85	6.65		19.50	24
0200	1-3/8" x 7-1/4" blank, super duct		110.40	.145		30	8.90		38.90	46.50
0400	7/8" or 1-3/8" insert type, 24" OC, 1-3/8" x 3-1/8", std.		128.80	.124		20	7.60		27.60	33.50
0600	1-3/8" x 7-1/4", super duct		92	.174		35	10.65		45.65	54.50
0800	Junction box, single duct, 1 level, 3-1/8"	1 Elec	3.68	2.170	Ea.	445	133		578	690
0820	3-1/8" x 7-1/4"		3.68	2.170		575	133		708	835
0840	2 level, 3-1/8" upper & lower		2.94	2.720		520	167		687	820

26 05 43.20 Underfloor Duct	Crew	Daily Output	Labor-Hours	Unit	Material	2020 Bare Costs Labor	2020 Bare Costs Equipment	Total	Total Incl O&P
0860	3-1/8" upper, 7-1/4" lower	1 Elec	2.48	3.220	Ea.	510	198		708	855
0880	Carpet pan for above		73.60	.109		370	6.65		376.65	420
0900	Terrazzo pan for above		61.64	.130		880	7.95		887.95	980
1000	Junction box, single duct, 1 level, 7-1/4"		2.48	3.220		520	198		718	865
1020	2 level, 7-1/4" upper & lower		2.48	3.220		595	198		793	950
1040	2 duct, two 3-1/8" upper & lower		2.94	2.720		805	167		972	1,125
1200	1 level, 2 duct, 3-1/8"		2.94	2.720		590	167		757	900
1220	Carpet pan for above boxes		73.60	.109		365	6.65		371.65	410
1240	Terrazzo pan for above boxes		61.64	.130		850	7.95		857.95	945
1260	Junction box, 1 level, two 3-1/8" x one 3-1/8" + one 7-1/4"		2.12	3.780		995	232		1,227	1,450
1280	2 level, two 3-1/8" upper, one 3-1/8" + one 7-1/4" lower		1.84	4.350		1,100	267		1,367	1,600
1300	Carpet pan for above boxes		73.60	.109		365	6.65		371.65	410
1320	Terrazzo pan for above boxes		61.64	.130		850	7.95		857.95	945
1400	Junction box, 1 level, 2 duct, 7-1/4"		2.12	3.780		1,475	232		1,707	1,975
1420	Two 3-1/8" + one 7-1/4"		1.84	4.350		1,500	267		1,767	2,050
1440	Carpet pan for above		73.60	.109		365	6.65		371.65	410
1460	Terrazzo pan for above		61.64	.130		850	7.95		857.95	945
1580	Junction box, 1 level, one 3-1/8" + one 7-1/4" x same		2.12	3.780		1,000	232		1,232	1,450
1600	Triple duct, 3-1/8"		2.12	3.780		1,000	232		1,232	1,450
1700	Junction box, 1 level, one 3-1/8" + two 7-1/4"		1.84	4.350		1,675	267		1,942	2,250
1720	Carpet pan for above		73.60	.109		365	6.65		371.65	410
1740	Terrazzo pan for above		61.64	.130		850	7.95		857.95	945
1800	Insert to conduit adapter, 3/4" & 1"		29.44	.272		36.50	16.65		53.15	65.50
2000	Support, single cell		24.84	.322		54.50	19.75		74.25	89.50
2200	Super duct		14.72	.543		55	33.50		88.50	110
2400	Double cell		14.72	.543		55	33.50		88.50	111
2600	Triple cell		10.12	.791		64	48.50		112.50	143
2800	Vertical elbow, standard duct		9.20	.870		98.50	53.50		152	188
3000	Super duct		7.36	1.090		98.50	67		165.50	208
3200	Cabinet connector, standard duct		29.44	.272		78.50	16.65		95.15	112
3400	Super duct		24.84	.322		76.50	19.75		96.25	114
3600	Conduit adapter, 1" to 1-1/4"		29.44	.272		73.50	16.65		90.15	106
3800	2" to 1-1/4"		24.84	.322		88.50	19.75		108.25	127
4000	Outlet, low tension (tele, computer, etc.)		7.36	1.090		104	67		171	214
4200	High tension, receptacle (120 volt)		7.36	1.090		106	67		173	217
4300	End closure, standard duct		147.20	.054		4.44	3.33		7.77	9.85
4310	Super duct		147.20	.054		8.40	3.33		11.73	14.20
4350	Elbow, horiz., standard duct		23.92	.334		252	20.50		272.50	310
4360	Super duct		23.92	.334		240	20.50		260.50	295
4380	Elbow, offset, standard duct		23.92	.334		98.50	20.50		119	139
4390	Super duct		23.92	.334		100	20.50		120.50	141
4400	Marker screw assembly for inserts		46	.174		19.40	10.65		30.05	37.50
4410	Y take off, standard duct		23.92	.334		153	20.50		173.50	200
4420	Super duct		23.92	.334		153	20.50		173.50	200
4430	Box opening plug, standard duct		147.20	.054		18.70	3.33		22.03	25.50
4440	Super duct		147.20	.054		18.70	3.33		22.03	25.50
4450	Sleeve coupling, standard duct		147.20	.054		49.50	3.33		52.83	59.50
4460	Super duct		147.20	.054		49.50	3.33		52.83	59.50
4470	Conduit adapter, standard duct, 3/4"		29.44	.272		80	16.65		96.65	113
4480	1" or 1-1/4"		29.44	.272		76	16.65		92.65	109
4500	1-1/2"		29.44	.272		74	16.65		90.65	107

26 05 83 – Wiring Connections

26 05 83.10 Motor Connections	Crew	Daily Output	Labor-Hours	Unit	Material	2020 Bare Costs Labor	2020 Bare Costs Equipment	Total	Total Incl O&P
0010 **MOTOR CONNECTIONS**									
0020 Flexible conduit and fittings, 115 volt, 1 phase, up to 1 HP motor	1 Elec	7.36	1.090	Ea.	5.90	67		72.90	106
0050 2 HP motor		5.98	1.340		10.40	82		92.40	133
0100 3 HP motor		5.06	1.580		9.20	97		106.20	154
0110 230 volt, 3 phase, 3 HP motor		6.24	1.280		7.15	78.50		85.65	125
0112 5 HP motor		5.03	1.590		6.05	97.50		103.55	152
0114 7-1/2 HP motor		4.24	1.890		9.35	116		125.35	183
0120 10 HP motor		3.86	2.070		17.40	127		144.40	208
0150 15 HP motor		3.04	2.640		17.40	162		179.40	260
0200 25 HP motor		2.48	3.220		29.50	198		227.50	325
0400 50 HP motor		2.02	3.950		54.50	242		296.50	420
0600 100 HP motor		1.38	5.800		121	355		476	665
1500 460 volt, 5 HP motor, 3 phase		7.36	1.090		6.35	67		73.35	106
1520 10 HP motor		7.36	1.090		6.35	67		73.35	106
1530 25 HP motor		5.52	1.450		11.25	89		100.25	144
1540 30 HP motor		5.52	1.450		11.25	89		100.25	144
1550 40 HP motor		4.60	1.740		16.70	107		123.70	177
1560 50 HP motor		4.60	1.740		22.50	107		129.50	184
1570 60 HP motor		3.50	2.290		25	140		165	237
1580 75 HP motor		3.22	2.480		31.50	152		183.50	262
1590 100 HP motor		2.30	3.480		53.50	214		267.50	380
1600 125 HP motor		1.84	4.350		58.50	267		325.50	460
1610 150 HP motor		1.66	4.830		60	296		356	505
1620 200 HP motor		1.38	5.800		99.50	355		454.50	640
2005 460 volt, 5 HP motor, 3 phase, w/sealtite		7.36	1.090		9.30	67		76.30	110
2010 10 HP motor		7.36	1.090		9.30	67		76.30	110
2015 25 HP motor		5.52	1.450		17	89		106	151
2020 30 HP motor		5.52	1.450		17	89		106	151
2025 40 HP motor		4.60	1.740		30	107		137	193
2030 50 HP motor		4.60	1.740		34.50	107		141.50	197
2035 60 HP motor		3.50	2.290		38.50	140		178.50	251
2040 75 HP motor		3.22	2.480		45	152		197	277
2045 100 HP motor		2.30	3.480		80	214		294	410
2055 150 HP motor		1.66	4.830		86.50	296		382.50	535
2060 200 HP motor		1.38	5.800		273	355		628	830

26 05 90 – Residential Applications

26 05 90.10 Residential Wiring

	Crew	Daily Output	Labor-Hours	Unit	Material	2020 Bare Costs Labor	2020 Bare Costs Equipment	Total	Total Incl O&P
0010 **RESIDENTIAL WIRING**									
0020 20' avg. runs and #14/2 wiring incl. unless otherwise noted									
1000 Service & panel, includes 24' SE-AL cable, service eye, meter,									
1010 Socket, panel board, main bkr., ground rod, 15 or 20 amp									
1020 1-pole circuit breakers, and misc. hardware									
1100 100 amp, with 10 branch breakers	1 Elec	1.09	7.310	Ea.	365	450		815	1,075
1110 With PVC conduit and wire		.85	9.450		405	580		985	1,300
1120 With RGS conduit and wire		.67	11.910		625	730		1,355	1,800
1150 150 amp, with 14 branch breakers		.95	8.440		885	520		1,405	1,750
1170 With PVC conduit and wire		.75	10.600		960	650		1,610	2,025
1180 With RGS conduit and wire		.62	12.980		1,325	795		2,120	2,650
1200 200 amp, with 18 branch breakers	2 Elec	1.66	9.660		1,125	595		1,720	2,100
1220 With PVC conduit and wire		1.34	11.910		1,200	730		1,930	2,425
1230 With RGS conduit and wire		1.14	14.030		1,650	860		2,510	3,075
1800 Lightning surge suppressor	1 Elec	29.44	.272		93.50	16.65		110.15	128

26 05 90.10 Residential Wiring	Crew	Daily Output	Labor-Hours	Unit	Material	2020 Bare Costs Labor	Equipment	Total	Total Incl O&P
2000 Switch devices									
2100 Single pole, 15 amp, ivory, with a 1-gang box, cover plate,									
2110 Type NM (Romex) cable	1 Elec	15.73	.509	Ea.	17.10	31		48.10	65.50
2120 Type MC cable		13.16	.608		29	37.50		66.50	87
2130 EMT & wire		5.25	1.520		40	93.50		133.50	183
2150 3-way, #14/3, type NM cable		13.39	.598		11.10	36.50		47.60	67
2170 Type MC cable		11.33	.706		27	43.50		70.50	94
2180 EMT & wire		4.60	1.740		33.50	107		140.50	196
2200 4-way, #14/3, type NM cable		13.39	.598		20	36.50		56.50	76.50
2220 Type MC cable		11.33	.706		36	43.50		79.50	104
2230 EMT & wire		4.60	1.740		42.50	107		149.50	206
2250 S.P., 20 amp, #12/2, type NM cable		12.26	.652		13.55	40		53.55	74.50
2270 Type MC cable		10.52	.761		24	46.50		70.50	96
2280 EMT & wire		4.46	1.790		38.50	110		148.50	206
2290 S.P. rotary dimmer, 600 W, no wiring		15.64	.512		35	31.50		66.50	85
2300 S.P. rotary dimmer, 600 W, type NM cable		13.39	.598		39	36.50		75.50	97.50
2320 Type MC cable		11.33	.706		51	43.50		94.50	121
2330 EMT & wire		4.60	1.740		63	107		170	229
2350 3-way rotary dimmer, type NM cable		12.26	.652		26.50	40		66.50	88.50
2370 Type MC cable		10.52	.761		38	46.50		84.50	112
2380 EMT & wire	▼	4.46	1.790	▼	50.50	110		160.50	220
2400 Interval timer wall switch, 20 amp, 1-30 min., #12/2									
2410 Type NM cable	1 Elec	13.39	.598	Ea.	65	36.50		101.50	126
2420 Type MC cable		11.33	.706		71	43.50		114.50	143
2430 EMT & wire	▼	4.60	1.740	▼	89.50	107		196.50	258
2500 Decorator style									
2510 S.P., 15 amp, type NM cable	1 Elec	15.73	.509	Ea.	23	31		54	72
2520 Type MC cable		13.16	.608		35	37.50		72.50	94
2530 EMT & wire		5.25	1.520		46	93.50		139.50	190
2550 3-way, #14/3, type NM cable		13.39	.598		17.25	36.50		53.75	73.50
2570 Type MC cable		11.33	.706		33	43.50		76.50	101
2580 EMT & wire		4.60	1.740		39.50	107		146.50	203
2600 4-way, #14/3, type NM cable		13.39	.598		26	36.50		62.50	83.50
2620 Type MC cable		11.33	.706		42	43.50		85.50	111
2630 EMT & wire		4.60	1.740		48.50	107		155.50	213
2650 S.P., 20 amp, #12/2, type NM cable		12.26	.652		19.65	40		59.65	81
2670 Type MC cable		10.52	.761		30	46.50		76.50	103
2680 EMT & wire		4.46	1.790		44.50	110		154.50	213
2700 S.P., slide dimmer, type NM cable		15.73	.509		41.50	31		72.50	92
2720 Type MC cable		13.16	.608		53	37.50		90.50	114
2730 EMT & wire		5.25	1.520		65.50	93.50		159	211
2750 S.P., touch dimmer, type NM cable		15.73	.509		58.50	31		89.50	111
2770 Type MC cable		13.16	.608		70	37.50		107.50	133
2780 EMT & wire		5.25	1.520		82.50	93.50		176	230
2800 3-way touch dimmer, type NM cable		12.26	.652		54.50	40		94.50	120
2820 Type MC cable		10.52	.761		66.50	46.50		113	143
2830 EMT & wire	▼	4.46	1.790	▼	79	110		189	251
3000 Combination devices									
3100 S.P. switch/15 amp recpt., ivory, 1-gang box, plate									
3110 Type NM cable	1 Elec	10.52	.761	Ea.	24	46.50		70.50	96
3120 Type MC cable		9.20	.870		36	53.50		89.50	119
3130 EMT & wire		4.05	1.980		48	121		169	234
3150 S.P. switch/pilot light, type NM cable		10.52	.761		25.50	46.50		72	97.50

For customer support on your Electrical Change Order Costs with RSMeans data, call 800.448.8182.

Pre-Installation Change Order Costs

26 05 90.10 Residential Wiring	Crew	Daily Output	Labor-Hours	Unit	Material	2020 Bare Costs Labor	Equipment	Total	Total Incl O&P
3170	Type MC cable	1 Elec	9.20	.870	Ea.	37.50	53.50		91	121
3180	EMT & wire		4.08	1.960		49.50	120		169.50	234
3190	2-S.P. switches, 2-#14/2, no wiring		12.88	.621		14.30	38		52.30	72
3200	2-S.P. switches, 2-#14/2, type NM cables		9.20	.870		26.50	53.50		80	109
3220	Type MC cable		8.18	.978		42.50	60		102.50	137
3230	EMT & wire		3.77	2.120		50.50	130		180.50	250
3250	3-way switch/15 amp recpt., #14/3, type NM cable		9.20	.870		33	53.50		86.50	116
3270	Type MC cable		8.18	.978		48.50	60		108.50	143
3280	EMT & wire		3.77	2.120		55	130		185	255
3300	2-3 way switches, 2-#14/3, type NM cables		8.18	.978		41.50	60		101.50	136
3320	Type MC cable		7.36	1.090		66	67		133	172
3330	EMT & wire		3.68	2.170		62.50	133		195.50	267
3350	S.P. switch/20 amp recpt., #12/2, type NM cable		9.20	.870		44	53.50		97.50	128
3370	Type MC cable		8.18	.978		50	60		110	145
3380	EMT & wire	▼	3.77	2.120	▼	68.50	130		198.50	270
3400	Decorator style									
3410	S.P. switch/15 amp recpt., type NM cable	1 Elec	10.52	.761	Ea.	30	46.50		76.50	103
3420	Type MC cable		9.20	.870		42	53.50		95.50	126
3430	EMT & wire		4.05	1.980		54.50	121		175.50	241
3450	S.P. switch/pilot light, type NM cable		10.52	.761		31.50	46.50		78	105
3470	Type MC cable		9.20	.870		43.50	53.50		97	128
3480	EMT & wire		4.05	1.980		56	121		177	243
3500	2-S.P. switches, 2-#14/2, type NM cables		9.20	.870		32.50	53.50		86	115
3520	Type MC cable		8.18	.978		48.50	60		108.50	143
3530	EMT & wire		3.77	2.120		56.50	130		186.50	256
3550	3-way/15 amp recpt., #14/3, type NM cable		9.20	.870		39	53.50		92.50	123
3570	Type MC cable		8.18	.978		54.50	60		114.50	150
3580	EMT & wire		3.77	2.120		61	130		191	262
3650	2-3 way switches, 2-#14/3, type NM cables		8.18	.978		48	60		108	142
3670	Type MC cable		7.36	1.090		72	67		139	179
3680	EMT & wire		3.68	2.170		68.50	133		201.50	274
3700	S.P. switch/20 amp recpt., #12/2, type NM cable		9.20	.870		50	53.50		103.50	135
3720	Type MC cable		8.18	.978		56	60		116	151
3730	EMT & wire	▼	3.77	2.120	▼	74.50	130		204.50	276
4000	Receptacle devices									
4010	Duplex outlet, 15 amp recpt., ivory, 1-gang box, plate									
4015	Type NM cable	1 Elec	13.39	.598	Ea.	9.80	36.50		46.30	65.50
4020	Type MC cable		11.33	.706		21.50	43.50		65	88
4030	EMT & wire		4.90	1.630		32.50	100		132.50	185
4050	With #12/2, type NM cable		11.33	.706		11.55	43.50		55.05	77
4070	Type MC cable		9.82	.815		22	50		72	98.50
4080	EMT & wire		4.33	1.850		36	114		150	209
4100	20 amp recpt., #12/2, type NM cable		11.33	.706		21.50	43.50		65	88
4120	Type MC cable		9.82	.815		32	50		82	110
4130	EMT & wire	▼	4.33	1.850	▼	46	114		160	220
4140	For GFI see Section 26 05 90.10 line 4300 below									
4150	Decorator style, 15 amp recpt., type NM cable	1 Elec	13.39	.598	Ea.	15.90	36.50		52.40	72
4170	Type MC cable		11.33	.706		27.50	43.50		71	95
4180	EMT & wire		4.90	1.630		38.50	100		138.50	192
4200	With #12/2, type NM cable		11.33	.706		17.65	43.50		61.15	84
4220	Type MC cable		9.82	.815		28	50		78	106
4230	EMT & wire		4.33	1.850		42.50	114		156.50	216
4250	20 amp recpt., #12/2, type NM cable		11.33	.706		27.50	43.50		71	95

26 05 Common Work Results for Electrical

26 05 90 – Residential Applications

26 05 90.10 Residential Wiring	Crew	Daily Output	Labor-Hours	Unit	Material	2020 Bare Costs Labor	Equipment	Total	Total Incl O&P
4270 Type MC cable	1 Elec	9.82	.815	Ea.	38	50		88	117
4280 EMT & wire		4.33	1.850		52.50	114		166.50	227
4300 GFI, 15 amp recpt., type NM cable		11.33	.706		23	43.50		66.50	89.50
4320 Type MC cable		9.82	.815		34.50	50		84.50	113
4330 EMT & wire		4.33	1.850		45.50	114		159.50	219
4350 GFI with #12/2, type NM cable		9.82	.815		24.50	50		74.50	102
4370 Type MC cable		8.46	.945		35	58		93	125
4380 EMT & wire		3.87	2.070		49.50	127		176.50	244
4400 20 amp recpt., #12/2, type NM cable		9.82	.815		59.50	50		109.50	140
4420 Type MC cable		8.46	.945		70	58		128	164
4430 EMT & wire		3.87	2.070		84	127		211	282
4500 Weather-proof cover for above receptacles, add	↓	29.44	.272	↓	2.22	16.65		18.87	27.50
4550 Air conditioner outlet, 20 amp-240 volt recpt.									
4560 30' of #12/2, 2 pole circuit breaker									
4570 Type NM cable	1 Elec	9.20	.870	Ea.	67	53.50		120.50	154
4580 Type MC cable		8.28	.966		79	59.50		138.50	176
4590 EMT & wire		3.68	2.170		92	133		225	299
4600 Decorator style, type NM cable		9.20	.870		72.50	53.50		126	159
4620 Type MC cable		8.28	.966		84.50	59.50		144	182
4630 EMT & wire	↓	3.68	2.170	↓	97.50	133		230.50	305
4650 Dryer outlet, 30 amp-240 volt recpt., 20' of #10/3									
4660 2 pole circuit breaker									
4670 Type NM cable	1 Elec	5.90	1.360	Ea.	60	83.50		143.50	190
4680 Type MC cable		5.25	1.520		68.50	93.50		162	215
4690 EMT & wire	↓	3.20	2.500	↓	80.50	153		233.50	315
4700 Range outlet, 50 amp-240 volt recpt., 30' of #8/3									
4710 Type NM cable	1 Elec	3.87	2.070	Ea.	90.50	127		217.50	289
4720 Type MC cable		3.68	2.170		146	133		279	360
4730 EMT & wire		2.72	2.940		115	180		295	395
4750 Central vacuum outlet, type NM cable		5.89	1.360		63.50	83.50		147	194
4770 Type MC cable		5.25	1.520		77.50	93.50		171	225
4780 EMT & wire	↓	3.20	2.500	↓	97	153		250	335
4800 30 amp-110 volt locking recpt., #10/2 circ. bkr.									
4810 Type NM cable	1 Elec	5.70	1.400	Ea.	73.50	86		159.50	209
4820 Type MC cable		4.97	1.610		91.50	99		190.50	247
4830 EMT & wire	↓	2.94	2.720	↓	108	167		275	365
4900 Low voltage outlets									
4910 Telephone recpt., 20' of 4/C phone wire	1 Elec	23.92	.334	Ea.	9.70	20.50		30.20	41
4920 TV recpt., 20' of RG59U coax wire, F type connector	"	14.72	.543	"	19.35	33.50		52.85	71
4950 Door bell chime, transformer, 2 buttons, 60' of bellwire									
4970 Economy model	1 Elec	10.58	.756	Ea.	63	46.50		109.50	139
4980 Custom model		10.58	.756		122	46.50		168.50	203
4990 Luxury model, 3 buttons	↓	8.74	.915	↓	207	56		263	310
6000 Lighting outlets									
6050 Wire only (for fixture), type NM cable	1 Elec	29.44	.272	Ea.	6.55	16.65		23.20	32
6070 Type MC cable		22.08	.362		12.45	22		34.45	46.50
6080 EMT & wire		9.20	.870		22.50	53.50		76	105
6100 Box (4"), and wire (for fixture), type NM cable		23	.348		16.55	21.50		38.05	50
6120 Type MC cable		18.40	.435		22.50	26.50		49	64
6130 EMT & wire	↓	10.12	.791		32.50	48.50		81	108
6200 Fixtures (use with line 6050 or 6100 above)									
6210 Canopy style, economy grade	1 Elec	36.80	.217	Ea.	24.50	13.35		37.85	47
6220 Custom grade	↓	36.80	.217	↓	58	13.35		71.35	84

Pre-Installation Change Order Costs

26 05 90.10 Residential Wiring	Crew	Daily Output	Labor-Hours	Unit	Material	2020 Bare Costs Labor	Equipment	Total	Total Incl O&P
6250	Dining room chandelier, economy grade	1 Elec	17.48	.458	Ea.	90	28		118	141
6260	Custom grade		17.48	.458		355	28		383	430
6270	Luxury grade		13.80	.580		1,425	35.50		1,460.50	1,625
6310	Kitchen fixture (fluorescent), economy grade		27.60	.290		79.50	17.80		97.30	114
6320	Custom grade		23	.348		162	21.50		183.50	210
6350	Outdoor, wall mounted, economy grade		27.60	.290		33.50	17.80		51.30	63
6360	Custom grade		27.60	.290		130	17.80		147.80	170
6370	Luxury grade		23	.348		272	21.50		293.50	330
6410	Outdoor PAR floodlights, 1 lamp, 150 watt		18.40	.435		30.50	26.50		57	73
6420	2 lamp, 150 watt each		18.40	.435		49.50	26.50		76	94
6425	Motion sensing, 2 lamp, 150 watt each		18.40	.435		120	26.50		146.50	172
6430	For infrared security sensor, add		29.44	.272		104	16.65		120.65	140
6450	Outdoor, quartz-halogen, 300 watt flood		18.40	.435		44.50	26.50		71	88
6600	Recessed downlight, round, pre-wired, 50 or 75 watt trim		27.60	.290		75	17.80		92.80	109
6610	With shower light trim		27.60	.290		103	17.80		120.80	140
6620	With wall washer trim		25.76	.311		103	19.05		122.05	143
6630	With eye-ball trim		25.76	.311		93.50	19.05		112.55	132
6700	Porcelain lamp holder		36.80	.217		3.04	13.35		16.39	23
6710	With pull switch		36.80	.217		11.75	13.35		25.10	33
6750	Fluorescent strip, 2-20 watt tube, wrap around diffuser, 24"		22.08	.362		50	22		72	88
6760	1-34 watt tube, 48"		22.08	.362		134	22		156	180
6770	2-34 watt tubes, 48"		18.40	.435		176	26.50		202.50	233
6800	Bathroom heat lamp, 1-250 watt		25.76	.311		37	19.05		56.05	69
6810	2-250 watt lamps		25.76	.311		70.50	19.05		89.55	107
6820	For timer switch, see Section 26 05 90.10 line 2400									
6900	Outdoor post lamp, incl. post, fixture, 35' of #14/2									
6910	Type NM cable	1 Elec	3.22	2.480	Ea.	355	152		507	620
6920	Photo-eye, add		24.84	.322		30	19.75		49.75	62.50
6950	Clock dial time switch, 24 hr., w/enclosure, type NM cable		10.52	.761		79.50	46.50		126	157
6970	Type MC cable		10.12	.791		91.50	48.50		140	173
6980	EMT & wire		4.46	1.790		102	110		212	277
7000	Alarm systems									
7050	Smoke detectors, box, #14/3, type NM cable	1 Elec	13.39	.598	Ea.	37	36.50		73.50	95
7070	Type MC cable		11.33	.706		48.50	43.50		92	118
7080	EMT & wire		4.60	1.740		55	107		162	220
7090	For relay output to security system, add					11.25			11.25	12.40
8000	Residential equipment									
8050	Disposal hook-up, incl. switch, outlet box, 3' of flex									
8060	20 amp-1 pole circ. bkr., and 25' of #12/2									
8070	Type NM cable	1 Elec	9.20	.870	Ea.	32	53.50		85.50	115
8080	Type MC cable		7.36	1.090		43	67		110	147
8090	EMT & wire		4.60	1.740		60	107		167	226
8100	Trash compactor or dishwasher hook-up, incl. outlet box,									
8110	3' of flex, 15 amp-1 pole circ. bkr., and 25' of #14/2									
8120	Type NM cable	1 Elec	9.20	.870	Ea.	17.65	53.50		71.15	99
8130	Type MC cable		7.36	1.090		31	67		98	134
8140	EMT & wire		4.60	1.740		45.50	107		152.50	209
8150	Hot water sink dispenser hook-up, use line 8100									
8200	Vent/exhaust fan hook-up, type NM cable	1 Elec	29.44	.272	Ea.	6.55	16.65		23.20	32
8220	Type MC cable		22.08	.362		12.45	22		34.45	46.50
8230	EMT & wire		9.20	.870		22.50	53.50		76	105
8250	Bathroom vent fan, 50 CFM (use with above hook-up)									
8260	Economy model	1 Elec	13.80	.580	Ea.	20.50	35.50		56	76

26 05 90.10 Residential Wiring		Crew	Daily Output	Labor-Hours	Unit	Material	2020 Bare Costs Labor	Equipment	Total	Total Incl O&P
8270	Low noise model	1 Elec	13.80	.580	Ea.	52.50	35.50		88	111
8280	Custom model	↓	11.04	.725	↓	129	44.50		173.50	208
8300	Bathroom or kitchen vent fan, 110 CFM									
8310	Economy model	1 Elec	13.80	.580	Ea.	73.50	35.50		109	134
8320	Low noise model	"	13.80	.580	"	106	35.50		141.50	170
8350	Paddle fan, variable speed (w/o lights)									
8360	Economy model (AC motor)	1 Elec	9.20	.870	Ea.	149	53.50		202.50	244
8362	With light kit		9.20	.870		193	53.50		246.50	293
8370	Custom model (AC motor)		9.20	.870		380	53.50		433.50	500
8372	With light kit		9.20	.870		425	53.50		478.50	550
8380	Luxury model (DC motor)		7.36	1.090		345	67		412	480
8382	With light kit		7.36	1.090		390	67		457	525
8390	Remote speed switch for above, add	↓	11.04	.725	↓	44.50	44.50		89	115
8500	Whole house exhaust fan, ceiling mount, 36", variable speed									
8510	Remote switch, incl. shutters, 20 amp-1 pole circ. bkr.									
8520	30' of #12/2, type NM cable	1 Elec	3.68	2.170	Ea.	1,525	133		1,658	1,875
8530	Type MC cable		3.22	2.480		1,550	152		1,702	1,925
8540	EMT & wire	↓	2.76	2.900	↓	1,550	178		1,728	2,000
8600	Whirlpool tub hook-up, incl. timer switch, outlet box									
8610	3' of flex, 20 amp-1 pole GFI circ. bkr.									
8620	30' of #12/2, type NM cable	1 Elec	4.60	1.740	Ea.	142	107		249	315
8630	Type MC cable		3.86	2.070		150	127		277	355
8640	EMT & wire	↓	3.13	2.560	↓	164	157		321	415
8650	Hot water heater hook-up, incl. 1-2 pole circ. bkr., box;									
8660	3' of flex, 20' of #10/2, type NM cable	1 Elec	4.60	1.740	Ea.	32	107		139	195
8670	Type MC cable		3.86	2.070		46	127		173	240
8680	EMT & wire	↓	3.13	2.560	↓	52.50	157		209.50	292
9000	Heating/air conditioning									
9050	Furnace/boiler hook-up, incl. firestat, local on-off switch									
9060	Emergency switch, and 40' of type NM cable	1 Elec	3.68	2.170	Ea.	63	133		196	267
9070	Type MC cable		3.22	2.480		79	152		231	315
9080	EMT & wire	↓	1.38	5.800	↓	103	355		458	645
9100	Air conditioner hook-up, incl. local 60 amp disc. switch									
9110	3' sealtite, 40 amp, 2 pole circuit breaker									
9130	40' of #8/2, type NM cable	1 Elec	3.22	2.480	Ea.	158	152		310	400
9140	Type MC cable		2.76	2.900		233	178		411	520
9150	EMT & wire	↓	1.20	6.690	↓	204	410		614	835
9200	Heat pump hook-up, 1-40 & 1-100 amp 2 pole circ. bkr.									
9210	Local disconnect switch, 3' sealtite									
9220	40' of #8/2 & 30' of #3/2									
9230	Type NM cable	1 Elec	1.20	6.690	Ea.	575	410		985	1,250
9240	Type MC cable		.99	8.050		605	495		1,100	1,400
9250	EMT & wire	↓	.86	9.250	↓	585	565		1,150	1,500
9500	Thermostat hook-up, using low voltage wire									
9520	Heating only, 25' of #18-3	1 Elec	22.08	.362	Ea.	7.45	22		29.45	41
9530	Heating/cooling, 25' of #18-4	"	18.40	.435	"	9.45	26.50		35.95	50

For customer support on your Electrical Change Order Costs with RSMeans data, call 800.448.8182.

Pre-Installation Change Order Costs

26 09 13 – Electrical Power Monitoring

26 09 13.10 Switchboard Instruments

	Crew	Daily Output	Labor-Hours	Unit	Material	2020 Bare Costs Labor	Equipment	Total	Total Incl O&P
0010 **SWITCHBOARD INSTRUMENTS** 3 phase, 4 wire									
0100 AC indicating, ammeter & switch	1 Elec	7.36	1.090	Ea.	2,925	67		2,992	3,325
0200 Voltmeter & switch		7.36	1.090		3,300	67		3,367	3,725
0300 Wattmeter		7.36	1.090		4,150	67		4,217	4,650
0400 AC recording, ammeter		3.68	2.170		7,375	133		7,508	8,325
0500 Voltmeter		3.68	2.170		7,375	133		7,508	8,325
0600 Ground fault protection, zero sequence		2.48	3.220		6,525	198		6,723	7,475
0700 Ground return path		2.48	3.220		6,525	198		6,723	7,475
0800 3 current transformers, 5 to 800 amp		1.84	4.350		3,025	267		3,292	3,750
0900 1,000 to 1,500 amp		1.20	6.690		4,375	410		4,785	5,400
1200 2,000 to 4,000 amp		.92	8.700		5,150	535		5,685	6,475
1300 Fused potential transformer, maximum 600 volt		7.36	1.090		1,150	67		1,217	1,350

26 09 13.20 Voltage Monitor Systems

	Crew	Daily Output	Labor-Hours	Unit	Material	2020 Bare Costs Labor	Equipment	Total	Total Incl O&P
0010 **VOLTAGE MONITOR SYSTEMS** (test equipment)									
0100 AC voltage monitor system, 120/240 V, one-channel				Ea.	2,875			2,875	3,175
0110 Modem adapter					360			360	395
0120 Add-on detector only					1,500			1,500	1,675
0150 AC voltage remote monitor sys., 3 channel, 120, 230, or 480 V					5,225			5,225	5,750
0160 With internal modem					5,525			5,525	6,075
0170 Combination temperature and humidity probe					810			810	890
0180 Add-on detector only					3,800			3,800	4,175
0190 With internal modem					4,125			4,125	4,550

26 09 23 – Lighting Control Devices

26 09 23.10 Energy Saving Lighting Devices

		Crew	Daily Output	Labor-Hours	Unit	Material	2020 Bare Costs Labor	Equipment	Total	Total Incl O&P
0010 **ENERGY SAVING LIGHTING DEVICES**										
0100 Occupancy sensors, passive infrared ceiling mounted	G	1 Elec	6.44	1.240	Ea.	82.50	76		158.50	204
0110 Ultrasonic ceiling mounted	G		6.44	1.240		116	76		192	241
0120 Dual technology ceiling mounted	G		5.98	1.340		149	82		231	286
0150 Automatic wall switches	G		22.08	.362		75.50	22		97.50	116
0160 Daylighting sensor, manual control, ceiling mounted	G		6.44	1.240		186	76		262	315
0170 Remote and dimming control with remote controller	G		5.98	1.340		222	82		304	365
0200 Passive infrared ceiling mounted			5.98	1.340		38.50	82		120.50	165
0400 Remote power pack	G		9.20	.870		38.50	53.50		92	122
0450 Photoelectric control, S.P.S.T. 120 V	G		7.36	1.090		27	67		94	130
0500 S.P.S.T. 208 V/277 V	G		7.36	1.090		28.50	67		95.50	131
0550 D.P.S.T. 120 V	G		5.52	1.450		207	89		296	360
0600 D.P.S.T. 208 V/277 V	G		5.52	1.450		234	89		323	390
0650 S.P.D.T. 208 V/277 V	G		5.52	1.450		219	89		308	370

26 09 26 – Lighting Control Panelboards

26 09 26.10 Lighting Control Relay Panel

	Crew	Daily Output	Labor-Hours	Unit	Material	2020 Bare Costs Labor	Equipment	Total	Total Incl O&P
0010 **LIGHTING CONTROL RELAY PANEL** with timeclock									
0100 4 Relay	1 Elec	2.30	3.480	Ea.	1,125	214		1,339	1,575
0110 8 Relay		2.12	3.780		1,525	232		1,757	2,050
0120 16 Relay		1.66	4.830		1,775	296		2,071	2,400
0130 24 Relay		1.38	5.800		2,675	355		3,030	3,475
0140 48 Relay		.92	8.700		1,825	535		2,360	2,800
0200 Room Controller, switching only									
0210 1 Relay	1 Elec	2.76	2.900	Ea.	1,650	178		1,828	2,100
0220 2 Relay		2.76	2.900		1,575	178		1,753	2,000
0230 3 Relay		2.76	2.900		1,900	178		2,078	2,375
0240 Dimming									

26 09 Instrumentation and Control for Electrical Systems

26 09 26 – Lighting Control Panelboards

26 09 26.10 Lighting Control Relay Panel	Crew	Daily Output	Labor-Hours	Unit	Material	2020 Bare Costs Labor	Equipment	Total	Total Incl O&P
0250	1 Relay	1 Elec	2.76	2.900	Ea.	1,475	178		1,653	1,900
0260	2 Relay		2.76	2.900		1,550	178		1,728	1,975
0270	3 Relay		2.76	2.900		1,900	178		2,078	2,375

26 12 Medium-Voltage Transformers

26 12 19 – Pad-Mounted, Liquid-Filled, Medium-Voltage Transformers

26 12 19.10 Transformer, Oil-Filled

		Crew	Daily Output	Labor-Hours	Unit	Material	2020 Bare Costs Labor	Equipment	Total	Total Incl O&P
0010	**TRANSFORMER, OIL-FILLED** primary delta or Y,									
0050	Pad mounted 5 kV or 15 kV, with taps, 277/480 V secondary, 3 phase									
0100	150 kVA	R-3	.60	33.440	Ea.	9,875	2,050	315	12,240	14,300
0110	225 kVA		.51	39.530		16,900	2,425	375	19,700	22,600
0200	300 kVA		.41	48.310		14,800	2,950	455	18,205	21,100
0300	500 kVA		.37	54.350		21,900	3,325	515	25,740	29,600
0400	750 kVA		.35	57.210		26,600	3,500	540	30,640	35,000
0500	1,000 kVA		.24	83.610		31,500	5,100	790	37,390	43,100
0600	1,500 kVA		.21	94.520		37,400	5,775	895	44,070	50,500
0700	2,000 kVA		.18	108		47,200	6,600	1,025	54,825	63,000
0710	2,500 kVA		.17	114		57,000	6,975	1,075	65,050	74,500
0720	3,000 kVA		.16	127		69,000	7,750	1,200	77,950	89,000
0800	3,750 kVA		.15	135		91,000	8,250	1,275	100,525	114,000
1990	Pole mounted distribution type, single phase									
2000	13.8 kV primary, 120/240 V secondary, 10 kVA	R-15	6.85	7	Ea.	1,150	420	40.50	1,610.50	1,950
2010	50 kVA		3.40	14.100		2,250	845	82	3,177	3,825
2020	100 kVA		2.53	18.970		4,000	1,150	110	5,260	6,225
2030	167 kVA		1.98	24.270		6,225	1,450	141	7,816	9,175
2900	2,400 V primary, 120/240 V secondary, 10 kVA		6.85	7		1,075	420	40.50	1,535.50	1,850
2910	15 kVA		6.16	7.790		1,300	470	45.50	1,815.50	2,175
2920	25 kVA		5.52	8.700		1,675	520	50.50	2,245.50	2,675
2930	37.5 kVA		3.96	12.130		2,050	730	70.50	2,850.50	3,400
2940	50 kVA		3.40	14.100		2,350	845	82	3,277	3,925
2950	75 kVA		2.76	17.390		3,550	1,050	101	4,701	5,550
2960	100 kVA		2.53	18.970		3,575	1,150	110	4,835	5,750

26 12 19.20 Transformer, Liquid-Filled

		Crew	Daily Output	Labor-Hours	Unit	Material	2020 Bare Costs Labor	Equipment	Total	Total Incl O&P
0010	**TRANSFORMER, LIQUID-FILLED** Pad mounted									
0020	5 kV or 15 kV primary, 277/480 volt secondary, 3 phase									
0050	225 kVA	R-3	.51	39.530	Ea.	14,000	2,425	375	16,800	19,500
0100	300 kVA		.41	48.310		16,700	2,950	455	20,105	23,300
0200	500 kVA		.37	54.350		21,100	3,325	515	24,940	28,700
0250	750 kVA		.35	57.210		27,200	3,500	540	31,240	35,700
0300	1,000 kVA		.24	83.610		31,600	5,100	790	37,490	43,300
0350	1,500 kVA		.21	94.520		36,900	5,775	895	43,570	50,000
0400	2,000 kVA		.18	108		45,700	6,600	1,025	53,325	61,000
0450	2,500 kVA		.17	114		55,500	6,975	1,075	63,550	72,500

Pre-Installation Change Order Costs

26 13 16 – Medium-Voltage Fusible Interrupter Switchgear

26 13 16.10 Switchgear	Crew	Daily Output	Labor-Hours	Unit	Material	2020 Bare Costs Labor	Equipment	Total	Total Incl O&P
0010 **SWITCHGEAR**, Incorporate switch with cable connections, transformer,									
0100 & low voltage section									
0200 Load interrupter switch, 600 amp, 2 position									
0300 NEMA 1, 4.8 kV, 300 kVA & below w/CLF fuses	R-3	.37	54.350	Ea.	20,200	3,325	515	24,040	27,700
0400 400 kVA & above w/CLF fuses		.35	57.210		23,700	3,500	540	27,740	31,900
0500 Non fusible		.38	53.020		18,100	3,250	500	21,850	25,300
0600 13.8 kV, 300 kVA & below w/CLF fuses		.35	57.210		29,800	3,500	540	33,840	38,500
0700 400 kVA & above w/CLF fuses		.33	60.390		29,800	3,700	570	34,070	38,800
0800 Non fusible	▼	.37	54.350		22,100	3,325	515	25,940	29,800
0900 Cable lugs for 2 feeders 4.8 kV or 13.8 kV	1 Elec	7.36	1.090		710	67		777	880
1000 Pothead, one 3 conductor or three 1 conductor		3.68	2.170		3,400	133		3,533	3,950
1100 Two 3 conductor or six 1 conductor		1.84	4.350		6,725	267		6,992	7,800
1200 Key interlocks	▼	7.36	1.090	▼	785	67		852	965
1300 Lightning arresters, distribution class (no charge)									
1400 Intermediate class or line type 4.8 kV	1 Elec	2.48	3.220	Ea.	3,825	198		4,023	4,500
1500 13.8 kV		1.84	4.350		5,075	267		5,342	5,975
1600 Station class, 4.8 kV		2.48	3.220		6,550	198		6,748	7,500
1700 13.8 kV	▼	1.84	4.350		11,300	267		11,567	12,800
1800 Transformers, 4,800 volts to 480/277 volts, 75 kVA	R-3	.63	31.970		19,900	1,950	300	22,150	25,100
1900 112.5 kVA		.60	33.440		24,300	2,050	315	26,665	30,200
2000 150 kVA		.52	38.140		27,700	2,325	360	30,385	34,400
2100 225 kVA		.44	45.290		31,800	2,775	430	35,005	39,600
2200 300 kVA		.38	53.020		35,600	3,250	500	39,350	44,500
2300 500 kVA		.33	60.390		46,800	3,700	570	51,070	57,500
2400 750 kVA		.27	74.960		53,000	4,575	710	58,285	66,000
2500 13,800 volts to 480/277 volts, 75 kVA		.56	35.640		28,100	2,175	335	30,610	34,500
2600 112.5 kVA		.51	39.530		37,300	2,425	375	40,100	45,000
2700 150 kVA		.45	44.370		37,600	2,700	420	40,720	45,900
2800 225 kVA		.38	53.020		43,400	3,250	500	47,150	53,000
2900 300 kVA		.34	58.750		44,300	3,600	555	48,455	55,000
3000 500 kVA		.29	70.130		49,100	4,275	665	54,040	61,000
3100 750 kVA	▼	.24	83.610		54,000	5,100	790	59,890	68,000
3200 Forced air cooling & temperature alarm	1 Elec	.92	8.700	▼	4,350	535		4,885	5,575
3300 Low voltage components									
3400 Maximum panel height 49-1/2", single or twin row									
3500 Breaker heights, type FA or FH, 6"									
3600 type KA or KH, 8"									
3700 type LA, 11"									
3800 type MA, 14"									
3900 Breakers, 2 pole, 15 to 60 amp, type FA	1 Elec	5.15	1.550	Ea.	750	95		845	960
4000 70 to 100 amp, type FA		3.86	2.070		960	127		1,087	1,250
4100 15 to 60 amp, type FH		5.15	1.550		1,125	95		1,220	1,400
4200 70 to 100 amp, type FH		3.86	2.070		1,475	127		1,602	1,825
4300 125 to 225 amp, type KA		3.13	2.560		1,400	157		1,557	1,775
4400 125 to 225 amp, type KH		3.13	2.560		2,350	157		2,507	2,825
4500 125 to 400 amp, type LA		2.30	3.480		2,775	214		2,989	3,375
4600 125 to 600 amp, type MA		1.66	4.830		5,750	296		6,046	6,775
4700 700 & 800 amp, type MA		1.38	5.800		7,225	355		7,580	8,475
4800 3 pole, 15 to 60 amp, type FA		4.88	1.640		900	101		1,001	1,150
4900 70 to 100 amp, type FA		3.68	2.170		1,125	133		1,258	1,450
5000 15 to 60 amp, type FH		4.88	1.640		1,525	101		1,626	1,825
5100 70 to 100 amp, type FH		3.68	2.170		1,775	133		1,908	2,150
5200 125 to 225 amp, type KA	▼	2.94	2.720	▼	2,150	167		2,317	2,625

26 13 Medium-Voltage Switchgear

26 13 16 – Medium-Voltage Fusible Interrupter Switchgear

26 13 16.10 Switchgear	Crew	Daily Output	Labor-Hours	Unit	Material	2020 Bare Costs Labor	Equipment	Total	Total Incl O&P
5300	125 to 225 amp, type KH	1 Elec	2.94	2.720	Ea.	3,900	167		4,067	4,525
5400	125 to 400 amp, type LA		2.12	3.780		3,575	232		3,807	4,300
5500	125 to 600 amp, type MA		1.47	5.430		5,150	335		5,485	6,175
5600	700 & 800 amp, type MA	▼	1.20	6.690	▼	6,400	410		6,810	7,650

26 22 Low-Voltage Transformers

26 22 13 – Low-Voltage Distribution Transformers

26 22 13.10 Transformer, Dry-Type

		Crew	Daily Output	Labor-Hours	Unit	Material	2020 Bare Costs Labor	Equipment	Total	Total Incl O&P
0010	**TRANSFORMER, DRY-TYPE**									
0050	Single phase, 240/480 volt primary, 120/240 volt secondary									
0100	1 kVA	1 Elec	1.84	4.350	Ea.	390	267		657	820
0300	2 kVA		1.47	5.430		700	335		1,035	1,275
0500	3 kVA		1.29	6.210		845	380		1,225	1,500
0700	5 kVA	▼	1.10	7.250		1,300	445		1,745	2,075
0900	7.5 kVA	2 Elec	2.02	7.910		1,250	485		1,735	2,100
1100	10 kVA		1.47	10.870		1,625	665		2,290	2,800
1300	15 kVA		1.10	14.490		2,075	890		2,965	3,600
1500	25 kVA		.92	17.390		2,475	1,075		3,550	4,325
1700	37.5 kVA		.74	21.740		3,325	1,325		4,650	5,650
1900	50 kVA		.64	24.840		3,125	1,525		4,650	5,700
2100	75 kVA	▼	.60	26.760		5,225	1,650		6,875	8,200
2110	100 kVA	R-3	.83	24.150		6,825	1,475	228	8,528	9,950
2120	167 kVA	"	.74	27.170		11,300	1,650	257	13,207	15,200
2190	480 V primary, 120/240 V secondary, nonvent., 15 kVA	2 Elec	1.10	14.490		1,625	890		2,515	3,100
2200	25 kVA		.83	19.320		2,950	1,175		4,125	5,025
2210	37 kVA		.69	23.190		3,375	1,425		4,800	5,850
2220	50 kVA		.60	26.760		3,975	1,650		5,625	6,825
2230	75 kVA		.55	28.990		5,275	1,775		7,050	8,475
2240	100 kVA		.46	34.780		6,875	2,125		9,000	10,800
2250	Low operating temperature (80°C), 25 kVA		.92	17.390		5,150	1,075		6,225	7,275
2260	37 kVA		.74	21.740		5,550	1,325		6,875	8,075
2270	50 kVA		.64	24.840		7,250	1,525		8,775	10,300
2280	75 kVA		.60	26.760		11,500	1,650		13,150	15,100
2290	100 kVA	▼	.51	31.620	▼	12,000	1,950		13,950	16,100
2300	3 phase, 480 volt primary, 120/208 volt secondary									
2310	Ventilated, 3 kVA	1 Elec	.92	8.700	Ea.	1,225	535		1,760	2,150
2700	6 kVA		.74	10.870		1,225	665		1,890	2,350
2900	9 kVA	▼	.64	12.420		1,450	760		2,210	2,700
3100	15 kVA	2 Elec	1.01	15.810		1,900	970		2,870	3,525
3300	30 kVA		.83	19.320		1,800	1,175		2,975	3,750
3500	45 kVA		.74	21.740		1,850	1,325		3,175	4,025
3700	75 kVA	▼	.64	24.840		2,500	1,525		4,025	5,025
3900	112.5 kVA	R-3	.83	24.150		4,075	1,475	228	5,778	6,925
4100	150 kVA		.78	25.580		5,000	1,575	242	6,817	8,100
4300	225 kVA		.60	33.440		8,975	2,050	315	11,340	13,300
4500	300 kVA		.51	39.530		9,325	2,425	375	12,125	14,300
4700	500 kVA		.41	48.310		18,200	2,950	455	21,605	24,900
4800	750 kVA		.32	62.110		22,100	3,800	585	26,485	30,600
4820	1,000 kVA	▼	.29	67.930		26,000	4,150	640	30,790	35,600
4850	K-4 rated, 15 kVA	2 Elec	1.01	15.810		3,550	970		4,520	5,350
4855	30 kVA		.83	19.320		5,825	1,175		7,000	8,175

For customer support on your Electrical Change Order Costs with RSMeans data, call 800.448.8182. **Pre-Installation Change Order Costs**

26 22 13 – Low-Voltage Distribution Transformers

26 22 13.10 Transformer, Dry-Type	Crew	Daily Output	Labor-Hours	Unit	Material	2020 Bare Costs Labor	Equipment	Total	Total Incl O&P
4860	45 kVA	2 Elec	.74	21.740	Ea.	7,450	1,325		8,775	10,200
4865	75 kVA	↓	.64	24.840		10,200	1,525		11,725	13,500
4870	112.5 kVA	R-3	.83	24.150		12,300	1,475	228	14,003	16,000
4875	150 kVA		.78	25.580		16,100	1,575	242	17,917	20,300
4880	225 kVA		.60	33.440		22,400	2,050	315	24,765	28,000
4885	300 kVA		.51	39.530		30,700	2,425	375	33,500	37,800
4890	500 kVA		.41	48.310		42,900	2,950	455	46,305	52,000
4900	K-13 rated, 15 kVA	2 Elec	1.01	15.810		4,025	970		4,995	5,875
4905	30 kVA		.83	19.320		5,775	1,175		6,950	8,150
4910	45 kVA		.74	21.740		6,975	1,325		8,300	9,625
4915	75 kVA	↓	.64	24.840		10,600	1,525		12,125	14,000
4920	112.5 kVA	R-3	.83	24.150		13,900	1,475	228	15,603	17,800
4925	150 kVA		.78	25.580		28,800	1,575	242	30,617	34,200
4930	225 kVA		.60	33.440		25,000	2,050	315	27,365	30,900
4935	300 kVA		.51	39.530		32,900	2,425	375	35,700	40,200
4940	500 kVA		.41	48.310	↓	56,000	2,950	455	59,405	66,500
5020	480 volt primary, 120/208 volt secondary									
5030	Nonventilated, 15 kVA	2 Elec	1.01	15.810	Ea.	3,450	970		4,420	5,225
5040	30 kVA		.74	21.740		5,000	1,325		6,325	7,475
5050	45 kVA		.64	24.840		5,900	1,525		7,425	8,750
5060	75 kVA	↓	.60	26.760		11,300	1,650		12,950	15,000
5070	112.5 kVA	R-3	.78	25.580		20,600	1,575	242	22,417	25,300
5081	150 kVA		.78	25.580		21,500	1,575	242	23,317	26,200
5090	225 kVA		.55	36.230		26,500	2,225	340	29,065	32,800
5100	300 kVA	↓	.46	43.480		23,800	2,650	410	26,860	30,600
5200	Low operating temperature (80°C), 30 kVA	2 Elec	.83	19.320		6,475	1,175		7,650	8,900
5210	45 kVA		.74	21.740		5,625	1,325		6,950	8,175
5220	75 kVA	↓	.64	24.840		8,450	1,525		9,975	11,600
5230	112.5 kVA	R-3	.83	24.150		11,200	1,475	228	12,903	14,800
5240	150 kVA		.78	25.580		14,300	1,575	242	16,117	18,400
5250	225 kVA		.60	33.440		19,500	2,050	315	21,865	24,800
5260	300 kVA		.51	39.530		32,400	2,425	375	35,200	39,600
5270	500 kVA	↓	.41	48.310	↓	35,200	2,950	455	38,605	43,600
5380	3 phase, 5 kV primary, 277/480 volt secondary									
5400	High voltage, 112.5 kVA	R-3	.78	25.580	Ea.	18,200	1,575	242	20,017	22,600
5410	150 kVA		.60	33.440		21,000	2,050	315	23,365	26,500
5420	225 kVA		.51	39.530		24,700	2,425	375	27,500	31,200
5430	300 kVA		.41	48.310		30,500	2,950	455	33,905	38,500
5440	500 kVA		.32	62.110		39,400	3,800	585	43,785	49,600
5450	750 kVA		.29	67.930		61,000	4,150	640	65,790	74,000
5460	1,000 kVA		.28	72.460		71,500	4,425	685	76,610	86,000
5470	1,500 kVA		.25	80.520		83,000	4,925	760	88,685	99,500
5480	2,000 kVA		.23	86.960		97,500	5,325	820	103,645	116,000
5490	2,500 kVA		.18	108		110,000	6,600	1,025	117,625	132,000
5500	3,000 kVA	↓	.17	120	↓	143,500	7,325	1,125	151,950	170,000
5590	15 kV primary, 277/480 volt secondary									
5600	High voltage, 112.5 kVA	R-3	.78	25.580	Ea.	28,400	1,575	242	30,217	33,800
5610	150 kVA		.60	33.440		33,200	2,050	315	35,565	39,900
5620	225 kVA		.51	39.530		36,600	2,425	375	39,400	44,300
5630	300 kVA		.41	48.310		43,100	2,950	455	46,505	52,500
5640	500 kVA		.32	62.110		54,000	3,800	585	58,385	66,000
5650	750 kVA		.29	67.930		71,000	4,150	640	75,790	85,000
5660	1,000 kVA	↓	.28	72.460	↓	80,500	4,425	685	85,610	96,000

26 22 13.10 Transformer, Dry-Type		Crew	Daily Output	Labor-Hours	Unit	Material	2020 Bare Costs Labor	Equipment	Total	Total Incl O&P
5670	1,500 kVA	R-3	.25	80.520	Ea.	93,000	4,925	760	98,685	110,000
5680	2,000 kVA		.23	86.960		103,000	5,325	820	109,145	122,000
5690	2,500 kVA		.18	108		119,500	6,600	1,025	127,125	142,000
5700	3,000 kVA		.17	120		142,000	7,325	1,125	150,450	168,500
6000	2400 volt primary, 480 volt secondary, 300 kVA		.41	48.310		30,500	2,950	455	33,905	38,500
6010	500 kVA		.32	62.110		39,400	3,800	585	43,785	49,600
6020	750 kVA		.29	67.930		56,000	4,150	640	60,790	69,000
9300	Energy efficient transformer 3 ph									
9400	15 kVA, 480VAC delta, 208Y/120VAC	R-3	1.38	14.490	Ea.	1,975	885	137	2,997	3,650
9405	30 kVA, 480VAC delta, 208Y/120VAC		1.38	14.490		2,450	885	137	3,472	4,175
9410	45 kVA, 480VAC delta, 208Y/120VAC		1.38	14.490		3,425	885	137	4,447	5,250
9415	75 kVA, 480VAC delta, 208Y/120VAC		1.15	17.390		4,500	1,075	164	5,739	6,700
9420	112 kVA, 480VAC delta, 208Y/120VAC		1.15	17.390		6,700	1,075	164	7,939	9,125
9425	150 kVA, 480VAC delta, 208Y/120VAC		1.15	17.390		15,400	1,075	164	16,639	18,800
9430	225 kVA, 480VAC delta, 208Y/120VAC		1.15	17.390		10,700	1,075	164	11,939	13,600
9435	300 kVA, 480VAC delta, 208Y/120VAC		1.15	17.390		15,000	1,075	164	16,239	18,300

26 22 13.20 Isolating Panels

		Crew	Daily Output	Labor-Hours	Unit	Material	Labor	Equipment	Total	Total Incl O&P
0010	**ISOLATING PANELS** used with isolating transformers									
0020	For hospital applications									
0100	Critical care area, 8 circuit, 3 kVA	1 Elec	.53	14.990	Ea.	7,075	920		7,995	9,150
0200	5 kVA		.50	16.100		7,325	990		8,315	9,525
0400	7.5 kVA		.48	16.720		4,900	1,025		5,925	6,925
0600	10 kVA		.40	19.760		8,075	1,200		9,275	10,700
0800	Operating room power & lighting, 8 circuit, 3 kVA		.53	14.990		5,300	920		6,220	7,200
1000	5 kVA		.50	16.100		5,775	990		6,765	7,825
1200	7.5 kVA		.48	16.720		4,050	1,025		5,075	5,975
1400	10 kVA		.40	19.760		4,950	1,200		6,150	7,250
1600	X-ray systems, 15 kVA, 90 amp		.40	19.760		15,800	1,200		17,000	19,200
1800	25 kVA, 125 amp		.33	24.150		16,600	1,475		18,075	20,500

26 22 13.30 Isolating Transformer

		Crew	Daily Output	Labor-Hours	Unit	Material	Labor	Equipment	Total	Total Incl O&P
0010	**ISOLATING TRANSFORMER**									
0100	Single phase, 120/240 volt primary, 120/240 volt secondary									
0200	0.50 kVA	1 Elec	3.68	2.170	Ea.	410	133		543	650
0400	1 kVA		1.84	4.350		945	267		1,212	1,450
0600	2 kVA		1.47	5.430		1,525	335		1,860	2,175
0800	3 kVA		1.29	6.210		860	380		1,240	1,500
1000	5 kVA		1.10	7.250		1,125	445		1,570	1,900
1200	7.5 kVA		1.01	7.910		1,425	485		1,910	2,300
1400	10 kVA		.74	10.870		1,825	665		2,490	3,025
1600	15 kVA		.55	14.490		2,350	890		3,240	3,900
1800	25 kVA		.46	17.390		3,375	1,075		4,450	5,300
1810	37.5 kVA	2 Elec	.74	21.740		5,575	1,325		6,900	8,100
1820	75 kVA	"	.60	26.760		8,375	1,650		10,025	11,700
1830	3 phase, 120/240 V primary, 120/240 V secondary, 112.5 kVA	R-3	.83	24.150		11,200	1,475	228	12,903	14,800
1840	150 kVA		.78	25.580		14,200	1,575	242	16,017	18,300
1850	225 kVA		.60	33.440		19,800	2,050	315	22,165	25,200
1860	300 kVA		.51	39.530		26,400	2,425	375	29,200	33,000
1870	500 kVA		.41	48.310		43,900	2,950	455	47,305	53,000
1880	750 kVA		.32	62.110		44,400	3,800	585	48,785	55,000

Pre-Installation Change Order Costs

26 22 Low-Voltage Transformers

26 22 13 – Low-Voltage Distribution Transformers

26 22 13.90 Transformer Handling

	26 22 13.90 Transformer Handling	Crew	Daily Output	Labor-Hours	Unit	Material	2020 Bare Costs Labor	Equipment	Total	Total Incl O&P
0010	**TRANSFORMER HANDLING** Add to normal labor cost in restricted areas									
5000	Transformers									
5150	15 kVA, approximately 200 pounds	2 Elec	2.48	6.440	Ea.		395		395	590
5160	25 kVA, approximately 300 pounds		2.30	6.960			425		425	635
5170	37.5 kVA, approximately 400 pounds		2.12	7.560			465		465	690
5180	50 kVA, approximately 500 pounds		1.84	8.700			535		535	795
5190	75 kVA, approximately 600 pounds		1.66	9.660			595		595	880
5200	100 kVA, approximately 700 pounds	↓	1.47	10.870			665		665	995
5210	112.5 kVA, approximately 800 pounds	3 Elec	2.02	11.860			730		730	1,075
5220	125 kVA, approximately 900 pounds		1.84	13.040			800		800	1,200
5230	150 kVA, approximately 1,000 pounds		1.66	14.490			890		890	1,325
5240	167 kVA, approximately 1,200 pounds		1.47	16.300			1,000		1,000	1,500
5250	200 kVA, approximately 1,400 pounds		1.29	18.630			1,150		1,150	1,700
5260	225 kVA, approximately 1,600 pounds		1.20	20.070			1,225		1,225	1,825
5270	250 kVA, approximately 1,800 pounds		1.01	23.720			1,450		1,450	2,175
5280	300 kVA, approximately 2,000 pounds		.92	26.090			1,600		1,600	2,375
5290	500 kVA, approximately 3,000 pounds		.69	34.780			2,125		2,125	3,175
5300	600 kVA, approximately 3,500 pounds		.62	38.940			2,400		2,400	3,550
5310	750 kVA, approximately 4,000 pounds		.55	43.480			2,675		2,675	3,975
5320	1,000 kVA, approximately 5,000 pounds	↓	.46	52.170	↓		3,200		3,200	4,775

26 22 16 – Low-Voltage Buck-Boost Transformers

26 22 16.10 Buck-Boost Transformer

	26 22 16.10 Buck-Boost Transformer	Crew	Daily Output	Labor-Hours	Unit	Material	2020 Bare Costs Labor	Equipment	Total	Total Incl O&P
0010	**BUCK-BOOST TRANSFORMER**									
0100	Single phase, 120/240 V primary, 12/24 V secondary									
0200	0.10 kVA	1 Elec	7.36	1.090	Ea.	100	67		167	210
0400	0.25 kVA		5.24	1.530		170	94		264	325
0600	0.50 kVA		3.68	2.170		192	133		325	410
0800	0.75 kVA		2.85	2.810		305	172		477	595
1000	1.0 kVA		1.84	4.350		258	267		525	680
1200	1.5 kVA		1.66	4.830		350	296		646	825
1400	2.0 kVA		1.47	5.430		580	335		915	1,125
1600	3.0 kVA		1.29	6.210		560	380		940	1,175
1800	5.0 kVA	↓	1.10	7.250		1,100	445		1,545	1,875
2000	3 phase, 240 V primary, 208/120 V secondary, 15 kVA	2 Elec	2.21	7.250		2,150	445		2,595	3,025
2200	30 kVA		1.47	10.870		2,800	665		3,465	4,075
2400	45 kVA		1.29	12.420		3,375	760		4,135	4,825
2600	75 kVA	↓	1.10	14.490		4,075	890		4,965	5,800
2800	112.5 kVA	R-3	1.29	15.530		5,075	950	147	6,172	7,150
3000	150 kVA		1.01	19.760		6,725	1,200	187	8,112	9,400
3200	225 kVA		.92	21.740		8,775	1,325	205	10,305	11,900
3400	300 kVA	↓	.83	24.150	↓	11,900	1,475	228	13,603	15,600

26 24 Switchboards and Panelboards

26 24 13 – Switchboards

26 24 13.10 Incoming Switchboards	Crew	Daily Output	Labor-Hours	Unit	Material	2020 Bare Costs Labor	Equipment	Total	Total Incl O&P
0010 **INCOMING SWITCHBOARDS** main service section									
0100 Aluminum bus bars, not including CT's or PT's									
0200 No main disconnect, includes CT compartment									
0300 120/208 volt, 4 wire, 600 amp	2 Elec	.92	17.390	Ea.	4,625	1,075		5,700	6,675
0400 800 amp		.81	19.760		4,625	1,200		5,825	6,875
0500 1,000 amp		.74	21.740		5,550	1,325		6,875	8,075
0600 1,200 amp		.66	24.150		5,550	1,475		7,025	8,300
0700 1,600 amp		.61	26.350		5,550	1,625		7,175	8,500
0800 2,000 amp		.57	28.050		6,000	1,725		7,725	9,150
1000 3,000 amp		.52	31.060		7,900	1,900		9,800	11,500
1200 277/480 volt, 4 wire, 600 amp		.92	17.390		4,625	1,075		5,700	6,675
1300 800 amp		.81	19.760		4,625	1,200		5,825	6,875
1400 1,000 amp		.74	21.740		5,875	1,325		7,200	8,425
1500 1,200 amp		.66	24.150		5,875	1,475		7,350	8,650
1600 1,600 amp		.61	26.350		5,875	1,625		7,500	8,850
1700 2,000 amp		.57	28.050		6,000	1,725		7,725	9,150
1800 3,000 amp		.52	31.060		7,900	1,900		9,800	11,500
1900 4,000 amp	▼	.48	33.440	▼	9,775	2,050		11,825	13,900
2000 Fused switch & CT compartment									
2100 120/208 volt, 4 wire, 400 amp	2 Elec	1.03	15.530	Ea.	2,700	955		3,655	4,400
2200 600 amp		.86	18.500		3,200	1,125		4,325	5,225
2300 800 amp		.77	20.700		12,500	1,275		13,775	15,700
2400 1,200 amp		.63	25.580		16,300	1,575		17,875	20,200
2500 277/480 volt, 4 wire, 400 amp		1.05	15.260		3,800	935		4,735	5,575
2600 600 amp		.86	18.500		5,950	1,125		7,075	8,250
2700 800 amp		.77	20.700		12,500	1,275		13,775	15,700
2800 1,200 amp	▼	.63	25.580	▼	16,300	1,575		17,875	20,200
2900 Pressure switch & CT compartment									
3000 120/208 volt, 4 wire, 800 amp	2 Elec	.74	21.740	Ea.	11,200	1,325		12,525	14,400
3100 1,200 amp		.61	26.350		21,800	1,625		23,425	26,300
3200 1,600 amp		.57	28.050		23,100	1,725		24,825	28,000
3300 2,000 amp		.52	31.060		24,600	1,900		26,500	29,800
3310 2,500 amp		.46	34.780		30,300	2,125		32,425	36,500
3320 3,000 amp		.40	39.530		40,800	2,425		43,225	48,400
3330 4,000 amp		.37	43.480		52,500	2,675		55,175	61,500
3340 120/208 volt, 4 wire, 800 amp, with ground fault		.74	21.740		18,600	1,325		19,925	22,400
3350 1,200 amp, with ground fault		.61	26.350		24,000	1,625		25,625	28,800
3360 1,600 amp, with ground fault		.57	28.050		26,100	1,725		27,825	31,300
3370 2,000 amp, with ground fault		.52	31.060		28,100	1,900		30,000	33,800
3400 277/480 volt, 4 wire, 800 amp, with ground fault		.74	21.740		18,600	1,325		19,925	22,400
3600 1,200 amp, with ground fault		.61	26.350		24,000	1,625		25,625	28,800
4000 1,600 amp, with ground fault		.57	28.050		26,100	1,725		27,825	31,300
4200 2,000 amp, with ground fault	▼	.52	31.060	▼	28,100	1,900		30,000	33,800
4400 Circuit breaker, molded case & CT compartment									
4600 3 pole, 4 wire, 600 amp	2 Elec	.86	18.500	Ea.	9,675	1,125		10,800	12,300
4800 800 amp	▼	.77	20.700	▼	11,600	1,275		12,875	14,600
5000 1,200 amp	▼	.63	25.580	▼	15,800	1,575		17,375	19,700
5100 Copper bus bars, not incl. CT's or PT's, add, minimum					15%				

26 24 13.20 In Plant Distribution Switchboards

	Crew	Daily Output	Labor-Hours	Unit	Material	Labor	Equipment	Total	Total Incl O&P
0010 **IN PLANT DISTRIBUTION SWITCHBOARDS**									
0100 Main lugs only, to 600 volt, 3 pole, 3 wire, 200 amp	2 Elec	1.10	14.490	Ea.	1,250	890		2,140	2,700
0110 400 amp	▼	1.10	14.490	▼	1,250	890		2,140	2,700

For customer support on your Electrical Change Order Costs with RSMeans data, call 800.448.8182.

Pre-Installation Change Order Costs

26 24 Switchboards and Panelboards

26 24 13 – Switchboards

26 24 13.20 In Plant Distribution Switchboards	Crew	Daily Output	Labor-Hours	Unit	Material	2020 Bare Costs Labor	Equipment	Total	Total Incl O&P
0120	600 amp	2 Elec	1.10	14.490	Ea.	1,300	890		2,190	2,750
0130	800 amp		.99	16.100		1,400	990		2,390	3,025
0140	1,200 amp		.85	18.900		1,700	1,150		2,850	3,600
0150	1,600 amp		.79	20.220		2,225	1,250		3,475	4,300
0160	2,000 amp		.75	21.210		2,475	1,300		3,775	4,675
0250	To 480 volt, 3 pole, 4 wire, 200 amp		1.10	14.490		1,100	890		1,990	2,525
0260	400 amp		1.10	14.490		1,250	890		2,140	2,700
0270	600 amp		1.10	14.490		1,375	890		2,265	2,825
0280	800 amp		.99	16.100		1,500	990		2,490	3,125
0290	1,200 amp		.85	18.900		1,825	1,150		2,975	3,750
0300	1,600 amp		.79	20.220		2,100	1,250		3,350	4,150
0310	2,000 amp		.75	21.210		2,450	1,300		3,750	4,650
0400	Main circuit breaker, to 600 volt, 3 pole, 3 wire, 200 amp		1.10	14.490		3,425	890		4,315	5,100
0410	400 amp		1.05	15.260		3,450	935		4,385	5,200
0420	600 amp		1.01	15.810		4,350	970		5,320	6,225
0430	800 amp		.96	16.720		7,275	1,025		8,300	9,525
0440	1,200 amp		.81	19.760		9,525	1,200		10,725	12,300
0450	1,600 amp		.77	20.700		15,300	1,275		16,575	18,700
0460	2,000 amp		.74	21.740		16,300	1,325		17,625	19,900
0550	277/480 volt, 3 pole, 4 wire, 200 amp		1.10	14.490		3,625	890		4,515	5,300
0560	400 amp		1.05	15.260		3,625	935		4,560	5,375
0570	600 amp		1.01	15.810		4,525	970		5,495	6,425
0580	800 amp		.96	16.720		7,650	1,025		8,675	9,950
0590	1,200 amp		.81	19.760		9,875	1,200		11,075	12,700
0600	1,600 amp		.77	20.700		15,300	1,275		16,575	18,800
0610	2,000 amp		.74	21.740		16,400	1,325		17,725	20,000
0700	Main fusible switch w/fuse, 208/240 volt, 3 pole, 3 wire, 200 amp		1.10	14.490		3,775	890		4,665	5,500
0710	400 amp		1.05	15.260		3,750	935		4,685	5,525
0720	600 amp		1.01	15.810		4,600	970		5,570	6,525
0730	800 amp		.96	16.720		9,425	1,025		10,450	11,900
0740	1,200 amp		.81	19.760		11,000	1,200		12,200	13,900
0800	120/208, 120/240 volt, 3 pole, 4 wire, 200 amp		1.10	14.490		3,375	890		4,265	5,025
0810	400 amp		1.05	15.260		3,450	935		4,385	5,200
0820	600 amp		1.01	15.810		4,400	970		5,370	6,275
0830	800 amp		.96	16.720		6,975	1,025		8,000	9,200
0840	1,200 amp		.81	19.760		8,125	1,200		9,325	10,700
0900	480 or 600 volt, 3 pole, 3 wire, 200 amp		1.10	14.490		3,650	890		4,540	5,325
0910	400 amp		1.05	15.260		3,575	935		4,510	5,350
0920	600 amp		1.01	15.810		4,375	970		5,345	6,275
0930	800 amp		.96	16.720		6,750	1,025		7,775	8,950
0940	1,200 amp		.81	19.760		7,825	1,200		9,025	10,400
1000	277 or 480 volt, 3 pole, 4 wire, 200 amp		1.10	14.490		3,775	890		4,665	5,500
1010	400 amp		1.05	15.260		3,725	935		4,660	5,500
1020	600 amp		1.01	15.810		4,575	970		5,545	6,475
1030	800 amp		.96	16.720		7,000	1,025		8,025	9,225
1040	1,200 amp		.81	19.760		8,125	1,200		9,325	10,700
1120	1,600 amp		.70	22.880		14,800	1,400		16,200	18,400
1130	2,000 amp		.63	25.580		19,500	1,575		21,075	23,800
1150	Pressure switch, bolted, 3 pole, 208/240 volt, 3 wire, 800 amp		.88	18.120		11,000	1,100		12,100	13,800
1160	1,200 amp		.74	21.740		14,000	1,325		15,325	17,400
1170	1,600 amp		.70	22.880		16,000	1,400		17,400	19,700
1180	2,000 amp		.63	25.580		18,200	1,575		19,775	22,400
1200	120/208 or 120/240 volt, 3 pole, 4 wire, 800 amp		.88	18.120		8,675	1,100		9,775	11,200

26 24 13.20 In Plant Distribution Switchboards	Crew	Daily Output	Labor-Hours	Unit	Material	2020 Bare Costs Labor	Equipment	Total	Total Incl O&P
1210	1,200 amp	2 Elec	.74	21.740	Ea.	10,100	1,325		11,425	13,200
1220	1,600 amp		.70	22.880		16,000	1,400		17,400	19,700
1230	2,000 amp		.63	25.580		18,200	1,575		19,775	22,400
1300	480 or 600 volt, 3 wire, 800 amp		.88	18.120		11,000	1,100		12,100	13,800
1310	1,200 amp		.74	21.740		15,400	1,325		16,725	18,900
1320	1,600 amp		.70	22.880		17,300	1,400		18,700	21,100
1330	2,000 amp		.63	25.580		19,500	1,575		21,075	23,800
1400	277/480 volt, 4 wire, 800 amp		.88	18.120		11,000	1,100		12,100	13,800
1410	1,200 amp		.74	21.740		15,400	1,325		16,725	18,900
1420	1,600 amp		.70	22.880		17,300	1,400		18,700	21,100
1430	2,000 amp		.63	25.580		19,500	1,575		21,075	23,800
1500	Main ground fault protector, 1,200-2,000 amp		4.97	3.220		3,225	198		3,423	3,850
1600	Busway connection, 200 amp		4.97	3.220		535	198		733	885
1610	400 amp		4.23	3.780		535	232		767	935
1620	600 amp		3.68	4.350		535	267		802	985
1630	800 amp		2.94	5.430		535	335		870	1,075
1640	1,200 amp		2.39	6.690		535	410		945	1,200
1650	1,600 amp		2.21	7.250		1,100	445		1,545	1,875
1660	2,000 amp		1.84	8.700		1,100	535		1,635	2,025
1700	Shunt trip for remote operation, 200 amp		7.36	2.170		605	133		738	870
1710	400 amp		7.36	2.170		970	133		1,103	1,275
1720	600 amp		7.36	2.170		1,150	133		1,283	1,450
1730	800 amp		7.36	2.170		1,475	133		1,608	1,825
1740	1,200-2,000 amp		7.36	2.170		3,300	133		3,433	3,825
1800	Motor operated main breaker, 200 amp		7.36	2.170		3,625	133		3,758	4,175
1810	400 amp		7.36	2.170		3,650	133		3,783	4,225
1820	600 amp		7.36	2.170		3,375	133		3,508	3,925
1830	800 amp		7.36	2.170		3,400	133		3,533	3,950
1840	1,200-2,000 amp		7.36	2.170		3,175	133		3,308	3,675
1900	Current/potential transformer metering compartment, 200-800 amp		4.97	3.220		2,150	198		2,348	2,675
1940	1,200 amp		4.97	3.220		6,000	198		6,198	6,900
1950	1,600-2,000 amp		4.97	3.220		9,875	198		10,073	11,200
2000	With watt meter, 200-800 amp		3.68	4.350		9,275	267		9,542	10,600
2040	1,200 amp		3.68	4.350		10,800	267		11,067	12,200
2050	1,600-2,000 amp	▼	3.68	4.350		11,500	267		11,767	13,000
2100	Split bus, 60-200 amp	1 Elec	4.88	1.640		194	101		295	365
2130	400 amp	2 Elec	4.23	3.780		320	232		552	695
2140	600 amp		3.31	4.830		390	296		686	865
2150	800 amp		2.39	6.690		495	410		905	1,150
2170	1,200 amp	▼	1.84	8.700		565	535		1,100	1,425
2250	Contactor control, 60 amp	1 Elec	1.84	4.350		1,275	267		1,542	1,800
2260	100 amp		1.38	5.800		1,350	355		1,705	2,000
2270	200 amp	▼	.92	8.700		2,025	535		2,560	3,050
2280	400 amp	2 Elec	.92	17.390		6,550	1,075		7,625	8,800
2290	600 amp		.77	20.700		7,325	1,275		8,600	9,950
2300	800 amp		.66	24.150		8,200	1,475		9,675	11,200
2500	Modifier, two distribution sections, add		.74	21.740		2,850	1,325		4,175	5,100
2520	Three distribution sections, add		.37	43.480		5,475	2,675		8,150	10,000
2560	Auxiliary pull section, 20", add		1.84	8.700		1,025	535		1,560	1,925
2580	24", add		1.66	9.660		1,025	595		1,620	2,000
2600	30", add		1.47	10.870		1,025	665		1,690	2,125
2620	36", add		1.29	12.420		1,225	760		1,985	2,475
2640	Dog house, 12", add	▼	2.21	7.250	▼	212	445		657	895

26 24 Switchboards and Panelboards

26 24 13 – Switchboards

26 24 13.20 In Plant Distribution Switchboards

		Crew	Daily Output	Labor-Hours	Unit	Material	2020 Bare Costs Labor	Equipment	Total	Total Incl O&P
2660	18", add	2 Elec	1.84	8.700	Ea.	420	535		955	1,250
3000	Transition section between switchboard and transformer									
3050	or motor control center, 4 wire alum. bus, 600 amp	2 Elec	1.05	15.260	Ea.	2,175	935		3,110	3,800
3100	800 amp		.92	17.390		2,450	1,075		3,525	4,275
3150	1,000 amp		.81	19.760		2,750	1,200		3,950	4,825
3200	1,200 amp		.74	21.740		3,025	1,325		4,350	5,300
3250	1,600 amp		.66	24.150		3,575	1,475		5,050	6,125
3300	2,000 amp		.61	26.350		4,100	1,625		5,725	6,900
3350	2,500 amp		.57	28.050		4,775	1,725		6,500	7,800
3400	3,000 amp		.52	31.060		5,500	1,900		7,400	8,875
4000	Weatherproof construction, per vertical section	↓	1.62	9.880	↓	2,525	605		3,130	3,675

26 24 13.30 Distribution Switchboards Section

		Crew	Daily Output	Labor-Hours	Unit	Material	2020 Bare Costs Labor	Equipment	Total	Total Incl O&P
0010	**DISTRIBUTION SWITCHBOARDS SECTION**									
0100	Aluminum bus bars, not including breakers									
0160	Subfeed lug-rated at 60 amp	2 Elec	1.20	13.380	Ea.	1,200	820		2,020	2,550
0170	100 amp		1.16	13.800		1,525	845		2,370	2,925
0180	200 amp		1.10	14.490		1,450	890		2,340	2,925
0190	400 amp		1.01	15.810		1,425	970		2,395	3,025
0195	120/208 or 277/480 volt, 4 wire, 400 amp		1.01	15.810		1,425	970		2,395	3,025
0200	600 amp		.92	17.390		1,700	1,075		2,775	3,475
0300	800 amp		.81	19.760		2,175	1,200		3,375	4,200
0400	1,000 amp		.74	21.740		2,500	1,325		3,825	4,725
0500	1,200 amp		.66	24.150		2,950	1,475		4,425	5,450
0600	1,600 amp		.61	26.350		4,400	1,625		6,025	7,250
0700	2,000 amp		.57	28.050		5,500	1,725		7,225	8,600
0800	2,500 amp		.55	28.990		5,950	1,775		7,725	9,200
0900	3,000 amp		.52	31.060		7,850	1,900		9,750	11,500
0950	4,000 amp	↓	.48	33.440	↓	7,875	2,050		9,925	11,700

26 24 13.40 Switchboards Feeder Section

		Crew	Daily Output	Labor-Hours	Unit	Material	2020 Bare Costs Labor	Equipment	Total	Total Incl O&P
0010	**SWITCHBOARDS FEEDER SECTION** group mounted devices									
0030	Circuit breakers									
0160	FA frame, 15 to 60 amp, 240 volt, 1 pole	1 Elec	7.36	1.090	Ea.	130	67		197	243
0170	2 pole		6.44	1.240		380	76		456	535
0180	3 pole		4.88	1.640		580	101		681	790
0210	480 volt, 1 pole		7.36	1.090		181	67		248	299
0220	2 pole		6.44	1.240		455	76		531	620
0230	3 pole		4.88	1.640		585	101		686	795
0260	600 volt, 2 pole		6.44	1.240		885	76		961	1,100
0270	3 pole		4.88	1.640		535	101		636	735
0280	FA frame, 70 to 100 amp, 240 volt, 1 pole		6.44	1.240		229	76		305	365
0310	2 pole		4.60	1.740		615	107		722	835
0320	3 pole		3.68	2.170		595	133		728	855
0330	480 volt, 1 pole		6.44	1.240		345	76		421	495
0360	2 pole		4.60	1.740		585	107		692	805
0370	3 pole		3.68	2.170		690	133		823	960
0380	600 volt, 2 pole		4.60	1.740		665	107		772	890
0410	3 pole		3.68	2.170		830	133		963	1,125
0420	KA frame, 70 to 225 amp		2.94	2.720		1,525	167		1,692	1,925
0430	LA frame, 125 to 400 amp		2.12	3.780		3,900	232		4,132	4,650
0460	MA frame, 450 to 600 amp		1.47	5.430		6,200	335		6,535	7,325
0470	700 to 800 amp		1.20	6.690		8,075	410		8,485	9,475
0480	MAL frame, 1,000 amp	↓	.92	8.700	↓	8,375	535		8,910	10,000

26 24 13.40 Switchboards Feeder Section	Crew	Daily Output	Labor-Hours	Unit	Material	2020 Bare Costs Labor	Equipment	Total	Total Incl O&P
0490	PA frame, 1,200 amp	1 Elec	.74	10.870	Ea.	17,000	665		17,665	19,700
0500	Branch circuit, fusible switch, 600 volt, double 30/30 amp		3.68	2.170		855	133		988	1,150
0550	60/60 amp		2.94	2.720		875	167		1,042	1,225
0600	100/100 amp		2.48	3.220		1,100	198		1,298	1,525
0650	Single, 30 amp		4.88	1.640		790	101		891	1,025
0700	60 amp		4.32	1.850		815	114		929	1,075
0750	100 amp		3.68	2.170		1,300	133		1,433	1,650
0800	200 amp		2.48	3.220		1,350	198		1,548	1,775
0850	400 amp		2.12	3.780		2,475	232		2,707	3,075
0900	600 amp		1.66	4.830		3,025	296		3,321	3,775
0950	800 amp		1.20	6.690		5,075	410		5,485	6,175
1000	1,200 amp		.74	10.870		5,800	665		6,465	7,400
1080	Branch circuit, circuit breakers, high interrupting capacity									
1100	60 amp, 240, 480 or 600 volt, 1 pole	1 Elec	7.36	1.090	Ea.	535	67		602	690
1120	2 pole		6.44	1.240		1,325	76		1,401	1,575
1140	3 pole		4.88	1.640		660	101		761	875
1150	100 amp, 240, 480 or 600 volt, 1 pole		6.44	1.240		595	76		671	770
1160	2 pole		4.60	1.740		1,475	107		1,582	1,775
1180	3 pole		3.68	2.170		705	133		838	980
1200	225 amp, 240, 480 or 600 volt, 2 pole		3.22	2.480		2,275	152		2,427	2,725
1220	3 pole		2.94	2.720		2,200	167		2,367	2,675
1240	400 amp, 240, 480 or 600 volt, 2 pole		2.30	3.480		3,250	214		3,464	3,900
1260	3 pole		2.12	3.780		2,975	232		3,207	3,625
1280	600 amp, 240, 480 or 600 volt, 2 pole		1.66	4.830		5,275	296		5,571	6,250
1300	3 pole		1.47	5.430		3,625	335		3,960	4,500
1320	800 amp, 240, 480 or 600 volt, 2 pole		1.38	5.800		4,650	355		5,005	5,650
1340	3 pole		1.20	6.690		5,450	410		5,860	6,575
1360	1,000 amp, 240, 480 or 600 volt, 2 pole		1.01	7.910		7,325	485		7,810	8,800
1380	3 pole		.92	8.700		6,050	535		6,585	7,450
1400	1,200 amp, 240, 480 or 600 volt, 2 pole		.83	9.660		7,425	595		8,020	9,050
1420	3 pole		.74	10.870		7,675	665		8,340	9,425
1700	Fusible switch, 240 V, 60 amp, 2 pole		2.94	2.720		415	167		582	705
1720	3 pole		2.76	2.900		570	178		748	890
1740	100 amp, 2 pole		2.48	3.220		420	198		618	760
1760	3 pole		2.30	3.480		640	214		854	1,025
1780	200 amp, 2 pole		1.84	4.350		880	267		1,147	1,375
1800	3 pole		1.75	4.580		1,025	281		1,306	1,550
1820	400 amp, 2 pole		1.38	5.800		1,650	355		2,005	2,350
1840	3 pole		1.20	6.690		2,075	410		2,485	2,875
1860	600 amp, 2 pole		.92	8.700		2,350	535		2,885	3,375
1880	3 pole		.83	9.660		2,875	595		3,470	4,050
1900	240-600 V, 800 amp, 2 pole		.64	12.420		5,225	760		5,985	6,875
1920	3 pole		.55	14.490		6,400	890		7,290	8,375
2000	600 V, 60 amp, 2 pole		2.94	2.720		630	167		797	940
2040	100 amp, 2 pole		2.48	3.220		645	198		843	1,000
2080	200 amp, 2 pole		1.84	4.350		1,125	267		1,392	1,625
2120	400 amp, 2 pole		1.38	5.800		2,225	355		2,580	2,975
2160	600 amp, 2 pole		.92	8.700		2,700	535		3,235	3,775
2500	Branch circuit, circuit breakers, 60 amp, 600 volt, 3 pole		4.88	1.640		465	101		566	660
2520	240, 480 or 600 volt, 1 pole		7.36	1.090		97	67		164	207
2540	240 volt, 2 pole		6.44	1.240		173	76		249	305
2560	480 or 600 volt, 2 pole		6.44	1.240		355	76		431	505
2580	240 volt, 3 pole		4.88	1.640		410	101		511	600

26 24 13 – Switchboards

26 24 13.40 Switchboards Feeder Section	Crew	Daily Output	Labor-Hours	Unit	Material	2020 Bare Costs Labor	Equipment	Total	Total Incl O&P
2600	480 volt, 3 pole	1 Elec	4.88	1.640	Ea.	470	101		571	670
2620	100 amp, 600 volt, 2 pole		4.60	1.740		450	107		557	655
2640	3 pole		3.68	2.170		570	133		703	825
2660	480 volt, 2 pole		4.60	1.740		390	107		497	590
2680	240 volt, 2 pole		4.60	1.740		222	107		329	405
2700	3 pole		3.68	2.170		355	133		488	595
2720	480 volt, 3 pole		3.68	2.170		525	133		658	775
2740	225 amp, 240, 480 or 600 volt, 2 pole		3.22	2.480		650	152		802	940
2760	3 pole		2.94	2.720		680	167		847	1,000
2780	400 amp, 240, 480 or 600 volt, 2 pole		2.30	3.480		1,375	214		1,589	1,825
2800	3 pole		2.12	3.780		1,575	232		1,807	2,100
2820	600 amp, 240 or 480 volt, 2 pole		1.66	4.830		2,225	296		2,521	2,900
2840	3 pole		1.47	5.430		2,725	335		3,060	3,500
2860	800 amp, 240, 480 or 600 volt, 2 pole		1.38	5.800		3,350	355		3,705	4,200
2880	3 pole		1.20	6.690		3,900	410		4,310	4,900
2900	1,000 amp, 240, 480 or 600 volt, 2 pole		1.01	7.910		4,100	485		4,585	5,250
2920	480 or 600 volt, 3 pole		.92	8.700		4,700	535		5,235	5,975
2940	1,200 amp, 240, 480 or 600 volt, 2 pole		.83	9.660		5,775	595		6,370	7,225
2960	3 pole		.74	10.870		6,300	665		6,965	7,950
2980	600 volt, 3 pole		.74	10.870		6,500	665		7,165	8,150

26 24 16 – Panelboards

26 24 16.10 Load Centers

		Crew	Daily Output	Labor-Hours	Unit	Material	2020 Bare Costs Labor	Equipment	Total	Total Incl O&P
0010	**LOAD CENTERS** (residential type)									
0100	3 wire, 120/240 V, 1 phase, including 1 pole plug-in breakers									
0200	100 amp main lugs, indoor, 8 circuits	1 Elec	1.29	6.210	Ea.	101	380		481	675
0300	12 circuits		1.10	7.250		127	445		572	800
0400	Rainproof, 8 circuits		1.29	6.210		130	380		510	710
0500	12 circuits		1.10	7.250		156	445		601	830
0600	200 amp main lugs, indoor, 16 circuits	R-1A	1.66	9.660		218	535		753	1,025
0700	20 circuits		1.38	11.590		208	640		848	1,175
0800	24 circuits		1.20	13.380		266	740		1,006	1,400
0900	30 circuits		1.10	14.490		285	800		1,085	1,525
1000	40 circuits		.74	21.740		435	1,200		1,635	2,275
1200	Rainproof, 16 circuits		1.66	9.660		250	535		785	1,075
1300	20 circuits		1.38	11.590		325	640		965	1,325
1400	24 circuits		1.20	13.380		340	740		1,080	1,475
1500	30 circuits		1.10	14.490		380	800		1,180	1,625
1600	40 circuits		.74	21.740		490	1,200		1,690	2,350
1800	400 amp main lugs, indoor, 42 circuits		.66	24.150		1,200	1,325		2,525	3,300
1900	Rainproof, 42 circuits		.66	24.150		1,625	1,325		2,950	3,750
2200	Plug in breakers, 20 amp, 1 pole, 4 wire, 120/208 volts									
2210	125 amp main lugs, indoor, 12 circuits	1 Elec	1.10	7.250	Ea.	194	445		639	875
2300	18 circuits		.74	10.870		261	665		926	1,275
2400	Rainproof, 12 circuits		1.10	7.250		215	445		660	895
2500	18 circuits		.74	10.870		315	665		980	1,350
2600	200 amp main lugs, indoor, 24 circuits	R-1A	1.20	13.380		380	740		1,120	1,525
2700	30 circuits		1.10	14.490		370	800		1,170	1,600
2800	36 circuits		.92	17.390		380	960		1,340	1,850
2900	42 circuits		.74	21.740		540	1,200		1,740	2,400
3000	Rainproof, 24 circuits		1.20	13.380		287	740		1,027	1,425
3100	30 circuits		1.10	14.490		305	800		1,105	1,525
3200	36 circuits		.92	17.390		465	960		1,425	1,950

26 24 16 – Panelboards

26 24 16.10 Load Centers

		Crew	Daily Output	Labor-Hours	Unit	Material	2020 Bare Costs Labor	Equipment	Total	Total Incl O&P
3300	42 circuits	R-1A	.74	21.740	Ea.	615	1,200		1,815	2,475
3500	400 amp main lugs, indoor, 42 circuits		.66	24.150		1,250	1,325		2,575	3,350
3600	Rainproof, 42 circuits		.66	24.150		1,575	1,325		2,900	3,700
3700	Plug-in breakers, 20 amp, 1 pole, 3 wire, 120/240 volts									
3800	100 amp main breaker, indoor, 12 circuits	1 Elec	1.10	7.250	Ea.	159	445		604	835
3900	18 circuits	"	.74	10.870		197	665		862	1,200
4000	200 amp main breaker, indoor, 20 circuits	R-1A	1.38	11.590		279	640		919	1,250
4200	24 circuits		1.20	13.380		305	740		1,045	1,425
4300	30 circuits		1.10	14.490		330	800		1,130	1,550
4400	40 circuits		.83	19.320		390	1,075		1,465	2,025
4500	Rainproof, 20 circuits		1.38	11.590		335	640		975	1,325
4600	24 circuits		1.20	13.380		390	740		1,130	1,525
4700	30 circuits		1.10	14.490		440	800		1,240	1,675
4800	40 circuits		.83	19.320		560	1,075		1,635	2,225
5000	400 amp main breaker, indoor, 42 circuits		.66	24.150		2,975	1,325		4,300	5,250
5100	Rainproof, 42 circuits		.66	24.150		3,150	1,325		4,475	5,425
5300	Plug in breakers, 20 amp, 1 pole, 4 wire, 120/208 volts									
5400	200 amp main breaker, indoor, 30 circuits	R-1A	1.10	14.490	Ea.	420	800		1,220	1,650
5500	42 circuits		.74	21.740		490	1,200		1,690	2,350
5600	Rainproof, 30 circuits		1.10	14.490		895	800		1,695	2,175
5700	42 circuits		.74	21.740		575	1,200		1,775	2,425

26 24 16.20 Panelboard and Load Center Circuit Breakers

		Crew	Daily Output	Labor-Hours	Unit	Material	2020 Bare Costs Labor	Equipment	Total	Total Incl O&P
0010	**PANELBOARD AND LOAD CENTER CIRCUIT BREAKERS**									
0050	Bolt-on, 10,000 amp I.C., 120 volt, 1 pole									
0100	15-50 amp	1 Elec	9.20	.870	Ea.	18.60	53.50		72.10	100
0200	60 amp		7.36	1.090		20.50	67		87.50	122
0300	70 amp		7.36	1.090		28	67		95	130
0350	240 volt, 2 pole									
0400	15-50 amp	1 Elec	7.36	1.090	Ea.	56.50	67		123.50	162
0500	60 amp		6.90	1.160		37.50	71		108.50	147
0600	80-100 amp		4.60	1.740		111	107		218	281
0700	3 pole, 15-60 amp		5.70	1.400		135	86		221	277
0800	70 amp		4.60	1.740		172	107		279	350
0900	80-100 amp		3.31	2.420		205	148		353	445
1000	22,000 amp I.C., 240 volt, 2 pole, 70-225 amp		2.48	3.220		555	198		753	905
1100	3 pole, 70-225 amp		2.12	3.780		345	232		577	720
1200	14,000 amp I.C., 277 volts, 1 pole, 15-30 amp		7.36	1.090		37	67		104	141
1300	22,000 amp I.C., 480 volts, 2 pole, 70-225 amp		2.48	3.220		525	198		723	870
1400	3 pole, 70-225 amp		2.12	3.780		1,050	232		1,282	1,500
2000	Plug-in panel or load center, 120/240 volt, to 60 amp, 1 pole		11.04	.725		6.40	44.50		50.90	73
2010	2 pole		8.28	.966		25.50	59.50		85	117
2020	3 pole		6.90	1.160		97.50	71		168.50	213
2030	100 amp, 2 pole		5.52	1.450		105	89		194	248
2040	3 pole		4.14	1.930		118	118		236	305
2050	150-200 amp, 2 pole		2.76	2.900		256	178		434	545
2060	Plug-in tandem, 120/240 V, 2-15 A, 1 pole		10.12	.791		30	48.50		78.50	105
2070	1-15 A & 1-20 A		10.12	.791		18.35	48.50		66.85	92
2080	2-20 A		10.12	.791		18.40	48.50		66.90	92
2082	Arc fault circuit interrupter, 120/240 V, 1-15 A & 1-20 A, 1 pole		10.12	.791		57	48.50		105.50	135
2100	High interrupting capacity, 120/240 volt, plug-in, 30 amp, 1 pole		11.04	.725		24	44.50		68.50	92.50
2110	60 amp, 2 pole		8.28	.966		26	59.50		85.50	117
2120	3 pole		6.90	1.160		257	71		328	390

For customer support on your Electrical Change Order Costs with RSMeans data, call 800.448.8182.

Pre-Installation Change Order Costs

26 24 16.20 Panelboard and Load Center Circuit Breakers	Crew	Daily Output	Labor-Hours	Unit	Material	2020 Bare Costs Labor	Equipment	Total	Total Incl O&P
2130	100 amp, 2 pole	1 Elec	5.52	1.450	Ea.	138	89		227	284
2140	3 pole		4.14	1.930		430	118		548	645
2150	125 amp, 2 pole		2.76	2.900		870	178		1,048	1,225
2200	Bolt-on, 30 amp, 1 pole		9.20	.870		61.50	53.50		115	147
2210	60 amp, 2 pole		6.90	1.160		103	71		174	220
2220	3 pole		5.70	1.400		271	86		357	425
2230	100 amp, 2 pole		4.60	1.740		293	107		400	480
2240	3 pole		3.31	2.420		470	148		618	740
2300	Ground fault, 240 volt, 30 amp, 1 pole		6.44	1.240		92.50	76		168.50	215
2310	2 pole		5.52	1.450		170	89		259	320
2350	Key operated, 240 volt, 1 pole, 30 amp		6.44	1.240		135	76		211	262
2360	Switched neutral, 240 volt, 30 amp, 2 pole		5.52	1.450		40	89		129	176
2370	3 pole		5.06	1.580		65.50	97		162.50	216
2400	Shunt trip, for 240 volt breaker, 60 amp, 1 pole		3.68	2.170		81.50	133		214.50	288
2410	2 pole		3.22	2.480		81.50	152		233.50	315
2420	3 pole		2.76	2.900		81.50	178		259.50	355
2430	100 amp, 2 pole		2.76	2.900		81.50	178		259.50	355
2440	3 pole		2.30	3.480		81.50	214		295.50	410
2450	150 amp, 2 pole		1.84	4.350		235	267		502	655
2500	Auxiliary switch, for 240 volt breaker, 60 amp, 1 pole		3.68	2.170		90	133		223	297
2510	2 pole		3.22	2.480		90	152		242	325
2520	3 pole		2.76	2.900		90	178		268	365
2530	100 amp, 2 pole		2.76	2.900		90	178		268	365
2540	3 pole		2.30	3.480		90	214		304	420
2550	150 amp, 2 pole		1.84	4.350		123	267		390	530
2600	Panel or load center, 277/480 volt, plug-in, 30 amp, 1 pole		11.04	.725		71	44.50		115.50	145
2610	60 amp, 2 pole		8.28	.966		218	59.50		277.50	330
2620	3 pole		6.90	1.160		320	71		391	455
2650	Bolt-on, 60 amp, 2 pole		6.90	1.160		218	71		289	345
2660	3 pole		5.70	1.400		320	86		406	480
2700	I-line, 277/480 volt, 30 amp, 1 pole		7.36	1.090		68.50	67		135.50	175
2710	60 amp, 2 pole		6.90	1.160		241	71		312	370
2720	3 pole		5.70	1.400		375	86		461	540
2730	100 amp, 1 pole		6.90	1.160		161	71		232	283
2740	2 pole		4.60	1.740		375	107		482	570
2750	3 pole		3.22	2.480		440	152		592	710
2800	High interrupting capacity, 277/480 volt, plug-in, 30 amp, 1 pole		11.04	.725		420	44.50		464.50	525
2810	60 amp, 2 pole		8.28	.966		650	59.50		709.50	805
2820	3 pole		6.44	1.240		755	76		831	945
2830	Bolt-on, 30 amp, 1 pole		7.36	1.090		420	67		487	560
2840	60 amp, 2 pole		6.90	1.160		570	71		641	730
2850	3 pole		5.70	1.400		635	86		721	830
2900	I-line, 30 amp, 1 pole		7.36	1.090		420	67		487	560
2910	60 amp, 2 pole		6.90	1.160		585	71		656	750
2920	3 pole		5.70	1.400		665	86		751	860
2930	100 amp, 1 pole		6.90	1.160		680	71		751	850
2940	2 pole		4.60	1.740		475	107		582	685
2950	3 pole		3.31	2.420		735	148		883	1,025
2960	Shunt trip, 277/480 volt breaker, remote oper., 30 amp, 1 pole		3.68	2.170		440	133		573	685
2970	60 amp, 2 pole		3.22	2.480		645	152		797	935
2980	3 pole		2.76	2.900		730	178		908	1,075
2990	100 amp, 1 pole		3.22	2.480		525	152		677	805
3000	2 pole		2.76	2.900		730	178		908	1,075

Pre-Installation Change Order Costs   For customer support on your Electrical Change Order Costs with RSMeans data, call 800.448.8182.

26 24 16.20 Panelboard and Load Center Circuit Breakers	Crew	Daily Output	Labor-Hours	Unit	Material	2020 Bare Costs Labor	Equipment	Total	Total Incl O&P
3010	3 pole	1 Elec	2.30	3.480	Ea.	670	214		884	1,050
3050	Under voltage trip, 277/480 volt breaker, 30 amp, 1 pole		3.68	2.170		440	133		573	685
3060	60 amp, 2 pole		3.22	2.480		645	152		797	935
3070	3 pole		2.76	2.900		730	178		908	1,075
3080	100 amp, 1 pole		3.22	2.480		525	152		677	805
3090	2 pole		2.76	2.900		730	178		908	1,075
3100	3 pole		2.30	3.480		670	214		884	1,050
3150	Motor operated, 277/480 volt breaker, 30 amp, 1 pole		3.68	2.170		785	133		918	1,050
3160	60 amp, 2 pole		3.22	2.480		820	152		972	1,125
3170	3 pole		2.76	2.900		1,100	178		1,278	1,475
3180	100 amp, 1 pole		3.22	2.480		875	152		1,027	1,200
3190	2 pole		2.76	2.900		1,100	178		1,278	1,475
3200	3 pole		2.30	3.480		950	214		1,164	1,375
3250	Panelboard spacers, per pole		36.80	.217		3.79	13.35		17.14	24
5110	NEMA 1 enclosure only, 600V, 3 p, 14k AIC, 100A	▼	2.02	3.950	▼	160	242		402	535

26 24 16.30 Panelboards Commercial Applications

		Crew	Daily Output	Labor-Hours	Unit	Material	2020 Bare Costs Labor	Equipment	Total	Total Incl O&P
0010	**PANELBOARDS COMMERCIAL APPLICATIONS**									
0050	NQOD, w/20 amp 1 pole bolt-on circuit breakers									
0100	3 wire, 120/240 volts, 100 amp main lugs									
0150	10 circuits	1 Elec	.92	8.700	Ea.	935	535		1,470	1,825
0200	14 circuits		.81	9.880		1,050	605		1,655	2,050
0250	18 circuits		.69	11.590		1,150	710		1,860	2,300
0300	20 circuits	▼	.60	13.380		1,275	820		2,095	2,650
0350	225 amp main lugs, 24 circuits	2 Elec	1.10	14.490		1,450	890		2,340	2,925
0400	30 circuits		.83	19.320		1,675	1,175		2,850	3,625
0450	36 circuits		.74	21.740		1,925	1,325		3,250	4,075
0500	38 circuits		.66	24.150		2,075	1,475		3,550	4,500
0550	42 circuits	▼	.61	26.350		2,150	1,625		3,775	4,750
0600	4 wire, 120/208 volts, 100 amp main lugs, 12 circuits	1 Elec	.92	8.700		1,025	535		1,560	1,925
0650	16 circuits		.69	11.590		1,175	710		1,885	2,325
0700	20 circuits		.60	13.380		1,350	820		2,170	2,700
0750	24 circuits		.55	14.490		1,275	890		2,165	2,725
0800	30 circuits	▼	.49	16.410		1,675	1,000		2,675	3,350
0850	225 amp main lugs, 32 circuits	2 Elec	.83	19.320		1,875	1,175		3,050	3,850
0900	34 circuits		.77	20.700		1,925	1,275		3,200	4,000
0950	36 circuits		.74	21.740		1,975	1,325		3,300	4,125
1000	42 circuits		.63	25.580		2,200	1,575		3,775	4,750
1010	400 amp main lugs, 42 circs		.63	25.580		2,200	1,575		3,775	4,750
1040	225 amp main lugs, NEMA 7, 12 circuits		.92	17.390		5,925	1,075		7,000	8,125
1100	24 circuits	▼	.37	43.480	▼	7,175	2,675		9,850	11,900
1200	NEHB, w/20 amp, 1 pole bolt-on circuit breakers									
1250	4 wire, 277/480 volts, 100 amp main lugs, 12 circuits	1 Elec	.81	9.880	Ea.	1,475	605		2,080	2,525
1300	20 circuits	"	.55	14.490		2,200	890		3,090	3,750
1350	225 amp main lugs, 24 circuits	2 Elec	.83	19.320		2,500	1,175		3,675	4,525
1400	30 circuits		.74	21.740		2,975	1,325		4,300	5,250
1448	32 circuits		4.51	3.550		3,475	218		3,693	4,125
1450	36 circuits		.66	24.150		3,475	1,475		4,950	6,000
1500	42 circuits		.55	28.990		3,950	1,775		5,725	7,000
1510	225 amp main lugs, NEMA 7, 12 circuits		.83	19.320		6,175	1,175		7,350	8,550
1590	24 circuits	▼	.28	57.970	▼	8,275	3,550		11,825	14,400
1600	NQOD panel, w/20 amp, 1 pole, circuit breakers									
1650	3 wire, 120/240 volt with main circuit breaker									

26 24 16.30 Panelboards Commercial Applications		Crew	Daily Output	Labor-Hours	Unit	Material	2020 Bare Costs Labor	Equipment	Total	Total Incl O&P
1700	100 amp main, 12 circuits	1 Elec	.74	10.870	Ea.	1,275	665		1,940	2,400
1750	20 circuits	"	.55	14.490		1,625	890		2,515	3,100
1800	225 amp main, 30 circuits	2 Elec	.63	25.580		3,050	1,575		4,625	5,675
1801	225 amp main, 32 circuits		4.60	3.480		3,050	214		3,264	3,675
1850	42 circuits		.48	33.440		3,525	2,050		5,575	6,925
1900	400 amp main, 30 circuits		.50	32.210		4,225	1,975		6,200	7,600
1950	42 circuits		.46	34.780		4,725	2,125		6,850	8,375
2000	4 wire, 120/208 volts with main circuit breaker									
2050	100 amp main, 24 circuits	1 Elec	.43	18.500	Ea.	1,825	1,125		2,950	3,725
2100	30 circuits	"	.37	21.740		2,125	1,325		3,450	4,325
2200	225 amp main, 32 circuits	2 Elec	.66	24.150		3,550	1,475		5,025	6,125
2250	42 circuits		.52	31.060		4,125	1,900		6,025	7,350
2300	400 amp main, 42 circuits		.44	36.230		5,225	2,225		7,450	9,050
2350	600 amp main, 42 circuits		.37	43.480		7,725	2,675		10,400	12,500
2400	NEHB, with 20 amp, 1 pole circuit breaker									
2450	4 wire, 277/480 volts with main circuit breaker									
2500	100 amp main, 24 circuits	1 Elec	.39	20.700	Ea.	2,875	1,275		4,150	5,050
2550	30 circuits	"	.35	22.880		3,350	1,400		4,750	5,800
2600	225 amp main, 30 circuits	2 Elec	.66	24.150		4,200	1,475		5,675	6,825
2650	42 circuits		.52	31.060		5,175	1,900		7,075	8,525
2700	400 amp main, 42 circuits		.42	37.810		6,225	2,325		8,550	10,300
2750	600 amp main, 42 circuits		.35	45.770		8,525	2,800		11,325	13,600
2900	Note: the following line items don't include branch circuit breakers									
2910	For branch circuit breakers information, see Section 26 24 16.20									
3010	Main lug, no main breaker, 240 volt, 1 pole, 3 wire, 100 amp	1 Elec	2.12	3.780	Ea.	605	232		837	1,025
3020	225 amp	2 Elec	2.21	7.250		660	445		1,105	1,375
3030	400 amp	"	1.66	9.660		1,125	595		1,720	2,125
3060	3 pole, 3 wire, 100 amp	1 Elec	2.12	3.780		655	232		887	1,075
3070	225 amp	2 Elec	2.21	7.250		700	445		1,145	1,425
3080	400 amp		1.66	9.660		1,225	595		1,820	2,225
3090	600 amp		1.47	10.870		1,425	665		2,090	2,575
3110	3 pole, 4 wire, 100 amp	1 Elec	2.12	3.780		720	232		952	1,150
3120	225 amp	2 Elec	2.21	7.250		825	445		1,270	1,575
3130	400 amp		1.66	9.660		1,225	595		1,820	2,225
3140	600 amp		1.47	10.870		1,425	665		2,090	2,575
3160	480 volt, 3 pole, 3 wire, 100 amp	1 Elec	2.12	3.780		785	232		1,017	1,200
3170	225 amp	2 Elec	2.21	7.250		1,100	445		1,545	1,875
3180	400 amp		1.66	9.660		1,550	595		2,145	2,575
3190	600 amp		1.47	10.870		1,725	665		2,390	2,900
3210	277/480 volt, 3 pole, 4 wire, 100 amp	1 Elec	2.12	3.780		765	232		997	1,175
3220	225 amp	2 Elec	2.21	7.250		930	445		1,375	1,675
3230	400 amp		1.66	9.660		1,500	595		2,095	2,525
3240	600 amp		1.47	10.870		1,675	665		2,340	2,850
3260	Main circuit breaker, 240 volt, 1 pole, 3 wire, 100 amp	1 Elec	1.84	4.350		740	267		1,007	1,200
3270	225 amp	2 Elec	1.84	8.700		1,775	535		2,310	2,750
3280	400 amp	"	1.47	10.870		2,775	665		3,440	4,050
3310	3 pole, 3 wire, 100 amp	1 Elec	1.84	4.350		890	267		1,157	1,375
3320	225 amp	2 Elec	1.84	8.700		2,025	535		2,560	3,025
3330	400 amp		1.47	10.870		3,175	665		3,840	4,475
3360	120/208 volt, 3 pole, 4 wire, 100 amp		3.68	4.350		850	267		1,117	1,325
3370	225 amp		1.84	8.700		2,025	535		2,560	3,025
3380	400 amp		1.47	10.870		3,175	665		3,840	4,475
3410	480 volt, 3 pole, 3 wire, 100 amp	1 Elec	1.84	4.350		1,375	267		1,642	1,925

26 24 Switchboards and Panelboards

26 24 16 – Panelboards

26 24 16.30 Panelboards Commercial Applications	Crew	Daily Output	Labor-Hours	Unit	Material	2020 Bare Costs Labor	Equipment	Total	Total Incl O&P
3420	225 amp	2 Elec	1.84	8.700	Ea.	2,375	535		2,910	3,425
3430	400 amp		1.47	10.870		3,550	665		4,215	4,900
3460	277/480 volt, 3 pole, 4 wire, 100 amp		3.68	4.350		1,350	267		1,617	1,900
3470	225 amp		1.84	8.700		2,325	535		2,860	3,350
3480	400 amp	↓	1.47	10.870		3,575	665		4,240	4,925
3510	Main circuit breaker, HIC, 240 volt, 1 pole, 3 wire, 100 amp	1 Elec	1.84	4.350		1,325	267		1,592	1,850
3520	225 amp	2 Elec	1.84	8.700		3,450	535		3,985	4,600
3530	400 amp	"	1.47	10.870		4,625	665		5,290	6,075
3560	3 pole, 3 wire, 100 amp	1 Elec	1.84	4.350		1,475	267		1,742	2,025
3570	225 amp	2 Elec	1.84	8.700		3,900	535		4,435	5,075
3580	400 amp	"	1.47	10.870		5,150	665		5,815	6,675
3610	120/208 volt, 3 pole, 4 wire, 100 amp	1 Elec	1.84	4.350		1,475	267		1,742	2,025
3620	225 amp	2 Elec	1.84	8.700		3,900	535		4,435	5,075
3630	400 amp	"	1.47	10.870		5,150	665		5,815	6,675
3660	480 volt, 3 pole, 3 wire, 100 amp	1 Elec	1.84	4.350		2,200	267		2,467	2,800
3670	225 amp	2 Elec	1.84	8.700		4,325	535		4,860	5,550
3680	400 amp	"	1.47	10.870		5,500	665		6,165	7,050
3710	277/480 volt, 3 pole, 4 wire, 100 amp	1 Elec	1.84	4.350		2,075	267		2,342	2,700
3720	225 amp	2 Elec	1.84	8.700		4,125	535		4,660	5,325
3730	400 amp	"	1.47	10.870		5,450	665		6,115	7,000
3760	Main circuit breaker, shunt trip, 100 amp	1 Elec	1.10	7.250		1,100	445		1,545	1,850
3770	225 amp	2 Elec	1.47	10.870		2,500	665		3,165	3,750
3780	400 amp	"	1.29	12.420	↓	3,625	760		4,385	5,125

26 24 19 – Motor-Control Centers

26 24 19.20 Motor Control Center Components

		Crew	Daily Output	Labor-Hours	Unit	Material	2020 Bare Costs Labor	Equipment	Total	Total Incl O&P
0010	**MOTOR CONTROL CENTER COMPONENTS**									
0100	Starter, size 1, FVNR, NEMA 1, type A, fusible	1 Elec	2.48	3.220	Ea.	1,650	198		1,848	2,125
0120	Circuit breaker		2.48	3.220		1,800	198		1,998	2,275
0140	Type B, fusible		2.48	3.220		1,825	198		2,023	2,300
0160	Circuit breaker		2.48	3.220		1,975	198		2,173	2,475
0180	NEMA 12, type A, fusible		2.39	3.340		1,700	205		1,905	2,150
0200	Circuit breaker		2.39	3.340		1,825	205		2,030	2,325
0220	Type B, fusible		2.39	3.340		1,850	205		2,055	2,350
0240	Circuit breaker		2.39	3.340		2,025	205		2,230	2,525
0300	Starter, size 1, FVR, NEMA 1, type A, fusible		1.84	4.350		2,375	267		2,642	3,025
0320	Circuit breaker		1.84	4.350		2,375	267		2,642	3,025
0340	Type B, fusible		1.84	4.350		2,625	267		2,892	3,275
0360	Circuit breaker		1.84	4.350		2,625	267		2,892	3,275
0380	NEMA 12, type A, fusible		1.75	4.580		2,400	281		2,681	3,075
0400	Circuit breaker		1.75	4.580		2,400	281		2,681	3,075
0420	Type B, fusible		1.75	4.580		2,650	281		2,931	3,350
0440	Circuit breaker	↓	1.75	4.580	↓	2,650	281		2,931	3,350
0490	Starter size 1, 2 speed, separate winding									
0500	NEMA 1, type A, fusible	1 Elec	2.39	3.340	Ea.	3,125	205		3,330	3,750
0520	Circuit breaker		2.39	3.340		3,125	205		3,330	3,750
0540	Type B, fusible		2.39	3.340		3,450	205		3,655	4,100
0560	Circuit breaker		2.39	3.340		3,450	205		3,655	4,100
0580	NEMA 12, type A, fusible		2.30	3.480		3,200	214		3,414	3,850
0600	Circuit breaker		2.30	3.480		3,200	214		3,414	3,850
0620	Type B, fusible		2.30	3.480		3,500	214		3,714	4,175
0640	Circuit breaker	↓	2.30	3.480	↓	3,500	214		3,714	4,175
0650	Starter size 1, 2 speed, consequent pole									

Pre-Installation Change Order Costs

26 24 19.20 Motor Control Center Components	Crew	Daily Output	Labor-Hours	Unit	Material	2020 Bare Costs Labor	Equipment	Total	Total Incl O&P
0660	NEMA 1, type A, fusible	1 Elec	2.39	3.340	Ea.	3,125	205		3,330	3,750
0680	Circuit breaker		2.39	3.340		3,125	205		3,330	3,750
0700	Type B, fusible		2.39	3.340		3,450	205		3,655	4,100
0720	Circuit breaker		2.39	3.340		3,450	205		3,655	4,100
0740	NEMA 12, type A, fusible		2.30	3.480		3,200	214		3,414	3,850
0760	Circuit breaker		2.30	3.480		3,200	214		3,414	3,850
0780	Type B, fusible		2.30	3.480		3,475	214		3,689	4,150
0800	Circuit breaker	▼	2.30	3.480	▼	3,500	214		3,714	4,175
0810	Starter size 1, 2 speed, space only									
0820	NEMA 1, type A, fusible	1 Elec	14.72	.543	Ea.	705	33.50		738.50	825
0840	Circuit breaker		14.72	.543		705	33.50		738.50	825
0860	Type B, fusible		14.72	.543		705	33.50		738.50	825
0880	Circuit breaker		14.72	.543		705	33.50		738.50	825
0900	NEMA 12, type A, fusible		13.80	.580		735	35.50		770.50	860
0920	Circuit breaker		13.80	.580		735	35.50		770.50	860
0940	Type B, fusible		13.80	.580		735	35.50		770.50	860
0960	Circuit breaker	▼	13.80	.580		735	35.50		770.50	860
1100	Starter size 2, FVNR, NEMA 1, type A, fusible	2 Elec	3.68	4.350		1,875	267		2,142	2,475
1120	Circuit breaker		3.68	4.350		2,050	267		2,317	2,675
1140	Type B, fusible		3.68	4.350		2,075	267		2,342	2,675
1160	Circuit breaker		3.68	4.350		2,250	267		2,517	2,875
1180	NEMA 12, type A, fusible		3.50	4.580		1,900	281		2,181	2,525
1200	Circuit breaker		3.50	4.580		2,075	281		2,356	2,725
1220	Type B, fusible		3.50	4.580		2,075	281		2,356	2,725
1240	Circuit breaker		3.50	4.580		2,275	281		2,556	2,925
1300	FVR, NEMA 1, type A, fusible		2.94	5.430		3,175	335		3,510	4,000
1320	Circuit breaker		2.94	5.430		3,175	335		3,510	4,000
1340	Type B, fusible		2.94	5.430		3,500	335		3,835	4,350
1360	Circuit breaker		2.94	5.430		3,500	335		3,835	4,350
1380	NEMA type 12, type A, fusible		2.76	5.800		3,225	355		3,580	4,075
1400	Circuit breaker		2.76	5.800		3,225	355		3,580	4,075
1420	Type B, fusible		2.76	5.800		3,550	355		3,905	4,425
1440	Circuit breaker		2.76	5.800	▼	3,550	355		3,905	4,425
1490	Starter size 2, 2 speed, separate winding									
1500	NEMA 1, type A, fusible	2 Elec	3.50	4.580	Ea.	3,550	281		3,831	4,325
1520	Circuit breaker		3.50	4.580		3,550	281		3,831	4,325
1540	Type B, fusible		3.50	4.580		3,900	281		4,181	4,725
1560	Circuit breaker		3.50	4.580		3,900	281		4,181	4,725
1570	NEMA 12, type A, fusible		3.31	4.830		3,600	296		3,896	4,425
1580	Circuit breaker		3.31	4.830		3,600	296		3,896	4,425
1600	Type B, fusible		3.31	4.830		3,950	296		4,246	4,800
1620	Circuit breaker	▼	3.31	4.830	▼	3,950	296		4,246	4,800
1630	Starter size 2, 2 speed, consequent pole									
1640	NEMA 1, type A, fusible	2 Elec	3.50	4.580	Ea.	4,025	281		4,306	4,850
1660	Circuit breaker		3.50	4.580		4,025	281		4,306	4,850
1680	Type B, fusible		3.50	4.580		4,250	281		4,531	5,100
1700	Circuit breaker		3.50	4.580		4,250	281		4,531	5,100
1720	NEMA 12, type A, fusible		3.50	4.580		4,050	281		4,331	4,875
1740	Circuit breaker		3.31	4.830		4,075	296		4,371	4,950
1760	Type B, fusible		3.31	4.830		4,325	296		4,621	5,200
1780	Sircuit breaker	▼	3.31	4.830	▼	4,325	296		4,621	5,200
1830	Starter size 2, autotransformer									
1840	NEMA 1, type A, fusible	2 Elec	3.13	5.120	Ea.	6,275	315		6,590	7,375

26 24 19.20 Motor Control Center Components	Crew	Daily Output	Labor-Hours	Unit	Material	2020 Bare Costs Labor	Equipment	Total	Total Incl O&P
1860	Circuit breaker	2 Elec	3.13	5.120	Ea.	6,425	315		6,740	7,550
1880	Type B, fusible		3.13	5.120		6,850	315		7,165	8,025
1900	Circuit breaker		3.13	5.120		6,850	315		7,165	8,025
1920	NEMA 12, type A, fusible		2.94	5.430		6,400	335		6,735	7,525
1940	Circuit breaker		2.94	5.430		6,400	335		6,735	7,525
1960	Type B, fusible		2.94	5.430		6,975	335		7,310	8,175
1980	Circuit breaker		2.94	5.430		6,975	335		7,310	8,175
2030	Starter size 2, space only									
2040	NEMA 1, type A, fusible	1 Elec	14.72	.543	Ea.	705	33.50		738.50	825
2060	Circuit breaker		14.72	.543		705	33.50		738.50	825
2080	Type B, fusible		14.72	.543		705	33.50		738.50	825
2100	Circuit breaker		14.72	.543		705	33.50		738.50	825
2120	NEMA 12, type A, fusible		13.80	.580		735	35.50		770.50	860
2140	Circuit breaker		13.80	.580		735	35.50		770.50	860
2160	Type B, fusible		13.80	.580		735	35.50		770.50	860
2180	Circuit breaker		13.80	.580		735	35.50		770.50	860
2300	Starter size 3, FVNR, NEMA 1, type A, fusible	2 Elec	1.84	8.700		3,575	535		4,110	4,725
2320	Circuit breaker		1.84	8.700		3,175	535		3,710	4,300
2340	Type B, fusible		1.84	8.700		3,925	535		4,460	5,125
2360	Circuit breaker		1.84	8.700		3,500	535		4,035	4,650
2380	NEMA 12, type A, fusible		1.75	9.150		3,650	560		4,210	4,825
2400	Circuit breaker		1.75	9.150		3,225	560		3,785	4,375
2420	Type B, fusible		1.75	9.150		4,075	560		4,635	5,300
2440	Circuit breaker		1.75	9.150		3,550	560		4,110	4,750
2500	Starter size 3, FVR, NEMA 1, type A, fusible		1.47	10.870		4,875	665		5,540	6,350
2520	Circuit breaker		1.47	10.870		4,650	665		5,315	6,125
2540	Type B, fusible		1.47	10.870		5,300	665		5,965	6,825
2560	Circuit breaker		1.47	10.870		5,100	665		5,765	6,600
2580	NEMA 12, type A, fusible		1.38	11.590		4,975	710		5,685	6,525
2600	Circuit breaker		1.38	11.590		4,750	710		5,460	6,275
2620	Type B, fusible		1.38	11.590		5,400	710		6,110	7,000
2640	Circuit breaker		1.38	11.590		5,600	710		6,310	7,225
2690	Starter size 3, 2 speed, separate winding									
2700	NEMA 1, type A, fusible	2 Elec	1.84	8.700	Ea.	5,600	535		6,135	6,975
2720	Circuit breaker		1.84	8.700		4,950	535		5,485	6,250
2740	Type B, fusible		1.84	8.700		6,125	535		6,660	7,550
2760	Circuit breaker		1.84	8.700		5,425	535		5,960	6,750
2780	NEMA 12, type A, fusible		1.75	9.150		5,725	560		6,285	7,125
2800	Circuit breaker		1.75	9.150		5,050	560		5,610	6,375
2820	Type B, fusible		1.75	9.150		6,250	560		6,810	7,700
2840	Circuit breaker		1.75	9.150		5,525	560		6,085	6,900
2850	Starter size 3, 2 speed, consequent pole									
2860	NEMA 1, type A, fusible	2 Elec	1.84	8.700	Ea.	6,250	535		6,785	7,675
2880	Circuit breaker		1.84	8.700		5,600	535		6,135	6,950
2900	Type B, fusible		1.84	8.700		6,775	535		7,310	8,275
2920	Circuit breaker		1.84	8.700		6,075	535		6,610	7,475
2940	NEMA 12, type A, fusible		1.75	9.150		6,375	560		6,935	7,825
2960	Circuit breaker		1.75	9.150		5,700	560		6,260	7,100
2980	Type B, fusible		1.75	9.150		6,900	560		7,460	8,425
3000	Circuit breaker		1.75	9.150		6,175	560		6,735	7,600
3100	Starter size 3, autotransformer, NEMA 1, type A, fusible		1.47	10.870		8,250	665		8,915	10,100
3120	Circuit breaker		1.47	10.870		7,775	665		8,440	9,550
3140	Type B, fusible		1.47	10.870		8,500	665		9,165	10,300

For customer support on your Electrical Change Order Costs with RSMeans data, call 800.448.8182. **Pre-Installation Change Order Costs**

26 24 19 – Motor-Control Centers

26 24 19.20 Motor Control Center Components	Crew	Daily Output	Labor-Hours	Unit	Material	2020 Bare Costs Labor	Equipment	Total	Total Incl O&P
3160	Circuit breaker	2 Elec	1.47	10.870	Ea.	8,500	665		9,165	10,300
3180	NEMA 12, type A, fusible		1.38	11.590		7,925	710		8,635	9,775
3200	Circuit breaker		1.38	11.590		7,925	710		8,635	9,775
3220	Type B, fusible		1.38	11.590		8,650	710		9,360	10,600
3240	Circuit breaker		1.38	11.590		8,650	710		9,360	10,600
3260	Starter size 3, space only, NEMA 1, type A, fusible	1 Elec	13.80	.580		1,200	35.50		1,235.50	1,375
3280	Circuit breaker		13.80	.580		940	35.50		975.50	1,075
3300	Type B, fusible		13.80	.580		1,200	35.50		1,235.50	1,375
3320	Circuit breaker		13.80	.580		940	35.50		975.50	1,075
3340	NEMA 12, type A, fusible		12.88	.621		1,275	38		1,313	1,450
3360	Circuit breaker		12.88	.621		995	38		1,033	1,150
3380	Type B, fusible		12.88	.621		1,275	38		1,313	1,450
3400	Circuit breaker		12.88	.621		995	38		1,033	1,150
3500	Starter size 4, FVNR, NEMA 1, type A, fusible	2 Elec	1.47	10.870		4,650	665		5,315	6,125
3520	Circuit breaker		1.47	10.870		4,225	665		4,890	5,650
3540	Type B, fusible		1.47	10.870		5,100	665		5,765	6,600
3560	Circuit breaker		1.47	10.870		4,650	665		5,315	6,125
3580	NEMA 12, type A, fusible		1.38	11.590		4,775	710		5,485	6,300
3600	Circuit breaker		1.38	11.590		4,300	710		5,010	5,775
3620	Type B, fusible		1.38	11.590		5,200	710		5,910	6,775
3640	Circuit breaker		1.38	11.590		4,725	710		5,435	6,250
3700	Starter size 4, FVR, NEMA 1, type A, fusible		1.10	14.490		6,375	890		7,265	8,325
3720	Circuit breaker		1.10	14.490		5,750	890		6,640	7,625
3740	Type B, fusible		1.10	14.490		6,950	890		7,840	8,975
3760	Circuit breaker		1.10	14.490		6,325	890		7,215	8,275
3780	NEMA 12, type A, fusible		1.07	14.990		6,500	920		7,420	8,525
3800	Circuit breaker		1.07	14.990		5,825	920		6,745	7,775
3820	Type B, fusible		1.07	14.990		5,875	920		6,795	7,850
3840	Circuit breaker		1.07	14.990		6,400	920		7,320	8,425
3890	Starter size 4, 2 speed, separate windings									
3900	NEMA 1, type A, fusible	2 Elec	1.47	10.870	Ea.	8,025	665		8,690	9,825
3920	Circuit breaker		1.47	10.870		6,025	665		6,690	7,650
3940	Type B, fusible		1.47	10.870		8,825	665		9,490	10,700
3960	Motor Circuit breaker		1.47	10.870		6,600	665		7,265	8,275
3980	NEMA 12, type A, fusible		1.38	11.590		8,175	710		8,885	10,100
4000	Circuit breaker		1.38	11.590		6,125	710		6,835	7,800
4020	Type B, fusible		1.38	11.590		8,975	710		9,685	10,900
4040	Circuit breaker		1.38	11.590		6,725	710		7,435	8,425
4050	Starter size 4, 2 speed, consequent pole									
4060	NEMA 1, type A, fusible	2 Elec	1.47	10.870	Ea.	9,375	665		10,040	11,300
4080	Circuit breaker		1.47	10.870		6,975	665		7,640	8,675
4100	Type B, fusible		1.47	10.870		10,300	665		10,965	12,400
4120	Circuit breaker		1.47	10.870		7,650	665		8,315	9,425
4140	NEMA 12, type A, fusible		1.38	11.590		9,525	710		10,235	11,600
4160	Circuit breaker		1.38	11.590		7,075	710		7,785	8,850
4180	Type B, fusible		1.38	11.590		10,500	710		11,210	12,600
4200	Circuit breaker		1.38	11.590		7,750	710		8,460	9,575
4300	Starter size 4, autotransformer, NEMA 1, type A, fusible		1.20	13.380		9,250	820		10,070	11,400
4320	Circuit breaker		1.20	13.380		9,325	820		10,145	11,400
4340	Type B, fusible		1.20	13.380		10,200	820		11,020	12,400
4360	Circuit breaker		1.20	13.380		10,200	820		11,020	12,400
4380	NEMA 12, type A, fusible		1.14	14.030		9,400	860		10,260	11,600
4400	Circuit breaker		1.14	14.030		9,450	860		10,310	11,700

26 24 19.20 Motor Control Center Components	Crew	Daily Output	Labor-Hours	Unit	Material	2020 Bare Costs Labor	Equipment	Total	Total Incl O&P
4420	Type B, fusible	2 Elec	1.14	14.030	Ea.	10,300	860		11,160	12,700
4440	Circuit breaker		1.14	14.030		10,300	860		11,160	12,600
4500	Starter size 4, space only, NEMA 1, type A, fusible	1 Elec	12.88	.621		1,675	38		1,713	1,900
4520	Circuit breaker		12.88	.621		1,200	38		1,238	1,375
4540	Type B, fusible		12.88	.621		1,675	38		1,713	1,900
4560	Circuit breaker		12.88	.621		1,200	38		1,238	1,375
4580	NEMA 12, type A, fusible		11.96	.669		1,775	41		1,816	2,000
4600	Circuit breaker		11.96	.669		1,275	41		1,316	1,450
4620	Type B, fusible		11.96	.669		1,775	41		1,816	2,000
4640	Circuit breaker		11.96	.669		1,275	41		1,316	1,450
4800	Starter size 5, FVNR, NEMA 1, type A, fusible	2 Elec	.92	17.390		9,600	1,075		10,675	12,200
4820	Circuit breaker		.92	17.390		6,825	1,075		7,900	9,100
4840	Type B, fusible		.92	17.390		10,500	1,075		11,575	13,200
4860	Circuit breaker		.92	17.390		7,500	1,075		8,575	9,850
4880	NEMA 12, type A, fusible		.88	18.120		9,750	1,100		10,850	12,400
4900	Circuit breaker		.88	18.120		6,950	1,100		8,050	9,300
4920	Type B, fusible		.88	18.120		10,700	1,100		11,800	13,400
4940	Circuit breaker		.88	18.120		7,625	1,100		8,725	10,100
5000	Starter size 5, FVR, NEMA 1, type A, fusible		.74	21.740		14,200	1,325		15,525	17,600
5020	Circuit breaker		.74	21.740		11,100	1,325		12,425	14,200
5040	Type B, fusible		.74	21.740		15,500	1,325		16,825	19,100
5060	Circuit breaker		.74	21.740		12,200	1,325		13,525	15,400
5080	NEMA 12, type A, fusible		.70	22.880		14,400	1,400		15,800	18,000
5100	Circuit breaker		.70	22.880		11,300	1,400		12,700	14,500
5120	Type B, fusible		.70	22.880		15,800	1,400		17,200	19,500
5140	Circuit breaker		.70	22.880		12,400	1,400		13,800	15,700
5190	Starter size 5, 2 speed, separate windings									
5200	NEMA 1, type A, fusible	2 Elec	.92	17.390	Ea.	19,500	1,075		20,575	23,100
5220	Circuit breaker		.92	17.390		14,600	1,075		15,675	17,600
5240	Type B, fusible		.92	17.390		21,400	1,075		22,475	25,200
5260	Circuit breaker		.92	17.390		16,000	1,075		17,075	19,200
5280	NEMA 12, type A, fusible		.88	18.120		19,800	1,100		20,900	23,500
5300	Circuit breaker		.88	18.120		14,700	1,100		15,800	17,900
5320	Type B, fusible		.88	18.120		21,700	1,100		22,800	25,600
5340	Circuit breaker		.88	18.120		16,200	1,100		17,300	19,500
5400	Starter size 5, autotransformer, NEMA 1, type A, fusible		.64	24.840		16,000	1,525		17,525	19,900
5420	Circuit breaker		.64	24.840		13,300	1,525		14,825	16,900
5440	Type B, fusible		.64	24.840		17,500	1,525		19,025	21,600
5460	Circuit breaker		.64	24.840		14,600	1,525		16,125	18,300
5480	NEMA 12, type A, fusible		.63	25.580		16,200	1,575		17,775	20,100
5500	Circuit breaker		.63	25.580		13,400	1,575		14,975	17,000
5520	Type B, fusible		.63	25.580		17,700	1,575		19,275	21,800
5540	Circuit breakers		.63	25.580		14,700	1,575		16,275	18,500
5600	Starter size 5, space only, NEMA 1, type A, fusible	1 Elec	11.04	.725		2,150	44.50		2,194.50	2,425
5620	Circuit breaker		11.04	.725		2,150	44.50		2,194.50	2,425
5640	Type B, fusible		11.04	.725		2,150	44.50		2,194.50	2,425
5660	Circuit breaker		11.04	.725		1,400	44.50		1,444.50	1,625
5680	NEMA 12, type A, fusible		10.12	.791		2,275	48.50		2,323.50	2,575
5700	Circuit breaker		10.12	.791		1,500	48.50		1,548.50	1,725
5720	Type B, fusible		10.12	.791		2,275	48.50		2,323.50	2,575
5740	Sircuit breaker		10.12	.791		1,500	48.50		1,548.50	1,725
5800	Fuse, light contactor NEMA 1, type A, 30 amp		2.48	3.220		1,500	198		1,698	1,950
5820	60 amp		1.84	4.350		1,700	267		1,967	2,275

26 24 19.20 Motor Control Center Components	Crew	Daily Output	Labor-Hours	Unit	Material	2020 Bare Costs Labor	Equipment	Total	Total Incl O&P
5840	100 amp	1 Elec	.92	8.700	Ea.	3,250	535		3,785	4,375
5860	200 amp		.74	10.870		7,525	665		8,190	9,300
5880	Type B, 30 amp		2.48	3.220		1,650	198		1,848	2,100
5900	60 amp		1.84	4.350		1,825	267		2,092	2,425
5920	100 amp		.92	8.700		3,575	535		4,110	4,725
5940	200 amp	2 Elec	1.47	10.870		8,250	665		8,915	10,100
5960	NEMA 12, type A, 30 amp	1 Elec	2.39	3.340		1,525	205		1,730	1,975
5980	60 amp		1.75	4.580		1,725	281		2,006	2,325
6000	100 amp		.87	9.150		3,300	560		3,860	4,475
6020	200 amp	2 Elec	1.38	11.590		7,650	710		8,360	9,475
6040	Type B, 30 amp	1 Elec	2.39	3.340		1,675	205		1,880	2,125
6060	60 amp		1.75	4.580		1,850	281		2,131	2,475
6080	100 amp		.87	9.150		3,625	560		4,185	4,825
6100	200 amp	2 Elec	1.38	11.590		8,375	710		9,085	10,300
6200	Circuit breaker, light contactor NEMA 1, type A, 30 amp	1 Elec	2.48	3.220		1,650	198		1,848	2,100
6220	60 amp		1.84	4.350		1,875	267		2,142	2,450
6240	100 amp		.92	8.700		2,875	535		3,410	3,975
6260	200 amp	2 Elec	1.47	10.870		6,250	665		6,915	7,875
6280	Type B, 30 amp	1 Elec	2.48	3.220		1,800	198		1,998	2,275
6300	60 amp		1.84	4.350		2,075	267		2,342	2,675
6320	100 amp		.92	8.700		3,150	535		3,685	4,250
6340	200 amp	2 Elec	1.47	10.870		6,825	665		7,490	8,500
6360	NEMA 12, type A, 30 amp	1 Elec	2.39	3.340		1,675	205		1,880	2,125
6380	60 amp		1.75	4.580		1,900	281		2,181	2,500
6400	100 amp		.87	9.150		2,925	560		3,485	4,050
6420	200 amp	2 Elec	1.38	11.590		6,275	710		6,985	7,950
6440	Type B, 30 amp	1 Elec	2.39	3.340		1,950	205		2,155	2,450
6460	60 amp		1.75	4.580		2,100	281		2,381	2,725
6480	100 amp		.87	9.150		3,200	560		3,760	4,350
6500	200 amp	2 Elec	1.38	11.590		6,900	710		7,610	8,625
6600	Fusible switch, NEMA 1, type A, 30 amp	1 Elec	4.88	1.640		1,100	101		1,201	1,350
6620	60 amp		4.60	1.740		1,175	107		1,282	1,450
6640	100 amp		3.68	2.170		1,300	133		1,433	1,625
6660	200 amp		2.94	2.720		2,250	167		2,417	2,700
6680	400 amp	2 Elec	4.23	3.780		5,275	232		5,507	6,150
6700	600 amp		2.94	5.430		5,700	335		6,035	6,775
6720	800 amp		2.39	6.690		15,600	410		16,010	17,800
6740	NEMA 12, type A, 30 amp	1 Elec	4.78	1.670		1,125	102		1,227	1,400
6760	60 amp		4.51	1.770		1,200	109		1,309	1,475
6780	100 amp		3.59	2.230		1,325	137		1,462	1,650
6800	200 amp		2.85	2.810		2,300	172		2,472	2,775
6820	400 amp	2 Elec	4.05	3.950		5,375	242		5,617	6,275
6840	600 amp		2.76	5.800		5,875	355		6,230	6,975
6860	800 amp		2.21	7.250		15,800	445		16,245	18,100
6900	Circuit breaker, NEMA 1, type A, 30 amp	1 Elec	4.88	1.640		1,000	101		1,101	1,250
6920	60 amp		4.60	1.740		1,000	107		1,107	1,250
6940	100 amp		3.68	2.170		1,000	133		1,133	1,300
6960	225 amp		2.94	2.720		1,800	167		1,967	2,225
6980	400 amp	2 Elec	4.23	3.780		3,425	232		3,657	4,125
7000	600 amp		2.94	5.430		4,000	335		4,335	4,900
7020	800 amp		2.39	6.690		8,375	410		8,785	9,800
7040	NEMA 12, type A, 30 amp	1 Elec	4.78	1.670		1,050	102		1,152	1,300
7060	60 amp		4.51	1.770		1,050	109		1,159	1,300

26 24 19.20 Motor Control Center Components	Crew	Daily Output	Labor-Hours	Unit	Material	2020 Bare Costs Labor	Equipment	Total	Total Incl O&P
7080	100 amp	1 Elec	3.59	2.230	Ea.	1,050	137		1,187	1,350
7100	225 amp	↓	2.85	2.810		1,825	172		1,997	2,275
7120	400 amp	2 Elec	4.05	3.950		3,475	242		3,717	4,175
7140	600 amp		2.76	5.800		4,075	355		4,430	5,000
7160	800 amp		2.21	7.250		8,525	445		8,970	10,100
7300	Incoming line, main lug only, 600 amp, alum., NEMA 1		1.47	10.870		1,275	665		1,940	2,400
7320	NEMA 12		1.38	11.590		1,300	710		2,010	2,475
7340	Copper, NEMA 1		1.47	10.870		1,350	665		2,015	2,475
7360	800 amp, alum., NEMA 1		1.38	11.590		3,200	710		3,910	4,575
7380	NEMA 12		1.29	12.420		3,250	760		4,010	4,700
7400	Copper, NEMA 1		1.38	11.590		3,325	710		4,035	4,725
7420	1,200 amp, copper, NEMA 1		1.29	12.420		3,450	760		4,210	4,925
7440	Incoming line, fusible switch, 400 amp, alum., NEMA 1		1.10	14.490		4,400	890		5,290	6,175
7460	NEMA 12		1.01	15.810		4,500	970		5,470	6,400
7480	Copper, NEMA 1		1.10	14.490		4,475	890		5,365	6,250
7500	600 amp, alum., NEMA 1		1.01	15.810		5,375	970		6,345	7,375
7520	NEMA 12		.92	17.390		5,475	1,075		6,550	7,625
7540	Copper, NEMA 1		1.01	15.810		5,450	970		6,420	7,450
7560	Incoming line, circuit breaker, 225 amp, alum., NEMA 1		1.10	14.490		2,300	890		3,190	3,850
7580	NEMA 12		1.01	15.810		2,350	970		3,320	4,025
7600	Copper, NEMA 1		1.10	14.490		2,375	890		3,265	3,925
7620	400 amp, alum., NEMA 1		1.10	14.490		3,425	890		4,315	5,100
7640	NEMA 12		1.01	15.810		3,475	970		4,445	5,275
7660	Copper, NEMA 1		1.10	14.490		3,500	890		4,390	5,175
7680	600 amp, alum., NEMA 1		1.01	15.810		4,000	970		4,970	5,850
7700	NEMA 12		.92	17.390		4,075	1,075		5,150	6,075
7720	Copper, NEMA 1		1.01	15.810		4,075	970		5,045	5,925
7740	800 amp, copper, NEMA 1	↓	.83	19.320		8,375	1,175		9,550	11,000
7760	Incoming line, for copper bus, add					126			126	139
7780	For 65,000 amp bus bracing, add					188			188	206
7800	For NEMA 3R enclosure, add					5,900			5,900	6,475
7820	For NEMA 12 enclosure, add					162			162	179
7840	For 1/4" x 1" ground bus, add	1 Elec	14.72	.543		105	33.50		138.50	165
7860	For 1/4" x 2" ground bus, add	"	11.04	.725		105	44.50		149.50	181
7900	Main rating basic section, alum., NEMA 1, 800 amp	2 Elec	1.29	12.420		345	760		1,105	1,500
7920	1,200 amp	"	1.10	14.490		675	890		1,565	2,075
7940	For copper bus, add					505			505	555
7960	For 65,000 amp bus bracing, add					340			340	375
7980	For NEMA 3R enclosure, add					5,900			5,900	6,475
8000	For NEMA 12, enclosure, add					162			162	179
8020	For 1/4" x 1" ground bus, add	1 Elec	14.72	.543		105	33.50		138.50	165
8040	For 1/4" x 2" ground bus, add		11.04	.725		105	44.50		149.50	181
8060	Unit devices, pilot light, standard		14.72	.543		90	33.50		123.50	149
8080	Pilot light, push to test		14.72	.543		126	33.50		159.50	189
8100	Pilot light, standard, and push button		11.04	.725		217	44.50		261.50	305
8120	Pilot light, push to test, and push button		11.04	.725		253	44.50		297.50	345
8140	Pilot light, standard, and select switch		11.04	.725		217	44.50		261.50	305
8160	Pilot light, push to test, and select switch	↓	11.04	.725	↓	253	44.50		297.50	345

26 24 19 – Motor-Control Centers

26 24 19.30 Motor Control Center	Crew	Daily Output	Labor-Hours	Unit	Material	2020 Bare Costs Labor	Equipment	Total	Total Incl O&P
0010 **MOTOR CONTROL CENTER** Consists of starters & structures									
0050 Starters, class 1, type B, comb. MCP, FVNR, with									
0100 control transformer, 10 HP, size 1, 12" high	1 Elec	2.48	3.220	Ea.	1,800	198		1,998	2,275
0200 25 HP, size 2, 18" high	2 Elec	3.68	4.350		2,050	267		2,317	2,675
0300 50 HP, size 3, 24" high		1.84	8.700		3,175	535		3,710	4,300
0350 75 HP, size 4, 24" high		1.47	10.870		4,225	665		4,890	5,650
0400 100 HP, size 4, 30" high		1.29	12.420		5,575	760		6,335	7,250
0500 200 HP, size 5, 48" high		.92	17.390		8,325	1,075		9,400	10,800
0600 400 HP, size 6, 72" high	↓	.74	21.740	↓	17,100	1,325		18,425	20,800
0800 Structures, 600 amp, 22,000 rms, takes any									
0900 combination of starters up to 72" high	2 Elec	1.47	10.870	Ea.	1,925	665		2,590	3,100
1000 Back to back, 72" front & 66" back	"	1.10	14.490		2,625	890		3,515	4,200
1100 For copper bus, add per structure					278			278	305
1200 For NEMA 12, add per structure					160			160	176
1300 For 42,000 rms, add per structure					220			220	242
1400 For 100,000 rms, size 1 & 2, add					715			715	785
1500 Size 3, add					1,150			1,150	1,250
1600 Size 4, add					925			925	1,025
1700 For pilot lights, add per starter	1 Elec	14.72	.543		125	33.50		158.50	187
1800 For push button, add per starter		14.72	.543		125	33.50		158.50	187
1900 For auxiliary contacts, add per starter	↓	14.72	.543	↓	221	33.50		254.50	293

26 24 19.40 Motor Starters and Controls

26 24 19.40 Motor Starters and Controls	Crew	Daily Output	Labor-Hours	Unit	Material	2020 Bare Costs Labor	Equipment	Total	Total Incl O&P
0010 **MOTOR STARTERS AND CONTROLS**									
0050 Magnetic, FVNR, with enclosure and heaters, 480 volt									
0080 2 HP, size 00	1 Elec	3.22	2.480	Ea.	216	152		368	465
0100 5 HP, size 0		2.12	3.780		380	232		612	765
0200 10 HP, size 1	↓	1.47	5.430		288	335		623	810
0300 25 HP, size 2	2 Elec	2.02	7.910		540	485		1,025	1,325
0400 50 HP, size 3		1.66	9.660		880	595		1,475	1,850
0500 100 HP, size 4		1.10	14.490		1,950	890		2,840	3,475
0600 200 HP, size 5		.83	19.320		4,575	1,175		5,750	6,800
0610 400 HP, size 6	↓	.74	21.740		20,100	1,325		21,425	24,100
0620 NEMA 7, 5 HP, size 0	1 Elec	1.47	5.430		1,500	335		1,835	2,150
0630 10 HP, size 1	"	1.01	7.910		1,575	485		2,060	2,450
0640 25 HP, size 2	2 Elec	1.66	9.660		2,525	595		3,120	3,650
0650 50 HP, size 3		1.10	14.490		3,800	890		4,690	5,500
0660 100 HP, size 4		.83	19.320		6,125	1,175		7,300	8,525
0670 200 HP, size 5	↓	.46	34.780		14,600	2,125		16,725	19,300
0700 Combination, with motor circuit protectors, 5 HP, size 0	1 Elec	1.66	4.830		1,075	296		1,371	1,650
0800 10 HP, size 1	"	1.20	6.690		1,125	410		1,535	1,825
0900 25 HP, size 2	2 Elec	1.84	8.700		1,575	535		2,110	2,525
1000 50 HP, size 3		1.21	13.180		2,275	810		3,085	3,700
1200 100 HP, size 4	↓	.74	21.740		4,900	1,325		6,225	7,375
1220 NEMA 7, 5 HP, size 0	1 Elec	1.20	6.690		3,925	410		4,335	4,900
1230 10 HP, size 1	"	.92	8.700		4,000	535		4,535	5,225
1240 25 HP, size 2	2 Elec	1.21	13.180		5,350	810		6,160	7,100
1250 50 HP, size 3		.74	21.740		8,850	1,325		10,175	11,700
1260 100 HP, size 4		.55	28.990		13,800	1,775		15,575	17,900
1270 200 HP, size 5	↓	.37	43.480		30,000	2,675		32,675	36,900
1400 Combination, with fused switch, 5 HP, size 0	1 Elec	1.66	4.830		650	296		946	1,150
1600 10 HP, size 1	"	1.20	6.690		695	410		1,105	1,375
1800 25 HP, size 2	2 Elec	1.84	8.700	↓	1,125	535		1,660	2,050

26 24 19.40 Motor Starters and Controls		Crew	Daily Output	Labor-Hours	Unit	Material	2020 Bare Costs Labor	Equipment	Total	Total Incl O&P
2000	50 HP, size 3	2 Elec	1.21	13.180	Ea.	1,900	810		2,710	3,300
2200	100 HP, size 4	↓	.74	21.740		3,325	1,325		4,650	5,625
2610	NEMA 4, with start-stop push button, size 1	1 Elec	1.20	6.690		1,750	410		2,160	2,525
2620	Size 2	2 Elec	1.84	8.700		2,325	535		2,860	3,375
2630	Size 3		1.21	13.180		3,700	810		4,510	5,250
2640	Size 4	↓	.74	21.740	↓	5,675	1,325		7,000	8,225
2650	NEMA 4, FVNR, including control transformer									
2660	Size 1	2 Elec	2.39	6.690	Ea.	1,675	410		2,085	2,450
2670	Size 2		1.84	8.700		2,400	535		2,935	3,425
2680	Size 3		1.21	13.180		3,825	810		4,635	5,400
2690	Size 4	↓	.74	21.740		5,900	1,325		7,225	8,450
2710	Magnetic, FVR, control circuit transformer, NEMA 1, size 1	1 Elec	1.20	6.690		990	410		1,400	1,700
2720	Size 2	2 Elec	1.84	8.700		1,575	535		2,110	2,550
2730	Size 3		1.21	13.180		2,400	810		3,210	3,850
2740	Size 4	↓	.74	21.740		5,225	1,325		6,550	7,725
2760	NEMA 4, size 1	1 Elec	1.01	7.910		1,425	485		1,910	2,300
2770	Size 2	2 Elec	1.47	10.870		2,300	665		2,965	3,525
2780	Size 3		1.10	14.490		3,450	890		4,340	5,125
2790	Size 4	↓	.64	24.840		7,075	1,525		8,600	10,100
2820	NEMA 12, size 1	1 Elec	1.01	7.910		1,175	485		1,660	2,000
2830	Size 2	2 Elec	1.47	10.870		1,850	665		2,515	3,025
2840	Size 3		1.10	14.490		2,900	890		3,790	4,525
2850	Size 4	↓	.64	24.840		5,975	1,525		7,500	8,850
2870	Combination FVR, fused, w/control XFMR & PB, NEMA 1, size 1	1 Elec	.92	8.700		1,700	535		2,235	2,675
2880	Size 2	2 Elec	1.38	11.590		2,425	710		3,135	3,725
2890	Size 3		1.01	15.810		3,625	970		4,595	5,425
2900	Size 4	↓	.64	24.840		7,325	1,525		8,850	10,300
2910	NEMA 4, size 1	1 Elec	.83	9.660		2,525	595		3,120	3,650
2920	Size 2	2 Elec	1.29	12.420		3,700	760		4,460	5,175
2930	Size 3		.92	17.390		5,825	1,075		6,900	8,000
2940	Size 4	↓	.55	28.990		9,675	1,775		11,450	13,300
2950	NEMA 12, size 1	1 Elec	.92	8.700		1,925	535		2,460	2,925
2960	Size 2	2 Elec	1.29	12.420		2,725	760		3,485	4,125
2970	Size 3		.92	17.390		4,000	1,075		5,075	6,000
2980	Size 4	↓	.55	28.990		7,950	1,775		9,725	11,400
3010	Manual, single phase, w/pilot, 1 pole, 120 V, NEMA 1	1 Elec	5.89	1.360		100	83.50		183.50	234
3020	NEMA 4		3.68	2.170		400	133		533	640
3030	2 pole, 120/240 V, NEMA 1		5.89	1.360		135	83.50		218.50	273
3040	NEMA 4		3.68	2.170		360	133		493	600
3041	3 phase, 3 pole, 600 V, NEMA 1		5.06	1.580		271	97		368	440
3042	NEMA 4		3.22	2.480		465	152		617	735
3043	NEMA 12	↓	3.22	2.480		310	152		462	565
3070	Auxiliary contact, normally open				↓	106			106	116
3500	Magnetic FVNR with NEMA 12, enclosure & heaters, 480 volt									
3600	5 HP, size 0	1 Elec	2.02	3.950	Ea.	260	242		502	645
3700	10 HP, size 1	"	1.38	5.800		390	355		745	960
3800	25 HP, size 2	2 Elec	1.84	8.700		735	535		1,270	1,600
3900	50 HP, size 3		1.47	10.870		1,125	665		1,790	2,250
4000	100 HP, size 4		.92	17.390		2,700	1,075		3,775	4,575
4100	200 HP, size 5	↓	.74	21.740		6,475	1,325		7,800	9,100
4200	Combination, with motor circuit protectors, 5 HP, size 0	1 Elec	1.56	5.120		865	315		1,180	1,425
4300	10 HP, size 1	"	1.10	7.250		895	445		1,340	1,650
4400	25 HP, size 2	2 Elec	1.66	9.660	↓	1,350	595		1,945	2,350

26 24 19 – Motor-Control Centers

26 24 19.40 Motor Starters and Controls	Crew	Daily Output	Labor-Hours	Unit	Material	2020 Bare Costs Labor	Equipment	Total	Total Incl O&P
4500	50 HP, size 3	2 Elec	1.10	14.490	Ea.	2,175	890		3,065	3,725
4600	100 HP, size 4	▼	.68	23.500		4,925	1,450		6,375	7,575
4700	Combination, with fused switch, 5 HP, size 0	1 Elec	1.56	5.120		815	315		1,130	1,375
4800	10 HP, size 1	"	1.10	7.250		850	445		1,295	1,600
4900	25 HP, size 2	2 Elec	1.66	9.660		1,300	595		1,895	2,300
5000	50 HP, size 3		1.10	14.490		2,075	890		2,965	3,600
5100	100 HP, size 4	▼	.68	23.500	▼	4,225	1,450		5,675	6,775
5200	Factory installed controls, adders to size 0 thru 5									
5300	Start-stop push button	1 Elec	29.44	.272	Ea.	54.50	16.65		71.15	85
5400	Hand-off-auto-selector switch		29.44	.272		54.50	16.65		71.15	85
5500	Pilot light		29.44	.272		102	16.65		118.65	138
5600	Start-stop-pilot		29.44	.272		157	16.65		173.65	198
5700	Auxiliary contact, NO or NC		29.44	.272		75	16.65		91.65	108
5800	NO-NC		29.44	.272		150	16.65		166.65	190
5810	Magnetic FVR, NEMA 7, w/heaters, size 1	▼	.61	13.180		3,400	810		4,210	4,950
5830	Size 2	2 Elec	1.01	15.810		5,600	970		6,570	7,625
5840	Size 3		.64	24.840		9,025	1,525		10,550	12,200
5850	Size 4	▼	.55	28.990		10,200	1,775		11,975	13,900
5860	Combination w/circuit breakers, heaters, control XFMR PB, size 1	1 Elec	.55	14.490		1,800	890		2,690	3,300
5870	Size 2	2 Elec	.74	21.740		2,325	1,325		3,650	4,525
5880	Size 3		.46	34.780		3,100	2,125		5,225	6,575
5890	Size 4	▼	.37	43.480		6,125	2,675		8,800	10,700
5900	Manual, 240 volt, 0.75 HP motor	1 Elec	3.68	2.170		58	133		191	262
5910	2 HP motor		3.68	2.170		157	133		290	370
6000	Magnetic, 240 volt, 1 or 2 pole, 0.75 HP motor		3.68	2.170		217	133		350	435
6020	2 HP motor		3.68	2.170		240	133		373	460
6040	5 HP motor		2.76	2.900		345	178		523	645
6060	10 HP motor		2.12	3.780		855	232		1,087	1,275
6100	3 pole, 0.75 HP motor		2.76	2.900		216	178		394	505
6120	5 HP motor		2.12	3.780		293	232		525	670
6140	10 HP motor		1.47	5.430		550	335		885	1,100
6160	15 HP motor		1.47	5.430		550	335		885	1,100
6180	20 HP motor	▼	1.01	7.910		900	485		1,385	1,725
6200	25 HP motor	2 Elec	2.02	7.910		900	485		1,385	1,725
6210	30 HP motor		1.66	9.660		900	595		1,495	1,875
6220	40 HP motor		1.66	9.660		2,000	595		2,595	3,075
6230	50 HP motor		1.66	9.660		2,000	595		2,595	3,075
6240	60 HP motor		1.10	14.490		4,675	890		5,565	6,450
6250	75 HP motor		1.10	14.490		4,675	890		5,565	6,450
6260	100 HP motor		1.10	14.490		4,675	890		5,565	6,450
6270	125 HP motor		.83	19.320		13,100	1,175		14,275	16,200
6280	150 HP motor		.83	19.320		13,100	1,175		14,275	16,200
6290	200 HP motor	▼	.83	19.320		13,100	1,175		14,275	16,200
6400	Starter & nonfused disconnect, 240 volt, 1-2 pole, 0 .75 HP motor	1 Elec	1.84	4.350		270	267		537	690
6410	2 HP motor		1.84	4.350		293	267		560	715
6420	5 HP motor		1.66	4.830		395	296		691	875
6430	10 HP motor		1.29	6.210		925	380		1,305	1,600
6440	3 pole, 0.75 HP motor		1.47	5.430		269	335		604	790
6450	5 HP motor		1.29	6.210		345	380		725	945
6460	10 HP motor		1.01	7.910		620	485		1,105	1,400
6470	15 HP motor	▼	.92	8.700		620	535		1,155	1,475
6480	20 HP motor	2 Elec	1.38	11.590		1,075	710		1,785	2,225
6490	25 HP motor	▼	1.38	11.590		1,075	710		1,785	2,225

26 24 19 – Motor-Control Centers

26 24 19.40 Motor Starters and Controls		Crew	Daily Output	Labor-Hours	Unit	Material	2020 Bare Costs Labor	Equipment	Total	Total Incl O&P
6500	30 HP motor	2 Elec	1.20	13.380	Ea.	1,075	820		1,895	2,400
6510	40 HP motor		1.14	14.030		2,300	860		3,160	3,800
6520	50 HP motor		1.03	15.530		2,300	955		3,255	3,950
6530	60 HP motor		.83	19.320		4,975	1,175		6,150	7,250
6540	75 HP motor		.70	22.880		4,975	1,400		6,375	7,575
6550	100 HP motor		.64	24.840		4,975	1,525		6,500	7,750
6560	125 HP motor		.55	28.990		13,900	1,775		15,675	18,000
6570	150 HP motor		.48	33.440		13,900	2,050		15,950	18,400
6580	200 HP motor		.46	34.780		13,900	2,125		16,025	18,500
6600	Starter & fused disconnect, 240 volt, 1-2 pole, 0.75 HP motor	1 Elec	1.84	4.350		283	267		550	705
6610	2 HP motor		1.84	4.350		305	267		572	730
6620	5 HP motor		1.66	4.830		410	296		706	890
6630	10 HP motor		1.29	6.210		965	380		1,345	1,625
6640	3 pole, 0.75 HP motor		1.47	5.430		282	335		617	805
6650	5 HP motor		1.29	6.210		360	380		740	960
6660	10 HP motor		1.01	7.910		665	485		1,150	1,450
6690	15 HP motor		.92	8.700		665	535		1,200	1,525
6700	20 HP motor	2 Elec	1.47	10.870		1,100	665		1,765	2,200
6710	25 HP motor		1.47	10.870		1,100	665		1,765	2,200
6720	30 HP motor		1.29	12.420		1,100	760		1,860	2,325
6730	40 HP motor		1.10	14.490		2,400	890		3,290	3,975
6740	50 HP motor		1.10	14.490		2,400	890		3,290	3,975
6750	60 HP motor		.83	19.320		5,075	1,175		6,250	7,375
6760	75 HP motor		.83	19.320		5,075	1,175		6,250	7,375
6770	100 HP motor		.64	24.840		5,075	1,525		6,600	7,875
6780	125 HP motor		.50	32.210		14,200	1,975		16,175	18,600
6790	Combination starter & nonfusible disconnect									
6800	240 volt, 1-2 pole, 0.75 HP motor	1 Elec	1.84	4.350	Ea.	660	267		927	1,125
6810	2 HP motor		1.84	4.350		690	267		957	1,150
6820	5 HP motor		1.38	5.800		710	355		1,065	1,300
6830	10 HP motor		1.10	7.250		1,075	445		1,520	1,825
6840	3 pole, 0.75 HP motor		1.66	4.830		655	296		951	1,150
6850	5 HP motor		1.20	6.690		690	410		1,100	1,375
6860	10 HP motor		.92	8.700		1,075	535		1,610	1,975
6870	15 HP motor		.92	8.700		1,075	535		1,610	1,975
6880	20 HP motor	2 Elec	1.21	13.180		1,775	810		2,585	3,150
6890	25 HP motor		1.21	13.180		1,775	810		2,585	3,150
6900	30 HP motor		1.21	13.180		1,775	810		2,585	3,150
6910	40 HP motor		.74	21.740		3,375	1,325		4,700	5,700
6920	50 HP motor		.74	21.740		3,375	1,325		4,700	5,700
6930	60 HP motor		.64	24.840		7,550	1,525		9,075	10,600
6940	75 HP motor		.64	24.840		7,550	1,525		9,075	10,600
6950	100 HP motor		.64	24.840		7,550	1,525		9,075	10,600
6960	125 HP motor		.55	28.990		19,800	1,775		21,575	24,500
6970	150 HP motor		.55	28.990		19,800	1,775		21,575	24,500
6980	200 HP motor		.55	28.990		19,800	1,775		21,575	24,500
6990	Combination starter and fused disconnect									
7000	240 volt, 1-2 pole, 0.75 HP motor	1 Elec	1.84	4.350	Ea.	720	267		987	1,175
7010	2 HP motor		1.84	4.350		720	267		987	1,175
7020	5 HP motor		1.38	5.800		755	355		1,110	1,350
7030	10 HP motor		1.10	7.250		1,175	445		1,620	1,950
7040	3 pole, 0.75 HP motor		1.66	4.830		720	296		1,016	1,225
7050	5 HP motor		1.20	6.690		755	410		1,165	1,450

For customer support on your Electrical Change Order Costs with RSMeans data, call 800.448.8182. **Pre-Installation Change Order Costs**

26 24 19.40 Motor Starters and Controls		Crew	Daily Output	Labor-Hours	Unit	Material	2020 Bare Costs Labor	2020 Bare Costs Equipment	Total	Total Incl O&P
7060	10 HP motor	1 Elec	.92	8.700	Ea.	1,175	535		1,710	2,100
7070	15 HP motor	↓	.92	8.700		1,175	535		1,710	2,100
7080	20 HP motor	2 Elec	1.21	13.180		1,950	810		2,760	3,350
7090	25 HP motor		1.21	13.180		1,950	810		2,760	3,350
7100	30 HP motor		1.21	13.180		1,950	810		2,760	3,350
7110	40 HP motor		.74	21.740		3,725	1,325		5,050	6,075
7120	50 HP motor		.74	21.740		3,725	1,325		5,050	6,075
7130	60 HP motor		.74	21.740		8,300	1,325		9,625	11,100
7140	75 HP motor		.64	24.840		8,300	1,525		9,825	11,400
7150	100 HP motor		.64	24.840		8,300	1,525		9,825	11,400
7160	125 HP motor		.64	24.840		21,900	1,525		23,425	26,400
7170	150 HP motor		.55	28.990		21,900	1,775		23,675	26,800
7180	200 HP motor	↓	.55	28.990	↓	21,900	1,775		23,675	26,800
7190	Combination starter & circuit breaker disconnect									
7200	240 volt, 1-2 pole, 0.75 HP motor	1 Elec	1.84	4.350	Ea.	680	267		947	1,150
7210	2 HP motor		1.84	4.350		680	267		947	1,150
7220	5 HP motor		1.38	5.800		715	355		1,070	1,325
7230	10 HP motor		1.10	7.250		1,100	445		1,545	1,850
7240	3 pole, 0.75 HP motor		1.66	4.830		705	296		1,001	1,225
7250	5 HP motor		1.20	6.690		735	410		1,145	1,425
7260	10 HP motor		.92	8.700		1,125	535		1,660	2,025
7270	15 HP motor	↓	.92	8.700		1,125	535		1,660	2,025
7280	20 HP motor	2 Elec	1.21	13.180		1,900	810		2,710	3,275
7290	25 HP motor		1.21	13.180		1,900	810		2,710	3,275
7300	30 HP motor		1.21	13.180		1,900	810		2,710	3,275
7310	40 HP motor		.74	21.740		4,125	1,325		5,450	6,500
7320	50 HP motor		.74	21.740		4,125	1,325		5,450	6,500
7330	60 HP motor		.74	21.740		9,525	1,325		10,850	12,500
7340	75 HP motor		.64	24.840		9,525	1,525		11,050	12,800
7350	100 HP motor		.64	24.840		9,525	1,525		11,050	12,800
7360	125 HP motor		.64	24.840		20,600	1,525		22,125	25,000
7370	150 HP motor		.55	28.990		20,600	1,775		22,375	25,400
7380	200 HP motor	↓	.55	28.990	↓	20,600	1,775		22,375	25,400
7400	Magnetic FVNR with enclosure & heaters, 2 pole,									
7410	230 volt, 1 HP, size 00	1 Elec	3.68	2.170	Ea.	197	133		330	415
7420	2 HP, size 0		3.68	2.170		219	133		352	440
7430	3 HP, size 1		2.76	2.900		252	178		430	540
7440	5 HP, size 1P		2.76	2.900		320	178		498	620
7450	115 volt, 1/3 HP, size 00		3.68	2.170		197	133		330	415
7460	1 HP, size 0		3.68	2.170		219	133		352	440
7470	2 HP, size 1		2.76	2.900		252	178		430	540
7480	3 HP, size 1P	↓	2.76	2.900		315	178		493	610
7500	3 pole, 480 volt, 600 HP, size 7	2 Elec	.64	24.840	↓	17,400	1,525		18,925	21,500
7590	Magnetic FVNR with heater, NEMA 1									
7600	600 volt, 3 pole, 5 HP motor	1 Elec	2.12	3.780	Ea.	261	232		493	630
7610	10 HP motor	"	1.47	5.430		293	335		628	820
7620	25 HP motor	2 Elec	2.02	7.910		550	485		1,035	1,325
7630	30 HP motor		1.66	9.660		900	595		1,495	1,875
7640	40 HP motor		1.66	9.660		900	595		1,495	1,875
7650	50 HP motor		1.66	9.660		900	595		1,495	1,875
7660	60 HP motor		1.10	14.490		2,000	890		2,890	3,525
7670	75 HP motor		1.10	14.490		2,425	890		3,315	4,000
7680	100 HP motor	↓	1.10	14.490	↓	2,800	890		3,690	4,400

26 24 19.40 Motor Starters and Controls		Crew	Daily Output	Labor-Hours	Unit	Material	2020 Bare Costs Labor	Equipment	Total	Total Incl O&P
7690	125 HP motor	2 Elec	.83	19.320	Ea.	4,575	1,175		5,750	6,800
7700	150 HP motor		.83	19.320		4,775	1,175		5,950	7,025
7710	200 HP motor		.83	19.320		5,150	1,175		6,325	7,450
7750	Starter & nonfused disconnect, 600 volt, 3 pole, 5 HP motor	1 Elec	1.29	6.210		355	380		735	955
7760	10 HP motor	"	1.01	7.910		390	485		875	1,150
7770	25 HP motor	2 Elec	1.38	11.590		645	710		1,355	1,750
7780	30 HP motor		1.20	13.380		1,075	820		1,895	2,400
7790	40 HP motor		1.20	13.380		1,075	820		1,895	2,400
7800	50 HP motor		1.20	13.380		1,075	820		1,895	2,400
7810	60 HP motor		.85	18.900		2,275	1,150		3,425	4,225
7820	75 HP motor		.85	18.900		2,700	1,150		3,850	4,675
7830	100 HP motor		.77	20.700		3,075	1,275		4,350	5,275
7840	125 HP motor		.64	24.840		4,975	1,525		6,500	7,750
7850	150 HP motor		.64	24.840		5,175	1,525		6,700	7,975
7860	200 HP motor		.55	28.990		5,550	1,775		7,325	8,750
7870	Starter & fused disconnect, 600 volt, 3 pole, 5 HP motor	1 Elec	1.29	6.210		445	380		825	1,050
7880	10 HP motor	"	1.01	7.910		475	485		960	1,250
7890	25 HP motor	2 Elec	1.38	11.590		735	710		1,445	1,850
7900	30 HP motor		1.20	13.380		1,125	820		1,945	2,450
7910	40 HP motor		1.20	13.380		1,125	820		1,945	2,450
7920	50 HP motor		1.20	13.380		1,125	820		1,945	2,450
7930	60 HP motor		.85	18.900		2,400	1,150		3,550	4,375
7940	75 HP motor		.85	18.900		2,825	1,150		3,975	4,825
7950	100 HP motor		.77	20.700		3,200	1,275		4,475	5,425
7960	125 HP motor		.64	24.840		5,150	1,525		6,675	7,925
7970	150 HP motor		.64	24.840		5,350	1,525		6,875	8,150
7980	200 HP motor		.55	28.990		5,725	1,775		7,500	8,950
7990	Combination starter and nonfusible disconnect									
8000	600 volt, 3 pole, 5 HP motor	1 Elec	1.66	4.830	Ea.	680	296		976	1,200
8010	10 HP motor	"	1.20	6.690		915	410		1,325	1,600
8020	25 HP motor	2 Elec	1.84	8.700		1,100	535		1,635	2,025
8030	30 HP motor		1.21	13.180		1,825	810		2,635	3,225
8040	40 HP motor		1.21	13.180		1,825	810		2,635	3,225
8050	50 HP motor		1.21	13.180		2,200	810		3,010	3,625
8060	60 HP motor		.74	21.740		3,500	1,325		4,825	5,825
8070	75 HP motor		.74	21.740		3,500	1,325		4,825	5,825
8080	100 HP motor		.74	21.740		4,350	1,325		5,675	6,775
8090	125 HP motor		.64	24.840		8,050	1,525		9,575	11,200
8100	150 HP motor		.64	24.840		9,375	1,525		10,900	12,600
8110	200 HP motor		.64	24.840		8,450	1,525		9,975	11,600
8140	Combination starter and fused disconnect									
8150	600 volt, 3 pole, 5 HP motor	1 Elec	1.66	4.830	Ea.	720	296		1,016	1,225
8160	10 HP motor	"	1.20	6.690		785	410		1,195	1,475
8170	25 HP motor	2 Elec	1.84	8.700		1,275	535		1,810	2,200
8180	30 HP motor		1.21	13.180		2,000	810		2,810	3,400
8190	40 HP motor		1.21	13.180		2,150	810		2,960	3,550
8200	50 HP motor		1.21	13.180		2,150	810		2,960	3,550
8210	60 HP motor		.74	21.740		3,750	1,325		5,075	6,100
8220	75 HP motor		.74	21.740		3,750	1,325		5,075	6,100
8230	100 HP motor		.74	21.740		3,750	1,325		5,075	6,100
8240	125 HP motor		.64	24.840		8,300	1,525		9,825	11,400
8250	150 HP motor		.64	24.840		8,300	1,525		9,825	11,400
8260	200 HP motor		.64	24.840		8,300	1,525		9,825	11,400

For customer support on your Electrical Change Order Costs with RSMeans data, call 800.448.8182. **Pre-Installation Change Order Costs**

26 24 19.40 Motor Starters and Controls	Crew	Daily Output	Labor-Hours	Unit	Material	2020 Bare Costs Labor	Equipment	Total	Total Incl O&P
8290 Combination starter & circuit breaker disconnect									
8300 600 volt, 3 pole, 5 HP motor	1 Elec	1.66	4.830	Ea.	1,025	296		1,321	1,575
8310 10 HP motor	"	1.20	6.690		1,000	410		1,410	1,700
8320 25 HP motor	2 Elec	1.84	8.700		1,500	535		2,035	2,450
8330 30 HP motor		1.21	13.180		1,900	810		2,710	3,275
8340 40 HP motor		1.21	13.180		1,900	810		2,710	3,275
8350 50 HP motor		1.21	13.180		1,975	810		2,785	3,350
8360 60 HP motor		.74	21.740		4,125	1,325		5,450	6,500
8370 75 HP motor		.74	21.740		4,125	1,325		5,450	6,500
8380 100 HP motor		.74	21.740		4,275	1,325		5,600	6,675
8390 125 HP motor		.64	24.840		9,525	1,525		11,050	12,800
8400 150 HP motor		.64	24.840		9,125	1,525		10,650	12,300
8410 200 HP motor	▼	.64	24.840	▼	9,875	1,525		11,400	13,100
8430 Starter & circuit breaker disconnect									
8440 600 volt, 3 pole, 5 HP motor	1 Elec	1.29	6.210	Ea.	845	380		1,225	1,500
8450 10 HP motor	"	1.01	7.910		875	485		1,360	1,675
8460 25 HP motor	2 Elec	1.38	11.590		1,125	710		1,835	2,300
8470 30 HP motor		1.20	13.380		1,600	820		2,420	3,000
8480 40 HP motor		1.20	13.380		1,600	820		2,420	3,000
8490 50 HP motor		1.20	13.380		1,600	820		2,420	3,000
8500 60 HP motor		.85	18.900		3,875	1,150		5,025	5,975
8510 75 HP motor		.85	18.900		4,300	1,150		5,450	6,450
8520 100 HP motor		.77	20.700		4,675	1,275		5,950	7,050
8530 125 HP motor		.64	24.840		6,450	1,525		7,975	9,375
8540 150 HP motor		.64	24.840		6,650	1,525		8,175	9,575
8550 200 HP motor	▼	.55	28.990		8,350	1,775		10,125	11,800
8900 240 volt, 1-2 pole, 0.75 HP motor	1 Elec	1.84	4.350		800	267		1,067	1,275
8910 2 HP motor		1.84	4.350		820	267		1,087	1,300
8920 5 HP motor		1.66	4.830		925	296		1,221	1,475
8930 10 HP motor		1.29	6.210		1,550	380		1,930	2,300
8950 3 pole, 0.75 HP motor		1.47	5.430		795	335		1,130	1,375
8970 5 HP motor		1.29	6.210		875	380		1,255	1,525
8980 10 HP motor		1.01	7.910		1,250	485		1,735	2,100
8990 15 HP motor	▼	.92	8.700		1,250	535		1,785	2,175
9100 20 HP motor	2 Elec	1.38	11.590		1,700	710		2,410	2,925
9110 25 HP motor		1.38	11.590		1,700	710		2,410	2,925
9120 30 HP motor		1.20	13.380		1,700	820		2,520	3,100
9130 40 HP motor		1.14	14.030		3,875	860		4,735	5,525
9140 50 HP motor		1.03	15.530		3,875	955		4,830	5,675
9150 60 HP motor		.83	19.320		7,875	1,175		9,050	10,400
9160 75 HP motor		.70	22.880		7,875	1,400		9,275	10,800
9170 100 HP motor		.64	24.840		7,875	1,525		9,400	10,900
9180 125 HP motor		.55	28.990		17,700	1,775		19,475	22,200
9190 150 HP motor		.48	33.440		17,700	2,050		19,750	22,600
9200 200 HP motor	▼	.46	34.780	▼	17,700	2,125		19,825	22,700

26 25 13.10 Aluminum Bus Duct	Crew	Daily Output	Labor-Hours	Unit	Material	2020 Bare Costs Labor	Equipment	Total	Total Incl O&P
0010 **ALUMINUM BUS DUCT** 10 ft. long									
0050 Indoor 3 pole 4 wire, plug-in, straight section, 225 amp	2 Elec	40.48	.395	L.F.	172	24.50		196.50	225
0100 400 amp		33.12	.483		216	29.50		245.50	281
0150 600 amp		29.44	.543		298	33.50		331.50	380
0200 800 amp		23.92	.669		345	41		386	440
0250 1,000 amp		22.08	.725		435	44.50		479.50	545
0300 1,350 amp		20.24	.791		305	48.50		353.50	405
0310 1,600 amp		16.56	.966		330	59.50		389.50	450
0320 2,000 amp		14.72	1.090		425	67		492	570
0330 2,500 amp		12.88	1.240		460	76		536	620
0340 3,000 amp		11.04	1.450		595	89		684	785
0350 Feeder, 600 amp		31.28	.512		113	31.50		144.50	172
0400 800 amp		25.76	.621		430	38		468	525
0450 1,000 amp		23.92	.669		535	41		576	645
0455 1,200 amp		23	.696		650	42.50		692.50	780
0500 1,350 amp		22.08	.725		236	44.50		280.50	325
0550 1,600 amp		18.40	.870		900	53.50		953.50	1,075
0600 2,000 amp		16.56	.966		1,100	59.50		1,159.50	1,300
0620 2,500 amp		12.88	1.240		1,350	76		1,426	1,600
0630 3,000 amp		11.04	1.450		1,575	89		1,664	1,850
0640 4,000 amp		9.20	1.740		680	107		787	910
0650 Elbow, 225 amp		4.05	3.950	Ea.	835	242		1,077	1,275
0700 400 amp		3.50	4.580		840	281		1,121	1,350
0750 600 amp		3.13	5.120		845	315		1,160	1,400
0800 800 amp		2.76	5.800		875	355		1,230	1,500
0850 1,000 amp		2.58	6.210		1,125	380		1,505	1,825
0870 1,200 amp		2.48	6.440		1,400	395		1,795	2,150
0900 1,350 amp		2.39	6.690		1,350	410		1,760	2,075
0950 1,600 amp		2.21	7.250		1,425	445		1,870	2,225
1000 2,000 amp		1.84	8.700		1,575	535		2,110	2,525
1020 2,500 amp		1.66	9.660		1,850	595		2,445	2,900
1030 3,000 amp		1.47	10.870		2,125	665		2,790	3,350
1040 4,000 amp		1.29	12.420		3,450	760		4,210	4,925
1100 Cable tap box end, 225 amp		3.31	4.830		1,775	296		2,071	2,400
1150 400 amp		2.94	5.430		2,175	335		2,510	2,900
1200 600 amp		2.39	6.690		1,475	410		1,885	2,225
1250 800 amp		2.02	7.910		1,825	485		2,310	2,725
1300 1,000 amp		1.84	8.700		3,500	535		4,035	4,650
1320 1,200 amp		1.84	8.700		1,875	535		2,410	2,875
1350 1,350 amp		1.47	10.870		1,550	665		2,215	2,700
1400 1,600 amp		1.29	12.420		1,825	760		2,585	3,125
1450 2,000 amp		1.10	14.490		2,075	890		2,965	3,600
1460 2,500 amp		.92	17.390		2,475	1,075		3,550	4,300
1470 3,000 amp		.74	21.740		2,725	1,325		4,050	4,975
1480 4,000 amp		.55	28.990		3,225	1,775		5,000	6,200
1500 Switchboard stub, 225 amp		5.34	3		1,850	184		2,034	2,300
1550 400 amp		4.97	3.220		1,875	198		2,073	2,375
1600 600 amp		4.23	3.780		2,100	232		2,332	2,675
1650 800 amp		3.68	4.350		1,900	267		2,167	2,500
1700 1,000 amp		2.94	5.430		1,700	335		2,035	2,375
1720 1,200 amp		2.85	5.610		1,750	345		2,095	2,425
1750 1,350 amp		2.76	5.800		1,725	355		2,080	2,425
1800 1,600 amp		2.39	6.690		1,825	410		2,235	2,600

26 25 Low-Voltage Enclosed Bus Assemblies

26 25 13 – Low-Voltage Busways

26 25 13.10 Aluminum Bus Duct		Crew	Daily Output	Labor-Hours	Unit	Material	2020 Bare Costs Labor	Equipment	Total	Total Incl O&P
1850	2,000 amp	2 Elec	2.21	7.250	Ea.	1,950	445		2,395	2,800
1860	2,500 amp		2.02	7.910		2,175	485		2,660	3,100
1870	3,000 amp		1.84	8.700		2,275	535		2,810	3,300
1880	4,000 amp		1.66	9.660		2,525	595		3,120	3,650
1890	Tee fittings, 225 amp		2.94	5.430		895	335		1,230	1,475
1900	400 amp		2.58	6.210		895	380		1,275	1,550
1950	600 amp		2.39	6.690		895	410		1,305	1,600
2000	800 amp		2.21	7.250		950	445		1,395	1,700
2050	1,000 amp		2.02	7.910		1,000	485		1,485	1,825
2070	1,200 amp		1.93	8.280		1,175	510		1,685	2,050
2100	1,350 amp		1.84	8.700		1,625	535		2,160	2,600
2150	1,600 amp		1.47	10.870		1,975	665		2,640	3,150
2200	2,000 amp		1.10	14.490		2,175	890		3,065	3,725
2220	2,500 amp		.92	17.390		2,600	1,075		3,675	4,450
2230	3,000 amp		.74	21.740		2,975	1,325		4,300	5,250
2240	4,000 amp		.55	28.990		4,925	1,775		6,700	8,075
2300	Wall flange, 600 amp		18.40	.870		246	53.50		299.50	350
2310	800 amp		14.72	1.090		246	67		313	370
2320	1,000 amp		11.96	1.340		246	82		328	395
2325	1,200 amp		11.04	1.450		246	89		335	405
2330	1,350 amp		9.94	1.610		246	99		345	420
2340	1,600 amp		8.28	1.930		246	118		364	445
2350	2,000 amp		7.36	2.170		246	133		379	470
2360	2,500 amp		6.07	2.640		246	162		408	510
2370	3,000 amp		4.97	3.220		246	198		444	565
2380	4,000 amp		3.68	4.350		390	267		657	825
2390	5,000 amp		2.76	5.800		390	355		745	960
2400	Vapor barrier		7.36	2.170		460	133		593	705
2420	Roof flange kit		3.68	4.350		885	267		1,152	1,375
2600	Expansion fitting, 225 amp		9.20	1.740		1,400	107		1,507	1,675
2610	400 amp		7.36	2.170		1,400	133		1,533	1,725
2620	600 amp		5.52	2.900		1,400	178		1,578	1,800
2630	800 amp		4.23	3.780		1,675	232		1,907	2,200
2640	1,000 amp		3.68	4.350		1,800	267		2,067	2,375
2650	1,350 amp		3.31	4.830		2,075	296		2,371	2,750
2660	1,600 amp		2.94	5.430		2,500	335		2,835	3,250
2670	2,000 amp		2.58	6.210		2,775	380		3,155	3,625
2680	2,500 amp		2.21	7.250		3,350	445		3,795	4,350
2690	3,000 amp		1.84	8.700		3,875	535		4,410	5,050
2700	4,000 amp		1.47	10.870		5,100	665		5,765	6,600
2800	Reducer nonfused, 400 amp		7.36	2.170		820	133		953	1,100
2810	600 amp		5.52	2.900		820	178		998	1,175
2820	800 amp		4.23	3.780		985	232		1,217	1,425
2830	1,000 amp		3.68	4.350		1,150	267		1,417	1,675
2840	1,350 amp		3.31	4.830		1,500	296		1,796	2,100
2850	1,600 amp		2.94	5.430		2,050	335		2,385	2,750
2860	2,000 amp		2.58	6.210		2,350	380		2,730	3,150
2870	2,500 amp		2.21	7.250		2,950	445		3,395	3,900
2880	3,000 amp		1.84	8.700		3,400	535		3,935	4,550
2890	4,000 amp		1.47	10.870		4,550	665		5,215	6,000
2950	Reducer fuse included, 225 amp		4.05	3.950		2,400	242		2,642	3,000
2960	400 amp		3.86	4.140		2,450	254		2,704	3,075
2970	600 amp		3.31	4.830		2,900	296		3,196	3,625

Pre-Installation Change Order Costs   For customer support on your Electrical Change Order Costs with RSMeans data, call 800.448.8182.   155

26 25 13.10 Aluminum Bus Duct		Crew	Daily Output	Labor-Hours	Unit	Material	2020 Bare Costs Labor	Equipment	Total	Total Incl O&P
2980	800 amp	2 Elec	2.94	5.430	Ea.	4,600	335		4,935	5,550
2990	1,000 amp		2.76	5.800		5,250	355		5,605	6,300
3000	1,200 amp		2.58	6.210		5,250	380		5,630	6,350
3010	1,600 amp		2.02	7.910		12,000	485		12,485	13,900
3020	2,000 amp		1.66	9.660		13,300	595		13,895	15,500
3100	Reducer circuit breaker, 225 amp		4.05	3.950		2,375	242		2,617	2,950
3110	400 amp		3.86	4.140		2,875	254		3,129	3,550
3120	600 amp		3.31	4.830		4,100	296		4,396	4,950
3130	800 amp		2.94	5.430		4,800	335		5,135	5,775
3140	1,000 amp		2.76	5.800		5,450	355		5,805	6,525
3150	1,200 amp		2.58	6.210		6,550	380		6,930	7,775
3160	1,600 amp		2.02	7.910		9,675	485		10,160	11,400
3170	2,000 amp		1.66	9.660		10,600	595		11,195	12,600
3250	Reducer circuit breaker, 75,000 AIC, 225 amp		4.05	3.950		3,700	242		3,942	4,425
3260	400 amp		3.86	4.140		3,700	254		3,954	4,450
3270	600 amp		3.31	4.830		4,950	296		5,246	5,900
3280	800 amp		2.94	5.430		5,450	335		5,785	6,475
3290	1,000 amp		2.76	5.800		8,750	355		9,105	10,200
3300	1,200 amp		2.58	6.210		8,750	380		9,130	10,200
3310	1,600 amp		2.02	7.910		9,675	485		10,160	11,400
3320	2,000 amp		1.66	9.660		10,600	595		11,195	12,600
3400	Reducer circuit breaker CLF 225 amp		4.05	3.950		3,800	242		4,042	4,550
3410	400 amp		3.86	4.140		4,525	254		4,779	5,350
3420	600 amp		3.31	4.830		6,775	296		7,071	7,900
3430	800 amp		2.94	5.430		7,075	335		7,410	8,275
3440	1,000 amp		2.76	5.800		7,375	355		7,730	8,625
3450	1,200 amp		2.58	6.210		9,625	380		10,005	11,200
3460	1,600 amp		2.02	7.910		9,675	485		10,160	11,400
3470	2,000 amp		1.66	9.660	↓	10,600	595		11,195	12,600
3550	Ground bus added to bus duct, 225 amp		294.40	.054	L.F.	33.50	3.33		36.83	42
3560	400 amp		294.40	.054		33.50	3.33		36.83	42
3570	600 amp		257.60	.062		33.50	3.81		37.31	42.50
3580	800 amp		220.80	.072		33.50	4.45		37.95	43.50
3590	1,000 amp		184	.087		33.50	5.35		38.85	45
3600	1,350 amp		165.60	.097		33.50	5.95		39.45	46
3610	1,600 amp		147.20	.109		33.50	6.65		40.15	47
3620	2,000 amp		147.20	.109		33.50	6.65		40.15	47
3630	2,500 amp		128.80	.124		33.50	7.60		41.10	48.50
3640	3,000 amp		110.40	.145		33.50	8.90		42.40	50.50
3650	4,000 amp		92	.174		33.50	10.65		44.15	53
3810	High short circuit, 400 amp		33.12	.483		208	29.50		237.50	273
3820	600 amp		29.44	.543		360	33.50		393.50	445
3830	800 amp		23.92	.669		415	41		456	515
3840	1,000 amp		22.08	.725		520	44.50		564.50	640
3850	1,350 amp		20.24	.791		189	48.50		237.50	280
3860	1,600 amp		16.56	.966		217	59.50		276.50	330
3870	2,000 amp		14.72	1.090		255	67		322	380
3880	2,500 amp		12.88	1.240		425	76		501	580
3890	3,000 amp		11.04	1.450	↓	480	89		569	660
3920	Cross, 225 amp		5.15	3.110	Ea.	1,350	191		1,541	1,750
3930	400 amp		4.23	3.780		1,350	232		1,582	1,825
3940	600 amp		3.68	4.350		1,350	267		1,617	1,875
3950	800 amp	↓	3.13	5.120	↓	1,400	315		1,715	2,025

Pre-Installation Change Order Costs

26 25 13.10 Aluminum Bus Duct		Crew	Daily Output	Labor-Hours	Unit	Material	2020 Bare Costs Labor	Equipment	Total	Total Incl O&P
3960	1,000 amp	2 Elec	2.76	5.800	Ea.	1,475	355		1,830	2,150
3970	1,350 amp		2.58	6.210		2,400	380		2,780	3,225
3980	1,600 amp		2.02	7.910		2,850	485		3,335	3,875
3990	2,000 amp		1.66	9.660		3,125	595		3,720	4,325
4000	2,500 amp		1.47	10.870		3,700	665		4,365	5,075
4010	3,000 amp		1.10	14.490		4,275	890		5,165	6,025
4020	4,000 amp		.92	17.390		6,525	1,075		7,600	8,775
4040	Cable tap box center, 225 amp		3.31	4.830		2,975	296		3,271	3,725
4050	400 amp		2.94	5.430		3,375	335		3,710	4,225
4060	600 amp		2.39	6.690		4,475	410		4,885	5,500
4070	800 amp		2.02	7.910		4,925	485		5,410	6,125
4080	1,000 amp		1.84	8.700		6,200	535		6,735	7,625
4090	1,350 amp		1.47	10.870		1,450	665		2,115	2,600
4100	1,600 amp		1.29	12.420		1,625	760		2,385	2,925
4110	2,000 amp		1.10	14.490		1,875	890		2,765	3,375
4120	2,500 amp		.92	17.390		2,275	1,075		3,350	4,100
4130	3,000 amp		.74	21.740		2,500	1,325		3,825	4,725
4140	4,000 amp		.55	28.990		3,350	1,775		5,125	6,325
4500	Weatherproof 3 pole 4 wire, feeder, 600 amp		27.60	.580	L.F.	136	35.50		171.50	202
4520	800 amp		22.08	.725		159	44.50		203.50	241
4540	1,000 amp		20.24	.791		181	48.50		229.50	271
4550	1,200 amp		19.32	.828		251	51		302	350
4560	1,350 amp		18.40	.870		283	53.50		336.50	390
4580	1,600 amp		15.64	1.020		330	62.50		392.50	455
4600	2,000 amp		14.72	1.090		385	67		452	525
4620	2,500 amp		11.04	1.450		500	89		589	680
4640	3,000 amp		9.20	1.740		565	107		672	785
4660	4,000 amp		7.36	2.170		815	133		948	1,100
5000	Indoor 3 pole, 3 wire, feeder, 600 amp		36.80	.435		104	26.50		130.50	154
5010	800 amp		29.44	.543		355	33.50		388.50	440
5020	1,000 amp		27.60	.580		395	35.50		430.50	490
5025	1,200 amp		26.68	.600		460	37		497	565
5030	1,350 amp		25.76	.621		179	38		217	254
5040	1,600 amp		22.08	.725		755	44.50		799.50	895
5050	2,000 amp		18.40	.870		900	53.50		953.50	1,075
5060	2,500 amp		14.72	1.090		1,100	67		1,167	1,300
5070	3,000 amp		12.88	1.240		1,250	76		1,326	1,500
5080	4,000 amp		11.04	1.450		490	89		579	670
5200	Plug-in type, 225 amp		46	.348		139	21.50		160.50	185
5210	400 amp		38.64	.414		170	25.50		195.50	225
5220	600 amp		33.12	.483		211	29.50		240.50	276
5230	800 amp		27.60	.580		295	35.50		330.50	380
5240	1,000 amp		25.76	.621		335	38		373	420
5245	1,200 amp		24.84	.644		236	39.50		275.50	320
5250	1,350 amp		23.92	.669		189	41		230	269
5260	1,600 amp		18.40	.870		217	53.50		270.50	320
5270	2,000 amp		16.56	.966		255	59.50		314.50	370
5280	2,500 amp		14.72	1.090		350	67		417	485
5290	3,000 amp		12.88	1.240		415	76		491	570
5300	4,000 amp		11.04	1.450		500	89		589	680
5330	High short circuit, 400 amp		38.64	.414		204	25.50		229.50	262
5340	600 amp		33.12	.483		252	29.50		281.50	320
5350	800 amp		27.60	.580		355	35.50		390.50	450

26 25 13.10 Aluminum Bus Duct	Crew	Daily Output	Labor-Hours	Unit	Material	2020 Bare Costs Labor	Equipment	Total	Total Incl O&P
5360	1,000 amp	2 Elec	25.76	.621	L.F.	405	38		443	500
5370	1,350 amp		23.92	.669		189	41		230	269
5380	1,600 amp		18.40	.870		217	53.50		270.50	320
5390	2,000 amp		16.56	.966		255	59.50		314.50	370
5400	2,500 amp		14.72	1.090		400	67		467	540
5410	3,000 amp		12.88	1.240		415	76		491	570
5440	Elbow, 225 amp		4.60	3.480	Ea.	655	214		869	1,050
5450	400 amp		4.05	3.950		655	242		897	1,075
5460	600 amp		3.68	4.350		655	267		922	1,125
5470	800 amp		3.13	5.120		700	315		1,015	1,250
5480	1,000 amp		2.94	5.430		715	335		1,050	1,275
5485	1,200 amp		2.85	5.610		760	345		1,105	1,350
5490	1,350 amp		2.76	5.800		730	355		1,085	1,325
5500	1,600 amp		2.58	6.210		1,125	380		1,505	1,800
5510	2,000 amp		2.21	7.250		1,225	445		1,670	2,000
5520	2,500 amp		1.84	8.700		1,500	535		2,035	2,450
5530	3,000 amp		1.66	9.660		1,800	595		2,395	2,850
5540	4,000 amp		1.47	10.870		2,400	665		3,065	3,625
5560	Tee fittings, 225 amp		3.31	4.830		800	296		1,096	1,325
5570	400 amp		2.94	5.430		800	335		1,135	1,375
5580	600 amp		2.76	5.800		800	355		1,155	1,425
5590	800 amp		2.58	6.210		860	380		1,240	1,500
5600	1,000 amp		2.39	6.690		890	410		1,300	1,575
5605	1,200 amp		2.30	6.960		945	425		1,370	1,675
5610	1,350 amp		2.21	7.250		1,325	445		1,770	2,100
5620	1,600 amp		1.66	9.660		1,550	595		2,145	2,575
5630	2,000 amp		1.29	12.420		1,725	760		2,485	3,025
5640	2,500 amp		1.10	14.490		2,150	890		3,040	3,675
5650	3,000 amp		.92	17.390		2,525	1,075		3,600	4,375
5660	4,000 amp		.64	24.840		3,700	1,525		5,225	6,325
5680	Cross, 225 amp		5.89	2.720		1,300	167		1,467	1,675
5690	400 amp		4.97	3.220		1,300	198		1,498	1,725
5700	600 amp		4.23	3.780		1,300	232		1,532	1,775
5710	800 amp		3.68	4.350		1,375	267		1,642	1,900
5720	1,000 amp		3.31	4.830		1,400	296		1,696	2,000
5730	1,350 amp		2.94	5.430		2,100	335		2,435	2,825
5740	1,600 amp		2.39	6.690		2,450	410		2,860	3,300
5750	2,000 amp		2.02	7.910		2,675	485		3,160	3,675
5760	2,500 amp		1.66	9.660		3,250	595		3,845	4,450
5770	3,000 amp		1.29	12.420		3,875	760		4,635	5,375
5780	4,000 amp		1.10	14.490		5,375	890		6,265	7,250
5800	Expansion fitting, 225 amp		10.67	1.500		1,100	92		1,192	1,350
5810	400 amp		8.46	1.890		1,100	116		1,216	1,400
5820	600 amp		6.44	2.480		1,100	152		1,252	1,450
5830	800 amp		4.78	3.340		1,325	205		1,530	1,750
5840	1,000 amp		4.23	3.780		1,450	232		1,682	1,925
5850	1,350 amp		3.86	4.140		1,475	254		1,729	2,000
5860	1,600 amp		3.31	4.830		1,825	296		2,121	2,475
5870	2,000 amp		2.94	5.430		2,175	335		2,510	2,900
5880	2,500 amp		2.58	6.210		2,450	380		2,830	3,275
5890	3,000 amp		2.21	7.250		2,875	445		3,320	3,800
5900	4,000 amp		1.66	9.660		3,925	595		4,520	5,200
5940	Reducer, nonfused, 400 amp		8.46	1.890		705	116		821	950

Pre-Installation Change Order Costs

26 25 13.10 Aluminum Bus Duct		Crew	Daily Output	Labor-Hours	Unit	Material	2020 Bare Costs Labor	Equipment	Total	Total Incl O&P
5950	600 amp	2 Elec	6.44	2.480	Ea.	705	152		857	1,000
5960	800 amp		4.78	3.340		740	205		945	1,125
5970	1,000 amp		4.23	3.780		910	232		1,142	1,350
5980	1,350 amp		3.86	4.140		1,350	254		1,604	1,850
5990	1,600 amp		3.31	4.830		1,525	296		1,821	2,150
6000	2,000 amp		2.94	5.430		1,800	335		2,135	2,475
6010	2,500 amp		2.58	6.210		2,300	380		2,680	3,100
6020	3,000 amp		2.02	7.910		2,675	485		3,160	3,675
6030	4,000 amp		1.66	9.660		3,675	595		4,270	4,925
6050	Reducer, fuse included, 225 amp		4.60	3.480		1,850	214		2,064	2,350
6060	400 amp		4.42	3.620		2,450	222		2,672	3,025
6070	600 amp		3.86	4.140		3,125	254		3,379	3,825
6080	800 amp		3.31	4.830		4,850	296		5,146	5,800
6090	1,000 amp		3.13	5.120		5,300	315		5,615	6,300
6100	1,350 amp		2.94	5.430		9,675	335		10,010	11,200
6110	1,600 amp		2.39	6.690		11,500	410		11,910	13,200
6120	2,000 amp		1.84	8.700		13,300	535		13,835	15,400
6160	Reducer, circuit breaker, 225 amp		4.60	3.480		2,275	214		2,489	2,825
6170	400 amp		4.42	3.620		2,775	222		2,997	3,375
6180	600 amp		3.86	4.140		4,000	254		4,254	4,775
6190	800 amp		3.31	4.830		4,650	296		4,946	5,575
6200	1,000 amp		3.13	5.120		5,300	315		5,615	6,300
6210	1,350 amp		2.94	5.430		6,400	335		6,735	7,550
6220	1,600 amp		2.39	6.690		9,550	410		9,960	11,100
6230	2,000 amp		1.84	8.700		10,400	535		10,935	12,300
6270	Cable tap box center, 225 amp		3.86	4.140		1,025	254		1,279	1,500
6280	400 amp		3.31	4.830		1,025	296		1,321	1,575
6290	600 amp		2.76	5.800		1,025	355		1,380	1,650
6300	800 amp		2.39	6.690		1,125	410		1,535	1,825
6310	1,000 amp		2.21	7.250		1,200	445		1,645	1,950
6320	1,350 amp		1.66	9.660		1,250	595		1,845	2,275
6330	1,600 amp		1.47	10.870		1,425	665		2,090	2,550
6340	2,000 amp		1.29	12.420		1,625	760		2,385	2,900
6350	2,500 amp		1.10	14.490		2,025	890		2,915	3,550
6360	3,000 amp		.92	17.390		2,300	1,075		3,375	4,125
6370	4,000 amp		.64	24.840		2,725	1,525		4,250	5,250
6390	Cable tap box end, 225 amp		3.86	4.140		625	254		879	1,075
6400	400 amp		3.31	4.830		625	296		921	1,125
6410	600 amp		2.76	5.800		625	355		980	1,225
6420	800 amp		2.39	6.690		685	410		1,095	1,375
6430	1,000 amp		2.21	7.250		740	445		1,185	1,475
6435	1,200 amp		1.93	8.280		805	510		1,315	1,650
6440	1,350 amp		1.66	9.660		785	595		1,380	1,750
6450	1,600 amp		1.47	10.870		890	665		1,555	1,975
6460	2,000 amp		1.29	12.420		1,000	760		1,760	2,225
6470	2,500 amp		1.10	14.490		1,175	890		2,065	2,600
6480	3,000 amp		.92	17.390		1,375	1,075		2,450	3,125
6490	4,000 amp		.64	24.840		1,650	1,525		3,175	4,100
7000	Weatherproof 3 pole 3 wire, feeder, 600 amp		31.28	.512	L.F.	134	31.50		165.50	194
7020	800 amp		25.76	.621		148	38		186	219
7040	1,000 amp		23.92	.669		159	41		200	236
7050	1,200 amp		23	.696		185	42.50		227.50	268
7060	1,350 amp		22.08	.725		215	44.50		259.50	305

26 25 13 – Low-Voltage Busways

26 25 13.10 Aluminum Bus Duct

		Crew	Daily Output	Labor-Hours	Unit	Material	2020 Bare Costs Labor	Equipment	Total	Total Incl O&P
7080	1,600 amp	2 Elec	18.40	.870	L.F.	249	53.50		302.50	355
7100	2,000 amp		16.56	.966		294	59.50		353.50	415
7120	2,500 amp		12.88	1.240		410	76		486	565
7140	3,000 amp		11.04	1.450		485	89		574	665
7160	4,000 amp		9.20	1.740		590	107		697	810

26 25 13.20 Bus Duct

		Crew	Daily Output	Labor-Hours	Unit	Material	2020 Bare Costs Labor	Equipment	Total	Total Incl O&P
0010	**BUS DUCT** 100 amp and less, aluminum or copper, plug-in									
0080	Bus duct, 3 pole 3 wire, 100 amp	1 Elec	38.64	.207	L.F.	78	12.70		90.70	105
0110	Elbow		3.68	2.170	Ea.	132	133		265	345
0120	Tee		1.84	4.350		187	267		454	600
0130	Wall flange		7.36	1.090		27.50	67		94.50	130
0140	Ground kit		14.72	.543		61.50	33.50		95	117
0180	3 pole 4 wire, 100 amp		36.80	.217	L.F.	84.50	13.35		97.85	112
0200	Cable tap box		2.85	2.810	Ea.	258	172		430	540
0300	End closure		14.72	.543		34.50	33.50		68	87.50
0400	Elbow		3.68	2.170		216	133		349	435
0500	Tee		1.84	4.350		315	267		582	740
0600	Hangers		9.20	.870		24.50	53.50		78	107
0700	Circuit breakers, 15 to 50 amp, 1 pole		7.36	1.090		670	67		737	840
0800	15 to 60 amp, 2 pole		6.16	1.300		360	80		440	515
0900	3 pole		4.88	1.640		450	101		551	650
1000	60 to 100 amp, 1 pole		6.16	1.300		705	80		785	895
1100	70 to 100 amp, 2 pole		4.88	1.640		1,300	101		1,401	1,575
1200	3 pole		4.14	1.930		1,750	118		1,868	2,100
1220	Switch, nonfused, 3 pole, 4 wire		7.36	1.090		219	67		286	340
1240	Fused, 3 fuses, 4 wire, 30 amp		7.36	1.090		360	67		427	500
1260	60 amp		4.88	1.640		425	101		526	615
1280	100 amp		4.14	1.930		570	118		688	800
1300	Plug, fusible, 3 pole 250 volt, 30 amp		4.88	1.640		535	101		636	735
1310	60 amp		4.88	1.640		600	101		701	810
1320	100 amp		4.14	1.930		815	118		933	1,075
1330	3 pole 480 volt, 30 amp		4.88	1.640		540	101		641	745
1340	60 amp		4.88	1.640		585	101		686	790
1350	100 amp		4.14	1.930		845	118		963	1,100
1360	Circuit breaker, 3 pole 250 volt, 60 amp		4.88	1.640		960	101		1,061	1,200
1370	3 pole 480 volt, 100 amp		4.14	1.930		960	118		1,078	1,225
2000	Bus duct, 2 wire, 250 volt, 30 amp		55.20	.145	L.F.	8.35	8.90		17.25	22.50
2100	60 amp		46	.174		8.35	10.65		19	25
2200	300 volt, 30 amp		55.20	.145		8.35	8.90		17.25	22.50
2300	60 amp		46	.174		8.35	10.65		19	25
2400	3 wire, 250 volt, 30 amp		55.20	.145		10.80	8.90		19.70	25
2500	60 amp		46	.174		10.55	10.65		21.20	27.50
2600	480/277 volt, 30 amp		55.20	.145		10.80	8.90		19.70	25
2700	60 amp		46	.174		10.80	10.65		21.45	28
2750	End feed, 300 volt 2 wire max. 30 amp		5.52	1.450	Ea.	75.50	89		164.50	215
2800	60 amp		5.06	1.580		75.50	97		172.50	227
2850	30 amp miniature		5.52	1.450		75.50	89		164.50	215
2900	3 wire, 30 amp		5.52	1.450		95	89		184	236
2950	60 amp		5.06	1.580		95	97		192	248
3000	30 amp miniature		5.52	1.450		95	89		184	236
3050	Center feed, 300 volt 2 wire, 30 amp		5.52	1.450		104	89		193	247
3100	60 amp		5.06	1.580		104	97		201	259

For customer support on your Electrical Change Order Costs with RSMeans data, call 800.448.8182. **Pre-Installation Change Order Costs**

26 25 Low-Voltage Enclosed Bus Assemblies

26 25 13 – Low-Voltage Busways

26 25 13.20 Bus Duct	Crew	Daily Output	Labor-Hours	Unit	Material	Labor	Equipment	Total	Total Incl O&P
						2020 Bare Costs			
3150 3 wire, 30 amp	1 Elec	5.52	1.450	Ea.	119	89		208	262
3200 60 amp		5.06	1.580		119	97		216	274
3220 Elbow, 30 amp		5.52	1.450		40.50	89		129.50	177
3240 60 amp		5.06	1.580		40.50	97		137.50	189
3260 End cap		36.80	.217		9.90	13.35		23.25	31
3280 Strength beam, 10'		13.80	.580		24.50	35.50		60	80
3300 Hanger		22.08	.362		5.25	22		27.25	39
3320 Tap box, nonfusible		5.80	1.380		84.50	84.50		169	219
3340 Fusible switch 30 amp, 1 fuse		5.52	1.450		465	89		554	640
3360 2 fuse		5.52	1.450		485	89		574	665
3380 3 fuse		5.52	1.450		530	89		619	715
3400 Circuit breaker handle on cover, 1 pole		5.52	1.450		62	89		151	201
3420 2 pole		5.52	1.450		580	89		669	765
3440 3 pole		5.52	1.450		710	89		799	915
3460 Circuit breaker external operhandle, 1 pole		5.52	1.450		74	89		163	214
3480 2 pole		5.52	1.450		750	89		839	955
3500 3 pole		5.52	1.450		815	89		904	1,025
3520 Terminal plug only		14.72	.543		99.50	33.50		133	160
3540 Terminal with receptacle		14.72	.543		109	33.50		142.50	170
3560 Fixture plug		14.72	.543		73.50	33.50		107	131
4000 Copper bus duct, lighting, 2 wire 300 volt, 20 amp		64.40	.124	L.F.	7.55	7.60		15.15	19.65
4020 35 amp		55.20	.145		7.55	8.90		16.45	21.50
4040 50 amp		50.60	.158		7.55	9.70		17.25	23
4060 60 amp		46	.174		7.55	10.65		18.20	24
4080 3 wire 300 volt, 20 amp		64.40	.124		7.15	7.60		14.75	19.20
4100 35 amp		55.20	.145		6.95	8.90		15.85	21
4120 50 amp		50.60	.158		7.15	9.70		16.85	22.50
4140 60 amp		46	.174		7.15	10.65		17.80	24
4160 Feeder in box, end, 1 circuit		5.52	1.450	Ea.	101	89		190	243
4180 2 circuit		5.06	1.580		106	97		203	260
4200 Center, 1 circuit		5.52	1.450		140	89		229	286
4220 2 circuit		5.06	1.580		145	97		242	305
4240 End cap		36.80	.217		17.05	13.35		30.40	38.50
4260 Hanger, surface mount		22.08	.362		10.20	22		32.20	44
4280 Coupling		36.80	.217		13	13.35		26.35	34

26 25 13.30 Copper Bus Duct

26 25 13.30 Copper Bus Duct	Crew	Daily Output	Labor-Hours	Unit	Material	Labor	Equipment	Total	Total Incl O&P
0010 **COPPER BUS DUCT**									
0100 Weatherproof 3 pole 4 wire, feeder duct, 600 amp	2 Elec	22.08	.725	L.F.	315	44.50		359.50	410
0110 800 amp		16.56	.966		410	59.50		469.50	540
0120 1,000 amp		15.64	1.020		520	62.50		582.50	665
0125 1,200 amp		15.18	1.050		555	64.50		619.50	710
0130 1,350 amp		14.72	1.090		580	67		647	735
0140 1,600 amp		11.04	1.450		640	89		729	835
0150 2,000 amp		9.20	1.740		740	107		847	975
0160 2,500 amp		6.44	2.480		855	152		1,007	1,175
0170 3,000 amp		4.60	3.480		1,125	214		1,339	1,550
0180 4,000 amp		3.31	4.830		1,750	296		2,046	2,375
0200 Indoor 3 pole 4 wire, plug-in, bus duct high short circuit, 400 amp		29.44	.543		450	33.50		483.50	545
0210 600 amp		23.92	.669		515	41		556	630
0220 800 amp		18.40	.870		785	53.50		838.50	945
0230 1,000 amp		16.56	.966		905	59.50		964.50	1,075
0240 1,350 amp		14.72	1.090		525	67		592	675

26 25 13.30 Copper Bus Duct		Crew	Daily Output	Labor-Hours	Unit	Material	2020 Bare Costs Labor	Equipment	Total	Total Incl O&P
0250	1,600 amp	2 Elec	11.04	1.450	L.F.	590	89		679	780
0260	2,000 amp		9.20	1.740		750	107		857	985
0270	2,500 amp		7.36	2.170		925	133		1,058	1,225
0280	3,000 amp		5.52	2.900		945	178		1,123	1,325
0310	Cross, 225 amp		2.76	5.800	Ea.	3,350	355		3,705	4,225
0320	400 amp		2.58	6.210		3,350	380		3,730	4,275
0330	600 amp		2.39	6.690		3,350	410		3,760	4,300
0340	800 amp		2.02	7.910		3,650	485		4,135	4,725
0350	1,000 amp		1.84	8.700		4,075	535		4,610	5,275
0360	1,350 amp		1.66	9.660		4,525	595		5,120	5,850
0370	1,600 amp		1.56	10.230		5,025	630		5,655	6,450
0380	2,000 amp		1.47	10.870		8,250	665		8,915	10,100
0390	2,500 amp		1.29	12.420		10,000	760		10,760	12,100
0400	3,000 amp		1.10	14.490		9,750	890		10,640	12,000
0410	4,000 amp		.92	17.390		12,700	1,075		13,775	15,600
0430	Expansion fitting, 225 amp		4.97	3.220		2,150	198		2,348	2,675
0440	400 amp		4.23	3.780		2,425	232		2,657	3,025
0450	600 amp		3.68	4.350		3,025	267		3,292	3,725
0460	800 amp		3.13	5.120		3,550	315		3,865	4,375
0470	1,000 amp		2.76	5.800		4,075	355		4,430	5,025
0480	1,350 amp		2.58	6.210		4,125	380		4,505	5,100
0490	1,600 amp		2.39	6.690		5,750	410		6,160	6,925
0500	2,000 amp		2.02	7.910		6,600	485		7,085	8,000
0510	2,500 amp		1.66	9.660		8,050	595		8,645	9,725
0520	3,000 amp		1.47	10.870		8,050	665		8,715	9,850
0530	4,000 amp		1.10	14.490		10,400	890		11,290	12,700
0550	Reducer nonfused, 225 amp		4.97	3.220		1,925	198		2,123	2,400
0560	400 amp		4.23	3.780		1,925	232		2,157	2,450
0570	600 amp		3.68	4.350		1,925	267		2,192	2,500
0580	800 amp		3.13	5.120		2,325	315		2,640	3,025
0590	1,000 amp		2.76	5.800		2,800	355		3,155	3,600
0600	1,350 amp		2.58	6.210		4,325	380		4,705	5,325
0610	1,600 amp		2.39	6.690		4,975	410		5,385	6,075
0620	2,000 amp		2.02	7.910		5,975	485		6,460	7,300
0630	2,500 amp		1.66	9.660		7,550	595		8,145	9,175
0640	3,000 amp		1.47	10.870		8,225	665		8,890	10,000
0650	4,000 amp		1.10	14.490		10,700	890		11,590	13,100
0670	Reducer fuse included, 225 amp		4.05	3.950		4,100	242		4,342	4,875
0680	400 amp		3.86	4.140		5,150	254		5,404	6,025
0690	600 amp		3.31	4.830		6,350	296		6,646	7,425
0700	800 amp		2.94	5.430		8,950	335		9,285	10,300
0710	1,000 amp		2.76	5.800		11,200	355		11,555	12,800
0720	1,350 amp		2.58	6.210		17,800	380		18,180	20,200
0730	1,600 amp		2.02	7.910		25,400	485		25,885	28,700
0740	2,000 amp		1.66	9.660		28,600	595		29,195	32,400
0790	Reducer, circuit breaker, 225 amp		4.05	3.950		5,200	242		5,442	6,075
0800	400 amp		3.86	4.140		6,075	254		6,329	7,050
0810	600 amp		3.31	4.830		8,625	296		8,921	9,925
0820	800 amp		2.94	5.430		10,100	335		10,435	11,600
0830	1,000 amp		2.76	5.800		11,500	355		11,855	13,100
0840	1,350 amp		2.58	6.210		14,100	380		14,480	16,100
0850	1,600 amp		2.02	7.910		20,800	485		21,285	23,600
0860	2,000 amp		1.66	9.660		22,800	595		23,395	26,000

26 25 13.30 Copper Bus Duct

		Crew	Daily Output	Labor-Hours	Unit	Material	2020 Bare Costs Labor	Equipment	Total	Total Incl O&P
0910	Cable tap box, center, 225 amp	2 Elec	2.94	5.430	Ea.	2,075	335		2,410	2,775
0920	400 amp		2.39	6.690		2,075	410		2,485	2,875
0930	600 amp		2.02	7.910		2,075	485		2,560	3,000
0940	800 amp		1.84	8.700		2,300	535		2,835	3,325
0950	1,000 amp		1.47	10.870		2,475	665		3,140	3,725
0960	1,350 amp		1.29	12.420		3,025	760		3,785	4,450
0970	1,600 amp		1.10	14.490		3,375	890		4,265	5,025
0980	2,000 amp		.92	17.390		4,075	1,075		5,150	6,075
1040	2,500 amp		.74	21.740		4,800	1,325		6,125	7,275
1060	3,000 amp		.55	28.990		4,700	1,775		6,475	7,825
1080	4,000 amp		.37	43.480		5,925	2,675		8,600	10,500
1800	Weatherproof 3 pole 3 wire, feeder duct, 600 amp		25.76	.621	L.F.	298	38		336	385
1820	800 amp		20.24	.791		360	48.50		408.50	470
1840	1,000 amp		18.40	.870		405	53.50		458.50	525
1850	1,200 amp		17.48	.915		420	56		476	545
1860	1,350 amp		16.56	.966		585	59.50		644.50	735
1880	1,600 amp		12.88	1.240		670	76		746	850
1900	2,000 amp		11.04	1.450		855	89		944	1,075
1920	2,500 amp		7.36	2.170		1,075	133		1,208	1,375
1940	3,000 amp		5.52	2.900		1,100	178		1,278	1,475
1960	4,000 amp		3.68	4.350		1,450	267		1,717	2,000
2000	Indoor 3 pole 3 wire, feeder duct, 600 amp		29.44	.543		248	33.50		281.50	325
2010	800 amp		23.92	.669		620	41		661	745
2020	1,000 amp		22.08	.725		335	44.50		379.50	435
2025	1,200 amp		20.24	.791		445	48.50		493.50	560
2030	1,350 amp		18.40	.870		490	53.50		543.50	615
2040	1,600 amp		14.72	1.090		1,075	67		1,142	1,275
2050	2,000 amp		12.88	1.240		1,375	76		1,451	1,650
2060	2,500 amp		9.20	1.740		890	107		997	1,150
2070	3,000 amp		7.36	2.170		885	133		1,018	1,175
2080	4,000 amp		5.52	2.900		1,200	178		1,378	1,600
2090	5,000 amp		4.60	3.480		1,475	214		1,689	1,950
2200	Indoor 3 pole 3 wire, bus duct plug-in, 225 amp		42.32	.378		172	23		195	224
2210	400 amp		33.12	.483		263	29.50		292.50	335
2220	600 amp		27.60	.580		330	35.50		365.50	420
2230	800 amp		22.08	.725		500	44.50		544.50	615
2240	1,000 amp		18.40	.870		540	53.50		593.50	675
2250	1,350 amp		16.56	.966		850	59.50		909.50	1,025
2260	1,600 amp		12.88	1.240		890	76		966	1,100
2270	2,000 amp		11.04	1.450		750	89		839	955
2280	2,500 amp		9.20	1.740		925	107		1,032	1,175
2290	3,000 amp		7.36	2.170		945	133		1,078	1,250
2330	High short circuit, 400 amp		33.12	.483		325	29.50		354.50	400
2340	600 amp		27.60	.580		405	35.50		440.50	500
2350	800 amp		22.08	.725		585	44.50		629.50	710
2360	1,000 amp		18.40	.870		635	53.50		688.50	780
2370	1,350 amp		16.56	.966		950	59.50		1,009.50	1,150
2380	1,600 amp		12.88	1.240		1,025	76		1,101	1,250
2390	2,000 amp		11.04	1.450		750	89		839	955
2400	2,500 amp		9.20	1.740		925	107		1,032	1,175
2410	3,000 amp		7.36	2.170		945	133		1,078	1,250
2440	Elbows, 225 amp		4.23	3.780	Ea.	1,600	232		1,832	2,125
2450	400 amp		3.86	4.140		1,450	254		1,704	1,975

26 25 13.30 Copper Bus Duct		Crew	Daily Output	Labor-Hours	Unit	Material	2020 Bare Costs Labor	Equipment	Total	Total Incl O&P
2460	600 amp	2 Elec	3.31	4.830	Ea.	1,450	296		1,746	2,050
2470	800 amp		2.94	5.430		1,575	335		1,910	2,225
2480	1,000 amp		2.76	5.800		1,650	355		2,005	2,325
2485	1,200 amp		2.67	6		1,725	370		2,095	2,450
2490	1,350 amp		2.58	6.210		1,850	380		2,230	2,625
2500	1,600 amp		2.39	6.690		2,000	410		2,410	2,800
2510	2,000 amp		1.84	8.700		2,450	535		2,985	3,500
2520	2,500 amp		1.66	9.660		3,725	595		4,320	4,975
2530	3,000 amp		1.47	10.870		3,650	665		4,315	5,000
2540	4,000 amp		1.29	12.420		4,650	760		5,410	6,250
2560	Tee fittings, 225 amp		2.58	6.210		1,675	380		2,055	2,425
2570	400 amp		2.21	7.250		1,675	445		2,120	2,500
2580	600 amp		1.84	8.700		1,675	535		2,210	2,650
2590	800 amp		1.66	9.660		1,825	595		2,420	2,875
2600	1,000 amp		1.47	10.870		2,025	665		2,690	3,250
2605	1,200 amp		1.38	11.590		2,175	710		2,885	3,450
2610	1,350 amp		1.29	12.420		2,350	760		3,110	3,700
2620	1,600 amp		1.10	14.490		2,725	890		3,615	4,325
2630	2,000 amp		.92	17.390		4,550	1,075		5,625	6,600
2640	2,500 amp		.64	24.840		5,325	1,525		6,850	8,125
2650	3,000 amp		.55	28.990		5,225	1,775		7,000	8,400
2660	4,000 amp		.46	34.780		6,750	2,125		8,875	10,600
2680	Cross, 225 amp		3.31	4.830		2,550	296		2,846	3,250
2690	400 amp		2.94	5.430		2,550	335		2,885	3,300
2700	600 amp		2.76	5.800		2,550	355		2,905	3,325
2710	800 amp		2.39	6.690		3,050	410		3,460	3,950
2720	1,000 amp		2.21	7.250		3,200	445		3,645	4,150
2730	1,350 amp		2.02	7.910		3,775	485		4,260	4,875
2740	1,600 amp		1.84	8.700		4,050	535		4,585	5,250
2750	2,000 amp		1.66	9.660		6,400	595		6,995	7,925
2760	2,500 amp		1.47	10.870		7,450	665		8,115	9,200
2770	3,000 amp		1.29	12.420		7,275	760		8,035	9,125
2780	4,000 amp		.92	17.390		9,300	1,075		10,375	11,800
2800	Expansion fitting, 225 amp		5.89	2.720		2,275	167		2,442	2,750
2810	400 amp		4.97	3.220		5,025	198		5,223	5,825
2820	600 amp		4.23	3.780		2,275	232		2,507	2,850
2830	800 amp		3.68	4.350		2,700	267		2,967	3,375
2840	1,000 amp		3.31	4.830		2,975	296		3,271	3,700
2850	1,350 amp		2.94	5.430		3,100	335		3,435	3,925
2860	1,600 amp		2.76	5.800		3,400	355		3,755	4,275
2870	2,000 amp		2.39	6.690		4,350	410		4,760	5,375
2880	2,500 amp		2.02	7.910		6,075	485		6,560	7,400
2890	3,000 amp		1.66	9.660		6,050	595		6,645	7,525
2900	4,000 amp		1.29	12.420		7,725	760		8,485	9,625
2920	Reducer nonfused, 225 amp		5.89	2.720		1,575	167		1,742	1,975
2930	400 amp		4.97	3.220		1,575	198		1,773	2,025
2940	600 amp		4.23	3.780		1,575	232		1,807	2,075
2950	800 amp		3.68	4.350		1,850	267		2,117	2,425
2960	1,000 amp		3.31	4.830		2,075	296		2,371	2,725
2970	1,350 amp		2.94	5.430		2,800	335		3,135	3,575
2980	1,600 amp		2.76	5.800		3,175	355		3,530	4,000
2990	2,000 amp		2.39	6.690		3,800	410		4,210	4,775
3000	2,500 amp		2.02	7.910		5,575	485		6,060	6,850

 Pre-Installation Change Order Costs

26 25 13.30 Copper Bus Duct		Crew	Daily Output	Labor-Hours	Unit	Material	2020 Bare Costs Labor	Equipment	Total	Total Incl O&P
3010	3,000 amp	2 Elec	1.66	9.660	Ea.	6,025	595		6,620	7,500
3020	4,000 amp		1.29	12.420		7,825	760		8,585	9,725
3040	Reducer fuse included, 225 amp		4.60	3.480		3,725	214		3,939	4,425
3050	400 amp		4.42	3.620		4,975	222		5,197	5,775
3060	600 amp		3.86	4.140		5,925	254		6,179	6,875
3070	800 amp		3.31	4.830		8,425	296		8,721	9,725
3080	1,000 amp		3.13	5.120		9,975	315		10,290	11,500
3090	1,350 amp		2.94	5.430		10,400	335		10,735	11,900
3100	1,600 amp		2.39	6.690		24,200	410		24,610	27,200
3110	2,000 amp		1.84	8.700		27,400	535		27,935	31,000
3160	Reducer circuit breaker, 225 amp		4.60	3.480		4,775	214		4,989	5,575
3170	400 amp		4.42	3.620		5,850	222		6,072	6,750
3180	600 amp		3.86	4.140		8,375	254		8,629	9,600
3190	800 amp		3.31	4.830		9,800	296		10,096	11,200
3200	1,000 amp		3.13	5.120		11,100	315		11,415	12,800
3210	1,350 amp		2.94	5.430		13,800	335		14,135	15,600
3220	1,600 amp		2.39	6.690		20,500	410		20,910	23,200
3230	2,000 amp		1.84	8.700		22,400	535		22,935	25,500
3280	3 pole, 3 wire, cable tap box center, 225 amp		3.31	4.830		2,350	296		2,646	3,050
3290	400 amp		2.76	5.800		2,350	355		2,705	3,125
3300	600 amp		2.39	6.690		2,350	410		2,760	3,200
3310	800 amp		2.21	7.250		2,625	445		3,070	3,550
3320	1,000 amp		1.66	9.660		2,850	595		3,445	4,025
3330	1,350 amp		1.47	10.870		3,475	665		4,140	4,825
3340	1,600 amp		1.29	12.420		3,875	760		4,635	5,400
3350	2,000 amp		1.10	14.490		4,800	890		5,690	6,625
3360	2,500 amp		.92	17.390		5,725	1,075		6,800	7,900
3370	3,000 amp		.64	24.840		5,650	1,525		7,175	8,475
3380	4,000 amp		.46	34.780		7,150	2,125		9,275	11,100
3400	Cable tap box end, 225 amp		3.31	4.830		1,275	296		1,571	1,850
3410	400 amp		2.76	5.800		1,200	355		1,555	1,850
3420	600 amp		2.39	6.690		1,425	410		1,835	2,175
3430	800 amp		2.21	7.250		1,425	445		1,870	2,225
3440	1,000 amp		1.66	9.660		1,850	595		2,445	2,900
3445	1,200 amp		1.56	10.230		1,750	630		2,380	2,850
3450	1,350 amp		1.47	10.870		1,875	665		2,540	3,050
3460	1,600 amp		1.29	12.420		2,150	760		2,910	3,475
3470	2,000 amp		1.10	14.490		2,500	890		3,390	4,075
3480	2,500 amp		.92	17.390		2,975	1,075		4,050	4,875
3490	3,000 amp		.64	24.840		2,925	1,525		4,450	5,475
3500	4,000 amp	▼	.46	34.780		3,700	2,125		5,825	7,250
4600	Plug-in, fusible switch w/3 fuses, 3 pole, 250 volt, 30 amp	1 Elec	3.68	2.170		465	133		598	715
4610	60 amp		3.31	2.420		610	148		758	890
4620	100 amp	▼	2.48	3.220		875	198		1,073	1,250
4630	200 amp	2 Elec	2.94	5.430		1,475	335		1,810	2,100
4640	400 amp		1.29	12.420		3,825	760		4,585	5,325
4650	600 amp	▼	.83	19.320		5,300	1,175		6,475	7,600
4700	4 pole, 120/208 volt, 30 amp	1 Elec	3.59	2.230		640	137		777	905
4710	60 amp		3.22	2.480		690	152		842	980
4720	100 amp	▼	2.39	3.340		960	205		1,165	1,350
4730	200 amp	2 Elec	2.76	5.800		1,600	355		1,955	2,300
4740	400 amp		1.20	13.380		3,775	820		4,595	5,375
4750	600 amp	▼	.74	21.740		5,300	1,325		6,625	7,800

26 25 13.30 Copper Bus Duct		Crew	Daily Output	Labor-Hours	Unit	Material	2020 Bare Costs Labor	Equipment	Total	Total Incl O&P
4800	3 pole, 480 volt, 30 amp	1 Elec	3.68	2.170	Ea.	480	133		613	725
4810	60 amp		3.31	2.420		505	148		653	775
4820	100 amp	↓	2.48	3.220		855	198		1,053	1,225
4830	200 amp	2 Elec	2.94	5.430		1,475	335		1,810	2,125
4840	400 amp		1.29	12.420		3,425	760		4,185	4,875
4850	600 amp		.83	19.320		4,850	1,175		6,025	7,125
4860	800 amp		.61	26.350		20,100	1,625		21,725	24,500
4870	1,000 amp		.55	28.990		20,600	1,775		22,375	25,300
4880	1,200 amp		.46	34.780		22,400	2,125		24,525	27,900
4890	1,600 amp	↓	.40	39.530		23,600	2,425		26,025	29,600
4900	4 pole, 277/480 volt, 30 amp	1 Elec	3.59	2.230		690	137		827	965
4910	60 amp		3.22	2.480		735	152		887	1,025
4920	100 amp		2.39	3.340		1,075	205		1,280	1,475
4930	200 amp	2 Elec	2.76	5.800		2,150	355		2,505	2,875
4940	400 amp		1.20	13.380		4,050	820		4,870	5,675
4950	600 amp		.74	21.740		5,525	1,325		6,850	8,050
5050	800 amp		.55	28.990		18,200	1,775		19,975	22,700
5060	1,000 amp		.52	31.060		20,900	1,900		22,800	25,800
5070	1,200 amp		.44	36.230		21,100	2,225		23,325	26,500
5080	1,600 amp	↓	.39	41.410		25,100	2,550		27,650	31,400
5150	Fusible with starter, 3 pole 250 volt, 30 amp	1 Elec	3.22	2.480		3,600	152		3,752	4,175
5160	60 amp		2.94	2.720		3,800	167		3,967	4,425
5170	100 amp	↓	2.30	3.480		4,300	214		4,514	5,050
5180	200 amp	2 Elec	2.58	6.210		7,100	380		7,480	8,375
5200	3 pole 480 volt, 30 amp	1 Elec	3.22	2.480		3,600	152		3,752	4,175
5210	60 amp		2.94	2.720		3,800	167		3,967	4,425
5220	100 amp	↓	2.30	3.480		4,300	214		4,514	5,050
5230	200 amp	2 Elec	2.58	6.210		7,100	380		7,480	8,375
5300	Fusible with contactor, 3 pole 250 volt, 30 amp	1 Elec	3.22	2.480		3,500	152		3,652	4,075
5310	60 amp		2.94	2.720		4,450	167		4,617	5,150
5320	100 amp	↓	2.30	3.480		6,225	214		6,439	7,175
5330	200 amp	2 Elec	2.58	6.210		7,125	380		7,505	8,425
5400	3 pole 480 volt, 30 amp	1 Elec	3.22	2.480		3,775	152		3,927	4,375
5410	60 amp		2.94	2.720		5,325	167		5,492	6,100
5420	100 amp	↓	2.30	3.480		7,300	214		7,514	8,350
5430	200 amp	2 Elec	2.58	6.210		7,475	380		7,855	8,800
5450	Fusible with capacitor, 3 pole 250 volt, 30 amp	1 Elec	2.76	2.900		9,100	178		9,278	10,300
5460	60 amp		1.84	4.350		10,600	267		10,867	12,000
5500	3 pole 480 volt, 30 amp		2.76	2.900		7,625	178		7,803	8,650
5510	60 amp		1.84	4.350		9,525	267		9,792	10,900
5600	Circuit breaker, 3 pole, 250 volt, 60 amp		4.14	1.930		655	118		773	895
5610	100 amp		2.94	2.720		805	167		972	1,125
5650	4 pole, 120/208 volt, 60 amp		4.05	1.980		740	121		861	995
5660	100 amp		2.85	2.810		880	172		1,052	1,225
5700	3 pole, 4 wire 277/480 volt, 60 amp		3.96	2.020		1,000	124		1,124	1,275
5710	100 amp	↓	2.76	2.900		1,100	178		1,278	1,500
5720	225 amp	2 Elec	2.94	5.430		2,450	335		2,785	3,200
5730	400 amp		1.10	14.490		5,100	890		5,990	6,950
5740	600 amp		.88	18.120		6,875	1,100		7,975	9,200
5750	700 amp		.55	28.990		8,700	1,775		10,475	12,200
5760	800 amp		.55	28.990		8,700	1,775		10,475	12,200
5770	900 amp		.50	32.210		11,500	1,975		13,475	15,600
5780	1,000 amp	↓	.50	32.210		11,500	1,975		13,475	15,600

26 25 13.30 Copper Bus Duct		Crew	Daily Output	Labor-Hours	Unit	Material	2020 Bare Costs Labor	Equipment	Total	Total Incl O&P
5790	1,200 amp	2 Elec	.39	41.410	Ea.	13,800	2,550		16,350	19,000
5810	Circuit breaker w/HIC fuses, 3 pole 480 volt, 60 amp	1 Elec	4.05	1.980		1,250	121		1,371	1,550
5820	100 amp	"	2.85	2.810		1,375	172		1,547	1,775
5830	225 amp	2 Elec	3.13	5.120		4,400	315		4,715	5,300
5840	400 amp		1.29	12.420		6,975	760		7,735	8,800
5850	600 amp		.92	17.390		7,075	1,075		8,150	9,375
5860	700 amp		.59	27.170		9,375	1,675		11,050	12,800
5870	800 amp		.59	27.170		9,375	1,675		11,050	12,800
5880	900 amp		.52	31.060		20,300	1,900		22,200	25,100
5890	1,000 amp		.52	31.060		20,300	1,900		22,200	25,100
5950	3 pole 4 wire 277/480 volt, 60 amp	1 Elec	3.96	2.020		1,250	124		1,374	1,550
5960	100 amp	"	2.76	2.900		1,375	178		1,553	1,800
5970	225 amp	2 Elec	2.76	5.800		4,400	355		4,755	5,350
5980	400 amp		1.01	15.810		6,975	970		7,945	9,125
5990	600 amp		.86	18.500		7,075	1,125		8,200	9,475
6000	700 amp		.53	29.990		9,375	1,850		11,225	13,100
6010	800 amp		.53	29.990		9,375	1,850		11,225	13,100
6020	900 amp		.48	33.440		20,300	2,050		22,350	25,400
6030	1,000 amp		.48	33.440		20,300	2,050		22,350	25,400
6040	1,200 amp		.37	43.480		20,300	2,675		22,975	26,300
6100	Circuit breaker with starter, 3 pole 250 volt, 60 amp	1 Elec	2.94	2.720		1,875	167		2,042	2,325
6110	100 amp	"	2.30	3.480		2,900	214		3,114	3,525
6120	225 amp	2 Elec	2.76	5.800		4,125	355		4,480	5,050
6130	3 pole 480 volt, 60 amp	1 Elec	2.94	2.720		2,100	167		2,267	2,550
6140	100 amp	"	2.30	3.480		2,900	214		3,114	3,525
6150	225 amp	2 Elec	2.76	5.800		3,500	355		3,855	4,375
6200	Circuit breaker with contactor, 3 pole 250 volt, 60 amp	1 Elec	2.94	2.720		1,975	167		2,142	2,425
6210	100 amp	"	2.30	3.480		2,700	214		2,914	3,300
6220	225 amp	2 Elec	2.76	5.800		3,825	355		4,180	4,725
6250	3 pole 480 volt, 60 amp	1 Elec	2.94	2.720		1,975	167		2,142	2,425
6260	100 amp	"	2.30	3.480		2,700	214		2,914	3,300
6270	225 amp	2 Elec	2.76	5.800		3,600	355		3,955	4,475
6300	Circuit breaker with capacitor, 3 pole 250 volt, 60 amp	1 Elec	1.84	4.350		10,700	267		10,967	12,200
6310	3 pole 480 volt, 60 amp		1.84	4.350		11,700	267		11,967	13,300
6400	Add control transformer with pilot light to above starter		14.72	.543		590	33.50		623.50	700
6410	Switch, fusible, mechanically held contactor optional		14.72	.543		1,575	33.50		1,608.50	1,775
6430	Circuit breaker, mechanically held contactor optional		14.72	.543		1,575	33.50		1,608.50	1,775
6450	Ground neutralizer, 3 pole		14.72	.543		72	33.50		105.50	129

26 25 13.40 Copper Bus Duct

		Crew	Daily Output	Labor-Hours	Unit	Material	2020 Bare Costs Labor	Equipment	Total	Total Incl O&P
0010	**COPPER BUS DUCT** 10' long									
0050	Indoor 3 pole 4 wire, plug-in, straight section, 225 amp	2 Elec	36.80	.435	L.F.	222	26.50		248.50	285
1000	400 amp		29.44	.543		385	33.50		418.50	475
1500	600 amp		23.92	.669		455	41		496	560
2400	800 amp		18.40	.870		620	53.50		673.50	760
2450	1,000 amp		16.56	.966		305	59.50		364.50	425
2470	1,200 amp		15.64	1.020		395	62.50		457.50	530
2500	1,350 amp		14.72	1.090		415	67		482	560
2510	1,600 amp		11.04	1.450		470	89		559	650
2520	2,000 amp		9.20	1.740		595	107		702	815
2530	2,500 amp		7.36	2.170		735	133		868	1,000
2540	3,000 amp		5.52	2.900		855	178		1,033	1,200
2550	Feeder, 600 amp		25.76	.621		204	38		242	281

26 25 13.40 Copper Bus Duct		Crew	Daily Output	Labor-Hours	Unit	Material	2020 Bare Costs Labor	Equipment	Total	Total Incl O&P
2600	800 amp	2 Elec	20.24	.791	L.F.	755	48.50		803.50	900
2700	1,000 amp		18.40	.870		905	53.50		958.50	1,075
2750	1,200 amp		17.48	.915		1,100	56		1,156	1,275
2800	1,350 amp		16.56	.966		390	59.50		449.50	515
2900	1,600 amp		12.88	1.240		1,450	76		1,526	1,725
3000	2,000 amp		11.04	1.450		1,800	89		1,889	2,100
3010	2,500 amp		7.36	2.170		705	133		838	980
3020	3,000 amp		5.52	2.900		830	178		1,008	1,175
3030	4,000 amp		3.68	4.350		1,100	267		1,367	1,600
3040	5,000 amp		1.84	8.700		1,325	535		1,860	2,275
3100	Elbows, 225 amp		3.68	4.350	Ea.	1,575	267		1,842	2,150
3200	400 amp		3.31	4.830		1,575	296		1,871	2,200
3300	600 amp		2.94	5.430		1,575	335		1,910	2,250
3400	800 amp		2.58	6.210		1,725	380		2,105	2,475
3500	1,000 amp		2.39	6.690		1,925	410		2,335	2,700
3550	1,200 amp		2.30	6.960		1,875	425		2,300	2,700
3600	1,350 amp		2.21	7.250		1,900	445		2,345	2,725
3700	1,600 amp		2.02	7.910		2,075	485		2,560	3,000
3800	2,000 amp		1.66	9.660		2,550	595		3,145	3,675
3810	2,500 amp		1.47	10.870		4,025	665		4,690	5,425
3820	3,000 amp		1.29	12.420		4,500	760		5,260	6,075
3830	4,000 amp		1.10	14.490		5,825	890		6,715	7,725
3840	5,000 amp		.92	17.390		9,400	1,075		10,475	11,900
4000	End box, 225 amp		31.28	.512		171	31.50		202.50	235
4100	400 amp		29.44	.543		189	33.50		222.50	258
4200	600 amp		25.76	.621		189	38		227	265
4300	800 amp		23.92	.669		189	41		230	269
4400	1,000 amp		22.08	.725		181	44.50		225.50	265
4410	1,200 amp		21.16	.756		189	46.50		235.50	277
4500	1,350 amp		20.24	.791		180	48.50		228.50	270
4600	1,600 amp		18.40	.870		180	53.50		233.50	278
4700	2,000 amp		16.56	.966		221	59.50		280.50	330
4710	2,500 amp		14.72	1.090		221	67		288	345
4720	3,000 amp		12.88	1.240		213	76		289	345
4730	4,000 amp		11.04	1.450		258	89		347	415
4740	5,000 amp		9.20	1.740		258	107		365	445
4800	Cable tap box end, 225 amp		2.94	5.430		1,225	335		1,560	1,850
5000	400 amp		2.39	6.690		1,225	410		1,635	1,925
5100	600 amp		2.02	7.910		1,625	485		2,110	2,500
5200	800 amp		1.84	8.700		1,775	535		2,310	2,750
5300	1,000 amp		1.47	10.870		1,725	665		2,390	2,900
5350	1,200 amp		1.38	11.590		2,025	710		2,735	3,275
5400	1,350 amp		1.29	12.420		2,200	760		2,960	3,550
5500	1,600 amp		1.10	14.490		2,475	890		3,365	4,050
5600	2,000 amp		.92	17.390		2,775	1,075		3,850	4,650
5610	2,500 amp		.74	21.740		3,175	1,325		4,500	5,475
5620	3,000 amp		.55	28.990		3,700	1,775		5,475	6,725
5630	4,000 amp		.37	43.480		4,275	2,675		6,950	8,675
5640	5,000 amp		.18	86.960		5,200	5,325		10,525	13,700
5700	Switchboard stub, 225 amp		4.97	3.220		1,225	198		1,423	1,650
5800	400 amp		4.23	3.780		1,300	232		1,532	1,775
5900	600 amp		3.68	4.350		1,375	267		1,642	1,925
6000	800 amp		2.94	5.430		1,675	335		2,010	2,325

Pre-Installation Change Order Costs

26 25 13.40 Copper Bus Duct		Crew	Daily Output	Labor-Hours	Unit	Material	2020 Bare Costs Labor	Equipment	Total	Total Incl O&P
6100	1,000 amp	2 Elec	2.76	5.800	Ea.	1,950	355		2,305	2,650
6150	1,200 amp		2.58	6.210		2,275	380		2,655	3,075
6200	1,350 amp		2.39	6.690		2,400	410		2,810	3,250
6300	1,600 amp		2.21	7.250		2,700	445		3,145	3,625
6400	2,000 amp		1.84	8.700		3,275	535		3,810	4,400
6410	2,500 amp		1.66	9.660		4,000	595		4,595	5,275
6420	3,000 amp		1.47	10.870		4,500	665		5,165	5,950
6430	4,000 amp		1.29	12.420		5,850	760		6,610	7,575
6440	5,000 amp		1.10	14.490		7,200	890		8,090	9,225
6490	Tee fittings, 225 amp		2.21	7.250		2,275	445		2,720	3,150
6500	400 amp		1.84	8.700		2,275	535		2,810	3,300
6600	600 amp		1.66	9.660		2,275	595		2,870	3,375
6700	800 amp		1.47	10.870		2,350	665		3,015	3,575
6750	1,000 amp		1.29	12.420		2,575	760		3,335	3,975
6770	1,200 amp		1.20	13.380		2,900	820		3,720	4,425
6800	1,350 amp		1.10	14.490		3,075	890		3,965	4,700
7000	1,600 amp		.92	17.390		3,475	1,075		4,550	5,425
7100	2,000 amp		.74	21.740		4,125	1,325		5,450	6,525
7110	2,500 amp		.55	28.990		5,075	1,775		6,850	8,225
7120	3,000 amp		.46	34.780		5,700	2,125		7,825	9,450
7130	4,000 amp		.37	43.480		7,300	2,675		9,975	12,000
7140	5,000 amp		.18	86.960		8,625	5,325		13,950	17,500
7200	Plug-in fusible switches w/3 fuses, 600 volt, 3 pole, 30 amp	1 Elec	3.68	2.170		885	133		1,018	1,175
7300	60 amp		3.31	2.420		990	148		1,138	1,325
7400	100 amp		2.48	3.220		1,425	198		1,623	1,850
7500	200 amp	2 Elec	2.94	5.430		2,500	335		2,835	3,250
7600	400 amp		1.29	12.420		7,075	760		7,835	8,900
7700	600 amp		.83	19.320		8,400	1,175		9,575	11,000
7800	800 amp		.61	26.350		12,800	1,625		14,425	16,500
7900	1,200 amp		.46	34.780		24,100	2,125		26,225	29,700
7910	1,600 amp		.40	39.530		22,600	2,425		25,025	28,500
8000	Plug-in circuit breakers, molded case, 15 to 50 amp	1 Elec	4.05	1.980		835	121		956	1,100
8100	70 to 100 amp	"	2.85	2.810		930	172		1,102	1,275
8200	150 to 225 amp	2 Elec	3.13	5.120		2,525	315		2,840	3,250
8300	250 to 400 amp		1.29	12.420		4,425	760		5,185	5,975
8400	500 to 600 amp		.92	17.390		5,950	1,075		7,025	8,150
8500	700 to 800 amp		.59	27.170		7,350	1,675		9,025	10,600
8600	900 to 1,000 amp		.52	31.060		10,500	1,900		12,400	14,400
8700	1,200 amp		.40	39.530		12,600	2,425		15,025	17,500
8720	1,400 amp		.37	43.480		17,700	2,675		20,375	23,400
8730	1,600 amp		.37	43.480		19,400	2,675		22,075	25,300
8750	Circuit breakers, with current limiting fuse, 15 to 50 amp	1 Elec	4.05	1.980		1,675	121		1,796	2,025
8760	70 to 100 amp	"	2.85	2.810		1,975	172		2,147	2,425
8770	150 to 225 amp	2 Elec	3.13	5.120		4,250	315		4,565	5,150
8780	250 to 400 amp		1.29	12.420		6,575	760		7,335	8,375
8790	500 to 600 amp		.92	17.390		7,575	1,075		8,650	9,950
8800	700 to 800 amp		.59	27.170		12,500	1,675		14,175	16,200
8810	900 to 1,000 amp		.52	31.060		14,200	1,900		16,100	18,400
8850	Combination starter FVNR, fusible switch, NEMA size 0, 30 amp	1 Elec	1.84	4.350		2,250	267		2,517	2,875
8860	NEMA size 1, 60 amp		1.66	4.830		2,375	296		2,671	3,075
8870	NEMA size 2, 100 amp		1.20	6.690		3,000	410		3,410	3,900
8880	NEMA size 3, 200 amp	2 Elec	1.84	8.700		4,800	535		5,335	6,075
8900	Circuit breaker, NEMA size 0, 30 amp	1 Elec	1.84	4.350		2,325	267		2,592	2,950

Pre-Installation Change Order Costs   For customer support on your Electrical Change Order Costs with RSMeans data, call 800.448.8182.   169

26 25 13.40 Copper Bus Duct

		Crew	Daily Output	Labor-Hours	Unit	Material	2020 Bare Costs Labor	Equipment	Total	Total Incl O&P
8910	NEMA size 1, 60 amp	1 Elec	1.66	4.830	Ea.	2,400	296		2,696	3,100
8920	NEMA size 2, 100 amp		1.20	6.690		3,450	410		3,860	4,400
8930	NEMA size 3, 200 amp	2 Elec	1.84	8.700		4,375	535		4,910	5,625
8950	Combination contactor, fusible switch, NEMA size 0, 30 amp	1 Elec	1.84	4.350		1,325	267		1,592	1,850
8960	NEMA size 1, 60 amp		1.66	4.830		1,350	296		1,646	1,950
8970	NEMA size 2, 100 amp		1.20	6.690		2,025	410		2,435	2,825
8980	NEMA size 3, 200 amp	2 Elec	1.84	8.700		2,325	535		2,860	3,350
9000	Circuit breaker, NEMA size 0, 30 amp	1 Elec	1.84	4.350		1,500	267		1,767	2,050
9010	NEMA size 1, 60 amp		1.66	4.830		1,550	296		1,846	2,150
9020	NEMA size 2, 100 amp		1.20	6.690		2,400	410		2,810	3,225
9030	NEMA size 3, 200 amp	2 Elec	1.84	8.700		2,900	535		3,435	4,000
9050	Control transformer for above, NEMA size 0, 30 amp	1 Elec	7.36	1.090		256	67		323	380
9060	NEMA size 1, 60 amp		7.36	1.090		256	67		323	380
9070	NEMA size 2, 100 amp		6.44	1.240		355	76		431	505
9080	NEMA size 3, 200 amp	2 Elec	12.88	1.240		495	76		571	660
9100	Comb. fusible switch & lighting control, electrically held, 30 amp	1 Elec	1.84	4.350		1,050	267		1,317	1,550
9110	60 amp		1.66	4.830		1,500	296		1,796	2,100
9120	100 amp		1.20	6.690		1,925	410		2,335	2,725
9130	200 amp	2 Elec	1.84	8.700		4,750	535		5,285	6,050
9150	Mechanically held, 30 amp	1 Elec	1.84	4.350		1,300	267		1,567	1,850
9160	60 amp		1.66	4.830		1,950	296		2,246	2,600
9170	100 amp		1.20	6.690		2,500	410		2,910	3,350
9180	200 amp	2 Elec	1.84	8.700		5,100	535		5,635	6,425
9200	Ground bus added to bus duct, 225 amp		294.40	.054	L.F.	44.50	3.33		47.83	54
9210	400 amp		220.80	.072		44.50	4.45		48.95	55.50
9220	600 amp		220.80	.072		44.50	4.45		48.95	55.50
9230	800 amp		147.20	.109		52	6.65		58.65	67.50
9240	1,000 amp		147.20	.109		59.50	6.65		66.15	75.50
9250	1,350 amp		128.80	.124		85	7.60		92.60	105
9260	1,600 amp		110.40	.145		92	8.90		100.90	114
9270	2,000 amp		101.20	.158		120	9.70		129.70	146
9280	2,500 amp		92	.174		149	10.65		159.65	179
9290	3,000 amp		82.80	.193		170	11.85		181.85	205
9300	4,000 amp		73.60	.217		225	13.35		238.35	267
9310	5,000 amp		64.40	.248		270	15.25		285.25	320
9320	High short circuit bracing, add					18.60			18.60	20.50

26 25 13.60 Copper or Aluminum Bus Duct Fittings

		Crew	Daily Output	Labor-Hours	Unit	Material	2020 Bare Costs Labor	Equipment	Total	Total Incl O&P
0010	**COPPER OR ALUMINUM BUS DUCT FITTINGS**									
0100	Flange, wall, with vapor barrier, 225 amp	2 Elec	5.70	2.810	Ea.	845	172		1,017	1,175
0110	400 amp		5.52	2.900		795	178		973	1,150
0120	600 amp		5.34	3		845	184		1,029	1,200
0130	800 amp		4.97	3.220		845	198		1,043	1,225
0140	1,000 amp		4.60	3.480		845	214		1,059	1,250
0145	1,200 amp		4.42	3.620		845	222		1,067	1,250
0150	1,350 amp		4.23	3.780		845	232		1,077	1,275
0160	1,600 amp		3.86	4.140		845	254		1,099	1,300
0170	2,000 amp		3.68	4.350		845	267		1,112	1,325
0180	2,500 amp		3.31	4.830		845	296		1,141	1,375
0190	3,000 amp		2.94	5.430		845	335		1,180	1,425
0200	4,000 amp		2.39	6.690		820	410		1,230	1,500
0300	Roof, 225 amp		5.70	2.810		1,050	172		1,222	1,400
0310	400 amp		5.52	2.900		1,050	178		1,228	1,425

26 25 Low-Voltage Enclosed Bus Assemblies

26 25 13 – Low-Voltage Busways

26 25 13.60 Copper or Aluminum Bus Duct Fittings	Crew	Daily Output	Labor-Hours	Unit	Material	2020 Bare Costs Labor	Equipment	Total	Total Incl O&P
0320	600 amp	2 Elec	5.34	3	Ea.	1,050	184		1,234	1,425
0330	800 amp		4.97	3.220		1,050	198		1,248	1,450
0340	1,000 amp		4.60	3.480		1,050	214		1,264	1,475
0345	1,200 amp		4.42	3.620		1,050	222		1,272	1,475
0350	1,350 amp		4.23	3.780		1,050	232		1,282	1,500
0360	1,600 amp		3.86	4.140		1,050	254		1,304	1,525
0370	2,000 amp		3.68	4.350		1,050	267		1,317	1,550
0380	2,500 amp		3.31	4.830		1,050	296		1,346	1,600
0390	3,000 amp		2.94	5.430		1,050	335		1,385	1,650
0400	4,000 amp		2.39	6.690		1,050	410		1,460	1,750
0420	Support, floor mounted, 225 amp		18.40	.870		171	53.50		224.50	268
0430	400 amp		18.40	.870		171	53.50		224.50	268
0440	600 amp		16.56	.966		171	59.50		230.50	277
0450	800 amp		14.72	1.090		171	67		238	288
0460	1,000 amp		11.96	1.340		171	82		253	310
0465	1,200 amp		10.86	1.470		171	90		261	320
0470	1,350 amp		9.75	1.640		171	101		272	340
0480	1,600 amp		8.46	1.890		171	116		287	360
0490	2,000 amp		7.36	2.170		171	133		304	385
0500	2,500 amp		5.89	2.720		171	167		338	435
0510	3,000 amp		4.97	3.220		171	198		369	480
0520	4,000 amp		3.68	4.350		171	267		438	585
0540	Weather stop, 225 amp		11.04	1.450		530	89		619	710
0550	400 amp		9.20	1.740		530	107		637	740
0560	600 amp		8.28	1.930		530	118		648	755
0570	800 amp		7.36	2.170		530	133		663	780
0580	1,000 amp		5.89	2.720		530	167		697	830
0585	1,200 amp		5.43	2.950		530	181		711	850
0590	1,350 amp		4.97	3.220		530	198		728	875
0600	1,600 amp		4.23	3.780		530	232		762	925
0610	2,000 amp		3.68	4.350		530	267		797	975
0620	2,500 amp		2.94	5.430		530	335		865	1,075
0630	3,000 amp		2.39	6.690		530	410		940	1,200
0640	4,000 amp		1.84	8.700		530	535		1,065	1,375
0660	End closure, 225 amp		31.28	.512		198	31.50		229.50	265
0670	400 amp		29.44	.543		198	33.50		231.50	268
0680	600 amp		25.76	.621		198	38		236	275
0690	800 amp		23.92	.669		172	41		213	250
0700	1,000 amp		22.08	.725		198	44.50		242.50	284
0705	1,200 amp		21.16	.756		170	46.50		216.50	256
0710	1,350 amp		20.24	.791		170	48.50		218.50	259
0720	1,600 amp		18.40	.870		198	53.50		251.50	298
0730	2,000 amp		16.56	.966		277	59.50		336.50	395
0740	2,500 amp		14.72	1.090		253	67		320	380
0750	3,000 amp		12.88	1.240		253	76		329	390
0760	4,000 amp		11.04	1.450		255	89		344	410
0780	Switchboard stub, 3 pole 3 wire, 225 amp		5.52	2.900		1,025	178		1,203	1,400
0790	400 amp		4.78	3.340		1,025	205		1,230	1,425
0800	600 amp		4.23	3.780		1,025	232		1,257	1,475
0810	800 amp		3.31	4.830		1,250	296		1,546	1,825
0820	1,000 amp		3.13	5.120		1,450	315		1,765	2,075
0825	1,200 amp		2.94	5.430		1,450	335		1,785	2,100
0830	1,350 amp		2.76	5.800		1,725	355		2,080	2,425

26 25 13.60 Copper or Aluminum Bus Duct Fittings		Crew	Daily Output	Labor-Hours	Unit	Material	2020 Bare Costs Labor	Equipment	Total	Total Incl O&P
0840	1,600 amp	2 Elec	2.58	6.210	Ea.	2,050	380		2,430	2,850
0850	2,000 amp		2.21	7.250		2,400	445		2,845	3,275
0860	2,500 amp		1.84	8.700		2,925	535		3,460	4,025
0870	3,000 amp		1.66	9.660		3,325	595		3,920	4,525
0880	4,000 amp		1.47	10.870		4,275	665		4,940	5,725
0890	5,000 amp		1.29	12.420		5,300	760		6,060	6,975
0900	3 pole 4 wire, 225 amp		4.97	3.220		1,275	198		1,473	1,700
0910	400 amp		4.23	3.780		1,275	232		1,507	1,750
0920	600 amp		3.68	4.350		1,275	267		1,542	1,800
0930	800 amp		2.94	5.430		1,525	335		1,860	2,200
0940	1,000 amp		2.76	5.800		1,800	355		2,155	2,500
0950	1,350 amp		2.39	6.690		2,300	410		2,710	3,125
0960	1,600 amp		2.21	7.250		2,625	445		3,070	3,525
0970	2,000 amp		1.84	8.700		3,150	535		3,685	4,250
0980	2,500 amp		1.66	9.660		3,875	595		4,470	5,125
0990	3,000 amp		1.47	10.870		4,475	665		5,140	5,925
1000	4,000 amp		1.29	12.420		5,875	760		6,635	7,575
1050	Service head, weatherproof, 3 pole 3 wire, 225 amp		2.76	5.800		1,725	355		2,080	2,425
1060	400 amp		2.58	6.210		1,725	380		2,105	2,475
1070	600 amp		2.39	6.690		1,725	410		2,135	2,500
1080	800 amp		2.21	7.250		1,950	445		2,395	2,800
1090	1,000 amp		1.84	8.700		2,100	535		2,635	3,125
1100	1,350 amp		1.66	9.660		2,725	595		3,320	3,875
1110	1,600 amp		1.47	10.870		3,050	665		3,715	4,375
1120	2,000 amp		1.29	12.420		3,750	760		4,510	5,250
1130	2,500 amp		1.10	14.490		4,475	890		5,365	6,250
1140	3,000 amp		.83	19.320		5,175	1,175		6,350	7,475
1150	4,000 amp		.64	24.840		6,600	1,525		8,125	9,525
1200	3 pole 4 wire, 225 amp		2.39	6.690		1,950	410		2,360	2,750
1210	400 amp		2.21	7.250		1,950	445		2,395	2,800
1220	600 amp		2.02	7.910		1,950	485		2,435	2,875
1230	800 amp		1.84	8.700		2,250	535		2,785	3,275
1240	1,000 amp		1.56	10.230		2,600	630		3,230	3,775
1250	1,350 amp		1.38	11.590		3,100	710		3,810	4,450
1260	1,600 amp		1.29	12.420		3,400	760		4,160	4,850
1270	2,000 amp		1.10	14.490		4,500	890		5,390	6,275
1280	2,500 amp		.92	17.390		5,550	1,075		6,625	7,700
1290	3,000 amp		.74	21.740		6,550	1,325		7,875	9,175
1300	4,000 amp		.55	28.990		8,425	1,775		10,200	11,900
1350	Flanged end, 3 pole 3 wire, 225 amp		5.52	2.900		945	178		1,123	1,325
1360	400 amp		4.78	3.340		945	205		1,150	1,350
1370	600 amp		4.23	3.780		945	232		1,177	1,400
1380	800 amp		3.31	4.830		1,050	296		1,346	1,625
1390	1,000 amp		3.13	5.120		1,200	315		1,515	1,775
1395	1,200 amp		2.94	5.430		1,325	335		1,660	1,950
1400	1,350 amp		2.76	5.800		1,400	355		1,755	2,075
1410	1,600 amp		2.58	6.210		1,600	380		1,980	2,350
1420	2,000 amp		2.21	7.250		1,900	445		2,345	2,750
1430	2,500 amp		1.84	8.700		2,200	535		2,735	3,225
1440	3,000 amp		1.66	9.660		2,525	595		3,120	3,650
1450	4,000 amp		1.47	10.870		3,125	665		3,790	4,425
1500	3 pole 4 wire, 225 amp		4.97	3.220		1,100	198		1,298	1,500
1510	400 amp		4.23	3.780		1,100	232		1,332	1,550

For customer support on your Electrical Change Order Costs with RSMeans data, call 800.448.8182.

Pre-Installation Change Order Costs

26 25 13 – Low-Voltage Busways

26 25 13.60 Copper or Aluminum Bus Duct Fittings

		Crew	Daily Output	Labor-Hours	Unit	Material	2020 Bare Costs Labor	Equipment	Total	Total Incl O&P
1520	600 amp	2 Elec	3.68	4.350	Ea.	1,100	267		1,367	1,600
1530	800 amp		2.94	5.430		1,300	335		1,635	1,925
1540	1,000 amp		2.76	5.800		1,425	355		1,780	2,100
1545	1,200 amp		2.58	6.210		1,625	380		2,005	2,375
1550	1,350 amp		2.39	6.690		1,750	410		2,160	2,525
1560	1,600 amp		2.21	7.250		2,000	445		2,445	2,850
1570	2,000 amp		1.84	8.700		2,375	535		2,910	3,400
1580	2,500 amp		1.66	9.660		2,800	595		3,395	3,975
1590	3,000 amp		1.47	10.870		3,225	665		3,890	4,550
1600	4,000 amp		1.29	12.420		4,175	760		4,935	5,725
1650	Hanger, standard, 225 amp		58.88	.272		26.50	16.65		43.15	54
1660	400 amp		44.16	.362		26.50	22		48.50	62
1670	600 amp		36.80	.435		26.50	26.50		53	68.50
1680	800 amp		29.44	.543		26.50	33.50		60	78.50
1690	1,000 amp		22.08	.725		26.50	44.50		71	95
1695	1,200 amp		20.24	.791		26.50	48.50		75	101
1700	1,350 amp		18.40	.870		26.50	53.50		80	109
1710	1,600 amp		18.40	.870		26.50	53.50		80	109
1720	2,000 amp		16.56	.966		26.50	59.50		86	118
1730	2,500 amp		14.72	1.090		26.50	67		93.50	129
1740	3,000 amp		14.72	1.090		26.50	67		93.50	129
1750	4,000 amp		14.72	1.090		26.50	67		93.50	129
1800	Spring type, 225 amp		14.72	1.090		89	67		156	198
1810	400 amp		12.88	1.240		89	76		165	211
1820	600 amp		12.88	1.240		89	76		165	211
1830	800 amp		12.88	1.240		89	76		165	211
1840	1,000 amp		12.88	1.240		89	76		165	211
1845	1,200 amp		12.88	1.240		89	76		165	211
1850	1,350 amp		12.88	1.240		89	76		165	211
1860	1,600 amp		11.04	1.450		89	89		178	230
1870	2,000 amp		11.04	1.450		89	89		178	230
1880	2,500 amp		11.04	1.450		89	89		178	230
1890	3,000 amp		9.20	1.740		89	107		196	257
1900	4,000 amp		9.20	1.740		89	107		196	257

26 25 13.70 Feedrail

		Crew	Daily Output	Labor-Hours	Unit	Material	2020 Bare Costs Labor	Equipment	Total	Total Incl O&P
0010	**FEEDRAIL**, 12' mounting									
0050	Trolley busway, 3 pole									
0100	300 volt 60 amp, plain, 10' lengths	1 Elec	46	.174	L.F.	22	10.65		32.65	40
0300	Door track		46	.174		41.50	10.65		52.15	62
0500	Curved track		27.60	.290		16.35	17.80		34.15	44.50
0700	Coupling				Ea.	12.75			12.75	14.05
0900	Center feed	1 Elec	4.88	1.640		47	101		148	202
1100	End feed		4.88	1.640		52	101		153	207
1300	Hanger set		22.08	.362		2.89	22		24.89	36
3000	600 volt 100 amp, plain, 10' lengths		32.20	.248	L.F.	59	15.25		74.25	87.50
3300	Door track		32.20	.248	"	93	15.25		108.25	126
3700	Coupling				Ea.	53			53	58.50
4000	End cap	1 Elec	36.80	.217		67	13.35		80.35	93.50
4200	End feed		3.68	2.170		239	133		372	460
4500	Trolley, 600 volt, 20 amp		4.88	1.640		263	101		364	440
4700	30 amp		4.88	1.640		263	101		364	440
4900	Duplex, 40 amp		3.68	2.170		920	133		1,053	1,225

26 25 Low-Voltage Enclosed Bus Assemblies

26 25 13 – Low-Voltage Busways

26 25 13.70 Feedrail		Crew	Daily Output	Labor-Hours	Unit	Material	2020 Bare Costs Labor	Equipment	Total	Total Incl O&P
5000	60 amp	1 Elec	3.68	2.170	Ea.	895	133		1,028	1,175
5300	Fusible, 20 amp		3.68	2.170		540	133		673	795
5500	30 amp		3.68	2.170		540	133		673	795
5900	300 volt, 20 amp		4.88	1.640		249	101		350	425
6000	30 amp		4.88	1.640		325	101		426	510
6300	Fusible, 20 amp		4.32	1.850		350	114		464	555
6500	30 amp		4.32	1.850		490	114		604	710
7300	Busway, 250 volt 50 amp, 2 wire		64.40	.124	L.F.	20	7.60		27.60	34
7330	Coupling				Ea.	44			44	48.50
7340	Center feed	1 Elec	5.52	1.450		510	89		599	690
7350	End feed		5.52	1.450		124	89		213	268
7360	End cap		36.80	.217		29	13.35		42.35	52
7370	Hanger set		22.08	.362		3.12	22		25.12	36.50
7400	125/250 volt 50 amp, 3 wire		55.20	.145	L.F.	20.50	8.90		29.40	36
7430	Coupling		5.52	1.450	Ea.	61.50	89		150.50	200
7440	Center feed		5.52	1.450		505	89		594	685
7450	End feed		5.52	1.450		111	89		200	254
7460	End cap		36.80	.217		29.50	13.35		42.85	52.50
7470	Hanger set		22.08	.362		4.22	22		26.22	37.50
7480	Trolley, 250 volt, 2 pole, 20 amp		5.52	1.450		38.50	89		127.50	175
7490	30 amp		5.52	1.450		38.50	89		127.50	175
7500	125/250 volt, 3 pole, 20 amp		5.52	1.450		37.50	89		126.50	173
7510	30 amp		5.52	1.450		37.50	89		126.50	173
8000	Cleaning tools, 300 volt, dust remover					107			107	118
8100	Bus bar cleaner					201			201	221
8300	600 volt, dust remover, 60 amp					310			310	340
8400	100 amp					655			655	720
8600	Bus bar cleaner, 60 amp					805			805	885
8700	100 amp					805			805	885

26 27 Low-Voltage Distribution Equipment

26 27 13 – Electricity Metering

26 27 13.10 Meter Centers and Sockets

26 27 13.10	Meter Centers and Sockets	Crew	Daily Output	Labor-Hours	Unit	Material	2020 Bare Costs Labor	Equipment	Total	Total Incl O&P
0010	**METER CENTERS AND SOCKETS**									
0100	Sockets, single position, 4 terminal, 100 amp	1 Elec	2.94	2.720	Ea.	52.50	167		219.50	305
0200	150 amp		2.12	3.780		62.50	232		294.50	415
0300	200 amp		1.75	4.580		110	281		391	540
0400	Transformer rated, 20 amp		2.94	2.720		179	167		346	445
0500	Double position, 4 terminal, 100 amp		2.58	3.110		240	191		431	550
0600	150 amp		1.93	4.140		290	254		544	700
0700	200 amp		1.56	5.120		550	315		865	1,075
0800	Trans-socket, 13 terminal, 3 CT mounts, 400 amp		.92	8.700		1,225	535		1,760	2,150
0900	800 amp	2 Elec	1.10	14.490		1,525	890		2,415	3,000
1100	Meter centers and sockets, three phase, single pos, 7 terminal, 100 amp	1 Elec	2.58	3.110		136	191		327	435
1200	200 amp		1.93	4.140		239	254		493	645
1400	400 amp		1.56	5.120		825	315		1,140	1,375
2000	Meter center, main fusible switch, 1P 3W 120/240 V									
2030	400 amp	2 Elec	1.47	10.870	Ea.	855	665		1,520	1,925
2040	600 amp		1.01	15.810		1,300	970		2,270	2,900
2050	800 amp		.83	19.320		5,000	1,175		6,175	7,275
2060	Rainproof 1P 3W 120/240 V, 400 A		1.47	10.870		1,950	665		2,615	3,150

Pre-Installation Change Order Costs

26 27 13.10 Meter Centers and Sockets	Crew	Daily Output	Labor-Hours	Unit	Material	2020 Bare Costs Labor	2020 Bare Costs Equipment	Total	Total Incl O&P
2070	600 amp	2 Elec	1.01	15.810	Ea.	3,400	970		4,370	5,175
2080	800 amp		.83	19.320		5,325	1,175		6,500	7,625
2100	3P 4W 120/208 V, 400 amp		1.47	10.870		865	665		1,530	1,950
2110	600 amp		1.01	15.810		1,475	970		2,445	3,050
2120	800 amp		.83	19.320		2,300	1,175		3,475	4,300
2130	Rainproof 3P 4W 120/208 V, 400 amp		1.47	10.870		2,225	665		2,890	3,450
2140	600 amp		1.01	15.810		3,575	970		4,545	5,400
2150	800 amp	▼	.83	19.320	▼	7,750	1,175		8,925	10,300
2170	Main circuit breaker, 1P 3W 120/240 V									
2180	400 amp	2 Elec	1.47	10.870	Ea.	1,475	665		2,140	2,625
2190	600 amp		1.01	15.810		1,800	970		2,770	3,425
2200	800 amp		.83	19.320		2,775	1,175		3,950	4,850
2210	1,000 amp		.74	21.740		3,200	1,325		4,525	5,500
2220	1,200 amp		.70	22.880		4,300	1,400		5,700	6,825
2230	1,600 amp		.63	25.580		18,800	1,575		20,375	23,000
2240	Rainproof 1P 3W 120/240 V, 400 amp		1.47	10.870		3,000	665		3,665	4,300
2250	600 amp		1.01	15.810		4,700	970		5,670	6,625
2260	800 amp		.83	19.320		5,475	1,175		6,650	7,800
2270	1,000 amp		.74	21.740		7,550	1,325		8,875	10,300
2280	1,200 amp		.70	22.880		10,200	1,400		11,600	13,300
2300	3P 4W 120/208 V, 400 amp		1.47	10.870		3,350	665		4,015	4,700
2310	600 amp		1.01	15.810		5,600	970		6,570	7,600
2320	800 amp		.83	19.320		6,650	1,175		7,825	9,100
2330	1,000 amp		.74	21.740		8,725	1,325		10,050	11,600
2340	1,200 amp		.70	22.880		11,200	1,400		12,600	14,400
2350	1,600 amp		.63	25.580		22,900	1,575		24,475	27,500
2360	Rainproof 3P 4W 120/208 V, 400 amp		1.47	10.870		3,975	665		4,640	5,375
2370	600 amp		1.01	15.810		5,600	970		6,570	7,600
2380	800 amp		.83	19.320		6,650	1,175		7,825	9,100
2390	1,000 amp		.70	22.880		8,725	1,400		10,125	11,700
2400	1,200 amp	▼	.63	25.580	▼	11,200	1,575		12,775	14,600
2420	Main lugs terminal box, 1P 3W 120/240 V									
2430	800 amp	2 Elec	.86	18.500	Ea.	565	1,125		1,690	2,325
2440	1,200 amp		.66	24.150		595	1,475		2,070	2,850
2450	Rainproof 1P 3W 120/240 V, 225 amp		2.21	7.250		445	445		890	1,150
2460	800 amp		.86	18.500		655	1,125		1,780	2,425
2470	1,200 amp		.66	24.150		1,325	1,475		2,800	3,650
2500	3P 4W 120/208 V, 800 amp		.86	18.500		725	1,125		1,850	2,500
2510	1,200 amp		.66	24.150		1,475	1,475		2,950	3,825
2520	Rainproof 3P 4W 120/208 V, 225 amp		2.21	7.250		445	445		890	1,150
2530	800 amp		.86	18.500		725	1,125		1,850	2,500
2540	1,200 amp	▼	.66	24.150	▼	1,475	1,475		2,950	3,825
2590	Basic meter device									
2600	1P 3W 120/240 V 4 jaw 125A sockets, 3 meter	2 Elec	.92	17.390	Ea.	355	1,075		1,430	2,000
2610	4 meter		.83	19.320		535	1,175		1,710	2,375
2620	5 meter		.74	21.740		630	1,325		1,955	2,675
2630	6 meter		.55	28.990		570	1,775		2,345	3,275
2640	7 meter		.52	31.060		1,650	1,900		3,550	4,650
2650	8 meter		.48	33.440		1,800	2,050		3,850	5,025
2660	10 meter	▼	.44	36.230	▼	2,250	2,225		4,475	5,775
2680	Rainproof 1P 3W 120/240 V 4 jaw 125A sockets									
2690	3 meter	2 Elec	.92	17.390	Ea.	750	1,075		1,825	2,425
2700	4 meter	↓	.83	19.320	▼	900	1,175		2,075	2,775

26 27 13.10 Meter Centers and Sockets		Crew	Daily Output	Labor-Hours	Unit	Material	2020 Bare Costs Labor	Equipment	Total	Total Incl O&P
2710	6 meter	2 Elec	.55	28.990	Ea.	1,300	1,775		3,075	4,075
2720	7 meter		.52	31.060		1,650	1,900		3,550	4,650
2730	8 meter	▼	.48	33.440	▼	1,800	2,050		3,850	5,025
2750	1P 3W 120/240 V 4 jaw sockets									
2760	with 125A circuit breaker, 3 meter	2 Elec	.92	17.390	Ea.	1,400	1,075		2,475	3,150
2770	4 meter		.83	19.320		1,775	1,175		2,950	3,725
2780	5 meter		.74	21.740		2,225	1,325		3,550	4,400
2790	6 meter		.55	28.990		2,600	1,775		4,375	5,500
2800	7 meter		.52	31.060		3,175	1,900		5,075	6,325
2810	8 meter		.48	33.440		3,550	2,050		5,600	6,950
2820	10 meter	▼	.44	36.230	▼	4,425	2,225		6,650	8,175
2830	Rainproof 1P 3W 120/240 V 4 jaw sockets									
2840	with 125A circuit breaker, 3 meter	2 Elec	.92	17.390	Ea.	1,400	1,075		2,475	3,150
2850	4 meter		.83	19.320		1,775	1,175		2,950	3,725
2870	6 meter		.55	28.990		2,600	1,775		4,375	5,500
2880	7 meter		.52	31.060		3,175	1,900		5,075	6,325
2890	8 meter	▼	.48	33.440	▼	3,550	2,050		5,600	6,950
2920	1P 3W on 3P 4W 120/208 V system 5 jaw									
2930	125A sockets, 3 meter	2 Elec	.92	17.390	Ea.	750	1,075		1,825	2,425
2940	4 meter		.83	19.320		900	1,175		2,075	2,775
2950	5 meter		.74	21.740		1,125	1,325		2,450	3,200
2960	6 meter		.55	28.990		1,300	1,775		3,075	4,075
2970	7 meter		.52	31.060		1,650	1,900		3,550	4,650
2980	8 meter		.48	33.440		1,800	2,050		3,850	5,025
2990	10 meter	▼	.44	36.230	▼	2,250	2,225		4,475	5,775
3000	Rainproof 1P 3W on 3P 4W 120/208 V system									
3020	5 jaw 125A sockets, 3 meter	2 Elec	.92	17.390	Ea.	750	1,075		1,825	2,425
3030	4 meter		.83	19.320		900	1,175		2,075	2,775
3050	6 meter		.55	28.990		1,300	1,775		3,075	4,075
3060	7 meter		.52	31.060		1,650	1,900		3,550	4,650
3070	8 meter	▼	.48	33.440	▼	1,800	2,050		3,850	5,025
3090	1P 3W on 3P 4W 120/208 V system 5 jaw sockets									
3100	With 125A circuit breaker, 3 meter	2 Elec	.92	17.390	Ea.	1,400	1,075		2,475	3,150
3110	4 meter		.83	19.320		1,775	1,175		2,950	3,725
3120	5 meter		.74	21.740		2,225	1,325		3,550	4,400
3130	6 meter		.55	28.990		2,600	1,775		4,375	5,500
3140	7 meter		.52	31.060		3,175	1,900		5,075	6,325
3150	8 meter		.48	33.440		3,550	2,050		5,600	6,950
3160	10 meter	▼	.44	36.230	▼	4,425	2,225		6,650	8,175
3170	Rainproof 1P 3W on 3P 4W 120/208 V system									
3180	5 jaw sockets w/125A circuit breaker, 3 meter	2 Elec	.92	17.390	Ea.	1,400	1,075		2,475	3,150
3190	4 meter		.83	19.320		1,775	1,175		2,950	3,725
3210	6 meter		.55	28.990		2,600	1,775		4,375	5,500
3220	7 meter		.52	31.060		3,175	1,900		5,075	6,325
3230	8 meter	▼	.48	33.440	▼	3,550	2,050		5,600	6,950
3250	1P 3W 120/240 V 4 jaw sockets									
3260	with 200A circuit breaker, 3 meter	2 Elec	.92	17.390	Ea.	2,100	1,075		3,175	3,925
3270	4 meter		.83	19.320		2,850	1,175		4,025	4,900
3290	6 meter		.55	28.990		4,200	1,775		5,975	7,275
3300	7 meter		.52	31.060		4,950	1,900		6,850	8,275
3310	8 meter	▼	.52	31.060	▼	5,700	1,900		7,600	9,075
3330	Rainproof 1P 3W 120/240 V 4 jaw sockets									
3350	with 200A circuit breaker, 3 meter	2 Elec	.92	17.390	Ea.	2,100	1,075		3,175	3,925

For customer support on your Electrical Change Order Costs with RSMeans data, call 800.448.8182.

Pre-Installation Change Order Costs

26 27 13.10 Meter Centers and Sockets		Crew	Daily Output	Labor-Hours	Unit	Material	2020 Bare Costs Labor	Equipment	Total	Total Incl O&P
3360	4 meter	2 Elec	.83	19.320	Ea.	2,850	1,175		4,025	4,900
3380	6 meter		.55	28.990		4,200	1,775		5,975	7,275
3390	7 meter		.52	31.060		4,950	1,900		6,850	8,275
3400	8 meter		.48	33.440		5,700	2,050		7,750	9,300
3420	1P 3W on 3P 4W 120/208 V 5 jaw sockets									
3430	with 200A circuit breaker, 3 meter	2 Elec	.92	17.390	Ea.	2,100	1,075		3,175	3,925
3440	4 meter		.83	19.320		2,850	1,175		4,025	4,900
3460	6 meter		.55	28.990		4,225	1,775		6,000	7,300
3470	7 meter		.52	31.060		4,950	1,900		6,850	8,275
3480	8 meter		.48	33.440		5,700	2,050		7,750	9,300
3500	Rainproof 1P 3W on 3P 4W 120/208 V 5 jaw socket									
3510	with 200A circuit breaker, 3 meter	2 Elec	.92	17.390	Ea.	2,100	1,075		3,175	3,925
3520	4 meter		.83	19.320		2,850	1,175		4,025	4,900
3540	6 meter		.55	28.990		4,200	1,775		5,975	7,275
3550	7 meter		.52	31.060		4,950	1,900		6,850	8,275
3560	8 meter		.48	33.440		5,700	2,050		7,750	9,300
3600	Automatic circuit closing, add					81.50			81.50	89.50
3610	Manual circuit closing, add					93			93	102
3650	Branch meter device									
3660	3P 4W 208/120 or 240/120 V 7 jaw sockets									
3670	with 200A circuit breaker, 2 meter	2 Elec	.83	19.320	Ea.	3,525	1,175		4,700	5,650
3680	3 meter		.74	21.740		5,275	1,325		6,600	7,775
3690	4 meter		.64	24.840		7,025	1,525		8,550	10,000
3700	Main circuit breaker 42,000 rms, 400 amp		1.47	10.870		2,350	665		3,015	3,600
3710	600 amp		1.01	15.810		2,675	970		3,645	4,375
3720	800 amp		.83	19.320		3,700	1,175		4,875	5,850
3730	Rainproof main circ. breaker 42,000 rms, 400 amp		1.47	10.870		3,075	665		3,740	4,375
3740	600 amp		1.01	15.810		4,950	970		5,920	6,900
3750	800 amp		.83	19.320		6,650	1,175		7,825	9,100
3760	Main circuit breaker 65,000 rms, 400 amp		1.47	10.870		4,175	665		4,840	5,600
3770	600 amp		1.01	15.810		5,875	970		6,845	7,900
3780	800 amp		.83	19.320		6,650	1,175		7,825	9,100
3790	1,000 amp		.74	21.740		8,725	1,325		10,050	11,600
3800	1,200 amp		.70	22.880		11,200	1,400		12,600	14,400
3810	1,600 amp		.63	25.580		22,900	1,575		24,475	27,500
3820	Rainproof main circ. breaker 65,000 rms, 400 amp		1.47	10.870		4,175	665		4,840	5,600
3830	600 amp		1.01	15.810		5,875	970		6,845	7,900
3840	800 amp		.83	19.320		6,650	1,175		7,825	9,075
3850	1,000 amp		.74	21.740		8,725	1,325		10,050	11,600
3860	1,200 amp		.70	22.880		11,200	1,400		12,600	14,400
3880	Main circuit breaker 100,000 rms, 400 amp		1.47	10.870		4,175	665		4,840	5,600
3890	600 amp		1.01	15.810		5,875	970		6,845	7,900
3900	800 amp		.83	19.320		6,900	1,175		8,075	9,350
3910	Rainproof main circ. breaker 100,000 rms, 400 amp		1.47	10.870		4,175	665		4,840	5,600
3920	600 amp		1.01	15.810		5,875	970		6,845	7,900
3930	800 amp		.83	19.320		6,900	1,175		8,075	9,350
3940	Main lugs terminal box, 800 amp		.86	18.500		725	1,125		1,850	2,500
3950	1,600 amp		.66	24.150		2,825	1,475		4,300	5,300
3960	Rainproof, 800 amp		.86	18.500		725	1,125		1,850	2,500
3970	1,600 amp		.66	24.150		2,825	1,475		4,300	5,325

26 27 16.10 Cabinets	Crew	Daily Output	Labor-Hours	Unit	Material	2020 Bare Costs Labor	Equipment	Total	Total Incl O&P
0010 **CABINETS**									
7000 Cabinets, current transformer									
7050 Single door, 24" H x 24" W x 10" D	1 Elec	1.47	5.430	Ea.	215	335		550	730
7100 30" H x 24" W x 10" D		1.20	6.690		207	410		617	835
7150 36" H x 24" W x 10" D		1.01	7.910		510	485		995	1,275
7200 30" H x 30" W x 10" D		.92	8.700		282	535		817	1,100
7250 36" H x 30" W x 10" D		.83	9.660		375	595		970	1,300
7300 36" H x 36" W x 10" D		.74	10.870		385	665		1,050	1,425
7500 Double door, 48" H x 36" W x 10" D		.55	14.490		715	890		1,605	2,125
7550 24" H x 24" W x 12" D		.92	8.700		237	535		772	1,050
8000 NEMA 12, double door, floor mounted									
8020 54" H x 42" W x 8" D	2 Elec	5.52	2.900	Ea.	1,250	178		1,428	1,650
8040 60" H x 48" W x 8" D		4.97	3.220		1,700	198		1,898	2,175
8060 60" H x 48" W x 10" D		4.97	3.220		1,700	198		1,898	2,175
8080 60" H x 60" W x 10" D		4.60	3.480		1,975	214		2,189	2,500
8100 72" H x 60" W x 10" D		3.68	4.350		2,250	267		2,517	2,875
8120 72" H x 72" W x 10" D		3.13	5.120		2,025	315		2,340	2,700
8140 60" H x 48" W x 12" D		3.13	5.120		1,750	315		2,065	2,400
8160 60" H x 60" W x 12" D		2.94	5.430		1,975	335		2,310	2,675
8180 72" H x 60" W x 12" D		2.76	5.800		3,425	355		3,780	4,300
8200 72" H x 72" W x 12" D		2.76	5.800		3,825	355		4,180	4,725
8220 60" H x 48" W x 16" D		2.94	5.430		1,875	335		2,210	2,550
8240 72" H x 72" W x 16" D		2.39	6.690		2,650	410		3,060	3,525
8260 60" H x 48" W x 20" D		2.76	5.800		2,025	355		2,380	2,750
8280 72" H x 72" W x 20" D		2.02	7.910		2,875	485		3,360	3,900
8300 60" H x 48" W x 24" D		2.39	6.690		2,150	410		2,560	2,950
8320 72" H x 72" W x 24" D		1.84	8.700		3,025	535		3,560	4,125
8340 Pushbutton enclosure, oiltight									
8360 3-1/2" H x 3-1/4" W x 2-3/4" D, for 1 P.B.	1 Elec	11.04	.725	Ea.	54	44.50		98.50	126
8380 5-3/4" H x 3-1/4" W x 2-3/4" D, for 2 P.B.		10.12	.791		57	48.50		105.50	135
8400 8" H x 3-1/4" W x 2-3/4" D, for 3 P.B.		9.66	.828		66.50	51		117.50	149
8420 10-1/4" H x 3-1/4" W x 2-3/4" D, for 4 P.B.		9.66	.828		72	51		123	155
8460 12-1/2" H x 3-1/4" W x 3" D, for 5 P.B.		8.28	.966		133	59.50		192.50	235
8480 9-1/2" H x 6-1/4" W x 3" D, for 6 P.B.		7.82	1.020		91	62.50		153.50	193
8500 9-1/2" H x 8-1/2" W x 3" D, for 9 P.B.		7.36	1.090		112	67		179	224
8510 11-3/4" H x 8-1/2" W x 3" D, for 12 P.B.		6.44	1.240		118	76		194	243
8520 11-3/4" H x 10-3/4" W x 3" D, for 16 P.B.		5.98	1.340		125	82		207	259
8540 14" H x 10-3/4" W x 3" D, for 20 P.B.		4.60	1.740		203	107		310	380
8560 14" H x 13" W x 3" D, for 25 P.B.		4.14	1.930		221	118		339	420
8580 Sloping front pushbutton enclosures									
8600 3-1/2" H x 7-3/4" W x 4-7/8" D, for 3 P.B.	1 Elec	9.20	.870	Ea.	89.50	53.50		143	178
8620 7-1/4" H x 8-1/2" W x 6-3/4" D, for 6 P.B.		7.36	1.090		131	67		198	244
8640 9-1/2" H x 8-1/2" W x 7-7/8" D, for 9 P.B.		6.44	1.240		175	76		251	305
8660 11-1/4" H x 8-1/2" W x 9" D, for 12 P.B.		4.60	1.740		175	107		282	350
8680 11-3/4" H x 10" W x 9" D, for 16 P.B.		4.60	1.740		199	107		306	380
8700 11-3/4" H x 13" W x 9" D, for 20 P.B.		4.60	1.740		228	107		335	410
8720 14" H x 13" W x 10-1/8" D, for 25 P.B.		4.14	1.930		254	118		372	455
8740 Pedestals, not including P.B. enclosure or base									
8760 Straight column 4" x 4"	1 Elec	4.14	1.930	Ea.	238	118		356	440
8780 6" x 6"		3.68	2.170		370	133		503	610
8800 Angled column 4" x 4"		4.14	1.930		275	118		393	480
8820 6" x 6"		3.68	2.170		410	133		543	655

26 27 Low-Voltage Distribution Equipment

26 27 16 – Electrical Cabinets and Enclosures

26 27 16.10 Cabinets

		Crew	Daily Output	Labor-Hours	Unit	Material	2020 Bare Costs Labor	Equipment	Total	Total Incl O&P
8840	Pedestal, base 18" x 18"	1 Elec	9.20	.870	Ea.	141	53.50		194.50	235
8860	24" x 24"	↓	8.28	.966	↓	325	59.50		384.50	445
8900	Electronic rack enclosures									
8920	72" H x 19" W x 24" D	1 Elec	1.38	5.800	Ea.	2,675	355		3,030	3,475
8940	72" H x 23" W x 24" D		1.38	5.800		2,900	355		3,255	3,725
8960	72" H x 19" W x 30" D		1.20	6.690		2,950	410		3,360	3,825
8980	72" H x 19" W x 36" D		1.10	7.250		3,450	445		3,895	4,450
9000	72" H x 23" W x 36" D	↓	1.10	7.250	↓	3,800	445		4,245	4,825
9020	NEMA 12 & 4 enclosure panels									
9040	12" x 24"	1 Elec	18.40	.435	Ea.	55	26.50		81.50	101
9060	16" x 12"		18.40	.435		39	26.50		65.50	82.50
9080	20" x 16"		18.40	.435		57	26.50		83.50	102
9100	20" x 20"		17.48	.458		69	28		97	118
9120	24" x 20"		16.56	.483		89	29.50		118.50	142
9140	24" x 24"		15.64	.512		102	31.50		133.50	160
9160	30" x 20"		14.72	.543		106	33.50		139.50	167
9180	30" x 24"		14.72	.543		125	33.50		158.50	188
9200	36" x 24"		13.80	.580		150	35.50		185.50	218
9220	36" x 30"		13.80	.580		198	35.50		233.50	270
9240	42" x 24"		13.80	.580		178	35.50		213.50	249
9260	42" x 30"		12.88	.621		225	38		263	305
9280	42" x 36"		12.88	.621		266	38		304	350
9300	48" x 24"		12.88	.621		131	38		169	201
9320	48" x 30"		12.88	.621		171	38		209	245
9340	48" x 36"		11.96	.669		299	41		340	390
9360	60" x 36"	↓	11.04	.725	↓	370	44.50		414.50	475
9400	Wiring trough steel JIC, clamp cover									
9490	4" x 4", 12" long	1 Elec	11.04	.725	Ea.	85.50	44.50		130	160
9510	24" long		9.20	.870		109	53.50		162.50	200
9530	36" long		7.36	1.090		141	67		208	255
9540	48" long		6.44	1.240		157	76		233	285
9550	60" long		5.52	1.450		191	89		280	340
9560	6" x 6", 12" long		10.12	.791		119	48.50		167.50	203
9580	24" long		8.28	.966		155	59.50		214.50	259
9600	36" long		6.44	1.240		184	76		260	315
9610	48" long		5.52	1.450		234	89		323	390
9620	60" long	↓	4.60	1.740	↓	275	107		382	460

26 27 16.20 Cabinets and Enclosures

		Crew	Daily Output	Labor-Hours	Unit	Material	2020 Bare Costs Labor	Equipment	Total	Total Incl O&P
0010	**CABINETS AND ENCLOSURES** Nonmetallic									
0080	Enclosures fiberglass NEMA 4X									
0100	Wall mount, quick release latch door, 20" H x 16" W x 6" D	1 Elec	4.42	1.810	Ea.	910	111		1,021	1,175
0110	20" H x 20" W x 6" D		4.14	1.930		1,050	118		1,168	1,325
0120	24" H x 20" W x 6" D		3.86	2.070		1,125	127		1,252	1,450
0130	20" H x 16" W x 8" D		4.14	1.930		995	118		1,113	1,275
0140	20" H x 20" W x 8" D		3.86	2.070		1,100	127		1,227	1,425
0150	24" H x 24" W x 8" D		3.50	2.290		1,250	140		1,390	1,575
0160	30" H x 24" W x 8" D		2.94	2.720		1,425	167		1,592	1,825
0170	36" H x 30" W x 8" D		2.76	2.900		2,000	178		2,178	2,475
0180	20" H x 16" W x 10" D		3.22	2.480		1,200	152		1,352	1,550
0190	20" H x 20" W x 10" D		2.94	2.720		1,300	167		1,467	1,675
0200	24" H x 20" W x 10" D		2.76	2.900		1,375	178		1,553	1,775
0210	30" H x 24" W x 10" D	↓	2.58	3.110	↓	1,575	191		1,766	2,025

26 27 16.20 Cabinets and Enclosures		Crew	Daily Output	Labor-Hours	Unit	Material	2020 Bare Costs Labor	Equipment	Total	Total Incl O&P
0220	20" H x 16" W x 12" D	1 Elec	2.76	2.900	Ea.	1,300	178		1,478	1,700
0230	20" H x 20" W x 12" D		2.58	3.110		1,350	191		1,541	1,775
0240	24" H x 24" W x 12" D		2.39	3.340		1,475	205		1,680	1,925
0250	30" H x 24" W x 12" D		2.21	3.620		1,700	222		1,922	2,175
0260	36" H x 30" W x 12" D		2.02	3.950		2,300	242		2,542	2,875
0270	36" H x 36" W x 12" D		1.93	4.140		2,525	254		2,779	3,150
0280	48" H x 36" W x 12" D		1.84	4.350		2,975	267		3,242	3,675
0290	60" H x 36" W x 12" D		1.66	4.830		2,150	296		2,446	2,825
0300	30" H x 24" W x 16" D		1.29	6.210		1,950	380		2,330	2,725
0310	48" H x 36" W x 16" D		1.10	7.250		3,275	445		3,720	4,250
0320	60" H x 36" W x 16" D		.92	8.700		3,675	535		4,210	4,850
0480	Freestanding, one door, 72" H x 25" W x 25" D		.74	10.870		4,200	665		4,865	5,625
0490	Two doors with two panels, 72" H x 49" W x 24" D		.46	17.390		9,725	1,075		10,800	12,300
0500	Floor stand kits, for NEMA 4 & 12, 20" W or more, 6" H x 8" D		22.08	.362		145	22		167	192
0510	6" H x 10" D		22.08	.362		158	22		180	207
0520	6" H x 12" D		22.08	.362		180	22		202	230
0530	6" H x 18" D		22.08	.362		218	22		240	273
0540	12" H x 8" D		20.24	.395		185	24.50		209.50	239
0550	12" H x 10" D		20.24	.395		335	24.50		359.50	405
0560	12" H x 12" D		20.24	.395		345	24.50		369.50	415
0570	12" H x 16" D		20.24	.395		241	24.50		265.50	300
0580	12" H x 18" D		20.24	.395		262	24.50		286.50	325
0590	12" H x 20" D		20.24	.395		315	24.50		339.50	380
0600	18" H x 8" D		18.40	.435		216	26.50		242.50	278
0610	18" H x 10" D		18.40	.435		281	26.50		307.50	350
0620	18" H x 12" D		18.40	.435		299	26.50		325.50	370
0630	18" H x 16" D		18.40	.435		330	26.50		356.50	405
0640	24" H x 8" D		14.72	.543		269	33.50		302.50	345
0650	24" H x 10" D		14.72	.543		330	33.50		363.50	415
0660	24" H x 12" D		14.72	.543		355	33.50		388.50	440
0670	24" H x 16" D		14.72	.543		380	33.50		413.50	470
0680	Small, screw cover, 5-1/2" H x 4" W x 4-15/16" D		11.04	.725		146	44.50		190.50	227
0690	7-1/2" H x 4" W x 4-15/16" D		11.04	.725		95.50	44.50		140	171
0700	7-1/2" H x 6" W x 5-3/16" D		9.20	.870		177	53.50		230.50	274
0710	9-1/2" H x 6" W x 5-11/16" D		9.20	.870		184	53.50		237.50	283
0720	11-1/2" H x 8" W x 6-11/16" D		7.36	1.090		272	67		339	400
0730	13-1/2" H x 10" W x 7-3/16" D		6.44	1.240		330	76		406	480
0740	15-1/2" H x 12" W x 8-3/16" D		5.52	1.450		425	89		514	595
0750	17-1/2" H x 14" W x 8-11/16" D		4.60	1.740		500	107		607	710
0760	Screw cover with window, 6" H x 4" W x 5" D		11.04	.725		186	44.50		230.50	270
0770	8" H x 4" W x 5" D		10.12	.791		245	48.50		293.50	340
0780	8" H x 6" W x 5" D		10.12	.791		360	48.50		408.50	465
0790	10" H x 6" W x 6" D		9.20	.870		375	53.50		428.50	490
0800	12" H x 8" W x 7" D		7.36	1.090		480	67		547	625
0810	14" H x 10" W x 7" D		6.44	1.240		615	76		691	795
0820	16" H x 12" W x 8" D		5.52	1.450		730	89		819	935
0830	18" H x 14" W x 9" D		4.60	1.740		820	107		927	1,050
0840	Quick-release latch cover, 5-1/2" H x 4" W x 5" D		11.04	.725		191	44.50		235.50	277
0850	7-1/2" H x 4" W x 5" D		11.04	.725		196	44.50		240.50	281
0860	7-1/2" H x 6" W x 5-1/4" D		9.20	.870		227	53.50		280.50	330
0870	9-1/2" H x 6" W x 5-3/4" D		9.20	.870		140	53.50		193.50	234
0880	11-1/2" H x 8" W x 6-3/4" D		7.36	1.090		221	67		288	345
0890	13-1/2" H x 10" W x 7-1/4" D		6.44	1.240		251	76		327	390

For customer support on your Electrical Change Order Costs with RSMeans data, call 800.448.8182.

Pre-Installation Change Order Costs

26 27 16 – Electrical Cabinets and Enclosures

26 27 16.20 Cabinets and Enclosures	Crew	Daily Output	Labor-Hours	Unit	Material	2020 Bare Costs Labor	Equipment	Total	Total Incl O&P
0900 15-1/2" H x 12" W x 8-1/4" D	1 Elec	5.52	1.450	Ea.	560	89		649	750
0910 17-1/2" H x 14" W x 8-3/4" D		4.60	1.740		695	107		802	925
0920 Pushbutton, 1 hole 5-1/2" H x 4" W x 4-15/16" D		11.04	.725		545	44.50		589.50	665
0930 2 hole 7-1/2" H x 4" W x 4-15/16" D		10.12	.791		755	48.50		803.50	900
0940 4 hole 7-1/2" H x 6" W x 5-3/16" D		9.66	.828		530	51		581	660
0950 6 hole 9-1/2" H x 6" W x 5-11/16" D		8.28	.966		645	59.50		704.50	800
0960 8 hole 11-1/2" H x 8" W x 6-11/16" D		7.82	1.020		825	62.50		887.50	1,000
0970 12 hole 13-1/2" H x 10" W x 7-3/16" D		7.36	1.090		1,075	67		1,142	1,300
0980 20 hole 15-1/2" H x 12" W x 8-3/16" D		4.60	1.740		1,025	107		1,132	1,275
0990 30 hole 17-1/2" H x 14" W x 8-11/16" D	▼	4.14	1.930	▼	1,200	118		1,318	1,500
1450 Enclosures polyester NEMA 4X									
1460 Small, screw cover,									
1500 3-15/16" H x 3-15/16" W x 3-1/16" D	1 Elec	11.04	.725	Ea.	81	44.50		125.50	155
1510 5-3/16" H x 3-5/16" W x 3-1/16" D		11.04	.725		79.50	44.50		124	153
1520 5-7/8" H x 3-7/8" W x 4-3/16" D		11.04	.725		83.50	44.50		128	158
1530 5-7/8" H x 5-7/8" W x 4-3/16" D		11.04	.725		94	44.50		138.50	169
1540 7-5/8" H x 3-5/16" W x 3-1/16" D		11.04	.725		86.50	44.50		131	162
1550 10-3/16" H x 3-5/16" W x 3-1/16" D		9.20	.870		108	53.50		161.50	199
1560 Clear cover, 3-15/16" H x 3-15/16" W x 2-7/8" D		11.04	.725		91	44.50		135.50	166
1570 5-3/16" H x 3-5/16" W x 2-7/8" D		11.04	.725		97	44.50		141.50	173
1580 5-7/8" H x 3-7/8" W x 4" D		11.04	.725		120	44.50		164.50	199
1590 5-7/8" H x 5-7/8" W x 4" D		11.04	.725		148	44.50		192.50	229
1600 7-5/8" H x 3-5/16" W x 2-7/8" D		11.04	.725		109	44.50		153.50	186
1610 10-3/16" H x 3-5/16" W x 2-7/8" D		9.20	.870		137	53.50		190.50	230
1620 Pushbutton, 1 hole, 5-5/16" H x 3-5/16" W x 3-1/16" D		11.04	.725		74	44.50		118.50	148
1630 2 hole, 7-5/8" H x 3-5/16" W x 3-1/8" D		10.12	.791		79.50	48.50		128	160
1640 3 hole, 10-3/16" H x 3-5/16" W x 3-1/16" D		9.66	.828		94	51		145	179
8000 Wireway fiberglass, straight sect. screwcover, 12" L, 4" W x 4" D		36.80	.217		275	13.35		288.35	320
8010 6" W x 6" D		27.60	.290		380	17.80		397.80	445
8020 24" L, 4" W x 4" D		18.40	.435		350	26.50		376.50	425
8030 6" W x 6" D		13.80	.580		540	35.50		575.50	645
8040 36" L, 4" W x 4" D		12.24	.654		425	40		465	525
8050 6" W x 6" D		9.20	.870		675	53.50		728.50	820
8060 48" L, 4" W x 4" D		9.20	.870		515	53.50		568.50	650
8070 6" W x 6" D		6.90	1.160		840	71		911	1,025
8080 60" L, 4" W x 4" D		7.36	1.090		630	67		697	795
8090 6" W x 6" D		5.52	1.450		985	89		1,074	1,200
8100 Elbow, 90°, 4" W x 4" D		18.40	.435		275	26.50		301.50	340
8110 6" W x 6" D		16.56	.483		550	29.50		579.50	655
8120 Elbow, 45°, 4" W x 4" D		18.40	.435		273	26.50		299.50	340
8130 6" W x 6" D		16.56	.483		555	29.50		584.50	655
8140 Tee, 4" W x 4" D		14.72	.543		350	33.50		383.50	435
8150 6" W x 6" D		12.88	.621		650	38		688	770
8160 Cross, 4" W x 4" D		12.88	.621		380	38		418	475
8170 6" W x 6" D		11.04	.725		930	44.50		974.50	1,100
8180 Cut-off fitting, w/flange & adhesive, 4" W x 4" D		16.56	.483		217	29.50		246.50	282
8190 6" W x 6" D		14.72	.543		445	33.50		478.50	540
8200 Flexible ftng., hvy. neoprene coated nylon, 4" W x 4" D		18.40	.435		405	26.50		431.50	485
8210 6" W x 6" D		16.56	.483		585	29.50		614.50	690
8220 Closure plate, fiberglass, 4" W x 4" D		18.40	.435		77	26.50		103.50	124
8230 6" W x 6" D		16.56	.483		85.50	29.50		115	139
8240 Box connector, stainless steel type 304, 4" W x 4" D		18.40	.435		132	26.50		158.50	186
8250 6" W x 6" D	▼	16.56	.483	▼	151	29.50		180.50	210

26 27 Low-Voltage Distribution Equipment

26 27 16 – Electrical Cabinets and Enclosures

26 27 16.20 Cabinets and Enclosures		Crew	Daily Output	Labor-Hours	Unit	Material	2020 Bare Costs Labor	Equipment	Total	Total Incl O&P
8260	Hanger, 4" W x 4" D	1 Elec	92	.087	Ea.	33	5.35		38.35	44.50
8270	6" W x 6" D		73.60	.109		42	6.65		48.65	56.50
8280	Straight tube section fiberglass, 4" W x 4" D, 12" long		36.80	.217		248	13.35		261.35	293
8290	24" long		18.40	.435		300	26.50		326.50	370
8300	36" long		12.24	.654		275	40		315	360
8310	48" long		9.20	.870		370	53.50		423.50	485
8320	60" long		7.36	1.090		430	67		497	575
8330	120" long		3.68	2.170		685	133		818	950

26 27 19 – Multi-Outlet Assemblies

26 27 19.10 Wiring Duct

0010	**WIRING DUCT** Plastic									
1250	PVC, snap-in slots, adhesive backed									
1270	1-1/2" W x 2" H	2 Elec	110.40	.145	L.F.	6.30	8.90		15.20	20
1280	1-1/2" W x 3" H		110.40	.145		8.25	8.90		17.15	22.50
1290	1-1/2" W x 4" H		110.40	.145		8.80	8.90		17.70	23
1300	2" W x 1" H		110.40	.145		5.65	8.90		14.55	19.50
1310	2" W x 1-1/2" H		110.40	.145		6	8.90		14.90	19.85
1320	2" W x 2" H		110.40	.145		6	8.90		14.90	19.85
1340	2" W x 3" H		110.40	.145		8.25	8.90		17.15	22.50
1350	2" W x 4" H		110.40	.145		9.95	8.90		18.85	24
1360	2-1/2" W x 3" H		110.40	.145		8.95	8.90		17.85	23
1370	3" W x 1" H		101.20	.158		6.65	9.70		16.35	22
1390	3" W x 2" H		101.20	.158		6.95	9.70		16.65	22
1400	3" W x 3" H		101.20	.158		8.40	9.70		18.10	23.50
1410	3" W x 4" H		101.20	.158		10.50	9.70		20.20	26
1420	3" W x 5" H		101.20	.158		13.75	9.70		23.45	29.50
1430	4" W x 1-1/2" H		92	.174		9.70	10.65		20.35	26.50
1440	4" W x 2" H		92	.174		8.10	10.65		18.75	25
1450	4" W x 3" H		92	.174		9.20	10.65		19.85	26
1460	4" W x 4" H		92	.174		11.10	10.65		21.75	28
1470	4" W x 5" H		92	.174		15.60	10.65		26.25	33
1550	Cover, 1-1/2" W		184	.087		1.16	5.35		6.51	9.25
1560	2" W		184	.087		1.28	5.35		6.63	9.35
1570	2-1/2" W		184	.087		1.79	5.35		7.14	9.90
1580	3" W		184	.087		1.94	5.35		7.29	10.10
1590	4" W		184	.087		2.34	5.35		7.69	10.50

26 27 23 – Indoor Service Poles

26 27 23.40 Surface Raceway

0010	**SURFACE RACEWAY**									
0090	Metal, straight section									
0100	No. 500	1 Elec	92	.087	L.F.	1.56	5.35		6.91	9.65
0400	No. 1500, small pancake		82.80	.097		2.93	5.95		8.88	12.05
0600	No. 2000, base & cover, blank		82.80	.097		3.12	5.95		9.07	12.30
0610	Receptacle, 6" OC		36.80	.217		25.50	13.35		38.85	48
0620	12" OC		40.48	.198		16.60	12.10		28.70	36.50
0630	18" OC		42.32	.189		9.75	11.60		21.35	28
0650	30" OC		46	.174		5.85	10.65		16.50	22.50
0660	60" OC		46	.174		5.40	10.65		16.05	22
0670	No. 2400, base & cover, blank		73.60	.109		2.74	6.65		9.39	12.95
0680	Receptacle, 6" OC		38.64	.207		52	12.70		64.70	76
0690	12" OC		48.76	.164		35.50	10.05		45.55	54
0700	18" OC		50.60	.158		31.50	9.70		41.20	49

For customer support on your Electrical Change Order Costs with RSMeans data, call 800.448.8182. **Pre-Installation Change Order Costs**

26 27 23 – Indoor Service Poles

26 27 23.40 Surface Raceway	Crew	Daily Output	Labor-Hours	Unit	Material	2020 Bare Costs Labor	Equipment	Total	Total Incl O&P
0710	24" OC	1 Elec	52.44	.153	L.F.	12	9.35		21.35	27
0720	30" OC		54.28	.147		8.05	9.05		17.10	22.50
0730	60" OC		56.12	.143		6.45	8.75		15.20	20
0800	No. 3000, base & cover, blank		69	.116		5.80	7.10		12.90	17
0810	Receptacle, 6" OC		41.40	.193		52.50	11.85		64.35	75.50
0820	12" OC		57.04	.140		29.50	8.60		38.10	45.50
0830	18" OC		58.88	.136		24	8.35		32.35	39
0840	24" OC		60.72	.132		18	8.10		26.10	32
0850	30" OC		62.56	.128		16.20	7.85		24.05	29.50
0860	60" OC		64.40	.124		11.85	7.60		19.45	24.50
1000	No. 4000, base & cover, blank		59.80	.134		9.45	8.20		17.65	22.50
1010	Receptacle, 6" OC		37.72	.212		61	13		74	87
1020	12" OC		47.84	.167		44	10.25		54.25	64
1030	18" OC		49.68	.161		36	9.90		45.90	54
1040	24" OC		51.52	.155		29.50	9.55		39.05	46.50
1050	30" OC		53.36	.150		27	9.20		36.20	43.50
1060	60" OC		55.20	.145		20.50	8.90		29.40	36
1200	No. 6000, base & cover, blank		46	.174		16.65	10.65		27.30	34
1210	Receptacle, 6" OC		27.60	.290		75	17.80		92.80	109
1220	12" OC		34.04	.235		57	14.40		71.40	84.50
1230	18" OC		35.88	.223		48.50	13.70		62.20	73.50
1240	24" OC		37.72	.212		39.50	13		52.50	63
1250	30" OC		39.56	.202		37.50	12.40		49.90	60
1260	60" OC		41.40	.193	▼	29	11.85		40.85	49.50
2400	Fittings, elbows, No. 500		36.80	.217	Ea.	2.81	13.35		16.16	23
2800	Elbow cover, No. 2000		36.80	.217		5.10	13.35		18.45	25.50
2880	Tee, No. 500		38.64	.207		5.05	12.70		17.75	24.50
2900	No. 2000		24.84	.322		16.40	19.75		36.15	47.50
3000	Switch box, No. 500		14.72	.543		17.80	33.50		51.30	69
3400	Telephone outlet, No. 1500		14.72	.543		21	33.50		54.50	72.50
3600	Junction box, No. 1500	▼	14.72	.543	▼	13.85	33.50		47.35	64.50
3800	Plugmold wired sections, No. 2000									
4000	1 circuit, 6 outlets, 3' long	1 Elec	7.36	1.090	Ea.	54	67		121	159
4100	2 circuits, 8 outlets, 6' long		4.88	1.640		78.50	101		179.50	237
4110	Tele-power pole, alum, w/2 recept, 10'		3.68	2.170		300	133		433	530
4120	12'		3.54	2.260		355	139		494	595
4130	15'		3.40	2.350		450	144		594	710
4140	Steel, w/2 recept, 10'		3.68	2.170		185	133		318	400
4150	One phone fitting, 10'		3.68	2.170		201	133		334	420
4160	Alum, 4 outlets, 10'	▼	3.40	2.350	▼	435	144		579	690
4300	Overhead distribution systems, 125 volt									
4800	No. 2000, entrance end fitting	1 Elec	18.40	.435	Ea.	8.15	26.50		34.65	48.50
5000	Blank end fitting		36.80	.217		2.74	13.35		16.09	23
5200	Supporting clip		36.80	.217		1.62	13.35		14.97	21.50
5800	No. 3000, entrance end fitting		18.40	.435		12.90	26.50		39.40	53.50
6000	Blank end fitting		36.80	.217		3.58	13.35		16.93	24
6020	Internal elbow		18.40	.435		16.95	26.50		43.45	58
6030	External elbow		18.40	.435		24	26.50		50.50	66
6040	Device bracket		48.76	.164		5.75	10.05		15.80	21.50
6400	Hanger clamp		29.44	.272	▼	8.75	16.65		25.40	34.50
7000	No. 4000 base		82.80	.097	L.F.	6.15	5.95		12.10	15.65
7200	Divider		92	.087	"	1.16	5.35		6.51	9.25
7400	Entrance end fitting		14.72	.543	Ea.	30.50	33.50		64	83

26 27 23.40 Surface Raceway	Crew	Daily Output	Labor-Hours	Unit	Material	2020 Bare Costs Labor	Equipment	Total	Total Incl O&P
7600	Blank end fitting	1 Elec	36.80	.217	Ea.	8.45	13.35		21.80	29
7610	Recpt. & tele. cover		48.76	.164		14.75	10.05		24.80	31.50
7620	External elbow		14.72	.543		48.50	33.50		82	103
7630	Coupling		48.76	.164		7.40	10.05		17.45	23
7640	Divider clip & coupling		73.60	.109		1.46	6.65		8.11	11.55
7650	Panel connector		14.72	.543		30.50	33.50		64	83.50
7800	Take off connector		14.72	.543		102	33.50		135.50	162
8000	No. 6000, take off connector		14.72	.543		121	33.50		154.50	183
8100	Take off fitting		14.72	.543		89.50	33.50		123	148
8200	Hanger clamp	↓	29.44	.272		23	16.65		39.65	50
8230	Coupling					12.60			12.60	13.85
8240	One gang device plate	1 Elec	48.76	.164		11.60	10.05		21.65	28
8250	Two gang device plate		36.80	.217		14.20	13.35		27.55	35.50
8260	Blank end fitting		36.80	.217		12.65	13.35		26	34
8270	Combination elbow		12.88	.621		50	38		88	112
8300	Panel connector	↓	14.72	.543	↓	22	33.50		55.50	74
8500	Chan-L-Wire system installed in 1-5/8" x 1-5/8" strut. Strut									
8600	not incl., 30 amp, 4 wire, 3 phase	1 Elec	184	.043	L.F.	5.60	2.67		8.27	10.15
8700	Junction box		7.36	1.090	Ea.	37.50	67		104.50	141
8800	Insulating end cap		36.80	.217		10	13.35		23.35	31
8900	Strut splice plate		36.80	.217		13.25	13.35		26.60	34.50
9000	Tap		36.80	.217		27	13.35		40.35	50
9100	Fixture hanger	↓	55.20	.145		11.80	8.90		20.70	26.50
9200	Pulling tool				↓	101			101	111
9300	Non-metallic, straight section									
9310	7/16" x 7/8", base & cover, blank	1 Elec	147.20	.054	L.F.	2	3.33		5.33	7.15
9320	Base & cover w/adhesive		147.20	.054		1.71	3.33		5.04	6.85
9340	7/16" x 1-5/16", base & cover, blank		133.40	.060		2.20	3.68		5.88	7.90
9350	Base & cover w/adhesive		133.40	.060		2.26	3.68		5.94	8
9370	11/16" x 2-1/4", base & cover, blank		119.60	.067		3.02	4.10		7.12	9.40
9380	Base & cover w/adhesive		119.60	.067		3.24	4.10		7.34	9.65
9385	1-11/16" x 5-1/4", two compartment base & cover w/screws		73.60	.109	↓	10.40	6.65		17.05	21.50
9400	Fittings, elbows, 7/16" x 7/8"		46	.174	Ea.	1.97	10.65		12.62	18.05
9410	7/16" x 1-5/16"		41.40	.193		2.04	11.85		13.89	19.90
9420	11/16" x 2-1/4"		36.80	.217		2.21	13.35		15.56	22.50
9425	1-11/16" x 5-1/4"		25.76	.311		12.45	19.05		31.50	42
9430	Tees, 7/16" x 7/8"		32.20	.248		2.53	15.25		17.78	25.50
9440	7/16" x 1-5/16"		29.44	.272		2.59	16.65		19.24	28
9450	11/16" x 2-1/4"		27.60	.290		2.69	17.80		20.49	29.50
9455	1-11/16" x 5-1/4"		22.08	.362		19.95	22		41.95	55
9460	Cover clip, 7/16" x 7/8"		73.60	.109		.51	6.65		7.16	10.50
9470	7/16" x 1-5/16"		66.24	.121		.46	7.40		7.86	11.55
9480	11/16" x 2-1/4"		58.88	.136		.84	8.35		9.19	13.30
9484	1-11/16" x 5-1/4"		38.64	.207		2.94	12.70		15.64	22
9486	Wire clip, 1-11/16" x 5-1/4"		62.56	.128		.58	7.85		8.43	12.35
9490	Blank end, 7/16" x 7/8"		46	.174		.77	10.65		11.42	16.75
9500	7/16" x 1-5/16"		41.40	.193		.86	11.85		12.71	18.60
9510	11/16" x 2-1/4"		36.80	.217		1.28	13.35		14.63	21.50
9515	1-11/16" x 5-1/4"		34.96	.229		6.70	14.05		20.75	28.50
9520	Round fixture box, 5.5" diam. x 1"		23	.348		11.55	21.50		33.05	44.50
9530	Device box, 1 gang		27.60	.290		5.25	17.80		23.05	32.50
9540	2 gang	↓	23	.348	↓	7.65	21.50		29.15	40.50

26 27 26.10 Low Voltage Switching

		Crew	Daily Output	Labor-Hours	Unit	Material	2020 Bare Costs Labor	Equipment	Total	Total Incl O&P
0010	**LOW VOLTAGE SWITCHING**									
3600	Relays, 120 V or 277 V standard	1 Elec	11.04	.725	Ea.	48.50	44.50		93	120
3800	Flush switch, standard		36.80	.217		14.80	13.35		28.15	36
4000	Interchangeable		36.80	.217		19.30	13.35		32.65	41.50
4100	Surface switch, standard		36.80	.217		9.15	13.35		22.50	30
4200	Transformer 115 V to 25 V		11.04	.725		141	44.50		185.50	221
4400	Master control, 12 circuit, manual		3.68	2.170		153	133		286	365
4500	25 circuit, motorized		3.68	2.170		177	133		310	395
4600	Rectifier, silicon		11.04	.725		55.50	44.50		100	127
4800	Switchplates, 1 gang, 1, 2 or 3 switch, plastic		73.60	.109		5.80	6.65		12.45	16.30
5000	Stainless steel		73.60	.109		12.45	6.65		19.10	23.50
5400	2 gang, 3 switch, stainless steel		48.76	.164		26	10.05		36.05	43.50
5500	4 switch, plastic		48.76	.164		11.35	10.05		21.40	27.50
5600	2 gang, 4 switch, stainless steel		48.76	.164		24.50	10.05		34.55	42
5700	6 switch, stainless steel		48.76	.164		48.50	10.05		58.55	68.50
5800	3 gang, 9 switch, stainless steel		29.44	.272		74	16.65		90.65	107
5900	Receptacle, triple, 1 return, 1 feed		23.92	.334		51.50	20.50		72	87.50
6000	2 feed		18.40	.435		51.50	26.50		78	96.50
6100	Relay gang boxes, flush or surface, 6 gang		4.88	1.640		101	101		202	261
6200	12 gang		4.32	1.850		120	114		234	300
6400	18 gang		3.68	2.170		132	133		265	345
6500	Frame, to hold up to 6 relays		11.04	.725	▼	98.50	44.50		143	175
7200	Control wire, 2 conductor		5.80	1.380	C.L.F.	39	84.50		123.50	169
7400	3 conductor		4.60	1.740		37.50	107		144.50	201
7600	19 conductor		2.30	3.480		350	214		564	705
7800	26 conductor		1.84	4.350		480	267		747	920
8000	Weatherproof, 3 conductor	▼	4.60	1.740	▼	79	107		186	246

26 27 26.20 Wiring Devices Elements

		Crew	Daily Output	Labor-Hours	Unit	Material	2020 Bare Costs Labor	Equipment	Total	Total Incl O&P
0010	**WIRING DEVICES ELEMENTS**									
0200	Toggle switch, quiet type, single pole, 15 amp	1 Elec	36.80	.217	Ea.	.57	13.35		13.92	20.50
0500	20 amp		24.84	.322		3.81	19.75		23.56	33.50
0510	30 amp		21.16	.378		25.50	23		48.50	63
0530	Lock handle, 20 amp		24.84	.322		32.50	19.75		52.25	65.50
0540	Security key, 20 amp		23.92	.334		94	20.50		114.50	134
0550	Rocker, 15 amp		36.80	.217		3.10	13.35		16.45	23.50
0560	20 amp		24.84	.322		11.95	19.75		31.70	42.50
0600	3 way, 15 amp		21.16	.378		2.01	23		25.01	36.50
0800	20 amp		16.56	.483		3.89	29.50		33.39	48.50
0810	30 amp		8.28	.966		26.50	59.50		86	118
0830	Lock handle, 20 amp		16.56	.483		31.50	29.50		61	78.50
0840	Security key, 20 amp		15.64	.512		137	31.50		168.50	197
0850	Rocker, 15 amp		21.16	.378		9.10	23		32.10	44.50
0860	20 amp		16.56	.483		14.10	29.50		43.60	59.50
0900	4 way, 15 amp		13.80	.580		11.70	35.50		47.20	66
1000	20 amp		10.12	.791		52.50	48.50		101	130
1020	Lock handle, 20 amp		10.12	.791		43.50	48.50		92	120
1030	Rocker, 15 amp		13.80	.580		11	35.50		46.50	65
1040	20 amp		10.12	.791		21	48.50		69.50	95
1100	Toggle switch, quiet type, double pole, 15 amp		13.80	.580		15.35	35.50		50.85	70
1200	20 amp		10.12	.791		21	48.50		69.50	95.50
1210	30 amp		8.28	.966		37.50	59.50		97	130
1230	Lock handle, 20 amp	▼	10.12	.791	▼	38.50	48.50		87	114

26 27 26.20 Wiring Devices Elements	Crew	Daily Output	Labor-Hours	Unit	Material	2020 Bare Costs Labor	Equipment	Total	Total Incl O&P
1250	Security key, 20 amp	1 Elec	9.20	.870	Ea.	100	53.50		153.50	190
1420	Toggle switch quiet type, 1 pole, 2 throw center off, 15 amp		21.16	.378		49	23		72	88
1440	20 amp		16.56	.483		77.50	29.50		107	129
1460	2 pole, 2 throw center off, lock handle, 20 amp		10.12	.791		93.50	48.50		142	175
1480	1 pole, momentary contact, 15 amp		21.16	.378		22	23		45	59
1500	20 amp		16.56	.483		31.50	29.50		61	78.50
1520	Momentary contact, lock handle, 20 amp		16.56	.483		35	29.50		64.50	82.50
1650	Dimmer switch, 120 volt, incandescent, 600 watt, 1 pole G		14.72	.543		25	33.50		58.50	77
1700	600 watt, 3 way G		11.04	.725		12.20	44.50		56.70	79.50
1750	1,000 watt, 1 pole G		14.72	.543		48.50	33.50		82	103
1800	1,000 watt, 3 way G		11.04	.725		80.50	44.50		125	155
2000	1,500 watt, 1 pole G		10.12	.791		112	48.50		160.50	195
2100	2,000 watt, 1 pole G		7.36	1.090		162	67		229	278
2110	Fluorescent, 600 watt G		13.80	.580		122	35.50		157.50	187
2120	1,000 watt G		13.80	.580		170	35.50		205.50	240
2130	1,500 watt G		9.20	.870		315	53.50		368.50	425
2160	Explosion proof, toggle switch, wall, single pole 20 amp		4.88	1.640		296	101		397	475
2180	Receptacle, single outlet, 20 amp		4.88	1.640		680	101		781	895
2190	30 amp		3.68	2.170		1,075	133		1,208	1,400
2290	60 amp		2.30	3.480		890	214		1,104	1,300
2360	Plug, 20 amp		14.72	.543		217	33.50		250.50	289
2370	30 amp		11.04	.725		415	44.50		459.50	520
2380	60 amp		7.36	1.090		590	67		657	745
2410	Furnace, thermal cutoff switch with plate		23.92	.334		18.75	20.50		39.25	51
2460	Receptacle, duplex, 120 volt, grounded, 15 amp		36.80	.217		1.78	13.35		15.13	22
2470	20 amp		24.84	.322		11.75	19.75		31.50	42.50
2480	Ground fault interrupting, 15 amp		24.84	.322		14.95	19.75		34.70	46
2482	20 amp		24.84	.322		49.50	19.75		69.25	84
2486	Clock receptacle, 15 amp		36.80	.217		33.50	13.35		46.85	57
2490	Dryer, 30 amp		13.80	.580		5	35.50		40.50	58.50
2500	Range, 50 amp		10.12	.791		12.35	48.50		60.85	85.50
2530	Surge suppressor receptacle, duplex, 20 amp		24.84	.322		57.50	19.75		77.25	92.50
2532	Quad, 20 amp		18.40	.435		114	26.50		140.50	165
2540	Isolated ground receptacle, duplex, 20 amp		24.84	.322		30	19.75		49.75	62.50
2542	Quad, 20 amp		18.40	.435		37	26.50		63.50	80
2550	Simplex, 20 amp		24.84	.322		32.50	19.75		52.25	65.50
2560	Simplex, 30 amp		13.80	.580		42.50	35.50		78	99.50
2570	Cable reel w/receptacle 50' w/3#12, 120 V, 20 A	2 Elec	2.45	6.520		1,225	400		1,625	1,925
2600	Wall plates, stainless steel, 1 gang	1 Elec	73.60	.109		2.83	6.65		9.48	13.05
2800	2 gang		48.76	.164		4.82	10.05		14.87	20.50
3000	3 gang		29.44	.272		12.65	16.65		29.30	39
3100	4 gang		24.84	.322		12.20	19.75		31.95	43
3110	Brown plastic, 1 gang		73.60	.109		.42	6.65		7.07	10.40
3120	2 gang		48.76	.164		.83	10.05		10.88	15.90
3130	3 gang		29.44	.272		1.22	16.65		17.87	26.50
3140	4 gang		24.84	.322		3.20	19.75		22.95	33
3150	Brushed brass, 1 gang		73.60	.109		6.50	6.65		13.15	17.15
3160	Anodized aluminum, 1 gang		73.60	.109		3.33	6.65		9.98	13.60
3170	Switch cover, weatherproof, 1 gang		55.20	.145		5.80	8.90		14.70	19.60
3180	Vandal proof lock, 1 gang		55.20	.145		16.75	8.90		25.65	31.50
3200	Lampholder, keyless		23.92	.334		21.50	20.50		42	54
3400	Pullchain with receptacle		20.24	.395		25	24.50		49.50	63.50
3500	Pilot light, neon with jewel		24.84	.322		11.25	19.75		31	42

26 27 26 – Wiring Devices

26 27 26.20 Wiring Devices Elements	Crew	Daily Output	Labor-Hours	Unit	Material	2020 Bare Costs Labor	Equipment	Total	Total Incl O&P
3600	Receptacle, 20 amp, 250 volt, NEMA 6	1 Elec	24.84	.322	Ea.	26.50	19.75		46.25	58.50
3620	277 volt NEMA 7		24.84	.322		25.50	19.75		45.25	57.50
3640	125/250 volt NEMA 10		24.84	.322		25.50	19.75		45.25	57.50
3680	125/250 volt NEMA 14		23	.348		33	21.50		54.50	68.50
3700	3 pole, 250 volt NEMA 15		23	.348		35	21.50		56.50	70.50
3720	120/208 volt NEMA 18		23	.348		37.50	21.50		59	73
3740	30 amp, 125 volt NEMA 5		13.80	.580		21.50	35.50		57	76.50
3760	250 volt NEMA 6		13.80	.580		29	35.50		64.50	85
3780	277 volt NEMA 7		13.80	.580		34	35.50		69.50	90.50
3820	125/250 volt NEMA 14		12.88	.621		72.50	38		110.50	136
3840	3 pole, 250 volt NEMA 15		12.88	.621		66.50	38		104.50	130
3880	50 amp, 125 volt NEMA 5		10.12	.791		35	48.50		83.50	111
3900	250 volt NEMA 6		10.12	.791		33.50	48.50		82	109
3920	277 volt NEMA 7		10.12	.791		39.50	48.50		88	116
3960	125/250 volt NEMA 14		9.20	.870		104	53.50		157.50	194
3980	3 pole, 250 volt NEMA 15		9.20	.870		95.50	53.50		149	185
4020	60 amp, 125/250 volt, NEMA 14		7.36	1.090		103	67		170	214
4040	3 pole, 250 volt NEMA 15		7.36	1.090		121	67		188	233
4060	120/208 volt NEMA 18		7.36	1.090		92	67		159	201
4100	Receptacle locking, 20 amp, 125 volt NEMA L5		24.84	.322		27.50	19.75		47.25	60
4120	250 volt NEMA L6		24.84	.322		25.50	19.75		45.25	57.50
4140	277 volt NEMA L7		24.84	.322		29.50	19.75		49.25	62
4150	3 pole, 250 volt, NEMA L11		24.84	.322		24	19.75		43.75	56
4160	20 amp, 480 volt NEMA L8		24.84	.322		33	19.75		52.75	66
4180	600 volt NEMA L9		24.84	.322		41.50	19.75		61.25	75.50
4200	125/250 volt NEMA L10		24.84	.322		32	19.75		51.75	64.50
4230	125/250 volt NEMA L14		23	.348		34	21.50		55.50	69.50
4280	250 volt NEMA L15		23	.348		37.50	21.50		59	73.50
4300	480 volt NEMA L16		23	.348		31	21.50		52.50	66
4320	3 phase, 120/208 volt NEMA L18		23	.348		39.50	21.50		61	75.50
4340	277/480 volt NEMA L19		23	.348		41.50	21.50		63	77.50
4360	347/600 volt NEMA L20		23	.348		40.50	21.50		62	76.50
4380	120/208 volt NEMA L21		21.16	.378		42	23		65	80.50
4400	277/480 volt NEMA L22		21.16	.378		42.50	23		65.50	81
4420	347/600 volt NEMA L23		21.16	.378		52	23		75	92
4440	30 amp, 125 volt NEMA L5		13.80	.580		36	35.50		71.50	92.50
4460	250 volt NEMA L6		13.80	.580		38.50	35.50		74	95.50
4480	277 volt NEMA L7		13.80	.580		42	35.50		77.50	99
4500	480 volt NEMA L8		13.80	.580		39.50	35.50		75	96.50
4520	600 volt NEMA L9		13.80	.580		44.50	35.50		80	102
4540	125/250 volt NEMA L10		13.80	.580		49.50	35.50		85	108
4560	3 phase, 250 volt NEMA L11		13.80	.580		35	35.50		70.50	91.50
4620	125/250 volt NEMA L14		12.88	.621		55	38		93	117
4640	250 volt NEMA L15		12.88	.621		57	38		95	119
4660	480 volt NEMA L16		12.88	.621		59.50	38		97.50	122
4680	600 volt NEMA L17		12.88	.621		61.50	38		99.50	124
4700	120/208 volt NEMA L18		12.88	.621		61.50	38		99.50	125
4720	277/480 volt NEMA L19		12.88	.621		64	38		102	127
4740	347/600 volt NEMA L20		12.88	.621		67	38		105	131
4760	120/208 volt NEMA L21		11.96	.669		59.50	41		100.50	126
4780	277/480 volt NEMA L22		11.96	.669		60	41		101	127
4800	347/600 volt NEMA L23		11.96	.669		69	41		110	137
4840	Receptacle, corrosion resistant, 15 or 20 amp, 125 volt NEMA L5		24.84	.322		39	19.75		58.75	72

26 27 26.20 Wiring Devices Elements	Crew	Daily Output	Labor-Hours	Unit	Material	2020 Bare Costs Labor	Equipment	Total	Total Incl O&P
4860	250 volt NEMA L6	1 Elec	24.84	.322	Ea.	25	19.75		44.75	57
4870	Receptacle box assembly, cast aluminum, 60A, 4P5W, 3P, 120/208V, NEMA PR4		20.24	.395		810	24.50		834.50	925
4875	100A, 4P5W, 3P, 347/600V, NEMA PR4		20.24	.395		890	24.50		914.50	1,025
4900	Receptacle, cover plate, phenolic plastic, NEMA 5 & 6		73.60	.109		.68	6.65		7.33	10.70
4910	NEMA 7-23		73.60	.109		.74	6.65		7.39	10.75
4920	Stainless steel, NEMA 5 & 6		73.60	.109		2.99	6.65		9.64	13.25
4930	NEMA 7-23		73.60	.109		3.36	6.65		10.01	13.65
4940	Brushed brass NEMA 5 & 6		73.60	.109		6	6.65		12.65	16.55
4950	NEMA 7-23		73.60	.109		6.55	6.65		13.20	17.15
4960	Anodized aluminum, NEMA 5 & 6		73.60	.109		3.93	6.65		10.58	14.25
4970	NEMA 7-23		73.60	.109		6.50	6.65		13.15	17.10
4980	Weatherproof NEMA 7-23		55.20	.145		47	8.90		55.90	65
5000	Duplex receptacle, combo 15A/125V, 3 wire w/2-5V 0.7A, port USB, AL		24.84	.322		21	19.75		40.75	52.50
5002	15A/125V, 3 wire w/2-5V 0.7A, port USB, BK		24.84	.322		20	19.75		39.75	51.50
5004	15A/125V, 3 wire w/2-5V 0.7A, port USB, LA		24.84	.322		19.95	19.75		39.70	51.50
5006	15A/125V, 3 wire w/2-5V 0.7A, port USB, IV		24.84	.322		33	19.75		52.75	66
5008	15A/125V, 3 wire w/2-5V 0.7A, port USB, WH		24.84	.322		24.50	19.75		44.25	56.50
5010	Duplex receptacle, combo 15A/125V, 3 wire w/2-5V 2.1A, port USB, AL		24.84	.322		64	19.75		83.75	100
5012	15A/125V, 3 wire w/2-5V 2.1A, port USB, BK		24.84	.322		62.50	19.75		82.25	98.50
5014	15A/125V, 3 wire w/2-5V 2.1A, port USB, LA		24.84	.322		55	19.75		74.75	90
5016	15A/125V, 3 wire w/2-5V 2.1A, port USB, IV		24.84	.322		55.50	19.75		75.25	90.50
5018	15A/125V, 3 wire w/2-5V 2.1A, port USB, WH		24.84	.322		34	19.75		53.75	67
5020	Duplex receptacle, combo 20A/125V, 3 wire w/2-5V 2.1A, port USB, AL		24.84	.322		56.50	19.75		76.25	91.50
5022	20A/125V, 3 wire w/2-5V 2.1A, port USB, BK		24.84	.322		71	19.75		90.75	108
5024	20A/125V, 3 wire w/2-5V 2.1A, port USB, LA		24.84	.322		39	19.75		58.75	72
5026	20A/125V, 3 wire w/2-5V 2.1A, port USB, IV		24.84	.322		38.50	19.75		58.25	72
5028	20A/125V, 3 wire w/2-5V 2.1A, port USB, WH		24.84	.322		38.50	19.75		58.25	72
5100	Plug, 20 amp, 250 volt NEMA 6		27.60	.290		22	17.80		39.80	50.50
5110	277 volt NEMA 7		27.60	.290		24	17.80		41.80	53
5120	3 pole, 120/250 volt NEMA 10		23.92	.334		26.50	20.50		47	59.50
5130	125/250 volt NEMA 14		23.92	.334		55	20.50		75.50	91
5140	250 volt NEMA 15		23.92	.334		55.50	20.50		76	91.50
5150	120/208 volt NEMA 8		23.92	.334		63.50	20.50		84	101
5160	30 amp, 125 volt NEMA 5		11.96	.669		56	41		97	123
5170	250 volt NEMA 6		11.96	.669		55	41		96	122
5180	277 volt NEMA 7		11.96	.669		51	41		92	117
5190	125/250 volt NEMA 14		11.96	.669		63	41		104	130
5200	3 pole, 250 volt NEMA 15		11.04	.725		67	44.50		111.50	140
5210	50 amp, 125 volt NEMA 5		8.28	.966		84.50	59.50		144	182
5220	250 volt NEMA 6		8.28	.966		88.50	59.50		148	186
5230	277 volt NEMA 7		8.28	.966		94.50	59.50		154	193
5240	125/250 volt NEMA 14		8.28	.966		82	59.50		141.50	179
5250	3 pole, 250 volt NEMA 15		7.36	1.090		81	67		148	189
5260	60 amp, 125/250 volt NEMA 14		6.44	1.240		88.50	76		164.50	210
5270	3 pole, 250 volt NEMA 15		6.44	1.240		94	76		170	217
5280	120/208 volt NEMA 18		6.44	1.240		114	76		190	238
5300	Plug angle, 20 amp, 250 volt NEMA 6		27.60	.290		33	17.80		50.80	62.50
5310	30 amp, 125 volt NEMA 5		11.96	.669		61.50	41		102.50	129
5320	250 volt NEMA 6		11.96	.669		64	41		105	131
5330	277 volt NEMA 7		11.96	.669		75	41		116	144
5340	125/250 volt NEMA 14		11.96	.669		70	41		111	138
5350	3 pole, 250 volt NEMA 15		11.04	.725		75.50	44.50		120	149
5360	50 amp, 125 volt NEMA 5		8.28	.966		71	59.50		130.50	167

26 27 Low-Voltage Distribution Equipment

26 27 26 – Wiring Devices

26 27 26.20 Wiring Devices Elements	Crew	Daily Output	Labor-Hours	Unit	Material	2020 Bare Costs Labor	Equipment	Total	Total Incl O&P
5370 250 volt NEMA 6	1 Elec	8.28	.966	Ea.	67	59.50		126.50	162
5380 277 volt NEMA 7		8.28	.966		84	59.50		143.50	181
5390 125/250 volt NEMA 14		8.28	.966		85	59.50		144.50	182
5400 3 pole, 250 volt NEMA 15		7.36	1.090		90	67		157	199
5410 60 amp, 125/250 volt NEMA 14		6.44	1.240		101	76		177	224
5420 3 pole, 250 volt NEMA 15		6.44	1.240		103	76		179	226
5430 120/208 volt NEMA 18		6.44	1.240		106	76		182	230
5500 Plug, locking, 20 amp, 125 volt NEMA L5		27.60	.290		20	17.80		37.80	48.50
5510 250 volt NEMA L6		27.60	.290		20	17.80		37.80	48.50
5520 277 volt NEMA L7		27.60	.290		19.70	17.80		37.50	48
5530 480 volt NEMA L8		27.60	.290		22	17.80		39.80	50.50
5540 600 volt NEMA L9		27.60	.290		24	17.80		41.80	53
5550 3 pole, 125/250 volt NEMA L10		23.92	.334		25.50	20.50		46	58.50
5560 250 volt NEMA L11		23.92	.334		25.50	20.50		46	58.50
5570 480 volt NEMA L12		23.92	.334		29.50	20.50		50	63
5580 125/250 volt NEMA L14		23.92	.334		30	20.50		50.50	63.50
5590 250 volt NEMA L15		23.92	.334		37.50	20.50		58	71.50
5600 480 volt NEMA L16		23.92	.334		33.50	20.50		54	67.50
5610 4 pole, 120/208 volt NEMA L18		22.08	.362		36.50	22		58.50	73
5620 277/480 volt NEMA L19		22.08	.362		43.50	22		65.50	81
5630 347/600 volt NEMA L20		22.08	.362		38	22		60	74.50
5640 120/208 volt NEMA L21		22.08	.362		38.50	22		60.50	75.50
5650 277/480 volt NEMA L22		22.08	.362		41	22		63	78
5660 347/600 volt NEMA L23		22.08	.362		45.50	22		67.50	83
5670 30 amp, 125 volt NEMA L5		11.96	.669		30.50	41		71.50	94.50
5680 250 volt NEMA L6		11.96	.669		31.50	41		72.50	95.50
5690 277 volt NEMA L7		11.96	.669		31.50	41		72.50	95.50
5700 480 volt NEMA L8		11.96	.669		32.50	41		73.50	96.50
5710 600 volt NEMA L9		11.96	.669		33.50	41		74.50	98
5720 3 pole, 125/250 volt NEMA L10		10.12	.791		28	48.50		76.50	103
5730 250 volt NEMA L11		10.12	.791		27.50	48.50		76	103
5760 125/250 volt NEMA L14		10.12	.791		41	48.50		89.50	117
5770 250 volt NEMA L15		10.12	.791		41	48.50		89.50	117
5780 480 volt NEMA L16		10.12	.791		42.50	48.50		91	119
5790 600 volt NEMA L17		10.12	.791		42.50	48.50		91	119
5800 4 pole, 120/208 volt NEMA L18		9.20	.870		47.50	53.50		101	132
5810 120/208 volt NEMA L19		9.20	.870		48	53.50		101.50	133
5820 347/600 volt NEMA L20		9.20	.870		50.50	53.50		104	135
5830 120/208 volt NEMA L21		9.20	.870		44.50	53.50		98	129
5840 277/480 volt NEMA L22		9.20	.870		50.50	53.50		104	135
5850 347/600 volt NEMA L23		9.20	.870		51.50	53.50		105	137
6000 Connector, 20 amp, 250 volt NEMA 6		27.60	.290		35	17.80		52.80	65
6010 277 volt NEMA 7		27.60	.290		35	17.80		52.80	65
6020 3 pole, 120/250 volt NEMA 10		23.92	.334		39.50	20.50		60	74
6030 125/250 volt NEMA 14		23.92	.334		39.50	20.50		60	74
6040 250 volt NEMA 15		23.92	.334		40	20.50		60.50	74.50
6050 120/208 volt NEMA 18		23.92	.334		43.50	20.50		64	78.50
6060 30 amp, 125 volt NEMA 5		11.96	.669		69	41		110	137
6070 250 volt NEMA 6		11.96	.669		69	41		110	137
6080 277 volt NEMA 7		11.96	.669		69	41		110	137
6110 50 amp, 125 volt NEMA 5		8.28	.966		95	59.50		154.50	193
6120 250 volt NEMA 6		8.28	.966		95	59.50		154.50	193
6130 277 volt NEMA 7		8.28	.966		95	59.50		154.50	193

26 27 26.20 Wiring Devices Elements	Crew	Daily Output	Labor-Hours	Unit	Material	2020 Bare Costs Labor	Equipment	Total	Total Incl O&P
6200 Connector, locking, 20 amp, 125 volt NEMA L5	1 Elec	27.60	.290	Ea.	31	17.80		48.80	60.50
6210 250 volt NEMA L6		27.60	.290		31	17.80		48.80	60.50
6220 277 volt NEMA L7		27.60	.290		30	17.80		47.80	59.50
6230 480 volt NEMA L8		27.60	.290		35	17.80		52.80	65
6240 600 volt NEMA L9		27.60	.290		41	17.80		58.80	71.50
6250 3 pole, 125/250 volt NEMA L10		23.92	.334		40.50	20.50		61	75
6260 250 volt NEMA L11		23.92	.334		40.50	20.50		61	75
6280 125/250 volt NEMA L14		23.92	.334		42	20.50		62.50	76.50
6290 250 volt NEMA L15		23.92	.334		42	20.50		62.50	77
6300 480 volt NEMA L16		23.92	.334		45	20.50		65.50	80.50
6310 4 pole, 120/208 volt NEMA L18		22.08	.362		52.50	22		74.50	90.50
6320 277/480 volt NEMA L19		22.08	.362		54	22		76	92.50
6330 347/600 volt NEMA L20		22.08	.362		54.50	22		76.50	93
6340 120/208 volt NEMA L21		22.08	.362		62	22		84	101
6350 277/480 volt NEMA L22		22.08	.362		69.50	22		91.50	110
6360 347/600 volt NEMA L23		22.08	.362		78	22		100	119
6370 30 amp, 125 volt NEMA L5		11.96	.669		60.50	41		101.50	128
6380 250 volt NEMA L6		11.96	.669		61	41		102	128
6390 277 volt NEMA L7		11.96	.669		64.50	41		105.50	132
6400 480 volt NEMA L8		11.96	.669		64	41		105	131
6410 600 volt NEMA L9		11.96	.669		71.50	41		112.50	140
6420 3 pole, 125/250 volt NEMA L10		10.12	.791		78	48.50		126.50	158
6430 250 volt NEMA L11		10.12	.791		78	48.50		126.50	158
6460 125/250 volt NEMA L14		10.12	.791		84.50	48.50		133	165
6470 250 volt NEMA L15		10.12	.791		84	48.50		132.50	165
6480 480 volt NEMA L16		10.12	.791		90.50	48.50		139	172
6490 600 volt NEMA L17		10.12	.791		90	48.50		138.50	171
6500 4 pole, 120/208 volt NEMA L18		9.20	.870		97.50	53.50		151	187
6510 120/208 volt NEMA L19		9.20	.870		97	53.50		150.50	187
6520 347/600 volt NEMA L20		9.20	.870		99	53.50		152.50	189
6530 120/208 volt NEMA L21		9.20	.870		82.50	53.50		136	170
6540 277/480 volt NEMA L22		9.20	.870		88.50	53.50		142	177
6550 347/600 volt NEMA L23		9.20	.870		95	53.50		148.50	184
7000 Receptacle computer, 250 volt, 15 amp, 3 pole 4 wire		7.36	1.090		102	67		169	212
7010 20 amp, 2 pole 3 wire		7.36	1.090		122	67		189	234
7020 30 amp, 2 pole 3 wire		5.98	1.340		196	82		278	340
7030 30 amp, 3 pole 4 wire		5.98	1.340		212	82		294	355
7040 60 amp, 3 pole 4 wire		4.14	1.930		350	118		468	560
7050 100 amp, 3 pole 4 wire		2.76	2.900		435	178		613	745
7100 Connector computer, 250 volt, 15 amp, 3 pole 4 wire		24.84	.322		168	19.75		187.75	215
7110 20 amp, 2 pole 3 wire		24.84	.322		167	19.75		186.75	213
7120 30 amp, 2 pole 3 wire		13.80	.580		231	35.50		266.50	305
7130 30 amp, 3 pole 4 wire		13.80	.580		261	35.50		296.50	340
7140 60 amp, 3 pole 4 wire		7.36	1.090		440	67		507	585
7150 100 amp, 3 pole 4 wire		3.68	2.170		595	133		728	850
7200 Plug, computer, 250 volt, 15 amp, 3 pole 4 wire		24.84	.322		150	19.75		169.75	195
7210 20 amp, 2 pole 3 wire		24.84	.322		153	19.75		172.75	199
7220 30 amp, 2 pole 3 wire		13.80	.580		240	35.50		275.50	315
7230 30 amp, 3 pole 4 wire		13.80	.580		238	35.50		273.50	315
7240 60 amp, 3 pole 4 wire		7.36	1.090		355	67		422	495
7250 100 amp, 3 pole 4 wire		3.68	2.170		435	133		568	675
7300 Connector adapter to flexible conduit, 1/2"		55.20	.145		3.89	8.90		12.79	17.55
7310 3/4"		46	.174		4.97	10.65		15.62	21.50

26 27 26 – Wiring Devices

26 27 26.20 Wiring Devices Elements	Crew	Daily Output	Labor-Hours	Unit	Material	2020 Bare Costs Labor	Equipment	Total	Total Incl O&P
7320 1-1/4"	1 Elec	27.60	.290	Ea.	15.25	17.80		33.05	43.50
7330 1-1/2"		21.16	.378		20	23		43	56.50
8100 Pin/sleeve, 20A, 480V, DSN1, male inlet, 3 pole, 3 phase, 4 wire		14.72	.543		48.50	33.50		82	103
8110 20A, 125V, DSN20, male inlet, 3 pole, 3 phase, 3 wire		14.72	.543		39.50	33.50		73	93
8120 60A, 480V, DS60, male inlet, 3 pole, 3 phase, 3 wire		14.72	.543		325	33.50		358.50	410
8130 Pin/sleeve, 20A, 480V, DSN10, female RCPT, 3P, 3 phase, 4 wire		14.72	.543		92.50	33.50		126	152
8140 30A, 480V, DS30, female RCPT, 3P, 3 phase, 4 wire		14.72	.543		300	33.50		333.50	385
8500 Wiring device terminal strip, 2 pole, 12-2AWG, 300VAC, CSA, 600VAC, 10A		33.12	.242		3.44	14.80		18.24	26
8505 4 pole, 12-2AWG, 300VAC, CSA, 600VAC, 10A		33.12	.242		4.87	14.80		19.67	27.50
8510 6 pole, 12-2AWG, 300VAC, CSA, 600VAC, 10A		33.12	.242		5.50	14.80		20.30	28
8515 8 pole, 12-2AWG, 300VAC, CSA, 600VAC, 10A		27.60	.290		6.15	17.80		23.95	33.50
8520 10 pole, 12-2AWG, 300VAC, CSA, 600VAC, 10A		27.60	.290		6.80	17.80		24.60	34
8525 12 pole, 12-2AWG, 300VAC, CSA, 600VAC, 10A		27.60	.290		7.50	17.80		25.30	35
8530 Wiring device terminal strip jumper, 10A		33.12	.242		.84	14.80		15.64	23
8535 Wiring device terminal strip, 2 pole, 12-22AWG, 300VAC, CSA, 600VAC, 20A		31.28	.256		3.74	15.70		19.44	27.50
8540 4 pole, 12-22AWG, 300VAC, CSA, 600VAC, 20A		31.28	.256		5.10	15.70		20.80	29
8545 6 pole, 12-22AWG, 300VAC, CSA, 600VAC, 20A		31.28	.256		5.80	15.70		21.50	30
8550 8 pole, 12-22AWG, 300VAC, CSA, 600VAC, 20A		27.60	.290		6.45	17.80		24.25	33.50
8555 10 pole, 12-22AWG, 300VAC, CSA, 600VAC, 20A		27.60	.290		7.15	17.80		24.95	34.50
8560 12 pole, 12-22AWG, 300VAC, CSA, 600VAC, 20A		27.60	.290		7.85	17.80		25.65	35
8565 Wiring device terminal strip jumper, 20A		31.28	.256		1.03	15.70		16.73	24.50
8570 Wiring device terminal shorting type no cover, 4 pole, 6-10AWG, 600VAC, 50A		27.60	.290		19.40	17.80		37.20	48
8575 6 pole, 6-10AWG, 600VAC, 50A		27.60	.290		24.50	17.80		42.30	53.50
8580 8 pole, 6-10AWG, 600VAC, 50A		27.60	.290		28.50	17.80		46.30	58
8585 12 pole, 6-10AWG, 600VAC, 50A		27.60	.290		32	17.80		49.80	61.50
8590 Wiring device terminal 2 shorting pin 1/2 cover, 2 pole, 8-4AWG, 600V, 45A		27.60	.290		20.50	17.80		38.30	49
8595 4 pole, 8-4AWG, 600V, 45A		27.60	.290		25.50	17.80		43.30	54.50
8600 6 pole, 8-4AWG, 600V, 45A		27.60	.290		31.50	17.80		49.30	61.50
8605 8 pole, 8-4AWG, 600V, 45A		27.60	.290		45	17.80		62.80	76
8610 12 pole, 8-4AWG, 600V, 45A		27.60	.290		54	17.80		71.80	86
8615 Wiring device terminal 2 shorting pin W/cover, 4 pole, 8-4AWG, 600V, 45A		27.60	.290		34	17.80		51.80	64
8620 6 pole, 8-4AWG, 600V, 45A		27.60	.290		52	17.80		69.80	83.50
8625 8 pole, 8-4AWG, 600V, 45A		27.60	.290		54.50	17.80		72.30	86.50
8630 12 pole, 8-4AWG, 600V, 45A		27.60	.290		65	17.80		82.80	98

26 27 33 – Power Distribution Units

26 27 33.10 Power Distribution Unit Cabinet

0010 **POWER DISTRIBUTION UNIT CABINET**									
0100 Power distribution unit, single cabinet, 50 kVA output, 480/208 input	1 Elec	1.84	4.350	Ea.	9,200	267		9,467	10,500
0110 75 kVA output, 480/208 input		1.84	4.350		9,975	267		10,242	11,400
0120 100 kVA output, 480/208 input		1.84	4.350		10,900	267		11,167	12,400
0130 125 kVA output, 480/208 input		1.38	5.800		12,000	355		12,355	13,700
0140 150 kVA output, 480/208 input		1.38	5.800		14,500	355		14,855	16,500
0150 200 kVA output, 480/208 input		.92	8.700		16,000	535		16,535	18,400
0160 225 kVA output, 480/208 input		.92	8.700		16,900	535		17,435	19,400

26 27 33.20 Power Distribution Unit

0010 **POWER DISTRIBUTION UNIT**									
0050 3PH, 60 kVA, PDU									
0100 208V-208V/120V	2 Elec	1.84	8.700	Ea.	14,300	535		14,835	16,500
0110 480V-208V/120V		1.84	8.700		14,100	535		14,635	16,300
0120 600V-208V/120V		1.84	8.700		14,700	535		15,235	16,900
0125 3PH, 80 kVA, PDU									
0130 208V-208V/120V	2 Elec	1.84	8.700	Ea.	27,700	535		28,235	31,300

26 27 Low-Voltage Distribution Equipment

26 27 33 – Power Distribution Units

26 27 33.20 Power Distribution Unit	Crew	Daily Output	Labor-Hours	Unit	Material	2020 Bare Costs Labor	Equipment	Total	Total Incl O&P
0140	480V-208V/120V	2 Elec	1.84	8.700	Ea.	28,100	535		28,635	31,800
0150	600V-208V/120V		1.84	8.700		27,900	535		28,435	31,500
0200	3PH, 15 kVA, 480-208/120V, 208-208/120V		2.76	5.800		11,600	355		11,955	13,300
0210	30 kVA, 400-208/120V, 208-208/120V		2.76	5.800		19,200	355		19,555	21,600
0250	50 kVA, 480-208/120V, 208-208/120V		2.30	6.960		10,500	425		10,925	12,200
0260	75 kVA, 480-208/120V, 208-208/120V		2.30	6.960		11,800	425		12,225	13,600
0270	100 kVA, 480-208/120V, 208-208/120V		1.84	8.700		12,700	535		13,235	14,800
0280	125 kVA, 480-208/120V, 208-208/120V		1.84	8.700		13,700	535		14,235	15,900
0290	150 kVA, 480-208/120V, 208-208/120V		1.84	8.700		12,500	535		13,035	14,600
0300	200 kVA, 480-208/120V, 208-208/120V		1.38	11.590		13,800	710		14,510	16,300
0310	225 kVA, 480-208/120V, 208-208/120V		1.38	11.590		14,600	710		15,310	17,100

26 27 73 – Door Chimes

26 27 73.10 Doorbell System

		Crew	Daily Output	Labor-Hours	Unit	Material	2020 Bare Costs Labor	Equipment	Total	Total Incl O&P
0010	**DOORBELL SYSTEM**, incl. transformer, button & signal									
0100	6" bell	1 Elec	3.68	2.170	Ea.	158	133		291	370
0200	Buzzer		3.68	2.170		127	133		260	335
1000	Door chimes, 2 notes		14.72	.543		32	33.50		65.50	84.50
1020	with ambient light		11.04	.725		116	44.50		160.50	194
1100	Tube type, 3 tube system		11.04	.725		233	44.50		277.50	320
1180	4 tube system		9.20	.870		495	53.50		548.50	625
1900	For transformer & button, add		4.60	1.740		16.15	107		123.15	177
3000	For push button only		22.08	.362		.92	22		22.92	34
3200	Bell transformer		14.72	.543		24	33.50		57.50	76

26 28 Low-Voltage Circuit Protective Devices

26 28 13 – Fuses

26 28 13.10 Fuse Elements

		Crew	Daily Output	Labor-Hours	Unit	Material	2020 Bare Costs Labor	Equipment	Total	Total Incl O&P
0010	**FUSE ELEMENTS**									
0020	Cartridge, nonrenewable									
0050	250 volt, 30 amp	1 Elec	46	.174	Ea.	2.75	10.65		13.40	18.90
0100	60 amp		46	.174		5	10.65		15.65	21.50
0150	100 amp		36.80	.217		19.60	13.35		32.95	41.50
0200	200 amp		33.12	.242		44.50	14.80		59.30	71
0250	400 amp		27.60	.290		135	17.80		152.80	176
0300	600 amp		22.08	.362		236	22		258	292
0400	600 volt, 30 amp		36.80	.217		12.35	13.35		25.70	33.50
0450	60 amp		36.80	.217		18.40	13.35		31.75	40.50
0500	100 amp		33.12	.242		39.50	14.80		54.30	65
0550	200 amp		27.60	.290		73	17.80		90.80	107
0600	400 amp		22.08	.362		146	22		168	193
0650	600 amp		18.40	.435		231	26.50		257.50	294
0800	Dual element, time delay, 250 volt, 30 amp		46	.174		11.90	10.65		22.55	29
0850	60 amp		46	.174		13.90	10.65		24.55	31
0900	100 amp		36.80	.217		52	13.35		65.35	77.50
0950	200 amp		33.12	.242		113	14.80		127.80	146
1000	400 amp		27.60	.290		138	17.80		155.80	178
1050	600 amp		22.08	.362		227	22		249	283
1300	600 volt, 15 to 30 amp		36.80	.217		26	13.35		39.35	48.50
1350	35 to 60 amp		36.80	.217		38	13.35		51.35	61.50
1400	70 to 100 amp		33.12	.242		79.50	14.80		94.30	110

Pre-Installation Change Order Costs

	26 28 13.10 Fuse Elements	Crew	Daily Output	Labor-Hours	Unit	Material	2020 Bare Costs Labor	Equipment	Total	Total Incl O&P
1450	110 to 200 amp	1 Elec	27.60	.290	Ea.	157	17.80		174.80	200
1500	225 to 400 amp		22.08	.362		340	22		362	410
1550	600 amp		18.40	.435		435	26.50		461.50	515
1800	Class RK1, high capacity, 250 volt, 30 amp		46	.174		10.55	10.65		21.20	27.50
1850	60 amp		46	.174		20	10.65		30.65	38
1900	100 amp		36.80	.217		31	13.35		44.35	54
1950	200 amp		33.12	.242		80	14.80		94.80	110
2000	400 amp		27.60	.290		181	17.80		198.80	227
2050	600 amp		22.08	.362		340	22		362	410
2200	600 volt, 30 amp		36.80	.217		15.20	13.35		28.55	36.50
2250	60 amp		36.80	.217		28	13.35		41.35	50.50
2300	100 amp		33.12	.242		95.50	14.80		110.30	127
2350	200 amp		27.60	.290		135	17.80		152.80	175
2400	400 amp		22.08	.362		235	22		257	291
2450	600 amp		18.40	.435		485	26.50		511.50	575
2700	Class J, current limiting, 250 or 600 volt, 30 amp		36.80	.217		24	13.35		37.35	46
2750	60 amp		36.80	.217		37.50	13.35		50.85	61.50
2800	100 amp		33.12	.242		76.50	14.80		91.30	107
2850	200 amp		27.60	.290		117	17.80		134.80	156
2900	400 amp		22.08	.362		355	22		377	425
2950	600 amp		18.40	.435		370	26.50		396.50	450
3100	Class L, current limiting, 250 or 600 volt, 601 to 1,200 amp		14.72	.543		890	33.50		923.50	1,025
3150	1,500 to 1,600 amp		11.96	.669		860	41		901	1,000
3200	1,800 to 2,000 amp		9.20	.870		1,800	53.50		1,853.50	2,050
3250	2,500 amp		9.20	.870		1,475	53.50		1,528.50	1,675
3300	3,000 amp		7.36	1.090		2,350	67		2,417	2,700
3350	3,500 to 4,000 amp		7.36	1.090		2,275	67		2,342	2,600
3400	4,500 to 5,000 amp		6.16	1.300		2,200	80		2,280	2,550
3450	6,000 amp		5.24	1.530		6,550	94		6,644	7,375
3600	Plug, 120 volt, 1 to 10 amp		46	.174		5.35	10.65		16	22
3650	15 to 30 amp		46	.174		5	10.65		15.65	21.50
3700	Dual element 0.3 to 14 amp		46	.174		7.95	10.65		18.60	24.50
3750	15 to 30 amp		46	.174		7.90	10.65		18.55	24.50
3800	Fustat, 120 volt, 15 to 30 amp		46	.174		5.80	10.65		16.45	22.50
3850	0.3 to 14 amp		46	.174		6.85	10.65		17.50	23.50
3900	Adapters 0.3 to 10 amp		46	.174		7.50	10.65		18.15	24
3950	15 to 30 amp		46	.174		10.60	10.65		21.25	27.50
4000	F-frame current limiting fuse, 14 to 2 AWG, 3 ampere, 3P, aluminum terminal		36.80	.217		1,925	13.35		1,938.35	2,150
4010	7 ampere, 3P, aluminum terminal		36.80	.217		1,925	13.35		1,938.35	2,125
4020	15 AMP, 3P, aluminum terminal		36.80	.217		1,975	13.35		1,988.35	2,200
4030	30 AMP, 3P, aluminum terminal		36.80	.217		1,925	13.35		1,938.35	2,150
4040	50 AMP, 3P, aluminum terminal		36.80	.217		1,625	13.35		1,638.35	1,800
4050	F-frame current limiting fuse, 1 to 4/0 AWG, 100 AMP, 3P, aluminum terminal		36.80	.217		1,525	13.35		1,538.35	1,700
4060	150 AMP, 3P, aluminum terminal		36.80	.217		2,250	13.35		2,263.35	2,475

26 28 16 – Enclosed Switches and Circuit Breakers

26 28 16.10 Circuit Breakers

		Crew	Daily Output	Labor-Hours	Unit	Material	2020 Bare Costs Labor	Equipment	Total	Total Incl O&P
0010	**CIRCUIT BREAKERS** (in enclosure)									
0100	Enclosed (NEMA 1), 600 volt, 3 pole, 30 amp	1 Elec	2.94	2.720	Ea.	530	167		697	830
0200	60 amp		2.58	3.110		645	191		836	990
0400	100 amp		2.12	3.780		735	232		967	1,150
0500	200 amp		1.38	5.800		1,875	355		2,230	2,600
0600	225 amp		1.38	5.800		1,700	355		2,055	2,400

26 28 Low-Voltage Circuit Protective Devices

26 28 16 – Enclosed Switches and Circuit Breakers

26 28 16.10 Circuit Breakers		Crew	Daily Output	Labor-Hours	Unit	Material	2020 Bare Costs Labor	2020 Bare Costs Equipment	Total	Total Incl O&P
0700	400 amp	2 Elec	1.47	10.870	Ea.	2,900	665		3,565	4,200
0800	600 amp		1.10	14.490		4,200	890		5,090	5,950
1000	800 amp		.86	18.500		5,475	1,125		6,600	7,725
1200	1,000 amp		.77	20.700		6,925	1,275		8,200	9,500
1220	1,200 amp		.74	21.740		8,850	1,325		10,175	11,700
1240	1,600 amp		.66	24.150		16,100	1,475		17,575	19,900
1260	2,000 amp		.59	27.170		17,400	1,675		19,075	21,700
1400	1,200 amp with ground fault		.74	21.740		15,500	1,325		16,825	19,100
1600	1,600 amp with ground fault		.66	24.150		18,300	1,475		19,775	22,300
1800	2,000 amp with ground fault		.59	27.170		19,600	1,675		21,275	24,000
2000	Disconnect, 240 volt 3 pole, 5 HP motor	1 Elec	2.94	2.720		450	167		617	745
2020	10 HP motor		2.94	2.720		450	167		617	745
2040	15 HP motor		2.58	3.110		450	191		641	780
2060	20 HP motor		2.12	3.780		545	232		777	945
2080	25 HP motor		2.12	3.780		545	232		777	945
2100	30 HP motor		2.12	3.780		545	232		777	945
2120	40 HP motor		1.84	4.350		935	267		1,202	1,425
2140	50 HP motor		1.38	5.800		935	355		1,290	1,550
2160	60 HP motor		1.38	5.800		2,150	355		2,505	2,900
2180	75 HP motor	2 Elec	1.84	8.700		2,150	535		2,685	3,175
2200	100 HP motor		1.47	10.870		2,150	665		2,815	3,375
2220	125 HP motor		1.47	10.870		2,150	665		2,815	3,375
2240	150 HP motor		1.10	14.490		4,200	890		5,090	5,950
2260	200 HP motor		1.10	14.490		5,475	890		6,365	7,350
2300	Enclosed (NEMA 7), explosion proof, 600 volt 3 pole, 50 amp	1 Elec	2.12	3.780		1,925	232		2,157	2,475
2350	100 amp		1.38	5.800		2,000	355		2,355	2,725
2400	150 amp		.92	8.700		4,825	535		5,360	6,125
2450	250 amp	2 Elec	1.47	10.870		6,050	665		6,715	7,650
2500	400 amp	"	1.10	14.490		6,700	890		7,590	8,700

26 28 16.13 Circuit Breakers

0010	**CIRCUIT BREAKERS**									
0100	Circuit breaker current limiter, 225 to 400, ampere, 1 pole	1 Elec	44.16	.181	Ea.	835	11.10		846.10	930
0200	Circuit breaker current limiter, 200 to 500, ampere, 1 pole, thermal magnet		42.32	.189		690	11.60		701.60	775
0300	300 to 500, ampere, 1 pole, thermal magnet		42.32	.189		1,275	11.60		1,286.60	1,450
0400	Circuit breaker current limiter, 600 to 1000, ampere, 1 pole, tri pac		42.32	.189		1,750	11.60		1,761.60	1,950

26 28 16.20 Safety Switches

0010	**SAFETY SWITCHES**									
0100	General duty 240 volt, 3 pole NEMA 1, fusible, 30 amp	1 Elec	2.94	2.720	Ea.	71.50	167		238.50	325
0200	60 amp		2.12	3.780		122	232		354	480
0300	100 amp		1.75	4.580		210	281		491	650
0400	200 amp		1.20	6.690		450	410		860	1,100
0500	400 amp	2 Elec	1.66	9.660		1,175	595		1,770	2,175
0600	600 amp	"	1.10	14.490		2,300	890		3,190	3,850
0610	Nonfusible, 30 amp	1 Elec	2.94	2.720		57.50	167		224.50	310
0650	60 amp		2.12	3.780		76	232		308	430
0700	100 amp		1.75	4.580		178	281		459	615
0750	200 amp		1.20	6.690		330	410		740	970
0800	400 amp	2 Elec	1.66	9.660		895	595		1,490	1,875
0850	600 amp	"	1.10	14.490		1,850	890		2,740	3,350
1100	Heavy duty, 600 volt, 3 pole NEMA 1 nonfused									
1110	30 amp	1 Elec	2.94	2.720	Ea.	103	167		270	360
1500	60 amp		2.12	3.780		185	232		417	550

Pre-Installation Change Order Costs

26 28 16.20 Safety Switches		Crew	Daily Output	Labor-Hours	Unit	Material	2020 Bare Costs Labor	Equipment	Total	Total Incl O&P
1700	100 amp	1 Elec	1.75	4.580	Ea.	291	281		572	740
1900	200 amp		1.20	6.690		440	410		850	1,100
2100	400 amp	2 Elec	1.66	9.660		1,075	595		1,670	2,050
2300	600 amp		1.10	14.490		2,000	890		2,890	3,525
2500	800 amp		.86	18.500		3,600	1,125		4,725	5,675
2700	1,200 amp		.74	21.740		4,300	1,325		5,625	6,700
2900	Heavy duty, 240 volt, 3 pole NEMA 1 fusible									
2910	30 amp	1 Elec	2.94	2.720	Ea.	123	167		290	385
3000	60 amp		2.12	3.780		205	232		437	570
3300	100 amp		1.75	4.580		315	281		596	765
3500	200 amp		1.20	6.690		545	410		955	1,200
3700	400 amp	2 Elec	1.66	9.660		1,475	595		2,070	2,500
3900	600 amp		1.10	14.490		2,275	890		3,165	3,825
4100	800 amp		.86	18.500		5,650	1,125		6,775	7,900
4300	1,200 amp		.74	21.740		7,600	1,325		8,925	10,300
4340	2 pole fusible, 30 amp	1 Elec	3.22	2.480		93	152		245	330
4350	600 volt, 3 pole, fusible, 30 amp		2.94	2.720		199	167		366	465
4380	60 amp		2.12	3.780		243	232		475	610
4400	100 amp		1.75	4.580		440	281		721	905
4420	200 amp		1.20	6.690		625	410		1,035	1,300
4440	400 amp	2 Elec	1.66	9.660		1,675	595		2,270	2,725
4450	600 amp		1.10	14.490		2,975	890		3,865	4,600
4460	800 amp		.86	18.500		5,650	1,125		6,775	7,900
4480	1,200 amp		.74	21.740		7,600	1,325		8,925	10,300
4500	240 volt 3 pole NEMA 3R (no hubs), fusible									
4510	30 amp	1 Elec	2.85	2.810	Ea.	214	172		386	490
4700	60 amp		2.02	3.950		340	242		582	735
4900	100 amp		1.66	4.830		495	296		791	985
5100	200 amp		1.10	7.250		670	445		1,115	1,400
5300	400 amp	2 Elec	1.47	10.870		1,525	665		2,190	2,675
5500	600 amp	"	.92	17.390		3,075	1,075		4,150	4,975
5510	Heavy duty, 600 volt, 3 pole 3 ph. NEMA 3R fusible, 30 amp	1 Elec	2.85	2.810		335	172		507	625
5520	60 amp		2.02	3.950		410	242		652	810
5530	100 amp		1.66	4.830		600	296		896	1,100
5540	200 amp		1.10	7.250		835	445		1,280	1,575
5550	400 amp	2 Elec	1.47	10.870		2,025	665		2,690	3,225
5700	600 volt, 3 pole NEMA 3R nonfused									
5710	30 amp	1 Elec	2.85	2.810	Ea.	182	172		354	455
5900	60 amp		2.02	3.950		320	242		562	710
6100	100 amp		1.66	4.830		450	296		746	935
6300	200 amp		1.10	7.250		540	445		985	1,250
6500	400 amp	2 Elec	1.47	10.870		1,400	665		2,065	2,550
6700	600 amp	"	.92	17.390		2,925	1,075		4,000	4,825
6900	600 volt, 6 pole NEMA 3R nonfused, 30 amp	1 Elec	2.48	3.220		1,225	198		1,423	1,650
7100	60 amp		1.84	4.350		2,325	267		2,592	2,975
7300	100 amp		1.38	5.800		2,250	355		2,605	3,000
7500	200 amp		1.10	7.250		4,175	445		4,620	5,250
7600	600 volt, 3 pole NEMA 7 explosion proof nonfused									
7610	30 amp	1 Elec	2.02	3.950	Ea.	1,700	242		1,942	2,225
7620	60 amp		1.66	4.830		2,050	296		2,346	2,700
7630	100 amp		1.10	7.250		3,425	445		3,870	4,425
7640	200 amp		.74	10.870		7,375	665		8,040	9,100
7710	600 volt 6 pole, NEMA 3R fusible, 30 amp		2.48	3.220		1,825	198		2,023	2,300

26 28 16.20 Safety Switches

		Crew	Daily Output	Labor-Hours	Unit	Material	2020 Bare Costs Labor	Equipment	Total	Total Incl O&P
7900	60 amp	1 Elec	1.84	4.350	Ea.	2,575	267		2,842	3,225
8100	100 amp		1.38	5.800		2,050	355		2,405	2,775
8110	240 volt 3 pole, NEMA 12 fusible, 30 amp		2.85	2.810		310	172		482	595
8120	60 amp		2.02	3.950		545	242		787	960
8130	100 amp		1.66	4.830		660	296		956	1,175
8140	200 amp		1.10	7.250		740	445		1,185	1,475
8150	400 amp	2 Elec	1.47	10.870		2,025	665		2,690	3,225
8160	600 amp	"	.92	17.390		3,325	1,075		4,400	5,250
8180	600 volt 3 pole, NEMA 12 fusible, 30 amp	1 Elec	2.85	2.810		445	172		617	745
8190	60 amp		2.02	3.950		560	242		802	975
8200	100 amp		1.66	4.830		835	296		1,131	1,350
8210	200 amp		1.10	7.250		1,150	445		1,595	1,900
8220	400 amp	2 Elec	1.47	10.870		2,875	665		3,540	4,175
8230	600 amp	"	.92	17.390		4,775	1,075		5,850	6,850
8240	600 volt 3 pole, NEMA 12 nonfused, 30 amp	1 Elec	2.85	2.810		278	172		450	560
8250	60 amp		2.02	3.950		430	242		672	830
8260	100 amp		1.66	4.830		620	296		916	1,125
8270	200 amp		1.10	7.250		675	445		1,120	1,400
8280	400 amp	2 Elec	1.47	10.870		1,700	665		2,365	2,875
8290	600 amp	"	.92	17.390		3,475	1,075		4,550	5,425
8310	600 volt, 3 pole NEMA 4 fusible, 30 amp	1 Elec	2.76	2.900		930	178		1,108	1,300
8320	60 amp		2.02	3.950		1,100	242		1,342	1,575
8330	100 amp		1.66	4.830		2,275	296		2,571	2,950
8340	200 amp		1.10	7.250		3,000	445		3,445	3,950
8350	400 amp	2 Elec	1.47	10.870		5,975	665		6,640	7,575
8360	600 volt 3 pole NEMA 4 nonfused, 30 amp	1 Elec	2.76	2.900		825	178		1,003	1,175
8370	60 amp		2.02	3.950		985	242		1,227	1,425
8380	100 amp		1.66	4.830		3,400	296		3,696	4,175
8390	200 amp		1.10	7.250		2,775	445		3,220	3,700
8400	400 amp	2 Elec	1.47	10.870		5,100	665		5,765	6,600
8490	Motor starters, manual, single phase, NEMA 1	1 Elec	5.89	1.360		119	83.50		202.50	255
8500	NEMA 4		3.68	2.170		310	133		443	540
8700	NEMA 7		3.68	2.170		375	133		508	610
8900	NEMA 1 with pilot		5.89	1.360		148	83.50		231.50	287
8920	3 pole, NEMA 1, 230/460 volt, 5 HP, size 0		3.22	2.480		242	152		394	495
8940	10 HP, size 1		1.84	4.350		287	267		554	710
9010	Disc. switch, 600 volt 3 pole fusible, 30 amp, to 10 HP motor		2.94	2.720		420	167		587	710
9050	60 amp, to 30 HP motor		2.12	3.780		965	232		1,197	1,400
9070	100 amp, to 60 HP motor		1.75	4.580		965	281		1,246	1,475
9100	200 amp, to 125 HP motor		1.20	6.690		1,450	410		1,860	2,200
9110	400 amp, to 200 HP motor	2 Elec	1.66	9.660		3,650	595		4,245	4,875

26 28 16.40 Time Switches

		Crew	Daily Output	Labor-Hours	Unit	Material	2020 Bare Costs Labor	Equipment	Total	Total Incl O&P
0010	**TIME SWITCHES**									
0100	Single pole, single throw, 24 hour dial	1 Elec	3.68	2.170	Ea.	157	133		290	370
0200	24 hour dial with reserve power		3.31	2.420		810	148		958	1,100
0300	Astronomic dial		3.31	2.420		305	148		453	560
0400	Astronomic dial with reserve power		3.04	2.640		1,125	162		1,287	1,500
0500	7 day calendar dial		3.04	2.640		261	162		423	530
0600	7 day calendar dial with reserve power		2.94	2.720		254	167		421	530
0700	Photo cell 2,000 watt		7.36	1.090		34	67		101	137
1080	Load management device, 4 loads		1.84	4.350		1,375	267		1,642	1,900
1100	8 loads		.92	8.700		2,875	535		3,410	3,950

For customer support on your Electrical Change Order Costs with RSMeans data, call 800.448.8182.

Pre-Installation Change Order Costs

26 28 Low-Voltage Circuit Protective Devices

26 28 16 – Enclosed Switches and Circuit Breakers

26 28 16.50 Meter Socket Entry Hub	Crew	Daily Output	Labor-Hours	Unit	Material	2020 Bare Costs Labor	Equipment	Total	Total Incl O&P
0010 **METER SOCKET ENTRY HUB**									
0100 Meter socket entry hub closing cap, 3/4 to 4 inch	1 Elec	36.80	.217	Ea.	2.43	13.35		15.78	22.50
0110 Meter socket entry hub, 3/4 conduit		36.80	.217		20	13.35		33.35	42
0120 1 inch conduit		36.80	.217		21	13.35		34.35	43
0130 1.25 inch conduit		36.80	.217		19.50	13.35		32.85	41.50
0140 1.50 inch conduit		36.80	.217		19.60	13.35		32.95	41.50
0150 2 inch conduit		36.80	.217		20.50	13.35		33.85	42.50
0160 2.50 inch conduit		36.80	.217		62	13.35		75.35	88
0170 3 inch conduit		36.80	.217		93	13.35		106.35	122

26 29 Low-Voltage Controllers

26 29 13 – Enclosed Controllers

26 29 13.10 Contactors, AC

	Crew	Daily Output	Labor-Hours	Unit	Material	2020 Bare Costs Labor	Equipment	Total	Total Incl O&P
0010 **CONTACTORS, AC** Enclosed (NEMA 1)									
0050 Lighting, 600 volt 3 pole, electrically held									
0100 20 amp	1 Elec	3.68	2.170	Ea.	375	133		508	610
0200 30 amp		3.31	2.420		300	148		448	550
0300 60 amp		2.76	2.900		625	178		803	955
0400 100 amp		2.30	3.480		1,000	214		1,214	1,450
0500 200 amp		1.29	6.210		2,525	380		2,905	3,350
0600 300 amp	2 Elec	1.47	10.870		4,525	665		5,190	5,975
0800 600 volt 3 pole, mechanically held, 30 amp	1 Elec	3.31	2.420		590	148		738	870
0900 60 amp		2.76	2.900		1,175	178		1,353	1,575
1000 75 amp		2.58	3.110		1,500	191		1,691	1,925
1100 100 amp		2.30	3.480		1,700	214		1,914	2,200
1200 150 amp		1.84	4.350		3,275	267		3,542	4,000
1300 200 amp		1.29	6.210		5,650	380		6,030	6,800
1500 Magnetic with auxiliary contact, size 00, 9 amp		3.68	2.170		189	133		322	405
1600 Size 0, 18 amp		3.68	2.170		225	133		358	445
1700 Size 1, 27 amp		3.31	2.420		256	148		404	500
1800 Size 2, 45 amp		2.76	2.900		475	178		653	790
1900 Size 3, 90 amp		2.30	3.480		770	214		984	1,175
2000 Size 4, 135 amp		2.12	3.780		1,750	232		1,982	2,275
2100 Size 5, 270 amp	2 Elec	1.66	9.660		3,675	595		4,270	4,925
2200 Size 6, 540 amp		1.10	14.490		10,800	890		11,690	13,200
2300 Size 7, 810 amp		.92	17.390		14,500	1,075		15,575	17,600
2310 Size 8, 1,215 amp		.74	21.740		22,600	1,325		23,925	26,800
2500 Magnetic, 240 volt, 1-2 pole, 0.75 HP motor	1 Elec	3.68	2.170		152	133		285	365
2520 2 HP motor		3.31	2.420		201	148		349	440
2540 5 HP motor		2.30	3.480		415	214		629	775
2560 10 HP motor		1.29	6.210		680	380		1,060	1,325
2600 240 volt or less, 3 pole, 0.75 HP motor		3.68	2.170		152	133		285	365
2620 5 HP motor		3.31	2.420		189	148		337	430
2640 10 HP motor		3.31	2.420		219	148		367	460
2660 15 HP motor		2.30	3.480		440	214		654	800
2700 25 HP motor		2.30	3.480		440	214		654	800
2720 30 HP motor	2 Elec	2.58	6.210		730	380		1,110	1,375
2740 40 HP motor		2.58	6.210		730	380		1,110	1,375
2760 50 HP motor		1.47	10.870		730	665		1,395	1,800
2800 75 HP motor		1.47	10.870		1,725	665		2,390	2,875

26 29 13.10 Contactors, AC

		Crew	Daily Output	Labor-Hours	Unit	Material	2020 Bare Costs Labor	Equipment	Total	Total Incl O&P
2820	100 HP motor	2 Elec	.92	17.390	Ea.	1,725	1,075		2,800	3,475
2860	150 HP motor		.92	17.390		3,650	1,075		4,725	5,625
2880	200 HP motor	↓	.92	17.390		3,650	1,075		4,725	5,625
3000	600 volt, 3 pole, 5 HP motor	1 Elec	3.68	2.170		189	133		322	405
3020	10 HP motor		3.31	2.420		219	148		367	460
3040	25 HP motor		2.76	2.900		440	178		618	745
3100	50 HP motor	↓	2.30	3.480		730	214		944	1,125
3160	100 HP motor	2 Elec	2.58	6.210		1,725	380		2,105	2,450
3220	200 HP motor	"	1.47	10.870	↓	3,650	665		4,315	5,025

26 29 13.20 Control Stations

		Crew	Daily Output	Labor-Hours	Unit	Material	Labor	Equipment	Total	Total Incl O&P
0010	**CONTROL STATIONS**									
0050	NEMA 1, heavy duty, stop/start	1 Elec	7.36	1.090	Ea.	168	67		235	285
0100	Stop/start, pilot light		5.70	1.400		229	86		315	380
0200	Hand/off/automatic		5.70	1.400		124	86		210	264
0400	Stop/start/reverse		4.88	1.640		226	101		327	400
0500	NEMA 7, heavy duty, stop/start		5.52	1.450		610	89		699	805
0600	Stop/start, pilot light		3.68	2.170		745	133		878	1,025
0700	NEMA 7 or 9, 1 element		5.52	1.450		500	89		589	675
0800	2 element		5.52	1.450		640	89		729	835
0900	3 element		3.68	2.170		1,375	133		1,508	1,725
0910	Selector switch, 2 position		5.52	1.450		500	89		589	680
0920	3 position		3.68	2.170		500	133		633	750
0930	Oiltight, 1 element		7.36	1.090		124	67		191	236
0940	2 element		5.70	1.400		171	86		257	315
0950	3 element		4.88	1.640		157	101		258	320
0960	Selector switch, 2 position		5.70	1.400		125	86		211	265
0970	3 position	↓	4.88	1.640	↓	130	101		231	293

26 29 13.30 Control Switches

		Crew	Daily Output	Labor-Hours	Unit	Material	Labor	Equipment	Total	Total Incl O&P
0010	**CONTROL SWITCHES** Field installed									
6000	Push button 600 V 10A, momentary contact									
6150	Standard operator with colored button	1 Elec	31.28	.256	Ea.	21	15.70		36.70	46.50
6160	With single block 1NO 1NC		16.56	.483		46.50	29.50		76	95
6170	With double block 2NO 2NC		13.80	.580		67	35.50		102.50	127
6180	Std operator w/mushroom button 1-9/16" diam.	↓	31.28	.256	↓	42.50	15.70		58.20	70.50
6190	Std operator w/mushroom button 2-1/4" diam.									
6200	With single block 1NO 1NC	1 Elec	16.56	.483	Ea.	63	29.50		92.50	113
6210	With double block 2NO 2NC		13.80	.580		84.50	35.50		120	146
6500	Maintained contact, selector operator		31.28	.256		63	15.70		78.70	92.50
6510	With single block 1NO 1NC		16.56	.483		84.50	29.50		114	137
6520	With double block 2NO 2NC		13.80	.580		106	35.50		141.50	170
6560	Spring-return selector operator		31.28	.256		63	15.70		78.70	92.50
6570	With single block 1NO 1NC		16.56	.483		84.50	29.50		114	137
6580	With double block 2NO 2NC	↓	13.80	.580	↓	106	35.50		141.50	170
6620	Transformer operator w/illuminated									
6630	button 6 V #12 lamp	1 Elec	29.44	.272	Ea.	106	16.65		122.65	142
6640	With single block 1NO 1NC w/guard		14.72	.543		128	33.50		161.50	190
6650	With double block 2NO 2NC w/guard		11.96	.669		149	41		190	225
6690	Combination operator		31.28	.256		63	15.70		78.70	92.50
6700	With single block 1NO 1NC		16.56	.483		84.50	29.50		114	137
6710	With double block 2NO 2NC	↓	13.80	.580		106	35.50		141.50	170
9000	Indicating light unit, full voltage									
9010	110-125 V front mount	1 Elec	29.44	.272	Ea.	84.50	16.65		101.15	118

26 29 Low-Voltage Controllers

26 29 13 – Enclosed Controllers

26 29 13.30 Control Switches

		Crew	Daily Output	Labor-Hours	Unit	Material	2020 Bare Costs Labor	Equipment	Total	Total Incl O&P
9020	130 V resistor type	1 Elec	29.44	.272	Ea.	69.50	16.65		86.15	102
9030	6 V transformer type	▼	29.44	.272	▼	77	16.65		93.65	110

26 29 13.40 Relays

		Crew	Daily Output	Labor-Hours	Unit	Material	2020 Bare Costs Labor	Equipment	Total	Total Incl O&P
0010	**RELAYS** Enclosed (NEMA 1)									
0050	600 volt AC, 1 pole, 12 amp	1 Elec	4.88	1.640	Ea.	98	101		199	258
0100	2 pole, 12 amp		4.60	1.740		98	107		205	267
0200	4 pole, 10 amp		4.14	1.930		131	118		249	320
0500	250 volt DC, 1 pole, 15 amp		4.88	1.640		128	101		229	291
0600	2 pole, 10 amp		4.60	1.740		122	107		229	293
0700	4 pole, 4 amp	▼	4.14	1.930	▼	162	118		280	355

26 29 23 – Variable-Frequency Motor Controllers

26 29 23.10 Variable Frequency Drives/Adj. Frequency Drives

			Crew	Daily Output	Labor-Hours	Unit	Material	2020 Bare Costs Labor	Equipment	Total	Total Incl O&P
0010	**VARIABLE FREQUENCY DRIVES/ADJ. FREQUENCY DRIVES**										
0100	Enclosed (NEMA 1), 460 volt, for 3 HP motor size	G	1 Elec	.74	10.870	Ea.	1,825	665		2,490	3,000
0110	5 HP motor size	G		.74	10.870		2,050	665		2,715	3,250
0120	7.5 HP motor size	G		.62	12.980		2,450	795		3,245	3,850
0130	10 HP motor size	G	▼	.62	12.980		2,775	795		3,570	4,225
0140	15 HP motor size	G	2 Elec	.82	19.540		3,450	1,200		4,650	5,575
0150	20 HP motor size	G		.82	19.540		4,350	1,200		5,550	6,550
0160	25 HP motor size	G		.62	25.960		5,125	1,600		6,725	8,025
0170	30 HP motor size	G		.62	25.960		6,250	1,600		7,850	9,250
0180	40 HP motor size	G		.62	25.960		7,225	1,600		8,825	10,300
0190	50 HP motor size	G	▼	.49	32.810		9,650	2,025		11,675	13,600
0200	60 HP motor size	G	R-3	.52	38.820		11,700	2,375	365	14,440	16,800
0210	75 HP motor size	G		.52	38.820		13,700	2,375	365	16,440	19,000
0220	100 HP motor size	G		.46	43.480		16,800	2,650	410	19,860	22,800
0230	125 HP motor size	G		.46	43.480		18,200	2,650	410	21,260	24,400
0240	150 HP motor size	G		.46	43.480		23,600	2,650	410	26,660	30,400
0250	200 HP motor size	G	▼	.39	51.760		30,600	3,175	490	34,265	39,000
1100	Custom-engineered, 460 volt, for 3 HP motor size	G	1 Elec	.52	15.530		3,425	955		4,380	5,200
1110	5 HP motor size	G		.52	15.530		3,450	955		4,405	5,225
1120	7.5 HP motor size	G		.43	18.500		3,100	1,125		4,225	5,100
1130	10 HP motor size	G	▼	.43	18.500		3,250	1,125		4,375	5,300
1140	15 HP motor size	G	2 Elec	.57	28.050		4,700	1,725		6,425	7,725
1150	20 HP motor size	G		.57	28.050		4,350	1,725		6,075	7,325
1160	25 HP motor size	G		.43	37		5,375	2,275		7,650	9,300
1170	30 HP motor size	G		.43	37		6,625	2,275		8,900	10,700
1180	40 HP motor size	G		.43	37		7,925	2,275		10,200	12,100
1190	50 HP motor size	G	▼	.34	47		10,500	2,875		13,375	15,800
1200	60 HP motor size	G	R-3	.36	55.740		15,800	3,400	525	19,725	23,100
1210	75 HP motor size	G		.36	55.740		14,600	3,400	525	18,525	21,800
1220	100 HP motor size	G		.32	62.110		18,500	3,800	585	22,885	26,600
1230	125 HP motor size	G		.32	62.110		19,800	3,800	585	24,185	28,100
1240	150 HP motor size	G		.32	62.110		22,500	3,800	585	26,885	31,100
1250	200 HP motor size	G	▼	.27	74.960	▼	25,700	4,575	710	30,985	35,800
2000	For complex & special design systems to meet specific										
2010	requirements, obtain quote from vendor.										

26 31 13.50 Solar Energy - Photovoltaics		Crew	Daily Output	Labor-Hours	Unit	Material	2020 Bare Costs Labor	Equipment	Total	Total Incl O&P
0010	**SOLAR ENERGY - PHOTOVOLTAICS**									
0220	Alt. energy source, photovoltaic module, 6 watt, 15 V	G	1 Elec	7.36	1.090	Ea.	54	67		121	159
0230	10 watt, 16.3 V	G		7.36	1.090		119	67		186	231
0240	20 watt, 14.5 V	G		7.36	1.090		176	67		243	294
0250	36 watt, 17 V	G		7.36	1.090		175	67		242	293
0260	55 watt, 17 V	G		7.36	1.090		238	67		305	360
0270	75 watt, 17 V	G		7.36	1.090		400	67		467	540
0280	130 watt, 33 V	G		7.36	1.090		605	67		672	770
0290	140 watt, 33 V	G		7.36	1.090		495	67		562	645
0300	150 watt, 33 V	G		7.36	1.090		450	67		517	595
0310	DC to AC inverter for, 12 V, 2,000 watt	G		3.68	2.170		1,425	133		1,558	1,750
0320	12 V, 2,500 watt	G		3.68	2.170		1,100	133		1,233	1,425
0330	24 V, 2,500 watt	G		3.68	2.170		1,675	133		1,808	2,050
0340	12 V, 3,000 watt	G		2.76	2.900		1,325	178		1,503	1,725
0350	24 V, 3,000 watt	G		2.76	2.900		2,400	178		2,578	2,925
0360	24 V, 4,000 watt	G		1.84	4.350		3,675	267		3,942	4,450
0370	48 V, 4,000 watt	G		1.84	4.350		3,050	267		3,317	3,775
0380	48 V, 5,500 watt	G		1.84	4.350		3,000	267		3,267	3,700
0390	PV components, combiner box, 10 lug, NEMA 3R enclosure	G		3.68	2.170		270	133		403	495
0400	Fuse, 15 A for combiner box	G		36.80	.217		21.50	13.35		34.85	44
0410	Battery charger controller w/temperature sensor	G		3.68	2.170		450	133		583	695
0420	Digital readout panel, displays hours, volts, amps, etc.	G		3.68	2.170		219	133		352	440
0430	Deep cycle solar battery, 6 V, 180 Ah (C/20)	G		7.36	1.090		335	67		402	470
0440	Battery interconn, 15" AWG #2/0, sealed w/copper ring lugs	G		14.72	.543		16.60	33.50		50.10	68
0442	Battery interconn, 24" AWG #2/0, sealed w/copper ring lugs	G		14.72	.543		26.50	33.50		60	78.50
0444	Battery interconn, 60" AWG #2/0, sealed w/copper ring lugs	G		14.72	.543		53	33.50		86.50	108
0446	Batt temp computer probe, RJ11 jack, 15' cord	G		14.72	.543		22	33.50		55.50	73.50
0450	System disconnect, DC 175 amp circuit breaker	G		7.36	1.090		223	67		290	345
0460	Conduit box for inverter	G		7.36	1.090		63.50	67		130.50	169
0470	Low voltage disconnect	G		7.36	1.090		63.50	67		130.50	170
0480	Vented battery enclosure, wood	G	1 Carp	1.84	4.350		285	231		516	660
0490	PV rack system, roof, non-penetrating ballast, 1 panel	G	R-1A	28.06	.570		1,000	31.50		1,031.50	1,175
0500	Penetrating surface mount, on steel framing, 1 panel			4.72	3.390		58	187		245	345
0510	On wood framing, 1 panel			10.12	1.580		57	87.50		144.50	193
0520	With standoff, 1 panel			10.12	1.580		66	87.50		153.50	203
0530	Ground, ballast, fixed, 3 panel			18.86	.848		1,100	47		1,147	1,275
0540	4 panel			18.86	.848		1,500	47		1,547	1,725
0550	5 panel			18.86	.848		1,975	47		2,022	2,250
0560	6 panel			18.86	.848		2,275	47		2,322	2,575
0570	Adjustable, 3 panel			18.86	.848		1,175	47		1,222	1,350
0580	4 panel			18.86	.848		1,600	47		1,647	1,825
0590	5 panel			18.86	.848		2,100	47		2,147	2,400
0600	6 panel			18.86	.848		2,425	47		2,472	2,750
0610	Passive tracking, 1 panel			18.86	.848		685	47		732	820
0620	2 panel			18.86	.848		1,350	47		1,397	1,575
0630	3 panel			18.86	.848		1,950	47		1,997	2,225
0640	4 panel			18.86	.848		2,100	47		2,147	2,375
0650	6 panel			18.86	.848		2,300	47		2,347	2,600
0660	8 panel			18.86	.848		3,425	47		3,472	3,825
1020	Photovoltaic module, Thin film		1 Elec	7.36	1.090		176	67		243	294
1040	Photovoltaic module, Cadium telluride			7.36	1.090		238	67		305	360
1060	Photovoltaic module, Polycrystalline			7.36	1.090		176	67		243	294
1080	Photovoltaic module, Monocrystalline			7.36	1.090		605	67		672	770

26 32 Packaged Generator Assemblies

26 32 13 – Engine Generators

26 32 13.13 Diesel-Engine-Driven Generator Sets

		Crew	Daily Output	Labor-Hours	Unit	Material	2020 Bare Costs Labor	Equipment	Total	Total Incl O&P
0010	**DIESEL-ENGINE-DRIVEN GENERATOR SETS**									
2000	Diesel engine, including battery, charger,									
2010	muffler, & day tank, 30 kW	R-3	.51	39.530	Ea.	11,200	2,425	375	14,000	16,300
2100	50 kW		.39	51.760		21,800	3,175	490	25,465	29,300
2110	60 kW		.36	55.740		22,000	3,400	525	25,925	29,900
2200	75 kW		.32	62.110		25,300	3,800	585	29,685	34,100
2300	100 kW		.29	70.130		27,600	4,275	665	32,540	37,500
2400	125 kW		.27	74.960		29,000	4,575	710	34,285	39,500
2500	150 kW		.24	83.610		45,800	5,100	790	51,690	59,000
2501	Generator set, dsl eng in alum encl, incl btry, chgr, muf & day tank,150 kW		.24	83.610		41,500	5,100	790	47,390	54,000
2600	175 kW		.23	86.960		48,900	5,325	820	55,045	62,500
2700	200 kW		.22	90.580		49,900	5,525	855	56,280	64,000
2800	250 kW		.21	94.520		52,500	5,775	895	59,170	67,500
2850	275 kW		.20	98.810		58,000	6,050	935	64,985	74,000
2900	300 kW		.20	98.810		58,000	6,050	935	64,985	73,500
3000	350 kW		.18	108		67,000	6,600	1,025	74,625	85,000
3100	400 kW		.17	114		78,000	6,975	1,075	86,050	97,500
3200	500 kW		.17	120		106,500	7,325	1,125	114,950	129,000
3220	600 kW		.16	127		128,500	7,750	1,200	137,450	154,500
3230	650 kW	R-13	.35	120		170,500	7,125	675	178,300	199,000
3240	750 kW		.35	120		160,000	7,125	675	167,800	187,500
3250	800 kW		.33	126		156,500	7,475	710	164,685	184,000
3260	900 kW		.29	147		195,500	8,725	830	205,055	229,000
3270	1,000 kW		.25	169		205,000	10,000	950	215,950	241,500

26 32 13.16 Gas-Engine-Driven Generator Sets

		Crew	Daily Output	Labor-Hours	Unit	Material	2020 Bare Costs Labor	Equipment	Total	Total Incl O&P
0010	**GAS-ENGINE-DRIVEN GENERATOR SETS**									
0020	Gas or gasoline operated, includes battery,									
0050	charger, & muffler									
0200	7.5 kW	R-3	.76	26.190	Ea.	8,750	1,600	248	10,598	12,300
0300	11.5 kW		.65	30.620		12,400	1,875	289	14,564	16,700
0400	20 kW		.58	34.510		14,600	2,100	325	17,025	19,600
0500	35 kW		.51	39.530		17,400	2,425	375	20,200	23,100
0520	60 kW		.46	43.480		22,900	2,650	410	25,960	29,600
0600	80 kW	R-13	.37	114		28,500	6,750	640	35,890	42,200
0700	100 kW		.30	138		31,200	8,175	775	40,150	47,500
0800	125 kW		.26	163		64,000	9,675	920	74,595	86,000
0900	185 kW		.23	182		84,500	10,800	1,025	96,325	110,000

26 33 Battery Equipment

26 33 19 – Battery Units

26 33 19.10 Battery Units

		Crew	Daily Output	Labor-Hours	Unit	Material	2020 Bare Costs Labor	Equipment	Total	Total Incl O&P
0010	**BATTERY UNITS**									
0500	Salt water battery, 2.1 kWA, 24V, 750W, 30A, wired in crate, IP22 rated	1 Elec	14.95	.535	Ea.	955	33		988	1,100
0510	2.2 kWh, 48V, 800W, 17A, wired in crate, IP22 rated		14.95	.535		1,200	33		1,233	1,350
0520	25.9 kWh,48V, 11700W, 240A, wired in crate, IP2X rated		14.95	.535		14,200	33		14,233	15,600
0700	Nickel iron nife battery, deep cycle, 100Ah, 12V, 10 cells		13.11	.610		1,025	37.50		1,062.50	1,175
0710	24V, 10 cells		13.11	.610		2,175	37.50		2,212.50	2,450
0720	48V, 10 cells		13.11	.610		4,100	37.50		4,137.50	4,550
0730	Nickel iron nife battery, deep cycle, 300Ah, 12V, 10 cells		11.27	.710		3,025	43.50		3,068.50	3,400
0740	24V, 10 cells		11.27	.710		6,050	43.50		6,093.50	6,725
0750	48V, 10 cells		11.27	.710		12,700	43.50		12,743.50	14,000

26 33 Battery Equipment

26 33 53 – Static Uninterruptible Power Supply

26 33 53.10 Uninterruptible Power Supply/Conditioner Trans.	Crew	Daily Output	Labor-Hours	Unit	Material	2020 Bare Costs Labor	Equipment	Total	Total Incl O&P
0010 **UNINTERRUPTIBLE POWER SUPPLY/CONDITIONER TRANSFORMERS**									
0100 Volt. regulating, isolating transf., w/invert. & 10 min. battery pack									
0110 Single-phase, 120 V, 0.35 kVA	1 Elec	2.11	3.800	Ea.	1,150	233		1,383	1,600
0120 0.5 kVA		1.84	4.350		1,200	267		1,467	1,725
0130 For additional 55 min. battery, add to 0.35 kVA		2.11	3.800		675	233		908	1,075
0140 Add to 0.5 kVA		1.05	7.630		705	470		1,175	1,475
0150 Single-phase, 120 V, 0.75 kVA		.74	10.870		1,525	665		2,190	2,675
0160 1.0 kVA		.74	10.870		2,175	665		2,840	3,375
0170 1.5 kVA	2 Elec	1.05	15.260		3,725	935		4,660	5,500
0180 2 kVA	"	.82	19.540		4,050	1,200		5,250	6,225
0190 3 kVA	R-3	.58	34.510		5,000	2,100	325	7,425	9,000
0200 5 kVA		.39	51.760		7,100	3,175	490	10,765	13,100
0210 7.5 kVA		.30	65.880		9,075	4,025	625	13,725	16,700
0220 10 kVA		.26	77.640		12,200	4,750	735	17,685	21,400
0230 15 kVA		.20	98.810		15,700	6,050	935	22,685	27,200
0240 3 phase, 120/208 V input 120/208 V output, 20 kVA, incl 17 min. battery		.19	103		29,000	6,300	975	36,275	42,500
0242 30 kVA, incl 11 min. battery		.18	108		32,500	6,600	1,025	40,125	46,800
0250 40 kVA, incl 15 min. battery		.18	108		43,400	6,600	1,025	51,025	58,500
0260 480 V input 277/480 V output, 60 kVA, incl 6 min. battery		.17	114		48,800	6,975	1,075	56,850	65,000
0262 80 kVA, incl 4 min. battery		.17	114		58,000	6,975	1,075	66,050	75,500
0400 For additional 34 min./15 min. battery, add to 40 kVA		.66	30.450		15,100	1,850	288	17,238	19,800
0600 For complex & special design systems to meet specific									
0610 requirements, obtain quote from vendor									

26 35 Power Filters and Conditioners

26 35 13 – Capacitors

26 35 13.10 Capacitors Indoor

	Crew	Daily Output	Labor-Hours	Unit	Material	2020 Bare Costs Labor	Equipment	Total	Total Incl O&P
0010 **CAPACITORS INDOOR**									
0020 240 volts, single & 3 phase, 0.5 kVAR	1 Elec	2.48	3.220	Ea.	600	198		798	955
0100 1.0 kVAR		2.48	3.220		725	198		923	1,100
0150 2.5 kVAR		1.84	4.350		815	267		1,082	1,300
0200 5.0 kVAR		1.66	4.830		795	296		1,091	1,325
0250 7.5 kVAR		1.47	5.430		1,625	335		1,960	2,300
0300 10 kVAR		1.38	5.800		1,750	355		2,105	2,450
0350 15 kVAR		1.20	6.690		2,075	410		2,485	2,875
0400 20 kVAR		1.01	7.910		2,850	485		3,335	3,850
0450 25 kVAR		.92	8.700		3,350	535		3,885	4,475
1000 480 volts, single & 3 phase, 1 kVAR		2.48	3.220		545	198		743	895
1050 2 kVAR		2.48	3.220		630	198		828	990
1100 5 kVAR		1.84	4.350		795	267		1,062	1,275
1150 7.5 kVAR		1.84	4.350		855	267		1,122	1,325
1200 10 kVAR		1.84	4.350		1,200	267		1,467	1,725
1250 15 kVAR		1.84	4.350		1,525	267		1,792	2,075
1300 20 kVAR		1.47	5.430		1,600	335		1,935	2,250
1350 30 kVAR		1.38	5.800		1,575	355		1,930	2,275
1400 40 kVAR		1.10	7.250		2,275	445		2,720	3,175
1450 50 kVAR		1.01	7.910		2,875	485		3,360	3,900
2000 600 volts, single & 3 phase, 1 kVAR		2.48	3.220		565	198		763	915
2050 2 kVAR		2.48	3.220		650	198		848	1,000
2100 5 kVAR		1.84	4.350		795	267		1,062	1,275

For customer support on your Electrical Change Order Costs with RSMeans data, call 800.448.8182.

Pre-Installation Change Order Costs

26 35 Power Filters and Conditioners

26 35 13 – Capacitors

26 35 13.10 Capacitors Indoor		Crew	Daily Output	Labor-Hours	Unit	Material	2020 Bare Costs Labor	Equipment	Total	Total Incl O&P
2150	7.5 kVAR	1 Elec	1.84	4.350	Ea.	855	267		1,122	1,325
2200	10 kVAR		1.84	4.350		1,225	267		1,492	1,725
2250	15 kVAR		1.47	5.430		1,550	335		1,885	2,200
2300	20 kVAR		1.47	5.430		1,675	335		2,010	2,325
2350	25 kVAR		1.38	5.800		1,825	355		2,180	2,525
2400	35 kVAR		1.29	6.210		2,225	380		2,605	3,025
2450	50 kVAR	↓	1.20	6.690	↓	2,850	410		3,260	3,750

26 35 26 – Harmonic Filters

26 35 26.10 Computer Isolation Transformer

		Crew	Daily Output	Labor-Hours	Unit	Material	2020 Bare Costs Labor	Equipment	Total	Total Incl O&P
0010	**COMPUTER ISOLATION TRANSFORMER**									
0100	Computer grade									
0110	Single-phase, 120/240 V, 0.5 kVA	1 Elec	3.68	2.170	Ea.	455	133		588	700
0120	1.0 kVA		2.46	3.260		645	200		845	1,000
0130	2.5 kVA		1.84	4.350		985	267		1,252	1,475
0140	5 kVA	↓	1.05	7.630	↓	1,125	470		1,595	1,925

26 35 26.20 Computer Regulator Transformer

		Crew	Daily Output	Labor-Hours	Unit	Material	2020 Bare Costs Labor	Equipment	Total	Total Incl O&P
0010	**COMPUTER REGULATOR TRANSFORMER**									
0100	Ferro-resonant, constant voltage, variable transformer									
0110	Single-phase, 240 V, 0.5 kVA	1 Elec	2.46	3.260	Ea.	600	200		800	960
0120	1.0 kVA		1.84	4.350		825	267		1,092	1,300
0130	2.0 kVA		.92	8.700		1,425	535		1,960	2,350
0210	Plug-in unit 120 V, 0.14 kVA		7.36	1.090		350	67		417	480
0220	0.25 kVA		7.36	1.090		405	67		472	545
0230	0.5 kVA		7.36	1.090		600	67		667	760
0240	1.0 kVA		4.90	1.630		825	100		925	1,050
0250	2.0 kVA	↓	3.68	2.170	↓	1,425	133		1,558	1,750

26 35 26.30 Power Conditioner Transformer

		Crew	Daily Output	Labor-Hours	Unit	Material	2020 Bare Costs Labor	Equipment	Total	Total Incl O&P
0010	**POWER CONDITIONER TRANSFORMER**									
0100	Electronic solid state, buck-boost, transformer, w/tap switch									
0110	Single-phase, 115 V, 3.0 kVA, + or - 3% accuracy	2 Elec	1.47	10.870	Ea.	3,250	665		3,915	4,575
0120	208, 220, 230, or 240 V, 5.0 kVA, + or - 1.5% accuracy	3 Elec	1.47	16.300		4,200	1,000		5,200	6,125
0130	5.0 kVA, + or - 6% accuracy	2 Elec	1.05	15.260		3,775	935		4,710	5,550
0140	7.5 kVA, + or - 1.5% accuracy	3 Elec	1.38	17.390		5,350	1,075		6,425	7,475
0150	7.5 kVA, + or - 6% accuracy		1.47	16.300		4,450	1,000		5,450	6,400
0160	10.0 kVA, + or - 1.5% accuracy		1.22	19.610		7,125	1,200		8,325	9,625
0170	10.0 kVA, + or - 6% accuracy	↓	1.30	18.500	↓	6,050	1,125		7,175	8,350

26 35 26.40 Transient Voltage Suppressor Transformer

		Crew	Daily Output	Labor-Hours	Unit	Material	2020 Bare Costs Labor	Equipment	Total	Total Incl O&P
0010	**TRANSIENT VOLTAGE SUPPRESSOR TRANSFORMER**									
0110	Single-phase, 120 V, 1.8 kVA	1 Elec	3.68	2.170	Ea.	1,625	133		1,758	2,000
0120	3.6 kVA		3.68	2.170		2,625	133		2,758	3,075
0130	7.2 kVA		2.94	2.720		3,150	167		3,317	3,725
0150	240 V, 3.6 kVA		3.68	2.170		3,275	133		3,408	3,825
0160	7.2 kVA		3.68	2.170		3,825	133		3,958	4,400
0170	14.4 kVA		2.94	2.720		5,350	167		5,517	6,125
0210	Plug-in unit, 120 V, 1.8 kVA	↓	7.36	1.090	↓	1,125	67		1,192	1,350

26 35 53 – Voltage Regulators

26 35 53.10 Automatic Voltage Regulators

		Crew	Daily Output	Labor-Hours	Unit	Material	2020 Bare Costs Labor	Equipment	Total	Total Incl O&P
0010	**AUTOMATIC VOLTAGE REGULATORS**									
0100	Computer grade, solid state, variable transf. volt. regulator									
0110	Single-phase, 120 V, 8.6 kVA	2 Elec	1.22	13.080	Ea.	6,725	800		7,525	8,600
0120	17.3 kVA	↓	1.05	15.260		7,950	935		8,885	10,200

Pre-Installation Change Order Costs   For customer support on your Electrical Change Order Costs with RSMeans data, call 800.448.8182.

26 35 Power Filters and Conditioners

26 35 53 – Voltage Regulators

26 35 53.10 Automatic Voltage Regulators	Crew	Daily Output	Labor-Hours	Unit	Material	2020 Bare Costs Labor	Equipment	Total	Total Incl O&P
0130	208/240 V, 7.5/8.6 kVA	2 Elec	1.22	13.080	Ea.	6,725	800		7,525	8,600
0140	13.5/15.6 kVA		1.22	13.080		7,950	800		8,750	9,950
0150	27.0/31.2 kVA		1.05	15.260		10,000	935		10,935	12,400
0210	Two-phase, single control, 208/240 V, 15.0/17.3 kVA		1.05	15.260		7,950	935		8,885	10,200
0220	Individual phase control, 15.0/17.3 kVA	▼	1.05	15.260		7,950	935		8,885	10,200
0230	30.0/34.6 kVA	3 Elec	1.22	19.610		10,000	1,200		11,200	12,800
0310	Three-phase single control, 208/240 V, 26/30 kVA	2 Elec	.92	17.390		7,950	1,075		9,025	10,400
0320	380/480 V, 24/30 kVA	"	.92	17.390		7,950	1,075		9,025	10,400
0330	43/54 kVA	3 Elec	1.22	19.610		14,400	1,200		15,600	17,700
0340	Individual phase control, 208 V, 26 kVA	"	1.22	19.610		7,950	1,200		9,150	10,600
0350	52 kVA	R-3	.84	23.890		10,000	1,450	226	11,676	13,400
0360	340/480 V, 24/30 kVA	2 Elec	.92	17.390		7,950	1,075		9,025	10,400
0370	43/54 kVA	"	.92	17.390		10,000	1,075		11,075	12,600
0380	48/60 kVA	3 Elec	1.22	19.610		14,500	1,200		15,700	17,800
0390	86/108 kVA	R-3	.84	23.890	▼	16,100	1,450	226	17,776	20,200
0500	Standard grade, solid state, variable transformer volt. regulator									
0510	Single-phase, 115 V, 2.3 kVA	1 Elec	2.11	3.800	Ea.	2,125	233		2,358	2,700
0520	4.2 kVA		1.84	4.350		3,400	267		3,667	4,150
0530	6.6 kVA		1.05	7.630		4,175	470		4,645	5,300
0540	13.0 kVA	▼	1.05	7.630		7,225	470		7,695	8,650
0550	16.6 kVA	2 Elec	1.13	14.140		8,525	865		9,390	10,700
0610	230 V, 8.3 kVA		1.22	13.080		7,225	800		8,025	9,150
0620	21.4 kVA		1.13	14.140		8,525	865		9,390	10,700
0630	29.9 kVA		1.13	14.140		8,525	865		9,390	10,700
0710	460 V, 9.2 kVA		1.22	13.080		7,225	800		8,025	9,150
0720	20.7 kVA	▼	1.13	14.140		8,525	865		9,390	10,700
0810	Three-phase, 230 V, 13.1 kVA	3 Elec	1.30	18.500		7,225	1,125		8,350	9,650
0820	19.1 kVA		1.30	18.500		8,525	1,125		9,650	11,100
0830	25.1 kVA		1.13	21.210		8,525	1,300		9,825	11,300
0840	57.8 kVA	R-3	.87	22.880		15,500	1,400	216	17,116	19,300
0850	74.9 kVA	"	.84	23.890		15,500	1,450	226	17,176	19,400
0910	460 V, 14.3 kVA	3 Elec	1.30	18.500		7,225	1,125		8,350	9,650
0920	19.1 kVA		1.30	18.500		8,525	1,125		9,650	11,100
0930	27.9 kVA		1.13	21.210		8,525	1,300		9,825	11,300
0940	59.8 kVA	R-3	.92	21.740		15,500	1,325	205	17,030	19,200
0950	79.7 kVA		.87	22.880		17,300	1,400	216	18,916	21,400
0960	118 kVA	▼	.87	22.880	▼	18,200	1,400	216	19,816	22,400
1000	Laboratory grade, precision, electronic voltage regulator									
1110	Single-phase, 115 V, 0.5 kVA	1 Elec	2.11	3.800	Ea.	1,650	233		1,883	2,175
1120	1.0 kVA		1.84	4.350		1,750	267		2,017	2,325
1130	3.0 kVA	▼	.74	10.870		2,450	665		3,115	3,700
1140	6.0 kVA	2 Elec	1.34	11.910		4,550	730		5,280	6,100
1150	10.0 kVA	3 Elec	.92	26.090		5,925	1,600		7,525	8,875
1160	15.0 kVA	"	1.38	17.390		6,750	1,075		7,825	9,025
1210	230 V, 3.0 kVA	1 Elec	.74	10.870		2,775	665		3,440	4,050
1220	6.0 kVA	2 Elec	1.34	11.910		4,650	730		5,380	6,200
1230	10.0 kVA	3 Elec	1.57	15.260		6,175	935		7,110	8,200
1240	15.0 kVA	"	1.47	16.300	▼	6,975	1,000		7,975	9,175

26 35 53.30 Transient Suppressor/Voltage Regulator

0010	**TRANSIENT SUPPRESSOR/VOLTAGE REGULATOR** (without isolation)									
0110	Single-phase, 115 V, 1.0 kVA	1 Elec	2.46	3.260	Ea.	1,100	200		1,300	1,525
0120	2.0 kVA	▼	2.11	3.800	▼	1,525	233		1,758	2,025

For customer support on your Electrical Change Order Costs with RSMeans data, call 800.448.8182. **Pre-Installation Change Order Costs**

26 35 Power Filters and Conditioners

26 35 53 – Voltage Regulators

26 35 53.30 Transient Suppressor/Voltage Regulator		Crew	Daily Output	Labor-Hours	Unit	Material	2020 Bare Costs Labor	Equipment	Total	Total Incl O&P
0130	4.0 kVA	1 Elec	1.96	4.080	Ea.	1,875	250		2,125	2,425
0140	220 V, 1.0 kVA		2.46	3.260		1,100	200		1,300	1,525
0150	2.0 kVA		2.11	3.800		1,550	233		1,783	2,075
0160	4.0 kVA		1.96	4.080		1,975	250		2,225	2,550
0210	Plug-in unit, 120 V, 1.0 kVA		7.36	1.090		1,050	67		1,117	1,275
0220	2.0 kVA	▼	7.36	1.090	▼	1,500	67		1,567	1,750

26 36 Transfer Switches

26 36 13 – Manual Transfer Switches

26 36 13.10 Non-Automatic Transfer Switches

		Crew	Daily Output	Labor-Hours	Unit	Material	2020 Bare Costs Labor	Equipment	Total	Total Incl O&P
0010	**NON-AUTOMATIC TRANSFER SWITCHES** enclosed									
0100	Manual operated, 480 volt 3 pole, 30 amp	1 Elec	2.12	3.780	Ea.	1,100	232		1,332	1,575
0150	60 amp		1.75	4.580		2,025	281		2,306	2,650
0200	100 amp	▼	1.20	6.690		3,525	410		3,935	4,500
0250	200 amp	2 Elec	1.84	8.700		4,450	535		4,985	5,700
0300	400 amp		1.47	10.870		8,350	665		9,015	10,200
0350	600 amp	▼	.92	17.390		5,475	1,075		6,550	7,625
1000	250 volt 3 pole, 30 amp	1 Elec	2.12	3.780		1,050	232		1,282	1,500
1100	60 amp		1.75	4.580		1,700	281		1,981	2,275
1150	100 amp	▼	1.20	6.690		2,875	410		3,285	3,775
1200	200 amp	2 Elec	1.84	8.700		4,150	535		4,685	5,350
1300	600 amp	"	.92	17.390		10,500	1,075		11,575	13,100
1500	Electrically operated, 480 volt 3 pole, 60 amp	1 Elec	1.75	4.580		1,650	281		1,931	2,250
1600	100 amp	"	1.20	6.690		1,650	410		2,060	2,425
1650	200 amp	2 Elec	1.84	8.700		2,725	535		3,260	3,775
1700	400 amp		1.47	10.870		3,800	665		4,465	5,175
1750	600 amp	▼	.92	17.390		5,475	1,075		6,550	7,625
2000	250 volt 3 pole, 30 amp	1 Elec	2.12	3.780		1,800	232		2,032	2,325
2050	60 amp	"	1.75	4.580		2,150	281		2,431	2,775
2150	200 amp	2 Elec	1.84	8.700		3,525	535		4,060	4,675
2200	400 amp		1.47	10.870		4,925	665		5,590	6,400
2250	600 amp	▼	.92	17.390		7,075	1,075		8,150	9,400
2500	NEMA 3R, 480 volt 3 pole, 60 amp	1 Elec	1.66	4.830		2,550	296		2,846	3,250
2550	100 amp	"	1.10	7.250		3,875	445		4,320	4,900
2600	200 amp	2 Elec	1.66	9.660		5,300	595		5,895	6,700
2650	400 amp	"	1.29	12.420		4,150	760		4,910	5,700
2800	NEMA 3R, 250 volt 3 pole solid state, 100 amp	1 Elec	1.10	7.250		3,975	445		4,420	5,025
2850	150 amp	2 Elec	1.66	9.660		5,275	595		5,870	6,700
2900	250 volt 2 pole solid state, 100 amp	1 Elec	1.20	6.690		3,875	410		4,285	4,850
2950	150 amp	2 Elec	1.84	8.700	▼	5,150	535		5,685	6,450

26 36 23 – Automatic Transfer Switches

26 36 23.10 Automatic Transfer Switch Devices

		Crew	Daily Output	Labor-Hours	Unit	Material	2020 Bare Costs Labor	Equipment	Total	Total Incl O&P
0010	**AUTOMATIC TRANSFER SWITCH DEVICES**									
0015	Switches, enclosed 120/240 volt, 2 pole, 30 amp	1 Elec	2.21	3.620	Ea.	2,025	222		2,247	2,550
0020	70 amp		1.84	4.350		2,025	267		2,292	2,625
0030	100 amp	▼	1.24	6.440		2,025	395		2,420	2,825
0040	225 amp	2 Elec	1.93	8.280		2,975	510		3,485	4,025
0050	400 amp		1.56	10.230		4,525	630		5,155	5,925
0060	600 amp		.98	16.410		9,575	1,000		10,575	12,000
0070	800 amp	▼	.77	20.700	▼	11,300	1,275		12,575	14,300

26 36 Transfer Switches

26 36 23 – Automatic Transfer Switches

26 36 23.10 Automatic Transfer Switch Devices	Crew	Daily Output	Labor-Hours	Unit	Material	2020 Bare Costs Labor	Equipment	Total	Total Incl O&P
0100	Switches, enclosed 480 volt, 3 pole, 30 amp	1 Elec	2.12	3.780	Ea.	3,425	232		3,657	4,125
0200	60 amp		1.75	4.580		3,425	281		3,706	4,200
0300	100 amp		1.20	6.690		3,425	410		3,835	4,375
0400	150 amp	2 Elec	2.21	7.250		4,200	445		4,645	5,275
0500	225 amp		1.84	8.700		5,400	535		5,935	6,750
0600	260 amp		1.84	8.700		6,250	535		6,785	7,675
0700	400 amp		1.47	10.870		7,850	665		8,515	9,625
0800	600 amp		.92	17.390		11,300	1,075		12,375	14,000
0900	800 amp		.74	21.740		13,300	1,325		14,625	16,600
1000	1,000 amp		.70	22.880		18,500	1,400		19,900	22,500
1100	1,200 amp		.64	24.840		25,400	1,525		26,925	30,300
1200	1,600 amp		.55	28.990		28,900	1,775		30,675	34,500
1300	2,000 amp		.46	34.780		32,400	2,125		34,525	38,800
1600	Accessories, time delay on engine starting					284			284	310
1700	Adjustable time delay on retransfer					284			284	310
1800	Shunt trips for customer connections					505			505	555
1900	Maintenance select switch					115			115	127
2000	Auxiliary contact when normal fails					103			103	114
2100	Pilot light-emergency					115			115	127
2200	Pilot light-normal					115			115	127
2300	Auxiliary contact-closed on normal					133			133	147
2400	Auxiliary contact-closed on emergency					133			133	147
2500	Emergency source sensing, frequency relay					585			585	645

26 41 Facility Lightning Protection

26 41 13 – Lightning Protection for Structures

26 41 13.13 Lightning Protection for Buildings

		Crew	Daily Output	Labor-Hours	Unit	Material	2020 Bare Costs Labor	Equipment	Total	Total Incl O&P
0010	**LIGHTNING PROTECTION FOR BUILDINGS**									
0200	Air terminals & base, copper									
0400	3/8" diameter x 10" (to 75' high)	1 Elec	7.36	1.090	Ea.	26	67		93	128
0500	1/2" diameter x 12" (over 75' high)		7.36	1.090		30	67		97	133
0520	1/2" diameter x 24"		6.72	1.190		40	73		113	153
0540	1/2" diameter x 60"		6.16	1.300		73.50	80		153.50	200
1000	Aluminum, 1/2" diameter x 12" (to 75' high)		7.36	1.090		17.15	67		84.15	118
1020	1/2" diameter x 24"		6.72	1.190		18.95	73		91.95	130
1040	1/2" diameter x 60"		6.16	1.300		25	80		105	147
1100	5/8" diameter x 12" (over 75' high)		7.36	1.090		18.25	67		85.25	120
2000	Cable, copper, 220 lb. per thousand ft. (to 75' high)		294.40	.027	L.F.	3.18	1.67		4.85	6
2100	375 lb. per thousand ft. (over 75' high)		211.60	.038		5.75	2.32		8.07	9.80
2500	Aluminum, 101 lb. per thousand ft. (to 75' high)		257.60	.031		.91	1.91		2.82	3.84
2600	199 lb. per thousand ft. (over 75' high)		220.80	.036		1.33	2.22		3.55	4.77
3000	Arrester, 175 volt AC to ground		7.36	1.090	Ea.	163	67		230	279
3100	650 volt AC to ground		6.16	1.300	"	152	80		232	287

Pre-Installation Change Order Costs

26 51 13 – Interior Lighting Fixtures, Lamps, and Ballasts

26 51 13.10 Fixture Hangers	Crew	Daily Output	Labor-Hours	Unit	Material	2020 Bare Costs Labor	Equipment	Total	Total Incl O&P
0010 **FIXTURE HANGERS**									
0220 Box hub cover	1 Elec	29.44	.272	Ea.	4.13	16.65		20.78	29.50
0240 Canopy		11.04	.725		9.05	44.50		53.55	76
0260 Connecting block		36.80	.217		2.83	13.35		16.18	23
0280 Cushion hanger		14.72	.543		23.50	33.50		57	75.50
0300 Box hanger, with mounting strap		7.36	1.090		9.60	67		76.60	110
0320 Connecting block		36.80	.217		2.63	13.35		15.98	22.50
0340 Flexible, 1/2" diameter, 4" long		11.04	.725		15.75	44.50		60.25	83.50
0360 6" long		11.04	.725		17.10	44.50		61.60	85
0380 8" long		11.04	.725		18.95	44.50		63.45	87
0400 10" long		11.04	.725		20	44.50		64.50	88
0420 12" long		11.04	.725		21	44.50		65.50	89.50
0440 15" long		11.04	.725		22.50	44.50		67	90.50
0460 18" long		11.04	.725		26	44.50		70.50	94.50
0480 3/4" diameter, 4" long		9.20	.870		20	53.50		73.50	102
0500 6" long		9.20	.870		22	53.50		75.50	104
0520 8" long		9.20	.870		22.50	53.50		76	104
0540 10" long		9.20	.870		25	53.50		78.50	107
0560 12" long		9.20	.870		27	53.50		80.50	109
0580 15" long		9.20	.870		30	53.50		83.50	113
0600 18" long		9.20	.870		32.50	53.50		86	116

26 51 13.40 Interior HID Fixtures

	Crew	Daily Output	Labor-Hours	Unit	Material	2020 Bare Costs Labor	Equipment	Total	Total Incl O&P
0010 **INTERIOR HID FIXTURES** Incl. lamps and mounting hardware									
0700 High pressure sodium, recessed, round, 70 watt	1 Elec	3.22	2.480	Ea.	585	152		737	870
0720 100 watt		3.22	2.480		705	152		857	1,000
0740 150 watt		2.94	2.720		725	167		892	1,050
0760 Square, 70 watt		3.31	2.420		625	148		773	910
0780 100 watt		3.31	2.420		705	148		853	995
0820 250 watt		2.76	2.900		890	178		1,068	1,250
0840 1,000 watt	2 Elec	4.42	3.620		1,775	222		1,997	2,275
0860 Surface, round, 70 watt	1 Elec	2.76	2.900		960	178		1,138	1,325
0880 100 watt		2.76	2.900		985	178		1,163	1,350
0900 150 watt		2.48	3.220		955	198		1,153	1,350
0920 Square, 70 watt		2.76	2.900		830	178		1,008	1,175
0940 100 watt		2.76	2.900		865	178		1,043	1,225
0980 250 watt		2.30	3.480		695	214		909	1,075
1040 Pendent, round, 70 watt		2.76	2.900		980	178		1,158	1,350
1060 100 watt		2.76	2.900		860	178		1,038	1,200
1080 150 watt		2.48	3.220		960	198		1,158	1,350
1100 Square, 70 watt		2.76	2.900		970	178		1,148	1,350
1120 100 watt		2.76	2.900		985	178		1,163	1,350
1140 150 watt		2.48	3.220		1,000	198		1,198	1,425
1160 250 watt		2.30	3.480		1,375	214		1,589	1,825
1180 400 watt		2.21	3.620		1,450	222		1,672	1,900
1220 Wall, round, 70 watt		2.76	2.900		860	178		1,038	1,225
1240 100 watt		2.76	2.900		855	178		1,033	1,200
1260 150 watt		2.48	3.220		850	198		1,048	1,225
1300 Square, 70 watt		2.76	2.900		880	178		1,058	1,225
1320 100 watt		2.76	2.900		965	178		1,143	1,350
1340 150 watt		2.48	3.220		950	198		1,148	1,350
1360 250 watt		2.30	3.480		960	214		1,174	1,375
1380 400 watt	2 Elec	4.42	3.620		1,175	222		1,397	1,625

26 51 13 – Interior Lighting Fixtures, Lamps, and Ballasts

26 51 13.40 Interior HID Fixtures	Crew	Daily Output	Labor-Hours	Unit	Material	2020 Bare Costs Labor	Equipment	Total	Total Incl O&P
1400	1,000 watt	2 Elec	3.31	4.830	Ea.	1,775	296		2,071	2,400
1500	Metal halide, recessed, round, 175 watt	1 Elec	3.13	2.560		445	157		602	725
1520	250 watt	"	2.94	2.720		545	167		712	850
1540	400 watt	2 Elec	5.34	3		815	184		999	1,175
1580	Square, 175 watt	1 Elec	3.13	2.560		460	157		617	740
1640	Surface, round, 175 watt		2.67	3		600	184		784	935
1660	250 watt		2.48	3.220		1,225	198		1,423	1,650
1680	400 watt	2 Elec	4.42	3.620		1,225	222		1,447	1,675
1720	Square, 175 watt	1 Elec	2.67	3		655	184		839	995
1800	Pendent, round, 175 watt		2.67	3		875	184		1,059	1,225
1820	250 watt		2.48	3.220		1,325	198		1,523	1,775
1840	400 watt	2 Elec	4.42	3.620		1,225	222		1,447	1,675
1880	Square, 175 watt	1 Elec	2.67	3		500	184		684	825
1900	250 watt	"	2.48	3.220		740	198		938	1,100
1920	400 watt	2 Elec	4.42	3.620		1,250	222		1,472	1,700
1980	Wall, round, 175 watt	1 Elec	2.67	3		975	184		1,159	1,350
2000	250 watt	"	2.48	3.220		1,175	198		1,373	1,575
2020	400 watt	2 Elec	4.42	3.620		1,025	222		1,247	1,475
2060	Square, 175 watt	1 Elec	2.67	3		515	184		699	840
2080	250 watt	"	2.48	3.220		725	198		923	1,100
2100	400 watt	2 Elec	4.42	3.620		1,025	222		1,247	1,450
2800	High pressure sodium, vaporproof, recessed, 70 watt	1 Elec	3.22	2.480		765	152		917	1,075
2820	100 watt		3.22	2.480		780	152		932	1,075
2840	150 watt		2.94	2.720		800	167		967	1,125
2900	Surface, 70 watt		2.76	2.900		860	178		1,038	1,200
2920	100 watt		2.76	2.900		900	178		1,078	1,250
2940	150 watt		2.48	3.220		935	198		1,133	1,325
3000	Pendent, 70 watt		2.76	2.900		850	178		1,028	1,200
3020	100 watt		2.76	2.900		875	178		1,053	1,225
3040	150 watt		2.48	3.220		930	198		1,128	1,325
3100	Wall, 70 watt		2.76	2.900		915	178		1,093	1,275
3120	100 watt		2.76	2.900		955	178		1,133	1,325
3140	150 watt		2.48	3.220		995	198		1,193	1,400
3200	Metal halide, vaporproof, recessed, 175 watt		3.13	2.560		790	157		947	1,100
3220	250 watt		2.94	2.720		720	167		887	1,050
3240	400 watt	2 Elec	5.34	3		900	184		1,084	1,275
3260	1,000 watt	"	4.42	3.620		1,650	222		1,872	2,125
3280	Surface, 175 watt	1 Elec	2.67	3		1,150	184		1,334	1,525
3300	250 watt	"	2.48	3.220		1,025	198		1,223	1,425
3320	400 watt	2 Elec	4.42	3.620		1,275	222		1,497	1,725
3340	1,000 watt	"	3.31	4.830		1,850	296		2,146	2,500
3360	Pendent, 175 watt	1 Elec	2.67	3		1,250	184		1,434	1,650
3380	250 watt	"	2.48	3.220		1,050	198		1,248	1,450
3400	400 watt	2 Elec	4.42	3.620		1,250	222		1,472	1,700
3420	1,000 watt	"	3.31	4.830		2,025	296		2,321	2,675
3440	Wall, 175 watt	1 Elec	2.67	3		1,325	184		1,509	1,725
3460	250 watt	"	2.48	3.220		1,100	198		1,298	1,525
3480	400 watt	2 Elec	4.42	3.620		1,325	222		1,547	1,775
3500	1,000 watt	"	3.31	4.830		2,100	296		2,396	2,775

Pre-Installation Change Order Costs

26 51 13 – Interior Lighting Fixtures, Lamps, and Ballasts

26 51 13.50 Interior Lighting Fixtures		Crew	Daily Output	Labor-Hours	Unit	Material	2020 Bare Costs Labor	Equipment	Total	Total Incl O&P
0010	**INTERIOR LIGHTING FIXTURES** Including lamps, mounting									
0030	hardware and connections									
0100	Fluorescent, C.W. lamps, troffer, recess mounted in grid, RS									
0130	Grid ceiling mount									
0200	Acrylic lens, 1' W x 4' L, two 40 watt	1 Elec	5.24	1.530	Ea.	53.50	94		147.50	199
0210	1' W x 4' L, three 40 watt		4.97	1.610		60.50	99		159.50	214
0300	2' W x 2' L, two U40 watt		5.24	1.530		57.50	94		151.50	203
0400	2' W x 4' L, two 40 watt		4.88	1.640		56	101		157	212
0500	2' W x 4' L, three 40 watt		4.60	1.740		62	107		169	227
0600	2' W x 4' L, four 40 watt		4.32	1.850		64.50	114		178.50	240
0700	4' W x 4' L, four 40 watt	2 Elec	5.89	2.720		294	167		461	575
0800	4' W x 4' L, six 40 watt		5.70	2.810		310	172		482	595
0900	4' W x 4' L, eight 40 watt		5.34	3		350	184		534	660
0910	Acrylic lens, 1' W x 4' L, two 32 watt T8	G 1 Elec	5.24	1.530		76	94		170	224
0930	2' W x 2' L, two U32 watt T8	G	5.24	1.530		115	94		209	267
0940	2' W x 4' L, two 32 watt T8	G	4.88	1.640		87	101		188	246
0950	2' W x 4' L, three 32 watt T8	G	4.60	1.740		80.50	107		187.50	248
0960	2' W x 4' L, four 32 watt T8	G	4.32	1.850		75	114		189	252
1000	Surface mounted, RS									
1030	Acrylic lens with hinged & latched door frame									
1100	1' W x 4' L, two 40 watt	1 Elec	6.44	1.240	Ea.	68.50	76		144.50	189
1110	1' W x 4' L, three 40 watt		6.16	1.300		71	80		151	197
1200	2' W x 2' L, two U40 watt		6.44	1.240		73.50	76		149.50	194
1300	2' W x 4' L, two 40 watt		5.70	1.400		83.50	86		169.50	220
1400	2' W x 4' L, three 40 watt		5.24	1.530		85	94		179	234
1500	2' W x 4' L, four 40 watt		4.88	1.640		87	101		188	246
1501	2' W x 4' L, six 40 watt T8		4.78	1.670		87	102		189	249
1600	4' W x 4' L, four 40 watt	2 Elec	6.62	2.420		430	148		578	695
1700	4' W x 4' L, six 40 watt		6.07	2.640		465	162		627	755
1800	4' W x 4' L, eight 40 watt		5.70	2.810		485	172		657	790
1900	2' W x 8' L, four 40 watt		5.89	2.720		171	167		338	435
2000	2' W x 8' L, eight 40 watt		5.70	2.810		184	172		356	460
2010	Acrylic wrap around lens									
2020	6" W x 4' L, one 40 watt	1 Elec	7.36	1.090	Ea.	69.50	67		136.50	176
2030	6" W x 8' L, two 40 watt	2 Elec	7.36	2.170		76.50	133		209.50	282
2040	11" W x 4' L, two 40 watt	1 Elec	6.44	1.240		46.50	76		122.50	165
2050	11" W x 8' L, four 40 watt	2 Elec	6.07	2.640		77	162		239	325
2060	16" W x 4' L, four 40 watt	1 Elec	4.88	1.640		77	101		178	235
2070	16" W x 8' L, eight 40 watt	2 Elec	5.89	2.720		168	167		335	435
2080	2' W x 2' L, two U40 watt	1 Elec	6.44	1.240		94	76		170	217
2100	Strip fixture									
2200	4' long, one 40 watt, RS	1 Elec	7.82	1.020	Ea.	31.50	62.50		94	128
2300	4' long, two 40 watt, RS		7.36	1.090		49	67		116	154
2310	4' long, two 32 watt T8, RS	G	7.36	1.090		75	67		142	182
2400	4' long, one 40 watt, SL		7.36	1.090		54	67		121	159
2500	4' long, two 40 watt, SL		6.44	1.240		73	76		149	194
2580	8' long, one 60 watt T8, SL	G 2 Elec	12.33	1.300		113	80		193	243
2590	8' long, two 60 watt T8, SL	G	11.41	1.400		99.50	86		185.50	237
2600	8' long, one 75 watt, SL		12.33	1.300		57.50	80		137.50	182
2700	8' long, two 75 watt, SL		11.41	1.400		71	86		157	206
2800	4' long, two 60 watt, HO	1 Elec	6.16	1.300		112	80		192	242
2810	4' long, two 54 watt, T5HO	G "	6.16	1.300		176	80		256	310

26 51 13.50 Interior Lighting Fixtures		Crew	Daily Output	Labor-Hours	Unit	Material	2020 Bare Costs Labor	Equipment	Total	Total Incl O&P
2900	8' long, two 110 watt, HO	2 Elec	9.75	1.640	Ea.	110	101		211	271
2910	4' long, two 115 watt, VHO	1 Elec	5.98	1.340		145	82		227	281
2920	8' long, two 215 watt, VHO	2 Elec	9.57	1.670		157	102		259	325
2950	High bay pendent mounted, 16" W x 4' L, four 54 watt, T5HO G		8.19	1.950		298	120		418	510
2952	2' W x 4' L, six 54 watt, T5HO G		7.82	2.050		305	126		431	520
2954	2' W x 4' L, six 32 watt, T8 G		7.82	2.050		178	126		304	385
3000	Strip, pendent mounted, industrial, white porcelain enamel									
3100	4' long, two 40 watt, RS	1 Elec	5.24	1.530	Ea.	53	94		147	198
3110	4' long, two 32 watt T8, RS G		5.24	1.530		83	94		177	232
3200	4' long, two 60 watt, HO		4.60	1.740		82.50	107		189.50	250
3290	8' long, two 60 watt T8, SL G	2 Elec	8.10	1.980		121	121		242	315
3300	8' long, two 75 watt, SL		8.10	1.980		101	121		222	292
3400	8' long, two 110 watt, HO		7.36	2.170		130	133		263	340
3410	Acrylic finish, 4' long, two 40 watt, RS	1 Elec	5.24	1.530		87	94		181	236
3420	4' long, two 60 watt, HO		4.60	1.740		161	107		268	335
3430	4' long, two 115 watt, VHO		4.42	1.810		210	111		321	395
3440	8' long, two 75 watt, SL	2 Elec	8.10	1.980		172	121		293	370
3450	8' long, two 110 watt, HO		7.36	2.170		203	133		336	420
3460	8' long, two 215 watt, VHO		6.99	2.290		287	140		427	525
3470	Troffer, air handling, 2' W x 4' L with four 32 watt T8 G	1 Elec	3.68	2.170		119	133		252	330
3480	2' W x 2' L with two U32 watt T8 G		5.06	1.580		110	97		207	265
3490	Air connector insulated, 5" diameter		18.40	.435		69.50	26.50		96	116
3500	6" diameter		18.40	.435		68.50	26.50		95	115
3502	Troffer, direct/indirect, 2' W x 4' L with two 32 W T8 G		4.88	1.640		286	101		387	465
3510	Troffer parabolic lay-in, 1' W x 4' L with one 32 W T8 G		5.24	1.530		119	94		213	271
3520	1' W x 4' L with two 32 W T8 G		4.88	1.640		141	101		242	305
3525	2' W x 2' L with two U32 W T8 G		5.24	1.530		124	94		218	276
3530	2' W x 4' L with three 32 W T8 G		4.60	1.740		132	107		239	305
3531	Intr fxtr, fluor, troffer prismatic lay-in, 2' W x 4'l w/three 32 W T8		4.60	1.740		152	107		259	325
3535	Downlight, recess mounted G		7.36	1.090		148	67		215	263
3540	Wall wash reflector, recess mounted G		7.36	1.090		113	67		180	225
3550	Direct/indirect, 4' long, stl., pendent mtd. G		4.60	1.740		162	107		269	335
3560	4' long, alum., pendent mtd. G		4.60	1.740		335	107		442	525
3565	Prefabricated cove, 4' long, stl. continuous row G		4.60	1.740		210	107		317	390
3570	4' long, alum. continuous row G		4.60	1.740		345	107		452	540
3580	Wet location, recess mounted, 2' W x 4' L with two 32 watt T8 G		4.88	1.640		238	101		339	410
3590	Pendent mounted, 2' W x 4' L with two 32 watt T8 G		5.24	1.530		385	94		479	560
4000	Induction lamp, integral ballast, ceiling mounted									
4110	High bay, aluminum reflector, 160 watt	1 Elec	2.94	2.720	Ea.	920	167		1,087	1,250
4120	320 watt	"	2.76	2.900		1,650	178		1,828	2,100
4130	480 watt	2 Elec	5.34	3		2,500	184		2,684	3,025
4150	Low bay, aluminum reflector, 250 watt	1 Elec	2.94	2.720		935	167		1,102	1,275
4170	Garage, aluminum reflector, 80 watt		3.31	2.420		840	148		988	1,150
4180	Vandalproof, aluminum reflector, 100 watt		2.94	2.720		615	167		782	930
4220	Metal halide, integral ballast, ceiling, recess mounted									
4230	prismatic glass lens, floating door									
4240	2' W x 2' L, 250 watt	1 Elec	2.94	2.720	Ea.	300	167		467	580
4250	2' W x 2' L, 400 watt	2 Elec	5.34	3		370	184		554	680
4260	Surface mounted, 2' W x 2' L, 250 watt	1 Elec	2.48	3.220		340	198		538	670
4270	400 watt	2 Elec	4.42	3.620		405	222		627	775
4280	High bay, aluminum reflector,									
4290	Single unit, 400 watt	2 Elec	4.23	3.780	Ea.	425	232		657	810
4300	Single unit, 1,000 watt		3.68	4.350		610	267		877	1,075

For customer support on your Electrical Change Order Costs with RSMeans data, call 800.448.8182.

Pre-Installation Change Order Costs

26 51 13 - Interior Lighting Fixtures, Lamps, and Ballasts

26 51 13.50 Interior Lighting Fixtures	Crew	Daily Output	Labor-Hours	Unit	Material	2020 Bare Costs Labor	Equipment	Total	Total Incl O&P
4310	Twin unit, 400 watt	2 Elec	2.94	5.430	Ea.	820	335		1,155	1,400
4320	Low bay, aluminum reflector, 250W DX lamp	1 Elec	2.94	2.720		360	167		527	650
4330	400 watt lamp	2 Elec	4.60	3.480		545	214		759	920
4340	High pressure sodium integral ballast ceiling, recess mounted									
4350	prismatic glass lens, floating door									
4360	2' W x 2' L, 150 watt lamp	1 Elec	2.94	2.720	Ea.	400	167		567	690
4370	2' W x 2' L, 400 watt lamp	2 Elec	5.34	3		475	184		659	800
4380	Surface mounted, 2' W x 2' L, 150 watt lamp	1 Elec	2.48	3.220		485	198		683	830
4390	400 watt lamp	2 Elec	4.42	3.620		545	222		767	930
4400	High bay, aluminum reflector,									
4410	Single unit, 400 watt lamp	2 Elec	4.23	3.780	Ea.	390	232		622	775
4430	Single unit, 1,000 watt lamp	"	3.68	4.350		545	267		812	995
4440	Low bay, aluminum reflector, 150 watt lamp	1 Elec	2.94	2.720		340	167		507	620
4445	High bay H.I.D. quartz restrike	"	14.72	.543		169	33.50		202.50	235
4450	Incandescent, high hat can, round alzak reflector, prewired									
4470	100 watt	1 Elec	7.36	1.090	Ea.	68	67		135	174
4480	150 watt		7.36	1.090		104	67		171	214
4500	300 watt		6.16	1.300		241	80		321	385
4520	Round with reflector and baffles, 150 watt		7.36	1.090		54	67		121	159
4540	Round with concentric louver, 150 watt PAR		7.36	1.090		78.50	67		145.50	186
4600	Square glass lens with metal trim, prewired									
4630	100 watt	1 Elec	6.16	1.300	Ea.	55.50	80		135.50	180
4680	150 watt		6.16	1.300		97.50	80		177.50	226
4700	200 watt		6.16	1.300		97.50	80		177.50	226
4800	300 watt		5.24	1.530		146	94		240	300
4810	500 watt		4.60	1.740		286	107		393	475
4900	Ceiling/wall, surface mounted, metal cylinder, 75 watt		9.20	.870		51.50	53.50		105	136
4920	150 watt		9.20	.870		89	53.50		142.50	178
4930	300 watt		7.36	1.090		173	67		240	290
5000	500 watt		6.16	1.300		365	80		445	520
5010	Square, 100 watt		7.36	1.090		124	67		191	236
5020	150 watt		7.36	1.090		124	67		191	237
5030	300 watt		6.44	1.240		340	76		416	490
5040	500 watt		5.52	1.450		360	89		449	530
5200	Ceiling, surface mounted, opal glass drum									
5300	8", one 60 watt lamp	1 Elec	9.20	.870	Ea.	67	53.50		120.50	154
5400	10", two 60 watt lamps		7.36	1.090		74.50	67		141.50	182
5500	12", four 60 watt lamps		6.16	1.300		104	80		184	233
5510	Pendent, round, 100 watt		7.36	1.090		124	67		191	236
5520	150 watt		7.36	1.090		124	67		191	236
5530	300 watt		6.16	1.300		171	80		251	310
5540	500 watt		5.06	1.580		330	97		427	510
5550	Square, 100 watt		6.16	1.300		152	80		232	286
5560	150 watt		6.16	1.300		159	80		239	294
5570	300 watt		5.24	1.530		237	94		331	400
5580	500 watt		4.60	1.740		320	107		427	515
5600	Wall, round, 100 watt		7.36	1.090		66	67		133	172
5620	300 watt		7.36	1.090		133	67		200	247
5630	500 watt		6.16	1.300		390	80		470	550
5640	Square, 100 watt		7.36	1.090		123	67		190	236
5650	150 watt		7.36	1.090		108	67		175	219
5660	300 watt		6.44	1.240		167	76		243	297
5670	500 watt		5.52	1.450		278	89		367	435

26 51 13 – Interior Lighting Fixtures, Lamps, and Ballasts

26 51 13.50 Interior Lighting Fixtures		Crew	Daily Output	Labor-Hours	Unit	Material	2020 Bare Costs Labor	Equipment	Total	Total Incl O&P
6010	Vapor tight, incandescent, ceiling mounted, 200 watt	1 Elec	5.70	1.400	Ea.	78.50	86		164.50	215
6020	Recessed, 200 watt		6.16	1.300		127	80		207	259
6030	Pendent, 200 watt		6.16	1.300		78.50	80		158.50	206
6040	Wall, 200 watt		7.36	1.090		80.50	67		147.50	189
6100	Fluorescent, surface mounted, 2 lamps, 4' L, RS, 40 watt		2.94	2.720		119	167		286	380
6110	Industrial, 2 lamps, 4' L in tandem, 430 MA		2.02	3.950		212	242		454	595
6130	2 lamps, 4' L, 800 MA		1.75	4.580		181	281		462	620
6160	Pendent, indust, 2 lamps, 4' L in tandem, 430 MA		1.75	4.580		247	281		528	690
6170	2 lamps, 4' L, 430 MA		2.12	3.780		165	232		397	525
6180	2 lamps, 4' L, 800 MA		1.56	5.120		208	315		523	700
6850	Vandalproof, surface mounted, fluorescent, two 32 watt T8 G		2.94	2.720		278	167		445	555
6860	Incandescent, one 150 watt		7.36	1.090		94.50	67		161.50	204
6900	Mirror light, fluorescent, RS, acrylic enclosure, two 40 watt		7.36	1.090		115	67		182	227
6910	One 40 watt		7.36	1.090		99.50	67		166.50	209
6920	One 20 watt		11.04	.725		84	44.50		128.50	159
7000	Low bay, aluminum reflector, 70 watt, high pressure sodium		3.68	2.170		277	133		410	505
7010	250 watt		2.94	2.720		370	167		537	655
7020	400 watt	2 Elec	4.60	3.480		430	214		644	795
7500	Ballast replacement, by weight of ballast, to 15' high									
7520	Indoor fluorescent, less than 2 lb.	1 Elec	9.20	.870	Ea.	25	53.50		78.50	107
7540	Two 40W, watt reducer, 2 to 5 lb.		8.65	.925		71.50	57		128.50	163
7560	Two F96 slimline, over 5 lb.		7.36	1.090		111	67		178	222
7580	Vaportite ballast, less than 2 lb.		8.65	.925		25	57		82	112
7600	2 lb. to 5 lb.		8.19	.977		71.50	60		131.50	168
7620	Over 5 lb.		6.99	1.140		111	70		181	226
7630	Electronic ballast for two tubes		7.36	1.090		41.50	67		108.50	145
7640	Dimmable ballast one-lamp G		7.36	1.090		105	67		172	216
7650	Dimmable ballast two-lamp G		6.99	1.140		108	70		178	223
7690	Emergency ballast (factory installed in fixture)					152			152	167
7990	Decorator									
8000	Pendent RLM in colors, shallow dome, 12" diam., 100 watt	1 Elec	7.36	1.090	Ea.	80	67		147	188
8010	Regular dome, 12" diam., 100 watt		7.36	1.090		82.50	67		149.50	191
8020	16" diam., 200 watt		6.44	1.240		84	76		160	206
8030	18" diam., 300 watt		5.52	1.450		93.50	89		182.50	235
8100	Picture framing light		14.72	.543		98.50	33.50		132	158
8150	Miniature low voltage, recessed, pinhole		7.36	1.090		134	67		201	248
8160	Star		7.36	1.090		133	67		200	246
8170	Adjustable cone		7.36	1.090		169	67		236	286
8180	Eyeball		7.36	1.090		109	67		176	219
8190	Cone		7.36	1.090		129	67		196	242
8200	Coilex baffle		7.36	1.090		140	67		207	254
8210	Surface mounted, adjustable cylinder		7.36	1.090		140	67		207	254
8250	Chandeliers, incandescent									
8260	24" diam. x 42" high, 6 light candle	1 Elec	5.52	1.450	Ea.	440	89		529	610
8270	24" diam. x 42" high, 6 light candle w/glass shade		5.52	1.450		430	89		519	605
8280	17" diam. x 12" high, 8 light w/glass panels		7.36	1.090		282	67		349	410
8300	27" diam. x 29" high, 10 light bohemian lead crystal		3.68	2.170		460	133		593	705
8310	21" diam. x 9" high, 6 light sculptured ice crystal		7.36	1.090		420	67		487	560
8500	Accent lights, on floor or edge, 0.5 W low volt incandescent									
8520	incl. transformer & fastenings, based on 100' lengths									
8550	Lights in clear tubing, 12" OC	1 Elec	211.60	.038	L.F.	8.50	2.32		10.82	12.80
8560	6" OC		147.20	.054		11.10	3.33		14.43	17.15
8570	4" OC		119.60	.067		16.95	4.10		21.05	25

For customer support on your Electrical Change Order Costs with RSMeans data, call 800.448.8182.   **Pre-Installation Change Order Costs**

26 51 13 – Interior Lighting Fixtures, Lamps, and Ballasts

26 51 13.50 Interior Lighting Fixtures

		Crew	Daily Output	Labor-Hours	Unit	Material	2020 Bare Costs Labor	Equipment	Total	Total Incl O&P
8580	3" OC	1 Elec	115	.070	L.F.	18.85	4.27		23.12	27
8590	2" OC		92	.087		27.50	5.35		32.85	38
8600	Carpet, lights both sides 6" OC, in alum. extrusion		248.40	.032		25.50	1.98		27.48	31
8610	In bronze extrusion		248.40	.032		29	1.98		30.98	35
8620	Carpet-bare floor, lights 18" OC, in alum. extrusion		248.40	.032		20.50	1.98		22.48	25.50
8630	In bronze extrusion		248.40	.032		24	1.98		25.98	29.50
8640	Carpet edge-wall, lights 6" OC in alum. extrusion		248.40	.032		25.50	1.98		27.48	31
8650	In bronze extrusion		248.40	.032		29	1.98		30.98	35
8660	Bare floor, lights 18" OC, in alum. extrusion		276	.029		20.50	1.78		22.28	25
8670	In bronze extrusion		276	.029		24	1.78		25.78	29
8680	Bare floor conduit, alum. extrusion		276	.029		6.80	1.78		8.58	10.10
8690	In bronze extrusion		276	.029	↓	13.55	1.78		15.33	17.60
8700	Step edge to 36", lights 6" OC, in alum. extrusion		92	.087	Ea.	68.50	5.35		73.85	83
8710	In bronze extrusion		92	.087		71	5.35		76.35	86.50
8720	Step edge to 54", lights 6" OC, in alum. extrusion		92	.087		102	5.35		107.35	121
8730	In bronze extrusion		92	.087		108	5.35		113.35	127
8740	Step edge to 72", lights 6" OC, in alum. extrusion		92	.087		136	5.35		141.35	158
8750	In bronze extrusion		92	.087		149	5.35		154.35	172
8760	Connector, male		29.44	.272		2.53	16.65		19.18	28
8770	Female with pigtail		29.44	.272		5.30	16.65		21.95	31
8780	Clamps		368	.022		.51	1.33		1.84	2.55
8790	Transformers, 50 watt		7.36	1.090		100	67		167	210
8800	250 watt		3.68	2.170		320	133		453	550
8810	1,000 watt	↓	2.48	3.220	↓	350	198		548	680

26 51 13.55 Interior LED Fixtures

			Crew	Daily Output	Labor-Hours	Unit	Material	2020 Bare Costs Labor	Equipment	Total	Total Incl O&P
0010	**INTERIOR LED FIXTURES** Incl. lamps and mounting hardware										
0100	Downlight, recess mounted, 7.5" diameter, 25 watt	G	1 Elec	7.36	1.090	Ea.	340	67		407	475
0120	10" diameter, 36 watt	G		7.36	1.090		360	67		427	495
0160	cylinder, 10 watts	G		7.36	1.090		104	67		171	214
0180	20 watts	G		7.36	1.090		135	67		202	249
0900	Interior LED fixts, troffer, recess mounted, 2' x 2', 3,500K			7.82	1.020		129	62.50		191.50	234
0910	2' x 2', 4,000K			7.82	1.020		130	62.50		192.50	236
1000	Troffer, recess mounted, 2' x 4', 3,200 lumens	G		4.88	1.640		138	101		239	300
1010	4,800 lumens	G		4.60	1.740		150	107		257	325
1020	6,400 lumens	G		4.32	1.850		189	114		303	375
1100	Troffer retrofit lamp, 38 watt	G		19.32	.414		63.50	25.50		89	108
1110	60 watt	G		18.40	.435		156	26.50		182.50	211
1120	100 watt	G		16.56	.483		140	29.50		169.50	198
1200	Troffer, volumetric recess mounted, 2' x 2'	G		5.24	1.530		345	94		439	520
2000	Strip, surface mounted, one light bar 4' long, 3,500 K	G		7.82	1.020		305	62.50		367.50	435
2010	5,000 K	G		7.36	1.090		265	67		332	390
2020	Two light bar 4' long, 5,000 K	G		6.44	1.240		415	76		491	575
3000	Linear, suspended mounted, one light bar 4' long, 37 watt	G	↓	6.16	1.300		171	80		251	305
3010	One light bar 8' long, 74 watt	G	2 Elec	11.22	1.430		310	87.50		397.50	470
3020	Two light bar 4' long, 74 watt	G	1 Elec	5.24	1.530		335	94		429	505
3030	Two light bar 8' long, 148 watt	G	2 Elec	8.10	1.980		360	121		481	575
7000	Downlight, recess mtd., low profile, 4" diam., 9W, 3,000K		1 Elec	7.82	1.020		24.50	62.50		87	120
7010	4,000K			7.82	1.020		23	62.50		85.50	119
7020	5,000K		↓	7.82	1.020	↓	24	62.50		86.50	119

26 51 13 – Interior Lighting Fixtures, Lamps, and Ballasts

26 51 13.70 Residential Fixtures

	26 51 13.70 Residential Fixtures	Crew	Daily Output	Labor-Hours	Unit	Material	2020 Bare Costs Labor	Equipment	Total	Total Incl O&P
0010	**RESIDENTIAL FIXTURES**									
0400	Fluorescent, interior, surface, circline, 32 watt & 40 watt	1 Elec	18.40	.435	Ea.	164	26.50		190.50	220
0700	Shallow under cabinet, two 20 watt		14.72	.543		70.50	33.50		104	127
0900	Wall mounted, 4' L, two 32 watt T8, with baffle		9.20	.870		165	53.50		218.50	262
2000	Incandescent, exterior lantern, wall mounted, 60 watt		14.72	.543		61.50	33.50		95	117
2100	Post light, 150 W, with 7' post		3.68	2.170		291	133		424	520
2500	Lamp holder, weatherproof with 150 W PAR		14.72	.543		35.50	33.50		69	88.50
2550	With reflector and guard		11.04	.725		66	44.50		110.50	139
2600	Interior pendent, globe with shade, 150 W		18.40	.435		190	26.50		216.50	249

26 51 13.90 Ballast, Replacement HID

	26 51 13.90 Ballast, Replacement HID	Crew	Daily Output	Labor-Hours	Unit	Material	Labor	Equipment	Total	Total Incl O&P
0010	**BALLAST, REPLACEMENT HID**									
7510	Multi-tap 120/208/240/277 V									
7550	High pressure sodium, 70 watt	1 Elec	9.20	.870	Ea.	183	53.50		236.50	281
7560	100 watt		8.65	.925		130	57		187	228
7570	150 watt		8.28	.966		205	59.50		264.50	315
7580	250 watt		7.82	1.020		305	62.50		367.50	430
7590	400 watt		6.44	1.240		269	76		345	410
7600	1,000 watt		5.52	1.450		287	89		376	445
7610	Metal halide, 175 watt		7.36	1.090		97	67		164	207
7620	250 watt		7.36	1.090		128	67		195	241
7630	400 watt		6.44	1.240		158	76		234	287
7640	1,000 watt		5.52	1.450		258	89		347	415
7650	1,500 watt		4.60	1.740		245	107		352	430

26 51 19 – LED Interior Lighting

26 51 19.10 LED Interior Lighting

	26 51 19.10 LED Interior Lighting	Crew	Daily Output	Labor-Hours	Unit	Material	Labor	Equipment	Total	Total Incl O&P
0010	**LED INTERIOR LIGHTING**									
7000	Interior LED fixts, tape rope kit, 120V, 3' strip, 4W, w/5' cord, 3,000K	1 Elec	22.08	.362	Ea.	7.25	22		29.25	41
7010	5,000K		22.08	.362		10.95	22		32.95	45
7020	Interior LED fixts, tape rope kit, 120V, 6' strip, 8W, w/5' cord, 3,000K		21.16	.378		15.35	23		38.35	51.50
7030	5,000K		21.16	.378		22	23		45	59
7040	Interior LED fixts, tape rope kit, 120V, 13' strip, 15W, w/5' cord, 3,000K		20.24	.395		26	24.50		50.50	64.50
7050	5,000K		20.24	.395		28.50	24.50		53	67
7060	Interior LED fixts, tape rope kit, 120V, 20' strip, 22W, w/5' cord, 3,000K		19.32	.414		41	25.50		66.50	83
7070	5,000K		19.32	.414		41	25.50		66.50	83
7080	Interior LED fixts, tape rope kit, 120V, 33' strip, 37W, w/5' cord, 3,000K		18.40	.435		60.50	26.50		87	106
7090	5,000K		18.40	.435		60.50	26.50		87	106
7100	Interior LED fixts, tape rope kit, 120V, 50' strip, 55W, w/5' cord, 3,000K		17.48	.458		91.50	28		119.50	143
7110	5,000K		17.48	.458		131	28		159	186
7120	Interior LED tape light, starstrand, dimmable, 12V, super star, 2", 3,000K		23	.348		11.25	21.50		32.75	44.50
7130	Ultra star, 2", 3,000K		23	.348		14.05	21.50		35.55	47.50
7140	Rainbow star, 2", 3,000K		23	.348		11.25	21.50		32.75	44.50
7150	Interior LED tape light, starstrand, dimmable, 24V, elite star, 4", 35K		23	.348		9.80	21.50		31.30	43
7160	12", 35K		23	.348		21	21.50		42.50	55.50
7170	60", 35K		21.16	.378		98	23		121	143
7180	120", 35K		19.32	.414		123	25.50		148.50	173
7190	240", 35K		16.56	.483		297	29.50		326.50	370
7200	Interior LED tape lighting channel , 36", AL flex, starstrand, elite star		44.16	.181		21	11.10		32.10	39.50

Pre-Installation Change Order Costs

26 52 Safety Lighting

26 52 13 – Emergency and Exit Lighting

26 52 13.10 Emergency Lighting and Battery Units

		Crew	Daily Output	Labor-Hours	Unit	Material	2020 Bare Costs Labor	Equipment	Total	Total Incl O&P
0010	**EMERGENCY LIGHTING AND BATTERY UNITS**									
0300	Emergency light units, battery operated									
0350	Twin sealed beam light, 25 W, 6 V each									
0500	Lead battery operated	1 Elec	3.68	2.170	Ea.	140	133		273	350
0700	Nickel cadmium battery operated		3.68	2.170		330	133		463	565
0780	Additional remote mount, sealed beam, 25 W 6 V		24.56	.326		29	20		49	62
0781	Additional remote mount, sealed beam, 25 W 6 V		24.56	.326		29	20		49	62
0790	Twin sealed beam light, 25 W 6 V each		24.56	.326		62.50	20		82.50	98.50
0900	Self-contained fluorescent lamp pack	↓	9.20	.870	↓	181	53.50		234.50	279

26 52 13.16 Exit Signs

		Crew	Daily Output	Labor-Hours	Unit	Material	2020 Bare Costs Labor	Equipment	Total	Total Incl O&P
0010	**EXIT SIGNS**									
0080	Exit light ceiling or wall mount, incandescent, single face	1 Elec	7.36	1.090	Ea.	73	67		140	180
0100	Double face		6.16	1.300		50.50	80		130.50	175
0120	Explosion proof		3.50	2.290		585	140		725	850
0150	Fluorescent, single face		7.36	1.090		92.50	67		159.50	202
0160	Double face		6.16	1.300		76.50	80		156.50	204
0200	LED standard, single face	G	7.36	1.090		49.50	67		116.50	154
0220	Double face	G	6.16	1.300		52	80		132	176
0230	LED vandal-resistant, single face	G	6.69	1.200		218	73.50		291.50	350
0240	LED w/battery unit, single face	G	4.05	1.980		192	121		313	390
0260	Double face	G	3.68	2.170		223	133		356	445
0262	LED w/battery unit, vandal-resistant, single face	G	4.05	1.980		258	121		379	465
0270	Combination emergency light units and exit sign		3.68	2.170		179	133		312	395
0290	LED retrofit kits	G	55.20	.145		52.50	8.90		61.40	71
1500	Exit sign, 12 V, 1 face, remote end mounted (Type 602A1)	R-19	16.56	1.210		73	74.50		147.50	192
1780	With emergency battery, explosion proof	"	7.08	2.820	↓	4,675	173		4,848	5,375

26 54 Classified Location Lighting

26 54 13 – Incandescent Classified Location Lighting

26 54 13.20 Explosion Proof

		Crew	Daily Output	Labor-Hours	Unit	Material	2020 Bare Costs Labor	Equipment	Total	Total Incl O&P
0010	**EXPLOSION PROOF**, incl. lamps, mounting hardware and connections									
6310	Metal halide with ballast, ceiling, surface mounted, 175 watt	1 Elec	2.67	3	Ea.	1,400	184		1,584	1,825
6320	250 watt	"	2.48	3.220		1,675	198		1,873	2,150
6330	400 watt	2 Elec	4.42	3.620		1,800	222		2,022	2,300
6340	Ceiling, pendent mounted, 175 watt	1 Elec	2.39	3.340		1,325	205		1,530	1,750
6350	250 watt	"	2.21	3.620		1,600	222		1,822	2,100
6360	400 watt	2 Elec	3.86	4.140		1,725	254		1,979	2,275
6370	Wall, surface mounted, 175 watt	1 Elec	2.67	3		1,500	184		1,684	1,925
6380	250 watt	"	2.48	3.220		1,775	198		1,973	2,250
6390	400 watt	2 Elec	4.42	3.620		1,900	222		2,122	2,425
6400	High pressure sodium, ceiling surface mounted, 70 watt	1 Elec	2.76	2.900		2,275	178		2,453	2,800
6410	100 watt		2.76	2.900		2,375	178		2,553	2,875
6420	150 watt		2.48	3.220		2,500	198		2,698	3,050
6430	Pendent mounted, 70 watt		2.48	3.220		2,175	198		2,373	2,700
6440	100 watt		2.48	3.220		2,275	198		2,473	2,800
6450	150 watt		2.21	3.620		2,050	222		2,272	2,575
6460	Wall mounted, 70 watt		2.76	2.900		2,450	178		2,628	2,950
6470	100 watt		2.76	2.900		2,575	178		2,753	3,100
6480	150 watt		2.48	3.220		2,600	198		2,798	3,150
6510	Incandescent, ceiling mounted, 200 watt		3.68	2.170		1,650	133		1,783	2,000
6520	Pendent mounted, 200 watt		3.22	2.480	↓	1,400	152		1,552	1,775

26 54 Classified Location Lighting

26 54 13 – Incandescent Classified Location Lighting

26 54 13.20 Explosion Proof	Crew	Daily Output	Labor-Hours	Unit	Material	2020 Bare Costs Labor	Equipment	Total	Total Incl O&P
6530	Wall mounted, 200 watt	1 Elec	3.68	2.170	Ea.	1,625	133		1,758	1,975
6600	Fluorescent, RS, 4' long, ceiling mounted, two 40 watt		2.48	3.220		4,975	198		5,173	5,775
6610	Three 40 watt		2.02	3.950		7,200	242		7,442	8,275
6620	Four 40 watt		1.75	4.580		9,250	281		9,531	10,600
6630	Pendent mounted, two 40 watt		2.12	3.780		5,750	232		5,982	6,675
6640	Three 40 watt		1.75	4.580		8,175	281		8,456	9,425
6650	Four 40 watt		1.56	5.120		10,800	315		11,115	12,300

26 55 Special Purpose Lighting

26 55 33.10 Warning Beacons

		Crew	Daily Output	Labor-Hours	Unit	Material	2020 Bare Costs Labor	Equipment	Total	Total Incl O&P
0010	**WARNING BEACONS**									
0015	Surface mount with colored or clear lens									
0100	Rotating beacon									
0110	120V, 40 watt halogen	1 Elec	3.22	2.480	Ea.	118	152		270	355
0120	24V, 20 watt halogen	"	3.22	2.480	"	279	152		431	530
0200	Steady beacon									
0210	120V, 40 watt halogen	1 Elec	3.22	2.480	Ea.	120	152		272	360
0220	24V, 20 watt		3.22	2.480		124	152		276	365
0230	12V DC, incandescent		3.22	2.480		125	152		277	365
0300	Flashing beacon									
0310	120V, 40 watt halogen	1 Elec	3.22	2.480	Ea.	117	152		269	355
0320	24V, 20 watt halogen		3.22	2.480		118	152		270	355
0410	12V DC with two 6V lantern batteries		6.44	1.240		119	76		195	244

26 55 59 – Display Lighting

26 55 59.10 Track Lighting

		Crew	Daily Output	Labor-Hours	Unit	Material	2020 Bare Costs Labor	Equipment	Total	Total Incl O&P
0010	**TRACK LIGHTING**									
0080	Track, 1 circuit, 4' section	1 Elec	6.16	1.300	Ea.	43	80		123	167
0100	8' section	2 Elec	9.75	1.640		76	101		177	234
0200	12' section	"	8.10	1.980		112	121		233	305
0300	3 circuits, 4' section	1 Elec	6.16	1.300		117	80		197	247
0400	8' section	2 Elec	9.75	1.640		130	101		231	293
0500	12' section	"	8.10	1.980		168	121		289	365
1000	Feed kit, surface mounting	1 Elec	14.72	.543		15.85	33.50		49.35	67
1100	End cover		22.08	.362		8.30	22		30.30	42
1200	Feed kit, stem mounting, 1 circuit		14.72	.543		52.50	33.50		86	107
1300	3 circuit		14.72	.543		52.50	33.50		86	107
2000	Electrical joiner, for continuous runs, 1 circuit		29.44	.272		37.50	16.65		54.15	66
2100	3 circuit		29.44	.272		76.50	16.65		93.15	110
2200	Fixtures, spotlight, 75 W PAR halogen		14.72	.543		48	33.50		81.50	102
2210	50 W MR16 halogen		14.72	.543		175	33.50		208.50	242
3000	Wall washer, 250 W tungsten halogen		14.72	.543		134	33.50		167.50	197
3100	Low voltage, 25/50 W, 1 circuit		14.72	.543		135	33.50		168.50	198
3120	3 circuit		14.72	.543		193	33.50		226.50	263

26 55 63 – Detention Lighting

26 55 63.10 Detention Lighting Fixtures

		Crew	Daily Output	Labor-Hours	Unit	Material	2020 Bare Costs Labor	Equipment	Total	Total Incl O&P
0010	**DETENTION LIGHTING FIXTURES**									
3000	Fluorescent vandal resistant light fixture, high abuse troffer	1 Elec	4.97	1.610	Ea.	244	99		343	415
3010	50" L x 8" W x 4" H		4.97	1.610		216	99		315	385
3020	32W		4.97	1.610		310	99		409	485
3030	T8/32W		4.97	1.610		325	99		424	500

Pre-Installation Change Order Costs

26 55 Special Purpose Lighting

26 55 63 – Detention Lighting

26 55 63.10 Detention Lighting Fixtures	Crew	Daily Output	Labor-Hours	Unit	Material	2020 Bare Costs Labor	Equipment	Total	Total Incl O&P
3040	No lamp	1 Elec	5.15	1.550	Ea.	370	95		465	545
3050	Lamp included		4.97	1.610		425	99		524	615
3060	16 ga. cold rolled steel		4.97	1.610		340	99		439	520
3070	16 ga. cold rolled steel, ceiling/wall mount		4.97	1.610		345	99		444	525
3080	w/lamp, white		4.97	1.610		269	99		368	445
3090	wo/lamp, bronze		5.15	1.550		245	95		340	410
3100	w/lamp 6" L x 4-1/2" W x 8-1/2" H		4.97	1.610		85.50	99		184.50	241
3110	w/lamp 5-7/8" W x 4-1/2" D x 8-1/2" H clear		4.97	1.610		67	99		166	221
3120	w/lamp white		4.97	1.610		540	99		639	740
3130	Tall wallpack, w/lamp, bronze		4.97	1.610		128	99		227	288
3140	Tall wallpack, w/lamp, white		4.97	1.610		129	99		228	289

26 56 Exterior Lighting

26 56 13 – Lighting Poles and Standards

26 56 13.10 Lighting Poles

		Crew	Daily Output	Labor-Hours	Unit	Material	2020 Bare Costs Labor	Equipment	Total	Total Incl O&P
0010	**LIGHTING POLES**									
0100	Exterior, light poles, concrete, 30' above 5' below, 13.5" Base, 5.5" Tip	2 Elec	4.23	3.780	Ea.	1,575	232		1,807	2,075
0110	39' above 6' below, 15.5" Base, 5.25" Tip		4.23	3.780		1,850	232		2,082	2,400
0120	43' above 7' below, 17.25" Base, 6.5" Tip		4.23	3.780		1,875	232		2,107	2,425
0130	43' above 7' below, 19.5" Base, 8.25" Tip		4.23	3.780		1,900	232		2,132	2,450
2800	Light poles, anchor base									
2820	not including concrete bases									
2840	Aluminum pole, 8' high	1 Elec	3.68	2.170	Ea.	810	133		943	1,100
2850	10' high		3.68	2.170		855	133		988	1,150
2860	12' high		3.50	2.290		890	140		1,030	1,200
2870	14' high		3.13	2.560		920	157		1,077	1,225
2880	16' high		2.76	2.900		1,025	178		1,203	1,400
3000	20' high	R-3	2.67	7.500		1,100	460	71	1,631	1,975
3200	30' high		2.39	8.360		2,075	510	79	2,664	3,125
3400	35' high		2.12	9.450		2,275	580	89.50	2,944.50	3,450
3600	40' high		1.84	10.870		2,750	665	103	3,518	4,125
3800	Bracket arms, 1 arm	1 Elec	7.36	1.090		141	67		208	255
4000	2 arms		7.36	1.090		280	67		347	410
4200	3 arms		4.88	1.640		425	101		526	615
4400	4 arms		4.42	1.810		565	111		676	790
4500	Steel pole, galvanized, 8' high		3.50	2.290		675	140		815	950
4510	10' high		3.40	2.350		700	144		844	990
4520	12' high		3.13	2.560		755	157		912	1,075
4530	14' high		2.85	2.810		805	172		977	1,150
4540	16' high		2.67	3		855	184		1,039	1,225
4550	18' high		2.48	3.220		900	198		1,098	1,275
4600	20' high	R-3	2.39	8.360		1,275	510	79	1,864	2,250
4800	30' high		2.12	9.450		1,275	580	89.50	1,944.50	2,350
5000	35' high		2.02	9.880		1,550	605	93.50	2,248.50	2,725
5200	40' high		1.56	12.790		1,725	780	121	2,626	3,175
5400	Bracket arms, 1 arm	1 Elec	7.36	1.090		214	67		281	335
5600	2 arms		7.36	1.090		310	67		377	440
5800	3 arms		4.88	1.640		215	101		316	385
6000	4 arms		4.88	1.640		350	101		451	535
6100	Fiberglass pole, 1 or 2 fixtures, 20' high	R-3	3.68	5.430		860	330	51.50	1,241.50	1,500
6200	30' high		3.31	6.040		980	370	57	1,407	1,700

26 56 Exterior Lighting

26 56 13 - Lighting Poles and Standards

26 56 13.10 Lighting Poles

		Crew	Daily Output	Labor-Hours	Unit	Material	2020 Bare Costs Labor	Equipment	Total	Total Incl O&P
6300	35' high	R-3	2.94	6.790	Ea.	1,675	415	64	2,154	2,550
6400	40' high		2.58	7.760		1,850	475	73.50	2,398.50	2,800
6420	Wood pole, 4-1/2" x 5-1/8", 8' high	1 Elec	5.52	1.450		385	89		474	555
6430	10' high		5.52	1.450		450	89		539	630
6440	12' high		5.24	1.530		570	94		664	765
6450	15' high		4.60	1.740		665	107		772	890
6460	20' high		3.68	2.170		810	133		943	1,100
6461	Light poles,anchor base,w/o conc base, pwdr ct stl, 16' H	2 Elec	2.85	5.610		855	345		1,200	1,450
6462	20' high	R-3	2.67	7.500		1,275	460	71	1,806	2,175
6463	30' high		2.12	9.450		1,275	580	89.50	1,944.50	2,350
6464	35' high		2.21	9.060		1,550	555	85.50	2,190.50	2,650
6465	25' high		2.48	8.050		1,275	490	76	1,841	2,225
6470	Light pole conc base, max 6' buried, 2' exposed, 18" diam., average cost	C-6	5.52	8.700		190	380	9.75	579.75	790
7300	Transformer bases, not including concrete bases									
7320	Maximum pole size, steel, 40' high	1 Elec	1.84	4.350	Ea.	1,600	267		1,867	2,175
7340	Cast aluminum, 30' high		2.76	2.900		850	178		1,028	1,200
7350	40' high		2.30	3.480		1,300	214		1,514	1,750

26 56 19 - LED Exterior Lighting

2000	Exterior fixtures, LED roadway, type 2, 50W, neutral slipfitter bronze	2 Elec	4.78	3.340	Ea.	660	205		865	1,025
2010	3, 50W, neutral slipfitter bronze		4.78	3.340		635	205		840	1,000
2020	4, 50W, neutral slipfitter bronze		4.78	3.340		635	205		840	1,000
2030	Exterior fixtures, LED roadway, type 2, 78W, neutral slipfitter bronze		4.78	3.340		630	205		835	995
2040	3, 78W, neutral slipfitter bronze		4.78	3.340		675	205		880	1,050
2050	4, 78W, neutral slipfitter bronze		4.78	3.340		660	205		865	1,025
2060	Exterior fixtures, LED roadway, type 2, 105W, neutral slipfitter bronze		4.78	3.340		820	205		1,025	1,200
2070	3, 105W, neutral slipfitter bronze		4.78	3.340		785	205		990	1,175
2080	4, 105W, neutral slipfitter bronze		4.78	3.340		730	205		935	1,100
2090	Exterior fixtures, LED roadway, type 2, 125W, neutral slipfitter bronze		4.78	3.340		790	205		995	1,175

26 56 21 - HID Exterior Lighting

26 56 21.20 Roadway Luminaire

0010	**ROADWAY LUMINAIRE**									
2650	Roadway area luminaire, low pressure sodium, 135 watt	1 Elec	1.84	4.350	Ea.	835	267		1,102	1,325
2700	180 watt	"	1.84	4.350		995	267		1,262	1,500
2750	Metal halide, 400 watt	2 Elec	4.05	3.950		690	242		932	1,125
2760	1,000 watt		3.68	4.350		775	267		1,042	1,250
2780	High pressure sodium, 400 watt		4.05	3.950		825	242		1,067	1,275
2790	1,000 watt		3.68	4.350		940	267		1,207	1,425

26 56 23 - Area Lighting

26 56 23.10 Exterior Fixtures

0010	**EXTERIOR FIXTURES** With lamps									
0200	Wall mounted, incandescent, 100 watt	1 Elec	7.36	1.090	Ea.	51.50	67		118.50	156
0400	Quartz, 500 watt		4.88	1.640		62	101		163	219
0420	1,500 watt		3.86	2.070		108	127		235	305
1100	Wall pack, low pressure sodium, 35 watt		3.68	2.170		205	133		338	425
1150	55 watt		3.68	2.170		241	133		374	465
1160	High pressure sodium, 70 watt		3.68	2.170		197	133		330	415
1170	150 watt		3.68	2.170		218	133		351	435
1175	High pressure sodium, 250 watt		3.68	2.170		218	133		351	435
1180	Metal halide, 175 watt		3.68	2.170		221	133		354	440
1190	250 watt		3.68	2.170		251	133		384	475
1195	400 watt		3.68	2.170		410	133		543	655

Pre-Installation Change Order Costs

26 56 Exterior Lighting

26 56 23 – Area Lighting

26 56 23.10 Exterior Fixtures	Crew	Daily Output	Labor-Hours	Unit	Material	2020 Bare Costs Labor	Equipment	Total	Total Incl O&P
1250	Induction lamp, 40 watt	1 Elec	3.68	2.170	Ea.	495	133		628	745
1260	80 watt		3.68	2.170		590	133		723	845
1278	LED, poly lens, 26 watt		3.68	2.170		310	133		443	540
1280	110 watt		3.68	2.170		830	133		963	1,100
1500	LED, glass lens, 13 watt		3.68	2.170		310	133		443	540

26 56 26 – Landscape Lighting

26 56 26.20 Landscape Fixtures

		Crew	Daily Output	Labor-Hours	Unit	Material	2020 Bare Costs Labor	Equipment	Total	Total Incl O&P
0010	**LANDSCAPE FIXTURES**									
7380	Landscape recessed uplight, incl. housing, ballast, transformer									
7390	& reflector, not incl. conduit, wire, trench									
7420	Incandescent, 250 watt	1 Elec	4.60	1.740	Ea.	665	107		772	890
7440	Quartz, 250 watt		4.60	1.740		630	107		737	855
7460	500 watt		3.68	2.170		650	133		783	915

26 56 26.50 Landscape LED Fixtures

		Crew	Daily Output	Labor-Hours	Unit	Material	2020 Bare Costs Labor	Equipment	Total	Total Incl O&P
0010	**LANDSCAPE LED FIXTURES**									
0100	12 volt alum bullet hooded-BLK	1 Elec	4.60	1.740	Ea.	107	107		214	277
0200	12 volt alum bullet hooded-BRZ		4.60	1.740		107	107		214	277
0300	12 volt alum bullet hooded-GRN		4.60	1.740		107	107		214	277
1000	12 volt alum large bullet hooded-BLK		4.60	1.740		79	107		186	246
1100	12 volt alum large bullet hooded-BRZ		4.60	1.740		79	107		186	246
1200	12 volt alum large bullet hooded-GRN		4.60	1.740		79	107		186	246
2000	12 volt large bullet landscape light fixture		4.60	1.740		79	107		186	246
2100	12 volt alum light large bullet		4.60	1.740		79	107		186	246
2200	12 volt alum bullet light		4.60	1.740		79	107		186	246

26 56 33 – Walkway Lighting

26 56 33.10 Walkway Luminaire

		Crew	Daily Output	Labor-Hours	Unit	Material	2020 Bare Costs Labor	Equipment	Total	Total Incl O&P
0010	**WALKWAY LUMINAIRE**									
6500	Bollard light, lamp & ballast, 42" high with polycarbonate lens									
6800	Metal halide, 175 watt	1 Elec	2.76	2.900	Ea.	985	178		1,163	1,350
6900	High pressure sodium, 70 watt		2.76	2.900		950	178		1,128	1,325
7000	100 watt		2.76	2.900		950	178		1,128	1,325
7100	150 watt		2.76	2.900		925	178		1,103	1,300
7200	Incandescent, 150 watt		2.76	2.900		675	178		853	1,000
7810	Walkway luminaire, square 16", metal halide 250 watt		2.48	3.220		770	198		968	1,150
7820	High pressure sodium, 70 watt		2.76	2.900		880	178		1,058	1,225
7830	100 watt		2.76	2.900		895	178		1,073	1,250
7840	150 watt		2.76	2.900		895	178		1,073	1,250
7850	200 watt		2.76	2.900		900	178		1,078	1,250
7910	Round 19", metal halide, 250 watt		2.48	3.220		1,250	198		1,448	1,675
7920	High pressure sodium, 70 watt		2.76	2.900		1,375	178		1,553	1,775
7930	100 watt		2.76	2.900		1,375	178		1,553	1,775
7940	150 watt		2.76	2.900		1,375	178		1,553	1,775
7950	250 watt		2.48	3.220		1,300	198		1,498	1,725
8000	Sphere 14" opal, incandescent, 200 watt		3.68	2.170		355	133		488	590
8020	Sphere 18" opal, incandescent, 300 watt		3.22	2.480		430	152		582	695
8040	Sphere 16" clear, high pressure sodium, 70 watt		2.76	2.900		740	178		918	1,075
8050	100 watt		2.76	2.900		790	178		968	1,125
8100	Cube 16" opal, incandescent, 300 watt		3.22	2.480		470	152		622	740
8120	High pressure sodium, 70 watt		2.76	2.900		685	178		863	1,025
8130	100 watt		2.76	2.900		705	178		883	1,050
8230	Lantern, high pressure sodium, 70 watt		2.76	2.900		615	178		793	940

26 56 33 – Walkway Lighting

26 56 33.10 Walkway Luminaire

		Crew	Daily Output	Labor-Hours	Unit	Material	2020 Bare Costs Labor	Equipment	Total	Total Incl O&P
8240	100 watt	1 Elec	2.76	2.900	Ea.	660	178		838	995
8250	150 watt		2.76	2.900		620	178		798	945
8260	250 watt		2.48	3.220		865	198		1,063	1,250
8270	Incandescent, 300 watt		3.22	2.480		455	152		607	730
8330	Reflector 22" w/globe, high pressure sodium, 70 watt		2.76	2.900		555	178		733	875
8340	100 watt		2.76	2.900		565	178		743	885
8350	150 watt		2.76	2.900		570	178		748	895
8360	250 watt		2.48	3.220		730	198		928	1,100
0600	LED bollard pole, 6" diam. x 36" H, cast alum, surf. mtd., 8W, 6,000K, 120V		7.82	1.020		385	62.50		447.50	520
0610	6" diam. x 42" H, cast alum, surf. mtd., 12W, 5,100K,120V		7.82	1.020		565	62.50		627.50	715
0620	6" diam. x 42" H, cast alum, surf. mtd., 18W, 5,100K,120V		7.82	1.020		645	62.50		707.50	805
0630	6" diam. x 42" H, cast alum, surf. mtd., 24W, 5,100K,120V		7.82	1.020		505	62.50		567.50	650
0640	9.5" square x 42" H, cast resin, surf. mtd., A19, 9W, 120V		7.82	1.020		610	62.50		672.50	765
0650	9.75" diam. x 42" H, cast resin, side mount, dome, 9W, 120V		7.82	1.020		695	62.50		757.50	860
0660	9.75" diam. x 43" H, cast resin, burial base, dome, 9W, 120V		7.82	1.020		695	62.50		757.50	860
0670	8" diam. x 36" H, concrete security base, w/rebar, 8W, HID		7.36	1.090		1,025	67		1,092	1,225
0680	10" diam. x 49" H, concrete base, w/Sch 40 Steel pipe, 8W		6.21	1.290		1,175	79		1,254	1,400
0690	12" diam. x 48" H, concrete base, w/Sch 40 Steel pipe, 8W		6.21	1.290		1,325	79		1,404	1,600
0700	8" diam. x 36" H, concrete, cast aluminum dome, 90 degree, 4.1W		6.44	1.240		1,025	76		1,101	1,275
0710	8" diam. x 36" H, concrete, cast aluminum dome, 180 degree, 4.1W		6.44	1.240		1,275	76		1,351	1,525
0720	8" diam. x 48" H, concrete, cast aluminum dome, 360 degree, 4.1W		6.44	1.240		1,575	76		1,651	1,850

26 56 36 – Flood Lighting

26 56 36.20 Floodlights

		Crew	Daily Output	Labor-Hours	Unit	Material	2020 Bare Costs Labor	Equipment	Total	Total Incl O&P
0010	**FLOODLIGHTS** with ballast and lamp,									
1290	floor mtd, mount with swivel bracket									
1300	Induction lamp, 40 watt	1 Elec	2.76	2.900	Ea.	565	178		743	885
1310	80 watt		2.76	2.900		750	178		928	1,100
1320	150 watt		2.76	2.900		1,350	178		1,528	1,775
1400	Pole mounted, pole not included									
1950	Metal halide, 175 watt	1 Elec	2.48	3.220	Ea.	213	198		411	530
2000	400 watt	2 Elec	4.05	3.950		219	242		461	600
2200	1,000 watt		3.68	4.350		900	267		1,167	1,375
2210	1,500 watt		3.40	4.700		425	288		713	895
2250	Low pressure sodium, 55 watt	1 Elec	2.48	3.220		460	198		658	800
2270	90 watt		1.84	4.350		620	267		887	1,075
2290	180 watt		1.84	4.350		680	267		947	1,150
2340	High pressure sodium, 70 watt		2.48	3.220		254	198		452	575
2360	100 watt		2.48	3.220		261	198		459	580
2380	150 watt		2.48	3.220		293	198		491	620
2400	400 watt	2 Elec	4.05	3.950		320	242		562	715
2600	1,000 watt	"	3.68	4.350		635	267		902	1,100

For customer support on your Electrical Change Order Costs with RSMeans data, call 800.448.8182.

Pre-Installation Change Order Costs

26 61 13 – Lighting Accessories

26 61 13.30 Fixture Whips		Crew	Daily Output	Labor-Hours	Unit	Material	2020 Bare Costs Labor	Equipment	Total	Total Incl O&P
0010	**FIXTURE WHIPS**									
0080	3/8" Greenfield, 2 connectors, 6' long									
0100	TFFN wire, three #18	1 Elec	29.44	.272	Ea.	9.20	16.65		25.85	35
0150	Four #18		25.76	.311		9.90	19.05		28.95	39.50
0200	Three #16		29.44	.272		9.50	16.65		26.15	35.50
0250	Four #16		25.76	.311		10.35	19.05		29.40	40
0300	THHN wire, three #14		29.44	.272		6.55	16.65		23.20	32
0350	Four #14		25.76	.311		9.15	19.05		28.20	38.50
0360	Three #12		29.44	.272		12.60	16.65		29.25	39

26 61 23 – Lamps Applications

26 61 23.10 Lamps

		Crew	Daily Output	Labor-Hours	Unit	Material	2020 Bare Costs Labor	Equipment	Total	Total Incl O&P
0010	**LAMPS**									
0080	Fluorescent, rapid start, cool white, 2' long, 20 watt	1 Elec	.92	8.700	C	365	535		900	1,200
0100	4' long, 40 watt		.83	9.660		305	595		900	1,225
0120	3' long, 30 watt		.83	9.660		445	595		1,040	1,375
0125	3' long, 25 watt energy saver [G]		.83	9.660		1,425	595		2,020	2,425
0150	U-40 watt		.74	10.870		1,250	665		1,915	2,375
0155	U-34 watt energy saver [G]		.74	10.870		12.95	665		677.95	1,000
0170	4' long, 34 watt energy saver [G]		.83	9.660		870	595		1,465	1,825
0176	2' long, T8, 17 watt energy saver [G]		.92	8.700		385	535		920	1,225
0178	3' long, T8, 25 watt energy saver [G]		.83	9.660		410	595		1,005	1,325
0180	4' long, T8, 32 watt energy saver [G]		.83	9.660		237	595		832	1,150
0200	Slimline, 4' long, 40 watt		.83	9.660		1,150	595		1,745	2,125
0210	4' long, 30 watt energy saver [G]		.83	9.660		1,150	595		1,745	2,125
0300	8' long, 75 watt		.74	10.870		1,125	665		1,790	2,225
0350	8' long, 60 watt energy saver [G]		.74	10.870		435	665		1,100	1,475
0400	High output, 4' long, 60 watt		.83	9.660		670	595		1,265	1,625
0410	8' long, 95 watt energy saver [G]		.74	10.870		665	665		1,330	1,725
0500	8' long, 110 watt		.74	10.870		665	665		1,330	1,725
0512	2' long, T5, 14 watt energy saver [G]		.92	8.700		196	535		731	1,000
0514	3' long, T5, 21 watt energy saver [G]		.83	9.660		190	595		785	1,100
0516	4' long, T5, 28 watt energy saver [G]		.83	9.660		190	595		785	1,100
0517	4' long, T5, 54 watt energy saver [G]		.83	9.660		565	595		1,160	1,500
0520	Very high output, 4' long, 110 watt		.83	9.660		1,775	595		2,370	2,825
0525	8' long, 195 watt energy saver [G]		.64	12.420		1,525	760		2,285	2,800
0550	8' long, 215 watt		.64	12.420		1,450	760		2,210	2,725
0554	Full spectrum, 4' long, 60 watt		.83	9.660		725	595		1,320	1,675
0556	6' long, 85 watt		.83	9.660		815	595		1,410	1,775
0558	8' long, 110 watt		.74	10.870		2,150	665		2,815	3,375
0560	Twin tube compact lamp [G]		.83	9.660		475	595		1,070	1,400
0570	Double twin tube compact lamp [G]		.74	10.870		1,050	665		1,715	2,150
0600	Mercury vapor, mogul base, deluxe white, 100 watt		.28	28.990		5,675	1,775		7,450	8,900
0650	175 watt		.28	28.990		3,000	1,775		4,775	5,950
0700	250 watt		.28	28.990		5,500	1,775		7,275	8,700
0800	400 watt		.28	28.990		5,050	1,775		6,825	8,200
0900	1,000 watt		.18	43.480		14,000	2,675		16,675	19,400
1000	Metal halide, mogul base, 175 watt		.28	28.990		1,150	1,775		2,925	3,925
1100	250 watt		.28	28.990		1,800	1,775		3,575	4,650
1200	400 watt		.28	28.990		2,200	1,775		3,975	5,075
1300	1,000 watt		.18	43.480		4,300	2,675		6,975	8,725
1320	1,000 watt, 125,000 initial lumens		.18	43.480		17,700	2,675		20,375	23,500
1330	1,500 watt		.18	43.480		3,925	2,675		6,600	8,275

26 61 Lighting Systems and Accessories

26 61 23 – Lamps Applications

26 61 23.10 Lamps

	26 61 23.10 Lamps	Crew	Daily Output	Labor-Hours	Unit	Material	2020 Bare Costs Labor	Equipment	Total	Total Incl O&P
1350	High pressure sodium, 70 watt	1 Elec	.28	28.990	C	1,875	1,775		3,650	4,700
1360	100 watt		.28	28.990		1,925	1,775		3,700	4,775
1370	150 watt		.28	28.990		1,800	1,775		3,575	4,625
1380	250 watt		.28	28.990		2,150	1,775		3,925	5,025
1400	400 watt		.28	28.990		1,800	1,775		3,575	4,625
1450	1,000 watt		.18	43.480		4,950	2,675		7,625	9,425
1500	Low pressure sodium, 35 watt		.28	28.990		17,400	1,775		19,175	21,900
1550	55 watt		.28	28.990		19,900	1,775		21,675	24,600
1600	90 watt		.28	28.990		7,000	1,775		8,775	10,400
1650	135 watt		.18	43.480		28,900	2,675		31,575	35,800
1700	180 watt		.18	43.480		43,300	2,675		45,975	51,500
1750	Quartz line, clear, 500 watt		1.01	7.910		820	485		1,305	1,625
1760	1,500 watt		.18	43.480		2,325	2,675		5,000	6,525
1762	Spot, MR 16, 50 watt		1.20	6.690		1,150	410		1,560	1,875
1770	Tungsten halogen, T4, 400 watt		1.01	7.910		4,225	485		4,710	5,375
1775	T3, 1,200 watt		.28	28.990		5,400	1,775		7,175	8,575
1778	PAR 30, 50 watt		1.20	6.690		1,175	410		1,585	1,875
1780	PAR 38, 90 watt		1.20	6.690		10,200	410		10,610	11,800
1800	Incandescent, interior, A21, 100 watt		1.47	5.430		2,975	335		3,310	3,775
1900	A21, 150 watt		1.47	5.430		17,600	335		17,935	19,800
2000	A23, 200 watt		1.47	5.430		355	335		690	885
2200	PS 35, 300 watt		1.47	5.430		1,050	335		1,385	1,675
2210	PS 35, 500 watt		1.47	5.430		1,650	335		1,985	2,325
2230	PS 52, 1,000 watt		1.20	6.690		2,675	410		3,085	3,525
2240	PS 52, 1,500 watt		1.20	6.690		7,250	410		7,660	8,575
2300	R30, 75 watt		1.20	6.690		765	410		1,175	1,450
2400	R40, 100 watt		1.20	6.690		765	410		1,175	1,450
2500	Exterior, PAR 38, 75 watt		1.20	6.690		2,000	410		2,410	2,800
2600	PAR 38, 150 watt		1.20	6.690		2,250	410		2,660	3,075
2700	PAR 46, 200 watt		1.01	7.910		4,100	485		4,585	5,250
2800	PAR 56, 300 watt		1.01	7.910		2,800	485		3,285	3,825
3000	Guards, fluorescent lamp, 4' long		92	.087	Ea.	15.40	5.35		20.75	25
3200	8' long		82.80	.097	"	31	5.95		36.95	43

26 61 23.55 LED Lamps

	LED LAMPS	Crew	Daily Output	Labor-Hours	Unit	Material	2020 Bare Costs Labor	Equipment	Total	Total Incl O&P
0010	**LED LAMPS**									
0100	LED lamp, interior, shape A60, equal to 60 W G	1 Elec	147.20	.054	Ea.	20.50	3.33		23.83	27.50
0110	7 W LED decorative c, ca, f, g shape		147.20	.054		36.50	3.33		39.83	45
0205	LED lamp, interior, globe		147.20	.054		17.10	3.33		20.43	24
0210	2.2 W LED LMP		147.20	.054		70	3.33		73.33	82
0220	2.2 W LED replacement decorative lamp		147.20	.054		9.65	3.33		12.98	15.55
0230	3.5 W LED replacement decorative lamp		147.20	.054		11.20	3.33		14.53	17.30
0240	4.5 W 120V LED replacement decorative lamp		147.20	.054		13.45	3.33		16.78	19.75
0250	4.5 W 120V LED, 2700k replacement decorative lamp		147.20	.054		11.85	3.33		15.18	18
0260	4.9 W 120V LED, 2700k replacement decorative lamp candelabra base		147.20	.054		10.65	3.33		13.98	16.65
0270	4.9 W 120V LED, 3000k replacement decorative lamp candelabra base		147.20	.054		9.95	3.33		13.28	15.90
0280	5 W LED PAR 20 parabolic reflector lamp FL		128.80	.062		63.50	3.81		67.31	75.50
0300	Globe earth, equal to 100 W G		128.80	.062		29.50	3.81		33.31	38
0305	7 W LED reflector lamp 3000k DIM		128.80	.062		61.50	3.81		65.31	73
0310	10 W omni LED warm white light bulb E26 medium base 120 volt card		128.80	.062		18.45	3.81		22.26	26
0315	7 W LED reflector lamp WFL 2700k DIM		128.80	.062		61.50	3.81		65.31	73
0320	9 W omni LED warm white light bulb E26 medium base 120 volt card		128.80	.062		9.85	3.81		13.66	16.45
0500	8 W LED, A19 lamp, equal to 40 W		128.80	.062		6.25	3.81		10.06	12.50

For customer support on your Electrical Change Order Costs with RSMeans data, call 800.448.8182.

Pre-Installation Change Order Costs

26 61 23 – Lamps Applications

26 61 23.55 LED Lamps		Crew	Daily Output	Labor-Hours	Unit	Material	2020 Bare Costs Labor	Equipment	Total	Total Incl O&P
0505	9 W LED, A19 lamp, 5000K dimmable, equal to 60 W	1 Elec	128.80	.062	Ea.	7.80	3.81		11.61	14.25
0510	10.5 W LED, A19 lamp, dimmable, equal to 60 W		128.80	.062		8.90	3.81		12.71	15.40
0515	10 W LED, A19 lamp, frosted, dimmable, equal to 60 W		128.80	.062		14.20	3.81		18.01	21.50
0520	10 W LED, A19 lamp, omni-directional, dimmable, equal to 60 W		128.80	.062		8.30	3.81		12.11	14.80
0525	12 W LED, A19 lamp, dimmable, equal to 75 W		128.80	.062		12.05	3.81		15.86	18.90
1100	MR16, 3 W, replacement of halogen lamp 25 W [G]		119.60	.067		20.50	4.10		24.60	28.50
1200	6 W replacement of halogen lamp 45 W [G]		119.60	.067		21.50	4.10		25.60	29.50
2100	10 W, PAR20, equal to 60 W [G]		119.60	.067		28.50	4.10		32.60	37.50
2200	15 W, PAR30, equal to 100 W [G]		119.60	.067		54	4.10		58.10	65
2210	50 W LED lamp		119.60	.067		605	4.10		609.10	670
2220	11 Watt reflector dimmable warm white LED light bulb with medium base		119.60	.067		44	4.10		48.10	54.50
2221	12 W A-Line LED lamp DIM		119.60	.067		83	4.10		87.10	97
2225	13 Watt reflector LED warm white e26 with medium base 120 volt box		119.60	.067		41.50	4.10		45.60	51.50
2226	13 W br30 LED lamp		119.60	.067		72	4.10		76.10	85
2227	15 W 120V br30 inc LED lamp, 2700k		119.60	.067		67	4.10		71.10	80
2228	15 W 120V br30 inc LED lamp, 4000k		119.60	.067		63.50	4.10		67.60	76
2230	3 Watt dimmable warm white decorative LED lamp with medium base		147.20	.054		13	3.33		16.33	19.25
2240	.43 W night light LED daylight bulb E12 candelabra base 120 volt 2 pack		147.20	.054		5	3.33		8.33	10.45
2250	15 W omni-directional LED warm white e26 medium base 120 volt box		147.20	.054		42.50	3.33		45.83	52
2251	11 W omni-directional LED warm white e26 medium base 120 volt box		147.20	.054		30.50	3.33		33.83	38.50
2252	10 W omni-directional LED warm white e26 medium base 120 volt box		147.20	.054		29.50	3.33		32.83	37.50
2253	7 W omni A19 LED warm white e26 medium base 120 volt box		147.20	.054		22.50	3.33		25.83	29.50
2255	8 PAR 20 parabolic reflector 2700 LED lamp		147.20	.054		20	3.33		23.33	27
2256	8 PAR 20 parabolic reflector 3000 LED lamp		147.20	.054		19.80	3.33		23.13	27
2260	12 W PAR 38 120V LED 15 degree directional lamp		147.20	.054		215	3.33		218.33	241
2270	12 W PAR 38 120V LED 25 degree directional lamp		147.20	.054		33	3.33		36.33	41.50
2280	12 W PAR 38 120V LED 40 degree directional lamp		147.20	.054		215	3.33		218.33	241
2285	16 W PAR 38 120V 2700k LED parabolic reflector lamp		147.20	.054		32.50	3.33		35.83	40.50
2290	16 W PAR 38 120V 3000k LED parabolic reflector lamp		147.20	.054		24.50	3.33		27.83	32
3000	15 W PAR 30 LED daylight E26 medium base 120V box		147.20	.054		63.50	3.33		66.83	75
3100	17 W LED 3000k PAR 38 100 W replacement		147.20	.054		49.50	3.33		52.83	59.50
3110	17 W LED PAR 38 100 W replacement		147.20	.054		37	3.33		40.33	46
3120	17 W LED PAR 38 100 W replacement parabolic reflector		147.20	.054		39	3.33		42.33	47.50
3130	24 W LED T8 PW straight fluorescent lamp		8.18	.978		305	60		365	425
3135	Linear fluorescent LED lamp 120V		8.18	.978		141	60		201	245
3200	30 W LED 2700K recessed 8055E PAR 38 high power		147.20	.054		141	3.33		144.33	160
3210	30 W LED 4200K 120 degree 8055E PAR 38 high power		147.20	.054		142	3.33		145.33	161
3220	30 W LED 5700K 8055E PAR 38 high power		147.20	.054		143	3.33		146.33	162
3230	50 W LED 2700K 8045M PAR 38 277V high power		147.20	.054		360	3.33		363.33	400
3240	50 W LED 4200K 8045M PAR 38 277V high power retro fit		147.20	.054		191	3.33		194.33	215
3250	50 W LED 5700K 8045M PAR 38 277V high power retro fit		147.20	.054		360	3.33		363.33	400
8000	10 W LED PAR 30/fl 10 pk		147.20	.054		102	3.33		105.33	117
8010	Gen 3 PAR 30 15 W short neck power LED 120 VAC E26 80 +cri 300k dimm		147.20	.054		13.70	3.33		17.03	20
8020	12 PAR 30 2700K parabolic reflector LED lamp		147.20	.054		44.50	3.33		47.83	54
8030	12 PAR 30 3000K parabolic reflector LED lamp		147.20	.054		34	3.33		37.33	42.50
8040	3500K LED advantage T8 9 W 800LM 2ft linear 2 BD frosted		63.48	.126		11.10	7.75		18.85	23.50
8050	3500K LED litespan T8 9 W 900LM 2ft linear frosted		63.48	.126		14	7.75		21.75	27
8060	4000K LED advantage T8 9 W 800LM 2ft linear 2BD frosted		63.48	.126		13.30	7.75		21.05	26
8070	4000K LED litespan T8 9 W 900LM 2ft linear frosted		63.48	.126		14.65	7.75		22.40	27.50
8080	5000K LED advantage T8 9 W 800LM 2ft linear 2BD frosted		63.48	.126		11.90	7.75		19.65	24.50
8090	5000K LED litespan T8 9 W 900LM 2ft linear frosted		63.48	.126		11.65	7.75		19.40	24.50
8100	3500K LED advantage T8 18 W 1600LM 4ft linear 2 BD frosted		63.48	.126		11	7.75		18.75	23.50
8105	18 W LED 4ft T8 4000K frost 1600L linear lamp		63.48	.126		19.60	7.75		27.35	33

26 61 Lighting Systems and Accessories

26 61 23 – Lamps Applications

26 61 23.55 LED Lamps

		Crew	Daily Output	Labor-Hours	Unit	Material	2020 Bare Costs Labor	Equipment	Total	Total Incl O&P
8108	18 W LED 48 inch T8 4100K 1890LM linear lamp	1 Elec	63.48	.126	Ea.	60.50	7.75		68.25	78
8110	3500K LED advantage T8 18 W 1800LM 4ft linear 2 BD frosted		63.48	.126		31.50	7.75		39.25	46
8120	4000K LED advantage T8 18 W 1600LM 4ft linear 2 BD frosted		63.48	.126		21.50	7.75		29.25	35
8130	4000K LED litespan T8 18 W 1800LM 4ft linear frosted		63.48	.126		22.50	7.75		30.25	36
8140	5000K LED advantage T8 18 W 1600LM 4ft linear 2BD frosted		63.48	.126		14	7.75		21.75	27
8150	5000K LED litespan T8 18 W 1800LM 4ft linear 2BD frosted		63.48	.126		22.50	7.75		30.25	36
8200	16.5 T8 3000 IF-6U U-shape fluorescent LED lamp		59.80	.134		23	8.20		31.20	37
8210	16.5 T8 3500 IF-6U U-shape fluorescent LED lamp		59.80	.134		23.50	8.20		31.70	37.50
8220	16.5 T8 4000 IF-6U U-shape fluorescent LED lamp		59.80	.134		24	8.20		32.20	38.50
8230	16.5 T8 5000 IF-6U U-shape fluorescent LED lamp		59.80	.134		22.50	8.20		30.70	37
8240	18 W 6 inch T8 4100K U-shape frosted fluorescent LED lamp		55.20	.145		12.70	8.90		21.60	27.50
8250	18 W 6 inch T8 5000K U-shape frosted fluorescent LED lamp		55.20	.145		12.70	8.90		21.60	27.50
8260	Circular 12 W linear fluorescent LED 2700K MOD lamp		57.41	.139		320	8.55		328.55	365
8270	Circular 12 W linear fluorescent LED 3000K MOD lamp		57.41	.139		320	8.55		328.55	365
8280	Circular 12 W linear fluorescent LED 3500K MOD lamp		57.41	.139		289	8.55		297.55	335
8290	Circular 18 W linear fluorescent LED 3000K MOD lamp		57.41	.139		395	8.55		403.55	450
8300	Circular 18 W linear fluorescent LED 3500K MOD lamp		57.41	.139		435	8.55		443.55	495
8310	Circular 18 W linear fluorescent LED 5000K MOD lamp		57.41	.139		395	8.55		403.55	450
8320	Circular 18 W linear fluorescent LED 4100K MOD lamp		57.41	.139		420	8.55		428.55	480

26 71 Electrical Machines

26 71 13 – Motors Applications

26 71 13.10 Handling

		Crew	Daily Output	Labor-Hours	Unit	Material	2020 Bare Costs Labor	Equipment	Total	Total Incl O&P
0010	**HANDLING** Add to normal labor cost for restricted areas									
5000	Motors									
5100	1/2 HP, 23 pounds	1 Elec	3.68	2.170	Ea.		133		133	198
5110	3/4 HP, 28 pounds		3.68	2.170			133		133	198
5120	1 HP, 33 pounds		3.68	2.170			133		133	198
5130	1-1/2 HP, 44 pounds		2.94	2.720			167		167	248
5140	2 HP, 56 pounds		2.76	2.900			178		178	265
5150	3 HP, 71 pounds		2.12	3.780			232		232	345
5160	5 HP, 82 pounds		1.75	4.580			281		281	420
5170	7-1/2 HP, 124 pounds		1.38	5.800			355		355	530
5180	10 HP, 144 pounds		1.10	7.250			445		445	660
5190	15 HP, 185 pounds		.92	8.700			535		535	795
5200	20 HP, 214 pounds	2 Elec	1.38	11.590			710		710	1,050
5210	25 HP, 266 pounds		1.29	12.420			760		760	1,125
5220	30 HP, 310 pounds		1.10	14.490			890		890	1,325
5230	40 HP, 400 pounds		.92	17.390			1,075		1,075	1,600
5240	50 HP, 450 pounds		.83	19.320			1,175		1,175	1,775
5250	75 HP, 680 pounds		.74	21.740			1,325		1,325	1,975
5260	100 HP, 870 pounds	3 Elec	.92	26.090			1,600		1,600	2,375
5270	125 HP, 940 pounds		.74	32.610			2,000		2,000	2,975
5280	150 HP, 1,200 pounds		.64	37.270			2,275		2,275	3,400
5290	175 HP, 1,300 pounds		.55	43.480			2,675		2,675	3,975
5300	200 HP, 1,400 pounds		.46	52.170			3,200		3,200	4,775

26 71 13.20 Motors

		Crew	Daily Output	Labor-Hours	Unit	Material	2020 Bare Costs Labor	Equipment	Total	Total Incl O&P
0010	**MOTORS** 230/460 V, 60 HZ									
0050	Dripproof, premium efficiency, 1.15 service factor									
0060	1,800 RPM, 1/4 HP	1 Elec	4.90	1.630	Ea.	280	100		380	460
0070	1/3 HP		4.90	1.630		247	100		347	420

Pre-Installation Change Order Costs

26 71 13 – Motors Applications

26 71 13.20 Motors		Crew	Daily Output	Labor-Hours	Unit	Material	2020 Bare Costs Labor	Equipment	Total	Total Incl O&P
0080	1/2 HP	1 Elec	4.90	1.630	Ea.	207	100		307	375
0090	3/4 HP		4.90	1.630		320	100		420	500
0100	1 HP		4.14	1.930		345	118		463	555
0150	2 HP		4.14	1.930		370	118		488	585
0200	3 HP		4.14	1.930		805	118		923	1,050
0250	5 HP		4.14	1.930		595	118		713	830
0300	7.5 HP		3.86	2.070		985	127		1,112	1,275
0350	10 HP		3.68	2.170		1,175	133		1,308	1,500
0400	15 HP		2.94	2.720		1,600	167		1,767	2,025
0450	20 HP	2 Elec	4.78	3.340		2,125	205		2,330	2,625
0500	25 HP		4.60	3.480		2,575	214		2,789	3,150
0550	30 HP		4.42	3.620		2,725	222		2,947	3,300
0600	40 HP		3.68	4.350		3,625	267		3,892	4,400
0650	50 HP		2.94	5.430		3,800	335		4,135	4,700
0700	60 HP		2.58	6.210		4,725	380		5,105	5,775
0750	75 HP		2.21	7.250		5,825	445		6,270	7,050
0800	100 HP	3 Elec	2.48	9.660		5,950	595		6,545	7,425
0850	125 HP		1.93	12.420		7,175	760		7,935	9,000
0900	150 HP		1.66	14.490		9,800	890		10,690	12,100
0950	200 HP		1.38	17.390		10,800	1,075		11,875	13,500
1000	1,200 RPM, 1 HP	1 Elec	4.14	1.930		480	118		598	705
1050	2 HP		4.14	1.930		585	118		703	820
1100	3 HP		4.14	1.930		735	118		853	985
1150	5 HP		4.14	1.930		955	118		1,073	1,225
1200	3,600 RPM, 2 HP		4.14	1.930		515	118		633	740
1250	3 HP		4.14	1.930		575	118		693	810
1300	5 HP		4.14	1.930		600	118		718	835
1350	Totally enclosed, premium efficiency 1.15 service factor									
1360	1,800 RPM, 1/4 HP	1 Elec	4.90	1.630	Ea.	395	100		495	585
1370	1/3 HP		4.90	1.630		277	100		377	455
1380	1/2 HP		4.90	1.630		385	100		485	575
1390	3/4 HP		4.90	1.630		430	100		530	620
1400	1 HP		4.14	1.930		685	118		803	930
1450	2 HP		4.14	1.930		670	118		788	910
1500	3 HP		4.14	1.930		695	118		813	940
1550	5 HP		4.14	1.930		810	118		928	1,075
1600	7.5 HP		3.86	2.070		1,025	127		1,152	1,325
1650	10 HP		3.68	2.170		1,400	133		1,533	1,750
1700	15 HP		2.94	2.720		2,500	167		2,667	3,000
1750	20 HP	2 Elec	4.78	3.340		2,450	205		2,655	3,000
1800	25 HP		4.60	3.480		3,875	214		4,089	4,575
1850	30 HP		4.42	3.620		3,575	222		3,797	4,250
1900	40 HP		3.68	4.350		4,050	267		4,317	4,850
1950	50 HP		2.94	5.430		4,775	335		5,110	5,750
2000	60 HP		2.58	6.210		6,150	380		6,530	7,350
2050	75 HP		2.21	7.250		7,850	445		8,295	9,275
2100	100 HP	3 Elec	2.48	9.660		8,775	595		9,370	10,600
2150	125 HP		1.93	12.420		12,300	760		13,060	14,700
2200	150 HP		1.66	14.490		12,900	890		13,790	15,400
2250	200 HP		1.38	17.390		16,500	1,075		17,575	19,700
2300	1,200 RPM, 1 HP	1 Elec	4.14	1.930		475	118		593	700
2350	2 HP		4.14	1.930		600	118		718	835
2400	3 HP		4.14	1.930		780	118		898	1,025

Pre-Installation Change Order Costs   For customer support on your Electrical Change Order Costs with RSMeans data, call 800.448.8182.

26 71 13 – Motors Applications

26 71 13.20 Motors		Crew	Daily Output	Labor-Hours	Unit	Material	2020 Bare Costs Labor	Equipment	Total	Total Incl O&P
2450	5 HP	1 Elec	4.14	1.930	Ea.	970	118		1,088	1,250
2500	3,600 RPM, 2 HP		4.14	1.930		420	118		538	635
2550	3 HP		4.14	1.930		555	118		673	785
2600	5 HP	↓	4.14	1.930	↓	700	118		818	945

26 71 13.40 Motors Explosion Proof

		Crew	Daily Output	Labor-Hours	Unit	Material	2020 Bare Costs Labor	Equipment	Total	Total Incl O&P
0010	**MOTORS EXPLOSION PROOF**, 208-230/460 V, 60 HZ									
0020	1,800 RPM, 1/4 HP	1 Elec	4.60	1.740	Ea.	650	107		757	875
0030	1/3 HP		4.60	1.740		515	107		622	730
0040	1/2 HP		4.60	1.740		700	107		807	930
0050	3/4 HP		4.90	1.630		1,225	100		1,325	1,500
0060	1 HP		3.86	2.070		760	127		887	1,025
0070	2 HP		3.86	2.070		720	127		847	980
0080	3 HP		3.86	2.070		1,025	127		1,152	1,325
0090	5 HP		3.86	2.070		1,000	127		1,127	1,300
0100	7.5 HP		3.68	2.170		1,250	133		1,383	1,575
0110	10 HP		3.40	2.350		1,475	144		1,619	1,850
0120	15 HP	↓	2.94	2.720		1,875	167		2,042	2,300
0130	20 HP	2 Elec	4.60	3.480		2,275	214		2,489	2,825
0140	25 HP		4.42	3.620		2,700	222		2,922	3,300
0150	30 HP		4.05	3.950		3,050	242		3,292	3,700
0160	40 HP		3.50	4.580		5,000	281		5,281	5,900
0170	50 HP		2.94	5.430		5,100	335		5,435	6,100
0180	60 HP		2.39	6.690		8,425	410		8,835	9,850
0190	75 HP	↓	1.84	8.700		10,300	535		10,835	12,200
0200	100 HP	3 Elec	2.30	10.430		13,500	640		14,140	15,900
0210	125 HP		2.02	11.860		17,200	730		17,930	20,000
0220	150 HP		1.47	16.300		20,400	1,000		21,400	23,900
0230	200 HP	↓	1.10	21.740		25,000	1,325		26,325	29,500
1000	1,200 RPM, 1 HP	1 Elec	3.86	2.070		575	127		702	825
1010	2 HP		3.86	2.070		665	127		792	920
1020	3 HP		3.86	2.070		905	127		1,032	1,175
1030	5 HP		3.86	2.070		1,225	127		1,352	1,550
2000	3,600 RPM, 1/4 HP		4.60	1.740		490	107		597	700
2010	1/3 HP		4.60	1.740		515	107		622	725
2020	1/2 HP		4.60	1.740		650	107		757	875
2030	3/4 HP		4.60	1.740		690	107		797	915
2040	1 HP		3.86	2.070		745	127		872	1,000
2050	2 HP		3.86	2.070		935	127		1,062	1,225
2060	3 HP		3.86	2.070		1,025	127		1,152	1,325
2070	5 HP		3.86	2.070		1,375	127		1,502	1,700
2080	7.5 HP		3.68	2.170		1,550	133		1,683	1,900
2090	10 HP		3.40	2.350		1,800	144		1,944	2,200
2100	15 HP	↓	2.94	2.720		2,425	167		2,592	2,925
2110	20 HP	2 Elec	4.60	3.480		3,000	214		3,214	3,625
2120	25 HP		4.42	3.620		3,700	222		3,922	4,400
2130	30 HP		4.05	3.950		4,325	242		4,567	5,100
2140	40 HP		3.50	4.580		5,475	281		5,756	6,450
2150	50 HP		2.94	5.430		5,050	335		5,385	6,050
2160	60 HP		2.39	6.690		8,500	410		8,910	9,950
2170	75 HP	↓	1.84	8.700		10,400	535		10,935	12,200
2180	100 HP	3 Elec	2.30	10.430		14,500	640		15,140	16,900
2190	125 HP	↓	2.02	11.860	↓	17,800	730		18,530	20,700

For customer support on your Electrical Change Order Costs with RSMeans data, call 800.448.8182.   **Pre-Installation Change Order Costs**

26 71 Electrical Machines

26 71 13 – Motors Applications

26 71 13.40 Motors Explosion Proof		Crew	Daily Output	Labor-Hours	Unit	Material	2020 Bare Costs Labor	Equipment	Total	Total Incl O&P
2200	150 HP	3 Elec	1.47	16.300	Ea.	21,700	1,000		22,700	25,400
2210	200 HP		1.10	21.740		27,700	1,325		29,025	32,500
2220	250 HP	↓	1.10	21.740	↓	22,400	1,325		23,725	26,600

Division Notes

	CREW	DAILY OUTPUT	LABOR-HOURS	UNIT	BARE COSTS				TOTAL INCL O&P
					MAT.	LABOR	EQUIP.	TOTAL	

Estimating Tips
27 20 00 Data Communications
27 30 00 Voice Communications
27 40 00 Audio-Video Communications

- When estimating material costs for special systems, it is always prudent to obtain manufacturers' quotations for equipment prices and special installation requirements that may affect the total cost.

- For cost modifications for elevated tray installation, add the percentages to labor according to the height of the installation and only to the quantities exceeding the different height levels, not to the total tray quantities. Refer to subdivision 26 01 02.20 for labor adjustment factors.

- Do not overlook the costs for equipment used in the installation. If scissor lifts and boom lifts are available in the field, contractors may use them in lieu of the proposed ladders and rolling staging.

Reference Numbers
Reference numbers are shown at the beginning of some major classifications. These numbers refer to related items in the Reference Section. The reference information may be an estimating procedure, an alternate pricing method, or technical information.

Note: Not all subdivisions listed here necessarily appear. ∎

Same Data. Simplified.

Enjoy the convenience and efficiency of accessing your costs anywhere:

- **Skip the multiplier** by setting your location
- **Quickly search,** edit, favorite and share costs
- **Stay on top of price changes** with automatic updates

Discover more at rsmeans.com/online

27 05 Common Work Results for Communications

27 05 29 – Hangers and Supports for Communications Systems

27 05 29.10 Cable Support

27 05 29.10 Cable Support	Crew	Daily Output	Labor-Hours	Unit	Material	2020 Bare Costs Labor	Equipment	Total	Total Incl O&P
0010 **CABLE SUPPORT**									
0110 J-hook, single tier, single sided, 1" diam.	1 Elec	63.08	.127	Ea.	2.41	7.80		10.21	14.25
0120 1-1/2" diam.		62.38	.128		3.14	7.85		10.99	15.15
0130 2" diam.		61.92	.129		3.64	7.95		11.59	15.80
0140 4" diam.		61.46	.130		7.20	8		15.20	19.80
0330 Double tier, single sided, 2" diam.		61.64	.130		13	7.95		20.95	26
0340 4" diam.		60.90	.131		18.50	8.05		26.55	32.50
0430 Double sided, 2" diam.		60.54	.132		20	8.10		28.10	34
0440 4" diam.		60.17	.133		32.50	8.15		40.65	47.50
0530 Triple tier, single sided, 2" diam.		61.09	.131		16.95	8.05		25	30.50
0540 4" diam.		60.58	.132		25.50	8.10		33.60	40
0630 Double sided, 2" diam.		59.89	.134		23	8.20		31.20	37.50
0640 4" diam.	↓	59.52	.134	↓	38.50	8.25		46.75	55

27 11 Communications Equipment Room Fittings

27 11 16 – Communications Cabinets, Racks, Frames and Enclosures

27 11 16.10 Public Phone

27 11 16.10 Public Phone	Crew	Daily Output	Labor-Hours	Unit	Material	2020 Bare Costs Labor	Equipment	Total	Total Incl O&P
0010 **PUBLIC PHONE**									
7600 Telephone with wood backboard									
7620 Single door, 12" H x 12" W x 4" D	1 Elec	4.88	1.640	Ea.	114	101		215	276
7650 18" H x 12" W x 4" D		4.32	1.850		140	114		254	320
7700 24" H x 12" W x 4" D		3.86	2.070		211	127		338	420
7720 18" H x 18" W x 4" D		3.86	2.070		185	127		312	395
7750 24" H x 18" W x 4" D		3.68	2.170		262	133		395	485
7780 36" H x 36" W x 4" D		3.31	2.420		253	148		401	500
7800 24" H x 24" W x 6" D		3.31	2.420		350	148		498	610
7820 30" H x 24" W x 6" D		2.94	2.720		370	167		537	660
7850 30" H x 30" W x 6" D		2.48	3.220		565	198		763	915
7880 36" H x 30" W x 6" D		2.30	3.480		600	214		814	980
7900 48" H x 36" W x 6" D		2.02	3.950		960	242		1,202	1,400
7920 Double door, 48" H x 36" W x 6" D	↓	1.84	4.350	↓	1,325	267		1,592	1,850

27 11 16.20 Rack Mount Cabinet

27 11 16.20 Rack Mount Cabinet	Crew	Daily Output	Labor-Hours	Unit	Material	2020 Bare Costs Labor	Equipment	Total	Total Incl O&P
0010 **RACK MOUNT CABINET**									
0100 80" H x 24" W x 40" D									
0110 No sides	1 Elec	8.83	.906	Ea.	1,500	55.50		1,555.50	1,725
0120 One side		7.36	1.090		1,600	67		1,667	1,850
0130 Two sides	↓	7.23	1.110	↓	1,725	68		1,793	1,975
0200 80" H x 24" W x 42" D									
0210 No sides	1 Elec	8.83	.906	Ea.	1,525	55.50		1,580.50	1,775
0220 One side		7.36	1.090		1,575	67		1,642	1,825
0230 Two sides	↓	7.23	1.110	↓	1,600	68		1,668	1,850
0300 80" H x 24" W x 48" D									
0310 No sides	1 Elec	8.83	.906	Ea.	1,750	55.50		1,805.50	2,000
0320 One side		7.36	1.090		1,725	67		1,792	2,000
0330 Two sides	↓	7.23	1.110	↓	1,825	68		1,893	2,100
0400 80" H x 30" W x 40" D									
0410 No sides	1 Elec	8.83	.906	Ea.	1,625	55.50		1,680.50	1,875
0420 One side		7.36	1.090		1,700	67		1,767	1,975
0430 Two sides	↓	7.23	1.110	↓	1,775	68		1,843	2,050
0500 80" H x 30" W x 42" D									
0510 No sides	1 Elec	8.83	.906	Ea.	1,675	55.50		1,730.50	1,900

27 11 16 – Communications Cabinets, Racks, Frames and Enclosures

27 11 16.20 Rack Mount Cabinet		Crew	Daily Output	Labor-Hours	Unit	Material	2020 Bare Costs Labor	Equipment	Total	Total Incl O&P
0520	One side	1 Elec	7.36	1.090	Ea.	1,700	67		1,767	1,975
0530	Two sides	↓	7.23	1.110	↓	1,725	68		1,793	2,000
0600	80" H x 30" W x 48" D									
0610	No sides	1 Elec	8.83	.906	Ea.	1,825	55.50		1,880.50	2,100
0620	One side		7.36	1.090		1,925	67		1,992	2,225
0630	Two sides	↓	7.23	1.110	↓	2,050	68		2,118	2,350
0700	80" H x 32" W x 32" D									
0710	No sides	1 Elec	8.83	.906	Ea.	1,600	55.50		1,655.50	1,850
0720	One side		7.36	1.090		1,650	67		1,717	1,900
0730	Two sides	↓	7.23	1.110	↓	1,675	68		1,743	1,925
0800	80" H x 32" W x 40" D									
0810	No sides	1 Elec	8.83	.906	Ea.	1,775	55.50		1,830.50	2,025
0820	One side		7.36	1.090		1,850	67		1,917	2,125
0830	Two sides	↓	7.23	1.110	↓	1,925	68		1,993	2,200
0900	80" H x 32" W x 42" D									
0910	No sides	1 Elec	8.83	.906	Ea.	1,775	55.50		1,830.50	2,025
0920	One side		7.36	1.090		1,800	67		1,867	2,100
0930	Two sides	↓	7.23	1.110	↓	1,850	68		1,918	2,125
1000	80" H x 32" W x 48" D									
1010	No sides	1 Elec	8.83	.906	Ea.	1,975	55.50		2,030.50	2,250
1020	One side		7.36	1.090		2,075	67		2,142	2,400
1030	Two sides	↓	7.23	1.110	↓	1,475	68		1,543	1,725
1200	Freestanding cabinet with front and rear doors									
1300	84" H x 19" W x 24" D, gray									
1310	No sides	1 Elec	8.83	.906	Ea.	1,850	55.50		1,905.50	2,100
1320	One side		8.83	.906		2,775	55.50		2,830.50	3,150
1330	Two sides	↓	8.83	.906	↓	2,775	55.50		2,830.50	3,150
1400	84" H x 19" W x 24" D, white									
1410	No sides	1 Elec	8.83	.906	Ea.	2,350	55.50		2,405.50	2,650
1420	One side		8.83	.906		2,775	55.50		2,830.50	3,150
1430	Two sides	↓	8.83	.906	↓	2,775	55.50		2,830.50	3,150
1500	84" H x 19" W x 24" D, black									
1510	No sides	1 Elec	8.83	.906	Ea.	1,575	55.50		1,630.50	1,800
1520	One side		8.83	.906		2,650	55.50		2,705.50	3,000
1530	Two sides		8.83	.906		2,600	55.50		2,655.50	2,925
1600	84" H x 28" W x 34" D front glass, rear louvered door	↓	8.83	.906	↓	1,700	55.50		1,755.50	1,950

27 11 19 – Communications Termination Blocks and Patch Panels

27 11 19.10 Termination Blocks and Patch Panels

		Crew	Daily Output	Labor-Hours	Unit	Material	Labor	Equipment	Total	Total Incl O&P
0010	**TERMINATION BLOCKS AND PATCH PANELS**									
2960	Patch panel, RJ-45/110 type, 24 ports	2 Elec	5.52	2.900	Ea.	183	178		361	465
3000	48 ports	3 Elec	5.52	4.350		320	267		587	745
3040	96 ports	"	3.68	6.520		515	400		915	1,150
3100	Punch down termination per port	1 Elec	98.44	.081	↓		4.99		4.99	7.40

27 13 Communications Backbone Cabling

27 13 23 – Communications Optical Fiber Backbone Cabling

27 13 23.13 Communications Optical Fiber

		Crew	Daily Output	Labor-Hours	Unit	Material	2020 Bare Costs Labor	Equipment	Total	Total Incl O&P
0010	**COMMUNICATIONS OPTICAL FIBER**									
0040	Specialized tools & techniques cause installation costs to vary.									
0070	Fiber optic, cable, bulk simplex, single mode	1 Elec	7.36	1.090	C.L.F.	28	67		95	130
0080	Multi mode		7.36	1.090		36	67		103	139
0090	4 strand, single mode		6.75	1.180		46.50	72.50		119	159
0095	Multi mode		6.75	1.180		60.50	72.50		133	175
0100	12 strand, single mode		6.14	1.300		103	80		183	233
0105	Multi mode		6.14	1.300		119	80		199	250
0150	Jumper				Ea.	41.50			41.50	46
0200	Pigtail					43.50			43.50	48
0300	Connector	1 Elec	22.08	.362		31.50	22		53.50	68
0350	Finger splice		29.44	.272		45	16.65		61.65	74.50
0400	Transceiver (low cost bi-directional)		7.36	1.090		550	67		617	705
0450	Rack housing, 4 rack spaces, 12 panels (144 fibers)		1.84	4.350		735	267		1,002	1,200
1000	Cable, 62.5 microns, direct burial, 4 fiber	R-15	1104	.043	L.F.	1.12	2.61	.25	3.98	5.40
1020	Indoor, 2 fiber	R-19	920	.022		.53	1.34		1.87	2.57
1040	Outdoor, aerial/duct	"	1536.40	.013		.78	.80		1.58	2.05
1060	50 microns, direct burial, 8 fiber	R-22	3680	.010		1.56	.57		2.13	2.57
1080	12 fiber		3680	.010		2.92	.57		3.49	4.06
1100	Indoor, 12 fiber		698.28	.053		2.41	3		5.41	7.10
1120	Connectors, 62.5 micron cable, transmission	R-19	36.80	.543	Ea.	17.65	33.50		51.15	69
1140	Cable splice		36.80	.543		21	33.50		54.50	72.50
1160	125 micron cable, transmission		14.72	1.360		18.55	83.50		102.05	145
1180	Receiver, 1.2 mile range		18.40	1.090		340	67		407	470
1200	1.9 mile range		18.40	1.090		299	67		366	430
1220	6.2 mile range		4.60	4.350		390	267		657	830
1240	Transmitter, 1.2 mile range		18.40	1.090		380	67		447	515
1260	1.9 mile range		18.40	1.090		340	67		407	475
1280	6.2 mile range		4.60	4.350		520	267		787	975
1300	Modem, 1.2 mile range		4.60	4.350		220	267		487	640
1320	6.2 mile range		4.60	4.350		390	267		657	830
1340	1.9 mile range, 12 channel		4.60	4.350		2,475	267		2,742	3,125
1360	Repeater, 1.2 mile range		9.20	2.170		445	133		578	685
1380	1.9 mile range		9.20	2.170		590	133		723	850
1400	6.2 mile range		4.60	4.350		1,100	267		1,367	1,600
1420	1.2 mile range, digital		4.60	4.350		545	267		812	1,000
2040	Fiber optic cable, 48 strand, single mode, steel armor, material		1.54	13.020	M.L.F.	4,700	800		5,500	6,375

27 15 Communications Horizontal Cabling

27 15 01 – Communications Horizontal Cabling Applications

27 15 01.19 Fire Alarm Communications Conductors & Cables

		Crew	Daily Output	Labor-Hours	Unit	Material	2020 Bare Costs Labor	Equipment	Total	Total Incl O&P
0010	**FIRE ALARM COMMUNICATIONS CONDUCTORS AND CABLES**									
1500	Fire alarm FEP teflon 150 V to 200°C									
1550	#22, 1 pair	1 Elec	9.20	.870	C.L.F.	94	53.50		147.50	183
1600	2 pair		7.36	1.090		96	67		163	205
1650	4 pair		6.44	1.240		350	76		426	500
1700	6 pair		5.52	1.450		495	89		584	670
1750	8 pair		5.06	1.580		840	97		937	1,075
1800	10 pair		4.60	1.740		775	107		882	1,000
1850	#18, 1 pair	2 Elec	14.72	1.090		67.50	67		134.50	174
1900	2 pair		11.96	1.340		224	82		306	370

Pre-Installation Change Order Costs

27 15 Communications Horizontal Cabling

27 15 01 – Communications Horizontal Cabling Applications

27 15 01.19 Fire Alarm Communications Conductors & Cables	Crew	Daily Output	Labor-Hours	Unit	Material	2020 Bare Costs Labor	Equipment	Total	Total Incl O&P
1950	4 pair	2 Elec	8.83	1.810	C.L.F.	206	111		317	390
2000	6 pair		7.36	2.170		440	133		573	685
2050	8 pair		6.44	2.480		320	152		472	580
2100	10 pair		5.52	2.900		480	178		658	795

27 15 01.23 Audio-Video Communications Horizontal Cabling

		Crew	Daily Output	Labor-Hours	Unit	Material	Labor	Equipment	Total	Total Incl O&P
0010	**AUDIO-VIDEO COMMUNICATIONS HORIZONTAL CABLING**									
3000	S-video cable, MD4, 1M Pro interconnects, 3'	1 Elec	34.96	.229	Ea.	34.50	14.05		48.55	59
3010	6'		34.96	.229		28.50	14.05		42.55	52.50
3020	9'		34.96	.229		45.50	14.05		59.55	71
3030	16'		33.12	.242		49	14.80		63.80	76
3040	22'		33.12	.242		57.50	14.80		72.30	85
3300	HDMI cable, coaxial, HDMI/HDMI, Non plenum, 3', 24 kt terminations		36.80	.217		13.20	13.35		26.55	34.50
3310	6', 24 kt terminations		36.80	.217		15.40	13.35		28.75	37
3320	HDMI cable, coaxial, HDMI/HDMI metal, Non plenum, 3', 24 kt terminations		36.80	.217		15.05	13.35		28.40	36.50
3330	6', 24 kt terminations		36.80	.217		19.65	13.35		33	41.50
3340	15', 24 kt terminations		34.96	.229		27.50	14.05		41.55	51
3350	25', 24 kt terminations		34.96	.229		35.50	14.05		49.55	60.50

27 15 10 – Special Communications Cabling

27 15 10.23 Sound and Video Cables and Fittings

		Crew	Daily Output	Labor-Hours	Unit	Material	Labor	Equipment	Total	Total Incl O&P
0010	**SOUND AND VIDEO CABLES & FITTINGS**									
0900	TV antenna lead-in, 300 ohm, #20-2 conductor	1 Elec	6.44	1.240	C.L.F.	26.50	76		102.50	142
0950	Coaxial, feeder outlet		6.44	1.240		24.50	76		100.50	140
1000	Coaxial, main riser		5.52	1.450		17.80	89		106.80	152
1100	Sound, shielded with drain, #22-2 conductor		7.36	1.090		10.20	67		77.20	111
1150	#22-3 conductor		6.90	1.160		14.90	71		85.90	122
1200	#22-4 conductor		5.98	1.340		20.50	82		102.50	145
1250	Nonshielded, #22-2 conductor		9.20	.870		14.20	53.50		67.70	95
1300	#22-3 conductor		8.28	.966		25.50	59.50		85	117
1350	#22-4 conductor		7.36	1.090		28.50	67		95.50	131
1400	Microphone cable		7.36	1.090		33	67		100	136

27 15 13 – Communications Copper Horizontal Cabling

27 15 13.13 Communication Cables and Fittings

		Crew	Daily Output	Labor-Hours	Unit	Material	Labor	Equipment	Total	Total Incl O&P
0010	**COMMUNICATION CABLES AND FITTINGS**									
2200	Telephone twisted, PVC insulation, #22-2 conductor	1 Elec	9.20	.870	C.L.F.	12.45	53.50		65.95	93
2250	#22-3 conductor		8.28	.966		15.10	59.50		74.60	105
2300	#22-4 conductor		7.36	1.090		16.80	67		83.80	118
2350	#18-2 conductor		8.28	.966		25	59.50		84.50	116
2370	Telephone jack, eight pins		29.44	.272	Ea.	4.19	16.65		20.84	29.50
5000	High performance unshielded twisted pair (UTP)									
5100	Cable, category 3, #24, 2 pair solid, PVC jacket	1 Elec	9.20	.870	C.L.F.	11	53.50		64.50	91.50
5200	4 pair solid, PVC jacket		6.44	1.240		16	76		92	131
5300	25 pair solid, PVC jacket		2.76	2.900		90.50	178		268.50	365
5400	2 pair solid, plenum		9.20	.870		13.45	53.50		66.95	94.50
5500	4 pair solid, plenum		6.44	1.240		14.90	76		90.90	129
5600	25 pair solid, plenum		2.76	2.900		133	178		311	410
5700	4 pair stranded, PVC jacket		6.44	1.240		16.20	76		92.20	131
7000	Category 5, #24, 4 pair solid, PVC jacket		6.44	1.240		11.55	76		87.55	126
7100	4 pair solid, plenum		6.44	1.240		17.50	76		93.50	132
7200	4 pair stranded, PVC jacket		6.44	1.240		30	76		106	146
7210	Category 5e, #24, 4 pair solid, PVC jacket		6.44	1.240		15.60	76		91.60	130
7212	4 pair solid, plenum		6.44	1.240		21.50	76		97.50	137

27 15 13 – Communications Copper Horizontal Cabling

27 15 13.13 Communication Cables and Fittings		Crew	Daily Output	Labor-Hours	Unit	Material	2020 Bare Costs Labor	Equipment	Total	Total Incl O&P
7214	4 pair stranded, PVC jacket	1 Elec	6.44	1.240	C.L.F.	22	76		98	137
7240	Category 6, #24, 4 pair solid, PVC jacket		6.44	1.240		16.65	76		92.65	131
7242	4 pair solid, plenum		6.44	1.240		14	76		90	128
7244	4 pair stranded, PVC jacket		6.44	1.240	▼	15.45	76		91.45	130
7300	Connector, RJ45, category 5		73.60	.109	Ea.	.76	6.65		7.41	10.80
7302	Shielded RJ45, category 5		66.24	.121		1.73	7.40		9.13	12.95
7310	Jack, UTP RJ45, category 3		66.24	.121		.44	7.40		7.84	11.55
7312	Category 5		59.80	.134		6.25	8.20		14.45	19.10
7314	Category 5e		59.80	.134		4.98	8.20		13.18	17.70
7316	Category 6		59.80	.134		3.44	8.20		11.64	16
7322	Jack, shielded RJ45, category 5		55.20	.145		6.60	8.90		15.50	20.50
7324	Category 5e		55.20	.145		6.60	8.90		15.50	20.50
7326	Category 6		55.20	.145		6.90	8.90		15.80	21
7400	Voice/data expansion module, category 5e		7.36	1.090		47.50	67		114.50	152
7401	Modular jack, cat 6 keystone, RJ45, office white		15.18	.527		7.20	32.50		39.70	56
7402	RJ45, green		15.18	.527		7.10	32.50		39.60	56
7404	RJ45, grey		15.18	.527		8.05	32.50		40.55	57
7408	RJ45, orange		15.18	.527		7.40	32.50		39.90	56
7420	5M, RJ45, 568B, cord set		15.18	.527		42.50	32.50		75	94.50
7422	3M, RJ45, 568B, cord set		15.18	.527		38.50	32.50		71	90.50
7424	1M, RJ45, 568B, cord set		15.18	.527		33.50	32.50		66	85
7450	M12 TO RJ45, 2M ultra-loch D-mode, cord set	▼	14.26	.561	▼	48.50	34.50		83	105
8000	Multipair unshielded non-plenum cable, 150 V PVC jacket									
8002	#22, 2 pair	1 Elec	7.73	1.040	C.L.F.	18.30	64		82.30	115
8003	3 pair		7.36	1.090		30	67		97	133
8004	4 pair		6.72	1.190		68.50	73		141.50	184
8006	6 pair		5.70	1.400		41	86		127	174
8008	8 pair		5.24	1.530		95.50	94		189.50	245
8010	10 pair		4.88	1.640		75	101		176	233
8012	12 pair		3.86	2.070		189	127		316	395
8015	15 pair	▼	3.50	2.290		244	140		384	480
8020	20 pair	2 Elec	6.07	2.640		283	162		445	550
8025	25 pair		5.52	2.900		345	178		523	645
8030	30 pair		5.15	3.110		570	191		761	910
8040	40 pair		4.78	3.340		500	205		705	855
8050	50 pair	▼	4.51	3.550	▼	710	218		928	1,100
8100	Multipair unshielded non-plenum cable, 300 V PVC jacket									
8102	#20, 2 pair	1 Elec	6.72	1.190	C.L.F.	28	73		101	140
8103	3 pair		6.16	1.300		37.50	80		117.50	161
8104	4 pair		5.52	1.450		42.50	89		131.50	179
8106	6 pair		4.88	1.640		91.50	101		192.50	251
8108	8 pair	▼	4.05	1.980		105	121		226	297
8110	10 pair	2 Elec	6.72	2.380		116	146		262	345
8112	12 pair		5.70	2.810		144	172		316	415
8115	15 pair	▼	5.43	2.950		181	181		362	470
8201	#18, 1 pair	1 Elec	7.36	1.090		88	67		155	196
8202	2 pair		5.98	1.340		93.50	82		175.50	225
8203	3 pair		5.15	1.550		163	95		258	320
8204	4 pair		4.60	1.740		212	107		319	390
8206	6 pair		4.05	1.980		299	121		420	510
8208	8 pair	▼	3.68	2.170		297	133		430	525
8215	15 pair	2 Elec	4.60	3.480	▼	770	214		984	1,175
8300	Multipair shielded non-plenum cable, 300 V PVC jacket									

For customer support on your Electrical Change Order Costs with RSMeans data, call 800.448.8182.   **Pre-Installation Change Order Costs**

27 15 Communications Horizontal Cabling

27 15 13 – Communications Copper Horizontal Cabling

27 15 13.13 Communication Cables and Fittings

		Crew	Daily Output	Labor-Hours	Unit	Material	2020 Bare Costs Labor	Equipment	Total	Total Incl O&P
8303	#22, 3 pair	1 Elec	6.16	1.300	C.L.F.	71	80		151	197
8306	6 pair		5.24	1.530		136	94		230	290
8309	9 pair	▼	4.60	1.740		168	107		275	345
8312	12 pair	2 Elec	7.36	2.170		530	133		663	780
8315	15 pair		6.72	2.380		640	146		786	920
8317	17 pair		6.26	2.560		765	157		922	1,075
8319	19 pair		5.89	2.720		765	167		932	1,100
8327	27 pair	▼	5.43	2.950		1,350	181		1,531	1,750
8402	#20, 2 pair	1 Elec	6.16	1.300		128	80		208	259
8403	3 pair		5.70	1.400		154	86		240	297
8406	6 pair		4.05	1.980		385	121		506	605
8409	9 pair	▼	3.31	2.420		335	148		483	585
8412	12 pair	2 Elec	5.43	2.950		635	181		816	970
8415	15 pair	"	5.24	3.050		585	187		772	920
8502	#18, 2 pair	1 Elec	5.70	1.400		144	86		230	286
8503	3 pair		4.88	1.640		225	101		326	400
8504	4 pair		4.32	1.850		265	114		379	460
8506	6 pair		3.68	2.170		420	133		553	660
8509	9 pair	▼	3.13	2.560		640	157		797	935
8515	15 pair	2 Elec	4.42	3.620	▼	1,075	222		1,297	1,500

27 15 33 – Communications Coaxial Horizontal Cabling

27 15 33.10 Coaxial Cable and Fittings

		Crew	Daily Output	Labor-Hours	Unit	Material	2020 Bare Costs Labor	Equipment	Total	Total Incl O&P
0010	**COAXIAL CABLE & FITTINGS**									
3500	Coaxial connectors, 50 ohm impedance quick disconnect									
3540	BNC plug, for RG A/U #58 cable	1 Elec	38.64	.207	Ea.	3.35	12.70		16.05	22.50
3550	RG A/U #59 cable		38.64	.207		3.24	12.70		15.94	22.50
3560	RG A/U #62 cable		38.64	.207		3.28	12.70		15.98	22.50
3600	BNC jack, for RG A/U #58 cable		38.64	.207		3.74	12.70		16.44	23
3610	RG A/U #59 cable		38.64	.207		4.27	12.70		16.97	23.50
3620	RG A/U #62 cable		38.64	.207		4.15	12.70		16.85	23.50
3660	BNC panel jack, for RG A/U #58 cable		36.80	.217		9.55	13.35		22.90	30.50
3670	RG A/U #59 cable		36.80	.217		7.85	13.35		21.20	28.50
3680	RG A/U #62 cable		36.80	.217		7.85	13.35		21.20	28.50
3720	BNC bulkhead jack, for RG A/U #58 cable		36.80	.217		7.75	13.35		21.10	28.50
3730	RG A/U #59 cable		36.80	.217		7.75	13.35		21.10	28.50
3740	RG A/U #62 cable		36.80	.217	▼	7.75	13.35		21.10	28.50
3850	Coaxial cable, RG A/U #58, 50 ohm		7.36	1.090	C.L.F.	58.50	67		125.50	164
3860	RG A/U #59, 75 ohm		7.36	1.090		47	67		114	152
3870	RG A/U #62, 93 ohm		7.36	1.090		57	67		124	163
3875	RG 6/U, 75 ohm		7.36	1.090		42	67		109	146
3950	Fire rated, RG A/U #58, 50 ohm		7.36	1.090		113	67		180	224
3960	RG A/U #59, 75 ohm		7.36	1.090		149	67		216	263
3970	RG A/U #62, 93 ohm	▼	7.36	1.090	▼	131	67		198	244

27 15 43 – Communications Faceplates and Connectors

27 15 43.13 Communication Outlets

		Crew	Daily Output	Labor-Hours	Unit	Material	2020 Bare Costs Labor	Equipment	Total	Total Incl O&P
0010	**COMMUNICATION OUTLETS**									
0100	Voice/data devices not included									
0120	Voice/data outlets, single opening	1 Elec	44.16	.181	Ea.	7.60	11.10		18.70	25
0140	Two jack openings		44.16	.181		2.45	11.10		13.55	19.25
0160	One jack & one 3/4" round opening		44.16	.181		7.40	11.10		18.50	24.50
0180	One jack & one twinaxial opening		44.16	.181		7.55	11.10		18.65	25
0200	One jack & one connector cabling opening	▼	44.16	.181	▼	7.40	11.10		18.50	24.50

27 15 Communications Horizontal Cabling

27 15 43 – Communications Faceplates and Connectors

27 15 43.13 Communication Outlets		Crew	Daily Output	Labor-Hours	Unit	Material	2020 Bare Costs Labor	Equipment	Total	Total Incl O&P
0220	Two 3/8" coaxial openings	1 Elec	44.16	.181	Ea.	7.40	11.10		18.50	24.50
0300	Data outlets, single opening		44.16	.181		7.40	11.10		18.50	24.50
0320	One 25-pin subminiature opening		44.16	.181		7.40	11.10		18.50	24.50
1000	Voice/data wall plate, plastic, 1 gang, 1-port		66.24	.121		2.62	7.40		10.02	13.95
1020	2-port		66.24	.121		2.78	7.40		10.18	14.10
1040	3-port		66.24	.121		2.33	7.40		9.73	13.60
1060	4-port		66.24	.121		2.63	7.40		10.03	13.95
1080	6-port		66.24	.121		2.25	7.40		9.65	13.55
1100	2 gang, 6-port		44.16	.181		7.30	11.10		18.40	24.50
1120	Voice/data wall plate, stainless steel, 1 gang, 1-port		66.24	.121		8.60	7.40		16	20.50
1140	2-port		66.24	.121		8.20	7.40		15.60	20
1160	3-port		66.24	.121		7.95	7.40		15.35	19.80
1180	4-port		66.24	.121		8.25	7.40		15.65	20
1200	2 gang, 6-port	▼	44.16	.181	▼	12.65	11.10		23.75	30.50

27 21 Data Communications Network Equipment

27 21 29 – Data Communications Switches and Hubs

27 21 29.10 Switching and Routing Equipment

		Crew	Daily Output	Labor-Hours	Unit	Material	2020 Bare Costs Labor	Equipment	Total	Total Incl O&P
0010	**SWITCHING AND ROUTING EQUIPMENT**									
1100	Network hub, dual speed, 24 ports, includes cabinet	3 Elec	.61	39.530	Ea.	960	2,425		3,385	4,650
1300	Network switch, 50/60 HZ, 8 port, multi-platform, analog KVM	1 Elec	6.30	1.270		1,175	78		1,253	1,400
1310	16 port, multi-platform, analog KVM		5.52	1.450		1,400	89		1,489	1,650
1320	Network switch, 0x2x16,CAT5, analog KVM		5.52	1.450		1,225	89		1,314	1,450
1330	2x1x16, digital KVM, w/VM		14.72	.543		3,850	33.50		3,883.50	4,275
1340	2x1x32, digital KVM, w/VM		14.72	.543		4,575	33.50		4,608.50	5,075
1350	8x1x32, digital KVM, w/VM		14.72	.543		5,125	33.50		5,158.50	5,675
1500	KVM,1 RMU 16 port, no keyboard		14.72	.543		1,550	33.50		1,583.50	1,750
1510	KVM,1 RMU, 17" LCD, 16 port, US keyboard		13.94	.574		2,975	35		3,010	3,325
1520	KVM,1 RMU, 17" LCD, 16 port, UK keyboard		13.94	.574		2,975	35		3,010	3,325
2000	10/100/1000 Mbps, 24 ports		12.88	.621		620	38		658	735
2040	10/100/1000 Mbps, 48 ports		11.04	.725		1,400	44.50		1,444.50	1,625
2050	10/100/1000 Mbps, 5 port, industrial ethernet type		29.44	.272		330	16.65		346.65	390
2060	10/100/1000 Mbps, 6 port, industrial ethernet type		29.44	.272		1,300	16.65		1,316.65	1,450
2070	10/100/1000 Mbps, 10 port, X307-3		25.76	.311		3,075	19.05		3,094.05	3,425
2080	10/100/1000 Mbps, 10 port, X308-2		25.76	.311		2,850	19.05		2,869.05	3,175
3000	10/100/1000 Mbps, 20 port, front ports		18.40	.435		4,425	26.50		4,451.50	4,925
3010	10/100/1000 Mbps, 20 port, rear ports		18.40	.435		4,425	26.50		4,451.50	4,925
3040	KVM, 10/100/1000/10000 Mbps, 28 port, rear ports		14.72	.543		9,150	33.50		9,183.50	10,100
3050	KVM, 10/100/1000/10000 Mbps, 28 port, front ports	▼	14.72	.543		10,700	33.50		10,733.50	11,800
3070	KVM, 10/100/1000/10000 Mbps, 52 port, front ports	2 Elec	11.04	1.450		11,300	89		11,389	12,500
3080	KVM, 10/100/1000/10000 Mbps, 52 port, rear ports	"	11.04	1.450	▼	11,300	89		11,389	12,500

For customer support on your Electrical Change Order Costs with RSMeans data, call 800.448.8182. **Pre-Installation Change Order Costs**

27 32 Voice Communications Terminal Equipment

27 32 26 – Ring-Down Emergency Telephones

27 32 26.10 Emergency Phone Stations

27 32 26.10 Emergency Phone Stations	Crew	Daily Output	Labor-Hours	Unit	Material	2020 Bare Costs Labor	2020 Bare Costs Equipment	Total	Total Incl O&P
0010 **EMERGENCY PHONE STATIONS**									
0100 Hands free emergency speaker phone w/LED strobe	1 Elec	6.64	1.200	Ea.	1,025	73.50		1,098.50	1,225
0110 Wall/pole mounted emergency phone station w/LED strobe	"	4.23	1.890		1,825	116		1,941	2,175
2000 Call station, 30", pole mounted emergency phone, 120V/ VoIP, blue beacon	2 Elec	4.60	3.480		1,100	214		1,314	1,525
2010 Call station, 36"		4.60	3.480		1,575	214		1,789	2,050
2020 Call station, 5'-8"		4.42	3.620		2,650	222		2,872	3,250
2030 Call station, 9'		4.42	3.620		2,675	222		2,897	3,275

27 32 36 – TTY Equipment

27 32 36.10 TTY Telephone Equipment

27 32 36.10 TTY Telephone Equipment	Crew	Daily Output	Labor-Hours	Unit	Material	2020 Bare Costs Labor	2020 Bare Costs Equipment	Total	Total Incl O&P
0010 **TTY TELEPHONE EQUIPMENT**									
1620 Telephone, TTY, compact, pocket type				Ea.	355			355	390
1630 Advanced, desk type	2 Elec	18.40	.870		495	53.50		548.50	625
1640 Full-featured public, wall type	"	3.68	4.350		705	267		972	1,175

27 41 Audio-Video Systems

27 41 33 – Master Antenna Television Systems

27 41 33.10 TV Systems

27 41 33.10 TV Systems	Crew	Daily Output	Labor-Hours	Unit	Material	2020 Bare Costs Labor	2020 Bare Costs Equipment	Total	Total Incl O&P
0010 **TV SYSTEMS,** not including rough-in wires, cables & conduits									
0100 Master TV antenna system									
0200 VHF reception & distribution, 12 outlets	1 Elec	5.52	1.450	Outlet	126	89		215	271
0400 30 outlets		9.20	.870		254	53.50		307.50	360
0600 100 outlets		11.96	.669		295	41		336	385
0650 UHF reception, 100 outlets		11.96	.669		169	41		210	247
0800 VHF & UHF reception & distribution, 12 outlets		5.52	1.450		247	89		336	405
1000 30 outlets		9.20	.870		169	53.50		222.50	266
1200 100 outlets		11.96	.669		168	41		209	246
1400 School and deluxe systems, 12 outlets		2.21	3.620		325	222		547	690
1600 30 outlets		3.68	2.170		286	133		419	515
1800 80 outlets		4.88	1.640		275	101		376	450
1900 Amplifier		3.68	2.170	Ea.	820	133		953	1,100

27 42 Electronic Digital Systems

27 42 13 – Point of Sale Systems

27 42 13.10 Bar Code Scanner

27 42 13.10 Bar Code Scanner	Crew	Daily Output	Labor-Hours	Unit	Material	2020 Bare Costs Labor	2020 Bare Costs Equipment	Total	Total Incl O&P
0010 **BAR CODE SCANNER**									
0100 CR3600, w/B4 battery, handle	1 Elec	5.52	1.450	Outlet	1,025	89		1,114	1,250
0110 Palm		11.04	.725		900	44.50		944.50	1,050
0120 Batch, B4 batt, handle, chgstn, 3' USB		8.28	.966		1,050	59.50		1,109.50	1,250
0130 Palm, chgstn, 3' USB		11.04	.725		945	44.50		989.50	1,125
0140 CR2500, w/H2 handle, no cable		9.20	.870		690	53.50		743.50	835
0150 Core unit, palm, no cable		14.72	.543		640	33.50		673.50	755

27 51 16 – Public Address Systems

27 51 16.10 Public Address System	Crew	Daily Output	Labor-Hours	Unit	Material	2020 Bare Costs Labor	Equipment	Total	Total Incl O&P
0010 **PUBLIC ADDRESS SYSTEM**									
0100 Conventional, office	1 Elec	4.90	1.630	Speaker	171	100		271	335
0200 Industrial	"	2.48	3.220	"	330	198		528	655

27 51 19 – Sound Masking Systems

27 51 19.10 Sound System

27 51 19.10 Sound System	Crew	Daily Output	Labor-Hours	Unit	Material	Labor	Equipment	Total	Total Incl O&P
0010 **SOUND SYSTEM**, not including rough-in wires, cables & conduits									
0100 Components, projector outlet	1 Elec	7.36	1.090	Ea.	61.50	67		128.50	168
0200 Microphone		3.68	2.170		115	133		248	325
0400 Speakers, ceiling or wall		7.36	1.090		157	67		224	273
0600 Trumpets		3.68	2.170		291	133		424	520
0800 Privacy switch		7.36	1.090		117	67		184	228
1000 Monitor panel		3.68	2.170		520	133		653	770
1200 Antenna, AM/FM		3.68	2.170		155	133		288	370
1400 Volume control		7.36	1.090		60	67		127	166
1600 Amplifier, 250 W		.92	8.700		1,400	535		1,935	2,350
1800 Cabinets		.92	8.700		1,125	535		1,660	2,050
2000 Intercom, 30 station capacity, master station	2 Elec	1.84	8.700		2,400	535		2,935	3,450
2020 10 station capacity	"	3.68	4.350		1,475	267		1,742	2,025
2200 Remote station	1 Elec	7.36	1.090		217	67		284	340
2400 Intercom outlets		7.36	1.090		128	67		195	240
2600 Handset		3.68	2.170		425	133		558	665
2800 Emergency call system, 12 zones, annunciator		1.20	6.690		1,275	410		1,685	2,000
3000 Bell		4.88	1.640		132	101		233	295
3200 Light or relay		7.36	1.090		66	67		133	172
3400 Transformer		3.68	2.170		291	133		424	520
3600 House telephone, talking station		1.47	5.430		625	335		960	1,175
3800 Press to talk, release to listen		4.88	1.640		145	101		246	310
4000 System-on button					87			87	95.50
4200 Door release	1 Elec	3.68	2.170		155	133		288	370
4400 Combination speaker and microphone		7.36	1.090		265	67		332	390
4600 Termination box		2.94	2.720		83.50	167		250.50	340
4800 Amplifier or power supply		4.88	1.640		955	101		1,056	1,200
5000 Vestibule door unit		14.72	.543	Name	161	33.50		194.50	227
5200 Strip cabinet		24.84	.322	Ea.	330	19.75		349.75	395
5400 Directory		14.72	.543		156	33.50		189.50	221
6000 Master door, button buzzer type, 100 unit	2 Elec	.50	32.210		1,600	1,975		3,575	4,700
6020 200 unit		.28	57.970		3,000	3,550		6,550	8,600
6040 300 unit		.18	86.960		4,575	5,325		9,900	13,000
6060 Transformer	1 Elec	7.36	1.090		40.50	67		107.50	144
6080 Door opener		4.88	1.640		58.50	101		159.50	214
6100 Buzzer with door release and plate		3.68	2.170		58.50	133		191.50	262
6200 Intercom type, 100 unit	2 Elec	.50	32.210		1,975	1,975		3,950	5,125
6220 200 unit		.28	57.970		3,875	3,550		7,425	9,550
6240 300 unit		.18	86.960		5,825	5,325		11,150	14,400
6260 Amplifier	1 Elec	1.84	4.350		291	267		558	715
6280 Speaker with door release	"	3.68	2.170		87	133		220	294

Pre-Installation Change Order Costs

27 52 Healthcare Communications and Monitoring Systems

27 52 23 – Nurse Call/Code Blue Systems

27 52 23.10 Nurse Call Systems

	27 52 23.10 Nurse Call Systems	Crew	Daily Output	Labor-Hours	Unit	Material	2020 Bare Costs Labor	Equipment	Total	Total Incl O&P
0010	**NURSE CALL SYSTEMS**									
0100	Single bedside call station	1 Elec	7.36	1.090	Ea.	182	67		249	300
0200	Ceiling speaker station		7.36	1.090		90	67		157	199
0400	Emergency call station		7.36	1.090		92.50	67		159.50	202
0600	Pillow speaker		7.36	1.090		223	67		290	345
0800	Double bedside call station		3.68	2.170		169	133		302	385
1000	Duty station		3.68	2.170		137	133		270	350
1200	Standard call button		7.36	1.090		125	67		192	237
1400	Lights, corridor, dome or zone indicator		7.36	1.090		64.50	67		131.50	171
1600	Master control station for 20 stations	2 Elec	.60	26.760	Total	3,725	1,650		5,375	6,550

27 53 Distributed Systems

27 53 13 – Clock Systems

27 53 13.50 Clock Equipments

	27 53 13.50 Clock Equipments	Crew	Daily Output	Labor-Hours	Unit	Material	2020 Bare Costs Labor	Equipment	Total	Total Incl O&P
0010	**CLOCK EQUIPMENTS**, not including wires & conduits									
0100	Time system components, master controller	1 Elec	.30	26.350	Ea.	1,575	1,625		3,200	4,125
0200	Program bell		7.36	1.090		114	67		181	225
0400	Combination clock & speaker		2.94	2.720		200	167		367	470
0600	Frequency generator		1.84	4.350		2,725	267		2,992	3,375
0800	Job time automatic stamp recorder		3.68	2.170		560	133		693	815
1600	Master time clock system, clocks & bells, 20 room	4 Elec	.18	173		6,025	10,600		16,625	22,400
1800	50 room	"	.07	434		12,700	26,600		39,300	53,500
1900	Time clock	1 Elec	2.94	2.720		375	167		542	665
2000	100 cards in & out, 1 color					10.45			10.45	11.50
2200	2 colors					9.90			9.90	10.90
2800	Metal rack for 25 cards	1 Elec	6.44	1.240		50	76		126	168
4000	Wireless time systems component, master controller	"	1.84	4.350		630	267		897	1,075
4010	For transceiver and antenna, see Section 28 47 12.10									
4100	Wireless analog clock w/battery operated	1 Elec	7.36	1.090	Ea.	110	67		177	221
4200	Wireless digital clock w/battery operated	"	7.36	1.090	"	425	67		492	565

Division Notes

	CREW	DAILY OUTPUT	LABOR-HOURS	UNIT	BARE COSTS				TOTAL INCL O&P
					MAT.	LABOR	EQUIP.	TOTAL	

Estimating Tips

- When estimating material costs for electronic safety and security systems, it is always prudent to obtain manufacturers' quotations for equipment prices and special installation requirements that may affect the total cost.

- Fire alarm systems consist of control panels, annunciator panels, batteries with rack, charger, and fire alarm actuating and indicating devices. Some fire alarm systems include speakers, telephone lines, door closer controls, and other components. Be careful not to overlook the costs related to installation for these items. Also be aware of costs for integrated automation instrumentation and terminal devices, control equipment, control wiring, and programming. Insurance underwriters may have specific requirements for the type of materials to be installed or design requirements based on the hazard to be protected. Local jurisdictions may have requirements not covered by code. It is advisable to be aware of any special conditions.

- Security equipment includes items such as CCTV, access control, and other detection and identification systems to perform alert and alarm functions. Be sure to consider the costs related to installation for this security equipment, such as for integrated automation instrumentation and terminal devices, control equipment, control wiring, and programming.

Reference Numbers

Reference numbers are shown at the beginning of some major classifications. These numbers refer to related items in the Reference Section. The reference information may be an estimating procedure, an alternate pricing method, or technical information.

Same Data. Simplified.

Enjoy the convenience and efficiency of accessing your costs anywhere:

- **Skip the multiplier** by setting your location
- **Quickly search,** edit, favorite and share costs
- **Stay on top of price changes** with automatic updates

Discover more at rsmeans.com/online

Note: Trade Service, in part, has been used as a reference source for some of the material prices used in Division 28.

28 05 19 – Storage Appliances for Electronic Safety and Security

28 05 19.11 Digital Video Recorder (DVR)	Crew	Daily Output	Labor-Hours	Unit	Material	2020 Bare Costs Labor	Equipment	Total	Total Incl O&P
0010 **DIGITAL VIDEO RECORDERS**									
0100 Pentaplex hybrid, internet protocol, and hard drive									
0200 4 channel	1 Elec	1.22	6.540	Ea.	1,800	400		2,200	2,575
0300 8 channel		.92	8.700		3,925	535		4,460	5,125
0400 16 channel	↓	.92	8.700	↓	3,400	535		3,935	4,550

28 23 Video Management System

28 23 13 – Video Management System Interfaces

28 23 13.10 Closed Circuit Television System

		Crew	Daily Output	Labor-Hours	Unit	Material	2020 Bare Costs Labor	Equipment	Total	Total Incl O&P
0010	**CLOSED CIRCUIT TELEVISION SYSTEM**									
2000	Surveillance, one station (camera & monitor)	2 Elec	2.39	6.690	Total	665	410		1,075	1,350
2200	For additional camera stations, add	1 Elec	2.48	3.220	Ea.	320	198		518	645
2400	Industrial quality, one station (camera & monitor)	2 Elec	2.39	6.690	Total	1,650	410		2,060	2,425
2600	For additional camera stations, add	1 Elec	2.48	3.220	Ea.	690	198		888	1,050
2610	For low light, add		2.48	3.220		525	198		723	870
2620	For very low light, add		2.48	3.220		3,450	198		3,648	4,100
2800	For weatherproof camera station, add		1.20	6.690		540	410		950	1,200
3000	For pan and tilt, add		1.20	6.690		1,575	410		1,985	2,325
3200	For zoom lens - remote control, add		1.84	4.350		1,800	267		2,067	2,375
3400	Extended zoom lens		1.84	4.350		5,025	267		5,292	5,925
3410	For automatic iris for low light, add	↓	1.84	4.350	↓	1,025	267		1,292	1,525
3600	Educational TV studio, basic 3 camera system, black & white,									
3800	electrical & electronic equip. only	4 Elec	.74	43.480	Total	8,275	2,675		10,950	13,100
4000	Full console		.26	124		25,300	7,600		32,900	39,200
4100	As above, but color system		.26	124		50,500	7,600		58,100	67,000
4120	Full console	↓	.11	289	↓	190,000	17,700		207,700	235,500
4200	For film chain, black & white, add	1 Elec	.92	8.700	Ea.	12,500	535		13,035	14,500
4250	Color, add		.23	34.780		8,925	2,125		11,050	13,000
4400	For video recorders, add	↓	.92	8.700		2,450	535		2,985	3,475
4600	Premium	4 Elec	.37	86.960	↓	16,000	5,325		21,325	25,600

28 23 23 – Video Surveillance Systems Infrastructure

28 23 23.50 Video Surveillance Equipments

		Crew	Daily Output	Labor-Hours	Unit	Material	2020 Bare Costs Labor	Equipment	Total	Total Incl O&P
0010	**VIDEO SURVEILLANCE EQUIPMENTS**									
0200	Video cameras, wireless, hidden in exit signs, clocks, etc., incl. receiver	1 Elec	2.76	2.900	Ea.	171	178		349	455
0210	Accessories for video recorder, single camera		2.76	2.900		195	178		373	480
0220	For multiple cameras		2.76	2.900		1,675	178		1,853	2,125
0230	Video cameras, wireless, for under vehicle searching, complete		1.84	4.350		15,200	267		15,467	17,100
0400	Internet protocol network camera, day/night, color & power supply		2.39	3.340		1,250	205		1,455	1,675
0500	Monitor, color flat screen, liquid crystal display (LCD), 15"		2.48	3.220		535	198		733	885
0520	17"		2.48	3.220		470	198		668	815
0540	19"	↓	2.48	3.220		590	198		788	940

Pre-Installation Change Order Costs

28 31 Intrusion Detection

28 31 16 – Intrusion Detection Systems Infrastructure

28 31 16.50 Intrusion Detection	Crew	Daily Output	Labor-Hours	Unit	Material	2020 Bare Costs Labor	Equipment	Total	Total Incl O&P
0010 **INTRUSION DETECTION**, not including wires & conduits									
0100 Burglar alarm, battery operated, mechanical trigger	1 Elec	3.68	2.170	Ea.	289	133		422	520
0200 Electrical trigger		3.68	2.170		345	133		478	580
0400 For outside key control, add		7.36	1.090		89.50	67		156.50	198
0600 For remote signaling circuitry, add		7.36	1.090		143	67		210	257
0800 Card reader, flush type, standard		2.48	3.220		725	198		923	1,100
1000 Multi-code		2.48	3.220		1,250	198		1,448	1,675
1010 Card reader, proximity type		2.48	3.220		300	198		498	625
1200 Door switches, hinge switch		4.88	1.640		64.50	101		165.50	221
1400 Magnetic switch		4.88	1.640		105	101		206	266
1600 Exit control locks, horn alarm		3.68	2.170		218	133		351	440
1800 Flashing light alarm		3.68	2.170		247	133		380	470
2000 Indicating panels, 1 channel	▼	2.48	3.220		244	198		442	565
2200 10 channel	2 Elec	2.94	5.430		1,175	335		1,510	1,800
2400 20 channel		1.84	8.700		2,525	535		3,060	3,575
2600 40 channel	▼	1.05	15.260		4,650	935		5,585	6,525
2800 Ultrasonic motion detector, 12 V	1 Elec	2.12	3.780		200	232		432	565
3000 Infrared photoelectric detector		3.68	2.170		133	133		266	345
3200 Passive infrared detector		3.68	2.170		237	133		370	460
3400 Glass break alarm switch		7.36	1.090		85.50	67		152.50	194
3420 Switchmats, 30" x 5'		4.88	1.640		116	101		217	278
3440 30" x 25'		3.68	2.170		212	133		345	430
3460 Police connect panel		3.68	2.170		283	133		416	510
3480 Telephone dialer		4.88	1.640		400	101		501	590
3500 Alarm bell		3.68	2.170		106	133		239	315
3520 Siren		3.68	2.170		152	133		285	365
3540 Microwave detector, 10' to 200'		1.84	4.350		585	267		852	1,025
3560 10' to 350'	▼	1.84	4.350	▼	1,775	267		2,042	2,350

28 33 Security Monitoring and Control

28 33 11 – Electronic Structural Monitoring Systems

28 33 11.10 Load Cell Units

	Crew	Daily Output	Labor-Hours	Unit	Material	2020 Bare Costs Labor	Equipment	Total	Total Incl O&P
0010 **LOAD CELL UNITS**									
0100 Load cell, stainless steel, 0.5/1 ton rated, mounting unit	2 Elec	1.22	13.080	Ea.	665	800		1,465	1,925
0110 2/3 to 5/5 ton rated, mounting unit		1.10	14.490		705	890		1,595	2,100
0120 0.5/1/2/3.5/5 ton, self aligning bottom base		1.53	10.480		195	645		840	1,175
0130 Combo mounting unit, 10/25 ton rated		.79	20.220		1,825	1,250		3,075	3,850
0140 40/60 ton rated		.61	26.350		2,975	1,625		4,600	5,675
0150 60kg/3m, rated		1.69	9.450		1,550	580		2,130	2,575
0160 130kg/3m, rated		1.43	11.220		1,475	690		2,165	2,625
0170 280 kg, rated	▼	1.12	14.260	▼	1,475	875		2,350	2,900

28 46 Fire Detection and Alarm

28 46 11 – Fire Sensors and Detectors

28 46 11.21 Carbon-Monoxide Detection Sensors	Crew	Daily Output	Labor-Hours	Unit	Material	2020 Bare Costs Labor	Equipment	Total	Total Incl O&P
0010 **CARBON-MONOXIDE DETECTION SENSORS**									
8400 Smoke and carbon monoxide alarm battery operated photoelectric low profile	1 Elec	22.08	.362	Ea.	51.50	22		73.50	89.50
8410 low profile photoelectric battery powered		22.08	.362		40	22		62	77.50
8420 photoelectric low profile sealed lithium		22.08	.362		70.50	22		92.50	111
8430 Photoelectric low profile sealed lithium smoke and CO with voice combo		22.08	.362		50.50	22		72.50	88.50
8500 Carbon monoxide sensor, wall mount 1Mod 1 relay output smoke & heat		22.08	.362		490	22		512	575
8510 1Mod, 1relay no display, smoke & heat		22.08	.362		475	22		497	560
8530 1Mod 1 relay no HW logo smoke & heat		22.08	.362		515	22		537	600
8700 Carbon monoxide detector, battery operated, wall mounted		14.72	.543		43	33.50		76.50	97
8710 Hardwired, wall and ceiling mounted		7.36	1.090		84.50	67		151.50	193
8720 Duct mounted	▼	7.36	1.090		350	67		417	485
8730 Continuous air monitoring system, PC based, incl. remote sensors, min.	2 Elec	.23	69.570		31,600	4,275		35,875	41,100
8740 Maximum	"	.09	173	▼	56,500	10,600		67,100	78,000

28 46 11.27 Other Sensors

	Crew	Daily Output	Labor-Hours	Unit	Material	2020 Bare Costs Labor	Equipment	Total	Total Incl O&P
0010 **OTHER SENSORS**									
5200 Smoke detector, ceiling type	1 Elec	5.70	1.400	Ea.	120	86		206	261
5240 Smoke detector, addressable type		5.52	1.450		224	89		313	380
5400 Duct type		2.94	2.720		295	167		462	575
5420 Duct addressable type		2.94	2.720		520	167		687	825
8300 Smoke alarm with integrated strobe light 120 V, 16DB 60 fpm flash rate		14.72	.543		119	33.50		152.50	181
8310 Photoelectric smoke detector with strobe 120 V, 90 DB ceiling mount		11.04	.725		227	44.50		271.50	315
8320 120 V, 90 DB wall mount		11.04	.725		202	44.50		246.50	288
8330 120 V, 9 VDC, 90 DB ceiling mount		11.04	.725		201	44.50		245.50	288
8340 120 V, 9 VDC backup 90 DB wall mount	▼	11.04	.725		202	44.50		246.50	288
8440 Fire alarm beam detector, motorized reflective, infrared optical beam	2 Elec	1.47	10.870		750	665		1,415	1,825
8450 Fire alarm beam detector, motorized reflective, IR/UV optical beam	"	1.47	10.870	▼	1,025	665		1,690	2,125

28 46 11.50 Fire and Heat Detectors

	Crew	Daily Output	Labor-Hours	Unit	Material	2020 Bare Costs Labor	Equipment	Total	Total Incl O&P
0010 **FIRE & HEAT DETECTORS**									
5000 Detector, rate of rise	1 Elec	7.36	1.090	Ea.	52	67		119	157
5010 Heat addressable type		6.67	1.200		225	73.50		298.50	355
5100 Fixed temp fire alarm		6.44	1.240		42	76		118	160
5200 125 V SRF MT-135F		6.67	1.200		33.50	73.50		107	147
5300 MT-195F		6.67	1.200		34	73.50		107.50	148
5400 104/85 DB, indoor/outdoor, ceiling/wall, red		9.43	.848		73	52		125	158
5410 Ceiling/wall, white		9.43	.848		71.50	52		123.50	157
5420 102/98 DB, lumin 110cd, indoor/wall, red		9.43	.848		113	52		165	203
5430 Lumin15/75cd, indoor/wall, red		9.43	.848		110	52		162	199
5440 HI/LO DB, 24 V, fire marking, red	▼	9.43	.848	▼	81	52		133	167

28 46 20 – Fire Alarm

28 46 20.50 Alarm Panels and Devices

0010 **ALARM PANELS AND DEVICES**, not including wires & conduits	Crew	Daily Output	Labor-Hours	Unit	Material	2020 Bare Costs Labor	Equipment	Total	Total Incl O&P
2200 Intercom remote station	1 Elec	7.36	1.090	Ea.	89.50	67		156.50	198
2400 Intercom outlet		7.36	1.090		64.50	67		131.50	171
2600 Sound system, intercom handset		7.36	1.090		485	67		552	635
3590 Signal device, beacon		7.36	1.090		202	67		269	320
3594 2 zone	▼	1.24	6.440		310	395		705	930
3600 4 zone	2 Elec	1.84	8.700		410	535		945	1,250
3610 3 zone		1.84	8.700		640	535		1,175	1,500
3800 8 zone		.92	17.390		875	1,075		1,950	2,575
3810 5 zone		1.38	11.590		700	710		1,410	1,825
3900 10 zone		1.15	13.910		1,025	855		1,880	2,425
4000 12 zone	▼	.61	26.070		2,350	1,600		3,950	4,950

For customer support on your Electrical Change Order Costs with RSMeans data, call 800.448.8182.   **Pre-Installation Change Order Costs**

28 46 Fire Detection and Alarm

28 46 20 – Fire Alarm

28 46 20.50 Alarm Panels and Devices	Crew	Daily Output	Labor-Hours	Unit	Material	2020 Bare Costs Labor	Equipment	Total	Total Incl O&P
4020	Alarm device, tamper, flow	1 Elec	7.36	1.090	Ea.	230	67		297	355
4025	Fire alarm, loop expander card		14.72	.543		680	33.50		713.50	800
4050	Actuating device	↓	7.36	1.090		340	67		407	470
4160	Alarm control panel, addressable w/o voice, up to 200 points	2 Elec	1.05	15.220		4,575	935		5,510	6,425
4170	addressable w/voice, up to 400 points	"	.67	23.920		9,700	1,475		11,175	12,900
4175	Addressable interface device	1 Elec	6.67	1.200		149	73.50		222.50	274
4200	Battery and rack		3.68	2.170		460	133		593	710
4400	Automatic charger		7.36	1.090		525	67		592	680
4600	Signal bell		7.36	1.090		66.50	67		133.50	173
4610	Fire alarm signal bell 10" red 20-24 V P		7.36	1.090		154	67		221	269
4800	Trouble buzzer or manual station		7.36	1.090		88.50	67		155.50	197
5425	Duct smoke and heat detector 2 wire		7.36	1.090		125	67		192	238
5430	Fire alarm duct detector controller		2.76	2.900		254	178		432	545
5435	Fire alarm duct detector sensor kit		7.36	1.090		75	67		142	182
5440	Remote test station for smoke detector duct type		4.88	1.640		54.50	101		155.50	210
5460	Remote fire alarm indicator light		4.88	1.640		25	101		126	178
5600	Strobe and horn		4.88	1.640		137	101		238	300
5610	Strobe and horn (ADA type)		4.88	1.640		165	101		266	330
5620	Visual alarm (ADA type)		6.16	1.300		119	80		199	250
5800	electric bell		6.16	1.300		54	80		134	179
6000	Door holder, electro-magnetic		3.68	2.170		92	133		225	299
6200	Combination holder and closer		2.94	2.720		146	167		313	410
6600	Drill switch		7.36	1.090		400	67		467	540
6800	Master box		2.48	3.220		6,675	198		6,873	7,625
7000	Break glass station		7.36	1.090		58	67		125	163
7010	Break glass station, addressable		6.67	1.200		171	73.50		244.50	298
7800	Remote annunciator, 8 zone lamp	↓	1.66	4.830		212	296		508	675
8000	12 zone lamp	2 Elec	2.39	6.690		415	410		825	1,075
8200	16 zone lamp	"	2.02	7.910	↓	370	485		855	1,125

28 47 Mass Notification

28 47 12 – Notification Systems

28 47 12.10 Mass Notification System

		Crew	Daily Output	Labor-Hours	Unit	Material	2020 Bare Costs Labor	Equipment	Total	Total Incl O&P
0010	**MASS NOTIFICATION SYSTEM**									
0100	Wireless command center, 10,000 devices	2 Elec	1.22	13.080	Ea.	2,325	800		3,125	3,750
0200	Option, email notification					2,100			2,100	2,325
0210	Remote device supervision & monitor					2,650			2,650	2,900
0300	Antenna VHF or UHF, for medium range	1 Elec	3.68	2.170		93	133		226	300
0310	For high-power transmitter		1.84	4.350		1,125	267		1,392	1,650
0400	Transmitter, 25 watt		3.68	2.170		1,750	133		1,883	2,125
0410	40 watt		2.45	3.270		2,025	201		2,226	2,525
0420	100 watt		1.22	6.540		6,600	400		7,000	7,850
0500	Wireless receiver/control module for speaker		7.36	1.090		286	67		353	415
0600	Desktop paging controller, stand alone		3.68	2.170		850	133		983	1,125
0700	Auxiliary level inputs	↓	7.69	1.040	↓	103	64		167	208

28 52 Detention Security Systems

28 52 11 – Detention Monitoring and Control Systems

28 52 11.10 Detention Control Systems	Crew	Daily Output	Labor-Hours	Unit	Material	2020 Bare Costs Labor	Equipment	Total	Total Incl O&P
0010 **DETENTION CONTROL SYSTEMS**									
1000 Desk top control systems for 10 doors, 10 intercoms, and 10 lights	2 Elec	.28	57.970	Ea.	63,000	3,550		66,550	74,500
1020 Push button control panel systems for 10 doors, 10 intercoms, and 10 lights	"	.29	54.350	"	56,500	3,325		59,825	67,000

Pre-Installation Change Order Costs

Estimating Tips
33 10 00 Water Utilities
33 30 00 Sanitary Sewerage Utilities
33 40 00 Storm Drainage Utilities

- Never assume that the water, sewer, and drainage lines will go in at the early stages of the project. Consider the site access needs before dividing the site in half with open trenches, loose pipe, and machinery obstructions. Always inspect the site to establish that the site drawings are complete. Check off all existing utilities on your drawings as you locate them. Be especially careful with underground utilities because appurtenances are sometimes buried during regrading or repaving operations. If you find any discrepancies, mark up the site plan for further research. Differing site conditions can be very costly if discovered later in the project.

- See also Section 33 01 00 for restoration of pipe where removal/replacement may be undesirable. Use of new types of piping materials can reduce the overall project cost. Owners/design engineers should consider the installing contractor as a valuable source of current information on utility products and local conditions that could lead to significant cost savings.

Reference Numbers
Reference numbers are shown at the beginning of some major classifications. These numbers refer to related items in the Reference Section. The reference information may be an estimating procedure, an alternate pricing method, or technical information.

Note: Not all subdivisions listed here necessarily appear. ■

Same Data. Simplified.

Enjoy the convenience and efficiency of accessing your costs anywhere:

- **Skip the multiplier** by setting your location
- **Quickly search,** edit, favorite and share costs
- **Stay on top of price changes** with automatic updates

Discover more at rsmeans.com/online

33 71 Electrical Utility Transmission and Distribution

33 71 13 – Electrical Utility Towers

33 71 13.23 Steel Electrical Utility Towers

33 71 13.23 Steel Electrical Utility Towers	Crew	Daily Output	Labor-Hours	Unit	Material	2020 Bare Costs Labor	Equipment	Total	Total Incl O&P
0010 **STEEL ELECTRICAL UTILITY TOWERS**									
0500 Towers: material handling and spotting	R-7	20.76	2.310	Ton		101	8	109	162
0540 Steel tower erection	R-5	7.04	12.500		2,300	670	168	3,138	3,700
0550 Lace and box	"	6.53	13.470		2,300	720	181	3,201	3,800
0600 Special towers: material handling and spotting	R-7	11.33	4.240			185	14.65	199.65	296
0640 Special steel structure erection	R-6	6	14.670		2,875	785	289	3,949	4,650
0650 Special steel lace and box	"	5.79	15.210		2,875	815	300	3,990	4,700

33 71 13.80 Transmission Line Right of Way

33 71 13.80 Transmission Line Right of Way	Crew	Daily Output	Labor-Hours	Unit	Material	Labor	Equipment	Total	Total Incl O&P
0010 **TRANSMISSION LINE RIGHT OF WAY**									
0100 Clearing right of way	B-87	6.14	6.520	Acre		350	390	740	955
0200 Restoration & seeding	B-10D	3.68	3.260	"	495	169	525	1,189	1,375

33 71 16 – Electrical Utility Poles

33 71 16.23 Steel Electrical Utility Poles

33 71 16.23 Steel Electrical Utility Poles	Crew	Daily Output	Labor-Hours	Unit	Material	Labor	Equipment	Total	Total Incl O&P
0010 **STEEL ELECTRICAL UTILITY POLES**									
6000 Digging holes in earth, average	R-5	23.13	3.800	Ea.		204	51	255	360
6010 In rock, average	"	4.15	21.210	"		1,125	285	1,410	2,025
6020 Formed plate pole structure									
6030 Material handling and spotting	R-7	2.21	21.740	Ea.		950	75	1,025	1,525
6040 Erect steel plate pole	R-5	1.79	49.050		11,400	2,625	660	14,685	17,300
6050 Guys, anchors and hardware for pole, in earth		6.48	13.590		695	730	183	1,608	2,050
6060 In rock		16.52	5.330		830	286	71.50	1,187.50	1,425
6070 Foundations for line poles									
6080 Excavation, in earth	R-5	124.55	.707	C.Y.		38	9.50	47.50	67.50
6090 In rock		18.40	4.780			256	64	320	455
6110 Concrete foundations		10.12	8.700		164	465	117	746	1,000

33 71 16.33 Wood Electrical Utility Poles

33 71 16.33 Wood Electrical Utility Poles	Crew	Daily Output	Labor-Hours	Unit	Material	Labor	Equipment	Total	Total Incl O&P
0010 **WOOD ELECTRICAL UTILITY POLES**									
0011 Excludes excavation, backfill and cast-in-place concrete									
6200 Wood, class 3 Douglas Fir, penta-treated, 20'	R-3	2.85	7.010	Ea.	256	430	66	752	995
6600 30'		2.39	8.360		320	510	79	909	1,200
7000 40'		2.12	9.450		630	580	89.50	1,299.50	1,650
7200 45'		1.56	12.790		900	780	121	1,801	2,300
7400 Cross arms with hardware & insulators									
7600 4' long	1 Elec	2.30	3.480	Ea.	156	214		370	490
7800 5' long		2.21	3.620		168	222		390	515
8000 6' long		2.02	3.950		190	242		432	570
9000 Disposal of pole & hardware surplus material	R-7	19.20	2.500	Mile		109	8.65	117.65	175
9100 Disposal of crossarms & hardware surplus material	"	36.80	1.300	"		56.50	4.50	61	91

33 71 19 – Electrical Underground Ducts and Manholes

33 71 19.15 Underground Ducts and Manholes

33 71 19.15 Underground Ducts and Manholes	Crew	Daily Output	Labor-Hours	Unit	Material	Labor	Equipment	Total	Total Incl O&P
0010 **UNDERGROUND DUCTS AND MANHOLES**									
0011 Not incl. excavation, backfill, or concrete in slab and duct bank									
1000 Direct burial									
1010 PVC, schedule 40, w/coupling, 1/2" diameter	1 Elec	312.80	.026	L.F.	.41	1.57		1.98	2.79
1020 3/4" diameter		266.80	.030		.49	1.84		2.33	3.28
1030 1" diameter		239.20	.033		.80	2.05		2.85	3.93
1040 1-1/2" diameter		193.20	.041		1.19	2.54		3.73	5.10
1050 2" diameter		165.60	.048		1.55	2.96		4.51	6.10
1060 3" diameter	2 Elec	220.80	.072		2.86	4.45		7.31	9.75
1070 4" diameter		147.20	.109		3.79	6.65		10.44	14.10
1080 5" diameter		110.40	.145		5.65	8.90		14.55	19.45

For customer support on your Electrical Change Order Costs with RSMeans data, call 800.448.8182.

Pre-Installation Change Order Costs

33 71 19.15 Underground Ducts and Manholes	Crew	Daily Output	Labor-Hours	Unit	Material	2020 Bare Costs Labor	Equipment	Total	Total Incl O&P
1090	6" diameter	2 Elec	82.80	.193	L.F.	7.40	11.85		19.25	26
1110	Elbows, 1/2" diameter	1 Elec	44.16	.181	Ea.	.70	11.10		11.80	17.30
1120	3/4" diameter		34.96	.229		.72	14.05		14.77	22
1130	1" diameter		29.44	.272		1.45	16.65		18.10	26.50
1140	1-1/2" diameter		19.32	.414		3.25	25.50		28.75	41.50
1150	2" diameter		14.72	.543		4.25	33.50		37.75	54
1160	3" diameter		11.04	.725		14.70	44.50		59.20	82
1170	4" diameter		8.28	.966		27.50	59.50		87	119
1180	5" diameter		7.36	1.090		41	67		108	145
1190	6" diameter		4.60	1.740		42	107		149	206
1210	Adapters, 1/2" diameter		47.84	.167		.31	10.25		10.56	15.65
1220	3/4" diameter		39.56	.202		.44	12.40		12.84	18.95
1230	1" diameter		35.88	.223		.47	13.70		14.17	21
1240	1-1/2" diameter		32.20	.248		.85	15.25		16.10	23.50
1250	2" diameter		23.92	.334		1.67	20.50		22.17	32.50
1260	3" diameter		18.40	.435		3.42	26.50		29.92	43.50
1270	4" diameter		12.88	.621		5.75	38		43.75	63
1280	5" diameter		11.04	.725		11.60	44.50		56.10	79
1290	6" diameter		8.28	.966		22.50	59.50		82	114
1340	Bell end & cap, 1-1/2" diameter		32.20	.248		5.65	15.25		20.90	28.50
1350	Bell end & plug, 2" diameter		23.92	.334		7.05	20.50		27.55	38.50
1360	3" diameter		18.40	.435		9.90	26.50		36.40	50.50
1370	4" diameter		12.88	.621		11.85	38		49.85	69.50
1380	5" diameter		11.04	.725		19.15	44.50		63.65	87
1390	6" diameter		8.28	.966		20.50	59.50		80	111
1450	Base spacer, 2" diameter		51.52	.155		1.64	9.55		11.19	16
1460	3" diameter		42.32	.189		2.14	11.60		13.74	19.60
1470	4" diameter		37.72	.212		2.14	13		15.14	21.50
1480	5" diameter		34.04	.235		2.39	14.40		16.79	24
1490	6" diameter		31.28	.256		2.98	15.70		18.68	27
1550	Intermediate spacer, 2" diameter		55.20	.145		1.50	8.90		10.40	14.90
1560	3" diameter		42.32	.189		2.11	11.60		13.71	19.55
1570	4" diameter		37.72	.212		1.91	13		14.91	21.50
1580	5" diameter		34.04	.235		2.33	14.40		16.73	24
1590	6" diameter		31.28	.256	L.F.	3.10	15.70		18.80	27
4010	PVC, schedule 80, w/coupling, 1/2" diameter		197.80	.040	L.F.	1.24	2.48		3.72	5.05
4020	3/4" diameter		165.60	.048		1.69	2.96		4.65	6.25
4030	1" diameter		133.40	.060		2.29	3.68		5.97	8
4040	1-1/2" diameter		110.40	.072		3.78	4.45		8.23	10.75
4050	2" diameter		92	.087		5.15	5.35		10.50	13.60
4060	3" diameter	2 Elec	119.60	.134		10.75	8.20		18.95	24
4070	4" diameter		82.80	.193		15.40	11.85		27.25	34.50
4080	5" diameter		64.40	.248		21.50	15.25		36.75	46.50
4090	6" diameter		46	.348		30.50	21.50		52	65.50
4110	Elbows, 1/2" diameter	1 Elec	26.68	.300	Ea.	2.96	18.40		21.36	31
4120	3/4" diameter		21.16	.378		6.85	23		29.85	42
4130	1" diameter		18.40	.435		8.15	26.50		34.65	48.50
4140	1-1/2" diameter		14.72	.543		13.70	33.50		47.20	64.50
4150	2" diameter		11.04	.725		18.30	44.50		62.80	86
4160	3" diameter		8.28	.966		46	59.50		105.50	139
4170	4" diameter		6.44	1.240		80.50	76		156.50	202
4180	5" diameter		5.52	1.450		201	89		290	355
4190	6" diameter		3.68	2.170		87	133		220	294

33 71 19.15 Underground Ducts and Manholes

		Crew	Daily Output	Labor-Hours	Unit	Material	2020 Bare Costs Labor	Equipment	Total	Total Incl O&P
4210	Adapter, 1/2" diameter	1 Elec	35.88	.223	Ea.	.31	13.70		14.01	21
4220	3/4" diameter		30.36	.264		.44	16.15		16.59	24.50
4230	1" diameter		26.68	.300		.47	18.40		18.87	28
4240	1-1/2" diameter		23.92	.334		.85	20.50		21.35	31.50
4250	2" diameter		21.16	.378		1.67	23		24.67	36.50
4260	3" diameter		16.56	.483		3.42	29.50		32.92	48
4270	4" diameter		11.96	.669		5.75	41		46.75	67.50
4280	5" diameter		10.12	.791		11.60	48.50		60.10	85
4290	6" diameter		7.36	1.090		22.50	67		89.50	125
4310	Bell end & cap, 1-1/2" diameter		23.92	.334		5.65	20.50		26.15	36.50
4320	Bell end & plug, 2" diameter		21.16	.378		7.05	23		30.05	42.50
4330	3" diameter		16.56	.483		9.90	29.50		39.40	55
4340	4" diameter		11.96	.669		11.85	41		52.85	74
4350	5" diameter		10.12	.791		19.15	48.50		67.65	93
4360	6" diameter		7.36	1.090		20.50	67		87.50	122
4370	Base spacer, 2" diameter		38.64	.207		1.64	12.70		14.34	20.50
4380	3" diameter		30.36	.264		2.14	16.15		18.29	26.50
4390	4" diameter		26.68	.300		2.14	18.40		20.54	30
4400	5" diameter		23.92	.334		2.39	20.50		22.89	33
4410	6" diameter		23	.348		2.98	21.50		24.48	35.50
4420	Intermediate spacer, 2" diameter		41.40	.193		1.50	11.85		13.35	19.30
4430	3" diameter		31.28	.256		2.11	15.70		17.81	26
4440	4" diameter		28.52	.281		1.91	17.20		19.11	27.50
4450	5" diameter		25.76	.311		2.33	19.05		21.38	31
4460	6" diameter		23	.348		3.10	21.50		24.60	35.50

33 71 19.17 Electric and Telephone Underground

		Crew	Daily Output	Labor-Hours	Unit	Material	2020 Bare Costs Labor	Equipment	Total	Total Incl O&P
0010	**ELECTRIC AND TELEPHONE UNDERGROUND**									
0011	Not including excavation									
0200	backfill and cast in place concrete									
0400	Hand holes, precast concrete, with concrete cover									
0600	2' x 2' x 3' deep	R-3	2.21	9.060	Ea.	545	555	85.50	1,185.50	1,525
0800	3' x 3' x 3' deep		1.75	11.440		705	700	108	1,513	1,950
1000	4' x 4' x 4' deep		1.29	15.530		1,600	950	147	2,697	3,325
1200	Manholes, precast with iron racks & pulling irons, C.I. frame									
1400	and cover, 4' x 6' x 7' deep	B-13	1.84	30.430	Ea.	3,000	1,400	315	4,715	5,750
1600	6' x 8' x 7' deep		1.75	32.040		3,375	1,475	330	5,180	6,275
1800	6' x 10' x 7' deep		1.66	33.820		3,775	1,550	350	5,675	6,850
4200	Underground duct, banks ready for concrete fill, min. of 7.5"									
4400	between conduits, center to center									
4580	PVC, type EB, 1 @ 2" diameter	2 Elec	441.60	.036	L.F.	1.34	2.22		3.56	4.78
4600	2 @ 2" diameter		220.80	.072		2.68	4.45		7.13	9.55
4800	4 @ 2" diameter		110.40	.145		5.35	8.90		14.25	19.15
4900	1 @ 3" diameter		368	.043		1.94	2.67		4.61	6.10
5000	2 @ 3" diameter		184	.087		3.88	5.35		9.23	12.20
5200	4 @ 3" diameter		92	.174		7.75	10.65		18.40	24.50
5300	1 @ 4" diameter		294.40	.054		2.23	3.33		5.56	7.40
5400	2 @ 4" diameter		147.20	.109		4.46	6.65		11.11	14.85
5600	4 @ 4" diameter		73.60	.217		8.90	13.35		22.25	29.50
5800	6 @ 4" diameter		49.68	.322		13.40	19.75		33.15	44
5810	1 @ 5" diameter		239.20	.067		2.51	4.10		6.61	8.85
5820	2 @ 5" diameter		119.60	.134		5	8.20		13.20	17.70
5840	4 @ 5" diameter		64.40	.248		10.05	15.25		25.30	33.50

33 71 19.17 Electric and Telephone Underground		Crew	Daily Output	Labor-Hours	Unit	Material	2020 Bare Costs Labor	Equipment	Total	Total Incl O&P
5860	6 @ 5" diameter	2 Elec	46	.348	L.F.	15.05	21.50		36.55	48.50
5870	1 @ 6" diameter		184	.087		4.31	5.35		9.66	12.70
5880	2 @ 6" diameter		92	.174		8.60	10.65		19.25	25.50
5900	4 @ 6" diameter		46	.348		17.20	21.50		38.70	51
5920	6 @ 6" diameter		27.60	.580		26	35.50		61.50	81.50
6200	Rigid galvanized steel, 2 @ 2" diameter		165.60	.097		18.85	5.95		24.80	30
6400	4 @ 2" diameter		82.80	.193		37.50	11.85		49.35	59
6800	2 @ 3" diameter		92	.174		34.50	10.65		45.15	54
7000	4 @ 3" diameter		46	.348		69	21.50		90.50	108
7200	2 @ 4" diameter		64.40	.248		55	15.25		70.25	82.50
7400	4 @ 4" diameter		31.28	.512		110	31.50		141.50	167
7600	6 @ 4" diameter		20.24	.791		164	48.50		212.50	253
7620	2 @ 5" diameter		55.20	.290		86.50	17.80		104.30	122
7640	4 @ 5" diameter		27.60	.580		173	35.50		208.50	244
7660	6 @ 5" diameter		16.56	.966		260	59.50		319.50	375
7680	2 @ 6" diameter		36.80	.435		132	26.50		158.50	185
7700	4 @ 6" diameter		18.40	.870		263	53.50		316.50	370
7720	6 @ 6" diameter		12.88	1.240		395	76		471	550
7800	For cast-in-place concrete, add									
7810	Under 1 C.Y.	C-6	14.72	3.260	C.Y.	247	143	3.65	393.65	490
7820	1 C.Y. to 5 C.Y.		17.66	2.720		224	119	3.05	346.05	430
7830	Over 5 C.Y.		22.08	2.170		185	95	2.43	282.43	350
7850	For reinforcing rods, add									
7860	#4 to #7	2 Rodm	1.01	15.810	Ton	1,350	890		2,240	2,850
7870	#8 to #14	"	1.38	11.590	"	1,350	655		2,005	2,475
8000	Fittings, PVC type EB, elbow, 2" diameter	1 Elec	14.72	.543	Ea.	19.05	33.50		52.55	70.50
8200	3" diameter		12.88	.621		22	38		60	80.50
8400	4" diameter		11.04	.725		48	44.50		92.50	119
8420	5" diameter		9.20	.870		147	53.50		200.50	242
8440	6" diameter		8.28	.966		229	59.50		288.50	340
8500	Coupling, 2" diameter					1.02			1.02	1.12
8600	3" diameter					4.97			4.97	5.45
8700	4" diameter					5.40			5.40	5.90
8720	5" diameter					10.25			10.25	11.30
8740	6" diameter					28.50			28.50	31.50
8800	Adapter, 2" diameter	1 Elec	23.92	.334		1.21	20.50		21.71	32
9000	3" diameter		18.40	.435		3.68	26.50		30.18	43.50
9200	4" diameter		14.72	.543		4.03	33.50		37.53	54
9220	5" diameter		11.96	.669		10.25	41		51.25	72.50
9240	6" diameter		9.20	.870		15	53.50		68.50	96
9400	End bell, 2" diameter		14.72	.543		2.76	33.50		36.26	52.50
9600	3" diameter		12.88	.621		4.90	38		42.90	62
9800	4" diameter		11.04	.725		5.55	44.50		50.05	72
9810	5" diameter		9.20	.870		9.75	53.50		63.25	90
9820	6" diameter		7.36	1.090		12.15	67		79.15	113
9830	5° angle coupling, 2" diameter		23.92	.334		9.35	20.50		29.85	41
9840	3" diameter		18.40	.435		31	26.50		57.50	73.50
9850	4" diameter		14.72	.543		23	33.50		56.50	75
9860	5" diameter		11.96	.669		20	41		61	83
9870	6" diameter		9.20	.870		31	53.50		84.50	114
9880	Expansion joint, 2" diameter		14.72	.543		40	33.50		73.50	93.50
9890	3" diameter		16.56	.483		84	29.50		113.50	137
9900	4" diameter		11.04	.725		123	44.50		167.50	201

33 71 Electrical Utility Transmission and Distribution

33 71 19 – Electrical Underground Ducts and Manholes

33 71 19.17 Electric and Telephone Underground

		Crew	Daily Output	Labor-Hours	Unit	Material	2020 Bare Costs Labor	Equipment	Total	Total Incl O&P
9910	5" diameter	1 Elec	9.20	.870	Ea.	244	53.50		297.50	350
9920	6" diameter	↓	7.36	1.090		212	67		279	335
9930	Heat bender, 2" diameter					585			585	645
9940	6" diameter					1,800			1,800	1,975
9950	Cement, quart				↓	18.65			18.65	20.50
9960	Nylon polyethylene pull rope, 1/4"	2 Elec	1840	.009	L.F.	.22	.53		.75	1.03

33 71 23 – Insulators and Fittings

33 71 23.16 Post Insulators

		Crew	Daily Output	Labor-Hours	Unit	Material	2020 Bare Costs Labor	Equipment	Total	Total Incl O&P
0010	**POST INSULATORS**									
7400	Insulators, pedestal type	R-11	103.04	.543	Ea.		31.50	6.20	37.70	54.50
7490	See also line 33 71 39.13 1000									

33 71 26 – Transmission and Distribution Equipment

33 71 26.13 Capacitor Banks

		Crew	Daily Output	Labor-Hours	Unit	Material	2020 Bare Costs Labor	Equipment	Total	Total Incl O&P
0010	**CAPACITOR BANKS**									
1300	Station capacitors									
1350	Synchronous, 13 to 26 kV	R-11	2.86	19.570	MVAR	7,550	1,150	223	8,923	10,200
1360	46 kV		3.06	18.280		9,625	1,075	208	10,908	12,400
1370	69 kV		3.51	15.980		9,475	935	182	10,592	12,000
1380	161 kV		5.99	9.350		8,875	545	106	9,526	10,700
1390	500 kV		9.54	5.870		7,700	345	67	8,112	9,050
1450	Static, 13 to 26 kV		2.86	19.570		6,400	1,150	223	7,773	9,000
1460	46 kV		2.77	20.220		8,100	1,175	230	9,505	10,900
1470	69 kV		3.51	15.980		7,850	935	182	8,967	10,300
1480	161 kV		5.99	9.350		7,300	545	106	7,951	8,950
1490	500 kV		9.54	5.870	↓	6,650	345	67	7,062	7,900
1600	Voltage regulators, 13 to 26 kV	↓	.69	81.160	Ea.	290,000	4,725	925	295,650	327,000

33 71 26.23 Current Transformers

		Crew	Daily Output	Labor-Hours	Unit	Material	2020 Bare Costs Labor	Equipment	Total	Total Incl O&P
0010	**CURRENT TRANSFORMERS**									
4050	Current transformers, 13 to 26 kV	R-11	12.88	4.350	Ea.	3,575	254	49.50	3,878.50	4,350
4060	46 kV		8.58	6.520		10,400	380	74	10,854	12,100
4070	69 kV		6.44	8.700		10,800	510	99	11,409	12,800
4080	161 kV	↓	1.72	32.550	↓	35,000	1,900	370	37,270	41,700

33 71 26.26 Potential Transformers

		Crew	Daily Output	Labor-Hours	Unit	Material	2020 Bare Costs Labor	Equipment	Total	Total Incl O&P
0010	**POTENTIAL TRANSFORMERS**									
4100	Potential transformers, 13 to 26 kV	R-11	10.30	5.430	Ea.	5,100	315	62	5,477	6,150
4110	46 kV		7.36	7.610		10,500	445	86.50	11,031.50	12,300
4120	69 kV		5.72	9.790		11,100	570	111	11,781	13,200
4130	161 kV		2.06	27.170		24,000	1,575	310	25,885	29,100
4140	500 kV	↓	1.29	43.480	↓	71,500	2,525	495	74,520	83,500

33 71 26.28 Overhead Fuse Cutouts

		Crew	Daily Output	Labor-Hours	Unit	Material	2020 Bare Costs Labor	Equipment	Total	Total Incl O&P
0010	**OVERHEAD FUSE CUTOUTS**									
1000	Cutout Fuse Assemblies									
1100	15.0 kV, 110 kV BIL, 100 amp, type C, silicone	R-8	5.52	8.700	Ea.	121	470	50	641	900
1200	200 amp		5.52	8.700		197	470	50	717	985
1300	300 amp	↓	5.52	8.700	↓	235	470	50	755	1,025
2000	Cutout Fuse Links									
2100	Removeable button head, type K, 100 amp	R-3	9.20	2.170	Ea.	11.55	133	20.50	165.05	233
2140	140 amp	"	9.20	2.170	"	26	133	20.50	179.50	249
3000	Cutout Fuse Hardware									
3100	Tri-mount bracket, aluminum	R-3	9.20	2.170	Ea.	230	133	20.50	383.50	475
3200	Arrester, surge, 12.7 kV		9.20	2.170		61.50	133	20.50	215	288

 Pre-Installation Change Order Costs

33 71 Electrical Utility Transmission and Distribution

33 71 26 – Transmission and Distribution Equipment

33 71 26.28 Overhead Fuse Cutouts	Crew	Daily Output	Labor-Hours	Unit	Material	2020 Bare Costs Labor	Equipment	Total	Total Incl O&P
3300	Clamp, hot-line tap, bronze, 5"	R-3	9.20	2.170	Ea.	23	133	20.50	176.50	246
3400	Clamp, stirrup, bronze, 1-13/16"		9.20	2.170		16.85	133	20.50	170.35	239
3500	Insulator, dead-end, clevis tongue, 12-1/2"	↓	9.20	2.170	↓	29	133	20.50	182.50	253

33 71 39 – High-Voltage Wiring

33 71 39.13 Overhead High-Voltage Wiring

	33 71 39.13 Overhead High-Voltage Wiring	Crew	Daily Output	Labor-Hours	Unit	Material	2020 Bare Costs Labor	Equipment	Total	Total Incl O&P
0010	**OVERHEAD HIGH-VOLTAGE WIRING**									
0100	Conductors, primary circuits									
0110	Material handling and spotting	R-5	9	9.780	W.Mile		525	131	656	930
0120	For river crossing, add		10.12	8.700			465	117	582	830
0150	Conductors, per wire, 210 to 636 kcmil		1.80	48.800		13,300	2,625	655	16,580	19,300
0160	795 to 954 kcmil		1.72	51.150		27,200	2,750	685	30,635	34,800
0170	1,000 to 1,600 kcmil		1.35	65.070		49,800	3,500	875	54,175	61,000
0180	Over 1,600 kcmil		1.24	70.850		70,000	3,800	950	74,750	84,000
0200	For river crossing, add, 210 to 636 kcmil		1.14	77.140			4,125	1,025	5,150	7,350
0220	795 to 954 kcmil		1	87.750			4,700	1,175	5,875	8,350
0230	1,000 to 1,600 kcmil		.89	98.610			5,275	1,325	6,600	9,375
0240	Over 1,600 kcmil	↓	.80	109	↓		5,850	1,475	7,325	10,400
0300	Joints and dead ends	R-8	5.52	8.700	Ea.	1,900	470	50	2,420	2,875
0400	Sagging	R-5	6.75	13.040	W.Mile		700	175	875	1,250
0500	Clipping, per structure, 69 kV	R-10	8.83	5.430	Ea.		315	43.50	358.50	520
0510	161 kV		4.90	9.790			565	78	643	930
0520	345 to 500 kV	↓	2.33	20.620			1,200	164	1,364	1,950
0600	Make and install jumpers, per structure, 69 kV	R-8	2.94	16.300		515	885	94	1,494	2,000
0620	161 kV		1.10	43.480		1,025	2,350	251	3,626	4,925
0640	345 to 500 kV		.29	163		1,725	8,850	940	11,515	16,200
0700	Spacers	R-10	63.08	.761		103	44	6.05	153.05	186
0720	For river crossings, add	"	55.20	.870	↓		50.50	6.95	57.45	82.50
0800	Installing pulling line (500 kV only)	R-9	1.33	47.980	W.Mile	875	2,425	208	3,508	4,850
0810	Disposal of surplus material, high voltage conductors	R-7	6.40	7.500	Mile		325	26	351	525
0820	With trailer mounted reel stands	"	12.61	3.810	"		166	13.20	179.20	267
0900	Insulators and hardware, primary circuits									
0920	Material handling and spotting, 69 kV	R-7	441.60	.109	Ea.		4.74	.38	5.12	7.60
0930	161 kV		630.85	.076			3.32	.26	3.58	5.35
0950	345 to 500 kV	↓	883.20	.054			2.37	.19	2.56	3.80
1000	Disk insulators, 69 kV	R-5	809.60	.109		105	5.85	1.46	112.31	125
1020	161 kV		899.56	.098		120	5.25	1.31	126.56	141
1040	345 to 500 kV	↓	1012	.087	↓	120	4.66	1.17	125.83	140
1060	See Section 33 71 23.16 for pin or pedestal insulator									
1100	Install disk insulator at river crossing, add									
1110	69 kV	R-5	539.74	.163	Ea.		8.75	2.19	10.94	15.50
1120	161 kV		809.60	.109			5.85	1.46	7.31	10.35
1140	345 to 500 kV	↓	809.60	.109	↓		5.85	1.46	7.31	10.35
1150	Disposal of surplus material, high voltage insulators	R-7	38.40	1.250	Mile		54.50	4.33	58.83	87.50
1300	Overhead ground wire installation									
1320	Material handling and spotting	R-7	5.20	9.230	W.Mile		400	32	432	645
1340	Overhead ground wire	R-5	1.62	54.350		3,800	2,925	730	7,455	9,350
1350	At river crossing, add		1.08	81.750	↓		4,375	1,100	5,475	7,775
1360	Disposal of surplus material, grounding wire	↓	38.40	2.290	Mile		123	31	154	218
1400	Installing conductors, underbuilt circuits									
1420	Material handling and spotting	R-7	5.20	9.230	W.Mile		400	32	432	645
1440	Conductors, per wire, 210 to 636 kcmil	R-5	1.80	48.800		13,300	2,625	655	16,580	19,300
1450	795 to 954 kcmil	↓	1.72	51.150	↓	27,200	2,750	685	30,635	34,800

33 71 39.13 Overhead High-Voltage Wiring	Crew	Daily Output	Labor-Hours	Unit	Material	2020 Bare Costs Labor	Equipment	Total	Total Incl O&P
1460	1,000 to 1,600 kcmil	R-5	1.35	65.070	W.Mile	49,800	3,500	875	54,175	61,000
1470	Over 1,600 kcmil		1.24	70.850		70,000	3,800	950	74,750	84,000
1500	Joints and dead ends	R-8	5.52	8.700	Ea.	1,900	470	50	2,420	2,875
1550	Sagging	R-5	8.10	10.870	W.Mile		585	146	731	1,025
1600	Clipping, per structure, 69 kV	R-10	8.83	5.430	Ea.		315	43.50	358.50	520
1620	161 kV		4.90	9.790			565	78	643	930
1640	345 to 500 kV		2.33	20.620			1,200	164	1,364	1,950
1700	Making and installing jumpers, per structure, 69 kV	R-8	5.40	8.890		515	485	51.50	1,051.50	1,350
1720	161 kV		.88	54.350		1,025	2,950	315	4,290	5,900
1740	345 to 500 kV		.29	163		1,725	8,850	940	11,515	16,200
1800	Spacers	R-10	88.32	.543		103	31.50	4.33	138.83	165
1810	Disposal of surplus material, conductors & hardware	R-7	6.40	7.500	Mile		325	26	351	525
2000	Insulators and hardware for underbuilt circuits									
2100	Material handling and spotting	R-7	1104	.043	Ea.		1.90	.15	2.05	3.05
2150	Disk insulators, 69 kV	R-8	552	.087		105	4.72	.50	110.22	123
2160	161 kV		631.12	.076		120	4.13	.44	124.57	139
2170	345 to 500 kV		736	.065		120	3.54	.38	123.92	138
2180	Disposal of surplus material, insulators & hardware	R-7	38.40	1.250	Mile		54.50	4.33	58.83	87.50
2300	Sectionalizing switches, 69 kV	R-5	1.16	75.910	Ea.	27,000	4,075	1,025	32,100	36,900
2310	161 kV		.74	119		30,500	6,375	1,600	38,475	44,800
2500	Protective devices		5.06	17.390		8,825	930	234	9,989	11,400
2600	Clearance poles, 8 poles per mile									
2650	In earth, 69 kV	R-5	1.07	82.460	Mile	7,250	4,425	1,100	12,775	15,800
2660	161 kV	"	.59	149		11,900	8,000	2,000	21,900	27,300
2670	345 to 500 kV	R-6	.44	199		14,300	10,700	3,925	28,925	36,000
2800	In rock, 69 kV	R-5	.63	138		7,250	7,400	1,850	16,500	21,100
2820	161 kV	"	.32	273		11,900	14,600	3,675	30,175	39,000
2840	345 to 500 kV	R-6	.22	398		14,300	21,300	7,850	43,450	56,500

33 72 Utility Substations
33 72 26 – Substation Bus Assemblies
33 72 26.13 Aluminum Substation Bus Assemblies

		Crew	Daily Output	Labor-Hours	Unit	Material	Labor	Equipment	Total	Total Incl O&P
0010	**ALUMINUM SUBSTATION BUS ASSEMBLIES**									
7300	Bus	R-11	542.80	.103	Lb.	3.30	6	1.17	10.47	13.90

33 72 33 – Control House Equipment
33 72 33.43 Substation Backup Batteries

0010	**SUBSTATION BACKUP BATTERIES**									
9120	Battery chargers	R-11	10.30	5.430	Ea.	4,350	315	62	4,727	5,325
9200	Control batteries	"	12.88	4.350	K.A.H.	540	254	49.50	843.50	1,025

33 72 53 – Shunt Reactors
33 72 53.13 Reactors

0010	**REACTORS**									
8150	Reactors and resistors, 13 to 26 kV	R-11	25.76	2.170	Ea.	4,375	127	24.50	4,526.50	5,025
8160	46 kV		3.97	14.120		13,500	825	161	14,486	16,300
8170	69 kV		2.58	21.740		21,500	1,275	247	23,022	25,900
8180	161 kV		2.06	27.170		24,000	1,575	310	25,885	29,100
8190	500 kV		.07	760		93,000	44,400	8,650	146,050	178,000

33 73 Utility Transformers

33 73 23 – Dry-Type Utility Transformers

33 73 23.20 Primary Transformers

33 73 23.20 Primary Transformers	Crew	Daily Output	Labor-Hours	Unit	Material	2020 Bare Costs Labor	2020 Bare Costs Equipment	Total	Total Incl O&P
0010 **PRIMARY TRANSFORMERS**									
1000 Main conversion equipment									
1050 Power transformers, 13 to 26 kV	R-11	1.58	35.390	MVA	27,100	2,075	405	29,580	33,300
1060 46 kV		3.22	17.390		25,400	1,025	198	26,623	29,600
1070 69 kV		2.86	19.570		21,600	1,150	223	22,973	25,700
1080 110 kV		3.03	18.500		20,600	1,075	211	21,886	24,400
1090 161 kV		3.97	14.120		19,100	825	161	20,086	22,400
1100 500 kV		6.44	8.700		19,000	510	99	19,609	21,800
1200 Grounding transformers		2.86	19.570	Ea.	121,000	1,150	223	122,373	135,000

33 75 High-Voltage Switchgear and Protection Devices

33 75 13 – Air High-Voltage Circuit Breaker

33 75 13.13 High-Voltage Circuit Breaker, Air

	Crew	Daily Output	Labor-Hours	Unit	Material	Labor	Equipment	Total	Total Incl O&P
0010 **HIGH-VOLTAGE CIRCUIT BREAKER, AIR**									
2100 Air circuit breakers, 13 to 26 kV	R-11	.52	108	Ea.	68,000	6,300	1,225	75,525	85,500
2110 161 kV	"	.22	253	"	290,500	14,800	2,875	308,175	344,500

33 75 16 – Oil High-Voltage Circuit Breaker

33 75 16.13 Oil Circuit Breaker

	Crew	Daily Output	Labor-Hours	Unit	Material	Labor	Equipment	Total	Total Incl O&P
0010 **OIL CIRCUIT BREAKER**									
2000 Power circuit breakers									
2050 Oil circuit breakers, 13 to 26 kV	R-11	1.03	54.350	Ea.	68,000	3,175	620	71,795	80,000
2060 46 kV		.69	81.160		98,500	4,725	925	104,150	116,500
2070 69 kV		.41	135		228,500	7,875	1,525	237,900	265,000
2080 161 kV		.15	380		335,500	22,200	4,325	362,025	407,000
2090 500 kV		.06	1014		1,254,000	59,000	11,500	1,324,500	1,480,000

33 75 19 – Gas High-Voltage Circuit Breaker

33 75 19.13 Gas Circuit Breaker

	Crew	Daily Output	Labor-Hours	Unit	Material	Labor	Equipment	Total	Total Incl O&P
0010 **GAS CIRCUIT BREAKER**									
2150 Gas circuit breakers, 13 to 26 kV	R-11	.52	108	Ea.	254,500	6,300	1,225	262,025	290,500
2160 161 kV		.07	760		333,500	44,400	8,650	386,550	442,500
2170 500 kV		.04	1521		1,173,500	89,000	17,300	1,279,800	1,442,500

33 75 23 – Vacuum High-Voltage Circuit Breaker

33 75 23.13 Vacuum Circuit Breaker

	Crew	Daily Output	Labor-Hours	Unit	Material	Labor	Equipment	Total	Total Incl O&P
0010 **VACUUM CIRCUIT BREAKER**									
2000 Power circuit breakers									
2200 Vacuum circuit breakers, 13 to 26 kV	R-11	.52	108	Ea.	61,500	6,300	1,225	69,025	78,500

33 75 36 – High-Voltage Utility Fuses

33 75 36.13 Fuses

	Crew	Daily Output	Labor-Hours	Unit	Material	Labor	Equipment	Total	Total Incl O&P
0010 **FUSES**									
8250 Fuses, 13 to 26 kV	R-11	17.18	3.260	Ea.	2,475	190	37	2,702	3,025
8260 46 kV		10.30	5.430		2,825	315	62	3,202	3,675
8270 69 kV		7.36	7.610		3,050	445	86.50	3,581.50	4,100
8280 161 kV		4.30	13.030		3,700	760	148	4,608	5,375

33 75 High-Voltage Switchgear and Protection Devices

33 75 39 – High-Voltage Surge Arresters

33 75 39.13 Surge Arresters

		Crew	Daily Output	Labor-Hours	Unit	Material	2020 Bare Costs Labor	Equipment	Total	Total Incl O&P
0010	**SURGE ARRESTERS**									
8000	Protective equipment									
8050	Lightning arresters, 13 to 26 kV	R-11	17.18	3.260	Ea.	1,850	190	37	2,077	2,375
8060	46 kV		12.88	4.350		4,750	254	49.50	5,053.50	5,650
8070	69 kV		10.30	5.430		6,325	315	62	6,702	7,500
8080	161 kV		5.15	10.870		8,800	635	124	9,559	10,800
8090	500 kV		1.29	43.480		30,100	2,525	495	33,120	37,400

33 75 53 – High-Voltage Switches

33 75 53.13 Switches

		Crew	Daily Output	Labor-Hours	Unit	Material	2020 Bare Costs Labor	Equipment	Total	Total Incl O&P
0010	**SWITCHES**									
3000	Disconnecting switches									
3050	Gang operated switches									
3060	Manual operation, 13 to 26 kV	R-11	1.52	36.890	Ea.	17,000	2,150	420	19,570	22,400
3070	46 kV		1.03	54.350		28,800	3,175	620	32,595	37,100
3080	69 kV		.74	76.090		31,000	4,450	865	36,315	41,700
3090	161 kV		.52	108		37,200	6,300	1,225	44,725	51,500
3100	500 kV		.13	434		99,500	25,300	4,950	129,750	152,000
3110	Motor operation, 161 kV		.47	119		56,000	6,950	1,350	64,300	73,500
3120	500 kV		.26	217		146,000	12,700	2,475	161,175	182,000
3250	Circuit switches, 161 kV		.38	148		105,000	8,625	1,675	115,300	130,500
3300	Single pole switches									
3350	Disconnecting switches, 13 to 26 kV	R-11	25.76	2.170	Ea.	15,300	127	24.50	15,451.50	17,000
3360	46 kV		7.36	7.610		27,100	445	86.50	27,631.50	30,600
3370	69 kV		5.15	10.870		28,800	635	124	29,559	32,800
3380	161 kV		2.58	21.740		106,500	1,275	247	108,022	119,000
3390	500 kV		.20	276		302,000	16,100	3,150	321,250	360,000
3450	Grounding switches, 46 kV		5.15	10.870		40,300	635	124	41,059	45,500
3460	69 kV		3.43	16.320		40,700	950	186	41,836	46,400
3470	161 kV		2.06	27.170		43,500	1,575	310	45,385	50,500
3480	500 kV		.57	98.180		55,000	5,725	1,125	61,850	70,500

33 78 Substation Converter Stations

33 78 33 – Converter Stations

33 78 33.46 Substation Converter Stations

		Crew	Daily Output	Labor-Hours	Unit	Material	2020 Bare Costs Labor	Equipment	Total	Total Incl O&P
0010	**SUBSTATION CONVERTER STATIONS**									
9000	Station service equipment									
9100	Conversion equipment									
9110	Station service transformers	R-11	5.15	10.870	Ea.	105,000	635	124	105,759	116,500

Pre-Installation Change Order Costs

33 79 Site Grounding

33 79 83 – Site Grounding Conductors

33 79 83.13 Grounding Wire, Bar, and Rod	Crew	Daily Output	Labor-Hours	Unit	Material	2020 Bare Costs Labor	Equipment	Total	Total Incl O&P
0010 **GROUNDING WIRE, BAR, AND ROD**									
7500 Grounding systems	R-11	257.60	.217	Lb.	11.65	12.70	2.47	26.82	34.50

Division Notes

	CREW	DAILY OUTPUT	LABOR-HOURS	UNIT	BARE COSTS				TOTAL INCL O&P
					MAT.	LABOR	EQUIP.	TOTAL	

Estimating Tips
34 11 00 Rail Tracks
This subdivision includes items that may involve either repair of existing or construction of new railroad tracks. Additional preparation work, such as the roadbed earthwork, would be found in Division 31. Additional new construction siding and turnouts are found in Subdivision 34 72. Maintenance of railroads is found under 34 01 23 Operation and Maintenance of Railways.

34 40 00 Traffic Signals
This subdivision includes traffic signal systems. Other traffic control devices such as traffic signs are found in Subdivision 10 14 53 Traffic Signage.

34 70 00 Vehicle Barriers
This subdivision includes security vehicle barriers, guide and guard rails, crash barriers, and delineators. The actual maintenance and construction of concrete and asphalt pavement are found in Division 32.

Reference Numbers
Reference numbers are shown at the beginning of some major classifications. These numbers refer to related items in the Reference Section. The reference information may be an estimating procedure, an alternate pricing method, or technical information.

Note: Not all subdivisions listed here necessarily appear. ■

Same Data. Simplified.

Enjoy the convenience and efficiency of accessing your costs anywhere:

■ **Skip the multiplier** by setting your location

■ **Quickly search,** edit, favorite and share costs

■ **Stay on top of price changes** with automatic updates

Discover more at rsmeans.com/online

34 43 Airfield Signaling and Control Equipment

34 43 13 – Airfield Signals and Lighting

34 43 13.16 Airfield Runway and Taxiway Inset Lighting	Crew	Daily Output	Labor-Hours	Unit	Material	2020 Bare Costs Labor	Equipment	Total	Total Incl O&P
0010 **AIRFIELD RUNWAY AND TAXIWAY INSET LIGHTING**									
0100 Runway centerline, bidir., semi-flush, 200 W, w/shallow insert base	R-22	11.41	3.270	Ea.	2,350	184		2,534	2,875
0120 Flush, 200 W, w/shallow insert base		11.41	3.270		2,125	184		2,309	2,625
0130 for mounting in base housing		17.15	2.170		1,075	122		1,197	1,375
0150 Touchdown zone light, unidirectional, 200 W, w/shallow insert base		11.41	3.270		1,925	184		2,109	2,400
0160 115 W		11.41	3.270		1,825	184		2,009	2,300
0180 Unidirectional, 200 W, for mounting in base housing		17.11	2.180		840	122		962	1,100
0190 115 W		17.11	2.180		905	122		1,027	1,175
0210 Runway edge & threshold light, bidir., 200 W, for base housing		8.61	4.330		1,150	243		1,393	1,600
0240 Threshold & approach light, unidir., 200 W, for base housing		8.61	4.330		650	243		893	1,075
0260 Runway edge, bidirectional, 2-115 W, for base housing		11.41	3.270		1,725	184		1,909	2,175
0280 Runway threshold & end, bidir., 2-115 W, for base housing		11.41	3.270		1,900	184		2,084	2,350
0370 45 W, flush, for mounting in base housing		17.15	2.170		925	122		1,047	1,200
0380 115 W		17.15	2.170		950	122		1,072	1,225

34 43 23 – Weather Observation Equipment

34 43 23.16 Airfield Wind Cones

	Crew	Daily Output	Labor-Hours	Unit	Material	2020 Bare Costs Labor	Equipment	Total	Total Incl O&P
0010 **AIRFIELD WIND CONES**									
1200 Wind cone, 12' lighted assembly, rigid, w/obstruction light	R-21	1.25	26.210	Ea.	11,900	1,600	91.50	13,591.50	15,500
1210 Without obstruction light		1.40	23.460		9,950	1,450	82	11,482	13,100
1220 Unlighted assembly, w/obstruction light		1.55	21.220		10,800	1,300	74	12,174	13,900
1230 Without obstruction light		1.69	19.380		9,950	1,200	67.50	11,217.50	12,700
1240 Wind cone slip fitter, 2-1/2" pipe		20.09	1.630		132	100	5.70	237.70	300
1250 Wind cone sock, 12' x 3', cotton		6.04	5.430		720	335	18.95	1,073.95	1,300
1260 Nylon		6.04	5.430		690	335	18.95	1,043.95	1,275

For customer support on your Electrical Change Order Costs with RSMeans data, call 800.448.8182. **Pre-Installation Change Order Costs**

Estimating Tips

- When estimating costs for the installation of electrical power generation equipment, factors to review include access to the job site, access and setting up at the installation site, required connections, uncrating pads, anchors, leveling, final assembly of the components, and temporary protection from physical damage, such as environmental exposure.

- Be aware of the costs of equipment supports, concrete pads, and vibration isolators. Cross-reference them against other trades' specifications. Also, review site and structural drawings for items that must be included in the estimates.

- It is important to include items that are not documented in the plans and specifications but must be priced. These items include, but are not limited to, testing, dust protection, roof penetration, core drilling concrete floors and walls, patching, cleanup, and final adjustments. Add a contingency or allowance for utility company fees for power hookups, if needed.

- The project size and scope of electrical power generation equipment will have a significant impact on cost. The intent of RSMeans cost data is to provide a benchmark cost so that owners, engineers, and electrical contractors will have a comfortable number with which to start a project. Additionally, there are many websites available to use for research and to obtain a vendor's quote to finalize costs.

Reference Numbers

Reference numbers are shown at the beginning of some major classifications. These numbers refer to related items in the Reference Section. The reference information may be an estimating procedure, an alternate pricing method, or technical information.

Same Data. Simplified.

Enjoy the convenience and efficiency of accessing your costs anywhere:

- **Skip the multiplier** by setting your location
- **Quickly search,** edit, favorite and share costs
- **Stay on top of price changes** with automatic updates

Discover more at rsmeans.com/online

Note: Trade Service, in part, has been used as a reference source for some of the material prices used in Division 48.

48 13 13 – Hydroelectric Power Plant Water Turbines

48 13 13.10 Water Turbines and Components	Crew	Daily Output	Labor-Hours	Unit	Material	2020 Bare Costs Labor	2020 Bare Costs Equipment	Total	Total Incl O&P
0010 **WATER TURBINES & COMPONENTS**									
0500 Water turbine, 100-500 watt, 3P, 48VDC, 54"W x 24"H, 260LBS	2 Elec	1.84	8.700	Ea.	10,200	535		10,735	12,000
0600 Water turbine, 1500-2000 watt, 3P, 130-200VAC to DC, 95LBS		2.76	5.800		4,025	355		4,380	4,950
0501 Wind turbine marine rated, 160 W, 12 V, DC		2.02	7.910		1,175	485		1,660	2,025
0510 160 W, 24 V, DC		2.02	7.910		1,200	485		1,685	2,050
0520 160 W, 48 V, DC		2.02	7.910		1,175	485		1,660	2,025
1001 Wind turbine, 300 W, 12 V, DC		2.02	7.910		900	485		1,385	1,725
1011 300 W, 24 V, DC		2.02	7.910		915	485		1,400	1,725
1020 300 W, 48 V, DC		2.02	7.910		910	485		1,395	1,725
1030 Wind turbine, 400 W, 12 V, DC		2.02	7.910		910	485		1,395	1,725
1040 400 W, 24 V, DC		2.02	7.910		910	485		1,395	1,725
1050 400 W, 48 V, DC		2.02	7.910		915	485		1,400	1,725
3300 Wind turbine/charge control, 2,900-7,900 W, 5.5" H x 11' W, 180 lb.		1.84	8.700		16,600	535		17,135	19,100
3320 Wind turbine/charge control, 4,000-10,800 W, 5.5" H x 13' W, 240 lb.		1.84	8.700		27,500	535		28,035	31,100
3401 Wind turbine tower, 2 stage steel, 8' H x 6' W		1.84	8.700		2,075	535		2,610	3,100
4000 Anemometer, standard AC or 9V, 2-digit LCD, wood case, +/-2% of Input	2 Elec	1.84	8.700	Ea.	177	535		712	990

Pre-Installation Change Order Costs

Post-Installation Costs

Post-Installation
Change Order Costs

This second cost data section contains prices for change order work that occurs after the original installation is nearly or entirely complete. These costs are generally higher due to additional complications, such as the need for immediate crew changes and/or increases, and rush material orders.

As with Pre-Installation Change Orders, the Post-Installation Change Order costs allow for the additional expense of the actual change order work, but do not include demolition or the disposal of any removed materials. Such additional work must be determined for each project.

For change order work that is to be performed before the original installation is substantially complete, see the Pre-Installation section.

For more information on the numbering system used to organize this data and for a detailed explanation of the components that comprise this cost information, see "How RSMeans Unit Price Data Works."

The unit prices in Post-Installation Change Orders are for change order additions to the base contract.

In most cases, change orders resulting in a deduction or a credit reflect only the bare costs. The contractor retains the anticipated overhead and profit based on the original bid.

Same Data. Simplified.

Enjoy the convenience and efficiency of accessing your costs anywhere:

- **Skip the multiplier** by setting your location
- **Quickly search,** edit, favorite and share costs
- **Stay on top of price changes** with automatic updates

Discover more at rsmeans.com/online

Estimating Tips
01 20 00 Price and Payment Procedures

- Allowances that should be added to estimates to cover contingencies and job conditions that are not included in the national average material and labor costs are shown in Section 01 21.

- When estimating historic preservation projects (depending on the condition of the existing structure and the owner's requirements), a 15–20% contingency or allowance is recommended, regardless of the stage of the drawings.

01 30 00 Administrative Requirements

- Before determining a final cost estimate, it is good practice to review all the items listed in Subdivisions 01 31 and 01 32 to make final adjustments for items that may need customizing to specific job conditions.

- Requirements for initial and periodic submittals can represent a significant cost to the General Requirements of a job. Thoroughly check the submittal specifications when estimating a project to determine any costs that should be included.

01 40 00 Quality Requirements

- All projects will require some degree of quality control. This cost is not included in the unit cost of construction listed in each division. Depending upon the terms of the contract, the various costs of inspection and testing can be the responsibility of either the owner or the contractor. Be sure to include the required costs in your estimate.

01 50 00 Temporary Facilities and Controls

- Barricades, access roads, safety nets, scaffolding, security, and many more requirements for the execution of a safe project are elements of direct cost. These costs can easily be overlooked when preparing an estimate. When looking through the major classifications of this subdivision, determine which items apply to each division in your estimate.

- Construction equipment rental costs can be found in the Reference Section in Section 01 54 33. Operators' wages are not included in equipment rental costs.

- Equipment mobilization and demobilization costs are not included in equipment rental costs and must be considered separately.

- The cost of small tools provided by the installing contractor for his workers is covered in the "Overhead" column on the "Installing Contractor's Overhead and Profit" table that lists labor trades, base rates, and markups. Therefore, it is included in the "Total Incl. O&P" cost of any unit price line item.

01 70 00 Execution and Closeout Requirements

- When preparing an estimate, thoroughly read the specifications to determine the requirements for Contract Closeout. Final cleaning, record documentation, operation and maintenance data, warranties and bonds, and spare parts and maintenance materials can all be elements of cost for the completion of a contract. Do not overlook these in your estimate.

Reference Numbers

Reference numbers are shown at the beginning of some major classifications. These numbers refer to related items in the Reference Section. The reference information may be an estimating procedure, an alternate pricing method, or technical information.

Note: Not all subdivisions listed here necessarily appear. ∎

Same Data. Simplified.

Enjoy the convenience and efficiency of accessing your costs anywhere:

- **Skip the multiplier** by setting your location
- **Quickly search,** edit, favorite and share costs
- **Stay on top of price changes** with automatic updates

Discover more at rsmeans.com/online

01 21 Allowances

01 21 53 – Factors Allowance

01 21 53.60 Security Factors	Crew	Daily Output	Labor-Hours	Unit	Material	2020 Bare Costs Labor	Equipment	Total	Total Incl O&P
0010 **SECURITY FACTORS**									
0100 Additional costs due to security requirements									
0110 Daily search of personnel, supplies, equipment and vehicles									
0120 Physical search, inventory and doc of assets, at entry				Costs		30%			
0130 At entry and exit						50%			
0140 Physical search, at entry						6.25%			
0150 At entry and exit						12.50%			
0160 Electronic scan search, at entry						2%			
0170 At entry and exit						4%			
0180 Visual inspection only, at entry						.25%			
0190 At entry and exit						.50%			
0200 ID card or display sticker only, at entry						.12%			
0210 At entry and exit				↓		.25%			
0220 Day 1 as described below, then visual only for up to 5 day job duration									
0230 Physical search, inventory and doc of assets, at entry				Costs		5%			
0240 At entry and exit						10%			
0250 Physical search, at entry						1.25%			
0260 At entry and exit						2.50%			
0270 Electronic scan search, at entry						.42%			
0280 At entry and exit				↓		.83%			
0290 Day 1 as described below, then visual only for 6-10 day job duration									
0300 Physical search, inventory and doc of assets, at entry				Costs		2.50%			
0310 At entry and exit						5%			
0320 Physical search, at entry						.63%			
0330 At entry and exit						1.25%			
0340 Electronic scan search, at entry						.21%			
0350 At entry and exit				↓		.42%			
0360 Day 1 as described below, then visual only for 11-20 day job duration									
0370 Physical search, inventory and doc of assets, at entry				Costs		1.25%			
0380 At entry and exit						2.50%			
0390 Physical search, at entry						.31%			
0400 At entry and exit						.63%			
0410 Electronic scan search, at entry						.10%			
0420 At entry and exit				↓		.21%			
0430 Beyond 20 days, costs are negligible									
0440 Escort required to be with tradesperson during work effort				Costs		6.25%			

01 51 Temporary Utilities

01 51 13 – Temporary Electricity

01 51 13.50 Temporary Power Equip (Pro-Rated Per Job)

01 51 13.50 Temporary Power Equip (Pro-Rated Per Job)	Crew	Daily Output	Labor-Hours	Unit	Material	2020 Bare Costs Labor	Equipment	Total	Total Incl O&P
0010 **TEMPORARY POWER EQUIP (PRO-RATED PER JOB)**									
0020 Service, overhead feed, 3 use									
0030 100 Amp	1 Elec	1.05	7.620	Ea.	430	465		895	1,175
0040 200 Amp		.84	9.520		470	585		1,055	1,375
0050 400 Amp		.63	12.700		830	780		1,610	2,075
0060 600 Amp	↓	.42	19.050	↓	1,200	1,175		2,375	3,075
0100 Underground feed, 3 use									
0110 100 Amp	1 Elec	1.68	4.760	Ea.	445	292		737	925
0120 200 Amp		.97	8.280		520	510		1,030	1,325
0130 400 Amp		.84	9.520		875	585		1,460	1,825
0140 600 Amp	↓	.63	12.700	↓	1,075	780		1,855	2,325

Post-Installation Change Order Costs

01 51 13.50 Temporary Power Equip (Pro-Rated Per Job)	Crew	Daily Output	Labor-Hours	Unit	Material	2020 Bare Costs Labor	Equipment	Total	Total Incl O&P
0150	800 Amp	1 Elec	.42	19.050	Ea.	1,525	1,175		2,700	3,425
0160	1000 Amp		.29	27.210		1,700	1,675		3,375	4,325
0170	1200 Amp		.21	38.100		1,875	2,325		4,200	5,525
0180	2000 Amp		.17	47.620		2,250	2,925		5,175	6,825
0200	Transformers, 3 use									
0210	30 kVA	1 Elec	.84	9.520	Ea.	1,925	585		2,510	3,000
0220	45 kVA		.63	12.700		2,450	780		3,230	3,850
0230	75 kVA		.42	19.050		3,725	1,175		4,900	5,850
0240	112.5 kVA		.34	23.810		4,100	1,450		5,550	6,675
0250	Feeder, PVC, CU wire in trench									
0260	60 Amp	1 Elec	80.64	.099	L.F.	2.73	6.10		8.83	12.05
0270	100 Amp		71.40	.112		4.96	6.85		11.81	15.70
0280	200 Amp		49.56	.161		12	9.90		21.90	28
0290	400 Amp		35.28	.227		32	13.90		45.90	55.50
0300	Feeder, PVC, aluminum wire in trench									
0310	60 Amp	1 Elec	80.64	.099	L.F.	2.67	6.10		8.77	12
0320	100 Amp		71.40	.112		3.19	6.85		10.04	13.75
0330	200 Amp		49.56	.161		8.65	9.90		18.55	24.50
0340	400 Amp		35.28	.227		15.75	13.90		29.65	38
0350	Feeder, EMT, CU wire									
0360	60 Amp	1 Elec	75.60	.106	L.F.	3.10	6.50		9.60	13.05
0370	100 Amp		67.20	.119		5.75	7.30		13.05	17.20
0380	200 Amp		50.40	.159		12.85	9.75		22.60	28.50
0390	400 Amp		29.40	.272		32.50	16.70		49.20	60.50
0400	Feeder, EMT, aluminum wire									
0410	60 Amp	1 Elec	75.60	.106	L.F.	4.59	6.50		11.09	14.70
0420	100 Amp		67.20	.119		5.05	7.30		12.35	16.45
0430	200 Amp		50.40	.159		9.25	9.75		19	24.50
0440	400 Amp		29.40	.272		17.30	16.70		34	44
0500	Equipment, 3 use									
0510	Spider box, 50 Amp	1 Elec	6.72	1.190	Ea.	785	73		858	975
0520	Lighting cord, 100'		6.72	1.190		126	73		199	248
0530	Light stanchion		6.72	1.190		72	73		145	189
0540	Temporary cords, 100', 3 use									
0550	Feeder cord, 50 Amp	1 Elec	13.44	.595	Ea.	237	36.50		273.50	315
0560	Feeder cord, 100 Amp		10.08	.794		1,200	48.50		1,248.50	1,400
0570	Tap cord, 50 Amp		10.08	.794		630	48.50		678.50	770
0580	Tap cord, 100 Amp		5.04	1.590		1,100	97.50		1,197.50	1,350
0590	Temporary cords, 50', 3 use									
0600	Feeder cord, 50 Amp	1 Elec	13.44	.595	Ea.	305	36.50		341.50	390
0610	Feeder cord, 100 Amp		10.08	.794		625	48.50		673.50	765
0620	Tap cord, 50 Amp		10.08	.794		305	48.50		353.50	410
0630	Tap cord, 100 Amp		5.04	1.590		615	97.50		712.50	825
0700	Connections									
0710	Compressor or pump									
0720	30 Amp	1 Elec	5.88	1.360	Ea.	17	83.50		100.50	143
0730	60 Amp		4.45	1.800		30	110		140	198
0740	100 Amp		3.36	2.380		80	146		226	305
0750	Tower crane									
0760	60 Amp	1 Elec	3.78	2.120	Ea.	30	130		160	228
0770	100 Amp	"	2.52	3.170	"	80	194		274	380
0780	Manlift									
0790	Single	1 Elec	2.52	3.170	Ea.	34.50	194		228.50	330

01 51 Temporary Utilities

01 51 13 – Temporary Electricity

01 51 13.50 Temporary Power Equip (Pro-Rated Per Job)	Crew	Daily Output	Labor-Hours	Unit	Material	2020 Bare Costs Labor	Equipment	Total	Total Incl O&P
0800	Double	1 Elec	1.68	4.760	Ea.	76	292		368	520
0810	Welder with disconnect									
0820	50 Amp	1 Elec	4.20	1.900	Ea.	205	117		322	400
0830	100 Amp		3.19	2.510		395	154		549	660
0840	200 Amp		2.10	3.810		735	234		969	1,150
0850	400 Amp		.84	9.520		1,575	585		2,160	2,600
0860	Office trailer									
0870	60 Amp	1 Elec	3.78	2.120	Ea.	55	130		185	255
0880	100 Amp		2.52	3.170		83.50	194		277.50	380
0890	200 Amp		1.68	4.760		292	292		584	755

Post-Installation Change Order Costs

Estimating Tips
General

- The items in this division are usually priced per square foot or each. Most of these items are purchased by the owner and installed by the contractor. Do not assume the items in Division 12 will be purchased and installed by the contractor. Check the specifications for responsibilities and include receiving, storage, installation, and mechanical and electrical hookups in the appropriate divisions.

- Some items in this division require some type of support system that is not usually furnished with the item. Examples of these systems include blocking for the attachment of casework and heavy drapery rods. The required blocking must be added to the estimate in the appropriate division.

Reference Numbers

Reference numbers are shown at the beginning of some major classifications. These numbers refer to related items in the Reference Section. The reference information may be an estimating procedure, an alternate pricing method, or technical information.

Same Data. Simplified.

Enjoy the convenience and efficiency of accessing your costs anywhere:

- **Skip the multiplier** by setting your location
- **Quickly search,** edit, favorite and share costs
- **Stay on top of price changes** with automatic updates

Discover more at rsmeans.com/online

12 46 Furnishing Accessories

12 46 19 – Clocks

12 46 19.50 Wall Clocks	Crew	Daily Output	Labor-Hours	Unit	Material	2020 Bare Costs Labor	Equipment	Total	Total Incl O&P
0010 **WALL CLOCKS**									
0080 12" diameter, single face	1 Elec	6.72	1.190	Ea.	109	73		182	229
0100 Double face	"	5.21	1.540	"	161	94.50		255.50	320

For customer support on your Electrical Change Order Costs with RSMeans data, call 800.448.8182.

Post-Installation Change Order Costs

Estimating Tips
General

- The items and systems in this division are usually estimated, purchased, supplied, and installed as a unit by one or more subcontractors. The estimator must ensure that all parties are operating from the same set of specifications and assumptions, and that all necessary items are estimated and will be provided. Many times the complex items and systems are covered, but the more common ones, such as excavation or a crane, are overlooked for the very reason that everyone assumes nobody could miss them. The estimator should be the central focus and be able to ensure that all systems are complete.

- It is important to consider factors such as site conditions, weather, shape and size of building, as well as labor availability as they may impact the overall cost of erecting special structures and systems included in this division.

- Another area where problems can develop in this division is at the interface between systems.

The estimator must ensure, for instance, that anchor bolts, nuts, and washers are estimated and included for the air-supported structures and pre-engineered buildings to be bolted to their foundations. Utility supply is a common area where essential items or pieces of equipment can be missed or overlooked because each subcontractor may feel it is another's responsibility. The estimator should also be aware of certain items which may be supplied as part of a package but installed by others, and ensure that the installing contractor's estimate includes the cost of installation. Conversely, the estimator must also ensure that items are not costed by two different subcontractors, resulting in an inflated overall estimate.

13 30 00 Special Structures

- The foundations and floor slab, as well as rough mechanical and electrical, should be estimated, as this work is required for the assembly and erection of the structure. Generally, as noted in the data set, the pre-engineered building comes

as a shell. Pricing is based on the size and structural design parameters stated in the reference section. Additional features, such as windows and doors with their related structural framing, must also be included by the estimator. Here again, the estimator must have a clear understanding of the scope of each portion of the work and all the necessary interfaces.

Reference Numbers

Reference numbers are shown at the beginning of some major classifications. These numbers refer to related items in the Reference Section. The reference information may be an estimating procedure, an alternate pricing method, or technical information.

Note: Not all subdivisions listed here necessarily appear. ■

Same Data. Simplified.

Enjoy the convenience and efficiency of accessing your costs anywhere:

- **Skip the multiplier** by setting your location
- **Quickly search,** edit, favorite and share costs
- **Stay on top of price changes** with automatic updates

Discover more at rsmeans.com/online

13 47 13.16 Cathodic Prot. for Underground Storage Tanks	Crew	Daily Output	Labor-Hours	Unit	Material	2020 Bare Costs Labor	Equipment	Total	Total Incl O&P
0010 **CATHODIC PROTECTION FOR UNDERGROUND STORAGE TANKS**									
1000 Anodes, magnesium type, 9 #	R-15	15.54	3.090	Ea.	38.50	186	17.95	242.45	340
1010 17 #		10.92	4.400		81.50	264	25.50	371	515
1020 32 #		8.40	5.710		123	345	33	501	680
1030 48 #	▼	6.05	7.940		160	475	46	681	935
1100 Graphite type w/epoxy cap, 3" x 60" (32 #)	R-22	7.06	5.280		148	297		445	605
1110 4" x 80" (68 #)		5.04	7.400		246	415		661	890
1120 6" x 72" (80 #)		4.37	8.530		1,525	480		2,005	2,400
1130 6" x 36" (45 #)	▼	8.06	4.620		770	259		1,029	1,225
2000 Rectifiers, silicon type, air cooled, 28 V/10 A	R-19	2.94	6.800		2,325	420		2,745	3,175
2010 20 V/20 A		2.94	6.800		2,325	420		2,745	3,200
2100 Oil immersed, 28 V/10 A		2.52	7.940		2,875	490		3,365	3,900
2110 20 V/20 A	▼	2.52	7.940	▼	3,250	490		3,740	4,300
3000 Anode backfill, coke breeze	R-22	3234	.012	Lb.	.30	.65		.95	1.29
4000 Cable, HMWPE, No. 8		2.02	18.490	M.L.F.	605	1,050		1,655	2,225
4010 No. 6		2.02	18.490		820	1,050		1,870	2,450
4020 No. 4		2.02	18.490		1,225	1,050		2,275	2,900
4030 No. 2		2.02	18.490		1,900	1,050		2,950	3,650
4040 No. 1		1.85	20.170		2,575	1,125		3,700	4,525
4050 No. 1/0		1.85	20.170		3,350	1,125		4,475	5,350
4060 No. 2/0		1.85	20.170		5,250	1,125		6,375	7,450
4070 No. 4/0	▼	1.68	22.190	▼	7,225	1,250		8,475	9,800
5000 Test station, 7 terminal box, flush curb type w/lockable cover	R-19	10.08	1.980	Ea.	79.50	122		201.50	269
5010 Reference cell, 2" diam. PVC conduit, cplg., plug, set flush	"	4.03	4.960	"	158	305		463	630

Estimating Tips
22 10 00 Plumbing
Piping and Pumps

This subdivision is primarily basic pipe and related materials. The pipe may be used by any of the mechanical disciplines, i.e., plumbing, fire protection, heating, and air conditioning.

Note: CPVC plastic piping approved for fire protection is located in 21 11 13.

- The labor adjustment factors listed in Subdivision 22 01 02.20 apply throughout Divisions 21, 22, and 23. CAUTION: the correct percentage may vary for the same items. For example, the percentage add for the basic pipe installation should be based on the maximum height that the installer must install for that particular section. If the pipe is to be located 14' above the floor but it is suspended on threaded rod from beams, the bottom flange of which is 18' high (4' rods), then the height is actually 18' and the add is 20%. The pipe cover, however, does not have to go above the 14' and so the add should be 10%.

- Most pipe is priced first as straight pipe with a joint (coupling, weld, etc.) every 10' and a hanger usually every 10'. There are exceptions with hanger spacing such as for cast iron pipe (5')

and plastic pipe (3 per 10'). Following each type of pipe there are several lines listing sizes and the amount to be subtracted to delete couplings and hangers. This is for pipe that is to be buried or supported together on trapeze hangers. The reason that the couplings are deleted is that these runs are usually long, and frequently longer lengths of pipe are used. By deleting the couplings, the estimator is expected to look up and add back the correct reduced number of couplings.

- When preparing an estimate, it may be necessary to approximate the fittings. Fittings usually run between 25% and 50% of the cost of the pipe. The lower percentage is for simpler runs, and the higher number is for complex areas, such as mechanical rooms.

- For historic restoration projects, the systems must be as invisible as possible, and pathways must be sought for pipes, conduit, and ductwork. While installations in accessible spaces (such as basements and attics) are relatively straightforward to estimate, labor costs may be more difficult to determine when delivery systems must be concealed.

22 40 00 Plumbing Fixtures

- Plumbing fixture costs usually require two lines: the fixture itself and its "rough-in, supply, and waste."

- In the Assemblies Section (Plumbing D2010) for the desired fixture, the System Components Group at the center of the page shows the fixture on the first line. The rest of the list (fittings, pipe, tubing, etc.) will total up to what we refer to in the Unit Price section as "Rough-in, supply, waste, and vent." Note that for most fixtures we allow a nominal 5' of tubing to reach from the fixture to a main or riser.

- Remember that gas- and oil-fired units need venting.

Reference Numbers

Reference numbers are shown at the beginning of some major classifications. These numbers refer to related items in the Reference Section. The reference information may be an estimating procedure, an alternate pricing method, or technical information.

Note: Not all subdivisions listed here necessarily appear. ∎

Same Data. Simplified.

Enjoy the convenience and efficiency of accessing your costs anywhere:

- **Skip the multiplier** by setting your location
- **Quickly search,** edit, favorite and share costs
- **Stay on top of price changes** with automatic updates

Discover more at rsmeans.com/online

---

22 05 29.10 Hangers & Supp. for Plumb'g/HVAC Pipe/Equip.	Crew	Daily Output	Labor-Hours	Unit	Material	2020 Bare Costs Labor	Equipment	Total	Total Incl O&P
0010 **HANGERS AND SUPPORTS FOR PLUMB'G/HVAC PIPE/EQUIP.**									
0011 TYPE numbers per MSS-SP58									
2650 Rods, carbon steel									
2660 Continuous thread									
2670 1/4" thread size	1 Plum	120.96	.066	L.F.	2.43	4.26		6.69	9.05
2680 3/8" thread size		120.96	.066		2.59	4.26		6.85	9.25
2690 1/2" thread size		120.96	.066		4.07	4.26		8.33	10.90
2700 5/8" thread size		120.96	.066		5.75	4.26		10.01	12.75
2710 3/4" thread size		120.96	.066		10.60	4.26		14.86	18.10
2720 7/8" thread size	↓	120.96	.066		13.35	4.26		17.61	21
2721 1" thread size	Q-1	134.40	.119		22.50	6.90		29.40	35.50
2722 1-1/8" thread size	"	100.80	.159	↓	24.50	9.20		33.70	41
2730 For galvanized, add					40%				
8800 Wire cable support system									
8810 Cable with hook terminal and locking device									
8830 2 mm (.079") diam. cable (100 lb. cap.)									
8840 1 m (3.3') length, with hook	1 Shee	80.64	.099	Ea.	4.68	6.20		10.88	14.55
8850 2 m (6.6') length, with hook		70.56	.113		5.45	7.05		12.50	16.70
8860 3 m (9.9') length, with hook	↓	60.48	.132	↓	5.35	8.25		13.60	18.40
8900 3 mm (.118") diam. cable (200 lb. cap.)									
8910 1 m (3.3') length, with hook	1 Shee	80.64	.099	Ea.	11.80	6.20		18	22.50
8920 2 m (6.6') length, with hook		70.56	.113		13.05	7.05		20.10	25
8930 3 m (9.9') length, with hook	↓	60.48	.132	↓	14.20	8.25		22.45	28
9000 Cable system accessories									
9010 Anchor bolt, 3/8", with nut	1 Shee	117.60	.068	Ea.	1.36	4.24		5.60	7.95
9020 Air duct corner protector		134.40	.060		1.81	3.71		5.52	7.60
9030 Air duct support attachment	↓	117.60	.068	↓	2.18	4.24		6.42	8.85
9040 Flange clip, hammer-on style									
9044 For flange thickness 3/32"-9/64", 160 lb. cap.	1 Shee	151.20	.053	Ea.	.53	3.30		3.83	5.60
9048 For flange thickness 1/8"-1/4", 200 lb. cap.		134.40	.060		.38	3.71		4.09	6
9052 For flange thickness 5/16"-1/2", 200 lb. cap.		126	.063		.78	3.96		4.74	6.85
9056 For flange thickness 9/16"-3/4", 200 lb. cap.	↓	117.60	.068	↓	1.07	4.24		5.31	7.60

Estimating Tips

The labor adjustment factors listed in Subdivision 22 01 02.20 also apply to Division 23.

23 10 00 Facility Fuel Systems

- The prices in this subdivision for above- and below-ground storage tanks do not include foundations or hold-down slabs, unless noted. The estimator should refer to Divisions 3 and 31 for foundation system pricing. In addition to the foundations, required tank accessories, such as tank gauges, leak detection devices, and additional manholes and piping, must be added to the tank prices.

23 50 00 Central Heating Equipment

- When estimating the cost of an HVAC system, check to see who is responsible for providing and installing the temperature control system. It is possible to overlook controls, assuming that they would be included in the electrical estimate.
- When looking up a boiler, be careful on specified capacity. Some

manufacturers rate their products on output while others use input.
- Include HVAC insulation for pipe, boiler, and duct (wrap and liner).
- Be careful when looking up mechanical items to get the correct pressure rating and connection type (thread, weld, flange).

23 70 00 Central HVAC Equipment

- Combination heating and cooling units are sized by the air conditioning requirements. (See Reference No. R236000-20 for the preliminary sizing guide.)
- A ton of air conditioning is nominally 400 CFM.
- Rectangular duct is taken off by the linear foot for each size, but its cost is usually estimated by the pound. Remember that SMACNA standards now base duct on internal pressure.
- Prefabricated duct is estimated and purchased like pipe: straight sections and fittings.
- Note that cranes or other lifting equipment are not included on any

lines in Division 23. For example, if a crane is required to lift a heavy piece of pipe into place high above a gym floor, or to put a rooftop unit on the roof of a four-story building, etc., it must be added. Due to the potential for extreme variation—from nothing additional required to a major crane or helicopter—we feel that including a nominal amount for "lifting contingency" would be useless and detract from the accuracy of the estimate. When using equipment rental cost data from RSMeans, do not forget to include the cost of the operator(s).

Reference Numbers

Reference numbers are shown at the beginning of some major classifications. These numbers refer to related items in the Reference Section. The reference information may be an estimating procedure, an alternate pricing method, or technical information.

Note: Not all subdivisions listed here necessarily appear. ■

Same Data. Simplified.

Enjoy the convenience and efficiency of accessing your costs anywhere:
- **Skip the multiplier** by setting your location
- **Quickly search,** edit, favorite and share costs
- **Stay on top of price changes** with automatic updates

Discover more at rsmeans.com/online

23 83 Radiant Heating Units

23 83 33 – Electric Radiant Heaters

23 83 33.10 Electric Heating	Crew	Daily Output	Labor-Hours	Unit	Material	2020 Bare Costs Labor	Equipment	Total	Total Incl O&P
0010 **ELECTRIC HEATING**, not incl. conduit or feed wiring									
1100 Rule of thumb: Baseboard units, including control	1 Elec	3.70	2.160	kW	116	133		249	325
1300 Baseboard heaters, 2' long, 350 watt		6.72	1.190	Ea.	29	73		102	141
1400 3' long, 750 watt		6.72	1.190		35	73		108	148
1600 4' long, 1,000 watt		5.63	1.420		40.50	87		127.50	175
1800 5' long, 935 watt		4.79	1.670		44	102		146	202
2000 6' long, 1,500 watt		4.20	1.900		57	117		174	237
2200 7' long, 1,310 watt		3.70	2.160		61	133		194	265
2400 8' long, 2,000 watt		3.36	2.380		70	146		216	294
2600 9' long, 1,680 watt		3.02	2.650		86.50	163		249.50	335
2800 10' long, 1,875 watt		2.77	2.890		162	177		339	440
2950 Wall heaters with fan, 120 to 277 volt									
3160 Recessed, residential, 750 watt	1 Elec	5.04	1.590	Ea.	94.50	97.50		192	249
3170 1,000 watt		5.04	1.590		94.50	97.50		192	249
3180 1,250 watt		4.20	1.900		107	117		224	292
3190 1,500 watt		3.36	2.380		107	146		253	335
3210 2,000 watt		3.36	2.380		99	146		245	325
3230 2,500 watt		2.94	2.720		236	167		403	510
3240 3,000 watt		2.52	3.170		365	194		559	695
3250 4,000 watt		2.27	3.530		370	217		587	725
3260 Commercial, 750 watt		5.04	1.590		211	97.50		308.50	375
3270 1,000 watt		5.04	1.590		211	97.50		308.50	380
3280 1,250 watt		4.20	1.900		211	117		328	405
3290 1,500 watt		3.36	2.380		195	146		341	430
3300 2,000 watt		3.36	2.380		211	146		357	450
3310 2,500 watt		2.94	2.720		211	167		378	480
3320 3,000 watt		2.52	3.170		425	194		619	760
3330 4,000 watt		2.27	3.530		500	217		717	870
3600 Thermostats, integral		13.44	.595		29	36.50		65.50	86.50
3800 Line voltage, 1 pole		6.72	1.190		16.30	73		89.30	127
3810 2 pole		6.72	1.190		23.50	73		96.50	135
4000 Heat trace system, 400 degree									
4020 115 V, 2.5 watts/L.F.	1 Elec	445.20	.018	L.F.	11	1.10		12.10	13.75
4030 5 watts/L.F.		445.20	.018		11	1.10		12.10	13.75
4050 10 watts/L.F.		445.20	.018		9.20	1.10		10.30	11.75
4060 208 V, 5 watts/L.F.		445.20	.018		11	1.10		12.10	13.75
4080 480 V, 8 watts/L.F.		445.20	.018		11	1.10		12.10	13.75
4200 Heater raceway									
4260 Heat transfer cement									
4280 1 gallon				Ea.	64.50			64.50	71
4300 5 gallon				"	252			252	277
4320 Cable tie									
4340 3/4" pipe size	1 Elec	394.80	.020	Ea.	.02	1.24		1.26	1.87
4360 1" pipe size		372.96	.021		.02	1.32		1.34	1.98
4380 1-1/4" pipe size		336	.024		.02	1.46		1.48	2.20
4400 1-1/2" pipe size		298.20	.027		.02	1.65		1.67	2.47
4420 2" pipe size		268.80	.030		.05	1.83		1.88	2.78
4440 3" pipe size		134.40	.060		.05	3.65		3.70	5.50
4460 4" pipe size		84	.095		.06	5.85		5.91	8.75
4480 Thermostat NEMA 3R, 22 amp, 0-150 degree, 10' cap.		6.72	1.190		193	73		266	320
4500 Thermostat NEMA 4X, 25 amp, 40 degree, 5-1/2' cap.		5.88	1.360		248	83.50		331.50	395
4520 Thermostat NEMA 4X, 22 amp, 25-325 degree, 10' cap.		5.88	1.360		660	83.50		743.50	850
4540 Thermostat NEMA 4X, 22 amp, 15-140 degree		5.04	1.590		560	97.50		657.50	760

For customer support on your Electrical Change Order Costs with RSMeans data, call 800.448.8182. **Post-Installation Change Order Costs**

23 83 33.10 Electric Heating	Crew	Daily Output	Labor-Hours	Unit	Material	2020 Bare Costs Labor	Equipment	Total	Total Incl O&P
4580	Thermostat NEMA 4, 7, 9, 22 amp, 25-325 degree, 10' cap.	1 Elec	3.02	2.650	Ea.	810	163		973	1,125
4600	Thermostat NEMA 4, 7, 9, 22 amp, 15-140 degree		2.52	3.170		815	194		1,009	1,175
4720	Fiberglass application tape, 36 yard roll		9.24	.866		75	53		128	162
5000	Radiant heating ceiling panels, 2' x 4', 500 watt		13.44	.595		405	36.50		441.50	500
5050	750 watt		13.44	.595		430	36.50		466.50	530
5200	For recessed plaster frame, add		26.88	.298		138	18.25		156.25	179
5300	Infrared quartz heaters, 120 volts, 1,000 watt		5.63	1.420		350	87		437	515
5350	1,500 watt		4.20	1.900		335	117		452	545
5400	240 volts, 1,500 watt		4.20	1.900		405	117		522	620
5450	2,000 watt		3.36	2.380		335	146		481	585
5500	3,000 watt		2.52	3.170		330	194		524	655
5550	4,000 watt		2.18	3.660		380	225		605	755
5570	Modulating control		.67	11.900		141	730		871	1,225
5600	Unit heaters, heavy duty, with fan & mounting bracket									
5650	Single phase, 208-240-277 volt, 3 kW	1 Elec	5.04	1.590	Ea.	705	97.50		802.50	920
5750	5 kW		4.62	1.730		800	106		906	1,050
5800	7 kW		4.20	1.900		1,375	117		1,492	1,675
5850	10 kW		3.36	2.380		1,100	146		1,246	1,425
5950	15 kW		3.19	2.510		1,850	154		2,004	2,250
6000	480 volt, 3 kW		5.04	1.590		470	97.50		567.50	660
6020	4 kW		4.87	1.640		700	101		801	920
6040	5 kW		4.62	1.730		710	106		816	940
6060	7 kW		4.20	1.900		1,150	117		1,267	1,450
6080	10 kW		3.36	2.380		1,100	146		1,246	1,450
6100	13 kW		3.19	2.510		1,950	154		2,104	2,375
6120	15 kW		3.11	2.570		2,150	158		2,308	2,600
6140	20 kW		2.94	2.720		3,075	167		3,242	3,650
6300	3 phase, 208-240 volt, 5 kW		4.62	1.730		575	106		681	795
6320	7 kW		4.20	1.900		940	117		1,057	1,200
6340	10 kW		3.36	2.380		805	146		951	1,100
6360	15 kW		3.11	2.570		1,600	158		1,758	1,975
6380	20 kW		2.94	2.720		2,500	167		2,667	3,000
6400	25 kW		2.77	2.890		3,725	177		3,902	4,375
6500	480 volt, 5 kW		4.62	1.730		650	106		756	875
6520	7 kW		4.20	1.900		930	117		1,047	1,200
6540	10 kW		3.36	2.380		860	146		1,006	1,150
6560	13 kW		3.19	2.510		1,650	154		1,804	2,050
6580	15 kW		3.11	2.570		1,475	158		1,633	1,850
6600	20 kW		2.94	2.720		2,325	167		2,492	2,800
6620	25 kW		2.77	2.890		3,450	177		3,627	4,075
6630	30 kW		2.52	3.170		3,025	194		3,219	3,625
6640	40 kW		1.68	4.760		4,050	292		4,342	4,875
6650	50 kW		1.34	5.950		6,675	365		7,040	7,900
6800	Vertical discharge heaters, with fan									
6820	Single phase, 208-240-277 volt, 10 kW	1 Elec	3.36	2.380	Ea.	860	146		1,006	1,150
6840	15 kW		3.11	2.570		1,650	158		1,808	2,025
6900	3 phase, 208-240 volt, 10 kW		3.36	2.380		1,050	146		1,196	1,400
6920	15 kW		3.11	2.570		1,600	158		1,758	2,000
6940	20 kW		2.94	2.720		2,925	167		3,092	3,475
6960	25 kW		2.77	2.890		3,175	177		3,352	3,775
6980	30 kW		2.52	3.170		4,350	194		4,544	5,075
7000	40 kW		1.68	4.760		4,325	292		4,617	5,175
7020	50 kW		1.34	5.950		6,950	365		7,315	8,200

23 83 Radiant Heating Units

23 83 33 – Electric Radiant Heaters

23 83 33.10 Electric Heating	Crew	Daily Output	Labor-Hours	Unit	Material	2020 Bare Costs Labor	Equipment	Total	Total Incl O&P
7100	480 volt, 10 kW	1 Elec	3.36	2.380	Ea.	1,200	146		1,346	1,550
7120	15 kW		3.11	2.570		1,950	158		2,108	2,375
7140	20 kW		2.94	2.720		2,650	167		2,817	3,175
7160	25 kW		2.77	2.890		3,450	177		3,627	4,075
7180	30 kW		2.52	3.170		3,800	194		3,994	4,475
7200	40 kW		1.68	4.760		3,925	292		4,217	4,750
7410	Sill height convector heaters, 5" high x 2' long, 500 watt		5.63	1.420		365	87		452	535
7420	3' long, 750 watt		5.46	1.470		435	90		525	610
7430	4' long, 1,000 watt		5.21	1.540		500	94.50		594.50	690
7440	5' long, 1,250 watt		4.62	1.730		570	106		676	785
7450	6' long, 1,500 watt		4.03	1.980		645	121		766	890
7460	8' long, 2,000 watt		3.02	2.650		875	163		1,038	1,200
7470	10' long, 2,500 watt	↓	2.52	3.170	↓	1,100	194		1,294	1,500
7900	Cabinet convector heaters, 240 volt, three phase,									
7920	3' long, 2,000 watt	1 Elec	4.45	1.800	Ea.	2,300	110		2,410	2,700
7940	3,000 watt		4.45	1.800		2,400	110		2,510	2,825
7960	4,000 watt		4.45	1.800		2,475	110		2,585	2,875
7980	6,000 watt		3.86	2.070		2,675	127		2,802	3,150
8000	8,000 watt		3.86	2.070		2,750	127		2,877	3,225
8020	4' long, 4,000 watt		3.86	2.070		2,500	127		2,627	2,950
8040	6,000 watt		3.36	2.380		2,600	146		2,746	3,075
8060	8,000 watt		3.36	2.380		2,675	146		2,821	3,150
8080	10,000 watt	↓	3.36	2.380	↓	2,700	146		2,846	3,200
8100	Available also in 208 or 277 volt									
8200	Cabinet unit heaters, 120 to 277 volt, 1 pole,									
8220	wall mounted, 2 kW	1 Elec	3.86	2.070	Ea.	2,075	127		2,202	2,500
8230	3 kW		3.86	2.070		2,075	127		2,202	2,475
8240	4 kW		3.70	2.160		2,150	133		2,283	2,575
8250	5 kW		3.70	2.160		2,275	133		2,408	2,700
8260	6 kW		3.53	2.270		2,250	139		2,389	2,675
8270	8 kW		3.36	2.380		2,300	146		2,446	2,750
8280	10 kW		3.19	2.510		2,350	154		2,504	2,800
8290	12 kW		2.94	2.720		2,575	167		2,742	3,100
8300	13.5 kW		2.44	3.280		3,650	201		3,851	4,325
8310	16 kW		2.27	3.530		3,025	217		3,242	3,675
8320	20 kW		1.93	4.140		3,825	254		4,079	4,575
8330	24 kW		1.60	5.010		4,000	305		4,305	4,875
8350	Recessed, 2 kW		3.70	2.160		2,275	133		2,408	2,700
8370	3 kW		3.70	2.160		2,325	133		2,458	2,750
8380	4 kW		3.53	2.270		2,200	139		2,339	2,625
8390	5 kW		3.53	2.270		2,300	139		2,439	2,725
8400	6 kW		3.36	2.380		2,350	146		2,496	2,800
8410	8 kW		3.19	2.510		2,525	154		2,679	3,025
8420	10 kW		2.94	2.720		3,150	167		3,317	3,725
8430	12 kW		2.44	3.280		3,175	201		3,376	3,800
8440	13.5 kW		2.27	3.530		3,050	217		3,267	3,675
8450	16 kW		1.93	4.140		3,250	254		3,504	3,950
8460	20 kW		1.60	5.010		3,775	305		4,080	4,625
8470	24 kW		1.34	5.950		3,725	365		4,090	4,650
8490	Ceiling mounted, 2 kW		2.69	2.980		2,100	183		2,283	2,575
8510	3 kW		2.69	2.980		2,200	183		2,383	2,700
8520	4 kW		2.52	3.170		2,250	194		2,444	2,775
8530	5 kW	↓	2.52	3.170	↓	2,325	194		2,519	2,850

For customer support on your Electrical Change Order Costs with RSMeans data, call 800.448.8182. **Post-Installation Change Order Costs**

23 83 Radiant Heating Units

23 83 33 – Electric Radiant Heaters

23 83 33.10 Electric Heating	Crew	Daily Output	Labor-Hours	Unit	Material	2020 Bare Costs Labor	Equipment	Total	Total Incl O&P
8540	6 kW	1 Elec	2.35	3.400	Ea.	2,325	209		2,534	2,875
8550	8 kW		2.02	3.970		2,450	244		2,694	3,075
8560	10 kW		1.85	4.330		3,050	266		3,316	3,750
8570	12 kW		1.68	4.760		2,625	292		2,917	3,325
8580	13.5 kW		1.26	6.350		2,550	390		2,940	3,375
8590	16 kW		1.09	7.330		2,675	450		3,125	3,600
8600	20 kW		.76	10.580		3,700	650		4,350	5,050
8610	24 kW		.50	15.870		3,725	975		4,700	5,550
8630	208 to 480 V, 3 pole									
8650	Wall mounted, 2 kW	1 Elec	3.86	2.070	Ea.	2,025	127		2,152	2,450
8670	3 kW		3.86	2.070		2,375	127		2,502	2,825
8680	4 kW		3.70	2.160		2,425	133		2,558	2,875
8690	5 kW		3.70	2.160		2,500	133		2,633	2,950
8700	6 kW		3.53	2.270		2,525	139		2,664	2,975
8710	8 kW		3.36	2.380		2,625	146		2,771	3,100
8720	10 kW		3.19	2.510		2,775	154		2,929	3,275
8730	12 kW		2.94	2.720		2,825	167		2,992	3,350
8740	13.5 kW		2.44	3.280		2,825	201		3,026	3,400
8750	16 kW		2.27	3.530		2,925	217		3,142	3,550
8760	20 kW		1.93	4.140		4,475	254		4,729	5,300
8770	24 kW		1.60	5.010		4,200	305		4,505	5,050
8790	Recessed, 2 kW		3.70	2.160		2,250	133		2,383	2,675
8810	3 kW		3.70	2.160		2,300	133		2,433	2,750
8820	4 kW		3.53	2.270		2,400	139		2,539	2,850
8830	5 kW		3.53	2.270		2,425	139		2,564	2,875
8840	6 kW		3.36	2.380		2,350	146		2,496	2,800
8850	8 kW		3.19	2.510		2,425	154		2,579	2,900
8860	10 kW		2.94	2.720		2,450	167		2,617	2,950
8870	12 kW		2.44	3.280		2,750	201		2,951	3,300
8880	13.5 kW		2.27	3.530		2,375	217		2,592	2,950
8890	16 kW		1.93	4.140		5,175	254		5,429	6,075
8900	20 kW		1.60	5.010		4,300	305		4,605	5,175
8920	24 kW		1.34	5.950		4,350	365		4,715	5,350
8940	Ceiling mount, 2 kW		2.69	2.980		2,325	183		2,508	2,825
8950	3 kW		2.69	2.980		2,150	183		2,333	2,625
8960	4 kW		2.52	3.170		2,475	194		2,669	3,025
8970	5 kW		2.52	3.170		2,175	194		2,369	2,675
8980	6 kW		2.35	3.400		2,475	209		2,684	3,025
8990	8 kW		2.02	3.970		2,550	244		2,794	3,200
9000	10 kW		1.85	4.330		2,350	266		2,616	3,000
9020	13.5 kW		1.26	6.350		3,150	390		3,540	4,050
9030	16 kW		1.09	7.330		4,300	450		4,750	5,425
9040	20 kW		.76	10.580		4,300	650		4,950	5,700
9060	24 kW		.50	15.870		5,325	975		6,300	7,300

Division Notes

	CREW	DAILY OUTPUT	LABOR-HOURS	UNIT	BARE COSTS				TOTAL INCL O&P
					MAT.	LABOR	EQUIP.	TOTAL	

Estimating Tips

- When estimating material costs for special systems, it is always prudent to obtain manufacturers' quotations for equipment prices and special installation requirements that may affect the total cost.

- For cost modifications for elevated tray installation, add the percentages to labor according to the height of the installation and only to the quantities exceeding the different height levels, not to the total tray quantities. Refer to subdivision 26 01 02.20 for labor adjustment factors.

- Do not overlook the costs for equipment used in the installation.

Reference Numbers

Reference numbers are shown at the beginning of some major classifications. These numbers refer to related items in the Reference Section. The reference information may be an estimating procedure, an alternate pricing method, or technical information.

Same Data. Simplified.

Enjoy the convenience and efficiency of accessing your costs anywhere:

- **Skip the multiplier** by setting your location
- **Quickly search,** edit, favorite and share costs
- **Stay on top of price changes** with automatic updates

Discover more at rsmeans.com/online

Note: Trade Service, in part, has been used as a reference source for some of the material prices used in Division 25.

25 05 Common Work Results for Integrated Automation

25 05 28 – Pathways for Integrated Automation

25 05 28.39 Surface Raceways for Integrated Automation	Crew	Daily Output	Labor-Hours	Unit	Material	2020 Bare Costs Labor	Equipment	Total	Total Incl O&P
0010 **SURFACE RACEWAYS FOR INTEGRATED AUTOMATION**									
1000 Din rail, G type, 15 mm, steel, galvanized, 3', for termination blocks	1 Elec	40.32	.198	Ea.	12.45	12.15		24.60	32
1010 32 mm, steel, galvanized, 3', for termination blocks		40.32	.198		16.20	12.15		28.35	36
1020 35 mm, steel, galvanized, 3', for termination blocks		40.32	.198		14.55	12.15		26.70	34
1030 Din rail, rail mounted, 35 mm, aluminum, 3', for termination blocks		40.32	.198		20.50	12.15		32.65	40.50
1040 35 mm, aluminum, 6', for termination blocks		40.32	.198		25.50	12.15		37.65	46.50
2000 Terminal block, Din mounted, 35 mm, medium screw, connector end bracket		40.32	.198		2.19	12.15		14.34	20.50
2010 35 mm, large screw, connector end bracket		40.32	.198		3.43	12.15		15.58	22
2020 Large screw, connector end bracket		40.32	.198		4.88	12.15		17.03	23.50
3000 Terminal block, feed through, 35 mm, Din mounted, 600V, 20A		40.32	.198		1	12.15		13.15	19.20
3010 35 mm, Din mounted, 600V, 30A		40.32	.198		1.29	12.15		13.44	19.50
3020 Terminal block, fuse holder, LED indicator, Din mounted, 110V, 6A		40.32	.198		9.50	12.15		21.65	28.50
3030 Din mounted, 220V, 6A		40.32	.198		7.80	12.15		19.95	26.50
3040 Din mounted, 300V, 6A		40.32	.198		7.65	12.15		19.80	26.50

25 36 Integrated Automation Instrumentation & Terminal Devices

25 36 19 – Integrated Automation Current Sensors

25 36 19.10 Integrated Automation Current Sensors

	Crew	Daily Output	Labor-Hours	Unit	Material	2020 Bare Costs Labor	Equipment	Total	Total Incl O&P
0010 **INTEGRATED AUTOMATION CURRENT SENSORS**									
1000 Current limiter inrush, 100 to 480 VAC, 10A, power supply	1 Elec	40.32	.198	Ea.	86.50	12.15		98.65	114

Post-Installation Change Order Costs

Estimating Tips
26 05 00 Common Work Results for Electrical

- Conduit should be taken off in three main categories—power distribution, branch power, and branch lighting—so the estimator can concentrate on systems and components, therefore making it easier to ensure all items have been accounted for.

- For cost modifications for elevated conduit installation, add the percentages to labor according to the height of installation and only to the quantities exceeding the different height levels, not to the total conduit quantities. Refer to subdivision 26 01 02.20 for labor adjustment factors.

- Remember that aluminum wiring of equal ampacity is larger in diameter than copper and may require larger conduit.

- If more than three wires at a time are being pulled, deduct percentages from the labor hours of that grouping of wires.

- When taking off grounding systems, identify separately the type and size of wire, and list each unique type of ground connection.

- The estimator should take the weights of materials into consideration when completing a takeoff. Topics to consider include: How will the materials be supported? What methods of support are available? How high will the support structure have to reach? Will the final support structure be able to withstand the total burden? Is the support material included or separate from the fixture, equipment, and material specified?

- Do not overlook the costs for equipment used in the installation. If scaffolding or highlifts are available in the field, contractors may use them in lieu of the proposed ladders and rolling staging.

26 20 00 Low-Voltage Electrical Transmission

- Supports and concrete pads may be shown on drawings for the larger equipment, or the support system may be only a piece of plywood for the back of a panelboard. In either case, they must be included in the costs.

26 40 00 Electrical and Cathodic Protection

- When taking off cathodic protection systems, identify the type and size of cable, and list each unique type of anode connection.

26 50 00 Lighting

- Fixtures should be taken off room by room using the fixture schedule, specifications, and the ceiling plan. For large concentrations of lighting fixtures in the same area, deduct the percentages from labor hours.

Reference Numbers

Reference numbers are shown at the beginning of some major classifications. These numbers refer to related items in the Reference Section. The reference information may be an estimating procedure, an alternate pricing method, or technical information.

Note: Not all subdivisions listed here necessarily appear. ∎

Same Data. Simplified.

Enjoy the convenience and efficiency of accessing your costs anywhere:

- **Skip the multiplier** by setting your location
- **Quickly search,** edit, favorite and share costs
- **Stay on top of price changes** with automatic updates

Discover more at rsmeans.com/online

26 01 Operation and Maintenance of Electrical Systems

26 01 02 – Labor Adjustment

26 01 02.20 Labor Adjustment Factors	Crew	Daily Output	Labor-Hours	Unit	Material	2020 Bare Costs Labor	Equipment	Total	Total Incl O&P
0010 **LABOR ADJUSTMENT FACTORS** (For Div. 26, 27 and 28)									
0100 Subtract from labor for Economy of Scale for Wire									
0110 4-5 wires						25%			
0120 6-10 wires						30%			
0130 11-15 wires						35%			
0140 over 15 wires						40%			
0150 Labor adjustment factors (For Div. 26, 27, 28 and 48)									
0200 Labor factors: The below are reasonable suggestions, but									
0210 each project must be evaluated for its own peculiarities, and									
0220 the adjustments be increased or decreased depending on the									
0230 severity of the special conditions.									
1000 Add to labor for elevated installation (above floor level)									
1010 10' to 14.5' high						10%			
1020 15' to 19.5' high						20%			
1030 20' to 24.5' high						25%			
1040 25' to 29.5' high						35%			
1050 30' to 34.5' high						40%			
1060 35' to 39.5' high						50%			
1070 40' and higher						55%			
2000 Add to labor for crawl space									
2010 3' high						40%			
2020 4' high						30%			
3000 Add to labor for multi-story building									
3100 For new construction (No elevator available)									
3110 Add for floors 3 thru 10						5%			
3120 Add for floors 11 thru 15						10%			
3130 Add for floors 16 thru 20						15%			
3140 Add for floors 21 thru 30						20%			
3150 Add for floors 31 and up						30%			
3200 For existing structure (Elevator available)									
3210 Add for work on floor 3 and above						2%			
4000 Add to labor for working in existing occupied buildings									
4010 Hospital						35%			
4020 Office building						25%			
4030 School						20%			
4040 Factory or warehouse						15%			
4050 Multi-dwelling						15%			
5000 Add to labor, miscellaneous									
5010 Cramped shaft						35%			
5020 Congested area						15%			
5030 Excessive heat or cold						30%			
6000 Labor factors: the above are reasonable suggestions, but									
6100 each project should be evaluated for its own peculiarities									
6200 Other factors to be considered are:									
6210 Movement of material and equipment through finished areas						10%			
6220 Equipment room min security direct access w/authorization						15%			
6230 Attic space						25%			
6240 No service road						25%			
6250 Poor unloading/storage area, no hydraulic lifts or jacks						20%			
6260 Congested site area/heavy traffic						20%			
7000 Correctional facilities (no compounding division 1 adjustment factors)									
7010 Minimum security w/facilities escort						30%			
7020 Medium security w/facilities and correctional officer escort						40%			

For customer support on your Electrical Change Order Costs with RSMeans data, call 800.448.8182.   **Post-Installation Change Order Costs**

26 01 Operation and Maintenance of Electrical Systems

26 01 02 – Labor Adjustment

26 01 02.20 Labor Adjustment Factors	Crew	Daily Output	Labor-Hours	Unit	Material	2020 Bare Costs Labor	Equipment	Total	Total Incl O&P
7030	Max security w/facilities & correctional officer escort (no inmate contact)						50%			

26 05 Common Work Results for Electrical

26 05 13 – Medium-Voltage Cables

26 05 13.10 Cable Terminations

		Crew	Daily Output	Labor-Hours	Unit	Material	2020 Bare Costs Labor	Equipment	Total	Total Incl O&P
0010	**CABLE TERMINATIONS**, 5 kV to 35 kV									
0100	Indoor, insulation diameter range 0.64" to 1.08"									
0300	Padmount, 5 kV	1 Elec	6.72	1.190	Ea.	87.50	73		160.50	205
0400	10 kV		5.38	1.490		109	91.50		200.50	255
0500	15 kV		5.04	1.590		142	97.50		239.50	300
0600	25 kV		4.70	1.700		210	104		314	385
0700	Insulation diameter range 1.05" to 1.8"									
0800	Padmount, 5 kV	1 Elec	6.72	1.190	Ea.	36.50	73		109.50	150
0900	10 kV		5.04	1.590		163	97.50		260.50	325
1000	15 kV		4.70	1.700		169	104		273	340
1100	25 kV		4.45	1.800		242	110		352	430
1200	Insulation diameter range 1.53" to 2.32"									
1300	Padmount, 5 kV	1 Elec	6.22	1.290	Ea.	152	79		231	285
1400	10 kV		4.70	1.700		175	104		279	350
1500	15 kV		4.45	1.800		194	110		304	380
1600	25 kV		4.20	1.900		315	117		432	525
1700	Outdoor systems, #4 stranded to 1/0 stranded									
1800	5 kV	1 Elec	6.22	1.290	Ea.	108	79		187	236
1900	15 kV		4.45	1.800		120	110		230	296
2000	25 kV		4.20	1.900		178	117		295	370
2100	35 kV		4.03	1.980		325	121		446	540
2200	#1 solid to 4/0 stranded, 5 kV		5.80	1.380		146	84.50		230.50	286
2300	15 kV		4.20	1.900		184	117		301	375
2400	25 kV		4.03	1.980		254	121		375	460
2500	35 kV		3.86	2.070		355	127		482	580
2600	2/0 solid to 350 kcmil stranded, 5 kV		5.38	1.490		193	91.50		284.50	350
2700	15 kV		4.03	1.980		210	121		331	410
2800	25 kV		3.86	2.070		294	127		421	515
2900	35 kV		3.70	2.160		405	133		538	640
3000	400 kcmil compact to 750 kcmil stranded, 5 kV		5.04	1.590		175	97.50		272.50	335
3100	15 kV		3.86	2.070		231	127		358	445
3200	25 kV		3.70	2.160		335	133		468	565
3300	35 kV		3.53	2.270		340	139		479	575
3400	1,000 kcmil, 5 kV		4.70	1.700		250	104		354	430
3500	15 kV		3.70	2.160		390	133		523	620
3600	25 kV		3.53	2.270		340	139		479	575
3700	35 kV		3.36	2.380		470	146		616	735

26 05 13.16 Medium-Voltage, Single Cable

		Crew	Daily Output	Labor-Hours	Unit	Material	2020 Bare Costs Labor	Equipment	Total	Total Incl O&P
0010	**MEDIUM-VOLTAGE, SINGLE CABLE** Splicing & terminations not included									
0040	Copper, XLP shielding, 5 kV, #6	2 Elec	3.70	4.330	C.L.F.	205	266		471	620
0050	#4		3.70	4.330		266	266		532	690
0100	#2		3.36	4.760		305	292		597	770
0200	#1		3.36	4.760		405	292		697	885
0400	1/0		3.19	5.010		400	305		705	900
0600	2/0		3.02	5.290		490	325		815	1,025
0800	4/0		2.69	5.950		655	365		1,020	1,275

26 05 Common Work Results for Electrical

26 05 13 – Medium-Voltage Cables

26 05 13.16 Medium-Voltage, Single Cable		Crew	Daily Output	Labor-Hours	Unit	Material	2020 Bare Costs Labor	Equipment	Total	Total Incl O&P
1000	250 kcmil	3 Elec	3.78	6.350	C.L.F.	785	390		1,175	1,450
1200	350 kcmil		3.28	7.330		1,025	450		1,475	1,800
1400	500 kcmil	▼	3.02	7.940		1,300	485		1,785	2,175
1600	15 kV, ungrounded neutral, #1	2 Elec	3.36	4.760		440	292		732	920
1800	1/0		3.19	5.010		525	305		830	1,050
2000	2/0		3.02	5.290		600	325		925	1,150
2200	4/0	▼	2.69	5.950		795	365		1,160	1,425
2400	250 kcmil	3 Elec	3.78	6.350		880	390		1,270	1,550
2600	350 kcmil		3.28	7.330		1,100	450		1,550	1,900
2800	500 kcmil	▼	3.02	7.940		1,250	485		1,735	2,100
3000	25 kV, grounded neutral, 1/0	2 Elec	3.02	5.290		730	325		1,055	1,275
3200	2/0		2.86	5.600		800	345		1,145	1,400
3400	4/0	▼	2.52	6.350		1,000	390		1,390	1,675
3600	250 kcmil	3 Elec	3.53	6.800		1,250	415		1,665	2,000
3800	350 kcmil		3.02	7.940		1,475	485		1,960	2,350
3900	500 kcmil	▼	2.77	8.660		1,725	530		2,255	2,700
4000	35 kV, grounded neutral, 1/0	2 Elec	2.86	5.600		715	345		1,060	1,300
4200	2/0		2.69	5.950		905	365		1,270	1,550
4400	4/0	▼	2.35	6.800		1,150	415		1,565	1,875
4600	250 kcmil	3 Elec	3.28	7.330		1,325	450		1,775	2,150
4800	350 kcmil		2.77	8.660		1,600	530		2,130	2,575
5000	500 kcmil	▼	2.52	9.520		1,900	585		2,485	2,950
5050	Aluminum, XLP shielding, 5 kV, #2	2 Elec	4.20	3.810		291	234		525	670
5070	#1		3.70	4.330		300	266		566	725
5090	1/0		3.36	4.760		350	292		642	820
5100	2/0		3.19	5.010		390	305		695	890
5150	4/0	▼	3.02	5.290		460	325		785	995
5200	250 kcmil	3 Elec	4.03	5.950		560	365		925	1,150
5220	350 kcmil		3.78	6.350		660	390		1,050	1,300
5240	500 kcmil		3.28	7.330		845	450		1,295	1,600
5260	750 kcmil	▼	3.02	7.940		1,125	485		1,610	1,950
5300	15 kV aluminum, XLP, #1	2 Elec	3.70	4.330		375	266		641	805
5320	1/0		3.36	4.760		385	292		677	860
5340	2/0		3.19	5.010		415	305		720	920
5360	4/0	▼	3.02	5.290		510	325		835	1,050
5380	250 kcmil	3 Elec	4.03	5.950		610	365		975	1,225
5400	350 kcmil		3.78	6.350		685	390		1,075	1,325
5420	500 kcmil		3.28	7.330		945	450		1,395	1,725
5440	750 kcmil	▼	3.02	7.940	▼	1,250	485		1,735	2,100

26 05 19 – Low-Voltage Electrical Power Conductors and Cables

26 05 19.13 Undercarpet Electrical Power Cables

		Crew	Daily Output	Labor-Hours	Unit	Material	Labor	Equipment	Total	Total Incl O&P
0010	**UNDERCARPET ELECTRICAL POWER CABLES**									
0020	Power System									
0100	Cable flat, 3 conductor, #12, w/attached bottom shield	1 Elec	824.88	.010	L.F.	5.90	.60		6.50	7.35
0200	Shield, top, steel		1485.12	.005	"	5.30	.33		5.63	6.35
0250	Splice, 3 conductor		40.32	.198	Ea.	16.10	12.15		28.25	36
0300	Top shield		80.64	.099		1.49	6.10		7.59	10.70
0350	Tap		33.60	.238		20.50	14.60		35.10	44.50
0400	Insulating patch, splice, tap & end		40.32	.198		51.50	12.15		63.65	75
0450	Fold		193.20	.041			2.54		2.54	3.78
0500	Top shield, tap & fold		80.64	.099		1.49	6.10		7.59	10.70
0700	Transition, block assembly		64.68	.124	▼	75.50	7.60		83.10	94.50

For customer support on your Electrical Change Order Costs with RSMeans data, call 800.448.8182.

Post-Installation Change Order Costs

26 05 19 – Low-Voltage Electrical Power Conductors and Cables

26 05 19.13 Undercarpet Electrical Power Cables	Crew	Daily Output	Labor-Hours	Unit	Material	2020 Bare Costs Labor	Equipment	Total	Total Incl O&P
0750	Receptacle frame & base	1 Elec	26.88	.298	Ea.	41.50	18.25		59.75	72.50
0800	Cover receptacle		100.80	.079		3.49	4.87		8.36	11.10
0850	Cover blank		134.40	.060		4.11	3.65		7.76	10
0860	Receptacle, direct connected, single		21	.381		88.50	23.50		112	133
0870	Dual		13.44	.595		145	36.50		181.50	215
0880	Combination high & low, tension		17.64	.454		107	28		135	160
0900	Box, floor with cover		16.80	.476		89.50	29		118.50	142
0920	Floor service w/barrier		3.36	2.380		254	146		400	495
1000	Wall, surface, with cover		16.80	.476		59	29		88	109
1100	Wall, flush, with cover		16.80	.476	▼	41.50	29		70.50	89
1450	Cable flat, 5 conductor #12, w/attached bottom shield		672	.012	L.F.	9.65	.73		10.38	11.75
1550	Shield, top, steel		1485.12	.005	"	9.60	.33		9.93	11.05
1600	Splice, 5 conductor		40.32	.198	Ea.	26	12.15		38.15	46.50
1650	Top shield		80.64	.099		1.49	6.10		7.59	10.70
1700	Tap		40.32	.198		34.50	12.15		46.65	56
1750	Insulating patch, splice tap & end		69.72	.115		60.50	7.05		67.55	77
1800	Transition, block assembly		64.68	.124		54	7.60		61.60	71
1850	Box, wall, flush with cover		16.80	.476	▼	55	29		84	104
1900	Cable flat, 4 conductor #12		783.72	.010	L.F.	6.80	.63		7.43	8.40
1950	3 conductor #10		824.88	.010		6.55	.60		7.15	8.10
1960	4 conductor #10		783.72	.010		8.60	.63		9.23	10.40
1970	5 conductor #10	▼	742.56	.011	▼	10.45	.66		11.11	12.50
2500	Telephone System									
2510	Transition fitting wall box, surface	1 Elec	20.16	.397	Ea.	66.50	24.50		91	110
2520	Flush		20.16	.397		66.50	24.50		91	110
2530	Flush, for PC board		20.16	.397		66.50	24.50		91	110
2540	Floor service box	▼	3.36	2.380		262	146		408	505
2550	Cover, surface					24			24	26.50
2560	Flush					24			24	26.50
2570	Flush for PC board					24			24	26.50
2700	Floor fitting w/duplex jack & cover	1 Elec	17.64	.454		60.50	28		88.50	108
2720	Low profile		44.52	.180		19.95	11		30.95	38.50
2740	Miniature w/duplex jack		44.52	.180		37.50	11		48.50	58
2760	25 pair kit		17.64	.454		56	28		84	103
2780	Low profile		44.52	.180		20.50	11		31.50	39
2800	Call director kit for 5 cable		15.96	.501		104	31		135	161
2820	4 pair kit		15.96	.501		106	31		137	163
2840	3 pair kit		15.96	.501		132	31		163	191
2860	Comb. 25 pair & 3 conductor power		17.64	.454		104	28		132	157
2880	5 conductor power		17.64	.454		102	28		130	154
2900	PC board, 8 per 3 pair		135.24	.059		77.50	3.63		81.13	90.50
2920	6 per 4 pair		135.24	.059		77.50	3.63		81.13	90.50
2940	3 pair adapter		135.24	.059		70.50	3.63		74.13	83
2950	Plug		64.68	.124		3.53	7.60		11.13	15.20
2960	Couplers		269.64	.030	▼	9.95	1.82		11.77	13.65
3000	Bottom shield for 25 pair cable		3712.80	.002	L.F.	.95	.13		1.08	1.24
3020	4 pair		3712.80	.002		.45	.13		.58	.70
3040	Top shield for 25 pair cable		3712.80	.002	▼	.95	.13		1.08	1.24
3100	Cable assembly, double-end, 50', 25 pair		9.91	.807	Ea.	273	49.50		322.50	375
3110	3 pair		19.82	.404		86.50	25		111.50	132
3120	4 pair		19.82	.404	▼	104	25		129	152
3140	Bulk 3 pair		1237.32	.006	L.F.	1.47	.40		1.87	2.21
3160	4 pair	▼	1237.32	.006	"	1.83	.40		2.23	2.60

26 05 19.13 Undercarpet Electrical Power Cables

		Crew	Daily Output	Labor-Hours	Unit	Material	2020 Bare Costs Labor	Equipment	Total	Total Incl O&P
3500	Data System									
3520	Cable 25 conductor w/connection 40', 75 ohm	1 Elec	12.18	.657	Ea.	94	40.50		134.50	164
3530	Single lead		18.48	.433		257	26.50		283.50	325
3540	Dual lead		18.48	.433		273	26.50		299.50	340
3560	Shields same for 25 conductor as 25 pair telephone									
3570	Single & dual, none required									
3590	BNC coax connectors, plug	1 Elec	33.60	.238	Ea.	16.80	14.60		31.40	40.50
3600	TNC coax connectors, plug	"	33.60	.238	"	16.90	14.60		31.50	40.50
3700	Cable-bulk									
3710	Single lead	1 Elec	1237.32	.006	L.F.	4.07	.40		4.47	5.05
3720	Dual lead	"	1237.32	.006	"	5.90	.40		6.30	7.10
3730	Hand tool crimp				Ea.	570			570	625
3740	Hand tool notch				"	22.50			22.50	25
3750	Boxes & floor fitting same as telephone									
3790	Data cable notching, 90°	1 Elec	81.48	.098	Ea.		6		6	8.95
3800	180°		50.40	.159			9.75		9.75	14.50
8100	Drill floor		134.40	.060		4.86	3.65		8.51	10.80
8200	Marking floor		1344	.006	L.F.		.37		.37	.54
8300	Tape, hold down		5376	.001	"	.20	.09		.29	.36
8350	Tape primer, 500' per can		80.64	.099	Ea.	54	6.10		60.10	68.50
8400	Tool, splicing				"	184			184	202

26 05 19.20 Armored Cable

		Crew	Daily Output	Labor-Hours	Unit	Material	2020 Bare Costs Labor	Equipment	Total	Total Incl O&P
0010	**ARMORED CABLE**									
0050	600 volt, copper (BX), #14, 2 conductor, solid	1 Elec	2.02	3.970	C.L.F.	49	244		293	420
0100	3 conductor, solid		1.85	4.330		82	266		348	485
0120	4 conductor, solid		1.68	4.760		116	292		408	565
0150	#12, 2 conductor, solid		1.93	4.140		51.50	254		305.50	435
0200	3 conductor, solid		1.68	4.760		86	292		378	530
0220	4 conductor, solid		1.51	5.290		119	325		444	615
0250	#10, 2 conductor, solid		1.68	4.760		102	292		394	545
0300	3 conductor, solid		1.34	5.950		136	365		501	695
0320	4 conductor, solid		1.18	6.800		210	415		625	850
0340	#8, 2 conductor, stranded		1.26	6.350		320	390		710	935
0350	3 conductor, stranded		1.09	7.330		320	450		770	1,025
0370	4 conductor, stranded		.92	8.660		460	530		990	1,300
0380	#6, 2 conductor, stranded		1.09	7.330		315	450		765	1,025
0400	3 conductor with PVC jacket, in cable tray, #6		2.60	3.070		665	188		853	1,000
0450	#4	2 Elec	4.54	3.530		815	217		1,032	1,225
0500	#2		3.86	4.140		1,000	254		1,254	1,475
0550	#1		3.36	4.760		1,275	292		1,567	1,825
0600	1/0		3.02	5.290		1,375	325		1,700	1,975
0650	2/0		2.86	5.600		1,400	345		1,745	2,050
0700	3/0		2.69	5.950		2,175	365		2,540	2,950
0750	4/0		2.52	6.350		2,625	390		3,015	3,475
0800	250 kcmil	3 Elec	3.02	7.940		3,300	485		3,785	4,350
0850	350 kcmil		2.77	8.660		4,100	530		4,630	5,325
0900	500 kcmil		2.52	9.520		5,675	585		6,260	7,125
0910	4 conductor with PVC jacket, in cable tray, #6	1 Elec	2.27	3.530		1,725	217		1,942	2,225
0920	#4	2 Elec	3.86	4.140		1,875	254		2,129	2,450
0930	#2		3.36	4.760		1,500	292		1,792	2,075
0940	#1		3.02	5.290		1,900	325		2,225	2,575
0950	1/0		2.86	5.600		1,775	345		2,120	2,450

26 05 19 – Low-Voltage Electrical Power Conductors and Cables

26 05 19.20 Armored Cable		Crew	Daily Output	Labor-Hours	Unit	Material	2020 Bare Costs Labor	Equipment	Total	Total Incl O&P
0960	2/0	2 Elec	2.69	5.950	C.L.F.	2,175	365		2,540	2,950
0970	3/0		2.52	6.350		2,875	390		3,265	3,750
0980	4/0		2.02	7.940		2,625	485		3,110	3,600
0990	250 kcmil	3 Elec	2.77	8.660		4,000	530		4,530	5,200
1000	350 kcmil		2.52	9.520		5,375	585		5,960	6,775
1010	500 kcmil		2.27	10.580		7,450	650		8,100	9,175
1050	5 kV, copper, 3 conductor with PVC jacket,									
1060	non-shielded, in cable tray, #4	2 Elec	319.20	.050	L.F.	9.20	3.08		12.28	14.70
1100	#2		302.40	.053		11.95	3.25		15.20	18
1200	#1		252	.063		15.20	3.90		19.10	22.50
1400	1/0		243.60	.066		17.55	4.03		21.58	25.50
1600	2/0		218.40	.073		20.50	4.49		24.99	29
2000	4/0		201.60	.079		27	4.87		31.87	37
2100	250 kcmil	3 Elec	277.20	.087		37	5.30		42.30	49
2150	350 kcmil		264.60	.091		45.50	5.55		51.05	58.50
2200	500 kcmil		226.80	.106		67.50	6.50		74	83.50
2400	15 kV, copper, 3 conductor with PVC jacket galv., steel armored									
2500	grounded neutral, in cable tray, #2	2 Elec	252	.063	L.F.	20	3.90		23.90	28
2600	#1		235.20	.068		20.50	4.17		24.67	28.50
2800	1/0		218.40	.073		24	4.49		28.49	33
2900	2/0		184.80	.087		31	5.30		36.30	42.50
3000	4/0		159.60	.100		35.50	6.15		41.65	48
3100	250 kcmil	3 Elec	226.80	.106		39.50	6.50		46	53
3150	350 kcmil		201.60	.119		46.50	7.30		53.80	62
3200	500 kcmil		176.40	.136		62.50	8.35		70.85	81
3400	15 kV, copper, 3 conductor with PVC jacket,									
3450	ungrounded neutral, in cable tray, #2	2 Elec	218.40	.073	L.F.	21	4.49		25.49	29.50
3500	#1		193.20	.083		23	5.10		28.10	33
3600	1/0		168	.095		26.50	5.85		32.35	37.50
3700	2/0		159.60	.100		32.50	6.15		38.65	44.50
3800	4/0		134.40	.119		39	7.30		46.30	54
4000	250 kcmil	3 Elec	176.40	.136		45.50	8.35		53.85	62.50
4050	350 kcmil		163.80	.147		60	9		69	79.50
4100	500 kcmil		151.20	.159		73	9.75		82.75	94.50
4200	600 volt, aluminum, 3 conductor in cable tray with PVC jacket									
4300	#2	2 Elec	453.60	.035	L.F.	4.96	2.16		7.12	8.65
4400	#1		386.40	.041		5.90	2.54		8.44	10.25
4500	1/0		336	.048		7.35	2.92		10.27	12.40
4600	2/0		302.40	.053		7.40	3.25		10.65	13
4700	3/0		285.60	.056		8.70	3.44		12.14	14.65
4800	4/0		268.80	.060		10.55	3.65		14.20	17.05
4900	250 kcmil	3 Elec	378	.063		12.65	3.90		16.55	19.70
5000	350 kcmil		302.40	.079		15.10	4.87		19.97	24
5200	500 kcmil		277.20	.087		18.75	5.30		24.05	28.50
5300	750 kcmil		239.40	.100		24	6.15		30.15	35.50
5400	600 volt, aluminum, 4 conductor in cable tray with PVC jacket									
5410	#2	2 Elec	436.80	.037	L.F.	5.90	2.25		8.15	9.80
5430	#1		369.60	.043		7.20	2.66		9.86	11.85
5450	1/0		319.20	.050		8.40	3.08		11.48	13.85
5470	2/0		285.60	.056		8.60	3.44		12.04	14.55
5480	3/0		268.80	.060		10.10	3.65		13.75	16.55
5500	4/0		252	.063		11.85	3.90		15.75	18.80
5520	250 kcmil	3 Elec	352.80	.068		12.70	4.17		16.87	20

26 05 19.20 Armored Cable	Crew	Daily Output	Labor-Hours	Unit	Material	2020 Bare Costs Labor	Equipment	Total	Total Incl O&P
5540	350 kcmil	3 Elec	277.20	.087	L.F.	16.40	5.30		21.70	26
5560	500 kcmil		252	.095		20.50	5.85		26.35	31
5580	750 kcmil		226.80	.106		28.50	6.50		35	40.50
5600	5 kV, aluminum, unshielded in cable tray, #2 with PVC jacket	2 Elec	319.20	.050		7.40	3.08		10.48	12.75
5700	#1 with PVC jacket		302.40	.053		8.30	3.25		11.55	14
5800	1/0 with PVC jacket		252	.063		8.50	3.90		12.40	15.15
6000	2/0 with PVC jacket		243.60	.066		8.70	4.03		12.73	15.55
6200	3/0 with PVC jacket		218.40	.073		10.45	4.49		14.94	18.20
6300	4/0 with PVC jacket		201.60	.079		12.25	4.87		17.12	21
6400	250 kcmil with PVC jacket	3 Elec	277.20	.087		13.45	5.30		18.75	22.50
6500	350 kcmil with PVC jacket		264.60	.091		15.75	5.55		21.30	25.50
6600	500 kcmil with PVC jacket		252	.095		18.65	5.85		24.50	29
6800	750 kcmil with PVC jacket		226.80	.106		23	6.50		29.50	35
6900	15 kV, aluminum, shielded-grounded, #2 with PVC jacket	2 Elec	268.80	.060		16.90	3.65		20.55	24
7000	#1 with PVC jacket		252	.063		17.40	3.90		21.30	25
7200	1/0 with PVC jacket		235.20	.068		18.75	4.17		22.92	26.50
7300	2/0 with PVC jacket		218.40	.073		19	4.49		23.49	27.50
7400	3/0 with PVC jacket		201.60	.079		21.50	4.87		26.37	31
7500	4/0 with PVC jacket		184.80	.087		22	5.30		27.30	32
7600	250 kcmil with PVC jacket	3 Elec	252	.095		24	5.85		29.85	35
7700	350 kcmil with PVC jacket		226.80	.106		28	6.50		34.50	40.50
7800	500 kcmil with PVC jacket		201.60	.119		34	7.30		41.30	48
8000	750 kcmil with PVC jacket		171.36	.140		40	8.60		48.60	57
8200	15 kV, aluminum, shielded-ungrounded, #1 with PVC jacket	2 Elec	210	.076		21	4.67		25.67	30.50
8300	1/0 with PVC jacket		193.20	.083		22	5.10		27.10	31.50
8400	2/0 with PVC jacket		176.40	.091		24	5.55		29.55	35
8500	3/0 with PVC jacket		168	.095		24.50	5.85		30.35	35.50
8600	4/0 with PVC jacket		159.60	.100		26.50	6.15		32.65	38
8700	250 kcmil with PVC jacket	3 Elec	226.80	.106		28.50	6.50		35	41
8800	350 kcmil with PVC jacket		201.60	.119		32.50	7.30		39.80	47
8900	500 kcmil with PVC jacket		176.40	.136		39.50	8.35		47.85	55.50
8950	750 kcmil with PVC jacket		146.16	.164		49	10.05		59.05	69
9010	600 volt, copper (MC) steel clad, #14, 2 wire	R-1A	6.85	2.330	C.L.F.	52	129		181	249
9020	3 wire		6.06	2.640		96.50	146		242.50	325
9030	4 wire		5.62	2.850		250	157		407	510
9040	#12, 2 wire		6.06	2.640		71.50	146		217.50	296
9050	3 wire		5.62	2.850		84.50	157		241.50	325
9060	4 wire		5.07	3.150		145	174		319	420
9070	#10, 2 wire		5.07	3.150		129	174		303	400
9080	3 wire		4.75	3.370		165	186		351	460
9090	4 wire		4.39	3.640		260	201		461	585
9100	#8, 2 wire, stranded		3.31	4.830		201	267		468	615
9110	3 wire, stranded		2.86	5.600		231	310		541	715
9120	4 wire, stranded		2.35	6.800		380	375		755	975
9130	#6, 2 wire, stranded		2.90	5.520		279	305		584	760
9200	600 volt, copper (MC) aluminum clad, #14, 2 wire		6.85	2.330		51	129		180	248
9210	3 wire		6.06	2.640		93	146		239	320
9220	4 wire		5.62	2.850		115	157		272	360
9230	#12, 2 wire		6.06	2.640		52	146		198	275
9240	3 wire		5.62	2.850		90	157		247	335
9250	4 wire		5.07	3.150		185	174		359	460
9260	#10, 2 wire		5.07	3.150		111	174		285	380
9270	3 wire		4.75	3.370		143	186		329	435

Post-Installation Change Order Costs

26 05 19.20 Armored Cable

		Crew	Daily Output	Labor-Hours	Unit	Material	2020 Bare Costs Labor	Equipment	Total	Total Incl O&P
9280	4 wire	R-1A	4.39	3.640	C.L.F.	271	201		472	595
9600	Alum (MC) aluminum clad, #6, 3 conductor w/#6 grnd		2.31	6.930		292	385		677	890
9610	4 conductor w/#6 grnd	▼	1.92	8.320		261	460		721	970
9620	#4, 3 conductor w/#6 grnd	R-1B	2.63	9.130		249	485		734	1,000
9630	4 conductor w/#6 grnd		2.24	10.700		269	570		839	1,150
9640	#2, 3 conductor w/#4 grnd		2.18	11.030		445	585		1,030	1,375
9650	4 conductor w/#4 grnd		1.86	12.930		500	690		1,190	1,575
9660	#1, 3 conductor w/#4 grnd		1.81	13.290		370	705		1,075	1,450
9670	4 conductor w/#4 grnd		1.55	15.440		680	820		1,500	1,975
9680	1/0, 3 conductor w/#4 grnd		1.81	13.230		420	705		1,125	1,525
9690	4 conductor w/#4 grnd		1.45	16.520		510	880		1,390	1,875
9700	2/0, 3 conductor w/#4 grnd		1.60	14.960		690	795		1,485	1,925
9710	4 conductor w/#4 grnd		1.29	18.670		490	995		1,485	2,025
9720	3/0, 3 conductor w/#4 grnd		1.47	16.330		885	870		1,755	2,275
9730	4 conductor w/#4 grnd		1.18	20.410		610	1,075		1,685	2,300
9740	4/0, 3 conductor w/#2 grnd		1.29	18.550		990	985		1,975	2,575
9750	4 conductor w/#2 grnd	▼	1.03	23.230		430	1,225		1,655	2,325
9760	250 kcmil, 3 conductor w/#1 grnd	R-1C	1.60	20.050		1,125	1,100	64	2,289	2,950
9770	4 conductor w/#1 grnd		1.28	25.060		1,525	1,375	79.50	2,979.50	3,825
9780	350 kcmil, 3 conductor w/1/0 grnd		1.48	21.650		1,500	1,200	69	2,769	3,500
9790	4 conductor w/1/0 grnd		1.18	27.020		1,150	1,500	86	2,736	3,600
9800	500 kcmil, 3 conductor w/#1 grnd		1.38	23.230		1,900	1,275	74	3,249	4,075
9810	4 conductor w/2/0 grnd		1.10	29.080		1,575	1,600	92.50	3,267.50	4,225
9840	750 kcmil, 3 conductor w/1/0 grnd		1.10	29.080		3,025	1,600	92.50	4,717.50	5,825
9850	4 conductor w/3/0 grnd	▼	.87	36.630	▼	3,125	2,025	116	5,266	6,575

26 05 19.25 Cable Connectors

		Crew	Daily Output	Labor-Hours	Unit	Material	2020 Bare Costs Labor	Equipment	Total	Total Incl O&P
0010	**CABLE CONNECTORS**									
0100	600 volt, nonmetallic, #14-2 wire	1 Elec	134.40	.060	Ea.	2.14	3.65		5.79	7.80
0200	#14-3 wire to #12-2 wire		111.72	.072		2.14	4.39		6.53	8.90
0300	#12-3 wire to #10-2 wire		95.76	.084		2.14	5.15		7.29	10
0400	#10-3 wire to #14-4 and #12-4 wire		84	.095		2.14	5.85		7.99	11.05
0500	#8-3 wire to #10-4 wire		67.20	.119		4.05	7.30		11.35	15.35
0600	#6-3 wire		33.60	.238		5.20	14.60		19.80	27.50
0800	SER, 3 #8 insulated + 1 #8 ground		26.88	.298		3.94	18.25		22.19	31.50
0900	3 #6 + 1 #6 ground		20.16	.397		4.92	24.50		29.42	42
1000	3 #4 + 1 #6 ground		18.48	.433		6.15	26.50		32.65	46.50
1100	3 #2 + 1 #4 ground		16.80	.476		13.20	29		42.20	58
1200	3 1/0 + 1 #2 ground		15.12	.529		26.50	32.50		59	77.50
1400	3 2/0 + 1 #1 ground		13.44	.595		37	36.50		73.50	95
1600	3 4/0 + 1 #2/0 ground		11.76	.680		36.50	41.50		78	103
1800	600 volt, armored, #14-2 wire		67.20	.119		1.58	7.30		8.88	12.65
2200	#14-4, #12-3 and #10-2 wire		33.60	.238		.99	14.60		15.59	23
2400	#12-4, #10-3 and #8-2 wire		26.88	.298		4.35	18.25		22.60	32
2600	#8-3 and #10-4 wire		21.84	.366		5.15	22.50		27.65	39
2650	#8-4 wire		18.48	.433		7.65	26.50		34.15	48
2700	PVC jacket connector, #6-3 wire, #6-4 wire		13.44	.595		7.50	36.50		44	62.50
2800	#4-3 wire, #4-4 wire		13.44	.595		7.50	36.50		44	62.50
2900	#2-3 wire		10.08	.794		7.50	48.50		56	80.50
3000	#1-3 wire, #2-4 wire		10.08	.794		16.95	48.50		65.45	91
3200	1/0-3 wire		9.24	.866		16.95	53		69.95	97.50
3400	2/0-3 wire, 1/0-4 wire		8.40	.952		16.95	58.50		75.45	106
3500	3/0-3 wire, 2/0-4 wire		7.56	1.060		22	65		87	122

26 05 19.25 Cable Connectors

		Crew	Daily Output	Labor-Hours	Unit	Material	2020 Bare Costs Labor	Equipment	Total	Total Incl O&P
3600	4/0-3 wire, 3/0-4 wire	1 Elec	5.88	1.360	Ea.	22	83.50		105.50	149
3800	250 kcmil-3 wire, 4/0-4 wire		5.04	1.590		43	97.50		140.50	192
4000	350 kcmil-3 wire, 250 kcmil-4 wire		4.20	1.900		43	117		160	221
4100	350 kcmil-4 wire		3.36	2.380		248	146		394	490
4200	500 kcmil-3 wire		3.36	2.380		248	146		394	490
4250	500 kcmil-4 wire, 750 kcmil-3 wire		2.94	2.720		274	167		441	550
4300	750 kcmil-4 wire		2.52	3.170		274	194		468	590
4400	5 kV, armored, #4		6.72	1.190		83.50	73		156.50	201
4600	#2		6.72	1.190		83.50	73		156.50	201
4800	#1		6.72	1.190		109	73		182	229
5000	1/0		5.38	1.490		135	91.50		226.50	284
5200	2/0		4.45	1.800		135	110		245	310
5500	4/0		3.36	2.380		181	146		327	415
5600	250 kcmil		3.02	2.650		181	163		344	440
5650	350 kcmil		2.69	2.980		214	183		397	505
5700	500 kcmil		2.10	3.810		279	234		513	655
5720	750 kcmil		1.85	4.330		335	266		601	765
5750	1,000 kcmil		1.68	4.760		400	292		692	875
5800	15 kV, armored, #1		3.36	2.380		174	146		320	410
5900	1/0		3.36	2.380		217	146		363	455
6000	3/0		3.02	2.650		320	163		483	590
6100	4/0		2.86	2.800		350	172		522	640
6200	250 kcmil		2.69	2.980		440	183		623	750
6300	350 kcmil		2.27	3.530		395	217		612	750
6400	500 kcmil		1.68	4.760		565	292		857	1,050

26 05 19.30 Cable Splicing

		Crew	Daily Output	Labor-Hours	Unit	Material	2020 Bare Costs Labor	Equipment	Total	Total Incl O&P
0010	**CABLE SPLICING** URD or similar, ideal conditions									
0100	#6 stranded to #1 stranded, 5 kV	1 Elec	3.36	2.380	Ea.	166	146		312	400
0120	15 kV		3.02	2.650		224	163		387	490
0140	25 kV		2.69	2.980		272	183		455	570
0200	#1 stranded to 4/0 stranded, 5 kV		3.02	2.650		171	163		334	430
0210	15 kV		2.69	2.980		245	183		428	540
0220	25 kV		2.35	3.400		272	209		481	610
0300	4/0 stranded to 500 kcmil stranded, 5 kV		2.77	2.890		171	177		348	450
0310	15 kV		2.44	3.280		405	201		606	745
0320	25 kV		2.10	3.810		465	234		699	860
0400	500 kcmil, 5 kV		2.69	2.980		228	183		411	525
0410	15 kV		2.35	3.400		405	209		614	755
0420	25 kV		1.93	4.140		465	254		719	890
0500	600 kcmil, 5 kV		2.44	3.280		228	201		429	550
0510	15 kV		2.02	3.970		430	244		674	835
0520	25 kV		1.68	4.760		495	292		787	980
0600	750 kcmil, 5 kV		2.18	3.660		228	225		453	585
0610	15 kV		1.85	4.330		430	266		696	865
0620	25 kV		1.60	5.010		495	305		800	1,000
0700	1,000 kcmil, 5 kV		1.93	4.140		228	254		482	630
0710	15 kV		1.60	5.010		465	305		770	975
0720	25 kV		1.34	5.950		475	365		840	1,075

26 05 19.35 Cable Terminations

		Crew	Daily Output	Labor-Hours	Unit	Material	2020 Bare Costs Labor	Equipment	Total	Total Incl O&P
0010	**CABLE TERMINATIONS**									
0015	Wire connectors, screw type, #22 to #14	1 Elec	218.40	.037	Ea.	.08	2.25		2.33	3.44
0020	#18 to #12		201.60	.040		.09	2.43		2.52	3.72

26 05 19.35 Cable Terminations		Crew	Daily Output	Labor-Hours	Unit	Material	2020 Bare Costs Labor	Equipment	Total	Total Incl O&P
0035	#16 to #10	1 Elec	193.20	.041	Ea.	.34	2.54		2.88	4.15
0040	#14 to #8		176.40	.045		.43	2.78		3.21	4.61
0045	#12 to #6		151.20	.053		.61	3.25		3.86	5.50
0050	Terminal lugs, solderless, #16 to #10		42	.190		.48	11.70		12.18	17.95
0100	#8 to #4		25.20	.317		.96	19.50		20.46	30
0150	#2 to #1		18.48	.433		1.03	26.50		27.53	40.50
0200	1/0 to 2/0		13.44	.595		2.05	36.50		38.55	57
0250	3/0		10.08	.794		6.05	48.50		54.55	79
0300	4/0		9.24	.866		5.45	53		58.45	85
0350	250 kcmil		7.56	1.060		4.07	65		69.07	101
0400	350 kcmil		5.88	1.360		5.55	83.50		89.05	130
0450	500 kcmil		5.04	1.590		10.20	97.50		107.70	156
0500	600 kcmil		4.87	1.640		11.90	101		112.90	163
0550	750 kcmil		4.37	1.830		13.20	112		125.20	182
0600	Split bolt connectors, tapped, #6		13.44	.595		4.73	36.50		41.23	59.50
0650	#4		11.76	.680		4.43	41.50		45.93	67
0700	#2		10.08	.794		8.20	48.50		56.70	81.50
0750	#1		9.24	.866		9.20	53		62.20	89
0800	1/0		8.40	.952		9.20	58.50		67.70	97
0850	2/0		7.56	1.060		16	65		81	115
0900	3/0		6.05	1.320		29.50	81		110.50	154
1000	4/0		5.38	1.490		36	91.50		127.50	176
1100	250 kcmil		4.79	1.670		36	102		138	193
1200	300 kcmil		4.45	1.800		44	110		154	212
1400	350 kcmil		3.86	2.070		44	127		171	237
1500	500 kcmil	▽	3.36	2.380	▽	101	146		247	330
1600	Crimp 1 hole lugs, copper or aluminum, 600 volt									
1620	#14	1 Elec	50.40	.159	Ea.	.77	9.75		10.52	15.35
1630	#12		42	.190		1.30	11.70		13	18.85
1640	#10		37.80	.212		1.30	13		14.30	21
1780	#8		30.24	.265		1.96	16.25		18.21	26
1800	#6		25.20	.317		2.28	19.50		21.78	31.50
2000	#4		22.68	.353		5.25	21.50		26.75	38
2200	#2		20.16	.397		3.61	24.50		28.11	40.50
2400	#1		16.80	.476		6.50	29		35.50	50.50
2500	1/0		14.70	.544		7.85	33.50		41.35	58
2600	2/0		12.60	.635		10.55	39		49.55	69.50
2800	3/0		10.08	.794		10.30	48.50		58.80	84
3000	4/0		9.24	.866		8.35	53		61.35	88
3200	250 kcmil		7.56	1.060		9.35	65		74.35	107
3400	300 kcmil		6.72	1.190		14.40	73		87.40	125
3500	350 kcmil		5.88	1.360		15.20	83.50		98.70	141
3600	400 kcmil		5.46	1.470		20	90		110	156
3800	500 kcmil		5.04	1.590		26.50	97.50		124	174
4000	600 kcmil		4.87	1.640		35	101		136	189
4200	700 kcmil		4.62	1.730		42.50	106		148.50	205
4400	750 kcmil	▽	4.37	1.830	▽	42.50	112		154.50	214
4500	Crimp 2-way connectors, copper or alum., 600 volt									
4510	#14	1 Elec	50.40	.159	Ea.	3.29	9.75		13.04	18.10
4520	#12		42	.190		3.50	11.70		15.20	21.50
4530	#10		37.80	.212		3.69	13		16.69	23.50
4540	#8		22.68	.353		3.73	21.50		25.23	36
4600	#6	▽	21	.381		5.30	23.50		28.80	41

Post-Installation Change Order Costs   For customer support on your Electrical Change Order Costs with RSMeans data, call 800.448.8182.

26 05 19.35 Cable Terminations

		Crew	Daily Output	Labor-Hours	Unit	Material	2020 Bare Costs Labor	2020 Bare Costs Equipment	Total	Total Incl O&P
4800	#4	1 Elec	19.32	.414	Ea.	5.15	25.50		30.65	43.50
5000	#2		16.80	.476		7.35	29		36.35	51.50
5200	#1		13.44	.595		7.40	36.50		43.90	62.50
5400	1/0		10.92	.733		7.40	45		52.40	75
5420	2/0		10.08	.794		10.20	48.50		58.70	84
5440	3/0		9.24	.866		9.60	53		62.60	89.50
5460	4/0		8.40	.952		10	58.50		68.50	98
5480	250 kcmil		7.56	1.060		22	65		87	121
5500	300 kcmil		7.14	1.120		21	68.50		89.50	125
5520	350 kcmil		6.72	1.190		19.95	73		92.95	131
5540	400 kcmil		6.13	1.300		26	80		106	148
5560	500 kcmil		5.21	1.540		31	94.50		125.50	175
5580	600 kcmil		4.62	1.730		62	106		168	226
5600	700 kcmil		3.78	2.120		65	130		195	266
5620	750 kcmil		3.36	2.380		72	146		218	296
7000	Compression equipment adapter, aluminum wire, #6		25.20	.317		9.80	19.50		29.30	40
7020	#4		22.68	.353		9.95	21.50		31.45	43
7040	#2		20.16	.397		9.20	24.50		33.70	46.50
7060	#1		16.80	.476		10.50	29		39.50	55
7080	1/0		15.12	.529		17.45	32.50		49.95	67.50
7100	2/0		12.60	.635		18.35	39		57.35	78
7140	4/0		9.24	.866		21.50	53		74.50	103
7160	250 kcmil		7.56	1.060		23	65		88	123
7180	300 kcmil		6.72	1.190		24	73		97	136
7200	350 kcmil		5.88	1.360		33	83.50		116.50	161
7220	400 kcmil		5.46	1.470		39	90		129	177
7240	500 kcmil		5.04	1.590		39.50	97.50		137	189
7260	600 kcmil		4.87	1.640		45	101		146	200
7280	750 kcmil	▼	4.37	1.830		54.50	112		166.50	227
8000	Compression tool, hand					2,000			2,000	2,200
8100	Hydraulic					2,500			2,500	2,750
8500	Hydraulic dies				▼	465			465	510

26 05 19.50 Mineral Insulated Cable

		Crew	Daily Output	Labor-Hours	Unit	Material	2020 Bare Costs Labor	2020 Bare Costs Equipment	Total	Total Incl O&P
0010	**MINERAL INSULATED CABLE** 600 volt									
0100	1 conductor, #12	1 Elec	1.34	5.950	C.L.F.	425	365		790	1,000
0200	#10		1.34	5.950		645	365		1,010	1,250
0400	#8		1.26	6.350		805	390		1,195	1,475
0500	#6	▼	1.18	6.800		855	415		1,270	1,550
0600	#4	2 Elec	2.02	7.940		1,275	485		1,760	2,125
0800	#2		1.85	8.660		1,725	530		2,255	2,700
0900	#1		1.76	9.070		2,000	555		2,555	3,025
1000	1/0		1.68	9.520		2,325	585		2,910	3,425
1100	2/0		1.60	10.030		2,925	615		3,540	4,150
1200	3/0		1.51	10.580		3,525	650		4,175	4,850
1400	4/0	▼	1.34	11.900		4,075	730		4,805	5,550
1410	250 kcmil	3 Elec	2.02	11.900		4,575	730		5,305	6,125
1420	350 kcmil		1.64	14.650		5,275	900		6,175	7,150
1430	500 kcmil	▼	1.64	14.650		6,650	900		7,550	8,650
1500	2 conductor, #12	1 Elec	1.18	6.800		1,125	415		1,540	1,850
1600	#10		1.01	7.940		1,375	485		1,860	2,225
1800	#8		.92	8.660		1,575	530		2,105	2,525
2000	#6	▼	.88	9.070		2,150	555		2,705	3,200

   Post-Installation Change Order Costs

26 05 19.50 Mineral Insulated Cable

	26 05 19.50 Mineral Insulated Cable	Crew	Daily Output	Labor-Hours	Unit	Material	2020 Bare Costs Labor	Equipment	Total	Total Incl O&P
2100	#4	2 Elec	1.68	9.520	C.L.F.	3,050	585		3,635	4,250
2200	3 conductor, #12	1 Elec	1.01	7.940		1,300	485		1,785	2,150
2400	#10		.92	8.660		1,475	530		2,005	2,425
2600	#8		.88	9.070		1,975	555		2,530	3,000
2800	#6		.84	9.520		2,600	585		3,185	3,725
3000	#4	2 Elec	1.51	10.580		3,425	650		4,075	4,725
3100	4 conductor, #12	1 Elec	1.01	7.940		1,375	485		1,860	2,225
3200	#10		.92	8.660		1,650	530		2,180	2,600
3400	#8		.84	9.520		2,300	585		2,885	3,400
3600	#6		.76	10.580		3,125	650		3,775	4,425
3620	7 conductor, #12		.92	8.660		1,875	530		2,405	2,875
3640	#10		.84	9.520		2,575	585		3,160	3,725
3800	Terminations, 600 volt, 1 conductor, #12		6.72	1.190	Ea.	22	73		95	133
4000	#10		6.38	1.250		22	76.50		98.50	138
4100	#8		6.13	1.300		22	80		102	143
4200	#6		5.63	1.420		22	87		109	154
4400	#4		5.21	1.540		22	94.50		116.50	165
4600	#2		4.79	1.670		33	102		135	190
4800	#1		4.45	1.800		33	110		143	201
5000	1/0		4.20	1.900		33	117		150	211
5100	2/0		3.95	2.030		33	125		158	222
5200	3/0		3.61	2.210		33	136		169	239
5400	4/0		3.36	2.380		73.50	146		219.50	298
5410	250 kcmil		3.36	2.380		73.50	146		219.50	298
5420	350 kcmil		3.36	2.380		126	146		272	355
5430	500 kcmil		3.36	2.380		126	146		272	355
5500	2 conductor, #12		5.63	1.420		22	87		109	154
5600	#10		5.38	1.490		33	91.50		124.50	173
5800	#8		5.21	1.540		33	94.50		127.50	178
6000	#6		4.79	1.670		33	102		135	190
6200	#4		4.45	1.800		73.50	110		183.50	245
6400	3 conductor, #12		4.79	1.670		33	102		135	190
6500	#10		4.62	1.730		33	106		139	195
6600	#8		4.37	1.830		33	112		145	204
6800	#6		4.03	1.980		33	121		154	218
7200	#4		3.86	2.070		73.50	127		200.50	270
7400	4 conductor, #12		3.86	2.070		37	127		164	230
7500	#10		3.70	2.160		37	133		170	238
7600	#8		3.53	2.270		37	139		176	248
8400	#6		3.36	2.380		77	146		223	300
8500	7 conductor, #12		2.94	2.720		36.50	167		203.50	289
8600	#10		2.52	3.170		79.50	194		273.50	380
8800	Crimping tool, plier type					92.50			92.50	102
9000	Stripping tool					350			350	385
9200	Hand vise					104			104	114

26 05 19.55 Non-Metallic Sheathed Cable

	26 05 19.55 Non-Metallic Sheathed Cable	Crew	Daily Output	Labor-Hours	Unit	Material	2020 Bare Costs Labor	Equipment	Total	Total Incl O&P
0010	**NON-METALLIC SHEATHED CABLE** 600 volt									
0100	Copper with ground wire (Romex)									
0150	#14, 2 conductor	1 Elec	2.27	3.530	C.L.F.	20	217		237	345
0200	3 conductor		2.02	3.970		29	244		273	395
0220	4 conductor		1.85	4.330		46.50	266		312.50	445
0250	#12, 2 conductor		2.10	3.810		29	234		263	380

26 05 19.55 Non-Metallic Sheathed Cable	Crew	Daily Output	Labor-Hours	Unit	Material	2020 Bare Costs Labor	2020 Bare Costs Equipment	Total	Total Incl O&P
0300	3 conductor	1 Elec	1.85	4.330	C.L.F.	45	266		311	445
0320	4 conductor		1.68	4.760		71.50	292		363.50	515
0350	#10, 2 conductor		1.85	4.330		49.50	266		315.50	450
0400	3 conductor		1.51	5.290		73	325		398	565
0420	4 conductor		1.34	5.950		115	365		480	670
0430	#8, 2 conductor		1.34	5.950		96.50	365		461.50	650
0450	3 conductor		1.26	6.350		128	390		518	720
0500	#6, 3 conductor		1.18	6.800		197	415		612	835
0520	#4, 3 conductor	2 Elec	2.02	7.940		440	485		925	1,200
0540	#2, 3 conductor	"	1.85	8.660		1,000	530		1,530	1,900
0550	SE type SER aluminum cable, 3 RHW and									
0600	1 bare neutral, 3 #8 & 1 #8	1 Elec	1.34	5.950	C.L.F.	69	365		434	620
0650	3 #6 & 1 #6	"	1.18	6.800		78	415		493	705
0700	3 #4 & 1 #6	2 Elec	2.02	7.940		86.50	485		571.50	820
0750	3 #2 & 1 #4		1.85	8.660		158	530		688	965
0800	3 #1/0 & 1 #2		1.68	9.520		184	585		769	1,075
0850	3 #2/0 & 1 #1		1.51	10.580		219	650		869	1,200
0900	3 #4/0 & 1 #2/0		1.34	11.900		320	730		1,050	1,425
1000	URD - triplex underground distribution cable, alum. 2 #4 + #4 neutral		2.35	6.800		92.50	415		507.50	720
1010	2 #2 + #4 neutral		2.23	7.190		117	440		557	785
1020	2 #2 + #2 neutral		2.14	7.470		116	460		576	810
1030	2 1/0 + #2 neutral		2.02	7.940		143	485		628	885
1040	2 1/0 + 1/0 neutral		1.93	8.280		162	510		672	935
1050	2 2/0 + #1 neutral		1.85	8.660		161	530		691	965
1060	2 2/0 + 2/0 neutral		1.76	9.070		169	555		724	1,025
1070	2 3/0 + 1/0 neutral		1.64	9.770		214	600		814	1,125
1080	2 3/0 + 3/0 neutral		1.64	9.770		218	600		818	1,125
1090	2 4/0 + 2/0 neutral		1.55	10.300		207	630		837	1,175
1100	2 4/0 + 4/0 neutral		1.55	10.300		248	630		878	1,225
1450	UF underground feeder cable, copper with ground, #14, 2 conductor	1 Elec	3.36	2.380		39.50	146		185.50	261
1500	#12, 2 conductor		2.94	2.720		37	167		204	289
1550	#10, 2 conductor		2.52	3.170		58.50	194		252.50	355
1600	#14, 3 conductor		2.94	2.720		40.50	167		207.50	293
1650	#12, 3 conductor		2.52	3.170		54.50	194		248.50	350
1700	#10, 3 conductor		2.10	3.810		88.50	234		322.50	445
1710	#8, 3 conductor		1.68	4.760		153	292		445	605
1720	#6, 3 conductor		1.51	5.290		235	325		560	745
2400	SEU service entrance cable, copper 2 conductors, #8 + #8 neutral		1.26	6.350		109	390		499	700
2600	#6 + #8 neutral		1.09	7.330		152	450		602	840
2800	#6 + #6 neutral		1.09	7.330		171	450		621	860
3000	#4 + #6 neutral	2 Elec	1.85	8.660		244	530		774	1,050
3200	#4 + #4 neutral		1.85	8.660		257	530		787	1,075
3400	#3 + #5 neutral		1.76	9.070		350	555		905	1,225
3600	#3 + #3 neutral		1.76	9.070		335	555		890	1,200
3800	#2 + #4 neutral		1.68	9.520		395	585		980	1,300
4000	#1 + #1 neutral		1.60	10.030		415	615		1,030	1,375
4200	1/0 + 1/0 neutral		1.51	10.580		935	650		1,585	2,000
4400	2/0 + 2/0 neutral		1.43	11.200		895	685		1,580	2,000
4600	3/0 + 3/0 neutral		1.34	11.900		995	730		1,725	2,175
4620	4/0 + 4/0 neutral		1.22	13.140		1,950	805		2,755	3,350
4800	Aluminum 2 conductors, #8 + #8 neutral	1 Elec	1.34	5.950		61.50	365		426.50	615
5000	#6 + #6 neutral	"	1.18	6.800		62.50	415		477.50	690
5100	#4 + #6 neutral	2 Elec	2.10	7.620		77	465		542	780

Post-Installation Change Order Costs

26 05 19 – Low-Voltage Electrical Power Conductors and Cables

26 05 19.55 Non-Metallic Sheathed Cable

		Crew	Daily Output	Labor-Hours	Unit	Material	2020 Bare Costs Labor	Equipment	Total	Total Incl O&P
5200	#4 + #4 neutral	2 Elec	2.02	7.940	C.L.F.	89	485		574	825
5300	#2 + #4 neutral		1.93	8.280		116	510		626	880
5400	#2 + #2 neutral		1.85	8.660		119	530		649	920
5450	1/0 + #2 neutral		1.76	9.070		177	555		732	1,025
5500	1/0 + 1/0 neutral		1.68	9.520		187	585		772	1,075
5550	2/0 + #1 neutral		1.60	10.030		197	615		812	1,125
5600	2/0 + 2/0 neutral		1.51	10.580		198	650		848	1,175
5800	3/0 + 1/0 neutral		1.43	11.200		267	685		952	1,325
6000	3/0 + 3/0 neutral		1.43	11.200		218	685		903	1,275
6200	4/0 + 2/0 neutral		1.34	11.900		218	730		948	1,325
6400	4/0 + 4/0 neutral	↓	1.34	11.900	↓	242	730		972	1,350
6500	Service entrance cap for copper SEU									
6600	100 amp	1 Elec	10.08	.794	Ea.	8.65	48.50		57.15	82
6700	150 amp		8.40	.952		12.90	58.50		71.40	101
6800	200 amp	↓	6.72	1.190	↓	19.60	73		92.60	131

26 05 19.70 Portable Cord

		Crew	Daily Output	Labor-Hours	Unit	Material	2020 Bare Costs Labor	Equipment	Total	Total Incl O&P
0010	**PORTABLE CORD** 600 volt									
0100	Type SO, #18, 2 conductor	1 Elec	823.20	.010	L.F.	.39	.60		.99	1.31
0110	3 conductor		823.20	.010		.37	.60		.97	1.30
0120	#16, 2 conductor		705.60	.011		.46	.70		1.16	1.55
0130	3 conductor		705.60	.011		4.75	.70		5.45	6.30
0140	4 conductor		705.60	.011		.58	.70		1.28	1.68
0240	#14, 2 conductor		705.60	.011		.58	.70		1.28	1.68
0250	3 conductor		705.60	.011		.62	.70		1.32	1.72
0260	4 conductor		705.60	.011		1.22	.70		1.92	2.38
0280	#12, 2 conductor		705.60	.011		.80	.70		1.50	1.92
0290	3 conductor		705.60	.011		.99	.70		1.69	2.13
0320	#10, 2 conductor		642.60	.012		1.09	.76		1.85	2.34
0330	3 conductor		642.60	.012		1.61	.76		2.37	2.91
0340	4 conductor		642.60	.012		1.75	.76		2.51	3.06
0360	#8, 2 conductor		466.20	.017		1.88	1.05		2.93	3.64
0370	3 conductor		453.60	.018		2.27	1.08		3.35	4.10
0380	4 conductor		441	.018		2.37	1.11		3.48	4.26
0400	#6, 2 conductor		441	.018		2.86	1.11		3.97	4.81
0410	3 conductor		411.60	.019		3.39	1.19		4.58	5.50
0420	4 conductor	↓	348.60	.023		3.58	1.41		4.99	6.05
0440	#4, 2 conductor	2 Elec	697.20	.023		4.20	1.41		5.61	6.70
0450	3 conductor		588	.027		4.40	1.67		6.07	7.35
0460	4 conductor		554.40	.029		5.85	1.77		7.62	9.10
0480	#2, 2 conductor		378	.042		5.50	2.60		8.10	9.90
0490	3 conductor		294	.054		7.90	3.34		11.24	13.65
0500	4 conductor	↓	235.20	.068	↓	10.10	4.17		14.27	17.30
2000	See 26 27 26.20 for Wiring Devices Elements									

26 05 19.75 Modular Flexible Wiring System

		Crew	Daily Output	Labor-Hours	Unit	Material	2020 Bare Costs Labor	Equipment	Total	Total Incl O&P
0010	**MODULAR FLEXIBLE WIRING SYSTEM**									
0020	Commercial system grid ceiling									
0100	Conversion Module	1 Elec	26.88	.298	Ea.	9	18.25		27.25	37
0120	Fixture cable, for fixture to fixture, 3 conductor, 15' long		33.60	.238		37	14.60		51.60	62.50
0150	Extender cable, 3 conductor, 15' long		40.32	.198		36	12.15		48.15	58
0200	Switch drop, 1 level, 9' long		26.88	.298		39	18.25		57.25	70
0220	2 level, 9' long		26.88	.298		38	18.25		56.25	69
0250	Power tee, 9' long	↓	26.88	.298	↓	31.50	18.25		49.75	61.50

26 05 19.75 Modular Flexible Wiring System	Crew	Daily Output	Labor-Hours	Unit	Material	2020 Bare Costs Labor	Equipment	Total	Total Incl O&P
1020 Industrial system open ceiling									
1100 Converter, interface between hardwiring and modular wiring	1 Elec	26.88	.298	Ea.	22	18.25		40.25	51
1120 Fixture cable, for fixture to fixture, 3 conductor, 21' long		26.88	.298		126	18.25		144.25	165
1125 3 conductor, 25' long		20.16	.397		162	24.50		186.50	215
1130 3 conductor, 31' long		15.96	.501		143	31		174	204
1150 Fixture cord drop, 10' long	▼	33.60	.238	▼	39	14.60		53.60	64.50

26 05 19.90 Wire	Crew	Daily Output	Labor-Hours	Unit	Material	2020 Bare Costs Labor	Equipment	Total	Total Incl O&P
0010 **WIRE**, normal installation conditions in wireway, conduit, cable tray									
0020 600 volt, copper type THW, solid, #14	1 Elec	10.92	.733	C.L.F.	7.50	45		52.50	75.50
0030 #12		9.24	.866		11.85	53		64.85	92
0040 #10		8.40	.952		19.15	58.50		77.65	108
0050 Stranded, #14		10.92	.733		9.10	45		54.10	77
0100 #12		9.24	.866		16.50	53		69.50	97
0120 #10		8.40	.952		26	58.50		84.50	116
0140 #8		6.72	1.190		41.50	73		114.50	155
0160 #6	▼	5.46	1.470		69	90		159	210
0180 #4	2 Elec	8.90	1.800		105	110		215	280
0200 #3		8.40	1.900		141	117		258	330
0220 #2		7.56	2.120		163	130		293	375
0240 #1		6.72	2.380		208	146		354	445
0260 1/0		5.54	2.890		271	177		448	560
0280 2/0		4.87	3.280		470	201		671	820
0300 3/0		4.20	3.810		335	234		569	715
0350 4/0	▼	3.70	4.330		540	266		806	990
0400 250 kcmil	3 Elec	5.04	4.760		585	292		877	1,075
0420 300 kcmil		4.79	5.010		825	305		1,130	1,375
0450 350 kcmil		4.54	5.290		800	325		1,125	1,375
0480 400 kcmil		4.28	5.600		1,450	345		1,795	2,075
0490 500 kcmil		4.03	5.950		1,150	365		1,515	1,825
0500 600 kcmil		3.28	7.330		1,525	450		1,975	2,350
0510 750 kcmil		2.77	8.660		1,375	530		1,905	2,325
0520 1,000 kcmil	▼	2.27	10.580		2,250	650		2,900	3,425
0540 600 volt, aluminum type THHN, stranded, #6	1 Elec	6.72	1.190		24.50	73		97.50	136
0560 #4	2 Elec	10.92	1.470		30	90		120	167
0580 #2		8.90	1.800		58	110		168	228
0600 #1		7.56	2.120		68.50	130		198.50	270
0620 1/0		6.72	2.380		96.50	146		242.50	325
0640 2/0		6.05	2.650		98.50	163		261.50	350
0680 3/0		5.54	2.890		123	177		300	400
0700 4/0	▼	5.21	3.070		157	188		345	455
0720 250 kcmil	3 Elec	7.31	3.280		165	201		366	480
0740 300 kcmil		6.80	3.530		228	217		445	570
0760 350 kcmil		6.30	3.810		231	234		465	605
0780 400 kcmil		5.80	4.140		262	254		516	670
0800 500 kcmil		5.04	4.760		296	292		588	760
0850 600 kcmil		4.79	5.010		375	305		680	875
0880 700 kcmil		4.28	5.600		460	345		805	1,025
0900 750 kcmil		4.03	5.950		450	365		815	1,050
0910 1,000 kcmil	▼	3.18	7.560		800	465		1,265	1,575
0920 600 volt, copper type THWN-THHN, solid, #14	1 Elec	10.92	.733		8	45		53	76
0940 #12		9.24	.866		12.05	53		65.05	92.50
0960 #10	▼	8.40	.952	▼	18.70	58.50		77.20	108

Post-Installation Change Order Costs

26 05 Common Work Results for Electrical

26 05 19 – Low-Voltage Electrical Power Conductors and Cables

	26 05 19.90 Wire	Crew	Daily Output	Labor-Hours	Unit	Material	2020 Bare Costs Labor	Equipment	Total	Total Incl O&P
1000	Stranded, #14	1 Elec	10.92	.733	C.L.F.	9.25	45		54.25	77
1200	#12		9.24	.866		12.75	53		65.75	93
1250	#10		8.40	.952		24.50	58.50		83	114
1300	#8		6.72	1.190		32.50	73		105.50	145
1350	#6		5.46	1.470		53	90		143	193
1400	#4	2 Elec	8.90	1.800		84	110		194	256
1450	#3		8.40	1.900		105	117		222	290
1500	#2		7.56	2.120		130	130		260	335
1550	#1		6.72	2.380		166	146		312	400
1600	1/0		5.54	2.890		240	177		417	530
1650	2/0		4.87	3.280		249	201		450	575
1700	3/0		4.20	3.810		310	234		544	690
2000	4/0		3.70	4.330		395	266		661	825
2200	250 kcmil	3 Elec	5.04	4.760		460	292		752	940
2400	300 kcmil		4.79	5.010		620	305		925	1,150
2600	350 kcmil		4.54	5.290		645	325		970	1,200
2700	400 kcmil		4.28	5.600		725	345		1,070	1,300
2800	500 kcmil		4.03	5.950		930	365		1,295	1,575
2802	600 kcmil		3.28	7.330		935	450		1,385	1,700
2804	750 kcmil		2.77	8.660		1,825	530		2,355	2,825
2805	1,000 kcmil		1.63	14.730		2,950	905		3,855	4,600
2900	600 volt, copper type XHHW, solid, #14	1 Elec	10.92	.733		13.25	45		58.25	81.50
2920	#12		9.24	.866		21	53		74	102
2940	#10		8.40	.952		32.50	58.50		91	123
3000	Stranded, #14		10.92	.733		11.65	45		56.65	80
3020	#12		9.24	.866		16.45	53		69.45	97
3040	#10		8.40	.952		24.50	58.50		83	114
3060	#8		6.72	1.190		40.50	73		113.50	154
3080	#6		5.46	1.470		65.50	90		155.50	206
3100	#4	2 Elec	8.90	1.800		92.50	110		202.50	266
3120	#2		7.56	2.120		144	130		274	355
3140	#1		6.72	2.380		190	146		336	425
3160	1/0		5.54	2.890		236	177		413	525
3180	2/0		4.87	3.280		295	201		496	625
3200	3/0		4.20	3.810		370	234		604	755
3220	4/0		3.70	4.330		460	266		726	900
3240	250 kcmil	3 Elec	5.04	4.760		540	292		832	1,025
3260	300 kcmil		4.79	5.010		640	305		945	1,175
3280	350 kcmil		4.54	5.290		700	325		1,025	1,250
3300	400 kcmil		4.28	5.600		860	345		1,205	1,450
3320	500 kcmil		4.03	5.950		1,000	365		1,365	1,650
3340	600 kcmil		3.28	7.330		1,250	450		1,700	2,050
3360	750 kcmil		2.77	8.660		2,200	530		2,730	3,225
3380	1,000 kcmil		2.02	11.900		2,925	730		3,655	4,300
5020	600 volt, aluminum type XHHW, stranded, #6	1 Elec	6.72	1.190		31	73		104	144
5040	#4	2 Elec	10.92	1.470		38.50	90		128.50	177
5060	#2		8.90	1.800		48	110		158	217
5080	#1		7.56	2.120		76.50	130		206.50	279
5100	1/0		6.72	2.380		91.50	146		237.50	315
5120	2/0		6.05	2.650		112	163		275	365
5140	3/0		5.54	2.890		134	177		311	410
5160	4/0		5.21	3.070		121	188		309	415
5180	250 kcmil	3 Elec	7.31	3.280		188	201		389	505

26 05 19 – Low-Voltage Electrical Power Conductors and Cables

26 05 19.90 Wire	Crew	Daily Output	Labor-Hours	Unit	Material	2020 Bare Costs Labor	Equipment	Total	Total Incl O&P
5200	300 kcmil	3 Elec	6.80	3.530	C.L.F.	250	217		467	595
5220	350 kcmil		6.30	3.810		265	234		499	640
5240	400 kcmil		5.80	4.140		297	254		551	705
5260	500 kcmil		5.04	4.760		325	292		617	790
5280	600 kcmil		4.79	5.010		420	305		725	920
5300	700 kcmil		4.54	5.290		475	325		800	1,000
5320	750 kcmil		4.28	5.600		450	345		795	1,000
5340	1,000 kcmil		3.02	7.940		710	485		1,195	1,500
5390	600 volt, copper type XLPE-USE (RHW), solid, #14	1 Elec	10.08	.794		14.25	48.50		62.75	88
5400	#12		9.24	.866		21.50	53		74.50	103
5420	#10		8.40	.952		28	58.50		86.50	118
5440	Stranded, #14		10.92	.733		18.05	45		63.05	87
5460	#12		9.24	.866		21.50	53		74.50	103
5480	#10		8.40	.952		33.50	58.50		92	124
5500	#8		6.72	1.190		40.50	73		113.50	154
5520	#6		5.46	1.470		68	90		158	209
5540	#4	2 Elec	8.90	1.800		104	110		214	278
5560	#2		7.56	2.120		168	130		298	380
5580	#1		6.72	2.380		226	146		372	465
5600	1/0		5.54	2.890		283	177		460	575
5620	2/0		4.87	3.280		370	201		571	710
5640	3/0		4.20	3.810		465	234		699	860
5660	4/0		3.70	4.330		505	266		771	950
5680	250 kcmil	3 Elec	5.04	4.760		610	292		902	1,100
5700	300 kcmil		4.79	5.010		700	305		1,005	1,225
5720	350 kcmil		4.54	5.290		785	325		1,110	1,350
5740	400 kcmil		4.28	5.600		920	345		1,265	1,500
5760	500 kcmil		4.03	5.950		1,000	365		1,365	1,650
5780	600 kcmil		3.28	7.330		1,500	450		1,950	2,325
5800	750 kcmil		2.77	8.660		2,325	530		2,855	3,350
5820	1,000 kcmil		2.27	10.580		3,125	650		3,775	4,425
5840	600 volt, aluminum type XLPE-USE (RHW), stranded, #6	1 Elec	6.72	1.190		38.50	73		111.50	151
5860	#4	2 Elec	10.92	1.470		45	90		135	184
5880	#2		8.90	1.800		63.50	110		173.50	234
5900	#1		7.56	2.120		90	130		220	293
5920	1/0		6.72	2.380		106	146		252	335
5940	2/0		6.05	2.650		124	163		287	380
5960	3/0		5.54	2.890		146	177		323	425
5980	4/0		5.21	3.070		162	188		350	460
6000	250 kcmil	3 Elec	7.31	3.280		220	201		421	540
6020	300 kcmil		6.80	3.530		285	217		502	635
6040	350 kcmil		6.30	3.810		289	234		523	670
6060	400 kcmil		5.80	4.140		350	254		604	765
6080	500 kcmil		5.04	4.760		390	292		682	865
6100	600 kcmil		4.79	5.010		500	305		805	1,000
6110	700 kcmil		4.54	5.290		545	325		870	1,075
6120	750 kcmil		4.28	5.600		585	345		930	1,150

26 05 23 – Control-Voltage Electrical Power Cables

26 05 23.10 Control Cable

		Crew	Daily Output	Labor-Hours	Unit	Material	Labor	Equipment	Total	Total Incl O&P
0010	**CONTROL CABLE**									
0020	600 volt, copper, #14 THWN wire with PVC jacket, 2 wires	1 Elec	7.56	1.060	C.L.F.	53.50	65		118.50	156
0030	3 wires		6.72	1.190		63	73		136	178

For customer support on your Electrical Change Order Costs with RSMeans data, call 800.448.8182. **Post-Installation Change Order Costs**

26 05 23 – Control-Voltage Electrical Power Cables

26 05 23.10 Control Cable	Crew	Daily Output	Labor-Hours	Unit	Material	2020 Bare Costs Labor	Equipment	Total	Total Incl O&P
0100	4 wires	1 Elec	5.88	1.360	C.L.F.	65	83.50		148.50	196
0150	5 wires		5.46	1.470		89	90		179	232
0200	6 wires		5.04	1.590		120	97.50		217.50	277
0300	8 wires		4.45	1.800		167	110		277	350
0400	10 wires		4.03	1.980		199	121		320	400
0500	12 wires		3.61	2.210		256	136		392	485
0600	14 wires		3.19	2.510		271	154		425	525
0700	16 wires		2.94	2.720		310	167		477	590
0800	18 wires		2.77	2.890		360	177		537	660
0810	19 wires		2.60	3.070		335	188		523	650
0900	20 wires		2.52	3.170		355	194		549	680
1000	22 wires	▼	2.35	3.400	▼	430	209		639	780

26 05 23.20 Special Wires and Fittings

		Crew	Daily Output	Labor-Hours	Unit	Material	2020 Bare Costs Labor	Equipment	Total	Total Incl O&P
0010	**SPECIAL WIRES & FITTINGS**									
0100	Fixture TFFN 600 volt 90°C stranded, #18	1 Elec	10.92	.733	C.L.F.	11.25	45		56.25	79.50
0150	#16		10.92	.733		14.80	45		59.80	83.50
0500	Thermostat, jacket non-plenum, twisted, #18-2 conductor		6.72	1.190		12.20	73		85.20	122
0550	#18-3 conductor		5.88	1.360		15.20	83.50		98.70	141
0600	#18-4 conductor		5.46	1.470		24	90		114	161
0650	#18-5 conductor		5.04	1.590		29.50	97.50		127	178
0700	#18-6 conductor		4.62	1.730		33.50	106		139.50	195
0750	#18-7 conductor		4.20	1.900		43.50	117		160.50	222
0800	#18-8 conductor		4.03	1.980		42.50	121		163.50	228
2460	Tray cable, type TC, copper, #16-2 conductor		7.90	1.010		37	62		99	134
2464	#16-3 conductor		7.06	1.130		61.50	69.50		131	171
2468	#16-4 conductor		6.13	1.300		65.50	80		145.50	191
2472	#16-5 conductor		5.63	1.420		51	87		138	186
2476	#16-7 conductor		4.62	1.730		70	106		176	235
2480	#16-9 conductor	▼	4.20	1.900		87.50	117		204.50	271
2484	#16-12 conductor	2 Elec	7.39	2.160		152	133		285	365
2488	#16-15 conductor		6.05	2.650		152	163		315	410
2492	#16-19 conductor		5.54	2.890		188	177		365	470
2496	#16-25 conductor	▼	5.38	2.980		265	183		448	565
2500	#14-2 conductor	1 Elec	7.56	1.060		42.50	65		107.50	144
2520	#14-3 conductor		6.72	1.190		41	73		114	154
2540	#14-4 conductor		5.88	1.360		58.50	83.50		142	189
2560	#14-5 conductor		5.46	1.470		89	90		179	232
2564	#14-7 conductor	▼	4.45	1.800		110	110		220	285
2568	#14-9 conductor	2 Elec	8.06	1.980		127	121		248	320
2572	#14-12 conductor		7.22	2.210		177	136		313	395
2576	#14-15 conductor		6.05	2.650		209	163		372	470
2578	#14-19 conductor		5.21	3.070		263	188		451	570
2582	#14-25 conductor	▼	4.45	3.590		350	220		570	715
2590	#12-2 conductor	1 Elec	7.06	1.130		62	69.50		131.50	172
2592	#12-3 conductor		6.38	1.250		67.50	76.50		144	188
2594	#12-4 conductor		5.54	1.440		126	88.50		214.50	271
2596	#12-5 conductor	▼	5.21	1.540		137	94.50		231.50	292
2598	#12-7 conductor	2 Elec	8.74	1.830		144	112		256	325
2602	#12-9 conductor		7.39	2.160		260	133		393	485
2604	#12-12 conductor		6.55	2.440		264	150		414	515
2606	#12-15 conductor		5.71	2.800		315	172		487	600
2608	#12-19 conductor	▼	5.04	3.170	▼	385	194		579	710

26 05 23.20 Special Wires and Fittings	Crew	Daily Output	Labor-Hours	Unit	Material	2020 Bare Costs Labor	Equipment	Total	Total Incl O&P
2610	#12-25 conductor	3 Elec	6.47	3.710	C.L.F.	520	228		748	915
2618	#10-2 conductor	1 Elec	6.72	1.190		101	73		174	220
2622	#10-3 conductor	"	6.13	1.300		101	80		181	231
2624	#10-4 conductor	2 Elec	10.75	1.490		192	91.50		283.50	345
2626	#10-5 conductor		9.91	1.610		163	99		262	325
2628	#10-7 conductor		7.90	2.030		235	125		360	445
2630	#10-9 conductor		7.06	2.270		298	139		437	535
2632	#10-12 conductor		6.05	2.650		330	163		493	605
2640	300 V, copper braided shield, PVC jacket									
2650	2 conductor #18 stranded	1 Elec	5.88	1.360	C.L.F.	36	83.50		119.50	164
2660	3 conductor #18 stranded		5.04	1.590		53.50	97.50		151	204
2670	4 conductor #18 stranded		5.04	1.590		194	97.50		291.50	360
3000	Strain relief grip for cable									
3050	Cord, top, #12-3	1 Elec	33.60	.238	Ea.	12.85	14.60		27.45	36
3060	#12-4		33.60	.238		12.85	14.60		27.45	36
3070	#12-5		32.76	.244		15.20	15		30.20	39
3100	#10-3		32.76	.244		15.20	15		30.20	39
3110	#10-4		31.92	.251		15.20	15.40		30.60	39.50
3120	#10-5		31.92	.251		18.35	15.40		33.75	43
3200	Bottom, #12-3		33.60	.238		30.50	14.60		45.10	55.50
3210	#12-4		33.60	.238		30.50	14.60		45.10	55.50
3220	#12-5		32.76	.244		32	15		47	57.50
3230	#10-3		32.76	.244		32	15		47	57.50
3300	#10-4		31.92	.251		32	15.40		47.40	58
3310	#10-5		31.92	.251		40	15.40		55.40	67
3400	Cable ties, standard, 4" length		159.60	.050		.18	3.08		3.26	4.78
3410	7" length		134.40	.060		.22	3.65		3.87	5.70
3420	14.5" length		75.60	.106		.49	6.50		6.99	10.20
3430	Heavy, 14.5" length		67.20	.119		.71	7.30		8.01	11.70
3500	Cable gland, nylon, 1/4" NPT thread		50.40	.159		2.11	9.75		11.86	16.80
3510	Cable gland, nylon, 3/8" NPT thread		50.40	.159		2.43	9.75		12.18	17.15
3520	Cable gland, nylon, 1/2" NPT thread		50.40	.159		3.04	9.75		12.79	17.85
3530	Cable gland, nylon, 3/4" NPT thread		50.40	.159		4.59	9.75		14.34	19.55
3540	Cable gland, nylon, 1" NPT thread		50.40	.159		7.25	9.75		17	22.50
3550	Cable gland, nylon, 1-1/4" NPT thread		50.40	.159		8.55	9.75		18.30	24
3970	U-cable guards									
3980	.75" x 5' U-cable guard, 14 gauge	1 Elec	20.16	.397	Ea.	11.30	24.50		35.80	49
3990	.75" x 8' U-cable guard, 14 gauge		20.16	.397		21	24.50		45.50	59.50
4000	1-1/8" x 5' U-cable guard		20.16	.397		16.40	24.50		40.90	54.50
4010	1-1/8" x 8' U-cable guard		20.16	.397		19.30	24.50		43.80	57.50
4020	2-3/16" x 5' U-cable guard		20.16	.397		33.50	24.50		58	73
4030	2-3/16" x 8' U-cable guard		20.16	.397		32.50	24.50		57	72.50
4040	3-3/16" x 5' U-cable guard		20.16	.397		37	24.50		61.50	77.50
4050	3-3/16" x 8' U-cable guard		20.16	.397		48	24.50		72.50	89.50
4060	4" x 8' U-cable guard		20.16	.397		42	24.50		66.50	82.50
4070	2" x 9' flanged-cable guard		16.80	.476		98.50	29		127.50	152
4080	3" x 9' flanged-cable guard		16.80	.476		125	29		154	182
4090	3.5" x 9' flanged-cable guard		16.80	.476		151	29		180	210
5000	2" x 5' flanged-cable guard, extension		16.80	.476		66.50	29		95.50	117
5010	3" x 5' flanged-cable guard, extension		16.80	.476		91	29		120	144
5020	.05" x 8' U-cable guard, black plastic		21.84	.366		5.55	22.50		28.05	39.50
5030	.75" x 8' U-cable guard, black plastic		21.84	.366		7.50	22.50		30	42
5040	1" x 8' U-cable guard, black plastic		21.84	.366		11.95	22.50		34.45	46.50

For customer support on your Electrical Change Order Costs with RSMeans data, call 800.448.8182.

Post-Installation Change Order Costs

26 05 26.80 Grounding	Crew	Daily Output	Labor-Hours	Unit	Material	2020 Bare Costs Labor	Equipment	Total	Total Incl O&P
0010 **GROUNDING**									
0030 Rod, copper clad, 8' long, 1/2" diameter	1 Elec	4.62	1.730	Ea.	26	106		132	187
0040 5/8" diameter		4.62	1.730		26	106		132	187
0050 3/4" diameter		4.45	1.800		41	110		151	209
0080 10' long, 1/2" diameter		4.03	1.980		28	121		149	212
0090 5/8" diameter		3.86	2.070		25.50	127		152.50	218
0100 3/4" diameter		3.70	2.160		61	133		194	265
0130 15' long, 3/4" diameter		3.36	2.380		63.50	146		209.50	287
0150 Coupling, bronze, 1/2" diameter					5.80			5.80	6.40
0160 5/8" diameter					6.25			6.25	6.90
0170 3/4" diameter					16.95			16.95	18.65
0190 Drive studs, 1/2" diameter					14.65			14.65	16.15
0210 5/8" diameter					23.50			23.50	25.50
0220 3/4" diameter					26.50			26.50	29
0230 Clamp, bronze, 1/2" diameter	1 Elec	26.88	.298		6.05	18.25		24.30	33.50
0240 5/8" diameter		26.88	.298		5.35	18.25		23.60	33
0250 3/4" diameter		26.88	.298		6.50	18.25		24.75	34
0260 Wire ground bare armored, #8-1 conductor		1.68	4.760	C.L.F.	85.50	292		377.50	530
0270 #6-1 conductor		1.51	5.290		96.50	325		421.50	590
0280 #4-1 conductor		1.34	5.950		168	365		533	730
0320 Bare copper wire, #14 solid		11.76	.680		8.15	41.50		49.65	71
0330 #12		10.92	.733		14.80	45		59.80	83.50
0340 #10		10.08	.794		20.50	48.50		69	95
0350 #8		9.24	.866		29	53		82	111
0360 #6		8.40	.952		45	58.50		103.50	137
0370 #4		6.72	1.190		113	73		186	233
0380 #2		4.20	1.900		146	117		263	335
0390 Bare copper wire, stranded, #8		9.24	.866		46	53		99	130
0400 #6		8.40	.952		48.50	58.50		107	141
0450 #4	2 Elec	13.44	1.190		98.50	73		171.50	217
0600 #2		8.40	1.900		102	117		219	286
0650 #1		7.56	2.120		149	130		279	360
0700 1/0		6.72	2.380		180	146		326	415
0750 2/0		6.05	2.650		291	163		454	560
0800 3/0		5.54	2.890		355	177		532	655
1000 4/0		4.79	3.340		485	205		690	840
1200 250 kcmil	3 Elec	6.05	3.970		505	244		749	920
1210 300 kcmil		5.54	4.330		645	266		911	1,100
1220 350 kcmil		5.04	4.760		630	292		922	1,125
1230 400 kcmil		4.79	5.010		695	305		1,000	1,225
1240 500 kcmil		4.28	5.600		905	345		1,250	1,500
1260 750 kcmil		3.02	7.940		1,500	485		1,985	2,375
1270 1,000 kcmil		2.52	9.520		1,800	585		2,385	2,850
1360 Bare aluminum, stranded, #6	1 Elec	7.56	1.060		16.75	65		81.75	115
1370 #4	2 Elec	13.44	1.190		36.50	73		109.50	149
1380 #2		10.92	1.470		52.50	90		142.50	192
1390 #1		8.90	1.800		45.50	110		155.50	215
1400 1/0		7.56	2.120		58	130		188	258
1410 2/0		6.72	2.380		89.50	146		235.50	315
1420 3/0		6.05	2.650		99	163		262	350
1430 4/0		5.54	2.890		126	177		303	400
1440 250 kcmil	3 Elec	7.81	3.070		132	188		320	425

26 05 26.80 Grounding		Crew	Daily Output	Labor-Hours	Unit	Material	2020 Bare Costs Labor	Equipment	Total	Total Incl O&P
1450	300 kcmil	3 Elec	7.31	3.280	C.L.F.	139	201		340	455
1460	400 kcmil		6.30	3.810		183	234		417	550
1470	500 kcmil		5.80	4.140		222	254		476	625
1480	600 kcmil		5.04	4.760		264	292		556	725
1490	700 kcmil		4.79	5.010		284	305		589	775
1500	750 kcmil		4.28	5.600		315	345		660	855
1510	1,000 kcmil		4.03	5.950		400	365		765	985
1800	Water pipe ground clamps, heavy duty									
2000	Bronze, 1/2" to 1" diameter	1 Elec	6.72	1.190	Ea.	33	73		106	146
2100	1-1/4" to 2" diameter		6.72	1.190		35.50	73		108.50	148
2200	2-1/2" to 3" diameter		5.04	1.590		69.50	97.50		167	222
2400	Grounding, exothermic welding									
2500	Exothermic welding reusable mold, cable to cable, parallel, vertical	1 Elec	6.72	1.190	Ea.	390	73		463	535
2505	Splice single		6.72	1.190		335	73		408	480
2510	Splice single to double		6.72	1.190		390	73		463	540
2520	Cable to cable, termination, Tee		6.72	1.190		380	73		453	530
2530	Cable to rod, termination, Tee		6.72	1.190		330	73		403	470
2535	Cable to rebar over, termination, Tee		6.72	1.190		111	73		184	231
2540	Cable to rod, termination, 90 Deg		6.72	1.190		380	73		453	530
2550	Cable to verticle flat steel		6.72	1.190		385	73		458	535
2555	Cable to 45 Deg		6.72	1.190		390	73		463	540
2730	Exothermic weld, 4/0 wire to 1" ground rod		5.88	1.360		10.90	83.50		94.40	136
2740	4/0 wire to building steel		5.88	1.360		10.90	83.50		94.40	136
2750	4/0 wire to motor frame		5.88	1.360		10.90	83.50		94.40	136
2760	4/0 wire to 4/0 wire		5.88	1.360		10.90	83.50		94.40	136
2770	4/0 wire to #4 wire		5.88	1.360		10.90	83.50		94.40	136
2780	4/0 wire to #8 wire		5.88	1.360		10.90	83.50		94.40	136
2790	Mold, reusable, for above					162			162	179
2800	Brazed connections, #6 wire	1 Elec	10.08	.794		18.55	48.50		67.05	93
3000	#2 wire		8.40	.952		25	58.50		83.50	115
3100	3/0 wire		6.72	1.190		37	73		110	150
3200	4/0 wire		5.88	1.360		42	83.50		125.50	171
3400	250 kcmil wire		4.20	1.900		49	117		166	228
3600	500 kcmil wire		3.36	2.380		60.50	146		206.50	284
3700	Insulated ground wire, copper #14		10.92	.733	C.L.F.	8.90	45		53.90	77
3710	#12		9.24	.866		12.20	53		65.20	92.50
3720	#10		8.40	.952		23.50	58.50		82	113
3730	#8		6.72	1.190		31.50	73		104.50	144
3740	#6		5.46	1.470		51	90		141	190
3750	#4	2 Elec	8.90	1.800		80.50	110		190.50	253
3770	#2		7.56	2.120		125	130		255	330
3780	#1		6.72	2.380		159	146		305	390
3790	1/0		5.54	2.890		230	177		407	515
3800	2/0		4.87	3.280		239	201		440	565
3810	3/0		4.20	3.810		297	234		531	675
3820	4/0		3.70	4.330		375	266		641	810
3830	250 kcmil	3 Elec	5.04	4.760		440	292		732	915
3840	300 kcmil		4.79	5.010		590	305		895	1,100
3850	350 kcmil		4.54	5.290		620	325		945	1,175
3860	400 kcmil		4.28	5.600		695	345		1,040	1,275
3870	500 kcmil		4.03	5.950		890	365		1,255	1,525
3880	600 kcmil		3.28	7.330		1,475	450		1,925	2,300
3890	750 kcmil		2.77	8.660		1,325	530		1,855	2,250

Post-Installation Change Order Costs

26 05 26 – Grounding and Bonding for Electrical Systems

26 05 26.80 Grounding

		Crew	Daily Output	Labor-Hours	Unit	Material	2020 Bare Costs Labor	2020 Bare Costs Equipment	Total	Total Incl O&P
3900	1,000 kcmil	3 Elec	2.27	10.580	C.L.F.	2,150	650		2,800	3,325
3960	Insulated ground wire, aluminum, #6	1 Elec	6.72	1.190		23.50	73		96.50	135
3970	#4	2 Elec	10.92	1.470		28.50	90		118.50	166
3980	#2		8.90	1.800		55.50	110		165.50	225
3990	#1		7.56	2.120		65.50	130		195.50	267
4000	1/0		6.72	2.380		92.50	146		238.50	320
4010	2/0		6.05	2.650		94.50	163		257.50	345
4020	3/0		5.54	2.890		118	177		295	395
4030	4/0		5.21	3.070		151	188		339	445
4040	250 kcmil	3 Elec	7.31	3.280		158	201		359	475
4050	300 kcmil		6.80	3.530		219	217		436	560
4060	350 kcmil		6.30	3.810		221	234		455	595
4070	400 kcmil		5.80	4.140		251	254		505	655
4080	500 kcmil		5.04	4.760		283	292		575	745
4090	600 kcmil		4.79	5.010		360	305		665	855
4100	700 kcmil		4.28	5.600		440	345		785	995
4110	750 kcmil		4.03	5.950		435	365		800	1,025
5000	Copper electrolytic ground rod system									
5010	Includes augering hole, mixing bentonite clay,									
5020	Installing rod, and terminating ground wire									
5100	Straight vertical type, 2" diam.									
5120	8.5' long, clamp connection	1 Elec	2.24	3.570	Ea.	720	219		939	1,125
5130	With exothermic weld connection		1.64	4.880		730	299		1,029	1,250
5140	10' long		1.97	4.050		710	248		958	1,150
5150	With exothermic weld connection		1.50	5.350		890	330		1,220	1,475
5160	12' long		1.81	4.410		1,025	271		1,296	1,525
5170	With exothermic weld connection		1.40	5.700		1,125	350		1,475	1,750
5180	20' long		1.46	5.470		1,400	335		1,735	2,025
5190	With exothermic weld connection		1.18	6.800		1,375	415		1,790	2,150
5195	40' long with exothermic weld connection	2 Elec	1.68	9.520		3,050	585		3,635	4,250
5200	L-shaped, 2" diam.									
5220	4' vert. x 10' horiz., clamp connection	1 Elec	4.48	1.790	Ea.	1,325	110		1,435	1,650
5230	With exothermic weld connection	"	2.59	3.090	"	1,325	190		1,515	1,725
5300	Protective box at grade level, with breather slots									
5320	Round 12" long, fiberlyte	1 Elec	26.88	.298	Ea.	70	18.25		88.25	104
5330	Concrete	"	13.44	.595		112	36.50		148.50	178
5400	Bentonite clay, 50# bag, 1 per 10' of rod					56			56	61.50
5500	Equipotential earthing bar	1 Elec	1.68	4.760		138	292		430	585
7000	Exothermic welding kit, multi vertical	"	6.72	1.190		720	73		793	900

26 05 29 – Hangers and Supports for Electrical Systems

26 05 29.20 Hangers

		Crew	Daily Output	Labor-Hours	Unit	Material	2020 Bare Costs Labor	2020 Bare Costs Equipment	Total	Total Incl O&P
0010	**HANGERS**									
0015	See section 22 05 29.10 for additional items									
0030	Conduit supports									
0050	Strap w/2 holes, rigid steel conduit									
0100	1/2" diameter	1 Elec	394.80	.020	Ea.	.11	1.24		1.35	1.97
0150	3/4" diameter		369.60	.022		.13	1.33		1.46	2.12
0200	1" diameter		336	.024		.18	1.46		1.64	2.37
0300	1-1/4" diameter		298.20	.027		.29	1.65		1.94	2.77
0350	1-1/2" diameter		268.80	.030		.38	1.83		2.21	3.13
0400	2" diameter		223.44	.036		.50	2.20		2.70	3.82
0500	2-1/2" diameter		134.40	.060		.78	3.65		4.43	6.30

26 05 29.20 Hangers		Crew	Daily Output	Labor-Hours	Unit	Material	2020 Bare Costs Labor	Equipment	Total	Total Incl O&P
0550	3" diameter	1 Elec	111.72	.072	Ea.	1.04	4.39		5.43	7.70
0600	3-1/2" diameter		84	.095		2.73	5.85		8.58	11.70
0650	4" diameter		67.20	.119		1.05	7.30		8.35	12.05
0700	EMT, 1/2" diameter		394.80	.020		.20	1.24		1.44	2.07
0800	3/4" diameter		369.60	.022		.19	1.33		1.52	2.19
0850	1" diameter		336	.024		.36	1.46		1.82	2.58
0900	1-1/4" diameter		298.20	.027		.74	1.65		2.39	3.26
0950	1-1/2" diameter		268.80	.030		.99	1.83		2.82	3.81
1000	2" diameter		223.44	.036		1	2.20		3.20	4.37
1100	2-1/2" diameter		134.40	.060		1.49	3.65		5.14	7.10
1150	3" diameter		111.72	.072		1.85	4.39		6.24	8.60
1200	3-1/2" diameter		84	.095		4.11	5.85		9.96	13.20
1250	4" diameter		67.20	.119		2.96	7.30		10.26	14.15
1400	Hanger, with bolt, 1/2" diameter		168	.048		.44	2.92		3.36	4.83
1450	3/4" diameter		159.60	.050		.46	3.08		3.54	5.10
1500	1" diameter		147.84	.054		.49	3.32		3.81	5.50
1550	1-1/4" diameter		134.40	.060		.61	3.65		4.26	6.10
1600	1-1/2" diameter		117.60	.068		.81	4.17		4.98	7.10
1650	2" diameter		109.20	.073		1.41	4.49		5.90	8.25
1700	2-1/2" diameter		84	.095		1.81	5.85		7.66	10.70
1750	3" diameter		53.76	.149		2.38	9.15		11.53	16.20
1800	3-1/2" diameter		42	.190		2.34	11.70		14.04	19.95
1850	4" diameter		33.60	.238		3.60	14.60		18.20	26
1900	Riser clamps, conduit, 1/2" diameter		33.60	.238		8.95	14.60		23.55	32
1950	3/4" diameter		30.24	.265		8.55	16.25		24.80	33.50
2000	1" diameter		25.20	.317		7.75	19.50		27.25	37.50
2100	1-1/4" diameter		22.68	.353		11.75	21.50		33.25	45
2150	1-1/2" diameter		22.68	.353		10.95	21.50		32.45	44
2200	2" diameter		16.80	.476		11.30	29		40.30	56
2250	2-1/2" diameter		16.80	.476		12.30	29		41.30	57
2300	3" diameter		15.12	.529		13.75	32.50		46.25	63.50
2350	3-1/2" diameter		15.12	.529		12.75	32.50		45.25	62.50
2400	4" diameter		11.76	.680		16.55	41.50		58.05	80
2500	Threaded rod, painted, 1/4" diameter		218.40	.037	L.F.	1.96	2.25		4.21	5.50
2600	3/8" diameter		168	.048		7.40	2.92		10.32	12.45
2700	1/2" diameter		117.60	.068		8.45	4.17		12.62	15.45
2800	5/8" diameter		84	.095		10.65	5.85		16.50	20.50
2900	3/4" diameter		50.40	.159		7.80	9.75		17.55	23
2940	Couplings painted, 1/4" diameter				C	390			390	425
2960	3/8" diameter					570			570	625
2970	1/2" diameter					485			485	530
2980	5/8" diameter					955			955	1,050
2990	3/4" diameter					1,200			1,200	1,325
3000	Nuts, galvanized, 1/4" diameter					13.05			13.05	14.35
3050	3/8" diameter					21			21	23
3100	1/2" diameter					58			58	64
3150	5/8" diameter					222			222	244
3200	3/4" diameter					157			157	172
3250	Washers, galvanized, 1/4" diameter					17.60			17.60	19.35
3300	3/8" diameter					20			20	22
3350	1/2" diameter					51.50			51.50	56.50
3400	5/8" diameter					102			102	112
3450	3/4" diameter					178			178	196

Post-Installation Change Order Costs

26 05 29.20 Hangers	Crew	Daily Output	Labor-Hours	Unit	Material	2020 Bare Costs Labor	Equipment	Total	Total Incl O&P
3500	Lock washers, galvanized, 1/4" diameter				C	11.40			11.40	12.55
3550	3/8" diameter					19.50			19.50	21.50
3600	1/2" diameter					29			29	32
3650	5/8" diameter					51.50			51.50	56.50
3700	3/4" diameter					101			101	111
3710	304 stainless steel, 1' long, 12GA, 1-5/8" x 1-5/8"	1 Elec	42	.190	Ea.	22.50	11.70		34.20	42.50
3715	304 stainless steel, 1'6" long, 12GA, 1-5/8" x 1-5/8"		28	.286		30.50	17.55		48.05	59.50
3720	304 stainless steel, 2' long, 12GA, 1-5/8" x 1-5/8"		21	.381		47	23.50		70.50	87
3725	304 stainless steel, 3' long, 12GA, 1-5/8" x 1-5/8"		14	.571		60.50	35		95.50	119
3730	304 stainless steel, 4' long, 12GA, 1-5/8" x 1-5/8"		10.50	.762		75	46.50		121.50	152
3735	304 stainless steel, 5' long, 12GA, 1-5/8" x 1-5/8"		8.40	.952		113	58.50		171.50	212
3740	304 stainless steel, 10' long, 12GA, 1-5/8" x 1-5/8"		4.20	1.900		171	117		288	365
3745	Aluminum, 1' length, 12GA, 1-5/8" x 1-5/8"		42	.190		6.85	11.70		18.55	25
3750	Aluminum, 1'6" length, 12GA, 1-5/8" x 1-5/8"		28	.286		10.15	17.55		27.70	37
3755	Aluminum, 2' length, 12GA, 1-5/8" x 1-5/8"		21	.381		12.80	23.50		36.30	49
3760	Aluminum, 3' length, 12GA, 1-5/8" x 1-5/8"		14	.571		19.95	35		54.95	74
3765	Aluminum, 4' length, 12GA, 1-5/8" x 1-5/8"		10.50	.762		27	46.50		73.50	99
3770	Aluminum, 5' length, 12GA, 1-5/8" x 1-5/8"		8.40	.952		31.50	58.50		90	122
3775	Aluminum, 10' length, 12GA, 1-5/8" x 1-5/8"		4.20	1.900		57.50	117		174.50	238
3800	Channels, steel, 3/4" x 1-1/2", 14 ga.		67.20	.119	L.F.	3.73	7.30		11.03	15
3900	1-1/2" x 1-1/2", 12 ga.		58.80	.136		4.60	8.35		12.95	17.50
4000	1-7/8" x 1-1/2"		50.40	.159		26.50	9.75		36.25	43.50
4100	3" x 1-1/2"		42	.190		25	11.70		36.70	45
4200	Spring nuts, long, 1/4"		100.80	.079	Ea.	.81	4.87		5.68	8.15
4250	3/8"		84	.095		.94	5.85		6.79	9.75
4300	1/2"		67.20	.119		1.04	7.30		8.34	12.05
4350	Spring nuts, short, 1/4"		100.80	.079		1.70	4.87		6.57	9.10
4400	3/8"		84	.095		2.13	5.85		7.98	11.05
4450	1/2"		67.20	.119		2.26	7.30		9.56	13.40
4500	Closure strip		168	.048	L.F.	4.58	2.92		7.50	9.40
4550	End cap		50.40	.159	Ea.	1.51	9.75		11.26	16.15
4600	End connector 3/4" conduit		33.60	.238		8.95	14.60		23.55	32
4650	Junction box, 1 channel		13.44	.595		46.50	36.50		83	106
4700	2 channel		11.76	.680		64.50	41.50		106	133
4750	3 channel		10.08	.794		81.50	48.50		130	162
4800	4 channel		8.40	.952		107	58.50		165.50	205
4850	Splice plate		33.60	.238		11.75	14.60		26.35	35
4900	Continuous concrete insert, 1-1/2" deep, 1' long		13.44	.595		25	36.50		61.50	82
4950	2' long		11.76	.680		30	41.50		71.50	95
5000	3' long		10.08	.794		30	48.50		78.50	106
5050	4' long		8.40	.952		43	58.50		101.50	135
5100	6' long		6.72	1.190		64.50	73		137.50	180
5150	3/4" deep, 1' long		13.44	.595		25.50	36.50		62	82.50
5200	2' long		11.76	.680		25.50	41.50		67	90
5250	3' long		10.08	.794		27	48.50		75.50	102
5300	4' long		8.40	.952		43.50	58.50		102	135
5350	6' long		6.72	1.190		65.50	73		138.50	181
5400	90° angle fitting 2-1/8" x 2-1/8"		50.40	.159		4.73	9.75		14.48	19.70
5450	Supports, suspension rod type, small		50.40	.159		77	9.75		86.75	99
5500	Large		33.60	.238		82	14.60		96.60	112
5550	Beam clamp, small		50.40	.159		35.50	9.75		45.25	53.50
5600	Large		33.60	.238		32	14.60		46.60	57
5650	U-support, small		50.40	.159		6.60	9.75		16.35	22

26 05 29.20 Hangers		Crew	Daily Output	Labor-Hours	Unit	Material	2020 Bare Costs Labor	Equipment	Total	Total Incl O&P
5700	Large	1 Elec	33.60	.238	Ea.	13.55	14.60		28.15	37
5750	Concrete insert, cast, for up to 1/2" threaded rod		13.44	.595		14.75	36.50		51.25	70.50
5800	Beam clamp, 1/4" clamp, for 1/4" threaded drop rod		26.88	.298		3.40	18.25		21.65	30.50
5900	3/8" clamp, for 3/8" threaded drop rod		26.88	.298		3.98	18.25		22.23	31.50
6000	Strap, rigid conduit, 1/2" diameter		453.60	.018		1.50	1.08		2.58	3.26
6050	3/4" diameter		369.60	.022		1.56	1.33		2.89	3.70
6100	1" diameter		352.80	.023		1.86	1.39		3.25	4.12
6150	1-1/4" diameter		336	.024		2.19	1.46		3.65	4.59
6200	1-1/2" diameter		336	.024		2.53	1.46		3.99	4.96
6250	2" diameter		224.28	.036		2.89	2.19		5.08	6.45
6300	2-1/2" diameter		224.28	.036		2.78	2.19		4.97	6.30
6350	3" diameter		134.40	.060		3.58	3.65		7.23	9.40
6400	3-1/2" diameter		111.72	.072		3.95	4.39		8.34	10.90
6450	4" diameter		84	.095		4.63	5.85		10.48	13.80
6500	5" diameter		67.20	.119		11.35	7.30		18.65	23.50
6550	6" diameter		50.40	.159		38.50	9.75		48.25	57
6600	EMT, 1/2" diameter		453.60	.018		1.26	1.08		2.34	3
6650	3/4" diameter		369.60	.022		1.33	1.33		2.66	3.44
6700	1" diameter		352.80	.023		1.39	1.39		2.78	3.60
6750	1-1/4" diameter		336	.024		1.76	1.46		3.22	4.12
6800	1-1/2" diameter		336	.024		2.25	1.46		3.71	4.66
6850	2" diameter		224.28	.036		2.15	2.19		4.34	5.65
6900	2-1/2" diameter		224.28	.036		2.31	2.19		4.50	5.80
6950	3" diameter		134.40	.060		2.93	3.65		6.58	8.65
6970	3-1/2" diameter		111.72	.072		3.90	4.39		8.29	10.85
6990	4" diameter		84	.095		4.36	5.85		10.21	13.50
7000	Clip, 1 hole for rigid conduit, 1/2" diameter		420	.019		.08	1.17		1.25	1.82
7050	3/4" diameter		394.80	.020		.93	1.24		2.17	2.87
7100	1" diameter		369.60	.022		2.26	1.33		3.59	4.47
7150	1-1/4" diameter		336	.024		2.08	1.46		3.54	4.46
7200	1-1/2" diameter		298.20	.027		2.58	1.65		4.23	5.30
7250	2" diameter		268.80	.030		1.99	1.83		3.82	4.91
7300	2-1/2" diameter		223.44	.036		6	2.20		8.20	9.85
7350	3" diameter		134.40	.060		5.10	3.65		8.75	11.05
7400	3-1/2" diameter		111.72	.072		7.55	4.39		11.94	14.85
7450	4" diameter		84	.095		15.80	5.85		21.65	26
7500	5" diameter		67.20	.119		121	7.30		128.30	145
7550	6" diameter		50.40	.159		266	9.75		275.75	310
7600	Hammer on purlin clip, 1/4" hole, 1/16" flange		50.40	.159		.76	9.75		10.51	15.35
7610	1/4" hole threaded, 1/4" flange, rod hanger		50.40	.159		1.70	9.75		11.45	16.35
7620	3/8" hole threaded, 1/4" flange, rod hanger		50.40	.159		1.78	9.75		11.53	16.45
7630	Z purlin clip, 1/4" & 3/8" flange threaded rod hanger		50.40	.159		1.36	9.75		11.11	16
7640	1/2" rigid, 3/4" conduit, 1" flange		50.40	.159		1.79	9.75		11.54	16.45
7650	1" conduit, 1" flange		50.40	.159		2.03	9.75		11.78	16.75
7660	1-1/4" conduit, 1" flange		50.40	.159		2.39	9.75		12.14	17.15
7670	Push-in type conduit clip 3/4" EMT, with 9/32" hole		50.40	.159		.58	9.75		10.33	15.15
7680	3/4" EMT, 1/2" rigid		50.40	.159		1.56	9.75		11.31	16.20
7690	Push-in type tandem conduit clip 3/4" - 1" EMT, 1/2" - 3/4" rigid		50.40	.159		1.66	9.75		11.41	16.35
7820	Conduit hangers, with bolt & 12" rod, 1/2" diameter		126	.063		7.80	3.90		11.70	14.40
7830	3/4" diameter		121.80	.066		7.85	4.03		11.88	14.60
7840	1" diameter		113.40	.071		7.85	4.33		12.18	15.10
7850	1-1/4" diameter		100.80	.079		8	4.87		12.87	16.05
7860	1-1/2" diameter		92.40	.087		8.20	5.30		13.50	16.90

For customer support on your Electrical Change Order Costs with RSMeans data, call 800.448.8182.

Post-Installation Change Order Costs

26 05 Common Work Results for Electrical

26 05 29 – Hangers and Supports for Electrical Systems

26 05 29.20 Hangers

		Crew	Daily Output	Labor-Hours	Unit	Material	2020 Bare Costs Labor	Equipment	Total	Total Incl O&P
7870	2" diameter	1 Elec	84	.095	Ea.	9.85	5.85		15.70	19.50
7880	2-1/2" diameter		67.20	.119		10.25	7.30		17.55	22
7890	3" diameter		50.40	.159		10.80	9.75		20.55	26.50
7900	3-1/2" diameter		37.80	.212		10.75	13		23.75	31
7910	4" diameter		29.40	.272		12.05	16.70		28.75	38.50
7920	5" diameter		25.20	.317		14.25	19.50		33.75	44.50
7930	6" diameter		21	.381		15.70	23.50		39.20	52.50
7950	Jay clamp, 1/2" diameter		26.88	.298		9.40	18.25		27.65	37.50
7960	3/4" diameter		26.88	.298		12.35	18.25		30.60	40.50
7970	1" diameter		26.88	.298		13.75	18.25		32	42
7980	1-1/4" diameter		25.20	.317		13.65	19.50		33.15	44
7990	1-1/2" diameter		25.20	.317		15.25	19.50		34.75	46
8000	2" diameter		25.20	.317		22.50	19.50		42	54
8010	2-1/2" diameter		23.52	.340		28	21		49	62
8020	3" diameter		23.52	.340		28	21		49	62
8030	3-1/2" diameter		21	.381		78.50	23.50		102	122
8040	4" diameter		21	.381		78.50	23.50		102	122
8050	5" diameter		16.80	.476		141	29		170	199
8060	6" diameter		13.44	.595		269	36.50		305.50	350
8070	Channels, 3/4" x 1-1/2" w/12" rods for 1/2" to 1" conduit		25.20	.317		12.95	19.50		32.45	43.50
8080	1-1/2" x 1-1/2" w/12" rods for 1-1/4" to 2" conduit		23.52	.340		13.15	21		34.15	45.50
8090	1-1/2" x 1-1/2" w/12" rods for 2-1/2" to 4" conduit		21.84	.366		16.40	22.50		38.90	51.50
8100	1-1/2" x 1-7/8" w/12" rods for 5" to 6" conduit		20.16	.397		61	24.50		85.50	104
8110	Beam clamp, conduit, plastic coated steel, 1/2" diam.		25.20	.317		43	19.50		62.50	76.50
8120	3/4" diameter		25.20	.317		42.50	19.50		62	76
8130	1" diameter		25.20	.317		50	19.50		69.50	84
8140	1-1/4" diameter		23.52	.340		62	21		83	99
8150	1-1/2" diameter		23.52	.340		73.50	21		94.50	112
8160	2" diameter		23.52	.340		73.50	21		94.50	112
8170	2-1/2" diameter		21.84	.366		101	22.50		123.50	145
8180	3" diameter		21.84	.366		112	22.50		134.50	158
8190	3-1/2" diameter		19.32	.414		84.50	25.50		110	131
8200	4" diameter		19.32	.414		116	25.50		141.50	166
8210	5" diameter		15.12	.529		250	32.50		282.50	325
8220	Channels, plastic coated									
8250	3/4" x 1-1/2", w/12" rods for 1/2" to 1" conduit	1 Elec	23.52	.340	Ea.	43.50	21		64.50	79
8260	1-1/2" x 1-1/2", w/12" rods for 1-1/4" to 2" conduit		21.84	.366		54.50	22.50		77	93.50
8270	1-1/2" x 1-1/2", w/12" rods for 2-1/2" to 3-1/2" conduit		20.16	.397		59	24.50		83.50	102
8280	1-1/2" x 1-7/8", w/12" rods for 4" to 5" conduit		18.48	.433		98	26.50		124.50	148
8290	1-1/2" x 1-7/8", w/12" rods for 6" conduit		16.80	.476		108	29		137	162
8320	Conduit hangers, plastic coated steel, with bolt & 12" rod, 1/2" diam.		117.60	.068		28	4.17		32.17	36.50
8330	3/4" diameter		113.40	.071		28.50	4.33		32.83	38
8340	1" diameter		105	.076		29.50	4.67		34.17	39.50
8350	1-1/4" diameter		92.40	.087		31	5.30		36.30	42
8360	1-1/2" diameter		84	.095		34.50	5.85		40.35	46.50
8370	2" diameter		75.60	.106		38.50	6.50		45	51.50
8380	2-1/2" diameter		58.80	.136		45.50	8.35		53.85	62.50
8390	3" diameter		42	.190		55	11.70		66.70	78
8400	3-1/2" diameter		29.40	.272		50	16.70		66.70	80.50
8410	4" diameter		21	.381		81.50	23.50		105	125
8420	5" diameter		16.80	.476		89.50	29		118.50	142
9000	Parallel type, conduit beam clamp, 1/2"		26.88	.298		7.75	18.25		26	35.50
9010	3/4"		26.88	.298		8.95	18.25		27.20	37

Post-Installation Change Order Costs   For customer support on your Electrical Change Order Costs with RSMeans data, call 800.448.8182.   309

26 05 29.20 Hangers

		Crew	Daily Output	Labor-Hours	Unit	Material	2020 Bare Costs Labor	Equipment	Total	Total Incl O&P
9020	1"	1 Elec	26.88	.298	Ea.	9.05	18.25		27.30	37
9030	1-1/4"		25.20	.317		12.90	19.50		32.40	43
9040	1-1/2"		25.20	.317		15.90	19.50		35.40	46.50
9050	2"		25.20	.317		18.95	19.50		38.45	50
9060	2-1/2"		23.52	.340		24	21		45	57.50
9070	3"		23.52	.340		29	21		50	63
9090	4"		21	.381		19.05	23.50		42.55	56
9110	Right angle, conduit beam clamp, 1/2"		26.88	.298		3.15	18.25		21.40	30.50
9120	3/4"		26.88	.298		3.39	18.25		21.64	30.50
9130	1"		26.88	.298		3.05	18.25		21.30	30.50
9140	1-1/4"		25.20	.317		4.66	19.50		24.16	34
9150	1-1/2"		25.20	.317		4.76	19.50		24.26	34.50
9160	2"		25.20	.317		6.95	19.50		26.45	36.50
9170	2-1/2"		23.52	.340		8.55	21		29.55	40.50
9180	3"		23.52	.340		9.35	21		30.35	41.50
9190	3-1/2"		21	.381		12.95	23.50		36.45	49.50
9200	4"		21	.381		14.40	23.50		37.90	51
9230	Adjustable, conduit hanger, 1/2"		26.88	.298		9.45	18.25		27.70	37.50
9240	3/4"		26.88	.298		7	18.25		25.25	34.50
9250	1"		26.88	.298		9.10	18.25		27.35	37
9260	1-1/4"		25.20	.317		9.90	19.50		29.40	40
9270	1-1/2"		25.20	.317		11.80	19.50		31.30	42
9280	2"		25.20	.317		4.91	19.50		24.41	34.50
9290	2-1/2"		23.52	.340		28	21		49	62
9300	3"		23.52	.340		33	21		54	67.50
9310	3-1/2"		21	.381		33	23.50		56.50	71
9320	4"		21	.381		39	23.50		62.50	78
9330	5"		16.80	.476		47	29		76	95.50
9340	6"		13.44	.595		59	36.50		95.50	120
9350	Combination conduit hanger, 3/8"		26.88	.298		18.45	18.25		36.70	47.50
9360	Adjustable flange, 3/8"		26.88	.298		21.50	18.25		39.75	51

26 05 29.30 Fittings and Channel Support

		Crew	Daily Output	Labor-Hours	Unit	Material	2020 Bare Costs Labor	Equipment	Total	Total Incl O&P
0010	**FITTINGS & CHANNEL SUPPORT**									
0020	Rooftop channel support									
0200	2-7/8" L x 1-5/8" W, 12 ga. pre galv., dbl. base	1 Elec	38.64	.207	Ea.	4.78	12.70		17.48	24
0210	2-7/8" L x 1-5/8" W, 12 ga. hot dip galv., dbl. base		38.64	.207		2.78	12.70		15.48	22
0220	2-7/16" L x 1-5/8" W, 12 ga. pre galv., sngl base		40.32	.198		2.29	12.15		14.44	20.50
0230	2-7/16" L x 1-5/8" W, 12 ga. hot dip galv., sngl base		40.32	.198		8.15	12.15		20.30	27
0240	13/16" L x 1-5/8" W, 14 ga. pre galv., sngl base		47.04	.170		1.95	10.45		12.40	17.70
0250	13/16" L x 1-5/8" W, 14 ga. hot dip galv., sngl base		47.04	.170		7.95	10.45		18.40	24.50
0260	1-5/8" L x 1-5/8" W, 12 ga. pre galv., sngl base		45.36	.176		2.11	10.80		12.91	18.40
0270	1-5/8" L x 1-5/8" W, 12 ga. hot dip galv., sngl base		45.36	.176		2.24	10.80		13.04	18.55
0280	1-5/8" x 28", 12 ga. pre galv., dbl. H block base		35.28	.227		4.84	13.90		18.74	26
0290	1-5/8" x 36", 12 ga. pre galv., dbl. H block base		33.60	.238		5.10	14.60		19.70	27.50
0300	1-5/8" x 42", 12 ga. pre galv., dbl. H block base		28.56	.280		5.75	17.20		22.95	32
0310	1-5/8" x 50", 12 ga. pre galv., dbl. H block base		31.92	.251		6.15	15.40		21.55	30
0320	1-5/8" x 60", 12 ga. pre galv., dbl. H block base		30.24	.265		6.75	16.25		23	31.50
0330	13/16" x 8" H, threaded rod pre galv., H block base		29.40	.272		3.24	16.70		19.94	28.50
0340	13/16" x 12" H, threaded rod pre galv., H block base		29.06	.275		4.05	16.90		20.95	29.50
0350	1-5/8" x 16" H, threaded rod dbl. H block base		28.56	.280		6.05	17.20		23.25	32
0360	1-5/8" x 12" H, pre galv., dbl. H block base		29.06	.275		11.60	16.90		28.50	38
0370	1-5/8" x 24"H, pre galv., dbl. H block base		28.56	.280		12.25	17.20		29.45	39

For customer support on your Electrical Change Order Costs with RSMeans data, call 800.448.8182.

Post-Installation Change Order Costs

26 05 29 – Hangers and Supports for Electrical Systems

26 05 29.30 Fittings and Channel Support	Crew	Daily Output	Labor-Hours	Unit	Material	2020 Bare Costs Labor	Equipment	Total	Total Incl O&P
0380	3-1/4" x 24" H, pre galv., dbl. H block base	1 Elec	26.88	.298	Ea.	28	18.25		46.25	57.50
0390	3-1/4" x 36" H, pre galv., dbl. H block base	↓	25.20	.317	↓	30.50	19.50		50	62.50

26 05 33 – Raceway and Boxes for Electrical Systems

26 05 33.13 Conduit

	26 05 33.13 Conduit	Crew	Daily Output	Labor-Hours	Unit	Material	2020 Bare Costs Labor	Equipment	Total	Total Incl O&P
0010	**CONDUIT** To 10' high, includes 2 terminations, 2 elbows,									
0020	11 beam clamps, and 11 couplings per 100 L.F.									
0300	Aluminum, 1/2" diameter	1 Elec	84	.095	L.F.	2.18	5.85		8.03	11.10
0500	3/4" diameter		75.60	.106		3.22	6.50		9.72	13.20
0700	1" diameter		67.20	.119		4.29	7.30		11.59	15.60
1000	1-1/4" diameter		58.80	.136		5.25	8.35		13.60	18.20
1030	1-1/2" diameter		54.60	.147		6.40	9		15.40	20.50
1050	2" diameter		50.40	.159		8.70	9.75		18.45	24
1070	2-1/2" diameter	↓	42	.190		11	11.70		22.70	29.50
1100	3" diameter	2 Elec	75.60	.212		15.40	13		28.40	36.50
1130	3-1/2" diameter		67.20	.238		19.40	14.60		34	43.50
1140	4" diameter		58.80	.272		23	16.70		39.70	50.50
1150	5" diameter		42	.381		43.50	23.50		67	82.50
1160	6" diameter	↓	33.60	.476	↓	76.50	29		105.50	128
1161	Field bends, 45° to 90°, 1/2" diameter	1 Elec	44.52	.180	Ea.		11		11	16.40
1162	3/4" diameter		39.48	.203			12.45		12.45	18.50
1163	1" diameter		36.96	.216			13.30		13.30	19.75
1164	1-1/4" diameter		19.32	.414			25.50		25.50	38
1165	1-1/2" diameter		17.64	.454			28		28	41.50
1166	2" diameter		13.44	.595			36.50		36.50	54.50
1170	Elbows, 1/2" diameter		33.60	.238		5.85	14.60		20.45	28.50
1200	3/4" diameter		26.88	.298		7.65	18.25		25.90	35.50
1230	1" diameter		23.52	.340		10.85	21		31.85	43
1250	1-1/4" diameter		20.16	.397		23	24.50		47.50	61.50
1270	1-1/2" diameter		16.80	.476		30	29		59	77
1300	2" diameter		13.44	.595		33.50	36.50		70	91
1330	2-1/2" diameter		10.08	.794		64	48.50		112.50	143
1350	3" diameter		6.72	1.190		104	73		177	224
1370	3-1/2" diameter		5.04	1.590		154	97.50		251.50	315
1400	4" diameter		4.20	1.900		177	117		294	370
1410	5" diameter		3.36	2.380		535	146		681	800
1420	6" diameter		2.10	3.810		735	234		969	1,150
1430	Couplings, 1/2" diameter		268.80	.030		2.14	1.83		3.97	5.05
1450	3/4" diameter		161.28	.050		3.24	3.04		6.28	8.10
1470	1" diameter		134.40	.060		4.20	3.65		7.85	10.05
1500	1-1/4" diameter		100.80	.079		4.50	4.87		9.37	12.20
1530	1-1/2" diameter		80.64	.099		5.90	6.10		12	15.55
1550	2" diameter		67.20	.119		8.50	7.30		15.80	20.50
1570	2-1/2" diameter		57.96	.138		14.50	8.45		22.95	28.50
1600	3" diameter		53.76	.149		24.50	9.15		33.65	40.50
1630	3-1/2" diameter		47.04	.170		27.50	10.45		37.95	46
1650	4" diameter		44.52	.180		41	11		52	61.50
1670	5" diameter		40.32	.198		101	12.15		113.15	129
1690	6" diameter	↓	38.64	.207	↓	236	12.70		248.70	279
1691	See note on line 26 05 33.13 9995									
1750	Rigid galvanized steel, 1/2" diameter	1 Elec	75.60	.106	L.F.	3.32	6.50		9.82	13.30
1770	3/4" diameter		67.20	.119		6.15	7.30		13.45	17.70
1800	1" diameter	↓	54.60	.147	↓	9.25	9		18.25	23.50

26 05 33.13 Conduit		Crew	Daily Output	Labor-Hours	Unit	Material	2020 Bare Costs Labor	Equipment	Total	Total Incl O&P
1830	1-1/4" diameter	1 Elec	50.40	.159	L.F.	6.40	9.75		16.15	21.50
1850	1-1/2" diameter		46.20	.173		11.20	10.60		21.80	28
1870	2" diameter		37.80	.212		11.75	13		24.75	32.50
1900	2-1/2" diameter		29.40	.272		17.10	16.70		33.80	44
1930	3" diameter	2 Elec	42	.381		19	23.50		42.50	56
1950	3-1/2" diameter		36.96	.433		24.50	26.50		51	66
1970	4" diameter		33.60	.476		29.50	29		58.50	76
1980	5" diameter		25.20	.635		47	39		86	110
1990	6" diameter		16.80	.952		72.50	58.50		131	167
1991	Field bends, 45° to 90°, 1/2" diameter	1 Elec	36.96	.216	Ea.		13.30		13.30	19.75
1992	3/4" diameter		33.60	.238			14.60		14.60	22
1993	1" diameter		30.24	.265			16.25		16.25	24
1994	1-1/4" diameter		15.96	.501			31		31	46
1995	1-1/2" diameter		15.12	.529			32.50		32.50	48.50
1996	2" diameter		10.92	.733			45		45	67
2000	Elbows, 1/2" diameter		26.88	.298		4.64	18.25		22.89	32
2030	3/4" diameter		23.52	.340		5.35	21		26.35	37
2050	1" diameter		20.16	.397		7.05	24.50		31.55	44.50
2070	1-1/4" diameter		15.12	.529		13.60	32.50		46.10	63.50
2100	1-1/2" diameter		13.44	.595		23	36.50		59.50	79.50
2130	2" diameter		10.08	.794		33	48.50		81.50	109
2150	2-1/2" diameter		6.72	1.190		36	73		109	149
2170	3" diameter		5.04	1.590		47	97.50		144.50	197
2200	3-1/2" diameter		3.53	2.270		78	139		217	293
2220	4" diameter		3.36	2.380		88.50	146		234.50	315
2230	5" diameter		2.94	2.720		269	167		436	545
2240	6" diameter		1.68	4.760		375	292		667	850
2250	Couplings, 1/2" diameter		224.28	.036		1.25	2.19		3.44	4.63
2270	3/4" diameter		134.40	.060		1.84	3.65		5.49	7.45
2300	1" diameter		111.72	.072		3.20	4.39		7.59	10.05
2330	1-1/4" diameter		84	.095		3.10	5.85		8.95	12.10
2350	1-1/2" diameter		67.20	.119		6.55	7.30		13.85	18.10
2370	2" diameter		56.28	.142		8.55	8.70		17.25	22.50
2400	2-1/2" diameter		47.88	.167		12.60	10.25		22.85	29
2430	3" diameter		44.52	.180		16.25	11		27.25	34.50
2450	3-1/2" diameter		39.48	.203		21	12.45		33.45	41.50
2470	4" diameter		36.96	.216		35	13.30		48.30	58.50
2480	5" diameter		33.60	.238		40.50	14.60		55.10	67
2490	6" diameter		31.92	.251		67	15.40		82.40	97
2491	See note on line 26 05 33.13 9995									
2500	Steel, intermediate conduit (IMC), 1/2" diameter	1 Elec	84	.095	L.F.	2.29	5.85		8.14	11.20
2530	3/4" diameter		75.60	.106		3.17	6.50		9.67	13.15
2550	1" diameter		58.80	.136		4.28	8.35		12.63	17.15
2570	1-1/4" diameter		54.60	.147		4.74	9		13.74	18.60
2600	1-1/2" diameter		50.40	.159		6.90	9.75		16.65	22
2630	2" diameter		42	.190		8.20	11.70		19.90	26.50
2650	2-1/2" diameter		33.60	.238		10.85	14.60		25.45	34
2670	3" diameter	2 Elec	50.40	.317		16.25	19.50		35.75	47
2700	3-1/2" diameter		45.36	.353		22.50	21.50		44	57
2730	4" diameter		42	.381		14.45	23.50		37.95	51
2731	Field bends, 45° to 90°, 1/2" diameter	1 Elec	36.96	.216	Ea.		13.30		13.30	19.75
2732	3/4" diameter		33.60	.238			14.60		14.60	22
2733	1" diameter		30.24	.265			16.25		16.25	24

Post-Installation Change Order Costs

26 05 33.13 Conduit		Crew	Daily Output	Labor-Hours	Unit	Material	2020 Bare Costs Labor	Equipment	Total	Total Incl O&P
2734	1-1/4" diameter	1 Elec	15.96	.501	Ea.		31		31	46
2735	1-1/2" diameter		15.12	.529			32.50		32.50	48.50
2736	2" diameter		10.92	.733			45		45	67
2750	Elbows, 1/2" diameter		26.88	.298		6.55	18.25		24.80	34
2770	3/4" diameter		23.52	.340		14.10	21		35.10	46.50
2800	1" diameter		20.16	.397		15.95	24.50		40.45	54
2830	1-1/4" diameter		15.12	.529		21.50	32.50		54	72
2850	1-1/2" diameter		13.44	.595		46	36.50		82.50	105
2870	2" diameter		10.08	.794		72.50	48.50		121	153
2900	2-1/2" diameter		6.72	1.190		48	73		121	162
2930	3" diameter		5.04	1.590		66.50	97.50		164	219
2950	3-1/2" diameter		3.53	2.270		236	139		375	465
2970	4" diameter		3.36	2.380		88	146		234	315
3000	Couplings, 1/2" diameter		246.12	.033		1.25	1.99		3.24	4.34
3030	3/4" diameter		147.84	.054		1.84	3.32		5.16	6.95
3050	1" diameter		123.48	.065		3.20	3.97		7.17	9.40
3070	1-1/4" diameter		92.40	.087		3.10	5.30		8.40	11.30
3100	1-1/2" diameter		73.92	.108		6.55	6.65		13.20	17.10
3130	2" diameter		61.32	.130		8.55	8		16.55	21.50
3150	2-1/2" diameter		52.92	.151		12.60	9.25		21.85	27.50
3170	3" diameter		49.56	.161		16.25	9.90		26.15	32.50
3200	3-1/2" diameter		43.68	.183		21	11.25		32.25	40
3230	4" diameter		41.16	.194		35	11.90		46.90	56.50
3231	See note on line 26 05 33.13 9995									
4100	Rigid steel, plastic coated, 40 mil thick									
4130	1/2" diameter	1 Elec	67.20	.119	L.F.	11.65	7.30		18.95	24
4150	3/4" diameter		58.80	.136		12	8.35		20.35	25.50
4170	1" diameter		46.20	.173		15.05	10.60		25.65	32.50
4200	1-1/4" diameter		42	.190		21	11.70		32.70	40.50
4230	1-1/2" diameter		37.80	.212		23	13		36	44.50
4250	2" diameter		29.40	.272		27.50	16.70		44.20	55
4270	2-1/2" diameter		21	.381		45.50	23.50		69	85
4300	3" diameter	2 Elec	36.96	.433		50	26.50		76.50	94.50
4330	3-1/2" diameter		33.60	.476		60	29		89	110
4350	4" diameter		30.24	.529		74.50	32.50		107	131
4370	5" diameter		25.20	.635		145	39		184	218
4400	Elbows, 1/2" diameter	1 Elec	23.52	.340	Ea.	24	21		45	57.50
4430	3/4" diameter		20.16	.397		21.50	24.50		46	60.50
4450	1" diameter		15.12	.529		25.50	32.50		58	76.50
4470	1-1/4" diameter		13.44	.595		37	36.50		73.50	95.50
4500	1-1/2" diameter		10.08	.794		38.50	48.50		87	115
4530	2" diameter		6.72	1.190		50	73		123	164
4550	2-1/2" diameter		5.04	1.590		103	97.50		200.50	258
4570	3" diameter		3.53	2.270		161	139		300	385
4600	3-1/2" diameter		3.36	2.380		253	146		399	495
4630	4" diameter		3.19	2.510		235	154		389	485
4650	5" diameter		2.94	2.720		595	167		762	905
4680	Couplings, 1/2" diameter		178.92	.045		6.95	2.74		9.69	11.75
4700	3/4" diameter		107.52	.074		6.80	4.56		11.36	14.25
4730	1" diameter		89.88	.089		8.25	5.45		13.70	17.20
4750	1-1/4" diameter		67.20	.119		12.15	7.30		19.45	24.50
4770	1-1/2" diameter		53.76	.149		12.10	9.15		21.25	27
4800	2" diameter		44.52	.180		17.20	11		28.20	35.50

26 05 33.13 Conduit		Crew	Daily Output	Labor-Hours	Unit	Material	2020 Bare Costs Labor	Equipment	Total	Total Incl O&P
4830	2-1/2" diameter	1 Elec	38.64	.207	Ea.	49	12.70		61.70	73
4850	3" diameter		36.12	.221		46.50	13.60		60.10	71
4870	3-1/2" diameter		31.92	.251		69.50	15.40		84.90	99.50
4900	4" diameter		30.24	.265		69	16.25		85.25	100
4950	5" diameter	↓	26.88	.298	↓	256	18.25		274.25	310
4951	See note on line 26 05 33.13 9995									
5000	Electric metallic tubing (EMT), 1/2" diameter	1 Elec	142.80	.056	L.F.	.91	3.44		4.35	6.10
5020	3/4" diameter		109.20	.073		1.29	4.49		5.78	8.10
5040	1" diameter		96.60	.083		2.16	5.10		7.26	9.95
5060	1-1/4" diameter		84	.095		3.52	5.85		9.37	12.55
5080	1-1/2" diameter		75.60	.106		4.16	6.50		10.66	14.20
5100	2" diameter		67.20	.119		5.25	7.30		12.55	16.65
5120	2-1/2" diameter	↓	50.40	.159		5.80	9.75		15.55	21
5140	3" diameter	2 Elec	84	.190		7.40	11.70		19.10	25.50
5160	3-1/2" diameter		75.60	.212		9.25	13		22.25	29.50
5180	4" diameter	↓	67.20	.238	↓	16.25	14.60		30.85	40
5200	Field bends, 45° to 90°, 1/2" diameter	1 Elec	74.76	.107	Ea.		6.55		6.55	9.80
5220	3/4" diameter		67.20	.119			7.30		7.30	10.90
5240	1" diameter		61.32	.130			8		8	11.90
5260	1-1/4" diameter		31.92	.251			15.40		15.40	23
5280	1-1/2" diameter		30.24	.265			16.25		16.25	24
5300	2" diameter		21.84	.366			22.50		22.50	33.50
5320	Offsets, 1/2" diameter		54.60	.147			9		9	13.40
5340	3/4" diameter		52.08	.154			9.40		9.40	14.05
5360	1" diameter		44.52	.180			11		11	16.40
5380	1-1/4" diameter		25.20	.317			19.50		19.50	29
5400	1-1/2" diameter		23.52	.340			21		21	31
5420	2" diameter		16.80	.476			29		29	43.50
5700	Elbows, 1" diameter		33.60	.238		3.91	14.60		18.51	26.50
5720	1-1/4" diameter		26.88	.298		5.25	18.25		23.50	33
5740	1-1/2" diameter		20.16	.397		6.10	24.50		30.60	43
5760	2" diameter		16.80	.476		8.75	29		37.75	53
5780	2-1/2" diameter		10.08	.794		21.50	48.50		70	96
5800	3" diameter		7.56	1.060		29.50	65		94.50	130
5820	3-1/2" diameter		5.88	1.360		40.50	83.50		124	169
5840	4" diameter		5.04	1.590		47.50	97.50		145	197
6200	Couplings, set screw, steel, 1/2" diameter		197.40	.041		1.93	2.49		4.42	5.85
6220	3/4" diameter		157.92	.051		2.74	3.11		5.85	7.65
6240	1" diameter		131.88	.061		4.18	3.72		7.90	10.15
6260	1-1/4" diameter		99.12	.081		8.95	4.95		13.90	17.20
6280	1-1/2" diameter		78.96	.101		1.96	6.20		8.16	11.40
6300	2" diameter		65.52	.122		17.40	7.50		24.90	30.50
6320	2-1/2" diameter		56.28	.142		7.65	8.70		16.35	21.50
6340	3" diameter		49.56	.161		7.20	9.90		17.10	22.50
6360	3-1/2" diameter		39.48	.203		10.05	12.45		22.50	29.50
6380	4" diameter	↓	36.12	.221	↓	70	13.60		83.60	97
6381	See note on line 26 05 33.13 9995									
6500	Box connectors, set screw, steel, 1/2" diameter	1 Elec	100.80	.079	Ea.	.43	4.87		5.30	7.75
6520	3/4" diameter		92.40	.087		.71	5.30		6.01	8.70
6540	1" diameter		75.60	.106		1.33	6.50		7.83	11.10
6560	1-1/4" diameter		58.80	.136		2.38	8.35		10.73	15.05
6580	1-1/2" diameter		50.40	.159		3.96	9.75		13.71	18.85
6600	2" diameter	↓	42	.190	↓	4.72	11.70		16.42	22.50

For customer support on your Electrical Change Order Costs with RSMeans data, call 800.448.8182. **Post-Installation Change Order Costs**

26 05 33.13 Conduit	Crew	Daily Output	Labor-Hours	Unit	Material	2020 Bare Costs Labor	Equipment	Total	Total Incl O&P
6620	2-1/2" diameter	1 Elec	30.24	.265	Ea.	9.50	16.25		25.75	34.50
6640	3" diameter		22.68	.353		41.50	21.50		63	77.50
6680	3-1/2" diameter		17.64	.454		12.85	28		40.85	55.50
6700	4" diameter		13.44	.595		15.35	36.50		51.85	71.50
6740	Insulated box connectors, set screw, steel, 1/2" diameter		100.80	.079		.82	4.87		5.69	8.15
6760	3/4" diameter		92.40	.087		1.81	5.30		7.11	9.90
6780	1" diameter		75.60	.106		3.48	6.50		9.98	13.50
6800	1-1/4" diameter		58.80	.136		4.14	8.35		12.49	17
6820	1-1/2" diameter		50.40	.159		5.95	9.75		15.70	21
6840	2" diameter		42	.190		7.95	11.70		19.65	26
6860	2-1/2" diameter		30.24	.265		36	16.25		52.25	63.50
6880	3" diameter		22.68	.353		30.50	21.50		52	65.50
6900	3-1/2" diameter		17.64	.454		58	28		86	105
6920	4" diameter		13.44	.595		43	36.50		79.50	102
7000	EMT to conduit adapters, 1/2" diameter (compression)		58.80	.136		4.19	8.35		12.54	17.05
7020	3/4" diameter		50.40	.159		6.55	9.75		16.30	21.50
7040	1" diameter		42	.190		10.35	11.70		22.05	29
7060	1-1/4" diameter		33.60	.238		20	14.60		34.60	44.50
7080	1-1/2" diameter		25.20	.317		24	19.50		43.50	55.50
7100	2" diameter		21	.381		37.50	23.50		61	76.50
7200	EMT to Greenfield adapters, 1/2" to 3/8" diameter (compression)		75.60	.106		2.59	6.50		9.09	12.50
7220	1/2" diameter		75.60	.106		5.85	6.50		12.35	16.05
7240	3/4" diameter		67.20	.119		7.10	7.30		14.40	18.75
7260	1" diameter		58.80	.136		18.65	8.35		27	33
7270	1-1/4" diameter		50.40	.159		23.50	9.75		33.25	40.50
7280	1-1/2" diameter		42	.190		26.50	11.70		38.20	46.50
7290	2" diameter		33.60	.238		39.50	14.60		54.10	65.50
7400	EMT, LB, LR or LL fittings with covers, 1/2" diameter, set screw		20.16	.397		5.80	24.50		30.30	43
7420	3/4" diameter		16.80	.476		8.10	29		37.10	52.50
7440	1" diameter		13.44	.595		12.05	36.50		48.55	68
7450	1-1/4" diameter		10.92	.733		17.95	45		62.95	87
7460	1-1/2" diameter		9.24	.866		24	53		77	106
7470	2" diameter		7.56	1.060		39.50	65		104.50	140
7600	EMT, "T" fittings with covers, 1/2" diameter, set screw		13.44	.595		6.75	36.50		43.25	62
7620	3/4" diameter		12.60	.635		8.70	39		47.70	67.50
7640	1" diameter		10.08	.794		11	48.50		59.50	84.50
7650	1-1/4" diameter		9.24	.866		16.35	53		69.35	97
7660	1-1/2" diameter		8.40	.952		21	58.50		79.50	110
7670	2" diameter		6.72	1.190		28.50	73		101.50	141
8000	EMT, expansion fittings, no jumper, 1/2" diameter		20.16	.397		102	24.50		126.50	149
8020	3/4" diameter		16.80	.476		150	29		179	209
8040	1" diameter		13.44	.595		161	36.50		197.50	232
8060	1-1/4" diameter		10.92	.733		183	45		228	268
8080	1-1/2" diameter		9.24	.866		242	53		295	345
8100	2" diameter		7.56	1.060		405	65		470	540
8110	2-1/2" diameter		5.88	1.360		490	83.50		573.50	665
8120	3" diameter		5.04	1.590		665	97.50		762.50	875
8140	4" diameter		4.20	1.900		945	117		1,062	1,225
8200	Split adapter, 1/2" diameter		92.40	.087		3.92	5.30		9.22	12.20
8210	3/4" diameter		75.60	.106		3.77	6.50		10.27	13.80
8220	1" diameter		58.80	.136		5.85	8.35		14.20	18.90
8230	1-1/4" diameter		50.40	.159		9.95	9.75		19.70	25.50
8240	1-1/2" diameter		42	.190		13.90	11.70		25.60	32.50

26 05 33.13 Conduit		Crew	Daily Output	Labor-Hours	Unit	Material	2020 Bare Costs Labor	Equipment	Total	Total Incl O&P
8250	2" diameter	1 Elec	30.24	.265	Ea.	39	16.25		55.25	67
8300	1 hole clips, 1/2" diameter		420	.019		.50	1.17		1.67	2.29
8320	3/4" diameter		394.80	.020		.23	1.24		1.47	2.10
8340	1" diameter		372.96	.021		.34	1.32		1.66	2.33
8360	1-1/4" diameter		336	.024		.48	1.46		1.94	2.71
8380	1-1/2" diameter		298.20	.027		1.19	1.65		2.84	3.76
8400	2" diameter		268.80	.030		1.75	1.83		3.58	4.65
8420	2-1/2" diameter		223.44	.036		4.12	2.20		6.32	7.80
8440	3" diameter		134.40	.060		4.62	3.65		8.27	10.55
8460	3-1/2" diameter		111.72	.072		8.95	4.39		13.34	16.35
8480	4" diameter		84	.095		8.85	5.85		14.70	18.40
8500	Clamp back spacers, 1/2" diameter		420	.019		.97	1.17		2.14	2.81
8510	3/4" diameter		394.80	.020		1.21	1.24		2.45	3.18
8520	1" diameter		372.96	.021		2.27	1.32		3.59	4.45
8530	1-1/4" diameter		336	.024		4.73	1.46		6.19	7.40
8540	1-1/2" diameter		298.20	.027		9.15	1.65		10.80	12.55
8550	2" diameter		268.80	.030		6.55	1.83		8.38	9.90
8560	2-1/2" diameter		223.44	.036		12.30	2.20		14.50	16.75
8570	3" diameter		134.40	.060		32	3.65		35.65	41
8580	3-1/2" diameter		111.72	.072		50.50	4.39		54.89	62
8590	4" diameter		84	.095		123	5.85		128.85	144
8600	Offset connectors, 1/2" diameter		33.60	.238		3.34	14.60		17.94	25.50
8610	3/4" diameter		26.88	.298		4.37	18.25		22.62	32
8620	1" diameter		20.16	.397		6.05	24.50		30.55	43
8650	90° pulling elbows, female, 1/2" diameter, with gasket		20.16	.397		9.35	24.50		33.85	47
8660	3/4" diameter		16.80	.476		16.60	29		45.60	62
8700	Couplings, compression, 1/2" diameter, steel		65.52	.122		1.57	7.50		9.07	12.90
8710	3/4" diameter		56.28	.142		1.70	8.70		10.40	14.85
8720	1" diameter		49.56	.161		2.77	9.90		12.67	17.80
8730	1-1/4" diameter		39.48	.203		6.10	12.45		18.55	25
8740	1-1/2" diameter		31.92	.251		9	15.40		24.40	33
8750	2" diameter		26.04	.307		16.15	18.85		35	46
8760	2-1/2" diameter		20.16	.397		45.50	24.50		70	86.50
8770	3" diameter		17.64	.454		33	28		61	78
8780	3-1/2" diameter		15.96	.501		133	31		164	192
8790	4" diameter		13.44	.595		81.50	36.50		118	144
8791	See note on line 26 05 33.13 9995									
8800	Box connectors, compression, 1/2" diam., steel	1 Elec	100.80	.079	Ea.	2.94	4.87		7.81	10.50
8810	3/4" diameter		92.40	.087		4.03	5.30		9.33	12.35
8820	1" diameter		75.60	.106		5.70	6.50		12.20	15.90
8830	1-1/4" diameter		58.80	.136		14.40	8.35		22.75	28.50
8840	1-1/2" diameter		50.40	.159		17.60	9.75		27.35	34
8850	2" diameter		42	.190		24	11.70		35.70	44
8860	2-1/2" diameter		30.24	.265		57.50	16.25		73.75	87
8870	3" diameter		22.68	.353		77.50	21.50		99	118
8880	3-1/2" diameter		17.64	.454		121	28		149	175
8890	4" diameter		13.44	.595		122	36.50		158.50	189
8900	Box connectors, insulated compression, 1/2" diam., steel		100.80	.079		2.14	4.87		7.01	9.60
8910	3/4" diameter		92.40	.087		2.44	5.30		7.74	10.60
8920	1" diameter		75.60	.106		4.51	6.50		11.01	14.60
8930	1-1/4" diameter		58.80	.136		9.40	8.35		17.75	23
8940	1-1/2" diameter		50.40	.159		13.90	9.75		23.65	30
8950	2" diameter		42	.190		20.50	11.70		32.20	40

26 05 33.13 Conduit		Crew	Daily Output	Labor-Hours	Unit	Material	2020 Bare Costs Labor	Equipment	Total	Total Incl O&P
8960	2-1/2" diameter	1 Elec	30.24	.265	Ea.	79	16.25		95.25	111
8970	3" diameter		22.68	.353		70.50	21.50		92	110
8980	3-1/2" diameter		17.64	.454		120	28		148	174
8990	4" diameter		13.44	.595		106	36.50		142.50	171
9100	PVC, schedule 40, 1/2" diameter		159.60	.050	L.F.	1.04	3.08		4.12	5.75
9110	3/4" diameter		121.80	.066		1.23	4.03		5.26	7.35
9120	1" diameter		105	.076		1.85	4.67		6.52	9
9130	1-1/4" diameter		92.40	.087		2.38	5.30		7.68	10.50
9140	1-1/2" diameter		84	.095		2.86	5.85		8.71	11.85
9150	2" diameter		75.60	.106		4.03	6.50		10.53	14.10
9160	2-1/2" diameter		54.60	.147		5.50	9		14.50	19.50
9170	3" diameter	2 Elec	92.40	.173		5.80	10.60		16.40	22
9180	3-1/2" diameter		84	.190		7.65	11.70		19.35	26
9190	4" diameter		75.60	.212		6.90	13		19.90	27
9200	5" diameter		58.80	.272		10	16.70		26.70	36
9210	6" diameter		50.40	.317		15	19.50		34.50	45.50
9220	Elbows, 1/2" diameter	1 Elec	42	.190	Ea.	.70	11.70		12.40	18.15
9225	3/4" diameter		35.28	.227		.72	13.90		14.62	21.50
9230	1" diameter		29.40	.272		1.45	16.70		18.15	26.50
9235	1-1/4" diameter		23.52	.340		3.26	21		24.26	34.50
9240	1-1/2" diameter		16.80	.476		3.25	29		32.25	47
9245	2" diameter	R-1A	30.58	.523		4.25	29		33.25	47.50
9250	2-1/2" diameter		22.43	.713		9.30	39.50		48.80	69
9255	3" diameter		19.24	.832		14.70	46		60.70	84.50
9260	3-1/2" diameter		18.65	.858		22.50	47.50		70	95
9265	4" diameter		15.29	1.050		27.50	58		85.50	117
9270	5" diameter		10.16	1.570		41	86.50		127.50	175
9275	6" diameter		9.32	1.720		42	95		137	188
9312	Couplings, 1/2" diameter	1 Elec	42	.190		.13	11.70		11.83	17.55
9314	3/4" diameter		35.28	.227		.37	13.90		14.27	21
9316	1" diameter		29.40	.272		.30	16.70		17	25.50
9318	1-1/4" diameter		23.52	.340		.42	21		21.42	31.50
9320	1-1/2" diameter		16.80	.476		.65	29		29.65	44
9322	2" diameter	R-1A	30.58	.523		1.51	29		30.51	44.50
9324	2-1/2" diameter		22.43	.713		2.16	39.50		41.66	61
9326	3" diameter		19.24	.832		1.78	46		47.78	70.50
9328	3-1/2" diameter		18.65	.858		2.46	47.50		49.96	73
9330	4" diameter		15.29	1.050		3.18	58		61.18	90
9332	5" diameter		10.16	1.570		8.40	86.50		94.90	138
9334	6" diameter		9.32	1.720		12.20	95		107.20	154
9335	See note on line 26 05 33.13 9995									
9340	Field bends, 45° & 90°, 1/2" diameter	1 Elec	37.80	.212	Ea.		13		13	19.35
9350	3/4" diameter		33.60	.238			14.60		14.60	22
9360	1" diameter		29.40	.272			16.70		16.70	25
9370	1-1/4" diameter		26.88	.298			18.25		18.25	27
9380	1-1/2" diameter		22.68	.353			21.50		21.50	32
9390	2" diameter		16.80	.476			29		29	43.50
9400	2-1/2" diameter		13.44	.595			36.50		36.50	54.50
9410	3" diameter		10.92	.733			45		45	67
9420	3-1/2" diameter		10.08	.794			48.50		48.50	72.50
9430	4" diameter		8.40	.952			58.50		58.50	87
9440	5" diameter		7.56	1.060			65		65	97
9450	6" diameter		6.72	1.190			73		73	109

26 05 33.13 Conduit		Crew	Daily Output	Labor-Hours	Unit	Material	2020 Bare Costs Labor	Equipment	Total	Total Incl O&P
9460	PVC adapters, 1/2" diameter	1 Elec	42	.190	Ea.	.31	11.70		12.01	17.75
9470	3/4" diameter		35.28	.227		.44	13.90		14.34	21
9480	1" diameter		31.92	.251		.47	15.40		15.87	23.50
9490	1-1/4" diameter		29.40	.272		.91	16.70		17.61	26
9500	1-1/2" diameter		26.88	.298		.85	18.25		19.10	28
9510	2" diameter		22.68	.353		1.67	21.50		23.17	34
9520	2-1/2" diameter		19.32	.414		2.41	25.50		27.91	40.50
9530	3" diameter		15.12	.529		3.42	32.50		35.92	52.50
9540	3-1/2" diameter		10.92	.733		6.25	45		51.25	74
9550	4" diameter		9.24	.866		5.75	53		58.75	85.50
9560	5" diameter		6.72	1.190		11.60	73		84.60	122
9570	6" diameter		5.04	1.590		22.50	97.50		120	170
9580	PVC-LB, LR or LL fittings & covers									
9590	1/2" diameter	1 Elec	16.80	.476	Ea.	2.82	29		31.82	46.50
9600	3/4" diameter		13.44	.595		5.75	36.50		42.25	61
9610	1" diameter		10.08	.794		5.55	48.50		54.05	78.50
9620	1-1/4" diameter		7.56	1.060		10.40	65		75.40	108
9630	1-1/2" diameter		5.88	1.360		12.30	83.50		95.80	138
9640	2" diameter		5.04	1.590		17.80	97.50		115.30	165
9650	2-1/2" diameter		5.04	1.590		96.50	97.50		194	251
9660	3" diameter		4.20	1.900		51	117		168	230
9670	3-1/2" diameter		3.36	2.380		68	146		214	292
9680	4" diameter		2.52	3.170		74	194		268	370
9690	PVC-tee fitting & cover									
9700	1/2"	1 Elec	11.76	.680	Ea.	6.35	41.50		47.85	69
9710	3/4"		10.92	.733		6.85	45		51.85	74.50
9720	1"		8.40	.952		6.40	58.50		64.90	94
9730	1-1/4"		7.56	1.060		13.15	65		78.15	111
9740	1-1/2"		6.72	1.190		16.05	73		89.05	127
9750	2"		5.88	1.360		21.50	83.50		105	148
9760	PVC-reducers, 3/4" x 1/2" diameter					1.21			1.21	1.33
9770	1" x 1/2" diameter					3.83			3.83	4.21
9780	1" x 3/4" diameter					3.36			3.36	3.70
9790	1-1/4" x 3/4" diameter					2.88			2.88	3.17
9800	1-1/4" x 1" diameter					5.20			5.20	5.70
9810	1-1/2" x 1-1/4" diameter					4.13			4.13	4.54
9820	2" x 1-1/4" diameter					5.40			5.40	5.95
9830	2-1/2" x 2" diameter					22			22	24
9840	3" x 2" diameter					25.50			25.50	28
9850	4" x 3" diameter					27.50			27.50	30.50
9860	Cement, quart					18.65			18.65	20.50
9870	Gallon					85			85	93.50
9880	Heat bender, to 6" diameter					1,800			1,800	1,975
9900	Add to labor for higher elevated installation									
9905	10' to 14.5' high, add						10%			
9910	15' to 20' high, add						20%			
9920	20' to 25' high, add						25%			
9930	25' to 30' high, add						35%			
9940	30' to 35' high, add						40%			
9950	35' to 40' high, add						50%			
9960	Over 40' high, add						55%			
9995	Do not include labor when adding couplings to a fitting installation R260533-30									

26 05 33.14 Conduit

26 05 33.14 Conduit		Crew	Daily Output	Labor-Hours	Unit	Material	2020 Bare Costs Labor	Equipment	Total	Total Incl O&P
0010	**CONDUIT** To 10' high, includes 11 couplings per 100'									
0200	Electric metallic tubing, 1/2" diameter	1 Elec	365.40	.022	L.F.	.70	1.34		2.04	2.78
0220	3/4" diameter		212.52	.038		1.05	2.31		3.36	4.59
0240	1" diameter		173.88	.046		1.80	2.82		4.62	6.20
0260	1-1/4" diameter		145.32	.055		3.13	3.38		6.51	8.50
0280	1-1/2" diameter		128.52	.062		3.64	3.82		7.46	9.70
0300	2" diameter		109.20	.073		4.62	4.49		9.11	11.80
0320	2-1/2" diameter		77.28	.104		4.74	6.35		11.09	14.65
0340	3" diameter	2 Elec	124.32	.129		5.50	7.90		13.40	17.80
0360	3-1/2" diameter		112.56	.142		7.70	8.70		16.40	21.50
0380	4" diameter		95.76	.167		14.45	10.25		24.70	31
0500	Steel rigid galvanized, 1/2" diameter	1 Elec	122.64	.065		2.23	4		6.23	8.40
0520	3/4" diameter		105	.076		4.74	4.67		9.41	12.15
0540	1" diameter		78.12	.102		7.60	6.30		13.90	17.75
0560	1-1/4" diameter		73.92	.108		4.65	6.65		11.30	15
0580	1-1/2" diameter		67.20	.119		9.05	7.30		16.35	21
0600	2" diameter		54.60	.147		8.60	9		17.60	23
0620	2-1/2" diameter		40.32	.198		14.65	12.15		26.80	34
0640	3" diameter	2 Elec	53.76	.298		16.15	18.25		34.40	45
0660	3-1/2" diameter		50.40	.317		21	19.50		40.50	52
0680	4" diameter		43.68	.366		25.50	22.50		48	61.50
0700	5" diameter		42	.381		31.50	23.50		55	69.50
0720	6" diameter		40.32	.397		48.50	24.50		73	89.50
1000	Steel intermediate conduit (IMC), 1/2" diameter	1 Elec	130.20	.061		1.14	3.77		4.91	6.85
1010	3/4" diameter		109.20	.073		1.39	4.49		5.88	8.25
1020	1" diameter		84	.095		2.15	5.85		8	11.05
1030	1-1/4" diameter		78.12	.102		2.44	6.30		8.74	12.05
1040	1-1/2" diameter		71.40	.112		3.69	6.85		10.54	14.30
1050	2" diameter		58.80	.136		4.21	8.35		12.56	17.10
1060	2-1/2" diameter		44.52	.180		7.55	11		18.55	24.50
1070	3" diameter	2 Elec	67.20	.238		11.60	14.60		26.20	35
1080	3-1/2" diameter		58.80	.272		13.90	16.70		30.60	40.50
1090	4" diameter		50.40	.317		9.30	19.50		28.80	39.50

26 05 33.15 Conduit Nipples

26 05 33.15 Conduit Nipples		Crew	Daily Output	Labor-Hours	Unit	Material	2020 Bare Costs Labor	Equipment	Total	Total Incl O&P
0010	**CONDUIT NIPPLES** With locknuts and bushings									
0100	Aluminum, 1/2" diameter, close	1 Elec	30.24	.265	Ea.	8.45	16.25		24.70	33.50
0120	1-1/2" long		30.24	.265		7.95	16.25		24.20	33
0140	2" long		30.24	.265		9.85	16.25		26.10	35
0160	2-1/2" long		30.24	.265		12.45	16.25		28.70	37.50
0180	3" long		30.24	.265		10.70	16.25		26.95	36
0200	3-1/2" long		30.24	.265		8.60	16.25		24.85	33.50
0220	4" long		30.24	.265		9	16.25		25.25	34
0240	5" long		30.24	.265		11.75	16.25		28	37
0260	6" long		30.24	.265		12.40	16.25		28.65	37.50
0280	8" long		30.24	.265		12.20	16.25		28.45	37.50
0300	10" long		30.24	.265		13.35	16.25		29.60	38.50
0320	12" long		30.24	.265		22	16.25		38.25	48.50
0340	3/4" diameter, close		26.88	.298		11.60	18.25		29.85	40
0360	1-1/2" long		26.88	.298		11.60	18.25		29.85	40
0380	2" long		26.88	.298		11.60	18.25		29.85	40
0400	2-1/2" long		26.88	.298		13.75	18.25		32	42
0420	3" long		26.88	.298		13.80	18.25		32.05	42

26 05 33.15 Conduit Nipples		Crew	Daily Output	Labor-Hours	Unit	Material	2020 Bare Costs Labor	Equipment	Total	Total Incl O&P
0440	3-1/2" long	1 Elec	26.88	.298	Ea.	18.90	18.25		37.15	48
0460	4" long		26.88	.298		14.25	18.25		32.50	42.50
0480	5" long		26.88	.298		15.75	18.25		34	44.50
0500	6" long		26.88	.298		15.55	18.25		33.80	44
0520	8" long		26.88	.298		15.80	18.25		34.05	44.50
0540	10" long		26.88	.298		17.90	18.25		36.15	46.50
0560	12" long		26.88	.298		24.50	18.25		42.75	54
0580	1" diameter, close		22.68	.353		16.50	21.50		38	50
0600	2" long		22.68	.353		24.50	21.50		46	59
0620	2-1/2" long		22.68	.353		17.45	21.50		38.95	51
0640	3" long		22.68	.353		20.50	21.50		42	55
0660	3-1/2" long		22.68	.353		30	21.50		51.50	65
0680	4" long		22.68	.353		23.50	21.50		45	57.50
0700	5" long		22.68	.353		23	21.50		44.50	57
0720	6" long		22.68	.353		22.50	21.50		44	56.50
0740	8" long		22.68	.353		33.50	21.50		55	69
0760	10" long		22.68	.353		60	21.50		81.50	98.50
0780	12" long		22.68	.353		56.50	21.50		78	94
0800	1-1/4" diameter, close		19.32	.414		25.50	25.50		51	66
0820	2" long		19.32	.414		33	25.50		58.50	74
0840	2-1/2" long		19.32	.414		23.50	25.50		49	64
0860	3" long		19.32	.414		29	25.50		54.50	70
0880	3-1/2" long		19.32	.414		29	25.50		54.50	70
0900	4" long		19.32	.414		26	25.50		51.50	66.50
0920	5" long		19.32	.414		33	25.50		58.50	74.50
0940	6" long		19.32	.414		37	25.50		62.50	79
0960	8" long		19.32	.414		57.50	25.50		83	102
0980	10" long		19.32	.414		35.50	25.50		61	77
1000	12" long		19.32	.414		47.50	25.50		73	90
1020	1-1/2" diameter, close		16.80	.476		33.50	29		62.50	80.50
1040	2" long		16.80	.476		35.50	29		64.50	82.50
1060	2-1/2" long		16.80	.476		36	29		65	83
1080	3" long		16.80	.476		35.50	29		64.50	82.50
1100	3-1/2" long		16.80	.476		35	29		64	82
1120	4" long		16.80	.476		35	29		64	82
1140	5" long		16.80	.476		42.50	29		71.50	90.50
1160	6" long		16.80	.476		44	29		73	92
1180	8" long		16.80	.476		43.50	29		72.50	91.50
1200	10" long		16.80	.476		51	29		80	99.50
1220	12" long		16.80	.476		65	29		94	115
1240	2" diameter, close		15.12	.529		48	32.50		80.50	101
1260	2-1/2" long		15.12	.529		46.50	32.50		79	99.50
1280	3" long		15.12	.529		49	32.50		81.50	103
1300	3-1/2" long		15.12	.529		49.50	32.50		82	103
1320	4" long		15.12	.529		51.50	32.50		84	106
1340	5" long		15.12	.529		57	32.50		89.50	112
1360	6" long		15.12	.529		52.50	32.50		85	107
1380	8" long		15.12	.529		76	32.50		108.50	132
1400	10" long		15.12	.529		76	32.50		108.50	132
1420	12" long		15.12	.529		91	32.50		123.50	149
1440	2-1/2" diameter, close		12.60	.635		106	39		145	174
1460	3" long		12.60	.635		98	39		137	166
1480	3-1/2" long		12.60	.635		101	39		140	169

For customer support on your Electrical Change Order Costs with RSMeans data, call 800.448.8182.

Post-Installation Change Order Costs

26 05 Common Work Results for Electrical

26 05 33 – Raceway and Boxes for Electrical Systems

26 05 33.15 Conduit Nipples		Crew	Daily Output	Labor-Hours	Unit	Material	2020 Bare Costs Labor	Equipment	Total	Total Incl O&P
1500	4" long	1 Elec	12.60	.635	Ea.	112	39		151	181
1520	5" long		12.60	.635		142	39		181	215
1540	6" long		12.60	.635		107	39		146	176
1560	8" long		12.60	.635		112	39		151	181
1580	10" long		12.60	.635		140	39		179	212
1600	12" long		12.60	.635		126	39		165	196
1620	3" diameter, close		10.08	.794		113	48.50		161.50	198
1640	3" long		10.08	.794		119	48.50		167.50	204
1660	3-1/2" long		10.08	.794		143	48.50		191.50	231
1680	4" long		10.08	.794		148	48.50		196.50	236
1700	5" long		10.08	.794		113	48.50		161.50	197
1720	6" long		10.08	.794		124	48.50		172.50	209
1740	8" long		10.08	.794		155	48.50		203.50	243
1760	10" long		10.08	.794		186	48.50		234.50	278
1780	12" long		10.08	.794		179	48.50		227.50	270
1800	3-1/2" diameter, close		9.24	.866		249	53		302	355
1820	4" long		9.24	.866		283	53		336	390
1840	5" long		9.24	.866		257	53		310	360
1860	6" long		9.24	.866		270	53		323	375
1880	8" long		9.24	.866		251	53		304	355
1900	10" long		9.24	.866		284	53		337	390
1920	12" long		9.24	.866		360	53		413	475
1940	4" diameter, close		7.56	1.060		279	65		344	400
1960	4" long		7.56	1.060		300	65		365	425
1980	5" long		7.56	1.060		300	65		365	425
2000	6" long		7.56	1.060		355	65		420	485
2020	8" long		7.56	1.060		345	65		410	475
2040	10" long		7.56	1.060		360	65		425	490
2060	12" long		7.56	1.060		330	65		395	455
2080	5" diameter, close		5.88	1.360		560	83.50		643.50	740
2100	5" long		5.88	1.360		655	83.50		738.50	845
2120	6" long		5.88	1.360		660	83.50		743.50	850
2140	8" long		5.88	1.360		705	83.50		788.50	905
2160	10" long		5.88	1.360		590	83.50		673.50	775
2180	12" long		5.88	1.360		610	83.50		693.50	800
2200	6" diameter, close		5.04	1.590		850	97.50		947.50	1,075
2220	5" long		5.04	1.590		910	97.50		1,007.50	1,150
2240	6" long		5.04	1.590		835	97.50		932.50	1,050
2260	8" long		5.04	1.590		990	97.50		1,087.50	1,225
2280	10" long		5.04	1.590		1,050	97.50		1,147.50	1,300
2300	12" long		5.04	1.590		1,075	97.50		1,172.50	1,350
2320	Rigid galvanized steel, 1/2" diameter, close		26.88	.298		3.03	18.25		21.28	30.50
2340	1-1/2" long		26.88	.298		3.24	18.25		21.49	30.50
2360	2" long		26.88	.298		3.34	18.25		21.59	30.50
2380	2-1/2" long		26.88	.298		3.49	18.25		21.74	31
2400	3" long		26.88	.298		3.64	18.25		21.89	31
2420	3-1/2" long		26.88	.298		3.78	18.25		22.03	31
2440	4" long		26.88	.298		3.93	18.25		22.18	31.50
2460	5" long		26.88	.298		4.14	18.25		22.39	31.50
2480	6" long		26.88	.298		4.58	18.25		22.83	32
2500	8" long		26.88	.298		6.45	18.25		24.70	34
2520	10" long		26.88	.298		7.50	18.25		25.75	35.50
2540	12" long		26.88	.298		7.90	18.25		26.15	35.50

26 05 33.15 Conduit Nipples		Crew	Daily Output	Labor-Hours	Unit	Material	2020 Bare Costs Labor	Equipment	Total	Total Incl O&P
2560	3/4" diameter, close	1 Elec	22.68	.353	Ea.	4.39	21.50		25.89	37
2580	2" long		22.68	.353		4.64	21.50		26.14	37
2600	2-1/2" long		22.68	.353		4.80	21.50		26.30	37.50
2620	3" long		22.68	.353		4.96	21.50		26.46	37.50
2640	3-1/2" long		22.68	.353		5.05	21.50		26.55	37.50
2660	4" long		22.68	.353		5.30	21.50		26.80	38
2680	5" long		22.68	.353		5.60	21.50		27.10	38
2700	6" long		22.68	.353		6.05	21.50		27.55	38.50
2720	8" long		22.68	.353		7.95	21.50		29.45	41
2740	10" long		22.68	.353		9.55	21.50		31.05	42.50
2760	12" long		22.68	.353		9.70	21.50		31.20	42.50
2780	1" diameter, close		19.32	.414		6.80	25.50		32.30	45.50
2800	2" long		19.32	.414		6.95	25.50		32.45	45.50
2820	2-1/2" long		19.32	.414		7.10	25.50		32.60	46
2840	3" long		19.32	.414		7.40	25.50		32.90	46
2860	3-1/2" long		19.32	.414		7.85	25.50		33.35	46.50
2880	4" long		19.32	.414		8	25.50		33.50	47
2900	5" long		19.32	.414		8.40	25.50		33.90	47.50
2920	6" long		19.32	.414		8.75	25.50		34.25	47.50
2940	8" long		19.32	.414		11.05	25.50		36.55	50
2960	10" long		19.32	.414		12.85	25.50		38.35	52
2980	12" long		19.32	.414		13.90	25.50		39.40	53.50
3000	1-1/4" diameter, close		16.80	.476		9.35	29		38.35	54
3020	2" long		16.80	.476		9.60	29		38.60	54
3040	3" long		16.80	.476		10.10	29		39.10	54.50
3060	3-1/2" long		16.80	.476		10.45	29		39.45	55
3080	4" long		16.80	.476		10.75	29		39.75	55.50
3100	5" long		16.80	.476		11.40	29		40.40	56
3120	6" long		16.80	.476		11.95	29		40.95	56.50
3140	8" long		16.80	.476		15.10	29		44.10	60
3160	10" long		16.80	.476		17.30	29		46.30	62.50
3180	12" long		16.80	.476		18.90	29		47.90	64.50
3200	1-1/2" diameter, close		15.12	.529		13.55	32.50		46.05	63.50
3220	2" long		15.12	.529		13.75	32.50		46.25	63.50
3240	2-1/2" long		15.12	.529		14.15	32.50		46.65	64
3260	3" long		15.12	.529		14.40	32.50		46.90	64.50
3280	3-1/2" long		15.12	.529		15	32.50		47.50	65
3300	4" long		15.12	.529		15.40	32.50		47.90	65.50
3320	5" long		15.12	.529		16.05	32.50		48.55	66
3340	6" long		15.12	.529		17.30	32.50		49.80	67.50
3360	8" long		15.12	.529		21	32.50		53.50	71.50
3380	10" long		15.12	.529		23	32.50		55.50	74
3400	12" long		15.12	.529		24	32.50		56.50	75
3420	2" diameter, close		13.44	.595		17.50	36.50		54	74
3440	2-1/2" long		13.44	.595		18.10	36.50		54.60	74.50
3460	3" long		13.44	.595		18.85	36.50		55.35	75
3480	3-1/2" long		13.44	.595		19.55	36.50		56.05	76
3500	4" long		13.44	.595		20	36.50		56.50	76.50
3520	5" long		13.44	.595		21	36.50		57.50	78
3540	6" long		13.44	.595		22	36.50		58.50	79
3560	8" long		13.44	.595		26	36.50		62.50	83
3580	10" long		13.44	.595		28.50	36.50		65	85.50
3600	12" long		13.44	.595		30.50	36.50		67	88

26 05 33.15 Conduit Nipples		Crew	Daily Output	Labor-Hours	Unit	Material	2020 Bare Costs Labor	Equipment	Total	Total Incl O&P
3620	2-1/2" diameter, close	1 Elec	10.92	.733	Ea.	60.50	45		105.50	134
3640	3" long		10.92	.733		64.50	45		109.50	138
3660	3-1/2" long		10.92	.733		66.50	45		111.50	140
3680	4" long		10.92	.733		67	45		112	141
3700	5" long		10.92	.733		69.50	45		114.50	144
3720	6" long		10.92	.733		71.50	45		116.50	146
3740	8" long		10.92	.733		77	45		122	152
3760	10" long		10.92	.733		80.50	45		125.50	156
3780	12" long		10.92	.733		85.50	45		130.50	161
3800	3" diameter, close		10.08	.794		60.50	48.50		109	139
3820	3" long		10.08	.794		66	48.50		114.50	145
3900	3-1/2" long		10.08	.794		62.50	48.50		111	142
3920	4" long		10.08	.794		64.50	48.50		113	144
3940	5" long		10.08	.794		66.50	48.50		115	146
3960	6" long		10.08	.794		69.50	48.50		118	149
3980	8" long		10.08	.794		75.50	48.50		124	156
4000	10" long		10.08	.794		80.50	48.50		129	161
4020	12" long		10.08	.794		88	48.50		136.50	170
4040	3-1/2" diameter, close		8.40	.952		105	58.50		163.50	202
4060	4" long		8.40	.952		108	58.50		166.50	206
4080	5" long		8.40	.952		111	58.50		169.50	209
4100	6" long		8.40	.952		114	58.50		172.50	213
4120	8" long		8.40	.952		120	58.50		178.50	219
4140	10" long		8.40	.952		126	58.50		184.50	226
4160	12" long		8.40	.952		133	58.50		191.50	234
4180	4" diameter, close		6.72	1.190		167	73		240	293
4200	4" long		6.72	1.190		169	73		242	295
4220	5" long		6.72	1.190		172	73		245	298
4240	6" long		6.72	1.190		174	73		247	300
4260	8" long		6.72	1.190		178	73		251	305
4280	10" long		6.72	1.190		183	73		256	310
4300	12" long		6.72	1.190		188	73		261	315
4320	5" diameter, close		5.04	1.590		240	97.50		337.50	410
4340	5" long		5.04	1.590		254	97.50		351.50	425
4360	6" long		5.04	1.590		259	97.50		356.50	430
4380	8" long		5.04	1.590		261	97.50		358.50	430
4400	10" long		5.04	1.590		280	97.50		377.50	455
4420	12" long		5.04	1.590		298	97.50		395.50	475
4440	6" diameter, close		4.20	1.900		420	117		537	635
4460	5" long		4.20	1.900		440	117		557	660
4480	6" long		4.20	1.900		450	117		567	670
4500	8" long		4.20	1.900		460	117		577	680
4520	10" long		4.20	1.900		480	117		597	700
4540	12" long		4.20	1.900		490	117		607	715
4560	Plastic coated, 40 mil thick, 1/2" diameter, 2" long		26.88	.298		28.50	18.25		46.75	58.50
4580	2-1/2" long		26.88	.298		25	18.25		43.25	54.50
4600	3" long		26.88	.298		25.50	18.25		43.75	55
4680	3-1/2" long		26.88	.298		30	18.25		48.25	60
4700	4" long		26.88	.298		27.50	18.25		45.75	57
4720	5" long		26.88	.298		30.50	18.25		48.75	60.50
4740	6" long		26.88	.298		26	18.25		44.25	55.50
4760	8" long		26.88	.298		36	18.25		54.25	66.50
4780	10" long		26.88	.298		39.50	18.25		57.75	70.50

26 05 33 – Raceway and Boxes for Electrical Systems

26 05 33.15 Conduit Nipples		Crew	Daily Output	Labor-Hours	Unit	Material	2020 Bare Costs Labor	Equipment	Total	Total Incl O&P
4800	12" long	1 Elec	26.88	.298	Ea.	44	18.25		62.25	75.50
4820	3/4" diameter, 2" long		21.84	.366		30.50	22.50		53	67
4840	2-1/2" long		21.84	.366		29	22.50		51.50	65.50
4860	3" long		21.84	.366		32.50	22.50		55	69.50
4880	3-1/2" long		21.84	.366		31.50	22.50		54	68
4900	4" long		21.84	.366		34	22.50		56.50	70.50
4920	5" long		21.84	.366		33	22.50		55.50	69.50
4940	6" long		21.84	.366		40.50	22.50		63	78.50
4960	8" long		21.84	.366		36	22.50		58.50	73
4980	10" long		21.84	.366		33	22.50		55.50	70
5000	12" long		21.84	.366		31.50	22.50		54	68
5020	1" diameter, 2" long		18.48	.433		33	26.50		59.50	76
5040	2-1/2" long		18.48	.433		32	26.50		58.50	75
5060	3" long		18.48	.433		36.50	26.50		63	80
5080	3-1/2" long		18.48	.433		37	26.50		63.50	80
5100	4" long		18.48	.433		31	26.50		57.50	73.50
5120	5" long		18.48	.433		29.50	26.50		56	72
5140	6" long		18.48	.433		34.50	26.50		61	77.50
5160	8" long		18.48	.433		40	26.50		66.50	83.50
5180	10" long		18.48	.433		42.50	26.50		69	86.50
5200	12" long		18.48	.433		39	26.50		65.50	82
5220	1-1/4" diameter, 2" long		15.12	.529		37.50	32.50		70	89.50
5240	2-1/2" long		15.12	.529		41	32.50		73.50	93.50
5260	3" long		15.12	.529		42	32.50		74.50	94.50
5280	3-1/2" long		15.12	.529		45	32.50		77.50	98
5300	4" long		15.12	.529		51	32.50		83.50	105
5320	5" long		15.12	.529		46	32.50		78.50	99
5340	6" long		15.12	.529		45	32.50		77.50	98.50
5360	8" long		15.12	.529		48	32.50		80.50	102
5380	10" long		15.12	.529		54	32.50		86.50	108
5400	12" long		15.12	.529		53.50	32.50		86	107
5420	1-1/2" diameter, 2" long		13.44	.595		46	36.50		82.50	105
5440	2-1/2" long		13.44	.595		46	36.50		82.50	106
5460	3" long		13.44	.595		44	36.50		80.50	103
5480	3-1/2" long		13.44	.595		48.50	36.50		85	108
5500	4" long		13.44	.595		41.50	36.50		78	100
5520	5" long		13.44	.595		41.50	36.50		78	100
5540	6" long		13.44	.595		49.50	36.50		86	109
5560	8" long		13.44	.595		58	36.50		94.50	119
5580	10" long		13.44	.595		77.50	36.50		114	140
5600	12" long		13.44	.595		70	36.50		106.50	132
5620	2" diameter, 2-1/2" long		11.76	.680		47.50	41.50		89	114
5640	3" long		11.76	.680		51.50	41.50		93	119
5660	3-1/2" long		11.76	.680		59	41.50		100.50	127
5680	4" long		11.76	.680		58.50	41.50		100	127
5700	5" long		11.76	.680		63.50	41.50		105	132
5720	6" long		11.76	.680		61	41.50		102.50	129
5740	8" long		11.76	.680		73	41.50		114.50	143
5760	10" long		11.76	.680		95.50	41.50		137	167
5780	12" long		11.76	.680		93.50	41.50		135	165
5800	2-1/2" diameter, 3-1/2" long		10.08	.794		124	48.50		172.50	210
5820	4" long		10.08	.794		127	48.50		175.50	213
5840	5" long		10.08	.794		152	48.50		200.50	240

For customer support on your Electrical Change Order Costs with RSMeans data, call 800.448.8182.

Post-Installation Change Order Costs

26 05 33 – Raceway and Boxes for Electrical Systems

26 05 33.15 Conduit Nipples		Crew	Daily Output	Labor-Hours	Unit	Material	2020 Bare Costs Labor	Equipment	Total	Total Incl O&P
5860	6" long	1 Elec	10.08	.794	Ea.	126	48.50		174.50	212
5880	8" long		10.08	.794		167	48.50		215.50	256
5900	10" long		10.08	.794		182	48.50		230.50	273
5920	12" long		10.08	.794		184	48.50		232.50	275
5940	3" diameter, 3-1/2" long		9.24	.866		141	53		194	234
5960	4" long		9.24	.866		126	53		179	218
5980	5" long		9.24	.866		147	53		200	240
6000	6" long		9.24	.866		161	53		214	256
6020	8" long		9.24	.866		198	53		251	297
6040	10" long		9.24	.866		222	53		275	325
6060	12" long		9.24	.866		218	53		271	320
6080	3-1/2" diameter, 4" long		7.56	1.060		199	65		264	315
6100	5" long		7.56	1.060		201	65		266	320
6120	6" long		7.56	1.060		186	65		251	300
6140	8" long		7.56	1.060		244	65		309	365
6160	10" long		7.56	1.060		284	65		349	410
6180	12" long		7.56	1.060		325	65		390	455
6200	4" diameter, 4" long		6.30	1.270		260	78		338	400
6220	5" long		6.30	1.270		278	78		356	420
6240	6" long		6.30	1.270		300	78		378	445
6260	8" long		6.30	1.270		320	78		398	470
6280	10" long		6.30	1.270		390	78		468	545
6300	12" long		6.30	1.270		435	78		513	595
6320	5" diameter, 5" long		4.62	1.730		265	106		371	450
6340	6" long		4.62	1.730		325	106		431	520
6360	8" long		4.62	1.730		340	106		446	530
6380	10" long		4.62	1.730		355	106		461	550
6400	12" long		4.62	1.730		390	106		496	590
6420	6" diameter, 5" long		3.78	2.120		515	130		645	765
6440	6" long		3.78	2.120		530	130		660	780
6460	8" long		3.78	2.120		555	130		685	805
6480	10" long		3.78	2.120		580	130		710	830
6500	12" long		3.78	2.120		610	130		740	865

26 05 33.16 Boxes for Electrical Systems

		Crew	Daily Output	Labor-Hours	Unit	Material	2020 Bare Costs Labor	Equipment	Total	Total Incl O&P
0010	**BOXES FOR ELECTRICAL SYSTEMS**									
0020	Pressed steel, octagon, 4"	1 Elec	16.80	.476	Ea.	3.70	29		32.70	47.50
0040	For Romex or BX		16.80	.476		7.15	29		36.15	51.50
0050	For Romex or BX, with bracket		16.80	.476		10.65	29		39.65	55.50
0060	Covers, blank		53.76	.149		1.09	9.15		10.24	14.80
0100	Extension rings		33.60	.238		5.75	14.60		20.35	28.50
0150	Square, 4"		16.80	.476		8.60	29		37.60	53
0160	For Romex or BX		16.80	.476		15.05	29		44.05	60
0170	For Romex or BX, with bracket		16.80	.476		11.40	29		40.40	56
0200	Extension rings		33.60	.238		6.05	14.60		20.65	28.50
0220	2-1/8" deep, 1" KO		16.80	.476		4.21	29		33.21	48
0250	Covers, blank		53.76	.149		.96	9.15		10.11	14.65
0260	Raised device		53.76	.149		1.96	9.15		11.11	15.75
0300	Plaster rings		53.76	.149		2.08	9.15		11.23	15.90
0350	Square, 4-11/16"		16.80	.476		5.50	29		34.50	49.50
0370	2-1/8" deep, 3/4" to 1-1/4" KO		16.80	.476		7.35	29		36.35	51.50
0400	Extension rings		33.60	.238		13.25	14.60		27.85	36.50
0450	Covers, blank		44.52	.180		1.41	11		12.41	17.95

26 05 33.16 Boxes for Electrical Systems	Crew	Daily Output	Labor-Hours	Unit	Material	2020 Bare Costs Labor	Equipment	Total	Total Incl O&P
0460	Raised device	1 Elec	44.52	.180	Ea.	9.05	11		20.05	26.50
0500	Plaster rings		44.52	.180		7.60	11		18.60	25
0550	Handy box		22.68	.353		3.15	21.50		24.65	35.50
0560	Covers, device		53.76	.149		1.26	9.15		10.41	15
0600	Extension rings		45.36	.176		4.50	10.80		15.30	21
0650	Switchbox		22.68	.353		7	21.50		28.50	39.50
0660	Romex or BX		22.68	.353		9.80	21.50		31.30	43
0670	with bracket		22.68	.353		9	21.50		30.50	42
0680	Partition, metal		22.68	.353		4.20	21.50		25.70	36.50
0700	Masonry, 1 gang, 2-1/2" deep		22.68	.353		10.45	21.50		31.95	43.50
0710	3-1/2" deep		22.68	.353		10.65	21.50		32.15	43.50
0750	2 gang, 2-1/2" deep		16.80	.476		22	29		51	68
0760	3-1/2" deep		16.80	.476		14.70	29		43.70	59.50
0800	3 gang, 2-1/2" deep		10.92	.733		23	45		68	92.50
0850	4 gang, 2-1/2" deep		8.40	.952		31.50	58.50		90	122
0860	5 gang, 2-1/2" deep		7.56	1.060		46	65		111	148
0870	6 gang, 2-1/2" deep		6.72	1.190		76.50	73		149.50	194
0880	Masonry thru-the-wall, 1 gang, 4" block		13.44	.595		44.50	36.50		81	104
0890	6" block		13.44	.595		58	36.50		94.50	119
0900	8" block		13.44	.595		77.50	36.50		114	140
0920	2 gang, 6" block		13.44	.595		87	36.50		123.50	150
0940	Bar hanger with 3/8" stud, for wood and masonry boxes		44.52	.180		6.65	11		17.65	23.50
0950	Concrete, set flush, 4" deep		16.80	.476		10.25	29		39.25	55
1000	Plate with 3/8" stud		67.20	.119		10.40	7.30		17.70	22.50
1100	Concrete, floor, 1 gang		4.45	1.800		140	110		250	315
1150	2 gang		3.36	2.380		209	146		355	445
1200	3 gang		2.27	3.530		335	217		552	690
1250	For duplex receptacle, pedestal mounted, add		20.16	.397		146	24.50		170.50	198
1270	Flush mounted, add		22.68	.353		43.50	21.50		65	79.50
1300	For telephone, pedestal mounted, add		25.20	.317		141	19.50		160.50	184
1350	Carpet flange, 1 gang		44.52	.180		72	11		83	95.50
1400	Cast, 1 gang, FS (2" deep), 1/2" hub		10.08	.794		31.50	48.50		80	108
1410	3/4" hub		10.08	.794		36.50	48.50		85	113
1420	FD (2-11/16" deep), 1/2" hub		10.08	.794		23.50	48.50		72	98
1430	3/4" hub		10.08	.794		27.50	48.50		76	103
1450	2 gang, FS, 1/2" hub		8.40	.952		62.50	58.50		121	156
1460	3/4" hub		8.40	.952		56.50	58.50		115	149
1470	FD, 1/2" hub		8.40	.952		76	58.50		134.50	171
1480	3/4" hub		8.40	.952		69	58.50		127.50	163
1500	3 gang, FS, 3/4" hub		7.56	1.060		101	65		166	208
1510	Switch cover, 1 gang, FS		53.76	.149		8.05	9.15		17.20	22.50
1520	2 gang		44.52	.180		12.60	11		23.60	30.50
1530	Duplex receptacle cover, 1 gang, FS		53.76	.149		8.70	9.15		17.85	23
1540	2 gang, FS		44.52	.180		13.25	11		24.25	31
1542	Weatherproof blank cover, 1 gang		53.76	.149		1.41	9.15		10.56	15.15
1544	2 gang		44.52	.180		2.85	11		13.85	19.55
1550	Weatherproof switch cover, 1 gang		53.76	.149		7.35	9.15		16.50	21.50
1554	2 gang		44.52	.180		12.20	11		23.20	30
1600	Weatherproof receptacle cover, 1 gang		53.76	.149		8.95	9.15		18.10	23.50
1604	2 gang		44.52	.180		11.20	11		22.20	29
1620	Weatherproof receptacle cover, tamper resistant, 1 gang		48.72	.164		17	10.05		27.05	33.50
1624	2 gang		40.32	.198		33	12.15		45.15	54.50
1750	FSC, 1 gang, 1/2" hub		9.24	.866		34.50	53		87.50	117

Post-Installation Change Order Costs

26 05 33 – Raceway and Boxes for Electrical Systems

26 05 33.16 Boxes for Electrical Systems	Crew	Daily Output	Labor-Hours	Unit	Material	2020 Bare Costs Labor	Equipment	Total	Total Incl O&P
1760	3/4" hub	1 Elec	9.24	.866	Ea.	42	53		95	126
1770	2 gang, 1/2" hub		7.56	1.060		63.50	65		128.50	167
1780	3/4" hub		7.56	1.060		69	65		134	173
1790	FDC, 1 gang, 1/2" hub		9.24	.866		37	53		90	120
1800	3/4" hub		9.24	.866		47.50	53		100.50	131
1810	2 gang, 1/2" hub		7.56	1.060		83	65		148	189
1820	3/4" hub		7.56	1.060		82.50	65		147.50	188
1850	Weatherproof in-use cover, 1 gang		53.76	.149		30.50	9.15		39.65	47
1870	2 gang		44.52	.180		38.50	11		49.50	59
2000	Poke-thru fitting, fire rated, for 3-3/4" floor		5.71	1.400		212	86		298	360
2040	For 7" floor		5.71	1.400		209	86		295	360
2100	Pedestal, 15 amp, duplex receptacle & blank plate		4.41	1.810		190	111		301	375
2120	Duplex receptacle and telephone plate		4.41	1.810		190	111		301	375
2140	Pedestal, 20 amp, duplex recept. & phone plate		4.20	1.900		191	117		308	385
2160	Telephone plate, both sides		4.41	1.810		179	111		290	360
2200	Abandonment plate		26.88	.298		55.50	18.25		73.75	88

26 05 33.17 Outlet Boxes, Plastic

		Crew	Daily Output	Labor-Hours	Unit	Material	2020 Bare Costs Labor	Equipment	Total	Total Incl O&P
0010	**OUTLET BOXES, PLASTIC**									
0050	4" diameter, round with 2 mounting nails	1 Elec	21	.381	Ea.	3.61	23.50		27.11	39
0100	Bar hanger mounted		21	.381		6.30	23.50		29.80	42
0200	4", square with 2 mounting nails		21	.381		6.95	23.50		30.45	42.50
0300	Plaster ring		53.76	.149		2.63	9.15		11.78	16.50
0400	Switch box with 2 mounting nails, 1 gang		25.20	.317		4.01	19.50		23.51	33.50
0500	2 gang		21	.381		4.94	23.50		28.44	40.50
0600	3 gang		16.80	.476		6.90	29		35.90	51
0700	Old work box		25.20	.317		5.75	19.50		25.25	35.50
1400	PVC, FSS, 1 gang, 1/2" hub		11.76	.680		16.35	41.50		57.85	80
1410	3/4" hub		11.76	.680		14.25	41.50		55.75	77.50
1420	FD, 1 gang for variable terminations		11.76	.680		9.90	41.50		51.40	73
1450	FS, 2 gang for variable terminations		10.08	.794		16.85	48.50		65.35	91
1480	Weatherproof blank cover, FS, 1 gang		53.76	.149		4.81	9.15		13.96	18.90
1500	2 gang		44.52	.180		6.40	11		17.40	23.50
1510	Weatherproof switch cover, FS, 1 gang		53.76	.149		10.25	9.15		19.40	25
1520	2 gang		44.52	.180		21	11		32	39.50
1530	Weatherproof duplex receptacle cover, FS, 1 gang		53.76	.149		16.10	9.15		25.25	31.50
1540	2 gang		44.52	.180		17	11		28	35
1750	FSC, 1 gang, 1/2" hub		10.92	.733		10.10	45		55.10	78
1760	3/4" hub		10.92	.733		10.60	45		55.60	78.50
1770	FSC, 2 gang, 1/2" hub		9.24	.866		18.80	53		71.80	99.50
1780	3/4" hub		9.24	.866		14.55	53		67.55	95
1790	FDC, 1 gang, 1/2" hub		10.92	.733		13.10	45		58.10	81.50
1800	3/4" hub		10.92	.733		13.30	45		58.30	81.50
1810	Weatherproof, T box w/3 holes		11.76	.680		13.70	41.50		55.20	77
1820	4" diameter round w/5 holes		11.76	.680		12.85	41.50		54.35	76
1850	In-use cover, 1 gang		53.76	.149		9.05	9.15		18.20	23.50
1870	2 gang		44.52	.180		12.90	11		23.90	30.50

26 05 33.18 Pull Boxes

		Crew	Daily Output	Labor-Hours	Unit	Material	2020 Bare Costs Labor	Equipment	Total	Total Incl O&P
0010	**PULL BOXES**									
0100	Steel, pull box, NEMA 1, type SC, 6" W x 6" H x 4" D	1 Elec	6.72	1.190	Ea.	12.30	73		85.30	123
0180	8" W x 6" H x 4" D		6.72	1.190		29	73		102	141
0200	8" W x 8" H x 4" D		6.72	1.190		16.15	73		89.15	127
0210	10" W x 10" H x 4" D		5.88	1.360		31.50	83.50		115	159

26 05 33.18 Pull Boxes		Crew	Daily Output	Labor-Hours	Unit	Material	2020 Bare Costs Labor	Equipment	Total	Total Incl O&P
0220	12" W x 12" H x 4" D	1 Elec	5.46	1.470	Ea.	29.50	90		119.50	166
0230	15" W x 15" H x 4" D		4.37	1.830		54.50	112		166.50	227
0240	18" W x 18" H x 4" D		3.70	2.160		49.50	133		182.50	251
0250	6" W x 6" H x 6" D		6.72	1.190		18.35	73		91.35	129
0260	8" W x 8" H x 6" D		6.30	1.270		19	78		97	137
0270	10" W x 10" H x 6" D		4.62	1.730		25.50	106		131.50	186
0300	10" W x 12" H x 6" D		4.45	1.800		38	110		148	206
0310	12" W x 12" H x 6" D		4.37	1.830		31	112		143	201
0320	15" W x 15" H x 6" D		3.86	2.070		49	127		176	243
0330	18" W x 18" H x 6" D		3.53	2.270		58.50	139		197.50	272
0340	24" W x 24" H x 6" D		2.69	2.980		117	183		300	400
0350	12" W x 12" H x 8" D		4.20	1.900		38	117		155	216
0360	15" W x 15" H x 8" D		3.78	2.120		57	130		187	257
0370	18" W x 18" H x 8" D		3.36	2.380		76	146		222	300
0380	24" W x 18" H x 6" D		3.11	2.570		145	158		303	395
0400	16" W x 20" H x 8" D		3.36	2.380		103	146		249	330
0500	20" W x 24" H x 8" D		2.69	2.980		138	183		321	425
0510	24" W x 24" H x 8" D		2.52	3.170		131	194		325	435
0600	24" W x 36" H x 8" D		2.27	3.530		191	217		408	530
0610	30" W x 30" H x 8" D		2.27	3.530		252	217		469	595
0620	36" W x 36" H x 8" D		1.68	4.760		270	292		562	730
0630	24" W x 24" H x 10" D		2.10	3.810		232	234		466	605
0650	Pull box, hinged, NEMA 1, 6" W x 6" H x 4" D		6.72	1.190		21	73		94	133
0660	8" W x 8" H x 4" D		6.72	1.190		29.50	73		102.50	142
0670	10" W x 10" H x 4" D		5.88	1.360		47	83.50		130.50	176
0680	12" W x 12" H x 4" D		5.04	1.590		49.50	97.50		147	200
0690	15" W x 15" H x 4" D		4.37	1.830		55	112		167	228
0700	18" W x 18" H x 4" D		3.70	2.160		63	133		196	266
0710	6" W x 6" H x 6" D		6.72	1.190		24.50	73		97.50	136
0720	8" W x 8" H x 6" D		6.30	1.270		33.50	78		111.50	153
0730	10" W x 10" H x 6" D		4.62	1.730		44	106		150	207
0740	12" W x 12" H x 6" D		4.37	1.830		56	112		168	229
0800	12" W x 16" H x 6" D		3.95	2.030		61.50	125		186.50	253
0820	18" W x 18" H x 6" D		3.53	2.270		97	139		236	315
1000	20" W x 20" H x 6" D		3.02	2.650		134	163		297	390
1010	24" W x 24" H x 6" D		2.69	2.980		204	183		387	495
1020	12" W x 12" H x 8" D		4.20	1.900		101	117		218	285
1030	15" W x 15" H x 8" D		3.78	2.120		115	130		245	320
1040	18" W x 18" H x 8" D		3.36	2.380		162	146		308	395
1200	20" W x 20" H x 8" D		2.69	2.980		203	183		386	495
1210	24" W x 24" H x 8" D		2.52	3.170		229	194		423	540
1220	30" W x 30" H x 8" D		2.27	3.530		330	217		547	685
1400	24" W x 36" H x 8" D		2.27	3.530		320	217		537	675
1600	24" W x 42" H x 8" D		1.68	4.760		475	292		767	955
1610	36" W x 36" H x 8" D		1.68	4.760		420	292		712	895
2100	Pull box, NEMA 3R, type SC, raintight & weatherproof									
2150	6" L x 6" W x 6" D	1 Elec	8.40	.952	Ea.	21.50	58.50		80	111
2200	8" L x 6" W x 6" D		6.72	1.190		31	73		104	143
2250	10" L x 6" W x 6" D		5.88	1.360		40	83.50		123.50	168
2300	12" L x 12" W x 6" D		4.20	1.900		84	117		201	267
2350	16" L x 16" W x 6" D		3.78	2.120		115	130		245	320
2400	20" L x 20" W x 6" D		3.36	2.380		108	146		254	335
2450	24" L x 18" W x 8" D		2.52	3.170		191	194		385	500

For customer support on your Electrical Change Order Costs with RSMeans data, call 800.448.8182.

Post-Installation Change Order Costs

26 05 33.18 Pull Boxes		Crew	Daily Output	Labor-Hours	Unit	Material	2020 Bare Costs Labor	Equipment	Total	Total Incl O&P
2500	24" L x 24" W x 10" D	1 Elec	2.10	3.810	Ea.	360	234		594	745
2550	30" L x 24" W x 12" D		1.68	4.760		495	292		787	980
2600	36" L x 36" W x 12" D		1.26	6.350		865	390		1,255	1,525
2800	Cast iron, pull boxes for surface mounting									
3000	NEMA 4, watertight & dust tight									
3050	6" L x 6" W x 6" D	1 Elec	3.36	2.380	Ea.	345	146		491	595
3100	8" L x 6" W x 6" D		2.69	2.980		520	183		703	845
3150	10" L x 6" W x 6" D		2.10	3.810		555	234		789	960
3200	12" L x 12" W x 6" D		1.93	4.140		930	254		1,184	1,400
3250	16" L x 16" W x 6" D		1.09	7.330		1,050	450		1,500	1,825
3300	20" L x 20" W x 6" D		.67	11.900		1,200	730		1,930	2,400
3350	24" L x 18" W x 8" D		.59	13.610		3,250	835		4,085	4,825
3400	24" L x 24" W x 10" D		.42	19.050		5,900	1,175		7,075	8,250
3450	30" L x 24" W x 12" D		.34	23.810		8,425	1,450		9,875	11,400
3500	36" L x 36" W x 12" D		.17	47.620		10,300	2,925		13,225	15,800
3510	NEMA 4 clamp cover, 6" L x 6" W x 4" D		3.36	2.380		340	146		486	590
3520	8" L x 6" W x 4" D		3.36	2.380		415	146		561	670
4000	NEMA 7, explosion proof									
4050	6" L x 6" W x 6" D	1 Elec	1.68	4.760	Ea.	935	292		1,227	1,450
4100	8" L x 6" W x 6" D		1.51	5.290		1,325	325		1,650	1,925
4150	10" L x 6" W x 6" D		1.34	5.950		1,650	365		2,015	2,350
4200	12" L x 12" W x 6" D		.84	9.520		3,000	585		3,585	4,175
4250	16" L x 14" W x 6" D		.50	15.870		4,375	975		5,350	6,275
4300	18" L x 18" W x 8" D		.42	19.050		8,100	1,175		9,275	10,700
4350	24" L x 18" W x 8" D		.34	23.810		9,575	1,450		11,025	12,700
4400	24" L x 24" W x 10" D		.25	31.750		14,000	1,950		15,950	18,300
4450	30" L x 24" W x 12" D		.17	47.620		17,900	2,925		20,825	24,100
5000	NEMA 9, dust tight 6" L x 6" W x 6" D		2.69	2.980		515	183		698	840
5050	8" L x 6" W x 6" D		2.27	3.530		615	217		832	995
5100	10" L x 6" W x 6" D		1.68	4.760		815	292		1,107	1,325
5150	12" L x 12" W x 6" D		1.34	5.950		1,575	365		1,940	2,275
5200	16" L x 16" W x 6" D		.84	9.520		2,600	585		3,185	3,725
5250	18" L x 18" W x 8" D		.59	13.610		4,050	835		4,885	5,700
5300	24" L x 18" W x 8" D		.50	15.870		5,750	975		6,725	7,775
5350	24" L x 24" W x 10" D		.34	23.810		7,725	1,450		9,175	10,700
5400	30" L x 24" W x 12" D		.25	31.750		11,600	1,950		13,550	15,700
6000	J.I.C. wiring boxes, NEMA 12, dust tight & drip tight									
6050	6" L x 8" W x 4" D	1 Elec	8.40	.952	Ea.	97	58.50		155.50	193
6100	8" L x 10" W x 4" D		6.72	1.190		161	73		234	286
6150	12" L x 14" W x 6" D		4.45	1.800		122	110		232	298
6200	14" L x 16" W x 6" D		3.95	2.030		158	125		283	360
6250	16" L x 20" W x 6" D		3.70	2.160		410	133		543	650
6300	24" L x 30" W x 6" D		2.69	2.980		525	183		708	845
6350	24" L x 30" W x 8" D		2.44	3.280		610	201		811	970
6400	24" L x 36" W x 8" D		2.27	3.530		500	217		717	870
6450	24" L x 42" W x 8" D		1.93	4.140		675	254		929	1,125
6500	24" L x 48" W x 8" D		1.68	4.760		740	292		1,032	1,250

26 05 33.23 Wireway

		Crew	Daily Output	Labor-Hours	Unit	Material	2020 Bare Costs Labor	Equipment	Total	Total Incl O&P
0010	**WIREWAY** to 10' high									
0020	For higher elevations, see Section 26 05 36.40									
0100	NEMA 1, screw cover w/fittings and supports, 2-1/2" x 2-1/2"	1 Elec	37.80	.212	L.F.	13.70	13		26.70	34.50
0200	4" x 4"	"	33.60	.238		14.75	14.60		29.35	38.50

26 05 33.23 Wireway		Crew	Daily Output	Labor-Hours	Unit	Material	2020 Bare Costs Labor	Equipment	Total	Total Incl O&P
0400	6" x 6"	2 Elec	50.40	.317	L.F.	26	19.50		45.50	57.50
0600	8" x 8"		33.60	.476		31	29		60	77.50
0620	10" x 10"		25.20	.635		61.50	39		100.50	126
0640	12" x 12"		16.80	.952		70.50	58.50		129	165
0800	Elbows, 90°, 2-1/2"	1 Elec	20.16	.397	Ea.	35.50	24.50		60	75.50
1000	4"		16.80	.476		41.50	29		70.50	89.50
1200	6"		15.12	.529		46	32.50		78.50	99
1400	8"		13.44	.595		73.50	36.50		110	135
1420	10"		10.08	.794		96.50	48.50		145	179
1440	12"		8.40	.952		136	58.50		194.50	237
1500	Elbows, 45°, 2-1/2"		20.16	.397		37	24.50		61.50	77
1510	4"		16.80	.476		43.50	29		72.50	91.50
1520	6"		15.12	.529		50	32.50		82.50	104
1530	8"		13.44	.595		73.50	36.50		110	135
1540	10"		10.08	.794		97.50	48.50		146	180
1550	12"		8.40	.952		185	58.50		243.50	291
1600	"T" box, 2-1/2"		15.12	.529		41	32.50		73.50	93.50
1800	4"		13.44	.595		54	36.50		90.50	114
2000	6"		11.76	.680		60	41.50		101.50	128
2200	8"		10.08	.794		102	48.50		150.50	186
2220	10"		8.40	.952		126	58.50		184.50	225
2240	12"		6.72	1.190		179	73		252	305
2300	Cross, 2-1/2"		13.44	.595		45.50	36.50		82	105
2310	4"		11.76	.680		55.50	41.50		97	123
2320	6"		10.08	.794		68.50	48.50		117	148
2400	Panel adapter, 2-1/2"		20.16	.397		7.60	24.50		32.10	45
2600	4"		16.80	.476		9.70	29		38.70	54
2800	6"		15.12	.529		13.90	32.50		46.40	64
3000	8"		13.44	.595		20	36.50		56.50	76.50
3020	10"		11.76	.680		31	41.50		72.50	96
3040	12"		10.08	.794		45.50	48.50		94	123
3200	Reducer, 4" to 2-1/2"		20.16	.397		13.75	24.50		38.25	51.50
3400	6" to 4"		16.80	.476		30.50	29		59.50	77.50
3600	8" to 6"		15.12	.529		37.50	32.50		70	89.50
3620	10" to 8"		13.44	.595		48	36.50		84.50	108
3640	12" to 10"		11.76	.680		62.50	41.50		104	131
3780	End cap, 2-1/2"		20.16	.397		5.75	24.50		30.25	43
3800	4"		16.80	.476		6.65	29		35.65	51
4000	6"		15.12	.529		7.85	32.50		40.35	57
4200	8"		13.44	.595		10.05	36.50		46.55	65.50
4220	10"		11.76	.680		17.70	41.50		59.20	81.50
4240	12"		10.08	.794		23.50	48.50		72	98
4300	U-connector, 2-1/2"		168	.048		4.91	2.92		7.83	9.75
4320	4"		168	.048		5.90	2.92		8.82	10.85
4340	6"		151.20	.053		6.90	3.25		10.15	12.45
4360	8"		142.80	.056		12.05	3.44		15.49	18.35
4380	10"		126	.063		17.20	3.90		21.10	24.50
4400	12"		109.20	.073		22	4.49		26.49	31
4420	Hanger, 2-1/2"		84	.095		13.50	5.85		19.35	23.50
4430	4"		84	.095		13.50	5.85		19.35	23.50
4440	6"		67.20	.119		16.05	7.30		23.35	28.50
4450	8"		54.60	.147		27	9		36	43
4460	10"		42	.190		49	11.70		60.70	71.50

26 05 33.23 Wireway		Crew	Daily Output	Labor-Hours	Unit	Material	2020 Bare Costs Labor	Equipment	Total	Total Incl O&P
4470	12"	1 Elec	33.60	.238	Ea.	71.50	14.60		86.10	101
4475	NEMA 3R, screw cover w/fittings and supports, 4" x 4"	▼	30.24	.265	L.F.	18.35	16.25		34.60	44
4480	6" x 6"	2 Elec	46.20	.346		20.50	21.50		42	54
4485	8" x 8"		30.24	.529		36	32.50		68.50	88
4490	12" x 12"	▼	15.12	1.060		66.50	65		131.50	170
4500	Hinged cover, with fittings and supports, 2-1/2" x 2-1/2"	1 Elec	50.40	.159		22.50	9.75		32.25	39.50
4520	4" x 4"	"	37.80	.212		15.20	13		28.20	36
4540	6" x 6"	2 Elec	67.20	.238		23	14.60		37.60	47.50
4560	8" x 8"		50.40	.317		40.50	19.50		60	74
4580	10" x 10"		42	.381		52.50	23.50		76	93
4600	12" x 12"	▼	20.16	.794	▼	93.50	48.50		142	176
4700	Elbows 90°, 2-1/2" x 2-1/2"	1 Elec	26.88	.298	Ea.	51.50	18.25		69.75	83.50
4720	4"		22.68	.353		84	21.50		105.50	125
4730	6"		19.32	.414		65	25.50		90.50	109
4740	8"		15.12	.529		78.50	32.50		111	135
4750	10"		11.76	.680		122	41.50		163.50	197
4760	12"		10.08	.794		198	48.50		246.50	291
4800	Tee box, hinged cover, 2-1/2" x 2-1/2"		19.32	.414		67.50	25.50		93	113
4810	4"		16.80	.476		97	29		126	151
4820	6"		15.12	.529		75	32.50		107.50	131
4830	8"		13.44	.595		178	36.50		214.50	251
4840	10"		10.08	.794		181	48.50		229.50	272
4860	12"		8.40	.952		235	58.50		293.50	345
4880	Cross box, hinged cover, 2-1/2" x 2-1/2"		15.12	.529		68	32.50		100.50	124
4900	4"		13.44	.595		96.50	36.50		133	161
4920	6"		10.92	.733		118	45		163	197
4940	8"		9.24	.866		169	53		222	265
4960	10"		8.40	.952		267	58.50		325.50	380
4980	12"		7.56	1.060	▼	259	65		324	380
5000	NEMA 12, hinged cover, 2-1/2" x 2-1/2"		33.60	.238	L.F.	27	14.60		41.60	51.50
5020	4" x 4"	▼	29.40	.272		30.50	16.70		47.20	58.50
5040	6" x 6"	2 Elec	50.40	.317		45.50	19.50		65	79
5060	8" x 8"	"	42	.381	▼	61	23.50		84.50	102
5120	Elbows 90°, flanged, 2-1/2" x 2-1/2"	1 Elec	19.32	.414	Ea.	110	25.50		135.50	159
5140	4"		16.80	.476		138	29		167	196
5160	6"		15.12	.529		165	32.50		197.50	230
5180	8"		12.60	.635		153	39		192	226
5240	Tee box, flanged, 2-1/2" x 2-1/2"		15.12	.529		143	32.50		175.50	206
5260	4"		13.44	.595		187	36.50		223.50	261
5280	6"		12.60	.635		151	39		190	224
5300	8"		10.92	.733		212	45		257	300
5360	Cross box, flanged, 2-1/2" x 2-1/2"		12.60	.635		188	39		227	265
5380	4"		10.92	.733		155	45		200	237
5400	6"		10.08	.794		200	48.50		248.50	293
5420	8"		8.40	.952		250	58.50		308.50	360
5480	Flange gasket, 2-1/2"		134.40	.060		2.83	3.65		6.48	8.55
5500	4"		67.20	.119		3.87	7.30		11.17	15.15
5520	6"		44.52	.180		6.25	11		17.25	23.50
5530	8"	▼	33.60	.238	▼	6.85	14.60		21.45	29.50

26 05 33.25 Conduit Fittings for Rigid Galvanized Steel	Crew	Daily Output	Labor-Hours	Unit	Material	2020 Bare Costs Labor	Equipment	Total	Total Incl O&P
0010 **CONDUIT FITTINGS FOR RIGID GALVANIZED STEEL**									
0050 Standard, locknuts, 1/2" diameter				Ea.	.24			.24	.26
0100 3/4" diameter					.48			.48	.53
0300 1" diameter					.74			.74	.82
0500 1-1/4" diameter					.95			.95	1.04
0700 1-1/2" diameter					1.58			1.58	1.74
1000 2" diameter					2.22			2.22	2.44
1030 2-1/2" diameter					5.45			5.45	6
1050 3" diameter					7.75			7.75	8.55
1070 3-1/2" diameter					13.40			13.40	14.70
1100 4" diameter					18.15			18.15	19.95
1110 5" diameter					37			37	40.50
1120 6" diameter					64.50			64.50	71
1130 Bushings, plastic, 1/2" diameter	1 Elec	33.60	.238		.11	14.60		14.71	22
1150 3/4" diameter		26.88	.298		.12	18.25		18.37	27
1170 1" diameter		23.52	.340		.24	21		21.24	31.50
1200 1-1/4" diameter		20.16	.397		.30	24.50		24.80	37
1230 1-1/2" diameter		15.12	.529		.42	32.50		32.92	49
1250 2" diameter		12.60	.635		.74	39		39.74	59
1270 2-1/2" diameter		10.92	.733		1.38	45		46.38	68.50
1300 3" diameter		10.08	.794		1.61	48.50		50.11	74.50
1330 3-1/2" diameter		9.24	.866		1.92	53		54.92	81
1350 4" diameter		7.56	1.060		2.12	65		67.12	99.50
1360 5" diameter		5.88	1.360		14.85	83.50		98.35	140
1370 6" diameter		4.20	1.900		54.50	117		171.50	234
1390 Steel, 1/2" diameter		33.60	.238		.73	14.60		15.33	23
1400 3/4" diameter		26.88	.298		1	18.25		19.25	28
1430 1" diameter		23.52	.340		1.56	21		22.56	32.50
1450 Steel insulated, 1-1/4" diameter		20.16	.397		5.85	24.50		30.35	43
1470 1-1/2" diameter		15.12	.529		7.15	32.50		39.65	56.50
1500 2" diameter		12.60	.635		8.50	39		47.50	67.50
1530 2-1/2" diameter		10.92	.733		20	45		65	89
1550 3" diameter		10.08	.794		23.50	48.50		72	98
1570 3-1/2" diameter		9.24	.866		30.50	53		83.50	113
1600 4" diameter		7.56	1.060		38	65		103	139
1610 5" diameter		5.88	1.360		79.50	83.50		163	211
1620 6" diameter		4.20	1.900		177	117		294	370
1630 Sealing locknuts, 1/2" diameter		33.60	.238		2.08	14.60		16.68	24.50
1650 3/4" diameter		26.88	.298		1.90	18.25		20.15	29
1670 1" diameter		23.52	.340		5.50	21		26.50	37
1700 1-1/4" diameter		20.16	.397		6	24.50		30.50	43
1730 1-1/2" diameter		15.12	.529		6.85	32.50		39.35	56
1750 2" diameter		12.60	.635		8.70	39		47.70	67.50
1760 Grounding bushing, insulated, 1/2" diameter		26.88	.298		5.90	18.25		24.15	33.50
1770 3/4" diameter		23.52	.340		8.15	21		29.15	40
1780 1" diameter		16.80	.476		12.05	29		41.05	57
1800 1-1/4" diameter		15.12	.529		14.10	32.50		46.60	64
1830 1-1/2" diameter		13.44	.595		16.40	36.50		52.90	72.50
1850 2" diameter		10.92	.733		15	45		60	83.50
1870 2-1/2" diameter		10.08	.794		35.50	48.50		84	112
1900 3" diameter		9.24	.866		37	53		90	120
1930 3-1/2" diameter		7.56	1.060		30	65		95	130

Post-Installation Change Order Costs

26 05 33.25 Conduit Fittings for Rigid Galvanized Steel	Crew	Daily Output	Labor-Hours	Unit	Material	2020 Bare Costs Labor	Equipment	Total	Total Incl O&P
1950	4" diameter	1 Elec	6.72	1.190	Ea.	70	73		143	186
1960	5" diameter		5.04	1.590		91.50	97.50		189	246
1970	6" diameter		3.36	2.380		132	146		278	360
1990	Coupling, with set screw, 1/2" diameter		42	.190		5.15	11.70		16.85	23
2000	3/4" diameter		33.60	.238		6.45	14.60		21.05	29
2030	1" diameter		29.40	.272		10.35	16.70		27.05	36.50
2050	1-1/4" diameter		23.52	.340		14	21		35	46.50
2070	1-1/2" diameter		19.32	.414		16.20	25.50		41.70	56
2090	2" diameter		16.80	.476		37.50	29		66.50	84.50
2100	2-1/2" diameter		15.12	.529		84.50	32.50		117	141
2110	3" diameter		12.60	.635		106	39		145	175
2120	3-1/2" diameter		10.08	.794		183	48.50		231.50	274
2130	4" diameter		8.40	.952		210	58.50		268.50	320
2140	5" diameter		7.56	1.060		292	65		357	415
2150	6" diameter		6.72	1.190		500	73		573	660
2160	Box connector with set screw, plain, 1/2" diameter		58.80	.136		2.66	8.35		11.01	15.40
2170	3/4" diameter		50.40	.159		4.73	9.75		14.48	19.70
2180	1" diameter		42	.190		8.40	11.70		20.10	26.50
2190	Insulated, 1-1/4" diameter		33.60	.238		12.10	14.60		26.70	35.50
2200	1-1/2" diameter		25.20	.317		17	19.50		36.50	47.50
2210	2" diameter		16.80	.476		35.50	29		64.50	82.50
2220	2-1/2" diameter		15.12	.529		104	32.50		136.50	164
2230	3" diameter		12.60	.635		168	39		207	242
2240	3-1/2" diameter		10.08	.794		207	48.50		255.50	300
2250	4" diameter		8.40	.952		216	58.50		274.50	325
2260	5" diameter		7.56	1.060		475	65		540	615
2270	6" diameter		6.72	1.190		485	73		558	645
2280	LB, LR or LL fittings & covers, 1/2" diameter		13.44	.595		8.80	36.50		45.30	64
2290	3/4" diameter		10.92	.733		10.60	45		55.60	78.50
2300	1" diameter		9.24	.866		15.90	53		68.90	96.50
2330	1-1/4" diameter		6.72	1.190		31	73		104	143
2350	1-1/2" diameter		5.04	1.590		30.50	97.50		128	179
2370	2" diameter		4.20	1.900		73	117		190	255
2380	2-1/2" diameter		3.36	2.380		136	146		282	365
2390	3" diameter		2.94	2.720		179	167		346	445
2400	3-1/2" diameter		2.52	3.170		220	194		414	530
2410	4" diameter		2.10	3.810		266	234		500	645
2420	T fittings, with cover, 1/2" diameter		10.08	.794		12.25	48.50		60.75	86
2430	3/4" diameter		9.24	.866		13.35	53		66.35	93.50
2440	1" diameter		7.56	1.060		20.50	65		85.50	120
2450	1-1/4" diameter		5.04	1.590		26.50	97.50		124	174
2470	1-1/2" diameter		4.20	1.900		46.50	117		163.50	225
2500	2" diameter		3.36	2.380		48	146		194	270
2510	2-1/2" diameter		2.94	2.720		111	167		278	370
2520	3" diameter		2.52	3.170		146	194		340	450
2530	3-1/2" diameter		2.10	3.810		270	234		504	645
2540	4" diameter		1.68	4.760		291	292		583	755
2550	Nipples chase, plain, 1/2" diameter		33.60	.238		.46	14.60		15.06	22.50
2560	3/4" diameter		26.88	.298		1.79	18.25		20.04	29
2570	1" diameter		23.52	.340		2.57	21		23.57	34
2600	Insulated, 1-1/4" diameter		20.16	.397		9.50	24.50		34	47
2630	1-1/2" diameter		15.12	.529		14.85	32.50		47.35	65
2650	2" diameter		12.60	.635		16.65	39		55.65	76.50

26 05 33.25 Conduit Fittings for Rigid Galvanized Steel	Crew	Daily Output	Labor-Hours	Unit	Material	2020 Bare Costs Labor	Equipment	Total	Total Incl O&P
2660	2-1/2" diameter	1 Elec	10.08	.794	Ea.	49	48.50		97.50	127
2670	3" diameter		8.40	.952		47	58.50		105.50	139
2680	3-1/2" diameter		7.56	1.060		75	65		140	180
2690	4" diameter		6.72	1.190		103	73		176	222
2700	5" diameter		5.88	1.360		335	83.50		418.50	495
2710	6" diameter		5.04	1.590		530	97.50		627.50	725
2720	Nipples offset, plain, 1/2" diameter		33.60	.238		6.90	14.60		21.50	29.50
2730	3/4" diameter		26.88	.298		6.90	18.25		25.15	34.50
2740	1" diameter		20.16	.397		10.45	24.50		34.95	48
2750	Insulated, 1-1/4" diameter		16.80	.476		8.70	29		37.70	53
2760	1-1/2" diameter		15.12	.529		10.60	32.50		43.10	60
2770	2" diameter		13.44	.595		15	36.50		51.50	71
2780	3" diameter		11.76	.680		99	41.50		140.50	171
2850	Coupling, expansion, 1/2" diameter		10.08	.794		59.50	48.50		108	138
2880	3/4" diameter		8.40	.952		70	58.50		128.50	164
2900	1" diameter		6.72	1.190		92	73		165	210
2920	1-1/4" diameter		5.38	1.490		116	91.50		207.50	263
2940	1-1/2" diameter		4.45	1.800		158	110		268	340
2960	2" diameter		3.86	2.070		231	127		358	445
2980	2-1/2" diameter		3.02	2.650		385	163		548	665
3000	3" diameter		2.52	3.170		450	194		644	785
3020	3-1/2" diameter		2.35	3.400		670	209		879	1,050
3040	4" diameter		2.02	3.970		805	244		1,049	1,250
3060	5" diameter		1.68	4.760		1,275	292		1,567	1,825
3080	6" diameter		1.51	5.290		2,175	325		2,500	2,850
3100	Expansion deflection, 1/2" diameter		10.08	.794		315	48.50		363.50	425
3120	3/4" diameter		10.08	.794		355	48.50		403.50	465
3140	1" diameter		8.40	.952		415	58.50		473.50	545
3160	1-1/4" diameter		5.38	1.490		660	91.50		751.50	860
3180	1-1/2" diameter		4.45	1.800		795	110		905	1,050
3200	2" diameter		3.86	2.070		940	127		1,067	1,225
3220	2-1/2" diameter		3.02	2.650		1,050	163		1,213	1,400
3240	3" diameter		2.52	3.170		1,600	194		1,794	2,050
3260	3-1/2" diameter		2.35	3.400		1,575	209		1,784	2,025
3280	4" diameter		2.02	3.970		1,800	244		2,044	2,350
3300	5" diameter		1.68	4.760		3,425	292		3,717	4,200
3320	6" diameter		1.51	5.290		4,925	325		5,250	5,875
3340	Ericson, 1/2" diameter		13.44	.595		4.54	36.50		41.04	59.50
3360	3/4" diameter		11.76	.680		5.50	41.50		47	68
3380	1" diameter		9.24	.866		9.20	53		62.20	89
3400	1-1/4" diameter		6.72	1.190		17.70	73		90.70	128
3420	1-1/2" diameter		5.88	1.360		23	83.50		106.50	149
3440	2" diameter		4.20	1.900		43	117		160	221
3460	2-1/2" diameter		3.36	2.380		77	146		223	300
3480	3" diameter		2.94	2.720		126	167		293	385
3500	3-1/2" diameter		2.52	3.170		165	194		359	470
3520	4" diameter		2.27	3.530		193	217		410	530
3540	5" diameter		2.10	3.810		395	234		629	785
3560	6" diameter		1.93	4.140		505	254		759	935
3580	Split, 1/2" diameter		26.88	.298		5.05	18.25		23.30	32.50
3600	3/4" diameter		22.68	.353		6.20	21.50		27.70	39
3620	1" diameter		16.80	.476		13.05	29		42.05	58
3640	1-1/4" diameter		13.44	.595		16.90	36.50		53.40	73

26 05 33 – Raceway and Boxes for Electrical Systems

26 05 33.25 Conduit Fittings for Rigid Galvanized Steel		Crew	Daily Output	Labor-Hours	Unit	Material	2020 Bare Costs Labor	Equipment	Total	Total Incl O&P
3660	1-1/2" diameter	1 Elec	11.76	.680	Ea.	17.90	41.50		59.40	81.50
3680	2" diameter		10.08	.794		36.50	48.50		85	113
3700	2-1/2" diameter		8.40	.952		94.50	58.50		153	191
3720	3" diameter		7.56	1.060		145	65		210	257
3740	3-1/2" diameter		6.72	1.190		226	73		299	360
3760	4" diameter		5.88	1.360		220	83.50		303.50	365
3780	5" diameter		5.04	1.590		445	97.50		542.50	635
3800	6" diameter		4.20	1.900		595	117		712	825
4600	Reducing bushings, 3/4" to 1/2" diameter		45.36	.176		2.94	10.80		13.74	19.35
4620	1" to 3/4" diameter		38.64	.207		3.41	12.70		16.11	22.50
4640	1-1/4" to 1" diameter		33.60	.238		5.70	14.60		20.30	28.50
4660	1-1/2" to 1-1/4" diameter		30.24	.265		7.05	16.25		23.30	32
4680	2" to 1-1/2" diameter		26.88	.298		15.60	18.25		33.85	44
4740	2-1/2" to 2" diameter		25.20	.317		19.35	19.50		38.85	50.50
4760	3" to 2-1/2" diameter		23.52	.340		23.50	21		44.50	57
4800	Through-wall seal, 1/2" diameter		6.72	1.190		79	73		152	196
4820	3/4" diameter		6.30	1.270		340	78		418	485
4840	1" diameter		5.46	1.470		330	90		420	495
4860	1-1/4" diameter		4.62	1.730		345	106		451	540
4880	1-1/2" diameter		4.20	1.900		495	117		612	720
4900	2" diameter		3.53	2.270		350	139		489	590
4920	2-1/2" diameter		2.94	2.720		630	167		797	940
4940	3" diameter		2.52	3.170		670	194		864	1,025
4960	3-1/2" diameter		2.10	3.810		1,300	234		1,534	1,775
4980	4" diameter		1.68	4.760		1,000	292		1,292	1,525
5000	5" diameter		1.26	6.350		990	390		1,380	1,675
5020	6" diameter		.84	9.520		990	585		1,575	1,975
5100	Cable supports, 2 or more wires									
5120	1-1/2" diameter	1 Elec	6.72	1.190	Ea.	196	73		269	325
5140	2" diameter		5.04	1.590		246	97.50		343.50	415
5160	2-1/2" diameter		3.36	2.380		264	146		410	510
5180	3" diameter		2.94	2.720		340	167		507	625
5200	3-1/2" diameter		2.18	3.660		490	225		715	875
5220	4" diameter		1.68	4.760		560	292		852	1,050
5240	5" diameter		1.26	6.350		770	390		1,160	1,425
5260	6" diameter		.84	9.520		1,225	585		1,810	2,225
5280	Service entrance cap, 1/2" diameter		13.44	.595		6.45	36.50		42.95	61.50
5300	3/4" diameter		10.92	.733		7.05	45		52.05	75
5320	1" diameter		8.40	.952		5.75	58.50		64.25	93.50
5340	1-1/4" diameter		6.72	1.190		5.80	73		78.80	115
5360	1-1/2" diameter		5.46	1.470		11.65	90		101.65	147
5380	2" diameter		4.62	1.730		25.50	106		131.50	186
5400	2-1/2" diameter		3.36	2.380		98.50	146		244.50	325
5420	3" diameter		2.86	2.800		107	172		279	375
5440	3-1/2" diameter		2.52	3.170		169	194		363	475
5460	4" diameter		2.27	3.530		224	217		441	565
5750	90° pull elbows steel, female, 1/2" diameter		13.44	.595		9.50	36.50		46	65
5760	3/4" diameter		10.92	.733		11.95	45		56.95	80
5780	1" diameter		9.24	.866		17.75	53		70.75	98.50
5800	1-1/4" diameter		6.72	1.190		28.50	73		101.50	141
5820	1-1/2" diameter		5.04	1.590		39.50	97.50		137	189
5840	2" diameter		4.20	1.900		64	117		181	245
6000	Explosion proof, flexible coupling									

26 05 33.25 Conduit Fittings for Rigid Galvanized Steel	Crew	Daily Output	Labor-Hours	Unit	Material	2020 Bare Costs Labor	Equipment	Total	Total Incl O&P
6010	1/2" diameter, 4" long	1 Elec	10.08	.794	Ea.	161	48.50		209.50	250
6020	6" long		10.08	.794		147	48.50		195.50	234
6050	12" long		10.08	.794		198	48.50		246.50	291
6070	18" long		10.08	.794		255	48.50		303.50	355
6090	24" long		10.08	.794		345	48.50		393.50	455
6110	30" long		10.08	.794		325	48.50		373.50	435
6130	36" long		10.08	.794		365	48.50		413.50	480
6140	3/4" diameter, 4" long		8.40	.952		170	58.50		228.50	274
6150	6" long		8.40	.952		180	58.50		238.50	285
6180	12" long		8.40	.952		234	58.50		292.50	345
6200	18" long		8.40	.952		305	58.50		363.50	420
6220	24" long		8.40	.952		375	58.50		433.50	495
6240	30" long		8.40	.952		375	58.50		433.50	495
6260	36" long		8.40	.952		460	58.50		518.50	590
6270	1" diameter, 6" long		6.72	1.190		340	73		413	485
6300	12" long		6.72	1.190		435	73		508	590
6320	18" long		6.72	1.190		495	73		568	650
6340	24" long		6.72	1.190		740	73		813	920
6360	30" long		6.72	1.190		900	73		973	1,100
6380	36" long		6.72	1.190		805	73		878	995
6390	1-1/4" diameter, 12" long		5.38	1.490		590	91.50		681.50	785
6410	18" long		5.38	1.490		725	91.50		816.50	935
6430	24" long		5.38	1.490		955	91.50		1,046.50	1,175
6450	30" long		5.38	1.490		975	91.50		1,066.50	1,200
6470	36" long		5.38	1.490		1,625	91.50		1,716.50	1,900
6480	1-1/2" diameter, 12" long		4.45	1.800		780	110		890	1,025
6500	18" long		4.45	1.800		940	110		1,050	1,200
6520	24" long		4.45	1.800		1,150	110		1,260	1,425
6540	30" long		4.45	1.800		1,750	110		1,860	2,100
6560	36" long		4.45	1.800		1,450	110		1,560	1,775
6570	2" diameter, 12" long		3.86	2.070		1,050	127		1,177	1,350
6590	18" long		3.86	2.070		1,350	127		1,477	1,675
6610	24" long		3.86	2.070		1,450	127		1,577	1,800
6630	30" long		3.86	2.070		1,725	127		1,852	2,075
6650	36" long		3.86	2.070		1,950	127		2,077	2,350
7000	Close up plug, 1/2" diameter, explosion proof		33.60	.238		2.98	14.60		17.58	25.50
7010	3/4" diameter		26.88	.298		3.50	18.25		21.75	31
7020	1" diameter		23.52	.340		4.13	21		25.13	35.50
7030	1-1/4" diameter		20.16	.397		4.60	24.50		29.10	41.50
7040	1-1/2" diameter		15.12	.529		6.10	32.50		38.60	55
7050	2" diameter		12.60	.635		10.35	39		49.35	69.50
7060	2-1/2" diameter		10.92	.733		15.85	45		60.85	84.50
7070	3" diameter		10.08	.794		24	48.50		72.50	99
7080	3-1/2" diameter		9.24	.866		28	53		81	110
7090	4" diameter		7.56	1.060		48	65		113	150
7091	Elbow female, 45°, 1/2"		13.44	.595		18.90	36.50		55.40	75.50
7092	3/4"		10.92	.733		24.50	45		69.50	94
7093	1"		9.24	.866		30	53		83	112
7094	1-1/4"		6.72	1.190		37.50	73		110.50	151
7095	1-1/2"		5.04	1.590		31	97.50		128.50	179
7096	2"		4.20	1.900		47	117		164	226
7097	2-1/2"		3.78	2.120		123	130		253	330
7098	3"		3.53	2.270		140	139		279	360

For customer support on your Electrical Change Order Costs with RSMeans data, call 800.448.8182.

Post-Installation Change Order Costs

26 05 33.25 Conduit Fittings for Rigid Galvanized Steel	Crew	Daily Output	Labor-Hours	Unit	Material	2020 Bare Costs Labor	Equipment	Total	Total Incl O&P
7099	3-1/2"	1 Elec	3.36	2.380	Ea.	213	146		359	450
7100	4"		3.19	2.510		261	154		415	515
7101	90°, 1/2"		13.44	.595		18.50	36.50		55	75
7102	3/4"		10.92	.733		22	45		67	91
7103	1"		9.24	.866		27	53		80	109
7104	1-1/4"		6.72	1.190		40.50	73		113.50	154
7105	1-1/2"		5.04	1.590		61	97.50		158.50	212
7106	2"		4.20	1.900		115	117		232	300
7107	2-1/2"		3.78	2.120		204	130		334	420
7110	Elbows 90°, long male & female, 1/2" diameter, explosion proof		13.44	.595		27	36.50		63.50	84.50
7120	3/4" diameter		10.92	.733		30	45		75	100
7130	1" diameter		9.24	.866		40.50	53		93.50	124
7140	1-1/4" diameter		6.72	1.190		62.50	73		135.50	178
7150	1-1/2" diameter		5.04	1.590		52.50	97.50		150	203
7160	2" diameter		4.20	1.900		70.50	117		187.50	252
7170	Capped elbow, 1/2" diameter, explosion proof		9.24	.866		27	53		80	109
7180	3/4" diameter		6.72	1.190		36.50	73		109.50	149
7190	1" diameter		5.04	1.590		39.50	97.50		137	189
7200	1-1/4" diameter		4.20	1.900		86	117		203	269
7210	Pulling elbow, 1/2" diameter, explosion proof		9.24	.866		168	53		221	264
7220	3/4" diameter		6.72	1.190		231	73		304	365
7230	1" diameter		5.04	1.590		293	97.50		390.50	470
7240	1-1/4" diameter		4.20	1.900		380	117		497	595
7250	1-1/2" diameter		4.20	1.900		465	117		582	690
7260	2" diameter		3.36	2.380		545	146		691	815
7270	2-1/2" diameter		2.94	2.720		920	167		1,087	1,250
7280	3" diameter		2.52	3.170		1,425	194		1,619	1,875
7290	3-1/2" diameter		2.10	3.810		1,950	234		2,184	2,500
7300	4" diameter		1.85	4.330		1,925	266		2,191	2,525
7310	LB conduit body, 1/2" diameter		9.24	.866		71.50	53		124.50	158
7320	3/4" diameter		6.72	1.190		94.50	73		167.50	213
7330	T conduit body, 1/2" diameter		7.56	1.060		72	65		137	177
7340	3/4" diameter		5.04	1.590		90.50	97.50		188	245
7350	Explosion proof, round box w/cover, 3 threaded hubs, 1/2" diameter		6.72	1.190		70.50	73		143.50	187
7351	3/4" diameter		6.72	1.190		94.50	73		167.50	213
7352	1" diameter		6.30	1.270		101	78		179	227
7353	1-1/4" diameter		5.88	1.360		106	83.50		189.50	240
7354	1-1/2" diameter		5.88	1.360		285	83.50		368.50	440
7355	2" diameter		5.04	1.590		284	97.50		381.50	455
7356	Round box w/cover & mtng flange, 3 threaded hubs, 1/2" diameter		6.72	1.190		73.50	73		146.50	190
7357	3/4" diameter		6.72	1.190		67	73		140	183
7358	4 threaded hubs, 1" diameter		5.88	1.360		80	83.50		163.50	212
7400	Unions, 1/2" diameter		16.80	.476		14.55	29		43.55	59.50
7410	3/4" - 1/2" diameter		13.44	.595		23	36.50		59.50	80
7420	3/4" diameter		13.44	.595		24.50	36.50		61	81.50
7430	1" diameter		11.76	.680		45	41.50		86.50	112
7440	1-1/4" diameter		10.08	.794		67.50	48.50		116	147
7450	1-1/2" diameter		8.40	.952		85	58.50		143.50	181
7460	2" diameter		7.14	1.120		112	68.50		180.50	225
7480	2-1/2" diameter		6.72	1.190		169	73		242	295
7490	3" diameter		5.88	1.360		248	83.50		331.50	395
7500	3-1/2" diameter		5.04	1.590		480	97.50		577.50	670
7510	4" diameter		4.20	1.900		440	117		557	660

26 05 33.25 Conduit Fittings for Rigid Galvanized Steel		Crew	Daily Output	Labor-Hours	Unit	Material	2020 Bare Costs Labor	Equipment	Total	Total Incl O&P
7680	Reducer, 3/4" to 1/2"	1 Elec	45.36	.176	Ea.	6.50	10.80		17.30	23.50
7690	1" to 1/2"		38.64	.207		6.20	12.70		18.90	25.50
7700	1" to 3/4"		38.64	.207		7.80	12.70		20.50	27.50
7710	1-1/4" to 3/4"		33.60	.238		10.05	14.60		24.65	33
7720	1-1/4" to 1"		33.60	.238		12	14.60		26.60	35.50
7730	1-1/2" to 1"		30.24	.265		15.35	16.25		31.60	41
7740	1-1/2" to 1-1/4"		30.24	.265		15.45	16.25		31.70	41
7750	2" to 3/4"		26.88	.298		22	18.25		40.25	51
7760	2" to 1-1/4"		26.88	.298		24.50	18.25		42.75	54
7770	2" to 1-1/2"		26.88	.298		31.50	18.25		49.75	61.50
7780	2-1/2" to 1-1/2"		25.20	.317		27	19.50		46.50	58.50
7790	3" to 2"		25.20	.317		25.50	19.50		45	57.50
7800	3-1/2" to 2-1/2"		23.52	.340		81.50	21		102.50	121
7810	4" to 3"		23.52	.340		84	21		105	123
7820	Sealing fitting, vertical/horizontal, 1/2" diameter		12.18	.657		23	40.50		63.50	85.50
7830	3/4" diameter		11.17	.716		27	44		71	95
7840	1" diameter		9.58	.835		34	51.50		85.50	114
7850	1-1/4" diameter		8.40	.952		33.50	58.50		92	124
7860	1-1/2" diameter		7.39	1.080		54	66.50		120.50	158
7870	2" diameter		6.72	1.190		66.50	73		139.50	183
7880	2-1/2" diameter		5.63	1.420		98.50	87		185.50	238
7890	3" diameter		4.79	1.670		134	102		236	300
7900	3-1/2" diameter		3.95	2.030		350	125		475	570
7910	4" diameter		3.36	2.380		585	146		731	860
7920	Sealing hubs, 1" by 1-1/2"		10.08	.794		34.50	48.50		83	110
7930	1-1/4" by 2"		8.40	.952		65	58.50		123.50	159
7940	1-1/2" by 2"		7.56	1.060		89.50	65		154.50	196
7950	2" by 2-1/2"		6.72	1.190		119	73		192	240
7960	3" by 4"		5.88	1.360		213	83.50		296.50	360
7970	4" by 5"		5.04	1.590		455	97.50		552.50	645
7980	Drain, 1/2"		26.88	.298		132	18.25		150.25	172
7990	Breather, 1/2"		26.88	.298		142	18.25		160.25	183
7991	Explosion proof sealant compound, hub, fittings, 60 min set time, 1lb. pail		40.32	.198		26.50	12.15		38.65	47
7992	5 lb. pail		40.32	.198		75	12.15		87.15	101
7993	2 oz. tube		42	.190		81	11.70		92.70	107
7994	6 oz. tube		42	.190		149	11.70		160.70	181
7995	2.0 oz. cartridge		42	.190		53	11.70		64.70	76
7996	6.0 oz. cartridge		42	.190		104	11.70		115.70	132
7997	2.0 oz. box		42	.190		35	11.70		46.70	55.50
7998	8.0 oz. box		42	.190		107	11.70		118.70	134
7999	1.0 lb. box		42	.190		176	11.70		187.70	211
8000	Plastic coated 40 mil thick									
8010	LB, LR or LL conduit body w/cover, 1/2" diameter	1 Elec	10.92	.733	Ea.	65	45		110	139
8020	3/4" diameter		9.24	.866		84	53		137	172
8030	1" diameter		6.72	1.190		100	73		173	219
8040	1-1/4" diameter		5.04	1.590		148	97.50		245.50	310
8050	1-1/2" diameter		4.20	1.900		155	117		272	345
8060	2" diameter		3.78	2.120		263	130		393	485
8070	2-1/2" diameter		3.36	2.380		520	146		666	785
8080	3" diameter		2.94	2.720		515	167		682	815
8090	3-1/2" diameter		2.52	3.170		935	194		1,129	1,325
8100	4" diameter		2.10	3.810		1,075	234		1,309	1,525
8150	T conduit body with cover, 1/2" diameter		9.24	.866		77	53		130	164

 Post-Installation Change Order Costs

26 05 33.25 Conduit Fittings for Rigid Galvanized Steel	Crew	Daily Output	Labor-Hours	Unit	Material	2020 Bare Costs Labor	Equipment	Total	Total Incl O&P
8160	3/4" diameter	1 Elec	7.56	1.060	Ea.	98	65		163	205
8170	1" diameter		5.04	1.590		115	97.50		212.50	272
8180	1-1/4" diameter		4.20	1.900		172	117		289	365
8190	1-1/2" diameter		3.78	2.120		196	130		326	410
8200	2" diameter		3.36	2.380		251	146		397	495
8210	2-1/2" diameter		2.94	2.720		485	167		652	780
8220	3" diameter		2.52	3.170		660	194		854	1,025
8230	3-1/2" diameter		2.10	3.810		1,150	234		1,384	1,600
8240	4" diameter		1.68	4.760		1,100	292		1,392	1,650
8300	FS conduit body, 1 gang, 3/4" diameter		9.24	.866		83.50	53		136.50	171
8310	1" diameter		8.40	.952		70.50	58.50		129	165
8350	2 gang, 3/4" diameter		7.56	1.060		127	65		192	237
8360	1" diameter		6.72	1.190		184	73		257	310
8400	Duplex receptacle cover		53.76	.149		62	9.15		71.15	82
8410	Switch cover		53.76	.149		74.50	9.15		83.65	95
8420	Switch, vaportight cover		44.52	.180		227	11		238	266
8430	Blank, cover		53.76	.149		49	9.15		58.15	67.50
8520	FSC conduit body, 1 gang, 3/4" diameter		8.40	.952		81.50	58.50		140	177
8530	1" diameter		7.56	1.060		84.50	65		149.50	190
8550	2 gang, 3/4" diameter		6.72	1.190		142	73		215	265
8560	1" diameter		5.88	1.360		196	83.50		279.50	340
8590	Conduit hubs, 1/2" diameter		15.12	.529		48.50	32.50		81	102
8600	3/4" diameter		13.44	.595		55	36.50		91.50	115
8610	1" diameter		11.76	.680		71	41.50		112.50	140
8620	1-1/4" diameter		10.08	.794		85.50	48.50		134	167
8630	1-1/2" diameter		8.40	.952		94.50	58.50		153	191
8640	2" diameter		7.39	1.080		133	66.50		199.50	245
8650	2-1/2" diameter		7.14	1.120		183	68.50		251.50	305
8660	3" diameter		6.72	1.190		285	73		358	425
8670	3-1/2" diameter		6.30	1.270		400	78		478	555
8680	4" diameter		5.88	1.360		390	83.50		473.50	555
8690	5" diameter		5.04	1.590		470	97.50		567.50	665
8700	Plastic coated 40 mil thick									
8710	Pipe strap, stamped 1 hole, 1/2" diameter	1 Elec	394.80	.020	Ea.	17.40	1.24		18.64	21
8720	3/4" diameter		369.60	.022		15.45	1.33		16.78	19
8730	1" diameter		336	.024		16.60	1.46		18.06	20.50
8740	1-1/4" diameter		298.20	.027		27	1.65		28.65	32
8750	1-1/2" diameter		268.80	.030		27.50	1.83		29.33	32.50
8760	2" diameter		223.44	.036		36.50	2.20		38.70	43.50
8770	2-1/2" diameter		168	.048		38.50	2.92		41.42	46.50
8780	3" diameter		111.72	.072		52.50	4.39		56.89	64
8790	3-1/2" diameter		92.40	.087		120	5.30		125.30	140
8800	4" diameter		75.60	.106		124	6.50		130.50	147
8810	5" diameter		58.80	.136		170	8.35		178.35	199
8840	Clamp back spacers, 3/4" diameter		369.60	.022		24	1.33		25.33	28
8850	1" diameter		336	.024		32.50	1.46		33.96	38
8860	1-1/4" diameter		298.20	.027		31.50	1.65		33.15	37
8870	1-1/2" diameter		268.80	.030		47	1.83		48.83	54
8880	2" diameter		223.44	.036		71.50	2.20		73.70	82
8900	3" diameter		111.72	.072		106	4.39		110.39	123
8920	4" diameter		75.60	.106		86	6.50		92.50	104
8950	Touch-up plastic coating, spray, 12 oz.					80			80	88
8960	Sealing fittings, 1/2" diameter	1 Elec	9.24	.866		79.50	53		132.50	167

26 05 Common Work Results for Electrical

26 05 33 – Raceway and Boxes for Electrical Systems

26 05 33.25 Conduit Fittings for Rigid Galvanized Steel

		Crew	Daily Output	Labor-Hours	Unit	Material	2020 Bare Costs Labor	Equipment	Total	Total Incl O&P
8970	3/4" diameter	1 Elec	7.56	1.060	Ea.	120	65		185	229
8980	1" diameter		6.30	1.270		91	78		169	216
8990	1-1/4" diameter		5.46	1.470		137	90		227	285
9000	1-1/2" diameter		4.62	1.730		168	106		274	345
9010	2" diameter		4.03	1.980		192	121		313	390
9020	2-1/2" diameter		3.36	2.380		350	146		496	600
9030	3" diameter		2.94	2.720		350	167		517	635
9040	3-1/2" diameter		2.52	3.170		860	194		1,054	1,225
9050	4" diameter		2.10	3.810		1,325	234		1,559	1,825
9060	5" diameter		1.43	5.600		2,425	345		2,770	3,175
9070	Unions, 1/2" diameter		15.12	.529		64	32.50		96.50	119
9080	3/4" diameter		12.60	.635		82	39		121	149
9090	1" diameter		10.92	.733		85	45		130	161
9100	1-1/4" diameter		9.24	.866		147	53		200	241
9110	1-1/2" diameter		7.98	1		222	61.50		283.50	335
9120	2" diameter		6.72	1.190		263	73		336	400
9130	2-1/2" diameter		6.30	1.270		345	78		423	495
9140	3" diameter		5.71	1.400		385	86		471	555
9150	3-1/2" diameter		4.87	1.640		580	101		681	790
9160	4" diameter		4.03	1.980		710	121		831	960
9170	5" diameter		3.36	2.380		1,125	146		1,271	1,450

26 05 33.30 Electrical Nonmetallic Tubing (ENT)

0010	ELECTRICAL NONMETALLIC TUBING (ENT)									
0050	Flexible, 1/2" diameter	1 Elec	226.80	.035	L.F.	.63	2.16		2.79	3.92
0100	3/4" diameter		193.20	.041		.81	2.54		3.35	4.67
0200	1" diameter		121.80	.066		1.69	4.03		5.72	7.85
0210	1-1/4" diameter		105	.076		1.63	4.67		6.30	8.75
0220	1-1/2" diameter		84	.095		2.83	5.85		8.68	11.80
0230	2" diameter		63	.127		3.39	7.80		11.19	15.35
0300	Connectors, to outlet box, 1/2" diameter		193.20	.041	Ea.	1.17	2.54		3.71	5.05
0310	3/4" diameter		176.40	.045		2.21	2.78		4.99	6.55
0320	1" diameter		168	.048		2.76	2.92		5.68	7.40
0400	Couplings, to conduit, 1/2" diameter		121.80	.066		1.15	4.03		5.18	7.25
0410	3/4" diameter		109.20	.073		1.33	4.49		5.82	8.15
0420	1" diameter		105	.076		2.52	4.67		7.19	9.70

26 05 33.33 Raceway/Boxes for Utility Substations

0010	RACEWAY/BOXES FOR UTILITY SUBSTATIONS									
7000	Conduit, conductors, and insulators									
7100	Conduit, metallic	R-11	470.40	.119	Lb.	3.47	6.95	1.35	11.77	15.65
7110	Non-metallic	"	672	.083	"	9.05	4.86	.95	14.86	18.25
7190	See Section 26 05 33									
7200	Wire and cable	R-11	588	.095	Lb.	9.95	5.55	1.08	16.58	20.50
7290	See Section 26 05 19									

26 05 33.35 Flexible Metallic Conduit

0010	FLEXIBLE METALLIC CONDUIT									
0050	Steel, 3/8" diameter	1 Elec	168	.048	L.F.	.44	2.92		3.36	4.83
0100	1/2" diameter		168	.048		.52	2.92		3.44	4.92
0200	3/4" diameter		134.40	.060		.70	3.65		4.35	6.20
0250	1" diameter		84	.095		1.38	5.85		7.23	10.20
0300	1-1/4" diameter		58.80	.136		1.60	8.35		9.95	14.20
0350	1-1/2" diameter		42	.190		2.93	11.70		14.63	20.50
0370	2" diameter		33.60	.238		3.71	14.60		18.31	26

26 05 33.35 Flexible Metallic Conduit	Crew	Daily Output	Labor-Hours	Unit	Material	2020 Bare Costs Labor	Equipment	Total	Total Incl O&P
0380	2-1/2" diameter	1 Elec	25.20	.317	L.F.	4.03	19.50		23.53	33.50
0390	3" diameter	2 Elec	42	.381		6.90	23.50		30.40	42.50
0400	3-1/2" diameter		33.60	.476		7.75	29		36.75	52
0410	4" diameter		25.20	.635		8.90	39		47.90	68
0420	Connectors, plain, 3/8" diameter	1 Elec	84	.095	Ea.	2.14	5.85		7.99	11.05
0430	1/2" diameter		67.20	.119		2.32	7.30		9.62	13.45
0440	3/4" diameter		58.80	.136		2.44	8.35		10.79	15.15
0450	1" diameter		42	.190		5.05	11.70		16.75	23
0452	1-1/4" diameter		37.80	.212		6.55	13		19.55	26.50
0454	1-1/2" diameter		33.60	.238		9.60	14.60		24.20	32.50
0456	2" diameter		23.52	.340		13.90	21		34.90	46.50
0458	2-1/2" diameter		21	.381		23	23.50		46.50	60.50
0460	3" diameter		16.80	.476		32.50	29		61.50	79.50
0462	3-1/2" diameter		13.44	.595		105	36.50		141.50	171
0464	4" diameter		10.92	.733		116	45		161	195
0490	Insulated, 1" diameter		33.60	.238		9.40	14.60		24	32.50
0500	1-1/4" diameter		33.60	.238		15.15	14.60		29.75	38.50
0550	1-1/2" diameter		26.88	.298		28	18.25		46.25	57.50
0600	2" diameter		19.32	.414		30	25.50		55.50	70.50
0610	2-1/2" diameter		16.80	.476		66	29		95	116
0620	3" diameter		14.28	.560		134	34.50		168.50	199
0630	3-1/2" diameter		10.92	.733		325	45		370	420
0640	4" diameter		8.40	.952		345	58.50		403.50	465
0650	Connectors 90°, plain, 3/8" diameter		67.20	.119		3.37	7.30		10.67	14.60
0660	1/2" diameter		50.40	.159		6.15	9.75		15.90	21.50
0700	3/4" diameter		42	.190		8.55	11.70		20.25	27
0750	1" diameter		33.60	.238		13.85	14.60		28.45	37
0790	Insulated, 1" diameter		33.60	.238		19.95	14.60		34.55	44
0800	1-1/4" diameter		25.20	.317		27	19.50		46.50	59
0850	1-1/2" diameter		19.32	.414		53	25.50		78.50	96.50
0900	2" diameter		15.12	.529		77	32.50		109.50	134
0910	2-1/2" diameter		13.44	.595		194	36.50		230.50	269
0920	3" diameter		11.76	.680		254	41.50		295.50	340
0930	3-1/2" diameter		9.24	.866		660	53		713	805
0940	4" diameter		6.72	1.190		825	73		898	1,025
0960	Couplings, to flexible conduit, 1/2" diameter		42	.190		1.10	11.70		12.80	18.60
0970	3/4" diameter		33.60	.238		1.84	14.60		16.44	24
0980	1" diameter		29.40	.272		2.99	16.70		19.69	28.50
0990	1-1/4" diameter		23.52	.340		7.25	21		28.25	39
1000	1-1/2" diameter		19.32	.414		8.60	25.50		34.10	47.50
1010	2" diameter		16.80	.476		17.10	29		46.10	62.50
1020	2-1/2" diameter		15.12	.529		45	32.50		77.50	98.50
1030	3" diameter		12.60	.635		114	39		153	183
1032	Aluminum, 3/8" diameter		176.40	.045	L.F.	.52	2.78		3.30	4.71
1034	1/2" diameter		176.40	.045		.70	2.78		3.48	4.91
1036	3/4" diameter		138.60	.058		1	3.54		4.54	6.35
1038	1" diameter		88.20	.091		1.79	5.55		7.34	10.25
1040	1-1/4" diameter		63	.127		2.08	7.80		9.88	13.90
1042	1-1/2" diameter		44.52	.180		3.64	11		14.64	20.50
1044	2" diameter		35.28	.227		5.25	13.90		19.15	26.50
1046	2-1/2" diameter		26.88	.298		5.05	18.25		23.30	32.50
1048	3" diameter	2 Elec	44.52	.359		8.85	22		30.85	43
1050	3-1/2" diameter		35.28	.454		7.20	28		35.20	49.50

26 05 33.35 Flexible Metallic Conduit	Crew	Daily Output	Labor-Hours	Unit	Material	2020 Bare Costs Labor	Equipment	Total	Total Incl O&P
1052	4" diameter	2 Elec	26.88	.595	L.F.	10.60	36.50		47.10	66
1070	Sealtite, 3/8" diameter	1 Elec	117.60	.068		.77	4.17		4.94	7.05
1080	1/2" diameter		117.60	.068		1.30	4.17		5.47	7.65
1090	3/4" diameter		84	.095		1.82	5.85		7.67	10.70
1100	1" diameter		58.80	.136		2.68	8.35		11.03	15.40
1200	1-1/4" diameter		42	.190		3.58	11.70		15.28	21.50
1300	1-1/2" diameter		33.60	.238		3.98	14.60		18.58	26.50
1400	2" diameter		25.20	.317		5.15	19.50		24.65	34.50
1410	2-1/2" diameter		22.68	.353		9.20	21.50		30.70	42
1420	3" diameter	2 Elec	42	.381		11.70	23.50		35.20	48
1440	4" diameter	"	25.20	.635		24	39		63	84.50
1490	Connectors, plain, 3/8" diameter	1 Elec	58.80	.136	Ea.	2.76	8.35		11.11	15.50
1500	1/2" diameter		58.80	.136		3.62	8.35		11.97	16.45
1700	3/4" diameter		42	.190		5.20	11.70		16.90	23
1900	1" diameter		33.60	.238		8.30	14.60		22.90	31
1910	Insulated, 1" diameter		33.60	.238		11.40	14.60		26	34.50
2000	1-1/4" diameter		26.88	.298		13.55	18.25		31.80	42
2100	1-1/2" diameter		22.68	.353		25	21.50		46.50	59.50
2200	2" diameter		16.80	.476		58	29		87	107
2210	2-1/2" diameter		12.60	.635		124	39		163	194
2220	3" diameter		10.08	.794		176	48.50		224.50	267
2240	4" diameter		6.72	1.190		232	73		305	365
2290	Connectors, 90°, 3/8" diameter		58.80	.136		5.95	8.35		14.30	19
2300	1/2" diameter		58.80	.136		7.95	8.35		16.30	21
2400	3/4" diameter		42	.190		8.65	11.70		20.35	27
2600	1" diameter		33.60	.238		16.95	14.60		31.55	40.50
2790	Insulated, 1" diameter		33.60	.238		14.25	14.60		28.85	37.50
2800	1-1/4" diameter		26.88	.298		16.60	18.25		34.85	45.50
3000	1-1/2" diameter		22.68	.353		29	21.50		50.50	64
3100	2" diameter		16.80	.476		31	29		60	77.50
3110	2-1/2" diameter		11.76	.680		147	41.50		188.50	224
3120	3" diameter		9.24	.866		415	53		468	535
3140	4" diameter		5.88	1.360		196	83.50		279.50	340
4300	Coupling sealtite to rigid, 1/2" diameter		16.80	.476		4.98	29		33.98	49
4500	3/4" diameter		15.12	.529		7	32.50		39.50	56
4800	1" diameter		11.76	.680		7.95	41.50		49.45	71
4900	1-1/4" diameter		10.08	.794		19.50	48.50		68	94
5000	1-1/2" diameter		9.24	.866		24	53		77	106
5100	2" diameter		8.40	.952		38	58.50		96.50	129
5110	2-1/2" diameter		7.98	1		177	61.50		238.50	287
5120	3" diameter		7.56	1.060		180	65		245	295
5130	3-1/2" diameter		7.56	1.060		208	65		273	325
5140	4" diameter		7.14	1.120		237	68.50		305.50	365

26 05 33.95 Cutting and Drilling

		Crew	Daily Output	Labor-Hours	Unit	Material	2020 Bare Costs Labor	Equipment	Total	Total Incl O&P
0010	**CUTTING AND DRILLING**									
0100	Hole drilling to 10' high, concrete wall									
0110	8" thick, 1/2" pipe size	R-31	10.08	.794	Ea.	.27	48.50	6.20	54.97	79.50
0120	3/4" pipe size		10.08	.794		.27	48.50	6.20	54.97	79.50
0130	1" pipe size		7.98	1		.35	61.50	7.80	69.65	100
0140	1-1/4" pipe size		7.98	1		.35	61.50	7.80	69.65	100
0150	1-1/2" pipe size		7.98	1		.35	61.50	7.80	69.65	100
0160	2" pipe size		3.70	2.160		.53	133	16.85	150.38	216

 Post-Installation Change Order Costs

26 05 33.95 Cutting and Drilling	Crew	Daily Output	Labor-Hours	Unit	Material	2020 Bare Costs Labor	Equipment	Total	Total Incl O&P
0170	2-1/2" pipe size	R-31	3.70	2.160	Ea.	.53	133	16.85	150.38	216
0180	3" pipe size		3.70	2.160		.53	133	16.85	150.38	216
0190	3-1/2" pipe size		2.77	2.890		.60	177	22.50	200.10	290
0200	4" pipe size		2.77	2.890		.60	177	22.50	200.10	290
0500	12" thick, 1/2" pipe size		7.90	1.010		.40	62	7.90	70.30	102
0520	3/4" pipe size		7.90	1.010		.40	62	7.90	70.30	102
0540	1" pipe size		6.13	1.300		.52	80	10.15	90.67	131
0560	1-1/4" pipe size		6.13	1.300		.52	80	10.15	90.67	131
0570	1-1/2" pipe size		6.13	1.300		.52	80	10.15	90.67	131
0580	2" pipe size		3.02	2.650		.80	163	20.50	184.30	266
0590	2-1/2" pipe size		3.02	2.650		.80	163	20.50	184.30	266
0600	3" pipe size		3.02	2.650		.80	163	20.50	184.30	266
0610	3-1/2" pipe size		2.35	3.400		.90	209	26.50	236.40	340
0630	4" pipe size		2.10	3.810		.90	234	30	264.90	385
0650	16" thick, 1/2" pipe size		6.38	1.250		.54	76.50	9.75	86.79	125
0670	3/4" pipe size		5.88	1.360		.54	83.50	10.60	94.64	136
0690	1" pipe size		5.04	1.590		.69	97.50	12.40	110.59	159
0710	1-1/4" pipe size		4.62	1.730		.69	106	13.50	120.19	174
0730	1-1/2" pipe size		4.62	1.730		.69	106	13.50	120.19	174
0750	2" pipe size		2.52	3.170		1.06	194	25	220.06	320
0770	2-1/2" pipe size		2.27	3.530		1.06	217	27.50	245.56	350
0790	3" pipe size		2.10	3.810		1.06	234	30	265.06	385
0810	3-1/2" pipe size		1.93	4.140		1.20	254	32.50	287.70	415
0830	4" pipe size		1.68	4.760		1.20	292	37	330.20	475
0850	20" thick, 1/2" pipe size		5.38	1.490		.67	91.50	11.65	103.82	150
0870	3/4" pipe size		5.04	1.590		.67	97.50	12.40	110.57	159
0890	1" pipe size		4.20	1.900		.86	117	14.85	132.71	191
0910	1-1/4" pipe size		4.03	1.980		.86	121	15.45	137.31	199
0930	1-1/2" pipe size		3.86	2.070		.86	127	16.15	144.01	208
0950	2" pipe size		2.27	3.530		1.33	217	27.50	245.83	350
0970	2-1/2" pipe size		2.02	3.970		1.33	244	31	276.33	400
0990	3" pipe size		1.85	4.330		1.33	266	34	301.33	435
1010	3-1/2" pipe size		1.68	4.760		1.50	292	37	330.50	480
1030	4" pipe size		1.43	5.600		1.50	345	43.50	390	560
1050	24" thick, 1/2" pipe size		4.62	1.730		.81	106	13.50	120.31	174
1070	3/4" pipe size		4.28	1.870		.81	115	14.60	130.41	188
1090	1" pipe size		3.61	2.210		1.04	136	17.25	154.29	222
1110	1-1/4" pipe size		3.36	2.380		1.04	146	18.60	165.64	239
1130	1-1/2" pipe size		3.36	2.380		1.04	146	18.60	165.64	239
1150	2" pipe size		2.02	3.970		1.59	244	31	276.59	400
1170	2-1/2" pipe size		1.85	4.330		1.59	266	34	301.59	435
1190	3" pipe size		1.68	4.760		1.59	292	37	330.59	480
1210	3-1/2" pipe size		1.51	5.290		1.80	325	41.50	368.30	530
1230	4" pipe size		1.26	6.350		1.80	390	49.50	441.30	635
1500	Brick wall, 8" thick, 1/2" pipe size		15.12	.529		.27	32.50	4.13	36.90	53.50
1520	3/4" pipe size		15.12	.529		.27	32.50	4.13	36.90	53.50
1540	1" pipe size		11.17	.716		.35	44	5.60	49.95	72
1560	1-1/4" pipe size		11.17	.716		.35	44	5.60	49.95	72
1580	1-1/2" pipe size		11.17	.716		.35	44	5.60	49.95	72
1600	2" pipe size		4.79	1.670		.53	102	13.05	115.58	168
1620	2-1/2" pipe size		4.79	1.670		.53	102	13.05	115.58	168
1640	3" pipe size		4.79	1.670		.53	102	13.05	115.58	168
1660	3-1/2" pipe size		3.70	2.160		.60	133	16.85	150.45	216

26 05 33.95 Cutting and Drilling		Crew	Daily Output	Labor-Hours	Unit	Material	2020 Bare Costs Labor	Equipment	Total	Total Incl O&P
1680	4" pipe size	R-31	3.36	2.380	Ea.	.60	146	18.60	165.20	238
1700	12" thick, 1/2" pipe size		12.18	.657		.40	40.50	5.15	46.05	66
1720	3/4" pipe size		12.18	.657		.40	40.50	5.15	46.05	66
1740	1" pipe size		9.24	.866		.52	53	6.75	60.27	87
1760	1-1/4" pipe size		9.24	.866		.52	53	6.75	60.27	87
1780	1-1/2" pipe size		9.24	.866		.52	53	6.75	60.27	87
1800	2" pipe size		4.20	1.900		.80	117	14.85	132.65	191
1820	2-1/2" pipe size		4.20	1.900		.80	117	14.85	132.65	191
1840	3" pipe size		4.20	1.900		.80	117	14.85	132.65	191
1860	3-1/2" pipe size		3.19	2.510		.90	154	19.60	174.50	251
1880	4" pipe size		2.77	2.890		.90	177	22.50	200.40	290
1900	16" thick, 1/2" pipe size		10.33	.774		.54	47.50	6.05	54.09	77.50
1920	3/4" pipe size		10.33	.774		.54	47.50	6.05	54.09	77.50
1940	1" pipe size		7.81	1.020		.69	62.50	7.95	71.14	103
1960	1-1/4" pipe size		7.81	1.020		.69	62.50	7.95	71.14	103
1980	1-1/2" pipe size		7.81	1.020		.69	62.50	7.95	71.14	103
2000	2" pipe size		3.70	2.160		1.06	133	16.85	150.91	217
2010	2-1/2" pipe size		3.70	2.160		1.06	133	16.85	150.91	217
2030	3" pipe size		3.70	2.160		1.06	133	16.85	150.91	217
2050	3-1/2" pipe size		2.77	2.890		1.20	177	22.50	200.70	290
2070	4" pipe size		2.52	3.170		1.20	194	25	220.20	320
2090	20" thick, 1/2" pipe size		8.99	.890		.67	54.50	6.95	62.12	90
2110	3/4" pipe size		8.99	.890		.67	54.50	6.95	62.12	90
2130	1" pipe size		6.72	1.190		.86	73	9.30	83.16	120
2150	1-1/4" pipe size		6.72	1.190		.86	73	9.30	83.16	120
2170	1-1/2" pipe size		6.72	1.190		.86	73	9.30	83.16	120
2190	2" pipe size		3.36	2.380		1.33	146	18.60	165.93	239
2210	2-1/2" pipe size		3.36	2.380		1.33	146	18.60	165.93	239
2230	3" pipe size		3.36	2.380		1.33	146	18.60	165.93	239
2250	3-1/2" pipe size		2.52	3.170		1.50	194	25	220.50	320
2270	4" pipe size		2.27	3.530		1.50	217	27.50	246	350
2290	24" thick, 1/2" pipe size		7.90	1.010		.81	62	7.90	70.71	102
2310	3/4" pipe size		7.90	1.010		.81	62	7.90	70.71	102
2330	1" pipe size		5.96	1.340		1.04	82	10.45	93.49	135
2350	1-1/4" pipe size		5.96	1.340		1.04	82	10.45	93.49	135
2370	1-1/2" pipe size		5.96	1.340		1.04	82	10.45	93.49	135
2390	2" pipe size		3.02	2.650		1.59	163	20.50	185.09	267
2410	2-1/2" pipe size		3.02	2.650		1.59	163	20.50	185.09	267
2430	3" pipe size		3.02	2.650		1.59	163	20.50	185.09	267
2450	3-1/2" pipe size		2.35	3.400		1.80	209	26.50	237.30	340
2470	4" pipe size		2.10	3.810		1.80	234	30	265.80	385
3000	Knockouts to 8' high, metal boxes & enclosures									
3020	With hole saw, 1/2" pipe size	1 Elec	44.52	.180	Ea.		11		11	16.40
3040	3/4" pipe size		39.48	.203			12.45		12.45	18.50
3050	1" pipe size		33.60	.238			14.60		14.60	22
3060	1-1/4" pipe size		30.24	.265			16.25		16.25	24
3070	1-1/2" pipe size		26.88	.298			18.25		18.25	27
3080	2" pipe size		22.68	.353			21.50		21.50	32
3090	2-1/2" pipe size		16.80	.476			29		29	43.50
4010	3" pipe size		13.44	.595			36.50		36.50	54.50
4030	3-1/2" pipe size		10.92	.733			45		45	67
4050	4" pipe size		9.24	.866			53		53	79

   Post-Installation Change Order Costs

26 05 36 – Cable Trays for Electrical Systems

26 05 36.10 Cable Tray Ladder Type	Crew	Daily Output	Labor-Hours	Unit	Material	2020 Bare Costs Labor	Equipment	Total	Total Incl O&P
0010 **CABLE TRAY LADDER TYPE** w/ftngs. & supports, 4" dp., to 15' elev.									
0100 For higher elevations, see Section 26 05 36.40									
0160 Galvanized steel tray									
0170 4" rung spacing, 6" wide	2 Elec	82.32	.194	L.F.	14.90	11.90		26.80	34
0180 9" wide		77.28	.207		16.30	12.70		29	37
0200 12" wide		72.24	.221		17.95	13.60		31.55	40
0400 18" wide		68.88	.232		21	14.25		35.25	44
0600 24" wide		65.52	.244		24	15		39	49
0650 30" wide		57.12	.280		30.50	17.20		47.70	59
0800 6" rung spacing, 6" wide		84	.190		13.60	11.70		25.30	32.50
0850 9" wide		78.96	.203		14.90	12.45		27.35	35
0860 12" wide		73.92	.216		15.65	13.30		28.95	37
0870 18" wide		70.56	.227		17.90	13.90		31.80	40
0880 24" wide		67.20	.238		19.95	14.60		34.55	44
0890 30" wide		58.80	.272		23.50	16.70		40.20	51
0910 9" rung spacing, 6" wide		85.68	.187		12.65	11.45		24.10	31
0920 9" wide		82.32	.194		13	11.90		24.90	32
0930 12" wide		78.96	.203		14.90	12.45		27.35	35
0940 18" wide		75.60	.212		16.85	13		29.85	38
0950 24" wide		72.24	.221		19.15	13.60		32.75	41
0960 30" wide		67.20	.238		21	14.60		35.60	45
0980 12" rung spacing, 6" wide		89.04	.180		12.45	11		23.45	30
0990 9" wide		87.36	.183		12.75	11.25		24	31
1000 12" wide		84	.190		13.50	11.70		25.20	32.50
1010 18" wide		80.64	.198		14.50	12.15		26.65	34
1020 24" wide		78.96	.203		15.55	12.45		28	35.50
1030 30" wide		73.92	.216		17.65	13.30		30.95	39
1041 18" rung spacing, 6" wide		90.72	.176		12.40	10.80		23.20	30
1042 9" wide		89.04	.180		12.55	11		23.55	30
1043 12" wide		85.68	.187		13.35	11.45		24.80	32
1044 18" wide		82.32	.194		14.05	11.90		25.95	33
1045 24" wide		80.64	.198		14.50	12.15		26.65	34
1046 30" wide		75.60	.212		15.25	13		28.25	36
1050 Elbows horiz. 9" rung spacing, 90°, 12" radius, 6" wide		8.06	1.980	Ea.	57.50	121		178.50	245
1060 9" wide		7.06	2.270		61	139		200	274
1070 12" wide		6.38	2.510		66	154		220	300
1080 18" wide		5.21	3.070		87	188		275	375
1090 24" wide		4.54	3.530		97.50	217		314.50	425
1100 30" wide		4.03	3.970		129	244		373	505
1120 90°, 24" radius, 6" wide		7.73	2.070		119	127		246	320
1130 9" wide		6.72	2.380		122	146		268	350
1140 12" wide		6.05	2.650		125	163		288	380
1150 18" wide		4.87	3.280		136	201		337	450
1160 24" wide		4.20	3.810		146	234		380	510
1170 30" wide		3.70	4.330		159	266		425	570
1190 90°, 36" radius, 6" wide		7.39	2.160		131	133		264	340
1200 9" wide		6.38	2.510		138	154		292	380
1210 12" wide		5.71	2.800		145	172		317	415
1220 18" wide		4.54	3.530		166	217		383	500
1230 24" wide		3.86	4.140		181	254		435	580
1240 30" wide		3.36	4.760		218	292		510	675
1260 45°, 12" radius, 6" wide		11.09	1.440		44.50	88.50		133	181

26 05 36.10 Cable Tray Ladder Type		Crew	Daily Output	Labor-Hours	Unit	Material	2020 Bare Costs Labor	Equipment	Total	Total Incl O&P
1270	9" wide	2 Elec	9.24	1.730	Ea.	46	106		152	209
1280	12" wide		8.06	1.980		48	121		169	234
1290	18" wide		6.38	2.510		50.50	154		204.50	285
1300	24" wide		5.21	3.070		65.50	188		253.50	350
1310	30" wide		4.54	3.530		83.50	217		300.50	410
1330	45°, 24" radius, 6" wide		10.75	1.490		56	91.50		147.50	198
1340	9" wide		8.90	1.800		60	110		170	230
1350	12" wide		7.73	2.070		67	127		194	263
1360	18" wide		6.05	2.650		69	163		232	320
1370	24" wide		4.87	3.280		83.50	201		284.50	390
1380	30" wide		4.20	3.810		99.50	234		333.50	460
1400	45°, 36" radius, 6" wide		10.42	1.540		80.50	94.50		175	230
1410	9" wide		8.57	1.870		85.50	115		200.50	265
1420	12" wide		7.39	2.160		87	133		220	293
1430	18" wide		5.71	2.800		94	172		266	360
1440	24" wide		4.54	3.530		99.50	217		316.50	430
1450	30" wide	↓	3.86	4.140	↓	129	254		383	520
1470	Elbows horizontal, 4" rung spacing, use 9" rung x 1.50									
1480	6" rung spacing, use 9" rung x 1.20									
1490	12" rung spacing, use 9" rung x 0.93									
1500	Elbows vertical, 9" rung spacing, 90°, 12" radius, 6" wide	2 Elec	8.06	1.980	Ea.	79	121		200	268
1510	9" wide		7.06	2.270		80.50	139		219.50	296
1520	12" wide		6.38	2.510		83.50	154		237.50	320
1530	18" wide		5.21	3.070		87	188		275	375
1540	24" wide		4.54	3.530		90.50	217		307.50	420
1550	30" wide		4.03	3.970		97.50	244		341.50	470
1570	24" radius, 6" wide		7.73	2.070		112	127		239	310
1580	9" wide		6.72	2.380		115	146		261	345
1590	12" wide		6.05	2.650		119	163		282	370
1600	18" wide		4.87	3.280		125	201		326	440
1610	24" wide		4.20	3.810		132	234		366	495
1620	30" wide		3.70	4.330		138	266		404	545
1640	36" radius, 6" wide		7.39	2.160		150	133		283	360
1650	9" wide		6.38	2.510		155	154		309	400
1660	12" wide		5.71	2.800		157	172		329	430
1670	18" wide		4.54	3.530		171	217		388	510
1680	24" wide		3.86	4.140		173	254		427	570
1690	30" wide	↓	3.36	4.760	↓	195	292		487	650
1710	Elbows vertical, 4" rung spacing, use 9" rung x 1.25									
1720	6" rung spacing, use 9" rung x 1.15									
1730	12" rung spacing, use 9" rung x 0.90									
1740	Tee horizontal, 9" rung spacing, 12" radius, 6" wide	2 Elec	4.20	3.810	Ea.	122	234		356	485
1750	9" wide		3.86	4.140		131	254		385	525
1760	12" wide		3.70	4.330		138	266		404	545
1770	18" wide		3.36	4.760		157	292		449	610
1780	24" wide		3.02	5.290		185	325		510	690
1790	30" wide		2.86	5.600		221	345		566	755
1810	24" radius, 6" wide		3.86	4.140		190	254		444	590
1820	9" wide		3.53	4.540		199	279		478	635
1830	12" wide		3.36	4.760		206	292		498	660
1840	18" wide		3.02	5.290		223	325		548	730
1850	24" wide		2.69	5.950		242	365		607	810
1860	30" wide	↓	2.52	6.350		263	390		653	870

Post-Installation Change Order Costs

26 05 36.10 Cable Tray Ladder Type		Crew	Daily Output	Labor-Hours	Unit	Material	2020 Bare Costs Labor	Equipment	Total	Total Incl O&P
1880	36" radius, 6" wide	2 Elec	3.53	4.540	Ea.	284	279		563	725
1890	9" wide		3.19	5.010		296	305		601	785
1900	12" wide		3.02	5.290		315	325		640	830
1910	18" wide		2.69	5.950		340	365		705	915
1920	24" wide		2.35	6.800		380	415		795	1,050
1930	30" wide		2.18	7.330		415	450		865	1,125
1980	Tee vertical, 9" rung spacing, 12" radius, 6" wide		4.54	3.530		207	217		424	550
1990	9" wide		4.37	3.660		209	225		434	565
2000	12" wide		4.20	3.810		211	234		445	580
2010	18" wide		3.86	4.140		213	254		467	615
2020	24" wide		3.70	4.330		218	266		484	635
2030	30" wide		3.36	4.760		232	292		524	690
2050	24" radius, 6" wide		4.20	3.810		375	234		609	760
2060	9" wide		4.03	3.970		380	244		624	780
2070	12" wide		3.86	4.140		385	254		639	800
2080	18" wide		3.53	4.540		395	279		674	850
2090	24" wide		3.36	4.760		405	292		697	880
2100	30" wide		3.02	5.290		415	325		740	945
2120	36" radius, 6" wide		3.86	4.140		695	254		949	1,150
2130	9" wide		3.70	4.330		710	266		976	1,175
2140	12" wide		3.53	4.540		715	279		994	1,200
2150	18" wide		3.19	5.010		725	305		1,030	1,250
2160	24" wide		3.02	5.290		735	325		1,060	1,300
2170	30" wide		2.69	5.950		760	365		1,125	1,375
2190	Tee, 4" rung spacing, use 9" rung x 1.30									
2200	6" rung spacing, use 9" rung x 1.20									
2210	12" rung spacing, use 9" rung x 0.90									
2220	Cross horizontal, 9" rung spacing, 12" radius, 6" wide	2 Elec	3.36	4.760	Ea.	150	292		442	600
2230	9" wide		3.19	5.010		158	305		463	635
2240	12" wide		3.02	5.290		173	325		498	675
2250	18" wide		2.86	5.600		188	345		533	715
2260	24" wide		2.52	6.350		209	390		599	810
2270	30" wide		2.35	6.800		253	415		668	900
2290	24" radius, 6" wide		3.02	5.290		293	325		618	805
2300	9" wide		2.86	5.600		300	345		645	840
2310	12" wide		2.69	5.950		310	365		675	885
2320	18" wide		2.52	6.350		330	390		720	945
2330	24" wide		2.18	7.330		350	450		800	1,050
2340	30" wide		2.02	7.940		375	485		860	1,125
2360	36" radius, 6" wide		2.69	5.950		410	365		775	995
2370	9" wide		2.52	6.350		425	390		815	1,050
2380	12" wide		2.35	6.800		445	415		860	1,100
2390	18" wide		2.18	7.330		490	450		940	1,200
2400	24" wide		1.85	8.660		545	530		1,075	1,400
2410	30" wide		1.68	9.520		590	585		1,175	1,525
2430	Cross horizontal, 4" rung spacing, use 9" rung x 1.30									
2440	6" rung spacing, use 9" rung x 1.20									
2450	12" rung spacing, use 9" rung x 0.90									
2460	Reducer, 9" to 6" wide tray	2 Elec	10.92	1.470	Ea.	83.50	90		173.50	226
2470	12" to 9" wide tray		10.08	1.590		85.50	97.50		183	239
2480	18" to 12" wide tray		8.74	1.830		85.50	112		197.50	261
2490	24" to 18" wide tray		7.56	2.120		87	130		217	290
2500	30" to 24" wide tray		6.72	2.380		89	146		235	315

26 05 36.10 Cable Tray Ladder Type		Crew	Daily Output	Labor-Hours	Unit	Material	2020 Bare Costs Labor	Equipment	Total	Total Incl O&P
2510	36" to 30" wide tray	2 Elec	5.88	2.720	Ea.	99.50	167		266.50	355
2511	Reducer, 18" to 6" wide tray		8.74	1.830		89	112		201	265
2512	24" to 12" wide tray		7.56	2.120		90.50	130		220.50	294
2513	30" to 18" wide tray		6.72	2.380		93	146		239	320
2514	30" to 12" wide tray		6.72	2.380		98.50	146		244.50	325
2515	36" to 24" wide tray		5.88	2.720		101	167		268	360
2516	36" to 18" wide tray		5.88	2.720		101	167		268	360
2517	36" to 12" wide tray		5.88	2.720		101	167		268	360
2520	Dropout or end plate, 6" wide		26.88	.595		9.35	36.50		45.85	65
2530	9" wide		23.52	.680		11	41.50		52.50	74
2540	12" wide		21.84	.733		12.05	45		57.05	80.50
2550	18" wide		18.48	.866		14.10	53		67.10	94.50
2560	24" wide		16.80	.952		17.10	58.50		75.60	106
2570	30" wide		15.12	1.060		18.70	65		83.70	118
2590	Tray connector		40.32	.397	▼	16.55	24.50		41.05	54.50
3200	Aluminum tray, 4" deep, 6" rung spacing, 6" wide		112.56	.142	L.F.	15.75	8.70		24.45	30.50
3210	9" wide		107.52	.149		16.75	9.15		25.90	32
3220	12" wide		104.16	.154		17.65	9.40		27.05	33.50
3230	18" wide		95.76	.167		19.70	10.25		29.95	37
3240	24" wide		89.04	.180		23	11		34	41.50
3250	30" wide		84	.190		25	11.70		36.70	45.50
3270	9" rung spacing, 6" wide		117.60	.136		12.80	8.35		21.15	26.50
3280	9" wide		112.56	.142		13.50	8.70		22.20	28
3290	12" wide		109.20	.147		13.70	9		22.70	28.50
3300	18" wide		102.48	.156		16	9.60		25.60	32
3310	24" wide		97.44	.164		18.60	10.05		28.65	35.50
3320	30" wide		90.72	.176		20.50	10.80		31.30	38.50
3340	12" rung spacing, 6" wide		122.64	.130		13.65	8		21.65	27
3350	9" wide		117.60	.136		13.80	8.35		22.15	27.50
3360	12" wide		112.56	.142		14.15	8.70		22.85	28.50
3370	18" wide		107.52	.149		15.10	9.15		24.25	30
3380	24" wide		104.16	.154		16.75	9.40		26.15	32.50
3390	30" wide		95.76	.167		17.95	10.25		28.20	35
3401	18" rung spacing, 6" wide		126	.127		13.55	7.80		21.35	26.50
3402	9" wide tray		120.96	.132		13.70	8.10		21.80	27
3403	12" wide tray		117.60	.136		14.15	8.35		22.50	28
3404	18" wide tray		112.56	.142		15	8.70		23.70	29.50
3405	24" wide tray		109.20	.147		16.15	9		25.15	31
3406	30" wide tray		100.80	.159	▼	16.75	9.75		26.50	33
3410	Elbows horiz., 9" rung spacing, 90°, 12" radius, 6" wide		8.06	1.980	Ea.	60.50	121		181.50	248
3420	9" wide		7.06	2.270		62.50	139		201.50	276
3430	12" wide		6.38	2.510		74	154		228	310
3440	18" wide		5.21	3.070		83.50	188		271.50	370
3450	24" wide		4.54	3.530		103	217		320	435
3460	30" wide		4.03	3.970		109	244		353	485
3480	24" radius, 6" wide		7.73	2.070		108	127		235	310
3490	9" wide		6.72	2.380		112	146		258	340
3500	12" wide		6.05	2.650		118	163		281	370
3510	18" wide		4.87	3.280		128	201		329	440
3520	24" wide		4.20	3.810		141	234		375	505
3530	30" wide		3.70	4.330		153	266		419	565
3550	90°, 36" radius, 6" wide		7.39	2.160		119	133		252	325
3560	9" wide		6.38	2.510	▼	122	154		276	365

Post-Installation Change Order Costs

26 05 36 – Cable Trays for Electrical Systems

26 05 36.10 Cable Tray Ladder Type		Crew	Daily Output	Labor-Hours	Unit	Material	2020 Bare Costs Labor	Equipment	Total	Total Incl O&P
3570	12" wide	2 Elec	5.71	2.800	Ea.	136	172		308	405
3580	18" wide		4.54	3.530		153	217		370	490
3590	24" wide		3.86	4.140		172	254		426	570
3600	30" wide		3.36	4.760		190	292		482	645
3620	45°, 12" radius, 6" wide		11.09	1.440		40	88.50		128.50	176
3630	9" wide		9.24	1.730		41.50	106		147.50	204
3640	12" wide		8.06	1.980		43	121		164	229
3650	18" wide		6.38	2.510		51	154		205	285
3660	24" wide		5.21	3.070		63	188		251	350
3670	30" wide		4.54	3.530		72.50	217		289.50	400
3690	45°, 24" radius, 6" wide		10.75	1.490		53.50	91.50		145	195
3700	9" wide		8.90	1.800		60.50	110		170.50	231
3710	12" wide		7.73	2.070		62	127		189	258
3720	18" wide		6.05	2.650		66.50	163		229.50	315
3730	24" wide		4.87	3.280		77	201		278	385
3740	30" wide		4.20	3.810		92	234		326	450
3760	45°, 36" radius, 6" wide		10.42	1.540		72.50	94.50		167	221
3770	9" wide		8.57	1.870		74	115		189	253
3780	12" wide		7.39	2.160		75.50	133		208.50	280
3790	18" wide		5.71	2.800		83	172		255	350
3800	24" wide		4.54	3.530		102	217		319	435
3810	30" wide		3.86	4.140		108	254		362	500
3830	Elbows horizontal, 4" rung spacing, use 9" rung x 1.50									
3840	6" rung spacing, use 9" rung x 1.20									
3850	12" rung spacing, use 9" rung x 0.93									
3860	Elbows vertical, 9" rung spacing, 90°, 12" radius, 6" wide	2 Elec	8.06	1.980	Ea.	84.50	121		205.50	274
3870	9" wide		7.06	2.270		89	139		228	305
3880	12" wide		6.38	2.510		90.50	154		244.50	330
3890	18" wide		5.21	3.070		93.50	188		281.50	385
3900	24" wide		4.54	3.530		99.50	217		316.50	430
3910	30" wide		4.03	3.970		104	244		348	480
3930	24" radius, 6" wide		7.73	2.070		92	127		219	290
3940	9" wide		6.72	2.380		95	146		241	320
3950	12" wide		6.05	2.650		98	163		261	350
3960	18" wide		4.87	3.280		104	201		305	415
3970	24" wide		4.20	3.810		110	234		344	470
3980	30" wide		3.70	4.330		116	266		382	520
4000	36" radius, 6" wide		7.39	2.160		142	133		275	355
4010	9" wide		6.38	2.510		145	154		299	390
4020	12" wide		5.71	2.800		154	172		326	425
4030	18" wide		4.54	3.530		160	217		377	495
4040	24" wide		3.86	4.140		166	254		420	565
4050	30" wide		3.36	4.760		179	292		471	630
4070	Elbows vertical, 4" rung spacing, use 9" rung x 1.25									
4080	6" rung spacing, use 9" rung x 1.15									
4090	12" rung spacing, use 9" rung x 0.90									
4100	Tee horizontal, 9" rung spacing, 12" radius, 6" wide	2 Elec	4.20	3.810	Ea.	94	234		328	455
4110	9" wide		3.86	4.140		101	254		355	490
4120	12" wide		3.70	4.330		96	266		362	500
4130	18" wide		3.53	4.540		122	279		401	550
4140	24" wide		3.36	4.760		133	292		425	580
4150	30" wide		3.02	5.290		156	325		481	655
4170	24" radius, 6" wide		3.86	4.140		160	254		414	555

26 05 36.10 Cable Tray Ladder Type		Crew	Daily Output	Labor-Hours	Unit	Material	2020 Bare Costs Labor	Equipment	Total	Total Incl O&P
4180	9" wide	2 Elec	3.53	4.540	Ea.	169	279		448	600
4190	12" wide		3.36	4.760		173	292		465	625
4200	18" wide		3.19	5.010		190	305		495	670
4210	24" wide		3.02	5.290		207	325		532	715
4220	30" wide		2.69	5.950		225	365		590	795
4240	36" radius, 6" wide		3.53	4.540		227	279		506	665
4250	9" wide		3.19	5.010		233	305		538	715
4260	12" wide		3.02	5.290		255	325		580	765
4270	18" wide		2.86	5.600		268	345		613	805
4280	24" wide		2.69	5.950		296	365		661	870
4290	30" wide		2.35	6.800		375	415		790	1,025
4310	Tee vertical, 9" rung spacing, 12" radius, 6" wide		4.54	3.530		210	217		427	550
4320	9" wide		4.37	3.660		218	225		443	575
4330	12" wide		4.20	3.810		218	234		452	590
4340	18" wide		3.86	4.140		224	254		478	625
4350	24" wide		3.70	4.330		242	266		508	660
4360	30" wide		3.53	4.540		246	279		525	685
4380	24" radius, 6" wide		4.20	3.810		355	234		589	745
4390	9" wide		4.03	3.970		360	244		604	765
4400	12" wide		3.86	4.140		370	254		624	785
4410	18" wide		3.53	4.540		375	279		654	830
4420	24" wide		3.36	4.760		385	292		677	860
4430	30" wide		3.19	5.010		395	305		700	895
4450	36" radius, 6" wide		3.86	4.140		785	254		1,039	1,250
4460	9" wide		3.70	4.330		800	266		1,066	1,275
4470	12" wide		3.53	4.540		835	279		1,114	1,325
4480	18" wide		3.19	5.010		840	305		1,145	1,375
4490	24" wide		3.02	5.290		860	325		1,185	1,425
4500	30" wide	▼	2.86	5.600	▼	870	345		1,215	1,475
4520	Tees, 4" rung spacing, use 9" rung x 1.30									
4530	6" rung spacing, use 9" rung x 1.20									
4540	12" rung spacing, use 9" rung x 0.90									
4550	Cross horizontal, 9" rung spacing, 12" radius, 6" wide	2 Elec	3.70	4.330	Ea.	130	266		396	540
4560	9" wide		3.53	4.540		145	279		424	575
4570	12" wide		3.36	4.760		139	292		431	590
4580	18" wide		3.02	5.290		178	325		503	680
4590	24" wide		2.86	5.600		184	345		529	710
4600	30" wide		2.52	6.350		216	390		606	820
4620	24" radius, 6" wide		3.36	4.760		250	292		542	710
4630	9" wide		3.19	5.010		259	305		564	745
4640	12" wide		3.02	5.290		267	325		592	780
4650	18" wide		2.69	5.950		280	365		645	855
4660	24" wide		2.52	6.350		305	390		695	915
4670	30" wide		2.18	7.330		325	450		775	1,025
4690	36" radius, 6" wide		3.02	5.290		296	325		621	810
4700	9" wide		2.86	5.600		310	345		655	850
4710	12" wide		2.69	5.950		325	365		690	905
4720	18" wide		2.35	6.800		350	415		765	1,000
4730	24" wide		2.18	7.330		430	450		880	1,150
4740	30" wide	▼	1.85	8.660	▼	475	530		1,005	1,325
4760	Cross horizontal, 4" rung spacing, use 9" rung x 1.30									
4770	6" rung spacing, use 9" rung x 1.20									
4780	12" rung spacing, use 9" rung x 0.90									

26 05 36.10 Cable Tray Ladder Type		Crew	Daily Output	Labor-Hours	Unit	Material	2020 Bare Costs Labor	Equipment	Total	Total Incl O&P
4790	Reducer, 9" to 6" wide tray	2 Elec	13.44	1.190	Ea.	69.50	73		142.50	186
4800	12" to 9" wide tray		11.76	1.360		72.50	83.50		156	204
4810	18" to 12" wide tray		10.42	1.540		72.50	94.50		167	221
4820	24" to 18" wide tray		8.90	1.800		74	110		184	246
4830	30" to 24" wide tray		7.73	2.070		75.50	127		202.50	272
4840	36" to 30" wide tray		6.72	2.380		80	146		226	305
4841	Reducer, 18" to 6" wide tray		10.42	1.540		72.50	94.50		167	221
4842	24" to 12" wide tray		8.90	1.800		74	110		184	246
4843	30" to 18" wide tray		7.73	2.070		75.50	127		202.50	272
4844	30" to 12" wide tray		7.73	2.070		75.50	127		202.50	272
4845	36" to 24" wide tray		6.72	2.380		80	146		226	305
4846	36" to 18" wide tray		6.72	2.380		80	146		226	305
4847	36" to 12" wide tray		6.72	2.380		80	146		226	305
4850	Dropout or end plate, 6" wide		26.88	.595		9.10	36.50		45.60	64.50
4860	9" wide tray		23.52	.680		9.85	41.50		51.35	73
4870	12" wide tray		21.84	.733		10.60	45		55.60	78.50
4880	18" wide tray		18.48	.866		14.30	53		67.30	94.50
4890	24" wide tray		16.80	.952		16.75	58.50		75.25	105
4900	30" wide tray		15.12	1.060		19.85	65		84.85	119
4920	Tray connector		40.32	.397		13.30	24.50		37.80	51
8000	Elbow 36" radius horiz., 60°, 6" wide tray		8.90	1.800		101	110		211	275
8010	9" wide tray		7.56	2.120		110	130		240	315
8020	12" wide tray		6.55	2.440		113	150		263	345
8030	18" wide tray		5.21	3.070		120	188		308	410
8040	24" wide tray		4.20	3.810		141	234		375	505
8050	30" wide tray		3.70	4.330		144	266		410	555
8060	30°, 6" wide tray		11.76	1.360		77	83.50		160.50	209
8070	9" wide tray		9.58	1.670		81.50	102		183.50	243
8080	12" wide tray		8.23	1.940		83	119		202	269
8090	18" wide tray		6.22	2.570		87.50	158		245.50	330
8100	24" wide tray		4.87	3.280		107	201		308	415
8110	30" wide tray		4.03	3.970		110	244		354	485
8120	Adjustable, 6" wide tray		10.42	1.540		102	94.50		196.50	254
8130	9" wide tray		8.57	1.870		108	115		223	290
8140	12" wide tray		7.39	2.160		110	133		243	320
8150	18" wide tray		5.71	2.800		120	172		292	390
8160	24" wide tray		4.54	3.530		127	217		344	460
8170	30" wide tray		3.86	4.140		142	254		396	535
8180	Wye 36" radius horiz., 45°, 6" wide tray		3.86	4.140		111	254		365	500
8190	9" wide tray		3.70	4.330		117	266		383	525
8200	12" wide tray		3.53	4.540		122	279		401	550
8210	18" wide tray		3.19	5.010		138	305		443	610
8220	24" wide tray		3.02	5.290		154	325		479	655
8230	30" wide tray		2.86	5.600		172	345		517	700
8240	Elbow 36" radius vert. in/outside, 60°, 6" wide tray		8.90	1.800		111	110		221	286
8250	9" wide tray		7.56	2.120		116	130		246	320
8260	12" wide tray		6.55	2.440		117	150		267	350
8270	18" wide tray		5.21	3.070		122	188		310	415
8280	24" wide tray		4.20	3.810		129	234		363	490
8290	30" wide tray		3.70	4.330		138	266		404	545
8300	45°, 6" wide tray		10.42	1.540		100	94.50		194.50	251
8310	9" wide tray		8.57	1.870		106	115		221	288
8320	12" wide tray		7.39	2.160		108	133		241	315

Post-Installation Change Order Costs   For customer support on your Electrical Change Order Costs with RSMeans data, call 800.448.8182.

26 05 36.10 Cable Tray Ladder Type		Crew	Daily Output	Labor-Hours	Unit	Material	2020 Bare Costs Labor	Equipment	Total	Total Incl O&P
8330	18" wide tray	2 Elec	5.71	2.800	Ea.	111	172		283	380
8340	24" wide tray		4.54	3.530		117	217		334	450
8350	30" wide tray		3.86	4.140		120	254		374	510
8360	30°, 6" wide tray		11.76	1.360		88	83.50		171.50	221
8370	9" wide tray		9.58	1.670		89.50	102		191.50	252
8380	12" wide tray		8.23	1.940		95.50	119		214.50	282
8390	18" wide tray		6.22	2.570		97	158		255	340
8400	24" wide tray		4.87	3.280		99.50	201		300.50	410
8410	30" wide tray		4.03	3.970		105	244		349	480
8660	Adjustable, 6" wide tray		10.42	1.540		100	94.50		194.50	251
8670	9" wide tray		8.57	1.870		106	115		221	288
8680	12" wide tray		7.39	2.160		108	133		241	315
8690	18" wide tray		5.71	2.800		120	172		292	390
8700	24" wide tray		4.54	3.530		127	217		344	460
8710	30" wide tray		3.86	4.140		141	254		395	535
8720	Cross, vertical, 24" radius, 6" wide tray		3.02	5.290		730	325		1,055	1,275
8730	9" wide tray		2.86	5.600		740	345		1,085	1,325
8740	12" wide tray		2.69	5.950		765	365		1,130	1,375
8750	18" wide tray		2.35	6.800		800	415		1,215	1,500
8760	24" wide tray		2.18	7.330		825	450		1,275	1,575
8770	30" wide tray		1.85	8.660		875	530		1,405	1,750
9200	Splice plate	1 Elec	40.32	.198	Pr.	13.95	12.15		26.10	33.50
9210	Expansion joint		40.32	.198		16.75	12.15		28.90	36.50
9220	Horizontal hinged		40.32	.198		13.95	12.15		26.10	33.50
9230	Vertical hinged		40.32	.198		16.75	12.15		28.90	36.50
9240	Ladder hanger, vertical		23.52	.340	Ea.	4.18	21		25.18	35.50
9250	Ladder to channel connector		20.16	.397		48.50	24.50		73	90
9260	Ladder to box connector, 30" wide		15.96	.501		48.50	31		79.50	99.50
9270	24" wide		16.80	.476		48.50	29		77.50	97
9280	18" wide		17.64	.454		47	28		75	93
9290	12" wide		18.48	.433		44	26.50		70.50	88
9300	9" wide		19.32	.414		39.50	25.50		65	81
9310	6" wide		20.16	.397		31	24.50		55.50	71
9320	Ladder floor flange		20.16	.397		31.50	24.50		56	71
9330	Cable roller for tray, 30" wide		8.40	.952		275	58.50		333.50	390
9340	24" wide		9.24	.866		234	53		287	335
9350	18" wide		10.08	.794		223	48.50		271.50	320
9360	12" wide		10.92	.733		175	45		220	260
9370	9" wide		11.76	.680		155	41.50		196.50	233
9380	6" wide		12.60	.635		126	39		165	197
9390	Pulley, single wheel		10.08	.794		288	48.50		336.50	390
9400	Triple wheel		8.40	.952		575	58.50		633.50	715
9440	Nylon cable tie, 14" long		67.20	.119		.57	7.30		7.87	11.55
9450	Ladder, hold down clamp		50.40	.159		8.45	9.75		18.20	24
9460	Cable clamp		50.40	.159		9	9.75		18.75	24.50
9470	Wall bracket, 30" wide tray		15.96	.501		51	31		82	102
9480	24" wide tray		16.80	.476		39.50	29		68.50	87
9490	18" wide tray		17.64	.454		35.50	28		63.50	80.50
9500	12" wide tray		18.48	.433		33	26.50		59.50	76
9510	9" wide tray		19.32	.414		29	25.50		54.50	70
9520	6" wide tray		20.16	.397		28.50	24.50		53	67.50

Post-Installation Change Order Costs

26 05 36 – Cable Trays for Electrical Systems

26 05 36.20 Cable Tray Solid Bottom	Crew	Daily Output	Labor-Hours	Unit	Material	2020 Bare Costs Labor	Equipment	Total	Total Incl O&P
0010 **CABLE TRAY SOLID BOTTOM** w/ftngs. & supports, 3" deep, to 15' high									
0200 For higher elevations, see Section 26 05 36.40									
0220 Galvanized steel, tray, 6" wide	2 Elec	100.80	.159	L.F.	11.80	9.75		21.55	27.50
0240 12" wide		84	.190		15.15	11.70		26.85	34
0260 18" wide		58.80	.272		18.40	16.70		35.10	45.50
0280 24" wide		50.40	.317		21.50	19.50		41	52.50
0300 30" wide		42	.381		25.50	23.50		49	63.50
0340 Elbow horizontal, 90°, 12" radius, 6" wide		8.06	1.980	Ea.	89.50	121		210.50	280
0360 12" wide		5.71	2.800		100	172		272	365
0370 18" wide		4.54	3.530		115	217		332	445
0380 24" wide		3.70	4.330		140	266		406	550
0390 30" wide		3.19	5.010		167	305		472	645
0420 24" radius, 6" wide		7.73	2.070		128	127		255	330
0440 12" wide		5.38	2.980		148	183		331	435
0450 18" wide		4.20	3.810		172	234		406	540
0460 24" wide		3.36	4.760		200	292		492	655
0470 30" wide		2.86	5.600		234	345		579	770
0500 36" radius, 6" wide		7.39	2.160		186	133		319	400
0520 12" wide		5.04	3.170		210	194		404	520
0530 18" wide		3.86	4.140		257	254		511	665
0540 24" wide		3.02	5.290		278	325		603	790
0550 30" wide		2.52	6.350		335	390		725	950
0580 Elbow vertical, 90°, 12" radius, 6" wide		8.06	1.980		109	121		230	300
0600 12" wide		5.71	2.800		119	172		291	385
0610 18" wide		4.54	3.530		134	217		351	470
0620 24" wide		3.70	4.330		136	266		402	545
0630 30" wide		3.19	5.010		145	305		450	620
0670 24" radius, 6" wide		7.73	2.070		153	127		280	360
0690 12" wide		5.38	2.980		169	183		352	460
0700 18" wide		4.20	3.810		183	234		417	550
0710 24" wide		3.36	4.760		198	292		490	655
0720 30" wide		2.86	5.600		210	345		555	740
0750 36" radius, 6" wide		7.39	2.160		210	133		343	430
0770 12" wide		5.54	2.890		234	177		411	520
0780 18" wide		3.86	4.140		257	254		511	665
0790 24" wide		3.02	5.290		281	325		606	795
0800 30" wide		2.52	6.350		305	390		695	920
0840 Tee horizontal, 12" radius, 6" wide		4.20	3.810		126	234		360	490
0860 12" wide		3.36	4.760		136	292		428	585
0870 18" wide		2.86	5.600		162	345		507	690
0880 24" wide		2.35	6.800		181	415		596	820
0890 30" wide		2.18	7.330		207	450		657	900
0940 24" radius, 6" wide		3.86	4.140		195	254		449	595
0960 12" wide		3.02	5.290		228	325		553	735
0970 18" wide		2.52	6.350		253	390		643	860
0980 24" wide		2.02	7.940		345	485		830	1,100
0990 30" wide		1.85	8.660		370	530		900	1,200
1020 36" radius, 6" wide		3.53	4.540		305	279		584	755
1040 12" wide		2.69	5.950		345	365		710	920
1050 18" wide		2.18	7.330		370	450		820	1,075
1060 24" wide		1.85	8.660		480	530		1,010	1,325
1070 30" wide		1.68	9.520		520	585		1,105	1,450

26 05 36 – Cable Trays for Electrical Systems

26 05 36.20 Cable Tray Solid Bottom	Crew	Daily Output	Labor-Hours	Unit	Material	2020 Bare Costs Labor	Equipment	Total	Total Incl O&P
1100 Tee vertical, 12" radius, 6" wide	2 Elec	4.20	3.810	Ea.	207	234		441	580
1120 12" wide		3.36	4.760		216	292		508	670
1130 18" wide		3.02	5.290		219	325		544	725
1140 24" wide		2.86	5.600		240	345		585	775
1150 30" wide		2.52	6.350		257	390		647	865
1180 24" radius, 6" wide		3.86	4.140		305	254		559	715
1200 12" wide		3.02	5.290		320	325		645	835
1210 18" wide		2.69	5.950		335	365		700	915
1220 24" wide		2.52	6.350		350	390		740	965
1230 30" wide		2.18	7.330		385	450		835	1,100
1260 36" radius, 6" wide		3.53	4.540		475	279		754	940
1280 12" wide		2.69	5.950		485	365		850	1,075
1290 18" wide		2.35	6.800		520	415		935	1,200
1300 24" wide		2.18	7.330		540	450		990	1,250
1310 30" wide		1.85	8.660		630	530		1,160	1,475
1340 Cross horizontal, 12" radius, 6" wide		3.36	4.760		150	292		442	600
1360 12" wide		2.86	5.600		164	345		509	690
1370 18" wide		2.35	6.800		191	415		606	830
1380 24" wide		2.02	7.940		216	485		701	960
1390 30" wide		1.68	9.520		240	585		825	1,125
1420 24" radius, 6" wide		3.02	5.290		274	325		599	785
1440 12" wide		2.52	6.350		305	390		695	920
1450 18" wide		2.02	7.940		345	485		830	1,100
1460 24" wide		1.68	9.520		440	585		1,025	1,350
1470 30" wide		1.51	10.580		475	650		1,125	1,500
1500 36" radius, 6" wide		2.69	5.950		450	365		815	1,050
1520 12" wide		2.18	7.330		495	450		945	1,200
1530 18" wide		1.68	9.520		540	585		1,125	1,450
1540 24" wide		1.51	10.580		635	650		1,285	1,675
1550 30" wide		1.34	11.900		690	730		1,420	1,825
1580 Drop out or end plate, 6" wide		26.88	.595		18.30	36.50		54.80	74.50
1600 12" wide		21.84	.733		23	45		68	92
1610 18" wide		18.48	.866		24.50	53		77.50	106
1620 24" wide		16.80	.952		29	58.50		87.50	119
1630 30" wide		15.12	1.060		32	65		97	133
1660 Reducer, 12" to 6" wide		10.08	1.590		88	97.50		185.50	242
1680 18" to 12" wide		8.90	1.800		89.50	110		199.50	263
1700 18" to 6" wide		8.90	1.800		89.50	110		199.50	263
1720 24" to 18" wide		7.73	2.070		93	127		220	291
1740 24" to 12" wide		7.73	2.070		93	127		220	291
1760 30" to 24" wide		6.72	2.380		100	146		246	325
1780 30" to 18" wide		6.72	2.380		100	146		246	325
1800 30" to 12" wide		6.72	2.380		100	146		246	325
1820 36" to 30" wide		6.05	2.650		103	163		266	355
1840 36" to 24" wide		6.05	2.650		103	163		266	355
1860 36" to 18" wide		6.05	2.650		103	163		266	355
1880 36" to 12" wide		6.05	2.650		103	163		266	355
2000 Aluminum tray, 6" wide		126	.127	L.F.	10.95	7.80		18.75	23.50
2020 12" wide		109.20	.147		14.40	9		23.40	29
2030 18" wide		84	.190		18.05	11.70		29.75	37.50
2040 24" wide		75.60	.212		23	13		36	44.50
2050 30" wide		58.80	.272		27	16.70		43.70	54.50
2080 Elbow horizontal, 90°, 12" radius, 6" wide		8.06	1.980	Ea.	88.50	121		209.50	278

For customer support on your Electrical Change Order Costs with RSMeans data, call 800.448.8182.

Post-Installation Change Order Costs

26 05 36 – Cable Trays for Electrical Systems

26 05 36.20 Cable Tray Solid Bottom	Crew	Daily Output	Labor-Hours	Unit	Material	2020 Bare Costs Labor	Equipment	Total	Total Incl O&P
2100	12" wide	2 Elec	6.38	2.510	Ea.	99.50	154		253.50	340
2110	18" wide		5.71	2.800		122	172		294	390
2120	24" wide		4.87	3.280		139	201		340	455
2130	30" wide		4.20	3.810		170	234		404	535
2160	24" radius, 6" wide		7.73	2.070		126	127		253	330
2180	12" wide		6.05	2.650		148	163		311	405
2190	18" wide		5.38	2.980		170	183		353	460
2200	24" wide		4.54	3.530		209	217		426	550
2210	30" wide		3.86	4.140		240	254		494	645
2240	36" radius, 6" wide		7.39	2.160		186	133		319	400
2260	12" wide		5.71	2.800		223	172		395	500
2270	18" wide		5.04	3.170		265	194		459	580
2280	24" wide		4.20	3.810		284	234		518	660
2290	30" wide		3.53	4.540		320	279		599	770
2320	Elbow vertical, 90°, 12" radius, 6" wide		8.06	1.980		105	121		226	296
2340	12" wide		6.38	2.510		109	154		263	350
2350	18" wide		5.71	2.800		123	172		295	390
2360	24" wide		4.87	3.280		130	201		331	445
2370	30" wide		4.20	3.810		138	234		372	500
2400	24" radius, 6" wide		7.73	2.070		148	127		275	350
2420	12" wide		6.05	2.650		162	163		325	420
2430	18" wide		5.38	2.980		174	183		357	465
2440	24" wide		4.54	3.530		188	217		405	525
2450	30" wide		3.86	4.140		200	254		454	600
2480	36" radius, 6" wide		7.39	2.160		196	133		329	415
2500	12" wide		5.71	2.800		220	172		392	500
2510	18" wide		5.04	3.170		240	194		434	555
2520	24" wide		4.20	3.810		249	234		483	625
2530	30" wide		3.53	4.540		276	279		555	720
2560	Tee horizontal, 12" radius, 6" wide		4.20	3.810		137	234		371	500
2580	12" wide		3.70	4.330		166	266		432	575
2590	18" wide		3.36	4.760		186	292		478	640
2600	24" wide		3.02	5.290		208	325		533	715
2610	30" wide		2.52	6.350		240	390		630	845
2640	24" radius, 6" wide		3.86	4.140		225	254		479	630
2660	12" wide		3.36	4.760		253	292		545	715
2670	18" wide		3.02	5.290		286	325		611	800
2680	24" wide		2.52	6.350		350	390		740	965
2690	30" wide		2.02	7.940		400	485		885	1,175
2720	36" radius, 6" wide		3.53	4.540		365	279		644	815
2740	12" wide		3.02	5.290		400	325		725	925
2750	18" wide		2.69	5.950		460	365		825	1,050
2760	24" wide		2.18	7.330		540	450		990	1,250
2770	30" wide		1.68	9.520		590	585		1,175	1,525
2800	Tee vertical, 12" radius, 6" wide		4.20	3.810		195	234		429	565
2820	12" wide		3.70	4.330		199	266		465	615
2830	18" wide		3.53	4.540		207	279		486	640
2840	24" wide		3.36	4.760		220	292		512	675
2850	30" wide		3.02	5.290		240	325		565	750
2880	24" radius, 6" wide		3.86	4.140		286	254		540	695
2900	12" wide		3.36	4.760		305	292		597	770
2910	18" wide		3.19	5.010		325	305		630	815
2920	24" wide		3.02	5.290		350	325		675	870

26 05 36 – Cable Trays for Electrical Systems

	26 05 36.20 Cable Tray Solid Bottom	Crew	Daily Output	Labor-Hours	Unit	Material	2020 Bare Costs Labor	Equipment	Total	Total Incl O&P
2930	30" wide	2 Elec	2.69	5.950	Ea.	385	365		750	970
2960	36" radius, 6" wide		3.53	4.540		450	279		729	910
2980	12" wide		2.86	5.600		470	345		815	1,025
2990	18" wide		2.86	5.600		495	345		840	1,050
3000	24" wide		2.69	5.950		510	365		875	1,100
3010	30" wide		2.35	6.800		550	415		965	1,225
3040	Cross horizontal, 12" radius, 6" wide		3.70	4.330		174	266		440	585
3060	12" wide		3.36	4.760		196	292		488	650
3070	18" wide		2.86	5.600		223	345		568	755
3080	24" wide		2.35	6.800		253	415		668	900
3090	30" wide		2.18	7.330		286	450		736	985
3120	24" radius, 6" wide		3.36	4.760		305	292		597	770
3140	12" wide		3.02	5.290		350	325		675	870
3150	18" wide		2.52	6.350		380	390		770	1,000
3160	24" wide		2.02	7.940		460	485		945	1,225
3170	30" wide		1.85	8.660		510	530		1,040	1,350
3200	36" radius, 6" wide		3.02	5.290		530	325		855	1,075
3220	12" wide		2.69	5.950		560	365		925	1,150
3230	18" wide		2.18	7.330		620	450		1,070	1,350
3240	24" wide		1.68	9.520		675	585		1,260	1,625
3250	30" wide		1.51	10.580		730	650		1,380	1,775
3280	Dropout, or end plate, 6" wide		26.88	.595		17.45	36.50		53.95	73.50
3300	12" wide		21.84	.733		19.75	45		64.75	88.50
3310	18" wide		18.48	.866		23.50	53		76.50	105
3320	24" wide		16.80	.952		26	58.50		84.50	116
3330	30" wide		15.12	1.060		30.50	65		95.50	131
3380	Reducer, 12" to 6" wide		11.76	1.360		85.50	83.50		169	218
3400	18" to 12" wide		10.08	1.590		88.50	97.50		186	242
3420	18" to 6" wide		10.08	1.590		88.50	97.50		186	242
3440	24" to 18" wide		8.90	1.800		93.50	110		203.50	267
3460	24" to 12" wide		8.90	1.800		93.50	110		203.50	267
3480	30" to 24" wide		7.73	2.070		102	127		229	300
3500	30" to 18" wide		7.73	2.070		102	127		229	300
3520	30" to 12" wide		7.73	2.070		102	127		229	300
3540	36" to 30" wide		6.72	2.380		107	146		253	335
3560	36" to 24" wide		6.72	2.380		109	146		255	335
3580	36" to 18" wide		6.72	2.380		109	146		255	335
3600	36" to 12" wide		6.72	2.380		111	146		257	340

26 05 36.30 Cable Tray Trough

		Crew	Daily Output	Labor-Hours	Unit	Material	2020 Bare Costs Labor	Equipment	Total	Total Incl O&P
0010	**CABLE TRAY TROUGH** vented, w/ftngs. & supports, 6" deep, to 10' high									
0020	For higher elevations, see Section 26 05 36.40									
0200	Galvanized steel, tray, 6" wide	2 Elec	75.60	.212	L.F.	15.55	13		28.55	36.50
0240	12" wide		67.20	.238		14.20	14.60		28.80	37.50
0260	18" wide		58.80	.272		21.50	16.70		38.20	48.50
0280	24" wide		50.40	.317		48.50	19.50		68	82.50
0300	30" wide		42	.381		47.50	23.50		71	87.50
0340	Elbow horizontal, 90°, 12" radius, 6" wide		6.38	2.510	Ea.	124	154		278	365
0360	12" wide		4.70	3.400		147	209		356	470
0370	18" wide		3.70	4.330		170	266		436	580
0380	24" wide		3.02	5.290		210	325		535	715
0390	30" wide		2.69	5.950		237	365		602	805
0420	24" radius, 6" wide		6.05	2.650		183	163		346	445

For customer support on your Electrical Change Order Costs with RSMeans data, call 800.448.8182. **Post-Installation Change Order Costs**

26 05 36.30 Cable Tray Trough		Crew	Daily Output	Labor-Hours	Unit	Material	2020 Bare Costs Labor	Equipment	Total	Total Incl O&P
0440	12" wide	2 Elec	4.37	3.660	Ea.	212	225		437	570
0450	18" wide		3.36	4.760		255	292		547	715
0460	24" wide		2.69	5.950		288	365		653	860
0470	30" wide		2.35	6.800		335	415		750	985
0500	36" radius, 6" wide		5.71	2.800		239	172		411	520
0520	12" wide		4.03	3.970		272	244		516	665
0530	18" wide		3.02	5.290		325	325		650	845
0540	24" wide		2.35	6.800		335	415		750	990
0550	30" wide		2.02	7.940		375	485		860	1,150
0580	Elbow vertical, 90°, 12" radius, 6" wide		6.38	2.510		136	154		290	380
0600	12" wide		4.70	3.400		151	209		360	475
0610	18" wide		3.70	4.330		152	266		418	565
0620	24" wide		3.02	5.290		175	325		500	680
0630	30" wide		2.69	5.950		179	365		544	740
0660	24" radius, 6" wide		6.05	2.650		194	163		357	455
0680	12" wide		4.37	3.660		205	225		430	560
0690	18" wide		3.36	4.760		228	292		520	685
0700	24" wide		2.69	5.950		235	365		600	805
0710	30" wide		2.35	6.800		264	415		679	910
0740	36" radius, 6" wide		5.71	2.800		262	172		434	545
0760	12" wide		4.03	3.970		279	244		523	670
0770	18" wide		3.02	5.290		310	325		635	825
0780	24" wide		2.35	6.800		330	415		745	985
0790	30" wide		2.02	7.940		340	485		825	1,100
0820	Tee horizontal, 12" radius, 6" wide		3.36	4.760		158	292		450	610
0840	12" wide		2.69	5.950		177	365		542	740
0850	18" wide		2.35	6.800		200	415		615	840
0860	24" wide		2.02	7.940		232	485		717	980
0870	30" wide		1.85	8.660		264	530		794	1,075
0900	24" radius, 6" wide		3.02	5.290		262	325		587	775
0920	12" wide		2.35	6.800		286	415		701	935
0930	18" wide		2.02	7.940		320	485		805	1,075
0940	24" wide		1.68	9.520		415	585		1,000	1,325
0950	30" wide		1.51	10.580		450	650		1,100	1,450
0980	36" radius, 6" wide		2.69	5.950		375	365		740	960
1000	12" wide		2.02	7.940		430	485		915	1,200
1010	18" wide		1.68	9.520		470	585		1,055	1,375
1020	24" wide		1.34	11.900		565	730		1,295	1,700
1030	30" wide		1.18	13.610		600	835		1,435	1,900
1060	Tee vertical, 12" radius, 6" wide		3.36	4.760		249	292		541	710
1080	12" wide		2.69	5.950		253	365		618	825
1090	18" wide		2.52	6.350		260	390		650	865
1100	24" wide		2.35	6.800		274	415		689	920
1110	30" wide		2.18	7.330		290	450		740	990
1140	24" radius, 6" wide		3.02	5.290		335	325		660	855
1160	12" wide		2.35	6.800		350	415		765	1,000
1170	18" wide		2.18	7.330		370	450		820	1,075
1180	24" wide		2.02	7.940		395	485		880	1,150
1190	30" wide		1.85	8.660		415	530		945	1,250
1220	36" radius, 6" wide		2.69	5.950		545	365		910	1,150
1240	12" wide		2.02	7.940		560	485		1,045	1,350
1250	18" wide		1.85	8.660		580	530		1,110	1,425
1260	24" wide		1.68	9.520		610	585		1,195	1,550

26 05 Common Work Results for Electrical

26 05 36 – Cable Trays for Electrical Systems

26 05 36.30 Cable Tray Trough	Crew	Daily Output	Labor-Hours	Unit	Material	2020 Bare Costs Labor	Equipment	Total	Total Incl O&P
1270	30" wide	2 Elec	1.51	10.580	Ea.	710	650		1,360	1,750
1300	Cross horizontal, 12" radius, 6" wide		2.69	5.950		196	365		561	760
1320	12" wide		2.35	6.800		198	415		613	840
1330	18" wide		2.02	7.940		222	485		707	970
1340	24" wide		1.68	9.520		233	585		818	1,125
1350	30" wide		1.51	10.580		262	650		912	1,250
1380	24" radius, 6" wide		2.35	6.800		310	415		725	960
1400	12" wide		2.02	7.940		325	485		810	1,075
1410	18" wide		1.68	9.520		355	585		940	1,250
1420	24" wide		1.34	11.900		450	730		1,180	1,575
1430	30" wide		1.18	13.610		500	835		1,335	1,800
1460	36" radius, 6" wide		2.02	7.940		530	485		1,015	1,300
1480	12" wide		1.68	9.520		545	585		1,130	1,475
1490	18" wide		1.34	11.900		560	730		1,290	1,700
1500	24" wide		1.01	15.870		670	975		1,645	2,175
1510	30" wide		.84	19.050		740	1,175		1,915	2,550
1540	Dropout or end plate, 6" wide		21.84	.733		23.50	45		68.50	93
1560	12" wide		18.48	.866		29	53		82	111
1580	18" wide		16.80	.952		31.50	58.50		90	122
1600	24" wide		15.12	1.060		36.50	65		101.50	137
1620	30" wide		13.44	1.190		39	73		112	152
1660	Reducer, 12" to 6" wide		7.90	2.030		95.50	125		220.50	290
1680	18" to 12" wide		7.06	2.270		99	139		238	315
1700	18" to 6" wide		7.06	2.270		99	139		238	315
1720	24" to 18" wide		6.05	2.650		106	163		269	360
1740	24" to 12" wide		6.05	2.650		106	163		269	360
1760	30" to 24" wide		5.38	2.980		110	183		293	395
1780	30" to 18" wide		5.38	2.980		110	183		293	395
1800	30" to 12" wide		5.38	2.980		110	183		293	395
1820	36" to 30" wide		4.87	3.280		117	201		318	430
1840	36" to 24" wide		4.87	3.280		117	201		318	430
1860	36" to 18" wide		4.87	3.280		119	201		320	430
1880	36" to 12" wide		4.87	3.280		119	201		320	430
2000	Aluminum, tray, vented, 6" wide		100.80	.159	L.F.	16.10	9.75		25.85	32.50
2010	9" wide		92.40	.173		18.25	10.60		28.85	36
2020	12" wide		84	.190		23.50	11.70		35.20	43.50
2030	18" wide		75.60	.212		24.50	13		37.50	46.50
2040	24" wide		67.20	.238		29	14.60		43.60	53.50
2050	30" wide		58.80	.272		38.50	16.70		55.20	67
2080	Elbow horiz., 90°, 12" radius, 6" wide		6.38	2.510	Ea.	102	154		256	340
2090	9" wide		5.88	2.720		110	167		277	370
2100	12" wide		5.21	3.070		117	188		305	410
2110	18" wide		4.70	3.400		137	209		346	460
2120	24" wide		3.86	4.140		158	254		412	555
2130	30" wide		3.36	4.760		195	292		487	650
2160	24" radius, 6" wide		6.05	2.650		147	163		310	405
2180	12" wide		4.87	3.280		173	201		374	490
2190	18" wide		4.37	3.660		198	225		423	555
2200	24" wide		3.53	4.540		226	279		505	665
2210	30" wide		3.02	5.290		259	325		584	770
2240	36" radius, 6" wide		5.71	2.800		212	172		384	490
2260	12" wide		4.54	3.530		241	217		458	585
2270	18" wide		4.03	3.970		275	244		519	665

For customer support on your Electrical Change Order Costs with RSMeans data, call 800.448.8182.

Post-Installation Change Order Costs

26 05 36 – Cable Trays for Electrical Systems

26 05 36.30 Cable Tray Trough	Crew	Daily Output	Labor-Hours	Unit	Material	2020 Bare Costs Labor	Equipment	Total	Total Incl O&P
2280	24" wide	2 Elec	3.19	5.010	Ea.	300	305		605	790
2290	30" wide		2.86	5.600		355	345		700	900
2320	Elbow vertical, 90°, 12" radius, 6" wide		6.38	2.510		123	154		277	365
2330	9" wide		5.88	2.720		132	167		299	395
2340	12" wide		5.21	3.070		133	188		321	425
2350	18" wide		4.70	3.400		137	209		346	460
2360	24" wide		3.86	4.140		151	254		405	545
2370	30" wide		3.36	4.760		158	292		450	610
2400	24" radius, 6" wide		6.05	2.650		167	163		330	425
2420	12" wide		4.87	3.280		181	201		382	500
2430	18" wide		4.37	3.660		197	225		422	550
2440	24" wide		3.53	4.540		199	279		478	635
2450	30" wide		3.02	5.290		211	325		536	715
2480	36" radius, 6" wide		5.71	2.800		211	172		383	490
2500	12" wide		4.54	3.530		225	217		442	570
2510	18" wide		4.03	3.970		253	244		497	645
2520	24" wide		3.19	5.010		273	305		578	760
2530	30" wide		2.86	5.600		300	345		645	840
2560	Tee horizontal, 12" radius, 6" wide		3.36	4.760		154	292		446	605
2570	9" wide		3.19	5.010		158	305		463	635
2580	12" wide		3.02	5.290		173	325		498	675
2590	18" wide		2.69	5.950		199	365		564	765
2600	24" wide		2.35	6.800		229	415		644	870
2610	30" wide		2.02	7.940		246	485		731	995
2640	24" radius, 6" wide		3.02	5.290		235	325		560	745
2660	12" wide		2.69	5.950		267	365		632	840
2670	18" wide		2.35	6.800		294	415		709	945
2680	24" wide		2.02	7.940		370	485		855	1,125
2690	30" wide		1.68	9.520		395	585		980	1,300
2720	36" radius, 6" wide		2.69	5.950		380	365		745	960
2740	12" wide		2.35	6.800		435	415		850	1,100
2750	18" wide		2.02	7.940		490	485		975	1,275
2760	24" wide		1.68	9.520		575	585		1,160	1,500
2770	30" wide		1.34	11.900		610	730		1,340	1,750
2800	Tee vertical, 12" radius, 6" wide		3.36	4.760		217	292		509	675
2810	9" wide		3.19	5.010		218	305		523	700
2820	12" wide		3.02	5.290		234	325		559	745
2830	18" wide		2.86	5.600		238	345		583	770
2840	24" wide		2.69	5.950		241	365		606	810
2850	30" wide		2.52	6.350		253	390		643	860
2880	24" radius, 6" wide		3.02	5.290		305	325		630	820
2900	12" wide		2.69	5.950		330	365		695	910
2910	18" wide		2.52	6.350		355	390		745	970
2920	24" wide		2.35	6.800		375	415		790	1,025
2930	30" wide		2.18	7.330		405	450		855	1,125
2960	36" radius, 6" wide		2.69	5.950		465	365		830	1,050
2980	12" wide		2.35	6.800		490	415		905	1,150
2990	18" wide		2.18	7.330		510	450		960	1,225
3000	24" wide		2.02	7.940		545	485		1,030	1,325
3010	30" wide		1.85	8.660		590	530		1,120	1,425
3040	Cross horizontal, 12" radius, 6" wide		3.02	5.290		192	325		517	695
3050	9" wide		2.86	5.600		207	345		552	735
3060	12" wide		2.69	5.950		212	365		577	780

26 05 36.30 Cable Tray Trough		Crew	Daily Output	Labor-Hours	Unit	Material	2020 Bare Costs Labor	Equipment	Total	Total Incl O&P
3070	18" wide	2 Elec	2.35	6.800	Ea.	222	415		637	865
3080	24" wide		2.02	7.940		248	485		733	995
3090	30" wide		1.85	8.660		315	530		845	1,125
3120	24" radius, 6" wide		2.69	5.950		345	365		710	925
3140	12" wide		2.35	6.800		380	415		795	1,025
3150	18" wide		2.02	7.940		405	485		890	1,175
3160	24" wide		1.68	9.520		470	585		1,055	1,400
3170	30" wide		1.51	10.580		510	650		1,160	1,525
3200	36" radius, 6" wide		2.35	6.800		600	415		1,015	1,275
3220	12" wide		2.02	7.940		630	485		1,115	1,425
3230	18" wide		1.68	9.520		685	585		1,270	1,625
3240	24" wide		1.34	11.900		795	730		1,525	1,950
3250	30" wide		1.18	13.610		860	835		1,695	2,200
3280	Dropout, or end plate, 6" wide		21.84	.733		19.65	45		64.65	88.50
3300	12" wide		18.48	.866		23.50	53		76.50	105
3310	18" wide		16.80	.952		28	58.50		86.50	118
3320	24" wide		15.12	1.060		34	65		99	134
3330	30" wide		13.44	1.190		36	73		109	149
3370	Reducer, 9" to 6" wide		10.08	1.590		96.50	97.50		194	251
3380	12" to 6" wide		9.58	1.670		101	102		203	264
3390	12" to 9" wide		9.58	1.670		101	102		203	264
3400	18" to 12" wide		8.06	1.980		108	121		229	300
3420	18" to 6" wide		8.06	1.980		108	121		229	300
3430	18" to 9" wide		8.06	1.980		108	121		229	300
3440	24" to 18" wide		7.06	2.270		116	139		255	335
3460	24" to 12" wide		7.06	2.270		116	139		255	335
3470	24" to 9" wide		7.06	2.270		117	139		256	335
3475	24" to 6" wide		7.06	2.270		117	139		256	335
3480	30" to 24" wide		6.05	2.650		119	163		282	370
3500	30" to 18" wide		6.05	2.650		119	163		282	370
3520	30" to 12" wide		6.05	2.650		121	163		284	375
3540	36" to 30" wide		5.38	2.980		123	183		306	405
3560	36" to 24" wide		5.38	2.980		123	183		306	405
3580	36" to 18" wide		5.38	2.980		123	183		306	405
3600	36" to 12" wide		5.38	2.980		123	183		306	405
3610	Elbow horizontal, 60°, 12" radius, 6" wide		6.55	2.440		84	150		234	315
3620	9" wide		6.05	2.650		91.50	163		254.50	345
3630	12" wide		5.38	2.980		101	183		284	385
3640	18" wide		4.87	3.280		114	201		315	425
3650	24" wide		4.03	3.970		133	244		377	510
3680	Elbow horizontal, 45°, 12" radius, 6" wide		6.72	2.380		75.50	146		221.50	300
3690	9" wide		6.22	2.570		76.50	158		234.50	320
3700	12" wide		5.54	2.890		81	177		258	355
3710	18" wide		5.04	3.170		91	194		285	390
3720	24" wide		4.20	3.810		106	234		340	465
3750	Elbow horizontal, 30°, 12" radius, 6" wide		6.89	2.320		63	142		205	282
3760	9" wide		6.38	2.510		66	154		220	300
3770	12" wide		5.71	2.800		72	172		244	335
3780	18" wide		5.21	3.070		76.50	188		264.50	365
3790	24" wide		4.37	3.660		84	225		309	425
3820	Elbow vertical, 60° in/outside, 12" radius, 6" wide		6.55	2.440		104	150		254	335
3830	9" wide		6.05	2.650		106	163		269	360
3840	12" wide		5.38	2.980		107	183		290	390

Post-Installation Change Order Costs

26 05 36.30 Cable Tray Trough		Crew	Daily Output	Labor-Hours	Unit	Material	2020 Bare Costs Labor	Equipment	Total	Total Incl O&P
3850	18" wide	2 Elec	4.87	3.280	Ea.	111	201		312	420
3860	24" wide		4.03	3.970		116	244		360	490
3890	Elbow vertical, 45° in/outside, 12" radius, 6" wide		6.72	2.380		88.50	146		234.50	315
3900	9" wide		6.22	2.570		85.50	158		243.50	330
3910	12" wide		5.54	2.890		91.50	177		268.50	365
3920	18" wide		5.04	3.170		94.50	194		288.50	395
3930	24" wide		4.20	3.810		104	234		338	465
3960	Elbow vertical, 30° in/outside, 12" radius, 6" wide		6.89	2.320		72	142		214	292
3970	9" wide		6.38	2.510		76.50	154		230.50	315
3980	12" wide		5.71	2.800		78	172		250	340
3990	18" wide		5.21	3.070		81	188		269	370
4000	24" wide		4.37	3.660		82.50	225		307.50	425
4250	Reducer, left or right hand, 24" to 18" wide		7.06	2.270		110	139		249	330
4260	24" to 12" wide		7.06	2.270		110	139		249	330
4270	24" to 9" wide		7.06	2.270		110	139		249	330
4280	24" to 6" wide		7.06	2.270		111	139		250	330
4290	18" to 12" wide		8.06	1.980		103	121		224	294
4300	18" to 9" wide		8.06	1.980		103	121		224	294
4310	18" to 6" wide		8.06	1.980		103	121		224	294
4320	12" to 9" wide		9.58	1.670		98.50	102		200.50	261
4330	12" to 6" wide		9.58	1.670		98.50	102		200.50	261
4340	9" to 6" wide	▼	10.08	1.590		95.50	97.50		193	250
4350	Splice plate	1 Elec	40.32	.198		9.10	12.15		21.25	28
4360	Splice plate, expansion joint		40.32	.198		9.35	12.15		21.50	28.50
4370	Splice plate, hinged, horizontal		40.32	.198		7.50	12.15		19.65	26.50
4380	Vertical		40.32	.198		10.70	12.15		22.85	30
4390	Trough, hanger, vertical		23.52	.340		30	21		51	64
4400	Box connector, 24" wide		16.80	.476		36	29		65	83
4410	18" wide		17.64	.454		34	28		62	78.50
4420	12" wide		18.48	.433		32.50	26.50		59	75
4430	9" wide		19.32	.414		31	25.50		56.50	72
4440	6" wide		20.16	.397		29.50	24.50		54	69
4450	Floor flange		20.16	.397		31	24.50		55.50	70.50
4460	Hold down clamp		50.40	.159		3.53	9.75		13.28	18.40
4520	Wall bracket, 24" wide tray		16.80	.476		35	29		64	82
4530	18" wide tray		17.64	.454		34	28		62	79
4540	12" wide tray		18.48	.433		18.25	26.50		44.75	59.50
4550	9" wide tray		19.32	.414		16.30	25.50		41.80	56
4560	6" wide tray	▼	20.16	.397		14.85	24.50		39.35	53
5000	Cable channel aluminum, vented, 1-1/4" deep, 4" wide, straight		67.20	.119	L.F.	8	7.30		15.30	19.70
5010	Elbow horizontal, 36" radius, 90°		4.20	1.900	Ea.	145	117		262	335
5020	60°		4.62	1.730		111	106		217	280
5030	45°		5.04	1.590		90.50	97.50		188	245
5040	30°		5.46	1.470		77.50	90		167.50	219
5050	Adjustable		5.04	1.590		73.50	97.50		171	226
5060	Elbow vertical, 36" radius, 90°		4.20	1.900		156	117		273	345
5070	60°		4.62	1.730		120	106		226	290
5080	45°		5.04	1.590		98.50	97.50		196	253
5090	30°		5.46	1.470		87.50	90		177.50	230
5100	Adjustable		5.04	1.590		73.50	97.50		171	226
5110	Splice plate, hinged, horizontal		40.32	.198		10.20	12.15		22.35	29.50
5120	Splice plate, hinged, vertical		40.32	.198		14.45	12.15		26.60	34
5130	Hanger, vertical	▼	23.52	.340	▼	15.95	21		36.95	48.50

26 05 Common Work Results for Electrical

26 05 36 – Cable Trays for Electrical Systems

26 05 36.30 Cable Tray Trough

		Crew	Daily Output	Labor-Hours	Unit	Material	2020 Bare Costs Labor	2020 Bare Costs Equipment	Total	Total Incl O&P
5140	Single	1 Elec	23.52	.340	Ea.	24.50	21		45.50	57.50
5150	Double		16.80	.476		25	29		54	71
5160	Channel to box connector		20.16	.397		33	24.50		57.50	73
5170	Hold down clip		67.20	.119		4.60	7.30		11.90	15.95
5180	Wall bracket, single		23.52	.340		11.80	21		32.80	44
5190	Double		16.80	.476		15.15	29		44.15	60
5200	Cable roller		13.44	.595		186	36.50		222.50	259
5210	Splice plate		40.32	.198		5.70	12.15		17.85	24.50

26 05 36.36 Cable Trays for Utility Substations

		Crew	Daily Output	Labor-Hours	Unit	Material	2020 Bare Costs Labor	2020 Bare Costs Equipment	Total	Total Incl O&P
0010	**CABLE TRAYS FOR UTILITY SUBSTATIONS**									
7700	Cable tray	R-11	33.60	1.670	L.F.	19.15	97.50	19	135.65	187
7790	See Section 26 05 36									

26 05 36.40 Cable Tray, Covers and Dividers

		Crew	Daily Output	Labor-Hours	Unit	Material	2020 Bare Costs Labor	2020 Bare Costs Equipment	Total	Total Incl O&P
0010	**CABLE TRAY, COVERS AND DIVIDERS** To 10' high									
0011	For higher elevations, see lines 9900 – 9960									
0100	Covers, ventilated galv. steel, straight, 6" wide tray size	2 Elec	436.80	.037	L.F.	8.20	2.25		10.45	12.35
0200	9" wide tray size		386.40	.041		8.70	2.54		11.24	13.40
0300	12" wide tray size		336	.048		14.35	2.92		17.27	20
0400	18" wide tray size		252	.063		34	3.90		37.90	43.50
0500	24" wide tray size		184.80	.087		58.50	5.30		63.80	72.50
0600	30" wide tray size		151.20	.106		26.50	6.50		33	38.50
1000	Elbow horizontal, 90°, 12" radius, 6" wide tray size		126	.127	Ea.	43.50	7.80		51.30	59.50
1020	9" wide tray size		107.52	.149		48	9.15		57.15	66.50
1040	12" wide tray size		90.72	.176		51	10.80		61.80	72
1060	18" wide tray size		70.56	.227		70.50	13.90		84.40	98
1080	24" wide tray size		55.44	.289		85	17.70		102.70	120
1100	30" wide tray size		50.40	.317		108	19.50		127.50	148
1160	24" radius, 6" wide tray size		114.24	.140		72.50	8.60		81.10	92.50
1180	9" wide tray size		97.44	.164		74	10.05		84.05	96.50
1200	12" wide tray size		80.64	.198		82.50	12.15		94.65	109
1220	18" wide tray size		63.84	.251		102	15.40		117.40	135
1240	24" wide tray size		50.40	.317		124	19.50		143.50	165
1260	30" wide tray size		43.68	.366		163	22.50		185.50	213
1320	36" radius, 6" wide tray size		100.80	.159		108	9.75		117.75	134
1340	9" wide tray size		87.36	.183		119	11.25		130.25	148
1360	12" wide tray size		70.56	.227		127	13.90		140.90	161
1380	18" wide tray size		60.48	.265		162	16.25		178.25	202
1400	24" wide tray size		43.68	.366		193	22.50		215.50	246
1420	30" wide tray size		38.64	.414		229	25.50		254.50	290
1480	Elbow horizontal, 45°, 12" radius, 6" wide tray size		126	.127		32.50	7.80		40.30	47.50
1500	9" wide tray size		107.52	.149		39	9.15		48.15	56.50
1520	12" wide tray size		90.72	.176		42.50	10.80		53.30	63
1540	18" wide tray size		73.92	.216		52	13.30		65.30	77
1560	24" wide tray size		63.84	.251		60.50	15.40		75.90	89.50
1580	30" wide tray size		55.44	.289		71.50	17.70		89.20	105
1640	24" radius, 6" wide tray size		114.24	.140		45.50	8.60		54.10	63
1660	9" wide tray size		97.44	.164		51	10.05		61.05	71
1680	12" wide tray size		80.64	.198		57.50	12.15		69.65	81
1700	18" wide tray size		67.20	.238		66.50	14.60		81.10	95
1720	24" wide tray size		58.80	.272		76	16.70		92.70	109
1740	30" wide tray size		50.40	.317		96.50	19.50		116	135
1800	36" radius, 6" wide tray size		100.80	.159		67	9.75		76.75	88

Post-Installation Change Order Costs

26 05 36 – Cable Trays for Electrical Systems

26 05 36.40 Cable Tray, Covers and Dividers		Crew	Daily Output	Labor-Hours	Unit	Material	2020 Bare Costs Labor	Equipment	Total	Total Incl O&P
1820	9" wide tray size	2 Elec	87.36	.183	Ea.	75.50	11.25		86.75	100
1840	12" wide tray size		70.56	.227		83	13.90		96.90	112
1860	18" wide tray size		63.84	.251		95.50	15.40		110.90	128
1880	24" wide tray size		52.08	.307		116	18.85		134.85	156
1900	30" wide tray size		43.68	.366		126	22.50		148.50	173
1960	Elbow vertical, 90°, 12" radius, 6" wide tray size		126	.127		38.50	7.80		46.30	54
1980	9" wide tray size		107.52	.149		39	9.15		48.15	56.50
2000	12" wide tray size		90.72	.176		42	10.80		52.80	62.50
2020	18" wide tray size		73.92	.216		48	13.30		61.30	73
2040	24" wide tray size		57.12	.280		49	17.20		66.20	79.50
2060	30" wide tray size		50.40	.317		54	19.50		73.50	88
2120	24" radius, 6" wide tray size		114.24	.140		47	8.60		55.60	65
2140	9" wide tray size		97.44	.164		52	10.05		62.05	72
2160	12" wide tray size		80.64	.198		54	12.15		66.15	77
2180	18" wide tray size		67.20	.238		71.50	14.60		86.10	101
2200	24" wide tray size		52.08	.307		80	18.85		98.85	116
2220	30" wide tray size		43.68	.366		88	22.50		110.50	131
2280	36" radius, 6" wide tray size		100.80	.159		53	9.75		62.75	73
2300	9" wide tray size		87.36	.183		68	11.25		79.25	92
2320	12" wide tray size		70.56	.227		75.50	13.90		89.40	104
2340	18" wide tray size		63.84	.251		95.50	15.40		110.90	128
2350	24" wide tray size		45.36	.353		106	21.50		127.50	148
2360	30" wide tray size		38.64	.414		127	25.50		152.50	177
2400	Tee horizontal, 12" radius, 6" wide tray size		77.28	.207		69.50	12.70		82.20	95.50
2410	9" wide tray size		67.20	.238		71.50	14.60		86.10	101
2420	12" wide tray size		57.12	.280		81	17.20		98.20	115
2430	18" wide tray size		50.40	.317		98	19.50		117.50	137
2440	24" wide tray size		43.68	.366		123	22.50		145.50	170
2460	30" wide tray size		30.24	.529		145	32.50		177.50	208
2500	24" radius, 6" wide tray size		73.92	.216		112	13.30		125.30	143
2510	9" wide tray size		63.84	.251		119	15.40		134.40	153
2520	12" wide tray size		53.76	.298		122	18.25		140.25	161
2530	18" wide tray size		47.04	.340		157	21		178	203
2540	24" wide tray size		40.32	.397		240	24.50		264.50	300
2560	30" wide tray size		26.88	.595		274	36.50		310.50	355
2600	36" radius, 6" wide tray size		70.56	.227		191	13.90		204.90	231
2610	9" wide tray size		60.48	.265		194	16.25		210.25	238
2620	12" wide tray size		50.40	.317		216	19.50		235.50	267
2630	18" wide tray size		43.68	.366		253	22.50		275.50	310
2640	24" wide tray size		36.96	.433		335	26.50		361.50	405
2660	30" wide tray size		23.52	.680		360	41.50		401.50	460
2700	Cross horizontal, 12" radius, 6" wide tray size		57.12	.280		104	17.20		121.20	140
2710	9" wide tray size		53.76	.298		110	18.25		128.25	148
2720	12" wide tray size		50.40	.317		122	19.50		141.50	164
2730	18" wide tray size		43.68	.366		145	22.50		167.50	193
2740	24" wide tray size		30.24	.529		171	32.50		203.50	237
2760	30" wide tray size		25.20	.635		194	39		233	272
2800	24" radius, 6" wide tray size		53.76	.298		187	18.25		205.25	233
2810	9" wide tray size		50.40	.317		203	19.50		222.50	252
2820	12" wide tray size		47.04	.340		219	21		240	272
2830	18" wide tray size		40.32	.397		262	24.50		286.50	325
2840	24" wide tray size		26.88	.595		320	36.50		356.50	405
2860	30" wide tray size		21.84	.733		350	45		395	450

26 05 36.40 Cable Tray, Covers and Dividers	Crew	Daily Output	Labor-Hours	Unit	Material	2020 Bare Costs Labor	Equipment	Total	Total Incl O&P
2900	36" radius, 6" wide tray size	2 Elec	50.40	.317	Ea.	320	19.50		339.50	380
2910	9" wide tray size		47.04	.340		320	21		341	380
2920	12" wide tray size		43.68	.366		345	22.50		367.50	415
2930	18" wide tray size		36.96	.433		390	26.50		416.50	470
2940	24" wide tray size		23.52	.680		490	41.50		531.50	600
2960	30" wide tray size		18.48	.866		525	53		578	660
3000	Reducer, 9" to 6" wide tray size		107.52	.149		44.50	9.15		53.65	62.50
3010	12" to 6" wide tray size		90.72	.176		46.50	10.80		57.30	67
3020	12" to 9" wide tray size		90.72	.176		46.50	10.80		57.30	67
3030	18" to 12" wide tray size		73.92	.216		49.50	13.30		62.80	74.50
3050	18" to 6" wide tray size		73.92	.216		49.50	13.30		62.80	74.50
3060	24" to 18" wide tray size		67.20	.238		69.50	14.60		84.10	98
3070	24" to 12" wide tray size		67.20	.238		63.50	14.60		78.10	92
3090	30" to 24" wide tray size		58.80	.272		73.50	16.70		90.20	106
3100	30" to 18" wide tray size		58.80	.272		73.50	16.70		90.20	106
3110	30" to 12" wide tray size		58.80	.272		63.50	16.70		80.20	95
3140	36" to 30" wide tray size		53.76	.298		80	18.25		98.25	115
3150	36" to 24" wide tray size		53.76	.298		80	18.25		98.25	115
3160	36" to 18" wide tray size		53.76	.298		80	18.25		98.25	115
3170	36" to 12" wide tray size		53.76	.298	▼	80	18.25		98.25	115
3250	Covers, aluminum, straight, 6" wide tray size		436.80	.037	L.F.	4.79	2.25		7.04	8.60
3270	9" wide tray size		386.40	.041		5.80	2.54		8.34	10.15
3290	12" wide tray size		336	.048		6.90	2.92		9.82	11.95
3310	18" wide tray size		268.80	.060		9.15	3.65		12.80	15.50
3330	24" wide tray size		218.40	.073		11.40	4.49		15.89	19.25
3350	30" wide tray size		168	.095	▼	12.60	5.85		18.45	22.50
3400	Elbow horizontal, 90°, 12" radius, 6" wide tray size		126	.127	Ea.	36	7.80		43.80	51
3410	9" wide tray size		107.52	.149		37.50	9.15		46.65	54.50
3420	12" wide tray size		90.72	.176		40.50	10.80		51.30	60.50
3430	18" wide tray size		73.92	.216		52.50	13.30		65.80	78
3440	24" wide tray size		58.80	.272		66.50	16.70		83.20	98
3460	30" wide tray size		53.76	.298		79.50	18.25		97.75	115
3500	24" radius, 6" wide tray size		114.24	.140		46.50	8.60		55.10	64
3510	9" wide tray size		97.44	.164		57.50	10.05		67.55	78
3520	12" wide tray size		80.64	.198		63	12.15		75.15	87.50
3530	18" wide tray size		67.20	.238		75	14.60		89.60	105
3540	24" wide tray size		53.76	.298		92.50	18.25		110.75	129
3560	30" wide tray size		47.04	.340		113	21		134	156
3600	36" radius, 6" wide tray size		100.80	.159		81	9.75		90.75	104
3610	9" wide tray size		87.36	.183		89	11.25		100.25	114
3620	12" wide tray size		70.56	.227		101	13.90		114.90	132
3630	18" wide tray size		63.84	.251		120	15.40		135.40	155
3640	24" wide tray size		47.04	.340		146	21		167	192
3660	30" wide tray size		42	.381		168	23.50		191.50	219
3700	Elbow horizontal, 45°, 12" radius, 6" wide tray size		126	.127		26.50	7.80		34.30	40.50
3710	9" wide tray size		107.52	.149		27	9.15		36.15	43
3720	12" wide tray size		90.72	.176		30.50	10.80		41.30	49.50
3730	18" wide tray size		73.92	.216		35	13.30		48.30	58
3740	24" wide tray size		67.20	.238		38.50	14.60		53.10	64.50
3760	30" wide tray size		58.80	.272		48.50	16.70		65.20	78.50
3800	24" radius, 6" wide tray size		114.24	.140		29	8.60		37.60	45
3810	9" wide tray size		97.44	.164		37	10.05		47.05	55.50
3820	12" wide tray size		80.64	.198		38.50	12.15		50.65	60

Post-Installation Change Order Costs

26 05 36.40 Cable Tray, Covers and Dividers		Crew	Daily Output	Labor- Hours	Unit	Material	2020 Bare Costs Labor	Equipment	Total	Total Incl O&P
3830	18" wide tray size	2 Elec	67.20	.238	Ea.	46.50	14.60		61.10	73
3840	24" wide tray size		60.48	.265		57.50	16.25		73.75	87
3860	30" wide tray size		53.76	.298		66	18.25		84.25	100
3900	36" radius, 6" wide tray size		100.80	.159		51	9.75		60.75	71
3910	9" wide tray size		87.36	.183		55.50	11.25		66.75	78
3920	12" wide tray size		70.56	.227		58.50	13.90		72.40	85
3930	18" wide tray size		63.84	.251		70.50	15.40		85.90	101
3940	24" wide tray size		53.76	.298		85.50	18.25		103.75	122
3960	30" wide tray size		47.04	.340		97.50	21		118.50	138
3970	36" wide tray size		42	.381		111	23.50		134.50	157
4000	Elbow vertical, 90°, 12" radius, 6" wide tray size		126	.127		30.50	7.80		38.30	45
4010	9" wide tray size		107.52	.149		30.50	9.15		39.65	47
4020	12" wide tray size		90.72	.176		33.50	10.80		44.30	52.50
4030	18" wide tray size		73.92	.216		38.50	13.30		51.80	62.50
4040	24" wide tray size		58.80	.272		40	16.70		56.70	69
4060	30" wide tray size		53.76	.298		41	18.25		59.25	72
4070	36" wide tray size		45.36	.353		50	21.50		71.50	87
4100	24" radius, 6" wide tray size		114.24	.140		35	8.60		43.60	51
4110	9" wide tray size		97.44	.164		38	10.05		48.05	56.50
4120	12" wide tray size		80.64	.198		41	12.15		53.15	63
4130	18" wide tray size		67.20	.238		49.50	14.60		64.10	76.50
4140	24" wide tray size		53.76	.298		54	18.25		72.25	86.50
4160	30" wide tray size		47.04	.340		66	21		87	104
4170	36" wide tray size		40.32	.397		72	24.50		96.50	116
4200	36" radius, 6" wide tray size		100.80	.159		39	9.75		48.75	57.50
4210	9" wide tray size		87.36	.183		48	11.25		59.25	70
4220	12" wide tray size		70.56	.227		55.50	13.90		69.40	81.50
4230	18" wide tray size		63.84	.251		67.50	15.40		82.90	97.50
4240	24" wide tray size		47.04	.340		79.50	21		100.50	119
4260	30" wide tray size		42	.381		101	23.50		124.50	146
4270	36" wide tray size		36.96	.433		108	26.50		134.50	159
4300	Tee horizontal, 12" radius, 6" wide tray size		90.72	.176		49.50	10.80		60.30	70.50
4310	9" wide tray size		73.92	.216		52.50	13.30		65.80	78
4320	12" wide tray size		67.20	.238		58.50	14.60		73.10	86&P
4330	18" wide tray size		57.12	.280		69.50	17.20		86.70	102
4340	24" wide tray size		47.04	.340		88	21		109	128
4360	30" wide tray size		36.96	.433		103	26.50		129.50	154
4370	36" wide tray size		30.24	.529		125	32.50		157.50	186
4400	24" radius, 6" wide tray size		80.64	.198		80.50	12.15		92.65	107
4410	9" wide tray size		67.20	.238		85.50	14.60		100.10	116
4420	12" wide tray size		60.48	.265		95.50	16.25		111.75	129
4430	18" wide tray size		50.40	.317		112	19.50		131.50	152
4440	24" wide tray size		40.32	.397		178	24.50		202.50	233
4460	30" wide tray size		33.60	.476		197	29		226	260
4470	36" wide tray size		26.88	.595		219	36.50		255.50	296
4500	36" radius, 6" wide tray size		73.92	.216		143	13.30		156.30	177
4510	9" wide tray size		60.48	.265		146	16.25		162.25	185
4520	12" wide tray size		53.76	.298		160	18.25		178.25	203
4530	18" wide tray size		47.04	.340		182	21		203	231
4540	24" wide tray size		36.96	.433		231	26.50		257.50	294
4560	30" wide tray size		30.24	.529		260	32.50		292.50	335
4570	36" wide tray size		23.52	.680		293	41.50		334.50	385
4600	Cross horizontal, 12" radius, 6" wide tray size		67.20	.238		75.50	14.60		90.10	106

26 05 36.40 Cable Tray, Covers and Dividers		Crew	Daily Output	Labor-Hours	Unit	Material	2020 Bare Costs Labor	Equipment	Total	Total Incl O&P
4610	9" wide tray size	2 Elec	60.48	.265	Ea.	80	16.25		96.25	112
4620	12" wide tray size		53.76	.298		89	18.25		107.25	125
4630	18" wide tray size		47.04	.340		108	21		129	150
4640	24" wide tray size		40.32	.397		125	24.50		149.50	174
4660	30" wide tray size		33.60	.476		148	29		177	207
4670	36" wide tray size		26.88	.595		171	36.50		207.50	243
4700	24" radius, 6" wide tray size		60.48	.265		141	16.25		157.25	180
4710	9" wide tray size		53.76	.298		151	18.25		169.25	194
4720	12" wide tray size		47.04	.340		163	21		184	210
4730	18" wide tray size		40.32	.397		190	24.50		214.50	246
4740	24" wide tray size		33.60	.476		226	29		255	293
4760	30" wide tray size		26.88	.595		265	36.50		301.50	345
4770	36" wide tray size		20.16	.794		287	48.50		335.50	390
4800	36" radius, 6" wide tray size		53.76	.298		231	18.25		249.25	281
4810	9" wide tray size		47.04	.340		241	21		262	296
4820	12" wide tray size		42	.381		267	23.50		290.50	330
4830	18" wide tray size		36.96	.433		293	26.50		319.50	365
4840	24" wide tray size		30.24	.529		360	32.50		392.50	450
4860	30" wide tray size		23.52	.680		415	41.50		456.50	520
4870	36" wide tray size		18.48	.866		455	53		508	580
4900	Reducer, 9" to 6" wide tray size		107.52	.149		38	9.15		47.15	55.50
4910	12" to 6" wide tray size		90.72	.176		39.50	10.80		50.30	59.50
4920	12" to 9" wide tray size		90.72	.176		39.50	10.80		50.30	59.50
4930	18" to 12" wide tray size		73.92	.216		43.50	13.30		56.80	68
4950	18" to 6" wide tray size		73.92	.216		43.50	13.30		56.80	68
4960	24" to 18" wide tray size		67.20	.238		56	14.60		70.60	83.50
4970	24" to 12" wide tray size		67.20	.238		48	14.60		62.60	75
4990	30" to 24" wide tray size		58.80	.272		58.50	16.70		75.20	89
5000	30" to 18" wide tray size		58.80	.272		58.50	16.70		75.20	89
5010	30" to 12" wide tray size		58.80	.272		58.50	16.70		75.20	89
5040	36" to 30" wide tray size		53.76	.298		64.50	18.25		82.75	98
5050	36" to 24" wide tray size		53.76	.298		64.50	18.25		82.75	98
5060	36" to 18" wide tray size		53.76	.298		64.50	18.25		82.75	98
5070	36" to 12" wide tray size		53.76	.298		64.50	18.25		82.75	98
5710	Tray cover hold down clamp	1 Elec	50.40	.159		10.35	9.75		20.10	26
8000	Divider strip, straight, galvanized, 3" deep		168	.048	L.F.	5.55	2.92		8.47	10.45
8020	4" deep		151.20	.053		6.75	3.25		10	12.30
8040	6" deep		134.40	.060		8.85	3.65		12.50	15.20
8060	Aluminum, straight, 3" deep		176.40	.045		5.55	2.78		8.33	10.25
8080	4" deep		159.60	.050		6.85	3.08		9.93	12.15
8100	6" deep		142.80	.056		8.75	3.44		12.19	14.75
8110	Divider strip, vertical fitting, 3" deep									
8120	12" radius, galvanized, 30°	1 Elec	23.52	.340	Ea.	26.50	21		47.50	60.50
8140	45°		22.68	.353		33.50	21.50		55	69
8160	60°		21.84	.366		35.50	22.50		58	72.50
8180	90°		21	.381		44.50	23.50		68	84
8200	Aluminum, 30°		24.36	.328		17.80	20		37.80	49.50
8220	45°		23.52	.340		21	21		42	54
8240	60°		22.68	.353		24.50	21.50		46	59
8260	90°		21.84	.366		31	22.50		53.50	67.50
8280	24" radius, galvanized, 30°		21	.381		40	23.50		63.50	79
8300	45°		20.16	.397		42.50	24.50		67	83.50
8320	60°		19.32	.414		54	25.50		79.50	97

Post-Installation Change Order Costs

26 05 36.40 Cable Tray, Covers and Dividers	Crew	Daily Output	Labor-Hours	Unit	Material	2020 Bare Costs Labor	2020 Bare Costs Equipment	Total	Total Incl O&P
8340	90°	1 Elec	18.48	.433	Ea.	73	26.50		99.50	120
8360	Aluminum, 30°		21.84	.366		28	22.50		50.50	64.50
8380	45°		21	.381		32.50	23.50		56	70.50
8400	60°		20.16	.397		40.50	24.50		65	81
8420	90°		19.32	.414		55	25.50		80.50	98.50
8440	36" radius, galvanized, 30°		18.48	.433		51	26.50		77.50	95.50
8460	45°		17.64	.454		59.50	28		87.50	107
8480	60°		16.80	.476		69.50	29		98.50	120
8500	90°		15.96	.501		96.50	31		127.50	152
8520	Aluminum, 30°		19.32	.414		42.50	25.50		68	85
8540	45°		18.48	.433		55	26.50		81.50	100
8560	60°		17.64	.454		71.50	28		99.50	120
8570	90°		16.80	.476		93.50	29		122.50	147
8590	Divider strip, vertical fitting, 4" deep									
8600	12" radius, galvanized, 30°	1 Elec	22.68	.353	Ea.	32	21.50		53.50	67
8610	45°		21.84	.366		37	22.50		59.50	74
8620	60°		21	.381		41.50	23.50		65	81
8630	90°		20.16	.397		51	24.50		75.50	92.50
8640	Aluminum, 30°		23.52	.340		24	21		45	57.50
8650	45°		22.68	.353		27.50	21.50		49	62.50
8660	60°		21.84	.366		31.50	22.50		54	68.50
8670	90°		21	.381		37.50	23.50		61	76
8680	24" radius, galvanized, 30°		20.16	.397		51	24.50		75.50	92.50
8690	45°		19.32	.414		64	25.50		89.50	109
8700	60°		18.48	.433		72.50	26.50		99	119
8710	90°		17.64	.454		96.50	28		124.50	148
8720	Aluminum, 30°		21	.381		37.50	23.50		61	76
8730	45°		20.16	.397		45.50	24.50		70	86.50
8740	60°		19.32	.414		54	25.50		79.50	97
8750	90°		18.48	.433		73.50	26.50		100	121
8760	36" radius, galvanized, 30°		19.32	.414		60.50	25.50		86	105
8770	45°		18.48	.433		69	26.50		95.50	116
8780	60°		17.64	.454		87	28		115	137
8790	90°		16.80	.476		116	29		145	171
8800	Aluminum, 30°		20.16	.397		58.50	24.50		83	101
8810	45°		19.32	.414		75	25.50		100.50	121
8820	60°		18.48	.433		91.50	26.50		118	141
8830	90°		17.64	.454		116	28		144	169
8840	Divider strip, vertical fitting, 6" deep									
8850	12" radius, galvanized, 30°	1 Elec	20.16	.397	Ea.	36.50	24.50		61	77
8860	45°		19.32	.414		41	25.50		66.50	83
8870	60°		18.48	.433		47	26.50		73.50	91.50
8880	90°		17.64	.454		59	28		87	106
8890	Aluminum, 30°		21	.381		27.50	23.50		51	65
8900	45°		20.16	.397		32.50	24.50		57	72
8910	60°		19.32	.414		34	25.50		59.50	75.50
8920	90°		18.48	.433		40	26.50		66.50	83.50
8930	24" radius, galvanized, 30°		19.32	.414		51	25.50		76.50	94
8940	45°		18.48	.433		64	26.50		90.50	110
8950	60°		17.64	.454		73	28		101	122
8960	90°		16.80	.476		96.50	29		125.50	150
8970	Aluminum, 30°		20.16	.397		38.50	24.50		63	78.50
8980	45°		19.32	.414		52	25.50		77.50	95.50

26 05 Common Work Results for Electrical

26 05 36 – Cable Trays for Electrical Systems

26 05 36.40 Cable Tray, Covers and Dividers

		Crew	Daily Output	Labor-Hours	Unit	Material	2020 Bare Costs Labor	Equipment	Total	Total Incl O&P
8990	60°	1 Elec	18.48	.433	Ea.	57.50	26.50		84	103
9000	90°		17.64	.454		78	28		106	128
9010	36" radius, galvanized, 30°		18.48	.433		59.50	26.50		86	105
9020	45°		17.64	.454		73	28		101	122
9030	60°		16.80	.476		96.50	29		125.50	150
9040	90°		15.96	.501		127	31		158	185
9050	Aluminum, 30°		19.32	.414		59.50	25.50		85	104
9060	45°		18.48	.433		79.50	26.50		106	127
9070	60°		17.64	.454		94	28		122	145
9080	90°		16.80	.476		113	29		142	168
9120	Divider strip, horizontal fitting, galvanized, 3" deep		27.72	.289		36.50	17.70		54.20	66.50
9130	4" deep		25.20	.317		40.50	19.50		60	74
9140	6" deep		22.68	.353		52	21.50		73.50	89
9150	Aluminum, 3" deep		29.40	.272		27.50	16.70		44.20	55
9160	4" deep		26.88	.298		30	18.25		48.25	60
9170	6" deep		24.36	.328		40	20		60	74
9300	Divider strip protector		252	.032	L.F.	2.96	1.95		4.91	6.15
9310	Fastener, ladder tray				Ea.	.52			.52	.57
9320	Trough or solid bottom tray				"	.39			.39	.43
9900	Add to labor for higher elevated installation									
9905	10' to 14.5' high, add						10%			
9910	15' to 20' high, add						20%			
9920	20' to 25' high, add						25%			
9930	25' to 30' high, add						35%			
9940	30' to 35' high, add						40%			
9950	35' to 40' high, add						50%			
9960	Over 40' high, add						55%			

26 05 39 – Underfloor Raceways for Electrical Systems

26 05 39.30 Conduit In Concrete Slab

		Crew	Daily Output	Labor-Hours	Unit	Material	2020 Bare Costs Labor	Equipment	Total	Total Incl O&P
0010	**CONDUIT IN CONCRETE SLAB** Including terminations,									
0020	fittings and supports									
3230	PVC, schedule 40, 1/2" diameter	1 Elec	226.80	.035	L.F.	.66	2.16		2.82	3.95
3250	3/4" diameter		193.20	.041		.70	2.54		3.24	4.55
3270	1" diameter		168	.048		.93	2.92		3.85	5.40
3300	1-1/4" diameter		142.80	.056		1.26	3.44		4.70	6.50
3330	1-1/2" diameter		117.60	.068		1.53	4.17		5.70	7.90
3350	2" diameter		100.80	.079		1.91	4.87		6.78	9.35
3370	2-1/2" diameter		75.60	.106		3.15	6.50		9.65	13.10
3400	3" diameter	2 Elec	134.40	.119		4.16	7.30		11.46	15.50
3430	3-1/2" diameter		100.80	.159		5.40	9.75		15.15	20.50
3440	4" diameter		84	.190		5.70	11.70		17.40	23.50
3450	5" diameter		67.20	.238		8.95	14.60		23.55	32
3460	6" diameter		50.40	.317		11.55	19.50		31.05	41.50
3530	Sweeps, 1" diameter, 30" radius	1 Elec	26.88	.298	Ea.	11.50	18.25		29.75	39.50
3550	1-1/4" diameter		20.16	.397		10.30	24.50		34.80	48
3570	1-1/2" diameter		17.64	.454		11.25	28		39.25	54
3600	2" diameter		15.12	.529		13.50	32.50		46	63.50
3630	2-1/2" diameter		11.76	.680		96	41.50		137.50	167
3650	3" diameter		8.40	.952		81.50	58.50		140	177
3670	3-1/2" diameter		6.72	1.190		62.50	73		135.50	178
3700	4" diameter		5.88	1.360		30	83.50		113.50	158
3710	5" diameter		5.04	1.590		66	97.50		163.50	218

Post-Installation Change Order Costs

26 05 39 – Underfloor Raceways for Electrical Systems

26 05 39.30 Conduit In Concrete Slab		Crew	Daily Output	Labor-Hours	Unit	Material	2020 Bare Costs Labor	Equipment	Total	Total Incl O&P
3730	Couplings, 1/2" diameter				Ea.	.14			.14	.16
3750	3/4" diameter					.18			.18	.20
3770	1" diameter					.28			.28	.30
3800	1-1/4" diameter					.46			.46	.50
3830	1-1/2" diameter					.52			.52	.57
3850	2" diameter					.70			.70	.77
3870	2-1/2" diameter					1.33			1.33	1.47
3900	3" diameter					2.11			2.11	2.32
3930	3-1/2" diameter					2.94			2.94	3.23
3950	4" diameter					3.29			3.29	3.62
3960	5" diameter					6.75			6.75	7.45
3970	6" diameter					10.60			10.60	11.70
4030	End bells, 1" diameter, PVC	1 Elec	50.40	.159		3.22	9.75		12.97	18.05
4050	1-1/4" diameter		44.52	.180		5.45	11		16.45	22.50
4100	1-1/2" diameter		40.32	.198		3.72	12.15		15.87	22
4150	2" diameter		28.56	.280		5.25	17.20		22.45	31.50
4170	2-1/2" diameter		22.68	.353		4.86	21.50		26.36	37.50
4200	3" diameter		16.80	.476		7.60	29		36.60	52
4250	3-1/2" diameter		13.44	.595		8.25	36.50		44.75	63.50
4300	4" diameter		11.76	.680		9.30	41.50		50.80	72.50
4310	5" diameter		10.08	.794		14.50	48.50		63	88.50
4320	6" diameter		7.56	1.060	↓	14.85	65		79.85	113
4350	Rigid galvanized steel, 1/2" diameter		168	.048	L.F.	2.80	2.92		5.72	7.45
4400	3/4" diameter		142.80	.056		5.35	3.44		8.79	10.95
4450	1" diameter		109.20	.073		8.25	4.49		12.74	15.80
4500	1-1/4" diameter		92.40	.087		5.45	5.30		10.75	13.90
4600	1-1/2" diameter		84	.095		10.05	5.85		15.90	19.80
4800	2" diameter	↓	75.60	.106	↓	9.85	6.50		16.35	20.50

26 05 39.40 Conduit In Trench

		Crew	Daily Output	Labor-Hours	Unit	Material	Labor	Equipment	Total	Total Incl O&P
0010	**CONDUIT IN TRENCH** Includes terminations and fittings									
0200	Rigid galvanized steel, 2" diameter	1 Elec	126	.063	L.F.	9.35	3.90		13.25	16.10
0400	2-1/2" diameter	"	84	.095		15.60	5.85		21.45	26
0600	3" diameter	2 Elec	134.40	.119		17.45	7.30		24.75	30
0800	3-1/2" diameter		117.60	.136		23	8.35		31.35	38
1000	4" diameter		84	.190		28	11.70		39.70	48
1200	5" diameter		67.20	.238		45	14.60		59.60	71.50
1400	6" diameter	↓	50.40	.317	↓	69	19.50		88.50	105

26 05 43 – Underground Ducts and Raceways for Electrical Systems

26 05 43.10 Trench Duct

		Crew	Daily Output	Labor-Hours	Unit	Material	Labor	Equipment	Total	Total Incl O&P
0010	**TRENCH DUCT** Steel with cover									
0020	Standard adjustable, depths to 4"									
0100	Straight, single compartment, 9" wide	2 Elec	33.60	.476	L.F.	289	29		318	365
0200	12" wide		26.88	.595		365	36.50		401.50	455
0400	18" wide		21.84	.733		395	45		440	500
0600	24" wide		18.48	.866		415	53		468	535
0700	27" wide		17.64	.907		215	55.50		270.50	320
0800	30" wide		16.80	.952		228	58.50		286.50	335
1000	36" wide		13.44	1.190		261	73		334	395
1020	Two compartment, 9" wide		31.92	.501		132	31		163	191
1030	12" wide		25.20	.635		157	39		196	231
1040	18" wide		20.16	.794		199	48.50		247.50	292
1050	24" wide	↓	16.80	.952	↓	247	58.50		305.50	360

26 05 43.10 Trench Duct	Crew	Daily Output	Labor-Hours	Unit	Material	2020 Bare Costs Labor	Equipment	Total	Total Incl O&P
1060	30" wide	2 Elec	15.12	1.060	L.F.	315	65		380	445
1070	36" wide		11.76	1.360		350	83.50		433.50	510
1090	Three compartment, 9" wide		30.24	.529		152	32.50		184.50	216
1100	12" wide		23.52	.680		175	41.50		216.50	255
1110	18" wide		18.48	.866		213	53		266	315
1120	24" wide		15.12	1.060		267	65		332	390
1130	30" wide		13.44	1.190		330	73		403	475
1140	36" wide		10.08	1.590		380	97.50		477.50	560
1200	Horizontal elbow, 9" wide		4.54	3.530	Ea.	455	217		672	820
1400	12" wide		3.86	4.140		440	254		694	865
1600	18" wide		3.36	4.760		625	292		917	1,125
1800	24" wide		2.69	5.950		895	365		1,260	1,525
1900	27" wide		2.52	6.350		1,075	390		1,465	1,775
2000	30" wide		2.18	7.330		1,175	450		1,625	1,975
2200	36" wide		2.02	7.940		1,575	485		2,060	2,450
2220	Two compartment, 9" wide		3.19	5.010		735	305		1,040	1,275
2230	12" wide		2.52	6.350		810	390		1,200	1,475
2240	18" wide		2.02	7.940		975	485		1,460	1,800
2250	24" wide		1.68	9.520		1,175	585		1,760	2,150
2260	30" wide		1.51	10.580		1,550	650		2,200	2,700
2270	36" wide		1.34	11.900		1,900	730		2,630	3,150
2290	Three compartment, 9" wide		3.02	5.290		725	325		1,050	1,275
2300	12" wide		2.35	6.800		810	415		1,225	1,500
2310	18" wide		1.85	8.660		985	530		1,515	1,875
2320	24" wide		1.51	10.580		1,250	650		1,900	2,350
2330	30" wide		1.34	11.900		1,625	730		2,355	2,850
2350	36" wide		1.18	13.610		1,975	835		2,810	3,425
2400	Vertical elbow, 9" wide		4.54	3.530		179	217		396	515
2600	12" wide		3.86	4.140		172	254		426	570
2800	18" wide		3.36	4.760		197	292		489	650
3000	24" wide		2.69	5.950		245	365		610	815
3100	27" wide		2.52	6.350		256	390		646	860
3200	30" wide		2.18	7.330		271	450		721	970
3400	36" wide		2.02	7.940		297	485		782	1,050
3600	Cross, 9" wide		3.36	4.760		745	292		1,037	1,250
3800	12" wide		2.69	5.950		785	365		1,150	1,400
4000	18" wide		2.18	7.330		940	450		1,390	1,700
4200	24" wide		1.85	8.660		1,175	530		1,705	2,075
4300	27" wide		1.85	8.660		1,325	530		1,855	2,250
4400	30" wide		1.68	9.520		1,450	585		2,035	2,475
4600	36" wide		1.51	10.580		1,825	650		2,475	2,975
4620	Two compartment, 9" wide		3.19	5.010		730	305		1,035	1,275
4630	12" wide		2.52	6.350		765	390		1,155	1,425
4640	18" wide		2.02	7.940		930	485		1,415	1,750
4650	24" wide		1.68	9.520		1,175	585		1,760	2,175
4660	30" wide		1.51	10.580		1,550	650		2,200	2,700
4670	36" wide		1.34	11.900		1,875	730		2,605	3,150
4690	Three compartment, 9" wide		3.02	5.290		740	325		1,065	1,300
4700	12" wide		2.35	6.800		850	415		1,265	1,550
4710	18" wide		1.85	8.660		1,000	530		1,530	1,900
4720	24" wide		1.51	10.580		1,250	650		1,900	2,350
4730	30" wide		1.34	11.900		1,625	730		2,355	2,875
4740	36" wide		1.18	13.610		2,000	835		2,835	3,450

Post-Installation Change Order Costs

26 05 43 – Underground Ducts and Raceways for Electrical Systems

26 05 43.10 Trench Duct	Crew	Daily Output	Labor-Hours	Unit	Material	2020 Bare Costs Labor	Equipment	Total	Total Incl O&P
4800	End closure, 9" wide	2 Elec	12.10	1.320	Ea.	44	81		125	169
5000	12" wide		10.08	1.590		50	97.50		147.50	200
5200	18" wide		8.40	1.900		77	117		194	259
5400	24" wide		6.72	2.380		101	146		247	330
5500	27" wide		5.88	2.720		117	167		284	375
5600	30" wide		5.54	2.890		156	177		333	435
5800	36" wide		4.87	3.280		151	201		352	465
6000	Tees, 9" wide		3.36	4.760		425	292		717	905
6200	12" wide		3.02	5.290		495	325		820	1,025
6400	18" wide		2.69	5.950		635	365		1,000	1,250
6600	24" wide		2.52	6.350		910	390		1,300	1,575
6700	27" wide		2.35	6.800		1,000	415		1,415	1,725
6800	30" wide		2.18	7.330		1,175	450		1,625	1,975
7000	36" wide		1.68	9.520		1,550	585		2,135	2,575
7020	Two compartment, 9" wide		3.19	5.010		495	305		800	1,000
7030	12" wide		2.86	5.600		530	345		875	1,100
7040	18" wide		2.52	6.350		710	390		1,100	1,350
7050	24" wide		2.35	6.800		960	415		1,375	1,675
7060	30" wide		2.02	7.940		1,300	485		1,785	2,150
7070	36" wide		1.60	10.030		1,625	615		2,240	2,725
7090	Three compartment, 9" wide		3.02	5.290		565	325		890	1,100
7100	12" wide		2.69	5.950		595	365		960	1,200
7110	18" wide		2.35	6.800		750	415		1,165	1,450
7120	24" wide		2.18	7.330		1,025	450		1,475	1,825
7130	30" wide		1.85	8.660		1,350	530		1,880	2,300
7140	36" wide		1.51	10.580		1,725	650		2,375	2,875
7200	Riser, and cabinet connector, 9" wide		4.54	3.530		186	217		403	525
7400	12" wide		3.86	4.140		217	254		471	620
7600	18" wide		3.36	4.760		228	292		520	685
7800	24" wide		2.69	5.950		325	365		690	900
7900	27" wide		2.52	6.350		300	390		690	910
8000	30" wide		2.18	7.330		370	450		820	1,075
8200	36" wide		1.68	9.520		430	585		1,015	1,350
8400	Insert assembly, cell to conduit adapter, 1-1/4"	1 Elec	13.44	.595		74	36.50		110.50	136
8500	Adjustable partition	"	268.80	.030	L.F.	26	1.83		27.83	31.50
8600	Depth of duct over 4", per 1", add					12.10			12.10	13.30
8700	Support post	1 Elec	201.60	.040		27.50	2.43		29.93	34
8800	Cover double tile trim, 2 sides					43			43	47
8900	4 sides					120			120	132
9160	Trench duct 3-1/2" x 4-1/2", add					11.50			11.50	12.65
9170	Trench duct 4" x 5", add					11.50			11.50	12.65
9200	For carpet trim, add					39			39	43
9210	For double carpet trim, add					119			119	131

26 05 43.20 Underfloor Duct

		Crew	Daily Output	Labor-Hours	Unit	Material	2020 Bare Costs Labor	Equipment	Total	Total Incl O&P
0010	**UNDERFLOOR DUCT**									
0100	Duct, 1-3/8" x 3-1/8" blank, standard	2 Elec	134.40	.119	L.F.	12.85	7.30		20.15	25
0200	1-3/8" x 7-1/4" blank, super duct		100.80	.159		30	9.75		39.75	47.50
0400	7/8" or 1-3/8" insert type, 24" OC, 1-3/8" x 3-1/8", std.		117.60	.136		20	8.35		28.35	34.50
0600	1-3/8" x 7-1/4", super duct		84	.190		35	11.70		46.70	56
0800	Junction box, single duct, 1 level, 3-1/8"	1 Elec	3.36	2.380	Ea.	445	146		591	705
0820	3-1/8" x 7-1/4"		3.36	2.380		575	146		721	850
0840	2 level, 3-1/8" upper & lower		2.69	2.980		520	183		703	840

26 05 43.20 Underfloor Duct	Crew	Daily Output	Labor-Hours	Unit	Material	2020 Bare Costs Labor	Equipment	Total	Total Incl O&P
0860 3-1/8" upper, 7-1/4" lower	1 Elec	2.27	3.530	Ea.	510	217		727	880
0880 Carpet pan for above		67.20	.119		370	7.30		377.30	420
0900 Terrazzo pan for above		56.28	.142		880	8.70		888.70	985
1000 Junction box, single duct, 1 level, 7-1/4"		2.27	3.530		520	217		737	890
1020 2 level, 7-1/4" upper & lower		2.27	3.530		595	217		812	975
1040 2 duct, two 3-1/8" upper & lower		2.69	2.980		805	183		988	1,150
1200 1 level, 2 duct, 3-1/8"		2.69	2.980		590	183		773	920
1220 Carpet pan for above boxes		67.20	.119		365	7.30		372.30	410
1240 Terrazzo pan for above boxes		56.28	.142		850	8.70		858.70	950
1260 Junction box, 1 level, two 3-1/8" x one 3-1/8" + one 7-1/4"		1.93	4.140		995	254		1,249	1,475
1280 2 level, two 3-1/8" upper, one 3-1/8" + one 7-1/4" lower		1.68	4.760		1,100	292		1,392	1,625
1300 Carpet pan for above boxes		67.20	.119		365	7.30		372.30	410
1320 Terrazzo pan for above boxes		56.28	.142		850	8.70		858.70	950
1400 Junction box, 1 level, 2 duct, 7-1/4"		1.93	4.140		1,475	254		1,729	2,000
1420 Two 3-1/8" + one 7-1/4"		1.68	4.760		1,500	292		1,792	2,075
1440 Carpet pan for above		67.20	.119		365	7.30		372.30	410
1460 Terrazzo pan for above		56.28	.142		850	8.70		858.70	950
1580 Junction box, 1 level, one 3-1/8" + one 7-1/4" x same		1.93	4.140		1,000	254		1,254	1,475
1600 Triple duct, 3-1/8"		1.93	4.140		1,000	254		1,254	1,475
1700 Junction box, 1 level, one 3-1/8" + two 7-1/4"		1.68	4.760		1,675	292		1,967	2,275
1720 Carpet pan for above		67.20	.119		365	7.30		372.30	410
1740 Terrazzo pan for above		56.28	.142		850	8.70		858.70	950
1800 Insert to conduit adapter, 3/4" & 1"		26.88	.298		36.50	18.25		54.75	67.50
2000 Support, single cell		22.68	.353		54.50	21.50		76	92
2200 Super duct		13.44	.595		55	36.50		91.50	115
2400 Double cell		13.44	.595		55	36.50		91.50	116
2600 Triple cell		9.24	.866		64	53		117	150
2800 Vertical elbow, standard duct		8.40	.952		98.50	58.50		157	195
3000 Super duct		6.72	1.190		98.50	73		171.50	217
3200 Cabinet connector, standard duct		26.88	.298		78.50	18.25		96.75	114
3400 Super duct		22.68	.353		76.50	21.50		98	117
3600 Conduit adapter, 1" to 1-1/4"		26.88	.298		73.50	18.25		91.75	108
3800 2" to 1-1/4"		22.68	.353		88.50	21.50		110	129
4000 Outlet, low tension (tele, computer, etc.)		6.72	1.190		104	73		177	223
4200 High tension, receptacle (120 volt)		6.72	1.190		106	73		179	226
4300 End closure, standard duct		134.40	.060		4.44	3.65		8.09	10.35
4310 Super duct		134.40	.060		8.40	3.65		12.05	14.70
4350 Elbow, horiz., standard duct		21.84	.366		252	22.50		274.50	310
4360 Super duct		21.84	.366		240	22.50		262.50	298
4380 Elbow, offset, standard duct		21.84	.366		98.50	22.50		121	142
4390 Super duct		21.84	.366		100	22.50		122.50	144
4400 Marker screw assembly for inserts		42	.190		19.40	11.70		31.10	39
4410 Y take off, standard duct		21.84	.366		153	22.50		175.50	203
4420 Super duct		21.84	.366		153	22.50		175.50	203
4430 Box opening plug, standard duct		134.40	.060		18.70	3.65		22.35	26
4440 Super duct		134.40	.060		18.70	3.65		22.35	26
4450 Sleeve coupling, standard duct		134.40	.060		49.50	3.65		53.15	60
4460 Super duct		134.40	.060		49.50	3.65		53.15	60
4470 Conduit adapter, standard duct, 3/4"		26.88	.298		80	18.25		98.25	115
4480 1" or 1-1/4"		26.88	.298		76	18.25		94.25	111
4500 1-1/2"		26.88	.298		74	18.25		92.25	109

26 05 Common Work Results for Electrical

26 05 83 – Wiring Connections

26 05 83.10 Motor Connections

		Crew	Daily Output	Labor-Hours	Unit	Material	2020 Bare Costs Labor	Equipment	Total	Total Incl O&P
0010	**MOTOR CONNECTIONS**									
0020	Flexible conduit and fittings, 115 volt, 1 phase, up to 1 HP motor	1 Elec	6.72	1.190	Ea.	5.90	73		78.90	116
0050	2 HP motor		5.46	1.470		10.40	90		100.40	145
0100	3 HP motor		4.62	1.730		9.20	106		115.20	168
0110	230 volt, 3 phase, 3 HP motor		5.70	1.400		7.15	86		93.15	136
0112	5 HP motor		4.59	1.740		6.05	107		113.05	166
0114	7-1/2 HP motor		3.87	2.070		9.35	127		136.35	199
0120	10 HP motor		3.53	2.270		17.40	139		156.40	226
0150	15 HP motor		2.77	2.890		17.40	177		194.40	283
0200	25 HP motor		2.27	3.530		29.50	217		246.50	355
0400	50 HP motor		1.85	4.330		54.50	266		320.50	455
0600	100 HP motor		1.26	6.350		121	390		511	715
1500	460 volt, 5 HP motor, 3 phase		6.72	1.190		6.35	73		79.35	116
1520	10 HP motor		6.72	1.190		6.35	73		79.35	116
1530	25 HP motor		5.04	1.590		11.25	97.50		108.75	157
1540	30 HP motor		5.04	1.590		11.25	97.50		108.75	157
1550	40 HP motor		4.20	1.900		16.70	117		133.70	192
1560	50 HP motor		4.20	1.900		22.50	117		139.50	199
1570	60 HP motor		3.19	2.510		25	154		179	257
1580	75 HP motor		2.94	2.720		31.50	167		198.50	283
1590	100 HP motor		2.10	3.810		53.50	234		287.50	410
1600	125 HP motor		1.68	4.760		58.50	292		350.50	500
1610	150 HP motor		1.51	5.290		60	325		385	550
1620	200 HP motor		1.26	6.350		99.50	390		489.50	690
2005	460 volt, 5 HP motor, 3 phase, w/sealtite		6.72	1.190		9.30	73		82.30	119
2010	10 HP motor		6.72	1.190		9.30	73		82.30	119
2015	25 HP motor		5.04	1.590		17	97.50		114.50	164
2020	30 HP motor		5.04	1.590		17	97.50		114.50	164
2025	40 HP motor		4.20	1.900		30	117		147	208
2030	50 HP motor		4.20	1.900		34.50	117		151.50	212
2035	60 HP motor		3.19	2.510		38.50	154		192.50	271
2040	75 HP motor		2.94	2.720		45	167		212	298
2045	100 HP motor		2.10	3.810		80	234		314	440
2055	150 HP motor		1.51	5.290		86.50	325		411.50	580
2060	200 HP motor		1.26	6.350		273	390		663	880

26 05 90 – Residential Applications

26 05 90.10 Residential Wiring

		Crew	Daily Output	Labor-Hours	Unit	Material	2020 Bare Costs Labor	Equipment	Total	Total Incl O&P
0010	**RESIDENTIAL WIRING**									
0020	20' avg. runs and #14/2 wiring incl. unless otherwise noted									
1000	Service & panel, includes 24' SE-AL cable, service eye, meter,									
1010	Socket, panel board, main bkr., ground rod, 15 or 20 amp									
1020	1-pole circuit breakers, and misc. hardware									
1100	100 amp, with 10 branch breakers	1 Elec	1	8	Ea.	365	490		855	1,125
1110	With PVC conduit and wire		.77	10.350		405	635		1,040	1,400
1120	With RGS conduit and wire		.61	13.050		625	800		1,425	1,900
1150	150 amp, with 14 branch breakers		.87	9.250		885	565		1,450	1,825
1170	With PVC conduit and wire		.69	11.610		960	710		1,670	2,100
1180	With RGS conduit and wire		.56	14.210		1,325	870		2,195	2,775
1200	200 amp, with 18 branch breakers	2 Elec	1.51	10.580		1,125	650		1,775	2,200
1220	With PVC conduit and wire		1.23	13.050		1,200	800		2,000	2,525
1230	With RGS conduit and wire		1.04	15.360		1,650	940		2,590	3,200
1800	Lightning surge suppressor	1 Elec	26.88	.298		93.50	18.25		111.75	130

Post-Installation Change Order Costs   For customer support on your Electrical Change Order Costs with RSMeans data, call 800.448.8182.

26 05 90.10 Residential Wiring	Crew	Daily Output	Labor-Hours	Unit	Material	2020 Bare Costs Labor	Equipment	Total	Total Incl O&P
2000	Switch devices									
2100	Single pole, 15 amp, ivory, with a 1-gang box, cover plate,									
2110	Type NM (Romex) cable	1 Elec	14.36	.557	Ea.	17.10	34		51.10	70
2120	Type MC cable		12.01	.666		29	41		70	92.50
2130	EMT & wire		4.80	1.670		40	102		142	197
2150	3-way, #14/3, type NM cable		12.22	.655		11.10	40		51.10	72.50
2170	Type MC cable		10.34	.774		27	47.50		74.50	100
2180	EMT & wire		4.20	1.900		33.50	117		150.50	211
2200	4-way, #14/3, type NM cable		12.22	.655		20	40		60	82
2220	Type MC cable		10.34	.774		36	47.50		83.50	110
2230	EMT & wire		4.20	1.900		42.50	117		159.50	221
2250	S.P., 20 amp, #12/2, type NM cable		11.20	.714		13.55	44		57.55	80.50
2270	Type MC cable		9.60	.833		24	51		75	103
2280	EMT & wire		4.07	1.960		38.50	120		158.50	221
2290	S.P. rotary dimmer, 600 W, no wiring		14.28	.560		35	34.50		69.50	89.50
2300	S.P. rotary dimmer, 600 W, type NM cable		12.22	.655		39	40		79	103
2320	Type MC cable		10.34	.774		51	47.50		98.50	127
2330	EMT & wire		4.20	1.900		63	117		180	244
2350	3-way rotary dimmer, type NM cable		11.20	.714		26.50	44		70.50	94.50
2370	Type MC cable		9.60	.833		38	51		89	118
2380	EMT & wire		4.07	1.960		50.50	120		170.50	235
2400	Interval timer wall switch, 20 amp, 1-30 min., #12/2									
2410	Type NM cable	1 Elec	12.22	.655	Ea.	65	40		105	132
2420	Type MC cable		10.34	.774		71	47.50		118.50	149
2430	EMT & wire		4.20	1.900		89.50	117		206.50	273
2500	Decorator style									
2510	S.P., 15 amp, type NM cable	1 Elec	14.36	.557	Ea.	23	34		57	76.50
2520	Type MC cable		12.01	.666		35	41		76	99.50
2530	EMT & wire		4.80	1.670		46	102		148	204
2550	3-way, #14/3, type NM cable		12.22	.655		17.25	40		57.25	79
2570	Type MC cable		10.34	.774		33	47.50		80.50	107
2580	EMT & wire		4.20	1.900		39.50	117		156.50	218
2600	4-way, #14/3, type NM cable		12.22	.655		26	40		66	89
2620	Type MC cable		10.34	.774		42	47.50		89.50	117
2630	EMT & wire		4.20	1.900		48.50	117		165.50	228
2650	S.P., 20 amp, #12/2, type NM cable		11.20	.714		19.65	44		63.65	87
2670	Type MC cable		9.60	.833		30	51		81	109
2680	EMT & wire		4.07	1.960		44.50	120		164.50	228
2700	S.P., slide dimmer, type NM cable		14.36	.557		41.50	34		75.50	96.50
2720	Type MC cable		12.01	.666		53	41		94	120
2730	EMT & wire		4.80	1.670		65.50	102		167.50	225
2750	S.P., touch dimmer, type NM cable		14.36	.557		58.50	34		92.50	116
2770	Type MC cable		12.01	.666		70	41		111	138
2780	EMT & wire		4.80	1.670		82.50	102		184.50	244
2800	3-way touch dimmer, type NM cable		11.20	.714		54.50	44		98.50	126
2820	Type MC cable		9.60	.833		66.50	51		117.50	149
2830	EMT & wire		4.07	1.960		79	120		199	266
3000	Combination devices									
3100	S.P. switch/15 amp recpt., ivory, 1-gang box, plate									
3110	Type NM cable	1 Elec	9.60	.833	Ea.	24	51		75	103
3120	Type MC cable		8.40	.952		36	58.50		94.50	127
3130	EMT & wire		3.70	2.160		48	133		181	250
3150	S.P. switch/pilot light, type NM cable		9.60	.833		25.50	51		76.50	104

Post-Installation Change Order Costs

26 05 90.10 Residential Wiring		Crew	Daily Output	Labor-Hours	Unit	Material	2020 Bare Costs Labor	Equipment	Total	Total Incl O&P
3170	Type MC cable	1 Elec	8.40	.952	Ea.	37.50	58.50		96	128
3180	EMT & wire		3.72	2.150		49.50	132		181.50	251
3190	2-S.P. switches, 2-#14/2, no wiring		11.76	.680		14.30	41.50		55.80	77.50
3200	2-S.P. switches, 2-#14/2, type NM cables		8.40	.952		26.50	58.50		85	116
3220	Type MC cable		7.47	1.070		42.50	65.50		108	145
3230	EMT & wire		3.44	2.320		50.50	142		192.50	268
3250	3-way switch/15 amp recpt., #14/3, type NM cable		8.40	.952		33	58.50		91.50	123
3270	Type MC cable		7.47	1.070		48.50	65.50		114	151
3280	EMT & wire		3.44	2.320		55	142		197	273
3300	2-3 way switches, 2-#14/3, type NM cables		7.47	1.070		41.50	65.50		107	144
3320	Type MC cable		6.72	1.190		66	73		139	182
3330	EMT & wire		3.36	2.380		62.50	146		208.50	286
3350	S.P. switch/20 amp recpt., #12/2, type NM cable		8.40	.952		44	58.50		102.50	136
3370	Type MC cable		7.47	1.070		50	65.50		115.50	153
3380	EMT & wire		3.44	2.320		68.50	142		210.50	288
3400	Decorator style									
3410	S.P. switch/15 amp recpt., type NM cable	1 Elec	9.60	.833	Ea.	30	51		81	110
3420	Type MC cable		8.40	.952		42	58.50		100.50	133
3430	EMT & wire		3.70	2.160		54.50	133		187.50	257
3450	S.P. switch/pilot light, type NM cable		9.60	.833		31.50	51		82.50	111
3470	Type MC cable		8.40	.952		43.50	58.50		102	135
3480	EMT & wire		3.70	2.160		56	133		189	259
3500	2-S.P. switches, 2-#14/2, type NM cables		8.40	.952		32.50	58.50		91	123
3520	Type MC cable		7.47	1.070		48.50	65.50		114	151
3530	EMT & wire		3.44	2.320		56.50	142		198.50	274
3550	3-way/15 amp recpt., #14/3, type NM cable		8.40	.952		39	58.50		97.50	130
3570	Type MC cable		7.47	1.070		54.50	65.50		120	158
3580	EMT & wire		3.44	2.320		61	142		203	280
3650	2-3 way switches, 2-#14/3, type NM cables		7.47	1.070		48	65.50		113.50	150
3670	Type MC cable		6.72	1.190		72	73		145	189
3680	EMT & wire		3.36	2.380		68.50	146		214.50	293
3700	S.P. switch/20 amp recpt., #12/2, type NM cable		8.40	.952		50	58.50		108.50	142
3720	Type MC cable		7.47	1.070		56	65.50		121.50	159
3730	EMT & wire		3.44	2.320		74.50	142		216.50	294
4000	Receptacle devices									
4010	Duplex outlet, 15 amp recpt., ivory, 1-gang box, plate									
4015	Type NM cable	1 Elec	12.22	.655	Ea.	9.80	40		49.80	71
4020	Type MC cable		10.34	.774		21.50	47.50		69	94
4030	EMT & wire		4.48	1.790		32.50	110		142.50	200
4050	With #12/2, type NM cable		10.34	.774		11.55	47.50		59.05	83
4070	Type MC cable		8.96	.893		22	55		77	106
4080	EMT & wire		3.96	2.020		36	124		160	225
4100	20 amp recpt., #12/2, type NM cable		10.34	.774		21.50	47.50		69	94
4120	Type MC cable		8.96	.893		32	55		87	117
4130	EMT & wire		3.96	2.020		46	124		170	236
4140	For GFI see Section 26 05 90.10 line 4300 below									
4150	Decorator style, 15 amp recpt., type NM cable	1 Elec	12.22	.655	Ea.	15.90	40		55.90	77.50
4170	Type MC cable		10.34	.774		27.50	47.50		75	101
4180	EMT & wire		4.48	1.790		38.50	110		148.50	207
4200	With #12/2, type NM cable		10.34	.774		17.65	47.50		65.15	90
4220	Type MC cable		8.96	.893		28	55		83	113
4230	EMT & wire		3.96	2.020		42.50	124		166.50	232
4250	20 amp recpt., #12/2, type NM cable		10.34	.774		27.50	47.50		75	101

Post-Installation Change Order Costs   For customer support on your Electrical Change Order Costs with RSMeans data, call 800.448.8182.

26 05 90.10 Residential Wiring		Crew	Daily Output	Labor-Hours	Unit	Material	2020 Bare Costs Labor	Equipment	Total	Total Incl O&P
4270	Type MC cable	1 Elec	8.96	.893	Ea.	38	55		93	124
4280	EMT & wire		3.96	2.020		52.50	124		176.50	243
4300	GFI, 15 amp recpt., type NM cable		10.34	.774		23	47.50		70.50	95.50
4320	Type MC cable		8.96	.893		34.50	55		89.50	120
4330	EMT & wire		3.96	2.020		45.50	124		169.50	235
4350	GFI with #12/2, type NM cable		8.96	.893		24.50	55		79.50	109
4370	Type MC cable		7.73	1.040		35	64		99	134
4380	EMT & wire		3.54	2.260		49.50	139		188.50	261
4400	20 amp recpt., #12/2, type NM cable		8.96	.893		59.50	55		114.50	147
4420	Type MC cable		7.73	1.040		70	64		134	172
4430	EMT & wire		3.54	2.260		84	139		223	299
4500	Weather-proof cover for above receptacles, add	↓	26.88	.298	↓	2.22	18.25		20.47	29.50
4550	Air conditioner outlet, 20 amp-240 volt recpt.									
4560	30' of #12/2, 2 pole circuit breaker									
4570	Type NM cable	1 Elec	8.40	.952	Ea.	67	58.50		125.50	161
4580	Type MC cable		7.56	1.060		79	65		144	184
4590	EMT & wire		3.36	2.380		92	146		238	320
4600	Decorator style, type NM cable		8.40	.952		72.50	58.50		131	167
4620	Type MC cable		7.56	1.060		84.50	65		149.50	190
4630	EMT & wire	↓	3.36	2.380	↓	97.50	146		243.50	325
4650	Dryer outlet, 30 amp-240 volt recpt., 20' of #10/3									
4660	2 pole circuit breaker									
4670	Type NM cable	1 Elec	5.38	1.490	Ea.	60	91.50		151.50	202
4680	Type MC cable		4.80	1.670		68.50	102		170.50	229
4690	EMT & wire	↓	2.92	2.740	↓	80.50	168		248.50	340
4700	Range outlet, 50 amp-240 volt recpt., 30' of #8/3									
4710	Type NM cable	1 Elec	3.54	2.260	Ea.	90.50	139		229.50	305
4720	Type MC cable		3.36	2.380		146	146		292	380
4730	EMT & wire		2.49	3.220		115	198		313	420
4750	Central vacuum outlet, type NM cable		5.38	1.490		63.50	91.50		155	206
4770	Type MC cable		4.80	1.670		77.50	102		179.50	239
4780	EMT & wire	↓	2.92	2.740	↓	97	168		265	355
4800	30 amp-110 volt locking recpt., #10/2 circ. bkr.									
4810	Type NM cable	1 Elec	5.21	1.540	Ea.	73.50	94.50		168	222
4820	Type MC cable		4.54	1.760		91.50	108		199.50	261
4830	EMT & wire	↓	2.69	2.980	↓	108	183		291	390
4900	Low voltage outlets									
4910	Telephone recpt., 20' of 4/C phone wire	1 Elec	21.84	.366	Ea.	9.70	22.50		32.20	44
4920	TV recpt., 20' of RG59U coax wire, F type connector	"	13.44	.595	"	19.35	36.50		55.85	76
4950	Door bell chime, transformer, 2 buttons, 60' of bellwire									
4970	Economy model	1 Elec	9.66	.828	Ea.	63	51		114	145
4980	Custom model		9.66	.828		122	51		173	210
4990	Luxury model, 3 buttons	↓	7.98	1	↓	207	61.50		268.50	320
6000	Lighting outlets									
6050	Wire only (for fixture), type NM cable	1 Elec	26.88	.298	Ea.	6.55	18.25		24.80	34
6070	Type MC cable		20.16	.397		12.45	24.50		36.95	50
6080	EMT & wire		8.40	.952		22.50	58.50		81	112
6100	Box (4"), and wire (for fixture), type NM cable		21	.381		16.55	23.50		40.05	53
6120	Type MC cable		16.80	.476		22.50	29		51.50	68
6130	EMT & wire	↓	9.24	.866	↓	32.50	53		85.50	115
6200	Fixtures (use with line 6050 or 6100 above)									
6210	Canopy style, economy grade	1 Elec	33.60	.238	Ea.	24.50	14.60		39.10	49
6220	Custom grade	↓	33.60	.238	↓	58	14.60		72.60	86

For customer support on your Electrical Change Order Costs with RSMeans data, call 800.448.8182. **Post-Installation Change Order Costs**

26 05 90.10 Residential Wiring	Crew	Daily Output	Labor-Hours	Unit	Material	2020 Bare Costs Labor	2020 Bare Costs Equipment	Total	Total Incl O&P
6250	Dining room chandelier, economy grade	1 Elec	15.96	.501	Ea.	90	31		121	145
6260	Custom grade		15.96	.501		355	31		386	435
6270	Luxury grade		12.60	.635		1,425	39		1,464	1,625
6310	Kitchen fixture (fluorescent), economy grade		25.20	.317		79.50	19.50		99	117
6320	Custom grade		21	.381		162	23.50		185.50	213
6350	Outdoor, wall mounted, economy grade		25.20	.317		33.50	19.50		53	65.50
6360	Custom grade		25.20	.317		130	19.50		149.50	172
6370	Luxury grade		21	.381		272	23.50		295.50	335
6410	Outdoor PAR floodlights, 1 lamp, 150 watt		16.80	.476		30.50	29		59.50	77
6420	2 lamp, 150 watt each		16.80	.476		49.50	29		78.50	98
6425	Motion sensing, 2 lamp, 150 watt each		16.80	.476		120	29		149	176
6430	For infrared security sensor, add		26.88	.298		104	18.25		122.25	142
6450	Outdoor, quartz-halogen, 300 watt flood		16.80	.476		44.50	29		73.50	92
6600	Recessed downlight, round, pre-wired, 50 or 75 watt trim		25.20	.317		75	19.50		94.50	112
6610	With shower light trim		25.20	.317		103	19.50		122.50	142
6620	With wall washer trim		23.52	.340		103	21		124	145
6630	With eye-ball trim		23.52	.340		93.50	21		114.50	134
6700	Porcelain lamp holder		33.60	.238		3.04	14.60		17.64	25.50
6710	With pull switch		33.60	.238		11.75	14.60		26.35	35
6750	Fluorescent strip, 2-20 watt tube, wrap around diffuser, 24"		20.16	.397		50	24.50		74.50	91.50
6760	1-34 watt tube, 48"		20.16	.397		134	24.50		158.50	184
6770	2-34 watt tubes, 48"		16.80	.476		176	29		205	237
6800	Bathroom heat lamp, 1-250 watt		23.52	.340		37	21		58	71.50
6810	2-250 watt lamps		23.52	.340		70.50	21		91.50	109
6820	For timer switch, see Section 26 05 90.10 line 2400									
6900	Outdoor post lamp, incl. post, fixture, 35' of #14/2									
6910	Type NM cable	1 Elec	2.94	2.720	Ea.	355	167		522	645
6920	Photo-eye, add		22.68	.353		30	21.50		51.50	65
6950	Clock dial time switch, 24 hr., w/enclosure, type NM cable		9.60	.833		79.50	51		130.50	164
6970	Type MC cable		9.24	.866		91.50	53		144.50	180
6980	EMT & wire		4.07	1.960		102	120		222	292
7000	Alarm systems									
7050	Smoke detectors, box, #14/3, type NM cable	1 Elec	12.22	.655	Ea.	37	40		77	101
7070	Type MC cable		10.34	.774		48.50	47.50		96	124
7080	EMT & wire		4.20	1.900		55	117		172	235
7090	For relay output to security system, add					11.25			11.25	12.40
8000	Residential equipment									
8050	Disposal hook-up, incl. switch, outlet box, 3' of flex									
8060	20 amp-1 pole circ. bkr., and 25' of #12/2									
8070	Type NM cable	1 Elec	8.40	.952	Ea.	32	58.50		90.50	122
8080	Type MC cable		6.72	1.190		43	73		116	157
8090	EMT & wire		4.20	1.900		60	117		177	241
8100	Trash compactor or dishwasher hook-up, incl. outlet box,									
8110	3' of flex, 15 amp-1 pole circ. bkr., and 25' of #14/2									
8120	Type NM cable	1 Elec	8.40	.952	Ea.	17.65	58.50		76.15	106
8130	Type MC cable		6.72	1.190		31	73		104	143
8140	EMT & wire		4.20	1.900		45.50	117		162.50	224
8150	Hot water sink dispenser hook-up, use line 8100									
8200	Vent/exhaust fan hook-up, type NM cable	1 Elec	26.88	.298	Ea.	6.55	18.25		24.80	34
8220	Type MC cable		20.16	.397		12.45	24.50		36.95	50
8230	EMT & wire		8.40	.952		22.50	58.50		81	112
8250	Bathroom vent fan, 50 CFM (use with above hook-up)									
8260	Economy model	1 Elec	12.60	.635	Ea.	20.50	39		59.50	81

26 05 90.10 Residential Wiring	Crew	Daily Output	Labor-Hours	Unit	Material	2020 Bare Costs Labor	Equipment	Total	Total Incl O&P
8270	Low noise model	1 Elec	12.60	.635	Ea.	52.50	39		91.50	116
8280	Custom model	↓	10.08	.794	↓	129	48.50		177.50	215
8300	Bathroom or kitchen vent fan, 110 CFM									
8310	Economy model	1 Elec	12.60	.635	Ea.	73.50	39		112.50	139
8320	Low noise model	"	12.60	.635	"	106	39		145	175
8350	Paddle fan, variable speed (w/o lights)									
8360	Economy model (AC motor)	1 Elec	8.40	.952	Ea.	149	58.50		207.50	251
8362	With light kit		8.40	.952		193	58.50		251.50	300
8370	Custom model (AC motor)		8.40	.952		380	58.50		438.50	505
8372	With light kit		8.40	.952		425	58.50		483.50	555
8380	Luxury model (DC motor)		6.72	1.190		345	73		418	490
8382	With light kit		6.72	1.190		390	73		463	535
8390	Remote speed switch for above, add	↓	10.08	.794	↓	44.50	48.50		93	122
8500	Whole house exhaust fan, ceiling mount, 36", variable speed									
8510	Remote switch, incl. shutters, 20 amp-1 pole circ. bkr.									
8520	30' of #12/2, type NM cable	1 Elec	3.36	2.380	Ea.	1,525	146		1,671	1,900
8530	Type MC cable		2.94	2.720		1,550	167		1,717	1,950
8540	EMT & wire	↓	2.52	3.170	↓	1,550	194		1,744	2,025
8600	Whirlpool tub hook-up, incl. timer switch, outlet box									
8610	3' of flex, 20 amp-1 pole GFI circ. bkr.									
8620	30' of #12/2, type NM cable	1 Elec	4.20	1.900	Ea.	142	117		259	330
8630	Type MC cable		3.53	2.270		150	139		289	370
8640	EMT & wire	↓	2.86	2.800	↓	164	172		336	435
8650	Hot water heater hook-up, incl. 1-2 pole circ. bkr., box;									
8660	3' of flex, 20' of #10/2, type NM cable	1 Elec	4.20	1.900	Ea.	32	117		149	210
8670	Type MC cable		3.53	2.270		46	139		185	258
8680	EMT & wire	↓	2.86	2.800	↓	52.50	172		224.50	315
9000	Heating/air conditioning									
9050	Furnace/boiler hook-up, incl. firestat, local on-off switch									
9060	Emergency switch, and 40' of type NM cable	1 Elec	3.36	2.380	Ea.	63	146		209	286
9070	Type MC cable		2.94	2.720		79	167		246	335
9080	EMT & wire	↓	1.26	6.350	↓	103	390		493	695
9100	Air conditioner hook-up, incl. local 60 amp disc. switch									
9110	3' sealtite, 40 amp, 2 pole circuit breaker									
9130	40' of #8/2, type NM cable	1 Elec	2.94	2.720	Ea.	158	167		325	420
9140	Type MC cable		2.52	3.170		233	194		427	545
9150	EMT & wire	↓	1.09	7.330	↓	204	450		654	895
9200	Heat pump hook-up, 1-40 & 1-100 amp 2 pole circ. bkr.									
9210	Local disconnect switch, 3' sealtite									
9220	40' of #8/2 & 30' of #3/2									
9230	Type NM cable	1 Elec	1.09	7.330	Ea.	575	450		1,025	1,300
9240	Type MC cable		.91	8.820		605	540		1,145	1,475
9250	EMT & wire	↓	.79	10.130	↓	585	620		1,205	1,575
9500	Thermostat hook-up, using low voltage wire									
9520	Heating only, 25' of #18-3	1 Elec	20.16	.397	Ea.	7.45	24.50		31.95	44.50
9530	Heating/cooling, 25' of #18-4	"	16.80	.476	"	9.45	29		38.45	54

26 09 Instrumentation and Control for Electrical Systems

26 09 13 – Electrical Power Monitoring

26 09 13.10 Switchboard Instruments

	26 09 13.10 Switchboard Instruments		Crew	Daily Output	Labor-Hours	Unit	Material	2020 Bare Costs Labor	2020 Bare Costs Equipment	Total	Total Incl O&P
0010	**SWITCHBOARD INSTRUMENTS** 3 phase, 4 wire										
0100	AC indicating, ammeter & switch		1 Elec	6.72	1.190	Ea.	2,925	73		2,998	3,325
0200	Voltmeter & switch			6.72	1.190		3,300	73		3,373	3,725
0300	Wattmeter			6.72	1.190		4,150	73		4,223	4,650
0400	AC recording, ammeter			3.36	2.380		7,375	146		7,521	8,350
0500	Voltmeter			3.36	2.380		7,375	146		7,521	8,350
0600	Ground fault protection, zero sequence			2.27	3.530		6,525	217		6,742	7,500
0700	Ground return path			2.27	3.530		6,525	217		6,742	7,500
0800	3 current transformers, 5 to 800 amp			1.68	4.760		3,025	292		3,317	3,775
0900	1,000 to 1,500 amp			1.09	7.330		4,375	450		4,825	5,475
1200	2,000 to 4,000 amp			.84	9.520		5,150	585		5,735	6,550
1300	Fused potential transformer, maximum 600 volt			6.72	1.190		1,150	73		1,223	1,350

26 09 13.20 Voltage Monitor Systems

	26 09 13.20 Voltage Monitor Systems		Crew	Daily Output	Labor-Hours	Unit	Material	2020 Bare Costs Labor	2020 Bare Costs Equipment	Total	Total Incl O&P
0010	**VOLTAGE MONITOR SYSTEMS** (test equipment)										
0100	AC voltage monitor system, 120/240 V, one-channel					Ea.	2,875			2,875	3,175
0110	Modem adapter						360			360	395
0120	Add-on detector only						1,500			1,500	1,675
0150	AC voltage remote monitor sys., 3 channel, 120, 230, or 480 V						5,225			5,225	5,750
0160	With internal modem						5,525			5,525	6,075
0170	Combination temperature and humidity probe						810			810	890
0180	Add-on detector only						3,800			3,800	4,175
0190	With internal modem						4,125			4,125	4,550

26 09 23 – Lighting Control Devices

26 09 23.10 Energy Saving Lighting Devices

	26 09 23.10 Energy Saving Lighting Devices			Crew	Daily Output	Labor-Hours	Unit	Material	2020 Bare Costs Labor	2020 Bare Costs Equipment	Total	Total Incl O&P
0010	**ENERGY SAVING LIGHTING DEVICES**											
0100	Occupancy sensors, passive infrared ceiling mounted		G	1 Elec	5.88	1.360	Ea.	82.50	83.50		166	215
0110	Ultrasonic ceiling mounted		G		5.88	1.360		116	83.50		199.50	252
0120	Dual technology ceiling mounted		G		5.46	1.470		149	90		239	298
0150	Automatic wall switches		G		20.16	.397		75.50	24.50		100	120
0160	Daylighting sensor, manual control, ceiling mounted		G		5.88	1.360		186	83.50		269.50	330
0170	Remote and dimming control with remote controller		G		5.46	1.470		222	90		312	380
0200	Passive infrared ceiling mounted				5.46	1.470		38.50	90		128.50	177
0400	Remote power pack		G		8.40	.952		38.50	58.50		97	130
0450	Photoelectric control, S.P.S.T. 120 V		G		6.72	1.190		27	73		100	139
0500	S.P.S.T. 208 V/277 V		G		6.72	1.190		28.50	73		101.50	141
0550	D.P.S.T. 120 V		G		5.04	1.590		207	97.50		304.50	370
0600	D.P.S.T. 208 V/277 V		G		5.04	1.590		234	97.50		331.50	405
0650	S.P.D.T. 208 V/277 V		G		5.04	1.590		219	97.50		316.50	385

26 09 26 – Lighting Control Panelboards

26 09 26.10 Lighting Control Relay Panel

	26 09 26.10 Lighting Control Relay Panel		Crew	Daily Output	Labor-Hours	Unit	Material	2020 Bare Costs Labor	2020 Bare Costs Equipment	Total	Total Incl O&P
0010	**LIGHTING CONTROL RELAY PANEL** with timeclock										
0100	4 Relay		1 Elec	2.10	3.810	Ea.	1,125	234		1,359	1,600
0110	8 Relay			1.93	4.140		1,525	254		1,779	2,075
0120	16 Relay			1.51	5.290		1,775	325		2,100	2,425
0130	24 Relay			1.26	6.350		2,675	390		3,065	3,525
0140	48 Relay			.84	9.520		1,825	585		2,410	2,875
0200	Room Controller, switching only										
0210	1 Relay		1 Elec	2.52	3.170	Ea.	1,650	194		1,844	2,125
0220	2 Relay			2.52	3.170		1,575	194		1,769	2,025
0230	3 Relay			2.52	3.170		1,900	194		2,094	2,400
0240	Dimming										

26 09 Instrumentation and Control for Electrical Systems

26 09 26 – Lighting Control Panelboards

26 09 26.10 Lighting Control Relay Panel		Crew	Daily Output	Labor-Hours	Unit	Material	2020 Bare Costs Labor	Equipment	Total	Total Incl O&P
0250	1 Relay	1 Elec	2.52	3.170	Ea.	1,475	194		1,669	1,925
0260	2 Relay		2.52	3.170		1,550	194		1,744	2,000
0270	3 Relay		2.52	3.170		1,900	194		2,094	2,400

26 12 Medium-Voltage Transformers

26 12 19 – Pad-Mounted, Liquid-Filled, Medium-Voltage Transformers

26 12 19.10 Transformer, Oil-Filled

		Crew	Daily Output	Labor-Hours	Unit	Material	Labor	Equipment	Total	Total Incl O&P
0010	**TRANSFORMER, OIL-FILLED** primary delta or Y,									
0050	Pad mounted 5 kV or 15 kV, with taps, 277/480 V secondary, 3 phase									
0100	150 kVA	R-3	.55	36.630	Ea.	9,875	2,250	345	12,470	14,600
0110	225 kVA		.46	43.290		16,900	2,650	410	19,960	23,000
0200	300 kVA		.38	52.910		14,800	3,225	500	18,525	21,600
0300	500 kVA		.34	59.520		21,900	3,650	560	26,110	30,100
0400	750 kVA		.32	62.660		26,600	3,825	590	31,015	35,600
0500	1,000 kVA		.22	91.580		31,500	5,600	865	37,965	43,900
0600	1,500 kVA		.19	103		37,400	6,300	975	44,675	51,500
0700	2,000 kVA		.17	119		47,200	7,275	1,125	55,600	64,000
0710	2,500 kVA		.16	125		57,000	7,650	1,175	65,825	75,500
0720	3,000 kVA		.14	140		69,000	8,550	1,325	78,875	90,500
0800	3,750 kVA		.13	148		91,000	9,050	1,400	101,450	115,500
1990	Pole mounted distribution type, single phase									
2000	13.8 kV primary, 120/240 V secondary, 10 kVA	R-15	6.26	7.670	Ea.	1,150	460	44.50	1,654.50	2,000
2010	50 kVA		3.11	15.440		2,250	925	89.50	3,264.50	3,950
2020	100 kVA		2.31	20.780		4,000	1,250	121	5,371	6,375
2030	167 kVA		1.81	26.580		6,225	1,600	154	7,979	9,400
2900	2,400 V primary, 120/240 V secondary, 10 kVA		6.26	7.670		1,075	460	44.50	1,579.50	1,900
2910	15 kVA		5.63	8.530		1,300	510	49.50	1,859.50	2,250
2920	25 kVA		5.04	9.520		1,675	570	55.50	2,300.50	2,750
2930	37.5 kVA		3.61	13.290		2,050	800	77	2,927	3,525
2940	50 kVA		3.11	15.440		2,350	925	89.50	3,364.50	4,050
2950	75 kVA		2.52	19.050		3,550	1,150	111	4,811	5,725
2960	100 kVA		2.31	20.780		3,575	1,250	121	4,946	5,900

26 12 19.20 Transformer, Liquid-Filled

		Crew	Daily Output	Labor-Hours	Unit	Material	Labor	Equipment	Total	Total Incl O&P
0010	**TRANSFORMER, LIQUID-FILLED** Pad mounted									
0020	5 kV or 15 kV primary, 277/480 volt secondary, 3 phase									
0050	225 kVA	R-3	.46	43.290	Ea.	14,000	2,650	410	17,060	19,900
0100	300 kVA		.38	52.910		16,700	3,225	500	20,425	23,800
0200	500 kVA		.34	59.520		21,100	3,650	560	25,310	29,200
0250	750 kVA		.32	62.660		27,200	3,825	590	31,615	36,300
0300	1,000 kVA		.22	91.580		31,600	5,600	865	38,065	44,100
0350	1,500 kVA		.19	103		36,900	6,300	975	44,175	51,000
0400	2,000 kVA		.17	119		45,700	7,275	1,125	54,100	62,000
0450	2,500 kVA		.16	125		55,500	7,650	1,175	64,325	73,500

For customer support on your Electrical Change Order Costs with RSMeans data, call 800.448.8182.

Post-Installation Change Order Costs

26 13 16.10 Switchgear	Crew	Daily Output	Labor-Hours	Unit	Material	2020 Bare Costs Labor	Equipment	Total	Total Incl O&P
0010 **SWITCHGEAR**, Incorporate switch with cable connections, transformer,									
0100 & low voltage section									
0200 Load interrupter switch, 600 amp, 2 position									
0300 NEMA 1, 4.8 kV, 300 kVA & below w/CLF fuses	R-3	.34	59.520	Ea.	20,200	3,650	560	24,410	28,200
0400 400 kVA & above w/CLF fuses		.32	62.660		23,700	3,825	590	28,115	32,500
0500 Non fusible		.34	58.070		18,100	3,550	550	22,200	25,800
0600 13.8 kV, 300 kVA & below w/CLF fuses		.32	62.660		29,800	3,825	590	34,215	39,100
0700 400 kVA & above w/CLF fuses		.30	66.140		29,800	4,050	625	34,475	39,400
0800 Non fusible		.34	59.520		22,100	3,650	560	26,310	30,300
0900 Cable lugs for 2 feeders 4.8 kV or 13.8 kV	1 Elec	6.72	1.190		710	73		783	890
1000 Pothead, one 3 conductor or three 1 conductor		3.36	2.380		3,400	146		3,546	3,975
1100 Two 3 conductor or six 1 conductor		1.68	4.760		6,725	292		7,017	7,825
1200 Key interlocks		6.72	1.190		785	73		858	975
1300 Lightning arresters, distribution class (no charge)									
1400 Intermediate class or line type 4.8 kV	1 Elec	2.27	3.530	Ea.	3,825	217		4,042	4,525
1500 13.8 kV		1.68	4.760		5,075	292		5,367	6,000
1600 Station class, 4.8 kV		2.27	3.530		6,550	217		6,767	7,525
1700 13.8 kV		1.68	4.760		11,300	292		11,592	12,800
1800 Transformers, 4,800 volts to 480/277 volts, 75 kVA	R-3	.57	35.010		19,900	2,150	330	22,380	25,500
1900 112.5 kVA		.55	36.630		24,300	2,250	345	26,895	30,500
2000 150 kVA		.48	41.770		27,700	2,550	395	30,645	34,700
2100 225 kVA		.40	49.600		31,800	3,025	470	35,295	40,000
2200 300 kVA		.34	58.070		35,600	3,550	550	39,700	45,000
2300 500 kVA		.30	66.140		46,800	4,050	625	51,475	58,000
2400 750 kVA		.24	82.100		53,000	5,025	775	58,800	67,000
2500 13,800 volts to 480/277 volts, 75 kVA		.51	39.030		28,100	2,375	370	30,845	34,900
2600 112.5 kVA		.46	43.290		37,300	2,650	410	40,360	45,400
2700 150 kVA		.41	48.590		37,600	2,975	460	41,035	46,300
2800 225 kVA		.34	58.070		43,400	3,550	550	47,500	53,500
2900 300 kVA		.31	64.350		44,300	3,925	610	48,835	55,500
3000 500 kVA		.26	76.800		49,100	4,700	725	54,525	62,000
3100 750 kVA		.22	91.580		54,000	5,600	865	60,465	69,000
3200 Forced air cooling & temperature alarm	1 Elec	.84	9.520		4,350	585		4,935	5,650
3300 Low voltage components									
3400 Maximum panel height 49-1/2", single or twin row									
3500 Breaker heights, type FA or FH, 6"									
3600 type KA or KH, 8"									
3700 type LA, 11"									
3800 type MA, 14"									
3900 Breakers, 2 pole, 15 to 60 amp, type FA	1 Elec	4.70	1.700	Ea.	750	104		854	975
4000 70 to 100 amp, type FA		3.53	2.270		960	139		1,099	1,250
4100 15 to 60 amp, type FH		4.70	1.700		1,125	104		1,229	1,400
4200 70 to 100 amp, type FH		3.53	2.270		1,475	139		1,614	1,825
4300 125 to 225 amp, type KA		2.86	2.800		1,400	172		1,572	1,800
4400 125 to 225 amp, type KH		2.86	2.800		2,350	172		2,522	2,850
4500 125 to 400 amp, type LA		2.10	3.810		2,775	234		3,009	3,400
4600 125 to 600 amp, type MA		1.51	5.290		5,750	325		6,075	6,800
4700 700 & 800 amp, type MA		1.26	6.350		7,225	390		7,615	8,525
4800 3 pole, 15 to 60 amp, type FA		4.45	1.800		900	110		1,010	1,150
4900 70 to 100 amp, type FA		3.36	2.380		1,125	146		1,271	1,475
5000 15 to 60 amp, type FH		4.45	1.800		1,525	110		1,635	1,850
5100 70 to 100 amp, type FH		3.36	2.380		1,775	146		1,921	2,175
5200 125 to 225 amp, type KA		2.69	2.980		2,150	183		2,333	2,650

26 13 Medium-Voltage Switchgear

26 13 16 – Medium-Voltage Fusible Interrupter Switchgear

26 13 16.10 Switchgear		Crew	Daily Output	Labor-Hours	Unit	Material	2020 Bare Costs Labor	Equipment	Total	Total Incl O&P
5300	125 to 225 amp, type KH	1 Elec	2.69	2.980	Ea.	3,900	183		4,083	4,550
5400	125 to 400 amp, type LA		1.93	4.140		3,575	254		3,829	4,325
5500	125 to 600 amp, type MA		1.34	5.950		5,150	365		5,515	6,225
5600	700 & 800 amp, type MA	▼	1.09	7.330	▼	6,400	450		6,850	7,725

26 22 Low-Voltage Transformers

26 22 13 – Low-Voltage Distribution Transformers

26 22 13.10 Transformer, Dry-Type

		Crew	Daily Output	Labor-Hours	Unit	Material	2020 Bare Costs Labor	Equipment	Total	Total Incl O&P
0010	**TRANSFORMER, DRY-TYPE**									
0050	Single phase, 240/480 volt primary, 120/240 volt secondary									
0100	1 kVA	1 Elec	1.68	4.760	Ea.	390	292		682	860
0300	2 kVA		1.34	5.950		700	365		1,065	1,325
0500	3 kVA		1.18	6.800		845	415		1,260	1,550
0700	5 kVA	▼	1.01	7.940		1,300	485		1,785	2,150
0900	7.5 kVA	2 Elec	1.85	8.660		1,250	530		1,780	2,175
1100	10 kVA		1.34	11.900		1,625	730		2,355	2,875
1300	15 kVA		1.01	15.870		2,075	975		3,050	3,725
1500	25 kVA		.84	19.050		2,475	1,175		3,650	4,475
1700	37.5 kVA		.67	23.810		3,325	1,450		4,775	5,850
1900	50 kVA		.59	27.210		3,125	1,675		4,800	5,900
2100	75 kVA	▼	.55	29.300		5,225	1,800		7,025	8,425
2110	100 kVA	R-3	.76	26.460		6,825	1,625	250	8,700	10,200
2120	167 kVA	"	.67	29.760		11,300	1,825	281	13,406	15,400
2190	480 V primary, 120/240 V secondary, nonvent., 15 kVA	2 Elec	1.01	15.870		1,625	975		2,600	3,225
2200	25 kVA		.76	21.160		2,950	1,300		4,250	5,175
2210	37 kVA		.63	25.400		3,375	1,550		4,925	6,050
2220	50 kVA		.55	29.300		3,975	1,800		5,775	7,050
2230	75 kVA		.50	31.750		5,275	1,950		7,225	8,725
2240	100 kVA		.42	38.100		6,875	2,325		9,200	11,100
2250	Low operating temperature (80°C), 25 kVA		.84	19.050		5,150	1,175		6,325	7,425
2260	37 kVA		.67	23.810		5,550	1,450		7,000	8,275
2270	50 kVA		.59	27.210		7,250	1,675		8,925	10,500
2280	75 kVA		.55	29.300		11,500	1,800		13,300	15,300
2290	100 kVA	▼	.46	34.630	▼	12,000	2,125		14,125	16,400
2300	3 phase, 480 volt primary, 120/208 volt secondary									
2310	Ventilated, 3 kVA	1 Elec	.84	9.520	Ea.	1,225	585		1,810	2,225
2700	6 kVA		.67	11.900		1,225	730		1,955	2,425
2900	9 kVA	▼	.59	13.610		1,450	835		2,285	2,825
3100	15 kVA	2 Elec	.92	17.320		1,900	1,075		2,975	3,650
3300	30 kVA		.76	21.160		1,800	1,300		3,100	3,900
3500	45 kVA		.67	23.810		1,850	1,450		3,300	4,225
3700	75 kVA	▼	.59	27.210		2,500	1,675		4,175	5,225
3900	112.5 kVA	R-3	.76	26.460		4,075	1,625	250	5,950	7,150
4100	150 kVA		.71	28.010		5,000	1,700	265	6,965	8,350
4300	225 kVA		.55	36.630		8,975	2,250	345	11,570	13,600
4500	300 kVA		.46	43.290		9,325	2,650	410	12,385	14,700
4700	500 kVA		.38	52.910		18,200	3,225	500	21,925	25,400
4800	750 kVA		.29	68.030		22,100	4,150	645	26,895	31,200
4820	1,000 kVA	▼	.27	74.400		26,000	4,550	705	31,255	36,300
4850	K-4 rated, 15 kVA	2 Elec	.92	17.320		3,550	1,075		4,625	5,475
4855	30 kVA	▼	.76	21.160		5,825	1,300		7,125	8,325

For customer support on your Electrical Change Order Costs with RSMeans data, call 800.448.8182. **Post-Installation Change Order Costs**

26 22 Low-Voltage Transformers

26 22 13 – Low-Voltage Distribution Transformers

26 22 13.10 Transformer, Dry-Type		Crew	Daily Output	Labor-Hours	Unit	Material	2020 Bare Costs Labor	Equipment	Total	Total Incl O&P
4860	45 kVA	2 Elec	.67	23.810	Ea.	7,450	1,450		8,900	10,400
4865	75 kVA	↓	.59	27.210		10,200	1,675		11,875	13,700
4870	112.5 kVA	R-3	.76	26.460		12,300	1,625	250	14,175	16,200
4875	150 kVA		.71	28.010		16,100	1,700	265	18,065	20,500
4880	225 kVA		.55	36.630		22,400	2,250	345	24,995	28,300
4885	300 kVA		.46	43.290		30,700	2,650	410	33,760	38,200
4890	500 kVA	↓	.38	52.910		42,900	3,225	500	46,625	52,500
4900	K-13 rated, 15 kVA	2 Elec	.92	17.320		4,025	1,075		5,100	6,000
4905	30 kVA		.76	21.160		5,775	1,300		7,075	8,300
4910	45 kVA		.67	23.810		6,975	1,450		8,425	9,825
4915	75 kVA	↓	.59	27.210		10,600	1,675		12,275	14,200
4920	112.5 kVA	R-3	.76	26.460		13,900	1,625	250	15,775	18,000
4925	150 kVA		.71	28.010		28,800	1,700	265	30,765	34,400
4930	225 kVA		.55	36.630		25,000	2,250	345	27,595	31,200
4935	300 kVA		.46	43.290		32,900	2,650	410	35,960	40,600
4940	500 kVA	↓	.38	52.910	↓	56,000	3,225	500	59,725	67,000
5020	480 volt primary, 120/208 volt secondary									
5030	Nonventilated, 15 kVA	2 Elec	.92	17.320	Ea.	3,450	1,075		4,525	5,350
5040	30 kVA		.67	23.810		5,000	1,450		6,450	7,675
5050	45 kVA		.59	27.210		5,900	1,675		7,575	8,950
5060	75 kVA	↓	.55	29.300		11,300	1,800		13,100	15,200
5070	112.5 kVA	R-3	.71	28.010		20,600	1,700	265	22,565	25,500
5081	150 kVA		.71	28.010		21,500	1,700	265	23,465	26,400
5090	225 kVA		.50	39.680		26,500	2,425	375	29,300	33,100
5100	300 kVA	↓	.42	47.620		23,800	2,900	450	27,150	31,000
5200	Low operating temperature (80°C), 30 kVA	2 Elec	.76	21.160		6,475	1,300		7,775	9,050
5210	45 kVA		.67	23.810		5,625	1,450		7,075	8,375
5220	75 kVA	↓	.59	27.210		8,450	1,675		10,125	11,800
5230	112.5 kVA	R-3	.76	26.460		11,200	1,625	250	13,075	15,000
5240	150 kVA		.71	28.010		14,300	1,700	265	16,265	18,600
5250	225 kVA		.55	36.630		19,500	2,250	345	22,095	25,100
5260	300 kVA		.46	43.290		32,400	2,650	410	35,460	40,000
5270	500 kVA	↓	.38	52.910	↓	35,200	3,225	500	38,925	44,100
5380	3 phase, 5 kV primary, 277/480 volt secondary									
5400	High voltage, 112.5 kVA	R-3	.71	28.010	Ea.	18,200	1,700	265	20,165	22,800
5410	150 kVA		.55	36.630		21,000	2,250	345	23,595	26,800
5420	225 kVA		.46	43.290		24,700	2,650	410	27,760	31,600
5430	300 kVA		.38	52.910		30,500	3,225	500	34,225	39,000
5440	500 kVA		.29	68.030		39,400	4,150	645	44,195	50,000
5450	750 kVA		.27	74.400		61,000	4,550	705	66,255	74,500
5460	1,000 kVA		.25	79.370		71,500	4,850	750	77,100	86,500
5470	1,500 kVA		.23	88.180		83,000	5,400	835	89,235	100,500
5480	2,000 kVA		.21	95.240		97,500	5,825	900	104,225	116,500
5490	2,500 kVA		.17	119		110,000	7,275	1,125	118,400	133,000
5500	3,000 kVA	↓	.15	132	↓	143,500	8,075	1,250	152,825	171,500
5590	15 kV primary, 277/480 volt secondary									
5600	High voltage, 112.5 kVA	R-3	.71	28.010	Ea.	28,400	1,700	265	30,365	34,000
5610	150 kVA		.55	36.630		33,200	2,250	345	35,795	40,200
5620	225 kVA		.46	43.290		36,600	2,650	410	39,660	44,700
5630	300 kVA		.38	52.910		43,100	3,225	500	46,825	53,000
5640	500 kVA		.29	68.030		54,000	4,150	645	58,795	66,500
5650	750 kVA		.27	74.400		71,000	4,550	705	76,255	85,500
5660	1,000 kVA	↓	.25	79.370		80,500	4,850	750	86,100	96,500

Post-Installation Change Order Costs For customer support on your Electrical Change Order Costs with RSMeans data, call 800.448.8182.

26 22 13.10 Transformer, Dry-Type	Crew	Daily Output	Labor-Hours	Unit	Material	2020 Bare Costs Labor	Equipment	Total	Total Incl O&P
5670	1,500 kVA	R-3	.23	88.180	Ea.	93,000	5,400	835	99,235	111,000
5680	2,000 kVA		.21	95.240		103,000	5,825	900	109,725	122,500
5690	2,500 kVA		.17	119		119,500	7,275	1,125	127,900	143,000
5700	3,000 kVA		.15	132		142,000	8,075	1,250	151,325	170,000
6000	2400 volt primary, 480 volt secondary, 300 kVA		.38	52.910		30,500	3,225	500	34,225	39,000
6010	500 kVA		.29	68.030		39,400	4,150	645	44,195	50,000
6020	750 kVA		.27	74.400		56,000	4,550	705	61,255	69,500
9300	Energy efficient transformer 3 ph									
9400	15 kVA, 480VAC delta, 208Y/120VAC	R-3	1.26	15.870	Ea.	1,975	970	150	3,095	3,800
9405	30 kVA, 480VAC delta, 208Y/120VAC		1.26	15.870		2,450	970	150	3,570	4,325
9410	45 kVA, 480VAC delta, 208Y/120VAC		1.26	15.870		3,425	970	150	4,545	5,400
9415	75 kVA, 480VAC delta, 208Y/120VAC		1.05	19.050		4,500	1,175	180	5,855	6,875
9420	112 kVA, 480VAC delta, 208Y/120VAC		1.05	19.050		6,700	1,175	180	8,055	9,300
9425	150 kVA, 480VAC delta, 208Y/120VAC		1.05	19.050		15,400	1,175	180	16,755	18,900
9430	225 kVA, 480VAC delta, 208Y/120VAC		1.05	19.050		10,700	1,175	180	12,055	13,700
9435	300 kVA, 480VAC delta, 208Y/120VAC		1.05	19.050		15,000	1,175	180	16,355	18,400

26 22 13.20 Isolating Panels

		Crew	Daily Output	Labor-Hours	Unit	Material	Labor	Equipment	Total	Total Incl O&P
0010	**ISOLATING PANELS** used with isolating transformers									
0020	For hospital applications									
0100	Critical care area, 8 circuit, 3 kVA	1 Elec	.49	16.420	Ea.	7,075	1,000		8,075	9,275
0200	5 kVA		.45	17.640		7,325	1,075		8,400	9,650
0400	7.5 kVA		.44	18.320		4,900	1,125		6,025	7,075
0600	10 kVA		.37	21.650		8,075	1,325		9,400	10,900
0800	Operating room power & lighting, 8 circuit, 3 kVA		.49	16.420		5,300	1,000		6,300	7,325
1000	5 kVA		.45	17.640		5,775	1,075		6,850	7,950
1200	7.5 kVA		.44	18.320		4,050	1,125		5,175	6,125
1400	10 kVA		.37	21.650		4,950	1,325		6,275	7,425
1600	X-ray systems, 15 kVA, 90 amp		.37	21.650		15,800	1,325		17,125	19,400
1800	25 kVA, 125 amp		.30	26.460		16,600	1,625		18,225	20,700

26 22 13.30 Isolating Transformer

		Crew	Daily Output	Labor-Hours	Unit	Material	Labor	Equipment	Total	Total Incl O&P
0010	**ISOLATING TRANSFORMER**									
0100	Single phase, 120/240 volt primary, 120/240 volt secondary									
0200	0.50 kVA	1 Elec	3.36	2.380	Ea.	410	146		556	665
0400	1 kVA		1.68	4.760		945	292		1,237	1,475
0600	2 kVA		1.34	5.950		1,525	365		1,890	2,225
0800	3 kVA		1.18	6.800		860	415		1,275	1,575
1000	5 kVA		1.01	7.940		1,125	485		1,610	1,975
1200	7.5 kVA		.92	8.660		1,425	530		1,955	2,375
1400	10 kVA		.67	11.900		1,825	730		2,555	3,100
1600	15 kVA		.50	15.870		2,350	975		3,325	4,025
1800	25 kVA		.42	19.050		3,375	1,175		4,550	5,450
1810	37.5 kVA	2 Elec	.67	23.810		5,575	1,450		7,025	8,300
1820	75 kVA	"	.55	29.300		8,375	1,800		10,175	11,900
1830	3 phase, 120/240 V primary, 120/240 V secondary, 112.5 kVA	R-3	.76	26.460		11,200	1,625	250	13,075	15,000
1840	150 kVA		.71	28.010		14,200	1,700	265	16,165	18,500
1850	225 kVA		.55	36.630		19,800	2,250	345	22,395	25,500
1860	300 kVA		.46	43.290		26,400	2,650	410	29,460	33,400
1870	500 kVA		.38	52.910		43,900	3,225	500	47,625	53,500
1880	750 kVA		.29	68.030		44,400	4,150	645	49,195	55,500

26 22 Low-Voltage Transformers

26 22 13 - Low-Voltage Distribution Transformers

26 22 13.90 Transformer Handling	Crew	Daily Output	Labor-Hours	Unit	Material	2020 Bare Costs Labor	Equipment	Total	Total Incl O&P
0010 **TRANSFORMER HANDLING** Add to normal labor cost in restricted areas									
5000 Transformers									
5150 15 kVA, approximately 200 pounds	2 Elec	2.27	7.050	Ea.		435		435	645
5160 25 kVA, approximately 300 pounds		2.10	7.620			465		465	695
5170 37.5 kVA, approximately 400 pounds		1.93	8.280			510		510	755
5180 50 kVA, approximately 500 pounds		1.68	9.520			585		585	870
5190 75 kVA, approximately 600 pounds		1.51	10.580			650		650	965
5200 100 kVA, approximately 700 pounds		1.34	11.900			730		730	1,075
5210 112.5 kVA, approximately 800 pounds	3 Elec	1.85	12.990			795		795	1,175
5220 125 kVA, approximately 900 pounds		1.68	14.290			875		875	1,300
5230 150 kVA, approximately 1,000 pounds		1.51	15.870			975		975	1,450
5240 167 kVA, approximately 1,200 pounds		1.34	17.860			1,100		1,100	1,625
5250 200 kVA, approximately 1,400 pounds		1.18	20.410			1,250		1,250	1,875
5260 225 kVA, approximately 1,600 pounds		1.09	21.980			1,350		1,350	2,000
5270 250 kVA, approximately 1,800 pounds		.92	25.970			1,600		1,600	2,375
5280 300 kVA, approximately 2,000 pounds		.84	28.570			1,750		1,750	2,600
5290 500 kVA, approximately 3,000 pounds		.63	38.100			2,325		2,325	3,475
5300 600 kVA, approximately 3,500 pounds		.56	42.640			2,625		2,625	3,900
5310 750 kVA, approximately 4,000 pounds		.50	47.620			2,925		2,925	4,350
5320 1,000 kVA, approximately 5,000 pounds		.42	57.140			3,500		3,500	5,225

26 22 16 - Low-Voltage Buck-Boost Transformers

26 22 16.10 Buck-Boost Transformer	Crew	Daily Output	Labor-Hours	Unit	Material	2020 Bare Costs Labor	Equipment	Total	Total Incl O&P
0010 **BUCK-BOOST TRANSFORMER**									
0100 Single phase, 120/240 V primary, 12/24 V secondary									
0200 0.10 kVA	1 Elec	6.72	1.190	Ea.	100	73		173	219
0400 0.25 kVA		4.79	1.670		170	102		272	340
0600 0.50 kVA		3.36	2.380		192	146		338	430
0800 0.75 kVA		2.60	3.070		305	188		493	620
1000 1.0 kVA		1.68	4.760		258	292		550	720
1200 1.5 kVA		1.51	5.290		350	325		675	870
1400 2.0 kVA		1.34	5.950		580	365		945	1,175
1600 3.0 kVA		1.18	6.800		560	415		975	1,225
1800 5.0 kVA		1.01	7.940		1,100	485		1,585	1,950
2000 3 phase, 240 V primary, 208/120 V secondary, 15 kVA	2 Elec	2.02	7.940		2,150	485		2,635	3,100
2200 30 kVA		1.34	11.900		2,800	730		3,530	4,150
2400 45 kVA		1.18	13.610		3,375	835		4,210	4,950
2600 75 kVA		1.01	15.870		4,075	975		5,050	5,925
2800 112.5 kVA	R-3	1.18	17.010		5,075	1,050	161	6,286	7,300
3000 150 kVA		.92	21.650		6,725	1,325	205	8,255	9,600
3200 225 kVA		.84	23.810		8,775	1,450	225	10,450	12,100
3400 300 kVA		.76	26.460		11,900	1,625	250	13,775	15,800

26 24 13.10 Incoming Switchboards	Crew	Daily Output	Labor-Hours	Unit	Material	2020 Bare Costs Labor	Equipment	Total	Total Incl O&P
0010 **INCOMING SWITCHBOARDS** main service section									
0100 Aluminum bus bars, not including CT's or PT's									
0200 No main disconnect, includes CT compartment									
0300 120/208 volt, 4 wire, 600 amp	2 Elec	.84	19.050	Ea.	4,625	1,175		5,800	6,825
0400 800 amp		.74	21.650		4,625	1,325		5,950	7,050
0500 1,000 amp		.67	23.810		5,550	1,450		7,000	8,275
0600 1,200 amp		.60	26.460		5,550	1,625		7,175	8,525
0700 1,600 amp		.55	28.860		5,550	1,775		7,325	8,725
0800 2,000 amp		.52	30.720		6,000	1,875		7,875	9,400
1000 3,000 amp		.47	34.010		7,900	2,075		9,975	11,800
1200 277/480 volt, 4 wire, 600 amp		.84	19.050		4,625	1,175		5,800	6,825
1300 800 amp		.74	21.650		4,625	1,325		5,950	7,050
1400 1,000 amp		.67	23.810		5,875	1,450		7,325	8,625
1500 1,200 amp		.60	26.460		5,875	1,625		7,500	8,875
1600 1,600 amp		.55	28.860		5,875	1,775		7,650	9,075
1700 2,000 amp		.52	30.720		6,000	1,875		7,875	9,400
1800 3,000 amp		.47	34.010		7,900	2,075		9,975	11,800
1900 4,000 amp	▼	.44	36.630	▼	9,775	2,250		12,025	14,200
2000 Fused switch & CT compartment									
2100 120/208 volt, 4 wire, 400 amp	2 Elec	.94	17.010	Ea.	2,700	1,050		3,750	4,525
2200 600 amp		.79	20.260		3,200	1,250		4,450	5,375
2300 800 amp		.71	22.680		12,500	1,400		13,900	15,900
2400 1,200 amp		.57	28.010		16,300	1,725		18,025	20,500
2500 277/480 volt, 4 wire, 400 amp		.96	16.710		3,800	1,025		4,825	5,700
2600 600 amp		.79	20.260		5,950	1,250		7,200	8,400
2700 800 amp		.71	22.680		12,500	1,400		13,900	15,900
2800 1,200 amp	▼	.57	28.010	▼	16,300	1,725		18,025	20,500
2900 Pressure switch & CT compartment									
3000 120/208 volt, 4 wire, 800 amp	2 Elec	.67	23.810	Ea.	11,200	1,450		12,650	14,600
3100 1,200 amp		.55	28.860		21,800	1,775		23,575	26,500
3200 1,600 amp		.52	30.720		23,100	1,875		24,975	28,200
3300 2,000 amp		.47	34.010		24,600	2,075		26,675	30,100
3310 2,500 amp		.42	38.100		30,300	2,325		32,625	36,800
3320 3,000 amp		.37	43.290		40,800	2,650		43,450	48,800
3330 4,000 amp		.34	47.620		52,500	2,925		55,425	62,000
3340 120/208 volt, 4 wire, 800 amp, with ground fault		.67	23.810		18,600	1,450		20,050	22,600
3350 1,200 amp, with ground fault		.55	28.860		24,000	1,775		25,775	29,000
3360 1,600 amp, with ground fault		.52	30.720		26,100	1,875		27,975	31,500
3370 2,000 amp, with ground fault		.47	34.010		28,100	2,075		30,175	34,100
3400 277/480 volt, 4 wire, 800 amp, with ground fault		.67	23.810		18,600	1,450		20,050	22,600
3600 1,200 amp, with ground fault		.55	28.860		24,000	1,775		25,775	29,000
4000 1,600 amp, with ground fault		.52	30.720		26,100	1,875		27,975	31,500
4200 2,000 amp, with ground fault	▼	.47	34.010	▼	28,100	2,075		30,175	34,100
4400 Circuit breaker, molded case & CT compartment									
4600 3 pole, 4 wire, 600 amp	2 Elec	.79	20.260	Ea.	9,675	1,250		10,925	12,500
4800 800 amp		.71	22.680		11,600	1,400		13,000	14,800
5000 1,200 amp	▼	.57	28.010	▼	15,800	1,725		17,525	20,000
5100 Copper bus bars, not incl. CT's or PT's, add, minimum					15%				

26 24 13.20 In Plant Distribution Switchboards

	Crew	Daily Output	Labor-Hours	Unit	Material	Labor	Equipment	Total	Total Incl O&P
0010 **IN PLANT DISTRIBUTION SWITCHBOARDS**									
0100 Main lugs only, to 600 volt, 3 pole, 3 wire, 200 amp	2 Elec	1.01	15.870	Ea.	1,250	975		2,225	2,825
0110 400 amp	▼	1.01	15.870	▼	1,250	975		2,225	2,825

26 24 Switchboards and Panelboards

26 24 13 – Switchboards

26 24 13.20 In Plant Distribution Switchboards	Crew	Daily Output	Labor-Hours	Unit	Material	2020 Bare Costs Labor	Equipment	Total	Total Incl O&P
0120	600 amp	2 Elec	1.01	15.870	Ea.	1,300	975		2,275	2,875
0130	800 amp		.91	17.640		1,400	1,075		2,475	3,150
0140	1,200 amp		.77	20.700		1,700	1,275		2,975	3,775
0150	1,600 amp		.72	22.150		2,225	1,350		3,575	4,475
0160	2,000 amp		.69	23.230		2,475	1,425		3,900	4,850
0250	To 480 volt, 3 pole, 4 wire, 200 amp		1.01	15.870		1,100	975		2,075	2,650
0260	400 amp		1.01	15.870		1,250	975		2,225	2,825
0270	600 amp		1.01	15.870		1,375	975		2,350	2,950
0280	800 amp		.91	17.640		1,500	1,075		2,575	3,250
0290	1,200 amp		.77	20.700		1,825	1,275		3,100	3,925
0300	1,600 amp		.72	22.150		2,100	1,350		3,450	4,325
0310	2,000 amp		.69	23.230		2,450	1,425		3,875	4,825
0400	Main circuit breaker, to 600 volt, 3 pole, 3 wire, 200 amp		1.01	15.870		3,425	975		4,400	5,225
0410	400 amp		.96	16.710		3,450	1,025		4,475	5,325
0420	600 amp		.92	17.320		4,350	1,075		5,425	6,350
0430	800 amp		.87	18.320		7,275	1,125		8,400	9,675
0440	1,200 amp		.74	21.650		9,525	1,325		10,850	12,500
0450	1,600 amp		.71	22.680		15,300	1,400		16,700	18,900
0460	2,000 amp		.67	23.810		16,300	1,450		17,750	20,100
0550	277/480 volt, 3 pole, 4 wire, 200 amp		1.01	15.870		3,625	975		4,600	5,425
0560	400 amp		.96	16.710		3,625	1,025		4,650	5,500
0570	600 amp		.92	17.320		4,525	1,075		5,600	6,550
0580	800 amp		.87	18.320		7,650	1,125		8,775	10,100
0590	1,200 amp		.74	21.650		9,875	1,325		11,200	12,900
0600	1,600 amp		.71	22.680		15,300	1,400		16,700	19,000
0610	2,000 amp		.67	23.810		16,400	1,450		17,850	20,200
0700	Main fusible switch w/fuse, 208/240 volt, 3 pole, 3 wire, 200 amp		1.01	15.870		3,775	975		4,750	5,625
0710	400 amp		.96	16.710		3,750	1,025		4,775	5,650
0720	600 amp		.92	17.320		4,600	1,075		5,675	6,650
0730	800 amp		.87	18.320		9,425	1,125		10,550	12,100
0740	1,200 amp		.74	21.650		11,000	1,325		12,325	14,100
0800	120/208, 120/240 volt, 3 pole, 4 wire, 200 amp		1.01	15.870		3,375	975		4,350	5,150
0810	400 amp		.96	16.710		3,450	1,025		4,475	5,325
0820	600 amp		.92	17.320		4,400	1,075		5,475	6,400
0830	800 amp		.87	18.320		6,975	1,125		8,100	9,350
0840	1,200 amp		.74	21.650		8,125	1,325		9,450	10,900
0900	480 or 600 volt, 3 pole, 3 wire, 200 amp		1.01	15.870		3,650	975		4,625	5,450
0910	400 amp		.96	16.710		3,575	1,025		4,600	5,475
0920	600 amp		.92	17.320		4,375	1,075		5,450	6,400
0930	800 amp		.87	18.320		6,750	1,125		7,875	9,100
0940	1,200 amp		.74	21.650		7,825	1,325		9,150	10,600
1000	277 or 480 volt, 3 pole, 4 wire, 200 amp		1.01	15.870		3,775	975		4,750	5,625
1010	400 amp		.96	16.710		3,725	1,025		4,750	5,625
1020	600 amp		.92	17.320		4,575	1,075		5,650	6,600
1030	800 amp		.87	18.320		7,000	1,125		8,125	9,375
1040	1,200 amp		.74	21.650		8,125	1,325		9,450	10,900
1120	1,600 amp		.64	25.060		14,800	1,525		16,325	18,600
1130	2,000 amp		.57	28.010		19,500	1,725		21,225	24,100
1150	Pressure switch, bolted, 3 pole, 208/240 volt, 3 wire, 800 amp		.81	19.840		11,000	1,225		12,225	13,900
1160	1,200 amp		.67	23.810		14,000	1,450		15,450	17,600
1170	1,600 amp		.64	25.060		16,000	1,525		17,525	19,900
1180	2,000 amp		.57	28.010		18,200	1,725		19,925	22,700
1200	120/208 or 120/240 volt, 3 pole, 4 wire, 800 amp		.81	19.840		8,675	1,225		9,900	11,400

Post-Installation Change Order Costs   For customer support on your Electrical Change Order Costs with RSMeans data, call 800.448.8182.

26 24 Switchboards and Panelboards

26 24 13 – Switchboards

26 24 13.20 In Plant Distribution Switchboards	Crew	Daily Output	Labor-Hours	Unit	Material	2020 Bare Costs Labor	Equipment	Total	Total Incl O&P
1210	1,200 amp	2 Elec	.67	23.810	Ea.	10,100	1,450		11,550	13,400
1220	1,600 amp		.64	25.060		16,000	1,525		17,525	19,900
1230	2,000 amp		.57	28.010		18,200	1,725		19,925	22,700
1300	480 or 600 volt, 3 wire, 800 amp		.81	19.840		11,000	1,225		12,225	13,900
1310	1,200 amp		.67	23.810		15,400	1,450		16,850	19,100
1320	1,600 amp		.64	25.060		17,300	1,525		18,825	21,300
1330	2,000 amp		.57	28.010		19,500	1,725		21,225	24,100
1400	277/480 volt, 4 wire, 800 amp		.81	19.840		11,000	1,225		12,225	13,900
1410	1,200 amp		.67	23.810		15,400	1,450		16,850	19,100
1420	1,600 amp		.64	25.060		17,300	1,525		18,825	21,300
1430	2,000 amp		.57	28.010		19,500	1,725		21,225	24,100
1500	Main ground fault protector, 1,200-2,000 amp		4.54	3.530		3,225	217		3,442	3,875
1600	Busway connection, 200 amp		4.54	3.530		535	217		752	910
1610	400 amp		3.86	4.140		535	254		789	970
1620	600 amp		3.36	4.760		535	292		827	1,025
1630	800 amp		2.69	5.950		535	365		900	1,125
1640	1,200 amp		2.18	7.330		535	450		985	1,250
1650	1,600 amp		2.02	7.940		1,100	485		1,585	1,950
1660	2,000 amp		1.68	9.520		1,100	585		1,685	2,100
1700	Shunt trip for remote operation, 200 amp		6.72	2.380		605	146		751	885
1710	400 amp		6.72	2.380		970	146		1,116	1,300
1720	600 amp		6.72	2.380		1,150	146		1,296	1,475
1730	800 amp		6.72	2.380		1,475	146		1,621	1,850
1740	1,200-2,000 amp		6.72	2.380		3,300	146		3,446	3,850
1800	Motor operated main breaker, 200 amp		6.72	2.380		3,625	146		3,771	4,200
1810	400 amp		6.72	2.380		3,650	146		3,796	4,250
1820	600 amp		6.72	2.380		3,375	146		3,521	3,950
1830	800 amp		6.72	2.380		3,400	146		3,546	3,975
1840	1,200-2,000 amp		6.72	2.380		3,175	146		3,321	3,700
1900	Current/potential transformer metering compartment, 200-800 amp		4.54	3.530		2,150	217		2,367	2,700
1940	1,200 amp		4.54	3.530		6,000	217		6,217	6,925
1950	1,600-2,000 amp		4.54	3.530		9,875	217		10,092	11,200
2000	With watt meter, 200-800 amp		3.36	4.760		9,275	292		9,567	10,600
2040	1,200 amp		3.36	4.760		10,800	292		11,092	12,200
2050	1,600-2,000 amp		3.36	4.760		11,500	292		11,792	13,000
2100	Split bus, 60-200 amp	1 Elec	4.45	1.800		194	110		304	375
2130	400 amp	2 Elec	3.86	4.140		320	254		574	730
2140	600 amp		3.02	5.290		390	325		715	910
2150	800 amp		2.18	7.330		495	450		945	1,225
2170	1,200 amp		1.68	9.520		565	585		1,150	1,500
2250	Contactor control, 60 amp	1 Elec	1.68	4.760		1,275	292		1,567	1,825
2260	100 amp		1.26	6.350		1,350	390		1,740	2,050
2270	200 amp		.84	9.520		2,025	585		2,610	3,125
2280	400 amp	2 Elec	.84	19.050		6,550	1,175		7,725	8,950
2290	600 amp		.71	22.680		7,325	1,400		8,725	10,100
2300	800 amp		.60	26.460		8,200	1,625		9,825	11,500
2500	Modifier, two distribution sections, add		.67	23.810		2,850	1,450		4,300	5,300
2520	Three distribution sections, add		.34	47.620		5,475	2,925		8,400	10,400
2560	Auxiliary pull section, 20", add		1.68	9.520		1,025	585		1,610	2,000
2580	24", add		1.51	10.580		1,025	650		1,675	2,100
2600	30", add		1.34	11.900		1,025	730		1,755	2,200
2620	36", add		1.18	13.610		1,225	835		2,060	2,600
2640	Dog house, 12", add		2.02	7.940		212	485		697	960

For customer support on your Electrical Change Order Costs with RSMeans data, call 800.448.8182.

Post-Installation Change Order Costs

26 24 Switchboards and Panelboards

26 24 13 – Switchboards

26 24 13.20 In Plant Distribution Switchboards	Crew	Daily Output	Labor-Hours	Unit	Material	2020 Bare Costs Labor	Equipment	Total	Total Incl O&P
2660	18", add	2 Elec	1.68	9.520	Ea.	420	585		1,005	1,325
3000	Transition section between switchboard and transformer									
3050	or motor control center, 4 wire alum. bus, 600 amp	2 Elec	.96	16.710	Ea.	2,175	1,025		3,200	3,925
3100	800 amp		.84	19.050		2,450	1,175		3,625	4,425
3150	1,000 amp		.74	21.650		2,750	1,325		4,075	5,000
3200	1,200 amp		.67	23.810		3,025	1,450		4,475	5,500
3250	1,600 amp		.60	26.460		3,575	1,625		5,200	6,350
3300	2,000 amp		.55	28.860		4,100	1,775		5,875	7,125
3350	2,500 amp		.52	30.720		4,775	1,875		6,650	8,050
3400	3,000 amp		.47	34.010		5,500	2,075		7,575	9,150
4000	Weatherproof construction, per vertical section		1.48	10.820		2,525	665		3,190	3,775

26 24 13.30 Distribution Switchboards Section

		Crew	Daily Output	Labor-Hours	Unit	Material	2020 Bare Costs Labor	Equipment	Total	Total Incl O&P
0010	**DISTRIBUTION SWITCHBOARDS SECTION**									
0100	Aluminum bus bars, not including breakers									
0160	Subfeed lug-rated at 60 amp	2 Elec	1.09	14.650	Ea.	1,200	900		2,100	2,675
0170	100 amp		1.06	15.120		1,525	930		2,455	3,050
0180	200 amp		1.01	15.870		1,450	975		2,425	3,050
0190	400 amp		.92	17.320		1,425	1,075		2,500	3,150
0195	120/208 or 277/480 volt, 4 wire, 400 amp		.92	17.320		1,425	1,075		2,500	3,150
0200	600 amp		.84	19.050		1,700	1,175		2,875	3,625
0300	800 amp		.74	21.650		2,175	1,325		3,500	4,375
0400	1,000 amp		.67	23.810		2,500	1,450		3,950	4,925
0500	1,200 amp		.60	26.460		2,950	1,625		4,575	5,675
0600	1,600 amp		.55	28.860		4,400	1,775		6,175	7,475
0700	2,000 amp		.52	30.720		5,500	1,875		7,375	8,850
0800	2,500 amp		.50	31.750		5,950	1,950		7,900	9,450
0900	3,000 amp		.47	34.010		7,850	2,075		9,925	11,800
0950	4,000 amp		.44	36.630		7,875	2,250		10,125	12,000

26 24 13.40 Switchboards Feeder Section

		Crew	Daily Output	Labor-Hours	Unit	Material	2020 Bare Costs Labor	Equipment	Total	Total Incl O&P
0010	**SWITCHBOARDS FEEDER SECTION** group mounted devices									
0030	Circuit breakers									
0160	FA frame, 15 to 60 amp, 240 volt, 1 pole	1 Elec	6.72	1.190	Ea.	130	73		203	252
0170	2 pole		5.88	1.360		380	83.50		463.50	545
0180	3 pole		4.45	1.800		580	110		690	805
0210	480 volt, 1 pole		6.72	1.190		181	73		254	310
0220	2 pole		5.88	1.360		455	83.50		538.50	630
0230	3 pole		4.45	1.800		585	110		695	810
0260	600 volt, 2 pole		5.88	1.360		885	83.50		968.50	1,100
0270	3 pole		4.45	1.800		535	110		645	750
0280	FA frame, 70 to 100 amp, 240 volt, 1 pole		5.88	1.360		229	83.50		312.50	375
0310	2 pole		4.20	1.900		615	117		732	850
0320	3 pole		3.36	2.380		595	146		741	870
0330	480 volt, 1 pole		5.88	1.360		345	83.50		428.50	505
0360	2 pole		4.20	1.900		585	117		702	820
0370	3 pole		3.36	2.380		690	146		836	975
0380	600 volt, 2 pole		4.20	1.900		665	117		782	905
0410	3 pole		3.36	2.380		830	146		976	1,125
0420	KA frame, 70 to 225 amp		2.69	2.980		1,525	183		1,708	1,950
0430	LA frame, 125 to 400 amp		1.93	4.140		3,900	254		4,154	4,675
0460	MA frame, 450 to 600 amp		1.34	5.950		6,200	365		6,565	7,375
0470	700 to 800 amp		1.09	7.330		8,075	450		8,525	9,550
0480	MAL frame, 1,000 amp		.84	9.520		8,375	585		8,960	10,100

26 24 Switchboards and Panelboards

26 24 13 — Switchboards

26 24 13.40 Switchboards Feeder Section	Crew	Daily Output	Labor-Hours	Unit	Material	2020 Bare Costs Labor	Equipment	Total	Total Incl O&P
0490	PA frame, 1,200 amp	1 Elec	.67	11.900	Ea.	17,000	730		17,730	19,800
0500	Branch circuit, fusible switch, 600 volt, double 30/30 amp		3.36	2.380		855	146		1,001	1,150
0550	60/60 amp		2.69	2.980		875	183		1,058	1,225
0600	100/100 amp		2.27	3.530		1,100	217		1,317	1,550
0650	Single, 30 amp		4.45	1.800		790	110		900	1,025
0700	60 amp		3.95	2.030		815	125		940	1,075
0750	100 amp		3.36	2.380		1,300	146		1,446	1,675
0800	200 amp		2.27	3.530		1,350	217		1,567	1,800
0850	400 amp		1.93	4.140		2,475	254		2,729	3,100
0900	600 amp		1.51	5.290		3,025	325		3,350	3,800
0950	800 amp		1.09	7.330		5,075	450		5,525	6,250
1000	1,200 amp	▼	.67	11.900	▼	5,800	730		6,530	7,475
1080	Branch circuit, circuit breakers, high interrupting capacity									
1100	60 amp, 240, 480 or 600 volt, 1 pole	1 Elec	6.72	1.190	Ea.	535	73		608	700
1120	2 pole		5.88	1.360		1,325	83.50		1,408.50	1,575
1140	3 pole		4.45	1.800		660	110		770	890
1150	100 amp, 240, 480 or 600 volt, 1 pole		5.88	1.360		595	83.50		678.50	780
1160	2 pole		4.20	1.900		1,475	117		1,592	1,800
1180	3 pole		3.36	2.380		705	146		851	995
1200	225 amp, 240, 480 or 600 volt, 2 pole		2.94	2.720		2,275	167		2,442	2,750
1220	3 pole		2.69	2.980		2,200	183		2,383	2,700
1240	400 amp, 240, 480 or 600 volt, 2 pole		2.10	3.810		3,250	234		3,484	3,925
1260	3 pole		1.93	4.140		2,975	254		3,229	3,650
1280	600 amp, 240, 480 or 600 volt, 2 pole		1.51	5.290		5,275	325		5,600	6,275
1300	3 pole		1.34	5.950		3,625	365		3,990	4,550
1320	800 amp, 240, 480 or 600 volt, 2 pole		1.26	6.350		4,650	390		5,040	5,700
1340	3 pole		1.09	7.330		5,450	450		5,900	6,650
1360	1,000 amp, 240, 480 or 600 volt, 2 pole		.92	8.660		7,325	530		7,855	8,875
1380	3 pole		.84	9.520		6,050	585		6,635	7,525
1400	1,200 amp, 240, 480 or 600 volt, 2 pole		.76	10.580		7,425	650		8,075	9,150
1420	3 pole		.67	11.900		7,675	730		8,405	9,500
1700	Fusible switch, 240 V, 60 amp, 2 pole		2.69	2.980		415	183		598	725
1720	3 pole		2.52	3.170		570	194		764	915
1740	100 amp, 2 pole		2.27	3.530		420	217		637	785
1760	3 pole		2.10	3.810		640	234		874	1,050
1780	200 amp, 2 pole		1.68	4.760		880	292		1,172	1,400
1800	3 pole		1.60	5.010		1,025	305		1,330	1,575
1820	400 amp, 2 pole		1.26	6.350		1,650	390		2,040	2,400
1840	3 pole		1.09	7.330		2,075	450		2,525	2,950
1860	600 amp, 2 pole		.84	9.520		2,350	585		2,935	3,450
1880	3 pole		.76	10.580		2,875	650		3,525	4,150
1900	240-600 V, 800 amp, 2 pole		.59	13.610		5,225	835		6,060	7,000
1920	3 pole		.50	15.870		6,400	975		7,375	8,500
2000	600 V, 60 amp, 2 pole		2.69	2.980		630	183		813	960
2040	100 amp, 2 pole		2.27	3.530		645	217		862	1,025
2080	200 amp, 2 pole		1.68	4.760		1,125	292		1,417	1,650
2120	400 amp, 2 pole		1.26	6.350		2,225	390		2,615	3,025
2160	600 amp, 2 pole		.84	9.520		2,700	585		3,285	3,850
2500	Branch circuit, circuit breakers, 60 amp, 600 volt, 3 pole		4.45	1.800		465	110		575	675
2520	240, 480 or 600 volt, 1 pole		6.72	1.190		97	73		170	216
2540	240 volt, 2 pole		5.88	1.360		173	83.50		256.50	315
2560	480 or 600 volt, 2 pole		5.88	1.360		355	83.50		438.50	515
2580	240 volt, 3 pole	▼	4.45	1.800	▼	410	110		520	615

For customer support on your Electrical Change Order Costs with RSMeans data, call 800.448.8182. **Post-Installation Change Order Costs**

26 24 Switchboards and Panelboards

26 24 13 – Switchboards

26 24 13.40 Switchboards Feeder Section

		Crew	Daily Output	Labor-Hours	Unit	Material	2020 Bare Costs Labor	Equipment	Total	Total Incl O&P
2600	480 volt, 3 pole	1 Elec	4.45	1.800	Ea.	470	110		580	685
2620	100 amp, 600 volt, 2 pole		4.20	1.900		450	117		567	670
2640	3 pole		3.36	2.380		570	146		716	840
2660	480 volt, 2 pole		4.20	1.900		390	117		507	605
2680	240 volt, 2 pole		4.20	1.900		222	117		339	420
2700	3 pole		3.36	2.380		355	146		501	610
2720	480 volt, 3 pole		3.36	2.380		525	146		671	790
2740	225 amp, 240, 480 or 600 volt, 2 pole		2.94	2.720		650	167		817	965
2760	3 pole		2.69	2.980		680	183		863	1,025
2780	400 amp, 240, 480 or 600 volt, 2 pole		2.10	3.810		1,375	234		1,609	1,850
2800	3 pole		1.93	4.140		1,575	254		1,829	2,125
2820	600 amp, 240 or 480 volt, 2 pole		1.51	5.290		2,225	325		2,550	2,925
2840	3 pole		1.34	5.950		2,725	365		3,090	3,550
2860	800 amp, 240, 480 or 600 volt, 2 pole		1.26	6.350		3,350	390		3,740	4,250
2880	3 pole		1.09	7.330		3,900	450		4,350	4,975
2900	1,000 amp, 240, 480 or 600 volt, 2 pole		.92	8.660		4,100	530		4,630	5,325
2920	480 or 600 volt, 3 pole		.84	9.520		4,700	585		5,285	6,050
2940	1,200 amp, 240, 480 or 600 volt, 2 pole		.76	10.580		5,775	650		6,425	7,325
2960	3 pole		.67	11.900		6,300	730		7,030	8,025
2980	600 volt, 3 pole		.67	11.900		6,500	730		7,230	8,225

26 24 16 – Panelboards

26 24 16.10 Load Centers

		Crew	Daily Output	Labor-Hours	Unit	Material	2020 Bare Costs Labor	Equipment	Total	Total Incl O&P
0010	**LOAD CENTERS** (residential type)									
0100	3 wire, 120/240 V, 1 phase, including 1 pole plug-in breakers									
0200	100 amp main lugs, indoor, 8 circuits	1 Elec	1.18	6.800	Ea.	101	415		516	730
0300	12 circuits		1.01	7.940		127	485		612	865
0400	Rainproof, 8 circuits		1.18	6.800		130	415		545	765
0500	12 circuits		1.01	7.940		156	485		641	895
0600	200 amp main lugs, indoor, 16 circuits	R-1A	1.51	10.580		218	585		803	1,100
0700	20 circuits		1.26	12.700		208	700		908	1,275
0800	24 circuits		1.09	14.650		266	810		1,076	1,500
0900	30 circuits		1.01	15.870		285	875		1,160	1,625
1000	40 circuits		.67	23.810		435	1,325		1,760	2,425
1200	Rainproof, 16 circuits		1.51	10.580		250	585		835	1,150
1300	20 circuits		1.26	12.700		325	700		1,025	1,400
1400	24 circuits		1.09	14.650		340	810		1,150	1,575
1500	30 circuits		1.01	15.870		380	875		1,255	1,725
1600	40 circuits		.67	23.810		490	1,325		1,815	2,500
1800	400 amp main lugs, indoor, 42 circuits		.60	26.460		1,200	1,450		2,650	3,500
1900	Rainproof, 42 circuits		.60	26.460		1,625	1,450		3,075	3,950
2200	Plug in breakers, 20 amp, 1 pole, 4 wire, 120/208 volts									
2210	125 amp main lugs, indoor, 12 circuits	1 Elec	1.01	7.940	Ea.	194	485		679	940
2300	18 circuits		.67	11.900		261	730		991	1,350
2400	Rainproof, 12 circuits		1.01	7.940		215	485		700	960
2500	18 circuits		.67	11.900		315	730		1,045	1,425
2600	200 amp main lugs, indoor, 24 circuits	R-1A	1.09	14.650		380	810		1,190	1,625
2700	30 circuits		1.01	15.870		370	875		1,245	1,700
2800	36 circuits		.84	19.050		380	1,050		1,430	2,000
2900	42 circuits		.67	23.810		540	1,325		1,865	2,550
3000	Rainproof, 24 circuits		1.09	14.650		287	810		1,097	1,525
3100	30 circuits		1.01	15.870		305	875		1,180	1,625
3200	36 circuits		.84	19.050		465	1,050		1,515	2,100

26 24 16 – Panelboards

26 24 16.10 Load Centers

26 24 16.10 Load Centers		Crew	Daily Output	Labor-Hours	Unit	Material	2020 Bare Costs Labor	2020 Bare Costs Equipment	Total	Total Incl O&P
3300	42 circuits	R-1A	.67	23.810	Ea.	615	1,325		1,940	2,625
3500	400 amp main lugs, indoor, 42 circuits		.60	26.460		1,250	1,450		2,700	3,550
3600	Rainproof, 42 circuits		.60	26.460		1,575	1,450		3,025	3,900
3700	Plug-in breakers, 20 amp, 1 pole, 3 wire, 120/240 volts									
3800	100 amp main breaker, indoor, 12 circuits	1 Elec	1.01	7.940	Ea.	159	485		644	900
3900	18 circuits	"	.67	11.900		197	730		927	1,300
4000	200 amp main breaker, indoor, 20 circuits	R-1A	1.26	12.700		279	700		979	1,350
4200	24 circuits		1.09	14.650		305	810		1,115	1,525
4300	30 circuits		1.01	15.870		330	875		1,205	1,650
4400	40 circuits		.76	21.160		390	1,175		1,565	2,175
4500	Rainproof, 20 circuits		1.26	12.700		335	700		1,035	1,425
4600	24 circuits		1.09	14.650		390	810		1,200	1,625
4700	30 circuits		1.01	15.870		440	875		1,315	1,775
4800	40 circuits		.76	21.160		560	1,175		1,735	2,375
5000	400 amp main breaker, indoor, 42 circuits		.60	26.460		2,975	1,450		4,425	5,450
5100	Rainproof, 42 circuits		.60	26.460		3,150	1,450		4,600	5,625
5300	Plug in breakers, 20 amp, 1 pole, 4 wire, 120/208 volts									
5400	200 amp main breaker, indoor, 30 circuits	R-1A	1.01	15.870	Ea.	420	875		1,295	1,750
5500	42 circuits		.67	23.810		490	1,325		1,815	2,500
5600	Rainproof, 30 circuits		1.01	15.870		895	875		1,770	2,275
5700	42 circuits		.67	23.810		575	1,325		1,900	2,575

26 24 16.20 Panelboard and Load Center Circuit Breakers

26 24 16.20 Panelboard and Load Center Circuit Breakers		Crew	Daily Output	Labor-Hours	Unit	Material	2020 Bare Costs Labor	2020 Bare Costs Equipment	Total	Total Incl O&P
0010	**PANELBOARD AND LOAD CENTER CIRCUIT BREAKERS**									
0050	Bolt-on, 10,000 amp I.C., 120 volt, 1 pole									
0100	15-50 amp	1 Elec	8.40	.952	Ea.	18.60	58.50		77.10	108
0200	60 amp		6.72	1.190		20.50	73		93.50	132
0300	70 amp		6.72	1.190		28	73		101	140
0350	240 volt, 2 pole									
0400	15-50 amp	1 Elec	6.72	1.190	Ea.	56.50	73		129.50	171
0500	60 amp		6.30	1.270		37.50	78		115.50	157
0600	80-100 amp		4.20	1.900		111	117		228	296
0700	3 pole, 15-60 amp		5.21	1.540		135	94.50		229.50	290
0800	70 amp		4.20	1.900		172	117		289	365
0900	80-100 amp		3.02	2.650		205	163		368	470
1000	22,000 amp I.C., 240 volt, 2 pole, 70-225 amp		2.27	3.530		555	217		772	930
1100	3 pole, 70-225 amp		1.93	4.140		345	254		599	755
1200	14,000 amp I.C., 277 volts, 1 pole, 15-30 amp		6.72	1.190		37	73		110	150
1300	22,000 amp I.C., 480 volts, 2 pole, 70-225 amp		2.27	3.530		525	217		742	895
1400	3 pole, 70-225 amp		1.93	4.140		1,050	254		1,304	1,525
2000	Plug-in panel or load center, 120/240 volt, to 60 amp, 1 pole		10.08	.794		6.40	48.50		54.90	79.50
2010	2 pole		7.56	1.060		25.50	65		90.50	125
2020	3 pole		6.30	1.270		97.50	78		175.50	223
2030	100 amp, 2 pole		5.04	1.590		105	97.50		202.50	261
2040	3 pole		3.78	2.120		118	130		248	325
2050	150-200 amp, 2 pole		2.52	3.170		256	194		450	570
2060	Plug-in tandem, 120/240 V, 2-15 A, 1 pole		9.24	.866		30	53		83	112
2070	1-15 A & 1-20 A		9.24	.866		18.35	53		71.35	99
2080	2-20 A		9.24	.866		18.40	53		71.40	99
2082	Arc fault circuit interrupter, 120/240 V, 1-15 A & 1-20 A, 1 pole		9.24	.866		57	53		110	142
2100	High interrupting capacity, 120/240 volt, plug-in, 30 amp, 1 pole		10.08	.794		24	48.50		72.50	99
2110	60 amp, 2 pole		7.56	1.060		26	65		91	126
2120	3 pole		6.30	1.270		257	78		335	400

26 24 16 – Panelboards

26 24 16.20 Panelboard and Load Center Circuit Breakers	Crew	Daily Output	Labor-Hours	Unit	Material	2020 Bare Costs Labor	Equipment	Total	Total Incl O&P
2130	100 amp, 2 pole	1 Elec	5.04	1.590	Ea.	138	97.50		235.50	297
2140	3 pole		3.78	2.120		430	130		560	665
2150	125 amp, 2 pole		2.52	3.170		870	194		1,064	1,250
2200	Bolt-on, 30 amp, 1 pole		8.40	.952		61.50	58.50		120	155
2210	60 amp, 2 pole		6.30	1.270		103	78		181	230
2220	3 pole		5.21	1.540		271	94.50		365.50	440
2230	100 amp, 2 pole		4.20	1.900		293	117		410	495
2240	3 pole		3.02	2.650		470	163		633	760
2300	Ground fault, 240 volt, 30 amp, 1 pole		5.88	1.360		92.50	83.50		176	226
2310	2 pole		5.04	1.590		170	97.50		267.50	330
2350	Key operated, 240 volt, 1 pole, 30 amp		5.88	1.360		135	83.50		218.50	273
2360	Switched neutral, 240 volt, 30 amp, 2 pole		5.04	1.590		40	97.50		137.50	189
2370	3 pole		4.62	1.730		65.50	106		171.50	230
2400	Shunt trip, for 240 volt breaker, 60 amp, 1 pole		3.36	2.380		81.50	146		227.50	305
2410	2 pole		2.94	2.720		81.50	167		248.50	340
2420	3 pole		2.52	3.170		81.50	194		275.50	380
2430	100 amp, 2 pole		2.52	3.170		81.50	194		275.50	380
2440	3 pole		2.10	3.810		81.50	234		315.50	440
2450	150 amp, 2 pole		1.68	4.760		235	292		527	695
2500	Auxiliary switch, for 240 volt breaker, 60 amp, 1 pole		3.36	2.380		90	146		236	315
2510	2 pole		2.94	2.720		90	167		257	345
2520	3 pole		2.52	3.170		90	194		284	390
2530	100 amp, 2 pole		2.52	3.170		90	194		284	390
2540	3 pole		2.10	3.810		90	234		324	450
2550	150 amp, 2 pole		1.68	4.760		123	292		415	570
2600	Panel or load center, 277/480 volt, plug-in, 30 amp, 1 pole		10.08	.794		71	48.50		119.50	151
2610	60 amp, 2 pole		7.56	1.060		218	65		283	335
2620	3 pole		6.30	1.270		320	78		398	465
2650	Bolt-on, 60 amp, 2 pole		6.30	1.270		218	78		296	355
2660	3 pole		5.21	1.540		320	94.50		414.50	490
2700	I-line, 277/480 volt, 30 amp, 1 pole		6.72	1.190		68.50	73		141.50	185
2710	60 amp, 2 pole		6.30	1.270		241	78		319	380
2720	3 pole		5.21	1.540		375	94.50		469.50	550
2730	100 amp, 1 pole		6.30	1.270		161	78		239	293
2740	2 pole		4.20	1.900		375	117		492	585
2750	3 pole		2.94	2.720		440	167		607	735
2800	High interrupting capacity, 277/480 volt, plug-in, 30 amp, 1 pole		10.08	.794		420	48.50		468.50	535
2810	60 amp, 2 pole		7.56	1.060		650	65		715	810
2820	3 pole		5.88	1.360		755	83.50		838.50	955
2830	Bolt-on, 30 amp, 1 pole		6.72	1.190		420	73		493	570
2840	60 amp, 2 pole		6.30	1.270		570	78		648	740
2850	3 pole		5.21	1.540		635	94.50		729.50	840
2900	I-line, 30 amp, 1 pole		6.72	1.190		420	73		493	570
2910	60 amp, 2 pole		6.30	1.270		585	78		663	760
2920	3 pole		5.21	1.540		665	94.50		759.50	870
2930	100 amp, 1 pole		6.30	1.270		680	78		758	860
2940	2 pole		4.20	1.900		475	117		592	700
2950	3 pole		3.02	2.650		735	163		898	1,050
2960	Shunt trip, 277/480 volt breaker, remote oper., 30 amp, 1 pole		3.36	2.380		440	146		586	700
2970	60 amp, 2 pole		2.94	2.720		645	167		812	960
2980	3 pole		2.52	3.170		730	194		924	1,100
2990	100 amp, 1 pole		2.94	2.720		525	167		692	830
3000	2 pole		2.52	3.170		730	194		924	1,100

Post-Installation Change Order Costs   For customer support on your Electrical Change Order Costs with RSMeans data, call 800.448.8182.

26 24 16.20 Panelboard and Load Center Circuit Breakers		Crew	Daily Output	Labor-Hours	Unit	Material	2020 Bare Costs Labor	Equipment	Total	Total Incl O&P
3010	3 pole	1 Elec	2.10	3.810	Ea.	670	234		904	1,100
3050	Under voltage trip, 277/480 volt breaker, 30 amp, 1 pole		3.36	2.380		440	146		586	700
3060	60 amp, 2 pole		2.94	2.720		645	167		812	960
3070	3 pole		2.52	3.170		730	194		924	1,100
3080	100 amp, 1 pole		2.94	2.720		525	167		692	830
3090	2 pole		2.52	3.170		730	194		924	1,100
3100	3 pole		2.10	3.810		670	234		904	1,100
3150	Motor operated, 277/480 volt breaker, 30 amp, 1 pole		3.36	2.380		785	146		931	1,075
3160	60 amp, 2 pole		2.94	2.720		820	167		987	1,150
3170	3 pole		2.52	3.170		1,100	194		1,294	1,500
3180	100 amp, 1 pole		2.94	2.720		875	167		1,042	1,225
3190	2 pole		2.52	3.170		1,100	194		1,294	1,500
3200	3 pole		2.10	3.810		950	234		1,184	1,400
3250	Panelboard spacers, per pole		33.60	.238		3.79	14.60		18.39	26
5110	NEMA 1 enclosure only, 600V, 3 p, 14k AIC, 100A		1.85	4.330		160	266		426	570

26 24 16.30 Panelboards Commercial Applications

		Crew	Daily Output	Labor-Hours	Unit	Material	2020 Bare Costs Labor	Equipment	Total	Total Incl O&P
0010	**PANELBOARDS COMMERCIAL APPLICATIONS**									
0050	NQOD, w/20 amp 1 pole bolt-on circuit breakers									
0100	3 wire, 120/240 volts, 100 amp main lugs									
0150	10 circuits	1 Elec	.84	9.520	Ea.	935	585		1,520	1,900
0200	14 circuits		.74	10.820		1,050	665		1,715	2,150
0250	18 circuits		.63	12.700		1,150	780		1,930	2,400
0300	20 circuits		.55	14.650		1,275	900		2,175	2,775
0350	225 amp main lugs, 24 circuits	2 Elec	1.01	15.870		1,450	975		2,425	3,050
0400	30 circuits		.76	21.160		1,675	1,300		2,975	3,775
0450	36 circuits		.67	23.810		1,925	1,450		3,375	4,275
0500	38 circuits		.60	26.460		2,075	1,625		3,700	4,725
0550	42 circuits		.55	28.860		2,150	1,775		3,925	4,975
0600	4 wire, 120/208 volts, 100 amp main lugs, 12 circuits	1 Elec	.84	9.520		1,025	585		1,610	2,000
0650	16 circuits		.63	12.700		1,175	780		1,955	2,425
0700	20 circuits		.55	14.650		1,350	900		2,250	2,825
0750	24 circuits		.50	15.870		1,275	975		2,250	2,850
0800	30 circuits		.45	17.970		1,675	1,100		2,775	3,500
0850	225 amp main lugs, 32 circuits	2 Elec	.76	21.160		1,875	1,300		3,175	4,000
0900	34 circuits		.71	22.680		1,925	1,400		3,325	4,175
0950	36 circuits		.67	23.810		1,975	1,450		3,425	4,325
1000	42 circuits		.57	28.010		2,200	1,725		3,925	4,975
1010	400 amp main lugs, 42 circs		.57	28.010		2,200	1,725		3,925	4,975
1040	225 amp main lugs, NEMA 7, 12 circuits		.84	19.050		5,925	1,175		7,100	8,275
1100	24 circuits		.34	47.620		7,175	2,925		10,100	12,200
1200	NEHB, w/20 amp, 1 pole bolt-on circuit breakers									
1250	4 wire, 277/480 volts, 100 amp main lugs, 12 circuits	1 Elec	.74	10.820	Ea.	1,475	665		2,140	2,625
1300	20 circuits	"	.50	15.870		2,200	975		3,175	3,875
1350	225 amp main lugs, 24 circuits	2 Elec	.76	21.160		2,500	1,300		3,800	4,675
1400	30 circuits		.67	23.810		2,975	1,450		4,425	5,450
1448	32 circuits		4.12	3.890		3,475	239		3,714	4,150
1450	36 circuits		.60	26.460		3,475	1,625		5,100	6,225
1500	42 circuits		.50	31.750		3,950	1,950		5,900	7,250
1510	225 amp main lugs, NEMA 7, 12 circuits		.76	21.160		6,175	1,300		7,475	8,700
1590	24 circuits		.25	63.490		8,275	3,900		12,175	14,900
1600	NQOD panel, w/20 amp, 1 pole, circuit breakers									
1650	3 wire, 120/240 volt with main circuit breaker									

For customer support on your Electrical Change Order Costs with RSMeans data, call 800.448.8182.   **Post-Installation Change Order Costs**

26 24 16.30 Panelboards Commercial Applications		Crew	Daily Output	Labor-Hours	Unit	Material	2020 Bare Costs Labor	Equipment	Total	Total Incl O&P
1700	100 amp main, 12 circuits	1 Elec	.67	11.900	Ea.	1,275	730		2,005	2,475
1750	20 circuits	"	.50	15.870		1,625	975		2,600	3,225
1800	225 amp main, 30 circuits	2 Elec	.57	28.010		3,050	1,725		4,775	5,900
1801	225 amp main, 32 circuits		4.20	3.810		3,050	234		3,284	3,700
1850	42 circuits		.44	36.630		3,525	2,250		5,775	7,225
1900	400 amp main, 30 circuits		.45	35.270		4,225	2,175		6,400	7,875
1950	42 circuits		.42	38.100		4,725	2,325		7,050	8,675
2000	4 wire, 120/208 volts with main circuit breaker									
2050	100 amp main, 24 circuits	1 Elec	.39	20.260	Ea.	1,825	1,250		3,075	3,875
2100	30 circuits	"	.34	23.810		2,125	1,450		3,575	4,525
2200	225 amp main, 32 circuits	2 Elec	.60	26.460		3,550	1,625		5,175	6,350
2250	42 circuits		.47	34.010		4,125	2,075		6,200	7,625
2300	400 amp main, 42 circuits		.40	39.680		5,225	2,425		7,650	9,375
2350	600 amp main, 42 circuits		.34	47.620		7,725	2,925		10,650	12,900
2400	NEHB, with 20 amp, 1 pole circuit breaker									
2450	4 wire, 277/480 volts with main circuit breaker									
2500	100 amp main, 24 circuits	1 Elec	.35	22.680	Ea.	2,875	1,400		4,275	5,225
2550	30 circuits	"	.32	25.060		3,350	1,525		4,875	6,000
2600	225 amp main, 30 circuits	2 Elec	.60	26.460		4,200	1,625		5,825	7,050
2650	42 circuits		.47	34.010		5,175	2,075		7,250	8,800
2700	400 amp main, 42 circuits		.39	41.410		6,225	2,550		8,775	10,600
2750	600 amp main, 42 circuits		.32	50.130		8,525	3,075		11,600	14,000
2900	Note: the following line items don't include branch circuit breakers									
2910	For branch circuit breakers information, see Section 26 24 16.20									
3010	Main lug, no main breaker, 240 volt, 1 pole, 3 wire, 100 amp	1 Elec	1.93	4.140	Ea.	605	254		859	1,050
3020	225 amp	2 Elec	2.02	7.940		660	485		1,145	1,450
3030	400 amp	"	1.51	10.580		1,125	650		1,775	2,225
3060	3 pole, 3 wire, 100 amp	1 Elec	1.93	4.140		655	254		909	1,100
3070	225 amp	2 Elec	2.02	7.940		700	485		1,185	1,500
3080	400 amp		1.51	10.580		1,225	650		1,875	2,325
3090	600 amp		1.34	11.900		1,425	730		2,155	2,650
3110	3 pole, 4 wire, 100 amp	1 Elec	1.93	4.140		720	254		974	1,175
3120	225 amp	2 Elec	2.02	7.940		825	485		1,310	1,625
3130	400 amp		1.51	10.580		1,225	650		1,875	2,325
3140	600 amp		1.34	11.900		1,425	730		2,155	2,650
3160	480 volt, 3 pole, 3 wire, 100 amp	1 Elec	1.93	4.140		785	254		1,039	1,250
3170	225 amp	2 Elec	2.02	7.940		1,100	485		1,585	1,950
3180	400 amp		1.51	10.580		1,550	650		2,200	2,675
3190	600 amp		1.34	11.900		1,725	730		2,455	2,975
3210	277/480 volt, 3 pole, 4 wire, 100 amp	1 Elec	1.93	4.140		765	254		1,019	1,225
3220	225 amp	2 Elec	2.02	7.940		930	485		1,415	1,750
3230	400 amp		1.51	10.580		1,500	650		2,150	2,625
3240	600 amp		1.34	11.900		1,675	730		2,405	2,925
3260	Main circuit breaker, 240 volt, 1 pole, 3 wire, 100 amp	1 Elec	1.68	4.760		740	292		1,032	1,250
3270	225 amp	2 Elec	1.68	9.520		1,775	585		2,360	2,825
3280	400 amp	"	1.34	11.900		2,775	730		3,505	4,125
3310	3 pole, 3 wire, 100 amp	1 Elec	1.68	4.760		890	292		1,182	1,425
3320	225 amp	2 Elec	1.68	9.520		2,025	585		2,610	3,100
3330	400 amp		1.34	11.900		3,175	730		3,905	4,550
3360	120/208 volt, 3 pole, 4 wire, 100 amp		3.36	4.760		850	292		1,142	1,375
3370	225 amp		1.68	9.520		2,025	585		2,610	3,100
3380	400 amp		1.34	11.900		3,175	730		3,905	4,550
3410	480 volt, 3 pole, 3 wire, 100 amp	1 Elec	1.68	4.760		1,375	292		1,667	1,950

26 24 16 – Panelboards

26 24 16.30 Panelboards Commercial Applications		Crew	Daily Output	Labor-Hours	Unit	Material	2020 Bare Costs Labor	Equipment	Total	Total Incl O&P
3420	225 amp	2 Elec	1.68	9.520	Ea.	2,375	585		2,960	3,500
3430	400 amp		1.34	11.900		3,550	730		4,280	4,975
3460	277/480 volt, 3 pole, 4 wire, 100 amp		3.36	4.760		1,350	292		1,642	1,925
3470	225 amp		1.68	9.520		2,325	585		2,910	3,425
3480	400 amp		1.34	11.900		3,575	730		4,305	5,000
3510	Main circuit breaker, HIC, 240 volt, 1 pole, 3 wire, 100 amp	1 Elec	1.68	4.760		1,325	292		1,617	1,875
3520	225 amp	2 Elec	1.68	9.520		3,450	585		4,035	4,675
3530	400 amp	"	1.34	11.900		4,625	730		5,355	6,150
3560	3 pole, 3 wire, 100 amp	1 Elec	1.68	4.760		1,475	292		1,767	2,050
3570	225 amp	2 Elec	1.68	9.520		3,900	585		4,485	5,150
3580	400 amp	"	1.34	11.900		5,150	730		5,880	6,750
3610	120/208 volt, 3 pole, 4 wire, 100 amp	1 Elec	1.68	4.760		1,475	292		1,767	2,050
3620	225 amp	2 Elec	1.68	9.520		3,900	585		4,485	5,150
3630	400 amp	"	1.34	11.900		5,150	730		5,880	6,750
3660	480 volt, 3 pole, 3 wire, 100 amp	1 Elec	1.68	4.760		2,200	292		2,492	2,825
3670	225 amp	2 Elec	1.68	9.520		4,325	585		4,910	5,625
3680	400 amp	"	1.34	11.900		5,500	730		6,230	7,125
3710	277/480 volt, 3 pole, 4 wire, 100 amp	1 Elec	1.68	4.760		2,075	292		2,367	2,725
3720	225 amp	2 Elec	1.68	9.520		4,125	585		4,710	5,400
3730	400 amp	"	1.34	11.900		5,450	730		6,180	7,075
3760	Main circuit breaker, shunt trip, 100 amp	1 Elec	1.01	7.940		1,100	485		1,585	1,925
3770	225 amp	2 Elec	1.34	11.900		2,500	730		3,230	3,825
3780	400 amp	"	1.18	13.610		3,625	835		4,460	5,250

26 24 19 – Motor-Control Centers

26 24 19.20 Motor Control Center Components

26 24 19.20 Motor Control Center Components		Crew	Daily Output	Labor-Hours	Unit	Material	2020 Bare Costs Labor	Equipment	Total	Total Incl O&P
0010	**MOTOR CONTROL CENTER COMPONENTS**									
0100	Starter, size 1, FVNR, NEMA 1, type A, fusible	1 Elec	2.27	3.530	Ea.	1,650	217		1,867	2,150
0120	Circuit breaker		2.27	3.530		1,800	217		2,017	2,300
0140	Type B, fusible		2.27	3.530		1,825	217		2,042	2,325
0160	Circuit breaker		2.27	3.530		1,975	217		2,192	2,500
0180	NEMA 12, type A, fusible		2.18	3.660		1,700	225		1,925	2,175
0200	Circuit breaker		2.18	3.660		1,825	225		2,050	2,350
0220	Type B, fusible		2.18	3.660		1,850	225		2,075	2,375
0240	Circuit breaker		2.18	3.660		2,025	225		2,250	2,550
0300	Starter, size 1, FVR, NEMA 1, type A, fusible		1.68	4.760		2,375	292		2,667	3,050
0320	Circuit breaker		1.68	4.760		2,375	292		2,667	3,050
0340	Type B, fusible		1.68	4.760		2,625	292		2,917	3,300
0360	Circuit breaker		1.68	4.760		2,625	292		2,917	3,300
0380	NEMA 12, type A, fusible		1.60	5.010		2,400	305		2,705	3,100
0400	Circuit breaker		1.60	5.010		2,400	305		2,705	3,100
0420	Type B, fusible		1.60	5.010		2,650	305		2,955	3,375
0440	Circuit breaker		1.60	5.010		2,650	305		2,955	3,375
0490	Starter size 1, 2 speed, separate winding									
0500	NEMA 1, type A, fusible	1 Elec	2.18	3.660	Ea.	3,125	225		3,350	3,775
0520	Circuit breaker		2.18	3.660		3,125	225		3,350	3,775
0540	Type B, fusible		2.18	3.660		3,450	225		3,675	4,125
0560	Circuit breaker		2.18	3.660		3,450	225		3,675	4,125
0580	NEMA 12, type A, fusible		2.10	3.810		3,200	234		3,434	3,875
0600	Circuit breaker		2.10	3.810		3,200	234		3,434	3,875
0620	Type B, fusible		2.10	3.810		3,500	234		3,734	4,200
0640	Circuit breaker		2.10	3.810		3,500	234		3,734	4,200
0650	Starter size 1, 2 speed, consequent pole									

26 24 19 – Motor-Control Centers

26 24 19.20 Motor Control Center Components	Crew	Daily Output	Labor-Hours	Unit	Material	2020 Bare Costs Labor	Equipment	Total	Total Incl O&P
0660	NEMA 1, type A, fusible	1 Elec	2.18	3.660	Ea.	3,125	225		3,350	3,775
0680	Circuit breaker		2.18	3.660		3,125	225		3,350	3,775
0700	Type B, fusible		2.18	3.660		3,450	225		3,675	4,125
0720	Circuit breaker		2.18	3.660		3,450	225		3,675	4,125
0740	NEMA 12, type A, fusible		2.10	3.810		3,200	234		3,434	3,875
0760	Circuit breaker		2.10	3.810		3,200	234		3,434	3,875
0780	Type B, fusible		2.10	3.810		3,475	234		3,709	4,175
0800	Circuit breaker	▼	2.10	3.810	▼	3,500	234		3,734	4,200
0810	Starter size 1, 2 speed, space only									
0820	NEMA 1, type A, fusible	1 Elec	13.44	.595	Ea.	705	36.50		741.50	830
0840	Circuit breaker		13.44	.595		705	36.50		741.50	830
0860	Type B, fusible		13.44	.595		705	36.50		741.50	830
0880	Circuit breaker		13.44	.595		705	36.50		741.50	830
0900	NEMA 12, type A, fusible		12.60	.635		735	39		774	865
0920	Circuit breaker		12.60	.635		735	39		774	865
0940	Type B, fusible		12.60	.635		735	39		774	865
0960	Circuit breaker	▼	12.60	.635		735	39		774	865
1100	Starter size 2, FVNR, NEMA 1, type A, fusible	2 Elec	3.36	4.760		1,875	292		2,167	2,500
1120	Circuit breaker		3.36	4.760		2,050	292		2,342	2,700
1140	Type B, fusible		3.36	4.760		2,075	292		2,367	2,700
1160	Circuit breaker		3.36	4.760		2,250	292		2,542	2,900
1180	NEMA 12, type A, fusible		3.19	5.010		1,900	305		2,205	2,550
1200	Circuit breaker		3.19	5.010		2,075	305		2,380	2,750
1220	Type B, fusible		3.19	5.010		2,075	305		2,380	2,750
1240	Circuit breaker		3.19	5.010		2,275	305		2,580	2,950
1300	FVR, NEMA 1, type A, fusible		2.69	5.950		3,175	365		3,540	4,050
1320	Circuit breaker		2.69	5.950		3,175	365		3,540	4,050
1340	Type B, fusible		2.69	5.950		3,500	365		3,865	4,400
1360	Circuit breaker		2.69	5.950		3,500	365		3,865	4,400
1380	NEMA type 12, type A, fusible		2.52	6.350		3,225	390		3,615	4,125
1400	Circuit breaker		2.52	6.350		3,225	390		3,615	4,125
1420	Type B, fusible		2.52	6.350		3,550	390		3,940	4,475
1440	Circuit breaker	▼	2.52	6.350	▼	3,550	390		3,940	4,475
1490	Starter size 2, 2 speed, separate winding									
1500	NEMA 1, type A, fusible	2 Elec	3.19	5.010	Ea.	3,550	305		3,855	4,350
1520	Circuit breaker		3.19	5.010		3,550	305		3,855	4,350
1540	Type B, fusible		3.19	5.010		3,900	305		4,205	4,750
1560	Circuit breaker		3.19	5.010		3,900	305		4,205	4,750
1570	NEMA 12, type A, fusible		3.02	5.290		3,600	325		3,925	4,450
1580	Circuit breaker		3.02	5.290		3,600	325		3,925	4,450
1600	Type B, fusible		3.02	5.290		3,950	325		4,275	4,825
1620	Circuit breaker	▼	3.02	5.290	▼	3,950	325		4,275	4,825
1630	Starter size 2, 2 speed, consequent pole									
1640	NEMA 1, type A, fusible	2 Elec	3.19	5.010	Ea.	4,025	305		4,330	4,875
1660	Circuit breaker		3.19	5.010		4,025	305		4,330	4,875
1680	Type B, fusible		3.19	5.010		4,250	305		4,555	5,125
1700	Circuit breaker		3.19	5.010		4,250	305		4,555	5,125
1720	NEMA 12, type A, fusible		3.19	5.010		4,050	305		4,355	4,900
1740	Circuit breaker		3.02	5.290		4,075	325		4,400	4,975
1760	Type B, fusible		3.02	5.290		4,325	325		4,650	5,225
1780	Circuit breaker	▼	3.02	5.290	▼	4,325	325		4,650	5,225
1830	Starter size 2, autotransformer									
1840	NEMA 1, type A, fusible	2 Elec	2.86	5.600	Ea.	6,275	345		6,620	7,400

Post-Installation Change Order Costs   For customer support on your Electrical Change Order Costs with RSMeans data, call 800.448.8182.

26 24 19.20 Motor Control Center Components	Crew	Daily Output	Labor-Hours	Unit	Material	2020 Bare Costs Labor	Equipment	Total	Total Incl O&P
1860 Circuit breaker	2 Elec	2.86	5.600	Ea.	6,425	345		6,770	7,575
1880 Type B, fusible		2.86	5.600		6,850	345		7,195	8,050
1900 Circuit breaker		2.86	5.600		6,850	345		7,195	8,050
1920 NEMA 12, type A, fusible		2.69	5.950		6,400	365		6,765	7,575
1940 Circuit breaker		2.69	5.950		6,400	365		6,765	7,575
1960 Type B, fusible		2.69	5.950		6,975	365		7,340	8,225
1980 Circuit breaker	↓	2.69	5.950	↓	6,975	365		7,340	8,225
2030 Starter size 2, space only									
2040 NEMA 1, type A, fusible	1 Elec	13.44	.595	Ea.	705	36.50		741.50	830
2060 Circuit breaker		13.44	.595		705	36.50		741.50	830
2080 Type B, fusible		13.44	.595		705	36.50		741.50	830
2100 Circuit breaker		13.44	.595		705	36.50		741.50	830
2120 NEMA 12, type A, fusible		12.60	.635		735	39		774	865
2140 Circuit breaker		12.60	.635		735	39		774	865
2160 Type B, fusible		12.60	.635		735	39		774	865
2180 Circuit breaker	↓	12.60	.635		735	39		774	865
2300 Starter size 3, FVNR, NEMA 1, type A, fusible	2 Elec	1.68	9.520		3,575	585		4,160	4,800
2320 Circuit breaker		1.68	9.520		3,175	585		3,760	4,375
2340 Type B, fusible		1.68	9.520		3,925	585		4,510	5,200
2360 Circuit breaker		1.68	9.520		3,500	585		4,085	4,725
2380 NEMA 12, type A, fusible		1.60	10.030		3,650	615		4,265	4,925
2400 Circuit breaker		1.60	10.030		3,225	615		3,840	4,475
2420 Type B, fusible		1.60	10.030		4,075	615		4,690	5,400
2440 Circuit breaker		1.60	10.030		3,550	615		4,165	4,850
2500 Starter size 3, FVR, NEMA 1, type A, fusible		1.34	11.900		4,875	730		5,605	6,425
2520 Circuit breaker		1.34	11.900		4,650	730		5,380	6,200
2540 Type B, fusible		1.34	11.900		5,300	730		6,030	6,900
2560 Circuit breaker		1.34	11.900		5,100	730		5,830	6,675
2580 NEMA 12, type A, fusible		1.26	12.700		4,975	780		5,755	6,625
2600 Circuit breaker		1.26	12.700		4,750	780		5,530	6,375
2620 Type B, fusible		1.26	12.700		5,400	780		6,180	7,100
2640 Circuit breaker	↓	1.26	12.700	↓	5,600	780		6,380	7,325
2690 Starter size 3, 2 speed, separate winding									
2700 NEMA 1, type A, fusible	2 Elec	1.68	9.520	Ea.	5,600	585		6,185	7,050
2720 Circuit breaker		1.68	9.520		4,950	585		5,535	6,325
2740 Type B, fusible		1.68	9.520		6,125	585		6,710	7,625
2760 Circuit breaker		1.68	9.520		5,425	585		6,010	6,825
2780 NEMA 12, type A, fusible		1.60	10.030		5,725	615		6,340	7,225
2800 Circuit breaker		1.60	10.030		5,050	615		5,665	6,475
2820 Type B, fusible		1.60	10.030		6,250	615		6,865	7,800
2840 Circuit breaker	↓	1.60	10.030	↓	5,525	615		6,140	7,000
2850 Starter size 3, 2 speed, consequent pole									
2860 NEMA 1, type A, fusible	2 Elec	1.68	9.520	Ea.	6,250	585		6,835	7,750
2880 Circuit breaker		1.68	9.520		5,600	585		6,185	7,025
2900 Type B, fusible		1.68	9.520		6,775	585		7,360	8,350
2920 Circuit breaker		1.68	9.520		6,075	585		6,660	7,550
2940 NEMA 12, type A, fusible		1.60	10.030		6,375	615		6,990	7,925
2960 Circuit breaker		1.60	10.030		5,700	615		6,315	7,200
2980 Type B, fusible		1.60	10.030		6,900	615		7,515	8,525
3000 Circuit breaker		1.60	10.030		6,175	615		6,790	7,700
3100 Starter size 3, autotransformer, NEMA 1, type A, fusible		1.34	11.900		8,250	730		8,980	10,200
3120 Circuit breaker		1.34	11.900		7,775	730		8,505	9,625
3140 Type B, fusible	↓	1.34	11.900	↓	8,500	730		9,230	10,400

For customer support on your Electrical Change Order Costs with RSMeans data, call 800.448.8182.   **Post-Installation Change Order Costs**

26 24 19.20 Motor Control Center Components	Crew	Daily Output	Labor-Hours	Unit	Material	2020 Bare Costs Labor	Equipment	Total	Total Incl O&P
3160	Circuit breaker	2 Elec	1.34	11.900	Ea.	8,500	730		9,230	10,400
3180	NEMA 12, type A, fusible		1.26	12.700		7,925	780		8,705	9,875
3200	Circuit breaker		1.26	12.700		7,925	780		8,705	9,875
3220	Type B, fusible		1.26	12.700		8,650	780		9,430	10,700
3240	Circuit breaker		1.26	12.700		8,650	780		9,430	10,700
3260	Starter size 3, space only, NEMA 1, type A, fusible	1 Elec	12.60	.635		1,200	39		1,239	1,375
3280	Circuit breaker		12.60	.635		940	39		979	1,075
3300	Type B, fusible		12.60	.635		1,200	39		1,239	1,375
3320	Circuit breaker		12.60	.635		940	39		979	1,075
3340	NEMA 12, type A, fusible		11.76	.680		1,275	41.50		1,316.50	1,450
3360	Circuit breaker		11.76	.680		995	41.50		1,036.50	1,150
3380	Type B, fusible		11.76	.680		1,275	41.50		1,316.50	1,450
3400	Circuit breaker		11.76	.680		995	41.50		1,036.50	1,150
3500	Starter size 4, FVNR, NEMA 1, type A, fusible	2 Elec	1.34	11.900		4,650	730		5,380	6,200
3520	Circuit breaker		1.34	11.900		4,225	730		4,955	5,725
3540	Type B, fusible		1.34	11.900		5,100	730		5,830	6,675
3560	Circuit breaker		1.34	11.900		4,650	730		5,380	6,200
3580	NEMA 12, type A, fusible		1.26	12.700		4,775	780		5,555	6,400
3600	Circuit breaker		1.26	12.700		4,300	780		5,080	5,875
3620	Type B, fusible		1.26	12.700		5,200	780		5,980	6,875
3640	Circuit breaker		1.26	12.700		4,725	780		5,505	6,350
3700	Starter size 4, FVR, NEMA 1, type A, fusible		1.01	15.870		6,375	975		7,350	8,450
3720	Circuit breaker		1.01	15.870		5,750	975		6,725	7,750
3740	Type B, fusible		1.01	15.870		6,950	975		7,925	9,100
3760	Circuit breaker		1.01	15.870		6,325	975		7,300	8,400
3780	NEMA 12, type A, fusible		.97	16.420		6,500	1,000		7,500	8,650
3800	Circuit breaker		.97	16.420		5,825	1,000		6,825	7,900
3820	Type B, fusible		.97	16.420		5,875	1,000		6,875	7,975
3840	Circuit breaker		.97	16.420		6,400	1,000		7,400	8,550
3890	Starter size 4, 2 speed, separate windings									
3900	NEMA 1, type A, fusible	2 Elec	1.34	11.900	Ea.	8,025	730		8,755	9,900
3920	Circuit breaker		1.34	11.900		6,025	730		6,755	7,725
3940	Type B, fusible		1.34	11.900		8,825	730		9,555	10,800
3960	Circuit breaker		1.34	11.900		6,600	730		7,330	8,350
3980	NEMA 12, type A, fusible		1.26	12.700		8,175	780		8,955	10,200
4000	Circuit breaker		1.26	12.700		6,125	780		6,905	7,900
4020	Type B, fusible		1.26	12.700		8,975	780		9,755	11,000
4040	Circuit breaker		1.26	12.700		6,725	780		7,505	8,525
4050	Starter size 4, 2 speed, consequent pole									
4060	NEMA 1, type A, fusible	2 Elec	1.34	11.900	Ea.	9,375	730		10,105	11,400
4080	Circuit breaker		1.34	11.900		6,975	730		7,705	8,750
4100	Type B, fusible		1.34	11.900		10,300	730		11,030	12,500
4120	Circuit breaker		1.34	11.900		7,650	730		8,380	9,500
4140	NEMA 12, type A, fusible		1.26	12.700		9,525	780		10,305	11,700
4160	Circuit breaker		1.26	12.700		7,075	780		7,855	8,950
4180	Type B, fusible		1.26	12.700		10,500	780		11,280	12,700
4200	Circuit breaker		1.26	12.700		7,750	780		8,530	9,675
4300	Starter size 4, autotransformer, NEMA 1, type A, fusible		1.09	14.650		9,250	900		10,150	11,600
4320	Circuit breaker		1.09	14.650		9,325	900		10,225	11,600
4340	Type B, fusible		1.09	14.650		10,200	900		11,100	12,600
4360	Circuit breaker		1.09	14.650		10,200	900		11,100	12,600
4380	NEMA 12, type A, fusible		1.04	15.360		9,400	940		10,340	11,700
4400	Circuit breaker		1.04	15.360		9,450	940		10,390	11,800

26 24 19.20 Motor Control Center Components	Crew	Daily Output	Labor-Hours	Unit	Material	2020 Bare Costs Labor	Equipment	Total	Total Incl O&P
4420	Type B, fusible	2 Elec	1.04	15.360	Ea.	10,300	940		11,240	12,800
4440	Circuit breaker		1.04	15.360		10,300	940		11,240	12,700
4500	Starter size 4, space only, NEMA 1, type A, fusible	1 Elec	11.76	.680		1,675	41.50		1,716.50	1,900
4520	Circuit breaker		11.76	.680		1,200	41.50		1,241.50	1,375
4540	Type B, fusible		11.76	.680		1,675	41.50		1,716.50	1,900
4560	Circuit breaker		11.76	.680		1,200	41.50		1,241.50	1,375
4580	NEMA 12, type A, fusible		10.92	.733		1,775	45		1,820	2,025
4600	Circuit breaker		10.92	.733		1,275	45		1,320	1,475
4620	Type B, fusible		10.92	.733		1,775	45		1,820	2,025
4640	Circuit breaker		10.92	.733		1,275	45		1,320	1,475
4800	Starter size 5, FVNR, NEMA 1, type A, fusible	2 Elec	.84	19.050		9,600	1,175		10,775	12,400
4820	Circuit breaker		.84	19.050		6,825	1,175		8,000	9,250
4840	Type B, fusible		.84	19.050		10,500	1,175		11,675	13,400
4860	Circuit breaker		.84	19.050		7,500	1,175		8,675	10,000
4880	NEMA 12, type A, fusible		.81	19.840		9,750	1,225		10,975	12,500
4900	Circuit breaker		.81	19.840		6,950	1,225		8,175	9,450
4920	Type B, fusible		.81	19.840		10,700	1,225		11,925	13,500
4940	Circuit breaker		.81	19.840		7,625	1,225		8,850	10,200
5000	Starter size 5, FVR, NEMA 1, type A, fusible		.67	23.810		14,200	1,450		15,650	17,800
5020	Circuit breaker		.67	23.810		11,100	1,450		12,550	14,400
5040	Type B, fusible		.67	23.810		15,500	1,450		16,950	19,300
5060	Circuit breaker		.67	23.810		12,200	1,450		13,650	15,600
5080	NEMA 12, type A, fusible		.64	25.060		14,400	1,525		15,925	18,200
5100	Circuit breaker		.64	25.060		11,300	1,525		12,825	14,700
5120	Type B, fusible		.64	25.060		15,800	1,525		17,325	19,700
5140	Circuit breaker		.64	25.060		12,400	1,525		13,925	15,900
5190	Starter size 5, 2 speed, separate windings									
5200	NEMA 1, type A, fusible	2 Elec	.84	19.050	Ea.	19,500	1,175		20,675	23,300
5220	Circuit breaker		.84	19.050		14,600	1,175		15,775	17,800
5240	Type B, fusible		.84	19.050		21,400	1,175		22,575	25,400
5260	Circuit breaker		.84	19.050		16,000	1,175		17,175	19,400
5280	NEMA 12, type A, fusible		.81	19.840		19,800	1,225		21,025	23,600
5300	Circuit breaker		.81	19.840		14,700	1,225		15,925	18,000
5320	Type B, fusible		.81	19.840		21,700	1,225		22,925	25,700
5340	Circuit breaker		.81	19.840		16,200	1,225		17,425	19,600
5400	Starter size 5, autotransformer, NEMA 1, type A, fusible		.59	27.210		16,000	1,675		17,675	20,100
5420	Circuit breaker		.59	27.210		13,300	1,675		14,975	17,100
5440	Type B, fusible		.59	27.210		17,500	1,675		19,175	21,800
5460	Circuit breaker		.59	27.210		14,600	1,675		16,275	18,500
5480	NEMA 12, type A, fusible		.57	28.010		16,200	1,725		17,925	20,400
5500	Circuit breaker		.57	28.010		13,400	1,725		15,125	17,300
5520	Type B, fusible		.57	28.010		17,700	1,725		19,425	22,100
5540	Circuit breakers		.57	28.010		14,700	1,725		16,425	18,800
5600	Starter size 5, space only, NEMA 1, type A, fusible	1 Elec	10.08	.794		2,150	48.50		2,198.50	2,425
5620	Circuit breaker		10.08	.794		2,150	48.50		2,198.50	2,425
5640	Type B, fusible		10.08	.794		2,150	48.50		2,198.50	2,425
5660	Circuit breaker		10.08	.794		1,400	48.50		1,448.50	1,625
5680	NEMA 12, type A, fusible		9.24	.866		2,275	53		2,328	2,575
5700	Circuit breaker		9.24	.866		1,500	53		1,553	1,725
5720	Type B, fusible		9.24	.866		2,275	53		2,328	2,575
5740	Circuit breaker		9.24	.866		1,500	53		1,553	1,725
5800	Fuse, light contactor NEMA 1, type A, 30 amp		2.27	3.530		1,500	217		1,717	1,975
5820	60 amp		1.68	4.760		1,700	292		1,992	2,300

26 24 19.20 Motor Control Center Components	Crew	Daily Output	Labor-Hours	Unit	Material	2020 Bare Costs Labor	Equipment	Total	Total Incl O&P
5840	100 amp	1 Elec	.84	9.520	Ea.	3,250	585		3,835	4,450
5860	200 amp		.67	11.900		7,525	730		8,255	9,375
5880	Type B, 30 amp		2.27	3.530		1,650	217		1,867	2,125
5900	60 amp		1.68	4.760		1,825	292		2,117	2,450
5920	100 amp	▼	.84	9.520		3,575	585		4,160	4,800
5940	200 amp	2 Elec	1.34	11.900		8,250	730		8,980	10,200
5960	NEMA 12, type A, 30 amp	1 Elec	2.18	3.660		1,525	225		1,750	2,000
5980	60 amp		1.60	5.010		1,725	305		2,030	2,350
6000	100 amp	▼	.80	10.030		3,300	615		3,915	4,575
6020	200 amp	2 Elec	1.26	12.700		7,650	780		8,430	9,575
6040	Type B, 30 amp	1 Elec	2.18	3.660		1,675	225		1,900	2,150
6060	60 amp		1.60	5.010		1,850	305		2,155	2,500
6080	100 amp	▼	.80	10.030		3,625	615		4,240	4,925
6100	200 amp	2 Elec	1.26	12.700		8,375	780		9,155	10,400
6200	Circuit breaker, light contactor NEMA 1, type A, 30 amp	1 Elec	2.27	3.530		1,650	217		1,867	2,125
6220	60 amp		1.68	4.760		1,875	292		2,167	2,475
6240	100 amp	▼	.84	9.520		2,875	585		3,460	4,050
6260	200 amp	2 Elec	1.34	11.900		6,250	730		6,980	7,950
6280	Type B, 30 amp	1 Elec	2.27	3.530		1,800	217		2,017	2,300
6300	60 amp		1.68	4.760		2,075	292		2,367	2,700
6320	100 amp	▼	.84	9.520		3,150	585		3,735	4,325
6340	200 amp	2 Elec	1.34	11.900		6,825	730		7,555	8,575
6360	NEMA 12, type A, 30 amp	1 Elec	2.18	3.660		1,675	225		1,900	2,150
6380	60 amp		1.60	5.010		1,900	305		2,205	2,525
6400	100 amp	▼	.80	10.030		2,925	615		3,540	4,150
6420	200 amp	2 Elec	1.26	12.700		6,275	780		7,055	8,050
6440	Type B, 30 amp	1 Elec	2.18	3.660		1,950	225		2,175	2,475
6460	60 amp		1.60	5.010		2,100	305		2,405	2,750
6480	100 amp	▼	.80	10.030		3,200	615		3,815	4,450
6500	200 amp	2 Elec	1.26	12.700		6,900	780		7,680	8,725
6600	Fusible switch, NEMA 1, type A, 30 amp	1 Elec	4.45	1.800		1,100	110		1,210	1,375
6620	60 amp		4.20	1.900		1,175	117		1,292	1,475
6640	100 amp		3.36	2.380		1,300	146		1,446	1,650
6660	200 amp	▼	2.69	2.980		2,250	183		2,433	2,725
6680	400 amp	2 Elec	3.86	4.140		5,275	254		5,529	6,175
6700	600 amp		2.69	5.950		5,700	365		6,065	6,825
6720	800 amp	▼	2.18	7.330		15,600	450		16,050	17,900
6740	NEMA 12, type A, 30 amp	1 Elec	4.37	1.830		1,125	112		1,237	1,425
6760	60 amp		4.12	1.940		1,200	119		1,319	1,500
6780	100 amp		3.28	2.440		1,325	150		1,475	1,675
6800	200 amp	▼	2.60	3.070		2,300	188		2,488	2,800
6820	400 amp	2 Elec	3.70	4.330		5,375	266		5,641	6,325
6840	600 amp		2.52	6.350		5,875	390		6,265	7,025
6860	800 amp	▼	2.02	7.940		15,800	485		16,285	18,100
6900	Circuit breaker, NEMA 1, type A, 30 amp	1 Elec	4.45	1.800		1,000	110		1,110	1,275
6920	60 amp		4.20	1.900		1,000	117		1,117	1,275
6940	100 amp		3.36	2.380		1,000	146		1,146	1,325
6960	225 amp	▼	2.69	2.980		1,800	183		1,983	2,250
6980	400 amp	2 Elec	3.86	4.140		3,425	254		3,679	4,150
7000	600 amp		2.69	5.950		4,000	365		4,365	4,950
7020	800 amp	▼	2.18	7.330		8,375	450		8,825	9,875
7040	NEMA 12, type A, 30 amp	1 Elec	4.37	1.830		1,050	112		1,162	1,325
7060	60 amp	▼	4.12	1.940		1,050	119		1,169	1,325

26 24 19.20 Motor Control Center Components	Crew	Daily Output	Labor-Hours	Unit	Material	2020 Bare Costs Labor	Equipment	Total	Total Incl O&P
7080	100 amp	1 Elec	3.28	2.440	Ea.	1,050	150		1,200	1,375
7100	225 amp	↓	2.60	3.070		1,825	188		2,013	2,300
7120	400 amp	2 Elec	3.70	4.330		3,475	266		3,741	4,225
7140	600 amp	↓	2.52	6.350		4,075	390		4,465	5,050
7160	800 amp		2.02	7.940		8,525	485		9,010	10,100
7300	Incoming line, main lug only, 600 amp, alum., NEMA 1		1.34	11.900		1,275	730		2,005	2,475
7320	NEMA 12		1.26	12.700		1,300	780		2,080	2,575
7340	Copper, NEMA 1		1.34	11.900		1,350	730		2,080	2,550
7360	800 amp, alum., NEMA 1		1.26	12.700		3,200	780		3,980	4,675
7380	NEMA 12		1.18	13.610		3,250	835		4,085	4,825
7400	Copper, NEMA 1		1.26	12.700		3,325	780		4,105	4,825
7420	1,200 amp, copper, NEMA 1		1.18	13.610		3,450	835		4,285	5,050
7440	Incoming line, fusible switch, 400 amp, alum., NEMA 1		1.01	15.870		4,400	975		5,375	6,300
7460	NEMA 12		.92	17.320		4,500	1,075		5,575	6,525
7480	Copper, NEMA 1		1.01	15.870		4,475	975		5,450	6,375
7500	600 amp, alum., NEMA 1		.92	17.320		5,375	1,075		6,450	7,500
7520	NEMA 12		.84	19.050		5,475	1,175		6,650	7,775
7540	Copper, NEMA 1		.92	17.320		5,450	1,075		6,525	7,575
7560	Incoming line, circuit breaker, 225 amp, alum., NEMA 1		1.01	15.870		2,300	975		3,275	3,975
7580	NEMA 12		.92	17.320		2,350	1,075		3,425	4,150
7600	Copper, NEMA 1		1.01	15.870		2,375	975		3,350	4,050
7620	400 amp, alum., NEMA 1		1.01	15.870		3,425	975		4,400	5,225
7640	NEMA 12		.92	17.320		3,475	1,075		4,550	5,400
7660	Copper, NEMA 1		1.01	15.870		3,500	975		4,475	5,300
7680	600 amp, alum., NEMA 1		.92	17.320		4,000	1,075		5,075	5,975
7700	NEMA 12		.84	19.050		4,075	1,175		5,250	6,225
7720	Copper, NEMA 1		.92	17.320		4,075	1,075		5,150	6,050
7740	800 amp, copper, NEMA 1	↓	.76	21.160		8,375	1,300		9,675	11,100
7760	Incoming line, for copper bus, add					126			126	139
7780	For 65,000 amp bus bracing, add					188			188	206
7800	For NEMA 3R enclosure, add					5,900			5,900	6,475
7820	For NEMA 12 enclosure, add					162			162	179
7840	For 1/4" x 1" ground bus, add	1 Elec	13.44	.595		105	36.50		141.50	170
7860	For 1/4" x 2" ground bus, add	"	10.08	.794		105	48.50		153.50	188
7900	Main rating basic section, alum., NEMA 1, 800 amp	2 Elec	1.18	13.610		345	835		1,180	1,625
7920	1,200 amp	"	1.01	15.870		675	975		1,650	2,200
7940	For copper bus, add					505			505	555
7960	For 65,000 amp bus bracing, add					340			340	375
7980	For NEMA 3R enclosure, add					5,900			5,900	6,475
8000	For NEMA 12, enclosure, add					162			162	179
8020	For 1/4" x 1" ground bus, add	1 Elec	13.44	.595		105	36.50		141.50	170
8040	For 1/4" x 2" ground bus, add		10.08	.794		105	48.50		153.50	188
8060	Unit devices, pilot light, standard		13.44	.595		90	36.50		126.50	154
8080	Pilot light, push to test		13.44	.595		126	36.50		162.50	194
8100	Pilot light, standard, and push button		10.08	.794		217	48.50		265.50	310
8120	Pilot light, push to test, and push button		10.08	.794		253	48.50		301.50	350
8140	Pilot light, standard, and select switch		10.08	.794		217	48.50		265.50	310
8160	Pilot light, push to test, and select switch	↓	10.08	.794	↓	253	48.50		301.50	350

For customer support on your Electrical Change Order Costs with RSMeans data, call 800.448.8182. **Post-Installation Change Order Costs**

26 24 19 – Motor-Control Centers

26 24 19.30 Motor Control Center	Crew	Daily Output	Labor-Hours	Unit	Material	2020 Bare Costs Labor	Equipment	Total	Total Incl O&P
0010 **MOTOR CONTROL CENTER** Consists of starters & structures									
0050 Starters, class 1, type B, comb. MCP, FVNR, with									
0100 control transformer, 10 HP, size 1, 12" high	1 Elec	2.27	3.530	Ea.	1,800	217		2,017	2,300
0200 25 HP, size 2, 18" high	2 Elec	3.36	4.760		2,050	292		2,342	2,700
0300 50 HP, size 3, 24" high		1.68	9.520		3,175	585		3,760	4,375
0350 75 HP, size 4, 24" high		1.34	11.900		4,225	730		4,955	5,725
0400 100 HP, size 4, 30" high		1.18	13.610		5,575	835		6,410	7,375
0500 200 HP, size 5, 48" high		.84	19.050		8,325	1,175		9,500	10,900
0600 400 HP, size 6, 72" high	↓	.67	23.810	↓	17,100	1,450		18,550	21,000
0800 Structures, 600 amp, 22,000 rms, takes any									
0900 combination of starters up to 72" high	2 Elec	1.34	11.900	Ea.	1,925	730		2,655	3,175
1000 Back to back, 72" front & 66" back	"	1.01	15.870		2,625	975		3,600	4,325
1100 For copper bus, add per structure					278			278	305
1200 For NEMA 12, add per structure					160			160	176
1300 For 42,000 rms, add per structure					220			220	242
1400 For 100,000 rms, size 1 & 2, add					715			715	785
1500 Size 3, add					1,150			1,150	1,250
1600 Size 4, add					925			925	1,025
1700 For pilot lights, add per starter	1 Elec	13.44	.595		125	36.50		161.50	192
1800 For push button, add per starter		13.44	.595		125	36.50		161.50	192
1900 For auxiliary contacts, add per starter	↓	13.44	.595	↓	221	36.50		257.50	298

26 24 19.40 Motor Starters and Controls

	Crew	Daily Output	Labor-Hours	Unit	Material	2020 Bare Costs Labor	Equipment	Total	Total Incl O&P
0010 **MOTOR STARTERS AND CONTROLS**									
0050 Magnetic, FVNR, with enclosure and heaters, 480 volt									
0080 2 HP, size 00	1 Elec	2.94	2.720	Ea.	216	167		383	485
0100 5 HP, size 0		1.93	4.140		380	254		634	800
0200 10 HP, size 1	↓	1.34	5.950		288	365		653	860
0300 25 HP, size 2	2 Elec	1.85	8.660		540	530		1,070	1,375
0400 50 HP, size 3		1.51	10.580		880	650		1,530	1,925
0500 100 HP, size 4		1.01	15.870		1,950	975		2,925	3,600
0600 200 HP, size 5		.76	21.160		4,575	1,300		5,875	6,950
0610 400 HP, size 6	↓	.67	23.810		20,100	1,450		21,550	24,300
0620 NEMA 7, 5 HP, size 0	1 Elec	1.34	5.950		1,500	365		1,865	2,200
0630 10 HP, size 1	"	.92	8.660		1,575	530		2,105	2,525
0640 25 HP, size 2	2 Elec	1.51	10.580		2,525	650		3,175	3,750
0650 50 HP, size 3		1.01	15.870		3,800	975		4,775	5,625
0660 100 HP, size 4		.76	21.160		6,125	1,300		7,425	8,675
0670 200 HP, size 5	↓	.42	38.100		14,600	2,325		16,925	19,600
0700 Combination, with motor circuit protectors, 5 HP, size 0	1 Elec	1.51	5.290		1,075	325		1,400	1,675
0800 10 HP, size 1	"	1.09	7.330		1,125	450		1,575	1,900
0900 25 HP, size 2	2 Elec	1.68	9.520		1,575	585		2,160	2,600
1000 50 HP, size 3		1.11	14.430		2,275	885		3,160	3,825
1200 100 HP, size 4	↓	.67	23.810		4,900	1,450		6,350	7,575
1220 NEMA 7, 5 HP, size 0	1 Elec	1.09	7.330		3,925	450		4,375	4,975
1230 10 HP, size 1	"	.84	9.520		4,000	585		4,585	5,300
1240 25 HP, size 2	2 Elec	1.11	14.430		5,350	885		6,235	7,225
1250 50 HP, size 3		.67	23.810		8,850	1,450		10,300	11,900
1260 100 HP, size 4		.50	31.750		13,800	1,950		15,750	18,100
1270 200 HP, size 5		.34	47.620		30,000	2,925		32,925	37,300
1400 Combination, with fused switch, 5 HP, size 0	1 Elec	1.51	5.290		650	325		975	1,200
1600 10 HP, size 1	"	1.09	7.330		695	450		1,145	1,425
1800 25 HP, size 2	2 Elec	1.68	9.520	↓	1,125	585		1,710	2,125

26 24 19.40 Motor Starters and Controls		Crew	Daily Output	Labor-Hours	Unit	Material	2020 Bare Costs Labor	Equipment	Total	Total Incl O&P
2000	50 HP, size 3	2 Elec	1.11	14.430	Ea.	1,900	885		2,785	3,425
2200	100 HP, size 4	↓	.67	23.810		3,325	1,450		4,775	5,825
2610	NEMA 4, with start-stop push button, size 1	1 Elec	1.09	7.330		1,750	450		2,200	2,600
2620	Size 2	2 Elec	1.68	9.520		2,325	585		2,910	3,450
2630	Size 3		1.11	14.430		3,700	885		4,585	5,375
2640	Size 4	↓	.67	23.810	↓	5,675	1,450		7,125	8,425
2650	NEMA 4, FVNR, including control transformer									
2660	Size 1	2 Elec	2.18	7.330	Ea.	1,675	450		2,125	2,525
2670	Size 2		1.68	9.520		2,400	585		2,985	3,500
2680	Size 3		1.11	14.430		3,825	885		4,710	5,525
2690	Size 4	↓	.67	23.810		5,900	1,450		7,350	8,650
2710	Magnetic, FVR, control circuit transformer, NEMA 1, size 1	1 Elec	1.09	7.330		990	450		1,440	1,775
2720	Size 2	2 Elec	1.68	9.520		1,575	585		2,160	2,625
2730	Size 3		1.11	14.430		2,400	885		3,285	3,975
2740	Size 4	↓	.67	23.810		5,225	1,450		6,675	7,925
2760	NEMA 4, size 1	1 Elec	.92	8.660		1,425	530		1,955	2,375
2770	Size 2	2 Elec	1.34	11.900		2,300	730		3,030	3,600
2780	Size 3		1.01	15.870		3,450	975		4,425	5,250
2790	Size 4	↓	.59	27.210		7,075	1,675		8,750	10,300
2820	NEMA 12, size 1	1 Elec	.92	8.660		1,175	530		1,705	2,075
2830	Size 2	2 Elec	1.34	11.900		1,850	730		2,580	3,100
2840	Size 3		1.01	15.870		2,900	975		3,875	4,650
2850	Size 4	↓	.59	27.210		5,975	1,675		7,650	9,050
2870	Combination FVR, fused, w/control XFMR & PB, NEMA 1, size 1	1 Elec	.84	9.520		1,700	585		2,285	2,750
2880	Size 2	2 Elec	1.26	12.700		2,425	780		3,205	3,825
2890	Size 3		.92	17.320		3,625	1,075		4,700	5,550
2900	Size 4	↓	.59	27.210		7,325	1,675		9,000	10,500
2910	NEMA 4, size 1	1 Elec	.76	10.580		2,525	650		3,175	3,750
2920	Size 2	2 Elec	1.18	13.610		3,700	835		4,535	5,300
2930	Size 3		.84	19.050		5,825	1,175		7,000	8,150
2940	Size 4	↓	.50	31.750		9,675	1,950		11,625	13,500
2950	NEMA 12, size 1	1 Elec	.84	9.520		1,925	585		2,510	3,000
2960	Size 2	2 Elec	1.18	13.610		2,725	835		3,560	4,250
2970	Size 3		.84	19.050		4,000	1,175		5,175	6,150
2980	Size 4	↓	.50	31.750		7,950	1,950		9,900	11,700
3010	Manual, single phase, w/pilot, 1 pole, 120 V, NEMA 1	1 Elec	5.38	1.490		100	91.50		191.50	246
3020	NEMA 4		3.36	2.380		400	146		546	655
3030	2 pole, 120/240 V, NEMA 1		5.38	1.490		135	91.50		226.50	285
3040	NEMA 4		3.36	2.380		360	146		506	615
3041	3 phase, 3 pole, 600 V, NEMA 1		4.62	1.730		271	106		377	455
3042	NEMA 4		2.94	2.720		465	167		632	760
3043	NEMA 12	↓	2.94	2.720		310	167		477	590
3070	Auxiliary contact, normally open				↓	106			106	116
3500	Magnetic FVNR with NEMA 12, enclosure & heaters, 480 volt									
3600	5 HP, size 0	1 Elec	1.85	4.330	Ea.	260	266		526	680
3700	10 HP, size 1	"	1.26	6.350		390	390		780	1,000
3800	25 HP, size 2	2 Elec	1.68	9.520		735	585		1,320	1,675
3900	50 HP, size 3		1.34	11.900		1,125	730		1,855	2,325
4000	100 HP, size 4		.84	19.050		2,700	1,175		3,875	4,725
4100	200 HP, size 5	↓	.67	23.810		6,475	1,450		7,925	9,300
4200	Combination, with motor circuit protectors, 5 HP, size 0	1 Elec	1.43	5.600		865	345		1,210	1,450
4300	10 HP, size 1	"	1.01	7.940		895	485		1,380	1,700
4400	25 HP, size 2	2 Elec	1.51	10.580	↓	1,350	650		2,000	2,450

For customer support on your Electrical Change Order Costs with RSMeans data, call 800.448.8182.
Post-Installation Change Order Costs

26 24 19.40 Motor Starters and Controls	Crew	Daily Output	Labor-Hours	Unit	Material	2020 Bare Costs Labor	Equipment	Total	Total Incl O&P
4500	50 HP, size 3	2 Elec	1.01	15.870	Ea.	2,175	975		3,150	3,850
4600	100 HP, size 4	↓	.62	25.740		4,925	1,575		6,500	7,775
4700	Combination, with fused switch, 5 HP, size 0	1 Elec	1.43	5.600		815	345		1,160	1,400
4800	10 HP, size 1	"	1.01	7.940		850	485		1,335	1,650
4900	25 HP, size 2	2 Elec	1.51	10.580		1,300	650		1,950	2,400
5000	50 HP, size 3		1.01	15.870		2,075	975		3,050	3,725
5100	100 HP, size 4	↓	.62	25.740	↓	4,225	1,575		5,800	6,975
5200	Factory installed controls, adders to size 0 thru 5									
5300	Start-stop push button	1 Elec	26.88	.298	Ea.	54.50	18.25		72.75	87
5400	Hand-off-auto-selector switch		26.88	.298		54.50	18.25		72.75	87
5500	Pilot light		26.88	.298		102	18.25		120.25	140
5600	Start-stop-pilot		26.88	.298		157	18.25		175.25	200
5700	Auxiliary contact, NO or NC		26.88	.298		75	18.25		93.25	110
5800	NO-NC		26.88	.298		150	18.25		168.25	192
5810	Magnetic FVR, NEMA 7, w/heaters, size 1	↓	.55	14.430		3,400	885		4,285	5,075
5830	Size 2	2 Elec	.92	17.320		5,600	1,075		6,675	7,750
5840	Size 3		.59	27.210		9,025	1,675		10,700	12,400
5850	Size 4	↓	.50	31.750		10,200	1,950		12,150	14,100
5860	Combination w/circuit breakers, heaters, control XFMR PB, size 1	1 Elec	.50	15.870		1,800	975		2,775	3,425
5870	Size 2	2 Elec	.67	23.810		2,325	1,450		3,775	4,725
5880	Size 3		.42	38.100		3,100	2,325		5,425	6,875
5890	Size 4	↓	.34	47.620		6,125	2,925		9,050	11,100
5900	Manual, 240 volt, 0.75 HP motor	1 Elec	3.36	2.380		58	146		204	281
5910	2 HP motor		3.36	2.380		157	146		303	390
6000	Magnetic, 240 volt, 1 or 2 pole, 0.75 HP motor		3.36	2.380		217	146		363	455
6020	2 HP motor		3.36	2.380		240	146		386	480
6040	5 HP motor		2.52	3.170		345	194		539	670
6060	10 HP motor		1.93	4.140		855	254		1,109	1,325
6100	3 pole, 0.75 HP motor		2.52	3.170		216	194		410	530
6120	5 HP motor		1.93	4.140		293	254		547	705
6140	10 HP motor		1.34	5.950		550	365		915	1,150
6160	15 HP motor		1.34	5.950		550	365		915	1,150
6180	20 HP motor	↓	.92	8.660		900	530		1,430	1,775
6200	25 HP motor	2 Elec	1.85	8.660		900	530		1,430	1,775
6210	30 HP motor		1.51	10.580		900	650		1,550	1,950
6220	40 HP motor		1.51	10.580		2,000	650		2,650	3,175
6230	50 HP motor		1.51	10.580		2,000	650		2,650	3,175
6240	60 HP motor		1.01	15.870		4,675	975		5,650	6,575
6250	75 HP motor		1.01	15.870		4,675	975		5,650	6,575
6260	100 HP motor		1.01	15.870		4,675	975		5,650	6,575
6270	125 HP motor		.76	21.160		13,100	1,300		14,400	16,300
6280	150 HP motor		.76	21.160		13,100	1,300		14,400	16,300
6290	200 HP motor	↓	.76	21.160		13,100	1,300		14,400	16,300
6400	Starter & nonfused disconnect, 240 volt, 1-2 pole, 0 .75 HP motor	1 Elec	1.68	4.760		270	292		562	730
6410	2 HP motor		1.68	4.760		293	292		585	755
6420	5 HP motor		1.51	5.290		395	325		720	920
6430	10 HP motor		1.18	6.800		925	415		1,340	1,650
6440	3 pole, 0.75 HP motor		1.34	5.950		269	365		634	840
6450	5 HP motor		1.18	6.800		345	415		760	1,000
6460	10 HP motor		.92	8.660		620	530		1,150	1,475
6470	15 HP motor	↓	.84	9.520		620	585		1,205	1,550
6480	20 HP motor	2 Elec	1.26	12.700		1,075	780		1,855	2,325
6490	25 HP motor	↓	1.26	12.700		1,075	780		1,855	2,325

26 24 19.40 Motor Starters and Controls		Crew	Daily Output	Labor-Hours	Unit	Material	2020 Bare Costs Labor	Equipment	Total	Total Incl O&P
6500	30 HP motor	2 Elec	1.09	14.650	Ea.	1,075	900		1,975	2,525
6510	40 HP motor		1.04	15.360		2,300	940		3,240	3,925
6520	50 HP motor		.94	17.010		2,300	1,050		3,350	4,075
6530	60 HP motor		.76	21.160		4,975	1,300		6,275	7,400
6540	75 HP motor		.64	25.060		4,975	1,525		6,500	7,775
6550	100 HP motor		.59	27.210		4,975	1,675		6,650	7,950
6560	125 HP motor		.50	31.750		13,900	1,950		15,850	18,200
6570	150 HP motor		.44	36.630		13,900	2,250		16,150	18,700
6580	200 HP motor	▼	.42	38.100		13,900	2,325		16,225	18,800
6600	Starter & fused disconnect, 240 volt, 1-2 pole, 0.75 HP motor	1 Elec	1.68	4.760		283	292		575	745
6610	2 HP motor		1.68	4.760		305	292		597	770
6620	5 HP motor		1.51	5.290		410	325		735	935
6630	10 HP motor		1.18	6.800		965	415		1,380	1,675
6640	3 pole, 0.75 HP motor		1.34	5.950		282	365		647	855
6650	5 HP motor		1.18	6.800		360	415		775	1,025
6660	10 HP motor		.92	8.660		665	530		1,195	1,525
6690	15 HP motor	▼	.84	9.520		665	585		1,250	1,600
6700	20 HP motor	2 Elec	1.34	11.900		1,100	730		1,830	2,275
6710	25 HP motor		1.34	11.900		1,100	730		1,830	2,275
6720	30 HP motor		1.18	13.610		1,100	835		1,935	2,450
6730	40 HP motor		1.01	15.870		2,400	975		3,375	4,100
6740	50 HP motor		1.01	15.870		2,400	975		3,375	4,100
6750	60 HP motor		.76	21.160		5,075	1,300		6,375	7,525
6760	75 HP motor		.76	21.160		5,075	1,300		6,375	7,525
6770	100 HP motor		.59	27.210		5,075	1,675		6,750	8,075
6780	125 HP motor	▼	.45	35.270	▼	14,200	2,175		16,375	18,800
6790	Combination starter & nonfusible disconnect									
6800	240 volt, 1-2 pole, 0.75 HP motor	1 Elec	1.68	4.760	Ea.	660	292		952	1,150
6810	2 HP motor		1.68	4.760		690	292		982	1,200
6820	5 HP motor		1.26	6.350		710	390		1,100	1,350
6830	10 HP motor		1.01	7.940		1,075	485		1,560	1,900
6840	3 pole, 0.75 HP motor		1.51	5.290		655	325		980	1,200
6850	5 HP motor		1.09	7.330		690	450		1,140	1,425
6860	10 HP motor		.84	9.520		1,075	585		1,660	2,050
6870	15 HP motor	▼	.84	9.520		1,075	585		1,660	2,050
6880	20 HP motor	2 Elec	1.11	14.430		1,775	885		2,660	3,275
6890	25 HP motor		1.11	14.430		1,775	885		2,660	3,275
6900	30 HP motor		1.11	14.430		1,775	885		2,660	3,275
6910	40 HP motor		.67	23.810		3,375	1,450		4,825	5,900
6920	50 HP motor		.67	23.810		3,375	1,450		4,825	5,900
6930	60 HP motor		.59	27.210		7,550	1,675		9,225	10,800
6940	75 HP motor		.59	27.210		7,550	1,675		9,225	10,800
6950	100 HP motor		.59	27.210		7,550	1,675		9,225	10,800
6960	125 HP motor		.50	31.750		19,800	1,950		21,750	24,700
6970	150 HP motor		.50	31.750		19,800	1,950		21,750	24,700
6980	200 HP motor	▼	.50	31.750	▼	19,800	1,950		21,750	24,700
6990	Combination starter and fused disconnect									
7000	240 volt, 1-2 pole, 0.75 HP motor	1 Elec	1.68	4.760	Ea.	720	292		1,012	1,225
7010	2 HP motor		1.68	4.760		720	292		1,012	1,225
7020	5 HP motor		1.26	6.350		755	390		1,145	1,400
7030	10 HP motor		1.01	7.940		1,175	485		1,660	2,025
7040	3 pole, 0.75 HP motor		1.51	5.290		720	325		1,045	1,275
7050	5 HP motor		1.09	7.330		755	450		1,205	1,500

For customer support on your Electrical Change Order Costs with RSMeans data, call 800.448.8182.

Post-Installation Change Order Costs

26 24 19 – Motor-Control Centers

26 24 19.40 Motor Starters and Controls		Crew	Daily Output	Labor-Hours	Unit	Material	2020 Bare Costs Labor	Equipment	Total	Total Incl O&P
7060	10 HP motor	1 Elec	.84	9.520	Ea.	1,175	585		1,760	2,175
7070	15 HP motor	↓	.84	9.520		1,175	585		1,760	2,175
7080	20 HP motor	2 Elec	1.11	14.430		1,950	885		2,835	3,475
7090	25 HP motor		1.11	14.430		1,950	885		2,835	3,475
7100	30 HP motor		1.11	14.430		1,950	885		2,835	3,475
7110	40 HP motor		.67	23.810		3,725	1,450		5,175	6,275
7120	50 HP motor		.67	23.810		3,725	1,450		5,175	6,275
7130	60 HP motor		.67	23.810		8,300	1,450		9,750	11,300
7140	75 HP motor		.59	27.210		8,300	1,675		9,975	11,600
7150	100 HP motor		.59	27.210		8,300	1,675		9,975	11,600
7160	125 HP motor		.59	27.210		21,900	1,675		23,575	26,600
7170	150 HP motor		.50	31.750		21,900	1,950		23,850	27,000
7180	200 HP motor	↓	.50	31.750	↓	21,900	1,950		23,850	27,000
7190	Combination starter & circuit breaker disconnect									
7200	240 volt, 1-2 pole, 0.75 HP motor	1 Elec	1.68	4.760	Ea.	680	292		972	1,175
7210	2 HP motor		1.68	4.760		680	292		972	1,175
7220	5 HP motor		1.26	6.350		715	390		1,105	1,375
7230	10 HP motor		1.01	7.940		1,100	485		1,585	1,925
7240	3 pole, 0.75 HP motor		1.51	5.290		705	325		1,030	1,250
7250	5 HP motor		1.09	7.330		735	450		1,185	1,475
7260	10 HP motor		.84	9.520		1,125	585		1,710	2,100
7270	15 HP motor	↓	.84	9.520		1,125	585		1,710	2,100
7280	20 HP motor	2 Elec	1.11	14.430		1,900	885		2,785	3,400
7290	25 HP motor		1.11	14.430		1,900	885		2,785	3,400
7300	30 HP motor		1.11	14.430		1,900	885		2,785	3,400
7310	40 HP motor		.67	23.810		4,125	1,450		5,575	6,700
7320	50 HP motor		.67	23.810		4,125	1,450		5,575	6,700
7330	60 HP motor		.67	23.810		9,525	1,450		10,975	12,700
7340	75 HP motor		.59	27.210		9,525	1,675		11,200	13,000
7350	100 HP motor		.59	27.210		9,525	1,675		11,200	13,000
7360	125 HP motor		.59	27.210		20,600	1,675		22,275	25,200
7370	150 HP motor		.50	31.750		20,600	1,950		22,550	25,600
7380	200 HP motor	↓	.50	31.750	↓	20,600	1,950		22,550	25,600
7400	Magnetic FVNR with enclosure & heaters, 2 pole,									
7410	230 volt, 1 HP, size 00	1 Elec	3.36	2.380	Ea.	197	146		343	435
7420	2 HP, size 0		3.36	2.380		219	146		365	460
7430	3 HP, size 1		2.52	3.170		252	194		446	565
7440	5 HP, size 1P		2.52	3.170		320	194		514	645
7450	115 volt, 1/3 HP, size 00		3.36	2.380		197	146		343	435
7460	1 HP, size 0		3.36	2.380		219	146		365	460
7470	2 HP, size 1		2.52	3.170		252	194		446	565
7480	3 HP, size 1P	↓	2.52	3.170		315	194		509	635
7500	3 pole, 480 volt, 600 HP, size 7	2 Elec	.59	27.210	↓	17,400	1,675		19,075	21,700
7590	Magnetic FVNR with heater, NEMA 1									
7600	600 volt, 3 pole, 5 HP motor	1 Elec	1.93	4.140	Ea.	261	254		515	665
7610	10 HP motor	"	1.34	5.950		293	365		658	870
7620	25 HP motor	2 Elec	1.85	8.660		550	530		1,080	1,400
7630	30 HP motor		1.51	10.580		900	650		1,550	1,950
7640	40 HP motor		1.51	10.580		900	650		1,550	1,950
7650	50 HP motor		1.51	10.580		900	650		1,550	1,950
7660	60 HP motor		1.01	15.870		2,000	975		2,975	3,650
7670	75 HP motor		1.01	15.870		2,425	975		3,400	4,125
7680	100 HP motor		1.01	15.870		2,800	975		3,775	4,525

26 24 19.40 Motor Starters and Controls		Crew	Daily Output	Labor-Hours	Unit	Material	2020 Bare Costs Labor	Equipment	Total	Total Incl O&P
7690	125 HP motor	2 Elec	.76	21.160	Ea.	4,575	1,300		5,875	6,950
7700	150 HP motor		.76	21.160		4,775	1,300		6,075	7,175
7710	200 HP motor	↓	.76	21.160		5,150	1,300		6,450	7,600
7750	Starter & nonfused disconnect, 600 volt, 3 pole, 5 HP motor	1 Elec	1.18	6.800		355	415		770	1,000
7760	10 HP motor	"	.92	8.660		390	530		920	1,225
7770	25 HP motor	2 Elec	1.26	12.700		645	780		1,425	1,850
7780	30 HP motor		1.09	14.650		1,075	900		1,975	2,525
7790	40 HP motor		1.09	14.650		1,075	900		1,975	2,525
7800	50 HP motor		1.09	14.650		1,075	900		1,975	2,525
7810	60 HP motor		.77	20.700		2,275	1,275		3,550	4,400
7820	75 HP motor		.77	20.700		2,700	1,275		3,975	4,850
7830	100 HP motor		.71	22.680		3,075	1,400		4,475	5,450
7840	125 HP motor		.59	27.210		4,975	1,675		6,650	7,950
7850	150 HP motor		.59	27.210		5,175	1,675		6,850	8,175
7860	200 HP motor	↓	.50	31.750		5,550	1,950		7,500	9,000
7870	Starter & fused disconnect, 600 volt, 3 pole, 5 HP motor	1 Elec	1.18	6.800		445	415		860	1,100
7880	10 HP motor	"	.92	8.660		475	530		1,005	1,325
7890	25 HP motor	2 Elec	1.26	12.700		735	780		1,515	1,950
7900	30 HP motor		1.09	14.650		1,125	900		2,025	2,575
7910	40 HP motor		1.09	14.650		1,125	900		2,025	2,575
7920	50 HP motor		1.09	14.650		1,125	900		2,025	2,575
7930	60 HP motor		.77	20.700		2,400	1,275		3,675	4,550
7940	75 HP motor		.77	20.700		2,825	1,275		4,100	5,000
7950	100 HP motor		.71	22.680		3,200	1,400		4,600	5,600
7960	125 HP motor		.59	27.210		5,150	1,675		6,825	8,125
7970	150 HP motor		.59	27.210		5,350	1,675		7,025	8,350
7980	200 HP motor	↓	.50	31.750	↓	5,725	1,950		7,675	9,200
7990	Combination starter and nonfusible disconnect									
8000	600 volt, 3 pole, 5 HP motor	1 Elec	1.51	5.290	Ea.	680	325		1,005	1,225
8010	10 HP motor	"	1.09	7.330		915	450		1,365	1,675
8020	25 HP motor	2 Elec	1.68	9.520		1,100	585		1,685	2,100
8030	30 HP motor		1.11	14.430		1,825	885		2,710	3,350
8040	40 HP motor		1.11	14.430		1,825	885		2,710	3,350
8050	50 HP motor		1.11	14.430		2,200	885		3,085	3,750
8060	60 HP motor		.67	23.810		3,500	1,450		4,950	6,025
8070	75 HP motor		.67	23.810		3,500	1,450		4,950	6,025
8080	100 HP motor		.67	23.810		4,350	1,450		5,800	6,975
8090	125 HP motor		.59	27.210		8,050	1,675		9,725	11,400
8100	150 HP motor		.59	27.210		9,375	1,675		11,050	12,800
8110	200 HP motor	↓	.59	27.210	↓	8,450	1,675		10,125	11,800
8140	Combination starter and fused disconnect									
8150	600 volt, 3 pole, 5 HP motor	1 Elec	1.51	5.290	Ea.	720	325		1,045	1,275
8160	10 HP motor	"	1.09	7.330		785	450		1,235	1,525
8170	25 HP motor	2 Elec	1.68	9.520		1,275	585		1,860	2,275
8180	30 HP motor		1.11	14.430		2,000	885		2,885	3,525
8190	40 HP motor		1.11	14.430		2,150	885		3,035	3,675
8200	50 HP motor		1.11	14.430		2,150	885		3,035	3,675
8210	60 HP motor		.67	23.810		3,750	1,450		5,200	6,300
8220	75 HP motor		.67	23.810		3,750	1,450		5,200	6,300
8230	100 HP motor		.67	23.810		3,750	1,450		5,200	6,300
8240	125 HP motor		.59	27.210		8,300	1,675		9,975	11,600
8250	150 HP motor		.59	27.210		8,300	1,675		9,975	11,600
8260	200 HP motor	↓	.59	27.210	↓	8,300	1,675		9,975	11,600

26 24 19.40 Motor Starters and Controls		Crew	Daily Output	Labor-Hours	Unit	Material	2020 Bare Costs Labor	Equipment	Total	Total Incl O&P
8290	Combination starter & circuit breaker disconnect									
8300	600 volt, 3 pole, 5 HP motor	1 Elec	1.51	5.290	Ea.	1,025	325		1,350	1,600
8310	10 HP motor	"	1.09	7.330		1,000	450		1,450	1,775
8320	25 HP motor	2 Elec	1.68	9.520		1,500	585		2,085	2,525
8330	30 HP motor		1.11	14.430		1,900	885		2,785	3,400
8340	40 HP motor		1.11	14.430		1,900	885		2,785	3,400
8350	50 HP motor		1.11	14.430		1,975	885		2,860	3,475
8360	60 HP motor		.67	23.810		4,125	1,450		5,575	6,700
8370	75 HP motor		.67	23.810		4,125	1,450		5,575	6,700
8380	100 HP motor		.67	23.810		4,275	1,450		5,725	6,875
8390	125 HP motor		.59	27.210		9,525	1,675		11,200	13,000
8400	150 HP motor		.59	27.210		9,125	1,675		10,800	12,500
8410	200 HP motor	▼	.59	27.210	▼	9,875	1,675		11,550	13,300
8430	Starter & circuit breaker disconnect									
8440	600 volt, 3 pole, 5 HP motor	1 Elec	1.18	6.800	Ea.	845	415		1,260	1,550
8450	10 HP motor	"	.92	8.660		875	530		1,405	1,750
8460	25 HP motor	2 Elec	1.26	12.700		1,125	780		1,905	2,400
8470	30 HP motor		1.09	14.650		1,600	900		2,500	3,125
8480	40 HP motor		1.09	14.650		1,600	900		2,500	3,125
8490	50 HP motor		1.09	14.650		1,600	900		2,500	3,125
8500	60 HP motor		.77	20.700		3,875	1,275		5,150	6,150
8510	75 HP motor		.77	20.700		4,300	1,275		5,575	6,625
8520	100 HP motor		.71	22.680		4,675	1,400		6,075	7,225
8530	125 HP motor		.59	27.210		6,450	1,675		8,125	9,575
8540	150 HP motor		.59	27.210		6,650	1,675		8,325	9,775
8550	200 HP motor	▼	.50	31.750		8,350	1,950		10,300	12,100
8900	240 volt, 1-2 pole, 0.75 HP motor	1 Elec	1.68	4.760		800	292		1,092	1,325
8910	2 HP motor		1.68	4.760		820	292		1,112	1,350
8920	5 HP motor		1.51	5.290		925	325		1,250	1,500
8930	10 HP motor		1.18	6.800		1,550	415		1,965	2,350
8950	3 pole, 0.75 HP motor		1.34	5.950		795	365		1,160	1,425
8970	5 HP motor		1.18	6.800		875	415		1,290	1,575
8980	10 HP motor		.92	8.660		1,250	530		1,780	2,175
8990	15 HP motor	▼	.84	9.520		1,250	585		1,835	2,250
9100	20 HP motor	2 Elec	1.26	12.700	Ea.	1,700	780		2,480	3,025
9110	25 HP motor		1.26	12.700		1,700	780		2,480	3,025
9120	30 HP motor		1.09	14.650		1,700	900		2,600	3,225
9130	40 HP motor		1.04	15.360		3,875	940		4,815	5,650
9140	50 HP motor		.94	17.010		3,875	1,050		4,925	5,800
9150	60 HP motor		.76	21.160		7,875	1,300		9,175	10,600
9160	75 HP motor		.64	25.060		7,875	1,525		9,400	11,000
9170	100 HP motor		.59	27.210		7,875	1,675		9,550	11,100
9180	125 HP motor		.50	31.750		17,700	1,950		19,650	22,400
9190	150 HP motor		.44	36.630		17,700	2,250		19,950	22,900
9200	200 HP motor	▼	.42	38.100	▼	17,700	2,325		20,025	23,000

26 25 13.10 Aluminum Bus Duct	Crew	Daily Output	Labor-Hours	Unit	Material	2020 Bare Costs Labor	Equipment	Total	Total Incl O&P
0010 **ALUMINUM BUS DUCT** 10 ft. long									
0050 Indoor 3 pole 4 wire, plug-in, straight section, 225 amp	2 Elec	36.96	.433	L.F.	172	26.50		198.50	229
0100 400 amp		30.24	.529		216	32.50		248.50	286
0150 600 amp		26.88	.595		298	36.50		334.50	385
0200 800 amp		21.84	.733		345	45		390	445
0250 1,000 amp		20.16	.794		435	48.50		483.50	555
0300 1,350 amp		18.48	.866		305	53		358	415
0310 1,600 amp		15.12	1.060		330	65		395	455
0320 2,000 amp		13.44	1.190		425	73		498	580
0330 2,500 amp		11.76	1.360		460	83.50		543.50	630
0340 3,000 amp		10.08	1.590		595	97.50		692.50	800
0350 Feeder, 600 amp		28.56	.560		113	34.50		147.50	176
0400 800 amp		23.52	.680		430	41.50		471.50	530
0450 1,000 amp		21.84	.733		535	45		580	650
0455 1,200 amp		21	.762		650	46.50		696.50	785
0500 1,350 amp		20.16	.794		236	48.50		284.50	335
0550 1,600 amp		16.80	.952		900	58.50		958.50	1,075
0600 2,000 amp		15.12	1.060		1,100	65		1,165	1,300
0620 2,500 amp		11.76	1.360		1,350	83.50		1,433.50	1,600
0630 3,000 amp		10.08	1.590		1,575	97.50		1,672.50	1,875
0640 4,000 amp		8.40	1.900	▼	680	117		797	925
0650 Elbow, 225 amp		3.70	4.330	Ea.	835	266		1,101	1,300
0700 400 amp		3.19	5.010		840	305		1,145	1,375
0750 600 amp		2.86	5.600		845	345		1,190	1,450
0800 800 amp		2.52	6.350		875	390		1,265	1,550
0850 1,000 amp		2.35	6.800		1,125	415		1,540	1,875
0870 1,200 amp		2.27	7.050		1,400	435		1,835	2,200
0900 1,350 amp		2.18	7.330		1,350	450		1,800	2,150
0950 1,600 amp		2.02	7.940		1,425	485		1,910	2,300
1000 2,000 amp		1.68	9.520		1,575	585		2,160	2,600
1020 2,500 amp		1.51	10.580		1,850	650		2,500	3,000
1030 3,000 amp		1.34	11.900		2,125	730		2,855	3,425
1040 4,000 amp		1.18	13.610		3,450	835		4,285	5,050
1100 Cable tap box end, 225 amp		3.02	5.290		1,775	325		2,100	2,425
1150 400 amp		2.69	5.950		2,175	365		2,540	2,950
1200 600 amp		2.18	7.330		1,475	450		1,925	2,300
1250 800 amp		1.85	8.660		1,825	530		2,355	2,800
1300 1,000 amp		1.68	9.520		3,500	585		4,085	4,725
1320 1,200 amp		1.68	9.520		1,875	585		2,460	2,950
1350 1,350 amp		1.34	11.900		1,550	730		2,280	2,775
1400 1,600 amp		1.18	13.610		1,825	835		2,660	3,250
1450 2,000 amp		1.01	15.870		2,075	975		3,050	3,725
1460 2,500 amp		.84	19.050		2,475	1,175		3,650	4,450
1470 3,000 amp		.67	23.810		2,725	1,450		4,175	5,175
1480 4,000 amp		.50	31.750		3,225	1,950		5,175	6,450
1500 Switchboard stub, 225 amp		4.87	3.280		1,850	201		2,051	2,325
1550 400 amp		4.54	3.530		1,875	217		2,092	2,400
1600 600 amp		3.86	4.140		2,100	254		2,354	2,700
1650 800 amp		3.36	4.760		1,900	292		2,192	2,525
1700 1,000 amp		2.69	5.950		1,700	365		2,065	2,425
1720 1,200 amp		2.60	6.140		1,750	375		2,125	2,475
1750 1,350 amp		2.52	6.350		1,725	390		2,115	2,475
1800 1,600 amp		2.18	7.330	▼	1,825	450		2,275	2,675

For customer support on your Electrical Change Order Costs with RSMeans data, call 800.448.8182.

Post-Installation Change Order Costs

26 25 13.10 Aluminum Bus Duct		Crew	Daily Output	Labor-Hours	Unit	Material	2020 Bare Costs Labor	Equipment	Total	Total Incl O&P
1850	2,000 amp	2 Elec	2.02	7.940	Ea.	1,950	485		2,435	2,875
1860	2,500 amp		1.85	8.660		2,175	530		2,705	3,175
1870	3,000 amp		1.68	9.520		2,275	585		2,860	3,375
1880	4,000 amp		1.51	10.580		2,525	650		3,175	3,750
1890	Tee fittings, 225 amp		2.69	5.950		895	365		1,260	1,525
1900	400 amp		2.35	6.800		895	415		1,310	1,600
1950	600 amp		2.18	7.330		895	450		1,345	1,650
2000	800 amp		2.02	7.940		950	485		1,435	1,775
2050	1,000 amp		1.85	8.660		1,000	530		1,530	1,900
2070	1,200 amp		1.76	9.070		1,175	555		1,730	2,125
2100	1,350 amp		1.68	9.520		1,625	585		2,210	2,675
2150	1,600 amp		1.34	11.900		1,975	730		2,705	3,225
2200	2,000 amp		1.01	15.870		2,175	975		3,150	3,850
2220	2,500 amp		.84	19.050		2,600	1,175		3,775	4,600
2230	3,000 amp		.67	23.810		2,975	1,450		4,425	5,450
2240	4,000 amp		.50	31.750		4,925	1,950		6,875	8,325
2300	Wall flange, 600 amp		16.80	.952		246	58.50		304.50	360
2310	800 amp		13.44	1.190		246	73		319	380
2320	1,000 amp		10.92	1.470		246	90		336	405
2325	1,200 amp		10.08	1.590		246	97.50		343.50	415
2330	1,350 amp		9.07	1.760		246	108		354	430
2340	1,600 amp		7.56	2.120		246	130		376	465
2350	2,000 amp		6.72	2.380		246	146		392	490
2360	2,500 amp		5.54	2.890		246	177		423	535
2370	3,000 amp		4.54	3.530		246	217		463	590
2380	4,000 amp		3.36	4.760		390	292		682	865
2390	5,000 amp		2.52	6.350		390	390		780	1,000
2400	Vapor barrier		6.72	2.380		460	146		606	720
2420	Roof flange kit		3.36	4.760		885	292		1,177	1,400
2600	Expansion fitting, 225 amp		8.40	1.900		1,400	117		1,517	1,700
2610	400 amp		6.72	2.380		1,400	146		1,546	1,750
2620	600 amp		5.04	3.170		1,400	194		1,594	1,825
2630	800 amp		3.86	4.140		1,675	254		1,929	2,225
2640	1,000 amp		3.36	4.760		1,800	292		2,092	2,400
2650	1,350 amp		3.02	5.290		2,075	325		2,400	2,775
2660	1,600 amp		2.69	5.950		2,500	365		2,865	3,300
2670	2,000 amp		2.35	6.800		2,775	415		3,190	3,675
2680	2,500 amp		2.02	7.940		3,350	485		3,835	4,425
2690	3,000 amp		1.68	9.520		3,875	585		4,460	5,125
2700	4,000 amp		1.34	11.900		5,100	730		5,830	6,675
2800	Reducer nonfused, 400 amp		6.72	2.380		820	146		966	1,125
2810	600 amp		5.04	3.170		820	194		1,014	1,200
2820	800 amp		3.86	4.140		985	254		1,239	1,450
2830	1,000 amp		3.36	4.760		1,150	292		1,442	1,700
2840	1,350 amp		3.02	5.290		1,500	325		1,825	2,125
2850	1,600 amp		2.69	5.950		2,050	365		2,415	2,800
2860	2,000 amp		2.35	6.800		2,350	415		2,765	3,200
2870	2,500 amp		2.02	7.940		2,950	485		3,435	3,975
2880	3,000 amp		1.68	9.520		3,400	585		3,985	4,625
2890	4,000 amp		1.34	11.900		4,550	730		5,280	6,075
2950	Reducer fuse included, 225 amp		3.70	4.330		2,400	266		2,666	3,050
2960	400 amp		3.53	4.540		2,450	279		2,729	3,125
2970	600 amp		3.02	5.290		2,900	325		3,225	3,650

26 25 13.10 Aluminum Bus Duct		Crew	Daily Output	Labor-Hours	Unit	Material	2020 Bare Costs Labor	Equipment	Total	Total Incl O&P
2980	800 amp	2 Elec	2.69	5.950	Ea.	4,600	365		4,965	5,600
2990	1,000 amp		2.52	6.350		5,250	390		5,640	6,350
3000	1,200 amp		2.35	6.800		5,250	415		5,665	6,400
3010	1,600 amp		1.85	8.660		12,000	530		12,530	14,000
3020	2,000 amp		1.51	10.580		13,300	650		13,950	15,600
3100	Reducer circuit breaker, 225 amp		3.70	4.330		2,375	266		2,641	3,000
3110	400 amp		3.53	4.540		2,875	279		3,154	3,600
3120	600 amp		3.02	5.290		4,100	325		4,425	4,975
3130	800 amp		2.69	5.950		4,800	365		5,165	5,825
3140	1,000 amp		2.52	6.350		5,450	390		5,840	6,575
3150	1,200 amp		2.35	6.800		6,550	415		6,965	7,825
3160	1,600 amp		1.85	8.660		9,675	530		10,205	11,500
3170	2,000 amp		1.51	10.580		10,600	650		11,250	12,700
3250	Reducer circuit breaker, 75,000 AIC, 225 amp		3.70	4.330		3,700	266		3,966	4,475
3260	400 amp		3.53	4.540		3,700	279		3,979	4,500
3270	600 amp		3.02	5.290		4,950	325		5,275	5,925
3280	800 amp		2.69	5.950		5,450	365		5,815	6,525
3290	1,000 amp		2.52	6.350		8,750	390		9,140	10,200
3300	1,200 amp		2.35	6.800		8,750	415		9,165	10,200
3310	1,600 amp		1.85	8.660		9,675	530		10,205	11,500
3320	2,000 amp		1.51	10.580		10,600	650		11,250	12,700
3400	Reducer circuit breaker CLF 225 amp		3.70	4.330		3,800	266		4,066	4,600
3410	400 amp		3.53	4.540		4,525	279		4,804	5,400
3420	600 amp		3.02	5.290		6,775	325		7,100	7,925
3430	800 amp		2.69	5.950		7,075	365		7,440	8,325
3440	1,000 amp		2.52	6.350		7,375	390		7,765	8,675
3450	1,200 amp		2.35	6.800		9,625	415		10,040	11,200
3460	1,600 amp		1.85	8.660		9,675	530		10,205	11,500
3470	2,000 amp		1.51	10.580		10,600	650		11,250	12,700
3550	Ground bus added to bus duct, 225 amp		268.80	.060	L.F.	33.50	3.65		37.15	42.50
3560	400 amp		268.80	.060		33.50	3.65		37.15	42.50
3570	600 amp		235.20	.068		33.50	4.17		37.67	43
3580	800 amp		201.60	.079		33.50	4.87		38.37	44.50
3590	1,000 amp		168	.095		33.50	5.85		39.35	45.50
3600	1,350 amp		151.20	.106		33.50	6.50		40	46.50
3610	1,600 amp		134.40	.119		33.50	7.30		40.80	48
3620	2,000 amp		134.40	.119		33.50	7.30		40.80	48
3630	2,500 amp		117.60	.136		33.50	8.35		41.85	49.50
3640	3,000 amp		100.80	.159		33.50	9.75		43.25	51.50
3650	4,000 amp		84	.190		33.50	11.70		45.20	54.50
3810	High short circuit, 400 amp		30.24	.529		208	32.50		240.50	278
3820	600 amp		26.88	.595		360	36.50		396.50	450
3830	800 amp		21.84	.733		415	45		460	520
3840	1,000 amp		20.16	.794		520	48.50		568.50	650
3850	1,350 amp		18.48	.866		189	53		242	287
3860	1,600 amp		15.12	1.060		217	65		282	335
3870	2,000 amp		13.44	1.190		255	73		328	390
3880	2,500 amp		11.76	1.360		425	83.50		508.50	590
3890	3,000 amp		10.08	1.590		480	97.50		577.50	675
3920	Cross, 225 amp		4.70	3.400	Ea.	1,350	209		1,559	1,775
3930	400 amp		3.86	4.140		1,350	254		1,604	1,850
3940	600 amp		3.36	4.760		1,350	292		1,642	1,900
3950	800 amp		2.86	5.600		1,400	345		1,745	2,050

Post-Installation Change Order Costs

26 25 Low-Voltage Enclosed Bus Assemblies

26 25 13 – Low-Voltage Busways

26 25 13.10 Aluminum Bus Duct	Crew	Daily Output	Labor-Hours	Unit	Material	2020 Bare Costs Labor	Equipment	Total	Total Incl O&P
3960	1,000 amp	2 Elec	2.52	6.350	Ea.	1,475	390		1,865	2,200
3970	1,350 amp		2.35	6.800		2,400	415		2,815	3,275
3980	1,600 amp		1.85	8.660		2,850	530		3,380	3,950
3990	2,000 amp		1.51	10.580		3,125	650		3,775	4,425
4000	2,500 amp		1.34	11.900		3,700	730		4,430	5,150
4010	3,000 amp		1.01	15.870		4,275	975		5,250	6,150
4020	4,000 amp		.84	19.050		6,525	1,175		7,700	8,925
4040	Cable tap box center, 225 amp		3.02	5.290		2,975	325		3,300	3,750
4050	400 amp		2.69	5.950		3,375	365		3,740	4,275
4060	600 amp		2.18	7.330		4,475	450		4,925	5,575
4070	800 amp		1.85	8.660		4,925	530		5,455	6,200
4080	1,000 amp		1.68	9.520		6,200	585		6,785	7,700
4090	1,350 amp		1.34	11.900		1,450	730		2,180	2,675
4100	1,600 amp		1.18	13.610		1,625	835		2,460	3,050
4110	2,000 amp		1.01	15.870		1,875	975		2,850	3,500
4120	2,500 amp		.84	19.050		2,275	1,175		3,450	4,250
4130	3,000 amp		.67	23.810		2,500	1,450		3,950	4,925
4140	4,000 amp		.50	31.750		3,350	1,950		5,300	6,575
4500	Weatherproof 3 pole 4 wire, feeder, 600 amp		25.20	.635	L.F.	136	39		175	207
4520	800 amp		20.16	.794		159	48.50		207.50	248
4540	1,000 amp		18.48	.866		181	53		234	278
4550	1,200 amp		17.64	.907		251	55.50		306.50	360
4560	1,350 amp		16.80	.952		283	58.50		341.50	395
4580	1,600 amp		14.28	1.120		330	68.50		398.50	460
4600	2,000 amp		13.44	1.190		385	73		458	535
4620	2,500 amp		10.08	1.590		500	97.50		597.50	695
4640	3,000 amp		8.40	1.900		565	117		682	800
4660	4,000 amp		6.72	2.380		815	146		961	1,100
5000	Indoor 3 pole, 3 wire, feeder, 600 amp		33.60	.476		104	29		133	158
5010	800 amp		26.88	.595		355	36.50		391.50	445
5020	1,000 amp		25.20	.635		395	39		434	495
5025	1,200 amp		24.36	.657		460	40.50		500.50	570
5030	1,350 amp		23.52	.680		179	41.50		220.50	259
5040	1,600 amp		20.16	.794		755	48.50		803.50	905
5050	2,000 amp		16.80	.952		900	58.50		958.50	1,075
5060	2,500 amp		13.44	1.190		1,100	73		1,173	1,300
5070	3,000 amp		11.76	1.360		1,250	83.50		1,333.50	1,500
5080	4,000 amp		10.08	1.590		490	97.50		587.50	685
5200	Plug-in type, 225 amp		42	.381		139	23.50		162.50	188
5210	400 amp		35.28	.454		170	28		198	229
5220	600 amp		30.24	.529		211	32.50		243.50	281
5230	800 amp		25.20	.635		295	39		334	385
5240	1,000 amp		23.52	.680		335	41.50		376.50	425
5245	1,200 amp		22.68	.705		236	43.50		279.50	325
5250	1,350 amp		21.84	.733		189	45		234	275
5260	1,600 amp		16.80	.952		217	58.50		275.50	325
5270	2,000 amp		15.12	1.060		255	65		320	375
5280	2,500 amp		13.44	1.190		350	73		423	495
5290	3,000 amp		11.76	1.360		415	83.50		498.50	580
5300	4,000 amp		10.08	1.590		500	97.50		597.50	695
5330	High short circuit, 400 amp		35.28	.454		204	28		232	266
5340	600 amp		30.24	.529		252	32.50		284.50	325
5350	800 amp		25.20	.635		355	39		394	455

Post-Installation Change Order Costs   For customer support on your Electrical Change Order Costs with RSMeans data, call 800.448.8182.

26 25 Low-Voltage Enclosed Bus Assemblies

26 25 13 – Low-Voltage Busways

26 25 13.10 Aluminum Bus Duct	Crew	Daily Output	Labor-Hours	Unit	Material	2020 Bare Costs Labor	2020 Bare Costs Equipment	Total	Total Incl O&P
5360	1,000 amp	2 Elec	23.52	.680	L.F.	405	41.50		446.50	505
5370	1,350 amp		21.84	.733		189	45		234	275
5380	1,600 amp		16.80	.952		217	58.50		275.50	325
5390	2,000 amp		15.12	1.060		255	65		320	375
5400	2,500 amp		13.44	1.190		400	73		473	550
5410	3,000 amp		11.76	1.360		415	83.50		498.50	580
5440	Elbow, 225 amp		4.20	3.810	Ea.	655	234		889	1,075
5450	400 amp		3.70	4.330		655	266		921	1,125
5460	600 amp		3.36	4.760		655	292		947	1,150
5470	800 amp		2.86	5.600		700	345		1,045	1,275
5480	1,000 amp		2.69	5.950		715	365		1,080	1,325
5485	1,200 amp		2.60	6.140		760	375		1,135	1,400
5490	1,350 amp		2.52	6.350		730	390		1,120	1,375
5500	1,600 amp		2.35	6.800		1,125	415		1,540	1,850
5510	2,000 amp		2.02	7.940		1,225	485		1,710	2,075
5520	2,500 amp		1.68	9.520		1,500	585		2,085	2,525
5530	3,000 amp		1.51	10.580		1,800	650		2,450	2,950
5540	4,000 amp		1.34	11.900		2,400	730		3,130	3,700
5560	Tee fittings, 225 amp		3.02	5.290		800	325		1,125	1,375
5570	400 amp		2.69	5.950		800	365		1,165	1,425
5580	600 amp		2.52	6.350		800	390		1,190	1,475
5590	800 amp		2.35	6.800		860	415		1,275	1,575
5600	1,000 amp		2.18	7.330		890	450		1,340	1,650
5605	1,200 amp		2.10	7.620		945	465		1,410	1,750
5610	1,350 amp		2.02	7.940		1,325	485		1,810	2,175
5620	1,600 amp		1.51	10.580		1,550	650		2,200	2,675
5630	2,000 amp		1.18	13.610		1,725	835		2,560	3,150
5640	2,500 amp		1.01	15.870		2,150	975		3,125	3,800
5650	3,000 amp		.84	19.050		2,525	1,175		3,700	4,525
5660	4,000 amp		.59	27.210		3,700	1,675		5,375	6,525
5680	Cross, 225 amp		5.38	2.980		1,300	183		1,483	1,700
5690	400 amp		4.54	3.530		1,300	217		1,517	1,750
5700	600 amp		3.86	4.140		1,300	254		1,554	1,800
5710	800 amp		3.36	4.760		1,375	292		1,667	1,925
5720	1,000 amp		3.02	5.290		1,400	325		1,725	2,025
5730	1,350 amp		2.69	5.950		2,100	365		2,465	2,875
5740	1,600 amp		2.18	7.330		2,450	450		2,900	3,375
5750	2,000 amp		1.85	8.660		2,675	530		3,205	3,750
5760	2,500 amp		1.51	10.580		3,250	650		3,900	4,550
5770	3,000 amp		1.18	13.610		3,875	835		4,710	5,500
5780	4,000 amp		1.01	15.870		5,375	975		6,350	7,375
5800	Expansion fitting, 225 amp		9.74	1.640		1,100	101		1,201	1,375
5810	400 amp		7.73	2.070		1,100	127		1,227	1,425
5820	600 amp		5.88	2.720		1,100	167		1,267	1,475
5830	800 amp		4.37	3.660		1,325	225		1,550	1,775
5840	1,000 amp		3.86	4.140		1,450	254		1,704	1,950
5850	1,350 amp		3.53	4.540		1,475	279		1,754	2,050
5860	1,600 amp		3.02	5.290		1,825	325		2,150	2,500
5870	2,000 amp		2.69	5.950		2,175	365		2,540	2,950
5880	2,500 amp		2.35	6.800		2,450	415		2,865	3,325
5890	3,000 amp		2.02	7.940		2,875	485		3,360	3,875
5900	4,000 amp		1.51	10.580		3,925	650		4,575	5,300
5940	Reducer, nonfused, 400 amp		7.73	2.070		705	127		832	965

Post-Installation Change Order Costs

26 25 Low-Voltage Enclosed Bus Assemblies

26 25 13 – Low-Voltage Busways

26 25 13.10 Aluminum Bus Duct		Crew	Daily Output	Labor-Hours	Unit	Material	2020 Bare Costs Labor	Equipment	Total	Total Incl O&P
5950	600 amp	2 Elec	5.88	2.720	Ea.	705	167		872	1,025
5960	800 amp		4.37	3.660		740	225		965	1,150
5970	1,000 amp		3.86	4.140		910	254		1,164	1,375
5980	1,350 amp		3.53	4.540		1,350	279		1,629	1,900
5990	1,600 amp		3.02	5.290		1,525	325		1,850	2,175
6000	2,000 amp		2.69	5.950		1,800	365		2,165	2,525
6010	2,500 amp		2.35	6.800		2,300	415		2,715	3,150
6020	3,000 amp		1.85	8.660		2,675	530		3,205	3,750
6030	4,000 amp		1.51	10.580		3,675	650		4,325	5,025
6050	Reducer, fuse included, 225 amp		4.20	3.810		1,850	234		2,084	2,375
6060	400 amp		4.03	3.970		2,450	244		2,694	3,075
6070	600 amp		3.53	4.540		3,125	279		3,404	3,875
6080	800 amp		3.02	5.290		4,850	325		5,175	5,825
6090	1,000 amp		2.86	5.600		5,300	345		5,645	6,325
6100	1,350 amp		2.69	5.950		9,675	365		10,040	11,200
6110	1,600 amp		2.18	7.330		11,500	450		11,950	13,300
6120	2,000 amp		1.68	9.520		13,300	585		13,885	15,500
6160	Reducer, circuit breaker, 225 amp		4.20	3.810		2,275	234		2,509	2,850
6170	400 amp		4.03	3.970		2,775	244		3,019	3,425
6180	600 amp		3.53	4.540		4,000	279		4,279	4,825
6190	800 amp		3.02	5.290		4,650	325		4,975	5,600
6200	1,000 amp		2.86	5.600		5,300	345		5,645	6,325
6210	1,350 amp		2.69	5.950		6,400	365		6,765	7,600
6220	1,600 amp		2.18	7.330		9,550	450		10,000	11,200
6230	2,000 amp		1.68	9.520		10,400	585		10,985	12,400
6270	Cable tap box center, 225 amp		3.53	4.540		1,025	279		1,304	1,550
6280	400 amp		3.02	5.290		1,025	325		1,350	1,600
6290	600 amp		2.52	6.350		1,025	390		1,415	1,700
6300	800 amp		2.18	7.330		1,125	450		1,575	1,900
6310	1,000 amp		2.02	7.940		1,200	485		1,685	2,025
6320	1,350 amp		1.51	10.580		1,250	650		1,900	2,375
6330	1,600 amp		1.34	11.900		1,425	730		2,155	2,625
6340	2,000 amp		1.18	13.610		1,625	835		2,460	3,025
6350	2,500 amp		1.01	15.870		2,025	975		3,000	3,675
6360	3,000 amp		.84	19.050		2,300	1,175		3,475	4,275
6370	4,000 amp		.59	27.210		2,725	1,675		4,400	5,450
6390	Cable tap box end, 225 amp		3.53	4.540		625	279		904	1,100
6400	400 amp		3.02	5.290		625	325		950	1,175
6410	600 amp		2.52	6.350		625	390		1,015	1,275
6420	800 amp		2.18	7.330		685	450		1,135	1,425
6430	1,000 amp		2.02	7.940		740	485		1,225	1,550
6435	1,200 amp		1.76	9.070		805	555		1,360	1,725
6440	1,350 amp		1.51	10.580		785	650		1,435	1,825
6450	1,600 amp		1.34	11.900		890	730		1,620	2,050
6460	2,000 amp		1.18	13.610		1,000	835		1,835	2,350
6470	2,500 amp		1.01	15.870		1,175	975		2,150	2,725
6480	3,000 amp		.84	19.050		1,375	1,175		2,550	3,275
6490	4,000 amp		.59	27.210		1,650	1,675		3,325	4,300
7000	Weatherproof 3 pole 3 wire, feeder, 600 amp		28.56	.560	L.F.	134	34.50		168.50	198
7020	800 amp		23.52	.680		148	41.50		189.50	224
7040	1,000 amp		21.84	.733		159	45		204	242
7050	1,200 amp		21	.762		185	46.50		231.50	274
7060	1,350 amp		20.16	.794		215	48.50		263.50	310

Post-Installation Change Order Costs   For customer support on your Electrical Change Order Costs with RSMeans data, call 800.448.8182.

26 25 13.10 Aluminum Bus Duct

		Crew	Daily Output	Labor-Hours	Unit	Material	2020 Bare Costs Labor	Equipment	Total	Total Incl O&P
7080	1,600 amp	2 Elec	16.80	.952	L.F.	249	58.50		307.50	360
7100	2,000 amp		15.12	1.060		294	65		359	420
7120	2,500 amp		11.76	1.360		410	83.50		493.50	575
7140	3,000 amp		10.08	1.590		485	97.50		582.50	680
7160	4,000 amp		8.40	1.900		590	117		707	825

26 25 13.20 Bus Duct

		Crew	Daily Output	Labor-Hours	Unit	Material	2020 Bare Costs Labor	Equipment	Total	Total Incl O&P
0010	**BUS DUCT** 100 amp and less, aluminum or copper, plug-in									
0080	Bus duct, 3 pole 3 wire, 100 amp	1 Elec	35.28	.227	L.F.	78	13.90		91.90	107
0110	Elbow		3.36	2.380	Ea.	132	146		278	360
0120	Tee		1.68	4.760		187	292		479	640
0130	Wall flange		6.72	1.190		27.50	73		100.50	140
0140	Ground kit		13.44	.595		61.50	36.50		98	122
0180	3 pole 4 wire, 100 amp		33.60	.238	L.F.	84.50	14.60		99.10	115
0200	Cable tap box		2.60	3.070	Ea.	258	188		446	565
0300	End closure		13.44	.595		34.50	36.50		71	92.50
0400	Elbow		3.36	2.380		216	146		362	455
0500	Tee		1.68	4.760		315	292		607	780
0600	Hangers		8.40	.952		24.50	58.50		83	114
0700	Circuit breakers, 15 to 50 amp, 1 pole		6.72	1.190		670	73		743	850
0800	15 to 60 amp, 2 pole		5.63	1.420		360	87		447	525
0900	3 pole		4.45	1.800		450	110		560	665
1000	60 to 100 amp, 1 pole		5.63	1.420		705	87		792	905
1100	70 to 100 amp, 2 pole		4.45	1.800		1,300	110		1,410	1,600
1200	3 pole		3.78	2.120		1,750	130		1,880	2,125
1220	Switch, nonfused, 3 pole, 4 wire		6.72	1.190		219	73		292	350
1240	Fused, 3 fuses, 4 wire, 30 amp		6.72	1.190		360	73		433	510
1260	60 amp		4.45	1.800		425	110		535	630
1280	100 amp		3.78	2.120		570	130		700	820
1300	Plug, fusible, 3 pole 250 volt, 30 amp		4.45	1.800		535	110		645	750
1310	60 amp		4.45	1.800		600	110		710	825
1320	100 amp		3.78	2.120		815	130		945	1,100
1330	3 pole 480 volt, 30 amp		4.45	1.800		540	110		650	760
1340	60 amp		4.45	1.800		585	110		695	805
1350	100 amp		3.78	2.120		845	130		975	1,125
1360	Circuit breaker, 3 pole 250 volt, 60 amp		4.45	1.800		960	110		1,070	1,225
1370	3 pole 480 volt, 100 amp		3.78	2.120		960	130		1,090	1,250
2000	Bus duct, 2 wire, 250 volt, 30 amp		50.40	.159	L.F.	8.35	9.75		18.10	23.50
2100	60 amp		42	.190		8.35	11.70		20.05	26.50
2200	300 volt, 30 amp		50.40	.159		8.35	9.75		18.10	23.50
2300	60 amp		42	.190		8.35	11.70		20.05	26.50
2400	3 wire, 250 volt, 30 amp		50.40	.159		10.80	9.75		20.55	26.50
2500	60 amp		42	.190		10.55	11.70		22.25	29
2600	480/277 volt, 30 amp		50.40	.159		10.80	9.75		20.55	26.50
2700	60 amp		42	.190		10.80	11.70		22.50	29.50
2750	End feed, 300 volt 2 wire max. 30 amp		5.04	1.590	Ea.	75.50	97.50		173	228
2800	60 amp		4.62	1.730		75.50	106		181.50	241
2850	30 amp miniature		5.04	1.590		75.50	97.50		173	228
2900	3 wire, 30 amp		5.04	1.590		95	97.50		192.50	249
2950	60 amp		4.62	1.730		95	106		201	262
3000	30 amp miniature		5.04	1.590		95	97.50		192.50	249
3050	Center feed, 300 volt 2 wire, 30 amp		5.04	1.590		104	97.50		201.50	260
3100	60 amp		4.62	1.730		104	106		210	273

26 25 Low-Voltage Enclosed Bus Assemblies

26 25 13 – Low-Voltage Busways

26 25 13.20 Bus Duct

		Crew	Daily Output	Labor-Hours	Unit	Material	2020 Bare Costs Labor	Equipment	Total	Total Incl O&P
3150	3 wire, 30 amp	1 Elec	5.04	1.590	Ea.	119	97.50		216.50	275
3200	60 amp		4.62	1.730		119	106		225	288
3220	Elbow, 30 amp		5.04	1.590		40.50	97.50		138	190
3240	60 amp		4.62	1.730		40.50	106		146.50	203
3260	End cap		33.60	.238		9.90	14.60		24.50	33
3280	Strength beam, 10'		12.60	.635		24.50	39		63.50	85
3300	Hanger		20.16	.397		5.25	24.50		29.75	42.50
3320	Tap box, nonfusible		5.29	1.510		84.50	92.50		177	231
3340	Fusible switch 30 amp, 1 fuse		5.04	1.590		465	97.50		562.50	655
3360	2 fuse		5.04	1.590		485	97.50		582.50	680
3380	3 fuse		5.04	1.590		530	97.50		627.50	730
3400	Circuit breaker handle on cover, 1 pole		5.04	1.590		62	97.50		159.50	214
3420	2 pole		5.04	1.590		580	97.50		677.50	780
3440	3 pole		5.04	1.590		710	97.50		807.50	930
3460	Circuit breaker external operhandle, 1 pole		5.04	1.590		74	97.50		171.50	227
3480	2 pole		5.04	1.590		750	97.50		847.50	970
3500	3 pole		5.04	1.590		815	97.50		912.50	1,050
3520	Terminal plug only		13.44	.595		99.50	36.50		136	165
3540	Terminal with receptacle		13.44	.595		109	36.50		145.50	175
3560	Fixture plug		13.44	.595		73.50	36.50		110	136
4000	Copper bus duct, lighting, 2 wire 300 volt, 20 amp		58.80	.136	L.F.	7.55	8.35		15.90	21
4020	35 amp		50.40	.159		7.55	9.75		17.30	23
4040	50 amp		46.20	.173		7.55	10.60		18.15	24
4060	60 amp		42	.190		7.55	11.70		19.25	25.50
4080	3 wire 300 volt, 20 amp		58.80	.136		7.15	8.35		15.50	20.50
4100	35 amp		50.40	.159		6.95	9.75		16.70	22
4120	50 amp		46.20	.173		7.15	10.60		17.75	23.50
4140	60 amp		42	.190		7.15	11.70		18.85	25.50
4160	Feeder in box, end, 1 circuit		5.04	1.590	Ea.	101	97.50		198.50	256
4180	2 circuit		4.62	1.730		106	106		212	274
4200	Center, 1 circuit		5.04	1.590		140	97.50		237.50	299
4220	2 circuit		4.62	1.730		145	106		251	315
4240	End cap		33.60	.238		17.05	14.60		31.65	41
4260	Hanger, surface mount		20.16	.397		10.20	24.50		34.70	47.50
4280	Coupling		33.60	.238	Ea.	13	14.60		27.60	36.50

26 25 13.30 Copper Bus Duct

		Crew	Daily Output	Labor-Hours	Unit	Material	2020 Bare Costs Labor	Equipment	Total	Total Incl O&P
0010	**COPPER BUS DUCT**									
0100	Weatherproof 3 pole 4 wire, feeder duct, 600 amp	2 Elec	20.16	.794	L.F.	315	48.50		363.50	420
0110	800 amp		15.12	1.060		410	65		475	545
0120	1,000 amp		14.28	1.120		520	68.50		588.50	670
0125	1,200 amp		13.86	1.150		555	70.50		625.50	720
0130	1,350 amp		13.44	1.190		580	73		653	745
0140	1,600 amp		10.08	1.590		640	97.50		737.50	850
0150	2,000 amp		8.40	1.900		740	117		857	990
0160	2,500 amp		5.88	2.720		855	167		1,022	1,200
0170	3,000 amp		4.20	3.810		1,125	234		1,359	1,575
0180	4,000 amp		3.02	5.290		1,750	325		2,075	2,400
0200	Indoor 3 pole 4 wire, plug-in, bus duct high short circuit, 400 amp		26.88	.595		450	36.50		486.50	550
0210	600 amp		21.84	.733		515	45		560	635
0220	800 amp		16.80	.952		785	58.50		843.50	950
0230	1,000 amp		15.12	1.060		905	65		970	1,100
0240	1,350 amp		13.44	1.190		525	73		598	685

Post-Installation Change Order Costs   For customer support on your Electrical Change Order Costs with RSMeans data, call 800.448.8182.   417

26 25 13.30 Copper Bus Duct		Crew	Daily Output	Labor-Hours	Unit	Material	2020 Bare Costs Labor	Equipment	Total	Total Incl O&P
0250	1,600 amp	2 Elec	10.08	1.590	L.F.	590	97.50		687.50	795
0260	2,000 amp		8.40	1.900		750	117		867	1,000
0270	2,500 amp		6.72	2.380		925	146		1,071	1,250
0280	3,000 amp		5.04	3.170		945	194		1,139	1,350
0310	Cross, 225 amp		2.52	6.350	Ea.	3,350	390		3,740	4,275
0320	400 amp		2.35	6.800		3,350	415		3,765	4,325
0330	600 amp		2.18	7.330		3,350	450		3,800	4,375
0340	800 amp		1.85	8.660		3,650	530		4,180	4,800
0350	1,000 amp		1.68	9.520		4,075	585		4,660	5,350
0360	1,350 amp		1.51	10.580		4,525	650		5,175	5,950
0370	1,600 amp		1.43	11.200		5,025	685		5,710	6,550
0380	2,000 amp		1.34	11.900		8,250	730		8,980	10,200
0390	2,500 amp		1.18	13.610		10,000	835		10,835	12,300
0400	3,000 amp		1.01	15.870		9,750	975		10,725	12,200
0410	4,000 amp		.84	19.050		12,700	1,175		13,875	15,800
0430	Expansion fitting, 225 amp		4.54	3.530		2,150	217		2,367	2,700
0440	400 amp		3.86	4.140		2,425	254		2,679	3,050
0450	600 amp		3.36	4.760		3,025	292		3,317	3,750
0460	800 amp		2.86	5.600		3,550	345		3,895	4,400
0470	1,000 amp		2.52	6.350		4,075	390		4,465	5,075
0480	1,350 amp		2.35	6.800		4,125	415		4,540	5,150
0490	1,600 amp		2.18	7.330		5,750	450		6,200	7,000
0500	2,000 amp		1.85	8.660		6,600	530		7,130	8,075
0510	2,500 amp		1.51	10.580		8,050	650		8,700	9,825
0520	3,000 amp		1.34	11.900		8,050	730		8,780	9,925
0530	4,000 amp		1.01	15.870		10,400	975		11,375	12,900
0550	Reducer nonfused, 225 amp		4.54	3.530		1,925	217		2,142	2,425
0560	400 amp		3.86	4.140		1,925	254		2,179	2,475
0570	600 amp		3.36	4.760		1,925	292		2,217	2,525
0580	800 amp		2.86	5.600		2,325	345		2,670	3,050
0590	1,000 amp		2.52	6.350		2,800	390		3,190	3,650
0600	1,350 amp		2.35	6.800		4,325	415		4,740	5,375
0610	1,600 amp		2.18	7.330		4,975	450		5,425	6,150
0620	2,000 amp		1.85	8.660		5,975	530		6,505	7,375
0630	2,500 amp		1.51	10.580		7,550	650		8,200	9,275
0640	3,000 amp		1.34	11.900		8,225	730		8,955	10,100
0650	4,000 amp		1.01	15.870		10,700	975		11,675	13,300
0670	Reducer fuse included, 225 amp		3.70	4.330		4,100	266		4,366	4,925
0680	400 amp		3.53	4.540		5,150	279		5,429	6,075
0690	600 amp		3.02	5.290		6,350	325		6,675	7,450
0700	800 amp		2.69	5.950		8,950	365		9,315	10,400
0710	1,000 amp		2.52	6.350		11,200	390		11,590	12,900
0720	1,350 amp		2.35	6.800		17,800	415		18,215	20,200
0730	1,600 amp		1.85	8.660		25,400	530		25,930	28,800
0740	2,000 amp		1.51	10.580		28,600	650		29,250	32,500
0790	Reducer, circuit breaker, 225 amp		3.70	4.330		5,200	266		5,466	6,125
0800	400 amp		3.53	4.540		6,075	279		6,354	7,100
0810	600 amp		3.02	5.290		8,625	325		8,950	9,950
0820	800 amp		2.69	5.950		10,100	365		10,465	11,600
0830	1,000 amp		2.52	6.350		11,500	390		11,890	13,200
0840	1,350 amp		2.35	6.800		14,100	415		14,515	16,100
0850	1,600 amp		1.85	8.660		20,800	530		21,330	23,700
0860	2,000 amp		1.51	10.580		22,800	650		23,450	26,100

For customer support on your Electrical Change Order Costs with RSMeans data, call 800.448.8182.

Post-Installation Change Order Costs

26 25 Low-Voltage Enclosed Bus Assemblies

26 25 13 – Low-Voltage Busways

26 25 13.30 Copper Bus Duct		Crew	Daily Output	Labor-Hours	Unit	Material	2020 Bare Costs Labor	Equipment	Total	Total Incl O&P
0910	Cable tap box, center, 225 amp	2 Elec	2.69	5.950	Ea.	2,075	365		2,440	2,825
0920	400 amp		2.18	7.330		2,075	450		2,525	2,950
0930	600 amp		1.85	8.660		2,075	530		2,605	3,075
0940	800 amp		1.68	9.520		2,300	585		2,885	3,400
0950	1,000 amp		1.34	11.900		2,475	730		3,205	3,800
0960	1,350 amp		1.18	13.610		3,025	835		3,860	4,575
0970	1,600 amp		1.01	15.870		3,375	975		4,350	5,150
0980	2,000 amp		.84	19.050		4,075	1,175		5,250	6,225
1040	2,500 amp		.67	23.810		4,800	1,450		6,250	7,475
1060	3,000 amp		.50	31.750		4,700	1,950		6,650	8,075
1080	4,000 amp		.34	47.620		5,925	2,925		8,850	10,900
1800	Weatherproof 3 pole 3 wire, feeder duct, 600 amp		23.52	.680	L.F.	298	41.50		339.50	390
1820	800 amp		18.48	.866		360	53		413	480
1840	1,000 amp		16.80	.952		405	58.50		463.50	530
1850	1,200 amp		15.96	1		420	61.50		481.50	550
1860	1,350 amp		15.12	1.060		585	65		650	740
1880	1,600 amp		11.76	1.360		670	83.50		753.50	860
1900	2,000 amp		10.08	1.590		855	97.50		952.50	1,100
1920	2,500 amp		6.72	2.380		1,075	146		1,221	1,400
1940	3,000 amp		5.04	3.170		1,100	194		1,294	1,500
1960	4,000 amp		3.36	4.760		1,450	292		1,742	2,025
2000	Indoor 3 pole 3 wire, feeder duct, 600 amp		26.88	.595		248	36.50		284.50	330
2010	800 amp		21.84	.733		620	45		665	750
2020	1,000 amp		20.16	.794		335	48.50		383.50	445
2025	1,200 amp		18.48	.866		445	53		498	570
2030	1,350 amp		16.80	.952		490	58.50		548.50	620
2040	1,600 amp		13.44	1.190		1,075	73		1,148	1,275
2050	2,000 amp		11.76	1.360		1,375	83.50		1,458.50	1,650
2060	2,500 amp		8.40	1.900		890	117		1,007	1,150
2070	3,000 amp		6.72	2.380		885	146		1,031	1,200
2080	4,000 amp		5.04	3.170		1,200	194		1,394	1,625
2090	5,000 amp		4.20	3.810		1,475	234		1,709	1,975
2200	Indoor 3 pole 3 wire, bus duct plug-in, 225 amp		38.64	.414		172	25.50		197.50	227
2210	400 amp		30.24	.529		263	32.50		295.50	340
2220	600 amp		25.20	.635		330	39		369	425
2230	800 amp		20.16	.794		500	48.50		548.50	625
2240	1,000 amp		16.80	.952		540	58.50		598.50	680
2250	1,350 amp		15.12	1.060		850	65		915	1,025
2260	1,600 amp		11.76	1.360		890	83.50		973.50	1,100
2270	2,000 amp		10.08	1.590		750	97.50		847.50	970
2280	2,500 amp		8.40	1.900		925	117		1,042	1,200
2290	3,000 amp		6.72	2.380		945	146		1,091	1,275
2330	High short circuit, 400 amp		30.24	.529		325	32.50		357.50	405
2340	600 amp		25.20	.635		405	39		444	505
2350	800 amp		20.16	.794		585	48.50		633.50	720
2360	1,000 amp		16.80	.952		635	58.50		693.50	785
2370	1,350 amp		15.12	1.060		950	65		1,015	1,150
2380	1,600 amp		11.76	1.360		1,025	83.50		1,108.50	1,250
2390	2,000 amp		10.08	1.590		750	97.50		847.50	970
2400	2,500 amp		8.40	1.900		925	117		1,042	1,200
2410	3,000 amp		6.72	2.380		945	146		1,091	1,275
2440	Elbows, 225 amp		3.86	4.140	Ea.	1,600	254		1,854	2,150
2450	400 amp		3.53	4.540		1,450	279		1,729	2,025

Post-Installation Change Order Costs

Post-Installation Change Order Costs   For customer support on your Electrical Change Order Costs with RSMeans data, call 800.448.8182.

26 25 Low-Voltage Enclosed Bus Assemblies

26 25 13 – Low-Voltage Busways

26 25 13.30 Copper Bus Duct		Crew	Daily Output	Labor-Hours	Unit	Material	2020 Bare Costs Labor	Equipment	Total	Total Incl O&P
2460	600 amp	2 Elec	3.02	5.290	Ea.	1,450	325		1,775	2,075
2470	800 amp		2.69	5.950		1,575	365		1,940	2,275
2480	1,000 amp		2.52	6.350		1,650	390		2,040	2,375
2485	1,200 amp		2.44	6.570		1,725	405		2,130	2,500
2490	1,350 amp		2.35	6.800		1,850	415		2,265	2,675
2500	1,600 amp		2.18	7.330		2,000	450		2,450	2,875
2510	2,000 amp		1.68	9.520		2,450	585		3,035	3,575
2520	2,500 amp		1.51	10.580		3,725	650		4,375	5,075
2530	3,000 amp		1.34	11.900		3,650	730		4,380	5,075
2540	4,000 amp		1.18	13.610		4,650	835		5,485	6,375
2560	Tee fittings, 225 amp		2.35	6.800		1,675	415		2,090	2,475
2570	400 amp		2.02	7.940		1,675	485		2,160	2,575
2580	600 amp		1.68	9.520		1,675	585		2,260	2,725
2590	800 amp		1.51	10.580		1,825	650		2,475	2,975
2600	1,000 amp		1.34	11.900		2,025	730		2,755	3,325
2605	1,200 amp		1.26	12.700		2,175	780		2,955	3,550
2610	1,350 amp		1.18	13.610		2,350	835		3,185	3,825
2620	1,600 amp		1.01	15.870		2,725	975		3,700	4,450
2630	2,000 amp		.84	19.050		4,550	1,175		5,725	6,750
2640	2,500 amp		.59	27.210		5,325	1,675		7,000	8,325
2650	3,000 amp		.50	31.750		5,225	1,950		7,175	8,650
2660	4,000 amp		.42	38.100		6,750	2,325		9,075	10,900
2680	Cross, 225 amp		3.02	5.290		2,550	325		2,875	3,275
2690	400 amp		2.69	5.950		2,550	365		2,915	3,350
2700	600 amp		2.52	6.350		2,550	390		2,940	3,375
2710	800 amp		2.18	7.330		3,050	450		3,500	4,025
2720	1,000 amp		2.02	7.940		3,200	485		3,685	4,225
2730	1,350 amp		1.85	8.660		3,775	530		4,305	4,950
2740	1,600 amp		1.68	9.520		4,050	585		4,635	5,325
2750	2,000 amp		1.51	10.580		6,400	650		7,050	8,025
2760	2,500 amp		1.34	11.900		7,450	730		8,180	9,275
2770	3,000 amp		1.18	13.610		7,275	835		8,110	9,250
2780	4,000 amp		.84	19.050		9,300	1,175		10,475	12,000
2800	Expansion fitting, 225 amp		5.38	2.980		2,275	183		2,458	2,775
2810	400 amp		4.54	3.530		5,025	217		5,242	5,850
2820	600 amp		3.86	4.140		2,275	254		2,529	2,875
2830	800 amp		3.36	4.760		2,700	292		2,992	3,400
2840	1,000 amp		3.02	5.290		2,975	325		3,300	3,725
2850	1,350 amp		2.69	5.950		3,100	365		3,465	3,975
2860	1,600 amp		2.52	6.350		3,400	390		3,790	4,325
2870	2,000 amp		2.18	7.330		4,350	450		4,800	5,450
2880	2,500 amp		1.85	8.660		6,075	530		6,605	7,475
2890	3,000 amp		1.51	10.580		6,050	650		6,700	7,625
2900	4,000 amp		1.18	13.610		7,725	835		8,560	9,750
2920	Reducer nonfused, 225 amp		5.38	2.980		1,575	183		1,758	2,000
2930	400 amp		4.54	3.530		1,575	217		1,792	2,050
2940	600 amp		3.86	4.140		1,575	254		1,829	2,100
2950	800 amp		3.36	4.760		1,850	292		2,142	2,450
2960	1,000 amp		3.02	5.290		2,075	325		2,400	2,750
2970	1,350 amp		2.69	5.950		2,800	365		3,165	3,625
2980	1,600 amp		2.52	6.350		3,175	390		3,565	4,050
2990	2,000 amp		2.18	7.330		3,800	450		4,250	4,850
3000	2,500 amp		1.85	8.660		5,575	530		6,105	6,925

For customer support on your Electrical Change Order Costs with RSMeans data, call 800.448.8182.

Post-Installation Change Order Costs

26 25 13.30 Copper Bus Duct		Crew	Daily Output	Labor-Hours	Unit	Material	2020 Bare Costs Labor	2020 Bare Costs Equipment	Total	Total Incl O&P
3010	3,000 amp	2 Elec	1.51	10.580	Ea.	6,025	650		6,675	7,600
3020	4,000 amp		1.18	13.610		7,825	835		8,660	9,850
3040	Reducer fuse included, 225 amp		4.20	3.810		3,725	234		3,959	4,450
3050	400 amp		4.03	3.970		4,975	244		5,219	5,825
3060	600 amp		3.53	4.540		5,925	279		6,204	6,925
3070	800 amp		3.02	5.290		8,425	325		8,750	9,750
3080	1,000 amp		2.86	5.600		9,975	345		10,320	11,500
3090	1,350 amp		2.69	5.950		10,400	365		10,765	11,900
3100	1,600 amp		2.18	7.330		24,200	450		24,650	27,300
3110	2,000 amp		1.68	9.520		27,400	585		27,985	31,100
3160	Reducer circuit breaker, 225 amp		4.20	3.810		4,775	234		5,009	5,600
3170	400 amp		4.03	3.970		5,850	244		6,094	6,800
3180	600 amp		3.53	4.540		8,375	279		8,654	9,650
3190	800 amp		3.02	5.290		9,800	325		10,125	11,300
3200	1,000 amp		2.86	5.600		11,100	345		11,445	12,800
3210	1,350 amp		2.69	5.950		13,800	365		14,165	15,600
3220	1,600 amp		2.18	7.330		20,500	450		20,950	23,300
3230	2,000 amp		1.68	9.520		22,400	585		22,985	25,600
3280	3 pole, 3 wire, cable tap box center, 225 amp		3.02	5.290		2,350	325		2,675	3,075
3290	400 amp		2.52	6.350		2,350	390		2,740	3,175
3300	600 amp		2.18	7.330		2,350	450		2,800	3,275
3310	800 amp		2.02	7.940		2,625	485		3,110	3,625
3320	1,000 amp		1.51	10.580		2,850	650		3,500	4,125
3330	1,350 amp		1.34	11.900		3,475	730		4,205	4,900
3340	1,600 amp		1.18	13.610		3,875	835		4,710	5,525
3350	2,000 amp		1.01	15.870		4,800	975		5,775	6,750
3360	2,500 amp		.84	19.050		5,725	1,175		6,900	8,050
3370	3,000 amp		.59	27.210		5,650	1,675		7,325	8,675
3380	4,000 amp		.42	38.100		7,150	2,325		9,475	11,400
3400	Cable tap box end, 225 amp		3.02	5.290		1,275	325		1,600	1,875
3410	400 amp		2.52	6.350		1,200	390		1,590	1,900
3420	600 amp		2.18	7.330		1,425	450		1,875	2,250
3430	800 amp		2.02	7.940		1,425	485		1,910	2,300
3440	1,000 amp		1.51	10.580		1,850	650		2,500	3,000
3445	1,200 amp		1.43	11.200		1,750	685		2,435	2,950
3450	1,350 amp		1.34	11.900		1,875	730		2,605	3,125
3460	1,600 amp		1.18	13.610		2,150	835		2,985	3,600
3470	2,000 amp		1.01	15.870		2,500	975		3,475	4,200
3480	2,500 amp		.84	19.050		2,975	1,175		4,150	5,025
3490	3,000 amp		.59	27.210		2,925	1,675		4,600	5,675
3500	4,000 amp		.42	38.100		3,700	2,325		6,025	7,550
4600	Plug-in, fusible switch w/3 fuses, 3 pole, 250 volt, 30 amp	1 Elec	3.36	2.380		465	146		611	730
4610	60 amp		3.02	2.650		610	163		773	910
4620	100 amp		2.27	3.530		875	217		1,092	1,275
4630	200 amp	2 Elec	2.69	5.950		1,475	365		1,840	2,150
4640	400 amp		1.18	13.610		3,825	835		4,660	5,450
4650	600 amp		.76	21.160		5,300	1,300		6,600	7,750
4700	4 pole, 120/208 volt, 30 amp	1 Elec	3.28	2.440		640	150		790	925
4710	60 amp		2.94	2.720		690	167		857	1,000
4720	100 amp		2.18	3.660		960	225		1,185	1,375
4730	200 amp	2 Elec	2.52	6.350		1,600	390		1,990	2,350
4740	400 amp		1.09	14.650		3,775	900		4,675	5,500
4750	600 amp		.67	23.810		5,300	1,450		6,750	8,000

26 25 Low-Voltage Enclosed Bus Assemblies

26 25 13 – Low-Voltage Busways

26 25 13.30 Copper Bus Duct		Crew	Daily Output	Labor-Hours	Unit	Material	2020 Bare Costs Labor	Equipment	Total	Total Incl O&P
4800	3 pole, 480 volt, 30 amp	1 Elec	3.36	2.380	Ea.	480	146		626	740
4810	60 amp		3.02	2.650		505	163		668	795
4820	100 amp		2.27	3.530		855	217		1,072	1,250
4830	200 amp	2 Elec	2.69	5.950		1,475	365		1,840	2,175
4840	400 amp		1.18	13.610		3,425	835		4,260	5,000
4850	600 amp		.76	21.160		4,850	1,300		6,150	7,275
4860	800 amp		.55	28.860		20,100	1,775		21,875	24,700
4870	1,000 amp		.50	31.750		20,600	1,950		22,550	25,500
4880	1,200 amp		.42	38.100		22,400	2,325		24,725	28,200
4890	1,600 amp		.37	43.290		23,600	2,650		26,250	30,000
4900	4 pole, 277/480 volt, 30 amp	1 Elec	3.28	2.440		690	150		840	985
4910	60 amp		2.94	2.720		735	167		902	1,050
4920	100 amp		2.18	3.660		1,075	225		1,300	1,500
4930	200 amp	2 Elec	2.52	6.350		2,150	390		2,540	2,925
4940	400 amp		1.09	14.650		4,050	900		4,950	5,800
4950	600 amp		.67	23.810		5,525	1,450		6,975	8,250
5050	800 amp		.50	31.750		18,200	1,950		20,150	22,900
5060	1,000 amp		.47	34.010		20,900	2,075		22,975	26,100
5070	1,200 amp		.40	39.680		21,100	2,425		23,525	26,800
5080	1,600 amp		.35	45.350		25,100	2,775		27,875	31,800
5150	Fusible with starter, 3 pole 250 volt, 30 amp	1 Elec	2.94	2.720		3,600	167		3,767	4,200
5160	60 amp		2.69	2.980		3,800	183		3,983	4,450
5170	100 amp		2.10	3.810		4,300	234		4,534	5,075
5180	200 amp	2 Elec	2.35	6.800		7,100	415		7,515	8,425
5200	3 pole 480 volt, 30 amp	1 Elec	2.94	2.720		3,600	167		3,767	4,200
5210	60 amp		2.69	2.980		3,800	183		3,983	4,450
5220	100 amp		2.10	3.810		4,300	234		4,534	5,075
5230	200 amp	2 Elec	2.35	6.800		7,100	415		7,515	8,425
5300	Fusible with contactor, 3 pole 250 volt, 30 amp	1 Elec	2.94	2.720		3,500	167		3,667	4,100
5310	60 amp		2.69	2.980		4,450	183		4,633	5,175
5320	100 amp		2.10	3.810		6,225	234		6,459	7,200
5330	200 amp	2 Elec	2.35	6.800		7,125	415		7,540	8,475
5400	3 pole 480 volt, 30 amp	1 Elec	2.94	2.720		3,775	167		3,942	4,400
5410	60 amp		2.69	2.980		5,325	183		5,508	6,125
5420	100 amp		2.10	3.810		7,300	234		7,534	8,375
5430	200 amp	2 Elec	2.35	6.800		7,475	415		7,890	8,850
5450	Fusible with capacitor, 3 pole 250 volt, 30 amp	1 Elec	2.52	3.170		9,100	194		9,294	10,300
5460	60 amp		1.68	4.760		10,600	292		10,892	12,000
5500	3 pole 480 volt, 30 amp		2.52	3.170		7,625	194		7,819	8,675
5510	60 amp		1.68	4.760		9,525	292		9,817	10,900
5600	Circuit breaker, 3 pole, 250 volt, 60 amp		3.78	2.120		655	130		785	915
5610	100 amp		2.69	2.980		805	183		988	1,150
5650	4 pole, 120/208 volt, 60 amp		3.70	2.160		740	133		873	1,000
5660	100 amp		2.60	3.070		880	188		1,068	1,250
5700	3 pole, 4 wire 277/480 volt, 60 amp		3.61	2.210		1,000	136		1,136	1,300
5710	100 amp		2.52	3.170		1,100	194		1,294	1,525
5720	225 amp	2 Elec	2.69	5.950		2,450	365		2,815	3,250
5730	400 amp		1.01	15.870		5,100	975		6,075	7,075
5740	600 amp		.81	19.840		6,875	1,225		8,100	9,350
5750	700 amp		.50	31.750		8,700	1,950		10,650	12,500
5760	800 amp		.50	31.750		8,700	1,950		10,650	12,500
5770	900 amp		.45	35.270		11,500	2,175		13,675	15,800
5780	1,000 amp		.45	35.270		11,500	2,175		13,675	15,800

For customer support on your Electrical Change Order Costs with RSMeans data, call 800.448.8182.

Post-Installation Change Order Costs

26 25 13.30 Copper Bus Duct

		Crew	Daily Output	Labor-Hours	Unit	Material	2020 Bare Costs Labor	Equipment	Total	Total Incl O&P
5790	1,200 amp	2 Elec	.35	45.350	Ea.	13,800	2,775		16,575	19,400
5810	Circuit breaker w/HIC fuses, 3 pole 480 volt, 60 amp	1 Elec	3.70	2.160		1,250	133		1,383	1,575
5820	100 amp	"	2.60	3.070		1,375	188		1,563	1,800
5830	225 amp	2 Elec	2.86	5.600		4,400	345		4,745	5,325
5840	400 amp		1.18	13.610		6,975	835		7,810	8,925
5850	600 amp		.84	19.050		7,075	1,175		8,250	9,525
5860	700 amp		.54	29.760		9,375	1,825		11,200	13,000
5870	800 amp		.54	29.760		9,375	1,825		11,200	13,000
5880	900 amp		.47	34.010		20,300	2,075		22,375	25,400
5890	1,000 amp		.47	34.010		20,300	2,075		22,375	25,400
5950	3 pole 4 wire 277/480 volt, 60 amp	1 Elec	3.61	2.210		1,250	136		1,386	1,575
5960	100 amp	"	2.52	3.170		1,375	194		1,569	1,825
5970	225 amp	2 Elec	2.52	6.350		4,400	390		4,790	5,400
5980	400 amp		.92	17.320		6,975	1,075		8,050	9,250
5990	600 amp		.79	20.260		7,075	1,250		8,325	9,625
6000	700 amp		.49	32.840		9,375	2,025		11,400	13,300
6010	800 amp		.49	32.840		9,375	2,025		11,400	13,300
6020	900 amp		.44	36.630		20,300	2,250		22,550	25,700
6030	1,000 amp		.44	36.630		20,300	2,250		22,550	25,700
6040	1,200 amp		.34	47.620		20,300	2,925		23,225	26,700
6100	Circuit breaker with starter, 3 pole 250 volt, 60 amp	1 Elec	2.69	2.980		1,875	183		2,058	2,350
6110	100 amp	"	2.10	3.810		2,900	234		3,134	3,550
6120	225 amp	2 Elec	2.52	6.350		4,125	390		4,515	5,100
6130	3 pole 480 volt, 60 amp	1 Elec	2.69	2.980		2,100	183		2,283	2,575
6140	100 amp	"	2.10	3.810		2,900	234		3,134	3,550
6150	225 amp	2 Elec	2.52	6.350		3,500	390		3,890	4,425
6200	Circuit breaker with contactor, 3 pole 250 volt, 60 amp	1 Elec	2.69	2.980		1,975	183		2,158	2,450
6210	100 amp	"	2.10	3.810		2,700	234		2,934	3,325
6220	225 amp	2 Elec	2.52	6.350		3,825	390		4,215	4,775
6250	3 pole 480 volt, 60 amp	1 Elec	2.69	2.980		1,975	183		2,158	2,450
6260	100 amp	"	2.10	3.810		2,700	234		2,934	3,325
6270	225 amp	2 Elec	2.52	6.350		3,600	390		3,990	4,525
6300	Circuit breaker with capacitor, 3 pole 250 volt, 60 amp	1 Elec	1.68	4.760		10,700	292		10,992	12,200
6310	3 pole 480 volt, 60 amp		1.68	4.760		11,700	292		11,992	13,300
6400	Add control transformer with pilot light to above starter		13.44	.595		590	36.50		626.50	705
6410	Switch, fusible, mechanically held contactor optional		13.44	.595		1,575	36.50		1,611.50	1,775
6430	Circuit breaker, mechanically held contactor optional		13.44	.595		1,575	36.50		1,611.50	1,775
6450	Ground neutralizer, 3 pole		13.44	.595		72	36.50		108.50	134

26 25 13.40 Copper Bus Duct

		Crew	Daily Output	Labor-Hours	Unit	Material	2020 Bare Costs Labor	Equipment	Total	Total Incl O&P
0010	**COPPER BUS DUCT** 10' long									
0050	Indoor 3 pole 4 wire, plug-in, straight section, 225 amp	2 Elec	33.60	.476	L.F.	222	29		251	289
1000	400 amp		26.88	.595		385	36.50		421.50	480
1500	600 amp		21.84	.733		455	45		500	565
2400	800 amp		16.80	.952		620	58.50		678.50	765
2450	1,000 amp		15.12	1.060		305	65		370	430
2470	1,200 amp		14.28	1.120		395	68.50		463.50	535
2500	1,350 amp		13.44	1.190		415	73		488	570
2510	1,600 amp		10.08	1.590		470	97.50		567.50	665
2520	2,000 amp		8.40	1.900		595	117		712	830
2530	2,500 amp		6.72	2.380		735	146		881	1,025
2540	3,000 amp		5.04	3.170		855	194		1,049	1,225
2550	Feeder, 600 amp		23.52	.680		204	41.50		245.50	286

26 25 13.40 Copper Bus Duct		Crew	Daily Output	Labor-Hours	Unit	Material	2020 Bare Costs Labor	Equipment	Total	Total Incl O&P
2600	800 amp	2 Elec	18.48	.866	L.F.	755	53		808	910
2700	1,000 amp		16.80	.952		905	58.50		963.50	1,075
2750	1,200 amp		15.96	1		1,100	61.50		1,161.50	1,300
2800	1,350 amp		15.12	1.060		390	65		455	520
2900	1,600 amp		11.76	1.360		1,450	83.50		1,533.50	1,725
3000	2,000 amp		10.08	1.590		1,800	97.50		1,897.50	2,125
3010	2,500 amp		6.72	2.380		705	146		851	995
3020	3,000 amp		5.04	3.170		830	194		1,024	1,200
3030	4,000 amp		3.36	4.760		1,100	292		1,392	1,625
3040	5,000 amp		1.68	9.520		1,325	585		1,910	2,350
3100	Elbows, 225 amp		3.36	4.760	Ea.	1,575	292		1,867	2,175
3200	400 amp		3.02	5.290		1,575	325		1,900	2,225
3300	600 amp		2.69	5.950		1,575	365		1,940	2,300
3400	800 amp		2.35	6.800		1,725	415		2,140	2,525
3500	1,000 amp		2.18	7.330		1,925	450		2,375	2,775
3550	1,200 amp		2.10	7.620		1,875	465		2,340	2,775
3600	1,350 amp		2.02	7.940		1,900	485		2,385	2,800
3700	1,600 amp		1.85	8.660		2,075	530		2,605	3,075
3800	2,000 amp		1.51	10.580		2,550	650		3,200	3,775
3810	2,500 amp		1.34	11.900		4,025	730		4,755	5,500
3820	3,000 amp		1.18	13.610		4,500	835		5,335	6,200
3830	4,000 amp		1.01	15.870		5,825	975		6,800	7,850
3840	5,000 amp		.84	19.050		9,400	1,175		10,575	12,100
4000	End box, 225 amp		28.56	.560		171	34.50		205.50	239
4100	400 amp		26.88	.595		189	36.50		225.50	263
4200	600 amp		23.52	.680		189	41.50		230.50	270
4300	800 amp		21.84	.733		189	45		234	275
4400	1,000 amp		20.16	.794		181	48.50		229.50	272
4410	1,200 amp		19.32	.828		189	51		240	284
4500	1,350 amp		18.48	.866		180	53		233	277
4600	1,600 amp		16.80	.952		180	58.50		238.50	285
4700	2,000 amp		15.12	1.060		221	65		286	340
4710	2,500 amp		13.44	1.190		221	73		294	350
4720	3,000 amp		11.76	1.360		213	83.50		296.50	360
4730	4,000 amp		10.08	1.590		258	97.50		355.50	430
4740	5,000 amp		8.40	1.900		258	117		375	460
4800	Cable tap box end, 225 amp		2.69	5.950		1,225	365		1,590	1,900
5000	400 amp		2.18	7.330		1,225	450		1,675	2,000
5100	600 amp		1.85	8.660		1,625	530		2,155	2,575
5200	800 amp		1.68	9.520		1,775	585		2,360	2,825
5300	1,000 amp		1.34	11.900		1,725	730		2,455	2,975
5350	1,200 amp		1.26	12.700		2,025	780		2,805	3,375
5400	1,350 amp		1.18	13.610		2,200	835		3,035	3,675
5500	1,600 amp		1.01	15.870		2,475	975		3,450	4,175
5600	2,000 amp		.84	19.050		2,775	1,175		3,950	4,800
5610	2,500 amp		.67	23.810		3,175	1,450		4,625	5,675
5620	3,000 amp		.50	31.750		3,700	1,950		5,650	6,975
5630	4,000 amp		.34	47.620		4,275	2,925		7,200	9,050
5640	5,000 amp		.17	95.240		5,200	5,850		11,050	14,400
5700	Switchboard stub, 225 amp		4.54	3.530		1,225	217		1,442	1,675
5800	400 amp		3.86	4.140		1,300	254		1,554	1,800
5900	600 amp		3.36	4.760		1,375	292		1,667	1,950
6000	800 amp		2.69	5.950		1,675	365		2,040	2,375

For customer support on your Electrical Change Order Costs with RSMeans data, call 800.448.8182.

Post-Installation Change Order Costs

26 25 13.40 Copper Bus Duct		Crew	Daily Output	Labor-Hours	Unit	Material	2020 Bare Costs Labor	Equipment	Total	Total Incl O&P
6100	1,000 amp	2 Elec	2.52	6.350	Ea.	1,950	390		2,340	2,700
6150	1,200 amp		2.35	6.800		2,275	415		2,690	3,125
6200	1,350 amp		2.18	7.330		2,400	450		2,850	3,325
6300	1,600 amp		2.02	7.940		2,700	485		3,185	3,700
6400	2,000 amp		1.68	9.520		3,275	585		3,860	4,475
6410	2,500 amp		1.51	10.580		4,000	650		4,650	5,375
6420	3,000 amp		1.34	11.900		4,500	730		5,230	6,025
6430	4,000 amp		1.18	13.610		5,850	835		6,685	7,700
6440	5,000 amp		1.01	15.870		7,200	975		8,175	9,350
6490	Tee fittings, 225 amp		2.02	7.940		2,275	485		2,760	3,225
6500	400 amp		1.68	9.520		2,275	585		2,860	3,375
6600	600 amp		1.51	10.580		2,275	650		2,925	3,475
6700	800 amp		1.34	11.900		2,350	730		3,080	3,650
6750	1,000 amp		1.18	13.610		2,575	835		3,410	4,100
6770	1,200 amp		1.09	14.650		2,900	900		3,800	4,550
6800	1,350 amp		1.01	15.870		3,075	975		4,050	4,825
7000	1,600 amp		.84	19.050		3,475	1,175		4,650	5,575
7100	2,000 amp		.67	23.810		4,125	1,450		5,575	6,725
7110	2,500 amp		.50	31.750		5,075	1,950		7,025	8,475
7120	3,000 amp		.42	38.100		5,700	2,325		8,025	9,750
7130	4,000 amp		.34	47.620		7,300	2,925		10,225	12,400
7140	5,000 amp	▼	.17	95.240		8,625	5,850		14,475	18,200
7200	Plug-in fusible switches w/3 fuses, 600 volt, 3 pole, 30 amp	1 Elec	3.36	2.380		885	146		1,031	1,175
7300	60 amp		3.02	2.650		990	163		1,153	1,350
7400	100 amp	▼	2.27	3.530		1,425	217		1,642	1,875
7500	200 amp	2 Elec	2.69	5.950		2,500	365		2,865	3,300
7600	400 amp		1.18	13.610		7,075	835		7,910	9,025
7700	600 amp		.76	21.160		8,400	1,300		9,700	11,200
7800	800 amp		.55	28.860		12,800	1,775		14,575	16,700
7900	1,200 amp		.42	38.100		24,100	2,325		26,425	30,000
7910	1,600 amp	▼	.37	43.290		22,600	2,650		25,250	28,900
8000	Plug-in circuit breakers, molded case, 15 to 50 amp	1 Elec	3.70	2.160		835	133		968	1,125
8100	70 to 100 amp	"	2.60	3.070		930	188		1,118	1,300
8200	150 to 225 amp	2 Elec	2.86	5.600		2,525	345		2,870	3,275
8300	250 to 400 amp		1.18	13.610		4,425	835		5,260	6,100
8400	500 to 600 amp		.84	19.050		5,950	1,175		7,125	8,300
8500	700 to 800 amp		.54	29.760		7,350	1,825		9,175	10,800
8600	900 to 1,000 amp		.47	34.010		10,500	2,075		12,575	14,700
8700	1,200 amp		.37	43.290		12,600	2,650		15,250	17,900
8720	1,400 amp		.34	47.620		17,700	2,925		20,625	23,800
8730	1,600 amp	▼	.34	47.620		19,400	2,925		22,325	25,700
8750	Circuit breakers, with current limiting fuse, 15 to 50 amp	1 Elec	3.70	2.160		1,675	133		1,808	2,050
8760	70 to 100 amp	"	2.60	3.070		1,975	188		2,163	2,450
8770	150 to 225 amp	2 Elec	2.86	5.600		4,250	345		4,595	5,175
8780	250 to 400 amp		1.18	13.610		6,575	835		7,410	8,500
8790	500 to 600 amp		.84	19.050		7,575	1,175		8,750	10,100
8800	700 to 800 amp		.54	29.760		12,500	1,825		14,325	16,400
8810	900 to 1,000 amp	▼	.47	34.010		14,200	2,075		16,275	18,700
8850	Combination starter FVNR, fusible switch, NEMA size 0, 30 amp	1 Elec	1.68	4.760		2,250	292		2,542	2,900
8860	NEMA size 1, 60 amp		1.51	5.290		2,375	325		2,700	3,100
8870	NEMA size 2, 100 amp	▼	1.09	7.330		3,000	450		3,450	3,975
8880	NEMA size 3, 200 amp	2 Elec	1.68	9.520		4,800	585		5,385	6,150
8900	Circuit breaker, NEMA size 0, 30 amp	1 Elec	1.68	4.760		2,325	292		2,617	2,975

26 25 Low-Voltage Enclosed Bus Assemblies

26 25 13 – Low-Voltage Busways

26 25 13.40 Copper Bus Duct		Crew	Daily Output	Labor-Hours	Unit	Material	2020 Bare Costs Labor	Equipment	Total	Total Incl O&P
8910	NEMA size 1, 60 amp	1 Elec	1.51	5.290	Ea.	2,400	325		2,725	3,125
8920	NEMA size 2, 100 amp	↓	1.09	7.330		3,450	450		3,900	4,475
8930	NEMA size 3, 200 amp	2 Elec	1.68	9.520		4,375	585		4,960	5,700
8950	Combination contactor, fusible switch, NEMA size 0, 30 amp	1 Elec	1.68	4.760		1,325	292		1,617	1,875
8960	NEMA size 1, 60 amp		1.51	5.290		1,350	325		1,675	1,975
8970	NEMA size 2, 100 amp	↓	1.09	7.330		2,025	450		2,475	2,900
8980	NEMA size 3, 200 amp	2 Elec	1.68	9.520		2,325	585		2,910	3,425
9000	Circuit breaker, NEMA size 0, 30 amp	1 Elec	1.68	4.760		1,500	292		1,792	2,075
9010	NEMA size 1, 60 amp		1.51	5.290		1,550	325		1,875	2,175
9020	NEMA size 2, 100 amp	↓	1.09	7.330		2,400	450		2,850	3,300
9030	NEMA size 3, 200 amp	2 Elec	1.68	9.520		2,900	585		3,485	4,075
9050	Control transformer for above, NEMA size 0, 30 amp	1 Elec	6.72	1.190		256	73		329	390
9060	NEMA size 1, 60 amp		6.72	1.190		256	73		329	390
9070	NEMA size 2, 100 amp	↓	5.88	1.360		355	83.50		438.50	515
9080	NEMA size 3, 200 amp	2 Elec	11.76	1.360		495	83.50		578.50	670
9100	Comb. fusible switch & lighting control, electrically held, 30 amp	1 Elec	1.68	4.760		1,050	292		1,342	1,575
9110	60 amp		1.51	5.290		1,500	325		1,825	2,125
9120	100 amp	↓	1.09	7.330		1,925	450		2,375	2,800
9130	200 amp	2 Elec	1.68	9.520		4,750	585		5,335	6,125
9150	Mechanically held, 30 amp	1 Elec	1.68	4.760		1,300	292		1,592	1,875
9160	60 amp		1.51	5.290		1,950	325		2,275	2,625
9170	100 amp	↓	1.09	7.330		2,500	450		2,950	3,425
9180	200 amp	2 Elec	1.68	9.520	↓	5,100	585		5,685	6,500
9200	Ground bus added to bus duct, 225 amp		268.80	.060	L.F.	44.50	3.65		48.15	54.50
9210	400 amp		201.60	.079		44.50	4.87		49.37	56.50
9220	600 amp		201.60	.079		44.50	4.87		49.37	56.50
9230	800 amp		134.40	.119		52	7.30		59.30	68.50
9240	1,000 amp		134.40	.119		59.50	7.30		66.80	76.50
9250	1,350 amp		117.60	.136		85	8.35		93.35	106
9260	1,600 amp		100.80	.159		92	9.75		101.75	116
9270	2,000 amp		92.40	.173		120	10.60		130.60	148
9280	2,500 amp		84	.190		149	11.70		160.70	180
9290	3,000 amp		75.60	.212		170	13		183	206
9300	4,000 amp		67.20	.238		225	14.60		239.60	269
9310	5,000 amp	↓	58.80	.272		270	16.70		286.70	320
9320	High short circuit bracing, add				↓	18.60			18.60	20.50

26 25 13.60 Copper or Aluminum Bus Duct Fittings

		Crew	Daily Output	Labor-Hours	Unit	Material	2020 Bare Costs Labor	Equipment	Total	Total Incl O&P
0010	**COPPER OR ALUMINUM BUS DUCT FITTINGS**									
0100	Flange, wall, with vapor barrier, 225 amp	2 Elec	5.21	3.070	Ea.	845	188		1,033	1,200
0110	400 amp		5.04	3.170		795	194		989	1,175
0120	600 amp		4.87	3.280		845	201		1,046	1,225
0130	800 amp		4.54	3.530		845	217		1,062	1,250
0140	1,000 amp		4.20	3.810		845	234		1,079	1,275
0145	1,200 amp		4.03	3.970		845	244		1,089	1,300
0150	1,350 amp		3.86	4.140		845	254		1,099	1,300
0160	1,600 amp		3.53	4.540		845	279		1,124	1,350
0170	2,000 amp		3.36	4.760		845	292		1,137	1,375
0180	2,500 amp		3.02	5.290		845	325		1,170	1,425
0190	3,000 amp		2.69	5.950		845	365		1,210	1,475
0200	4,000 amp		2.18	7.330		820	450		1,270	1,575
0300	Roof, 225 amp		5.21	3.070		1,050	188		1,238	1,425
0310	400 amp	↓	5.04	3.170	↓	1,050	194		1,244	1,450

26 25 13 – Low-Voltage Busways

26 25 13.60 Copper or Aluminum Bus Duct Fittings		Crew	Daily Output	Labor-Hours	Unit	Material	2020 Bare Costs Labor	Equipment	Total	Total Incl O&P
0320	600 amp	2 Elec	4.87	3.280	Ea.	1,050	201		1,251	1,450
0330	800 amp		4.54	3.530		1,050	217		1,267	1,475
0340	1,000 amp		4.20	3.810		1,050	234		1,284	1,500
0345	1,200 amp		4.03	3.970		1,050	244		1,294	1,525
0350	1,350 amp		3.86	4.140		1,050	254		1,304	1,525
0360	1,600 amp		3.53	4.540		1,050	279		1,329	1,575
0370	2,000 amp		3.36	4.760		1,050	292		1,342	1,575
0380	2,500 amp		3.02	5.290		1,050	325		1,375	1,625
0390	3,000 amp		2.69	5.950		1,050	365		1,415	1,700
0400	4,000 amp		2.18	7.330		1,050	450		1,500	1,825
0420	Support, floor mounted, 225 amp		16.80	.952		171	58.50		229.50	275
0430	400 amp		16.80	.952		171	58.50		229.50	275
0440	600 amp		15.12	1.060		171	65		236	285
0450	800 amp		13.44	1.190		171	73		244	297
0460	1,000 amp		10.92	1.470		171	90		261	320
0465	1,200 amp		9.91	1.610		171	99		270	335
0470	1,350 amp		8.90	1.800		171	110		281	350
0480	1,600 amp		7.73	2.070		171	127		298	375
0490	2,000 amp		6.72	2.380		171	146		317	405
0500	2,500 amp		5.38	2.980		171	183		354	460
0510	3,000 amp		4.54	3.530		171	217		388	510
0520	4,000 amp		3.36	4.760		171	292		463	625
0540	Weather stop, 225 amp		10.08	1.590		530	97.50		627.50	725
0550	400 amp		8.40	1.900		530	117		647	755
0560	600 amp		7.56	2.120		530	130		660	775
0570	800 amp		6.72	2.380		530	146		676	795
0580	1,000 amp		5.38	2.980		530	183		713	850
0585	1,200 amp		4.96	3.230		530	198		728	875
0590	1,350 amp		4.54	3.530		530	217		747	900
0600	1,600 amp		3.86	4.140		530	254		784	960
0610	2,000 amp		3.36	4.760		530	292		822	1,025
0620	2,500 amp		2.69	5.950		530	365		895	1,125
0630	3,000 amp		2.18	7.330		530	450		980	1,250
0640	4,000 amp		1.68	9.520		530	585		1,115	1,450
0660	End closure, 225 amp		28.56	.560		198	34.50		232.50	269
0670	400 amp		26.88	.595		198	36.50		234.50	273
0680	600 amp		23.52	.680		198	41.50		239.50	280
0690	800 amp		21.84	.733		172	45		217	256
0700	1,000 amp		20.16	.794		198	48.50		246.50	291
0705	1,200 amp		19.32	.828		170	51		221	263
0710	1,350 amp		18.48	.866		170	53		223	266
0720	1,600 amp		16.80	.952		198	58.50		256.50	305
0730	2,000 amp		15.12	1.060		277	65		342	400
0740	2,500 amp		13.44	1.190		253	73		326	390
0750	3,000 amp		11.76	1.360		253	83.50		336.50	405
0760	4,000 amp		10.08	1.590		255	97.50		352.50	425
0780	Switchboard stub, 3 pole 3 wire, 225 amp		5.04	3.170		1,025	194		1,219	1,425
0790	400 amp		4.37	3.660		1,025	225		1,250	1,450
0800	600 amp		3.86	4.140		1,025	254		1,279	1,500
0810	800 amp		3.02	5.290		1,250	325		1,575	1,850
0820	1,000 amp		2.86	5.600		1,450	345		1,795	2,100
0825	1,200 amp		2.69	5.950		1,450	365		1,815	2,150
0830	1,350 amp		2.52	6.350		1,725	390		2,115	2,475

26 25 Low-Voltage Enclosed Bus Assemblies

26 25 13 – Low-Voltage Busways

26 25 13.60 Copper or Aluminum Bus Duct Fittings	Crew	Daily Output	Labor-Hours	Unit	Material	2020 Bare Costs Labor	Equipment	Total	Total Incl O&P
0840	1,600 amp	2 Elec	2.35	6.800	Ea.	2,050	415		2,465	2,900
0850	2,000 amp		2.02	7.940		2,400	485		2,885	3,350
0860	2,500 amp		1.68	9.520		2,925	585		3,510	4,100
0870	3,000 amp		1.51	10.580		3,325	650		3,975	4,625
0880	4,000 amp		1.34	11.900		4,275	730		5,005	5,800
0890	5,000 amp		1.18	13.610		5,300	835		6,135	7,100
0900	3 pole 4 wire, 225 amp		4.54	3.530		1,275	217		1,492	1,725
0910	400 amp		3.86	4.140		1,275	254		1,529	1,775
0920	600 amp		3.36	4.760		1,275	292		1,567	1,825
0930	800 amp		2.69	5.950		1,525	365		1,890	2,250
0940	1,000 amp		2.52	6.350		1,800	390		2,190	2,550
0950	1,350 amp		2.18	7.330		2,300	450		2,750	3,200
0960	1,600 amp		2.02	7.940		2,625	485		3,110	3,600
0970	2,000 amp		1.68	9.520		3,150	585		3,735	4,325
0980	2,500 amp		1.51	10.580		3,875	650		4,525	5,225
0990	3,000 amp		1.34	11.900		4,475	730		5,205	6,000
1000	4,000 amp		1.18	13.610		5,875	835		6,710	7,700
1050	Service head, weatherproof, 3 pole 3 wire, 225 amp		2.52	6.350		1,725	390		2,115	2,475
1060	400 amp		2.35	6.800		1,725	415		2,140	2,525
1070	600 amp		2.18	7.330		1,725	450		2,175	2,575
1080	800 amp		2.02	7.940		1,950	485		2,435	2,875
1090	1,000 amp		1.68	9.520		2,100	585		2,685	3,200
1100	1,350 amp		1.51	10.580		2,725	650		3,375	3,975
1110	1,600 amp		1.34	11.900		3,050	730		3,780	4,450
1120	2,000 amp		1.18	13.610		3,750	835		4,585	5,375
1130	2,500 amp		1.01	15.870		4,475	975		5,450	6,375
1140	3,000 amp		.76	21.160		5,175	1,300		6,475	7,625
1150	4,000 amp		.59	27.210		6,600	1,675		8,275	9,725
1200	3 pole 4 wire, 225 amp		2.18	7.330		1,950	450		2,400	2,825
1210	400 amp		2.02	7.940		1,950	485		2,435	2,875
1220	600 amp		1.85	8.660		1,950	530		2,480	2,950
1230	800 amp		1.68	9.520		2,250	585		2,835	3,350
1240	1,000 amp		1.43	11.200		2,600	685		3,285	3,875
1250	1,350 amp		1.26	12.700		3,100	780		3,880	4,550
1260	1,600 amp		1.18	13.610		3,400	835		4,235	4,975
1270	2,000 amp		1.01	15.870		4,500	975		5,475	6,400
1280	2,500 amp		.84	19.050		5,550	1,175		6,725	7,850
1290	3,000 amp		.67	23.810		6,550	1,450		8,000	9,375
1300	4,000 amp		.50	31.750		8,425	1,950		10,375	12,200
1350	Flanged end, 3 pole 3 wire, 225 amp		5.04	3.170		945	194		1,139	1,350
1360	400 amp		4.37	3.660		945	225		1,170	1,375
1370	600 amp		3.86	4.140		945	254		1,199	1,425
1380	800 amp		3.02	5.290		1,050	325		1,375	1,650
1390	1,000 amp		2.86	5.600		1,200	345		1,545	1,800
1395	1,200 amp		2.69	5.950		1,325	365		1,690	2,000
1400	1,350 amp		2.52	6.350		1,400	390		1,790	2,125
1410	1,600 amp		2.35	6.800		1,600	415		2,015	2,400
1420	2,000 amp		2.02	7.940		1,900	485		2,385	2,825
1430	2,500 amp		1.68	9.520		2,200	585		2,785	3,300
1440	3,000 amp		1.51	10.580		2,525	650		3,175	3,750
1450	4,000 amp		1.34	11.900		3,125	730		3,855	4,500
1500	3 pole 4 wire, 225 amp		4.54	3.530		1,100	217		1,317	1,525
1510	400 amp		3.86	4.140		1,100	254		1,354	1,575

Post-Installation Change Order Costs

26 25 Low-Voltage Enclosed Bus Assemblies

26 25 13 – Low-Voltage Busways

26 25 13.60 Copper or Aluminum Bus Duct Fittings

		Crew	Daily Output	Labor-Hours	Unit	Material	2020 Bare Costs Labor	Equipment	Total	Total Incl O&P
1520	600 amp	2 Elec	3.36	4.760	Ea.	1,100	292		1,392	1,625
1530	800 amp		2.69	5.950		1,300	365		1,665	1,975
1540	1,000 amp		2.52	6.350		1,425	390		1,815	2,150
1545	1,200 amp		2.35	6.800		1,625	415		2,040	2,425
1550	1,350 amp		2.18	7.330		1,750	450		2,200	2,600
1560	1,600 amp		2.02	7.940		2,000	485		2,485	2,925
1570	2,000 amp		1.68	9.520		2,375	585		2,960	3,475
1580	2,500 amp		1.51	10.580		2,800	650		3,450	4,075
1590	3,000 amp		1.34	11.900		3,225	730		3,955	4,625
1600	4,000 amp		1.18	13.610		4,175	835		5,010	5,850
1650	Hanger, standard, 225 amp		53.76	.298		26.50	18.25		44.75	56
1660	400 amp		40.32	.397		26.50	24.50		51	65.50
1670	600 amp		33.60	.476		26.50	29		55.50	72.50
1680	800 amp		26.88	.595		26.50	36.50		63	83.50
1690	1,000 amp		20.16	.794		26.50	48.50		75	102
1695	1,200 amp		18.48	.866		26.50	53		79.50	108
1700	1,350 amp		16.80	.952		26.50	58.50		85	116
1710	1,600 amp		16.80	.952		26.50	58.50		85	116
1720	2,000 amp		15.12	1.060		26.50	65		91.50	126
1730	2,500 amp		13.44	1.190		26.50	73		99.50	138
1740	3,000 amp		13.44	1.190		26.50	73		99.50	138
1750	4,000 amp		13.44	1.190		26.50	73		99.50	138
1800	Spring type, 225 amp		13.44	1.190		89	73		162	207
1810	400 amp		11.76	1.360		89	83.50		172.50	222
1820	600 amp		11.76	1.360		89	83.50		172.50	222
1830	800 amp		11.76	1.360		89	83.50		172.50	222
1840	1,000 amp		11.76	1.360		89	83.50		172.50	222
1845	1,200 amp		11.76	1.360		89	83.50		172.50	222
1850	1,350 amp		11.76	1.360		89	83.50		172.50	222
1860	1,600 amp		10.08	1.590		89	97.50		186.50	243
1870	2,000 amp		10.08	1.590		89	97.50		186.50	243
1880	2,500 amp		10.08	1.590		89	97.50		186.50	243
1890	3,000 amp		8.40	1.900		89	117		206	272
1900	4,000 amp		8.40	1.900		89	117		206	272

26 25 13.70 Feedrail

		Crew	Daily Output	Labor-Hours	Unit	Material	2020 Bare Costs Labor	Equipment	Total	Total Incl O&P
0010	**FEEDRAIL**, 12' mounting									
0050	Trolley busway, 3 pole									
0100	300 volt 60 amp, plain, 10' lengths	1 Elec	42	.190	L.F.	22	11.70		33.70	41.50
0300	Door track		42	.190		41.50	11.70		53.20	63.50
0500	Curved track		25.20	.317		16.35	19.50		35.85	47
0700	Coupling				Ea.	12.75			12.75	14.05
0900	Center feed	1 Elec	4.45	1.800		47	110		157	216
1100	End feed		4.45	1.800		52	110		162	221
1300	Hanger set		20.16	.397		2.89	24.50		27.39	39.50
3000	600 volt 100 amp, plain, 10' lengths		29.40	.272	L.F.	59	16.70		75.70	90
3300	Door track		29.40	.272	"	93	16.70		109.70	128
3700	Coupling				Ea.	53			53	58.50
4000	End cap	1 Elec	33.60	.238		67	14.60		81.60	95.50
4200	End feed		3.36	2.380		239	146		385	480
4500	Trolley, 600 volt, 20 amp		4.45	1.800		263	110		373	455
4700	30 amp		4.45	1.800		263	110		373	455
4900	Duplex, 40 amp		3.36	2.380		920	146		1,066	1,250

26 25 Low-Voltage Enclosed Bus Assemblies

26 25 13 – Low-Voltage Busways

26 25 13.70 Feedrail

		Crew	Daily Output	Labor-Hours	Unit	Material	2020 Bare Costs Labor	2020 Bare Costs Equipment	Total	Total Incl O&P
5000	60 amp	1 Elec	3.36	2.380	Ea.	895	146		1,041	1,200
5300	Fusible, 20 amp		3.36	2.380		540	146		686	810
5500	30 amp		3.36	2.380		540	146		686	810
5900	300 volt, 20 amp		4.45	1.800		249	110		359	440
6000	30 amp		4.45	1.800		325	110		435	525
6300	Fusible, 20 amp		3.95	2.030		350	125		475	570
6500	30 amp		3.95	2.030		490	125		615	725
7300	Busway, 250 volt 50 amp, 2 wire		58.80	.136	L.F.	20	8.35		28.35	35
7330	Coupling				Ea.	44			44	48.50
7340	Center feed	1 Elec	5.04	1.590		510	97.50		607.50	705
7350	End feed		5.04	1.590		124	97.50		221.50	281
7360	End cap		33.60	.238		29	14.60		43.60	54
7370	Hanger set		20.16	.397		3.12	24.50		27.62	40
7400	125/250 volt 50 amp, 3 wire		50.40	.159	L.F.	20.50	9.75		30.25	37
7430	Coupling		5.04	1.590	Ea.	61.50	97.50		159	213
7440	Center feed		5.04	1.590		505	97.50		602.50	700
7450	End feed		5.04	1.590		111	97.50		208.50	267
7460	End cap		33.60	.238		29.50	14.60		44.10	54.50
7470	Hanger set		20.16	.397		4.22	24.50		28.72	41
7480	Trolley, 250 volt, 2 pole, 20 amp		5.04	1.590		38.50	97.50		136	188
7490	30 amp		5.04	1.590		38.50	97.50		136	188
7500	125/250 volt, 3 pole, 20 amp		5.04	1.590		37.50	97.50		135	186
7510	30 amp		5.04	1.590		37.50	97.50		135	186
8000	Cleaning tools, 300 volt, dust remover					107			107	118
8100	Bus bar cleaner					201			201	221
8300	600 volt, dust remover, 60 amp					310			310	340
8400	100 amp					655			655	720
8600	Bus bar cleaner, 60 amp					805			805	885
8700	100 amp					805			805	885

26 27 Low-Voltage Distribution Equipment

26 27 13 – Electricity Metering

26 27 13.10 Meter Centers and Sockets

		Crew	Daily Output	Labor-Hours	Unit	Material	2020 Bare Costs Labor	2020 Bare Costs Equipment	Total	Total Incl O&P
0010	**METER CENTERS AND SOCKETS**									
0100	Sockets, single position, 4 terminal, 100 amp	1 Elec	2.69	2.980	Ea.	52.50	183		235.50	330
0200	150 amp		1.93	4.140		62.50	254		316.50	450
0300	200 amp		1.60	5.010		110	305		415	580
0400	Transformer rated, 20 amp		2.69	2.980		179	183		362	470
0500	Double position, 4 terminal, 100 amp		2.35	3.400		240	209		449	575
0600	150 amp		1.76	4.540		290	279		569	735
0700	200 amp		1.43	5.600		550	345		895	1,125
0800	Trans-socket, 13 terminal, 3 CT mounts, 400 amp		.84	9.520		1,225	585		1,810	2,225
0900	800 amp	2 Elec	1.01	15.870		1,525	975		2,500	3,125
1100	Meter centers and sockets, three phase, single pos, 7 terminal, 100 amp	1 Elec	2.35	3.400		136	209		345	460
1200	200 amp		1.76	4.540		239	279		518	680
1400	400 amp		1.43	5.600		825	345		1,170	1,425
2000	Meter center, main fusible switch, 1P 3W 120/240 V									
2030	400 amp	2 Elec	1.34	11.900	Ea.	855	730		1,585	2,025
2040	600 amp		.92	17.320		1,300	1,075		2,375	3,025
2050	800 amp		.76	21.160		5,000	1,300		6,300	7,425
2060	Rainproof 1P 3W 120/240 V, 400 A		1.34	11.900		1,950	730		2,680	3,225

26 27 Low-Voltage Distribution Equipment

26 27 13 – Electricity Metering

26 27 13.10 Meter Centers and Sockets	Crew	Daily Output	Labor-Hours	Unit	Material	2020 Bare Costs Labor	Equipment	Total	Total Incl O&P
2070	600 amp	2 Elec	.92	17.320	Ea.	3,400	1,075		4,475	5,300
2080	800 amp		.76	21.160		5,325	1,300		6,625	7,775
2100	3P 4W 120/208 V, 400 amp		1.34	11.900		865	730		1,595	2,025
2110	600 amp		.92	17.320		1,475	1,075		2,550	3,175
2120	800 amp		.76	21.160		2,300	1,300		3,600	4,450
2130	Rainproof 3P 4W 120/208 V, 400 amp		1.34	11.900		2,225	730		2,955	3,525
2140	600 amp		.92	17.320		3,575	1,075		4,650	5,525
2150	800 amp		.76	21.160		7,750	1,300		9,050	10,500
2170	Main circuit breaker, 1P 3W 120/240 V									
2180	400 amp	2 Elec	1.34	11.900	Ea.	1,475	730		2,205	2,700
2190	600 amp		.92	17.320		1,800	1,075		2,875	3,550
2200	800 amp		.76	21.160		2,775	1,300		4,075	5,000
2210	1,000 amp		.67	23.810		3,200	1,450		4,650	5,700
2220	1,200 amp		.64	25.060		4,300	1,525		5,825	7,025
2230	1,600 amp		.57	28.010		18,800	1,725		20,525	23,300
2240	Rainproof 1P 3W 120/240 V, 400 amp		1.34	11.900		3,000	730		3,730	4,375
2250	600 amp		.92	17.320		4,700	1,075		5,775	6,750
2260	800 amp		.76	21.160		5,475	1,300		6,775	7,950
2270	1,000 amp		.67	23.810		7,550	1,450		9,000	10,500
2280	1,200 amp		.64	25.060		10,200	1,525		11,725	13,500
2300	3P 4W 120/208 V, 400 amp		1.34	11.900		3,350	730		4,080	4,775
2310	600 amp		.92	17.320		5,600	1,075		6,675	7,725
2320	800 amp		.76	21.160		6,650	1,300		7,950	9,250
2330	1,000 amp		.67	23.810		8,725	1,450		10,175	11,800
2340	1,200 amp		.64	25.060		11,200	1,525		12,725	14,600
2350	1,600 amp		.57	28.010		22,900	1,725		24,625	27,800
2360	Rainproof 3P 4W 120/208 V, 400 amp		1.34	11.900		3,975	730		4,705	5,450
2370	600 amp		.92	17.320		5,600	1,075		6,675	7,725
2380	800 amp		.76	21.160		6,650	1,300		7,950	9,250
2390	1,000 amp		.64	25.060		8,725	1,525		10,250	11,900
2400	1,200 amp		.57	28.010		11,200	1,725		12,925	14,900
2420	Main lugs terminal box, 1P 3W 120/240 V									
2430	800 amp	2 Elec	.79	20.260	Ea.	565	1,250		1,815	2,475
2440	1,200 amp		.60	26.460		595	1,625		2,220	3,075
2450	Rainproof 1P 3W 120/240 V, 225 amp		2.02	7.940		445	485		930	1,225
2460	800 amp		.79	20.260		655	1,250		1,905	2,575
2470	1,200 amp		.60	26.460		1,325	1,625		2,950	3,875
2500	3P 4W 120/208 V, 800 amp		.79	20.260		725	1,250		1,975	2,650
2510	1,200 amp		.60	26.460		1,475	1,625		3,100	4,050
2520	Rainproof 3P 4W 120/208 V, 225 amp		2.02	7.940		445	485		930	1,225
2530	800 amp		.79	20.260		725	1,250		1,975	2,650
2540	1,200 amp		.60	26.460		1,475	1,625		3,100	4,050
2590	Basic meter device									
2600	1P 3W 120/240 V 4 jaw 125A sockets, 3 meter	2 Elec	.84	19.050	Ea.	355	1,175		1,530	2,150
2610	4 meter		.76	21.160		535	1,300		1,835	2,525
2620	5 meter		.67	23.810		630	1,450		2,080	2,875
2630	6 meter		.50	31.750		570	1,950		2,520	3,525
2640	7 meter		.47	34.010		1,650	2,075		3,725	4,925
2650	8 meter		.44	36.630		1,800	2,250		4,050	5,325
2660	10 meter		.40	39.680		2,250	2,425		4,675	6,100
2680	Rainproof 1P 3W 120/240 V 4 jaw 125A sockets									
2690	3 meter	2 Elec	.84	19.050	Ea.	750	1,175		1,925	2,575
2700	4 meter		.76	21.160		900	1,300		2,200	2,925

26 27 13.10 Meter Centers and Sockets	Crew	Daily Output	Labor-Hours	Unit	Material	2020 Bare Costs Labor	Equipment	Total	Total Incl O&P
2710	6 meter	2 Elec	.50	31.750	Ea.	1,300	1,950		3,250	4,325
2720	7 meter		.47	34.010		1,650	2,075		3,725	4,925
2730	8 meter		.44	36.630		1,800	2,250		4,050	5,325
2750	1P 3W 120/240 V 4 jaw sockets									
2760	with 125A circuit breaker, 3 meter	2 Elec	.84	19.050	Ea.	1,400	1,175		2,575	3,300
2770	4 meter		.76	21.160		1,775	1,300		3,075	3,875
2780	5 meter		.67	23.810		2,225	1,450		3,675	4,600
2790	6 meter		.50	31.750		2,600	1,950		4,550	5,750
2800	7 meter		.47	34.010		3,175	2,075		5,250	6,600
2810	8 meter		.44	36.630		3,550	2,250		5,800	7,250
2820	10 meter		.40	39.680		4,425	2,425		6,850	8,500
2830	Rainproof 1P 3W 120/240 V 4 jaw sockets									
2840	with 125A circuit breaker, 3 meter	2 Elec	.84	19.050	Ea.	1,400	1,175		2,575	3,300
2850	4 meter		.76	21.160		1,775	1,300		3,075	3,875
2870	6 meter		.50	31.750		2,600	1,950		4,550	5,750
2880	7 meter		.47	34.010		3,175	2,075		5,250	6,600
2890	8 meter		.44	36.630		3,550	2,250		5,800	7,250
2920	1P 3W on 3P 4W 120/208 V system 5 jaw									
2930	125A sockets, 3 meter	2 Elec	.84	19.050	Ea.	750	1,175		1,925	2,575
2940	4 meter		.76	21.160		900	1,300		2,200	2,925
2950	5 meter		.67	23.810		1,125	1,450		2,575	3,400
2960	6 meter		.50	31.750		1,300	1,950		3,250	4,325
2970	7 meter		.47	34.010		1,650	2,075		3,725	4,925
2980	8 meter		.44	36.630		1,800	2,250		4,050	5,325
2990	10 meter		.40	39.680		2,250	2,425		4,675	6,100
3000	Rainproof 1P 3W on 3P 4W 120/208 V system									
3020	5 jaw 125A sockets, 3 meter	2 Elec	.84	19.050	Ea.	750	1,175		1,925	2,575
3030	4 meter		.76	21.160		900	1,300		2,200	2,925
3050	6 meter		.50	31.750		1,300	1,950		3,250	4,325
3060	7 meter		.47	34.010		1,650	2,075		3,725	4,925
3070	8 meter		.44	36.630		1,800	2,250		4,050	5,325
3090	1P 3W on 3P 4W 120/208 V system 5 jaw sockets									
3100	With 125A circuit breaker, 3 meter	2 Elec	.84	19.050	Ea.	1,400	1,175		2,575	3,300
3110	4 meter		.76	21.160		1,775	1,300		3,075	3,875
3120	5 meter		.67	23.810		2,225	1,450		3,675	4,600
3130	6 meter		.50	31.750		2,600	1,950		4,550	5,750
3140	7 meter		.47	34.010		3,175	2,075		5,250	6,600
3150	8 meter		.44	36.630		3,550	2,250		5,800	7,250
3160	10 meter		.40	39.680		4,425	2,425		6,850	8,500
3170	Rainproof 1P 3W on 3P 4W 120/208 V system									
3180	5 jaw sockets w/125A circuit breaker, 3 meter	2 Elec	.84	19.050	Ea.	1,400	1,175		2,575	3,300
3190	4 meter		.76	21.160		1,775	1,300		3,075	3,875
3210	6 meter		.50	31.750		2,600	1,950		4,550	5,750
3220	7 meter		.47	34.010		3,175	2,075		5,250	6,600
3230	8 meter		.44	36.630		3,550	2,250		5,800	7,250
3250	1P 3W 120/240 V 4 jaw sockets									
3260	with 200A circuit breaker, 3 meter	2 Elec	.84	19.050	Ea.	2,100	1,175		3,275	4,075
3270	4 meter		.76	21.160		2,850	1,300		4,150	5,050
3290	6 meter		.50	31.750		4,200	1,950		6,150	7,525
3300	7 meter		.47	34.010		4,950	2,075		7,025	8,550
3310	8 meter		.47	34.010		5,700	2,075		7,775	9,350
3330	Rainproof 1P 3W 120/240 V 4 jaw sockets									
3350	with 200A circuit breaker, 3 meter	2 Elec	.84	19.050	Ea.	2,100	1,175		3,275	4,075

For customer support on your Electrical Change Order Costs with RSMeans data, call 800.448.8182.

Post-Installation Change Order Costs

26 27 Low-Voltage Distribution Equipment

26 27 13 – Electricity Metering

26 27 13.10 Meter Centers and Sockets		Crew	Daily Output	Labor-Hours	Unit	Material	2020 Bare Costs Labor	Equipment	Total	Total Incl O&P
3360	4 meter	2 Elec	.76	21.160	Ea.	2,850	1,300		4,150	5,050
3380	6 meter		.50	31.750		4,200	1,950		6,150	7,525
3390	7 meter		.47	34.010		4,950	2,075		7,025	8,550
3400	8 meter		.44	36.630		5,700	2,250		7,950	9,600
3420	1P 3W on 3P 4W 120/208 V 5 jaw sockets									
3430	with 200A circuit breaker, 3 meter	2 Elec	.84	19.050	Ea.	2,100	1,175		3,275	4,075
3440	4 meter		.76	21.160		2,850	1,300		4,150	5,050
3460	6 meter		.50	31.750		4,225	1,950		6,175	7,550
3470	7 meter		.47	34.010		4,950	2,075		7,025	8,550
3480	8 meter		.44	36.630		5,700	2,250		7,950	9,600
3500	Rainproof 1P 3W on 3P 4W 120/208 V 5 jaw socket									
3510	with 200A circuit breaker, 3 meter	2 Elec	.84	19.050	Ea.	2,100	1,175		3,275	4,075
3520	4 meter		.76	21.160		2,850	1,300		4,150	5,050
3540	6 meter		.50	31.750		4,200	1,950		6,150	7,525
3550	7 meter		.47	34.010		4,950	2,075		7,025	8,550
3560	8 meter		.44	36.630		5,700	2,250		7,950	9,600
3600	Automatic circuit closing, add					81.50			81.50	89.50
3610	Manual circuit closing, add					93			93	102
3650	Branch meter device									
3660	3P 4W 208/120 or 240/120 V 7 jaw sockets									
3670	with 200A circuit breaker, 2 meter	2 Elec	.76	21.160	Ea.	3,525	1,300		4,825	5,800
3680	3 meter		.67	23.810		5,275	1,450		6,725	7,975
3690	4 meter		.59	27.210		7,025	1,675		8,700	10,200
3700	Main circuit breaker 42,000 rms, 400 amp		1.34	11.900		2,350	730		3,080	3,675
3710	600 amp		.92	17.320		2,675	1,075		3,750	4,500
3720	800 amp		.76	21.160		3,700	1,300		5,000	6,000
3730	Rainproof main circ. breaker 42,000 rms, 400 amp		1.34	11.900		3,075	730		3,805	4,450
3740	600 amp		.92	17.320		4,950	1,075		6,025	7,025
3750	800 amp		.76	21.160		6,650	1,300		7,950	9,250
3760	Main circuit breaker 65,000 rms, 400 amp		1.34	11.900		4,175	730		4,905	5,675
3770	600 amp		.92	17.320		5,875	1,075		6,950	8,025
3780	800 amp		.76	21.160		6,650	1,300		7,950	9,250
3790	1,000 amp		.67	23.810		8,725	1,450		10,175	11,800
3800	1,200 amp		.64	25.060		11,200	1,525		12,725	14,600
3810	1,600 amp		.57	28.010		22,900	1,725		24,625	27,800
3820	Rainproof main circ. breaker 65,000 rms, 400 amp		1.34	11.900		4,175	730		4,905	5,675
3830	600 amp		.92	17.320		5,875	1,075		6,950	8,025
3840	800 amp		.76	21.160		6,650	1,300		7,950	9,225
3850	1,000 amp		.67	23.810		8,725	1,450		10,175	11,800
3860	1,200 amp		.64	25.060		11,200	1,525		12,725	14,600
3880	Main circuit breaker 100,000 rms, 400 amp		1.34	11.900		4,175	730		4,905	5,675
3890	600 amp		.92	17.320		5,875	1,075		6,950	8,025
3900	800 amp		.76	21.160		6,900	1,300		8,200	9,500
3910	Rainproof main circ. breaker 100,000 rms, 400 amp		1.34	11.900		4,175	730		4,905	5,675
3920	600 amp		.92	17.320		5,875	1,075		6,950	8,025
3930	800 amp		.76	21.160		6,900	1,300		8,200	9,500
3940	Main lugs terminal box, 800 amp		.79	20.260		725	1,250		1,975	2,650
3950	1,600 amp		.60	26.460		2,825	1,625		4,450	5,525
3960	Rainproof, 800 amp		.79	20.260		725	1,250		1,975	2,650
3970	1,600 amp		.60	26.460		2,825	1,625		4,450	5,550

26 27 16.10 Cabinets	Crew	Daily Output	Labor-Hours	Unit	Material	2020 Bare Costs Labor	Equipment	Total	Total Incl O&P
0010 **CABINETS**									
7000 Cabinets, current transformer									
7050 Single door, 24" H x 24" W x 10" D	1 Elec	1.34	5.950	Ea.	215	365		580	780
7100 30" H x 24" W x 10" D		1.09	7.330		207	450		657	895
7150 36" H x 24" W x 10" D		.92	8.660		510	530		1,040	1,350
7200 30" H x 30" W x 10" D		.84	9.520		282	585		867	1,175
7250 36" H x 30" W x 10" D		.76	10.580		375	650		1,025	1,375
7300 36" H x 36" W x 10" D		.67	11.900		385	730		1,115	1,500
7500 Double door, 48" H x 36" W x 10" D		.50	15.870		715	975		1,690	2,250
7550 24" H x 24" W x 12" D	↓	.84	9.520	↓	237	585		822	1,125
8000 NEMA 12, double door, floor mounted									
8020 54" H x 42" W x 8" D	2 Elec	5.04	3.170	Ea.	1,250	194		1,444	1,675
8040 60" H x 48" W x 8" D		4.54	3.530		1,700	217		1,917	2,200
8060 60" H x 48" W x 10" D		4.54	3.530		1,700	217		1,917	2,200
8080 60" H x 60" W x 10" D		4.20	3.810		1,975	234		2,209	2,525
8100 72" H x 60" W x 10" D		3.36	4.760		2,250	292		2,542	2,900
8120 72" H x 72" W x 10" D		2.86	5.600		2,025	345		2,370	2,725
8140 60" H x 48" W x 12" D		2.86	5.600		1,750	345		2,095	2,425
8160 60" H x 60" W x 12" D		2.69	5.950		1,975	365		2,340	2,725
8180 72" H x 60" W x 12" D		2.52	6.350		3,425	390		3,815	4,350
8200 72" H x 72" W x 12" D		2.52	6.350		3,825	390		4,215	4,775
8220 60" H x 48" W x 16" D		2.69	5.950		1,875	365		2,240	2,600
8240 72" H x 72" W x 16" D		2.18	7.330		2,650	450		3,100	3,600
8260 60" H x 48" W x 20" D		2.52	6.350		2,025	390		2,415	2,800
8280 72" H x 72" W x 20" D		1.85	8.660		2,875	530		3,405	3,975
8300 60" H x 48" W x 24" D		2.18	7.330		2,150	450		2,600	3,025
8320 72" H x 72" W x 24" D	↓	1.68	9.520	↓	3,025	585		3,610	4,200
8340 Pushbutton enclosure, oiltight									
8360 3-1/2" H x 3-1/4" W x 2-3/4" D, for 1 P.B.	1 Elec	10.08	.794	Ea.	54	48.50		102.50	132
8380 5-3/4" H x 3-1/4" W x 2-3/4" D, for 2 P.B.		9.24	.866		57	53		110	142
8400 8" H x 3-1/4" W x 2-3/4" D, for 3 P.B.		8.82	.907		66.50	55.50		122	157
8420 10-1/4" H x 3-1/4" W x 2-3/4" D, for 4 P.B.		8.82	.907		72	55.50		127.50	162
8460 12-1/2" H x 3-1/4" W x 3" D, for 5 P.B.		7.56	1.060		133	65		198	243
8480 9-1/2" H x 6-1/4" W x 3" D, for 6 P.B.		7.14	1.120		91	68.50		159.50	202
8500 9-1/2" H x 8-1/2" W x 3" D, for 9 P.B.		6.72	1.190		112	73		185	233
8510 11-3/4" H x 8-1/2" W x 3" D, for 12 P.B.		5.88	1.360		118	83.50		201.50	254
8520 11-3/4" H x 10-3/4" W x 3" D, for 16 P.B.		5.46	1.470		125	90		215	271
8540 14" H x 10-3/4" W x 3" D, for 20 P.B.		4.20	1.900		203	117		320	395
8560 14" H x 13" W x 3" D, for 25 P.B.	↓	3.78	2.120	↓	221	130		351	435
8580 Sloping front pushbutton enclosures									
8600 3-1/2" H x 7-3/4" W x 4-7/8" D, for 3 P.B.	1 Elec	8.40	.952	Ea.	89.50	58.50		148	186
8620 7-1/4" H x 8-1/2" W x 6-3/4" D, for 6 P.B.		6.72	1.190		131	73		204	253
8640 9-1/2" H x 8-1/2" W x 7-7/8" D, for 9 P.B.		5.88	1.360		175	83.50		258.50	315
8660 11-1/4" H x 8-1/2" W x 9" D, for 12 P.B.		4.20	1.900		175	117		292	365
8680 11-3/4" H x 10" W x 9" D, for 16 P.B.		4.20	1.900		199	117		316	395
8700 11-3/4" H x 13" W x 9" D, for 20 P.B.		4.20	1.900		228	117		345	425
8720 14" H x 13" W x 10-1/8" D, for 25 P.B.	↓	3.78	2.120	↓	254	130		384	475
8740 Pedestals, not including P.B. enclosure or base									
8760 Straight column 4" x 4"	1 Elec	3.78	2.120	Ea.	238	130		368	455
8780 6" x 6"		3.36	2.380		370	146		516	625
8800 Angled column 4" x 4"		3.78	2.120		275	130		405	500
8820 6" x 6"	↓	3.36	2.380		410	146		556	670

26 27 16 – Electrical Cabinets and Enclosures

26 27 16.10 Cabinets

		Crew	Daily Output	Labor-Hours	Unit	Material	2020 Bare Costs Labor	Equipment	Total	Total Incl O&P
8840	Pedestal, base 18" x 18"	1 Elec	8.40	.952	Ea.	141	58.50		199.50	242
8860	24" x 24"	↓	7.56	1.060	↓	325	65		390	450
8900	Electronic rack enclosures									
8920	72" H x 19" W x 24" D	1 Elec	1.26	6.350	Ea.	2,675	390		3,065	3,525
8940	72" H x 23" W x 24" D		1.26	6.350		2,900	390		3,290	3,775
8960	72" H x 19" W x 30" D		1.09	7.330		2,950	450		3,400	3,900
8980	72" H x 19" W x 36" D		1.01	7.940		3,450	485		3,935	4,525
9000	72" H x 23" W x 36" D	↓	1.01	7.940	↓	3,800	485		4,285	4,900
9020	NEMA 12 & 4 enclosure panels									
9040	12" x 24"	1 Elec	16.80	.476	Ea.	55	29		84	105
9060	16" x 12"		16.80	.476		39	29		68	86.50
9080	20" x 16"		16.80	.476		57	29		86	106
9100	20" x 20"		15.96	.501		69	31		100	122
9120	24" x 20"		15.12	.529		89	32.50		121.50	147
9140	24" x 24"		14.28	.560		102	34.50		136.50	164
9160	30" x 20"		13.44	.595		106	36.50		142.50	172
9180	30" x 24"		13.44	.595		125	36.50		161.50	193
9200	36" x 24"		12.60	.635		150	39		189	223
9220	36" x 30"		12.60	.635		198	39		237	275
9240	42" x 24"		12.60	.635		178	39		217	254
9260	42" x 30"		11.76	.680		225	41.50		266.50	310
9280	42" x 36"		11.76	.680		266	41.50		307.50	355
9300	48" x 24"		11.76	.680		131	41.50		172.50	206
9320	48" x 30"		11.76	.680		171	41.50		212.50	250
9340	48" x 36"		10.92	.733		299	45		344	395
9360	60" x 36"	↓	10.08	.794	↓	370	48.50		418.50	485
9400	Wiring trough steel JIC, clamp cover									
9490	4" x 4", 12" long	1 Elec	10.08	.794	Ea.	85.50	48.50		134	167
9510	24" long		8.40	.952		109	58.50		167.50	207
9530	36" long		6.72	1.190		141	73		214	264
9540	48" long		5.88	1.360		157	83.50		240.50	296
9550	60" long		5.04	1.590		191	97.50		288.50	355
9560	6" x 6", 12" long		9.24	.866		119	53		172	210
9580	24" long		7.56	1.060		155	65		220	267
9600	36" long		5.88	1.360		184	83.50		267.50	325
9610	48" long		5.04	1.590		234	97.50		331.50	405
9620	60" long	↓	4.20	1.900		275	117		392	475

26 27 16.20 Cabinets and Enclosures

		Crew	Daily Output	Labor-Hours	Unit	Material	2020 Bare Costs Labor	Equipment	Total	Total Incl O&P
0010	**CABINETS AND ENCLOSURES** Nonmetallic									
0080	Enclosures fiberglass NEMA 4X									
0100	Wall mount, quick release latch door, 20" H x 16" W x 6" D	1 Elec	4.03	1.980	Ea.	910	121		1,031	1,175
0110	20" H x 20" W x 6" D		3.78	2.120		1,050	130		1,180	1,350
0120	24" H x 20" W x 6" D		3.53	2.270		1,125	139		1,264	1,450
0130	20" H x 16" W x 8" D		3.78	2.120		995	130		1,125	1,300
0140	20" H x 20" W x 8" D		3.53	2.270		1,100	139		1,239	1,425
0150	24" H x 24" W x 8" D		3.19	2.510		1,250	154		1,404	1,600
0160	30" H x 24" W x 8" D		2.69	2.980		1,425	183		1,608	1,850
0170	36" H x 30" W x 8" D		2.52	3.170		2,000	194		2,194	2,500
0180	20" H x 16" W x 10" D		2.94	2.720		1,200	167		1,367	1,575
0190	20" H x 20" W x 10" D		2.69	2.980		1,300	183		1,483	1,700
0200	24" H x 20" W x 10" D		2.52	3.170		1,375	194		1,569	1,800
0210	30" H x 24" W x 10" D	↓	2.35	3.400	↓	1,575	209		1,784	2,050

26 27 16.20 Cabinets and Enclosures		Crew	Daily Output	Labor-Hours	Unit	Material	2020 Bare Costs Labor	Equipment	Total	Total Incl O&P
0220	20" H x 16" W x 12" D	1 Elec	2.52	3.170	Ea.	1,300	194		1,494	1,725
0230	20" H x 20" W x 12" D		2.35	3.400		1,350	209		1,559	1,800
0240	24" H x 24" W x 12" D		2.18	3.660		1,475	225		1,700	1,950
0250	30" H x 24" W x 12" D		2.02	3.970		1,700	244		1,944	2,225
0260	36" H x 30" W x 12" D		1.85	4.330		2,300	266		2,566	2,925
0270	36" H x 36" W x 12" D		1.76	4.540		2,525	279		2,804	3,200
0280	48" H x 36" W x 12" D		1.68	4.760		2,975	292		3,267	3,700
0290	60" H x 36" W x 12" D		1.51	5.290		2,150	325		2,475	2,850
0300	30" H x 24" W x 16" D		1.18	6.800		1,950	415		2,365	2,775
0310	48" H x 36" W x 16" D		1.01	7.940		3,275	485		3,760	4,325
0320	60" H x 36" W x 16" D		.84	9.520		3,675	585		4,260	4,925
0480	Freestanding, one door, 72" H x 25" W x 25" D		.67	11.900		4,200	730		4,930	5,700
0490	Two doors with two panels, 72" H x 49" W x 24" D		.42	19.050		9,725	1,175		10,900	12,500
0500	Floor stand kits, for NEMA 4 & 12, 20" W or more, 6" H x 8" D		20.16	.397		145	24.50		169.50	196
0510	6" H x 10" D		20.16	.397		158	24.50		182.50	211
0520	6" H x 12" D		20.16	.397		180	24.50		204.50	234
0530	6" H x 18" D		20.16	.397		218	24.50		242.50	277
0540	12" H x 8" D		18.48	.433		185	26.50		211.50	243
0550	12" H x 10" D		18.48	.433		335	26.50		361.50	410
0560	12" H x 12" D		18.48	.433		345	26.50		371.50	420
0570	12" H x 16" D		18.48	.433		241	26.50		267.50	305
0580	12" H x 18" D		18.48	.433		262	26.50		288.50	330
0590	12" H x 20" D		18.48	.433		315	26.50		341.50	385
0600	18" H x 8" D		16.80	.476		216	29		245	282
0610	18" H x 10" D		16.80	.476		281	29		310	355
0620	18" H x 12" D		16.80	.476		299	29		328	375
0630	18" H x 16" D		16.80	.476		330	29		359	410
0640	24" H x 8" D		13.44	.595		269	36.50		305.50	350
0650	24" H x 10" D		13.44	.595		330	36.50		366.50	420
0660	24" H x 12" D		13.44	.595		355	36.50		391.50	445
0670	24" H x 16" D		13.44	.595		380	36.50		416.50	475
0680	Small, screw cover, 5-1/2" H x 4" W x 4-15/16" D		10.08	.794		146	48.50		194.50	234
0690	7-1/2" H x 4" W x 4-15/16" D		10.08	.794		95.50	48.50		144	178
0700	7-1/2" H x 6" W x 5-3/16" D		8.40	.952		177	58.50		235.50	281
0710	9-1/2" H x 6" W x 5-11/16" D		8.40	.952		184	58.50		242.50	290
0720	11-1/2" H x 8" W x 6-11/16" D		6.72	1.190		272	73		345	410
0730	13-1/2" H x 10" W x 7-3/16" D		5.88	1.360		330	83.50		413.50	490
0740	15-1/2" H x 12" W x 8-3/16" D		5.04	1.590		425	97.50		522.50	610
0750	17-1/2" H x 14" W x 8-11/16" D		4.20	1.900		500	117		617	725
0760	Screw cover with window, 6" H x 4" W x 5" D		10.08	.794		186	48.50		234.50	277
0770	8" H x 4" W x 5" D		9.24	.866		245	53		298	350
0780	8" H x 6" W x 5" D		9.24	.866		360	53		413	475
0790	10" H x 6" W x 6" D		8.40	.952		375	58.50		433.50	495
0800	12" H x 8" W x 7" D		6.72	1.190		480	73		553	635
0810	14" H x 10" W x 7" D		5.88	1.360		615	83.50		698.50	805
0820	16" H x 12" W x 8" D		5.04	1.590		730	97.50		827.50	950
0830	18" H x 14" W x 9" D		4.20	1.900		820	117		937	1,075
0840	Quick-release latch cover, 5-1/2" H x 4" W x 5" D		10.08	.794		191	48.50		239.50	284
0850	7-1/2" H x 4" W x 5" D		10.08	.794		196	48.50		244.50	288
0860	7-1/2" H x 6" W x 5-1/4" D		8.40	.952		227	58.50		285.50	335
0870	9-1/2" H x 6" W x 5-3/4" D		8.40	.952		140	58.50		198.50	241
0880	11-1/2" H x 8" W x 6-3/4" D		6.72	1.190		221	73		294	350
0890	13-1/2" H x 10" W x 7-1/4" D		5.88	1.360		251	83.50		334.50	400

For customer support on your Electrical Change Order Costs with RSMeans data, call 800.448.8182. **Post-Installation Change Order Costs**

26 27 16 – Electrical Cabinets and Enclosures

26 27 16.20 Cabinets and Enclosures		Crew	Daily Output	Labor-Hours	Unit	Material	2020 Bare Costs Labor	Equipment	Total	Total Incl O&P
0900	15-1/2" H x 12" W x 8-1/4" D	1 Elec	5.04	1.590	Ea.	560	97.50		657.50	765
0910	17-1/2" H x 14" W x 8-3/4" D		4.20	1.900		695	117		812	940
0920	Pushbutton, 1 hole 5-1/2" H x 4" W x 4-15/16" D		10.08	.794		545	48.50		593.50	675
0930	2 hole 7-1/2" H x 4" W x 4-15/16" D		9.24	.866		755	53		808	910
0940	4 hole 7-1/2" H x 6" W x 5-3/16" D		8.82	.907		530	55.50		585.50	670
0950	6 hole 9-1/2" H x 6" W x 5-11/16" D		7.56	1.060		645	65		710	805
0960	8 hole 11-1/2" H x 8" W x 6-11/16" D		7.14	1.120		825	68.50		893.50	1,000
0970	12 hole 13-1/2" H x 10" W x 7-3/16" D		6.72	1.190		1,075	73		1,148	1,300
0980	20 hole 15-1/2" H x 12" W x 8-3/16" D		4.20	1.900		1,025	117		1,142	1,300
0990	30 hole 17-1/2" H x 14" W x 8-11/16" D	▼	3.78	2.120	▼	1,200	130		1,330	1,525
1450	Enclosures polyester NEMA 4X									
1460	Small, screw cover,									
1500	3-15/16" H x 3-15/16" W x 3-1/16" D	1 Elec	10.08	.794	Ea.	81	48.50		129.50	162
1510	5-3/16" H x 3-5/16" W x 3-1/16" D		10.08	.794		79.50	48.50		128	160
1520	5-7/8" H x 3-7/8" W x 4-3/16" D		10.08	.794		83.50	48.50		132	165
1530	5-7/8" H x 5-7/8" W x 4-3/16" D		10.08	.794		94	48.50		142.50	176
1540	7-5/8" H x 3-5/16" W x 3-1/16" D		10.08	.794		86.50	48.50		135	168
1550	10-3/16" H x 3-5/16" W x 3-1/16" D		8.40	.952		108	58.50		166.50	206
1560	Clear cover, 3-15/16" H x 3-15/16" W x 2-7/8" D		10.08	.794		91	48.50		139.50	173
1570	5-3/16" H x 3-5/16" W x 2-7/8" D		10.08	.794		97	48.50		145.50	180
1580	5-7/8" H x 3-7/8" W x 4" D		10.08	.794		120	48.50		168.50	206
1590	5-7/8" H x 5-7/8" W x 4" D		10.08	.794		148	48.50		196.50	236
1600	7-5/8" H x 3-5/16" W x 2-7/8" D		10.08	.794		109	48.50		157.50	193
1610	10-3/16" H x 3-5/16" W x 2-7/8" D		8.40	.952		137	58.50		195.50	237
1620	Pushbutton, 1 hole, 5-5/16" H x 3-5/16" W x 3-1/16" D		10.08	.794		74	48.50		122.50	154
1630	2 hole, 7-5/8" H x 3-5/16" W x 3-1/8" D		9.24	.866		79.50	53		132.50	167
1640	3 hole, 10-3/16" H x 3-5/16" W x 3-1/16" D		8.82	.907		94	55.50		149.50	186
8000	Wireway fiberglass, straight sect. screwcover, 12" L, 4" W x 4" D		33.60	.238		275	14.60		289.60	320
8010	6" W x 6" D		25.20	.317		380	19.50		399.50	450
8020	24" L, 4" W x 4" D		16.80	.476		350	29		379	430
8030	6" W x 6" D		12.60	.635		540	39		579	650
8040	36" L, 4" W x 4" D		11.17	.716		425	44		469	530
8050	6" W x 6" D		8.40	.952		675	58.50		733.50	825
8060	48" L, 4" W x 4" D		8.40	.952		515	58.50		573.50	655
8070	6" W x 6" D		6.30	1.270		840	78		918	1,050
8080	60" L, 4" W x 4" D		6.72	1.190		630	73		703	805
8090	6" W x 6" D		5.04	1.590		985	97.50		1,082.50	1,225
8100	Elbow, 90°, 4" W x 4" D		16.80	.476		275	29		304	345
8110	6" W x 6" D		15.12	.529		550	32.50		582.50	660
8120	Elbow, 45°, 4" W x 4" D		16.80	.476		273	29		302	345
8130	6" W x 6" D		15.12	.529		555	32.50		587.50	660
8140	Tee, 4" W x 4" D		13.44	.595		350	36.50		386.50	440
8150	6" W x 6" D		11.76	.680		650	41.50		691.50	775
8160	Cross, 4" W x 4" D		11.76	.680		380	41.50		421.50	480
8170	6" W x 6" D		10.08	.794		930	48.50		978.50	1,100
8180	Cut-off fitting, w/flange & adhesive, 4" W x 4" D		15.12	.529		217	32.50		249.50	287
8190	6" W x 6" D		13.44	.595		445	36.50		481.50	545
8200	Flexible ftng., hvy. neoprene coated nylon, 4" W x 4" D		16.80	.476		405	29		434	490
8210	6" W x 6" D		15.12	.529		585	32.50		617.50	695
8220	Closure plate, fiberglass, 4" W x 4" D		16.80	.476		77	29		106	128
8230	6" W x 6" D		15.12	.529		85.50	32.50		118	143
8240	Box connector, stainless steel type 304, 4" W x 4" D		16.80	.476		132	29		161	190
8250	6" W x 6" D	▼	15.12	.529	▼	151	32.50		183.50	215

26 27 16 – Electrical Cabinets and Enclosures

26 27 16.20 Cabinets and Enclosures

		Crew	Daily Output	Labor-Hours	Unit	Material	2020 Bare Costs Labor	Equipment	Total	Total Incl O&P
8260	Hanger, 4" W x 4" D	1 Elec	84	.095	Ea.	33	5.85		38.85	45
8270	6" W x 6" D		67.20	.119		42	7.30		49.30	57.50
8280	Straight tube section fiberglass, 4" W x 4" D, 12" long		33.60	.238		248	14.60		262.60	295
8290	24" long		16.80	.476		300	29		329	375
8300	36" long		11.17	.716		275	44		319	365
8310	48" long		8.40	.952		370	58.50		428.50	490
8320	60" long		6.72	1.190		430	73		503	585
8330	120" long		3.36	2.380		685	146		831	965

26 27 19 – Multi-Outlet Assemblies

26 27 19.10 Wiring Duct

		Crew	Daily Output	Labor-Hours	Unit	Material	2020 Bare Costs Labor	Equipment	Total	Total Incl O&P
0010	**WIRING DUCT** Plastic									
1250	PVC, snap-in slots, adhesive backed									
1270	1-1/2" W x 2" H	2 Elec	100.80	.159	L.F.	6.30	9.75		16.05	21.50
1280	1-1/2" W x 3" H		100.80	.159		8.25	9.75		18	23.50
1290	1-1/2" W x 4" H		100.80	.159		8.80	9.75		18.55	24
1300	2" W x 1" H		100.80	.159		5.65	9.75		15.40	21
1310	2" W x 1-1/2" H		100.80	.159		6	9.75		15.75	21
1320	2" W x 2" H		100.80	.159		6	9.75		15.75	21
1340	2" W x 3" H		100.80	.159		8.25	9.75		18	23.50
1350	2" W x 4" H		100.80	.159		9.95	9.75		19.70	25.50
1360	2-1/2" W x 3" H		100.80	.159		8.95	9.75		18.70	24.50
1370	3" W x 1" H		92.40	.173		6.65	10.60		17.25	23
1390	3" W x 2" H		92.40	.173		6.95	10.60		17.55	23.50
1400	3" W x 3" H		92.40	.173		8.40	10.60		19	25
1410	3" W x 4" H		92.40	.173		10.50	10.60		21.10	27.50
1420	3" W x 5" H		92.40	.173		13.75	10.60		24.35	31
1430	4" W x 1-1/2" H		84	.190		9.70	11.70		21.40	28
1440	4" W x 2" H		84	.190		8.10	11.70		19.80	26.50
1450	4" W x 3" H		84	.190		9.20	11.70		20.90	27.50
1460	4" W x 4" H		84	.190		11.10	11.70		22.80	29.50
1470	4" W x 5" H		84	.190		15.60	11.70		27.30	34.50
1550	Cover, 1-1/2" W		168	.095		1.16	5.85		7.01	10
1560	2" W		168	.095		1.28	5.85		7.13	10.10
1570	2-1/2" W		168	.095		1.79	5.85		7.64	10.65
1580	3" W		168	.095		1.94	5.85		7.79	10.85
1590	4" W		168	.095		2.34	5.85		8.19	11.25

26 27 23 – Indoor Service Poles

26 27 23.40 Surface Raceway

		Crew	Daily Output	Labor-Hours	Unit	Material	2020 Bare Costs Labor	Equipment	Total	Total Incl O&P
0010	**SURFACE RACEWAY**									
0090	Metal, straight section									
0100	No. 500	1 Elec	84	.095	L.F.	1.56	5.85		7.41	10.40
0400	No. 1500, small pancake		75.60	.106		2.93	6.50		9.43	12.85
0600	No. 2000, base & cover, blank		75.60	.106		3.12	6.50		9.62	13.10
0610	Receptacle, 6" OC		33.60	.238		25.50	14.60		40.10	50
0620	12" OC		36.96	.216		16.60	13.30		29.90	38
0630	18" OC		38.64	.207		9.75	12.70		22.45	29.50
0650	30" OC		42	.190		5.85	11.70		17.55	24
0660	60" OC		42	.190		5.40	11.70		17.10	23.50
0670	No. 2400, base & cover, blank		67.20	.119		2.74	7.30		10.04	13.90
0680	Receptacle, 6" OC		35.28	.227		52	13.90		65.90	77.50
0690	12" OC		44.52	.180		35.50	11		46.50	55.50
0700	18" OC		46.20	.173		31.50	10.60		42.10	50.50

For customer support on your Electrical Change Order Costs with RSMeans data, call 800.448.8182.

Post-Installation Change Order Costs

26 27 23.40 Surface Raceway	Crew	Daily Output	Labor-Hours	Unit	Material	2020 Bare Costs Labor	Equipment	Total	Total Incl O&P
0710	24" OC	1 Elec	47.88	.167	L.F.	12	10.25		22.25	28.50
0720	30" OC		49.56	.161		8.05	9.90		17.95	23.50
0730	60" OC		51.24	.156		6.45	9.60		16.05	21.50
0800	No. 3000, base & cover, blank		63	.127		5.80	7.80		13.60	18
0810	Receptacle, 6" OC		37.80	.212		52.50	13		65.50	77.50
0820	12" OC		52.08	.154		29.50	9.40		38.90	46.50
0830	18" OC		53.76	.149		24	9.15		33.15	40
0840	24" OC		55.44	.144		18	8.85		26.85	33
0850	30" OC		57.12	.140		16.20	8.60		24.80	30.50
0860	60" OC		58.80	.136		11.85	8.35		20.20	25.50
1000	No. 4000, base & cover, blank		54.60	.147		9.45	9		18.45	24
1010	Receptacle, 6" OC		34.44	.232		61	14.25		75.25	88.50
1020	12" OC		43.68	.183		44	11.25		55.25	65.50
1030	18" OC		45.36	.176		36	10.80		46.80	55.50
1040	24" OC		47.04	.170		29.50	10.45		39.95	48
1050	30" OC		48.72	.164		27	10.05		37.05	45
1060	60" OC		50.40	.159		20.50	9.75		30.25	37
1200	No. 6000, base & cover, blank		42	.190		16.65	11.70		28.35	35.50
1210	Receptacle, 6" OC		25.20	.317		75	19.50		94.50	112
1220	12" OC		31.08	.257		57	15.80		72.80	86.50
1230	18" OC		32.76	.244		48.50	15		63.50	75.50
1240	24" OC		34.44	.232		39.50	14.25		53.75	64.50
1250	30" OC		36.12	.221		37.50	13.60		51.10	61.50
1260	60" OC		37.80	.212	▼	29	13		42	51.50
2400	Fittings, elbows, No. 500		33.60	.238	Ea.	2.81	14.60		17.41	25
2800	Elbow cover, No. 2000		33.60	.238		5.10	14.60		19.70	27.50
2880	Tee, No. 500		35.28	.227		5.05	13.90		18.95	26
2900	No. 2000		22.68	.353		16.40	21.50		37.90	50
3000	Switch box, No. 500		13.44	.595		17.80	36.50		54.30	74
3400	Telephone outlet, No. 1500		13.44	.595		21	36.50		57.50	77.50
3600	Junction box, No. 1500	▼	13.44	.595	▼	13.85	36.50		50.35	69.50
3800	Plugmold wired sections, No. 2000									
4000	1 circuit, 6 outlets, 3' long	1 Elec	6.72	1.190	Ea.	54	73		127	168
4100	2 circuits, 8 outlets, 6' long		4.45	1.800		78.50	110		188.50	251
4110	Tele-power pole, alum, w/2 recept, 10'		3.36	2.380		300	146		446	545
4120	12'		3.23	2.470		355	152		507	615
4130	15'		3.11	2.570		450	158		608	730
4140	Steel, w/2 recept, 10'		3.36	2.380		185	146		331	420
4150	One phone fitting, 10'		3.36	2.380		201	146		347	440
4160	Alum, 4 outlets, 10'	▼	3.11	2.570	▼	435	158		593	710
4300	Overhead distribution systems, 125 volt									
4800	No. 2000, entrance end fitting	1 Elec	16.80	.476	Ea.	8.15	29		37.15	52.50
5000	Blank end fitting		33.60	.238		2.74	14.60		17.34	25
5200	Supporting clip		33.60	.238		1.62	14.60		16.22	24
5800	No. 3000, entrance end fitting		16.80	.476		12.90	29		41.90	57.50
6000	Blank end fitting		33.60	.238		3.58	14.60		18.18	26
6020	Internal elbow		16.80	.476		16.95	29		45.95	62
6030	External elbow		16.80	.476		24	29		53	70
6040	Device bracket		44.52	.180		5.75	11		16.75	22.50
6400	Hanger clamp		26.88	.298	▼	8.75	18.25		27	36.50
7000	No. 4000 base		75.60	.106	L.F.	6.15	6.50		12.65	16.45
7200	Divider		84	.095	"	1.16	5.85		7.01	10
7400	Entrance end fitting		13.44	.595	Ea.	30.50	36.50		67	88

26 27 23.40 Surface Raceway	Crew	Daily Output	Labor-Hours	Unit	Material	2020 Bare Costs Labor	Equipment	Total	Total Incl O&P
7600	Blank end fitting	1 Elec	33.60	.238	Ea.	8.45	14.60		23.05	31.50
7610	Recpt. & tele. cover		44.52	.180		14.75	11		25.75	32.50
7620	External elbow		13.44	.595		48.50	36.50		85	108
7630	Coupling		44.52	.180		7.40	11		18.40	24.50
7640	Divider clip & coupling		67.20	.119		1.46	7.30		8.76	12.50
7650	Panel connector		13.44	.595		30.50	36.50		67	88.50
7800	Take off connector		13.44	.595		102	36.50		138.50	167
8000	No. 6000, take off connector		13.44	.595		121	36.50		157.50	188
8100	Take off fitting		13.44	.595		89.50	36.50		126	153
8200	Hanger clamp	↓	26.88	.298		23	18.25		41.25	52
8230	Coupling					12.60			12.60	13.85
8240	One gang device plate	1 Elec	44.52	.180		11.60	11		22.60	29
8250	Two gang device plate		33.60	.238		14.20	14.60		28.80	37.50
8260	Blank end fitting		33.60	.238		12.65	14.60		27.25	36
8270	Combination elbow		11.76	.680		50	41.50		91.50	117
8300	Panel connector	↓	13.44	.595	↓	22	36.50		58.50	79
8500	Chan-L-Wire system installed in 1-5/8" x 1-5/8" strut. Strut									
8600	not incl., 30 amp, 4 wire, 3 phase	1 Elec	168	.048	L.F.	5.60	2.92		8.52	10.55
8700	Junction box		6.72	1.190	Ea.	37.50	73		110.50	150
8800	Insulating end cap		33.60	.238		10	14.60		24.60	33
8900	Strut splice plate		33.60	.238		13.25	14.60		27.85	36.50
9000	Tap		33.60	.238		27	14.60		41.60	52
9100	Fixture hanger	↓	50.40	.159		11.80	9.75		21.55	27.50
9200	Pulling tool				↓	101			101	111
9300	Non-metallic, straight section									
9310	7/16" x 7/8", base & cover, blank	1 Elec	134.40	.060	L.F.	2	3.65		5.65	7.65
9320	Base & cover w/adhesive		134.40	.060		1.71	3.65		5.36	7.35
9340	7/16" x 1-5/16", base & cover, blank		121.80	.066		2.20	4.03		6.23	8.40
9350	Base & cover w/adhesive		121.80	.066		2.26	4.03		6.29	8.50
9370	11/16" x 2-1/4", base & cover, blank		109.20	.073		3.02	4.49		7.51	10
9380	Base & cover w/adhesive		109.20	.073		3.24	4.49		7.73	10.25
9385	1-11/16" x 5-1/4", two compartment base & cover w/screws		67.20	.119	↓	10.40	7.30		17.70	22.50
9400	Fittings, elbows, 7/16" x 7/8"		42	.190	Ea.	1.97	11.70		13.67	19.55
9410	7/16" x 1-5/16"		37.80	.212		2.04	13		15.04	21.50
9420	11/16" x 2-1/4"		33.60	.238		2.21	14.60		16.81	24.50
9425	1-11/16" x 5-1/4"		23.52	.340		12.45	21		33.45	44.50
9430	Tees, 7/16" x 7/8"		29.40	.272		2.53	16.70		19.23	28
9440	7/16" x 1-5/16"		26.88	.298		2.59	18.25		20.84	30
9450	11/16" x 2-1/4"		25.20	.317		2.69	19.50		22.19	32
9455	1-11/16" x 5-1/4"		20.16	.397		19.95	24.50		44.45	58.50
9460	Cover clip, 7/16" x 7/8"		67.20	.119		.51	7.30		7.81	11.45
9470	7/16" x 1-5/16"		60.48	.132		.46	8.10		8.56	12.60
9480	11/16" x 2-1/4"		53.76	.149		.84	9.15		9.99	14.50
9484	1-11/16" x 5-1/4"		35.28	.227		2.94	13.90		16.84	23.50
9486	Wire clip, 1-11/16" x 5-1/4"		57.12	.140		.58	8.60		9.18	13.45
9490	Blank end, 7/16" x 7/8"		42	.190		.77	11.70		12.47	18.25
9500	7/16" x 1-5/16"		37.80	.212		.86	13		13.86	20.50
9510	11/16" x 2-1/4"		33.60	.238		1.28	14.60		15.88	23.50
9515	1-11/16" x 5-1/4"		31.92	.251		6.70	15.40		22.10	30.50
9520	Round fixture box, 5.5" diam. x 1"		21	.381		11.55	23.50		35.05	47.50
9530	Device box, 1 gang		25.20	.317		5.25	19.50		24.75	35
9540	2 gang	↓	21	.381	↓	7.65	23.50		31.15	43.50

26 27 Low-Voltage Distribution Equipment

26 27 26 – Wiring Devices

26 27 26.10 Low Voltage Switching	Crew	Daily Output	Labor-Hours	Unit	Material	2020 Bare Costs Labor	Equipment	Total	Total Incl O&P
0010 **LOW VOLTAGE SWITCHING**									
3600 Relays, 120 V or 277 V standard	1 Elec	10.08	.794	Ea.	48.50	48.50		97	126
3800 Flush switch, standard		33.60	.238		14.80	14.60		29.40	38.50
4000 Interchangeable		33.60	.238		19.30	14.60		33.90	43.50
4100 Surface switch, standard		33.60	.238		9.15	14.60		23.75	32
4200 Transformer 115 V to 25 V		10.08	.794		141	48.50		189.50	228
4400 Master control, 12 circuit, manual		3.36	2.380		153	146		299	385
4500 25 circuit, motorized		3.36	2.380		177	146		323	410
4600 Rectifier, silicon		10.08	.794		55.50	48.50		104	134
4800 Switchplates, 1 gang, 1, 2 or 3 switch, plastic		67.20	.119		5.80	7.30		13.10	17.25
5000 Stainless steel		67.20	.119		12.45	7.30		19.75	24.50
5400 2 gang, 3 switch, stainless steel		44.52	.180		26	11		37	45
5500 4 switch, plastic		44.52	.180		11.35	11		22.35	29
5600 2 gang, 4 switch, stainless steel		44.52	.180		24.50	11		35.50	43.50
5700 6 switch, stainless steel		44.52	.180		48.50	11		59.50	70
5800 3 gang, 9 switch, stainless steel		26.88	.298		74	18.25		92.25	109
5900 Receptacle, triple, 1 return, 1 feed		21.84	.366		51.50	22.50		74	90.50
6000 2 feed		16.80	.476		51.50	29		80.50	101
6100 Relay gang boxes, flush or surface, 6 gang		4.45	1.800		101	110		211	275
6200 12 gang		3.95	2.030		120	125		245	315
6400 18 gang		3.36	2.380		132	146		278	365
6500 Frame, to hold up to 6 relays		10.08	.794		98.50	48.50		147	182
7200 Control wire, 2 conductor		5.29	1.510	C.L.F.	39	92.50		131.50	181
7400 3 conductor		4.20	1.900		37.50	117		154.50	216
7600 19 conductor		2.10	3.810		350	234		584	735
7800 26 conductor		1.68	4.760		480	292		772	960
8000 Weatherproof, 3 conductor		4.20	1.900		79	117		196	261

26 27 26.20 Wiring Devices Elements

26 27 26.20 Wiring Devices Elements	Crew	Daily Output	Labor-Hours	Unit	Material	2020 Bare Costs Labor	Equipment	Total	Total Incl O&P
0010 **WIRING DEVICES ELEMENTS**									
0200 Toggle switch, quiet type, single pole, 15 amp	1 Elec	33.60	.238	Ea.	.57	14.60		15.17	22.50
0500 20 amp		22.68	.353		3.81	21.50		25.31	36
0510 30 amp		19.32	.414		25.50	25.50		51	66.50
0530 Lock handle, 20 amp		22.68	.353		32.50	21.50		54	68
0540 Security key, 20 amp		21.84	.366		94	22.50		116.50	137
0550 Rocker, 15 amp		33.60	.238		3.10	14.60		17.70	25.50
0560 20 amp		22.68	.353		11.95	21.50		33.45	45
0600 3 way, 15 amp		19.32	.414		2.01	25.50		27.51	40
0800 20 amp		15.12	.529		3.89	32.50		36.39	53
0810 30 amp		7.56	1.060		26.50	65		91.50	127
0830 Lock handle, 20 amp		15.12	.529		31.50	32.50		64	83
0840 Security key, 20 amp		14.28	.560		137	34.50		171.50	201
0850 Rocker, 15 amp		19.32	.414		9.10	25.50		34.60	48
0860 20 amp		15.12	.529		14.10	32.50		46.60	64
0900 4 way, 15 amp		12.60	.635		11.70	39		50.70	71
1000 20 amp		9.24	.866		52.50	53		105.50	137
1020 Lock handle, 20 amp		9.24	.866		43.50	53		96.50	127
1030 Rocker, 15 amp		12.60	.635		11	39		50	70
1040 20 amp		9.24	.866		21	53		74	102
1100 Toggle switch, quiet type, double pole, 15 amp		12.60	.635		15.35	39		54.35	75
1200 20 amp		9.24	.866		21	53		74	103
1210 30 amp		7.56	1.060		37.50	65		102.50	138
1230 Lock handle, 20 amp		9.24	.866		38.50	53		91.50	121

26 27 Low-Voltage Distribution Equipment

26 27 26 – Wiring Devices

26 27 26.20 Wiring Devices Elements		Crew	Daily Output	Labor-Hours	Unit	Material	2020 Bare Costs Labor	Equipment	Total	Total Incl O&P
1250	Security key, 20 amp	1 Elec	8.40	.952	Ea.	100	58.50		158.50	197
1420	Toggle switch quiet type, 1 pole, 2 throw center off, 15 amp		19.32	.414		49	25.50		74.50	91.50
1440	20 amp		15.12	.529		77.50	32.50		110	134
1460	2 pole, 2 throw center off, lock handle, 20 amp		9.24	.866		93.50	53		146.50	182
1480	1 pole, momentary contact, 15 amp		19.32	.414		22	25.50		47.50	62.50
1500	20 amp		15.12	.529		31.50	32.50		64	83
1520	Momentary contact, lock handle, 20 amp		15.12	.529		35	32.50		67.50	87
1650	Dimmer switch, 120 volt, incandescent, 600 watt, 1 pole G		13.44	.595		25	36.50		61.50	82
1700	600 watt, 3 way G		10.08	.794		12.20	48.50		60.70	86
1750	1,000 watt, 1 pole G		13.44	.595		48.50	36.50		85	108
1800	1,000 watt, 3 way G		10.08	.794		80.50	48.50		129	161
2000	1,500 watt, 1 pole G		9.24	.866		112	53		165	202
2100	2,000 watt, 1 pole G		6.72	1.190		162	73		235	287
2110	Fluorescent, 600 watt G		12.60	.635		122	39		161	192
2120	1,000 watt G		12.60	.635		170	39		209	245
2130	1,500 watt G		8.40	.952		315	58.50		373.50	430
2160	Explosion proof, toggle switch, wall, single pole 20 amp		4.45	1.800		296	110		406	490
2180	Receptacle, single outlet, 20 amp		4.45	1.800		680	110		790	910
2190	30 amp		3.36	2.380		1,075	146		1,221	1,425
2290	60 amp		2.10	3.810		890	234		1,124	1,325
2360	Plug, 20 amp		13.44	.595		217	36.50		253.50	294
2370	30 amp		10.08	.794		415	48.50		463.50	530
2380	60 amp		6.72	1.190		590	73		663	755
2410	Furnace, thermal cutoff switch with plate		21.84	.366		18.75	22.50		41.25	54
2460	Receptacle, duplex, 120 volt, grounded, 15 amp		33.60	.238		1.78	14.60		16.38	24
2470	20 amp		22.68	.353		11.75	21.50		33.25	45
2480	Ground fault interrupting, 15 amp		22.68	.353		14.95	21.50		36.45	48.50
2482	20 amp		22.68	.353		49.50	21.50		71	86.50
2486	Clock receptacle, 15 amp		33.60	.238		33.50	14.60		48.10	59
2490	Dryer, 30 amp		12.60	.635		5	39		44	63.50
2500	Range, 50 amp		9.24	.866		12.35	53		65.35	92.50
2530	Surge suppressor receptacle, duplex, 20 amp		22.68	.353		57.50	21.50		79	95
2532	Quad, 20 amp		16.80	.476		114	29		143	169
2540	Isolated ground receptacle, duplex, 20 amp		22.68	.353		30	21.50		51.50	65
2542	Quad, 20 amp		16.80	.476		37	29		66	84
2550	Simplex, 20 amp		22.68	.353		32.50	21.50		54	68
2560	Simplex, 30 amp	↓	12.60	.635		42.50	39		81.50	105
2570	Cable reel w/receptacle 50' w/3#12, 120 V, 20 A	2 Elec	2.24	7.140		1,225	440		1,665	1,975
2600	Wall plates, stainless steel, 1 gang	1 Elec	67.20	.119		2.83	7.30		10.13	14
2800	2 gang		44.52	.180		4.82	11		15.82	21.50
3000	3 gang		26.88	.298		12.65	18.25		30.90	41
3100	4 gang		22.68	.353		12.20	21.50		33.70	45.50
3110	Brown plastic, 1 gang		67.20	.119		.42	7.30		7.72	11.35
3120	2 gang		44.52	.180		.83	11		11.83	17.30
3130	3 gang		26.88	.298		1.22	18.25		19.47	28.50
3140	4 gang		22.68	.353		3.20	21.50		24.70	35.50
3150	Brushed brass, 1 gang		67.20	.119		6.50	7.30		13.80	18.10
3160	Anodized aluminum, 1 gang		67.20	.119		3.33	7.30		10.63	14.55
3170	Switch cover, weatherproof, 1 gang		50.40	.159		5.80	9.75		15.55	21
3180	Vandal proof lock, 1 gang		50.40	.159		16.75	9.75		26.50	33
3200	Lampholder, keyless		21.84	.366		21.50	22.50		44	57
3400	Pullchain with receptacle		18.48	.433		25	26.50		51.50	67
3500	Pilot light, neon with jewel	↓	22.68	.353	↓	11.25	21.50		32.75	44.50

For customer support on your Electrical Change Order Costs with RSMeans data, call 800.448.8182. **Post-Installation Change Order Costs**

26 27 26.20 Wiring Devices Elements	Crew	Daily Output	Labor-Hours	Unit	Material	2020 Bare Costs Labor	Equipment	Total	Total Incl O&P
3600 Receptacle, 20 amp, 250 volt, NEMA 6	1 Elec	22.68	.353	Ea.	26.50	21.50		48	61
3620 277 volt NEMA 7		22.68	.353		25.50	21.50		47	60
3640 125/250 volt NEMA 10		22.68	.353		25.50	21.50		47	60
3680 125/250 volt NEMA 14		21	.381		33	23.50		56.50	71.50
3700 3 pole, 250 volt NEMA 15		21	.381		35	23.50		58.50	73.50
3720 120/208 volt NEMA 18		21	.381		37.50	23.50		61	76
3740 30 amp, 125 volt NEMA 5		12.60	.635		21.50	39		60.50	81.50
3760 250 volt NEMA 6		12.60	.635		29	39		68	90
3780 277 volt NEMA 7		12.60	.635		34	39		73	95.50
3820 125/250 volt NEMA 14		11.76	.680		72.50	41.50		114	142
3840 3 pole, 250 volt NEMA 15		11.76	.680		66.50	41.50		108	135
3880 50 amp, 125 volt NEMA 5		9.24	.866		35	53		88	118
3900 250 volt NEMA 6		9.24	.866		33.50	53		86.50	116
3920 277 volt NEMA 7		9.24	.866		39.50	53		92.50	123
3960 125/250 volt NEMA 14		8.40	.952		104	58.50		162.50	201
3980 3 pole, 250 volt NEMA 15		8.40	.952		95.50	58.50		154	192
4020 60 amp, 125/250 volt, NEMA 14		6.72	1.190		103	73		176	223
4040 3 pole, 250 volt NEMA 15		6.72	1.190		121	73		194	242
4060 120/208 volt NEMA 18		6.72	1.190		92	73		165	210
4100 Receptacle locking, 20 amp, 125 volt NEMA L5		22.68	.353		27.50	21.50		49	62.50
4120 250 volt NEMA L6		22.68	.353		25.50	21.50		47	60
4140 277 volt NEMA L7		22.68	.353		29.50	21.50		51	64.50
4150 3 pole, 250 volt, NEMA L11		22.68	.353		24	21.50		45.50	58.50
4160 20 amp, 480 volt NEMA L8		22.68	.353		33	21.50		54.50	68.50
4180 600 volt NEMA L9		22.68	.353		41.50	21.50		63	78
4200 125/250 volt NEMA L10		22.68	.353		32	21.50		53.50	67
4230 125/250 volt NEMA L14		21	.381		34	23.50		57.50	72.50
4280 250 volt NEMA L15		21	.381		37.50	23.50		61	76.50
4300 480 volt NEMA L16		21	.381		31	23.50		54.50	69
4320 3 phase, 120/208 volt NEMA L18		21	.381		39.50	23.50		63	78.50
4340 277/480 volt NEMA L19		21	.381		41.50	23.50		65	80.50
4360 347/600 volt NEMA L20		21	.381		40.50	23.50		64	79.50
4380 120/208 volt NEMA L21		19.32	.414		42	25.50		67.50	84
4400 277/480 volt NEMA L22		19.32	.414		42.50	25.50		68	84.50
4420 347/600 volt NEMA L23		19.32	.414		52	25.50		77.50	95.50
4440 30 amp, 125 volt NEMA L5		12.60	.635		36	39		75	97.50
4460 250 volt NEMA L6		12.60	.635		38.50	39		77.50	101
4480 277 volt NEMA L7		12.60	.635		42	39		81	104
4500 480 volt NEMA L8		12.60	.635		39.50	39		78.50	102
4520 600 volt NEMA L9		12.60	.635		44.50	39		83.50	107
4540 125/250 volt NEMA L10		12.60	.635		49.50	39		88.50	113
4560 3 phase, 250 volt NEMA L11		12.60	.635		35	39		74	96.50
4620 125/250 volt NEMA L14		11.76	.680		55	41.50		96.50	123
4640 250 volt NEMA L15		11.76	.680		57	41.50		98.50	125
4660 480 volt NEMA L16		11.76	.680		59.50	41.50		101	128
4680 600 volt NEMA L17		11.76	.680		61.50	41.50		103	130
4700 120/208 volt NEMA L18		11.76	.680		61.50	41.50		103	130
4720 277/480 volt NEMA L19		11.76	.680		64	41.50		105.50	133
4740 347/600 volt NEMA L20		11.76	.680		67	41.50		108.50	136
4760 120/208 volt NEMA L21		10.92	.733		59.50	45		104.50	132
4780 277/480 volt NEMA L22		10.92	.733		60	45		105	133
4800 347/600 volt NEMA L23		10.92	.733		69	45		114	143
4840 Receptacle, corrosion resistant, 15 or 20 amp, 125 volt NEMA L5		22.68	.353		39	21.50		60.50	74.50

26 27 26.20 Wiring Devices Elements	Crew	Daily Output	Labor-Hours	Unit	Material	2020 Bare Costs Labor	Equipment	Total	Total Incl O&P
4860	250 volt NEMA L6	1 Elec	22.68	.353	Ea.	25	21.50		46.50	59.50
4870	Receptacle box assembly, cast aluminum, 60A, 4P5W, 3P, 120/208V, NEMA PR4		18.48	.433		810	26.50		836.50	930
4875	100A, 4P5W, 3P, 347/600V, NEMA PR4		18.48	.433		890	26.50		916.50	1,025
4900	Receptacle, cover plate, phenolic plastic, NEMA 5 & 6		67.20	.119		.68	7.30		7.98	11.65
4910	NEMA 7-23		67.20	.119		.74	7.30		8.04	11.70
4920	Stainless steel, NEMA 5 & 6		67.20	.119		2.99	7.30		10.29	14.20
4930	NEMA 7-23		67.20	.119		3.36	7.30		10.66	14.60
4940	Brushed brass NEMA 5 & 6		67.20	.119		6	7.30		13.30	17.50
4950	NEMA 7-23		67.20	.119		6.55	7.30		13.85	18.10
4960	Anodized aluminum, NEMA 5 & 6		67.20	.119		3.93	7.30		11.23	15.20
4970	NEMA 7-23		67.20	.119		6.50	7.30		13.80	18.05
4980	Weatherproof NEMA 7-23		50.40	.159		47	9.75		56.75	66
5000	Duplex receptacle, combo 15A/125V, 3 wire w/2-5V 0.7A, port USB, AL		22.68	.353		21	21.50		42.50	55
5002	15A/125V, 3 wire w/2-5V 0.7A, port USB, BK		22.68	.353		20	21.50		41.50	54
5004	15A/125V, 3 wire w/2-5V 0.7A, port USB, LA		22.68	.353		19.95	21.50		41.45	54
5006	15A/125V, 3 wire w/2-5V 0.7A, port USB, IV		22.68	.353		33	21.50		54.50	68.50
5008	15A/125V, 3 wire w/2-5V 0.7A, port USB, WH		22.68	.353		24.50	21.50		46	59
5010	Duplex receptacle, combo 15A/125V, 3 wire w/2-5V 2.1A, port USB, AL		22.68	.353		64	21.50		85.50	103
5012	15A/125V, 3 wire w/2-5V 2.1A, port USB, BK		22.68	.353		62.50	21.50		84	101
5014	15A/125V, 3 wire w/2-5V 2.1A, port USB, LA		22.68	.353		55	21.50		76.50	92.50
5016	15A/125V, 3 wire w/2-5V 2.1A, port USB, IV		22.68	.353		55.50	21.50		77	93
5018	15A/125V, 3 wire w/2-5V 2.1A, port USB, WH		22.68	.353		34	21.50		55.50	69.50
5020	Duplex receptacle, combo 20A/125V, 3 wire w/2-5V 2.1A, port USB, AL		22.68	.353		56.50	21.50		78	94
5022	20A/125V, 3 wire w/2-5V 2.1A, port USB, BK		22.68	.353		71	21.50		92.50	110
5024	20A/125V, 3 wire w/2-5V 2.1A, port USB, LA		22.68	.353		39	21.50		60.50	74.50
5026	20A/125V, 3 wire w/2-5V 2.1A, port USB, IV		22.68	.353		38.50	21.50		60	74.50
5028	20A/125V, 3 wire w/2-5V 2.1A, port USB, WH		22.68	.353		38.50	21.50		60	74.50
5100	Plug, 20 amp, 250 volt NEMA 6		25.20	.317		22	19.50		41.50	53
5110	277 volt NEMA 7		25.20	.317		24	19.50		43.50	55.50
5120	3 pole, 120/250 volt NEMA 10		21.84	.366		26.50	22.50		49	62.50
5130	125/250 volt NEMA 14		21.84	.366		55	22.50		77.50	94
5140	250 volt NEMA 15		21.84	.366		55.50	22.50		78	94.50
5150	120/208 volt NEMA 8		21.84	.366		63.50	22.50		86	104
5160	30 amp, 125 volt NEMA 5		10.92	.733		56	45		101	129
5170	250 volt NEMA 6		10.92	.733		55	45		100	128
5180	277 volt NEMA 7		10.92	.733		51	45		96	123
5190	125/250 volt NEMA 14		10.92	.733		63	45		108	136
5200	3 pole, 250 volt NEMA 15		10.08	.794		67	48.50		115.50	147
5210	50 amp, 125 volt NEMA 5		7.56	1.060		84.50	65		149.50	190
5220	250 volt NEMA 6		7.56	1.060		88.50	65		153.50	195
5230	277 volt NEMA 7		7.56	1.060		94.50	65		159.50	201
5240	125/250 volt NEMA 14		7.56	1.060		82	65		147	188
5250	3 pole, 250 volt NEMA 15		6.72	1.190		81	73		154	198
5260	60 amp, 125/250 volt NEMA 14		5.88	1.360		88.50	83.50		172	221
5270	3 pole, 250 volt NEMA 15		5.88	1.360		94	83.50		177.50	228
5280	120/208 volt NEMA 18		5.88	1.360		114	83.50		197.50	249
5300	Plug angle, 20 amp, 250 volt NEMA 6		25.20	.317		33	19.50		52.50	65
5310	30 amp, 125 volt NEMA 5		10.92	.733		61.50	45		106.50	135
5320	250 volt NEMA 6		10.92	.733		64	45		109	137
5330	277 volt NEMA 7		10.92	.733		75	45		120	150
5340	125/250 volt NEMA 14		10.92	.733		70	45		115	144
5350	3 pole, 250 volt NEMA 15		10.08	.794		75.50	48.50		124	156
5360	50 amp, 125 volt NEMA 5		7.56	1.060		71	65		136	175

For customer support on your Electrical Change Order Costs with RSMeans data, call 800.448.8182.

Post-Installation Change Order Costs

26 27 26.20 Wiring Devices Elements	Crew	Daily Output	Labor-Hours	Unit	Material	2020 Bare Costs Labor	Equipment	Total	Total Incl O&P
5370	250 volt NEMA 6	1 Elec	7.56	1.060	Ea.	67	65		132	171
5380	277 volt NEMA 7		7.56	1.060		84	65		149	190
5390	125/250 volt NEMA 14		7.56	1.060		85	65		150	191
5400	3 pole, 250 volt NEMA 15		6.72	1.190		90	73		163	208
5410	60 amp, 125/250 volt NEMA 14		5.88	1.360		101	83.50		184.50	235
5420	3 pole, 250 volt NEMA 15		5.88	1.360		103	83.50		186.50	237
5430	120/208 volt NEMA 18		5.88	1.360		106	83.50		189.50	241
5500	Plug, locking, 20 amp, 125 volt NEMA L5		25.20	.317		20	19.50		39.50	51
5510	250 volt NEMA L6		25.20	.317		20	19.50		39.50	51
5520	277 volt NEMA L7		25.20	.317		19.70	19.50		39.20	50.50
5530	480 volt NEMA L8		25.20	.317		22	19.50		41.50	53
5540	600 volt NEMA L9		25.20	.317		24	19.50		43.50	55.50
5550	3 pole, 125/250 volt NEMA L10		21.84	.366		25.50	22.50		48	61.50
5560	250 volt NEMA L11		21.84	.366		25.50	22.50		48	61.50
5570	480 volt NEMA L12		21.84	.366		29.50	22.50		52	66
5580	125/250 volt NEMA L14		21.84	.366		30	22.50		52.50	66.50
5590	250 volt NEMA L15		21.84	.366		37.50	22.50		60	74.50
5600	480 volt NEMA L16		21.84	.366		33.50	22.50		56	70.50
5610	4 pole, 120/208 volt NEMA L18		20.16	.397		36.50	24.50		61	76.50
5620	277/480 volt NEMA L19		20.16	.397		43.50	24.50		68	84.50
5630	347/600 volt NEMA L20		20.16	.397		38	24.50		62.50	78
5640	120/208 volt NEMA L21		20.16	.397		38.50	24.50		63	79
5650	277/480 volt NEMA L22		20.16	.397		41	24.50		65.50	81.50
5660	347/600 volt NEMA L23		20.16	.397		45.50	24.50		70	86.50
5670	30 amp, 125 volt NEMA L5		10.92	.733		30.50	45		75.50	101
5680	250 volt NEMA L6		10.92	.733		31.50	45		76.50	102
5690	277 volt NEMA L7		10.92	.733		31.50	45		76.50	102
5700	480 volt NEMA L8		10.92	.733		32.50	45		77.50	103
5710	600 volt NEMA L9		10.92	.733		33.50	45		78.50	104
5720	3 pole, 125/250 volt NEMA L10		9.24	.866		28	53		81	110
5730	250 volt NEMA L11		9.24	.866		27.50	53		80.50	110
5760	125/250 volt NEMA L14		9.24	.866		41	53		94	124
5770	250 volt NEMA L15		9.24	.866		41	53		94	124
5780	480 volt NEMA L16		9.24	.866		42.50	53		95.50	126
5790	600 volt NEMA L17		9.24	.866		42.50	53		95.50	126
5800	4 pole, 120/208 volt NEMA L18		8.40	.952		47.50	58.50		106	139
5810	120/208 volt NEMA L19		8.40	.952		48	58.50		106.50	140
5820	347/600 volt NEMA L20		8.40	.952		50.50	58.50		109	143
5830	120/208 volt NEMA L21		8.40	.952		44.50	58.50		103	136
5840	277/480 volt NEMA L22		8.40	.952		50.50	58.50		109	143
5850	347/600 volt NEMA L23		8.40	.952		51.50	58.50		110	144
6000	Connector, 20 amp, 250 volt NEMA 6		25.20	.317		35	19.50		54.50	67.50
6010	277 volt NEMA 7		25.20	.317		35	19.50		54.50	67.50
6020	3 pole, 120/250 volt NEMA 10		21.84	.366		39.50	22.50		62	77
6030	125/250 volt NEMA 14		21.84	.366		39.50	22.50		62	77
6040	250 volt NEMA 15		21.84	.366		40	22.50		62.50	77.50
6050	120/208 volt NEMA 18		21.84	.366		43.50	22.50		66	81.50
6060	30 amp, 125 volt NEMA 5		10.92	.733		69	45		114	143
6070	250 volt NEMA 6		10.92	.733		69	45		114	143
6080	277 volt NEMA 7		10.92	.733		69	45		114	143
6110	50 amp, 125 volt NEMA 5		7.56	1.060		95	65		160	201
6120	250 volt NEMA 6		7.56	1.060		95	65		160	201
6130	277 volt NEMA 7		7.56	1.060		95	65		160	201

26 27 Low-Voltage Distribution Equipment

26 27 26 – Wiring Devices

26 27 26.20 Wiring Devices Elements	Crew	Daily Output	Labor-Hours	Unit	Material	2020 Bare Costs Labor	Equipment	Total	Total Incl O&P
6200	Connector, locking, 20 amp, 125 volt NEMA L5	1 Elec	25.20	.317	Ea.	31	19.50		50.50	63
6210	250 volt NEMA L6		25.20	.317		31	19.50		50.50	63
6220	277 volt NEMA L7		25.20	.317		30	19.50		49.50	62
6230	480 volt NEMA L8		25.20	.317		35	19.50		54.50	67.50
6240	600 volt NEMA L9		25.20	.317		41	19.50		60.50	74
6250	3 pole, 125/250 volt NEMA L10		21.84	.366		40.50	22.50		63	78
6260	250 volt NEMA L11		21.84	.366		40.50	22.50		63	78
6280	125/250 volt NEMA L14		21.84	.366		42	22.50		64.50	79.50
6290	250 volt NEMA L15		21.84	.366		42	22.50		64.50	80
6300	480 volt NEMA L16		21.84	.366		45	22.50		67.50	83.50
6310	4 pole, 120/208 volt NEMA L18		20.16	.397		52.50	24.50		77	94
6320	277/480 volt NEMA L19		20.16	.397		54	24.50		78.50	96
6330	347/600 volt NEMA L20		20.16	.397		54.50	24.50		79	96.50
6340	120/208 volt NEMA L21		20.16	.397		62	24.50		86.50	105
6350	277/480 volt NEMA L22		20.16	.397		69.50	24.50		94	113
6360	347/600 volt NEMA L23		20.16	.397		78	24.50		102.50	122
6370	30 amp, 125 volt NEMA L5		10.92	.733		60.50	45		105.50	134
6380	250 volt NEMA L6		10.92	.733		61	45		106	134
6390	277 volt NEMA L7		10.92	.733		64.50	45		109.50	138
6400	480 volt NEMA L8		10.92	.733		64	45		109	137
6410	600 volt NEMA L9		10.92	.733		71.50	45		116.50	146
6420	3 pole, 125/250 volt NEMA L10		9.24	.866		78	53		131	165
6430	250 volt NEMA L11		9.24	.866		78	53		131	165
6460	125/250 volt NEMA L14		9.24	.866		84.50	53		137.50	172
6470	250 volt NEMA L15		9.24	.866		84	53		137	172
6480	480 volt NEMA L16		9.24	.866		90.50	53		143.50	179
6490	600 volt NEMA L17		9.24	.866		90	53		143	178
6500	4 pole, 120/208 volt NEMA L18		8.40	.952		97.50	58.50		156	194
6510	120/208 volt NEMA L19		8.40	.952		97	58.50		155.50	194
6520	347/600 volt NEMA L20		8.40	.952		99	58.50		157.50	196
6530	120/208 volt NEMA L21		8.40	.952		82.50	58.50		141	178
6540	277/480 volt NEMA L22		8.40	.952		88.50	58.50		147	185
6550	347/600 volt NEMA L23		8.40	.952		95	58.50		153.50	191
7000	Receptacle computer, 250 volt, 15 amp, 3 pole 4 wire		6.72	1.190		102	73		175	221
7010	20 amp, 2 pole 3 wire		6.72	1.190		122	73		195	243
7020	30 amp, 2 pole 3 wire		5.46	1.470		196	90		286	350
7030	30 amp, 3 pole 4 wire		5.46	1.470		212	90		302	365
7040	60 amp, 3 pole 4 wire		3.78	2.120		350	130		480	580
7050	100 amp, 3 pole 4 wire		2.52	3.170		435	194		629	770
7100	Connector computer, 250 volt, 15 amp, 3 pole 4 wire		22.68	.353		168	21.50		189.50	217
7110	20 amp, 2 pole 3 wire		22.68	.353		167	21.50		188.50	215
7120	30 amp, 2 pole 3 wire		12.60	.635		231	39		270	310
7130	30 amp, 3 pole 4 wire		12.60	.635		261	39		300	345
7140	60 amp, 3 pole 4 wire		6.72	1.190		440	73		513	595
7150	100 amp, 3 pole 4 wire		3.36	2.380		595	146		741	865
7200	Plug, computer, 250 volt, 15 amp, 3 pole 4 wire		22.68	.353		150	21.50		171.50	197
7210	20 amp, 2 pole 3 wire		22.68	.353		153	21.50		174.50	201
7220	30 amp, 2 pole 3 wire		12.60	.635		240	39		279	320
7230	30 amp, 3 pole 4 wire		12.60	.635		238	39		277	320
7240	60 amp, 3 pole 4 wire		6.72	1.190		355	73		428	505
7250	100 amp, 3 pole 4 wire		3.36	2.380		435	146		581	690
7300	Connector adapter to flexible conduit, 1/2"		50.40	.159		3.89	9.75		13.64	18.80
7310	3/4"		42	.190		4.97	11.70		16.67	23

Post-Installation Change Order Costs

26 27 26.20 Wiring Devices Elements

		Crew	Daily Output	Labor-Hours	Unit	Material	2020 Bare Costs Labor	Equipment	Total	Total Incl O&P
7320	1-1/4"	1 Elec	25.20	.317	Ea.	15.25	19.50		34.75	46
7330	1-1/2"		19.32	.414		20	25.50		45.50	60
8100	Pin/sleeve, 20A, 480V, DSN1, male inlet, 3 pole, 3 phase, 4 wire		13.44	.595		48.50	36.50		85	108
8110	20A, 125V, DSN20, male inlet, 3 pole, 3 phase, 3 wire		13.44	.595		39.50	36.50		76	98
8120	60A, 480V, DS60, male inlet, 3 pole, 3 phase, 3 wire		13.44	.595		325	36.50		361.50	415
8130	Pin/sleeve, 20A, 480V, DSN10, female RCPT, 3P, 3 phase, 4 wire		13.44	.595		92.50	36.50		129	157
8140	30A, 480V, DS30, female RCPT, 3P, 3 phase, 4 wire		13.44	.595		300	36.50		336.50	390
8500	Wiring device terminal strip, 2 pole, 12-2AWG, 300VAC, CSA, 600VAC, 10A		30.24	.265		3.44	16.25		19.69	28
8505	4 pole, 12-2AWG, 300VAC, CSA, 600VAC, 10A		30.24	.265		4.87	16.25		21.12	29.50
8510	6 pole, 12-2AWG, 300VAC, CSA, 600VAC, 10A		30.24	.265		5.50	16.25		21.75	30
8515	8 pole, 12-2AWG, 300VAC, CSA, 600VAC, 10A		25.20	.317		6.15	19.50		25.65	36
8520	10 pole, 12-2AWG, 300VAC, CSA, 600VAC, 10A		25.20	.317		6.80	19.50		26.30	36.50
8525	12 pole, 12-2AWG, 300VAC, CSA, 600VAC, 10A		25.20	.317		7.50	19.50		27	37.50
8530	Wiring device terminal strip jumper, 10A		30.24	.265		.84	16.25		17.09	25
8535	Wiring device terminal strip, 2 pole, 12-22AWG, 300VAC, CSA, 600VAC, 20A		28.56	.280		3.74	17.20		20.94	29.50
8540	4 pole, 12-22AWG, 300VAC, CSA, 600VAC, 20A		28.56	.280		5.10	17.20		22.30	31
8545	6 pole, 12-22AWG, 300VAC, CSA, 600VAC, 20A		28.56	.280		5.80	17.20		23	32
8550	8 pole, 12-22AWG, 300VAC, CSA, 600VAC, 20A		25.20	.317		6.45	19.50		25.95	36
8555	10 pole, 12-22AWG, 300VAC, CSA, 600VAC, 20A		25.20	.317		7.15	19.50		26.65	37
8560	12 pole, 12-22AWG, 300VAC, CSA, 600VAC, 20A		25.20	.317		7.85	19.50		27.35	37.50
8565	Wiring device terminal strip jumper, 20A		28.56	.280		1.03	17.20		18.23	26.50
8570	Wiring device terminal shorting type no cover, 4 pole, 6-10AWG, 600VAC, 50A		25.20	.317		19.40	19.50		38.90	50.50
8575	6 pole, 6-10AWG, 600VAC, 50A		25.20	.317		24.50	19.50		44	56
8580	8 pole, 6-10AWG, 600VAC, 50A		25.20	.317		28.50	19.50		48	60.50
8585	12 pole, 6-10AWG, 600VAC, 50A		25.20	.317		32	19.50		51.50	64
8590	Wiring device terminal 2 shorting pin 1/2 cover, 2 pole, 8-4AWG, 600V, 45A		25.20	.317		20.50	19.50		40	51.50
8595	4 pole, 8-4AWG, 600V, 45A		25.20	.317		25.50	19.50		45	57
8600	6 pole, 8-4AWG, 600V, 45A		25.20	.317		31.50	19.50		51	64
8605	8 pole, 8-4AWG, 600V, 45A		25.20	.317		45	19.50		64.50	78.50
8610	12 pole, 8-4AWG, 600V, 45A		25.20	.317		54	19.50		73.50	88.50
8615	Wiring device terminal 2 shorting pin W/cover, 4 pole, 8-4AWG, 600V, 45A		25.20	.317		34	19.50		53.50	66.50
8620	6 pole, 8-4AWG, 600V, 45A		25.20	.317		52	19.50		71.50	86
8625	8 pole, 8-4AWG, 600V, 45A		25.20	.317		54.50	19.50		74	89
8630	12 pole, 8-4AWG, 600V, 45A		25.20	.317		65	19.50		84.50	101

26 27 33 – Power Distribution Units

26 27 33.10 Power Distribution Unit Cabinet

		Crew	Daily Output	Labor-Hours	Unit	Material	2020 Bare Costs Labor	Equipment	Total	Total Incl O&P
0010	**POWER DISTRIBUTION UNIT CABINET**									
0100	Power distribution unit, single cabinet, 50 kVA output, 480/208 input	1 Elec	1.68	4.760	Ea.	9,200	292		9,492	10,500
0110	75 kVA output, 480/208 input		1.68	4.760		9,975	292		10,267	11,400
0120	100 kVA output, 480/208 input		1.68	4.760		10,900	292		11,192	12,400
0130	125 kVA output, 480/208 input		1.26	6.350		12,000	390		12,390	13,800
0140	150 kVA output, 480/208 input		1.26	6.350		14,500	390		14,890	16,600
0150	200 kVA output, 480/208 input		.84	9.520		16,000	585		16,585	18,500
0160	225 kVA output, 480/208 input		.84	9.520		16,900	585		17,485	19,500

26 27 33.20 Power Distribution Unit

		Crew	Daily Output	Labor-Hours	Unit	Material	2020 Bare Costs Labor	Equipment	Total	Total Incl O&P
0010	**POWER DISTRIBUTION UNIT**									
0050	3PH, 60 kVA, PDU									
0100	208V-208V/120V	2 Elec	1.68	9.520	Ea.	14,300	585		14,885	16,600
0110	480V-208V/120V		1.68	9.520		14,100	585		14,685	16,400
0120	600V-208V/120V		1.68	9.520		14,700	585		15,285	17,000
0125	3PH, 80 kVA, PDU									
0130	208V-208V/120V	2 Elec	1.68	9.520	Ea.	27,700	585		28,285	31,400

26 27 Low-Voltage Distribution Equipment

26 27 33 – Power Distribution Units

26 27 33.20 Power Distribution Unit	Crew	Daily Output	Labor-Hours	Unit	Material	2020 Bare Costs Labor	Equipment	Total	Total Incl O&P
0140	480V-208V/120V	2 Elec	1.68	9.520	Ea.	28,100	585		28,685	31,900
0150	600V-208V/120V		1.68	9.520		27,900	585		28,485	31,600
0200	3PH, 15 kVA, 480-208/120V, 208-208/120V		2.52	6.350		11,600	390		11,990	13,400
0210	30 kVA, 400-208/120V, 208-208/120V		2.52	6.350		19,200	390		19,590	21,700
0250	50 kVA, 480-208/120V, 208-208/120V		2.10	7.620		10,500	465		10,965	12,300
0260	75 kVA, 480-208/120V, 208-208/120V		2.10	7.620		11,800	465		12,265	13,700
0270	100 kVA, 480-208/120V, 208-208/120V		1.68	9.520		12,700	585		13,285	14,900
0280	125 kVA, 480-208/120V, 208-208/120V		1.68	9.520		13,700	585		14,285	16,000
0290	150 kVA, 480-208/120V, 208-208/120V		1.68	9.520		12,500	585		13,085	14,700
0300	200 kVA, 480-208/120V, 208-208/120V		1.26	12.700		13,800	780		14,580	16,400
0310	225 kVA, 480-208/120V, 208-208/120V		1.26	12.700		14,600	780		15,380	17,200

26 27 73 – Door Chimes

26 27 73.10 Doorbell System

		Crew	Daily Output	Labor-Hours	Unit	Material	Labor	Equipment	Total	Total Incl O&P
0010	**DOORBELL SYSTEM**, incl. transformer, button & signal									
0100	6" bell	1 Elec	3.36	2.380	Ea.	158	146		304	390
0200	Buzzer		3.36	2.380		127	146		273	355
1000	Door chimes, 2 notes		13.44	.595		32	36.50		68.50	89.50
1020	with ambient light		10.08	.794		116	48.50		164.50	201
1100	Tube type, 3 tube system		10.08	.794		233	48.50		281.50	330
1180	4 tube system		8.40	.952		495	58.50		553.50	630
1900	For transformer & button, add		4.20	1.900		16.15	117		133.15	192
3000	For push button only		20.16	.397		.92	24.50		25.42	37.50
3200	Bell transformer		13.44	.595		24	36.50		60.50	81

26 28 Low-Voltage Circuit Protective Devices

26 28 13 – Fuses

26 28 13.10 Fuse Elements

		Crew	Daily Output	Labor-Hours	Unit	Material	Labor	Equipment	Total	Total Incl O&P
0010	**FUSE ELEMENTS**									
0020	Cartridge, nonrenewable									
0050	250 volt, 30 amp	1 Elec	42	.190	Ea.	2.75	11.70		14.45	20.50
0100	60 amp		42	.190		5	11.70		16.70	23
0150	100 amp		33.60	.238		19.60	14.60		34.20	43.50
0200	200 amp		30.24	.265		44.50	16.25		60.75	73
0250	400 amp		25.20	.317		135	19.50		154.50	178
0300	600 amp		20.16	.397		236	24.50		260.50	296
0400	600 volt, 30 amp		33.60	.238		12.35	14.60		26.95	35.50
0450	60 amp		33.60	.238		18.40	14.60		33	42.50
0500	100 amp		30.24	.265		39.50	16.25		55.75	67
0550	200 amp		25.20	.317		73	19.50		92.50	109
0600	400 amp		20.16	.397		146	24.50		170.50	197
0650	600 amp		16.80	.476		231	29		260	298
0800	Dual element, time delay, 250 volt, 30 amp		42	.190		11.90	11.70		23.60	30.50
0850	60 amp		42	.190		13.90	11.70		25.60	32.50
0900	100 amp		33.60	.238		52	14.60		66.60	79.50
0950	200 amp		30.24	.265		113	16.25		129.25	148
1000	400 amp		25.20	.317		138	19.50		157.50	180
1050	600 amp		20.16	.397		227	24.50		251.50	287
1300	600 volt, 15 to 30 amp		33.60	.238		26	14.60		40.60	50.50
1350	35 to 60 amp		33.60	.238		38	14.60		52.60	63.50
1400	70 to 100 amp		30.24	.265		79.50	16.25		95.75	112

For customer support on your Electrical Change Order Costs with RSMeans data, call 800.448.8182.   **Post-Installation Change Order Costs**

26 28 Low-Voltage Circuit Protective Devices

26 28 13 – Fuses

26 28 13.10 Fuse Elements	Crew	Daily Output	Labor-Hours	Unit	Material	2020 Bare Costs Labor	Equipment	Total	Total Incl O&P
1450	110 to 200 amp	1 Elec	25.20	.317	Ea.	157	19.50		176.50	202
1500	225 to 400 amp		20.16	.397		340	24.50		364.50	410
1550	600 amp		16.80	.476		435	29		464	520
1800	Class RK1, high capacity, 250 volt, 30 amp		42	.190		10.55	11.70		22.25	29
1850	60 amp		42	.190		20	11.70		31.70	39.50
1900	100 amp		33.60	.238		31	14.60		45.60	56
1950	200 amp		30.24	.265		80	16.25		96.25	112
2000	400 amp		25.20	.317		181	19.50		200.50	229
2050	600 amp		20.16	.397		340	24.50		364.50	410
2200	600 volt, 30 amp		33.60	.238		15.20	14.60		29.80	38.50
2250	60 amp		33.60	.238		28	14.60		42.60	52.50
2300	100 amp		30.24	.265		95.50	16.25		111.75	129
2350	200 amp		25.20	.317		135	19.50		154.50	177
2400	400 amp		20.16	.397		235	24.50		259.50	295
2450	600 amp		16.80	.476		485	29		514	580
2700	Class J, current limiting, 250 or 600 volt, 30 amp		33.60	.238		24	14.60		38.60	48
2750	60 amp		33.60	.238		37.50	14.60		52.10	63.50
2800	100 amp		30.24	.265		76.50	16.25		92.75	109
2850	200 amp		25.20	.317		117	19.50		136.50	158
2900	400 amp		20.16	.397		355	24.50		379.50	425
2950	600 amp		16.80	.476		370	29		399	455
3100	Class L, current limiting, 250 or 600 volt, 601 to 1,200 amp		13.44	.595		890	36.50		926.50	1,025
3150	1,500 to 1,600 amp		10.92	.733		860	45		905	1,000
3200	1,800 to 2,000 amp		8.40	.952		1,800	58.50		1,858.50	2,050
3250	2,500 amp		8.40	.952		1,475	58.50		1,533.50	1,675
3300	3,000 amp		6.72	1.190		2,350	73		2,423	2,700
3350	3,500 to 4,000 amp		6.72	1.190		2,275	73		2,348	2,600
3400	4,500 to 5,000 amp		5.63	1.420		2,200	87		2,287	2,550
3450	6,000 amp		4.79	1.670		6,550	102		6,652	7,375
3600	Plug, 120 volt, 1 to 10 amp		42	.190		5.35	11.70		17.05	23.50
3650	15 to 30 amp		42	.190		5	11.70		16.70	23
3700	Dual element 0.3 to 14 amp		42	.190		7.95	11.70		19.65	26
3750	15 to 30 amp		42	.190		7.90	11.70		19.60	26
3800	Fustat, 120 volt, 15 to 30 amp		42	.190		5.80	11.70		17.50	24
3850	0.3 to 14 amp		42	.190		6.85	11.70		18.55	25
3900	Adapters 0.3 to 10 amp		42	.190		7.50	11.70		19.20	25.50
3950	15 to 30 amp		42	.190		10.60	11.70		22.30	29
4000	F-frame current limiting fuse, 14 to 2 AWG, 3 ampere, 3P, aluminum terminal		33.60	.238		1,925	14.60		1,939.60	2,150
4010	7 ampere, 3P, aluminum terminal		33.60	.238		1,925	14.60		1,939.60	2,125
4020	15 AMP, 3P, aluminum terminal		33.60	.238		1,975	14.60		1,989.60	2,200
4030	30 AMP, 3P, aluminum terminal		33.60	.238		1,925	14.60		1,939.60	2,150
4040	50 AMP, 3P, aluminum terminal		33.60	.238		1,625	14.60		1,639.60	1,800
4050	F-frame current limiting fuse, 1 to 4/0 AWG, 100 AMP, 3P, aluminum terminal		33.60	.238		1,525	14.60		1,539.60	1,700
4060	150 AMP, 3P, aluminum terminal		33.60	.238		2,250	14.60		2,264.60	2,475

26 28 16 – Enclosed Switches and Circuit Breakers

26 28 16.10 Circuit Breakers

		Crew	Daily Output	Labor-Hours	Unit	Material	Labor	Equipment	Total	Total Incl O&P
0010	**CIRCUIT BREAKERS** (in enclosure)									
0100	Enclosed (NEMA 1), 600 volt, 3 pole, 30 amp	1 Elec	2.69	2.980	Ea.	530	183		713	850
0200	60 amp		2.35	3.400		645	209		854	1,025
0400	100 amp		1.93	4.140		735	254		989	1,200
0500	200 amp		1.26	6.350		1,875	390		2,265	2,650
0600	225 amp		1.26	6.350		1,700	390		2,090	2,450

26 28 16.10 Circuit Breakers

		Crew	Daily Output	Labor-Hours	Unit	Material	2020 Bare Costs Labor	Equipment	Total	Total Incl O&P
0700	400 amp	2 Elec	1.34	11.900	Ea.	2,900	730		3,630	4,275
0800	600 amp		1.01	15.870		4,200	975		5,175	6,075
1000	800 amp		.79	20.260		5,475	1,250		6,725	7,875
1200	1,000 amp		.71	22.680		6,925	1,400		8,325	9,675
1220	1,200 amp		.67	23.810		8,850	1,450		10,300	11,900
1240	1,600 amp		.60	26.460		16,100	1,625		17,725	20,100
1260	2,000 amp		.54	29.760		17,400	1,825		19,225	21,900
1400	1,200 amp with ground fault		.67	23.810		15,500	1,450		16,950	19,300
1600	1,600 amp with ground fault		.60	26.460		18,300	1,625		19,925	22,500
1800	2,000 amp with ground fault		.54	29.760		19,600	1,825		21,425	24,200
2000	Disconnect, 240 volt 3 pole, 5 HP motor	1 Elec	2.69	2.980		450	183		633	765
2020	10 HP motor		2.69	2.980		450	183		633	765
2040	15 HP motor		2.35	3.400		450	209		659	805
2060	20 HP motor		1.93	4.140		545	254		799	980
2080	25 HP motor		1.93	4.140		545	254		799	980
2100	30 HP motor		1.93	4.140		545	254		799	980
2120	40 HP motor		1.68	4.760		935	292		1,227	1,450
2140	50 HP motor		1.26	6.350		935	390		1,325	1,600
2160	60 HP motor		1.26	6.350		2,150	390		2,540	2,950
2180	75 HP motor	2 Elec	1.68	9.520		2,150	585		2,735	3,250
2200	100 HP motor		1.34	11.900		2,150	730		2,880	3,450
2220	125 HP motor		1.34	11.900		2,150	730		2,880	3,450
2240	150 HP motor		1.01	15.870		4,200	975		5,175	6,075
2260	200 HP motor		1.01	15.870		5,475	975		6,450	7,475
2300	Enclosed (NEMA 7), explosion proof, 600 volt 3 pole, 50 amp	1 Elec	1.93	4.140		1,925	254		2,179	2,500
2350	100 amp		1.26	6.350		2,000	390		2,390	2,775
2400	150 amp		.84	9.520		4,825	585		5,410	6,200
2450	250 amp	2 Elec	1.34	11.900		6,050	730		6,780	7,725
2500	400 amp	"	1.01	15.870		6,700	975		7,675	8,825

26 28 16.13 Circuit Breakers

		Crew	Daily Output	Labor-Hours	Unit	Material	2020 Bare Costs Labor	Equipment	Total	Total Incl O&P
0010	**CIRCUIT BREAKERS**									
0100	Circuit breaker current limiter, 225 to 400, ampere, 1 pole	1 Elec	40.32	.198	Ea.	835	12.15		847.15	935
0200	Circuit breaker current limiter, 200 to 500, ampere, 1 pole, thermal magnet		38.64	.207		690	12.70		702.70	780
0300	300 to 500, ampere, 1 pole, thermal magnet		38.64	.207		1,275	12.70		1,287.70	1,450
0400	Circuit breaker current limiter, 600 to 1000, ampere, 1 pole, tri pac		38.64	.207		1,750	12.70		1,762.70	1,950

26 28 16.20 Safety Switches

		Crew	Daily Output	Labor-Hours	Unit	Material	2020 Bare Costs Labor	Equipment	Total	Total Incl O&P
0010	**SAFETY SWITCHES**									
0100	General duty 240 volt, 3 pole NEMA 1, fusible, 30 amp	1 Elec	2.69	2.980	Ea.	71.50	183		254.50	350
0200	60 amp		1.93	4.140		122	254		376	515
0300	100 amp		1.60	5.010		210	305		515	690
0400	200 amp		1.09	7.330		450	450		900	1,175
0500	400 amp	2 Elec	1.51	10.580		1,175	650		1,825	2,275
0600	600 amp	"	1.01	15.870		2,300	975		3,275	3,975
0610	Nonfusible, 30 amp	1 Elec	2.69	2.980		57.50	183		240.50	335
0650	60 amp		1.93	4.140		76	254		330	465
0700	100 amp		1.60	5.010		178	305		483	655
0750	200 amp		1.09	7.330		330	450		780	1,025
0800	400 amp	2 Elec	1.51	10.580		895	650		1,545	1,950
0850	600 amp	"	1.01	15.870		1,850	975		2,825	3,475
1100	Heavy duty, 600 volt, 3 pole NEMA 1 nonfused									
1110	30 amp	1 Elec	2.69	2.980	Ea.	103	183		286	385
1500	60 amp		1.93	4.140		185	254		439	585

Post-Installation Change Order Costs

26 28 16.20 Safety Switches		Crew	Daily Output	Labor-Hours	Unit	Material	2020 Bare Costs Labor	Equipment	Total	Total Incl O&P
1700	100 amp	1 Elec	1.60	5.010	Ea.	291	305		596	780
1900	200 amp		1.09	7.330		440	450		890	1,150
2100	400 amp	2 Elec	1.51	10.580		1,075	650		1,725	2,150
2300	600 amp		1.01	15.870		2,000	975		2,975	3,650
2500	800 amp		.79	20.260		3,600	1,250		4,850	5,825
2700	1,200 amp		.67	23.810		4,300	1,450		5,750	6,900
2900	Heavy duty, 240 volt, 3 pole NEMA 1 fusible									
2910	30 amp	1 Elec	2.69	2.980	Ea.	123	183		306	405
3000	60 amp		1.93	4.140		205	254		459	605
3300	100 amp		1.60	5.010		315	305		620	805
3500	200 amp		1.09	7.330		545	450		995	1,275
3700	400 amp	2 Elec	1.51	10.580		1,475	650		2,125	2,600
3900	600 amp		1.01	15.870		2,275	975		3,250	3,950
4100	800 amp		.79	20.260		5,650	1,250		6,900	8,050
4300	1,200 amp		.67	23.810		7,600	1,450		9,050	10,500
4340	2 pole fusible, 30 amp	1 Elec	2.94	2.720		93	167		260	350
4350	600 volt, 3 pole, fusible, 30 amp		2.69	2.980		199	183		382	490
4380	60 amp		1.93	4.140		243	254		497	645
4400	100 amp		1.60	5.010		440	305		745	945
4420	200 amp		1.09	7.330		625	450		1,075	1,350
4440	400 amp	2 Elec	1.51	10.580		1,675	650		2,325	2,825
4450	600 amp		1.01	15.870		2,975	975		3,950	4,725
4460	800 amp		.79	20.260		5,650	1,250		6,900	8,050
4480	1,200 amp		.67	23.810		7,600	1,450		9,050	10,500
4500	240 volt 3 pole NEMA 3R (no hubs), fusible									
4510	30 amp	1 Elec	2.60	3.070	Ea.	214	188		402	515
4700	60 amp		1.85	4.330		340	266		606	770
4900	100 amp		1.51	5.290		495	325		820	1,025
5100	200 amp		1.01	7.940		670	485		1,155	1,450
5300	400 amp	2 Elec	1.34	11.900		1,525	730		2,255	2,750
5500	600 amp	"	.84	19.050		3,075	1,175		4,250	5,125
5510	Heavy duty, 600 volt, 3 pole 3 ph. NEMA 3R fusible, 30 amp	1 Elec	2.60	3.070		335	188		523	650
5520	60 amp		1.85	4.330		410	266		676	845
5530	100 amp		1.51	5.290		600	325		925	1,150
5540	200 amp		1.01	7.940		835	485		1,320	1,650
5550	400 amp	2 Elec	1.34	11.900		2,025	730		2,755	3,300
5700	600 volt, 3 pole NEMA 3R nonfused									
5710	30 amp	1 Elec	2.60	3.070	Ea.	182	188		370	480
5900	60 amp		1.85	4.330		320	266		586	745
6100	100 amp		1.51	5.290		450	325		775	980
6300	200 amp		1.01	7.940		540	485		1,025	1,325
6500	400 amp	2 Elec	1.34	11.900		1,400	730		2,130	2,625
6700	600 amp	"	.84	19.050		2,925	1,175		4,100	4,975
6900	600 volt, 6 pole NEMA 3R nonfused, 30 amp	1 Elec	2.27	3.530		1,225	217		1,442	1,675
7100	60 amp		1.68	4.760		2,325	292		2,617	3,000
7300	100 amp		1.26	6.350		2,250	390		2,640	3,050
7500	200 amp		1.01	7.940		4,175	485		4,660	5,325
7600	600 volt, 3 pole NEMA 7 explosion proof nonfused									
7610	30 amp	1 Elec	1.85	4.330	Ea.	1,700	266		1,966	2,275
7620	60 amp		1.51	5.290		2,050	325		2,375	2,725
7630	100 amp		1.01	7.940		3,425	485		3,910	4,500
7640	200 amp		.67	11.900		7,375	730		8,105	9,175
7710	600 volt 6 pole, NEMA 3R fusible, 30 amp		2.27	3.530		1,825	217		2,042	2,325

26 28 Low-Voltage Circuit Protective Devices

26 28 16 – Enclosed Switches and Circuit Breakers

26 28 16.20 Safety Switches		Crew	Daily Output	Labor-Hours	Unit	Material	2020 Bare Costs Labor	Equipment	Total	Total Incl O&P
7900	60 amp	1 Elec	1.68	4.760	Ea.	2,575	292		2,867	3,250
8100	100 amp		1.26	6.350		2,050	390		2,440	2,825
8110	240 volt 3 pole, NEMA 12 fusible, 30 amp		2.60	3.070		310	188		498	620
8120	60 amp		1.85	4.330		545	266		811	995
8130	100 amp		1.51	5.290		660	325		985	1,200
8140	200 amp	↓	1.01	7.940		740	485		1,225	1,550
8150	400 amp	2 Elec	1.34	11.900		2,025	730		2,755	3,300
8160	600 amp	"	.84	19.050		3,325	1,175		4,500	5,400
8180	600 volt 3 pole, NEMA 12 fusible, 30 amp	1 Elec	2.60	3.070		445	188		633	770
8190	60 amp		1.85	4.330		560	266		826	1,000
8200	100 amp		1.51	5.290		835	325		1,160	1,400
8210	200 amp	↓	1.01	7.940		1,150	485		1,635	1,975
8220	400 amp	2 Elec	1.34	11.900		2,875	730		3,605	4,250
8230	600 amp	"	.84	19.050		4,775	1,175		5,950	7,000
8240	600 volt 3 pole, NEMA 12 nonfused, 30 amp	1 Elec	2.60	3.070		278	188		466	585
8250	60 amp		1.85	4.330		430	266		696	865
8260	100 amp		1.51	5.290		620	325		945	1,175
8270	200 amp	↓	1.01	7.940		675	485		1,160	1,475
8280	400 amp	2 Elec	1.34	11.900		1,700	730		2,430	2,950
8290	600 amp	"	.84	19.050		3,475	1,175		4,650	5,575
8310	600 volt, 3 pole NEMA 4 fusible, 30 amp	1 Elec	2.52	3.170		930	194		1,124	1,325
8320	60 amp		1.85	4.330		1,100	266		1,366	1,625
8330	100 amp		1.51	5.290		2,275	325		2,600	2,975
8340	200 amp	↓	1.01	7.940		3,000	485		3,485	4,025
8350	400 amp	2 Elec	1.34	11.900		5,975	730		6,705	7,650
8360	600 volt 3 pole NEMA 4 nonfused, 30 amp	1 Elec	2.52	3.170		825	194		1,019	1,200
8370	60 amp		1.85	4.330		985	266		1,251	1,475
8380	100 amp		1.51	5.290		3,400	325		3,725	4,200
8390	200 amp	↓	1.01	7.940		2,775	485		3,260	3,775
8400	400 amp	2 Elec	1.34	11.900		5,100	730		5,830	6,675
8490	Motor starters, manual, single phase, NEMA 1	1 Elec	5.38	1.490		119	91.50		210.50	267
8500	NEMA 4		3.36	2.380		310	146		456	555
8700	NEMA 7		3.36	2.380		375	146		521	625
8900	NEMA 1 with pilot		5.38	1.490		148	91.50		239.50	299
8920	3 pole, NEMA 1, 230/460 volt, 5 HP, size 0		2.94	2.720		242	167		409	515
8940	10 HP, size 1		1.68	4.760		287	292		579	750
9010	Disc. switch, 600 volt 3 pole fusible, 30 amp, to 10 HP motor		2.69	2.980		420	183		603	730
9050	60 amp, to 30 HP motor		1.93	4.140		965	254		1,219	1,425
9070	100 amp, to 60 HP motor		1.60	5.010		965	305		1,270	1,500
9100	200 amp, to 125 HP motor	↓	1.09	7.330		1,450	450		1,900	2,275
9110	400 amp, to 200 HP motor	2 Elec	1.51	10.580	↓	3,650	650		4,300	4,975

26 28 16.40 Time Switches

		Crew	Daily Output	Labor-Hours	Unit	Material	2020 Bare Costs Labor	Equipment	Total	Total Incl O&P
0010	**TIME SWITCHES**									
0100	Single pole, single throw, 24 hour dial	1 Elec	3.36	2.380	Ea.	157	146		303	390
0200	24 hour dial with reserve power		3.02	2.650		810	163		973	1,125
0300	Astronomic dial		3.02	2.650		305	163		468	580
0400	Astronomic dial with reserve power		2.77	2.890		1,125	177		1,302	1,525
0500	7 day calendar dial		2.77	2.890		261	177		438	550
0600	7 day calendar dial with reserve power		2.69	2.980		254	183		437	550
0700	Photo cell 2,000 watt		6.72	1.190		34	73		107	147
1080	Load management device, 4 loads		1.68	4.760		1,375	292		1,667	1,925
1100	8 loads	↓	.84	9.520	↓	2,875	585		3,460	4,025

26 28 Low-Voltage Circuit Protective Devices

26 28 16 – Enclosed Switches and Circuit Breakers

26 28 16.50 Meter Socket Entry Hub		Crew	Daily Output	Labor-Hours	Unit	Material	2020 Bare Costs Labor	Equipment	Total	Total Incl O&P
0010	**METER SOCKET ENTRY HUB**									
0100	Meter socket entry hub closing cap, 3/4 to 4 inch	1 Elec	33.60	.238	Ea.	2.43	14.60		17.03	24.50
0110	Meter socket entry hub, 3/4 conduit		33.60	.238		20	14.60		34.60	44
0120	1 inch conduit		33.60	.238		21	14.60		35.60	45
0130	1.25 inch conduit		33.60	.238		19.50	14.60		34.10	43.50
0140	1.50 inch conduit		33.60	.238		19.60	14.60		34.20	43.50
0150	2 inch conduit		33.60	.238		20.50	14.60		35.10	44.50
0160	2.50 inch conduit		33.60	.238		62	14.60		76.60	90
0170	3 inch conduit		33.60	.238		93	14.60		107.60	124

26 29 Low-Voltage Controllers

26 29 13 – Enclosed Controllers

26 29 13.10 Contactors, AC

		Crew	Daily Output	Labor-Hours	Unit	Material	2020 Bare Costs Labor	Equipment	Total	Total Incl O&P
0010	**CONTACTORS, AC** Enclosed (NEMA 1)									
0050	Lighting, 600 volt 3 pole, electrically held									
0100	20 amp	1 Elec	3.36	2.380	Ea.	375	146		521	625
0200	30 amp		3.02	2.650		300	163		463	570
0300	60 amp		2.52	3.170		625	194		819	980
0400	100 amp		2.10	3.810		1,000	234		1,234	1,475
0500	200 amp		1.18	6.800		2,525	415		2,940	3,400
0600	300 amp	2 Elec	1.34	11.900		4,525	730		5,255	6,050
0800	600 volt 3 pole, mechanically held, 30 amp	1 Elec	3.02	2.650		590	163		753	890
0900	60 amp		2.52	3.170		1,175	194		1,369	1,600
1000	75 amp		2.35	3.400		1,500	209		1,709	1,950
1100	100 amp		2.10	3.810		1,700	234		1,934	2,225
1200	150 amp		1.68	4.760		3,275	292		3,567	4,025
1300	200 amp		1.18	6.800		5,650	415		6,065	6,850
1500	Magnetic with auxiliary contact, size 00, 9 amp		3.36	2.380		189	146		335	425
1600	Size 0, 18 amp		3.36	2.380		225	146		371	465
1700	Size 1, 27 amp		3.02	2.650		256	163		419	525
1800	Size 2, 45 amp		2.52	3.170		475	194		669	815
1900	Size 3, 90 amp		2.10	3.810		770	234		1,004	1,200
2000	Size 4, 135 amp		1.93	4.140		1,750	254		2,004	2,300
2100	Size 5, 270 amp	2 Elec	1.51	10.580		3,675	650		4,325	5,025
2200	Size 6, 540 amp		1.01	15.870		10,800	975		11,775	13,400
2300	Size 7, 810 amp		.84	19.050		14,500	1,175		15,675	17,800
2310	Size 8, 1,215 amp		.67	23.810		22,600	1,450		24,050	27,000
2500	Magnetic, 240 volt, 1-2 pole, 0.75 HP motor	1 Elec	3.36	2.380		152	146		298	385
2520	2 HP motor		3.02	2.650		201	163		364	465
2540	5 HP motor		2.10	3.810		415	234		649	805
2560	10 HP motor		1.18	6.800		680	415		1,095	1,375
2600	240 volt or less, 3 pole, 0.75 HP motor		3.36	2.380		152	146		298	385
2620	5 HP motor		3.02	2.650		189	163		352	450
2640	10 HP motor		3.02	2.650		219	163		382	485
2660	15 HP motor		2.10	3.810		440	234		674	830
2700	25 HP motor		2.10	3.810		440	234		674	830
2720	30 HP motor	2 Elec	2.35	6.800		730	415		1,145	1,425
2740	40 HP motor		2.35	6.800		730	415		1,145	1,425
2760	50 HP motor		1.34	11.900		730	730		1,460	1,875
2800	75 HP motor		1.34	11.900		1,725	730		2,455	2,950

26 29 13.10 Contactors, AC

		Crew	Daily Output	Labor-Hours	Unit	Material	2020 Bare Costs Labor	Equipment	Total	Total Incl O&P
2820	100 HP motor	2 Elec	.84	19.050	Ea.	1,725	1,175		2,900	3,625
2860	150 HP motor		.84	19.050		3,650	1,175		4,825	5,775
2880	200 HP motor		.84	19.050		3,650	1,175		4,825	5,775
3000	600 volt, 3 pole, 5 HP motor	1 Elec	3.36	2.380		189	146		335	425
3020	10 HP motor		3.02	2.650		219	163		382	485
3040	25 HP motor		2.52	3.170		440	194		634	770
3100	50 HP motor		2.10	3.810		730	234		964	1,150
3160	100 HP motor	2 Elec	2.35	6.800		1,725	415		2,140	2,500
3220	200 HP motor	"	1.34	11.900		3,650	730		4,380	5,100

26 29 13.20 Control Stations

		Crew	Daily Output	Labor-Hours	Unit	Material	2020 Bare Costs Labor	Equipment	Total	Total Incl O&P
0010	**CONTROL STATIONS**									
0050	NEMA 1, heavy duty, stop/start	1 Elec	6.72	1.190	Ea.	168	73		241	294
0100	Stop/start, pilot light		5.21	1.540		229	94.50		323.50	390
0200	Hand/off/automatic		5.21	1.540		124	94.50		218.50	277
0400	Stop/start/reverse		4.45	1.800		226	110		336	415
0500	NEMA 7, heavy duty, stop/start		5.04	1.590		610	97.50		707.50	820
0600	Stop/start, pilot light		3.36	2.380		745	146		891	1,025
0700	NEMA 7 or 9, 1 element		5.04	1.590		500	97.50		597.50	690
0800	2 element		5.04	1.590		640	97.50		737.50	850
0900	3 element		3.36	2.380		1,375	146		1,521	1,750
0910	Selector switch, 2 position		5.04	1.590		500	97.50		597.50	695
0920	3 position		3.36	2.380		500	146		646	765
0930	Oiltight, 1 element		6.72	1.190		124	73		197	245
0940	2 element		5.21	1.540		171	94.50		265.50	330
0950	3 element		4.45	1.800		157	110		267	335
0960	Selector switch, 2 position		5.21	1.540		125	94.50		219.50	278
0970	3 position		4.45	1.800		130	110		240	305

26 29 13.30 Control Switches

		Crew	Daily Output	Labor-Hours	Unit	Material	2020 Bare Costs Labor	Equipment	Total	Total Incl O&P
0010	**CONTROL SWITCHES** Field installed									
6000	Push button 600 V 10A, momentary contact									
6150	Standard operator with colored button	1 Elec	28.56	.280	Ea.	21	17.20		38.20	48.50
6160	With single block 1NO 1NC		15.12	.529		46.50	32.50		79	99.50
6170	With double block 2NO 2NC		12.60	.635		67	39		106	132
6180	Std operator w/mushroom button 1-9/16" diam.		28.56	.280		42.50	17.20		59.70	72.50
6190	Std operator w/mushroom button 2-1/4" diam.									
6200	With single block 1NO 1NC	1 Elec	15.12	.529	Ea.	63	32.50		95.50	118
6210	With double block 2NO 2NC		12.60	.635		84.50	39		123.50	151
6500	Maintained contact, selector operator		28.56	.280		63	17.20		80.20	94.50
6510	With single block 1NO 1NC		15.12	.529		84.50	32.50		117	142
6520	With double block 2NO 2NC		12.60	.635		106	39		145	175
6560	Spring-return selector operator		28.56	.280		63	17.20		80.20	94.50
6570	With single block 1NO 1NC		15.12	.529		84.50	32.50		117	142
6580	With double block 2NO 2NC		12.60	.635		106	39		145	175
6620	Transformer operator w/illuminated									
6630	button 6 V #12 lamp	1 Elec	26.88	.298	Ea.	106	18.25		124.25	144
6640	With single block 1NO 1NC w/guard		13.44	.595		128	36.50		164.50	195
6650	With double block 2NO 2NC w/guard		10.92	.733		149	45		194	231
6690	Combination operator		28.56	.280		63	17.20		80.20	94.50
6700	With single block 1NO 1NC		15.12	.529		84.50	32.50		117	142
6710	With double block 2NO 2NC		12.60	.635		106	39		145	175
9000	Indicating light unit, full voltage									
9010	110-125 V front mount	1 Elec	26.88	.298	Ea.	84.50	18.25		102.75	120

Post-Installation Change Order Costs

26 29 Low-Voltage Controllers

26 29 13 – Enclosed Controllers

26 29 13.30 Control Switches		Crew	Daily Output	Labor-Hours	Unit	Material	2020 Bare Costs Labor	Equipment	Total	Total Incl O&P
9020	130 V resistor type	1 Elec	26.88	.298	Ea.	69.50	18.25		87.75	104
9030	6 V transformer type	↓	26.88	.298	↓	77	18.25		95.25	112

26 29 13.40 Relays

		Crew	Daily Output	Labor-Hours	Unit	Material	Labor	Equipment	Total	Total Incl O&P
0010	**RELAYS** Enclosed (NEMA 1)									
0050	600 volt AC, 1 pole, 12 amp	1 Elec	4.45	1.800	Ea.	98	110		208	272
0100	2 pole, 12 amp		4.20	1.900		98	117		215	282
0200	4 pole, 10 amp		3.78	2.120		131	130		261	340
0500	250 volt DC, 1 pole, 15 amp		4.45	1.800		128	110		238	305
0600	2 pole, 10 amp		4.20	1.900		122	117		239	310
0700	4 pole, 4 amp	↓	3.78	2.120	↓	162	130		292	370

26 29 23 – Variable-Frequency Motor Controllers

26 29 23.10 Variable Frequency Drives/Adj. Frequency Drives

			Crew	Daily Output	Labor-Hours	Unit	Material	Labor	Equipment	Total	Total Incl O&P
0010	**VARIABLE FREQUENCY DRIVES/ADJ. FREQUENCY DRIVES**										
0100	Enclosed (NEMA 1), 460 volt, for 3 HP motor size	G	1 Elec	.67	11.900	Ea.	1,825	730		2,555	3,075
0110	5 HP motor size	G		.67	11.900		2,050	730		2,780	3,325
0120	7.5 HP motor size	G		.56	14.210		2,450	870		3,320	3,975
0130	10 HP motor size	G	↓	.56	14.210		2,775	870		3,645	4,350
0140	15 HP motor size	G	2 Elec	.75	21.400		3,450	1,325		4,775	5,750
0150	20 HP motor size	G		.75	21.400		4,350	1,325		5,675	6,725
0160	25 HP motor size	G		.56	28.430		5,125	1,750		6,875	8,250
0170	30 HP motor size	G		.56	28.430		6,250	1,750		8,000	9,475
0180	40 HP motor size	G		.56	28.430		7,225	1,750		8,975	10,600
0190	50 HP motor size	G	↓	.45	35.940		9,650	2,200		11,850	13,900
0200	60 HP motor size	G	R-3	.47	42.520		11,700	2,600	400	14,700	17,200
0210	75 HP motor size	G		.47	42.520		13,700	2,600	400	16,700	19,400
0220	100 HP motor size	G		.42	47.620		16,800	2,900	450	20,150	23,200
0230	125 HP motor size	G		.42	47.620		18,200	2,900	450	21,550	24,800
0240	150 HP motor size	G		.42	47.620		23,600	2,900	450	26,950	30,800
0250	200 HP motor size	G		.35	56.690		30,600	3,475	535	34,610	39,500
1100	Custom-engineered, 460 volt, for 3 HP motor size	G	1 Elec	.47	17.010		3,425	1,050		4,475	5,325
1110	5 HP motor size	G		.47	17.010		3,450	1,050		4,500	5,350
1120	7.5 HP motor size	G		.39	20.260		3,100	1,250		4,350	5,250
1130	10 HP motor size	G	↓	.39	20.260		3,250	1,250		4,500	5,450
1140	15 HP motor size	G	2 Elec	.52	30.720		4,700	1,875		6,575	7,975
1150	20 HP motor size	G		.52	30.720		4,350	1,875		6,225	7,575
1160	25 HP motor size	G		.39	40.530		5,375	2,475		7,850	9,625
1170	30 HP motor size	G		.39	40.530		6,625	2,475		9,100	11,000
1180	40 HP motor size	G		.39	40.530		7,925	2,475		10,400	12,400
1190	50 HP motor size	G	↓	.31	51.480		10,500	3,150		13,650	16,200
1200	60 HP motor size	G	R-3	.33	61.050		15,800	3,725	575	20,100	23,600
1210	75 HP motor size	G		.33	61.050		14,600	3,725	575	18,900	22,300
1220	100 HP motor size	G		.29	68.030		18,500	4,150	645	23,295	27,200
1230	125 HP motor size	G		.29	68.030		19,800	4,150	645	24,595	28,700
1240	150 HP motor size	G		.29	68.030		22,500	4,150	645	27,295	31,700
1250	200 HP motor size	G	↓	.24	82.100	↓	25,700	5,025	775	31,500	36,500
2000	For complex & special design systems to meet specific										
2010	requirements, obtain quote from vendor.										

26 31 13.50 Solar Energy - Photovoltaics		Crew	Daily Output	Labor-Hours	Unit	Material	2020 Bare Costs Labor	Equipment	Total	Total Incl O&P
0010	**SOLAR ENERGY - PHOTOVOLTAICS**									
0220	Alt. energy source, photovoltaic module, 6 watt, 15 V	G 1 Elec	6.72	1.190	Ea.	54	73		127	169
0230	10 watt, 16.3 V	G	6.72	1.190		119	73		192	240
0240	20 watt, 14.5 V	G	6.72	1.190		176	73		249	305
0250	36 watt, 17 V	G	6.72	1.190		175	73		248	300
0260	55 watt, 17 V	G	6.72	1.190		238	73		311	370
0270	75 watt, 17 V	G	6.72	1.190		400	73		473	550
0280	130 watt, 33 V	G	6.72	1.190		605	73		678	780
0290	140 watt, 33 V	G	6.72	1.190		495	73		568	655
0300	150 watt, 33 V	G	6.72	1.190		450	73		523	605
0310	DC to AC inverter for, 12 V, 2,000 watt	G	3.36	2.380		1,425	146		1,571	1,775
0320	12 V, 2,500 watt	G	3.36	2.380		1,100	146		1,246	1,450
0330	24 V, 2,500 watt	G	3.36	2.380		1,675	146		1,821	2,075
0340	12 V, 3,000 watt	G	2.52	3.170		1,325	194		1,519	1,750
0350	24 V, 3,000 watt	G	2.52	3.170		2,400	194		2,594	2,950
0360	24 V, 4,000 watt	G	1.68	4.760		3,675	292		3,967	4,475
0370	48 V, 4,000 watt	G	1.68	4.760		3,050	292		3,342	3,800
0380	48 V, 5,500 watt	G	1.68	4.760		3,000	292		3,292	3,725
0390	PV components, combiner box, 10 lug, NEMA 3R enclosure	G	3.36	2.380		270	146		416	515
0400	Fuse, 15 A for combiner box	G	33.60	.238		21.50	14.60		36.10	46
0410	Battery charger controller w/temperature sensor	G	3.36	2.380		450	146		596	710
0420	Digital readout panel, displays hours, volts, amps, etc.	G	3.36	2.380		219	146		365	460
0430	Deep cycle solar battery, 6 V, 180 Ah (C/20)	G	6.72	1.190		335	73		408	480
0440	Battery interconn, 15" AWG #2/0, sealed w/copper ring lugs	G	13.44	.595		16.60	36.50		53.10	73
0442	Battery interconn, 24" AWG #2/0, sealed w/copper ring lugs	G	13.44	.595		26.50	36.50		63	83.50
0444	Battery interconn, 60" AWG #2/0, sealed w/copper ring lugs	G	13.44	.595		53	36.50		89.50	113
0446	Batt temp computer probe, RJ11 jack, 15' cord	G	13.44	.595		22	36.50		58.50	78.50
0450	System disconnect, DC 175 amp circuit breaker	G	6.72	1.190		223	73		296	355
0460	Conduit box for inverter	G	6.72	1.190		63.50	73		136.50	179
0470	Low voltage disconnect	G	6.72	1.190		63.50	73		136.50	179
0480	Vented battery enclosure, wood	G 1 Carp	1.68	4.760		285	253		538	695
0490	PV rack system, roof, non-penetrating ballast, 1 panel	G R-1A	25.62	.625		1,000	34.50		1,034.50	1,175
0500	Penetrating surface mount, on steel framing, 1 panel		4.31	3.720		58	205		263	370
0510	On wood framing, 1 panel		9.24	1.730		57	95.50		152.50	205
0520	With standoff, 1 panel		9.24	1.730		66	95.50		161.50	215
0530	Ground, ballast, fixed, 3 panel		17.22	.929		1,100	51.50		1,151.50	1,275
0540	4 panel		17.22	.929		1,500	51.50		1,551.50	1,725
0550	5 panel		17.22	.929		1,975	51.50		2,026.50	2,250
0560	6 panel		17.22	.929		2,275	51.50		2,326.50	2,575
0570	Adjustable, 3 panel		17.22	.929		1,175	51.50		1,226.50	1,350
0580	4 panel		17.22	.929		1,600	51.50		1,651.50	1,825
0590	5 panel		17.22	.929		2,100	51.50		2,151.50	2,400
0600	6 panel		17.22	.929		2,425	51.50		2,476.50	2,750
0610	Passive tracking, 1 panel		17.22	.929		685	51.50		736.50	825
0620	2 panel		17.22	.929		1,350	51.50		1,401.50	1,575
0630	3 panel		17.22	.929		1,950	51.50		2,001.50	2,225
0640	4 panel		17.22	.929		2,100	51.50		2,151.50	2,375
0650	6 panel		17.22	.929		2,300	51.50		2,351.50	2,600
0660	8 panel		17.22	.929		3,425	51.50		3,476.50	3,825
1020	Photovoltaic module, Thin film	1 Elec	6.72	1.190		176	73		249	305
1040	Photovoltaic module, Cadium telluride		6.72	1.190		238	73		311	370
1060	Photovoltaic module, Polycrystalline		6.72	1.190		176	73		249	305
1080	Photovoltaic module, Monocrystalline		6.72	1.190		605	73		678	780

26 32 Packaged Generator Assemblies

26 32 13 – Engine Generators

26 32 13.13 Diesel-Engine-Driven Generator Sets	Crew	Daily Output	Labor-Hours	Unit	Material	2020 Bare Costs Labor	Equipment	Total	Total Incl O&P
0010 **DIESEL-ENGINE-DRIVEN GENERATOR SETS**									
2000 Diesel engine, including battery, charger,									
2010 muffler, & day tank, 30 kW	R-3	.46	43.290	Ea.	11,200	2,650	410	14,260	16,700
2100 50 kW		.35	56.690		21,800	3,475	535	25,810	29,800
2110 60 kW		.33	61.050		22,000	3,725	575	26,300	30,400
2200 75 kW		.29	68.030		25,300	4,150	645	30,095	34,700
2300 100 kW		.26	76.800		27,600	4,700	725	33,025	38,200
2400 125 kW		.24	82.100		29,000	5,025	775	34,800	40,200
2500 150 kW		.22	91.580		45,800	5,600	865	52,265	60,000
2501 Generator set, dsl eng in alum encl, incl btry, chgr, muf & day tank,150 kW		.22	91.580		41,500	5,600	865	47,965	55,000
2600 175 kW		.21	95.240		48,900	5,825	900	55,625	63,000
2700 200 kW		.20	99.210		49,900	6,075	940	56,915	65,000
2800 250 kW		.19	103		52,500	6,300	975	59,775	68,500
2850 275 kW		.18	108		58,000	6,600	1,025	65,625	75,000
2900 300 kW		.18	108		58,000	6,600	1,025	65,625	74,500
3000 350 kW		.17	119		67,000	7,275	1,125	75,400	86,000
3100 400 kW		.16	125		78,000	7,650	1,175	86,825	98,500
3200 500 kW		.15	132		106,500	8,075	1,250	115,825	130,500
3220 600 kW		.14	140		128,500	8,550	1,325	138,375	156,000
3230 650 kW	R-13	.32	131		170,500	7,775	740	179,015	200,000
3240 750 kW		.32	131		160,000	7,775	740	168,515	188,500
3250 800 kW		.30	138		156,500	8,175	775	165,450	185,000
3260 900 kW		.26	161		195,500	9,550	905	205,955	230,000
3270 1,000 kW		.23	185		205,000	11,000	1,050	217,050	243,000

26 32 13.16 Gas-Engine-Driven Generator Sets

	Crew	Daily Output	Labor-Hours	Unit	Material	2020 Bare Costs Labor	Equipment	Total	Total Incl O&P
0010 **GAS-ENGINE-DRIVEN GENERATOR SETS**									
0020 Gas or gasoline operated, includes battery,									
0050 charger, & muffler									
0200 7.5 kW	R-3	.70	28.690	Ea.	8,750	1,750	271	10,771	12,500
0300 11.5 kW		.60	33.530		12,400	2,050	315	14,765	17,000
0400 20 kW		.53	37.790		14,600	2,300	355	17,255	19,900
0500 35 kW		.46	43.290		17,400	2,650	410	20,460	23,500
0520 60 kW		.42	47.620		22,900	2,900	450	26,250	30,000
0600 80 kW	R-13	.34	125		28,500	7,400	705	36,605	43,200
0700 100 kW		.28	151		31,200	8,950	850	41,000	48,600
0800 125 kW		.24	178		64,000	10,600	1,000	75,600	87,500
0900 185 kW		.21	200		84,500	11,900	1,125	97,525	112,000

26 33 Battery Equipment

26 33 19 – Battery Units

26 33 19.10 Battery Units

	Crew	Daily Output	Labor-Hours	Unit	Material	2020 Bare Costs Labor	Equipment	Total	Total Incl O&P
0010 **BATTERY UNITS**									
0500 Salt water battery, 2.1 kWA, 24V, 750W, 30A, wired in crate, IP22 rated	1 Elec	13.65	.586	Ea.	955	36		991	1,100
0510 2.2 kWh, 48V, 800W, 17A, wired in crate, IP22 rated		13.65	.586		1,200	36		1,236	1,350
0520 25.9 kWh,48V, 11700W, 240A, wired in crate, IP2X rated		13.65	.586		14,200	36		14,236	15,700
0700 Nickel iron nife battery, deep cycle, 100Ah, 12V, 10 cells		11.97	.668		1,025	41		1,066	1,175
0710 24V, 10 cells		11.97	.668		2,175	41		2,216	2,450
0720 48V, 10 cells		11.97	.668		4,100	41		4,141	4,550
0730 Nickel iron nife battery, deep cycle, 300Ah, 12V, 10 cells		10.29	.777		3,025	47.50		3,072.50	3,400
0740 24V, 10 cells		10.29	.777		6,050	47.50		6,097.50	6,725
0750 48V, 10 cells		10.29	.777		12,700	47.50		12,747.50	14,000

Post-Installation Change Order Costs   For customer support on your Electrical Change Order Costs with RSMeans data, call 800.448.8182.

26 33 Battery Equipment

26 33 53 – Static Uninterruptible Power Supply

26 33 53.10 Uninterruptible Power Supply/Conditioner Trans.	Crew	Daily Output	Labor-Hours	Unit	Material	2020 Bare Costs Labor	2020 Bare Costs Equipment	Total	Total Incl O&P
0010 **UNINTERRUPTIBLE POWER SUPPLY/CONDITIONER TRANSFORMERS**									
0100 Volt. regulating, isolating transf., w/invert. & 10 min. battery pack									
0110 Single-phase, 120 V, 0.35 kVA	1 Elec	1.92	4.160	Ea.	1,150	255		1,405	1,625
0120 0.5 kVA		1.68	4.760		1,200	292		1,492	1,750
0130 For additional 55 min. battery, add to 0.35 kVA		1.92	4.160		675	255		930	1,125
0140 Add to 0.5 kVA		.96	8.350		705	510		1,215	1,550
0150 Single-phase, 120 V, 0.75 kVA		.67	11.900		1,525	730		2,255	2,750
0160 1.0 kVA		.67	11.900		2,175	730		2,905	3,450
0170 1.5 kVA	2 Elec	.96	16.710		3,725	1,025		4,750	5,625
0180 2 kVA	"	.75	21.400		4,050	1,325		5,375	6,400
0190 3 kVA	R-3	.53	37.790		5,000	2,300	355	7,655	9,350
0200 5 kVA		.35	56.690		7,100	3,475	535	11,110	13,600
0210 7.5 kVA		.28	72.150		9,075	4,400	680	14,155	17,300
0220 10 kVA		.24	85.030		12,200	5,200	805	18,205	22,100
0230 15 kVA		.18	108		15,700	6,600	1,025	23,325	28,200
0240 3 phase, 120/208 V input 120/208 V output, 20 kVA, incl 17 min. battery		.18	113		29,000	6,900	1,075	36,975	43,500
0242 30 kVA, incl 11 min. battery		.17	119		32,500	7,275	1,125	40,900	47,900
0250 40 kVA, incl 15 min. battery		.17	119		43,400	7,275	1,125	51,800	60,000
0260 480 V input 277/480 V output, 60 kVA, incl 6 min. battery		.16	125		48,800	7,650	1,175	57,625	66,000
0262 80 kVA, incl 4 min. battery		.16	125		58,000	7,650	1,175	66,825	76,500
0400 For additional 34 min./15 min. battery, add to 40 kVA		.60	33.350		15,100	2,050	315	17,465	20,100
0600 For complex & special design systems to meet specific									
0610 requirements, obtain quote from vendor									

26 35 Power Filters and Conditioners

26 35 13 – Capacitors

26 35 13.10 Capacitors Indoor

	Crew	Daily Output	Labor-Hours	Unit	Material	2020 Bare Costs Labor	2020 Bare Costs Equipment	Total	Total Incl O&P
0010 **CAPACITORS INDOOR**									
0020 240 volts, single & 3 phase, 0.5 kVAR	1 Elec	2.27	3.530	Ea.	600	217		817	980
0100 1.0 kVAR		2.27	3.530		725	217		942	1,125
0150 2.5 kVAR		1.68	4.760		815	292		1,107	1,325
0200 5.0 kVAR		1.51	5.290		795	325		1,120	1,350
0250 7.5 kVAR		1.34	5.950		1,625	365		1,990	2,350
0300 10 kVAR		1.26	6.350		1,750	390		2,140	2,500
0350 15 kVAR		1.09	7.330		2,075	450		2,525	2,950
0400 20 kVAR		.92	8.660		2,850	530		3,380	3,925
0450 25 kVAR		.84	9.520		3,350	585		3,935	4,550
1000 480 volts, single & 3 phase, 1 kVAR		2.27	3.530		545	217		762	920
1050 2 kVAR		2.27	3.530		630	217		847	1,025
1100 5 kVAR		1.68	4.760		795	292		1,087	1,300
1150 7.5 kVAR		1.68	4.760		855	292		1,147	1,375
1200 10 kVAR		1.68	4.760		1,200	292		1,492	1,750
1250 15 kVAR		1.68	4.760		1,525	292		1,817	2,100
1300 20 kVAR		1.34	5.950		1,600	365		1,965	2,300
1350 30 kVAR		1.26	6.350		1,575	390		1,965	2,325
1400 40 kVAR		1.01	7.940		2,275	485		2,760	3,250
1450 50 kVAR		.92	8.660		2,875	530		3,405	3,975
2000 600 volts, single & 3 phase, 1 kVAR		2.27	3.530		565	217		782	940
2050 2 kVAR		2.27	3.530		650	217		867	1,025
2100 5 kVAR		1.68	4.760		795	292		1,087	1,300

26 35 13 – Capacitors

26 35 13.10 Capacitors Indoor		Crew	Daily Output	Labor-Hours	Unit	Material	2020 Bare Costs Labor	Equipment	Total	Total Incl O&P
2150	7.5 kVAR	1 Elec	1.68	4.760	Ea.	855	292		1,147	1,375
2200	10 kVAR		1.68	4.760		1,225	292		1,517	1,750
2250	15 kVAR		1.34	5.950		1,550	365		1,915	2,250
2300	20 kVAR		1.34	5.950		1,675	365		2,040	2,375
2350	25 kVAR		1.26	6.350		1,825	390		2,215	2,575
2400	35 kVAR		1.18	6.800		2,225	415		2,640	3,075
2450	50 kVAR	↓	1.09	7.330	↓	2,850	450		3,300	3,825

26 35 26 – Harmonic Filters

26 35 26.10 Computer Isolation Transformer

		Crew	Daily Output	Labor-Hours	Unit	Material	Labor	Equipment	Total	Total Incl O&P
0010	**COMPUTER ISOLATION TRANSFORMER**									
0100	Computer grade									
0110	Single-phase, 120/240 V, 0.5 kVA	1 Elec	3.36	2.380	Ea.	455	146		601	715
0120	1.0 kVA		2.24	3.570		645	219		864	1,025
0130	2.5 kVA		1.68	4.760		985	292		1,277	1,500
0140	5 kVA	↓	.96	8.350	↓	1,125	510		1,635	2,000

26 35 26.20 Computer Regulator Transformer

		Crew	Daily Output	Labor-Hours	Unit	Material	Labor	Equipment	Total	Total Incl O&P
0010	**COMPUTER REGULATOR TRANSFORMER**									
0100	Ferro-resonant, constant voltage, variable transformer									
0110	Single-phase, 240 V, 0.5 kVA	1 Elec	2.24	3.570	Ea.	600	219		819	985
0120	1.0 kVA		1.68	4.760		825	292		1,117	1,350
0130	2.0 kVA		.84	9.520		1,425	585		2,010	2,425
0210	Plug-in unit 120 V, 0.14 kVA		6.72	1.190		350	73		423	490
0220	0.25 kVA		6.72	1.190		405	73		478	555
0230	0.5 kVA		6.72	1.190		600	73		673	770
0240	1.0 kVA		4.48	1.790		825	110		935	1,075
0250	2.0 kVA	↓	3.36	2.380	↓	1,425	146		1,571	1,775

26 35 26.30 Power Conditioner Transformer

		Crew	Daily Output	Labor-Hours	Unit	Material	Labor	Equipment	Total	Total Incl O&P
0010	**POWER CONDITIONER TRANSFORMER**									
0100	Electronic solid state, buck-boost, transformer, w/tap switch									
0110	Single-phase, 115 V, 3.0 kVA, + or - 3% accuracy	2 Elec	1.34	11.900	Ea.	3,250	730		3,980	4,650
0120	208, 220, 230, or 240 V, 5.0 kVA, + or - 1.5% accuracy	3 Elec	1.34	17.860		4,200	1,100		5,300	6,250
0130	5.0 kVA, + or - 6% accuracy	2 Elec	.96	16.710		3,775	1,025		4,800	5,675
0140	7.5 kVA, + or - 1.5% accuracy	3 Elec	1.26	19.050		5,350	1,175		6,525	7,625
0150	7.5 kVA, + or - 6% accuracy		1.34	17.860		4,450	1,100		5,550	6,525
0160	10.0 kVA, + or - 1.5% accuracy		1.12	21.480		7,125	1,325		8,450	9,775
0170	10.0 kVA, + or - 6% accuracy	↓	1.18	20.260		6,050	1,250		7,300	8,500

26 35 26.40 Transient Voltage Suppressor Transformer

		Crew	Daily Output	Labor-Hours	Unit	Material	Labor	Equipment	Total	Total Incl O&P
0010	**TRANSIENT VOLTAGE SUPPRESSOR TRANSFORMER**									
0110	Single-phase, 120 V, 1.8 kVA	1 Elec	3.36	2.380	Ea.	1,625	146		1,771	2,025
0120	3.6 kVA		3.36	2.380		2,625	146		2,771	3,100
0130	7.2 kVA		2.69	2.980		3,150	183		3,333	3,750
0150	240 V, 3.6 kVA		3.36	2.380		3,275	146		3,421	3,850
0160	7.2 kVA		3.36	2.380		3,825	146		3,971	4,425
0170	14.4 kVA		2.69	2.980		5,350	183		5,533	6,150
0210	Plug-in unit, 120 V, 1.8 kVA	↓	6.72	1.190	↓	1,125	73		1,198	1,350

26 35 53 – Voltage Regulators

26 35 53.10 Automatic Voltage Regulators

		Crew	Daily Output	Labor-Hours	Unit	Material	Labor	Equipment	Total	Total Incl O&P
0010	**AUTOMATIC VOLTAGE REGULATORS**									
0100	Computer grade, solid state, variable transf. volt. regulator									
0110	Single-phase, 120 V, 8.6 kVA	2 Elec	1.12	14.320	Ea.	6,725	880		7,605	8,700
0120	17.3 kVA	↓	.96	16.710	↓	7,950	1,025		8,975	10,300

26 35 53.10 Automatic Voltage Regulators		Crew	Daily Output	Labor-Hours	Unit	Material	2020 Bare Costs Labor	Equipment	Total	Total Incl O&P
0130	208/240 V, 7.5/8.6 kVA	2 Elec	1.12	14.320	Ea.	6,725	880		7,605	8,700
0140	13.5/15.6 kVA		1.12	14.320		7,950	880		8,830	10,100
0150	27.0/31.2 kVA		.96	16.710		10,000	1,025		11,025	12,500
0210	Two-phase, single control, 208/240 V, 15.0/17.3 kVA		.96	16.710		7,950	1,025		8,975	10,300
0220	Individual phase control, 15.0/17.3 kVA		.96	16.710		7,950	1,025		8,975	10,300
0230	30.0/34.6 kVA	3 Elec	1.12	21.480		10,000	1,325		11,325	13,000
0310	Three-phase single control, 208/240 V, 26/30 kVA	2 Elec	.84	19.050		7,950	1,175		9,125	10,500
0320	380/480 V, 24/30 kVA	"	.84	19.050		7,950	1,175		9,125	10,500
0330	43/54 kVA	3 Elec	1.12	21.480		14,400	1,325		15,725	17,900
0340	Individual phase control, 208 V, 26 kVA	"	1.12	21.480		7,950	1,325		9,275	10,700
0350	52 kVA	R-3	.76	26.160		10,000	1,600	247	11,847	13,600
0360	340/480 V, 24/30 kVA	2 Elec	.84	19.050		7,950	1,175		9,125	10,500
0370	43/54 kVA	"	.84	19.050		10,000	1,175		11,175	12,800
0380	48/60 kVA	3 Elec	1.12	21.480		14,500	1,325		15,825	18,000
0390	86/108 kVA	R-3	.76	26.160		16,100	1,600	247	17,947	20,400
0500	Standard grade, solid state, variable transformer volt. regulator									
0510	Single-phase, 115 V, 2.3 kVA	1 Elec	1.92	4.160	Ea.	2,125	255		2,380	2,725
0520	4.2 kVA		1.68	4.760		3,400	292		3,692	4,175
0530	6.6 kVA		.96	8.350		4,175	510		4,685	5,375
0540	13.0 kVA		.96	8.350		7,225	510		7,735	8,725
0550	16.6 kVA	2 Elec	1.03	15.490		8,525	950		9,475	10,800
0610	230 V, 8.3 kVA		1.12	14.320		7,225	880		8,105	9,250
0620	21.4 kVA		1.03	15.490		8,525	950		9,475	10,800
0630	29.9 kVA		1.03	15.490		8,525	950		9,475	10,800
0710	460 V, 9.2 kVA		1.12	14.320		7,225	880		8,105	9,250
0720	20.7 kVA		1.03	15.490		8,525	950		9,475	10,800
0810	Three-phase, 230 V, 13.1 kVA	3 Elec	1.18	20.260		7,225	1,250		8,475	9,800
0820	19.1 kVA		1.18	20.260		8,525	1,250		9,775	11,200
0830	25.1 kVA		1.03	23.230		8,525	1,425		9,950	11,500
0840	57.8 kVA	R-3	.80	25.060		15,500	1,525	237	17,262	19,500
0850	74.9 kVA	"	.76	26.160		15,500	1,600	247	17,347	19,600
0910	460 V, 14.3 kVA	3 Elec	1.18	20.260		7,225	1,250		8,475	9,800
0920	19.1 kVA		1.18	20.260		8,525	1,250		9,775	11,200
0930	27.9 kVA		1.03	23.230		8,525	1,425		9,950	11,500
0940	59.8 kVA	R-3	.84	23.810		15,500	1,450	225	17,175	19,400
0950	79.7 kVA		.80	25.060		17,300	1,525	237	19,062	21,600
0960	118 kVA		.80	25.060		18,200	1,525	237	19,962	22,600
1000	Laboratory grade, precision, electronic voltage regulator									
1110	Single-phase, 115 V, 0.5 kVA	1 Elec	1.92	4.160	Ea.	1,650	255		1,905	2,200
1120	1.0 kVA		1.68	4.760		1,750	292		2,042	2,350
1130	3.0 kVA		.67	11.900		2,450	730		3,180	3,775
1140	6.0 kVA	2 Elec	1.23	13.050		4,550	800		5,350	6,200
1150	10.0 kVA	3 Elec	.84	28.570		5,925	1,750		7,675	9,100
1160	15.0 kVA	"	1.26	19.050		6,750	1,175		7,925	9,175
1210	230 V, 3.0 kVA	1 Elec	.67	11.900		2,775	730		3,505	4,125
1220	6.0 kVA	2 Elec	1.23	13.050		4,650	800		5,450	6,300
1230	10.0 kVA	3 Elec	1.44	16.710		6,175	1,025		7,200	8,325
1240	15.0 kVA	"	1.34	17.860		6,975	1,100		8,075	9,300

26 35 53.30 Transient Suppressor/Voltage Regulator

		Crew	Daily Output	Labor-Hours	Unit	Material	Labor	Equipment	Total	Total Incl O&P
0010	**TRANSIENT SUPPRESSOR/VOLTAGE REGULATOR** (without isolation)									
0110	Single-phase, 115 V, 1.0 kVA	1 Elec	2.24	3.570	Ea.	1,100	219		1,319	1,550
0120	2.0 kVA		1.92	4.160		1,525	255		1,780	2,050

26 35 Power Filters and Conditioners

26 35 53 – Voltage Regulators

26 35 53.30 Transient Suppressor/Voltage Regulator	Crew	Daily Output	Labor-Hours	Unit	Material	2020 Bare Costs Labor	Equipment	Total	Total Incl O&P
0130	4.0 kVA	1 Elec	1.79	4.470	Ea.	1,875	274		2,149	2,450
0140	220 V, 1.0 kVA		2.24	3.570		1,100	219		1,319	1,550
0150	2.0 kVA		1.92	4.160		1,550	255		1,805	2,100
0160	4.0 kVA		1.79	4.470		1,975	274		2,249	2,575
0210	Plug-in unit, 120 V, 1.0 kVA		6.72	1.190		1,050	73		1,123	1,275
0220	2.0 kVA		6.72	1.190		1,500	73		1,573	1,750

26 36 Transfer Switches

26 36 13 – Manual Transfer Switches

26 36 13.10 Non-Automatic Transfer Switches

		Crew	Daily Output	Labor-Hours	Unit	Material	Labor	Equipment	Total	Total Incl O&P
0010	**NON-AUTOMATIC TRANSFER SWITCHES** enclosed									
0100	Manual operated, 480 volt 3 pole, 30 amp	1 Elec	1.93	4.140	Ea.	1,100	254		1,354	1,600
0150	60 amp		1.60	5.010		2,025	305		2,330	2,675
0200	100 amp		1.09	7.330		3,525	450		3,975	4,575
0250	200 amp	2 Elec	1.68	9.520		4,450	585		5,035	5,775
0300	400 amp		1.34	11.900		8,350	730		9,080	10,300
0350	600 amp		.84	19.050		5,475	1,175		6,650	7,775
1000	250 volt 3 pole, 30 amp	1 Elec	1.93	4.140		1,050	254		1,304	1,525
1100	60 amp		1.60	5.010		1,700	305		2,005	2,300
1150	100 amp		1.09	7.330		2,875	450		3,325	3,850
1200	200 amp	2 Elec	1.68	9.520		4,150	585		4,735	5,425
1300	600 amp	"	.84	19.050		10,500	1,175		11,675	13,300
1500	Electrically operated, 480 volt 3 pole, 60 amp	1 Elec	1.60	5.010		1,650	305		1,955	2,275
1600	100 amp	"	1.09	7.330		1,650	450		2,100	2,500
1650	200 amp	2 Elec	1.68	9.520		2,725	585		3,310	3,850
1700	400 amp		1.34	11.900		3,800	730		4,530	5,250
1750	600 amp		.84	19.050		5,475	1,175		6,650	7,775
2000	250 volt 3 pole, 30 amp	1 Elec	1.93	4.140		1,800	254		2,054	2,350
2050	60 amp	"	1.60	5.010		2,150	305		2,455	2,800
2150	200 amp	2 Elec	1.68	9.520		3,525	585		4,110	4,750
2200	400 amp		1.34	11.900		4,925	730		5,655	6,475
2250	600 amp		.84	19.050		7,075	1,175		8,250	9,550
2500	NEMA 3R, 480 volt 3 pole, 60 amp	1 Elec	1.51	5.290		2,550	325		2,875	3,275
2550	100 amp	"	1.01	7.940		3,875	485		4,360	4,975
2600	200 amp	2 Elec	1.51	10.580		5,300	650		5,950	6,800
2650	400 amp	"	1.18	13.610		4,150	835		4,985	5,825
2800	NEMA 3R, 250 volt 3 pole solid state, 100 amp	1 Elec	1.01	7.940		3,975	485		4,460	5,100
2850	150 amp	2 Elec	1.51	10.580		5,275	650		5,925	6,800
2900	250 volt 2 pole solid state, 100 amp	1 Elec	1.09	7.330		3,875	450		4,325	4,925
2950	150 amp	2 Elec	1.68	9.520		5,150	585		5,735	6,525

26 36 23 – Automatic Transfer Switches

26 36 23.10 Automatic Transfer Switch Devices

		Crew	Daily Output	Labor-Hours	Unit	Material	Labor	Equipment	Total	Total Incl O&P
0010	**AUTOMATIC TRANSFER SWITCH DEVICES**									
0015	Switches, enclosed 120/240 volt, 2 pole, 30 amp	1 Elec	2.02	3.970	Ea.	2,025	244		2,269	2,600
0020	70 amp		1.68	4.760		2,025	292		2,317	2,650
0030	100 amp		1.13	7.050		2,025	435		2,460	2,875
0040	225 amp	2 Elec	1.76	9.070		2,975	555		3,530	4,100
0050	400 amp		1.43	11.200		4,525	685		5,210	6,025
0060	600 amp		.89	17.970		9,575	1,100		10,675	12,200
0070	800 amp		.71	22.680		11,300	1,400		12,700	14,500

26 36 Transfer Switches

26 36 23 – Automatic Transfer Switches

26 36 23.10 Automatic Transfer Switch Devices	Crew	Daily Output	Labor-Hours	Unit	Material	2020 Bare Costs Labor	Equipment	Total	Total Incl O&P
0100	Switches, enclosed 480 volt, 3 pole, 30 amp	1 Elec	1.93	4.140	Ea.	3,425	254		3,679	4,150
0200	60 amp		1.60	5.010		3,425	305		3,730	4,225
0300	100 amp		1.09	7.330		3,425	450		3,875	4,450
0400	150 amp	2 Elec	2.02	7.940		4,200	485		4,685	5,350
0500	225 amp		1.68	9.520		5,400	585		5,985	6,825
0600	260 amp		1.68	9.520		6,250	585		6,835	7,750
0700	400 amp		1.34	11.900		7,850	730		8,580	9,700
0800	600 amp		.84	19.050		11,300	1,175		12,475	14,200
0900	800 amp		.67	23.810		13,300	1,450		14,750	16,800
1000	1,000 amp		.64	25.060		18,500	1,525		20,025	22,700
1100	1,200 amp		.59	27.210		25,400	1,675		27,075	30,500
1200	1,600 amp		.50	31.750		28,900	1,950		30,850	34,700
1300	2,000 amp		.42	38.100		32,400	2,325		34,725	39,100
1600	Accessories, time delay on engine starting					284			284	310
1700	Adjustable time delay on retransfer					284			284	310
1800	Shunt trips for customer connections					505			505	555
1900	Maintenance select switch					115			115	127
2000	Auxiliary contact when normal fails					103			103	114
2100	Pilot light-emergency					115			115	127
2200	Pilot light-normal					115			115	127
2300	Auxiliary contact-closed on normal					133			133	147
2400	Auxiliary contact-closed on emergency					133			133	147
2500	Emergency source sensing, frequency relay					585			585	645

26 41 Facility Lightning Protection

26 41 13 – Lightning Protection for Structures

26 41 13.13 Lightning Protection for Buildings

		Crew	Daily Output	Labor-Hours	Unit	Material	2020 Bare Costs Labor	Equipment	Total	Total Incl O&P
0010	**LIGHTNING PROTECTION FOR BUILDINGS**									
0200	Air terminals & base, copper									
0400	3/8" diameter x 10" (to 75' high)	1 Elec	6.72	1.190	Ea.	26	73		99	138
0500	1/2" diameter x 12" (over 75' high)		6.72	1.190		30	73		103	142
0520	1/2" diameter x 24"		6.13	1.300		40	80		120	163
0540	1/2" diameter x 60"		5.63	1.420		73.50	87		160.50	211
1000	Aluminum, 1/2" diameter x 12" (to 75' high)		6.72	1.190		17.15	73		90.15	128
1020	1/2" diameter x 24"		6.13	1.300		18.95	80		98.95	140
1040	1/2" diameter x 60"		5.63	1.420		25	87		112	158
1100	5/8" diameter x 12" (over 75' high)		6.72	1.190		18.25	73		91.25	129
2000	Cable, copper, 220 lb. per thousand ft. (to 75' high)		268.80	.030	L.F.	3.18	1.83		5.01	6.20
2100	375 lb. per thousand ft. (over 75' high)		193.20	.041		5.75	2.54		8.29	10.15
2500	Aluminum, 101 lb. per thousand ft. (to 75' high)		235.20	.034		.91	2.09		3	4.11
2600	199 lb. per thousand ft. (over 75' high)		201.60	.040		1.33	2.43		3.76	5.10
3000	Arrester, 175 volt AC to ground		6.72	1.190	Ea.	163	73		236	288
3100	650 volt AC to ground		5.63	1.420	"	152	87		239	298

Post-Installation Change Order Costs

26 51 13 – Interior Lighting Fixtures, Lamps, and Ballasts

26 51 13.10 Fixture Hangers		Crew	Daily Output	Labor-Hours	Unit	Material	2020 Bare Costs Labor	Equipment	Total	Total Incl O&P
0010	**FIXTURE HANGERS**									
0220	Box hub cover	1 Elec	26.88	.298	Ea.	4.13	18.25		22.38	31.50
0240	Canopy		10.08	.794		9.05	48.50		57.55	82.50
0260	Connecting block		33.60	.238		2.83	14.60		17.43	25
0280	Cushion hanger		13.44	.595		23.50	36.50		60	80.50
0300	Box hanger, with mounting strap		6.72	1.190		9.60	73		82.60	120
0320	Connecting block		33.60	.238		2.63	14.60		17.23	25
0340	Flexible, 1/2" diameter, 4" long		10.08	.794		15.75	48.50		64.25	90
0360	6" long		10.08	.794		17.10	48.50		65.60	91.50
0380	8" long		10.08	.794		18.95	48.50		67.45	93.50
0400	10" long		10.08	.794		20	48.50		68.50	94.50
0420	12" long		10.08	.794		21	48.50		69.50	96
0440	15" long		10.08	.794		22.50	48.50		71	97
0460	18" long		10.08	.794		26	48.50		74.50	101
0480	3/4" diameter, 4" long		8.40	.952		20	58.50		78.50	109
0500	6" long		8.40	.952		22	58.50		80.50	112
0520	8" long		8.40	.952		22.50	58.50		81	112
0540	10" long		8.40	.952		25	58.50		83.50	115
0560	12" long		8.40	.952		27	58.50		85.50	117
0580	15" long		8.40	.952		30	58.50		88.50	120
0600	18" long		8.40	.952		32.50	58.50		91	123

26 51 13.40 Interior HID Fixtures

26 51 13.40 Interior HID Fixtures		Crew	Daily Output	Labor-Hours	Unit	Material	2020 Bare Costs Labor	Equipment	Total	Total Incl O&P
0010	**INTERIOR HID FIXTURES** Incl. lamps and mounting hardware									
0700	High pressure sodium, recessed, round, 70 watt	1 Elec	2.94	2.720	Ea.	585	167		752	895
0720	100 watt		2.94	2.720		705	167		872	1,025
0740	150 watt		2.69	2.980		725	183		908	1,075
0760	Square, 70 watt		3.02	2.650		625	163		788	930
0780	100 watt		3.02	2.650		705	163		868	1,025
0820	250 watt		2.52	3.170		890	194		1,084	1,275
0840	1,000 watt	2 Elec	4.03	3.970		1,775	244		2,019	2,325
0860	Surface, round, 70 watt	1 Elec	2.52	3.170		960	194		1,154	1,350
0880	100 watt		2.52	3.170		985	194		1,179	1,375
0900	150 watt		2.27	3.530		955	217		1,172	1,375
0920	Square, 70 watt		2.52	3.170		830	194		1,024	1,200
0940	100 watt		2.52	3.170		865	194		1,059	1,250
0980	250 watt		2.10	3.810		695	234		929	1,125
1040	Pendent, round, 70 watt		2.52	3.170		980	194		1,174	1,375
1060	100 watt		2.52	3.170		860	194		1,054	1,225
1080	150 watt		2.27	3.530		960	217		1,177	1,375
1100	Square, 70 watt		2.52	3.170		970	194		1,164	1,375
1120	100 watt		2.52	3.170		985	194		1,179	1,375
1140	150 watt		2.27	3.530		1,000	217		1,217	1,450
1160	250 watt		2.10	3.810		1,375	234		1,609	1,850
1180	400 watt		2.02	3.970		1,450	244		1,694	1,950
1220	Wall, round, 70 watt		2.52	3.170		860	194		1,054	1,250
1240	100 watt		2.52	3.170		855	194		1,049	1,225
1260	150 watt		2.27	3.530		850	217		1,067	1,250
1300	Square, 70 watt		2.52	3.170		880	194		1,074	1,250
1320	100 watt		2.52	3.170		965	194		1,159	1,375
1340	150 watt		2.27	3.530		950	217		1,167	1,375
1360	250 watt		2.10	3.810		960	234		1,194	1,400
1380	400 watt	2 Elec	4.03	3.970		1,175	244		1,419	1,675

26 51 13.40 Interior HID Fixtures		Crew	Daily Output	Labor-Hours	Unit	Material	2020 Bare Costs Labor	Equipment	Total	Total Incl O&P
1400	1,000 watt	2 Elec	3.02	5.290	Ea.	1,775	325		2,100	2,425
1500	Metal halide, recessed, round, 175 watt	1 Elec	2.86	2.800		445	172		617	745
1520	250 watt	"	2.69	2.980		545	183		728	870
1540	400 watt	2 Elec	4.87	3.280		815	201		1,016	1,200
1580	Square, 175 watt	1 Elec	2.86	2.800		460	172		632	760
1640	Surface, round, 175 watt		2.44	3.280		600	201		801	960
1660	250 watt	↓	2.27	3.530		1,225	217		1,442	1,675
1680	400 watt	2 Elec	4.03	3.970		1,225	244		1,469	1,725
1720	Square, 175 watt	1 Elec	2.44	3.280		655	201		856	1,025
1800	Pendent, round, 175 watt		2.44	3.280		875	201		1,076	1,250
1820	250 watt	↓	2.27	3.530		1,325	217		1,542	1,800
1840	400 watt	2 Elec	4.03	3.970		1,225	244		1,469	1,725
1880	Square, 175 watt	1 Elec	2.44	3.280		500	201		701	850
1900	250 watt	"	2.27	3.530		740	217		957	1,125
1920	400 watt	2 Elec	4.03	3.970		1,250	244		1,494	1,750
1980	Wall, round, 175 watt	1 Elec	2.44	3.280		975	201		1,176	1,375
2000	250 watt	"	2.27	3.530		1,175	217		1,392	1,600
2020	400 watt	2 Elec	4.03	3.970		1,025	244		1,269	1,525
2060	Square, 175 watt	1 Elec	2.44	3.280		515	201		716	865
2080	250 watt	"	2.27	3.530		725	217		942	1,125
2100	400 watt	2 Elec	4.03	3.970		1,025	244		1,269	1,500
2800	High pressure sodium, vaporproof, recessed, 70 watt	1 Elec	2.94	2.720		765	167		932	1,100
2820	100 watt		2.94	2.720		780	167		947	1,100
2840	150 watt		2.69	2.980		800	183		983	1,150
2900	Surface, 70 watt		2.52	3.170		860	194		1,054	1,225
2920	100 watt		2.52	3.170		900	194		1,094	1,275
2940	150 watt		2.27	3.530		935	217		1,152	1,350
3000	Pendent, 70 watt		2.52	3.170		850	194		1,044	1,225
3020	100 watt		2.52	3.170		875	194		1,069	1,250
3040	150 watt		2.27	3.530		930	217		1,147	1,350
3100	Wall, 70 watt		2.52	3.170		915	194		1,109	1,300
3120	100 watt		2.52	3.170		955	194		1,149	1,350
3140	150 watt		2.27	3.530		995	217		1,212	1,425
3200	Metal halide, vaporproof, recessed, 175 watt		2.86	2.800		790	172		962	1,125
3220	250 watt	↓	2.69	2.980		720	183		903	1,050
3240	400 watt	2 Elec	4.87	3.280		900	201		1,101	1,300
3260	1,000 watt	"	4.03	3.970		1,650	244		1,894	2,175
3280	Surface, 175 watt	1 Elec	2.44	3.280		1,150	201		1,351	1,550
3300	250 watt	"	2.27	3.530		1,025	217		1,242	1,450
3320	400 watt	2 Elec	4.03	3.970		1,275	244		1,519	1,775
3340	1,000 watt	"	3.02	5.290		1,850	325		2,175	2,525
3360	Pendent, 175 watt	1 Elec	2.44	3.280		1,250	201		1,451	1,675
3380	250 watt	"	2.27	3.530		1,050	217		1,267	1,475
3400	400 watt	2 Elec	4.03	3.970		1,250	244		1,494	1,750
3420	1,000 watt	"	3.02	5.290		2,025	325		2,350	2,700
3440	Wall, 175 watt	1 Elec	2.44	3.280		1,325	201		1,526	1,750
3460	250 watt	"	2.27	3.530		1,100	217		1,317	1,550
3480	400 watt	2 Elec	4.03	3.970		1,325	244		1,569	1,825
3500	1,000 watt	"	3.02	5.290		2,100	325		2,425	2,800

26 51 Interior Lighting

26 51 13 – Interior Lighting Fixtures, Lamps, and Ballasts

26 51 13.50 Interior Lighting Fixtures		Crew	Daily Output	Labor-Hours	Unit	Material	2020 Bare Costs Labor	Equipment	Total	Total Incl O&P
0010	**INTERIOR LIGHTING FIXTURES** Including lamps, mounting									
0030	hardware and connections									
0100	Fluorescent, C.W. lamps, troffer, recess mounted in grid, RS									
0130	Grid ceiling mount									
0200	Acrylic lens, 1' W x 4' L, two 40 watt	1 Elec	4.79	1.670	Ea.	53.50	102		155.50	212
0210	1' W x 4' L, three 40 watt		4.54	1.760		60.50	108		168.50	228
0300	2' W x 2' L, two U40 watt		4.79	1.670		57.50	102		159.50	216
0400	2' W x 4' L, two 40 watt		4.45	1.800		56	110		166	226
0500	2' W x 4' L, three 40 watt		4.20	1.900		62	117		179	242
0600	2' W x 4' L, four 40 watt		3.95	2.030		64.50	125		189.50	256
0700	4' W x 4' L, four 40 watt	2 Elec	5.38	2.980		294	183		477	595
0800	4' W x 4' L, six 40 watt		5.21	3.070		310	188		498	620
0900	4' W x 4' L, eight 40 watt		4.87	3.280		350	201		551	685
0910	Acrylic lens, 1' W x 4' L, two 32 watt T8 [G]	1 Elec	4.79	1.670		76	102		178	237
0930	2' W x 2' L, two U32 watt T8 [G]		4.79	1.670		115	102		217	280
0940	2' W x 4' L, two 32 watt T8 [G]		4.45	1.800		87	110		197	260
0950	2' W x 4' L, three 32 watt T8 [G]		4.20	1.900		80.50	117		197.50	263
0960	2' W x 4' L, four 32 watt T8 [G]		3.95	2.030		75	125		200	268
1000	Surface mounted, RS									
1030	Acrylic lens with hinged & latched door frame									
1100	1' W x 4' L, two 40 watt	1 Elec	5.88	1.360	Ea.	68.50	83.50		152	200
1110	1' W x 4' L, three 40 watt		5.63	1.420		71	87		158	208
1200	2' W x 2' L, two U40 watt		5.88	1.360		73.50	83.50		157	205
1300	2' W x 4' L, two 40 watt		5.21	1.540		83.50	94.50		178	233
1400	2' W x 4' L, three 40 watt		4.79	1.670		85	102		187	247
1500	2' W x 4' L, four 40 watt		4.45	1.800		87	110		197	260
1501	2' W x 4' L, six 40 watt T8		4.37	1.830		87	112		199	263
1600	4' W x 4' L, four 40 watt	2 Elec	6.05	2.650		430	163		593	715
1700	4' W x 4' L, six 40 watt		5.54	2.890		465	177		642	780
1800	4' W x 4' L, eight 40 watt		5.21	3.070		485	188		673	815
1900	2' W x 8' L, four 40 watt		5.38	2.980		171	183		354	460
2000	2' W x 8' L, eight 40 watt		5.21	3.070		184	188		372	480
2010	Acrylic wrap around lens									
2020	6" W x 4' L, one 40 watt	1 Elec	6.72	1.190	Ea.	69.50	73		142.50	186
2030	6" W x 8' L, two 40 watt	2 Elec	6.72	2.380		76.50	146		222.50	300
2040	11" W x 4' L, two 40 watt	1 Elec	5.88	1.360		46.50	83.50		130	176
2050	11" W x 8' L, four 40 watt	2 Elec	5.54	2.890		77	177		254	350
2060	16" W x 4' L, four 40 watt	1 Elec	4.45	1.800		77	110		187	249
2070	16" W x 8' L, eight 40 watt	2 Elec	5.38	2.980		168	183		351	455
2080	2' W x 2' L, two U40 watt	1 Elec	5.88	1.360		94	83.50		177.50	228
2100	Strip fixture									
2200	4' long, one 40 watt, RS	1 Elec	7.14	1.120	Ea.	31.50	68.50		100	137
2300	4' long, two 40 watt, RS		6.72	1.190		49	73		122	163
2310	4' long, two 32 watt T8, RS [G]		6.72	1.190		75	73		148	192
2400	4' long, one 40 watt, SL		6.72	1.190		54	73		127	169
2500	4' long, two 40 watt, SL		5.88	1.360		73	83.50		156.50	205
2580	8' long, one 60 watt T8, SL [G]	2 Elec	11.26	1.420		113	87		200	254
2590	8' long, two 60 watt T8, SL [G]		10.42	1.540		99.50	94.50		194	250
2600	8' long, one 75 watt, SL		11.26	1.420		57.50	87		144.50	193
2700	8' long, two 75 watt, SL		10.42	1.540		71	94.50		165.50	219
2800	4' long, two 60 watt, HO	1 Elec	5.63	1.420		112	87		199	253
2810	4' long, two 54 watt, T5HO [G]	"	5.63	1.420		176	87		263	325

Post-Installation Change Order Costs   For customer support on your Electrical Change Order Costs with RSMeans data, call 800.448.8182.

465

26 51 13.50 Interior Lighting Fixtures		Crew	Daily Output	Labor-Hours	Unit	Material	2020 Bare Costs Labor	Equipment	Total	Total Incl O&P
2900	8' long, two 110 watt, HO	2 Elec	8.90	1.800	Ea.	110	110		220	285
2910	4' long, two 115 watt, VHO	1 Elec	5.46	1.470		145	90		235	293
2920	8' long, two 215 watt, VHO	2 Elec	8.74	1.830		157	112		269	340
2950	High bay pendent mounted, 16" W x 4' L, four 54 watt, T5HO [G]		7.48	2.140		298	131		429	525
2952	2' W x 4' L, six 54 watt, T5HO [G]		7.14	2.240		305	137		442	540
2954	2' W x 4' L, six 32 watt, T8 [G]		7.14	2.240		178	137		315	400
3000	Strip, pendent mounted, industrial, white porcelain enamel									
3100	4' long, two 40 watt, RS	1 Elec	4.79	1.670	Ea.	53	102		155	211
3110	4' long, two 32 watt T8, RS [G]		4.79	1.670		83	102		185	245
3200	4' long, two 60 watt, HO		4.20	1.900		82.50	117		199.50	265
3290	8' long, two 60 watt T8, SL [G]	2 Elec	7.39	2.160		121	133		254	330
3300	8' long, two 75 watt, SL		7.39	2.160		101	133		234	310
3400	8' long, two 110 watt, HO		6.72	2.380		130	146		276	360
3410	Acrylic finish, 4' long, two 40 watt, RS	1 Elec	4.79	1.670		87	102		189	249
3420	4' long, two 60 watt, HO		4.20	1.900		161	117		278	350
3430	4' long, two 115 watt, VHO		4.03	1.980		210	121		331	410
3440	8' long, two 75 watt, SL	2 Elec	7.39	2.160		172	133		305	385
3450	8' long, two 110 watt, HO		6.72	2.380		203	146		349	440
3460	8' long, two 215 watt, VHO		6.38	2.510		287	154		441	545
3470	Troffer, air handling, 2' W x 4' L with four 32 watt T8 [G]	1 Elec	3.36	2.380		119	146		265	350
3480	2' W x 2' L with two U32 watt T8 [G]		4.62	1.730		110	106		216	279
3490	Air connector insulated, 5" diameter		16.80	.476		69.50	29		98.50	120
3500	6" diameter		16.80	.476		68.50	29		97.50	119
3502	Troffer, direct/indirect, 2' W x 4' L with two 32 W T8 [G]		4.45	1.800		286	110		396	480
3510	Troffer parabolic lay-in, 1' W x 4' L with one 32 W T8 [G]		4.79	1.670		119	102		221	284
3520	1' W x 4' L with two 32 W T8 [G]		4.45	1.800		141	110		251	320
3525	2' W x 2' L with two U32 W T8 [G]		4.79	1.670		124	102		226	289
3530	2' W x 4' L with three 32 W T8 [G]		4.20	1.900		132	117		249	320
3531	Intr fxtr, fluor, troffer prismatic lay-in, 2' W x 4'l w/three 32 W T8		4.20	1.900		152	117		269	340
3535	Downlight, recess mounted [G]		6.72	1.190		148	73		221	272
3540	Wall wash reflector, recess mounted [G]		6.72	1.190		113	73		186	234
3550	Direct/indirect, 4' long, stl., pendent mtd. [G]		4.20	1.900		162	117		279	350
3560	4' long, alum., pendent mtd. [G]		4.20	1.900		335	117		452	540
3565	Prefabricated cove, 4' long, stl. continuous row [G]		4.20	1.900		210	117		327	405
3570	4' long, alum. continuous row [G]		4.20	1.900		345	117		462	555
3580	Wet location, recess mounted, 2' W x 4' L with two 32 watt T8 [G]		4.45	1.800		238	110		348	425
3590	Pendent mounted, 2' W x 4' L with two 32 watt T8 [G]		4.79	1.670		385	102		487	575
4000	Induction lamp, integral ballast, ceiling mounted									
4110	High bay, aluminum reflector, 160 watt	1 Elec	2.69	2.980	Ea.	920	183		1,103	1,275
4120	320 watt	"	2.52	3.170		1,650	194		1,844	2,125
4130	480 watt	2 Elec	4.87	3.280		2,500	201		2,701	3,050
4150	Low bay, aluminum reflector, 250 watt	1 Elec	2.69	2.980		935	183		1,118	1,300
4170	Garage, aluminum reflector, 80 watt		3.02	2.650		840	163		1,003	1,175
4180	Vandalproof, aluminum reflector, 100 watt		2.69	2.980		615	183		798	950
4220	Metal halide, integral ballast, ceiling, recess mounted									
4230	prismatic glass lens, floating door									
4240	2' W x 2' L, 250 watt	1 Elec	2.69	2.980	Ea.	300	183		483	600
4250	2' W x 2' L, 400 watt	2 Elec	4.87	3.280		370	201		571	705
4260	Surface mounted, 2' W x 2' L, 250 watt	1 Elec	2.27	3.530		340	217		557	695
4270	400 watt	2 Elec	4.03	3.970		405	244		649	810
4280	High bay, aluminum reflector,									
4290	Single unit, 400 watt	2 Elec	3.86	4.140	Ea.	425	254		679	845
4300	Single unit, 1,000 watt		3.36	4.760		610	292		902	1,100

Post-Installation Change Order Costs

26 51 Interior Lighting

26 51 13 - Interior Lighting Fixtures, Lamps, and Ballasts

26 51 13.50 Interior Lighting Fixtures		Crew	Daily Output	Labor-Hours	Unit	Material	2020 Bare Costs Labor	Equipment	Total	Total Incl O&P
4310	Twin unit, 400 watt	2 Elec	2.69	5.950	Ea.	820	365		1,185	1,450
4320	Low bay, aluminum reflector, 250W DX lamp	1 Elec	2.69	2.980		360	183		543	670
4330	400 watt lamp	2 Elec	4.20	3.810	↓	545	234		779	950
4340	High pressure sodium integral ballast ceiling, recess mounted									
4350	prismatic glass lens, floating door									
4360	2' W x 2' L, 150 watt lamp	1 Elec	2.69	2.980	Ea.	400	183		583	710
4370	2' W x 2' L, 400 watt lamp	2 Elec	4.87	3.280		475	201		676	825
4380	Surface mounted, 2' W x 2' L, 150 watt lamp	1 Elec	2.27	3.530		485	217		702	855
4390	400 watt lamp	2 Elec	4.03	3.970	↓	545	244		789	965
4400	High bay, aluminum reflector,									
4410	Single unit, 400 watt lamp	2 Elec	3.86	4.140	Ea.	390	254		644	810
4430	Single unit, 1,000 watt lamp	"	3.36	4.760		545	292		837	1,025
4440	Low bay, aluminum reflector, 150 watt lamp	1 Elec	2.69	2.980		340	183		523	640
4445	High bay H.I.D. quartz restrike	"	13.44	.595	↓	169	36.50		205.50	240
4450	Incandescent, high hat can, round alzak reflector, prewired									
4470	100 watt	1 Elec	6.72	1.190	Ea.	68	73		141	184
4480	150 watt		6.72	1.190		104	73		177	223
4500	300 watt		5.63	1.420		241	87		328	395
4520	Round with reflector and baffles, 150 watt		6.72	1.190		54	73		127	169
4540	Round with concentric louver, 150 watt PAR	↓	6.72	1.190	↓	78.50	73		151.50	196
4600	Square glass lens with metal trim, prewired									
4630	100 watt	1 Elec	5.63	1.420	Ea.	55.50	87		142.50	191
4680	150 watt		5.63	1.420		97.50	87		184.50	237
4700	200 watt		5.63	1.420		97.50	87		184.50	237
4800	300 watt		4.79	1.670		146	102		248	315
4810	500 watt		4.20	1.900		286	117		403	490
4900	Ceiling/wall, surface mounted, metal cylinder, 75 watt		8.40	.952		51.50	58.50		110	144
4920	150 watt		8.40	.952		89	58.50		147.50	185
4930	300 watt		6.72	1.190		173	73		246	299
5000	500 watt		5.63	1.420		365	87		452	530
5010	Square, 100 watt		6.72	1.190		124	73		197	245
5020	150 watt		6.72	1.190		124	73		197	246
5030	300 watt		5.88	1.360		340	83.50		423.50	500
5040	500 watt	↓	5.04	1.590	↓	360	97.50		457.50	545
5200	Ceiling, surface mounted, opal glass drum									
5300	8", one 60 watt lamp	1 Elec	8.40	.952	Ea.	67	58.50		125.50	161
5400	10", two 60 watt lamps		6.72	1.190		74.50	73		147.50	191
5500	12", four 60 watt lamps		5.63	1.420		104	87		191	244
5510	Pendent, round, 100 watt		6.72	1.190		124	73		197	245
5520	150 watt		6.72	1.190		124	73		197	245
5530	300 watt		5.63	1.420		171	87		258	320
5540	500 watt		4.62	1.730		330	106		436	525
5550	Square, 100 watt		5.63	1.420		152	87		239	297
5560	150 watt		5.63	1.420		159	87		246	305
5570	300 watt		4.79	1.670		237	102		339	415
5580	500 watt		4.20	1.900		320	117		437	530
5600	Wall, round, 100 watt		6.72	1.190		66	73		139	182
5620	300 watt		6.72	1.190		133	73		206	256
5630	500 watt		5.63	1.420		390	87		477	560
5640	Square, 100 watt		6.72	1.190		123	73		196	245
5650	150 watt		6.72	1.190		108	73		181	228
5660	300 watt		5.88	1.360		167	83.50		250.50	310
5670	500 watt	↓	5.04	1.590	↓	278	97.50		375.50	450

26 51 13.50 Interior Lighting Fixtures		Crew	Daily Output	Labor-Hours	Unit	Material	2020 Bare Costs Labor	Equipment	Total	Total Incl O&P
6010	Vapor tight, incandescent, ceiling mounted, 200 watt	1 Elec	5.21	1.540	Ea.	78.50	94.50		173	228
6020	Recessed, 200 watt		5.63	1.420		127	87		214	270
6030	Pendent, 200 watt		5.63	1.420		78.50	87		165.50	217
6040	Wall, 200 watt		6.72	1.190		80.50	73		153.50	198
6100	Fluorescent, surface mounted, 2 lamps, 4' L, RS, 40 watt		2.69	2.980		119	183		302	405
6110	Industrial, 2 lamps, 4' L in tandem, 430 MA		1.85	4.330		212	266		478	630
6130	2 lamps, 4' L, 800 MA		1.60	5.010		181	305		486	660
6160	Pendent, indust, 2 lamps, 4' L in tandem, 430 MA		1.60	5.010		247	305		552	730
6170	2 lamps, 4' L, 430 MA		1.93	4.140		165	254		419	560
6180	2 lamps, 4' L, 800 MA		1.43	5.600		208	345		553	740
6850	Vandalproof, surface mounted, fluorescent, two 32 watt T8 ⬛G		2.69	2.980		278	183		461	575
6860	Incandescent, one 150 watt		6.72	1.190		94.50	73		167.50	213
6900	Mirror light, fluorescent, RS, acrylic enclosure, two 40 watt		6.72	1.190		115	73		188	236
6910	One 40 watt		6.72	1.190		99.50	73		172.50	218
6920	One 20 watt		10.08	.794		84	48.50		132.50	165
7000	Low bay, aluminum reflector, 70 watt, high pressure sodium		3.36	2.380		277	146		423	520
7010	250 watt	⬇	2.69	2.980		370	183		553	675
7020	400 watt	2 Elec	4.20	3.810	⬇	430	234		664	825
7500	Ballast replacement, by weight of ballast, to 15' high									
7520	Indoor fluorescent, less than 2 lb.	1 Elec	8.40	.952	Ea.	25	58.50		83.50	115
7540	Two 40W, watt reducer, 2 to 5 lb.		7.90	1.010		71.50	62		133.50	171
7560	Two F96 slimline, over 5 lb.		6.72	1.190		111	73		184	231
7580	Vaportite ballast, less than 2 lb.		7.90	1.010		25	62		87	120
7600	2 lb. to 5 lb.		7.48	1.070		71.50	65.50		137	176
7620	Over 5 lb.		6.38	1.250		111	76.50		187.50	236
7630	Electronic ballast for two tubes		6.72	1.190		41.50	73		114.50	155
7640	Dimmable ballast one-lamp ⬛G		6.72	1.190		105	73		178	225
7650	Dimmable ballast two-lamp ⬛G	⬇	6.38	1.250		108	76.50		184.50	233
7690	Emergency ballast (factory installed in fixture)				⬇	152			152	167
7990	Decorator									
8000	Pendent RLM in colors, shallow dome, 12" diam., 100 watt	1 Elec	6.72	1.190	Ea.	80	73		153	197
8010	Regular dome, 12" diam., 100 watt		6.72	1.190		82.50	73		155.50	200
8020	16" diam., 200 watt		5.88	1.360		84	83.50		167.50	217
8030	18" diam., 300 watt		5.04	1.590		93.50	97.50		191	248
8100	Picture framing light		13.44	.595		98.50	36.50		135	163
8150	Miniature low voltage, recessed, pinhole		6.72	1.190		134	73		207	257
8160	Star		6.72	1.190		133	73		206	255
8170	Adjustable cone		6.72	1.190		169	73		242	295
8180	Eyeball		6.72	1.190		109	73		182	228
8190	Cone		6.72	1.190		129	73		202	251
8200	Coilex baffle		6.72	1.190		140	73		213	263
8210	Surface mounted, adjustable cylinder	⬇	6.72	1.190	⬇	140	73		213	263
8250	Chandeliers, incandescent									
8260	24" diam. x 42" high, 6 light candle	1 Elec	5.04	1.590	Ea.	440	97.50		537.50	625
8270	24" diam. x 42" high, 6 light candle w/glass shade		5.04	1.590		430	97.50		527.50	620
8280	17" diam. x 12" high, 8 light w/glass panels		6.72	1.190		282	73		355	420
8300	27" diam. x 29" high, 10 light bohemian lead crystal		3.36	2.380		460	146		606	720
8310	21" diam. x 9" high, 6 light sculptured ice crystal	⬇	6.72	1.190	⬇	420	73		493	570
8500	Accent lights, on floor or edge, 0.5 W low volt incandescent									
8520	incl. transformer & fastenings, based on 100' lengths									
8550	Lights in clear tubing, 12" OC	1 Elec	193.20	.041	L.F.	8.50	2.54		11.04	13.15
8560	6" OC		134.40	.060		11.10	3.65		14.75	17.65
8570	4" OC		109.20	.073		16.95	4.49		21.44	25.50

Post-Installation Change Order Costs

26 51 13 – Interior Lighting Fixtures, Lamps, and Ballasts

26 51 13.50 Interior Lighting Fixtures		Crew	Daily Output	Labor- Hours	Unit	Material	2020 Bare Costs Labor	Equipment	Total	Total Incl O&P
8580	3" OC	1 Elec	105	.076	L.F.	18.85	4.67		23.52	27.50
8590	2" OC		84	.095		27.50	5.85		33.35	38.50
8600	Carpet, lights both sides 6" OC, in alum. extrusion		226.80	.035		25.50	2.16		27.66	31
8610	In bronze extrusion		226.80	.035		29	2.16		31.16	35
8620	Carpet-bare floor, lights 18" OC, in alum. extrusion		226.80	.035		20.50	2.16		22.66	25.50
8630	In bronze extrusion		226.80	.035		24	2.16		26.16	29.50
8640	Carpet edge-wall, lights 6" OC in alum. extrusion		226.80	.035		25.50	2.16		27.66	31
8650	In bronze extrusion		226.80	.035		29	2.16		31.16	35
8660	Bare floor, lights 18" OC, in alum. extrusion		252	.032		20.50	1.95		22.45	25.50
8670	In bronze extrusion		252	.032		24	1.95		25.95	29.50
8680	Bare floor conduit, alum. extrusion		252	.032		6.80	1.95		8.75	10.35
8690	In bronze extrusion		252	.032	▼	13.55	1.95		15.50	17.85
8700	Step edge to 36", lights 6" OC, in alum. extrusion		84	.095	Ea.	68.50	5.85		74.35	83.50
8710	In bronze extrusion		84	.095		71	5.85		76.85	87
8720	Step edge to 54", lights 6" OC, in alum. extrusion		84	.095		102	5.85		107.85	122
8730	In bronze extrusion		84	.095		108	5.85		113.85	128
8740	Step edge to 72", lights 6" OC, in alum. extrusion		84	.095		136	5.85		141.85	159
8750	In bronze extrusion		84	.095		149	5.85		154.85	173
8760	Connector, male		26.88	.298		2.53	18.25		20.78	30
8770	Female with pigtail		26.88	.298		5.30	18.25		23.55	33
8780	Clamps		336	.024		.51	1.46		1.97	2.74
8790	Transformers, 50 watt		6.72	1.190		100	73		173	219
8800	250 watt		3.36	2.380		320	146		466	565
8810	1,000 watt	▼	2.27	3.530	▼	350	217		567	705

26 51 13.55 Interior LED Fixtures

			Crew	Daily Output	Labor- Hours	Unit	Material	2020 Bare Costs Labor	Equipment	Total	Total Incl O&P
0010	**INTERIOR LED FIXTURES** Incl. lamps and mounting hardware										
0100	Downlight, recess mounted, 7.5" diameter, 25 watt	G	1 Elec	6.72	1.190	Ea.	340	73		413	485
0120	10" diameter, 36 watt	G		6.72	1.190		360	73		433	505
0160	cylinder, 10 watts	G		6.72	1.190		104	73		177	223
0180	20 watts	G		6.72	1.190		135	73		208	258
0900	Interior LED fixts, troffer, recess mounted, 2' x 2', 3,500K			7.14	1.120		129	68.50		197.50	243
0910	2' x 2', 4,000K			7.14	1.120		130	68.50		198.50	245
1000	Troffer, recess mounted, 2' x 4', 3,200 lumens	G		4.45	1.800		138	110		248	315
1010	4,800 lumens	G		4.20	1.900		150	117		267	340
1020	6,400 lumens	G		3.95	2.030		189	125		314	395
1100	Troffer retrofit lamp, 38 watt	G		17.64	.454		63.50	28		91.50	111
1110	60 watt	G		16.80	.476		156	29		185	215
1120	100 watt	G		15.12	.529		140	32.50		172.50	203
1200	Troffer, volumetric recess mounted, 2' x 2'	G		4.79	1.670		345	102		447	535
2000	Strip, surface mounted, one light bar 4' long, 3,500 K	G		7.14	1.120		305	68.50		373.50	440
2010	5,000 K	G		6.72	1.190		265	73		338	400
2020	Two light bar 4' long, 5,000 K	G		5.88	1.360		415	83.50		498.50	585
3000	Linear, suspended mounted, one light bar 4' long, 37 watt	G	▼	5.63	1.420		171	87		258	320
3010	One light bar 8' long, 74 watt	G	2 Elec	10.25	1.560		310	95.50		405.50	485
3020	Two light bar 4' long, 74 watt	G	1 Elec	4.79	1.670		335	102		437	520
3030	Two light bar 8' long, 148 watt	G	2 Elec	7.39	2.160		360	133		493	590
7000	Downlight, recess mtd., low profile, 4" diam., 9W, 3,000K		1 Elec	7.14	1.120		24.50	68.50		93	129
7010	4,000K			7.14	1.120		23	68.50		91.50	128
7020	5,000K		▼	7.14	1.120	▼	24	68.50		92.50	128

26 51 13 – Interior Lighting Fixtures, Lamps, and Ballasts

26 51 13.70 Residential Fixtures	Crew	Daily Output	Labor-Hours	Unit	Material	2020 Bare Costs Labor	Equipment	Total	Total Incl O&P
0010 **RESIDENTIAL FIXTURES**									
0400 Fluorescent, interior, surface, circline, 32 watt & 40 watt	1 Elec	16.80	.476	Ea.	164	29		193	224
0700 Shallow under cabinet, two 20 watt		13.44	.595		70.50	36.50		107	132
0900 Wall mounted, 4' L, two 32 watt T8, with baffle		8.40	.952		165	58.50		223.50	269
2000 Incandescent, exterior lantern, wall mounted, 60 watt		13.44	.595		61.50	36.50		98	122
2100 Post light, 150 W, with 7' post		3.36	2.380		291	146		437	535
2500 Lamp holder, weatherproof with 150 W PAR		13.44	.595		35.50	36.50		72	93.50
2550 With reflector and guard		10.08	.794		66	48.50		114.50	145
2600 Interior pendent, globe with shade, 150 W	▼	16.80	.476	▼	190	29		219	253

26 51 13.90 Ballast, Replacement HID

	Crew	Daily Output	Labor-Hours	Unit	Material	2020 Bare Costs Labor	Equipment	Total	Total Incl O&P
0010 **BALLAST, REPLACEMENT HID**									
7510 Multi-tap 120/208/240/277 V									
7550 High pressure sodium, 70 watt	1 Elec	8.40	.952	Ea.	183	58.50		241.50	288
7560 100 watt		7.90	1.010		130	62		192	236
7570 150 watt		7.56	1.060		205	65		270	325
7580 250 watt		7.14	1.120		305	68.50		373.50	435
7590 400 watt		5.88	1.360		269	83.50		352.50	420
7600 1,000 watt		5.04	1.590		287	97.50		384.50	460
7610 Metal halide, 175 watt		6.72	1.190		97	73		170	216
7620 250 watt		6.72	1.190		128	73		201	250
7630 400 watt		5.88	1.360		158	83.50		241.50	298
7640 1,000 watt		5.04	1.590		258	97.50		355.50	430
7650 1,500 watt	▼	4.20	1.900	▼	245	117		362	445

26 51 19 – LED Interior Lighting

26 51 19.10 LED Interior Lighting

	Crew	Daily Output	Labor-Hours	Unit	Material	2020 Bare Costs Labor	Equipment	Total	Total Incl O&P
0010 **LED INTERIOR LIGHTING**									
7000 Interior LED fixts, tape rope kit, 120V, 3' strip, 4W, w/5' cord, 3,000K	1 Elec	20.16	.397	Ea.	7.25	24.50		31.75	44.50
7010 5,000K		20.16	.397		10.95	24.50		35.45	48.50
7020 Interior LED fixts, tape rope kit, 120V, 6' strip, 8W, w/5' cord, 3,000K		19.32	.414		15.35	25.50		40.85	55
7030 5,000K		19.32	.414		22	25.50		47.50	62.50
7040 Interior LED fixts, tape rope kit, 120V, 13' strip, 15W, w/5' cord, 3,000K		18.48	.433		26	26.50		52.50	68
7050 5,000K		18.48	.433		28.50	26.50		55	70.50
7060 Interior LED fixts, tape rope kit, 120V, 20' strip, 22W, w/5' cord, 3,000K		17.64	.454		41	28		69	86.50
7070 5,000K		17.64	.454		41	28		69	86.50
7080 Interior LED fixts, tape rope kit, 120V, 33' strip, 37W, w/5' cord, 3,000K		16.80	.476		60.50	29		89.50	110
7090 5,000K		16.80	.476		60.50	29		89.50	110
7100 Interior LED fixts, tape rope kit, 120V, 50' strip, 55W, w/5' cord, 3,000K		15.96	.501		91.50	31		122.50	147
7110 5,000K		15.96	.501		131	31		162	190
7120 Interior LED tape light, starstrand, dimmable, 12V, super star, 2", 3,000K		21	.381		11.25	23.50		34.75	47.50
7130 Ultra star, 2", 3,000K		21	.381		14.05	23.50		37.55	50.50
7140 Rainbow star, 2", 3,000K		21	.381		11.25	23.50		34.75	47.50
7150 Interior LED tape light, starstrand, dimmable, 24V, elite star, 4", 35K		21	.381		9.80	23.50		33.30	46
7160 12", 35K		21	.381		21	23.50		44.50	58.50
7170 60", 35K		19.32	.414		98	25.50		123.50	146
7180 120", 35K		17.64	.454		123	28		151	177
7190 240", 35K		15.12	.529		297	32.50		329.50	375
7200 Interior LED tape lighting channel , 36", AL flex, starstrand, elite star	▼	40.32	.198	▼	21	12.15		33.15	41

26 52 13.10 Emergency Lighting and Battery Units

		Crew	Daily Output	Labor-Hours	Unit	Material	2020 Bare Costs Labor	Equipment	Total	Total Incl O&P
0010	**EMERGENCY LIGHTING AND BATTERY UNITS**									
0300	Emergency light units, battery operated									
0350	Twin sealed beam light, 25 W, 6 V each									
0500	Lead battery operated	1 Elec	3.36	2.380	Ea.	140	146		286	370
0700	Nickel cadmium battery operated		3.36	2.380		330	146		476	580
0780	Additional remote mount, sealed beam, 25 W 6 V		22.43	.357		29	22		51	64.50
0781	Additional remote mount, sealed beam, 25 W 6 V		22.43	.357		29	22		51	64.50
0790	Twin sealed beam light, 25 W 6 V each		22.43	.357		62.50	22		84.50	101
0900	Self-contained fluorescent lamp pack	↓	8.40	.952	↓	181	58.50		239.50	286

26 52 13.16 Exit Signs

		Crew	Daily Output	Labor-Hours	Unit	Material	2020 Bare Costs Labor	Equipment	Total	Total Incl O&P
0010	**EXIT SIGNS**									
0080	Exit light ceiling or wall mount, incandescent, single face	1 Elec	6.72	1.190	Ea.	73	73		146	189
0100	Double face		5.63	1.420		50.50	87		137.50	186
0120	Explosion proof		3.19	2.510		585	154		739	870
0150	Fluorescent, single face		6.72	1.190		92.50	73		165.50	211
0160	Double face		5.63	1.420		76.50	87		163.50	215
0200	LED standard, single face	G	6.72	1.190		49.50	73		122.50	164
0220	Double face	G	5.63	1.420		52	87		139	187
0230	LED vandal-resistant, single face	G	6.11	1.310		218	80.50		298.50	360
0240	LED w/battery unit, single face	G	3.70	2.160		192	133		325	410
0260	Double face	G	3.36	2.380		223	146		369	460
0262	LED w/battery unit, vandal-resistant, single face	G	3.70	2.160		258	133		391	480
0270	Combination emergency light units and exit sign		3.36	2.380		179	146		325	415
0290	LED retrofit kits	G	50.40	.159		52.50	9.75		62.25	72
1500	Exit sign, 12 V, 1 face, remote end mounted (Type 602A1)	R-19	15.12	1.320		73	81		154	202
1780	With emergency battery, explosion proof	"	6.47	3.090	↓	4,675	190		4,865	5,400

26 54 Classified Location Lighting

26 54 13 – Incandescent Classified Location Lighting

26 54 13.20 Explosion Proof

		Crew	Daily Output	Labor-Hours	Unit	Material	2020 Bare Costs Labor	Equipment	Total	Total Incl O&P
0010	**EXPLOSION PROOF**, incl. lamps, mounting hardware and connections									
6310	Metal halide with ballast, ceiling, surface mounted, 175 watt	1 Elec	2.44	3.280	Ea.	1,400	201		1,601	1,850
6320	250 watt	"	2.27	3.530		1,675	217		1,892	2,175
6330	400 watt	2 Elec	4.03	3.970		1,800	244		2,044	2,350
6340	Ceiling, pendent mounted, 175 watt	1 Elec	2.18	3.660		1,325	225		1,550	1,775
6350	250 watt	"	2.02	3.970		1,600	244		1,844	2,150
6360	400 watt	2 Elec	3.53	4.540		1,725	279		2,004	2,325
6370	Wall, surface mounted, 175 watt	1 Elec	2.44	3.280		1,500	201		1,701	1,950
6380	250 watt	"	2.27	3.530		1,775	217		1,992	2,275
6390	400 watt	2 Elec	4.03	3.970		1,900	244		2,144	2,475
6400	High pressure sodium, ceiling surface mounted, 70 watt	1 Elec	2.52	3.170		2,275	194		2,469	2,825
6410	100 watt		2.52	3.170		2,375	194		2,569	2,900
6420	150 watt		2.27	3.530		2,500	217		2,717	3,075
6430	Pendent mounted, 70 watt		2.27	3.530		2,175	217		2,392	2,725
6440	100 watt		2.27	3.530		2,275	217		2,492	2,825
6450	150 watt		2.02	3.970		2,050	244		2,294	2,625
6460	Wall mounted, 70 watt		2.52	3.170		2,450	194		2,644	2,975
6470	100 watt		2.52	3.170		2,575	194		2,769	3,125
6480	150 watt		2.27	3.530		2,600	217		2,817	3,175
6510	Incandescent, ceiling mounted, 200 watt		3.36	2.380		1,650	146		1,796	2,025
6520	Pendent mounted, 200 watt	↓	2.94	2.720	↓	1,400	167		1,567	1,800

26 54 Classified Location Lighting

26 54 13 – Incandescent Classified Location Lighting

26 54 13.20 Explosion Proof		Crew	Daily Output	Labor-Hours	Unit	Material	2020 Bare Costs Labor	Equipment	Total	Total Incl O&P
6530	Wall mounted, 200 watt	1 Elec	3.36	2.380	Ea.	1,625	146		1,771	2,000
6600	Fluorescent, RS, 4' long, ceiling mounted, two 40 watt		2.27	3.530		4,975	217		5,192	5,800
6610	Three 40 watt		1.85	4.330		7,200	266		7,466	8,325
6620	Four 40 watt		1.60	5.010		9,250	305		9,555	10,700
6630	Pendent mounted, two 40 watt		1.93	4.140		5,750	254		6,004	6,700
6640	Three 40 watt		1.60	5.010		8,175	305		8,480	9,450
6650	Four 40 watt		1.43	5.600		10,800	345		11,145	12,300

26 55 Special Purpose Lighting

26 55 33.10 Warning Beacons

		Crew	Daily Output	Labor-Hours	Unit	Material	2020 Bare Costs Labor	Equipment	Total	Total Incl O&P
0010	**WARNING BEACONS**									
0015	Surface mount with colored or clear lens									
0100	Rotating beacon									
0110	120V, 40 watt halogen	1 Elec	2.94	2.720	Ea.	118	167		285	380
0120	24V, 20 watt halogen	"	2.94	2.720	"	279	167		446	555
0200	Steady beacon									
0210	120V, 40 watt halogen	1 Elec	2.94	2.720	Ea.	120	167		287	380
0220	24V, 20 watt		2.94	2.720		124	167		291	385
0230	12V DC, incandescent		2.94	2.720		125	167		292	385
0300	Flashing beacon									
0310	120V, 40 watt halogen	1 Elec	2.94	2.720	Ea.	117	167		284	375
0320	24V, 20 watt halogen		2.94	2.720		118	167		285	375
0410	12V DC with two 6V lantern batteries		5.88	1.360		119	83.50		202.50	255

26 55 59 – Display Lighting

26 55 59.10 Track Lighting

		Crew	Daily Output	Labor-Hours	Unit	Material	2020 Bare Costs Labor	Equipment	Total	Total Incl O&P
0010	**TRACK LIGHTING**									
0080	Track, 1 circuit, 4' section	1 Elec	5.63	1.420	Ea.	43	87		130	178
0100	8' section	2 Elec	8.90	1.800		76	110		186	248
0200	12' section	"	7.39	2.160		112	133		245	320
0300	3 circuits, 4' section	1 Elec	5.63	1.420		117	87		204	258
0400	8' section	2 Elec	8.90	1.800		130	110		240	305
0500	12' section	"	7.39	2.160		168	133		301	380
1000	Feed kit, surface mounting	1 Elec	13.44	.595		15.85	36.50		52.35	72
1100	End cover		20.16	.397		8.30	24.50		32.80	45.50
1200	Feed kit, stem mounting, 1 circuit		13.44	.595		52.50	36.50		89	112
1300	3 circuit		13.44	.595		52.50	36.50		89	112
2000	Electrical joiner, for continuous runs, 1 circuit		26.88	.298		37.50	18.25		55.75	68
2100	3 circuit		26.88	.298		76.50	18.25		94.75	112
2200	Fixtures, spotlight, 75 W PAR halogen		13.44	.595		48	36.50		84.50	107
2210	50 W MR16 halogen		13.44	.595		175	36.50		211.50	247
3000	Wall washer, 250 W tungsten halogen		13.44	.595		134	36.50		170.50	202
3100	Low voltage, 25/50 W, 1 circuit		13.44	.595		135	36.50		171.50	203
3120	3 circuit		13.44	.595		193	36.50		229.50	268

26 55 63 – Detention Lighting

26 55 63.10 Detention Lighting Fixtures

		Crew	Daily Output	Labor-Hours	Unit	Material	2020 Bare Costs Labor	Equipment	Total	Total Incl O&P
0010	**DETENTION LIGHTING FIXTURES**									
3000	Fluorescent vandal resistant light fixture, high abuse troffer	1 Elec	4.54	1.760	Ea.	244	108		352	430
3010	50" L x 8" W x 4" H		4.54	1.760		216	108		324	400
3020	32W		4.54	1.760		310	108		418	500
3030	T8/32W		4.54	1.760		325	108		433	515

For customer support on your Electrical Change Order Costs with RSMeans data, call 800.448.8182.

Post-Installation Change Order Costs

26 55 Special Purpose Lighting

26 55 63 – Detention Lighting

26 55 63.10 Detention Lighting Fixtures	Crew	Daily Output	Labor-Hours	Unit	Material	2020 Bare Costs Labor	Equipment	Total	Total Incl O&P
3040	No lamp	1 Elec	4.70	1.700	Ea.	370	104		474	560
3050	Lamp included		4.54	1.760		425	108		533	630
3060	16 ga. cold rolled steel		4.54	1.760		340	108		448	535
3070	16 ga. cold rolled steel, ceiling/wall mount		4.54	1.760		345	108		453	540
3080	w/lamp, white		4.54	1.760		269	108		377	455
3090	wo/lamp, bronze		4.70	1.700		245	104		349	425
3100	w/lamp 6" L x 4-1/2" W x 8-1/2" H		4.54	1.760		85.50	108		193.50	255
3110	w/lamp 5-7/8" W x 4-1/2" D x 8-1/2" H clear		4.54	1.760		67	108		175	235
3120	w/lamp white		4.54	1.760		540	108		648	755
3130	Tall wallpack, w/lamp, bronze		4.54	1.760		128	108		236	300
3140	Tall wallpack, w/lamp, white	↓	4.54	1.760	↓	129	108		237	305

26 56 Exterior Lighting

26 56 13 – Lighting Poles and Standards

26 56 13.10 Lighting Poles

		Crew	Daily Output	Labor-Hours	Unit	Material	2020 Bare Costs Labor	Equipment	Total	Total Incl O&P
0010	**LIGHTING POLES**									
0100	Exterior, light poles, concrete, 30' above 5' below, 13.5" Base, 5.5" Tip	2 Elec	3.86	4.140	Ea.	1,575	254		1,829	2,100
0110	39' above 6' below, 15.5" Base, 5.25" Tip		3.86	4.140		1,850	254		2,104	2,425
0120	43' above 7' below, 17.25" Base, 6.5" Tip		3.86	4.140		1,875	254		2,129	2,450
0130	43' above 7' below, 19.5" Base, 8.25" Tip	↓	3.86	4.140	↓	1,900	254		2,154	2,475
2800	Light poles, anchor base									
2820	not including concrete bases									
2840	Aluminum pole, 8' high	1 Elec	3.36	2.380	Ea.	810	146		956	1,100
2850	10' high		3.36	2.380		855	146		1,001	1,150
2860	12' high		3.19	2.510		890	154		1,044	1,200
2870	14' high		2.86	2.800		920	172		1,092	1,250
2880	16' high	↓	2.52	3.170		1,025	194		1,219	1,425
3000	20' high	R-3	2.44	8.210		1,100	500	77.50	1,677.50	2,025
3200	30' high		2.18	9.160		2,075	560	86.50	2,721.50	3,200
3400	35' high		1.93	10.350		2,275	635	98	3,008	3,550
3600	40' high	↓	1.68	11.900		2,750	725	112	3,587	4,225
3800	Bracket arms, 1 arm	1 Elec	6.72	1.190		141	73		214	264
4000	2 arms		6.72	1.700		280	73		353	420
4200	3 arms		4.45	1.800		425	110		535	630
4400	4 arms		4.03	1.980		565	121		686	805
4500	Steel pole, galvanized, 8' high		3.19	2.510		675	154		829	970
4510	10' high		3.11	2.570		700	158		858	1,000
4520	12' high		2.86	2.800		755	172		927	1,075
4530	14' high		2.60	3.070		805	188		993	1,175
4540	16' high		2.44	3.280		855	201		1,056	1,250
4550	18' high	↓	2.27	3.530		900	217		1,117	1,300
4600	20' high	R-3	2.18	9.160		1,275	560	86.50	1,921.50	2,325
4800	30' high		1.93	10.350		1,275	635	98	2,008	2,450
5000	35' high		1.85	10.820		1,550	660	102	2,312	2,825
5200	40' high	↓	1.43	14.010		1,725	855	132	2,712	3,300
5400	Bracket arms, 1 arm	1 Elec	6.72	1.190		214	73		287	345
5600	2 arms		6.72	1.190		310	73		383	450
5800	3 arms		4.45	1.800		215	110		325	400
6000	4 arms	↓	4.45	1.800		350	110		460	550
6100	Fiberglass pole, 1 or 2 fixtures, 20' high	R-3	3.36	5.950		860	365	56	1,281	1,550
6200	30' high		3.02	6.610		980	405	62.50	1,447.50	1,750

26 56 Exterior Lighting

26 56 13 – Lighting Poles and Standards

	26 56 13.10 Lighting Poles	Crew	Daily Output	Labor-Hours	Unit	Material	2020 Bare Costs Labor	Equipment	Total	Total Incl O&P
6300	35' high	R-3	2.69	7.440	Ea.	1,675	455	70.50	2,200.50	2,600
6400	40' high	↓	2.35	8.500		1,850	520	80.50	2,450.50	2,900
6420	Wood pole, 4-1/2" x 5-1/8", 8' high	1 Elec	5.04	1.590		385	97.50		482.50	570
6430	10' high		5.04	1.590		450	97.50		547.50	645
6440	12' high		4.79	1.670		570	102		672	780
6450	15' high		4.20	1.900		665	117		782	905
6460	20' high	↓	3.36	2.380		810	146		956	1,100
6461	Light poles, anchor base, w/o conc base, pwdr ct stl, 16' H	2 Elec	2.60	6.140		855	375		1,230	1,500
6462	20' high	R-3	2.44	8.210		1,275	500	77.50	1,852.50	2,225
6463	30' high		1.93	10.350		1,275	635	98	2,008	2,450
6464	35' high		2.02	9.920		1,550	605	93.50	2,248.50	2,725
6465	25' high	↓	2.27	8.820		1,275	540	83.50	1,898.50	2,300
6470	Light pole conc base, max 6' buried, 2' exposed, 18" diam., average cost	C-6	5.04	9.520	↓	190	415	10.65	615.65	845
7300	Transformer bases, not including concrete bases									
7320	Maximum pole size, steel, 40' high	1 Elec	1.68	4.760	Ea.	1,600	292		1,892	2,200
7340	Cast aluminum, 30' high		2.52	3.170		850	194		1,044	1,225
7350	40' high	↓	2.10	3.810	↓	1,300	234		1,534	1,775

26 56 19 – LED Exterior Lighting

		Crew	Daily Output	Labor-Hours	Unit	Material	Labor	Equipment	Total	Total Incl O&P
2000	Exterior fixtures, LED roadway, type 2, 50W, neutral slipfitter bronze	2 Elec	4.37	3.660	Ea.	660	225		885	1,050
2010	3, 50W, neutral slipfitter bronze		4.37	3.660		635	225		860	1,025
2020	4, 50W, neutral slipfitter bronze		4.37	3.660		635	225		860	1,025
2030	Exterior fixtures, LED roadway, type 2, 78W, neutral slipfitter bronze		4.37	3.660		630	225		855	1,025
2040	3, 78W, neutral slipfitter bronze		4.37	3.660		675	225		900	1,075
2050	4, 78W, neutral slipfitter bronze		4.37	3.660		660	225		885	1,050
2060	Exterior fixtures, LED roadway, type 2, 105W, neutral slipfitter bronze		4.37	3.660		820	225		1,045	1,225
2070	3, 105W, neutral slipfitter bronze		4.37	3.660		785	225		1,010	1,200
2080	4, 105W, neutral slipfitter bronze		4.37	3.660		730	225		955	1,150
2090	Exterior fixtures, LED roadway, type 2, 125W, neutral slipfitter bronze	↓	4.37	3.660	↓	790	225		1,015	1,200

26 56 21 – HID Exterior Lighting

26 56 21.20 Roadway Luminaire

		Crew	Daily Output	Labor-Hours	Unit	Material	Labor	Equipment	Total	Total Incl O&P
0010	**ROADWAY LUMINAIRE**									
2650	Roadway area luminaire, low pressure sodium, 135 watt	1 Elec	1.68	4.760	Ea.	835	292		1,127	1,350
2700	180 watt	"	1.68	4.760		995	292		1,287	1,525
2750	Metal halide, 400 watt	2 Elec	3.70	4.330		690	266		956	1,150
2760	1,000 watt		3.36	4.760		775	292		1,067	1,275
2780	High pressure sodium, 400 watt		3.70	4.330		825	266		1,091	1,300
2790	1,000 watt	↓	3.36	4.760	↓	940	292		1,232	1,450

26 56 23 – Area Lighting

26 56 23.10 Exterior Fixtures

		Crew	Daily Output	Labor-Hours	Unit	Material	Labor	Equipment	Total	Total Incl O&P
0010	**EXTERIOR FIXTURES** With lamps									
0200	Wall mounted, incandescent, 100 watt	1 Elec	6.72	1.190	Ea.	51.50	73		124.50	166
0400	Quartz, 500 watt		4.45	1.800		62	110		172	233
0420	1,500 watt		3.53	2.270		108	139		247	325
1100	Wall pack, low pressure sodium, 35 watt		3.36	2.380		205	146		351	445
1150	55 watt		3.36	2.380		241	146		387	480
1160	High pressure sodium, 70 watt		3.36	2.380		197	146		343	435
1170	150 watt		3.36	2.380		218	146		364	455
1175	High pressure sodium, 250 watt		3.36	2.380		218	146		364	455
1180	Metal halide, 175 watt		3.36	2.380		221	146		367	460
1190	250 watt		3.36	2.380		251	146		397	495
1195	400 watt	↓	3.36	2.380	↓	410	146		556	670

For customer support on your Electrical Change Order Costs with RSMeans data, call 800.448.8182. **Post-Installation Change Order Costs**

26 56 Exterior Lighting

26 56 23 – Area Lighting

26 56 23.10 Exterior Fixtures	Crew	Daily Output	Labor-Hours	Unit	Material	2020 Bare Costs Labor	Equipment	Total	Total Incl O&P
1250	Induction lamp, 40 watt	1 Elec	3.36	2.380	Ea.	495	146		641	760
1260	80 watt		3.36	2.380		590	146		736	860
1278	LED, poly lens, 26 watt		3.36	2.380		310	146		456	555
1280	110 watt		3.36	2.380		830	146		976	1,125
1500	LED, glass lens, 13 watt	▼	3.36	2.380	▼	310	146		456	555

26 56 26 – Landscape Lighting

26 56 26.20 Landscape Fixtures

		Crew	Daily Output	Labor-Hours	Unit	Material	Labor	Equipment	Total	Total Incl O&P
0010	**LANDSCAPE FIXTURES**									
7380	Landscape recessed uplight, incl. housing, ballast, transformer									
7390	& reflector, not incl. conduit, wire, trench									
7420	Incandescent, 250 watt	1 Elec	4.20	1.900	Ea.	665	117		782	905
7440	Quartz, 250 watt		4.20	1.900		630	117		747	870
7460	500 watt	▼	3.36	2.380	▼	650	146		796	930

26 56 26.50 Landscape LED Fixtures

		Crew	Daily Output	Labor-Hours	Unit	Material	Labor	Equipment	Total	Total Incl O&P
0010	**LANDSCAPE LED FIXTURES**									
0100	12 volt alum bullet hooded-BLK	1 Elec	4.20	1.900	Ea.	107	117		224	292
0200	12 volt alum bullet hooded-BRZ		4.20	1.900		107	117		224	292
0300	12 volt alum bullet hooded-GRN		4.20	1.900		107	117		224	292
1000	12 volt alum large bullet hooded-BLK		4.20	1.900		79	117		196	261
1100	12 volt alum large bullet hooded-BRZ		4.20	1.900		79	117		196	261
1200	12 volt alum large bullet hooded-GRN		4.20	1.900		79	117		196	261
2000	12 volt large bullet landscape light fixture		4.20	1.900		79	117		196	261
2100	12 volt alum light large bullet		4.20	1.900		79	117		196	261
2200	12 volt alum bullet light	▼	4.20	1.900	▼	79	117		196	261

26 56 33 – Walkway Lighting

26 56 33.10 Walkway Luminaire

		Crew	Daily Output	Labor-Hours	Unit	Material	Labor	Equipment	Total	Total Incl O&P
0010	**WALKWAY LUMINAIRE**									
6500	Bollard light, lamp & ballast, 42" high with polycarbonate lens									
6800	Metal halide, 175 watt	1 Elec	2.52	3.170	Ea.	985	194		1,179	1,375
6900	High pressure sodium, 70 watt		2.52	3.170		950	194		1,144	1,350
7000	100 watt		2.52	3.170		950	194		1,144	1,350
7100	150 watt		2.52	3.170		925	194		1,119	1,325
7200	Incandescent, 150 watt		2.52	3.170		675	194		869	1,025
7810	Walkway luminaire, square 16", metal halide 250 watt		2.27	3.530		770	217		987	1,175
7820	High pressure sodium, 70 watt		2.52	3.170		880	194		1,074	1,250
7830	100 watt		2.52	3.170		895	194		1,089	1,275
7840	150 watt		2.52	3.170		895	194		1,089	1,275
7850	200 watt		2.52	3.170		900	194		1,094	1,275
7910	Round 19", metal halide, 250 watt		2.27	3.530		1,250	217		1,467	1,700
7920	High pressure sodium, 70 watt		2.52	3.170		1,375	194		1,569	1,800
7930	100 watt		2.52	3.170		1,375	194		1,569	1,800
7940	150 watt		2.52	3.170		1,375	194		1,569	1,800
7950	250 watt		2.27	3.530		1,300	217		1,517	1,750
8000	Sphere 14" opal, incandescent, 200 watt		3.36	2.380		355	146		501	605
8020	Sphere 18" opal, incandescent, 300 watt		2.94	2.720		430	167		597	720
8040	Sphere 16" clear, high pressure sodium, 70 watt		2.52	3.170		740	194		934	1,100
8050	100 watt		2.52	3.170		790	194		984	1,150
8100	Cube 16" opal, incandescent, 300 watt		2.94	2.720		470	167		637	765
8120	High pressure sodium, 70 watt		2.52	3.170		685	194		879	1,050
8130	100 watt		2.52	3.170		705	194		899	1,075
8230	Lantern, high pressure sodium, 70 watt	▼	2.52	3.170	▼	615	194		809	965

26 56 Exterior Lighting

26 56 33 – Walkway Lighting

26 56 33.10 Walkway Luminaire

		Crew	Daily Output	Labor-Hours	Unit	Material	2020 Bare Costs Labor	2020 Bare Costs Equipment	Total	Total Incl O&P
8240	100 watt	1 Elec	2.52	3.170	Ea.	660	194		854	1,025
8250	150 watt		2.52	3.170		620	194		814	970
8260	250 watt		2.27	3.530		865	217		1,082	1,275
8270	Incandescent, 300 watt		2.94	2.720		455	167		622	755
8330	Reflector 22" w/globe, high pressure sodium, 70 watt		2.52	3.170		555	194		749	900
8340	100 watt		2.52	3.170		565	194		759	910
8350	150 watt		2.52	3.170		570	194		764	920
8360	250 watt		2.27	3.530		730	217		947	1,125
0600	LED bollard pole, 6" diam. x 36" H, cast alum, surf. mtd., 8W, 6,000K, 120V		7.14	1.120		385	68.50		453.50	525
0610	6" diam. x 42" H, cast alum, surf. mtd., 12W, 5,100K,120V		7.14	1.120		565	68.50		633.50	720
0620	6" diam. x 42" H, cast alum, surf. mtd., 18W, 5,100K,120V		7.14	1.120		645	68.50		713.50	810
0630	6" diam. x 42" H, cast alum, surf. mtd., 24W, 5,100K,120V		7.14	1.120		505	68.50		573.50	655
0640	9.5" square x 42" H, cast resin, surf. mtd., A19, 9W, 120V		7.14	1.120		610	68.50		678.50	770
0650	9.75" diam. x 42" H, cast resin, side mount, dome, 9W, 120V		7.14	1.120		695	68.50		763.50	865
0660	9.75" diam. x 43" H, cast resin, burial base, dome, 9W, 120V		7.14	1.120		695	68.50		763.50	865
0670	8" diam. x 36" H, concrete security base, w/rebar, 8W, HID		6.72	1.190		1,025	73		1,098	1,225
0680	10" diam. x 49" H, concrete base, w/Sch 40 Steel pipe, 8W		5.67	1.410		1,175	86.50		1,261.50	1,400
0690	12" diam. x 48" H, concrete base, w/Sch 40 Steel pipe, 8W		5.67	1.410		1,325	86.50		1,411.50	1,600
0700	8" diam. x 36" H, concrete, cast aluminum dome, 90 degree, 4.1W		5.88	1.360		1,025	83.50		1,108.50	1,275
0710	8" diam. x 36" H, concrete, cast aluminum dome, 180 degree, 4.1W		5.88	1.360		1,275	83.50		1,358.50	1,525
0720	8" diam. x 48" H, concrete, cast aluminum dome, 360 degree, 4.1W	↓	5.88	1.360	↓	1,575	83.50		1,658.50	1,850

26 56 36 – Flood Lighting

26 56 36.20 Floodlights

		Crew	Daily Output	Labor-Hours	Unit	Material	2020 Bare Costs Labor	2020 Bare Costs Equipment	Total	Total Incl O&P
0010	**FLOODLIGHTS** with ballast and lamp,									
1290	floor mtd, mount with swivel bracket									
1300	Induction lamp, 40 watt	1 Elec	2.52	3.170	Ea.	565	194		759	910
1310	80 watt		2.52	3.170		750	194		944	1,125
1320	150 watt	↓	2.52	3.170	↓	1,350	194		1,544	1,800
1400	Pole mounted, pole not included									
1950	Metal halide, 175 watt	1 Elec	2.27	3.530	Ea.	213	217		430	555
2000	400 watt	2 Elec	3.70	4.330		219	266		485	635
2200	1,000 watt		3.36	4.760		900	292		1,192	1,425
2210	1,500 watt	↓	3.11	5.150		425	315		740	935
2250	Low pressure sodium, 55 watt	1 Elec	2.27	3.530		460	217		677	825
2270	90 watt		1.68	4.760		620	292		912	1,125
2290	180 watt		1.68	4.760		680	292		972	1,175
2340	High pressure sodium, 70 watt		2.27	3.530		254	217		471	600
2360	100 watt		2.27	3.530		261	217		478	605
2380	150 watt	↓	2.27	3.530		293	217		510	645
2400	400 watt	2 Elec	3.70	4.330		320	266		586	750
2600	1,000 watt	"	3.36	4.760	↓	635	292		927	1,125

For customer support on your Electrical Change Order Costs with RSMeans data, call 800.448.8182.

Post-Installation Change Order Costs

26 61 13.30 Fixture Whips		Crew	Daily Output	Labor-Hours	Unit	Material	2020 Bare Costs Labor	Equipment	Total	Total Incl O&P
0010	**FIXTURE WHIPS**									
0080	3/8" Greenfield, 2 connectors, 6' long									
0100	TFFN wire, three #18	1 Elec	26.88	.298	Ea.	9.20	18.25		27.45	37
0150	Four #18		23.52	.340		9.90	21		30.90	42
0200	Three #16		26.88	.298		9.50	18.25		27.75	37.50
0250	Four #16		23.52	.340		10.35	21		31.35	42.50
0300	THHN wire, three #14		26.88	.298		6.55	18.25		24.80	34
0350	Four #14		23.52	.340		9.15	21		30.15	41
0360	Three #12		26.88	.298		12.60	18.25		30.85	41

26 61 23 – Lamps Applications

26 61 23.10 Lamps		Crew	Daily Output	Labor-Hours	Unit	Material	2020 Bare Costs Labor	Equipment	Total	Total Incl O&P
0010	**LAMPS**									
0080	Fluorescent, rapid start, cool white, 2' long, 20 watt	1 Elec	.84	9.520	C	365	585		950	1,275
0100	4' long, 40 watt		.76	10.580		305	650		955	1,300
0120	3' long, 30 watt		.76	10.580		445	650		1,095	1,450
0125	3' long, 25 watt energy saver G		.76	10.580		1,425	650		2,075	2,525
0150	U-40 watt		.67	11.900		1,250	730		1,980	2,450
0155	U-34 watt energy saver G		.67	11.900		12.95	730		742.95	1,100
0170	4' long, 34 watt energy saver G		.76	10.580		870	650		1,520	1,925
0176	2' long, T8, 17 watt energy saver G		.84	9.520		385	585		970	1,300
0178	3' long, T8, 25 watt energy saver G		.76	10.580		410	650		1,060	1,425
0180	4' long, T8, 32 watt energy saver G		.76	10.580		237	650		887	1,225
0200	Slimline, 4' long, 40 watt		.76	10.580		1,150	650		1,800	2,225
0210	4' long, 30 watt energy saver G		.76	10.580		1,150	650		1,800	2,225
0300	8' long, 75 watt		.67	11.900		1,125	730		1,855	2,300
0350	8' long, 60 watt energy saver G		.67	11.900		435	730		1,165	1,550
0400	High output, 4' long, 60 watt		.76	10.580		670	650		1,320	1,700
0410	8' long, 95 watt energy saver G		.67	11.900		665	730		1,395	1,800
0500	8' long, 110 watt		.67	11.900		665	730		1,395	1,800
0512	2' long, T5, 14 watt energy saver G		.84	9.520		196	585		781	1,075
0514	3' long, T5, 21 watt energy saver G		.76	10.580		190	650		840	1,175
0516	4' long, T5, 28 watt energy saver G		.76	10.580		190	650		840	1,175
0517	4' long, T5, 54 watt energy saver G		.76	10.580		565	650		1,215	1,575
0520	Very high output, 4' long, 110 watt		.76	10.580		1,775	650		2,425	2,925
0525	8' long, 195 watt energy saver G		.59	13.610		1,525	835		2,360	2,925
0550	8' long, 215 watt		.59	13.610		1,450	835		2,285	2,850
0554	Full spectrum, 4' long, 60 watt		.76	10.580		725	650		1,375	1,750
0556	6' long, 85 watt		.76	10.580		815	650		1,465	1,875
0558	8' long, 110 watt		.67	11.900		2,150	730		2,880	3,450
0560	Twin tube compact lamp G		.76	10.580		475	650		1,125	1,500
0570	Double twin tube compact lamp G		.67	11.900		1,050	730		1,780	2,225
0600	Mercury vapor, mogul base, deluxe white, 100 watt		.25	31.750		5,675	1,950		7,625	9,150
0650	175 watt		.25	31.750		3,000	1,950		4,950	6,200
0700	250 watt		.25	31.750		5,500	1,950		7,450	8,950
0800	400 watt		.25	31.750		5,050	1,950		7,000	8,450
0900	1,000 watt		.17	47.620		14,000	2,925		16,925	19,800
1000	Metal halide, mogul base, 175 watt		.25	31.750		1,150	1,950		3,100	4,175
1100	250 watt		.25	31.750		1,800	1,950		3,750	4,900
1200	400 watt		.25	31.750		2,200	1,950		4,150	5,325
1300	1,000 watt		.17	47.620		4,300	2,925		7,225	9,100
1320	1,000 watt, 125,000 initial lumens		.17	47.620		17,700	2,925		20,625	23,900
1330	1,500 watt		.17	47.620		3,925	2,925		6,850	8,650

26 61 23.10 Lamps	Crew	Daily Output	Labor-Hours	Unit	Material	2020 Bare Costs Labor	Equipment	Total	Total Incl O&P
1350	High pressure sodium, 70 watt	1 Elec	.25	31.750	C	1,875	1,950		3,825	4,950
1360	100 watt		.25	31.750		1,925	1,950		3,875	5,025
1370	150 watt		.25	31.750		1,800	1,950		3,750	4,875
1380	250 watt		.25	31.750		2,150	1,950		4,100	5,275
1400	400 watt		.25	31.750		1,800	1,950		3,750	4,875
1450	1,000 watt		.17	47.620		4,950	2,925		7,875	9,800
1500	Low pressure sodium, 35 watt		.25	31.750		17,400	1,950		19,350	22,100
1550	55 watt		.25	31.750		19,900	1,950		21,850	24,800
1600	90 watt		.25	31.750		7,000	1,950		8,950	10,600
1650	135 watt		.17	47.620		28,900	2,925		31,825	36,200
1700	180 watt		.17	47.620		43,300	2,925		46,225	52,000
1750	Quartz line, clear, 500 watt		.92	8.660		820	530		1,350	1,700
1760	1,500 watt		.17	47.620		2,325	2,925		5,250	6,900
1762	Spot, MR 16, 50 watt		1.09	7.330		1,150	450		1,600	1,950
1770	Tungsten halogen, T4, 400 watt		.92	8.660		4,225	530		4,755	5,450
1775	T3, 1,200 watt		.25	31.750		5,400	1,950		7,350	8,825
1778	PAR 30, 50 watt		1.09	7.330		1,175	450		1,625	1,950
1780	PAR 38, 90 watt		1.09	7.330		10,200	450		10,650	11,900
1800	Incandescent, interior, A21, 100 watt		1.34	5.950		2,975	365		3,340	3,825
1900	A21, 150 watt		1.34	5.950		17,600	365		17,965	19,800
2000	A23, 200 watt		1.34	5.950		355	365		720	935
2200	PS 35, 300 watt		1.34	5.950		1,050	365		1,415	1,725
2210	PS 35, 500 watt		1.34	5.950		1,650	365		2,015	2,375
2230	PS 52, 1,000 watt		1.09	7.330		2,675	450		3,125	3,600
2240	PS 52, 1,500 watt		1.09	7.330		7,250	450		7,700	8,650
2300	R30, 75 watt		1.09	7.330		765	450		1,215	1,500
2400	R40, 100 watt		1.09	7.330		765	450		1,215	1,500
2500	Exterior, PAR 38, 75 watt		1.09	7.330		2,000	450		2,450	2,875
2600	PAR 38, 150 watt		1.09	7.330		2,250	450		2,700	3,150
2700	PAR 46, 200 watt		.92	8.660		4,100	530		4,630	5,325
2800	PAR 56, 300 watt		.92	8.660		2,800	530		3,330	3,900
3000	Guards, fluorescent lamp, 4' long		84	.095	Ea.	15.40	5.85		21.25	25.50
3200	8' long		75.60	.106	"	31	6.50		37.50	43.50

26 61 23.55 LED Lamps

		Crew	Daily Output	Labor-Hours	Unit	Material	2020 Bare Costs Labor	Equipment	Total	Total Incl O&P
0010	**LED LAMPS**									
0100	LED lamp, interior, shape A60, equal to 60 W G	1 Elec	134.40	.060	Ea.	20.50	3.65		24.15	28
0110	7 W LED decorative c, ca, f, g shape		134.40	.060		36.50	3.65		40.15	45.50
0205	LED lamp, interior, globe		134.40	.060		17.10	3.65		20.75	24.50
0210	2.2 W LED LMP		134.40	.060		70	3.65		73.65	82.50
0220	2.2 W LED replacement decorative lamp		134.40	.060		9.65	3.65		13.30	16.05
0230	3.5 W LED replacement decorative lamp		134.40	.060		11.20	3.65		14.85	17.80
0240	4.5 W 120V LED replacement decorative lamp		134.40	.060		13.45	3.65		17.10	20.50
0250	4.5 W 120V LED, 2700k replacement decorative lamp		134.40	.060		11.85	3.65		15.50	18.50
0260	4.9 W 120V LED, 2700k replacement decorative lamp candelabra base		134.40	.060		10.65	3.65		14.30	17.15
0270	4.9 W 120V LED, 3000k replacement decorative lamp candelabra base		134.40	.060		9.95	3.65		13.60	16.40
0280	5 W LED PAR 20 parabolic reflector lamp FL		117.60	.068		63.50	4.17		67.67	76
0300	Globe earth, equal to 100 W G		117.60	.068		29.50	4.17		33.67	38.50
0305	7 W LED reflector lamp 3000k DIM		117.60	.068		61.50	4.17		65.67	73.50
0310	10 W omni LED warm white light bulb E26 medium base 120 volt card		117.60	.068		18.45	4.17		22.62	26.50
0315	7 W LED reflector lamp WFL 2700k DIM		117.60	.068		61.50	4.17		65.67	73.50
0320	9 W omni LED warm white light bulb E26 medium base 120 volt card		117.60	.068		9.85	4.17		14.02	17
0500	8 W LED, A19 lamp, equal to 40 W		117.60	.068		6.25	4.17		10.42	13.05

26 61 23.55 LED Lamps		Crew	Daily Output	Labor-Hours	Unit	Material	2020 Bare Costs Labor	Equipment	Total	Total Incl O&P
0505	9 W LED, A19 lamp, 5000K dimmable, equal to 60 W	1 Elec	117.60	.068	Ea.	7.80	4.17		11.97	14.80
0510	10.5 W LED, A19 lamp, dimmable, equal to 60 W		117.60	.068		8.90	4.17		13.07	15.95
0515	10 W LED, A19 lamp, frosted, dimmable, equal to 60 W		117.60	.068		14.20	4.17		18.37	22
0520	10 W LED, A19 lamp, omni-directional, dimmable, equal to 60 W		117.60	.068		8.30	4.17		12.47	15.35
0525	12 W LED, A19 lamp, dimmable, equal to 75 W		117.60	.068		12.05	4.17		16.22	19.45
1100	MR16, 3 W, replacement of halogen lamp 25 W [G]		109.20	.073		20.50	4.49		24.99	29
1200	6 W replacement of halogen lamp 45 W [G]		109.20	.073		21.50	4.49		25.99	30
2100	10 W, PAR20, equal to 60 W [G]		109.20	.073		28.50	4.49		32.99	38
2200	15 W, PAR30, equal to 100 W [G]		109.20	.073		54	4.49		58.49	65.50
2210	50 W LED lamp		109.20	.073		605	4.49		609.49	670
2220	11 Watt reflector dimmable warm white LED light bulb with medium base		109.20	.073		44	4.49		48.49	55
2221	12 W A-Line LED lamp DIM		109.20	.073		83	4.49		87.49	97.50
2225	13 Watt reflector LED warm white e26 with medium base 120 volt box		109.20	.073		41.50	4.49		45.99	52
2226	13 W br30 LED lamp		109.20	.073		72	4.49		76.49	85.50
2227	15 W 120V br30 inc LED lamp, 2700k		109.20	.073		67	4.49		71.49	80.50
2228	15 W 120V br30 inc LED lamp, 4000k		109.20	.073		63.50	4.49		67.99	76.50
2230	3 Watt dimmable warm white decorative LED lamp with medium base		134.40	.060		13	3.65		16.65	19.75
2240	.43 W night light LED daylight bulb E12 candelabra base 120 volt 2 pack		134.40	.060		5	3.65		8.65	10.95
2250	15 W omni-directional LED warm white e26 medium base 120 volt box		134.40	.060		42.50	3.65		46.15	52.50
2251	11 W omni-directional LED warm white e26 medium base 120 volt box		134.40	.060		30.50	3.65		34.15	39
2252	10 W omni-directional LED warm white e26 medium base 120 volt box		134.40	.060		29.50	3.65		33.15	38
2253	7 W omni A19 LED warm white e26 medium base 120 volt box		134.40	.060		22.50	3.65		26.15	30
2255	8 PAR 20 parabolic reflector 2700 LED lamp		134.40	.060		20	3.65		23.65	27.50
2256	8 PAR 20 parabolic reflector 3000 LED lamp		134.40	.060		19.80	3.65		23.45	27.50
2260	12 W PAR 38 120V LED 15 degree directional lamp		134.40	.060		215	3.65		218.65	241
2270	12 W PAR 38 120V LED 25 degree directional lamp		134.40	.060		33	3.65		36.65	42
2280	12 W PAR 38 120V LED 40 degree directional lamp		134.40	.060		215	3.65		218.65	241
2285	16 W PAR 38 120V 2700k LED parabolic reflector lamp		134.40	.060		32.50	3.65		36.15	41
2290	16 W PAR 38 120V 3000k LED parabolic reflector lamp		134.40	.060		24.50	3.65		28.15	32.50
3000	15 W PAR 30 LED daylight E26 medium base 120V box		134.40	.060		63.50	3.65		67.15	75.50
3100	17 W LED 3000k PAR 38 100 W replacement		134.40	.060		49.50	3.65		53.15	60
3110	17 W LED PAR 38 100 W replacement		134.40	.060		37	3.65		40.65	46.50
3120	17 W LED PAR 38 100 W replacement parabolic reflector		134.40	.060		39	3.65		42.65	48
3130	24 W LED T8 PW straight fluorescent lamp		7.47	1.070		305	65.50		370.50	435
3135	Linear fluorescent LED lamp 120V		7.47	1.070		141	65.50		206.50	253
3200	30 W LED 2700K recessed 8055E PAR 38 high power		134.40	.060		141	3.65		144.65	160
3210	30 W LED 4200K 120 degree 8055E PAR 38 high power		134.40	.060		142	3.65		145.65	161
3220	30 W LED 5700K 8055E PAR 38 high power		134.40	.060		143	3.65		146.65	162
3230	50 W LED 2700K 8045M PAR 38 277V high power		134.40	.060		360	3.65		363.65	400
3240	50 W LED 4200K 8045M PAR 38 277V high power retro fit		134.40	.060		191	3.65		194.65	215
3250	50 W LED 5700K 8045M PAR 38 277V high power retro fit		134.40	.060		360	3.65		363.65	400
8000	10 W LED PAR 30/fl 10 pk		134.40	.060		102	3.65		105.65	117
8010	Gen 3 PAR 30 15 W short neck power LED 120 VAC E26 80 +cri 300k dimm		134.40	.060		13.70	3.65		17.35	20.50
8020	12 PAR 30 2700K parabolic reflector LED lamp		134.40	.060		44.50	3.65		48.15	54.50
8030	12 PAR 30 3000K parabolic reflector LED lamp		134.40	.060		34	3.65		37.65	43
8040	3500K LED advantage T8 9 W 800LM 2ft linear 2 BD frosted		57.96	.138		11.10	8.45		19.55	25
8050	3500K LED litespan T8 9 W 900LM 2ft linear frosted		57.96	.138		14	8.45		22.45	28
8060	4000K LED advantage T8 9 W 800LM 2ft linear 2BD frosted		57.96	.138		13.30	8.45		21.75	27.50
8070	4000K LED litespan T8 9 W 900LM 2ft linear frosted		57.96	.138		14.65	8.45		23.10	28.50
8080	5000K LED advantage T8 9 W 800LM 2ft linear 2BD frosted		57.96	.138		11.90	8.45		20.35	25.50
8090	5000K LED litespan T8 9 W 900LM 2ft linear frosted		57.96	.138		11.65	8.45		20.10	25.50
8100	3500K LED advantage T8 18 W 1600LM 4ft linear 2 BD frosted		57.96	.138		11	8.45		19.45	24.50
8105	18 W LED 4ft T8 4000K frost 1600L linear lamp		57.96	.138		19.60	8.45		28.05	34

26 61 Lighting Systems and Accessories

26 61 23 – Lamps Applications

	26 61 23.55 LED Lamps	Crew	Daily Output	Labor-Hours	Unit	Material	2020 Bare Costs Labor	Equipment	Total	Total Incl O&P
8108	18 W LED 48 inch T8 4100K 1890LM linear lamp	1 Elec	57.96	.138	Ea.	60.50	8.45		68.95	79
8110	3500K LED advantage T8 18 W 1800LM 4ft linear 2 BD frosted		57.96	.138		31.50	8.45		39.95	47
8120	4000K LED advantage T8 18 W 1600LM 4ft linear 2 BD frosted		57.96	.138		21.50	8.45		29.95	36
8130	4000K LED litespan T8 18 W 1800LM 4ft linear frosted		57.96	.138		22.50	8.45		30.95	37
8140	5000K LED advantage T8 18 W 1600LM 4ft linear 2BD frosted		57.96	.138		14	8.45		22.45	28
8150	5000K LED litespan T8 18 W 1800LM 4ft linear 2BD frosted		57.96	.138		22.50	8.45		30.95	37
8200	16.5 T8 3000 IF-6U U-shape fluorescent LED lamp		54.60	.147		23	9		32	38.50
8210	16.5 T8 3500 IF-6U U-shape fluorescent LED lamp		54.60	.147		23.50	9		32.50	39
8220	16.5 T8 4000 IF-6U U-shape fluorescent LED lamp		54.60	.147		24	9		33	40
8230	16.5 T8 5000 IF-6U U-shape fluorescent LED lamp		54.60	.147		22.50	9		31.50	38.50
8240	18 W 6 inch T8 4100K U-shape frosted fluorescent LED lamp		50.40	.159		12.70	9.75		22.45	28.50
8250	18 W 6 inch T8 5000K U-shape frosted fluorescent LED lamp		50.40	.159		12.70	9.75		22.45	28.50
8260	Circular 12 W linear fluorescent LED 2700K MOD lamp		52.42	.153		320	9.35		329.35	365
8270	Circular 12 W linear fluorescent LED 3000K MOD lamp		52.42	.153		320	9.35		329.35	365
8280	Circular 12 W linear fluorescent LED 3500K MOD lamp		52.42	.153		289	9.35		298.35	335
8290	Circular 18 W linear fluorescent LED 3000K MOD lamp		52.42	.153		395	9.35		404.35	450
8300	Circular 18 W linear fluorescent LED 3500K MOD lamp		52.42	.153		435	9.35		444.35	495
8310	Circular 18 W linear fluorescent LED 5000K MOD lamp		52.42	.153		395	9.35		404.35	450
8320	Circular 18 W linear fluorescent LED 4100K MOD lamp		52.42	.153		420	9.35		429.35	480

26 71 Electrical Machines

26 71 13 – Motors Applications

26 71 13.10 Handling

		Crew	Daily Output	Labor-Hours	Unit	Material	2020 Bare Costs Labor	Equipment	Total	Total Incl O&P
0010	**HANDLING** Add to normal labor cost for restricted areas									
5000	Motors									
5100	1/2 HP, 23 pounds	1 Elec	3.36	2.380	Ea.		146		146	217
5110	3/4 HP, 28 pounds		3.36	2.380			146		146	217
5120	1 HP, 33 pounds		3.36	2.380			146		146	217
5130	1-1/2 HP, 44 pounds		2.69	2.980			183		183	272
5140	2 HP, 56 pounds		2.52	3.170			194		194	290
5150	3 HP, 71 pounds		1.93	4.140			254		254	380
5160	5 HP, 82 pounds		1.60	5.010			305		305	460
5170	7-1/2 HP, 124 pounds		1.26	6.350			390		390	580
5180	10 HP, 144 pounds		1.01	7.940			485		485	725
5190	15 HP, 185 pounds		.84	9.520			585		585	870
5200	20 HP, 214 pounds	2 Elec	1.26	12.700			780		780	1,150
5210	25 HP, 266 pounds		1.18	13.610			835		835	1,250
5220	30 HP, 310 pounds		1.01	15.870			975		975	1,450
5230	40 HP, 400 pounds		.84	19.050			1,175		1,175	1,750
5240	50 HP, 450 pounds		.76	21.160			1,300		1,300	1,925
5250	75 HP, 680 pounds		.67	23.810			1,450		1,450	2,175
5260	100 HP, 870 pounds	3 Elec	.84	28.570			1,750		1,750	2,600
5270	125 HP, 940 pounds		.67	35.710			2,200		2,200	3,250
5280	150 HP, 1,200 pounds		.59	40.820			2,500		2,500	3,725
5290	175 HP, 1,300 pounds		.50	47.620			2,925		2,925	4,350
5300	200 HP, 1,400 pounds		.42	57.140			3,500		3,500	5,225

26 71 13.20 Motors

		Crew	Daily Output	Labor-Hours	Unit	Material	2020 Bare Costs Labor	Equipment	Total	Total Incl O&P
0010	**MOTORS** 230/460 V, 60 HZ									
0050	Dripproof, premium efficiency, 1.15 service factor									
0060	1,800 RPM, 1/4 HP	1 Elec	4.48	1.790	Ea.	280	110		390	475
0070	1/3 HP		4.48	1.790		247	110		357	435

Post-Installation Change Order Costs

26 71 13.20 Motors		Crew	Daily Output	Labor-Hours	Unit	Material	2020 Bare Costs Labor	Equipment	Total	Total Incl O&P
0080	1/2 HP	1 Elec	4.48	1.790	Ea.	207	110		317	390
0090	3/4 HP		4.48	1.790		320	110		430	515
0100	1 HP		3.78	2.120		345	130		475	575
0150	2 HP		3.78	2.120		370	130		500	605
0200	3 HP		3.78	2.120		805	130		935	1,075
0250	5 HP		3.78	2.120		595	130		725	850
0300	7.5 HP		3.53	2.270		985	139		1,124	1,275
0350	10 HP		3.36	2.380		1,175	146		1,321	1,525
0400	15 HP	▼	2.69	2.980		1,600	183		1,783	2,050
0450	20 HP	2 Elec	4.37	3.660		2,125	225		2,350	2,650
0500	25 HP		4.20	3.810		2,575	234		2,809	3,175
0550	30 HP		4.03	3.970		2,725	244		2,969	3,350
0600	40 HP		3.36	4.760		3,625	292		3,917	4,425
0650	50 HP		2.69	5.950		3,800	365		4,165	4,750
0700	60 HP		2.35	6.800		4,725	415		5,140	5,825
0750	75 HP	▼	2.02	7.940		5,825	485		6,310	7,125
0800	100 HP	3 Elec	2.27	10.580		5,950	650		6,600	7,525
0850	125 HP		1.76	13.610		7,175	835		8,010	9,125
0900	150 HP		1.51	15.870		9,800	975		10,775	12,300
0950	200 HP	▼	1.26	19.050		10,800	1,175		11,975	13,700
1000	1,200 RPM, 1 HP	1 Elec	3.78	2.120		480	130		610	725
1050	2 HP		3.78	2.120		585	130		715	840
1100	3 HP		3.78	2.120		735	130		865	1,000
1150	5 HP		3.78	2.120		955	130		1,085	1,250
1200	3,600 RPM, 2 HP		3.78	2.120		515	130		645	760
1250	3 HP		3.78	2.120		575	130		705	830
1300	5 HP	▼	3.78	2.120	▼	600	130		730	855
1350	Totally enclosed, premium efficiency 1.15 service factor									
1360	1,800 RPM, 1/4 HP	1 Elec	4.48	1.790	Ea.	395	110		505	600
1370	1/3 HP		4.48	1.790		277	110		387	470
1380	1/2 HP		4.48	1.790		385	110		495	590
1390	3/4 HP		4.48	1.790		430	110		540	635
1400	1 HP		3.78	2.120		685	130		815	950
1450	2 HP		3.78	2.120		670	130		800	930
1500	3 HP		3.78	2.120		695	130		825	960
1550	5 HP		3.78	2.120		810	130		940	1,075
1600	7.5 HP		3.53	2.270		1,025	139		1,164	1,325
1650	10 HP		3.36	2.380		1,400	146		1,546	1,775
1700	15 HP	▼	2.69	2.980		2,500	183		2,683	3,025
1750	20 HP	2 Elec	4.37	3.660		2,450	225		2,675	3,025
1800	25 HP		4.20	3.810		3,875	234		4,109	4,600
1850	30 HP		4.03	3.970		3,575	244		3,819	4,300
1900	40 HP		3.36	4.760		4,050	292		4,342	4,875
1950	50 HP		2.69	5.950		4,775	365		5,140	5,800
2000	60 HP		2.35	6.800		6,150	415		6,565	7,400
2050	75 HP	▼	2.02	7.940		7,850	485		8,335	9,350
2100	100 HP	3 Elec	2.27	10.580		8,775	650		9,425	10,600
2150	125 HP		1.76	13.610		12,300	835		13,135	14,900
2200	150 HP		1.51	15.870		12,900	975		13,875	15,600
2250	200 HP	▼	1.26	19.050		16,500	1,175		17,675	19,900
2300	1,200 RPM, 1 HP	1 Elec	3.78	2.120		475	130		605	720
2350	2 HP		3.78	2.120		600	130		730	855
2400	3 HP	▼	3.78	2.120	▼	780	130		910	1,050

26 71 13.20 Motors

		Crew	Daily Output	Labor-Hours	Unit	Material	2020 Bare Costs Labor	Equipment	Total	Total Incl O&P
2450	5 HP	1 Elec	3.78	2.120	Ea.	970	130		1,100	1,275
2500	3,600 RPM, 2 HP		3.78	2.120		420	130		550	655
2550	3 HP		3.78	2.120		555	130		685	805
2600	5 HP		3.78	2.120		700	130		830	965

26 71 13.40 Motors Explosion Proof

		Crew	Daily Output	Labor-Hours	Unit	Material	2020 Bare Costs Labor	Equipment	Total	Total Incl O&P
0010	**MOTORS EXPLOSION PROOF**, 208-230/460 V, 60 HZ									
0020	1,800 RPM, 1/4 HP	1 Elec	4.20	1.900	Ea.	650	117		767	890
0030	1/3 HP		4.20	1.900		515	117		632	745
0040	1/2 HP		4.20	1.900		700	117		817	945
0050	3/4 HP		4.48	1.790		1,225	110		1,335	1,525
0060	1 HP		3.53	2.270		760	139		899	1,050
0070	2 HP		3.53	2.270		720	139		859	995
0080	3 HP		3.53	2.270		1,025	139		1,164	1,325
0090	5 HP		3.53	2.270		1,000	139		1,139	1,300
0100	7.5 HP		3.36	2.380		1,250	146		1,396	1,600
0110	10 HP		3.11	2.570		1,475	158		1,633	1,850
0120	15 HP		2.69	2.980		1,875	183		2,058	2,325
0130	20 HP	2 Elec	4.20	3.810		2,275	234		2,509	2,850
0140	25 HP		4.03	3.970		2,700	244		2,944	3,350
0150	30 HP		3.70	4.330		3,050	266		3,316	3,750
0160	40 HP		3.19	5.010		5,000	305		5,305	5,925
0170	50 HP		2.69	5.950		5,100	365		5,465	6,150
0180	60 HP		2.18	7.330		8,425	450		8,875	9,925
0190	75 HP		1.68	9.520		10,300	585		10,885	12,300
0200	100 HP	3 Elec	2.10	11.430		13,500	700		14,200	16,000
0210	125 HP		1.85	12.990		17,200	795		17,995	20,100
0220	150 HP		1.34	17.860		20,400	1,100		21,500	24,000
0230	200 HP		1.01	23.810		25,000	1,450		26,450	29,700
1000	1,200 RPM, 1 HP	1 Elec	3.53	2.270		575	139		714	840
1010	2 HP		3.53	2.270		665	139		804	935
1020	3 HP		3.53	2.270		905	139		1,044	1,200
1030	5 HP		3.53	2.270		1,225	139		1,364	1,550
2000	3,600 RPM, 1/4 HP		4.20	1.900		490	117		607	715
2010	1/3 HP		4.20	1.900		515	117		632	740
2020	1/2 HP		4.20	1.900		650	117		767	890
2030	3/4 HP		4.20	1.900		690	117		807	930
2040	1 HP		3.53	2.270		745	139		884	1,025
2050	2 HP		3.53	2.270		935	139		1,074	1,225
2060	3 HP		3.53	2.270		1,025	139		1,164	1,325
2070	5 HP		3.53	2.270		1,375	139		1,514	1,700
2080	7.5 HP		3.36	2.380		1,550	146		1,696	1,925
2090	10 HP		3.11	2.570		1,800	158		1,958	2,200
2100	15 HP		2.69	2.980		2,425	183		2,608	2,950
2110	20 HP	2 Elec	4.20	3.810		3,000	234		3,234	3,650
2120	25 HP		4.03	3.970		3,700	244		3,944	4,450
2130	30 HP		3.70	4.330		4,325	266		4,591	5,150
2140	40 HP		3.19	5.010		5,475	305		5,780	6,475
2150	50 HP		2.69	5.950		5,050	365		5,415	6,100
2160	60 HP		2.18	7.330		8,500	450		8,950	10,000
2170	75 HP		1.68	9.520		10,400	585		10,985	12,300
2180	100 HP	3 Elec	2.10	11.430		14,500	700		15,200	17,000
2190	125 HP		1.85	12.990		17,800	795		18,595	20,800

For customer support on your Electrical Change Order Costs with RSMeans data, call 800.448.8182.

Post-Installation Change Order Costs

26 71 13.40 Motors Explosion Proof	Crew	Daily Output	Labor-Hours	Unit	Material	2020 Bare Costs Labor	Equipment	Total	Total Incl O&P
2200	150 HP	3 Elec	1.34	17.860	Ea.	21,700	1,100		22,800	25,500
2210	200 HP		1.01	23.810		27,700	1,450		29,150	32,700
2220	250 HP	↓	1.01	23.810	↓	22,400	1,450		23,850	26,800

Division Notes

	CREW	DAILY OUTPUT	LABOR-HOURS	UNIT	BARE COSTS				TOTAL INCL O&P
					MAT.	LABOR	EQUIP.	TOTAL	

Estimating Tips
27 20 00 Data Communications
27 30 00 Voice Communications
27 40 00 Audio-Video Communications

- When estimating material costs for special systems, it is always prudent to obtain manufacturers' quotations for equipment prices and special installation requirements that may affect the total cost.

- For cost modifications for elevated tray installation, add the percentages to labor according to the height of the installation and only to the quantities exceeding the different height levels, not to the total tray quantities. Refer to subdivision 26 01 02.20 for labor adjustment factors.

- Do not overlook the costs for equipment used in the installation. If scissor lifts and boom lifts are available in the field, contractors may use them in lieu of the proposed ladders and rolling staging.

Reference Numbers
Reference numbers are shown at the beginning of some major classifications. These numbers refer to related items in the Reference Section. The reference information may be an estimating procedure, an alternate pricing method, or technical information.

Note: Not all subdivisions listed here necessarily appear. ■

Same Data. Simplified.

Enjoy the convenience and efficiency of accessing your costs anywhere:

- **Skip the multiplier** by setting your location
- **Quickly search,** edit, favorite and share costs
- **Stay on top of price changes** with automatic updates

Discover more at rsmeans.com/online

Note: Trade Service, in part, has been used as a reference source for some of the material prices used in Division 27.

27 05 Common Work Results for Communications

27 05 29 – Hangers and Supports for Communications Systems

27 05 29.10 Cable Support	Crew	Daily Output	Labor-Hours	Unit	Material	2020 Bare Costs Labor	Equipment	Total	Total Incl O&P
0010 **CABLE SUPPORT**									
0110 J-hook, single tier, single sided, 1" diam.	1 Elec	57.60	.139	Ea.	2.41	8.50		10.91	15.35
0120 1-1/2" diam.		56.95	.140		3.14	8.60		11.74	16.30
0130 2" diam.		56.53	.142		3.64	8.70		12.34	16.95
0140 4" diam.		56.11	.143		7.20	8.75		15.95	21
0330 Double tier, single sided, 2" diam.		56.28	.142		13	8.70		21.70	27.50
0340 4" diam.		55.61	.144		18.50	8.85		27.35	33.50
0430 Double sided, 2" diam.		55.27	.145		20	8.90		28.90	35
0440 4" diam.		54.94	.146		32.50	8.95		41.45	49
0530 Triple tier, single sided, 2" diam.		55.78	.143		16.95	8.80		25.75	32
0540 4" diam.		55.31	.145		25.50	8.85		34.35	41
0630 Double sided, 2" diam.		54.68	.146		23	9		32	39
0640 4" diam.	↓	54.35	.147	↓	38.50	9.05		47.55	56

27 11 Communications Equipment Room Fittings

27 11 16 – Communications Cabinets, Racks, Frames and Enclosures

27 11 16.10 Public Phone

	Crew	Daily Output	Labor-Hours	Unit	Material	2020 Bare Costs Labor	Equipment	Total	Total Incl O&P
0010 **PUBLIC PHONE**									
7600 Telephone with wood backboard									
7620 Single door, 12" H x 12" W x 4" D	1 Elec	4.45	1.800	Ea.	114	110		224	290
7650 18" H x 12" W x 4" D		3.95	2.030		140	125		265	340
7700 24" H x 12" W x 4" D		3.53	2.270		211	139		350	440
7720 18" H x 18" W x 4" D		3.53	2.270		185	139		324	410
7750 24" H x 18" W x 4" D		3.36	2.380		262	146		408	505
7780 36" H x 36" W x 4" D		3.02	2.650		253	163		416	520
7800 24" H x 24" W x 6" D		3.02	2.650		350	163		513	630
7820 30" H x 24" W x 6" D		2.69	2.980		370	183		553	680
7850 30" H x 30" W x 6" D		2.27	3.530		565	217		782	940
7880 36" H x 30" W x 6" D		2.10	3.810		600	234		834	1,000
7900 48" H x 36" W x 6" D		1.85	4.330		960	266		1,226	1,450
7920 Double door, 48" H x 36" W x 6" D	↓	1.68	4.760	↓	1,325	292		1,617	1,875

27 11 16.20 Rack Mount Cabinet

	Crew	Daily Output	Labor-Hours	Unit	Material	2020 Bare Costs Labor	Equipment	Total	Total Incl O&P
0010 **RACK MOUNT CABINET**									
0100 80" H x 24" W x 40" D									
0110 No sides	1 Elec	8.06	.992	Ea.	1,500	61		1,561	1,750
0120 One side		6.72	1.190		1,600	73		1,673	1,850
0130 Two sides	↓	6.60	1.210	↓	1,725	74		1,799	1,975
0200 80" H x 24" W x 42" D									
0210 No sides	1 Elec	8.06	.992	Ea.	1,525	61		1,586	1,800
0220 One side		6.72	1.190		1,575	73		1,648	1,825
0230 Two sides	↓	6.60	1.210		1,600	74		1,674	1,850
0300 80" H x 24" W x 48" D									
0310 No sides	1 Elec	8.06	.992	Ea.	1,750	61		1,811	2,025
0320 One side		6.72	1.190		1,725	73		1,798	2,000
0330 Two sides	↓	6.60	1.210		1,825	74		1,899	2,100
0400 80" H x 30" W x 40" D									
0410 No sides	1 Elec	8.06	.992	Ea.	1,625	61		1,686	1,900
0420 One side		6.72	1.190		1,700	73		1,773	1,975
0430 Two sides	↓	6.60	1.210		1,775	74		1,849	2,050
0500 80" H x 30" W x 42" D									
0510 No sides	1 Elec	8.06	.992	Ea.	1,675	61		1,736	1,925

Post-Installation Change Order Costs

27 11 Communications Equipment Room Fittings

27 11 16 – Communications Cabinets, Racks, Frames and Enclosures

27 11 16.20 Rack Mount Cabinet

		Crew	Daily Output	Labor-Hours	Unit	Material	2020 Bare Costs Labor	Equipment	Total	Total Incl O&P
0520	One side	1 Elec	6.72	1.190	Ea.	1,700	73		1,773	1,975
0530	Two sides	↓	6.60	1.210	↓	1,725	74		1,799	2,000
0600	80" H x 30" W x 48" D									
0610	No sides	1 Elec	8.06	.992	Ea.	1,825	61		1,886	2,125
0620	One side		6.72	1.190		1,925	73		1,998	2,225
0630	Two sides	↓	6.60	1.210	↓	2,050	74		2,124	2,350
0700	80" H x 32" W x 32" D									
0710	No sides	1 Elec	8.06	.992	Ea.	1,600	61		1,661	1,875
0720	One side		6.72	1.190		1,650	73		1,723	1,900
0730	Two sides	↓	6.60	1.210	↓	1,675	74		1,749	1,925
0800	80" H x 32" W x 40" D									
0810	No sides	1 Elec	8.06	.992	Ea.	1,775	61		1,836	2,050
0820	One side		6.72	1.190		1,850	73		1,923	2,125
0830	Two sides	↓	6.60	1.210	↓	1,925	74		1,999	2,200
0900	80" H x 32" W x 42" D									
0910	No sides	1 Elec	8.06	.992	Ea.	1,775	61		1,836	2,050
0920	One side		6.72	1.190		1,800	73		1,873	2,100
0930	Two sides	↓	6.60	1.210	↓	1,850	74		1,924	2,125
1000	80" H x 32" W x 48" D									
1010	No sides	1 Elec	8.06	.992	Ea.	1,975	61		2,036	2,275
1020	One side		6.72	1.190		2,075	73		2,148	2,400
1030	Two sides	↓	6.60	1.210	↓	1,475	74		1,549	1,725
1200	Freestanding cabinet with front and rear doors									
1300	84" H x 19" W x 24" D, gray									
1310	No sides	1 Elec	8.06	.992	Ea.	1,850	61		1,911	2,125
1320	One side		8.06	.992		2,775	61		2,836	3,175
1330	Two sides	↓	8.06	.992	↓	2,775	61		2,836	3,175
1400	84" H x 19" W x 24" D, white									
1410	No sides	1 Elec	8.06	.992	Ea.	2,350	61		2,411	2,675
1420	One side		8.06	.992		2,775	61		2,836	3,175
1430	Two sides	↓	8.06	.992	↓	2,775	61		2,836	3,175
1500	84" H x 19" W x 24" D, black									
1510	No sides	1 Elec	8.06	.992	Ea.	1,575	61		1,636	1,825
1520	One side		8.06	.992		2,650	61		2,711	3,025
1530	Two sides		8.06	.992		2,600	61		2,661	2,950
1600	84" H x 28" W x 34" D front glass, rear louvered door	↓	8.06	.992	↓	1,700	61		1,761	1,975

27 11 19 – Communications Termination Blocks and Patch Panels

27 11 19.10 Termination Blocks and Patch Panels

		Crew	Daily Output	Labor-Hours	Unit	Material	2020 Bare Costs Labor	Equipment	Total	Total Incl O&P
0010	**TERMINATION BLOCKS AND PATCH PANELS**									
2960	Patch panel, RJ-45/110 type, 24 ports	2 Elec	5.04	3.170	Ea.	183	194		377	490
3000	48 ports	3 Elec	5.04	4.760		320	292		612	785
3040	96 ports	"	3.36	7.140		515	440		955	1,225
3100	Punch down termination per port	1 Elec	89.88	.089	↓		5.45		5.45	8.15

27 13 Communications Backbone Cabling

27 13 23 – Communications Optical Fiber Backbone Cabling

27 13 23.13 Communications Optical Fiber

		Crew	Daily Output	Labor-Hours	Unit	Material	2020 Bare Costs Labor	Equipment	Total	Total Incl O&P
0010	**COMMUNICATIONS OPTICAL FIBER**									
0040	Specialized tools & techniques cause installation costs to vary.									
0070	Fiber optic, cable, bulk simplex, single mode	1 Elec	6.72	1.190	C.L.F.	28	73		101	140
0080	Multi mode		6.72	1.190		36	73		109	149
0090	4 strand, single mode		6.17	1.300		46.50	80		126.50	170
0095	Multi mode		6.17	1.300		60.50	80		140.50	186
0100	12 strand, single mode		5.60	1.430		103	87.50		190.50	245
0105	Multi mode		5.60	1.430		119	87.50		206.50	262
0150	Jumper				Ea.	41.50			41.50	46
0200	Pigtail					43.50			43.50	48
0300	Connector	1 Elec	20.16	.397		31.50	24.50		56	71.50
0350	Finger splice		26.88	.298		45	18.25		63.25	76.50
0400	Transceiver (low cost bi-directional)		6.72	1.190		550	73		623	715
0450	Rack housing, 4 rack spaces, 12 panels (144 fibers)		1.68	4.760		735	292		1,027	1,250
1000	Cable, 62.5 microns, direct burial, 4 fiber	R-15	1008	.048	L.F.	1.12	2.86	.28	4.26	5.80
1020	Indoor, 2 fiber	R-19	840	.024		.53	1.46		1.99	2.76
1040	Outdoor, aerial/duct	"	1402.80	.014		.78	.88		1.66	2.16
1060	50 microns, direct burial, 8 fiber	R-22	3360	.011		1.56	.62		2.18	2.65
1080	12 fiber		3360	.011		2.92	.62		3.54	4.14
1100	Indoor, 12 fiber		637.56	.058		2.41	3.28		5.69	7.55
1120	Connectors, 62.5 micron cable, transmission	R-19	33.60	.595	Ea.	17.65	36.50		54.15	74
1140	Cable splice		33.60	.595		21	36.50		57.50	77.50
1160	125 micron cable, transmission		13.44	1.490		18.55	91.50		110.05	157
1180	Receiver, 1.2 mile range		16.80	1.190		340	73		413	480
1200	1.9 mile range		16.80	1.190		299	73		372	440
1220	6.2 mile range		4.20	4.760		390	293		683	865
1240	Transmitter, 1.2 mile range		16.80	1.190		380	73		453	525
1260	1.9 mile range		16.80	1.190		340	73		413	485
1280	6.2 mile range		4.20	4.760		520	293		813	1,000
1300	Modem, 1.2 mile range		4.20	4.760		220	293		513	675
1320	6.2 mile range		4.20	4.760		390	293		683	865
1340	1.9 mile range, 12 channel		4.20	4.760		2,475	293		2,768	3,150
1360	Repeater, 1.2 mile range		8.40	2.380		445	146		591	705
1380	1.9 mile range		8.40	2.380		590	146		736	870
1400	6.2 mile range		4.20	4.760		1,100	293		1,393	1,625
1420	1.2 mile range, digital		4.20	4.760		545	293		838	1,025
2040	Fiber optic cable, 48 strand, single mode, steel armor, material		1.40	14.260	M.L.F.	4,700	875		5,575	6,475

27 15 Communications Horizontal Cabling

27 15 01 – Communications Horizontal Cabling Applications

27 15 01.19 Fire Alarm Communications Conductors & Cables

		Crew	Daily Output	Labor-Hours	Unit	Material	2020 Bare Costs Labor	Equipment	Total	Total Incl O&P
0010	**FIRE ALARM COMMUNICATIONS CONDUCTORS AND CABLES**									
1500	Fire alarm FEP teflon 150 V to 200°C									
1550	#22, 1 pair	1 Elec	8.40	.952	C.L.F.	94	58.50		152.50	190
1600	2 pair		6.72	1.190		96	73		169	214
1650	4 pair		5.88	1.360		350	83.50		433.50	510
1700	6 pair		5.04	1.590		495	97.50		592.50	685
1750	8 pair		4.62	1.730		840	106		946	1,075
1800	10 pair		4.20	1.900		775	117		892	1,025
1850	#18, 1 pair	2 Elec	13.44	1.190		67.50	73		140.50	184
1900	2 pair		10.92	1.470		224	90		314	380

For customer support on your Electrical Change Order Costs with RSMeans data, call 800.448.8182. **Post-Installation Change Order Costs**

27 15 Communications Horizontal Cabling

27 15 01 – Communications Horizontal Cabling Applications

27 15 01.19 Fire Alarm Communications Conductors & Cables	Crew	Daily Output	Labor-Hours	Unit	Material	2020 Bare Costs Labor	Equipment	Total	Total Incl O&P
1950	4 pair	2 Elec	8.06	1.980	C.L.F.	206	121		327	410
2000	6 pair		6.72	2.380		440	146		586	700
2050	8 pair		5.88	2.720		320	167		487	605
2100	10 pair		5.04	3.170		480	194		674	820

27 15 01.23 Audio-Video Communications Horizontal Cabling

		Crew	Daily Output	Labor-Hours	Unit	Material	2020 Bare Costs Labor	Equipment	Total	Total Incl O&P
0010	**AUDIO-VIDEO COMMUNICATIONS HORIZONTAL CABLING**									
3000	S-video cable, MD4, 1M Pro interconnects, 3'	1 Elec	31.92	.251	Ea.	34.50	15.40		49.90	61
3010	6'		31.92	.251		28.50	15.40		43.90	54.50
3020	9'		31.92	.251		45.50	15.40		60.90	73
3030	16'		30.24	.265		49	16.25		65.25	78
3040	22'		30.24	.265		57.50	16.25		73.75	87
3300	HDMI cable, coaxial, HDMI/HDMI, Non plenum, 3', 24 kt terminations		33.60	.238		13.20	14.60		27.80	36.50
3310	6', 24 kt terminations		33.60	.238		15.40	14.60		30	39
3320	HDMI cable, coaxial, HDMI/HDMI metal, Non plenum, 3', 24 kt terminations		33.60	.238		15.05	14.60		29.65	38.50
3330	6', 24 kt terminations		33.60	.238		19.65	14.60		34.25	43.50
3340	15', 24 kt terminations		31.92	.251		27.50	15.40		42.90	53
3350	25', 24 kt terminations		31.92	.251		35.50	15.40		50.90	62.50

27 15 10 – Special Communications Cabling

27 15 10.23 Sound and Video Cables and Fittings

		Crew	Daily Output	Labor-Hours	Unit	Material	2020 Bare Costs Labor	Equipment	Total	Total Incl O&P
0010	**SOUND AND VIDEO CABLES & FITTINGS**									
0900	TV antenna lead-in, 300 ohm, #20-2 conductor	1 Elec	5.88	1.360	C.L.F.	26.50	83.50		110	153
0950	Coaxial, feeder outlet		5.88	1.360		24.50	83.50		108	151
1000	Coaxial, main riser		5.04	1.590		17.80	97.50		115.30	165
1100	Sound, shielded with drain, #22-2 conductor		6.72	1.190		10.20	73		83.20	120
1150	#22-3 conductor		6.30	1.270		14.90	78		92.90	132
1200	#22-4 conductor		5.46	1.470		20.50	90		110.50	157
1250	Nonshielded, #22-2 conductor		8.40	.952		14.20	58.50		72.70	103
1300	#22-3 conductor		7.56	1.060		25.50	65		90.50	125
1350	#22-4 conductor		6.72	1.190		28.50	73		101.50	141
1400	Microphone cable		6.72	1.190		33	73		106	146

27 15 13 – Communications Copper Horizontal Cabling

27 15 13.13 Communication Cables and Fittings

		Crew	Daily Output	Labor-Hours	Unit	Material	2020 Bare Costs Labor	Equipment	Total	Total Incl O&P
0010	**COMMUNICATION CABLES AND FITTINGS**									
2200	Telephone twisted, PVC insulation, #22-2 conductor	1 Elec	8.40	.952	C.L.F.	12.45	58.50		70.95	101
2250	#22-3 conductor		7.56	1.060		15.10	65		80.10	114
2300	#22-4 conductor		6.72	1.190		16.80	73		89.80	127
2350	#18-2 conductor		7.56	1.060		25	65		90	125
2370	Telephone jack, eight pins		26.88	.298	Ea.	4.19	18.25		22.44	31.50
5000	High performance unshielded twisted pair (UTP)									
5100	Cable, category 3, #24, 2 pair solid, PVC jacket	1 Elec	8.40	.952	C.L.F.	11	58.50		69.50	99
5200	4 pair solid, PVC jacket		5.88	1.360		16	83.50		99.50	142
5300	25 pair solid, PVC jacket		2.52	3.170		90.50	194		284.50	390
5400	2 pair solid, plenum		8.40	.952		13.45	58.50		71.95	102
5500	4 pair solid, plenum		5.88	1.360		14.90	83.50		98.40	140
5600	25 pair solid, plenum		2.52	3.170		133	194		327	435
5700	4 pair stranded, PVC jacket		5.88	1.360		16.20	83.50		99.70	142
7000	Category 5, #24, 4 pair solid, PVC jacket		5.88	1.360		11.55	83.50		95.05	137
7100	4 pair solid, plenum		5.88	1.360		17.50	83.50		101	143
7200	4 pair stranded, PVC jacket		5.88	1.360		30	83.50		113.50	157
7210	Category 5e, #24, 4 pair solid, PVC jacket		5.88	1.360		15.60	83.50		99.10	141
7212	4 pair solid, plenum		5.88	1.360		21.50	83.50		105	148

27 15 13.13 Communication Cables and Fittings	Crew	Daily Output	Labor-Hours	Unit	Material	2020 Bare Costs Labor	Equipment	Total	Total Incl O&P
7214	4 pair stranded, PVC jacket	1 Elec	5.88	1.360	C.L.F.	22	83.50		105.50	148
7240	Category 6, #24, 4 pair solid, PVC jacket		5.88	1.360		16.65	83.50		100.15	142
7242	4 pair solid, plenum		5.88	1.360		14	83.50		97.50	139
7244	4 pair stranded, PVC jacket		5.88	1.360	↓	15.45	83.50		98.95	141
7300	Connector, RJ45, category 5		67.20	.119	Ea.	.76	7.30		8.06	11.75
7302	Shielded RJ45, category 5		60.48	.132		1.73	8.10		9.83	14
7310	Jack, UTP RJ45, category 3		60.48	.132		.44	8.10		8.54	12.60
7312	Category 5		54.60	.147		6.25	9		15.25	20.50
7314	Category 5e		54.60	.147		4.98	9		13.98	18.90
7316	Category 6		54.60	.147		3.44	9		12.44	17.20
7322	Jack, shielded RJ45, category 5		50.40	.159		6.60	9.75		16.35	22
7324	Category 5e		50.40	.159		6.60	9.75		16.35	22
7326	Category 6		50.40	.159		6.90	9.75		16.65	22
7400	Voice/data expansion module, category 5e		6.72	1.190		47.50	73		120.50	161
7401	Modular jack, cat 6 keystone, RJ45, office white		13.86	.577		7.20	35.50		42.70	60.50
7402	RJ45, green		13.86	.577		7.10	35.50		42.60	60.50
7404	RJ45, grey		13.86	.577		8.05	35.50		43.55	61.50
7408	RJ45, orange		13.86	.577		7.40	35.50		42.90	60.50
7420	5M, RJ45, 568B, cord set		13.86	.577		42.50	35.50		78	99
7422	3M, RJ45, 568B, cord set		13.86	.577		38.50	35.50		74	95
7424	1M, RJ45, 568B, cord set		13.86	.577		33.50	35.50		69	89.50
7450	M12 TO RJ45, 2M ultra-loch D-mode, cord set	↓	13.02	.614	↓	48.50	37.50		86	109
8000	Multipair unshielded non-plenum cable, 150 V PVC jacket									
8002	#22, 2 pair	1 Elec	7.06	1.130	C.L.F.	18.30	69.50		87.80	123
8003	3 pair		6.72	1.190		30	73		103	142
8004	4 pair		6.13	1.300		68.50	80		148.50	194
8006	6 pair		5.21	1.540		41	94.50		135.50	187
8008	8 pair		4.79	1.670		95.50	102		197.50	258
8010	10 pair		4.45	1.800		75	110		185	247
8012	12 pair		3.53	2.270		189	139		328	415
8015	15 pair	↓	3.19	2.510		244	154		398	500
8020	20 pair	2 Elec	5.54	2.890		283	177		460	575
8025	25 pair		5.04	3.170		345	194		539	670
8030	30 pair		4.70	3.400		570	209		779	935
8040	40 pair		4.37	3.660		500	225		725	885
8050	50 pair	↓	4.12	3.890	↓	710	239		949	1,150
8100	Multipair unshielded non-plenum cable, 300 V PVC jacket									
8102	#20, 2 pair	1 Elec	6.13	1.300	C.L.F.	28	80		108	150
8103	3 pair		5.63	1.420		37.50	87		124.50	172
8104	4 pair		5.04	1.590		42.50	97.50		140	192
8106	6 pair		4.45	1.800		91.50	110		201.50	265
8108	8 pair	↓	3.70	2.160		105	133		238	315
8110	10 pair	2 Elec	6.13	2.610		116	160		276	365
8112	12 pair		5.21	3.070		144	188		332	440
8115	15 pair	↓	4.96	3.230		181	198		379	495
8201	#18, 1 pair	1 Elec	6.72	1.190		88	73		161	206
8202	2 pair		5.46	1.470		93.50	90		183.50	237
8203	3 pair		4.70	1.700		163	104		267	335
8204	4 pair		4.20	1.900		212	117		329	405
8206	6 pair		3.70	2.160		299	133		432	525
8208	8 pair	↓	3.36	2.380		297	146		443	540
8215	15 pair	2 Elec	4.20	3.810	↓	770	234		1,004	1,200
8300	Multipair shielded non-plenum cable, 300 V PVC jacket									

27 15 13 – Communications Copper Horizontal Cabling

27 15 13.13 Communication Cables and Fittings

		Crew	Daily Output	Labor-Hours	Unit	Material	2020 Bare Costs Labor	Equipment	Total	Total Incl O&P
8303	#22, 3 pair	1 Elec	5.63	1.420	C.L.F.	71	87		158	208
8306	6 pair		4.79	1.670		136	102		238	305
8309	9 pair	↓	4.20	1.900		168	117		285	360
8312	12 pair	2 Elec	6.72	2.380		530	146		676	795
8315	15 pair		6.13	2.610		640	160		800	945
8317	17 pair		5.71	2.800		765	172		937	1,100
8319	19 pair		5.38	2.980		765	183		948	1,100
8327	27 pair	↓	4.96	3.230		1,350	198		1,548	1,775
8402	#20, 2 pair	1 Elec	5.63	1.420		128	87		215	270
8403	3 pair		5.21	1.540		154	94.50		248.50	310
8406	6 pair		3.70	2.160		385	133		518	620
8409	9 pair	↓	3.02	2.650		335	163		498	605
8412	12 pair	2 Elec	4.96	3.230		635	198		833	995
8415	15 pair	"	4.79	3.340		585	205		790	945
8502	#18, 2 pair	1 Elec	5.21	1.540		144	94.50		238.50	299
8503	3 pair		4.45	1.800		225	110		335	410
8504	4 pair		3.95	2.030		265	125		390	475
8506	6 pair		3.36	2.380		420	146		566	675
8509	9 pair	↓	2.86	2.800		640	172		812	955
8515	15 pair	2 Elec	4.03	3.970	↓	1,075	244		1,319	1,550

27 15 33 – Communications Coaxial Horizontal Cabling

27 15 33.10 Coaxial Cable and Fittings

		Crew	Daily Output	Labor-Hours	Unit	Material	2020 Bare Costs Labor	Equipment	Total	Total Incl O&P
0010	**COAXIAL CABLE & FITTINGS**									
3500	Coaxial connectors, 50 ohm impedance quick disconnect									
3540	BNC plug, for RG A/U #58 cable	1 Elec	35.28	.227	Ea.	3.35	13.90		17.25	24
3550	RG A/U #59 cable		35.28	.227		3.24	13.90		17.14	24
3560	RG A/U #62 cable		35.28	.227		3.28	13.90		17.18	24
3600	BNC jack, for RG A/U #58 cable		35.28	.227		3.74	13.90		17.64	24.50
3610	RG A/U #59 cable		35.28	.227		4.27	13.90		18.17	25
3620	RG A/U #62 cable		35.28	.227		4.15	13.90		18.05	25
3660	BNC panel jack, for RG A/U #58 cable		33.60	.238		9.55	14.60		24.15	32.50
3670	RG A/U #59 cable		33.60	.238		7.85	14.60		22.45	30.50
3680	RG A/U #62 cable		33.60	.238		7.85	14.60		22.45	30.50
3720	BNC bulkhead jack, for RG A/U #58 cable		33.60	.238		7.75	14.60		22.35	30.50
3730	RG A/U #59 cable		33.60	.238		7.75	14.60		22.35	30.50
3740	RG A/U #62 cable		33.60	.238	↓	7.75	14.60		22.35	30.50
3850	Coaxial cable, RG A/U #58, 50 ohm		6.72	1.190	C.L.F.	58.50	73		131.50	174
3860	RG A/U #59, 75 ohm		6.72	1.190		47	73		120	161
3870	RG A/U #62, 93 ohm		6.72	1.190		57	73		130	172
3875	RG 6/U, 75 ohm		6.72	1.190		42	73		115	156
3950	Fire rated, RG A/U #58, 50 ohm		6.72	1.190		113	73		186	233
3960	RG A/U #59, 75 ohm		6.72	1.190		149	73		222	272
3970	RG A/U #62, 93 ohm	↓	6.72	1.190	↓	131	73		204	253

27 15 43 – Communications Faceplates and Connectors

27 15 43.13 Communication Outlets

		Crew	Daily Output	Labor-Hours	Unit	Material	2020 Bare Costs Labor	Equipment	Total	Total Incl O&P
0010	**COMMUNICATION OUTLETS**									
0100	Voice/data devices not included									
0120	Voice/data outlets, single opening	1 Elec	40.32	.198	Ea.	7.60	12.15		19.75	26.50
0140	Two jack openings		40.32	.198		2.45	12.15		14.60	21
0160	One jack & one 3/4" round opening		40.32	.198		7.40	12.15		19.55	26.50
0180	One jack & one twinaxial opening		40.32	.198		7.55	12.15		19.70	26.50
0200	One jack & one connector cabling opening	↓	40.32	.198	↓	7.40	12.15		19.55	26.50

27 15 Communications Horizontal Cabling

27 15 43 – Communications Faceplates and Connectors

27 15 43.13 Communication Outlets		Crew	Daily Output	Labor-Hours	Unit	Material	2020 Bare Costs Labor	Equipment	Total	Total Incl O&P
0220	Two 3/8" coaxial openings	1 Elec	40.32	.198	Ea.	7.40	12.15		19.55	26.50
0300	Data outlets, single opening		40.32	.198		7.40	12.15		19.55	26.50
0320	One 25-pin subminiature opening		40.32	.198		7.40	12.15		19.55	26.50
1000	Voice/data wall plate, plastic, 1 gang, 1-port		60.48	.132		2.62	8.10		10.72	15
1020	2-port		60.48	.132		2.78	8.10		10.88	15.15
1040	3-port		60.48	.132		2.33	8.10		10.43	14.65
1060	4-port		60.48	.132		2.63	8.10		10.73	15
1080	6-port		60.48	.132		2.25	8.10		10.35	14.60
1100	2 gang, 6-port		40.32	.198		7.30	12.15		19.45	26
1120	Voice/data wall plate, stainless steel, 1 gang, 1-port		60.48	.132		8.60	8.10		16.70	21.50
1140	2-port		60.48	.132		8.20	8.10		16.30	21
1160	3-port		60.48	.132		7.95	8.10		16.05	21
1180	4-port		60.48	.132		8.25	8.10		16.35	21
1200	2 gang, 6-port	↓	40.32	.198	↓	12.65	12.15		24.80	32

27 21 Data Communications Network Equipment

27 21 29 – Data Communications Switches and Hubs

27 21 29.10 Switching and Routing Equipment

		Crew	Daily Output	Labor-Hours	Unit	Material	2020 Bare Costs Labor	Equipment	Total	Total Incl O&P
0010	**SWITCHING AND ROUTING EQUIPMENT**									
1100	Network hub, dual speed, 24 ports, includes cabinet	3 Elec	.55	43.290	Ea.	960	2,650		3,610	5,000
1300	Network switch, 50/60 HZ, 8 port, multi-platform, analog KVM	1 Elec	5.75	1.390		1,175	85.50		1,260.50	1,400
1310	16 port, multi-platform, analog KVM		5.04	1.590		1,400	97.50		1,497.50	1,675
1320	Network switch, 0x2x16,CAT5, analog KVM		5.04	1.590		1,225	97.50		1,322.50	1,475
1330	2x1x16, digital KVM, w/VM		13.44	.595		3,850	36.50		3,886.50	4,275
1340	2x1x32, digital KVM, w/VM		13.44	.595		4,575	36.50		4,611.50	5,075
1350	8x1x32, digital KVM, w/VM		13.44	.595		5,125	36.50		5,161.50	5,675
1500	KVM,1 RMU 16 port, no keyboard		13.44	.595		1,550	36.50		1,586.50	1,750
1510	KVM,1 RMU, 17" LCD, 16 port, US keyboard		12.73	.629		2,975	38.50		3,013.50	3,325
1520	KVM,1 RMU, 17" LCD, 16 port, UK keyboard		12.73	.629		2,975	38.50		3,013.50	3,325
2000	10/100/1000 Mbps, 24 ports		11.76	.680		620	41.50		661.50	740
2040	10/100/1000 Mbps, 48 ports		10.08	.794		1,400	48.50		1,448.50	1,625
2050	10/100/1000 Mbps, 5 port, industrial ethernet type		26.88	.298		330	18.25		348.25	390
2060	10/100/1000 Mbps, 6 port, industrial ethernet type		26.88	.298		1,300	18.25		1,318.25	1,450
2070	10/100/1000 Mbps, 10 port, X307-3		23.52	.340		3,075	21		3,096	3,425
2080	10/100/1000 Mbps, 10 port, X308-2		23.52	.340		2,850	21		2,871	3,175
3000	10/100/1000 Mbps, 20 port, front ports		16.80	.476		4,425	29		4,454	4,925
3010	10/100/1000 Mbps, 20 port, rear ports		16.80	.476		4,425	29		4,454	4,925
3040	KVM, 10/100/1000/10000 Mbps, 28 port, rear ports		13.44	.595		9,150	36.50		9,186.50	10,200
3050	KVM, 10/100/1000/10000 Mbps, 28 port, front ports	↓	13.44	.595		10,700	36.50		10,736.50	11,900
3070	KVM, 10/100/1000/10000 Mbps, 52 port, front ports	2 Elec	10.08	1.590		11,300	97.50		11,397.50	12,500
3080	KVM, 10/100/1000/10000 Mbps, 52 port, rear ports	"	10.08	1.590	↓	11,300	97.50		11,397.50	12,500

For customer support on your Electrical Change Order Costs with RSMeans data, call 800.448.8182.

Post-Installation Change Order Costs

27 32 Voice Communications Terminal Equipment

27 32 26 – Ring-Down Emergency Telephones

27 32 26.10 Emergency Phone Stations		Crew	Daily Output	Labor-Hours	Unit	Material	2020 Bare Costs Labor	2020 Bare Costs Equipment	Total	Total Incl O&P
0010	**EMERGENCY PHONE STATIONS**									
0100	Hands free emergency speaker phone w/LED strobe	1 Elec	6.07	1.320	Ea.	1,025	81		1,106	1,250
0110	Wall/pole mounted emergency phone station w/LED strobe	"	3.86	2.070		1,825	127		1,952	2,200
2000	Call station, 30", pole mounted emergency phone, 120V/ VoIP, blue beacon	2 Elec	4.20	3.810		1,100	234		1,334	1,550
2010	Call station, 36"		4.20	3.810		1,575	234		1,809	2,075
2020	Call station, 5'-8"		4.03	3.970		2,650	244		2,894	3,300
2030	Call station, 9'		4.03	3.970		2,675	244		2,919	3,325

27 32 36 – TTY Equipment

27 32 36.10 TTY Telephone Equipment		Crew	Daily Output	Labor-Hours	Unit	Material	2020 Bare Costs Labor	2020 Bare Costs Equipment	Total	Total Incl O&P
0010	**TTY TELEPHONE EQUIPMENT**									
1620	Telephone, TTY, compact, pocket type				Ea.	355			355	390
1630	Advanced, desk type	2 Elec	16.80	.952		495	58.50		553.50	630
1640	Full-featured public, wall type	"	3.36	4.760		705	292		997	1,200

27 41 Audio-Video Systems

27 41 33 – Master Antenna Television Systems

27 41 33.10 TV Systems

27 41 33.10 TV Systems		Crew	Daily Output	Labor-Hours	Unit	Material	2020 Bare Costs Labor	2020 Bare Costs Equipment	Total	Total Incl O&P
0010	**TV SYSTEMS,** not including rough-in wires, cables & conduits									
0100	Master TV antenna system									
0200	VHF reception & distribution, 12 outlets	1 Elec	5.04	1.590	Outlet	126	97.50		223.50	284
0400	30 outlets		8.40	.952		254	58.50		312.50	365
0600	100 outlets		10.92	.733		295	45		340	390
0650	UHF reception, 100 outlets		10.92	.733		169	45		214	253
0800	VHF & UHF reception & distribution, 12 outlets		5.04	1.590		247	97.50		344.50	415
1000	30 outlets		8.40	.952		169	58.50		227.50	273
1200	100 outlets		10.92	.733		168	45		213	252
1400	School and deluxe systems, 12 outlets		2.02	3.970		325	244		569	725
1600	30 outlets		3.36	2.380		286	146		432	530
1800	80 outlets		4.45	1.800		275	110		385	465
1900	Amplifier		3.36	2.380	Ea.	820	146		966	1,125

27 42 Electronic Digital Systems

27 42 13 – Point of Sale Systems

27 42 13.10 Bar Code Scanner

27 42 13.10 Bar Code Scanner		Crew	Daily Output	Labor-Hours	Unit	Material	2020 Bare Costs Labor	2020 Bare Costs Equipment	Total	Total Incl O&P
0010	**BAR CODE SCANNER**									
0100	CR3600, w/B4 battery, handle	1 Elec	5.04	1.590	Outlet	1,025	97.50		1,122.50	1,275
0110	Palm		10.08	.794		900	48.50		948.50	1,075
0120	Batch, B4 batt, handle, chgstn, 3' USB		7.56	1.060		1,050	65		1,115	1,250
0130	Palm, chgstn, 3' USB		10.08	.794		945	48.50		993.50	1,125
0140	CR2500, w/H2 handle, no cable		8.40	.952		690	58.50		748.50	840
0150	Core unit, palm, no cable		13.44	.595		640	36.50		676.50	760

27 51 Distributed Audio-Video Communications Systems

27 51 16 – Public Address Systems

27 51 16.10 Public Address System	Crew	Daily Output	Labor-Hours	Unit	Material	2020 Bare Costs Labor	Equipment	Total	Total Incl O&P
0010	**PUBLIC ADDRESS SYSTEM**									
0100	Conventional, office	1 Elec	4.48	1.790	Speaker	171	110		281	350
0200	Industrial	"	2.27	3.530	"	330	217		547	680

27 51 19 – Sound Masking Systems

27 51 19.10 Sound System

	27 51 19.10 Sound System	Crew	Daily Output	Labor-Hours	Unit	Material	Labor	Equipment	Total	Total Incl O&P
0010	**SOUND SYSTEM**, not including rough-in wires, cables & conduits									
0100	Components, projector outlet	1 Elec	6.72	1.190	Ea.	61.50	73		134.50	177
0200	Microphone		3.36	2.380		115	146		261	345
0400	Speakers, ceiling or wall		6.72	1.190		157	73		230	282
0600	Trumpets		3.36	2.380		291	146		437	535
0800	Privacy switch		6.72	1.190		117	73		190	237
1000	Monitor panel		3.36	2.380		520	146		666	785
1200	Antenna, AM/FM		3.36	2.380		155	146		301	390
1400	Volume control		6.72	1.190		60	73		133	175
1600	Amplifier, 250 W		.84	9.520		1,400	585		1,985	2,425
1800	Cabinets		.84	9.520		1,125	585		1,710	2,125
2000	Intercom, 30 station capacity, master station	2 Elec	1.68	9.520		2,400	585		2,985	3,525
2020	10 station capacity	"	3.36	4.760		1,475	292		1,767	2,050
2200	Remote station	1 Elec	6.72	1.190		217	73		290	350
2400	Intercom outlets		6.72	1.190		128	73		201	249
2600	Handset		3.36	2.380		425	146		571	680
2800	Emergency call system, 12 zones, annunciator		1.09	7.330		1,275	450		1,725	2,075
3000	Bell		4.45	1.800		132	110		242	310
3200	Light or relay		6.72	1.190		66	73		139	182
3400	Transformer		3.36	2.380		291	146		437	535
3600	House telephone, talking station		1.34	5.950		625	365		990	1,225
3800	Press to talk, release to listen	↓	4.45	1.800		145	110		255	325
4000	System-on button					87			87	95.50
4200	Door release	1 Elec	3.36	2.380		155	146		301	390
4400	Combination speaker and microphone		6.72	1.190		265	73		338	400
4600	Termination box		2.69	2.980		83.50	183		266.50	365
4800	Amplifier or power supply		4.45	1.800	↓	955	110		1,065	1,225
5000	Vestibule door unit		13.44	.595	Name	161	36.50		197.50	232
5200	Strip cabinet		22.68	.353	Ea.	330	21.50		351.50	395
5400	Directory	↓	13.44	.595		156	36.50		192.50	226
6000	Master door, button buzzer type, 100 unit	2 Elec	.45	35.270		1,600	2,175		3,775	4,975
6020	200 unit		.25	63.490		3,000	3,900		6,900	9,100
6040	300 unit	↓	.17	95.240		4,575	5,850		10,425	13,700
6060	Transformer	1 Elec	6.72	1.190		40.50	73		113.50	154
6080	Door opener		4.45	1.800		58.50	110		168.50	228
6100	Buzzer with door release and plate	↓	3.36	2.380		58.50	146		204.50	281
6200	Intercom type, 100 unit	2 Elec	.45	35.270		1,975	2,175		4,150	5,400
6220	200 unit		.25	63.490		3,875	3,900		7,775	10,100
6240	300 unit	↓	.17	95.240		5,825	5,850		11,675	15,100
6260	Amplifier	1 Elec	1.68	4.760		291	292		583	755
6280	Speaker with door release	"	3.36	2.380	↓	87	146		233	315

Post-Installation Change Order Costs

27 52 Healthcare Communications and Monitoring Systems

27 52 23 – Nurse Call/Code Blue Systems

27 52 23.10 Nurse Call Systems

	27 52 23.10 Nurse Call Systems	Crew	Daily Output	Labor-Hours	Unit	Material	2020 Bare Costs Labor	Equipment	Total	Total Incl O&P
0010	**NURSE CALL SYSTEMS**									
0100	Single bedside call station	1 Elec	6.72	1.190	Ea.	182	73		255	310
0200	Ceiling speaker station		6.72	1.190		90	73		163	208
0400	Emergency call station		6.72	1.190		92.50	73		165.50	211
0600	Pillow speaker		6.72	1.190		223	73		296	355
0800	Double bedside call station		3.36	2.380		169	146		315	405
1000	Duty station		3.36	2.380		137	146		283	370
1200	Standard call button		6.72	1.190		125	73		198	246
1400	Lights, corridor, dome or zone indicator		6.72	1.190		64.50	73		137.50	180
1600	Master control station for 20 stations	2 Elec	.55	29.300	Total	3,725	1,800		5,525	6,775

27 53 Distributed Systems

27 53 13 – Clock Systems

27 53 13.50 Clock Equipments

	27 53 13.50 Clock Equipments	Crew	Daily Output	Labor-Hours	Unit	Material	2020 Bare Costs Labor	Equipment	Total	Total Incl O&P
0010	**CLOCK EQUIPMENTS**, not including wires & conduits									
0100	Time system components, master controller	1 Elec	.28	28.860	Ea.	1,575	1,775		3,350	4,350
0200	Program bell		6.72	1.190		114	73		187	234
0400	Combination clock & speaker		2.69	2.980		200	183		383	490
0600	Frequency generator		1.68	4.760		2,725	292		3,017	3,400
0800	Job time automatic stamp recorder		3.36	2.380		560	146		706	830
1600	Master time clock system, clocks & bells, 20 room	4 Elec	.17	190		6,025	11,700		17,725	24,000
1800	50 room	"	.07	476		12,700	29,200		41,900	57,500
1900	Time clock	1 Elec	2.69	2.980		375	183		558	685
2000	100 cards in & out, 1 color					10.45			10.45	11.50
2200	2 colors					9.90			9.90	10.90
2800	Metal rack for 25 cards	1 Elec	5.88	1.360		50	83.50		133.50	179
4000	Wireless time systems component, master controller	"	1.68	4.760		630	292		922	1,125
4010	For transceiver and antenna, see Section 28 47 12.10									
4100	Wireless analog clock w/battery operated	1 Elec	6.72	1.190	Ea.	110	73		183	230
4200	Wireless digital clock w/battery operated	"	6.72	1.190	"	425	73		498	575

Division Notes

	CREW	DAILY OUTPUT	LABOR-HOURS	UNIT	BARE COSTS				TOTAL INCL O&P
					MAT.	LABOR	EQUIP.	TOTAL	

Estimating Tips

- When estimating material costs for electronic safety and security systems, it is always prudent to obtain manufacturers' quotations for equipment prices and special installation requirements that may affect the total cost.

- Fire alarm systems consist of control panels, annunciator panels, batteries with rack, charger, and fire alarm actuating and indicating devices. Some fire alarm systems include speakers, telephone lines, door closer controls, and other components. Be careful not to overlook the costs related to installation for these items. Also be aware of costs for integrated automation instrumentation and terminal devices, control equipment, control wiring, and programming. Insurance underwriters may have specific requirements for the type of materials to be installed or design requirements based on the hazard to be protected. Local jurisdictions may have requirements not covered by code. It is advisable to be aware of any special conditions.

- Security equipment includes items such as CCTV, access control, and other detection and identification systems to perform alert and alarm functions. Be sure to consider the costs related to installation for this security equipment, such as for integrated automation instrumentation and terminal devices, control equipment, control wiring, and programming.

Reference Numbers

Reference numbers are shown at the beginning of some major classifications. These numbers refer to related items in the Reference Section. The reference information may be an estimating procedure, an alternate pricing method, or technical information.

Same Data. Simplified.

Enjoy the convenience and efficiency of accessing your costs anywhere:

- **Skip the multiplier** by setting your location
- **Quickly search,** edit, favorite and share costs
- **Stay on top of price changes** with automatic updates

Discover more at rsmeans.com/online

Note: Trade Service, in part, has been used as a reference source for some of the material prices used in Division 28.

28 05 Common Work Results for Electronic Safety and Security

28 05 19 – Storage Appliances for Electronic Safety and Security

28 05 19.11 Digital Video Recorder (DVR)	Crew	Daily Output	Labor-Hours	Unit	Material	2020 Bare Costs Labor	Equipment	Total	Total Incl O&P
0010 **DIGITAL VIDEO RECORDERS**									
0100 Pentaplex hybrid, internet protocol, and hard drive									
0200 4 channel	1 Elec	1.12	7.160	Ea.	1,800	440		2,240	2,625
0300 8 channel		.84	9.520		3,925	585		4,510	5,200
0400 16 channel	↓	.84	9.520	↓	3,400	585		3,985	4,625

28 23 Video Management System

28 23 13 – Video Management System Interfaces

28 23 13.10 Closed Circuit Television System

	Crew	Daily Output	Labor-Hours	Unit	Material	2020 Bare Costs Labor	Equipment	Total	Total Incl O&P
0010 **CLOSED CIRCUIT TELEVISION SYSTEM**									
2000 Surveillance, one station (camera & monitor)	2 Elec	2.18	7.330	Total	665	450		1,115	1,400
2200 For additional camera stations, add	1 Elec	2.27	3.530	Ea.	320	217		537	670
2400 Industrial quality, one station (camera & monitor)	2 Elec	2.18	7.330	Total	1,650	450		2,100	2,500
2600 For additional camera stations, add	1 Elec	2.27	3.530	Ea.	690	217		907	1,075
2610 For low light, add		2.27	3.530		525	217		742	895
2620 For very low light, add		2.27	3.530		3,450	217		3,667	4,125
2800 For weatherproof camera station, add		1.09	7.330		540	450		990	1,275
3000 For pan and tilt, add		1.09	7.330		1,575	450		2,025	2,400
3200 For zoom lens - remote control, add		1.68	4.760		1,800	292		2,092	2,400
3400 Extended zoom lens		1.68	4.760		5,025	292		5,317	5,950
3410 For automatic iris for low light, add	↓	1.68	4.760	↓	1,025	292		1,317	1,550
3600 Educational TV studio, basic 3 camera system, black & white,									
3800 electrical & electronic equip. only	4 Elec	.67	47.620	Total	8,275	2,925		11,200	13,500
4000 Full console		.24	136		25,300	8,350		33,650	40,300
4100 As above, but color system		.24	136		50,500	8,350		58,850	68,000
4120 Full console	↓	.10	317	↓	190,000	19,400		209,400	238,000
4200 For film chain, black & white, add	1 Elec	.84	9.520	Ea.	12,500	585		13,085	14,600
4250 Color, add		.21	38.100		8,925	2,325		11,250	13,300
4400 For video recorders, add	↓	.84	9.520		2,450	585		3,035	3,550
4600 Premium	4 Elec	.34	95.240	↓	16,000	5,850		21,850	26,300

28 23 23 – Video Surveillance Systems Infrastructure

28 23 23.50 Video Surveillance Equipments

	Crew	Daily Output	Labor-Hours	Unit	Material	2020 Bare Costs Labor	Equipment	Total	Total Incl O&P
0010 **VIDEO SURVEILLANCE EQUIPMENTS**									
0200 Video cameras, wireless, hidden in exit signs, clocks, etc., incl. receiver	1 Elec	2.52	3.170	Ea.	171	194		365	480
0210 Accessories for video recorder, single camera		2.52	3.170		195	194		389	505
0220 For multiple cameras		2.52	3.170		1,675	194		1,869	2,150
0230 Video cameras, wireless, for under vehicle searching, complete		1.68	4.760		15,200	292		15,492	17,100
0400 Internet protocol network camera, day/night, color & power supply		2.18	3.660		1,250	225		1,475	1,700
0500 Monitor, color flat screen, liquid crystal display (LCD), 15"		2.27	3.530		535	217		752	910
0520 17"		2.27	3.530		470	217		687	840
0540 19"	↓	2.27	3.530	↓	590	217		807	965

For customer support on your Electrical Change Order Costs with RSMeans data, call 800.448.8182.

Post-Installation Change Order Costs

28 31 Intrusion Detection

28 31 16 – Intrusion Detection Systems Infrastructure

28 31 16.50 Intrusion Detection	Crew	Daily Output	Labor-Hours	Unit	Material	2020 Bare Costs Labor	Equipment	Total	Total Incl O&P
0010 **INTRUSION DETECTION**, not including wires & conduits									
0100 Burglar alarm, battery operated, mechanical trigger	1 Elec	3.36	2.380	Ea.	289	146		435	535
0200 Electrical trigger		3.36	2.380		345	146		491	595
0400 For outside key control, add		6.72	1.190		89.50	73		162.50	208
0600 For remote signaling circuitry, add		6.72	1.190		143	73		216	266
0800 Card reader, flush type, standard		2.27	3.530		725	217		942	1,125
1000 Multi-code		2.27	3.530		1,250	217		1,467	1,700
1010 Card reader, proximity type		2.27	3.530		300	217		517	650
1200 Door switches, hinge switch		4.45	1.800		64.50	110		174.50	235
1400 Magnetic switch		4.45	1.800		105	110		215	280
1600 Exit control locks, horn alarm		3.36	2.380		218	146		364	455
1800 Flashing light alarm		3.36	2.380		247	146		393	490
2000 Indicating panels, 1 channel	↓	2.27	3.530		244	217		461	590
2200 10 channel	2 Elec	2.69	5.950		1,175	365		1,540	1,850
2400 20 channel		1.68	9.520		2,525	585		3,110	3,650
2600 40 channel	↓	.96	16.710		4,650	1,025		5,675	6,650
2800 Ultrasonic motion detector, 12 V	1 Elec	1.93	4.140		200	254		454	600
3000 Infrared photoelectric detector		3.36	2.380		133	146		279	365
3200 Passive infrared detector		3.36	2.380		237	146		383	480
3400 Glass break alarm switch		6.72	1.190		85.50	73		158.50	204
3420 Switchmats, 30" x 5'		4.45	1.800		116	110		226	292
3440 30" x 25'		3.36	2.380		212	146		358	450
3460 Police connect panel		3.36	2.380		283	146		429	525
3480 Telephone dialer		4.45	1.800		400	110		510	605
3500 Alarm bell		3.36	2.380		106	146		252	335
3520 Siren		3.36	2.380		152	146		298	385
3540 Microwave detector, 10' to 200'		1.68	4.760		585	292		877	1,075
3560 10' to 350'	↓	1.68	4.760	↓	1,775	292		2,067	2,375

28 33 Security Monitoring and Control

28 33 11 – Electronic Structural Monitoring Systems

28 33 11.10 Load Cell Units

	Crew	Daily Output	Labor-Hours	Unit	Material	2020 Bare Costs Labor	Equipment	Total	Total Incl O&P
0010 **LOAD CELL UNITS**									
0100 Load cell, stainless steel, 0.5/1 ton rated, mounting unit	2 Elec	1.12	14.320	Ea.	665	880		1,545	2,025
0110 2/3 to 5/5 ton rated, mounting unit		1.01	15.870		705	975		1,680	2,225
0120 0.5/1/2/3.5/5 ton, self aligning bottom base		1.39	11.470		195	705		900	1,275
0130 Combo mounting unit, 10/25 ton rated		.72	22.150		1,825	1,350		3,175	4,025
0140 40/60 ton rated		.55	28.860		2,975	1,775		4,750	5,900
0150 60kg/3m, rated		1.55	10.350		1,550	635		2,185	2,650
0160 130kg/3m, rated		1.30	12.290		1,475	755		2,230	2,725
0170 280 kg, rated	↓	1.02	15.610	↓	1,475	960		2,435	3,025

28 46 11 – Fire Sensors and Detectors

28 46 11.21 Carbon-Monoxide Detection Sensors

		Crew	Daily Output	Labor-Hours	Unit	Material	2020 Bare Costs Labor	Equipment	Total	Total Incl O&P
0010	**CARBON-MONOXIDE DETECTION SENSORS**									
8400	Smoke and carbon monoxide alarm battery operated photoelectric low profile	1 Elec	20.16	.397	Ea.	51.50	24.50		76	93
8410	low profile photoelectric battery powered		20.16	.397		40	24.50		64.50	81
8420	photoelectric low profile sealed lithium		20.16	.397		70.50	24.50		95	114
8430	Photoelectric low profile sealed lithium smoke and CO with voice combo		20.16	.397		50.50	24.50		75	92
8500	Carbon monoxide sensor, wall mount 1Mod 1 relay output smoke & heat		20.16	.397		490	24.50		514.50	575
8510	1Mod, 1relay no display, smoke & heat		20.16	.397		475	24.50		499.50	560
8530	1Mod 1 relay no HW logo smoke & heat		20.16	.397		515	24.50		539.50	600
8700	Carbon monoxide detector, battery operated, wall mounted		13.44	.595		43	36.50		79.50	102
8710	Hardwired, wall and ceiling mounted		6.72	1.190		84.50	73		157.50	202
8720	Duct mounted		6.72	1.190		350	73		423	495
8730	Continuous air monitoring system, PC based, incl. remote sensors, min.	2 Elec	.21	76.190		31,600	4,675		36,275	41,700
8740	Maximum	"	.08	190		56,500	11,700		68,200	79,500

28 46 11.27 Other Sensors

		Crew	Daily Output	Labor-Hours	Unit	Material	2020 Bare Costs Labor	Equipment	Total	Total Incl O&P
0010	**OTHER SENSORS**									
5200	Smoke detector, ceiling type	1 Elec	5.21	1.540	Ea.	120	94.50		214.50	274
5240	Smoke detector, addressable type		5.04	1.590		224	97.50		321.50	390
5400	Duct type		2.69	2.980		295	183		478	595
5420	Duct addressable type		2.69	2.980		520	183		703	845
8300	Smoke alarm with integrated strobe light 120 V, 16DB 60 fpm flash rate		13.44	.595		119	36.50		155.50	186
8310	Photoelectric smoke detector with strobe 120 V, 90 DB ceiling mount		10.08	.794		227	48.50		275.50	320
8320	120 V, 90 DB wall mount		10.08	.794		202	48.50		250.50	295
8330	120 V, 9 VDC, 90 DB ceiling mount		10.08	.794		201	48.50		249.50	295
8340	120 V, 9 VDC backup 90 DB wall mount		10.08	.794		202	48.50		250.50	295
8440	Fire alarm beam detector, motorized reflective, infrared optical beam	2 Elec	1.34	11.900		750	730		1,480	1,900
8450	Fire alarm beam detector, motorized reflective, IR/UV optical beam	"	1.34	11.900		1,025	730		1,755	2,200

28 46 11.50 Fire and Heat Detectors

		Crew	Daily Output	Labor-Hours	Unit	Material	2020 Bare Costs Labor	Equipment	Total	Total Incl O&P
0010	**FIRE & HEAT DETECTORS**									
5000	Detector, rate of rise	1 Elec	6.72	1.190	Ea.	52	73		125	166
5010	Heat addressable type		6.09	1.310		225	80.50		305.50	365
5100	Fixed temp fire alarm		5.88	1.360		42	83.50		125.50	171
5200	125 V SRF MT-135F		6.09	1.310		33.50	80.50		114	157
5300	MT-195F		6.09	1.310		34	80.50		114.50	158
5400	104/85 DB, indoor/outdoor, ceiling/wall, red		8.61	.929		73	57		130	166
5410	Ceiling/wall, white		8.61	.929		71.50	57		128.50	164
5420	102/98 DB, lumin 110cd, indoor/wall, red		8.61	.929		113	57		170	210
5430	Lumin15/75cd, indoor/wall, red		8.61	.929		110	57		167	206
5440	HI/LO DB, 24 V, fire marking, red		8.61	.929		81	57		138	174

28 46 20 – Fire Alarm

28 46 20.50 Alarm Panels and Devices

		Crew	Daily Output	Labor-Hours	Unit	Material	2020 Bare Costs Labor	Equipment	Total	Total Incl O&P
0010	**ALARM PANELS AND DEVICES**, not including wires & conduits									
2200	Intercom remote station	1 Elec	6.72	1.190	Ea.	89.50	73		162.50	207
2400	Intercom outlet		6.72	1.190		64.50	73		137.50	180
2600	Sound system, intercom handset		6.72	1.190		485	73		558	645
3590	Signal device, beacon		6.72	1.190		202	73		275	330
3594	2 zone		1.13	7.050		310	435		745	985
3600	4 zone	2 Elec	1.68	9.520		410	585		995	1,325
3610	3 zone		1.68	9.520		640	585		1,225	1,575
3800	8 zone		.84	19.050		875	1,175		2,050	2,725
3810	5 zone		1.26	12.700		700	780		1,480	1,925
3900	10 zone		1.05	15.240		1,025	935		1,960	2,550
4000	12 zone		.56	28.560		2,350	1,750		4,100	5,175

For customer support on your Electrical Change Order Costs with RSMeans data, call 800.448.8182. **Post-Installation Change Order Costs**

28 46 Fire Detection and Alarm

28 46 20 – Fire Alarm

28 46 20.50 Alarm Panels and Devices

		Crew	Daily Output	Labor-Hours	Unit	Material	2020 Bare Costs Labor	Equipment	Total	Total Incl O&P
4020	Alarm device, tamper, flow	1 Elec	6.72	1.190	Ea.	230	73		303	360
4025	Fire alarm, loop expander card		13.44	.595		680	36.50		716.50	805
4050	Actuating device	↓	6.72	1.190		340	73		413	480
4160	Alarm control panel, addressable w/o voice, up to 200 points	2 Elec	.96	16.660		4,575	1,025		5,600	6,550
4170	addressable w/voice, up to 400 points	"	.61	26.200		9,700	1,600		11,300	13,100
4175	Addressable interface device	1 Elec	6.09	1.310		149	80.50		229.50	284
4200	Battery and rack		3.36	2.380		460	146		606	725
4400	Automatic charger		6.72	1.190		525	73		598	690
4600	Signal bell		6.72	1.190		66.50	73		139.50	182
4610	Fire alarm signal bell 10" red 20-24 V P		6.72	1.190		154	73		227	278
4800	Trouble buzzer or manual station		6.72	1.190		88.50	73		161.50	207
5425	Duct smoke and heat detector 2 wire		6.72	1.190		125	73		198	247
5430	Fire alarm duct detector controller		2.52	3.170		254	194		448	570
5435	Fire alarm duct detector sensor kit		6.72	1.190		75	73		148	192
5440	Remote test station for smoke detector duct type		4.45	1.800		54.50	110		164.50	224
5460	Remote fire alarm indicator light		4.45	1.800		25	110		135	192
5600	Strobe and horn		4.45	1.800		137	110		247	315
5610	Strobe and horn (ADA type)		4.45	1.800		165	110		275	345
5620	Visual alarm (ADA type)		5.63	1.420		119	87		206	261
5800	electric bell		5.63	1.420		54	87		141	190
6000	Door holder, electro-magnetic		3.36	2.380		92	146		238	320
6200	Combination holder and closer		2.69	2.980		146	183		329	430
6600	Drill switch		6.72	1.190		400	73		473	550
6800	Master box		2.27	3.530		6,675	217		6,892	7,650
7000	Break glass station		6.72	1.190		58	73		131	173
7010	Break glass station, addressable		6.09	1.310		171	80.50		251.50	310
7800	Remote annunciator, 8 zone lamp	↓	1.51	5.290		212	325		537	720
8000	12 zone lamp	2 Elec	2.18	7.330		415	450		865	1,125
8200	16 zone lamp	"	1.85	8.660	↓	370	530		900	1,200

28 47 Mass Notification

28 47 12 – Notification Systems

28 47 12.10 Mass Notification System

		Crew	Daily Output	Labor-Hours	Unit	Material	2020 Bare Costs Labor	Equipment	Total	Total Incl O&P
0010	**MASS NOTIFICATION SYSTEM**									
0100	Wireless command center, 10,000 devices	2 Elec	1.12	14.320	Ea.	2,325	880		3,205	3,850
0200	Option, email notification					2,100			2,100	2,325
0210	Remote device supervision & monitor					2,650			2,650	2,900
0300	Antenna VHF or UHF, for medium range	1 Elec	3.36	2.380		93	146		239	320
0310	For high-power transmitter		1.68	4.760		1,125	292		1,417	1,675
0400	Transmitter, 25 watt		3.36	2.380		1,750	146		1,896	2,150
0410	40 watt		2.23	3.580		2,025	220		2,245	2,550
0420	100 watt		1.12	7.160		6,600	440		7,040	7,900
0500	Wireless receiver/control module for speaker		6.72	1.190		286	73		359	425
0600	Desktop paging controller, stand alone		3.36	2.380		850	146		996	1,150
0700	Auxiliary level inputs	↓	7.02	1.140	↓	103	70		173	217

28 52 Detention Security Systems

28 52 11 – Detention Monitoring and Control Systems

28 52 11.10 Detention Control Systems	Crew	Daily Output	Labor-Hours	Unit	Material	2020 Bare Costs Labor	Equipment	Total	Total Incl O&P
0010 **DETENTION CONTROL SYSTEMS**									
1000 Desk top control systems for 10 doors, 10 intercoms, and 10 lights	2 Elec	.25	63.490	Ea.	63,000	3,900		66,900	75,000
1020 Push button control panel systems for 10 doors, 10 intercoms, and 10 lights	"	.27	59.520	"	56,500	3,650		60,150	67,500

For customer support on your Electrical Change Order Costs with RSMeans data, call 800.448.8182.

Post-Installation Change Order Costs

Estimating Tips
33 10 00 Water Utilities
33 30 00 Sanitary Sewerage Utilities
33 40 00 Storm Drainage Utilities

- Never assume that the water, sewer, and drainage lines will go in at the early stages of the project. Consider the site access needs before dividing the site in half with open trenches, loose pipe, and machinery obstructions. Always inspect the site to establish that the site drawings are complete. Check off all existing utilities on your drawings as you locate them. Be especially careful with underground utilities because appurtenances are sometimes buried during regrading or repaving operations. If you find any discrepancies, mark up the site plan for further research. Differing site conditions can be very costly if discovered later in the project.

- See also Section 33 01 00 for restoration of pipe where removal/replacement may be undesirable. Use of new types of piping materials can reduce the overall project cost. Owners/design engineers should consider the installing contractor as a valuable source of current information on utility products and local conditions that could lead to significant cost savings.

Reference Numbers
Reference numbers are shown at the beginning of some major classifications. These numbers refer to related items in the Reference Section. The reference information may be an estimating procedure, an alternate pricing method, or technical information.

Note: Not all subdivisions listed here necessarily appear. ■

Same Data. Simplified.

Enjoy the convenience and efficiency of accessing your costs anywhere:

- **Skip the multiplier** by setting your location
- **Quickly search,** edit, favorite and share costs
- **Stay on top of price changes** with automatic updates

Discover more at rsmeans.com/online

---

33 71 Electrical Utility Transmission and Distribution

33 71 13 – Electrical Utility Towers

33 71 13.23 Steel Electrical Utility Towers

		Crew	Daily Output	Labor-Hours	Unit	Material	2020 Bare Costs Labor	2020 Bare Costs Equipment	Total	Total Incl O&P
0010	**STEEL ELECTRICAL UTILITY TOWERS**									
0500	Towers: material handling and spotting	R-7	18.95	2.530	Ton		110	8.75	118.75	177
0540	Steel tower erection	R-5	6.43	13.690		2,300	735	184	3,219	3,825
0550	Lace and box	"	5.96	14.760		2,300	790	198	3,288	3,925
0600	Special towers: material handling and spotting	R-7	10.34	4.640			202	16.05	218.05	325
0640	Special steel structure erection	R-6	5.48	16.070		2,875	860	315	4,050	4,800
0650	Special steel lace and box	"	5.28	16.660	↓	2,875	895	330	4,100	4,850

33 71 13.80 Transmission Line Right of Way

		Crew	Daily Output	Labor-Hours	Unit	Material	2020 Bare Costs Labor	2020 Bare Costs Equipment	Total	Total Incl O&P
0010	**TRANSMISSION LINE RIGHT OF WAY**									
0100	Clearing right of way	B-87	5.60	7.140	Acre		385	430	815	1,050
0200	Restoration & seeding	B-10D	3.36	3.570	"	495	185	575	1,255	1,450

33 71 16 – Electrical Utility Poles

33 71 16.23 Steel Electrical Utility Poles

		Crew	Daily Output	Labor-Hours	Unit	Material	2020 Bare Costs Labor	2020 Bare Costs Equipment	Total	Total Incl O&P
0010	**STEEL ELECTRICAL UTILITY POLES**									
6000	Digging holes in earth, average	R-5	21.12	4.170	Ea.		224	56	280	395
6010	In rock, average	"	3.79	23.230	"		1,250	310	1,560	2,225
6020	Formed plate pole structure									
6030	Material handling and spotting	R-7	2.02	23.810	Ea.		1,050	82.50	1,132.50	1,675
6040	Erect steel plate pole	R-5	1.64	53.720		11,400	2,875	720	14,995	17,700
6050	Guys, anchors and hardware for pole, in earth		5.91	14.880		695	800	200	1,695	2,175
6060	In rock	↓	15.09	5.830	↓	830	315	78.50	1,223.50	1,475
6070	Foundations for line poles									
6080	Excavation, in earth	R-5	113.72	.774	C.Y.		41.50	10.40	51.90	73.50
6090	In rock		16.80	5.240	↓		281	70.50	351.50	500
6110	Concrete foundations	↓	9.24	9.520	↓	164	510	128	802	1,075

33 71 16.33 Wood Electrical Utility Poles

		Crew	Daily Output	Labor-Hours	Unit	Material	2020 Bare Costs Labor	2020 Bare Costs Equipment	Total	Total Incl O&P
0010	**WOOD ELECTRICAL UTILITY POLES**									
0011	Excludes excavation, backfill and cast-in-place concrete									
6200	Wood, class 3 Douglas Fir, penta-treated, 20'	R-3	2.60	7.680	Ea.	256	470	72.50	798.50	1,050
6600	30'		2.18	9.160		320	560	86.50	966.50	1,275
7000	40'		1.93	10.350		630	635	98	1,363	1,750
7200	45'	↓	1.43	14.010	↓	900	855	132	1,887	2,400
7400	Cross arms with hardware & insulators									
7600	4' long	1 Elec	2.10	3.810	Ea.	156	234		390	520
7800	5' long		2.02	3.970		168	244		412	550
8000	6' long	↓	1.85	4.330	↓	190	266		456	605
9000	Disposal of pole & hardware surplus material	R-7	17.53	2.740	Mile		119	9.50	128.50	191
9100	Disposal of crossarms & hardware surplus material	"	33.60	1.430	"		62.50	4.95	67.45	100

33 71 19 – Electrical Underground Ducts and Manholes

33 71 19.15 Underground Ducts and Manholes

		Crew	Daily Output	Labor-Hours	Unit	Material	2020 Bare Costs Labor	2020 Bare Costs Equipment	Total	Total Incl O&P
0010	**UNDERGROUND DUCTS AND MANHOLES**									
0011	Not incl. excavation, backfill, or concrete in slab and duct bank									
1000	Direct burial									
1010	PVC, schedule 40, w/coupling, 1/2" diameter	1 Elec	285.60	.028	L.F.	.41	1.72		2.13	3.01
1020	3/4" diameter		243.60	.033		.49	2.01		2.50	3.54
1030	1" diameter		218.40	.037		.80	2.25		3.05	4.23
1040	1-1/2" diameter		176.40	.045		1.19	2.78		3.97	5.45
1050	2" diameter	↓	151.20	.053		1.55	3.25		4.80	6.55
1060	3" diameter	2 Elec	201.60	.079		2.86	4.87		7.73	10.40
1070	4" diameter		134.40	.119		3.79	7.30		11.09	15.05
1080	5" diameter		100.80	.159	↓	5.65	9.75		15.40	20.50

Post-Installation Change Order Costs

33 71 19.15 Underground Ducts and Manholes	Crew	Daily Output	Labor-Hours	Unit	Material	2020 Bare Costs Labor	Equipment	Total	Total Incl O&P
1090	6" diameter	2 Elec	75.60	.212	L.F.	7.40	13		20.40	27.50
1110	Elbows, 1/2" diameter	1 Elec	40.32	.198	Ea.	.70	12.15		12.85	18.85
1120	3/4" diameter		31.92	.251		.72	15.40		16.12	24
1130	1" diameter		26.88	.298		1.45	18.25		19.70	28.50
1140	1-1/2" diameter		17.64	.454		3.25	28		31.25	45
1150	2" diameter		13.44	.595		4.25	36.50		40.75	59
1160	3" diameter		10.08	.794		14.70	48.50		63.20	88.50
1170	4" diameter		7.56	1.060		27.50	65		92.50	127
1180	5" diameter		6.72	1.190		41	73		114	155
1190	6" diameter		4.20	1.900		42	117		159	221
1210	Adapters, 1/2" diameter		43.68	.183		.31	11.25		11.56	17.10
1220	3/4" diameter		36.12	.221		.44	13.60		14.04	20.50
1230	1" diameter		32.76	.244		.47	15		15.47	23
1240	1-1/2" diameter		29.40	.272		.85	16.70		17.55	26
1250	2" diameter		21.84	.366		1.67	22.50		24.17	35.50
1260	3" diameter		16.80	.476		3.42	29		32.42	47.50
1270	4" diameter		11.76	.680		5.75	41.50		47.25	68.50
1280	5" diameter		10.08	.794		11.60	48.50		60.10	85.50
1290	6" diameter		7.56	1.060		22.50	65		87.50	122
1340	Bell end & cap, 1-1/2" diameter		29.40	.272		5.65	16.70		22.35	31
1350	Bell end & plug, 2" diameter		21.84	.366		7.05	22.50		29.55	41.50
1360	3" diameter		16.80	.476		9.90	29		38.90	54.50
1370	4" diameter		11.76	.680		11.85	41.50		53.35	75
1380	5" diameter		10.08	.794		19.15	48.50		67.65	93.50
1390	6" diameter		7.56	1.060		20.50	65		85.50	120
1450	Base spacer, 2" diameter		47.04	.170		1.64	10.45		12.09	17.35
1460	3" diameter		38.64	.207		2.14	12.70		14.84	21.50
1470	4" diameter		34.44	.232		2.14	14.25		16.39	23.50
1480	5" diameter		31.08	.257		2.39	15.80		18.19	26
1490	6" diameter		28.56	.280		2.98	17.20		20.18	29
1550	Intermediate spacer, 2" diameter		50.40	.159		1.50	9.75		11.25	16.15
1560	3" diameter		38.64	.207		2.11	12.70		14.81	21
1570	4" diameter		34.44	.232		1.91	14.25		16.16	23
1580	5" diameter		31.08	.257		2.33	15.80		18.13	26
1590	6" diameter		28.56	.280	▼	3.10	17.20		20.30	29
4010	PVC, schedule 80, w/coupling, 1/2" diameter		180.60	.044	L.F.	1.24	2.72		3.96	5.40
4020	3/4" diameter		151.20	.053		1.69	3.25		4.94	6.70
4030	1" diameter		121.80	.066		2.29	4.03		6.32	8.50
4040	1-1/2" diameter		100.80	.079		3.78	4.87		8.65	11.40
4050	2" diameter	▼	84	.095		5.15	5.85		11	14.35
4060	3" diameter	2 Elec	109.20	.147		10.75	9		19.75	25.50
4070	4" diameter		75.60	.212		15.40	13		28.40	36.50
4080	5" diameter		58.80	.272		21.50	16.70		38.20	49
4090	6" diameter	▼	42	.381	▼	30.50	23.50		54	68.50
4110	Elbows, 1/2" diameter	1 Elec	24.36	.328	Ea.	2.96	20		22.96	33.50
4120	3/4" diameter		19.32	.414		6.85	25.50		32.35	45.50
4130	1" diameter		16.80	.476		8.15	29		37.15	52.50
4140	1-1/2" diameter		13.44	.595		13.70	36.50		50.20	69.50
4150	2" diameter		10.08	.794		18.30	48.50		66.80	92.50
4160	3" diameter		7.56	1.060		46	65		111	148
4170	4" diameter		5.88	1.360		80.50	83.50		164	213
4180	5" diameter		5.04	1.590		201	97.50		298.50	365
4190	6" diameter		3.36	2.380	▼	87	146		233	315

33 71 19.15 Underground Ducts and Manholes

	Crew	Daily Output	Labor-Hours	Unit	Material	2020 Bare Costs Labor	Equipment	Total	Total Incl O&P
4210	Adapter, 1/2" diameter	1 Elec	32.76	.244	Ea.	.31	15		15.31	23
4220	3/4" diameter		27.72	.289		.44	17.70		18.14	27
4230	1" diameter		24.36	.328		.47	20		20.47	30.50
4240	1-1/2" diameter		21.84	.366		.85	22.50		23.35	34.50
4250	2" diameter		19.32	.414		1.67	25.50		27.17	40
4260	3" diameter		15.12	.529		3.42	32.50		35.92	52.50
4270	4" diameter		10.92	.733		5.75	45		50.75	73.50
4280	5" diameter		9.24	.866		11.60	53		64.60	92
4290	6" diameter		6.72	1.190		22.50	73		95.50	134
4310	Bell end & cap, 1-1/2" diameter		21.84	.366		5.65	22.50		28.15	39.50
4320	Bell end & plug, 2" diameter		19.32	.414		7.05	25.50		32.55	46
4330	3" diameter		15.12	.529		9.90	32.50		42.40	59.50
4340	4" diameter		10.92	.733		11.85	45		56.85	80
4350	5" diameter		9.24	.866		19.15	53		72.15	100
4360	6" diameter		6.72	1.190		20.50	73		93.50	132
4370	Base spacer, 2" diameter		35.28	.227		1.64	13.90		15.54	22.50
4380	3" diameter		27.72	.289		2.14	17.70		19.84	29
4390	4" diameter		24.36	.328		2.14	20		22.14	32.50
4400	5" diameter		21.84	.366		2.39	22.50		24.89	36
4410	6" diameter		21	.381		2.98	23.50		26.48	38.50
4420	Intermediate spacer, 2" diameter		37.80	.212		1.50	13		14.50	21
4430	3" diameter		28.56	.280		2.11	17.20		19.31	28
4440	4" diameter		26.04	.307		1.91	18.85		20.76	30
4450	5" diameter		23.52	.340		2.33	21		23.33	33.50
4460	6" diameter		21	.381		3.10	23.50		26.60	38.50

33 71 19.17 Electric and Telephone Underground

	Crew	Daily Output	Labor-Hours	Unit	Material	2020 Bare Costs Labor	Equipment	Total	Total Incl O&P
0010	**ELECTRIC AND TELEPHONE UNDERGROUND**									
0011	Not including excavation									
0200	backfill and cast in place concrete									
0400	Hand holes, precast concrete, with concrete cover									
0600	2' x 2' x 3' deep	R-3	2.02	9.920	Ea.	545	605	93.50	1,243.50	1,600
0800	3' x 3' x 3' deep		1.60	12.530		705	765	118	1,588	2,050
1000	4' x 4' x 4' deep		1.18	17.010		1,600	1,050	161	2,811	3,475
1200	Manholes, precast with iron racks & pulling irons, C.I. frame									
1400	and cover, 4' x 6' x 7' deep	B-13	1.68	33.330	Ea.	3,000	1,525	345	4,870	5,975
1600	6' x 8' x 7' deep		1.60	35.090		3,375	1,625	365	5,365	6,525
1800	6' x 10' x 7' deep		1.51	37.040		3,775	1,700	385	5,860	7,125
4200	Underground duct, banks ready for concrete fill, min. of 7.5"									
4400	between conduits, center to center									
4580	PVC, type EB, 1 @ 2" diameter	2 Elec	403.20	.040	L.F.	1.34	2.43		3.77	5.10
4600	2 @ 2" diameter		201.60	.079		2.68	4.87		7.55	10.20
4800	4 @ 2" diameter		100.80	.159		5.35	9.75		15.10	20.50
4900	1 @ 3" diameter		336	.048		1.94	2.92		4.86	6.50
5000	2 @ 3" diameter		168	.095		3.88	5.85		9.73	12.95
5200	4 @ 3" diameter		84	.190		7.75	11.70		19.45	26
5300	1 @ 4" diameter		268.80	.060		2.23	3.65		5.88	7.90
5400	2 @ 4" diameter		134.40	.119		4.46	7.30		11.76	15.80
5600	4 @ 4" diameter		67.20	.238		8.90	14.60		23.50	32
5800	6 @ 4" diameter		45.36	.353		13.40	21.50		34.90	46.50
5810	1 @ 5" diameter		218.40	.073		2.51	4.49		7	9.45
5820	2 @ 5" diameter		109.20	.147		5	9		14	18.90
5840	4 @ 5" diameter		58.80	.272		10.05	16.70		26.75	36

33 71 19 – Electrical Underground Ducts and Manholes

33 71 19.17 Electric and Telephone Underground		Crew	Daily Output	Labor-Hours	Unit	Material	2020 Bare Costs Labor	Equipment	Total	Total Incl O&P
5860	6 @ 5" diameter	2 Elec	42	.381	L.F.	15.05	23.50		38.55	51.50
5870	1 @ 6" diameter		168	.095		4.31	5.85		10.16	13.45
5880	2 @ 6" diameter		84	.190		8.60	11.70		20.30	27
5900	4 @ 6" diameter		42	.381		17.20	23.50		40.70	54
5920	6 @ 6" diameter		25.20	.635		26	39		65	86.50
6200	Rigid galvanized steel, 2 @ 2" diameter		151.20	.106		18.85	6.50		25.35	30.50
6400	4 @ 2" diameter		75.60	.212		37.50	13		50.50	61
6800	2 @ 3" diameter		84	.190		34.50	11.70		46.20	55.50
7000	4 @ 3" diameter		42	.381		69	23.50		92.50	111
7200	2 @ 4" diameter		58.80	.272		55	16.70		71.70	85
7400	4 @ 4" diameter		28.56	.560		110	34.50		144.50	171
7600	6 @ 4" diameter		18.48	.866		164	53		217	260
7620	2 @ 5" diameter		50.40	.317		86.50	19.50		106	125
7640	4 @ 5" diameter		25.20	.635		173	39		212	249
7660	6 @ 5" diameter		15.12	1.060		260	65		325	385
7680	2 @ 6" diameter		33.60	.476		132	29		161	189
7700	4 @ 6" diameter		16.80	.952		263	58.50		321.50	375
7720	6 @ 6" diameter	▼	11.76	1.360	▼	395	83.50		478.50	560
7800	For cast-in-place concrete, add									
7810	Under 1 C.Y.	C-6	13.44	3.570	C.Y.	247	156	4	407	510
7820	1 C.Y. to 5 C.Y.		16.13	2.980		224	130	3.34	357.34	445
7830	Over 5 C.Y.	▼	20.16	2.380	▼	185	104	2.67	291.67	365
7850	For reinforcing rods, add									
7860	#4 to #7	2 Rodm	.92	17.320	Ton	1,350	975		2,325	2,975
7870	#8 to #14	"	1.26	12.700	"	1,350	715		2,065	2,575
8000	Fittings, PVC type EB, elbow, 2" diameter	1 Elec	13.44	.595	Ea.	19.05	36.50		55.55	75.50
8200	3" diameter		11.76	.680		22	41.50		63.50	86
8400	4" diameter		10.08	.794		48	48.50		96.50	126
8420	5" diameter		8.40	.952		147	58.50		205.50	249
8440	6" diameter	▼	7.56	1.060		229	65		294	350
8500	Coupling, 2" diameter					1.02			1.02	1.12
8600	3" diameter					4.97			4.97	5.45
8700	4" diameter					5.40			5.40	5.90
8720	5" diameter					10.25			10.25	11.30
8740	6" diameter					28.50			28.50	31.50
8800	Adapter, 2" diameter	1 Elec	21.84	.366		1.21	22.50		23.71	35
9000	3" diameter		16.80	.476		3.68	29		32.68	47.50
9200	4" diameter		13.44	.595		4.03	36.50		40.53	59
9220	5" diameter		10.92	.733		10.25	45		55.25	78.50
9240	6" diameter		8.40	.952		15	58.50		73.50	104
9400	End bell, 2" diameter		13.44	.595		2.76	36.50		39.26	57.50
9600	3" diameter		11.76	.680		4.90	41.50		46.40	67.50
9800	4" diameter		10.08	.794		5.55	48.50		54.05	78.50
9810	5" diameter		8.40	.952		9.75	58.50		68.25	97.50
9820	6" diameter		6.72	1.190		12.15	73		85.15	122
9830	5° angle coupling, 2" diameter		21.84	.366		9.35	22.50		31.85	44
9840	3" diameter		16.80	.476		31	29		60	77.50
9850	4" diameter		13.44	.595		23	36.50		59.50	80
9860	5" diameter		10.92	.733		20	45		65	89
9870	6" diameter		8.40	.952		31	58.50		89.50	121
9880	Expansion joint, 2" diameter		13.44	.595		40	36.50		76.50	98.50
9890	3" diameter		15.12	.529		84	32.50		116.50	141
9900	4" diameter		10.08	.794		123	48.50		171.50	208

Post-Installation Change Order Costs   For customer support on your Electrical Change Order Costs with RSMeans data, call 800.448.8182.

33 71 19 – Electrical Underground Ducts and Manholes

33 71 19.17 Electric and Telephone Underground	Crew	Daily Output	Labor-Hours	Unit	Material	2020 Bare Costs Labor	Equipment	Total	Total Incl O&P
9910	5" diameter	1 Elec	8.40	.952	Ea.	244	58.50		302.50	355
9920	6" diameter		6.72	1.190		212	73		285	340
9930	Heat bender, 2" diameter					585			585	645
9940	6" diameter					1,800			1,800	1,975
9950	Cement, quart					18.65			18.65	20.50
9960	Nylon polyethylene pull rope, 1/4"	2 Elec	1680	.010	L.F.	.22	.58		.80	1.11

33 71 23 – Insulators and Fittings

33 71 23.16 Post Insulators

		Crew	Daily Output	Labor-Hours	Unit	Material	Labor	Equipment	Total	Total Incl O&P
0010	**POST INSULATORS**									
7400	Insulators, pedestal type	R-11	94.08	.595	Ea.		34.50	6.75	41.25	59.50
7490	See also line 33 71 39.13 1000									

33 71 26 – Transmission and Distribution Equipment

33 71 26.13 Capacitor Banks

		Crew	Daily Output	Labor-Hours	Unit	Material	Labor	Equipment	Total	Total Incl O&P
0010	**CAPACITOR BANKS**									
1300	Station capacitors									
1350	Synchronous, 13 to 26 kV	R-11	2.61	21.440	MVAR	7,550	1,250	244	9,044	10,400
1360	46 kV		2.80	20.020		9,625	1,175	228	11,028	12,600
1370	69 kV		3.20	17.500		9,475	1,025	199	10,699	12,100
1380	161 kV		5.47	10.240		8,875	600	117	9,592	10,800
1390	500 kV		8.71	6.430		7,700	375	73	8,148	9,125
1450	Static, 13 to 26 kV		2.61	21.440		6,400	1,250	244	7,894	9,200
1460	46 kV		2.53	22.150		8,100	1,300	252	9,652	11,100
1470	69 kV		3.20	17.500		7,850	1,025	199	9,074	10,400
1480	161 kV		5.47	10.240		7,300	600	117	8,017	9,050
1490	500 kV		8.71	6.430		6,650	375	73	7,098	7,975
1600	Voltage regulators, 13 to 26 kV		.63	88.890	Ea.	290,000	5,200	1,000	296,200	328,000

33 71 26.23 Current Transformers

		Crew	Daily Output	Labor-Hours	Unit	Material	Labor	Equipment	Total	Total Incl O&P
0010	**CURRENT TRANSFORMERS**									
4050	Current transformers, 13 to 26 kV	R-11	11.76	4.760	Ea.	3,575	278	54	3,907	4,400
4060	46 kV		7.84	7.150		10,400	415	81.50	10,896.50	12,100
4070	69 kV		5.88	9.520		10,800	555	108	11,463	12,800
4080	161 kV		1.57	35.650		35,000	2,075	405	37,480	42,000

33 71 26.26 Potential Transformers

		Crew	Daily Output	Labor-Hours	Unit	Material	Labor	Equipment	Total	Total Incl O&P
0010	**POTENTIAL TRANSFORMERS**									
4100	Potential transformers, 13 to 26 kV	R-11	9.41	5.950	Ea.	5,100	345	67.50	5,512.50	6,200
4110	46 kV		6.72	8.330		10,500	485	95	11,080	12,300
4120	69 kV		5.22	10.720		11,100	625	122	11,847	13,300
4130	161 kV		1.88	29.760		24,000	1,725	340	26,065	29,400
4140	500 kV		1.18	47.620		71,500	2,775	540	74,815	83,500

33 71 26.28 Overhead Fuse Cutouts

		Crew	Daily Output	Labor-Hours	Unit	Material	Labor	Equipment	Total	Total Incl O&P
0010	**OVERHEAD FUSE CUTOUTS**									
1000	Cutout Fuse Assemblies									
1100	15.0 kV, 110 kV BIL, 100 amp, type C, silicone	R-8	5.04	9.520	Ea.	121	515	55	691	970
1200	200 amp		5.04	9.520		197	515	55	767	1,050
1300	300 amp		5.04	9.520		235	515	55	805	1,100
2000	Cutout Fuse Links									
2100	Removeable button head, type K, 100 amp	R-3	8.40	2.380	Ea.	11.55	145	22.50	179.05	255
2140	140 amp	"	8.40	2.380	"	26	145	22.50	193.50	271
3000	Cutout Fuse Hardware									
3100	Tri-mount bracket, aluminum	R-3	8.40	2.380	Ea.	230	145	22.50	397.50	495
3200	Arrester, surge, 12.7 kV		8.40	2.380		61.50	145	22.50	229	310

Post-Installation Change Order Costs

33 71 Electrical Utility Transmission and Distribution

33 71 26 – Transmission and Distribution Equipment

33 71 26.28 Overhead Fuse Cutouts	Crew	Daily Output	Labor-Hours	Unit	Material	2020 Bare Costs Labor	Equipment	Total	Total Incl O&P
3300	Clamp, hot-line tap, bronze, 5"	R-3	8.40	2.380	Ea.	23	145	22.50	190.50	268
3400	Clamp, stirrup, bronze, 1-13/16"		8.40	2.380		16.85	145	22.50	184.35	261
3500	Insulator, dead-end, clevis tongue, 12-1/2"		8.40	2.380		29	145	22.50	196.50	274

33 71 39 – High-Voltage Wiring

33 71 39.13 Overhead High-Voltage Wiring

		Crew	Daily Output	Labor-Hours	Unit	Material	2020 Bare Costs Labor	Equipment	Total	Total Incl O&P
0010	**OVERHEAD HIGH-VOLTAGE WIRING**									
0100	Conductors, primary circuits									
0110	Material handling and spotting	R-5	8.22	10.710	W.Mile		575	144	719	1,025
0120	For river crossing, add		9.24	9.520			510	128	638	905
0150	Conductors, per wire, 210 to 636 kcmil		1.65	53.450		13,300	2,875	720	16,895	19,800
0160	795 to 954 kcmil		1.57	56.020		27,200	3,000	755	30,955	35,200
0170	1,000 to 1,600 kcmil		1.23	71.270		49,800	3,825	960	54,585	62,000
0180	Over 1,600 kcmil		1.13	77.600		70,000	4,150	1,050	75,200	84,500
0200	For river crossing, add, 210 to 636 kcmil		1.04	84.490			4,525	1,125	5,650	8,050
0220	795 to 954 kcmil		.92	96.110			5,150	1,300	6,450	9,150
0230	1,000 to 1,600 kcmil		.81	108			5,800	1,450	7,250	10,300
0240	Over 1,600 kcmil		.73	120			6,425	1,625	8,050	11,400
0300	Joints and dead ends	R-8	5.04	9.520	Ea.	1,900	515	55	2,470	2,925
0400	Sagging	R-5	6.16	14.290	W.Mile		765	192	957	1,350
0500	Clipping, per structure, 69 kV	R-10	8.06	5.950	Ea.		345	47.50	392.50	565
0510	161 kV		4.48	10.720			620	85.50	705.50	1,025
0520	345 to 500 kV		2.13	22.590			1,300	180	1,480	2,150
0600	Make and install jumpers, per structure, 69 kV	R-8	2.69	17.860		515	970	103	1,588	2,125
0620	161 kV		1.01	47.620		1,025	2,575	275	3,875	5,300
0640	345 to 500 kV		.27	178		1,725	9,675	1,025	12,425	17,500
0700	Spacers	R-10	57.60	.833		103	48	6.65	157.65	192
0720	For river crossings, add	"	50.40	.952			55	7.60	62.60	91
0800	Installing pulling line (500 kV only)	R-9	1.22	52.550	W.Mile	875	2,675	228	3,778	5,225
0810	Disposal of surplus material, high voltage conductors	R-7	5.85	8.210	Mile		360	28.50	388.50	575
0820	With trailer mounted reel stands	"	11.52	4.170	"		182	14.45	196.45	292
0900	Insulators and hardware, primary circuits									
0920	Material handling and spotting, 69 kV	R-7	403.20	.119	Ea.		5.20	.41	5.61	8.30
0930	161 kV		576	.083			3.63	.29	3.92	5.80
0950	345 to 500 kV		806.40	.060			2.60	.21	2.81	4.17
1000	Disk insulators, 69 kV	R-5	739.20	.119		105	6.40	1.60	113	126
1020	161 kV		821.34	.107		120	5.75	1.44	127.19	142
1040	345 to 500 kV		924	.095		120	5.10	1.28	126.38	141
1060	See Section 33 71 23.16 for pin or pedestal insulator									
1100	Install disk insulator at river crossing, add									
1110	69 kV	R-5	492.80	.179	Ea.		9.55	2.40	11.95	17
1120	161 kV		739.20	.119			6.40	1.60	8	11.30
1140	345 to 500 kV		739.20	.119			6.40	1.60	8	11.30
1150	Disposal of surplus material, high voltage insulators	R-7	35.06	1.370	Mile		59.50	4.74	64.24	95.50
1300	Overhead ground wire installation									
1320	Material handling and spotting	R-7	4.75	10.110	W.Mile		440	35	475	710
1340	Overhead ground wire	R-5	1.48	59.520		3,800	3,200	800	7,800	9,825
1350	At river crossing, add		.98	89.540			4,800	1,200	6,000	8,525
1360	Disposal of surplus material, grounding wire		35.06	2.510	Mile		135	33.50	168.50	239
1400	Installing conductors, underbuilt circuits									
1420	Material handling and spotting	R-7	4.75	10.110	W.Mile		440	35	475	710
1440	Conductors, per wire, 210 to 636 kcmil	R-5	1.65	53.450		13,300	2,875	720	16,895	19,800
1450	795 to 954 kcmil		1.57	56.020		27,200	3,000	755	30,955	35,200

33 71 39 – High-Voltage Wiring

33 71 39.13 Overhead High-Voltage Wiring	Crew	Daily Output	Labor-Hours	Unit	Material	2020 Bare Costs Labor	Equipment	Total	Total Incl O&P
1460	1,000 to 1,600 kcmil	R-5	1.23	71.270	W.Mile	49,800	3,825	960	54,585	62,000
1470	Over 1,600 kcmil		1.13	77.600		70,000	4,150	1,050	75,200	84,500
1500	Joints and dead ends	R-8	5.04	9.520	Ea.	1,900	515	55	2,470	2,925
1550	Sagging	R-5	7.39	11.900	W.Mile		640	160	800	1,125
1600	Clipping, per structure, 69 kV	R-10	8.06	5.950	Ea.		345	47.50	392.50	565
1620	161 kV		4.48	10.720			620	85.50	705.50	1,025
1640	345 to 500 kV		2.13	22.590			1,300	180	1,480	2,150
1700	Making and installing jumpers, per structure, 69 kV	R-8	4.93	9.730		515	530	56	1,101	1,425
1720	161 kV		.81	59.520		1,025	3,225	345	4,595	6,350
1740	345 to 500 kV		.27	178		1,725	9,675	1,025	12,425	17,500
1800	Spacers	R-10	80.64	.595		103	34.50	4.74	142.24	170
1810	Disposal of surplus material, conductors & hardware	R-7	5.85	8.210	Mile		360	28.50	388.50	575
2000	Insulators and hardware for underbuilt circuits									
2100	Material handling and spotting	R-7	1008	.048	Ea.		2.08	.16	2.24	3.33
2150	Disk insulators, 69 kV	R-8	504	.095		105	5.15	.55	110.70	123
2160	161 kV		576.24	.083		120	4.52	.48	125	139
2170	345 to 500 kV		672	.071		120	3.88	.41	124.29	138
2180	Disposal of surplus material, insulators & hardware	R-7	35.06	1.370	Mile		59.50	4.74	64.24	95.50
2300	Sectionalizing switches, 69 kV	R-5	1.06	83.140	Ea.	27,000	4,450	1,125	32,575	37,600
2310	161 kV		.67	130		30,500	6,975	1,750	39,225	45,800
2500	Protective devices		4.62	19.050		8,825	1,025	256	10,106	11,500
2600	Clearance poles, 8 poles per mile									
2650	In earth, 69 kV	R-5	.97	90.310	Mile	7,250	4,850	1,225	13,325	16,600
2660	161 kV	"	.54	163		11,900	8,750	2,200	22,850	28,600
2670	345 to 500 kV	R-6	.40	218		14,300	11,700	4,300	30,300	37,900
2800	In rock, 69 kV	R-5	.58	151		7,250	8,100	2,025	17,375	22,300
2820	161 kV	"	.29	299		11,900	16,000	4,025	31,925	41,500
2840	345 to 500 kV	R-6	.20	436		14,300	23,400	8,600	46,300	60,000

33 72 Utility Substations

33 72 26 – Substation Bus Assemblies

33 72 26.13 Aluminum Substation Bus Assemblies

		Crew	Daily Output	Labor-Hours	Unit	Material	Labor	Equipment	Total	Total Incl O&P
0010	**ALUMINUM SUBSTATION BUS ASSEMBLIES**									
7300	Bus	R-11	495.60	.113	Lb.	3.30	6.60	1.29	11.19	14.90

33 72 33 – Control House Equipment

33 72 33.43 Substation Backup Batteries

		Crew	Daily Output	Labor-Hours	Unit	Material	Labor	Equipment	Total	Total Incl O&P
0010	**SUBSTATION BACKUP BATTERIES**									
9120	Battery chargers	R-11	9.41	5.950	Ea.	4,350	345	67.50	4,762.50	5,375
9200	Control batteries	"	11.76	4.760	K.A.H.	540	278	54	872	1,075

33 72 53 – Shunt Reactors

33 72 53.13 Reactors

		Crew	Daily Output	Labor-Hours	Unit	Material	Labor	Equipment	Total	Total Incl O&P
0010	**REACTORS**									
8150	Reactors and resistors, 13 to 26 kV	R-11	23.52	2.380	Ea.	4,375	139	27	4,541	5,025
8160	46 kV		3.62	15.470		13,500	905	176	14,581	16,400
8170	69 kV		2.35	23.810		21,500	1,400	271	23,171	26,100
8180	161 kV		1.88	29.760		24,000	1,725	340	26,065	29,400
8190	500 kV		.07	833		93,000	48,600	9,475	151,075	185,500

For customer support on your Electrical Change Order Costs with RSMeans data, call 800.448.8182.

Post-Installation Change Order Costs

33 73 Utility Transformers

33 73 23 — Dry-Type Utility Transformers

33 73 23.20 Primary Transformers	Crew	Daily Output	Labor-Hours	Unit	Material	2020 Bare Costs Labor	Equipment	Total	Total Incl O&P
0010 **PRIMARY TRANSFORMERS**									
1000 Main conversion equipment									
1050 Power transformers, 13 to 26 kV	R-11	1.44	38.760	MVA	27,100	2,250	440	29,790	33,700
1060 46 kV		2.94	19.050		25,400	1,100	217	26,717	29,800
1070 69 kV		2.61	21.440		21,600	1,250	244	23,094	25,900
1080 110 kV		2.76	20.260		20,600	1,175	231	22,006	24,600
1090 161 kV		3.62	15.470		19,100	905	176	20,181	22,500
1100 500 kV		5.88	9.520	↓	19,000	555	108	19,663	21,800
1200 Grounding transformers	↓	2.61	21.440	Ea.	121,000	1,250	244	122,494	135,000

33 75 High-Voltage Switchgear and Protection Devices

33 75 13 — Air High-Voltage Circuit Breaker

33 75 13.13 High-Voltage Circuit Breaker, Air

	Crew	Daily Output	Labor-Hours	Unit	Material	Labor	Equipment	Total	Total Incl O&P
0010 **HIGH-VOLTAGE CIRCUIT BREAKER, AIR**									
2100 Air circuit breakers, 13 to 26 kV	R-11	.47	119	Ea.	68,000	6,950	1,350	76,300	86,500
2110 161 kV	"	.20	277	"	290,500	16,200	3,150	309,850	347,000

33 75 16 — Oil High-Voltage Circuit Breaker

33 75 16.13 Oil Circuit Breaker

	Crew	Daily Output	Labor-Hours	Unit	Material	Labor	Equipment	Total	Total Incl O&P
0010 **OIL CIRCUIT BREAKER**									
2000 Power circuit breakers									
2050 Oil circuit breakers, 13 to 26 kV	R-11	.94	59.520	Ea.	68,000	3,475	675	72,150	80,500
2060 46 kV		.63	88.890		98,500	5,200	1,000	104,700	117,500
2070 69 kV		.38	148		228,500	8,625	1,675	238,800	266,500
2080 161 kV		.13	416		335,500	24,300	4,725	364,525	410,500
2090 500 kV		.05	1111	↓	1,254,000	65,000	12,600	1,331,600	1,489,500

33 75 19 — Gas High-Voltage Circuit Breaker

33 75 19.13 Gas Circuit Breaker

	Crew	Daily Output	Labor-Hours	Unit	Material	Labor	Equipment	Total	Total Incl O&P
0010 **GAS CIRCUIT BREAKER**									
2150 Gas circuit breakers, 13 to 26 kV	R-11	.47	119	Ea.	254,500	6,950	1,350	262,800	291,500
2160 161 kV		.07	833		333,500	48,600	9,475	391,575	450,000
2170 500 kV	↓	.03	1666	↓	1,173,500	97,000	19,000	1,289,500	1,457,000

33 75 23 — Vacuum High-Voltage Circuit Breaker

33 75 23.13 Vacuum Circuit Breaker

	Crew	Daily Output	Labor-Hours	Unit	Material	Labor	Equipment	Total	Total Incl O&P
0010 **VACUUM CIRCUIT BREAKER**									
2000 Power circuit breakers									
2200 Vacuum circuit breakers, 13 to 26 kV	R-11	.47	119	Ea.	61,500	6,950	1,350	69,800	79,500

33 75 36 — High-Voltage Utility Fuses

33 75 36.13 Fuses

	Crew	Daily Output	Labor-Hours	Unit	Material	Labor	Equipment	Total	Total Incl O&P
0010 **FUSES**									
8250 Fuses, 13 to 26 kV	R-11	15.68	3.570	Ea.	2,475	208	40.50	2,723.50	3,050
8260 46 kV		9.41	5.950		2,825	345	67.50	3,237.50	3,725
8270 69 kV		6.72	8.330		3,050	485	95	3,630	4,175
8280 161 kV	↓	3.92	14.280	↓	3,700	835	163	4,698	5,500

33 75 High-Voltage Switchgear and Protection Devices

33 75 39 – High-Voltage Surge Arresters

33 75 39.13 Surge Arresters

33 75 39.13 Surge Arresters	Crew	Daily Output	Labor-Hours	Unit	Material	2020 Bare Costs Labor	2020 Bare Costs Equipment	Total	Total Incl O&P
0010 **SURGE ARRESTERS**									
8000 Protective equipment									
8050 Lightning arresters, 13 to 26 kV	R-11	15.68	3.570	Ea.	1,850	208	40.50	2,098.50	2,400
8060 46 kV		11.76	4.760		4,750	278	54	5,082	5,700
8070 69 kV		9.41	5.950		6,325	345	67.50	6,737.50	7,550
8080 161 kV		4.70	11.900		8,800	695	135	9,630	10,900
8090 500 kV		1.18	47.620		30,100	2,775	540	33,415	37,800

33 75 53 – High-Voltage Switches

33 75 53.13 Switches

33 75 53.13 Switches	Crew	Daily Output	Labor-Hours	Unit	Material	2020 Bare Costs Labor	2020 Bare Costs Equipment	Total	Total Incl O&P
0010 **SWITCHES**									
3000 Disconnecting switches									
3050 Gang operated switches									
3060 Manual operation, 13 to 26 kV	R-11	1.39	40.400	Ea.	17,000	2,350	460	19,810	22,700
3070 46 kV		.94	59.520		28,800	3,475	675	32,950	37,600
3080 69 kV		.67	83.330		31,000	4,875	950	36,825	42,400
3090 161 kV		.47	119		37,200	6,950	1,350	45,500	53,000
3100 500 kV		.12	476		99,500	27,800	5,425	132,725	156,500
3110 Motor operation, 161 kV		.43	130		56,000	7,575	1,475	65,050	74,500
3120 500 kV		.24	238		146,000	13,900	2,700	162,600	184,000
3250 Circuit switches, 161 kV		.34	162		105,000	9,450	1,850	116,300	131,500
3300 Single pole switches									
3350 Disconnecting switches, 13 to 26 kV	R-11	23.52	2.380	Ea.	15,300	139	27	15,466	17,000
3360 46 kV		6.72	8.330		27,100	485	95	27,680	30,600
3370 69 kV		4.70	11.900		28,800	695	135	29,630	32,900
3380 161 kV		2.35	23.810		106,500	1,400	271	108,171	119,500
3390 500 kV		.18	303		302,000	17,700	3,450	323,150	362,500
3450 Grounding switches, 46 kV		4.70	11.900		40,300	695	135	41,130	45,600
3460 69 kV		3.13	17.870		40,700	1,050	203	41,953	46,600
3470 161 kV		1.88	29.760		43,500	1,725	340	45,565	51,000
3480 500 kV		.52	107		55,000	6,250	1,225	62,475	71,000

33 78 Substation Converter Stations

33 78 33 – Converter Stations

33 78 33.46 Substation Converter Stations

33 78 33.46 Substation Converter Stations	Crew	Daily Output	Labor-Hours	Unit	Material	2020 Bare Costs Labor	2020 Bare Costs Equipment	Total	Total Incl O&P
0010 **SUBSTATION CONVERTER STATIONS**									
9000 Station service equipment									
9100 Conversion equipment									
9110 Station service transformers	R-11	4.70	11.900	Ea.	105,000	695	135	105,830	116,500

For customer support on your Electrical Change Order Costs with RSMeans data, call 800.448.8182.

Post-Installation Change Order Costs

33 79 83 – Site Grounding Conductors

33 79 83.13 Grounding Wire, Bar, and Rod	Crew	Daily Output	Labor-Hours	Unit	Material	2020 Bare Costs Labor	Equipment	Total	Total Incl O&P
0010 **GROUNDING WIRE, BAR, AND ROD**									
7500 Grounding systems	R-11	235.20	.238	Lb.	11.65	13.90	2.71	28.26	36.50

Division Notes

	CREW	DAILY OUTPUT	LABOR-HOURS	UNIT	BARE COSTS				TOTAL INCL O&P
					MAT.	LABOR	EQUIP.	TOTAL	

Estimating Tips
34 11 00 Rail Tracks
This subdivision includes items that may involve either repair of existing or construction of new railroad tracks. Additional preparation work, such as the roadbed earthwork, would be found in Division 31. Additional new construction siding and turnouts are found in Subdivision 34 72. Maintenance of railroads is found under 34 01 23 Operation and Maintenance of Railways.

34 40 00 Traffic Signals
This subdivision includes traffic signal systems. Other traffic control devices such as traffic signs are found in Subdivision 10 14 53 Traffic Signage.

34 70 00 Vehicle Barriers
This subdivision includes security vehicle barriers, guide and guard rails, crash barriers, and delineators. The actual maintenance and construction of concrete and asphalt pavement are found in Division 32.

Reference Numbers
Reference numbers are shown at the beginning of some major classifications. These numbers refer to related items in the Reference Section. The reference information may be an estimating procedure, an alternate pricing method, or technical information.

Note: Not all subdivisions listed here necessarily appear. ■

34 43 Airfield Signaling and Control Equipment

34 43 13 – Airfield Signals and Lighting

34 43 13.16 Airfield Runway and Taxiway Inset Lighting	Crew	Daily Output	Labor-Hours	Unit	Material	2020 Bare Costs Labor	Equipment	Total	Total Incl O&P
0010 **AIRFIELD RUNWAY AND TAXIWAY INSET LIGHTING**									
0100 Runway centerline, bidir., semi-flush, 200 W, w/shallow insert base	R-22	10.42	3.580	Ea.	2,350	201		2,551	2,900
0120 Flush, 200 W, w/shallow insert base		10.42	3.580		2,125	201		2,326	2,650
0130 for mounting in base housing		15.66	2.380		1,075	134		1,209	1,400
0150 Touchdown zone light, unidirectional, 200 W, w/shallow insert base		10.42	3.580		1,925	201		2,126	2,425
0160 115 W		10.42	3.580		1,825	201		2,026	2,325
0180 Unidirectional, 200 W, for mounting in base housing		15.62	2.390		840	134		974	1,125
0190 115 W		15.62	2.390		905	134		1,039	1,200
0210 Runway edge & threshold light, bidir., 200 W, for base housing		7.86	4.740		1,150	266		1,416	1,650
0240 Threshold & approach light, unidir., 200 W, for base housing		7.86	4.740		650	266		916	1,100
0260 Runway edge, bidirectional, 2-115 W, for base housing		10.42	3.580		1,725	201		1,926	2,200
0280 Runway threshold & end, bidir., 2-115 W, for base housing		10.42	3.580		1,900	201		2,101	2,375
0370 45 W, flush, for mounting in base housing		15.66	2.380		925	134		1,059	1,225
0380 115 W		15.66	2.380		950	134		1,084	1,250

34 43 23 – Weather Observation Equipment

34 43 23.16 Airfield Wind Cones

	Crew	Daily Output	Labor-Hours	Unit	Material	Labor	Equipment	Total	Total Incl O&P
0010 **AIRFIELD WIND CONES**									
1200 Wind cone, 12' lighted assembly, rigid, w/obstruction light	R-21	1.14	28.710	Ea.	11,900	1,750	100	13,750	15,700
1210 Without obstruction light		1.28	25.690		9,950	1,575	89.50	11,614.50	13,300
1220 Unlighted assembly, w/obstruction light		1.41	23.240		10,800	1,425	81	12,306	14,100
1230 Without obstruction light		1.55	21.220		9,950	1,300	74	11,324	12,900
1240 Wind cone slip fitter, 2-1/2" pipe		18.35	1.790		132	110	6.25	248.25	315
1250 Wind cone sock, 12' x 3', cotton		5.51	5.950		720	365	21	1,106	1,375
1260 Nylon		5.51	5.950		690	365	21	1,076	1,325

Post-Installation Change Order Costs

Estimating Tips

- When estimating costs for the installation of electrical power generation equipment, factors to review include access to the job site, access and setting up at the installation site, required connections, uncrating pads, anchors, leveling, final assembly of the components, and temporary protection from physical damage, such as environmental exposure.

- Be aware of the costs of equipment supports, concrete pads, and vibration isolators. Cross-reference them against other trades' specifications. Also, review site and structural drawings for items that must be included in the estimates.

- It is important to include items that are not documented in the plans and specifications but must be priced. These items include, but are not limited to, testing, dust protection, roof penetration, core drilling concrete floors and walls, patching, cleanup, and final adjustments. Add a contingency or allowance for utility company fees for power hookups, if needed.

- The project size and scope of electrical power generation equipment will have a significant impact on cost. The intent of RSMeans cost data is to provide a benchmark cost so that owners, engineers, and electrical contractors will have a comfortable number with which to start a project. Additionally, there are many websites available to use for research and to obtain a vendor's quote to finalize costs.

Reference Numbers

Reference numbers are shown at the beginning of some major classifications. These numbers refer to related items in the Reference Section. The reference information may be an estimating procedure, an alternate pricing method, or technical information.

Same Data. Simplified.

Enjoy the convenience and efficiency of accessing your costs anywhere:

- **Skip the multiplier** by setting your location
- **Quickly search,** edit, favorite and share costs
- **Stay on top of price changes** with automatic updates

Discover more at rsmeans.com/online

Note: Trade Service, in part, has been used as a reference source for some of the material prices used in Division 48.

48 13 Hydroelectric Plant Electrical Power Generation Equipment

48 13 13 – Hydroelectric Power Plant Water Turbines

48 13 13.10 Water Turbines and Components	Crew	Daily Output	Labor-Hours	Unit	Material	2020 Bare Costs Labor	Equipment	Total	Total Incl O&P
0010 **WATER TURBINES & COMPONENTS**									
0500 Water turbine, 100-500 watt, 3P, 48VDC, 54"W x 24"H, 260LBS	2 Elec	1.68	9.520	Ea.	10,200	585		10,785	12,100
0600 Water turbine, 1500-2000 watt, 3P, 130-200VAC to DC, 95LBS		2.52	6.350		4,025	390		4,415	5,000
0501 Wind turbine marine rated, 160 W, 12 V, DC		1.85	8.660		1,175	530		1,705	2,100
0510 160 W, 24 V, DC		1.85	8.660		1,200	530		1,730	2,125
0520 160 W, 48 V, DC		1.85	8.660		1,175	530		1,705	2,100
1001 Wind turbine, 300 W, 12 V, DC		1.85	8.660		900	530		1,430	1,775
1011 300 W, 24 V, DC		1.85	8.660		915	530		1,445	1,800
1020 300 W, 48 V, DC		1.85	8.660		910	530		1,440	1,800
1030 Wind turbine, 400 W, 12 V, DC		1.85	8.660		910	530		1,440	1,800
1040 400 W, 24 V, DC		1.85	8.660		910	530		1,440	1,800
1050 400 W, 48 V, DC		1.85	8.660		915	530		1,445	1,800
3300 Wind turbine/charge control, 2,900-7,900 W, 5.5" H x 11' W, 180 lb.		1.68	9.520		16,600	585		17,185	19,200
3320 Wind turbine/charge control, 4,000-10,800 W, 5.5" H x 13' W, 240 lb.		1.68	9.520		27,500	585		28,085	31,200
3401 Wind turbine tower, 2 stage steel, 8' H x 6' W		1.68	9.520		2,075	585		2,660	3,175
4000 Anemometer, standard AC or 9V, 2-digit LCD, wood case, +/-2% of Input	2 Elec	1.68	9.520	Ea.	177	585		762	1,075

For customer support on your Electrical Change Order Costs with RSMeans data, call 800.448.8182.   **Post-Installation Change Order Costs**

Reference Section

All the reference information is in one section, making it easy to find what you need to know . . . and easy to use the data set on a daily basis. This section is visually identified by a vertical black bar on the page edges.

In this Reference Section we've included Change Order Procedures and Considerations, with reference tables, explanations, and estimating information that support how we developed the unit price data; Crew Listings, a listing of all crews, equipment, and their costs; Historical Cost Indexes for cost comparisons over time; City Cost Indexes and Location Factors for adjusting costs to the region you are in; and an explanation of all the Abbreviations in the data set.

Table of Contents

Change Order Procedures and Considerations

Change orders, or "extras," in the construction process can be a burden to architects, engineers, contractors, or owners. On the other hand, change orders that are properly recognized and managed can ensure orderly, professional, and profitable progress for all who are involved in the project. There are many causes for change orders and/or change order requests. In all cases, change orders or change order requests should be addressed promptly and in a precise and prescribed manner. The following pages include information regarding change order pricing and procedures as well as information on how costs in this data set are developed.

The Causes of Change Orders

Change orders occur when the scope of work changes, whether by addition or deletion. Listed below are a few causes of change orders.

- **Unknown Concealed Conditions**
- **Completion Date Changes:** The owner changes the completion date for reasons unrelated to the construction process.
- **Design Discrepancies or Omissions**
- **Owner Changes:** The owner changes design criteria, scope of work, or project objectives during construction.

Procedures

Properly written contract documents must include the correct change order procedures for all parties—owner, designers, and contractors—to follow in order to avoid costly delays and litigation.

Being "in the right" is not always a sufficient or acceptable defense. The contract provisions requiring notification and documentation *must* be adhered to within a defined or reasonable time frame.

The fairest and simplest method of handling change orders is by a written proposal and acceptance by the owner's "authorized agent." A contractor would be wise to insist, prior to starting work on a project, that the owner identify the authorized agent who may sign or accept a change order proposal. The contractor should also know if there is a financial limit on the agent's authority.

Often, time is a critical factor when the need for a change arises. In such cases, the contractor might be directed to proceed on a "time and material" basis, rather than wait for the paperwork to be processed—a delay that could impede progress. In this situation the contractor still must follow the prescribed change order procedures, including but not limited to, notification and documentation.

All forms used for change orders should be dated and signed by the proper authority. Relying on memory can be a very costly mistake, especially if legal judgments are to be made and if certain field personnel are no longer available. See Figures R.10 through R.12 at the end of this reference section for examples of typical standard forms that can be used to document change order work (from *Means Forms for Building Construction Professionals*). A description and estimated value of the work are recorded on such forms. Completion of the form by the contractor, together with the signature of the owner or owner's agent, provide authorization for the extra work to proceed. For time and material change orders, a form should be used and signed DAILY describing the labor and material allocated to the change that day.

"Owners," or awarding authorities who do considerable and continual building construction (such as the federal government), realize the inevitability of change orders for numerous reasons, both predictable and unpredictable. As a result, the federal government, the American Institute of Architects (AIA), and many other engineering and contractor associations have developed their own standards and procedures to be followed by all parties to achieve contract continuance and timely completion, while being just and financially fair to all concerned.

In addition to the change order standards put forth by industry organizations, there are also many resources available on the subject.

Pricing Change Orders

When pricing change orders, regardless of their cause, the most significant factor is *when* the change occurs. The need for a change may be perceived in the field or requested by the architect/engineer *before* any of the actual installation has begun, or may evolve or appear while construction is progressing and the system in question is partially installed. In the latter cases, the original sequence of construction is disrupted, along with all contiguous and supporting systems.

Clearly, change orders cause the greatest impact when they occur after the installation has been completed and must be uncovered, or even replaced. Post-completion changes may be caused by necessary design changes, product failure, or changes in the owner's requirements that are not discovered until the building or the system begins to function.

Specified procedures of notification and record keeping must be adhered to and enforced regardless of the stage of construction: before, during or after installation.

For customer support on your Electrical Change Order Costs with RSMeans data, call 800.448.8182.

Some bid documents anticipate change orders by requiring that unit prices including overhead and profit percentages—for additional as well as deductible changes—be listed. Usually these unit prices do not take into account a ripple effect, or impact on other trades, and should be used for general guidance only.

It may not be possible to classify all changes according to the three previously mentioned time frames. Judgment and experience must be used to estimate those that occur between stages, as shown in Figure R.1.

This data set identifies two basic time frames for change orders: *Pre-installation Change Orders*, which occur before the start of construction, and *Post-installation Change Orders*, which involve reworking after the original installation. Change orders that occur during the construction phase may be priced according to the extent of work completed, using pre-installation or post-installation costs, depending upon percent of completion.

Factors that Affect Change Order Costs:

As an estimator begins to prepare a change order estimate, the following questions should be reviewed to determine their impact on the final price.

1. Is the change order work pre-installation or post-installation? (See Figure R.1).

2. How efficient is the existing crew at actual installation? (See Figure R.2).

3. Will you have to pay more or less for the new material required by the change order than you paid for the original purchase? (See Figure R.3).

4. How many electricians are required to install the change order work? (See Figure R.4).

5. How many foremen are needed to supervise the required crews? (See Figure R.5).

6. If the crews sizes are increased, what impact will that have on supervision requirements? (See Figure R.6).

7. What are the other impacts of increased crew size? (See Figure R.7).

8. As new journeymen, unfamiliar with the project, are brought onto the site, how long will it take them to become oriented to the project requirements? (See Figure R.8).

9. How much actual production can be gained by working overtime? (See Figure R.9).

10. Will this change delay the original completion schedule long enough so that the installation becomes subject to a new wage rate, dictated by the relevant labor contract?

11. Will this change delay the original completion schedule enough to put the installation into a new weather season, such as winter? For example, underground conduit banks were to be installed in September, but work was delayed until December. At this point, frost had penetrated the trench area, thereby changing the requirements of this task.

Handling

Change orders may increase not only material and installation costs, but also handling charges. For example, a change order may be processed in the pre-installation time frame, but after the major equipment has been ordered. In this case, the additional required equipment, such as motors and transformers, cannot be shipped with the original order. Arriving after the installation is in progress, the additional equipment must be handled and installed in restrictive conditions.

Productivity

Figures R.4 to R.9 provide information on labor as affected by change orders. Figure R.4 is a chart listing the labor requirements for the installation of various electrical components. Figure R.5 provides suggestions for efficient supervision ratios. Figures R.6, R.7, and R.8 illustrate the impact of increased crew size on productivity. Finally, the effects of overtime are analyzed in Figure R.9.

Efficiency Loss

During the course of a project, labor can be affected and productivity diminished for a variety of reasons. Change orders and resulting delays and inefficiencies frequently cause a decline in productivity.

For example, in order to maintain job progress, a new crew might be brought in just to handle the change order work. This increase in crew size may reduce the efficiency of the entire job crew due to overcrowding, dilution of supervision, and orientation and learning curve losses. In addition to the productivity drop, more tools, staging, and storage facilities may be required, thereby increasing job overhead and restricting the work space. Figures R.6, R.7, and R.8 are graphs showing the effects of increased crew size on additional supervision and crew orientation.

Construction Phase	Pre-installation	Installation	Post-installation
Source of Labor and Material Costs	Use Pre-installation Costs	Use Pre- or Post-installation Costs depending upon % of completion	Use Post-installation Costs

Figure R.1 Change Order Installation Phase

Notes:

1. Change order work that occurs during the installation varies in cost, according to how much of the installation has been completed. If little work has been done, the Pre-installation Change Order costs may be most appropriate. If, on the other hand, the installation is substantially complete, the Post-installation Change Order section costs will be most suitable.

2. Prices in Unit Price Change Order sections do not include demolition or the disposal of any removed materials or removal and/or replacement of non-electrical items.

3. The estimator must determine if unused or removed materials are in good enough condition to be returned to the manufacturer or supplier for credit, or if they can be salvaged and reused. Caution should be exercised here, as it is generally the case that once material has been delivered to the site, it is legally the owner's property. Therefore, approval to return such material must be obtained from the owner.

4. In pricing post-installation work, check to ensure that new labor rates have not been negotiated.

Activities (Productivity) Expressed as Percentages of a Workday			
Task	RSMeans Electrical Cost Data (for New Construction)	Pre-installation Change Orders	Post-Installation Change Orders
1. Study plans	3%	6%	6%
2. Material procurement	3%	3%	3%
3. Receiving and storing	3%	3%	3%
4. Mobilization	5%	5%	5%
5. Site Movement	5%	5%	8%
6. Layout and marking	8%	10%	12%
7. Actual installation	64%	59%	54%
8. Clean-up	3%	3%	3%
9. Breaks—non-productive	6%	6%	6%
Total	100%	100%	100%

Figure R.2 Change Order Installation Efficiency

The labor-hours expressed in *Electrical Costs with RSMeans Data* (for new construction) are based on average installation time, using an efficiency level of approximately 60-65%. For change order situations, adjustments to this efficiency level should reflect the daily labor-hour allocation for that particular occurrence.

If any of the specific percentages expressed in this chart (Figure R.2) do not apply to a particular project situation, then those percentage points should be reallocated to the appropriate task(s). Example: Using data for new construction, assume there is no new material being utilized. The percentages for Tasks 2 and 3 would therefore be reallocated to other tasks. If the time required for Tasks 2 and 3 can now be applied to installation, we can add the time allocated for *Material Procurement* and *Receiving and Storing* to the *Actual Installation* time for new construction, thereby increasing the *Actual Installation* percentage.

This chart shows that, due to reduced productivity, labor costs will be higher than those for new construction by 5% to 15% for pre-installation change orders and by 15% to 25% for post-installation change orders. Each job and change order is unique and must be examined individually. Many factors, covered elsewhere in this section, can have a significant impact on productivity and change order costs. All such factors should be considered in every case.

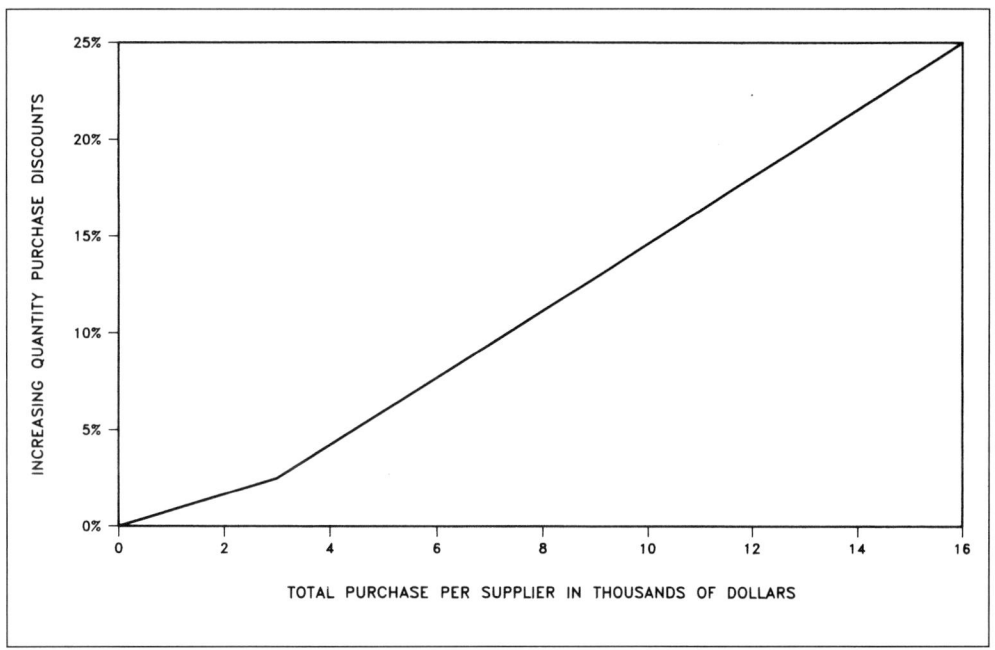

Figure R.3 Impact of Quantity on Change Order Material Purchasing

This chart shows the increasing quantity discounts as the power of the buyer increases. On a change order, however, it may be impossible to benefit from any large-quantity discounts. In addition, the contractor may forfeit the advantage of competitive pricing, as the following example shows.

Example:

A contractor bought over $20,000 worth of switchgear for an installation, and obtained the maximum discount for this purchase. After a period of time (before, during, or after the originally planned installation), it was determined that the project would require an additional matching section. The contractor had to purchase this section from the original supplier in order to ensure compatibility with the originally purchased equipment. For this second purchase, the supplier did not allow any discount, because of the small quantity purchased. Also the supplier is no longer in a competitive bid situation.

Material pricing for Pre-installation and Post-installation sections was developed based on small purchases. If large quantities of basic construction material are needed (i.e., conduit, wire boxes), the total cost of these materials should be modified. This can be done using Figure R.3 as a guide to quantity purchase discounts. Alternatively, material prices can be obtained from the current edition of *Electrical Costs with RSMeans Data*. Current prices for major pieces of equipment required for change orders should be determined by contacting equipment suppliers.

All major purchase equipment, such as switchgear, light fixtures, etc., were left at new construction pricing, as these items should be quoted for the individual requirements on each project.

Division 26 Electrical					
Material Unit	**Suggested Manpower**		**Material Unit**	**Suggested Manpower**	
Subdivision 26 05 19 Low-Voltage Electrical Power Conductors and Cables			**Subdivision 26 33 53 Static Uninterruptible Power Supply**		
Branch Conductors #14 gauge through #6	1 Electrician		Uninterruptible Power Supplies up to 1 kVA	1 Electrician	
Feeder Conductors #4 Gauge through Size 4/0	2 Electricians		from 1.5 to 15 kVA	2 Electricians	
Feeder Conductors 250 kcmil and up	3 Electricians				
			Subdivision 26 35 13 Capacitors		
			Capacitors	1 Electrician	
Subdivision 26 05 33 Raceway and Boxes					
Conduit 1/2" to 2-1/2" Diameter	1 Electrician		**Subdivision 26 35 26 Harmonic Filters**		
Conduit 3" Diameter and up	2 Electricians		Power Conditioners to 6 kVA	2 Electricians	
Duct	2 Electricians		Power Conditioners to 7.5 kVA and larger	3 Electricians	
Boxes and Devices	1 Electrician				
			Subdivision 26 36 13 Automatic Transfer Switches		
Subdivision 26 05 36 Cable Trays			Automatic Transfer Switches 40 through 100 Amp	1 Electrician	
Cable Trays	2 Electricians		150 Amp and up	2 Electricians	
Subdivision 26 05 90 Residential Wiring			**Subdivision 26 51 13 Interior Lighting Fixtures**		
Residential Wiring Device	1 Electrician		Fluorescent 4 Feet	1 Electrician	
			Fluorescent 8 Feet	2 Electricians	
Subdivision 26 22 13 Low-Voltage Distribution Transformers			High Intensity to 250 Watt	1 Electrician	
			High Intensity 400 Watt and larger	2 Electricians	
Transformers to 100 kVA	2 Electricians				
Transformers 112.5 kVA and larger	3 Electricians		**Subdivision 26 56 36 Flood Lighting**		
			Poles to 40 Feet High	2 Electricians	
Subdivision 26 24 13 Switchboards					
Switchgear Sections or Switchboard Sections	3 Electricians		**Subdivision 26 71 13 Motors**		
Switchgear Sections or Switchboard Make-Up	1 Electrician		Motors to 15 Horsepower	1 Electrician	
			Motors 20 to 75 Horsepower	2 Electricians	
Subdivision 26 24 16 Panelboards			Motors 100 to 200 Horsepower	3 Electricians	
Panelboards to 225 Amps	2 Electricians				
Panelboards Make-Up	1 Electrician				
Panelboards Interiors	1 Electrician				
Subdivision 26 25 13 Enclosed Bus Assemblies					
Bus Duct/Busway	2 Electricians				
Subdivision 26 32 13 Engine Generators					
Generator Sets	2 Electricians				

Division 27 Communications					
Material Unit	**Suggested Manpower**		**Material Unit**	**Suggested Manpower**	
Subdivision 27 51 19 Sound Masking System			**Subdivision 27 41 19 Portable Audio-Video Equipment**		
Sound System Devices	1 Electrician		Educational TV (3 Camera) System	4 Electricians	
Sound System	2 Electricians				
			Subdivision 27 53 13 Clock System		
Subdivision 27 52 23 Nurse Call Systems			Master Time Clock, 20 Rooms and up	4 Electricians	
Nurse Master Call System, 20 Stations and up	2 Electricians				

Division 28 Electronic Safety and Security					
Material Unit	**Suggested Manpower**		**Material Unit**	**Suggested Manpower**	
Subdivision 28 16 16 Intrusion Detection			**Subdivision 28 31 23 Fire Detection and Alarm**		
Detection Devices	1 Electrician		Detection Devices	1 Electrician	

Figure R.4 Manpower Required to Install Electrical Units

The preceding chart provides guidelines for the manpower needed to install various electrical units. The items are categorized according to their subdivision.

Crew	Crew Composition	
	Supervisory	Journeymen
3	1 Working Foreman	2 Electricians
4	1 Working Foreman	3 Electricians
5	1 Working Foreman	4 Electricians
6	1 Working Foreman	5 Electricians
7	1 Non-working Foreman	6 Electricians
8	1 Non-working Foreman	7 Electricians
9	1 Non-working Foreman 1 Working Foreman	7 Electricians
10	1 Non-Working Foreman 1 Working Foreman	8 Electricians
11	1 Non-working Foreman 1 Working Foreman	9 Electricians
12	1 Non-working (General) Foreman 2 Working Foremen	9 Electricians
13	1 Non-working (General) Foreman 2 Working Foremen	10 Electricians
14	1 Non-working (General) Foreman 1 Non-Working Foreman 1 Working Foreman	11 Electricians

Figure R.5 Efficient Ratios of Foremen to Crew

This chart is a guide for efficient ratios of foremen to electricians for various crew sizes. It should be noted that assigned supervision to an electrical work party or crew may be mandated by an existing labor agreement. The requirements may vary from one jurisdiction to another.

It is recommended that any crew larger than 14 be avoided, as an overall drop in efficiency or productivity tends to occur at this point. Should job conditions mandate this much manpower, one should consider establishing separate and more manageable crews under one non-working general foreman.

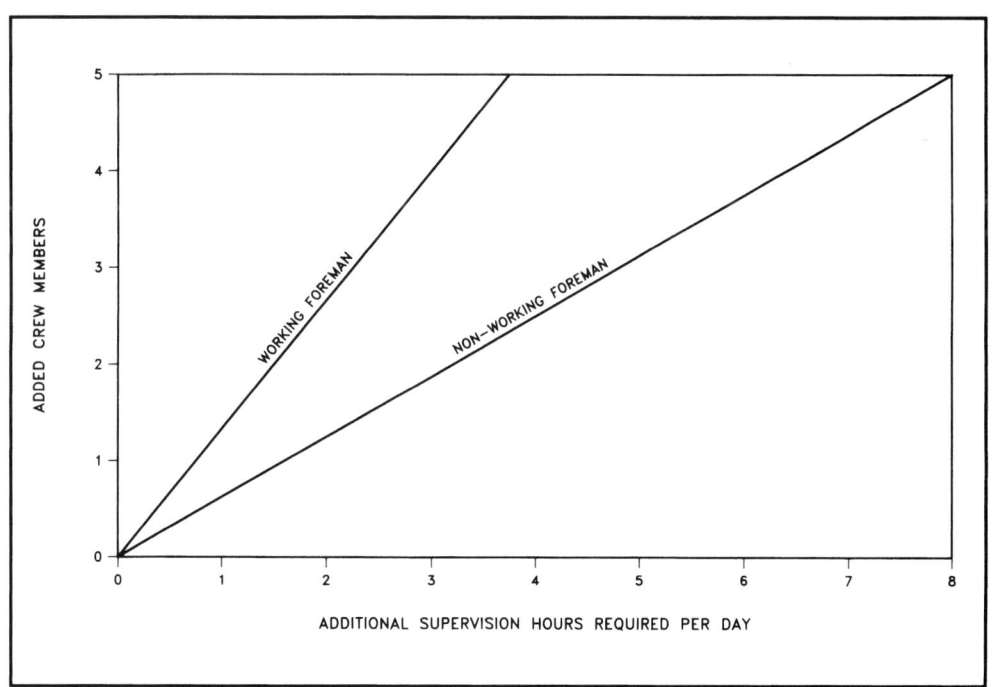

Figure R.6 Impact of Crew Size on Supervision

The graph above shows what impact increased crew size may have on the requirements for supervision. As the number of crew members grows, a greater number of foremen are required.

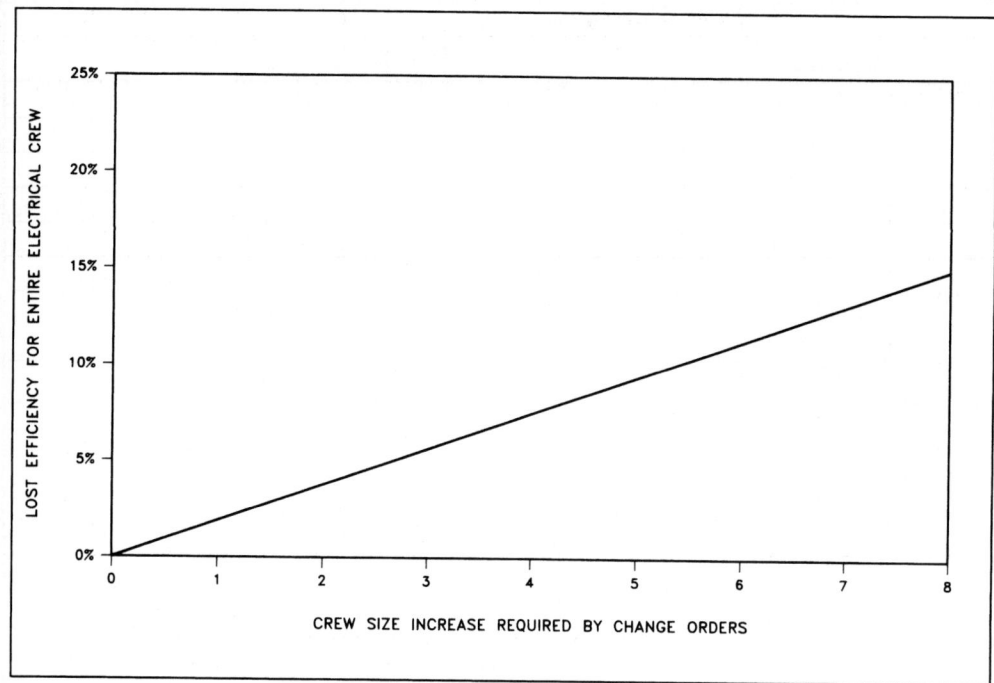

Figure R.7 Lost Efficiency Due to Increased Crew Size

This graph shows how change orders may negatively affect overall productivity by increasing crew size. The greater the increase in number of the crew, the more productivity drops. One of the factors that causes this productivity loss is overcrowding, producing restrictive conditions in the working space, and possibly, a shortage of any special tools and equipment required. Such factors affect not only the crew working on the elements directly involved in the change order, but other crews whose movement may also be hampered.

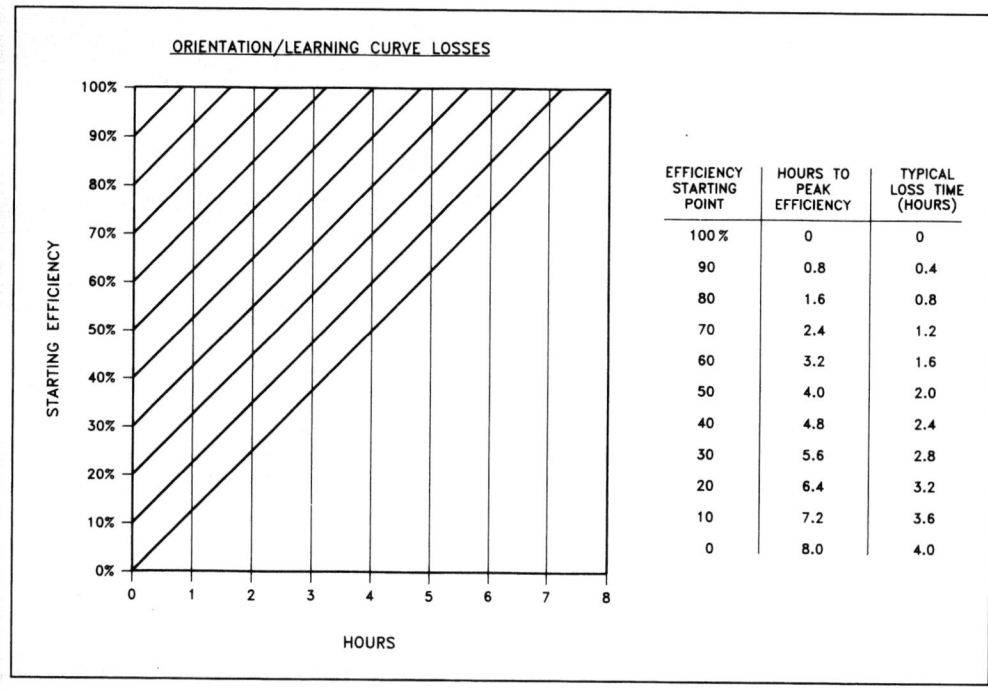

EFFICIENCY STARTING POINT	HOURS TO PEAK EFFICIENCY	TYPICAL LOSS TIME (HOURS)
100 %	0	0
90	0.8	0.4
80	1.6	0.8
70	2.4	1.2
60	3.2	1.6
50	4.0	2.0
40	4.8	2.4
30	5.6	2.8
20	6.4	3.2
10	7.2	3.6
0	8.0	4.0

Figure R.8 Time Required for Crew Orientation

This graph provides a means of estimating the amount of time it would take to bring a new crew, unfamiliar with a particular project, up to 100% efficiency. 100% represents the efficiency rate required to maintain scheduled progress. Looking at the graph, it can be seen that a crew that begins at 90% efficiency will attain 100% efficiency in less than one hour. A worker who begins the job at 40% efficiency will take almost 5 hours to reach 100%.

The three columns of numbers to the right of the graph list the following: The first column of percentages from 0-100 shows a range of starting efficiency levels of crews. The second column lists the time (in hours) required to bring each crew up to 100% efficiency. The third column lists the average loss time (in hours) that results from the learning period (required to reach 100% efficiency).

For customer support on your Electrical Change Order Costs with RSMeans data, call 800.448.8182.

Days per Week	Hours per Day	Production Efficiency					Payroll Cost Factors	
		1 Week	2 Weeks	3 Weeks	4 Weeks	Average 4 Weeks	@ 1-1/2 Times	@ 2 Times
5	8	100%	100%	100%	100%	100%	100%	100%
	9	100	100	95	90	96.25	105.6	111.1
	10	100	95	90	85	91.25	110.0	120.0
	11	95	90	75	65	81.25	113.6	127.3
	12	90	85	70	60	76.25	116.7	133.3
6	8	100	100	95	90	96.25	108.3	116.7
	9	100	95	90	85	92.50	113.0	125.9
	10	95	90	85	80	87.50	116.7	133.3
	11	95	85	70	65	78.75	119.7	139.4
	12	90	80	65	60	73.75	122.2	144.4
7	8	100	95	85	75	88.75	114.3	128.6
	9	95	90	80	70	83.75	118.3	136.5
	10	90	85	75	65	78.75	121.4	142.9
	11	85	80	65	60	72.50	124.0	148.1
	12	85	75	60	55	68.75	126.2	152.4

Figure R.9 Effects of Overtime

Caution: Under many labor agreements, Sundays and holidays are paid at a higher premium than the normal overtime rate.

The use of long-term overtime is counter-productive on almost any construction job; that is, the longer the period of overtime, the lower the actual production rate. Numerous studies have been conducted and, while they have resulted in slightly different numbers, all reach the same conclusion. Figure R.9 tabulates the effects of overtime work on efficiency.

As illustrated in Figure R.9, there can be a difference between the actual payroll cost per hour and the *effective* cost per hour for overtime work. This is due to the reduced production efficiency with the increase in weekly hours beyond 40. This difference between actual and effective cost results from overtime work over a prolonged period. Short-term overtime work does not result in as great a reduction in efficiency, and in such cases, effective cost may not vary significantly from the actual payroll cost. As the total hours per week are increased on a regular basis, more time is lost because of fatigue, lowered morale, and an increased accident rate.

As an example, assume a project where workers are working 6 days a week, 10 hours per day. From Figure R.9 (based on productivity studies), the average effective productive hours over a four-week period are:

$$.875 \times 60 = 52.5$$

Depending upon the locale and day of week, overtime hours may be paid at time and a half or double time. For time and a half, the overall (average) *actual* payroll cost (including regular and overtime hours) is determined as follows:

$$\frac{40 \text{ reg. hours.} + (20 \text{ overtime hrs.} \times 1.5)}{60 \text{ hrs.}} = 1.167$$

Based on 60 hours, the payroll cost per hour will be 116.7% of the normal rate at 40 hours per week. However, because the effective production (efficiency) for 60 hours is reduced to the equivalent of 52.5 hours, the effective cost of overtime is calculated as follows:

For time and a half:

$$\frac{40 \text{ reg. hrs.} + (20 \text{ overtime hrs.} \times 1.5)}{52.5 \text{ hrs.}} = 1.33$$

Installed cost will be 133% of the normal rate (for labor).

Thus, when figuring overtime, the actual cost per unit of work will be higher than the apparent overtime payroll dollar increase, due to the reduced productivity of the longer work week. These efficiency calculations are true only for those cost factors determined by hours worked. Costs that are applied weekly or monthly, such as equipment rentals, will not be similarly affected.

For customer support on your Electrical Change Order Costs with RSMeans data, call 800.448.8182.

527

**CONTRACT
CHANGE ORDER**

FROM

_____

_____

_____

_____

TO _____

_____

_____

_____

CHANGE ORDER NO.							
DATE							
PROJECT							
LOCATION							
JOB NO.							
ORIGINAL CONTRACT AMOUNT	$						
TOTAL PREVIOUS CONTRACT CHANGES							
TOTAL BEFORE THIS CHANGE ORDER							
AMOUNT OF THIS CHANGE ORDER							
REVISED CONTRACT TO DATE							

Gentlemen:

This CHANGE ORDER includes all Material, Labor and Equipment necessary to complete the following work and to adjust the total contract as indicated;

☐ the work below to be paid for at actual cost of Labor, Materials and Equipment plus _____ percent (_____%)

☐ the work below to be completed for the sum of_____

_____ dollars ($_____)

_____

_____

_____

_____

_____

_____

_____

_____

_____

_____

_____

_____

_____

_____

_____

_____

_____

_____

_____

_____

CHANGES APPROVED

The work covered by this order shall be performed under the same Terms and Conditions as that included in the original contract unless stated otherwise above.

By_____

By_____

Signed_____

By_____

Figure R.10

EXTRA
WORK ORDER

FROM

_____

_____

_____

_____

TO

_____

_____

_____

_____

EXTRA WORK ORDER NO. _____

DATE _____

PROJECT _____

LOCATION _____

JOB NO. _____

EXTRA WORK ORDER APPROVED

_____

BY _____

_____

BY _____

Gentlemen:

This EXTRA WORK ORDER includes all Material, Labor and Equipment necessary to complete the following work;

☐ the work below to be paid for at actual cost of Labor, Materials and Equipment plus _____ percent (_____%)

☐ the work below to be completed for the sum of_____dollars

($ _____)

DESCRIPTION

_____

_____

_____

_____

_____

_____

_____

_____

_____

_____

_____

_____

_____

_____

_____

_____

_____

_____

The work covered by this order shall be performed under the same Terms and Conditions as that included in the original contract unless stated otherwise above.

Signed_____

By_____

Figure R.11

For customer support on your Electrical Change Order Costs with RSMeans data, call 800.448.8182.

529

PROPOSAL

FROM

_____

_____

_____

_____

TO

_____

_____

_____

PROPOSAL NO. _____

DATE _____

PROJECT _____

LOCATION _____

CONSTRUCTION TO BEGIN _____

COMPLETION DATE _____

Gentlemen:

The undersigned proposes to furnish all materials and necessary equipment, and perform all labor necessary to complete the following work:

_____

_____

_____

_____

_____

_____

_____

_____

_____

_____

_____

_____

All of the above work to be completed in a substantial and workmanlike manner

☐ for the sum of_____dollars ($_____)

☐ to be paid for at actual cost of Labor, Materials and Equipment plus_____ percent (_____%)

Payments to be made as follows:_____

_____

_____The entire amount of the contract to be paid within_____ after completion.

Any alteration or deviation from the plans and specifications will be executed only upon written orders for same and will be added to or deducted from the sum quoted in this contract. All additional agreements must be in writing.

The Contractor agrees to carry Workers' Compensation and Public Liability Insurance and to pay all taxes on material and labor furnished under this contract as required by Federal laws and the laws of the State in which this work is performed.

Respectfully submitted,

Contractor _____

By_____

ACCEPTANCE

You are hereby authorized to furnish all material, equipment and labor required to complete the work described in the above proposal, for which the undersigned agrees to pay the amount stated in said proposal and according to the terms thereof.

Date _____ 20____   _____

_____

Figure R.12

Crew No.	Bare Costs		Incl. Subs O&P		Cost Per Labor-Hour	
Crew B-10D	Hr.	Daily	Hr.	Daily	Bare Costs	Incl. O&P
1 Equip. Oper. (medium)	$56.75	$454.00	$84.85	$678.80	$51.87	$77.65
.5 Laborer	42.10	168.40	63.25	253.00		
1 Dozer, 200 H.P.		1290.00		1419.00		
1 Sheepsft. Roller, Towed		430.15		473.17	143.35	157.68
12 L.H., Daily Totals		$2342.55		$2823.97	$195.21	$235.33
Crew B-13	Hr.	Daily	Hr.	Daily	Bare Costs	Incl. O&P
1 Labor Foreman (outside)	$44.10	$352.80	$66.25	$530.00	$46.04	$69.04
4 Laborers	42.10	1347.20	63.25	2024.00		
1 Equip. Oper. (crane)	59.20	473.60	88.50	708.00		
1 Equip. Oper. (oiler)	50.55	404.40	75.55	604.40		
1 Hyd. Crane, 25 Ton		581.70		639.87	10.39	11.43
56 L.H., Daily Totals		$3159.70		$4506.27	$56.42	$80.47
Crew B-87	Hr.	Daily	Hr.	Daily	Bare Costs	Incl. O&P
1 Laborer	$42.10	$336.80	$63.25	$506.00	$53.82	$80.53
4 Equip. Oper. (medium)	56.75	1816.00	84.85	2715.20		
2 Feller Bunchers, 100 H.P.		1607.10		1767.81		
1 Log Chipper, 22" Tree		770.10		847.11		
1 Dozer, 105 H.P.		609.70		670.67		
1 Chain Saw, Gas, 36" Long		46.65		51.31	75.84	83.42
40 L.H., Daily Totals		$5186.35		$6558.10	$129.66	$163.95
Crew C-6	Hr.	Daily	Hr.	Daily	Bare Costs	Incl. O&P
1 Labor Foreman (outside)	$44.10	$352.80	$66.25	$530.00	$43.74	$65.45
4 Laborers	42.10	1347.20	63.25	2024.00		
1 Cement Finisher	49.95	399.60	73.45	587.60		
2 Gas Engine Vibrators		53.10		58.41	1.11	1.22
48 L.H., Daily Totals		$2152.70		$3200.01	$44.85	$66.67
Crew Q-1	Hr.	Daily	Hr.	Daily	Bare Costs	Incl. O&P
1 Plumber	$64.45	$515.60	$96.45	$771.60	$58.00	$86.80
1 Plumber Apprentice	51.55	412.40	77.15	617.20		
16 L.H., Daily Totals		$928.00		$1388.80	$58.00	$86.80
Crew R-1A	Hr.	Daily	Hr.	Daily	Bare Costs	Incl. O&P
1 Electrician	$61.35	$490.80	$91.35	$730.80	$55.23	$82.22
1 Electrician Apprentice	49.10	392.80	73.10	584.80		
16 L.H., Daily Totals		$883.60		$1315.60	$55.23	$82.22
Crew R-1B	Hr.	Daily	Hr.	Daily	Bare Costs	Incl. O&P
1 Electrician	$61.35	$490.80	$91.35	$730.80	$53.18	$79.18
2 Electrician Apprentices	49.10	785.60	73.10	1169.60		
24 L.H., Daily Totals		$1276.40		$1900.40	$53.18	$79.18
Crew R-1C	Hr.	Daily	Hr.	Daily	Bare Costs	Incl. O&P
2 Electricians	$61.35	$981.60	$91.35	$1461.60	$55.23	$82.22
2 Electrician Apprentices	49.10	785.60	73.10	1169.60		
1 Portable cable puller, 8000 lb.		152.70		167.97	4.77	5.25
32 L.H., Daily Totals		$1919.90		$2799.17	$60.00	$87.47
Crew R-3	Hr.	Daily	Hr.	Daily	Bare Costs	Incl. O&P
1 Electrician Foreman	$61.85	$494.80	$92.10	$736.80	$61.12	$91.08
1 Electrician	61.35	490.80	91.35	730.80		
.5 Equip. Oper. (crane)	59.20	236.80	88.50	354.00		
.5 S.P. Crane, 4x4, 5 Ton		129.72		142.70	6.49	7.13
20 L.H., Daily Totals		$1352.13		$1964.30	$67.61	$98.21

Crew No.	Bare Costs		Incl. Subs O&P		Cost Per Labor-Hour	
Crew R-5	Hr.	Daily	Hr.	Daily	Bare Costs	Incl. O&P
1 Electrician Foreman	$61.85	$494.80	$92.10	$736.80	$53.61	$80.36
4 Electrician Linemen	61.35	1963.20	91.35	2923.20		
2 Electrician Operators	61.35	981.60	91.35	1461.60		
4 Electrician Groundmen	39.95	1278.40	60.95	1950.40		
1 Crew Truck		154.35		169.79		
1 Flatbed Truck, 20,000 GVW		199.45		219.40		
1 Pickup Truck, 3/4 Ton		109.90		120.89		
.2 Hyd. Crane, 55 Ton		196.30		215.93		
.2 Hyd. Crane, 12 Ton		94.35		103.79		
.2 Earth Auger, Truck-Mtd.		77.17		84.89		
1 Tractor w/Winch		368.50		405.35	13.64	15.00
88 L.H., Daily Totals		$5918.02		$8392.02	$67.25	$95.36
Crew R-6	Hr.	Daily	Hr.	Daily	Bare Costs	Incl. O&P
1 Electrician Foreman	$61.85	$494.80	$92.10	$736.80	$53.61	$80.36
4 Electrician Linemen	61.35	1963.20	91.35	2923.20		
2 Electrician Operators	61.35	981.60	91.35	1461.60		
4 Electrician Groundmen	39.95	1278.40	60.95	1950.40		
1 Crew Truck		154.35		169.79		
1 Flatbed Truck, 20,000 GVW		199.45		219.40		
1 Pickup Truck, 3/4 Ton		109.90		120.89		
.2 Hyd. Crane, 55 Ton		196.30		215.93		
.2 Hyd. Crane, 12 Ton		94.35		103.79		
.2 Earth Auger, Truck-Mtd.		77.17		84.89		
1 Tractor w/Winch		368.50		405.35		
3 Cable Trailers		648.60		713.46		
.5 Tensioning Rig		214.32		235.76		
.5 Cable Pulling Rig		1233.50		1356.85	37.46	41.21
88 L.H., Daily Totals		$8014.44		$10698.09	$91.07	$121.57
Crew R-7	Hr.	Daily	Hr.	Daily	Bare Costs	Incl. O&P
1 Electrician Foreman	$61.85	$494.80	$92.10	$736.80	$43.60	$66.14
5 Electrician Groundmen	39.95	1598.00	60.95	2438.00		
1 Crew Truck		154.35		169.79	3.22	3.54
48 L.H., Daily Totals		$2247.15		$3344.59	$46.82	$69.68
Crew R-8	Hr.	Daily	Hr.	Daily	Bare Costs	Incl. O&P
1 Electrician Foreman	$61.85	$494.80	$92.10	$736.80	$54.30	$81.34
3 Electrician Linemen	61.35	1472.40	91.35	2192.40		
2 Electrician Groundmen	39.95	639.20	60.95	975.20		
1 Pickup Truck, 3/4 Ton		109.90		120.89		
1 Crew Truck		154.35		169.79	5.51	6.06
48 L.H., Daily Totals		$2870.65		$4195.07	$59.81	$87.40
Crew R-9	Hr.	Daily	Hr.	Daily	Bare Costs	Incl. O&P
1 Electrician Foreman	$61.85	$494.80	$92.10	$736.80	$50.71	$76.24
1 Electrician Lineman	61.35	490.80	91.35	730.80		
2 Electrician Operators	61.35	981.60	91.35	1461.60		
4 Electrician Groundmen	39.95	1278.40	60.95	1950.40		
1 Pickup Truck, 3/4 Ton		109.90		120.89		
1 Crew Truck		154.35		169.79	4.13	4.54
64 L.H., Daily Totals		$3509.85		$5170.27	$54.84	$80.79
Crew R-10	Hr.	Daily	Hr.	Daily	Bare Costs	Incl. O&P
1 Electrician Foreman	$61.85	$494.80	$92.10	$736.80	$57.87	$86.41
4 Electrician Linemen	61.35	1963.20	91.35	2923.20		
1 Electrician Groundman	39.95	319.60	60.95	487.60		
1 Crew Truck		154.35		169.79		
3 Tram Cars		438.60		482.46	12.35	13.59
48 L.H., Daily Totals		$3370.55		$4799.85	$70.22	$100.00

Crew No.	Bare Costs		Incl. Subs O & P		Cost Per Labor-Hour	

Crew R-11	Hr.	Daily	Hr.	Daily	Bare Costs	Incl. O&P
1 Electrician Foreman	$61.85	$494.80	$92.10	$736.80	$58.36	$87.04
4 Electricians	61.35	1963.20	91.35	2923.20		
1 Equip. Oper. (crane)	59.20	473.60	88.50	708.00		
1 Common Laborer	42.10	336.80	63.25	506.00		
1 Crew Truck		154.35		169.79		
1 Hyd. Crane, 12 Ton		471.75		518.92	11.18	12.30
56 L.H., Daily Totals		$3894.50		$5562.71	$69.54	$99.33

Crew R-13	Hr.	Daily	Hr.	Daily	Bare Costs	Incl. O&P
1 Electrician Foreman	$61.85	$494.80	$92.10	$736.80	$59.29	$88.35
3 Electricians	61.35	1472.40	91.35	2192.40		
.25 Equip. Oper. (crane)	59.20	118.40	88.50	177.00		
1 Equipment Oiler	50.55	404.40	75.55	604.40		
.25 Hydraulic Crane, 33 Ton		236.65		260.32	5.63	6.20
42 L.H., Daily Totals		$2726.65		$3970.92	$64.92	$94.55

Crew R-15	Hr.	Daily	Hr.	Daily	Bare Costs	Incl. O&P
1 Electrician Foreman	$61.85	$494.80	$92.10	$736.80	$60.04	$89.45
4 Electricians	61.35	1963.20	91.35	2923.20		
1 Equipment Oper. (light)	53.00	424.00	79.20	633.60		
1 Telescoping Boom Lift, to 40'		283.15		311.46	5.90	6.49
48 L.H., Daily Totals		$3165.15		$4605.06	$65.94	$95.94

Crew R-18	Hr.	Daily	Hr.	Daily	Bare Costs	Incl. O&P
.25 Electrician Foreman	$61.85	$123.70	$92.10	$184.20	$53.85	$80.18
1 Electrician	61.35	490.80	91.35	730.80		
2 Electrician Apprentices	49.10	785.60	73.10	1169.60		
26 L.H., Daily Totals		$1400.10		$2084.60	$53.85	$80.18

Crew R-19	Hr.	Daily	Hr.	Daily	Bare Costs	Incl. O&P
.5 Electrician Foreman	$61.85	$247.40	$92.10	$368.40	$61.45	$91.50
2 Electricians	61.35	981.60	91.35	1461.60		
20 L.H., Daily Totals		$1229.00		$1830.00	$61.45	$91.50

Crew R-21	Hr.	Daily	Hr.	Daily	Bare Costs	Incl. O&P
1 Electrician Foreman	$61.85	$494.80	$92.10	$736.80	$61.36	$91.37
3 Electricians	61.35	1472.40	91.35	2192.40		
.1 Equip. Oper. (medium)	56.75	45.40	84.85	67.88		
.1 S.P. Crane, 4x4, 25 Ton		66.44		73.08	2.03	2.23
32.8 L.H., Daily Totals		$2079.04		$3070.16	$63.39	$93.60

Crew R-22	Hr.	Daily	Hr.	Daily	Bare Costs	Incl. O&P
.66 Electrician Foreman	$61.85	$326.57	$92.10	$486.29	$56.16	$83.62
2 Electricians	61.35	981.60	91.35	1461.60		
2 Electrician Apprentices	49.10	785.60	73.10	1169.60		
37.28 L.H., Daily Totals		$2093.77		$3117.49	$56.16	$83.62

Crew R-31	Hr.	Daily	Hr.	Daily	Bare Costs	Incl. O&P
1 Electrician	$61.35	$490.80	$91.35	$730.80	$61.35	$91.35
1 Core Drill, Electric, 2.5 H.P.		46.55		51.20	5.82	6.40
8 L.H., Daily Totals		$537.35		$782.01	$67.17	$97.75

For customer support on your Electrical Change Order Costs with RSMeans data, call 800.448.8182.

Historical Cost Indexes

The table below lists both the RSMeans® historical cost index based on Jan. 1, 1993 = 100 as well as the computed value of an index based on Jan. 1, 2020 costs. Since the Jan. 1, 2020 figure is estimated, space is left to write in the actual index figures as they become available through the quarterly *RSMeans Construction Cost Indexes*.

To compute the actual index based on Jan. 1, 2020 = 100, divide the historical cost index for a particular year by the actual Jan. 1, 2020 construction cost index. Space has been left to advance the index figures as the year progresses.

Year	Historical Cost Index Jan. 1, 1993 = 100		Current Index Based on Jan. 1, 2020 = 100		Year	Historical Cost Index Jan. 1, 1993 = 100	Current Index Based on Jan. 1, 2020 = 100		Year	Historical Cost Index Jan. 1, 1993 = 100	Current Index Based on Jan. 1, 2020 = 100	
	Est.	Actual	Est.	Actual		Actual	Est.	Actual		Actual	Est.	Actual
Oct 2020*					July 2005	151.6	63.4		July 1987	87.7	36.7	
July 2020*					2004	143.7	60.1		1986	84.2	35.2	
Apr 2020*					2003	132.0	55.2		1985	82.6	34.6	
Jan 2020*	239.1		100.0	100.0	2002	128.7	53.8		1984	82.0	34.3	
July 2019		232.2	97.1		2001	125.1	52.3		1983	80.2	33.5	
2018		222.9	93.2		2000	120.9	50.6		1982	76.1	31.8	
2017		213.6	89.3		1999	117.6	49.2		1981	70.0	29.3	
2016		207.3	86.7		1998	115.1	48.1		1980	62.9	26.3	
2015		206.2	86.2		1997	112.8	47.2		1979	57.8	24.2	
2014		204.9	85.7		1996	110.2	46.1		1978	53.5	22.4	
2013		201.2	84.1		1995	107.6	45.0		1977	49.5	20.7	
2012		194.6	81.4		1994	104.4	43.7		1976	46.9	19.6	
2011		191.2	80.0		1993	101.7	42.5		1975	44.8	18.7	
2010		183.5	76.7		1992	99.4	41.6		1974	41.4	17.3	
2009		180.1	75.3		1991	96.8	40.5		1973	37.7	15.8	
2008		180.4	75.4		1990	94.3	39.4		1972	34.8	14.6	
2007		169.4	70.8		1989	92.1	38.5		1971	32.1	13.4	
2006		162.0	67.8		1988	89.9	37.6		1970	28.7	12.0	

Adjustments to Costs

The "Historical Cost Index" can be used to convert national average building costs at a particular time to the approximate building costs for some other time.

Example:

Estimate and compare construction costs for different years in the same city.

To estimate the national average construction cost of a building in 1970, knowing that it cost $900,000 in 2020:

INDEX in 1970 = 28.7

INDEX in 2020 = 239.1

Note: The city cost indexes for Canada can be used to convert U.S. national averages to local costs in Canadian dollars.

Example:

To estimate and compare the cost of a building in Toronto, ON in 2020 with the known cost of $600,000 (US$) in New York, NY in 2020:

INDEX Toronto = 115.6

INDEX New York = 137.1

$$\frac{\text{INDEX Toronto}}{\text{INDEX New York}} \times \text{Cost New York} = \text{Cost Toronto}$$

$$\frac{115.6}{137.1} \times \$600,000 = .843 \times \$600,000 = \$505,908$$

The construction cost of the building in Toronto is $505,908 (CN$).

Time Adjustment Using the Historical Cost Indexes:

$$\frac{\text{Index for Year A}}{\text{Index for Year B}} \times \text{Cost in Year B} = \text{Cost in Year A}$$

$$\frac{\text{INDEX 1970}}{\text{INDEX 2020}} \times \text{Cost 2020} = \text{Cost 1970}$$

$$\frac{28.7}{239.1} \times \$900,000 = .120 \times \$900,000 = \$108,000$$

The construction cost of the building in 1970 was $108,000.

*Historical Cost Index updates and other resources are provided on the following website:
http://info.thegordiangroup.com/RSMeans.html

How to Use the City Cost Indexes

What you should know before you begin

RSMeans City Cost Indexes (CCI) are an extremely useful tool for when you want to compare costs from city to city and region to region.

This publication contains average construction cost indexes for 731 U.S. and Canadian cities covering over 930 three-digit zip code locations, as listed directly under each city.

Keep in mind that a City Cost Index number is a percentage ratio of a specific city's cost to the national average cost of the same item at a stated time period.

In other words, these index figures represent relative construction factors (or, if you prefer, multipliers) for material and installation costs, as well as the weighted average for Total In Place costs for each CSI MasterFormat division. Installation costs include both labor and equipment rental costs. When estimating equipment rental rates only for a specific location, use 01 54 33 EQUIPMENT RENTAL COSTS in the Reference Section.

The 30 City Average Index is the average of 30 major U.S. cities and serves as a national average.

Index figures for both material and installation are based on the 30 major city average of 100 and represent the cost relationship as of July 1, 2019. The index for each division is computed from representative material and labor quantities for that division. The weighted average for each city is a weighted total of the components listed above it. It does not include relative productivity between trades or cities.

As changes occur in local material prices, labor rates, and equipment rental rates (including fuel costs), the impact of these changes should be accurately measured by the change in the City Cost Index for each particular city (as compared to the 30 city average).

Therefore, if you know (or have estimated) building costs in one city today, you can easily convert those costs to expected building costs in another city.

In addition, by using the Historical Cost Index, you can easily convert national average building costs at a particular time to the approximate building costs for some other time. The City Cost Indexes can then be applied to calculate the costs for a particular city.

Quick calculations

Location Adjustment Using the City Cost Indexes:

$$\frac{\text{Index for City A}}{\text{Index for City B}} \times \text{Cost in City B} = \text{Cost in City A}$$

Time Adjustment for the National Average Using the Historical Cost Index:

$$\frac{\text{Index for Year A}}{\text{Index for Year B}} \times \text{Cost in Year B} = \text{Cost in Year A}$$

Adjustment from the National Average:

$$\frac{\text{Index for City A}}{100} \times \text{National Average Cost} = \text{Cost in City A}$$

Since each of the other RSMeans data sets contains many different items, any *one* item multiplied by the particular city index may give incorrect results. However, the larger the number of items compiled, the closer the results should be to actual costs for that particular city.

The City Cost Indexes for Canadian cities are calculated using Canadian material and equipment prices and labor rates in Canadian dollars. Therefore, indexes for Canadian cities can be used to convert U.S. national average prices to local costs in Canadian dollars.

How to use this section

1. Compare costs from city to city.

In using the RSMeans Indexes, remember that an index number is not a fixed number but a ratio: It's a percentage ratio of a building component's cost at any stated time to the national average cost of that same component at the same time period. Put in the form of an equation:

$$\frac{\text{Specific City Cost}}{\text{National Average Cost}} \times 100 = \text{City Index Number}$$

Therefore, when making cost comparisons between cities, do not subtract one city's index number from the index number of another city and read the result as a percentage difference. Instead, divide one city's index number by that of the other city. The resulting number may then be used as a multiplier to calculate cost differences from city to city.

The formula used to find cost differences between cities for the purpose of comparison is as follows:

$$\frac{\text{City A Index}}{\text{City B Index}} \times \text{City B Cost (Known)} = \text{City A Cost (Unknown)}$$

In addition, you can use RSMeans CCI to calculate and compare costs division by division between cities using the same basic formula. (Just be sure that you're comparing similar divisions.)

2. Compare a specific city's construction costs with the national average.

When you're studying construction location feasibility, it's advisable to compare a prospective project's cost index with an index of the national average cost.

For example, divide the weighted average index of construction costs of a specific city by that of the 30 City Average, which = 100.

$$\frac{\text{City Index}}{100} = \text{\% of National Average}$$

As a result, you get a ratio that indicates the relative cost of construction in that city in comparison with the national average.

3. Convert U.S. national average to actual costs in Canadian City.

$$\frac{\text{Index for Canadian City}}{100} \times \text{National Average Cost} = \text{Cost in Canadian City in \$ CAN}$$

4. Adjust construction cost data based on a national average.

When you use a source of construction cost data which is based on a national average (such as RSMeans cost data), it is necessary to adjust those costs to a specific location.

$$\frac{\text{City Index}}{100} \times \frac{\text{Cost Based on}}{\text{National Average Costs}} = \frac{\text{City Cost}}{\text{(Unknown)}}$$

5. When applying the City Cost Indexes to demolition projects, use the appropriate division installation index. For example, for removal of existing doors and windows, use the Division 8 (Openings) index.

What you might like to know about how we developed the Indexes

The information presented in the CCI is organized according to the Construction Specifications Institute (CSI) MasterFormat 2018 classification system.

To create a reliable index, RSMeans researched the building type most often constructed in the United States and Canada. Because it was concluded that no one type of building completely represented the building construction industry, nine different types of buildings were combined to create a composite model.

The exact material, labor, and equipment quantities are based on detailed analyses of these nine building types, and then each quantity is weighted in proportion to expected usage. These various material items, labor hours, and equipment rental rates are thus combined to form a composite building representing as closely as possible the actual usage of materials, labor, and equipment in the North American building construction industry.

The following structures were chosen to make up that composite model:

1. Factory, 1 story
2. Office, 2–4 stories
3. Store, Retail
4. Town Hall, 2–3 stories
5. High School, 2–3 stories
6. Hospital, 4–8 stories
7. Garage, Parking
8. Apartment, 1–3 stories
9. Hotel/Motel, 2–3 stories

For the purposes of ensuring the timeliness of the data, the components of the index for the composite model have been streamlined. They currently consist of:

- specific quantities of 66 commonly used construction materials;
- specific labor-hours for 21 building construction trades; and
- specific days of equipment rental for 6 types of construction equipment (normally used to install the 66 material items by the 21 trades.) Fuel costs and routine maintenance costs are included in the equipment cost.

Material and equipment price quotations are gathered quarterly from cities in the United States and Canada. These prices and the latest negotiated labor wage rates for 21 different building trades are used to compile the quarterly update of the City Cost Index.

The 30 major U.S. cities used to calculate the national average are:

Atlanta, GA	Memphis, TN
Baltimore, MD	Milwaukee, WI
Boston, MA	Minneapolis, MN
Buffalo, NY	Nashville, TN
Chicago, IL	New Orleans, LA
Cincinnati, OH	New York, NY
Cleveland, OH	Philadelphia, PA
Columbus, OH	Phoenix, AZ
Dallas, TX	Pittsburgh, PA
Denver, CO	St. Louis, MO
Detroit, MI	San Antonio, TX
Houston, TX	San Diego, CA
Indianapolis, IN	San Francisco, CA
Kansas City, MO	Seattle, WA
Los Angeles, CA	Washington, DC

What the CCI does not indicate

The weighted average for each city is a total of the divisional components weighted to reflect typical usage. It does not include the productivity variations between trades or cities.

In addition, the CCI does not take into consideration factors such as the following:

- managerial efficiency
- competitive conditions
- automation
- restrictive union practices
- unique local requirements
- regional variations due to specific building codes

UNITED STATES / ALABAMA

DIVISION		US 30 CITY AVERAGE MAT.	INST.	TOTAL	ANNISTON 362 MAT.	INST.	TOTAL	BIRMINGHAM 350-352 MAT.	INST.	TOTAL	BUTLER 369 MAT.	INST.	TOTAL	DECATUR 356 MAT.	INST.	TOTAL	DOTHAN 363 MAT.	INST.	TOTAL
015433	CONTRACTOR EQUIPMENT		100.0	100.0		101.9	101.9		104.7	104.7		99.5	99.5		101.9	101.9		99.5	99.5
0241, 31 - 34	SITE & INFRASTRUCTURE, DEMOLITION	100.0	100.0	100.0	90.9	89.5	89.9	91.2	94.5	93.5	105.4	84.4	90.9	84.5	88.1	87.0	102.9	84.8	90.4
0310	Concrete Forming & Accessories	100.0	100.0	100.0	85.6	68.7	71.2	90.6	69.7	72.8	82.3	68.8	70.8	90.9	64.6	66.9	90.3	69.1	72.2
0320	Concrete Reinforcing	100.0	100.0	100.0	88.8	71.2	80.3	94.2	71.3	83.1	93.9	70.5	82.6	88.1	67.7	78.2	93.9	71.3	83.0
0330	Cast-in-Place Concrete	100.0	100.0	100.0	82.3	67.7	76.9	103.8	69.3	90.9	80.3	67.8	75.6	101.9	67.2	89.0	80.3	67.7	75.6
03	CONCRETE	100.0	100.0	100.0	89.2	70.4	81.0	92.2	71.4	83.1	90.3	70.4	81.6	92.0	67.0	81.0	89.5	70.7	81.2
04	MASONRY	100.0	100.0	100.0	89.2	63.1	73.2	87.1	64.3	73.1	93.7	63.1	74.9	85.2	62.4	71.2	95.1	62.9	75.3
05	METALS	100.0	100.0	100.0	101.8	94.8	99.7	100.5	94.2	98.7	100.7	94.8	98.9	102.9	93.1	100.0	100.7	95.4	99.1
06	WOOD, PLASTICS & COMPOSITES	100.0	100.0	100.0	82.0	69.9	75.3	91.0	70.0	79.4	77.1	69.9	73.1	95.7	62.1	77.2	88.1	69.9	78.1
07	THERMAL & MOISTURE PROTECTION	100.0	100.0	100.0	96.0	63.0	81.6	94.0	67.6	82.5	96.1	65.9	82.9	93.1	64.5	80.7	96.0	65.7	82.8
08	OPENINGS	100.0	100.0	100.0	94.7	69.2	88.7	102.6	69.9	94.9	94.7	69.5	88.8	106.9	64.6	97.0	94.7	69.7	88.9
0920	Plaster & Gypsum Board	100.0	100.0	100.0	81.8	69.5	73.5	88.9	69.5	75.9	79.3	69.5	72.7	92.7	61.6	71.7	88.7	69.5	75.8
0950, 0980	Ceilings & Acoustic Treatment	100.0	100.0	100.0	77.1	69.5	72.0	84.8	69.5	74.5	77.1	69.5	72.0	83.2	61.6	68.6	77.1	69.5	72.0
0960	Flooring	100.0	100.0	100.0	82.0	69.1	78.3	99.6	69.1	91.0	86.5	69.1	81.6	89.9	69.1	84.1	91.6	69.1	85.3
0970, 0990	Wall Finishes & Painting/Coating	100.0	100.0	100.0	89.1	66.7	75.9	87.0	66.7	75.0	89.1	45.9	63.5	79.4	61.0	68.5	89.1	80.4	84.0
09	FINISHES	100.0	100.0	100.0	80.4	68.8	74.1	91.2	69.2	79.3	83.6	66.6	74.4	85.3	63.5	73.5	86.1	70.3	77.6
COVERS	DIVS. 10 - 14, 25, 28, 41, 43, 44, 46	100.0	100.0	100.0	100.0	74.0	94.2	100.0	83.6	96.3	100.0	74.0	94.2	100.0	72.9	94.0	100.0	74.0	94.2
21, 22, 23	FIRE SUPPRESSION, PLUMBING & HVAC	100.0	100.0	100.0	101.0	51.8	81.7	100.0	66.3	86.8	98.1	67.0	85.9	100.0	64.5	86.1	98.1	65.0	85.1
26, 27, 3370	ELECTRICAL, COMMUNICATIONS & UTIL.	100.0	100.0	100.0	97.7	59.9	79.1	98.4	59.7	79.3	99.8	61.7	81.0	94.5	63.8	79.4	98.6	76.1	87.5
MF2018	WEIGHTED AVERAGE	100.0	100.0	100.0	95.9	67.7	84.0	97.4	71.8	86.6	96.1	70.6	85.3	97.1	69.2	85.3	96.2	72.7	86.3

ALABAMA

DIVISION		EVERGREEN 364 MAT.	INST.	TOTAL	GADSDEN 359 MAT.	INST.	TOTAL	HUNTSVILLE 357-358 MAT.	INST.	TOTAL	JASPER 355 MAT.	INST.	TOTAL	MOBILE 365-366 MAT.	INST.	TOTAL	MONTGOMERY 360-361 MAT.	INST.	TOTAL
015433	CONTRACTOR EQUIPMENT		99.5	99.5		101.9	101.9		101.9	101.9		101.9	101.9		99.5	99.5		102.0	102.0
0241, 31 - 34	SITE & INFRASTRUCTURE, DEMOLITION	105.9	84.4	91.0	89.9	89.5	89.6	84.2	89.4	87.8	89.9	89.5	89.6	97.7	85.4	89.2	95.6	89.9	91.7
0310	Concrete Forming & Accessories	79.4	68.6	70.2	83.9	69.1	71.3	90.9	65.9	69.6	88.5	64.7	68.3	89.4	68.4	71.5	92.8	69.4	72.8
0320	Concrete Reinforcing	94.0	70.4	82.6	93.5	71.3	82.8	88.1	76.3	82.4	88.1	71.3	80.0	91.5	70.5	81.3	99.7	71.4	86.0
0330	Cast-in-Place Concrete	80.3	67.6	75.6	101.9	67.8	89.2	99.2	67.8	87.5	113.0	67.8	96.1	84.2	66.9	77.7	81.5	68.7	76.7
03	CONCRETE	90.7	70.2	81.7	96.5	70.7	85.1	90.8	70.1	81.7	100.1	68.7	86.3	85.5	69.9	78.7	84.1	71.1	78.4
04	MASONRY	93.8	62.9	74.8	83.8	63.1	71.1	86.3	63.3	72.2	81.4	63.1	70.2	92.3	61.4	73.3	89.4	63.2	73.3
05	METALS	100.7	94.6	98.9	100.7	95.2	99.1	102.9	96.7	101.0	100.7	95.1	99.0	102.8	94.9	100.5	101.7	94.3	99.5
06	WOOD, PLASTICS & COMPOSITES	73.8	69.9	71.6	86.7	69.9	77.4	95.7	65.3	79.0	93.0	64.1	77.1	86.7	69.9	77.4	90.3	70.0	79.1
07	THERMAL & MOISTURE PROTECTION	96.0	65.4	82.7	93.3	66.3	81.6	93.0	65.9	81.2	93.3	64.4	80.7	95.6	65.4	82.4	94.2	66.9	82.3
08	OPENINGS	94.7	69.3	88.8	103.4	69.7	95.6	106.6	68.5	97.7	103.4	66.6	94.8	97.3	69.5	90.8	96.2	69.8	90.0
0920	Plaster & Gypsum Board	78.3	69.5	72.4	85.1	69.5	74.6	92.7	64.8	73.9	89.5	63.6	72.1	85.6	69.5	74.8	85.9	69.5	74.9
0950, 0980	Ceilings & Acoustic Treatment	77.1	69.5	72.0	79.1	69.5	72.7	85.0	64.8	71.4	79.1	63.6	68.6	83.0	69.5	73.9	84.6	69.5	74.4
0960	Flooring	84.5	69.1	80.1	86.4	69.1	81.5	89.9	69.1	84.1	88.4	69.1	82.9	91.1	69.1	84.9	91.7	69.1	85.3
0970, 0990	Wall Finishes & Painting/Coating	89.1	45.9	63.5	79.4	57.4	66.4	79.4	63.7	70.1	79.4	66.7	71.9	92.3	45.9	64.8	91.1	66.7	76.7
09	FINISHES	82.9	66.5	74.1	82.9	67.8	74.7	85.7	65.9	75.0	84.0	65.4	73.9	86.2	66.2	75.4	88.3	68.9	77.8
COVERS	DIVS. 10 - 14, 25, 28, 41, 43, 44, 46	100.0	73.9	94.2	100.0	82.9	96.2	100.0	82.5	96.1	100.0	73.4	94.1	100.0	82.5	96.1	100.0	83.3	96.3
21, 22, 23	FIRE SUPPRESSION, PLUMBING & HVAC	98.1	59.8	83.0	102.3	62.3	86.6	100.0	66.9	87.0	102.3	65.7	87.9	100.0	61.4	84.8	100.0	64.3	86.0
26, 27, 3370	ELECTRICAL, COMMUNICATIONS & UTIL.	97.1	56.8	77.2	94.5	59.7	77.4	95.4	65.9	80.8	94.2	59.9	77.3	100.4	56.8	78.9	101.1	76.1	88.8
MF2018	WEIGHTED AVERAGE	95.8	68.3	84.2	97.3	70.2	85.9	97.1	71.7	86.4	97.8	69.8	86.0	96.7	68.7	84.9	96.3	73.1	86.5

ALABAMA / ALASKA

DIVISION		PHENIX CITY 368 MAT.	INST.	TOTAL	SELMA 367 MAT.	INST.	TOTAL	TUSCALOOSA 354 MAT.	INST.	TOTAL	ANCHORAGE 995-996 MAT.	INST.	TOTAL	FAIRBANKS 997 MAT.	INST.	TOTAL	JUNEAU 998 MAT.	INST.	TOTAL
015433	CONTRACTOR EQUIPMENT		99.5	99.5		99.5	99.5		101.9	101.9		110.5	110.5		112.9	112.9		110.5	110.5
0241, 31 - 34	SITE & INFRASTRUCTURE, DEMOLITION	109.8	85.5	93.0	102.7	85.5	90.8	84.8	89.5	88.0	118.4	121.7	120.7	117.8	124.5	122.5	133.6	121.7	125.4
0310	Concrete Forming & Accessories	85.6	67.6	70.2	83.5	69.1	71.2	90.8	69.2	72.4	120.5	116.2	116.8	124.7	115.0	116.4	125.7	116.2	117.6
0320	Concrete Reinforcing	93.9	66.9	80.9	93.9	71.3	83.0	88.1	71.3	80.0	148.4	120.7	135.0	147.2	120.7	134.4	138.9	120.7	130.1
0330	Cast-in-Place Concrete	80.3	67.9	75.7	80.3	67.8	75.6	103.4	67.9	90.2	105.1	117.5	109.7	109.7	115.6	111.9	116.1	117.5	116.6
03	CONCRETE	93.6	69.3	82.9	89.0	70.7	81.0	92.7	70.8	83.1	107.5	116.7	111.6	101.4	115.5	107.6	114.1	116.7	115.2
04	MASONRY	93.7	63.1	74.9	97.4	63.1	76.3	85.4	63.1	71.7	171.5	119.4	139.5	181.4	117.8	142.3	162.5	119.4	136.0
05	METALS	100.6	93.7	98.6	100.6	95.2	99.0	102.1	95.3	100.1	127.3	104.4	120.5	122.6	105.4	117.5	113.1	104.4	110.6
06	WOOD, PLASTICS & COMPOSITES	81.7	67.5	74.0	78.9	69.9	73.9	95.7	69.9	81.5	112.3	113.5	112.9	126.9	112.7	119.1	121.9	113.5	117.3
07	THERMAL & MOISTURE PROTECTION	96.5	66.1	83.2	95.9	65.9	82.8	93.1	66.3	81.4	176.9	116.9	150.8	186.2	115.8	155.6	188.4	116.9	157.3
08	OPENINGS	94.7	67.4	88.3	94.7	69.7	88.8	106.6	69.7	98.0	130.8	115.8	127.3	132.8	114.7	128.6	131.0	115.8	127.5
0920	Plaster & Gypsum Board	83.0	67.3	72.4	81.1	69.5	73.3	92.7	69.5	77.1	144.1	113.8	123.7	177.5	112.9	134.0	158.2	113.8	128.3
0950, 0980	Ceilings & Acoustic Treatment	77.1	67.3	70.5	77.1	69.5	72.0	85.0	69.5	74.6	131.3	113.8	119.5	123.9	112.9	116.5	136.1	113.8	121.0
0960	Flooring	88.4	69.1	83.0	87.0	69.1	81.9	89.9	69.1	84.1	118.3	119.4	118.6	116.7	119.4	117.4	125.0	119.4	123.5
0970, 0990	Wall Finishes & Painting/Coating	89.1	79.0	83.1	89.1	66.7	75.9	79.4	66.7	71.9	113.1	119.1	116.6	111.6	116.9	114.7	108.7	119.1	114.8
09	FINISHES	85.2	68.9	76.4	83.7	68.8	75.7	85.6	68.8	76.6	127.0	117.1	121.7	128.6	116.0	121.8	128.9	117.1	122.5
COVERS	DIVS. 10 - 14, 25, 28, 41, 43, 44, 46	100.0	82.7	96.1	100.0	70.4	93.4	100.0	82.9	96.2	100.0	114.1	103.2	100.0	113.6	103.0	100.0	114.1	103.2
21, 22, 23	FIRE SUPPRESSION, PLUMBING & HVAC	98.1	64.2	84.8	98.1	65.1	85.1	100.0	67.1	87.1	100.5	107.1	103.1	100.2	107.9	103.2	100.5	107.1	103.1
26, 27, 3370	ELECTRICAL, COMMUNICATIONS & UTIL.	99.2	65.4	82.6	98.3	76.1	87.3	95.0	59.7	77.6	117.2	109.4	113.3	124.2	109.4	116.9	110.2	109.4	109.8
MF2018	WEIGHTED AVERAGE	96.8	70.8	85.8	95.9	72.5	86.0	97.1	71.4	86.3	118.5	113.1	116.2	118.8	113.0	116.3	116.8	113.1	115.2

For customer support on your Electrical Change Order Costs with RSMeans data, call 800.448.8182.

City Cost Indexes

ALASKA / ARIZONA

DIVISION		ALASKA KETCHIKAN 999 MAT.	INST.	TOTAL	ARIZONA CHAMBERS 865 MAT.	INST.	TOTAL	FLAGSTAFF 860 MAT.	INST.	TOTAL	GLOBE 855 MAT.	INST.	TOTAL	KINGMAN 864 MAT.	INST.	TOTAL	MESA/TEMPE 852 MAT.	INST.	TOTAL
015433	CONTRACTOR EQUIPMENT		112.9	112.9		88.2	88.2		88.2	88.2		89.3	89.3		88.2	88.2		89.3	89.3
0241, 31 - 34	SITE & INFRASTRUCTURE, DEMOLITION	169.1	124.6	138.4	71.5	90.1	84.4	91.2	90.1	90.4	109.0	91.0	96.5	71.4	90.1	84.3	97.8	91.0	93.1
0310	Concrete Forming & Accessories	116.3	116.2	116.2	99.2	70.3	74.6	104.6	67.4	72.8	93.5	70.4	73.8	97.4	66.1	70.7	96.5	70.5	74.3
0320	Concrete Reinforcing	110.6	120.7	115.5	105.5	76.9	91.7	105.4	76.9	91.6	110.5	76.9	94.2	105.7	76.9	91.7	111.2	76.9	94.6
0330	Cast-in-Place Concrete	221.8	117.3	182.9	89.8	67.3	81.4	89.8	67.4	81.5	83.3	67.0	77.2	89.5	67.3	81.2	84.0	67.0	77.7
03	CONCRETE	170.9	116.7	147.1	93.4	70.3	83.3	114.1	69.0	94.3	101.9	70.3	88.0	93.1	68.4	82.3	93.1	70.4	83.1
04	MASONRY	188.6	119.4	146.1	96.0	59.4	73.5	96.1	59.4	73.6	101.7	59.3	75.7	96.0	59.4	73.5	101.9	59.3	75.8
05	METALS	122.8	105.4	117.7	103.0	71.6	93.8	103.5	71.7	94.2	105.8	72.5	96.0	103.7	71.6	94.3	106.2	72.6	96.3
06	WOOD, PLASTICS & COMPOSITES	117.1	113.5	115.1	101.8	73.0	85.9	108.2	68.8	86.5	92.7	73.1	81.9	96.4	67.3	80.4	96.7	73.1	83.7
07	THERMAL & MOISTURE PROTECTION	191.6	116.4	158.9	99.9	69.7	86.8	101.8	69.3	87.7	99.1	68.3	85.7	99.9	69.1	86.5	99.0	68.3	85.6
08	OPENINGS	128.6	115.8	125.6	105.8	69.3	97.3	105.9	70.1	97.4	94.9	69.4	88.9	106.0	67.9	97.1	94.9	69.4	89.0
0920	Plaster & Gypsum Board	161.0	113.8	129.2	92.8	72.5	79.1	96.5	68.2	77.5	88.2	72.5	77.6	84.2	66.6	72.4	91.7	72.5	78.7
0950, 0980	Ceilings & Acoustic Treatment	118.4	113.8	115.3	106.8	72.5	83.6	107.6	68.2	81.1	95.5	72.5	80.0	107.6	66.6	80.0	95.5	72.5	80.0
0960	Flooring	116.4	119.4	117.2	89.7	64.5	82.6	92.0	64.5	84.3	98.6	64.5	89.0	88.4	64.5	81.6	100.4	64.5	90.3
0970, 0990	Wall Finishes & Painting/Coating	111.6	119.1	116.0	87.9	58.6	70.5	87.9	58.6	70.5	90.6	58.6	71.6	87.9	58.6	70.5	90.6	58.6	71.6
09	FINISHES	129.2	117.1	122.7	91.3	68.1	78.8	94.5	65.7	78.9	96.6	68.2	81.2	89.9	64.8	76.3	96.4	68.2	81.2
COVERS	DIVS. 10 - 14, 25, 28, 41, 43, 44, 46	100.0	114.2	103.2	100.0	81.8	95.9	100.0	81.3	95.8	100.0	82.1	96.0	100.0	81.2	95.8	100.0	82.1	96.0
21, 22, 23	FIRE SUPPRESSION, PLUMBING & HVAC	98.1	107.2	101.7	97.7	76.8	89.5	100.3	77.0	91.1	96.6	76.9	88.9	97.7	76.8	89.5	100.1	77.0	91.1
26, 27, 3370	ELECTRICAL, COMMUNICATIONS & UTIL.	124.2	109.4	116.9	102.3	66.5	84.7	101.3	60.7	81.3	97.1	62.9	80.2	102.3	62.9	82.9	94.2	62.9	78.7
MF2018	WEIGHTED AVERAGE	128.4	113.4	122.1	98.3	71.8	87.1	102.4	70.5	88.9	99.5	71.4	87.6	98.3	70.4	86.5	98.7	71.5	87.2

ARIZONA / ARKANSAS

DIVISION		PHOENIX 850,853 MAT.	INST.	TOTAL	PRESCOTT 863 MAT.	INST.	TOTAL	SHOW LOW 859 MAT.	INST.	TOTAL	TUCSON 856 - 857 MAT.	INST.	TOTAL	BATESVILLE 725 MAT.	INST.	TOTAL	CAMDEN 717 MAT.	INST.	TOTAL
015433	CONTRACTOR EQUIPMENT		92.5	92.5		88.2	88.2		89.3	89.3		89.3	89.3		89.7	89.7		89.7	89.7
0241, 31 - 34	SITE & INFRASTRUCTURE, DEMOLITION	98.3	94.3	95.6	78.4	90.1	86.5	111.4	91.0	97.3	92.9	91.0	91.6	73.0	85.7	81.8	77.2	85.8	83.1
0310	Concrete Forming & Accessories	101.2	70.8	75.3	100.8	70.3	74.8	100.6	70.5	74.9	97.1	67.5	71.8	82.5	61.4	64.5	78.6	61.9	64.4
0320	Concrete Reinforcing	109.1	76.9	93.6	105.4	76.9	91.6	111.2	76.9	94.6	92.3	76.9	84.9	84.4	69.4	77.1	91.0	69.4	80.5
0330	Cast-in-Place Concrete	83.6	67.8	77.7	89.8	67.3	81.4	83.3	67.0	77.3	86.5	67.0	79.2	72.8	74.6	73.5	76.4	74.6	75.7
03	CONCRETE	95.5	70.8	84.6	99.0	70.3	86.4	104.2	70.3	89.3	91.6	69.0	81.7	73.2	68.1	70.9	75.7	68.3	72.4
04	MASONRY	94.8	62.1	74.7	96.1	59.4	73.6	101.8	59.3	75.7	88.9	59.3	70.7	94.1	60.8	73.6	104.7	60.8	77.7
05	METALS	107.8	73.7	97.8	103.5	71.6	94.2	105.6	72.5	95.9	107.0	72.6	96.9	94.7	76.0	89.2	101.1	76.1	93.8
06	WOOD, PLASTICS & COMPOSITES	102.1	72.0	85.5	103.3	73.0	86.6	101.2	73.1	85.7	96.9	69.0	81.6	91.1	63.5	75.9	89.2	64.0	75.4
07	THERMAL & MOISTURE PROTECTION	100.3	69.6	86.9	100.5	69.7	87.1	99.3	68.3	85.8	99.9	67.9	86.0	102.7	61.9	84.9	97.5	61.9	82.0
08	OPENINGS	100.7	71.4	93.9	105.9	69.3	97.4	94.2	71.1	88.8	91.4	70.2	86.5	100.0	61.0	90.9	103.1	61.3	93.4
0920	Plaster & Gypsum Board	99.4	71.2	80.4	93.3	72.5	79.3	93.5	72.5	79.4	96.3	68.2	77.4	73.7	62.9	66.5	84.7	63.5	70.4
0950, 0980	Ceilings & Acoustic Treatment	106.5	71.2	82.7	106.0	72.5	83.4	95.5	72.5	80.0	96.4	68.2	77.4	88.1	62.9	71.1	86.3	63.5	70.9
0960	Flooring	102.4	64.5	91.7	90.6	64.5	83.2	102.2	64.5	91.6	91.8	64.8	84.3	81.8	70.9	78.7	86.1	70.9	81.9
0970, 0990	Wall Finishes & Painting/Coating	95.3	57.2	72.8	87.9	58.6	70.5	90.6	58.6	71.6	91.3	58.6	71.9	92.2	54.5	69.9	88.8	54.5	68.5
09	FINISHES	101.0	68.1	83.2	92.0	68.1	79.1	98.6	68.2	82.2	94.5	65.9	79.0	76.3	62.6	68.9	79.8	62.9	70.7
COVERS	DIVS. 10 - 14, 25, 28, 41, 43, 44, 46	100.0	83.2	96.2	100.0	81.8	95.9	100.0	82.1	96.0	100.0	81.7	95.9	100.0	66.4	92.5	100.0	66.5	92.5
21, 22, 23	FIRE SUPPRESSION, PLUMBING & HVAC	100.0	78.5	91.6	100.3	76.8	91.0	96.6	76.9	88.9	100.1	74.8	90.2	96.5	51.2	78.7	96.4	57.5	81.2
26, 27, 3370	ELECTRICAL, COMMUNICATIONS & UTIL.	99.8	60.7	80.5	101.0	62.9	82.2	94.5	62.9	78.9	96.2	60.6	78.7	95.5	60.5	78.3	95.8	58.1	77.2
MF2018	WEIGHTED AVERAGE	100.6	72.3	88.6	99.9	71.3	87.8	99.7	71.5	87.8	97.7	70.2	86.1	91.6	63.6	79.8	94.0	64.7	81.6

ARKANSAS

DIVISION		FAYETTEVILLE 727 MAT.	INST.	TOTAL	FORT SMITH 729 MAT.	INST.	TOTAL	HARRISON 726 MAT.	INST.	TOTAL	HOT SPRINGS 719 MAT.	INST.	TOTAL	JONESBORO 724 MAT.	INST.	TOTAL	LITTLE ROCK 720 - 722 MAT.	INST.	TOTAL
015433	CONTRACTOR EQUIPMENT		89.7	89.7		89.7	89.7		89.7	89.7		89.7	89.7		112.7	112.7		92.4	92.4
0241, 31 - 34	SITE & INFRASTRUCTURE, DEMOLITION	72.5	85.7	81.6	77.7	85.5	83.1	77.7	85.7	83.3	80.2	85.7	84.0	96.7	102.4	100.7	85.3	90.3	88.8
0310	Concrete Forming & Accessories	78.4	61.8	64.3	95.7	61.6	66.7	86.5	61.8	65.4	76.3	61.8	64.0	85.7	61.9	65.4	93.1	62.4	66.9
0320	Concrete Reinforcing	84.4	62.1	73.6	85.4	66.3	76.2	84.0	69.4	76.9	89.3	69.4	79.7	81.5	65.1	73.6	90.4	69.5	80.3
0330	Cast-in-Place Concrete	72.9	74.6	73.5	83.3	75.5	80.4	80.8	74.5	78.5	78.1	74.6	76.8	79.4	75.8	78.0	80.7	76.9	79.3
03	CONCRETE	72.9	67.0	70.3	80.1	68.0	74.8	79.7	68.2	74.6	78.9	68.2	74.2	77.3	69.0	73.7	81.0	69.3	75.9
04	MASONRY	85.0	60.8	70.1	91.6	60.8	72.7	94.3	60.8	73.7	78.6	60.8	67.7	86.5	60.8	70.7	87.7	62.1	72.0
05	METALS	94.7	73.6	88.5	96.9	74.9	90.4	95.8	76.0	90.0	101.1	76.1	93.7	91.3	89.4	90.8	97.2	76.1	91.0
06	WOOD, PLASTICS & COMPOSITES	87.7	64.0	74.7	107.1	64.0	83.4	96.8	64.0	78.7	86.7	64.0	74.2	95.2	64.4	78.3	98.4	64.2	79.5
07	THERMAL & MOISTURE PROTECTION	103.5	61.9	85.4	103.8	61.7	85.5	103.0	61.9	85.1	97.8	61.9	82.2	108.4	61.9	88.2	98.5	63.1	83.1
08	OPENINGS	100.0	59.6	90.6	102.0	60.1	92.3	100.8	61.3	91.6	103.1	61.3	93.3	105.6	60.5	95.1	95.6	60.7	87.5
0920	Plaster & Gypsum Board	73.0	63.5	66.6	79.7	63.5	68.8	78.8	63.5	68.5	83.4	63.5	70.0	87.3	63.5	71.2	92.0	63.5	72.8
0950, 0980	Ceilings & Acoustic Treatment	88.1	63.5	71.5	89.8	63.5	72.0	89.8	63.5	72.0	86.3	63.5	70.9	92.1	63.5	72.8	87.3	63.5	71.2
0960	Flooring	78.9	70.9	76.7	88.3	70.9	83.4	84.1	70.9	80.4	85.1	70.9	81.1	59.3	70.9	62.6	89.2	81.0	86.9
0970, 0990	Wall Finishes & Painting/Coating	92.2	54.5	69.9	92.2	52.1	68.5	92.2	54.5	69.9	88.8	54.5	68.5	81.7	54.5	65.6	92.8	52.4	68.9
09	FINISHES	75.5	62.9	68.7	79.6	62.6	70.4	78.3	62.9	70.0	79.6	62.9	70.6	74.6	63.2	68.4	85.3	64.9	74.3
COVERS	DIVS. 10 - 14, 25, 28, 41, 43, 44, 46	100.0	66.1	92.4	100.0	77.7	95.0	100.0	58.5	93.0	100.0	66.2	92.5	100.0	67.0	92.6	100.0	78.2	95.1
21, 22, 23	FIRE SUPPRESSION, PLUMBING & HVAC	96.5	60.0	82.2	100.0	48.0	79.6	96.5	49.1	77.8	96.4	51.5	78.8	100.4	51.2	81.1	99.9	49.1	79.9
26, 27, 3370	ELECTRICAL, COMMUNICATIONS & UTIL.	90.0	53.9	72.2	93.1	57.0	75.3	94.2	57.5	76.1	97.7	65.2	81.7	99.0	62.3	81.0	100.1	66.7	83.7
MF2018	WEIGHTED AVERAGE	90.5	64.1	79.3	94.1	62.6	80.8	92.9	62.8	80.2	93.5	64.4	81.2	93.8	66.6	82.3	94.5	65.4	82.2

ARKANSAS / CALIFORNIA

DIVISION		PINE BLUFF 716 MAT.	INST.	TOTAL	RUSSELLVILLE 728 MAT.	INST.	TOTAL	TEXARKANA 718 MAT.	INST.	TOTAL	WEST MEMPHIS 723 MAT.	INST.	TOTAL	ALHAMBRA 917-918 MAT.	INST.	TOTAL	ANAHEIM 928 MAT.	INST.	TOTAL
015433	CONTRACTOR EQUIPMENT		89.7	89.7		89.7	89.7		90.9	90.9		112.7	112.7		95.4	95.4		99.2	99.2
0241, 31 - 34	SITE & INFRASTRUCTURE, DEMOLITION	82.3	85.6	84.6	74.2	85.7	82.2	92.9	87.8	89.3	103.1	102.8	102.9	100.8	105.3	103.9	97.7	105.5	103.1
0310	Concrete Forming & Accessories	76.1	61.9	64.0	83.1	61.7	64.8	81.4	61.5	64.5	90.6	62.2	66.4	114.4	139.5	135.8	102.9	139.7	134.3
0320	Concrete Reinforcing	90.9	69.4	80.5	84.9	69.4	77.4	90.5	69.3	80.2	81.5	63.3	72.7	105.0	133.6	118.9	95.2	133.5	113.8
0330	Cast-in-Place Concrete	78.1	74.6	76.8	76.3	74.5	75.6	85.2	74.4	81.2	83.2	75.9	80.5	83.3	127.8	99.9	86.7	131.0	103.2
03	CONCRETE	79.7	68.3	74.7	76.0	68.1	72.6	77.8	68.1	73.5	83.8	68.8	77.2	94.2	133.1	111.3	96.2	134.3	112.9
04	MASONRY	110.7	60.8	80.0	90.9	60.8	72.4	92.6	60.8	73.0	75.3	60.8	66.4	106.9	139.4	126.9	78.2	138.0	115.0
05	METALS	101.9	76.1	94.3	94.7	75.9	89.1	93.8	75.8	88.5	90.4	89.1	90.0	79.9	113.2	89.7	107.7	114.2	109.6
06	WOOD, PLASTICS & COMPOSITES	86.4	64.0	74.1	92.3	64.0	76.8	93.4	64.0	77.2	100.7	64.4	80.7	99.3	138.0	120.6	97.5	138.3	119.9
07	THERMAL & MOISTURE PROTECTION	97.9	62.0	82.3	103.7	61.9	85.5	98.5	61.9	82.6	108.9	61.9	88.4	104.2	131.8	116.2	108.9	134.9	120.2
08	OPENINGS	104.3	60.8	94.1	100.0	61.3	91.0	109.2	61.3	99.0	103.2	60.1	93.1	87.8	137.4	99.4	103.9	137.6	111.8
0920	Plaster & Gypsum Board	83.1	63.5	69.9	73.7	63.5	66.8	85.7	63.5	70.7	89.2	63.5	71.9	94.5	139.3	124.6	112.7	139.3	130.6
0950, 0980	Ceilings & Acoustic Treatment	86.3	63.5	70.9	88.1	63.5	71.5	89.6	63.5	72.0	90.2	63.5	72.2	111.0	139.3	130.1	109.3	139.3	129.5
0960	Flooring	84.9	70.9	80.9	81.3	70.9	78.4	86.9	43.4	74.7	61.6	70.9	64.2	104.8	121.1	109.4	99.2	121.1	105.3
0970, 0990	Wall Finishes & Painting/Coating	88.8	52.1	67.1	92.2	54.5	69.9	88.8	54.5	68.5	81.7	54.5	65.6	103.9	120.3	113.6	92.6	120.3	109.0
09	FINISHES	79.6	62.6	70.4	76.4	62.9	69.1	81.7	57.2	68.4	75.9	63.2	69.0	101.8	134.1	119.3	99.4	134.2	118.3
COVERS	DIVS. 10 - 14, 25, 28, 41, 43, 44, 46	100.0	77.7	95.0	100.0	66.5	92.5	100.0	66.1	92.5	100.0	67.0	92.6	100.0	119.6	104.4	100.0	120.2	104.4
21, 22, 23	FIRE SUPPRESSION, PLUMBING & HVAC	100.0	51.4	80.9	96.5	50.9	78.6	100.0	54.5	82.1	96.8	64.6	84.2	96.5	130.9	110.0	99.9	130.9	112.1
26, 27, 3370	ELECTRICAL, COMMUNICATIONS & UTIL.	95.9	57.5	77.0	93.1	57.5	75.6	97.6	57.5	77.9	100.5	65.3	83.1	121.0	129.2	125.0	91.5	113.3	102.2
MF2018	WEIGHTED AVERAGE	96.0	63.6	82.3	91.7	63.1	79.6	94.8	63.3	81.5	93.3	69.8	83.3	96.7	128.6	110.2	99.3	126.6	110.9

CALIFORNIA

DIVISION		BAKERSFIELD 932-933 MAT.	INST.	TOTAL	BERKELEY 947 MAT.	INST.	TOTAL	EUREKA 955 MAT.	INST.	TOTAL	FRESNO 936-938 MAT.	INST.	TOTAL	INGLEWOOD 903-905 MAT.	INST.	TOTAL	LONG BEACH 906-908 MAT.	INST.	TOTAL
015433	CONTRACTOR EQUIPMENT		98.7	98.7		100.2	100.2		97.1	97.1		97.3	97.3		96.9	96.9		96.9	96.9
0241, 31 - 34	SITE & INFRASTRUCTURE, DEMOLITION	96.1	106.8	103.5	109.8	106.0	107.2	109.0	102.7	104.7	98.9	103.2	101.8	87.7	102.0	97.6	94.7	102.0	99.7
0310	Concrete Forming & Accessories	105.8	139.2	134.3	113.0	168.1	160.0	111.2	154.1	147.8	102.9	153.3	145.8	107.3	139.9	135.1	102.0	139.9	134.3
0320	Concrete Reinforcing	99.8	133.4	116.0	90.7	135.2	112.2	103.7	136.6	119.6	83.3	134.2	107.9	99.8	133.7	116.2	99.0	133.7	115.8
0330	Cast-in-Place Concrete	88.1	130.2	103.8	110.4	133.5	119.0	94.1	129.6	107.3	95.5	129.2	108.0	81.5	130.2	99.6	92.7	130.2	106.7
03	CONCRETE	91.7	133.7	110.2	105.1	148.2	124.0	106.8	140.8	121.7	95.5	139.9	115.0	90.9	134.2	109.9	100.5	134.2	115.3
04	MASONRY	92.1	137.5	120.0	122.7	154.9	142.5	103.3	153.4	134.1	96.9	145.0	126.5	73.1	139.5	113.9	81.7	139.5	117.2
05	METALS	102.5	112.5	105.4	109.9	116.5	111.9	107.4	116.1	109.9	102.8	115.2	106.4	88.0	114.8	95.9	87.9	114.8	95.8
06	WOOD, PLASTICS & COMPOSITES	96.9	138.3	119.7	104.3	173.4	142.3	111.2	157.3	136.6	101.2	157.3	132.1	99.0	138.4	120.7	92.5	138.4	117.8
07	THERMAL & MOISTURE PROTECTION	108.5	124.3	115.4	112.2	155.7	131.2	112.6	151.2	129.4	98.9	132.9	113.7	107.2	133.4	118.6	107.5	133.4	118.8
08	OPENINGS	92.6	134.3	102.3	91.0	161.7	107.5	103.2	143.2	112.5	94.7	145.1	106.5	87.6	137.6	99.2	87.5	137.6	99.2
0920	Plaster & Gypsum Board	94.4	139.3	124.6	113.4	175.0	154.8	117.4	158.8	145.2	90.9	158.8	136.6	100.3	139.3	126.5	96.2	139.3	125.2
0950, 0980	Ceilings & Acoustic Treatment	100.7	139.3	126.8	112.6	175.0	151.0	113.3	158.8	144.0	98.1	158.8	139.0	114.0	139.3	131.1	114.0	139.3	131.1
0960	Flooring	102.2	121.1	107.5	124.0	143.3	129.4	103.2	143.3	114.4	100.4	121.5	106.3	114.7	121.1	116.5	111.5	121.1	114.2
0970, 0990	Wall Finishes & Painting/Coating	92.5	108.6	102.0	111.6	171.3	146.9	94.1	138.6	120.4	101.3	120.7	112.8	115.1	120.3	118.2	115.1	120.3	118.2
09	FINISHES	96.4	133.0	116.2	110.1	165.4	140.0	104.6	152.0	130.3	94.6	146.3	122.6	108.2	134.3	122.4	107.2	134.3	121.9
COVERS	DIVS. 10 - 14, 25, 28, 41, 43, 44, 46	100.0	117.7	103.9	100.0	134.0	107.6	100.0	130.8	106.9	100.0	130.8	106.9	100.0	120.5	104.6	100.0	120.5	104.6
21, 22, 23	FIRE SUPPRESSION, PLUMBING & HVAC	100.1	128.8	111.4	96.7	169.6	125.3	96.4	130.6	109.9	100.1	130.1	111.9	96.1	131.0	109.8	96.1	131.0	109.8
26, 27, 3370	ELECTRICAL, COMMUNICATIONS & UTIL.	107.6	108.1	107.8	99.9	161.8	130.4	98.4	124.9	111.5	97.0	107.4	102.2	98.5	129.2	113.6	98.3	129.2	113.5
MF2018	WEIGHTED AVERAGE	98.9	124.6	109.8	103.0	151.9	123.7	102.5	134.1	115.9	98.5	129.4	111.5	93.8	128.8	108.6	95.4	128.8	109.5

CALIFORNIA

DIVISION		LOS ANGELES 900-902 MAT.	INST.	TOTAL	MARYSVILLE 959 MAT.	INST.	TOTAL	MODESTO 953 MAT.	INST.	TOTAL	MOJAVE 935 MAT.	INST.	TOTAL	OAKLAND 946 MAT.	INST.	TOTAL	OXNARD 930 MAT.	INST.	TOTAL
015433	CONTRACTOR EQUIPMENT		103.3	103.3		97.1	97.1		97.1	97.1		97.3	97.3		100.2	100.2		96.3	96.3
0241, 31 - 34	SITE & INFRASTRUCTURE, DEMOLITION	93.7	109.2	104.4	105.6	102.6	103.5	100.6	102.6	102.0	91.9	103.4	99.8	115.2	106.0	108.8	99.2	101.6	100.8
0310	Concrete Forming & Accessories	105.4	140.1	135.0	101.6	153.6	145.9	98.1	153.7	145.5	113.6	139.3	135.5	102.3	168.2	158.4	105.5	139.8	134.7
0320	Concrete Reinforcing	100.7	133.7	116.6	103.7	134.1	118.4	107.4	134.1	120.3	100.4	133.4	116.4	92.8	136.9	114.1	98.6	133.5	115.5
0330	Cast-in-Place Concrete	84.1	131.4	101.7	105.2	129.2	114.1	94.2	129.3	107.2	82.6	130.1	100.2	104.8	133.5	115.4	96.1	130.4	108.9
03	CONCRETE	98.3	134.7	114.3	107.1	140.0	121.6	98.2	140.1	116.6	87.4	133.8	107.8	106.3	148.5	124.8	95.5	134.1	112.5
04	MASONRY	88.6	139.5	119.9	104.2	142.9	128.0	101.5	142.9	126.9	94.9	137.4	121.0	131.5	154.9	145.9	97.9	136.8	121.8
05	METALS	94.3	115.0	100.4	106.9	114.7	109.2	104.6	114.8	106.9	100.0	113.4	104.0	105.2	117.2	108.7	98.0	113.9	102.7
06	WOOD, PLASTICS & COMPOSITES	105.6	138.6	123.7	98.3	157.3	130.8	93.9	157.3	128.8	100.8	138.4	121.5	92.8	173.4	137.2	95.6	138.4	119.1
07	THERMAL & MOISTURE PROTECTION	103.4	133.9	116.7	112.0	138.9	123.7	111.6	139.5	123.7	105.6	123.7	113.5	110.8	155.7	130.4	108.6	133.3	119.3
08	OPENINGS	98.1	136.9	107.2	102.5	145.6	112.6	101.3	147.4	112.1	89.8	134.3	100.2	91.1	162.1	107.7	92.4	137.6	102.9
0920	Plaster & Gypsum Board	95.6	139.3	125.0	110.0	158.8	142.8	112.7	158.8	143.7	100.5	139.3	126.6	107.6	175.0	152.9	94.5	139.3	124.7
0950, 0980	Ceilings & Acoustic Treatment	121.4	139.3	133.5	112.5	158.8	143.7	109.3	158.8	142.6	99.4	139.3	126.3	103.8	175.0	151.8	100.8	139.3	126.8
0960	Flooring	112.5	122.2	115.3	98.8	119.5	104.6	99.2	133.0	108.7	101.3	121.1	106.8	117.0	143.3	124.4	93.8	121.1	101.5
0970, 0990	Wall Finishes & Painting/Coating	111.6	120.3	116.7	94.1	134.1	117.8	94.1	138.6	120.4	87.2	109.6	100.5	111.6	171.3	146.9	87.2	114.8	103.6
09	FINISHES	109.4	134.7	123.1	101.7	147.4	126.4	101.0	150.2	127.7	95.0	133.2	115.6	108.1	165.4	139.1	92.5	133.7	114.8
COVERS	DIVS. 10 - 14, 25, 28, 41, 43, 44, 46	100.0	121.0	104.7	100.0	130.8	106.9	100.0	130.8	106.9	100.0	114.8	104.0	100.0	134.0	107.6	100.0	120.4	104.5
21, 22, 23	FIRE SUPPRESSION, PLUMBING & HVAC	99.9	131.0	112.1	96.4	130.0	109.6	99.9	131.7	112.4	96.5	128.8	109.2	100.2	169.6	127.5	100.0	131.0	112.2
26, 27, 3370	ELECTRICAL, COMMUNICATIONS & UTIL.	96.9	131.0	113.7	95.1	120.7	107.7	97.5	109.9	103.6	95.5	108.1	101.7	99.1	161.8	130.0	101.4	117.9	109.5
MF2018	WEIGHTED AVERAGE	98.5	129.8	111.7	101.7	131.3	114.2	100.6	130.6	113.3	95.4	124.4	107.7	103.4	152.0	123.9	98.0	126.7	110.1

For customer support on your Electrical Change Order Costs with RSMeans data, call 800.448.8182.

City Cost Indexes

CALIFORNIA

	DIVISION	PALM SPRINGS 922 MAT.	INST.	TOTAL	PALO ALTO 943 MAT.	INST.	TOTAL	PASADENA 910-912 MAT.	INST.	TOTAL	REDDING 960 MAT.	INST.	TOTAL	RICHMOND 948 MAT.	INST.	TOTAL	RIVERSIDE 925 MAT.	INST.	TOTAL
015433	CONTRACTOR EQUIPMENT		98.1	98.1		100.2	100.2		95.4	95.4		97.1	97.1		100.2	100.2		98.1	98.1
0241, 31 - 34	SITE & INFRASTRUCTURE, DEMOLITION	89.3	103.6	99.2	106.1	106.0	106.0	97.5	105.3	102.9	122.2	102.6	108.7	114.3	106.0	108.5	96.2	103.6	101.3
0310	Concrete Forming & Accessories	99.8	136.1	130.7	100.5	168.2	158.2	103.4	139.6	134.2	105.0	153.6	146.5	115.5	167.8	160.1	103.5	139.7	134.4
0320	Concrete Reinforcing	109.0	133.5	120.9	90.7	136.7	113.0	106.0	133.6	119.3	137.8	134.1	136.0	90.7	136.7	112.9	105.9	133.5	119.3
0330	Cast-in-Place Concrete	82.7	131.0	100.6	93.4	133.5	108.4	79.1	127.8	97.2	107.2	129.3	115.4	107.4	133.3	117.1	89.9	131.0	105.2
03	CONCRETE	90.5	132.7	109.0	95.3	148.5	118.7	89.9	133.1	108.9	114.0	140.1	125.4	108.0	148.2	125.7	96.4	134.3	113.1
04	MASONRY	76.5	137.7	114.1	105.1	151.5	133.6	93.3	139.4	121.6	131.9	142.9	138.7	122.5	151.5	140.3	77.5	137.7	114.5
05	METALS	108.3	114.1	110.0	102.7	116.9	106.9	80.0	113.2	89.7	103.9	114.7	107.1	102.8	116.6	106.8	107.7	114.2	109.6
06	WOOD, PLASTICS & COMPOSITES	92.1	133.5	114.9	90.4	173.4	136.1	85.4	138.0	114.4	107.8	157.3	135.0	107.5	173.4	143.8	97.5	138.3	119.9
07	THERMAL & MOISTURE PROTECTION	108.1	134.3	119.5	110.0	155.0	129.6	103.8	131.8	116.0	128.2	138.9	132.9	110.6	154.8	129.8	109.0	134.8	120.2
08	OPENINGS	99.9	134.9	108.1	91.1	160.2	107.2	87.8	137.4	99.4	115.9	145.6	122.8	91.1	160.2	107.2	102.6	137.6	110.8
0920	Plaster & Gypsum Board	107.2	134.4	125.5	106.0	175.0	152.4	88.8	139.3	122.8	112.6	158.8	143.7	114.5	175.0	155.2	111.9	139.3	130.4
0950, 0980	Ceilings & Acoustic Treatment	106.0	134.4	125.1	102.0	175.0	151.2	111.0	139.3	130.1	148.8	158.8	155.5	102.0	175.0	151.2	114.2	139.3	131.1
0960	Flooring	101.5	121.1	107.0	115.6	143.3	123.4	98.5	121.1	104.9	87.8	128.6	99.3	126.5	143.3	131.3	102.9	121.1	108.0
0970, 0990	Wall Finishes & Painting/Coating	91.1	120.3	108.4	111.6	171.3	146.9	103.9	120.3	113.6	105.2	134.1	122.3	111.6	171.3	146.9	91.1	120.3	108.4
09	FINISHES	98.0	131.4	116.1	106.5	165.4	138.4	99.0	134.1	118.0	107.2	149.0	129.8	111.6	165.4	140.7	101.0	134.2	119.0
COVERS	DIVS. 10 - 14, 25, 28, 41, 43, 44, 46	100.0	119.6	104.4	100.0	134.0	107.6	100.0	119.6	104.4	100.0	108.8	106.9	100.0	133.8	107.5	100.0	119.3	104.3
21, 22, 23	FIRE SUPPRESSION, PLUMBING & HVAC	96.4	130.9	109.9	96.7	170.6	125.7	96.5	130.9	110.0	100.3	130.1	112.0	96.7	164.0	123.1	99.9	130.9	112.1
26, 27, 3370	ELECTRICAL, COMMUNICATIONS & UTIL.	94.6	110.8	102.6	99.0	176.6	137.2	117.5	129.2	123.2	101.3	120.8	110.9	99.6	133.2	116.2	91.3	114.4	102.7
MF2018	WEIGHTED AVERAGE	97.4	125.3	109.2	99.2	153.8	122.3	94.8	128.6	109.1	107.6	131.5	117.7	102.4	146.2	120.9	99.3	126.6	110.8

CALIFORNIA

	DIVISION	SACRAMENTO 942, 956-958 MAT.	INST.	TOTAL	SALINAS 939 MAT.	INST.	TOTAL	SAN BERNARDINO 923-924 MAT.	INST.	TOTAL	SAN DIEGO 919-921 MAT.	INST.	TOTAL	SAN FRANCISCO 940-941 MAT.	INST.	TOTAL	SAN JOSE 951 MAT.	INST.	TOTAL
015433	CONTRACTOR EQUIPMENT		99.0	99.0		97.3	97.3		98.1	98.1		102.2	102.2		109.8	109.8		98.7	98.7
0241, 31 - 34	SITE & INFRASTRUCTURE, DEMOLITION	95.6	111.3	106.4	112.3	103.2	106.0	76.1	103.6	95.1	105.7	107.8	107.2	116.4	113.1	114.1	131.6	97.9	108.3
0310	Concrete Forming & Accessories	100.5	156.0	147.8	108.9	156.6	149.6	107.1	139.7	134.9	103.8	129.1	125.4	106.1	168.3	159.1	103.9	167.9	158.5
0320	Concrete Reinforcing	86.0	134.4	109.4	99.1	134.6	116.3	105.9	133.5	119.3	102.6	133.4	117.5	105.6	132.1	118.4	94.7	135.0	114.2
0330	Cast-in-Place Concrete	88.6	130.2	104.1	95.0	129.6	107.8	62.1	131.0	87.8	90.5	124.5	103.2	116.9	133.3	123.0	109.4	132.4	118.0
03	CONCRETE	97.9	141.1	116.9	105.8	141.6	121.5	71.1	134.3	98.9	100.2	127.3	112.1	117.8	148.1	131.1	103.9	148.1	123.3
04	MASONRY	106.1	145.9	130.6	95.1	148.4	127.9	83.6	137.7	116.8	86.2	133.4	115.2	140.9	154.6	149.3	131.5	151.5	143.8
05	METALS	100.8	109.8	103.4	102.7	115.8	106.5	107.7	114.2	109.6	94.8	114.1	100.5	110.7	122.4	114.2	101.8	121.6	107.6
06	WOOD, PLASTICS & COMPOSITES	87.1	160.2	127.3	100.6	160.2	133.4	101.2	138.3	121.6	95.0	126.1	112.1	95.1	173.4	138.2	106.0	173.2	143.0
07	THERMAL & MOISTURE PROTECTION	122.2	142.5	131.0	106.2	145.4	123.3	107.1	134.8	119.2	108.2	121.5	114.0	116.5	155.7	133.5	108.0	155.1	128.5
08	OPENINGS	103.9	149.0	114.4	93.5	154.4	107.7	100.0	137.6	108.7	99.8	127.8	106.3	99.5	159.0	113.3	92.6	161.5	108.7
0920	Plaster & Gypsum Board	103.4	161.5	142.5	95.4	161.8	140.1	113.3	139.3	130.8	94.4	126.7	116.1	105.8	175.0	152.4	108.3	175.0	153.2
0950, 0980	Ceilings & Acoustic Treatment	102.0	161.5	142.1	99.4	161.8	141.5	109.3	139.3	129.5	121.1	126.7	124.8	111.7	175.0	154.4	109.3	175.0	153.6
0960	Flooring	115.8	128.6	119.4	95.9	143.3	109.2	105.0	121.1	109.5	108.9	121.1	112.3	117.5	143.3	124.7	92.4	143.3	106.7
0970, 0990	Wall Finishes & Painting/Coating	107.9	134.1	123.4	88.1	171.3	137.3	91.1	120.3	108.4	101.5	118.0	111.2	109.6	174.8	148.2	94.4	171.3	139.9
09	FINISHES	105.3	150.7	129.9	94.6	157.3	128.5	99.4	134.2	118.3	106.6	126.1	117.2	109.9	165.8	140.1	99.9	165.2	135.3
COVERS	DIVS. 10 - 14, 25, 28, 41, 43, 44, 46	100.0	131.7	107.1	100.0	131.2	107.0	100.0	117.7	103.9	100.0	115.7	103.5	100.0	128.4	106.3	100.0	133.4	107.5
21, 22, 23	FIRE SUPPRESSION, PLUMBING & HVAC	100.1	131.4	112.4	96.5	137.6	112.7	96.4	130.9	110.0	100.0	129.4	111.5	100.1	182.8	132.6	99.9	170.5	127.7
26, 27, 3370	ELECTRICAL, COMMUNICATIONS & UTIL.	94.8	120.8	107.6	96.5	132.1	114.0	94.6	112.6	103.5	103.8	104.3	104.0	99.2	182.8	140.4	100.7	176.6	138.1
MF2018	WEIGHTED AVERAGE	100.8	133.0	114.4	99.1	137.2	115.2	95.0	126.3	108.2	99.7	121.7	109.0	107.4	158.4	129.0	102.6	153.5	124.2

CALIFORNIA

	DIVISION	SAN LUIS OBISPO 934 MAT.	INST.	TOTAL	SAN MATEO 944 MAT.	INST.	TOTAL	SAN RAFAEL 949 MAT.	INST.	TOTAL	SANTA ANA 926-927 MAT.	INST.	TOTAL	SANTA BARBARA 931 MAT.	INST.	TOTAL	SANTA CRUZ 950 MAT.	INST.	TOTAL
015433	CONTRACTOR EQUIPMENT		97.3	97.3		100.2	100.2		99.4	99.4		98.1	98.1		97.3	97.3		98.7	98.7
0241, 31 - 34	SITE & INFRASTRUCTURE, DEMOLITION	104.4	103.4	103.7	112.1	106.0	107.9	107.6	111.2	110.1	87.8	103.6	98.7	99.1	103.4	102.1	131.3	97.7	108.1
0310	Concrete Forming & Accessories	115.3	139.7	136.1	106.3	168.2	159.0	111.5	167.8	159.5	107.4	139.7	134.9	106.1	139.7	134.8	103.9	156.9	149.0
0320	Concrete Reinforcing	100.4	133.4	116.4	90.7	136.9	113.0	91.3	135.1	112.5	109.6	133.5	121.2	98.6	133.4	115.4	117.1	134.6	125.6
0330	Cast-in-Place Concrete	102.1	130.2	112.6	104.0	133.5	115.0	121.1	132.3	125.2	79.3	131.0	98.6	95.8	130.2	108.6	108.7	131.4	117.1
03	CONCRETE	103.3	134.0	116.8	104.6	148.5	123.9	128.2	147.4	136.6	88.0	134.3	108.3	95.4	134.0	112.4	106.4	142.6	122.3
04	MASONRY	96.5	135.9	120.7	122.2	154.6	142.1	99.1	154.6	133.2	73.4	138.0	113.1	95.2	135.9	120.2	135.3	148.6	143.4
05	METALS	100.7	113.8	104.6	102.6	117.1	106.9	103.7	112.4	106.3	107.8	114.2	109.7	98.5	113.8	103.0	109.3	120.2	112.5
06	WOOD, PLASTICS & COMPOSITES	103.0	138.4	122.4	97.9	173.4	139.5	95.0	173.2	138.0	103.0	138.3	122.4	95.6	138.4	119.1	106.0	160.4	135.9
07	THERMAL & MOISTURE PROTECTION	106.4	132.6	117.8	110.4	156.4	130.4	115.0	155.4	132.5	108.4	134.9	119.9	105.8	132.6	117.5	107.6	147.9	125.1
08	OPENINGS	91.7	134.3	101.6	91.1	162.1	107.6	101.5	159.7	115.1	99.2	137.6	108.2	93.2	137.6	103.6	93.9	154.4	108.0
0920	Plaster & Gypsum Board	101.1	139.3	126.8	111.7	175.0	154.3	113.2	175.0	154.8	114.6	139.3	131.2	94.5	139.3	124.7	116.8	161.8	147.1
0950, 0980	Ceilings & Acoustic Treatment	99.4	139.3	126.3	102.0	175.0	151.2	109.3	175.0	153.6	109.3	139.3	129.5	100.8	139.3	126.8	110.9	161.8	145.2
0960	Flooring	102.0	121.1	107.4	119.5	143.3	126.2	132.6	143.3	135.6	105.6	121.1	110.0	95.0	121.1	102.4	96.6	143.3	109.7
0970, 0990	Wall Finishes & Painting/Coating	87.2	114.8	103.6	111.6	171.3	146.9	107.8	165.5	141.9	91.1	118.0	107.0	87.2	114.8	103.6	94.5	171.3	139.9
09	FINISHES	96.3	133.7	116.5	108.9	165.4	139.5	112.0	164.6	140.5	100.9	134.0	118.8	93.0	133.7	115.0	102.6	157.4	132.3
COVERS	DIVS. 10 - 14, 25, 28, 41, 43, 44, 46	100.0	126.5	105.9	100.0	134.0	107.6	100.0	133.3	107.4	100.0	120.2	104.5	100.0	119.5	104.3	100.0	131.5	107.0
21, 22, 23	FIRE SUPPRESSION, PLUMBING & HVAC	96.5	131.0	110.0	96.7	165.7	123.8	96.7	182.8	130.5	96.4	130.9	110.0	100.0	131.0	112.2	99.9	137.9	114.8
26, 27, 3370	ELECTRICAL, COMMUNICATIONS & UTIL.	95.5	110.8	103.1	99.0	168.0	133.0	96.2	124.9	110.3	94.6	111.9	103.1	94.5	113.7	103.9	99.8	132.1	115.7
MF2018	WEIGHTED AVERAGE	98.2	125.8	109.9	101.5	152.0	122.9	104.5	149.1	123.4	97.0	126.3	109.4	97.2	126.1	109.4	104.6	137.5	118.5

For customer support on your Electrical Change Order Costs with RSMeans data, call 800.448.8182.

539

CALIFORNIA / COLORADO

DIVISION		SANTA ROSA 954 MAT.	INST.	TOTAL	STOCKTON 952 MAT.	INST.	TOTAL	SUSANVILLE 961 MAT.	INST.	TOTAL	VALLEJO 945 MAT.	INST.	TOTAL	VAN NUYS 913-916 MAT.	INST.	TOTAL	ALAMOSA 811 MAT.	INST.	TOTAL
015433	CONTRACTOR EQUIPMENT		97.6	97.6		97.1	97.1		97.1	97.1		99.4	99.4		95.4	95.4		90.2	90.2
0241, 31 - 34	SITE & INFRASTRUCTURE, DEMOLITION	101.4	102.6	102.2	100.3	102.6	101.9	129.5	102.6	110.9	96.3	111.1	106.5	115.7	105.3	108.5	141.3	84.4	102.0
0310	Concrete Forming & Accessories	100.7	167.1	157.3	101.9	155.5	147.6	106.2	153.9	146.8	102.3	166.8	157.3	109.8	139.5	135.2	103.8	66.0	71.5
0320	Concrete Reinforcing	104.6	135.1	119.3	107.4	134.1	120.3	137.8	134.1	136.0	92.5	135.0	113.0	106.0	133.6	119.3	114.0	67.9	91.7
0330	Cast-in-Place Concrete	103.3	130.8	113.5	91.7	129.2	105.7	97.5	129.3	109.3	96.4	131.4	109.5	83.4	127.8	99.9	100.1	73.1	90.1
03	CONCRETE	107.1	146.9	124.6	97.3	140.9	116.5	117.0	140.2	127.2	102.4	146.6	121.8	104.4	133.1	117.0	112.5	69.4	93.6
04	MASONRY	101.8	153.5	133.6	101.4	142.9	126.9	130.4	142.9	138.1	77.4	153.5	124.1	106.9	139.4	126.9	130.1	59.5	86.7
05	METALS	108.1	117.7	110.9	103.9	114.7	107.0	102.9	114.7	106.4	103.7	111.7	106.0	79.2	113.2	89.2	105.2	78.5	97.3
06	WOOD, PLASTICS & COMPOSITES	93.6	173.0	137.3	99.5	159.9	132.8	109.6	157.6	136.0	84.8	173.2	133.4	94.1	138.0	118.3	98.5	67.4	81.4
07	THERMAL & MOISTURE PROTECTION	108.9	154.1	128.6	112.1	138.8	123.7	130.2	138.8	134.0	113.1	154.1	130.9	105.0	131.8	116.6	111.9	67.2	92.5
08	OPENINGS	100.8	161.4	114.9	101.3	147.0	112.0	116.8	145.8	123.6	103.2	159.7	116.3	87.7	137.4	99.3	96.3	68.5	89.8
0920	Plaster & Gypsum Board	109.2	175.0	153.5	112.7	161.5	145.5	113.1	159.1	144.1	107.5	175.0	152.9	92.8	139.3	124.1	80.8	66.5	71.2
0950, 0980	Ceilings & Acoustic Treatment	109.3	175.0	153.6	117.5	161.5	147.1	139.9	159.1	152.9	111.1	175.0	154.2	107.8	139.3	129.1	102.7	66.5	78.3
0960	Flooring	102.1	136.5	111.7	99.2	133.0	108.7	88.3	128.6	99.6	126.0	143.3	130.9	101.9	121.1	107.3	107.8	70.1	97.2
0970, 0990	Wall Finishes & Painting/Coating	91.1	165.5	135.2	94.1	134.1	117.8	105.2	134.1	122.3	108.7	165.5	142.3	103.9	120.3	113.6	101.7	76.9	87.0
09	FINISHES	100.0	163.0	134.1	102.6	151.3	129.0	106.7	149.2	129.7	108.4	164.4	138.7	101.2	134.1	119.0	99.8	67.5	82.3
COVERS	DIVS. 10 - 14, 25, 28, 41, 43, 44, 46	100.0	132.4	107.2	100.0	128.0	106.2	100.0	130.9	106.9	100.0	132.9	107.3	100.0	119.6	104.4	100.0	83.0	96.2
21, 22, 23	FIRE SUPPRESSION, PLUMBING & HVAC	96.4	182.2	130.1	99.9	131.7	112.4	96.8	130.1	109.9	100.2	147.2	118.7	96.5	130.9	110.0	96.5	72.4	87.0
26, 27, 3370	ELECTRICAL, COMMUNICATIONS & UTIL.	94.9	124.9	109.7	97.5	114.3	105.8	101.7	120.8	111.1	92.4	127.5	109.7	117.5	129.2	123.2	96.3	63.2	80.0
MF2018	WEIGHTED AVERAGE	101.2	148.4	121.2	100.7	131.3	113.7	107.3	131.6	117.5	100.2	141.6	117.7	97.8	128.6	110.8	103.6	70.3	89.5

COLORADO

DIVISION		BOULDER 803 MAT.	INST.	TOTAL	COLORADO SPRINGS 808-809 MAT.	INST.	TOTAL	DENVER 800-802 MAT.	INST.	TOTAL	DURANGO 813 MAT.	INST.	TOTAL	FORT COLLINS 805 MAT.	INST.	TOTAL	FORT MORGAN 807 MAT.	INST.	TOTAL
015433	CONTRACTOR EQUIPMENT		92.2	92.2		90.2	90.2		99.6	99.6		90.2	90.2		92.2	92.2		92.2	92.2
0241, 31 - 34	SITE & INFRASTRUCTURE, DEMOLITION	97.1	91.4	93.2	99.2	86.7	90.6	104.1	102.5	103.0	134.3	84.4	99.9	109.6	91.2	96.9	99.3	90.9	93.5
0310	Concrete Forming & Accessories	107.0	77.4	81.8	97.3	63.5	68.5	106.1	66.4	72.2	109.9	65.7	72.2	104.5	65.8	71.5	107.6	65.7	71.9
0320	Concrete Reinforcing	108.7	68.1	89.0	107.9	67.9	88.6	107.9	70.1	89.6	114.0	67.9	91.7	108.8	68.0	89.1	108.9	68.0	89.2
0330	Cast-in-Place Concrete	113.8	73.9	98.9	116.8	73.4	100.7	121.8	73.6	103.9	115.1	73.1	99.5	128.5	72.5	107.6	111.6	72.5	97.1
03	CONCRETE	105.2	74.9	91.9	108.7	68.4	91.0	111.4	70.2	93.3	114.2	69.3	94.5	116.3	69.2	95.6	103.7	69.1	88.5
04	MASONRY	101.6	65.4	79.4	103.7	62.3	78.3	104.9	64.1	79.9	117.5	60.6	82.5	118.9	63.1	84.6	116.7	63.1	83.8
05	METALS	94.4	77.9	89.6	97.5	77.7	91.7	99.9	79.0	93.8	105.2	78.4	97.3	95.7	77.7	90.4	94.2	77.8	89.4
06	WOOD, PLASTICS & COMPOSITES	107.1	81.4	93.0	96.6	63.6	78.5	108.4	67.1	85.7	108.4	67.4	85.8	104.6	67.0	83.9	107.1	67.0	85.0
07	THERMAL & MOISTURE PROTECTION	108.7	73.2	93.2	109.4	69.7	92.1	107.1	71.7	91.7	111.9	67.5	92.6	109.0	70.6	92.3	108.6	69.6	91.6
08	OPENINGS	95.4	76.2	90.9	99.5	66.4	91.8	102.5	68.8	94.6	103.1	68.5	95.0	95.3	68.3	89.0	95.3	68.2	89.0
0920	Plaster & Gypsum Board	116.3	81.3	92.8	100.3	62.9	75.2	111.7	66.6	81.4	95.3	66.5	75.9	110.3	66.6	80.9	116.3	66.5	82.8
0950, 0980	Ceilings & Acoustic Treatment	91.9	81.3	84.8	102.0	62.9	75.0	104.4	66.6	78.9	102.7	66.5	78.3	91.9	66.6	74.8	91.9	66.5	74.8
0960	Flooring	111.2	76.7	101.5	102.2	69.0	92.9	108.4	76.7	99.5	113.1	69.0	100.7	107.3	76.7	98.7	111.7	76.7	101.8
0970, 0990	Wall Finishes & Painting/Coating	97.1	76.9	85.1	96.8	76.9	85.0	103.4	76.9	87.7	101.7	76.9	87.0	97.1	76.9	85.1	97.1	76.9	85.1
09	FINISHES	100.6	77.7	88.2	97.4	65.3	80.0	102.5	68.9	84.3	102.4	67.4	83.5	99.2	68.6	82.6	100.7	68.5	83.3
COVERS	DIVS. 10 - 14, 25, 28, 41, 43, 44, 46	100.0	84.5	96.5	100.0	82.2	96.0	100.0	86.3	96.9	100.0	82.9	96.2	100.0	82.1	96.0	100.0	82.1	96.0
21, 22, 23	FIRE SUPPRESSION, PLUMBING & HVAC	96.5	74.8	87.9	100.1	72.2	89.1	99.9	73.8	89.7	96.5	58.7	81.6	100.0	71.7	88.9	96.5	73.5	87.4
26, 27, 3370	ELECTRICAL, COMMUNICATIONS & UTIL.	99.6	80.6	90.3	103.2	73.1	88.4	105.0	80.4	92.9	95.8	50.3	73.4	99.6	80.6	90.2	100.0	78.1	89.2
MF2018	WEIGHTED AVERAGE	98.7	77.0	89.6	101.2	71.5	88.6	103.0	75.4	91.3	103.9	65.6	87.8	102.2	73.5	90.1	99.3	73.5	88.4

COLORADO

DIVISION		GLENWOOD SPRINGS 816 MAT.	INST.	TOTAL	GOLDEN 804 MAT.	INST.	TOTAL	GRAND JUNCTION 815 MAT.	INST.	TOTAL	GREELEY 806 MAT.	INST.	TOTAL	MONTROSE 814 MAT.	INST.	TOTAL	PUEBLO 810 MAT.	INST.	TOTAL
015433	CONTRACTOR EQUIPMENT		93.3	93.3		92.2	92.2		93.3	93.3		92.2	92.2		91.7	91.7		90.2	90.2
0241, 31 - 34	SITE & INFRASTRUCTURE, DEMOLITION	150.9	91.9	110.2	110.4	91.4	97.3	133.7	92.1	105.0	95.9	91.2	92.7	144.0	87.9	105.3	125.2	84.5	97.1
0310	Concrete Forming & Accessories	100.6	65.5	70.7	99.7	66.7	71.5	108.7	66.6	72.8	102.4	65.5	71.2	100.1	65.5	70.6	106.1	63.7	70.0
0320	Concrete Reinforcing	112.8	68.0	91.2	108.9	68.0	89.2	113.2	67.8	91.2	108.7	68.0	89.0	112.7	67.8	91.0	109.0	67.9	89.1
0330	Cast-in-Place Concrete	100.1	72.1	89.6	111.7	73.9	97.6	110.8	73.5	96.9	107.3	72.5	94.3	100.1	72.6	89.8	99.4	73.7	89.8
03	CONCRETE	118.0	68.9	96.4	114.1	70.0	94.7	110.7	69.9	92.8	100.3	69.1	86.6	108.5	69.0	91.1	100.9	68.7	86.7
04	MASONRY	103.1	63.0	78.5	119.7	65.4	86.3	137.8	61.9	91.2	111.9	62.7	81.7	110.0	59.6	79.0	99.5	60.3	75.4
05	METALS	104.8	78.2	97.0	94.3	77.8	89.5	106.1	77.7	90.4	95.7	77.7	90.4	103.9	78.1	96.3	108.3	78.7	99.6
06	WOOD, PLASTICS & COMPOSITES	93.6	67.1	79.0	98.9	67.0	81.3	106.1	67.2	84.7	101.7	67.0	82.6	94.6	67.2	79.6	101.2	63.9	80.7
07	THERMAL & MOISTURE PROTECTION	111.8	68.2	92.8	109.8	71.6	93.2	110.9	69.3	92.8	108.2	70.4	91.8	112.0	67.2	92.5	110.4	67.8	91.8
08	OPENINGS	102.1	68.3	94.2	95.4	68.2	89.0	102.8	68.4	94.7	95.3	68.3	89.0	103.2	68.4	95.1	98.0	66.6	90.7
0920	Plaster & Gypsum Board	121.2	66.5	84.4	107.5	66.5	79.9	135.4	66.6	89.1	108.7	66.6	80.4	79.9	66.5	70.9	84.7	62.9	70.1
0950, 0980	Ceilings & Acoustic Treatment	101.9	66.5	78.0	91.9	66.5	74.8	101.9	66.6	78.1	91.9	66.6	74.8	102.7	66.5	78.3	110.0	62.9	78.3
0960	Flooring	106.9	76.7	98.4	104.9	76.7	97.0	112.4	70.1	100.5	106.2	73.1	96.9	110.4	73.1	100.0	109.1	73.1	99.0
0970, 0990	Wall Finishes & Painting/Coating	101.7	76.9	87.0	97.1	76.9	85.1	101.7	76.9	87.0	97.1	76.9	85.1	101.7	76.9	87.0	101.7	76.9	87.0
09	FINISHES	105.2	68.7	85.4	98.9	69.1	82.8	106.8	67.9	85.8	97.9	67.9	81.6	100.5	68.0	82.9	100.2	66.3	81.9
COVERS	DIVS. 10 - 14, 25, 28, 41, 43, 44, 46	100.0	82.3	96.1	100.0	82.9	96.2	100.0	83.1	96.2	100.0	82.1	96.0	100.0	82.6	96.1	100.0	82.9	96.2
21, 22, 23	FIRE SUPPRESSION, PLUMBING & HVAC	96.5	58.7	81.6	96.5	74.8	87.9	100.0	74.8	90.1	100.0	71.7	88.9	96.5	58.7	81.6	100.0	72.2	89.1
26, 27, 3370	ELECTRICAL, COMMUNICATIONS & UTIL.	93.4	52.3	73.1	100.0	80.6	90.4	95.5	54.6	75.3	99.6	80.6	90.2	95.5	50.3	73.2	96.4	63.3	80.0
MF2018	WEIGHTED AVERAGE	103.9	66.8	88.2	100.9	74.7	89.8	105.7	70.5	90.8	99.3	73.4	88.4	102.6	65.8	87.1	101.8	70.0	88.3

		COLORADO			CONNECTICUT														
		SALIDA			BRIDGEPORT			BRISTOL			HARTFORD			MERIDEN			NEW BRITAIN		
DIVISION		812			066			060			061			064			060		
		MAT.	INST.	TOTAL	MAT.	INST.	TOTAL	MAT.	INST.	TOTAL	MAT.	INST.	TOTAL	MAT.	INST.	TOTAL	MAT.	INST.	TOTAL
015433	CONTRACTOR EQUIPMENT		91.7	91.7		95.7	95.7		95.7	95.7		99.0	99.0		96.1	96.1		95.7	95.7
0241, 31 - 34	SITE & INFRASTRUCTURE, DEMOLITION	134.0	88.2	102.4	106.1	97.8	100.3	105.1	97.7	100.0	101.6	102.4	102.2	102.8	98.5	99.8	105.3	97.7	100.1
0310	Concrete Forming & Accessories	108.6	65.9	72.2	103.4	116.7	114.7	103.4	116.6	114.6	103.5	116.8	114.8	103.1	116.6	114.6	103.9	116.6	114.7
0320	Concrete Reinforcing	112.4	67.9	90.9	116.3	144.5	130.0	116.3	144.5	130.0	111.6	144.6	127.5	116.3	144.5	130.0	116.3	144.5	130.0
0330	Cast-in-Place Concrete	114.7	72.8	99.1	108.5	127.5	115.6	101.6	127.5	111.2	106.5	128.5	114.7	97.8	127.5	108.8	103.3	127.5	112.3
03	CONCRETE	109.1	69.3	91.6	104.1	124.5	113.0	100.9	124.4	111.2	101.1	124.8	111.5	99.1	124.4	110.2	101.7	124.4	111.7
04	MASONRY	138.6	60.6	90.7	109.4	130.9	122.6	100.8	130.9	119.3	100.5	131.0	119.2	100.4	130.9	119.2	102.5	130.9	120.0
05	METALS	103.6	78.6	96.2	98.1	116.6	103.5	98.1	116.5	103.5	103.1	116.0	106.9	95.4	116.5	101.6	94.5	116.5	100.9
06	WOOD, PLASTICS & COMPOSITES	102.4	67.2	83.0	105.0	114.2	110.1	105.0	114.2	110.1	96.2	114.4	106.2	105.0	114.2	110.1	105.0	114.2	110.1
07	THERMAL & MOISTURE PROTECTION	110.8	67.5	92.0	99.7	123.5	110.0	99.8	120.6	108.9	104.4	121.1	111.7	99.8	120.6	108.9	99.8	120.7	108.9
08	OPENINGS	96.3	68.4	89.8	96.4	120.6	102.0	96.4	120.6	102.0	98.1	120.7	103.4	98.5	120.6	103.7	96.4	120.6	102.0
0920	Plaster & Gypsum Board	80.2	66.5	71.0	115.7	114.3	114.8	115.7	114.3	114.8	102.5	114.3	110.5	117.3	114.3	115.3	115.7	114.3	114.8
0950, 0980	Ceilings & Acoustic Treatment	102.7	66.5	78.3	102.0	114.3	110.3	102.0	114.3	110.3	98.8	114.3	109.3	107.5	114.3	112.1	102.0	114.3	110.3
0960	Flooring	115.9	69.9	103.0	92.6	127.0	102.3	92.6	124.5	101.6	96.9	127.0	105.4	92.6	124.5	101.6	92.6	124.5	101.6
0970, 0990	Wall Finishes & Painting/Coating	101.7	76.9	87.0	92.0	125.5	111.8	92.0	129.1	113.9	98.1	129.1	116.4	92.0	129.1	113.9	92.0	129.1	113.9
09	FINISHES	101.1	67.4	82.8	94.7	118.9	107.8	94.7	118.8	107.8	96.2	119.4	108.8	96.0	118.8	108.4	94.7	118.8	107.8
COVERS	DIVS. 10 - 14, 25, 28, 41, 43, 44, 46	100.0	82.7	96.1	100.0	113.9	103.1	100.0	113.9	103.1	100.0	114.4	103.2	100.0	114.0	103.1	100.0	113.9	103.1
21, 22, 23	FIRE SUPPRESSION, PLUMBING & HVAC	96.5	73.6	87.5	100.1	118.6	107.4	100.1	118.6	107.4	100.0	118.6	107.3	96.6	118.6	105.2	100.1	118.6	107.4
26, 27, 3370	ELECTRICAL, COMMUNICATIONS & UTIL.	95.7	63.2	79.7	93.0	105.0	98.9	93.0	103.7	98.3	92.7	109.4	100.9	92.9	105.8	99.3	93.0	103.7	98.3
MF2018	WEIGHTED AVERAGE	103.1	70.9	89.5	99.3	117.0	106.8	98.5	116.7	106.0	99.6	118.0	107.3	97.2	117.0	105.6	98.1	116.7	106.0

		CONNECTICUT																	
		NEW HAVEN			NEW LONDON			NORWALK			STAMFORD			WATERBURY			WILLIMANTIC		
DIVISION		065			063			068			069			067			062		
		MAT.	INST.	TOTAL	MAT.	INST.	TOTAL	MAT.	INST.	TOTAL	MAT.	INST.	TOTAL	MAT.	INST.	TOTAL	MAT.	INST.	TOTAL
015433	CONTRACTOR EQUIPMENT		96.1	96.1		96.1	96.1		95.7	95.7		95.7	95.7		95.7	95.7		95.7	95.7
0241, 31 - 34	SITE & INFRASTRUCTURE, DEMOLITION	105.3	98.4	100.5	97.2	98.1	97.8	105.8	97.6	100.2	106.5	97.7	100.4	105.8	97.8	100.2	105.8	97.7	100.2
0310	Concrete Forming & Accessories	103.2	116.5	114.6	103.1	116.5	114.5	103.4	116.6	114.7	103.4	116.9	114.9	103.4	116.6	114.7	103.4	116.4	114.5
0320	Concrete Reinforcing	116.3	144.5	130.0	91.2	144.5	117.0	116.3	144.5	130.0	116.3	144.5	130.0	116.3	144.5	130.0	116.3	144.5	130.0
0330	Cast-in-Place Concrete	105.1	125.5	112.7	89.5	125.5	102.9	106.7	126.7	114.2	108.5	126.8	115.3	108.5	127.5	115.6	101.3	125.5	110.3
03	CONCRETE	116.3	123.7	119.6	89.2	123.7	104.4	103.3	124.1	112.4	104.1	124.3	113.0	104.1	124.4	113.0	100.8	123.6	110.8
04	MASONRY	101.1	130.9	119.4	99.3	130.9	118.7	100.6	130.1	118.7	101.4	130.1	119.0	101.4	130.9	119.5	100.6	130.9	119.3
05	METALS	94.7	116.4	101.1	94.4	116.4	100.9	98.1	116.5	103.5	98.1	116.8	103.6	98.1	116.5	103.5	97.9	116.3	103.3
06	WOOD, PLASTICS & COMPOSITES	105.0	114.2	110.1	105.0	114.2	110.1	105.0	114.2	110.1	105.0	114.2	110.1	105.0	114.2	110.1	105.0	114.2	110.1
07	THERMAL & MOISTURE PROTECTION	99.9	120.8	109.0	99.7	120.3	108.7	99.9	123.2	110.0	99.8	123.2	110.0	99.8	121.1	109.1	100.0	120.3	108.8
08	OPENINGS	96.4	120.6	102.0	98.7	120.6	103.8	96.4	120.6	102.0	96.4	120.6	102.0	96.4	120.6	102.0	98.8	120.6	103.9
0920	Plaster & Gypsum Board	115.7	114.3	114.8	115.7	114.3	114.8	115.7	114.3	114.8	115.7	114.3	114.8	115.7	114.3	114.8	115.7	114.3	114.8
0950, 0980	Ceilings & Acoustic Treatment	102.0	114.3	110.3	100.2	114.3	109.7	102.0	114.3	110.3	102.0	114.3	110.3	102.0	114.3	110.3	100.2	114.3	109.7
0960	Flooring	92.6	127.0	102.3	92.6	127.0	102.3	92.6	124.5	101.6	92.6	127.0	102.3	92.6	127.0	102.3	92.6	121.1	100.6
0970, 0990	Wall Finishes & Painting/Coating	92.0	125.5	111.8	92.0	129.1	113.9	92.0	125.5	111.8	92.0	125.5	111.8	92.0	129.1	113.9	92.0	129.1	113.9
09	FINISHES	94.7	118.9	107.8	93.9	119.3	107.6	94.7	118.5	107.6	94.8	118.9	107.8	94.6	119.3	107.9	94.5	118.3	107.3
COVERS	DIVS. 10 - 14, 25, 28, 41, 43, 44, 46	100.0	114.0	103.1	100.0	114.0	103.1	100.0	113.9	103.1	100.0	114.1	103.2	100.0	114.0	103.1	100.0	114.0	103.1
21, 22, 23	FIRE SUPPRESSION, PLUMBING & HVAC	100.1	118.6	107.4	96.6	118.6	105.2	100.1	118.6	107.4	100.1	118.6	107.4	100.1	118.6	107.4	100.1	118.4	107.3
26, 27, 3370	ELECTRICAL, COMMUNICATIONS & UTIL.	92.9	106.0	99.4	90.5	107.9	99.1	93.0	104.8	98.8	93.0	150.1	121.1	92.6	107.1	99.7	93.0	108.8	100.8
MF2018	WEIGHTED AVERAGE	99.9	117.0	107.1	95.2	117.2	104.5	98.8	116.7	106.4	99.0	123.2	109.2	98.9	117.2	106.6	98.7	117.1	106.5

		D.C.			DELAWARE									FLORIDA					
		WASHINGTON			DOVER			NEWARK			WILMINGTON			DAYTONA BEACH			FORT LAUDERDALE		
DIVISION		200 - 205			199			197			198			321			333		
		MAT.	INST.	TOTAL	MAT.	INST.	TOTAL	MAT.	INST.	TOTAL	MAT.	INST.	TOTAL	MAT.	INST.	TOTAL	MAT.	INST.	TOTAL
015433	CONTRACTOR EQUIPMENT		105.3	105.3		118.2	118.2		119.0	119.0		118.4	118.4		99.5	99.5		92.8	92.8
0241, 31 - 34	SITE & INFRASTRUCTURE, DEMOLITION	103.0	97.8	99.4	106.0	107.5	107.0	104.5	108.7	107.4	102.2	107.8	106.1	117.6	85.0	95.1	95.0	73.2	80.0
0310	Concrete Forming & Accessories	97.7	73.4	77.0	98.0	99.4	99.2	97.0	99.2	98.9	95.9	99.4	98.9	97.1	61.6	66.8	93.4	59.4	64.4
0320	Concrete Reinforcing	103.5	86.8	95.4	101.4	115.1	108.0	97.2	115.1	105.9	101.0	115.1	107.8	96.0	61.0	79.1	94.6	59.9	77.8
0330	Cast-in-Place Concrete	103.6	79.0	94.5	108.1	107.7	107.9	91.2	106.7	97.0	102.6	107.7	104.5	91.8	66.0	82.2	96.3	62.3	83.6
03	CONCRETE	106.0	78.6	94.0	100.4	106.0	102.9	93.4	105.7	98.8	97.7	106.0	101.4	90.5	64.8	79.2	93.4	62.3	79.7
04	MASONRY	97.6	87.8	91.6	99.0	98.5	98.7	101.1	98.5	99.5	94.4	98.5	96.9	86.4	60.9	70.7	87.6	54.4	67.2
05	METALS	103.9	95.6	101.5	102.9	121.9	108.5	104.7	123.1	110.1	103.1	121.9	108.6	101.9	88.9	98.1	96.9	88.2	94.4
06	WOOD, PLASTICS & COMPOSITES	95.4	72.2	82.6	92.3	97.6	95.2	91.0	97.4	94.5	85.7	97.6	92.2	93.2	60.1	75.0	78.3	60.6	68.6
07	THERMAL & MOISTURE PROTECTION	104.3	87.1	96.8	105.2	110.4	107.4	109.2	109.8	109.4	105.1	110.4	107.4	101.4	65.4	85.7	106.3	60.4	86.3
08	OPENINGS	100.7	74.6	94.6	91.2	107.4	95.0	91.2	107.3	94.9	88.6	107.4	93.0	92.9	59.4	85.1	94.2	59.5	86.1
0920	Plaster & Gypsum Board	104.3	71.4	82.2	96.0	97.4	96.9	96.7	97.4	97.2	97.3	97.4	97.4	92.7	59.5	70.4	107.2	60.0	75.5
0950, 0980	Ceilings & Acoustic Treatment	110.9	71.4	84.3	100.2	97.4	98.3	96.0	97.4	96.9	90.8	97.4	95.2	78.3	59.5	65.6	84.8	60.0	68.1
0960	Flooring	95.4	75.5	89.8	96.9	105.4	99.3	91.3	105.4	95.3	96.0	105.4	98.7	99.5	62.7	89.2	97.1	62.7	87.5
0970, 0990	Wall Finishes & Painting/Coating	103.7	73.1	85.6	92.6	119.1	108.3	88.2	119.1	106.4	95.3	119.1	109.4	103.8	61.4	78.7	96.5	58.3	73.9
09	FINISHES	98.5	72.8	84.6	95.9	101.6	99.0	89.7	101.6	96.1	94.2	101.6	98.2	91.5	61.1	75.1	91.3	59.4	74.1
COVERS	DIVS. 10 - 14, 25, 28, 41, 43, 44, 46	100.0	94.5	98.8	100.0	103.7	100.8	100.0	103.4	100.8	100.0	103.7	100.8	100.0	103.7	100.8	100.0	95.7	99.0
21, 22, 23	FIRE SUPPRESSION, PLUMBING & HVAC	100.0	88.4	95.4	100.0	119.4	107.6	100.3	119.3	107.7	100.1	119.4	107.7	99.9	75.6	90.4	100.0	66.7	86.9
26, 27, 3370	ELECTRICAL, COMMUNICATIONS & UTIL.	97.7	100.0	98.8	95.1	109.7	102.3	96.9	109.7	103.2	95.1	109.7	102.3	96.7	58.9	78.1	94.7	67.4	81.3
MF2018	WEIGHTED AVERAGE	101.1	87.5	95.4	99.0	109.6	103.5	98.3	109.7	103.1	97.9	109.7	102.9	97.2	69.5	85.5	96.1	66.4	83.6

For customer support on your Electrical Change Order Costs with RSMeans data, call 800.448.8182.

541

City Cost Indexes

FLORIDA

	DIVISION	FORT MYERS 339, 341			GAINESVILLE 326, 344			JACKSONVILLE 320, 322			LAKELAND 338			MELBOURNE 329			MIAMI 330 - 332, 340		
		MAT.	INST.	TOTAL	MAT.	INST.	TOTAL	MAT.	INST.	TOTAL	MAT.	INST.	TOTAL	MAT.	INST.	TOTAL	MAT.	INST.	TOTAL
015433	CONTRACTOR EQUIPMENT		99.5	99.5		99.5	99.5		99.5	99.5		99.5	99.5		99.5	99.5		94.9	94.9
0241, 31 - 34	SITE & INFRASTRUCTURE, DEMOLITION	106.7	85.2	91.9	127.1	84.8	97.9	117.6	85.2	95.2	108.7	85.2	92.5	125.6	85.1	97.7	96.5	78.9	84.3
0310	Concrete Forming & Accessories	89.6	63.8	67.6	92.5	57.1	62.4	96.9	60.2	65.6	86.4	63.8	67.1	93.7	61.7	66.4	99.3	60.2	66.0
0320	Concrete Reinforcing	95.7	76.1	86.2	101.8	61.5	82.3	96.0	61.3	79.2	98.0	76.4	87.6	97.1	65.6	81.9	101.6	60.0	81.5
0330	Cast-in-Place Concrete	100.4	66.0	87.6	105.4	65.7	90.6	92.7	65.8	82.7	102.7	66.1	89.1	110.7	66.0	94.1	93.1	63.9	82.2
03	CONCRETE	94.0	68.3	82.7	101.7	62.7	84.6	90.9	64.2	79.2	95.8	68.5	83.8	102.1	65.6	86.1	91.7	63.2	79.2
04	MASONRY	82.5	58.8	67.9	99.3	60.8	75.7	86.1	58.7	69.3	96.8	61.0	74.8	84.2	60.9	69.9	88.0	55.5	68.0
05	METALS	99.0	93.6	97.4	100.7	88.4	97.1	100.4	88.6	96.9	98.9	94.7	97.7	110.6	90.6	104.7	97.1	87.6	94.3
06	WOOD, PLASTICS & COMPOSITES	75.4	63.2	68.7	87.2	54.6	69.3	93.2	58.6	74.2	71.0	63.2	66.7	88.9	60.1	73.0	92.9	60.7	75.2
07	THERMAL & MOISTURE PROTECTION	106.1	63.2	87.4	101.8	62.5	84.7	101.7	62.4	84.6	106.0	63.8	87.6	101.9	64.1	85.5	106.1	61.4	86.6
08	OPENINGS	95.5	64.6	88.3	92.5	56.5	84.1	92.9	58.8	84.9	95.5	64.7	88.3	92.2	60.5	84.8	96.6	59.5	88.0
0920	Plaster & Gypsum Board	103.3	62.7	76.0	89.6	53.8	65.6	92.7	58.0	69.3	100.2	62.7	75.0	89.6	59.5	69.4	96.2	60.0	71.9
0950, 0980	Ceilings & Acoustic Treatment	79.7	62.7	68.2	72.9	53.8	60.1	78.3	58.0	64.6	79.7	62.7	68.2	77.5	59.5	65.4	85.6	60.0	68.3
0960	Flooring	94.1	77.2	89.4	97.0	62.7	87.4	99.5	62.7	89.2	92.2	62.7	83.9	97.2	62.7	87.5	98.6	62.7	88.5
0970, 0990	Wall Finishes & Painting/Coating	101.4	63.2	78.8	103.8	63.2	79.8	103.8	63.2	79.8	101.4	63.2	78.8	103.8	80.9	90.2	100.2	58.3	75.4
09	FINISHES	91.1	66.1	77.6	90.4	58.1	72.9	91.6	60.4	74.7	90.2	63.1	75.6	91.0	63.2	76.0	90.9	59.8	74.1
COVERS	DIVS. 10 - 14, 25, 28, 41, 43, 44, 46	100.0	80.0	95.5	100.0	80.8	95.7	100.0	78.0	95.1	100.0	80.0	95.5	100.0	81.4	95.9	100.0	81.5	95.9
21, 22, 23	FIRE SUPPRESSION, PLUMBING & HVAC	98.1	59.3	82.9	98.6	63.1	84.7	99.9	63.1	85.5	98.1	59.9	83.1	99.9	74.2	89.8	100.0	63.8	85.8
26, 27, 3370	ELECTRICAL, COMMUNICATIONS & UTIL.	96.6	60.0	78.6	97.0	58.6	78.0	96.4	61.9	79.4	95.0	58.5	77.0	97.7	62.7	80.5	98.4	77.9	88.3
MF2018	WEIGHTED AVERAGE	96.5	67.7	84.3	98.8	65.8	84.8	96.9	66.6	84.1	97.1	67.6	84.6	100.1	70.3	87.5	96.7	68.0	84.6

FLORIDA

	DIVISION	ORLANDO 327 - 328, 347			PANAMA CITY 324			PENSACOLA 325			SARASOTA 342			ST. PETERSBURG 337			TALLAHASSEE 323		
		MAT.	INST.	TOTAL	MAT.	INST.	TOTAL	MAT.	INST.	TOTAL	MAT.	INST.	TOTAL	MAT.	INST.	TOTAL	MAT.	INST.	TOTAL
015433	CONTRACTOR EQUIPMENT		102.0	102.0		99.5	99.5		99.5	99.5		99.5	99.5		99.5	99.5		102.0	102.0
0241, 31 - 34	SITE & INFRASTRUCTURE, DEMOLITION	115.8	89.6	97.7	131.4	85.2	99.5	131.3	85.0	99.3	119.0	85.2	95.7	110.4	85.0	92.9	110.2	89.8	96.1
0310	Concrete Forming & Accessories	101.8	61.5	67.4	96.2	66.2	70.6	94.2	61.6	66.4	93.7	63.7	68.1	92.5	60.8	65.5	99.0	62.2	67.6
0320	Concrete Reinforcing	104.9	65.6	85.9	100.3	67.4	84.4	102.8	67.4	85.7	96.6	76.2	86.7	98.0	76.0	87.4	98.4	61.5	80.5
0330	Cast-in-Place Concrete	111.6	66.7	94.9	97.4	65.9	85.7	120.4	65.3	99.9	105.1	66.1	90.6	103.8	65.8	89.7	92.3	66.7	82.8
03	CONCRETE	100.8	65.7	85.4	100.0	67.9	85.9	109.7	65.6	90.3	95.3	68.4	83.5	97.3	66.9	83.9	90.0	65.3	79.2
04	MASONRY	94.8	60.9	74.0	90.6	60.3	72.0	109.1	59.9	78.9	88.7	61.0	71.6	132.4	58.7	87.1	85.2	60.3	69.9
05	METALS	96.7	89.1	94.5	101.5	90.9	98.4	102.6	90.8	99.1	102.5	93.9	100.0	99.8	93.5	98.0	101.7	87.7	97.6
06	WOOD, PLASTICS & COMPOSITES	91.7	60.2	74.4	92.0	66.6	78.0	90.3	61.1	74.2	92.5	63.2	76.4	79.5	59.3	68.4	95.6	61.2	76.7
07	THERMAL & MOISTURE PROTECTION	109.2	66.0	90.4	102.0	64.1	85.5	101.9	63.2	85.1	100.1	63.8	84.3	106.2	62.2	87.1	96.7	64.0	82.4
08	OPENINGS	97.8	60.6	89.1	91.0	64.5	84.9	91.0	61.4	84.1	97.8	64.2	90.0	94.3	62.5	86.9	97.9	60.2	89.1
0920	Plaster & Gypsum Board	93.3	59.5	70.6	91.9	66.1	74.6	99.6	60.5	73.3	97.2	62.7	74.0	105.6	58.7	74.0	100.0	60.5	73.4
0950, 0980	Ceilings & Acoustic Treatment	89.7	59.5	69.3	77.5	66.1	69.8	77.5	60.5	66.0	83.9	62.7	69.6	81.6	58.7	66.1	86.2	60.5	68.9
0960	Flooring	95.7	62.7	86.4	99.1	72.5	91.6	95.1	62.7	85.9	103.5	54.9	89.9	96.1	60.8	86.2	97.6	61.7	87.5
0970, 0990	Wall Finishes & Painting/Coating	95.6	60.3	74.7	103.8	63.2	79.8	103.8	63.2	79.8	93.6	63.2	77.8	101.4	63.2	78.8	101.6	63.2	78.9
09	FINISHES	93.2	61.1	75.8	92.6	66.8	78.6	92.1	61.6	75.6	96.0	61.8	77.5	92.6	60.5	75.2	94.4	61.8	76.7
COVERS	DIVS. 10 - 14, 25, 28, 41, 43, 44, 46	100.0	81.7	95.9	100.0	80.6	95.7	100.0	79.7	95.5	100.0	80.0	95.5	100.0	79.6	95.4	100.0	81.5	95.9
21, 22, 23	FIRE SUPPRESSION, PLUMBING & HVAC	100.0	56.4	82.9	99.9	63.2	85.5	99.9	62.7	85.3	99.9	58.5	83.6	100.0	59.9	84.2	100.0	65.5	86.5
26, 27, 3370	ELECTRICAL, COMMUNICATIONS & UTIL.	98.4	63.1	81.0	95.5	58.6	77.3	99.0	50.7	75.2	96.9	58.5	78.0	95.0	60.6	78.0	103.7	58.6	81.4
MF2018	WEIGHTED AVERAGE	99.0	66.6	85.3	98.6	68.3	85.8	101.2	65.8	86.2	98.8	67.0	85.4	99.7	66.8	85.8	98.2	67.7	85.3

FLORIDA / GEORGIA

	DIVISION	TAMPA 335 - 336, 346			WEST PALM BEACH 334, 349			ALBANY 317, 398			ATHENS 306			ATLANTA 300 - 303, 399			AUGUSTA 308 - 309		
		MAT.	INST.	TOTAL	MAT.	INST.	TOTAL	MAT.	INST.	TOTAL	MAT.	INST.	TOTAL	MAT.	INST.	TOTAL	MAT.	INST.	TOTAL
015433	CONTRACTOR EQUIPMENT		99.5	99.5		92.8	92.8		93.7	93.7		92.6	92.6		96.6	96.6		92.6	92.6
0241, 31 - 34	SITE & INFRASTRUCTURE, DEMOLITION	110.9	87.0	94.4	91.6	73.2	78.9	107.3	76.7	86.1	102.8	90.8	94.5	99.7	96.0	97.1	95.9	91.9	93.2
0310	Concrete Forming & Accessories	95.4	64.1	68.7	96.7	59.1	64.7	90.6	66.5	70.1	90.0	43.6	50.4	94.3	72.3	75.5	90.9	73.0	75.6
0320	Concrete Reinforcing	94.6	76.5	85.8	97.3	57.5	78.1	96.5	71.3	84.3	94.5	63.5	79.5	93.9	71.3	83.0	94.9	70.7	83.2
0330	Cast-in-Place Concrete	101.5	66.2	88.3	91.5	62.2	80.6	94.6	69.1	85.1	109.7	69.5	94.7	113.1	71.1	97.5	103.6	70.0	91.1
03	CONCRETE	95.9	68.7	83.9	90.1	61.8	77.7	86.9	70.1	79.5	99.6	57.4	81.1	101.8	72.4	88.9	92.5	72.2	83.6
04	MASONRY	88.3	61.0	71.5	87.1	52.2	65.7	92.5	68.7	77.9	75.0	77.0	76.2	87.8	68.9	76.2	88.1	68.8	76.2
05	METALS	99.0	95.0	97.8	95.9	87.0	93.3	106.2	97.4	103.6	95.3	79.1	90.6	96.2	83.5	92.5	95.0	82.6	91.4
06	WOOD, PLASTICS & COMPOSITES	83.3	63.2	72.2	83.1	60.6	70.7	80.1	66.4	72.5	92.5	35.8	61.3	97.8	74.0	84.7	93.8	75.3	83.6
07	THERMAL & MOISTURE PROTECTION	106.5	63.8	87.9	106.0	60.6	86.2	100.6	68.0	86.4	95.4	68.3	83.6	96.8	73.8	86.8	95.1	71.4	84.8
08	OPENINGS	95.5	64.7	88.3	93.8	59.0	85.7	85.7	68.5	81.7	91.3	50.2	81.7	99.4	73.5	93.3	91.4	74.0	87.3
0920	Plaster & Gypsum Board	108.2	62.7	77.6	111.9	60.0	77.0	101.1	66.0	77.5	93.0	34.4	53.6	95.0	73.6	80.6	94.0	75.0	81.2
0950, 0980	Ceilings & Acoustic Treatment	84.8	62.7	69.9	79.7	60.0	66.4	77.5	66.0	69.8	96.6	34.4	54.7	89.7	73.6	78.8	97.5	75.0	82.4
0960	Flooring	97.1	62.7	87.5	98.9	54.9	86.6	99.4	70.8	91.4	96.2	81.9	92.2	99.1	70.8	91.1	96.4	70.8	89.2
0970, 0990	Wall Finishes & Painting/Coating	101.4	63.2	78.8	96.5	58.3	73.9	88.7	94.0	91.8	92.4	94.0	93.3	96.1	97.3	96.8	92.4	90.2	91.1
09	FINISHES	93.9	63.1	77.3	91.2	57.8	73.2	91.7	69.7	79.8	95.0	53.9	72.7	95.0	74.5	83.9	94.7	74.5	83.7
COVERS	DIVS. 10 - 14, 25, 28, 41, 43, 44, 46	100.0	80.0	95.6	100.0	80.7	95.7	100.0	85.9	96.8	100.0	82.7	96.1	100.0	87.2	97.2	100.0	87.0	97.1
21, 22, 23	FIRE SUPPRESSION, PLUMBING & HVAC	100.0	59.9	84.3	98.1	58.5	82.6	100.0	68.9	87.8	96.6	66.3	84.7	100.0	69.4	88.0	100.1	69.0	87.9
26, 27, 3370	ELECTRICAL, COMMUNICATIONS & UTIL.	94.7	63.1	79.1	95.7	67.4	81.8	95.5	61.9	79.0	97.8	63.9	81.1	97.6	71.5	84.7	96.7	67.3	83.1
MF2018	WEIGHTED AVERAGE	97.6	68.4	85.3	95.1	64.1	82.0	96.5	71.9	86.1	95.6	66.9	83.5	98.2	75.0	88.4	96.0	73.9	86.6

For customer support on your Electrical Change Order Costs with RSMeans data, call 800.448.8182.

GEORGIA

DIVISION		COLUMBUS 318 - 319			DALTON 307			GAINESVILLE 305			MACON 310 - 312			SAVANNAH 313 - 314			STATESBORO 304		
		MAT.	INST.	TOTAL	MAT.	INST.	TOTAL	MAT.	INST.	TOTAL	MAT.	INST.	TOTAL	MAT.	INST.	TOTAL	MAT.	INST.	TOTAL
015433	CONTRACTOR EQUIPMENT		93.7	93.7		107.6	107.6		92.6	92.6		103.2	103.2		96.7	96.7		95.5	95.5
0241, 31 - 34	SITE & INFRASTRUCTURE, DEMOLITION	107.2	76.7	86.2	102.6	95.8	97.9	102.5	90.7	94.3	108.6	91.0	96.4	107.0	82.0	89.7	103.7	78.0	85.9
0310	Concrete Forming & Accessories	90.5	66.6	70.1	83.2	64.5	67.3	93.3	40.7	48.5	90.3	66.6	70.1	93.4	72.0	75.2	78.5	52.2	56.1
0320	Concrete Reinforcing	96.6	71.3	84.4	94.1	60.9	78.0	94.4	63.4	79.4	97.8	71.3	85.0	104.1	70.7	88.0	93.7	71.2	82.8
0330	Cast-in-Place Concrete	94.2	68.8	84.8	106.6	68.0	92.2	115.3	68.7	98.0	93.0	68.3	83.8	98.3	69.5	87.6	109.5	68.6	94.3
03	CONCRETE	86.8	70.0	79.4	98.9	67.0	84.9	101.4	55.8	81.4	86.3	69.8	79.1	88.5	72.5	81.5	99.3	63.4	83.5
04	MASONRY	92.5	68.7	77.9	75.8	75.5	75.6	83.2	76.2	78.9	104.8	68.9	82.7	87.4	68.8	75.9	77.8	77.0	77.3
05	METALS	105.9	97.6	103.4	96.4	94.0	95.7	94.6	78.5	89.9	100.9	97.6	100.0	102.1	96.0	100.3	100.0	98.9	99.7
06	WOOD, PLASTICS & COMPOSITES	80.1	66.4	72.6	75.9	65.2	70.0	96.1	33.2	61.5	85.9	66.5	75.2	87.4	73.9	80.0	69.7	47.2	57.3
07	THERMAL & MOISTURE PROTECTION	100.5	68.9	86.8	97.3	70.1	85.5	95.4	67.7	83.3	99.0	70.6	86.6	98.5	70.4	86.3	95.9	67.1	83.4
08	OPENINGS	85.7	69.3	81.9	92.8	65.8	86.5	91.3	48.8	81.4	85.5	69.3	81.7	95.3	73.2	90.2	93.7	58.5	85.5
0920	Plaster & Gypsum Board	101.1	66.0	77.5	81.9	64.7	70.3	94.9	31.8	52.4	104.7	66.0	78.7	103.3	73.6	83.3	83.6	46.2	58.4
0950, 0980	Ceilings & Acoustic Treatment	77.5	66.0	69.8	110.2	64.7	79.5	96.6	31.8	52.9	72.8	66.0	68.3	87.3	73.6	78.0	105.8	46.2	65.6
0960	Flooring	99.4	70.8	91.4	97.0	81.9	92.8	97.8	81.9	93.3	77.9	70.8	75.9	95.6	70.8	88.6	114.6	81.9	105.4
0970, 0990	Wall Finishes & Painting/Coating	88.7	87.4	87.9	82.8	70.1	75.3	92.4	94.0	93.3	90.7	94.0	92.6	88.0	86.1	86.90	90.8	70.1	78.5
09	FINISHES	91.6	68.9	79.3	104.5	68.2	84.9	95.5	52.1	72.0	80.5	69.7	74.7	92.9	73.2	82.3	107.9	57.9	80.9
COVERS	DIVS. 10 - 14, 25, 28, 41, 43, 44, 46	100.0	85.9	96.9	100.0	85.3	96.7	100.0	35.5	85.6	100.0	86.0	96.9	100.0	87.0	97.1	100.0	83.8	96.4
21, 22, 23	FIRE SUPPRESSION, PLUMBING & HVAC	100.1	66.8	87.0	96.7	60.8	82.6	96.6	65.6	84.4	100.1	68.2	87.6	100.1	64.8	86.3	97.2	67.3	85.5
26, 27, 3370	ELECTRICAL, COMMUNICATIONS & UTIL.	95.7	67.7	81.9	106.9	62.2	84.9	97.8	70.8	84.5	94.2	62.7	78.6	99.1	63.7	81.7	98.3	63.4	81.1
MF2018	WEIGHTED AVERAGE	96.5	72.2	86.2	97.5	71.4	86.5	96.2	65.6	83.2	95.2	73.1	85.8	97.3	72.7	86.9	97.8	69.7	85.9

DIVISION		GEORGIA VALDOSTA 316			GEORGIA WAYCROSS 315			HAWAII HILO 967			HAWAII HONOLULU 968			HAWAII STATES & POSS., GUAM 969			IDAHO BOISE 836 - 837		
		MAT.	INST.	TOTAL	MAT.	INST.	TOTAL	MAT.	INST.	TOTAL	MAT.	INST.	TOTAL	MAT.	INST.	TOTAL	MAT.	INST.	TOTAL
015433	CONTRACTOR EQUIPMENT		93.7	93.7		93.7	93.7		97.8	97.8		99.3	99.3		160.5	160.5		94.6	94.6
0241, 31 - 34	SITE & INFRASTRUCTURE, DEMOLITION	116.9	76.7	89.1	113.7	77.0	88.3	142.5	103.1	115.3	150.0	106.6	120.0	186.3	98.5	125.7	86.8	92.0	90.4
0310	Concrete Forming & Accessories	81.7	41.6	47.5	83.5	64.0	66.8	108.2	123.8	121.5	120.0	123.7	123.2	110.5	51.5	60.2	100.2	81.1	83.9
0320	Concrete Reinforcing	98.7	59.5	79.8	98.7	59.8	79.9	143.4	127.0	135.5	166.8	127.0	147.6	253.7	27.4	144.2	107.6	79.5	94.0
0330	Cast-in-Place Concrete	92.6	68.7	83.7	104.7	68.6	91.2	180.3	123.6	159.1	145.0	123.5	137.0	156.3	95.7	133.7	90.6	96.6	92.8
03	CONCRETE	91.2	56.6	76.0	94.5	66.5	82.2	144.7	123.2	135.3	137.9	123.1	131.4	147.9	63.9	111.0	97.6	86.4	92.7
04	MASONRY	97.8	77.1	85.1	98.7	77.0	85.4	148.6	122.1	132.3	135.5	122.2	127.3	206.7	34.8	101.0	126.3	86.2	101.7
05	METALS	105.3	93.0	101.7	104.4	88.9	99.8	110.8	108.1	110.0	124.1	107.1	119.1	143.1	76.0	123.4	109.6	82.7	101.7
06	WOOD, PLASTICS & COMPOSITES	68.7	33.0	49.0	70.2	63.4	66.5	111.4	124.3	118.5	132.8	124.2	128.1	123.4	53.9	85.1	93.5	79.9	86.0
07	THERMAL & MOISTURE PROTECTION	100.8	66.6	86.0	100.6	69.1	86.9	128.7	119.3	124.6	146.3	120.0	134.9	150.7	58.2	110.4	101.4	86.6	94.9
08	OPENINGS	82.5	47.6	74.4	82.8	61.8	77.9	114.1	123.2	116.2	128.1	123.2	126.9	118.4	43.8	101.0	96.9	75.9	92.0
0920	Plaster & Gypsum Board	93.8	31.6	52.0	93.8	62.9	73.0	115.2	124.8	121.7	160.8	124.8	136.6	236.1	42.5	105.8	92.3	79.4	83.6
0950, 0980	Ceilings & Acoustic Treatment	75.9	31.6	46.0	74.0	62.9	66.5	136.6	124.8	128.6	144.8	124.8	131.3	257.2	42.5	112.4	102.8	79.4	87.0
0960	Flooring	93.0	83.5	90.3	94.3	81.9	90.8	103.8	139.3	113.8	121.3	139.3	126.4	121.2	39.5	98.2	92.9	90.7	92.3
0970, 0990	Wall Finishes & Painting/Coating	88.7	94.0	91.8	88.7	70.1	77.7	99.2	143.8	125.6	110.1	143.8	130.1	105.2	31.5	61.6	92.6	39.5	61.2
09	FINISHES	89.4	52.5	69.4	89.0	67.5	77.4	110.8	129.1	120.7	125.5	129.1	127.4	185.1	48.0	110.9	92.7	78.2	84.9
COVERS	DIVS. 10 - 14, 25, 28, 41, 43, 44, 46	100.0	82.2	96.0	100.0	85.5	96.8	100.0	113.7	103.1	100.0	113.6	103.0	100.0	66.0	92.4	100.0	88.6	97.4
21, 22, 23	FIRE SUPPRESSION, PLUMBING & HVAC	100.1	68.5	87.7	97.9	63.6	84.4	100.3	112.4	105.0	100.4	112.3	105.1	102.6	33.5	75.5	100.1	74.2	89.9
26, 27, 3370	ELECTRICAL, COMMUNICATIONS & UTIL.	94.1	57.8	76.2	97.7	63.4	80.8	108.1	124.6	116.2	109.8	124.5	117.1	158.1	35.9	97.9	97.2	68.2	82.9
MF2018	WEIGHTED AVERAGE	96.7	66.2	83.8	96.8	70.0	85.5	114.8	118.4	116.3	119.1	118.5	118.9	137.1	51.4	100.9	101.0	79.9	92.1

DIVISION		IDAHO COEUR D'ALENE 838			IDAHO IDAHO FALLS 834			IDAHO LEWISTON 835			IDAHO POCATELLO 832			IDAHO TWIN FALLS 833			ILLINOIS BLOOMINGTON 617		
		MAT.	INST.	TOTAL	MAT.	INST.	TOTAL	MAT.	INST.	TOTAL	MAT.	INST.	TOTAL	MAT.	INST.	TOTAL	MAT.	INST.	TOTAL
015433	CONTRACTOR EQUIPMENT		89.7	89.7		94.6	94.6		89.7	89.7		94.6	94.6		94.6	94.6		101.9	101.9
0241, 31 - 34	SITE & INFRASTRUCTURE, DEMOLITION	85.5	86.5	86.2	85.1	92.1	89.9	92.2	87.2	88.7	88.3	92.0	90.9	95.4	91.8	92.9	94.9	96.0	95.7
0310	Concrete Forming & Accessories	108.7	81.4	85.4	94.1	78.3	80.6	113.9	81.1	85.9	100.4	80.5	83.4	101.6	76.5	80.2	84.0	116.0	111.2
0320	Concrete Reinforcing	115.7	96.2	106.3	109.6	77.9	94.3	115.7	96.2	106.3	108.0	79.4	94.2	110.0	77.8	94.4	93.0	102.4	97.5
0330	Cast-in-Place Concrete	98.0	85.2	93.2	86.3	83.4	85.2	101.8	82.0	94.5	93.1	96.4	94.3	95.6	82.3	90.6	98.0	113.7	103.9
03	CONCRETE	103.9	85.3	95.7	89.6	80.2	85.5	107.5	84.1	97.2	96.7	86.0	92.0	104.6	79.0	93.4	91.9	113.6	101.4
04	MASONRY	128.0	87.3	103.0	121.5	87.2	100.4	128.5	84.7	101.6	124.0	86.2	100.8	126.7	82.8	99.7	110.2	121.7	117.3
05	METALS	103.0	88.1	98.6	118.1	81.6	107.4	102.4	88.2	98.3	118.2	82.0	107.6	118.2	81.4	107.4	95.2	123.1	103.4
06	WOOD, PLASTICS & COMPOSITES	96.7	79.8	87.4	87.2	76.3	81.2	102.6	79.8	90.1	93.5	79.9	86.0	94.6	76.3	84.5	81.7	113.0	99.0
07	THERMAL & MOISTURE PROTECTION	158.9	84.3	126.4	100.4	75.3	89.5	159.2	83.1	126.1	101.0	75.0	89.7	101.8	80.5	92.5	95.4	112.1	102.6
08	OPENINGS	113.8	75.9	105.0	99.8	71.3	93.2	106.7	79.6	100.4	97.6	68.8	90.9	100.6	62.4	91.7	91.3	118.7	97.7
0920	Plaster & Gypsum Board	167.0	79.4	108.1	80.3	75.7	77.2	168.9	79.4	108.7	81.9	79.4	80.2	84.0	75.7	78.4	89.1	113.5	105.5
0950, 0980	Ceilings & Acoustic Treatment	136.3	79.4	98.0	104.3	75.7	85.0	136.3	79.4	98.0	110.0	79.4	89.4	106.8	75.7	85.8	83.9	113.5	103.9
0960	Flooring	130.9	77.0	115.8	92.7	77.0	88.3	134.3	77.0	118.2	96.2	77.0	90.8	97.5	77.0	91.7	86.0	120.8	95.8
0970, 0990	Wall Finishes & Painting/Coating	111.2	73.8	89.1	92.6	39.5	61.2	111.2	71.3	87.6	92.5	39.5	61.1	92.6	39.5	61.2	87.6	134.8	115.5
09	FINISHES	158.3	79.5	115.7	90.9	74.0	81.8	159.7	78.9	116.0	93.7	75.9	84.1	94.5	72.9	82.8	86.1	119.0	103.9
COVERS	DIVS. 10 - 14, 25, 28, 41, 43, 44, 46	100.0	93.0	98.4	100.0	87.4	97.2	100.0	92.6	98.3	100.0	88.5	97.4	100.0	85.9	96.9	100.0	104.5	101.0
21, 22, 23	FIRE SUPPRESSION, PLUMBING & HVAC	99.4	85.5	94.0	101.1	81.8	93.5	100.9	85.1	94.6	100.0	73.6	89.6	100.0	72.2	89.1	96.5	105.4	100.0
26, 27, 3370	ELECTRICAL, COMMUNICATIONS & UTIL.	88.7	81.5	85.2	88.3	70.9	79.7	86.9	78.2	82.6	94.4	66.0	80.4	89.6	70.9	80.4	95.9	90.3	93.1
MF2018	WEIGHTED AVERAGE	108.1	84.4	98.1	100.5	79.9	91.8	108.2	83.5	97.8	102.0	78.4	92.0	103.2	76.8	92.1	95.0	109.5	101.1

For customer support on your Electrical Change Order Costs with RSMeans data, call 800.448.8182.

543

City Cost Indexes

ILLINOIS

DIVISION		CARBONDALE 629 MAT.	INST.	TOTAL	CENTRALIA 628 MAT.	INST.	TOTAL	CHAMPAIGN 618 - 619 MAT.	INST.	TOTAL	CHICAGO 606 - 608 MAT.	INST.	TOTAL	DECATUR 625 MAT.	INST.	TOTAL	EAST ST. LOUIS 620 - 622 MAT.	INST.	TOTAL
015433	CONTRACTOR EQUIPMENT		110.3	110.3		110.3	110.3		102.8	102.8		102.8	102.8		100.3	100.3		110.3	110.3
0241, 31 - 34	SITE & INFRASTRUCTURE, DEMOLITION	100.9	96.8	98.1	101.3	98.6	99.5	104.0	97.2	99.3	105.4	104.4	104.7	94.8	97.0	96.3	103.5	97.7	99.5
0310	Concrete Forming & Accessories	90.0	108.9	106.1	91.6	114.2	110.9	90.3	116.7	112.8	97.4	159.7	150.5	92.1	117.9	114.1	87.5	114.2	110.2
0320	Concrete Reinforcing	82.2	103.4	92.5	82.2	104.0	92.8	93.0	101.6	97.2	103.8	153.3	127.7	80.6	99.8	89.9	82.2	104.0	92.7
0330	Cast-in-Place Concrete	91.5	102.7	95.7	92.0	119.6	102.3	113.7	110.8	112.6	126.0	155.6	137.0	99.9	114.6	105.4	93.6	118.4	102.8
03	CONCRETE	79.9	107.3	91.9	80.4	115.8	96.0	104.1	112.7	107.9	108.0	156.5	129.3	90.3	114.3	100.8	81.4	115.4	96.3
04	MASONRY	75.0	111.5	97.4	75.0	122.9	104.5	133.1	125.4	128.4	105.3	163.4	141.0	70.9	122.8	102.8	75.3	123.0	104.6
05	METALS	100.9	130.4	109.6	100.9	133.5	110.5	95.2	120.3	102.6	95.9	146.8	110.8	104.9	120.7	109.5	102.1	133.5	111.3
06	WOOD, PLASTICS & COMPOSITES	87.7	105.8	97.7	90.2	111.2	101.7	88.5	114.0	102.5	100.0	158.4	132.1	88.7	116.8	104.2	84.8	111.2	99.4
07	THERMAL & MOISTURE PROTECTION	90.6	101.3	95.3	90.7	111.5	99.8	96.2	115.6	104.6	95.2	150.7	119.4	95.9	113.3	103.5	90.7	110.7	99.4
08	OPENINGS	87.3	116.3	94.1	87.3	119.2	94.8	91.8	116.9	97.7	100.9	169.3	116.9	98.2	118.2	102.9	87.4	118.4	94.6
0920	Plaster & Gypsum Board	93.7	106.2	102.1	95.0	111.6	106.2	91.0	114.5	106.8	101.3	160.0	140.8	96.5	117.4	110.6	92.2	111.6	105.3
0950, 0980	Ceilings & Acoustic Treatment	80.6	106.2	97.8	80.6	111.6	101.5	83.9	114.5	104.5	100.4	160.0	140.6	87.1	117.4	107.5	80.6	111.6	101.5
0960	Flooring	113.7	116.6	114.5	114.6	116.6	115.1	89.2	116.6	96.9	91.3	163.8	111.7	101.8	117.1	106.1	112.6	116.6	113.7
0970, 0990	Wall Finishes & Painting/Coating	103.1	105.0	104.2	103.1	111.6	108.2	87.6	114.9	103.7	96.7	170.3	140.2	95.5	117.8	108.7	103.1	111.6	108.2
09	FINISHES	92.1	110.7	102.1	92.5	114.4	104.4	88.1	116.9	103.7	95.8	162.4	131.8	92.2	118.3	106.3	91.6	114.4	103.9
COVERS	DIVS. 10 - 14, 25, 28, 41, 43, 44, 46	100.0	106.9	101.5	100.0	105.8	101.3	100.0	109.2	102.0	100.0	129.3	106.5	100.0	108.6	101.9	100.0	108.3	101.9
21, 22, 23	FIRE SUPPRESSION, PLUMBING & HVAC	96.3	106.8	100.4	96.3	97.8	96.9	96.5	107.0	100.6	99.9	137.4	114.6	99.9	98.8	99.5	99.9	99.2	99.6
26, 27, 3370	ELECTRICAL, COMMUNICATIONS & UTIL.	91.9	107.8	99.8	93.1	106.3	99.6	99.0	95.4	97.2	97.9	136.2	116.8	95.3	103.7	99.5	92.7	99.7	96.2
MF2018	WEIGHTED AVERAGE	92.5	109.6	99.7	92.8	111.2	100.5	98.5	110.5	103.6	100.1	145.6	119.3	96.6	110.1	102.3	93.8	110.4	100.9

ILLINOIS

DIVISION		EFFINGHAM 624 MAT.	INST.	TOTAL	GALESBURG 614 MAT.	INST.	TOTAL	JOLIET 604 MAT.	INST.	TOTAL	KANKAKEE 609 MAT.	INST.	TOTAL	LA SALLE 613 MAT.	INST.	TOTAL	NORTH SUBURBAN 600 - 603 MAT.	INST.	TOTAL
015433	CONTRACTOR EQUIPMENT		102.8	102.8		101.9	101.9		93.7	93.7		93.7	93.7		101.9	101.9		93.7	93.7
0241, 31 - 34	SITE & INFRASTRUCTURE, DEMOLITION	99.3	96.9	97.0	97.3	95.8	96.3	101.2	97.2	98.5	94.7	97.1	96.4	96.7	96.8	96.8	100.5	97.9	98.7
0310	Concrete Forming & Accessories	96.5	114.6	111.9	90.2	117.0	113.0	96.8	156.7	147.8	90.5	145.4	137.3	103.7	125.3	122.1	96.1	158.5	149.3
0320	Concrete Reinforcing	83.3	93.8	88.4	92.5	105.6	98.8	103.8	136.6	119.7	104.6	135.1	119.3	92.7	132.9	112.1	103.8	149.1	125.7
0330	Cast-in-Place Concrete	99.6	109.9	103.4	100.9	109.2	104.0	114.6	143.9	125.5	106.8	135.8	117.6	100.8	121.8	108.6	114.6	152.9	128.9
03	CONCRETE	91.0	109.9	99.3	94.8	113.0	102.8	101.0	147.9	121.6	95.0	139.7	114.7	95.6	126.1	109.0	101.0	154.1	124.3
04	MASONRY	79.1	115.3	101.3	110.4	122.4	117.8	104.5	156.7	136.6	100.6	147.0	129.1	110.4	129.7	122.3	101.2	158.8	136.6
05	METALS	101.9	114.9	105.7	95.2	123.6	103.5	94.1	134.7	106.0	94.1	133.4	105.6	95.3	142.0	109.0	95.1	141.6	108.8
06	WOOD, PLASTICS & COMPOSITES	91.0	114.0	103.6	88.3	115.4	103.2	96.6	157.7	130.2	90.0	144.7	120.1	103.8	123.2	114.5	95.4	158.2	130.0
07	THERMAL & MOISTURE PROTECTION	95.4	109.1	101.4	95.6	110.0	101.9	99.5	144.2	119.0	98.6	139.5	116.4	95.8	122.5	107.4	99.9	147.6	120.7
08	OPENINGS	93.3	114.5	98.2	91.3	115.9	97.0	99.3	161.7	113.9	92.4	156.2	107.3	91.3	133.3	101.1	99.4	167.9	115.4
0920	Plaster & Gypsum Board	96.3	114.5	108.5	91.0	115.9	107.8	93.7	159.5	138.0	92.5	146.1	128.6	98.3	124.0	115.6	97.2	160.0	139.5
0950, 0980	Ceilings & Acoustic Treatment	80.6	114.5	103.4	83.9	115.9	105.5	100.9	159.5	140.4	100.9	146.1	131.4	83.9	124.0	110.9	100.9	160.0	140.8
0960	Flooring	103.0	116.6	106.8	89.1	120.8	98.0	89.4	149.8	106.4	86.3	150.7	104.4	95.7	124.8	103.9	89.9	159.5	109.4
0970, 0990	Wall Finishes & Painting/Coating	95.5	109.3	103.7	87.6	98.5	94.0	89.0	163.0	132.8	89.0	135.7	116.6	87.6	133.0	114.5	91.0	163.9	134.1
09	FINISHES	91.3	115.1	104.2	87.4	116.3	103.1	91.6	157.4	127.2	90.0	145.8	120.2	90.4	124.7	109.0	92.2	160.1	129.0
COVERS	DIVS. 10 - 14, 25, 28, 41, 43, 44, 46	100.0	104.5	101.0	100.0	104.4	101.0	100.0	122.6	105.0	100.0	120.6	104.6	100.0	105.6	101.2	100.0	124.1	105.4
21, 22, 23	FIRE SUPPRESSION, PLUMBING & HVAC	96.4	104.5	99.6	96.5	105.7	100.1	100.0	130.0	111.8	96.5	129.6	109.5	96.5	128.4	109.0	99.9	136.2	114.1
26, 27, 3370	ELECTRICAL, COMMUNICATIONS & UTIL.	93.5	107.8	100.5	96.7	86.5	91.7	96.9	133.3	114.8	92.4	137.0	114.4	94.0	137.0	115.2	96.7	135.5	115.9
MF2018	WEIGHTED AVERAGE	95.1	109.1	101.0	95.7	108.5	101.1	98.3	138.8	115.4	95.0	134.9	111.9	95.9	127.0	109.0	98.3	142.9	117.1

ILLINOIS

DIVISION		PEORIA 615 - 616 MAT.	INST.	TOTAL	QUINCY 623 MAT.	INST.	TOTAL	ROCK ISLAND 612 MAT.	INST.	TOTAL	ROCKFORD 610 - 611 MAT.	INST.	TOTAL	SOUTH SUBURBAN 605 MAT.	INST.	TOTAL	SPRINGFIELD 626 - 627 MAT.	INST.	TOTAL
015433	CONTRACTOR EQUIPMENT		101.9	101.9		102.8	102.8		101.9	101.9		101.9	101.9		93.7	93.7		105.1	105.1
0241, 31 - 34	SITE & INFRASTRUCTURE, DEMOLITION	97.9	96.0	96.6	98.1	96.4	96.9	95.5	94.8	95.0	97.4	96.8	97.0	100.5	97.8	98.6	99.8	100.5	100.3
0310	Concrete Forming & Accessories	93.0	115.9	112.5	94.3	113.6	110.8	91.6	99.1	98.0	97.2	127.7	123.2	96.1	158.5	149.3	93.2	115.6	112.3
0320	Concrete Reinforcing	90.1	104.7	97.2	83.0	86.2	84.5	92.5	99.0	95.6	85.3	134.1	108.9	103.8	149.0	125.7	83.2	99.8	91.2
0330	Cast-in-Place Concrete	97.9	115.6	104.5	99.8	106.6	102.3	98.7	98.4	98.6	100.2	126.6	110.0	114.6	152.8	128.8	94.9	109.2	100.2
03	CONCRETE	92.0	114.6	102.0	90.7	107.1	97.9	92.7	100.0	95.9	92.7	129.0	108.7	101.0	154.0	124.3	88.5	111.2	98.5
04	MASONRY	109.7	121.5	116.9	98.9	113.2	107.7	110.2	100.2	104.1	85.6	142.3	120.4	101.2	158.7	136.6	80.7	122.2	106.2
05	METALS	97.9	124.2	105.6	101.9	112.6	105.1	95.2	118.5	102.1	97.9	141.5	110.7	95.1	141.4	108.7	102.5	119.1	107.3
06	WOOD, PLASTICS & COMPOSITES	96.4	113.2	105.6	88.6	114.0	102.6	90.0	97.5	94.1	96.4	123.6	111.3	95.4	158.2	130.0	90.0	114.1	103.3
07	THERMAL & MOISTURE PROTECTION	96.3	112.4	103.3	95.3	106.8	100.3	95.5	98.4	96.8	98.8	129.0	111.9	99.9	147.6	120.7	98.3	114.4	105.3
08	OPENINGS	96.6	119.5	102.0	94.0	112.1	98.2	91.3	103.3	94.1	96.6	136.5	105.9	99.4	167.9	115.4	99.2	116.4	103.2
0920	Plaster & Gypsum Board	95.5	113.7	107.7	95.0	114.5	108.1	91.0	97.6	95.5	95.5	124.4	114.9	97.2	160.0	139.5	99.9	114.5	109.7
0950, 0980	Ceilings & Acoustic Treatment	89.6	113.7	105.8	80.6	114.5	103.4	83.9	97.6	93.2	89.6	124.4	113.1	100.9	160.0	140.8	91.3	114.5	107.0
0960	Flooring	92.6	119.5	100.2	101.8	113.6	105.1	90.2	94.6	91.4	92.6	124.9	101.7	89.9	159.5	109.4	104.8	117.1	108.3
0970, 0990	Wall Finishes & Painting/Coating	87.6	134.8	115.5	95.5	111.5	105.0	87.6	95.5	92.3	87.6	142.7	120.2	91.0	163.9	134.1	98.0	111.5	106.0
09	FINISHES	90.2	118.7	105.6	90.7	114.2	103.4	87.7	97.6	93.0	90.2	128.5	110.9	92.2	160.1	129.0	98.0	115.9	107.7
COVERS	DIVS. 10 - 14, 25, 28, 41, 43, 44, 46	100.0	104.4	101.0	100.0	106.9	101.5	100.0	97.5	99.4	100.0	114.2	103.2	100.0	124.1	105.4	100.0	108.3	101.9
21, 22, 23	FIRE SUPPRESSION, PLUMBING & HVAC	100.0	100.2	100.0	96.4	102.9	98.9	96.5	97.5	96.9	100.1	116.1	106.4	99.9	136.2	114.1	99.9	103.7	101.4
26, 27, 3370	ELECTRICAL, COMMUNICATIONS & UTIL.	97.7	91.5	94.6	91.4	82.2	86.9	89.3	93.1	91.2	98.0	125.6	111.6	96.7	135.5	115.9	98.1	88.7	93.5
MF2018	WEIGHTED AVERAGE	97.6	108.8	102.3	95.7	104.2	99.3	94.6	99.5	96.7	96.6	125.5	108.8	98.3	142.8	117.1	97.5	108.3	102.1

For customer support on your Electrical Change Order Costs with RSMeans data, call 800.448.8182.

INDIANA

DIVISION		ANDERSON 460			BLOOMINGTON 474			COLUMBUS 472			EVANSVILLE 476 - 477			FORT WAYNE 467 - 468			GARY 463 - 464		
		MAT.	INST.	TOTAL	MAT.	INST.	TOTAL	MAT.	INST.	TOTAL	MAT.	INST.	TOTAL	MAT.	INST.	TOTAL	MAT.	INST.	TOTAL
015433	CONTRACTOR EQUIPMENT		94.5	94.5		81.5	81.5		81.5	81.5		110.3	110.3		94.5	94.5		94.5	94.5
0241, 31 - 34	SITE & INFRASTRUCTURE, DEMOLITION	98.7	89.0	92.0	86.5	87.7	87.3	83.2	87.4	86.1	91.9	115.1	107.9	99.8	88.8	92.2	99.3	92.8	94.8
0310	Concrete Forming & Accessories	94.9	77.4	80.0	100.8	82.1	84.9	95.0	80.2	82.4	94.3	78.6	80.9	92.9	73.7	76.5	95.0	111.1	108.7
0320	Concrete Reinforcing	104.9	83.0	94.3	90.4	86.9	88.7	90.8	86.9	88.9	99.0	81.0	90.3	104.9	77.9	91.9	104.9	115.8	110.2
0330	Cast-in-Place Concrete	106.3	74.9	94.6	101.3	75.7	91.7	100.8	76.5	91.8	96.8	83.2	91.7	113.0	75.2	98.9	111.1	112.2	111.5
03	CONCRETE	95.7	78.1	88.0	99.2	80.3	90.9	98.5	79.6	90.2	99.5	80.8	91.3	98.6	75.7	88.6	97.9	112.1	104.1
04	MASONRY	87.4	74.0	79.2	88.9	74.4	80.0	88.8	74.3	79.9	84.3	77.3	80.0	90.7	71.4	78.8	88.8	110.3	102.0
05	METALS	97.7	88.2	94.9	99.1	75.7	92.2	99.1	74.7	91.9	91.9	83.2	89.3	97.7	86.3	94.3	97.7	107.1	100.5
06	WOOD, PLASTICS & COMPOSITES	93.8	77.6	84.9	109.5	82.9	94.9	104.0	80.5	91.1	90.8	77.7	83.6	93.7	73.8	82.7	91.3	109.1	101.1
07	THERMAL & MOISTURE PROTECTION	109.2	75.0	94.4	95.8	77.9	88.0	95.2	78.7	88.0	100.0	81.7	92.1	108.9	76.8	94.9	107.7	106.2	107.0
08	OPENINGS	93.2	75.8	89.2	97.5	80.0	93.4	93.9	78.6	90.3	91.8	76.0	88.1	93.2	71.2	88.1	93.2	114.6	98.2
0920	Plaster & Gypsum Board	106.4	77.3	86.8	98.6	83.2	88.2	95.4	80.7	85.5	94.2	76.7	82.4	105.8	73.4	84.0	99.5	109.7	106.4
0950, 0980	Ceilings & Acoustic Treatment	89.8	77.3	81.4	77.9	83.2	81.5	77.9	80.7	79.8	81.6	76.7	78.3	89.8	73.4	78.7	89.8	109.7	103.2
0960	Flooring	93.5	75.0	88.3	99.6	83.2	95.0	94.4	83.2	91.2	94.1	71.4	87.7	93.5	71.2	87.2	93.5	110.5	98.3
0970, 0990	Wall Finishes & Painting/Coating	92.8	65.8	76.9	84.4	81.1	82.5	84.4	81.1	82.5	90.3	84.4	86.8	92.8	70.2	79.5	92.8	120.7	109.3
09	FINISHES	91.3	75.7	82.9	90.6	82.4	86.2	88.6	81.0	84.5	89.3	77.7	83.0	91.1	72.9	81.3	90.3	111.8	102.0
COVERS	DIVS. 10 - 14, 25, 28, 41, 43, 44, 46	100.0	87.4	97.2	100.0	88.3	97.4	100.0	88.0	97.3	100.0	91.2	98.0	100.0	87.2	97.1	100.0	107.9	101.8
21, 22, 23	FIRE SUPPRESSION, PLUMBING & HVAC	100.0	77.0	91.0	99.7	79.6	91.8	96.2	79.1	89.5	99.9	78.6	91.5	100.0	72.2	89.1	100.0		102.4
26, 27, 3370	ELECTRICAL, COMMUNICATIONS & UTIL.	87.7	83.2	85.5	99.9	86.6	93.3	99.1	86.7	93.0	95.7	81.4	88.7	88.4	74.9	81.7	99.5	110.1	104.7
MF2018	WEIGHTED AVERAGE	96.0	79.8	89.1	97.8	81.1	90.7	96.1	80.5	89.5	95.5	82.7	90.1	96.6	76.3	88.0	97.5	108.1	102.0

INDIANA

DIVISION		INDIANAPOLIS 461 - 462			KOKOMO 469			LAFAYETTE 479			LAWRENCEBURG 470			MUNCIE 473			NEW ALBANY 471		
		MAT.	INST.	TOTAL	MAT.	INST.	TOTAL	MAT.	INST.	TOTAL	MAT.	INST.	TOTAL	MAT.	INST.	TOTAL	MAT.	INST.	TOTAL
015433	CONTRACTOR EQUIPMENT		86.1	86.1		94.5	94.5		81.5	81.5		100.6	100.6		92.8	92.8		90.6	90.6
0241, 31 - 34	SITE & INFRASTRUCTURE, DEMOLITION	99.8	91.3	93.9	94.9	89.0	90.8	83.9	87.4	86.3	81.2	102.2	95.7	86.6	87.9	87.5	78.5	89.8	86.3
0310	Concrete Forming & Accessories	100.4	85.2	87.4	98.0	78.2	81.1	92.5	78.6	80.7	91.5	78.4	80.3	92.3	76.9	79.2	90.1	76.9	78.9
0320	Concrete Reinforcing	106.1	87.2	97.0	94.9	87.2	91.1	90.4	82.9	86.8	89.7	79.8	84.9	100.0	82.9	91.7	91.0	82.6	86.9
0330	Cast-in-Place Concrete	101.2	84.8	95.1	105.2	81.8	96.5	101.4	77.6	92.5	94.8	74.9	87.4	106.4	73.9	94.3	97.9	73.1	88.7
03	CONCRETE	99.0	84.8	92.8	92.5	81.6	87.7	98.7	78.7	89.9	91.8	78.0	85.7	97.4	77.6	88.7	97.3	76.8	88.3
04	MASONRY	89.8	78.7	83.0	87.1	77.2	81.0	94.0	74.0	81.7	73.8	74.2	74.1	90.5	74.1	80.4	80.2	71.1	74.6
05	METALS	94.7	76.0	89.2	94.2	89.8	92.9	97.4	73.7	90.5	93.9	86.2	91.7	100.8	88.0	97.0	96.0	82.2	91.9
06	WOOD, PLASTICS & COMPOSITES	100.1	86.0	92.4	96.8	77.4	86.1	101.2	79.2	89.1	89.1	78.3	83.2	102.6	77.1	88.6	91.2	77.6	83.7
07	THERMAL & MOISTURE PROTECTION	98.7	82.0	91.4	108.1	77.5	94.8	95.2	77.2	87.4	100.8	76.9	90.4	98.2	76.4	88.7	87.6	72.3	80.9
08	OPENINGS	103.9	81.9	98.8	88.4	76.9	85.7	92.4	76.7	88.8	93.6	75.0	89.3	91.0	75.6	87.4	91.4	76.3	87.9
0920	Plaster & Gypsum Board	96.8	85.8	89.4	111.0	77.1	88.2	93.3	79.4	83.9	71.7	78.4	76.2	94.2	77.3	82.8	92.0	77.5	82.2
0950, 0980	Ceilings & Acoustic Treatment	92.7	85.8	88.0	90.6	77.1	81.5	73.9	79.4	77.6	84.0	78.4	80.2	77.9	77.3	77.5	81.6	77.5	78.8
0960	Flooring	97.2	83.2	93.2	97.3	88.4	94.8	93.3	79.2	89.3	68.9	83.2	72.9	93.7	75.0	88.5	91.6	54.0	81.1
0970, 0990	Wall Finishes & Painting/Coating	96.5	81.1	87.4	92.8	68.8	78.6	84.4	81.4	82.6	85.2	73.0	78.0	84.4	65.8	73.4	90.3	66.3	76.1
09	FINISHES	94.9	84.6	89.4	93.1	79.1	85.5	87.2	79.2	82.8	78.8	78.9	78.8	87.9	75.4	81.1	88.5	71.4	79.2
COVERS	DIVS. 10 - 14, 25, 28, 41, 43, 44, 46	100.0	93.7	98.6	100.0	88.6	97.5	100.0	86.6	97.0	100.0	88.0	97.3	100.0	86.3	97.0	100.0	87.7	97.3
21, 22, 23	FIRE SUPPRESSION, PLUMBING & HVAC	100.1	79.9	92.2	96.5	79.3	89.7	96.2	76.2	88.3	97.0	75.9	88.7	99.7	76.9	90.7	96.4	78.6	89.4
26, 27, 3370	ELECTRICAL, COMMUNICATIONS & UTIL.	101.9	86.7	94.4	92.0	78.3	85.3	98.6	78.6	88.7	93.6	74.1	84.0	91.4	74.0	82.9	94.3	74.6	84.6
MF2018	WEIGHTED AVERAGE	98.7	83.2	92.1	94.2	81.1	88.6	95.8	78.1	88.3	92.5	79.6	87.0	96.2	78.2	88.6	93.9	77.3	86.9

INDIANA / IOWA

DIVISION		SOUTH BEND 465 - 466			TERRE HAUTE 478			WASHINGTON 475			BURLINGTON 526			CARROLL 514			CEDAR RAPIDS 522 - 524		
		MAT.	INST.	TOTAL	MAT.	INST.	TOTAL	MAT.	INST.	TOTAL	MAT.	INST.	TOTAL	MAT.	INST.	TOTAL	MAT.	INST.	TOTAL
015433	CONTRACTOR EQUIPMENT		108.2	108.2		110.3	110.3		110.3	110.3		98.8	98.8		98.8	98.8		95.7	95.7
0241, 31 - 34	SITE & INFRASTRUCTURE, DEMOLITION	97.7	94.2	95.3	93.4	115.4	108.6	93.2	115.7	108.7	98.0	91.8	93.7	87.1	92.6	90.9	99.9	91.4	94.0
0310	Concrete Forming & Accessories	95.1	76.0	78.9	95.0	77.1	79.7	96.0	80.5	82.8	96.5	94.2	94.5	84.1	83.9	83.9	102.5	83.9	86.6
0320	Concrete Reinforcing	104.1	83.6	94.2	99.0	83.0	91.3	91.6	82.6	87.2	94.7	97.8	96.2	95.4	85.8	90.8	95.3	80.2	88.0
0330	Cast-in-Place Concrete	108.2	78.4	97.1	93.7	78.3	88.0	102.0	85.0	95.6	107.3	54.8	87.8	107.3	82.2	97.9	107.6	83.4	98.6
03	CONCRETE	97.7	79.7	89.8	102.5	78.8	92.1	108.3	82.6	97.0	95.8	81.5	89.6	94.7	84.3	90.1	95.9	83.8	90.6
04	MASONRY	91.2	74.0	80.6	92.0	73.3	80.5	84.5	79.6	81.5	99.5	71.9	82.5	101.1	72.4	83.5	105.0	80.0	89.6
05	METALS	101.5	102.4	101.7	92.6	84.4	90.2	87.1	84.7	86.4	87.9	99.1	91.2	87.9	95.4	90.1	90.3	93.1	91.2
06	WOOD, PLASTICS & COMPOSITES	96.6	75.2	84.8	93.0	76.7	84.0	93.2	79.2	85.5	91.6	98.1	95.2	77.8	88.4	83.6	98.9	83.6	90.5
07	THERMAL & MOISTURE PROTECTION	103.2	78.7	92.5	100.1	79.1	91.0	100.0	83.5	92.8	104.4	77.1	92.5	104.7	78.8	93.4	105.5	80.1	94.4
08	OPENINGS	93.1	75.5	89.0	92.3	75.3	88.3	89.3	77.1	86.4	94.5	95.8	94.8	98.8	84.2	95.4	99.3	81.2	95.1
0920	Plaster & Gypsum Board	97.7	74.7	82.2	94.2	75.7	81.8	94.0	78.3	83.4	105.0	98.3	100.5	100.9	88.3	92.4	110.3	83.6	92.3
0950, 0980	Ceilings & Acoustic Treatment	94.2	74.7	81.1	81.6	75.7	77.6	76.7	78.3	77.8	99.0	98.3	98.5	99.0	88.3	91.8	101.4	83.6	89.4
0960	Flooring	90.7	86.2	89.5	94.1	75.7	88.9	95.1	83.2	91.7	93.7	69.1	86.8	88.0	79.5	85.6	107.9	84.3	101.3
0970, 0990	Wall Finishes & Painting/Coating	94.3	82.5	87.3	90.3	79.8	84.1	90.3	84.4	86.8	92.8	84.6	87.9	92.8	84.6	87.9	94.7	70.8	80.6
09	FINISHES	92.5	78.3	84.8	89.3	76.9	82.6	88.9	81.3	84.7	93.9	89.7	91.6	90.1	83.6	86.6	99.3	82.7	90.3
COVERS	DIVS. 10 - 14, 25, 28, 41, 43, 44, 46	100.0	88.7	97.5	100.0	88.7	97.5	100.0	91.8	98.2	100.0	93.4	98.5	100.0	89.1	97.6	100.0	93.4	98.5
21, 22, 23	FIRE SUPPRESSION, PLUMBING & HVAC	99.9	75.0	90.1	99.9	76.5	90.7	96.4	81.0	90.4	96.5	83.8	91.5	96.5	77.8	89.2	100.0	81.2	92.6
26, 27, 3370	ELECTRICAL, COMMUNICATIONS & UTIL.	101.3	83.9	92.7	94.1	83.5	88.8	94.5	83.6	89.2	101.1	72.2	86.9	101.8	76.9	89.5	98.8	80.1	89.5
MF2018	WEIGHTED AVERAGE	98.4	81.8	91.4	96.2	81.7	90.1	94.6	84.8	90.4	95.7	84.3	90.9	95.5	82.4	89.9	98.0	83.7	92.0

For customer support on your Electrical Change Order Costs with RSMeans data, call 800.448.8182.

545

City Cost Indexes

IOWA

DIVISION		COUNCIL BLUFFS 515			CRESTON 508			DAVENPORT 527 - 528			DECORAH 521			DES MOINES 500 - 503,509			DUBUQUE 520		
		MAT.	INST.	TOTAL	MAT.	INST.	TOTAL	MAT.	INST.	TOTAL	MAT.	INST.	TOTAL	MAT.	INST.	TOTAL	MAT.	INST.	TOTAL
015433	CONTRACTOR EQUIPMENT		95.0	95.0		98.8	98.8		98.8	98.8		98.8	98.8		102.4	102.4		94.5	94.5
0241, 31 - 34	SITE & INFRASTRUCTURE, DEMOLITION	104.5	88.5	93.4	92.9	93.6	93.4	98.4	94.6	95.8	96.6	91.6	93.2	98.3	99.3	99.0	97.7	88.7	91.5
0310	Concrete Forming & Accessories	83.5	73.6	75.0	79.2	86.7	85.6	102.0	95.9	96.8	94.0	71.8	75.1	95.6	89.5	90.4	84.9	81.5	82.0
0320	Concrete Reinforcing	97.3	79.8	88.8	96.3	85.9	91.3	95.3	99.0	97.1	94.7	85.2	90.1	101.0	101.6	101.3	94.0	79.9	87.2
0330	Cast-in-Place Concrete	111.8	79.1	99.6	114.6	85.7	103.8	103.6	95.2	100.5	104.4	76.0	93.8	97.6	92.1	95.6	105.3	83.0	97.0
03	CONCRETE	98.1	77.6	89.1	97.7	86.8	92.9	94.0	96.7	95.2	93.8	76.6	86.3	91.0	92.8	91.8	92.7	82.6	88.3
04	MASONRY	106.5	76.2	87.8	99.7	79.9	87.5	102.0	92.2	96.0	120.3	69.2	88.9	87.7	89.0	88.5	105.9	68.8	83.1
05	METALS	95.2	92.8	94.5	88.4	96.0	90.6	90.3	105.5	94.8	88.0	94.2	89.9	94.1	96.8	94.9	88.9	92.4	89.9
06	WOOD, PLASTICS & COMPOSITES	76.5	72.1	74.1	69.7	88.4	80.0	98.9	95.3	96.9	88.5	70.9	78.8	87.6	88.5	88.1	78.3	82.2	80.4
07	THERMAL & MOISTURE PROTECTION	104.8	74.8	91.8	106.2	82.2	95.8	104.9	91.6	99.1	104.6	69.9	89.5	98.5	87.6	93.8	105.1	76.5	92.6
08	OPENINGS	98.4	76.8	93.3	108.6	85.9	103.3	99.3	96.7	98.7	97.4	77.9	92.9	100.8	89.6	98.2	98.4	82.4	94.6
0920	Plaster & Gypsum Board	100.9	71.8	81.3	95.8	88.3	90.8	110.3	95.4	100.2	103.7	70.3	81.2	92.8	88.3	89.8	100.9	82.1	88.3
0950, 0980	Ceilings & Acoustic Treatment	99.0	71.8	80.6	90.2	88.3	88.9	101.4	95.4	97.3	99.0	70.3	79.6	93.4	88.3	90.0	99.0	82.1	87.6
0960	Flooring	87.0	84.3	86.2	81.3	69.1	77.8	96.1	89.4	94.2	93.3	69.1	86.5	95.2	94.5	95.0	99.1	69.1	90.7
0970, 0990	Wall Finishes & Painting/Coating	89.6	59.8	72.0	83.6	84.6	84.2	92.8	90.5	91.4	92.8	84.6	87.9	91.6	84.6	87.5	93.9	78.4	84.8
09	FINISHES	91.0	73.7	81.7	85.1	83.4	84.2	95.7	94.1	94.8	93.5	72.4	82.1	93.6	90.0	91.6	94.8	78.9	86.2
COVERS	DIVS. 10 - 14, 25, 28, 41, 43, 44, 46	100.0	91.4	98.1	100.0	91.6	98.1	100.0	96.9	99.3	100.0	88.6	97.5	100.0	95.6	99.0	100.0	92.3	98.3
21, 22, 23	FIRE SUPPRESSION, PLUMBING & HVAC	100.0	72.8	89.4	96.3	79.8	89.8	100.0	93.9	97.6	96.5	74.5	87.9	99.8	86.8	94.7	100.0	75.6	90.4
26, 27, 3370	ELECTRICAL, COMMUNICATIONS & UTIL.	103.9	82.7	93.4	93.5	76.9	85.3	96.9	87.6	92.3	98.8	47.2	73.4	105.1	84.3	94.8	102.5	76.9	89.9
MF2018	WEIGHTED AVERAGE	98.9	79.2	90.6	95.7	84.2	90.8	97.1	94.5	96.0	96.4	73.7	86.8	97.2	90.2	94.3	97.2	79.9	89.9

IOWA

DIVISION		FORT DODGE 505			MASON CITY 504			OTTUMWA 525			SHENANDOAH 516			SIBLEY 512			SIOUX CITY 510 - 511		
		MAT.	INST.	TOTAL	MAT.	INST.	TOTAL	MAT.	INST.	TOTAL	MAT.	INST.	TOTAL	MAT.	INST.	TOTAL	MAT.	INST.	TOTAL
015433	CONTRACTOR EQUIPMENT		98.8	98.8		98.8	98.8		94.5	94.5		95.0	95.0		98.8	98.8		98.8	98.8
0241, 31 - 34	SITE & INFRASTRUCTURE, DEMOLITION	101.4	90.5	93.9	101.5	91.5	94.6	98.0	86.7	90.2	102.6	88.6	92.9	107.7	91.6	96.6	109.6	93.3	98.3
0310	Concrete Forming & Accessories	79.8	78.0	78.3	83.6	71.4	73.2	91.9	87.1	87.8	85.1	77.2	78.3	85.7	37.8	44.9	102.5	75.5	79.4
0320	Concrete Reinforcing	96.3	85.2	90.9	96.2	85.2	90.9	94.7	98.0	96.3	97.3	85.8	91.7	97.3	85.1	91.4	95.3	100.8	98.0
0330	Cast-in-Place Concrete	107.6	41.9	83.1	107.6	71.2	94.0	108.0	63.9	91.6	108.1	82.6	98.6	105.9	55.5	87.2	106.6	89.3	100.2
03	CONCRETE	93.1	67.6	81.9	93.3	74.8	85.2	95.4	81.5	89.3	95.6	81.5	89.4	94.7	54.1	76.8	95.3	85.6	91.0
04	MASONRY	98.6	51.9	69.9	111.6	68.0	84.8	102.3	55.2	73.3	106.0	74.5	86.6	124.6	52.2	80.1	99.2	70.0	81.2
05	METALS	88.5	94.0	90.1	88.5	94.2	90.2	87.9	99.7	91.3	94.2	95.4	94.6	88.1	93.3	89.7	90.3	101.3	93.6
06	WOOD, PLASTICS & COMPOSITES	70.1	88.4	80.2	74.0	70.9	72.3	85.7	97.9	92.4	78.3	78.5	78.4	79.1	34.2	54.4	98.9	74.4	85.4
07	THERMAL & MOISTURE PROTECTION	105.5	65.2	88.0	105.0	70.5	90.0	105.3	70.3	90.1	104.0	73.3	90.7	104.3	55.2	82.9	104.9	73.4	91.2
08	OPENINGS	102.5	75.9	96.3	94.7	77.9	90.8	98.8	92.8	97.4	90.2	76.4	87.0	95.6	46.0	84.0	99.3	81.6	95.1
0920	Plaster & Gypsum Board	95.8	88.3	90.8	95.8	70.3	78.6	101.5	98.3	99.3	100.9	78.3	85.7	100.9	32.5	54.9	110.3	73.8	85.8
0950, 0980	Ceilings & Acoustic Treatment	90.2	88.3	88.9	90.2	70.3	76.8	99.0	98.3	98.5	99.0	78.3	85.0	99.0	32.5	54.2	101.4	73.8	82.8
0960	Flooring	82.6	69.1	78.8	84.6	69.1	80.2	102.2	69.1	92.9	87.7	73.9	83.8	88.9	69.1	83.3	96.1	73.3	89.7
0970, 0990	Wall Finishes & Painting/Coating	83.6	81.3	82.2	83.6	84.6	84.2	93.9	84.6	88.4	89.6	84.6	86.6	92.8	83.6	87.3	92.8	69.1	78.8
09	FINISHES	87.0	78.0	82.1	87.6	72.1	79.2	95.9	84.8	89.9	91.1	77.1	83.5	93.3	46.4	67.9	97.2	73.8	84.5
COVERS	DIVS. 10 - 14, 25, 28, 41, 43, 44, 46	100.0	84.8	96.6	100.0	88.2	97.4	100.0	87.3	97.2	100.0	88.2	97.4	100.0	78.9	95.3	100.0	91.4	98.1
21, 22, 23	FIRE SUPPRESSION, PLUMBING & HVAC	96.3	71.4	86.6	96.3	77.7	89.0	96.5	73.1	87.3	96.5	83.9	91.6	96.5	71.5	86.7	100.0	81.2	92.6
26, 27, 3370	ELECTRICAL, COMMUNICATIONS & UTIL.	99.8	69.4	84.8	98.9	47.2	73.4	100.9	71.0	86.2	98.8	78.4	88.7	98.8	47.2	73.4	98.8	74.0	86.6
MF2018	WEIGHTED AVERAGE	95.5	73.7	86.3	95.3	74.0	86.3	96.3	78.7	88.9	96.1	81.9	90.1	96.8	62.4	82.3	97.7	81.5	90.9

	IOWA						KANSAS												
DIVISION		SPENCER 513			WATERLOO 506 - 507			BELLEVILLE 669			COLBY 677			DODGE CITY 678			EMPORIA 668		
		MAT.	INST.	TOTAL	MAT.	INST.	TOTAL	MAT.	INST.	TOTAL	MAT.	INST.	TOTAL	MAT.	INST.	TOTAL	MAT.	INST.	TOTAL
015433	CONTRACTOR EQUIPMENT		98.8	98.8		98.8	98.8		103.7	103.7		103.7	103.7		103.7	103.7		101.9	101.9
0241, 31 - 34	SITE & INFRASTRUCTURE, DEMOLITION	107.8	90.4	95.7	106.9	92.7	97.1	111.3	92.0	98.0	106.7	92.7	97.0	109.4	91.7	97.2	103.1	89.7	93.8
0310	Concrete Forming & Accessories	91.8	37.5	45.5	94.4	68.7	72.5	94.8	54.3	60.3	97.5	61.2	66.6	91.1	61.1	65.5	86.2	70.0	72.4
0320	Concrete Reinforcing	97.3	85.1	91.4	96.9	80.3	88.9	98.9	102.6	100.7	97.1	102.6	99.8	94.7	102.3	98.4	97.5	102.8	100.1
0330	Cast-in-Place Concrete	105.9	65.5	90.9	115.2	84.0	103.6	121.0	83.5	107.1	119.5	87.2	107.5	121.6	86.9	108.7	117.0	87.3	106.0
03	CONCRETE	95.0	57.4	78.5	99.0	77.2	89.4	110.5	74.3	94.6	108.3	78.7	95.3	109.7	78.5	96.0	103.1	82.8	94.2
04	MASONRY	124.6	52.2	80.1	99.4	75.1	84.5	86.4	58.5	69.3	100.5	64.6	78.4	110.7	59.6	79.3	91.8	66.0	76.0
05	METALS	88.1	93.0	89.5	90.7	93.0	91.4	95.6	99.2	96.7	92.8	100.0	94.9	94.2	98.6	95.5	95.3	100.2	96.8
06	WOOD, PLASTICS & COMPOSITES	85.5	34.2	57.3	87.2	64.4	74.6	93.3	51.2	70.1	96.9	57.4	75.2	88.9	57.4	71.5	84.3	68.8	75.8
07	THERMAL & MOISTURE PROTECTION	105.3	55.8	83.8	105.3	77.2	93.0	91.4	62.1	78.7	97.8	65.4	83.7	97.8	66.6	84.2	89.5	75.4	83.3
08	OPENINGS	106.7	46.0	92.5	95.1	72.9	90.0	94.1	62.5	86.7	97.9	65.9	90.4	97.8	65.9	90.4	92.1	74.4	88.0
0920	Plaster & Gypsum Board	101.5	32.5	55.1	103.8	63.6	76.7	88.7	50.0	62.6	96.8	56.3	69.6	90.5	56.3	67.5	85.5	68.1	73.8
0950, 0980	Ceilings & Acoustic Treatment	99.0	32.5	54.2	91.8	63.6	72.8	76.5	50.0	58.6	77.3	56.3	63.2	77.3	56.3	63.2	76.5	68.1	70.8
0960	Flooring	91.6	69.1	85.3	89.6	78.9	86.6	92.9	68.4	86.0	87.6	68.4	82.2	83.8	68.4	79.4	88.3	68.4	82.7
0970, 0990	Wall Finishes & Painting/Coating	92.8	57.2	71.7	83.6	81.3	82.2	89.7	57.9	70.9	96.1	57.9	73.5	96.1	57.9	73.5	89.7	57.9	70.9
09	FINISHES	94.2	42.3	66.1	90.7	70.9	80.0	86.1	55.7	69.6	84.7	60.8	71.8	82.8	60.8	70.9	83.3	67.6	74.8
COVERS	DIVS. 10 - 14, 25, 28, 41, 43, 44, 46	100.0	78.9	95.3	100.0	91.0	98.0	100.0	83.3	96.3	100.0	86.1	96.9	100.0	86.0	96.9	100.0	85.9	96.9
21, 22, 23	FIRE SUPPRESSION, PLUMBING & HVAC	96.5	71.5	86.7	99.9	80.1	92.1	96.4	70.3	86.1	96.5	71.1	86.5	96.5	71.1	88.7	96.4	73.5	87.4
26, 27, 3370	ELECTRICAL, COMMUNICATIONS & UTIL.	100.4	47.2	74.2	95.4	61.6	78.7	104.1	62.7	83.7	95.5	67.9	81.9	92.9	67.9	80.6	101.4	66.9	84.4
MF2018	WEIGHTED AVERAGE	98.2	62.3	83.0	96.8	77.4	88.6	97.9	70.7	86.4	97.3	74.0	87.4	98.6	73.3	87.9	96.1	76.5	87.8

		KANSAS																
	DIVISION	FORT SCOTT			HAYS			HUTCHINSON			INDEPENDENCE			KANSAS CITY			LIBERAL		
		667			676			675			673			660 - 662			679		
		MAT.	INST.	TOTAL	MAT.	INST.	TOTAL	MAT.	INST.	TOTAL	MAT.	INST.	TOTAL	MAT.	INST.	TOTAL	MAT.	INST.	TOTAL
015433	CONTRACTOR EQUIPMENT		102.8	102.8		103.7	103.7		103.7	103.7		103.7	103.7		100.4	100.4		103.7	103.7
0241, 31 - 34	SITE & INFRASTRUCTURE, DEMOLITION	99.9	89.7	92.8	111.4	92.2	98.2	90.6	92.7	92.0	110.2	92.7	98.1	94.3	90.2	91.5	111.3	92.3	98.2
0310	Concrete Forming & Accessories	102.8	82.9	85.9	95.2	58.9	64.3	86.0	56.6	61.0	105.8	67.8	73.4	99.4	98.4	98.6	91.5	58.8	63.7
0320	Concrete Reinforcing	96.9	100.8	98.8	94.7	102.6	98.5	94.7	102.6	98.5	94.1	100.8	97.3	94.0	105.3	99.4	96.1	102.3	99.1
0330	Cast-in-Place Concrete	108.5	82.4	98.8	94.2	83.8	90.3	87.3	86.9	87.2	122.1	87.1	109.1	92.9	97.4	94.6	94.2	83.5	90.3
03	CONCRETE	98.4	86.6	93.2	99.8	76.5	89.6	83.0	76.5	80.2	110.8	81.3	97.8	90.7	99.7	94.6	101.7	76.3	90.5
04	MASONRY	92.7	56.1	70.2	109.5	58.6	78.2	100.2	64.5	78.3	97.8	64.5	77.3	93.5	98.1	96.3	108.2	53.6	74.6
05	METALS	95.3	99.0	96.4	92.3	99.8	94.5	92.1	99.0	94.1	92.0	98.8	94.0	103.1	106.2	104.0	92.6	98.7	94.4
06	WOOD, PLASTICS & COMPOSITES	103.8	90.5	96.5	93.9	57.4	73.8	84.0	51.4	66.0	107.3	65.9	84.5	99.6	98.5	99.0	89.6	57.4	71.9
07	THERMAL & MOISTURE PROTECTION	90.5	72.8	82.8	98.1	62.8	82.8	96.6	64.3	82.5	97.8	74.9	87.9	90.3	99.3	94.2	98.3	60.9	82.0
08	OPENINGS	92.1	85.9	90.6	97.8	65.9	90.4	97.8	62.6	89.6	95.9	70.2	89.9	93.2	97.4	94.2	97.9	65.9	90.4
0920	Plaster & Gypsum Board	91.2	90.4	90.7	94.3	56.3	68.7	89.2	50.2	63.0	105.0	65.1	78.1	84.8	98.6	94.1	91.4	56.3	67.8
0950, 0980	Ceilings & Acoustic Treatment	76.5	90.4	85.9	77.3	56.3	63.2	77.3	50.2	59.0	77.3	65.1	69.0	76.5	98.6	91.4	77.3	56.3	63.2
0960	Flooring	102.6	66.9	92.5	86.4	68.4	81.3	80.9	68.4	77.3	92.0	66.9	84.9	82.4	95.3	86.0	84.1	68.4	79.6
0970, 0990	Wall Finishes & Painting/Coating	91.3	77.5	83.1	96.1	57.9	73.5	96.1	57.9	73.5	96.1	57.9	73.5	97.1	102.7	100.4	96.1	57.9	73.5
09	FINISHES	88.5	80.7	84.3	84.5	59.3	70.9	80.3	57.3	67.9	87.3	66.1	75.8	83.6	98.3	91.5	83.7	59.3	70.5
COVERS	DIVS. 10 - 14, 25, 28, 41, 43, 44, 46	100.0	86.8	97.1	100.0	84.0	96.4	100.0	85.4	96.7	100.0	87.0	97.1	100.0	94.5	98.8	100.0	84.0	96.4
21, 22, 23	FIRE SUPPRESSION, PLUMBING & HVAC	96.4	67.4	85.0	96.5	67.4	85.1	96.5	71.1	86.5	96.5	70.4	86.3	99.9	99.1	99.6	96.5	69.3	85.8
26, 27, 3370	ELECTRICAL, COMMUNICATIONS & UTIL.	100.6	67.9	84.5	94.6	67.9	81.4	90.5	61.0	75.9	92.3	72.8	82.7	106.2	98.1	102.2	92.9	67.9	80.6
MF2018	WEIGHTED AVERAGE	96.0	77.1	88.0	96.5	72.0	86.1	92.5	71.9	83.8	97.2	76.0	88.2	97.3	98.8	97.8	96.5	71.7	86.0

		KANSAS									KENTUCKY								
	DIVISION	SALINA			TOPEKA			WICHITA			ASHLAND			BOWLING GREEN			CAMPTON		
		674			664 - 666			670 - 672			411 - 412			421 - 422			413 - 414		
		MAT.	INST.	TOTAL	MAT.	INST.	TOTAL	MAT.	INST.	TOTAL	MAT.	INST.	TOTAL	MAT.	INST.	TOTAL	MAT.	INST.	TOTAL
015433	CONTRACTOR EQUIPMENT		103.7	103.7		105.0	105.0		106.7	106.7		97.4	97.4		90.6	90.6		96.8	96.8
0241, 31 - 34	SITE & INFRASTRUCTURE, DEMOLITION	99.8	92.2	94.6	98.4	94.4	95.6	96.8	98.5	97.9	114.0	78.8	89.7	78.7	89.4	86.1	87.5	90.7	89.7
0310	Concrete Forming & Accessories	87.8	61.4	65.3	96.7	66.1	70.7	92.9	57.0	62.3	87.2	89.6	89.3	86.5	78.8	79.9	89.3	80.9	82.2
0320	Concrete Reinforcing	94.1	100.5	97.2	93.5	103.8	98.5	92.4	100.4	96.3	92.5	94.2	93.3	89.7	76.8	83.5	90.6	93.0	91.7
0330	Cast-in-Place Concrete	105.6	83.8	97.5	97.6	86.4	93.4	98.7	79.5	91.5	89.1	94.6	91.1	88.4	69.7	81.4	98.6	68.1	87.2
03	CONCRETE	96.8	77.3	88.2	92.3	80.8	87.3	92.4	73.7	84.2	93.1	93.3	93.2	91.6	75.7	84.6	95.0	78.8	87.9
04	MASONRY	125.4	54.9	82.1	87.7	64.3	73.3	97.5	54.1	70.8	90.7	92.0	91.5	92.7	69.2	78.3	89.7	54.7	68.2
05	METALS	94.0	98.9	95.5	99.5	100.0	99.6	96.2	97.5	96.6	95.1	107.7	98.8	96.7	82.8	92.6	96.0	89.8	94.2
06	WOOD, PLASTICS & COMPOSITES	85.4	60.7	71.8	99.9	65.2	80.8	98.7	54.4	74.3	73.4	87.7	81.3	85.7	80.6	82.9	84.2	88.3	86.5
07	THERMAL & MOISTURE PROTECTION	97.2	62.1	82.0	94.0	74.7	85.6	96.2	60.8	80.8	91.2	88.3	90.0	87.6	77.9	83.4	100.2	67.8	86.1
08	OPENINGS	97.8	67.3	90.7	103.5	74.1	96.7	101.3	63.8	92.5	90.6	89.0	90.2	91.4	77.1	88.1	92.6	87.4	91.4
0920	Plaster & Gypsum Board	89.2	59.8	69.4	99.5	64.2	75.7	95.7	53.1	67.0	59.3	87.6	78.4	87.6	80.6	82.9	87.6	87.6	87.6
0950, 0980	Ceilings & Acoustic Treatment	77.3	59.8	65.5	87.2	64.2	71.7	88.0	53.1	64.5	77.0	87.6	84.2	81.6	80.6	80.9	81.6	87.6	85.7
0960	Flooring	82.2	68.4	78.3	98.2	68.4	89.8	94.3	68.4	87.0	73.8	81.3	75.9	89.5	61.9	81.8	91.6	64.3	83.9
0970, 0990	Wall Finishes & Painting/Coating	96.1	57.9	73.5	92.8	67.7	78.0	94.9	57.9	73.0	91.6	90.4	90.9	90.3	67.6	76.9	90.3	53.9	68.8
09	FINISHES	81.5	61.3	70.6	93.9	65.7	78.6	90.3	57.7	72.7	76.1	88.1	82.6	87.1	74.6	80.3	87.9	75.9	81.4
COVERS	DIVS. 10 - 14, 25, 28, 41, 43, 44, 46	100.0	85.3	96.7	100.0	80.3	95.6	100.0	85.1	96.7	100.0	88.2	97.4	100.0	84.4	96.5	100.0	47.8	88.4
21, 22, 23	FIRE SUPPRESSION, PLUMBING & HVAC	100.0	70.5	88.5	100.0	73.2	89.5	99.8	71.0	88.5	96.2	85.0	91.8	99.9	77.0	90.9	96.4	76.4	88.6
26, 27, 3370	ELECTRICAL, COMMUNICATIONS & UTIL.	92.7	73.8	83.4	105.1	71.9	88.7	95.9	73.8	85.0	91.9	88.3	90.1	94.6	74.7	84.8	92.1	88.2	90.2
MF2018	WEIGHTED AVERAGE	97.2	73.5	87.2	98.6	76.5	89.3	97.0	72.6	86.7	93.2	89.7	91.7	94.6	77.2	87.3	94.4	78.0	87.5

		KENTUCKY																	
	DIVISION	CORBIN			COVINGTON			ELIZABETHTOWN			FRANKFORT			HAZARD			HENDERSON		
		407 - 409			410			427			406			417 - 418			424		
		MAT.	INST.	TOTAL	MAT.	INST.	TOTAL	MAT.	INST.	TOTAL	MAT.	INST.	TOTAL	MAT.	INST.	TOTAL	MAT.	INST.	TOTAL
015433	CONTRACTOR EQUIPMENT		96.8	96.8		100.6	100.6		90.6	90.6		100.1	100.1		96.8	96.8		110.3	110.3
0241, 31 - 34	SITE & INFRASTRUCTURE, DEMOLITION	91.6	91.2	91.3	82.7	101.9	95.9	72.9	89.3	84.3	90.0	96.8	94.7	85.3	91.8	89.7	81.7	114.6	104.5
0310	Concrete Forming & Accessories	85.0	76.0	77.3	84.9	70.8	72.8	81.3	72.1	73.5	102.0	78.5	82.0	85.8	81.8	82.4	92.3	77.1	79.3
0320	Concrete Reinforcing	89.6	92.3	90.9	89.3	78.6	84.1	90.2	83.7	87.0	94.0	84.0	89.2	91.0	92.5	91.7	89.8	84.4	87.2
0330	Cast-in-Place Concrete	91.0	72.5	84.1	94.3	79.6	88.9	79.9	67.6	75.3	91.1	77.0	85.8	94.8	69.6	85.4	78.1	84.5	80.5
03	CONCRETE	84.2	78.0	81.5	93.4	76.1	85.8	83.5	73.1	79.0	87.1	79.3	83.6	91.8	79.7	86.5	88.7	81.3	85.5
04	MASONRY	81.9	60.0	68.4	104.0	72.0	84.3	77.4	61.8	67.8	78.6	74.2	75.9	88.6	56.8	69.0	96.0	78.7	85.4
05	METALS	90.9	89.3	90.5	93.9	88.1	92.2	95.9	85.3	92.7	93.1	86.4	91.1	96.0	89.8	94.2	86.8	87.0	86.9
06	WOOD, PLASTICS & COMPOSITES	72.1	77.6	75.1	81.8	68.3	74.3	80.6	73.9	76.9	101.5	77.7	88.4	80.9	88.3	85.0	84.8	75.8	81.5
07	THERMAL & MOISTURE PROTECTION	103.8	69.5	88.9	101.1	72.1	88.5	87.0	68.5	79.0	101.8	75.7	90.5	100.1	69.5	86.8	99.3	82.8	92.1
08	OPENINGS	87.3	70.5	83.4	94.5	71.8	89.2	91.4	73.0	87.1	98.9	78.4	94.1	93.0	87.2	91.6	89.6	78.5	87.0
0920	Plaster & Gypsum Board	88.4	76.6	80.5	68.2	68.0	68.1	86.3	73.7	77.8	91.6	76.6	81.5	86.3	87.6	87.2	90.9	74.8	80.0
0950, 0980	Ceilings & Acoustic Treatment	74.7	76.6	76.0	84.0	68.0	73.2	81.6	73.7	76.2	90.5	76.6	81.1	81.6	87.6	85.7	76.7	74.8	75.4
0960	Flooring	88.4	64.3	81.6	66.3	83.5	71.1	86.9	73.8	83.2	98.9	78.6	93.2	89.8	64.3	82.6	93.2	76.0	88.4
0970, 0990	Wall Finishes & Painting/Coating	87.2	62.0	72.3	85.2	71.1	76.9	90.3	68.1	77.2	94.2	91.7	92.8	90.3	53.9	68.8	90.3	86.7	88.2
09	FINISHES	83.0	72.5	77.3	77.8	72.7	75.0	85.8	71.6	78.1	93.2	80.2	86.2	87.0	76.4	81.3	87.1	77.9	82.1
COVERS	DIVS. 10 - 14, 25, 28, 41, 43, 44, 46	100.0	89.8	97.7	100.0	85.9	96.8	100.0	70.0	93.3	100.0	58.5	90.8	100.0	48.5	88.5	100.0	55.8	90.1
21, 22, 23	FIRE SUPPRESSION, PLUMBING & HVAC	96.5	74.2	87.7	97.0	76.0	88.8	96.6	77.9	89.3	99.9	81.0	92.5	96.4	77.5	89.0	96.6	79.3	89.8
26, 27, 3370	ELECTRICAL, COMMUNICATIONS & UTIL.	90.0	88.2	89.1	95.6	72.4	84.2	91.9	77.9	85.0	100.6	77.9	89.4	92.1	88.2	90.2	94.0	77.9	86.1
MF2018	WEIGHTED AVERAGE	90.8	78.0	85.4	94.3	77.8	87.3	91.3	75.6	84.7	95.4	80.3	89.0	93.8	78.7	87.4	92.2	81.9	87.8

KENTUCKY

DIVISION		LEXINGTON 403 - 405			LOUISVILLE 400 - 402			OWENSBORO 423			PADUCAH 420			PIKEVILLE 415 - 416			SOMERSET 425 - 426		
		MAT.	INST.	TOTAL	MAT.	INST.	TOTAL	MAT.	INST.	TOTAL	MAT.	INST.	TOTAL	MAT.	INST.	TOTAL	MAT.	INST.	TOTAL
015433	CONTRACTOR EQUIPMENT		96.8	96.8		93.7	93.7		110.3	110.3		110.3	110.3		97.4	97.4		96.8	96.8
0241, 31 - 34	SITE & INFRASTRUCTURE, DEMOLITION	93.8	92.8	93.1	87.8	94.4	92.3	91.8	115.3	108.0	84.4	114.8	105.4	125.2	78.0	92.6	77.9	91.2	87.1
0310	Concrete Forming & Accessories	96.5	73.7	77.1	93.5	79.1	81.2	90.8	76.8	78.9	88.8	81.2	82.3	96.0	85.6	87.2	86.8	76.6	78.1
0320	Concrete Reinforcing	98.4	84.0	91.4	94.8	84.2	89.6	89.8	77.3	83.8	90.4	82.5	86.6	93.0	94.1	93.5	90.2	92.5	91.3
0330	Cast-in-Place Concrete	93.1	84.4	89.9	89.3	70.7	82.4	91.1	83.9	88.4	83.2	81.1	82.4	97.9	89.4	94.8	78.1	87.9	81.7
03	CONCRETE	86.9	79.7	83.8	85.8	77.4	82.1	100.7	79.8	91.5	93.3	81.7	88.2	106.8	89.7	99.3	78.6	83.7	80.8
04	MASONRY	80.6	71.1	74.8	78.7	72.0	74.6	88.8	78.6	82.5	91.5	79.2	83.9	88.5	83.2	85.3	83.3	63.2	71.0
05	METALS	93.3	86.7	91.3	95.5	85.5	92.5	88.3	84.7	87.3	85.3	87.5	86.0	95.0	107.5	98.7	95.9	90.0	94.2
06	WOOD, PLASTICS & COMPOSITES	87.3	71.6	78.6	88.8	80.8	84.4	86.3	75.8	80.5	84.0	81.4	82.6	83.3	87.7	85.7	81.5	77.6	79.3
07	THERMAL & MOISTURE PROTECTION	104.1	73.9	90.9	101.0	75.7	90.0	100.1	76.6	89.9	99.4	82.6	92.1	92.1	80.5	87.0	99.3	71.5	87.2
08	OPENINGS	87.5	74.6	84.5	88.8	76.8	86.0	89.6	76.7	86.6	89.0	81.1	87.1	91.2	85.7	89.9	92.1	77.1	88.6
0920	Plaster & Gypsum Board	97.7	70.4	79.4	93.7	80.6	84.9	89.3	74.8	79.5	88.6	80.6	83.2	63.1	87.6	79.6	86.3	76.6	79.8
0950, 0980	Ceilings & Acoustic Treatment	77.9	70.4	72.9	82.8	80.6	81.3	76.7	74.8	75.4	76.7	80.6	79.3	77.0	87.6	84.2	81.6	76.6	78.2
0960	Flooring	93.4	64.3	85.2	95.0	62.9	86.0	92.6	61.9	84.0	91.5	76.0	87.2	78.1	64.3	74.2	90.1	64.3	82.8
0970, 0990	Wall Finishes & Painting/Coating	87.2	79.7	82.8	93.9	68.1	78.7	90.3	86.7	88.2	90.3	73.1	80.2	91.6	91.7	91.7	90.3	68.1	77.2
09	FINISHES	86.4	72.0	78.6	90.8	74.9	82.2	87.3	74.7	80.5	86.5	79.4	82.7	78.7	82.6	80.8	86.4	73.3	79.3
COVERS	DIVS. 10 - 14, 25, 28, 41, 43, 44, 46	100.0	92.1	98.2	100.0	90.5	97.9	100.0	93.0	98.4	100.0	91.1	98.0	100.0	48.2	88.4	100.0	89.2	97.6
21, 22, 23	FIRE SUPPRESSION, PLUMBING & HVAC	100.0	76.4	90.8	99.9	80.0	92.1	99.9	77.1	91.0	96.6	78.2	89.4	96.2	82.3	90.7	96.6	74.0	87.7
26, 27, 3370	ELECTRICAL, COMMUNICATIONS & UTIL.	92.5	74.7	83.7	96.4	77.9	87.3	94.1	73.7	84.0	96.3	77.3	86.9	94.8	88.3	91.6	92.5	88.2	90.4
MF2018	WEIGHTED AVERAGE	93.1	78.1	86.7	93.9	79.6	87.8	94.7	81.0	88.6	92.4	83.2	88.6	95.8	85.4	91.4	91.6	79.5	86.5

LOUISIANA

DIVISION		ALEXANDRIA 713 - 714			BATON ROUGE 707 - 708			HAMMOND 704			LAFAYETTE 705			LAKE CHARLES 706			MONROE 712		
		MAT.	INST.	TOTAL	MAT.	INST.	TOTAL	MAT.	INST.	TOTAL	MAT.	INST.	TOTAL	MAT.	INST.	TOTAL	MAT.	INST.	TOTAL
015433	CONTRACTOR EQUIPMENT		90.9	90.9		91.9	91.9		89.3	89.3		89.3	89.3		88.8	88.8		90.9	90.9
0241, 31 - 34	SITE & INFRASTRUCTURE, DEMOLITION	98.7	87.7	91.1	102.1	91.2	94.6	101.2	85.7	90.5	102.5	87.9	92.4	103.2	87.1	92.0	98.7	87.6	91.1
0310	Concrete Forming & Accessories	77.8	60.0	62.6	99.0	71.8	75.8	79.0	54.9	58.5	95.6	67.8	71.9	96.4	67.8	72.0	77.4	59.4	62.1
0320	Concrete Reinforcing	92.2	54.3	73.9	93.3	54.3	74.5	92.1	54.3	73.8	93.4	54.3	74.5	93.4	54.3	74.5	91.2	54.3	73.3
0330	Cast-in-Place Concrete	89.0	66.2	80.5	88.8	73.1	82.9	87.6	62.6	78.3	87.1	66.1	79.3	91.7	66.1	82.2	89.0	64.1	79.7
03	CONCRETE	82.8	62.0	73.6	85.5	69.8	78.6	85.3	58.3	73.7	86.8	65.4	77.4	89.0	65.4	78.6	82.6	61.0	73.1
04	MASONRY	109.3	64.3	81.6	91.7	66.7	76.3	95.3	60.0	73.6	95.3	66.6	77.7	94.7	66.6	77.5	103.9	63.1	78.8
05	METALS	92.6	70.9	86.2	96.9	74.0	90.1	88.4	69.3	82.8	87.7	69.6	82.4	87.7	69.6	82.4	92.6	70.8	86.2
06	WOOD, PLASTICS & COMPOSITES	89.2	58.8	72.5	103.0	73.2	86.6	83.0	54.3	67.2	103.7	68.0	84.1	101.9	68.0	83.3	88.6	58.8	72.2
07	THERMAL & MOISTURE PROTECTION	99.0	65.6	84.5	97.7	69.2	85.3	96.7	62.9	82.0	97.3	67.4	84.3	97.1	67.2	84.1	99.0	64.8	84.1
08	OPENINGS	110.8	58.6	98.6	97.9	70.4	91.5	96.2	54.9	86.6	99.8	62.2	91.0	99.8	62.2	91.0	110.7	57.1	98.2
0920	Plaster & Gypsum Board	83.6	58.1	66.4	101.0	72.7	82.0	100.0	53.4	68.7	108.6	67.6	81.0	108.6	67.6	81.0	83.3	58.1	66.3
0950, 0980	Ceilings & Acoustic Treatment	87.1	58.1	67.6	92.9	72.7	79.3	96.9	53.4	67.6	94.4	67.6	76.3	95.3	67.6	76.6	87.1	58.1	67.6
0960	Flooring	85.1	68.2	80.3	93.0	68.2	86.0	89.9	68.2	83.8	98.2	68.2	89.8	98.2	68.2	89.8	84.7	68.2	80.1
0970, 0990	Wall Finishes & Painting/Coating	88.8	59.7	71.6	94.7	59.7	74.0	97.7	59.9	75.3	97.7	59.7	75.2	97.7	59.7	75.2	88.8	59.7	71.6
09	FINISHES	80.9	61.2	70.2	92.2	70.3	80.3	90.4	57.5	72.6	93.5	67.2	79.3	93.7	67.2	79.4	80.7	60.9	70.0
COVERS	DIVS. 10 - 14, 25, 28, 41, 43, 44, 46	100.0	79.1	95.3	100.0	84.2	96.5	100.0	80.7	95.7	100.0	83.3	96.3	100.0	83.3	96.3	100.0	78.9	95.3
21, 22, 23	FIRE SUPPRESSION, PLUMBING & HVAC	100.0	63.2	85.6	99.9	64.1	85.9	96.7	61.0	82.7	100.2	64.2	86.0	100.2	64.5	86.2	100.0	61.9	85.0
26, 27, 3370	ELECTRICAL, COMMUNICATIONS & UTIL.	94.3	61.4	78.1	101.7	58.0	80.2	97.0	69.6	83.5	98.1	63.8	81.2	97.7	65.9	82.0	95.9	57.3	76.9
MF2018	WEIGHTED AVERAGE	95.9	65.6	83.1	96.6	69.3	85.1	93.7	64.4	81.3	95.4	67.9	83.8	95.6	68.2	84.1	95.8	64.4	82.5

LOUISIANA / MAINE

DIVISION		NEW ORLEANS 700 - 701			SHREVEPORT 710 - 711			THIBODAUX 703			AUGUSTA 043			BANGOR 044			BATH 045		
		MAT.	INST.	TOTAL	MAT.	INST.	TOTAL	MAT.	INST.	TOTAL	MAT.	INST.	TOTAL	MAT.	INST.	TOTAL	MAT.	INST.	TOTAL
015433	CONTRACTOR EQUIPMENT		88.6	88.6		93.9	93.9		89.3	89.3		98.4	98.4		95.7	95.7		95.7	95.7
0241, 31 - 34	SITE & INFRASTRUCTURE, DEMOLITION	104.3	93.9	97.1	101.1	92.3	95.0	103.6	87.6	92.6	88.2	98.0	95.0	90.6	95.5	94.0	88.3	94.2	92.4
0310	Concrete Forming & Accessories	95.7	68.9	72.8	92.9	60.3	65.1	90.2	64.5	68.2	98.6	78.6	81.5	93.0	76.5	79.0	89.1	78.4	80.0
0320	Concrete Reinforcing	92.9	54.4	74.3	92.5	54.3	74.0	92.1	54.3	73.8	102.6	81.1	92.2	94.1	81.1	87.8	93.1	81.1	87.3
0330	Cast-in-Place Concrete	90.7	69.7	82.9	92.0	65.2	82.0	94.1	64.3	83.0	89.8	115.3	99.3	70.9	114.1	87.0	70.9	114.3	87.1
03	CONCRETE	94.4	66.6	82.2	86.2	61.7	75.4	90.6	63.2	78.5	95.5	92.3	94.1	88.5	91.0	89.6	88.6	91.9	90.1
04	MASONRY	99.8	62.6	77.0	94.5	63.4	75.4	119.9	62.8	84.8	101.0	94.1	96.8	115.8	93.7	102.2	122.5	94.1	105.0
05	METALS	99.1	62.6	88.4	96.7	70.2	88.9	88.4	69.5	82.8	105.4	92.3	101.6	93.7	93.0	93.5	92.1	92.9	92.3
06	WOOD, PLASTICS & COMPOSITES	100.3	71.1	84.2	98.0	59.7	76.9	90.5	65.5	76.8	93.5	75.6	83.6	90.7	73.0	80.9	85.1	75.4	79.8
07	THERMAL & MOISTURE PROTECTION	95.0	68.0	83.3	97.4	65.6	83.6	96.9	65.2	83.1	111.0	100.6	106.5	109.0	100.0	105.1	108.9	100.1	105.1
08	OPENINGS	98.8	64.2	90.7	104.7	56.6	93.5	100.7	56.9	90.5	102.7	78.5	97.1	96.4	77.1	91.9	96.4	78.4	92.2
0920	Plaster & Gypsum Board	96.5	70.5	79.0	91.5	58.8	69.5	101.6	65.0	77.0	109.7	74.5	86.0	116.8	71.9	86.6	112.0	74.5	86.8
0950, 0980	Ceilings & Acoustic Treatment	97.5	70.5	79.3	92.8	58.8	69.9	96.9	65.0	75.4	107.1	74.5	85.1	89.6	71.9	77.7	88.6	74.5	79.1
0960	Flooring	103.6	68.2	93.7	91.2	68.2	84.8	95.8	68.2	88.0	89.9	110.6	95.7	82.5	110.6	90.4	80.9	110.6	89.2
0970, 0990	Wall Finishes & Painting/Coating	103.6	60.6	78.2	87.7	59.7	71.1	98.9	59.9	75.8	96.0	88.9	91.8	90.6	88.9	89.6	90.6	88.9	89.6
09	FINISHES	96.6	68.2	81.2	87.7	61.5	73.5	92.6	64.8	77.6	95.8	84.7	89.8	89.7	83.0	86.1	88.3	84.6	86.3
COVERS	DIVS. 10 - 14, 25, 28, 41, 43, 44, 46	100.0	86.5	97.0	100.0	82.6	96.1	100.0	81.8	95.9	100.0	98.0	99.6	100.0	99.6	99.9	100.0	100.9	100.2
21, 22, 23	FIRE SUPPRESSION, PLUMBING & HVAC	100.1	62.5	85.4	99.9	62.8	85.3	96.7	62.5	83.2	100.0	74.3	89.9	100.3	74.1	90.0	96.7	74.3	87.9
26, 27, 3370	ELECTRICAL, COMMUNICATIONS & UTIL.	102.1	70.6	86.6	101.8	65.5	83.9	95.7	69.6	82.8	101.6	76.8	89.3	100.0	69.4	84.9	98.2	76.8	87.6
MF2018	WEIGHTED AVERAGE	99.0	68.5	86.1	97.1	66.3	84.1	96.0	67.0	83.8	100.4	85.6	94.2	97.1	83.9	91.5	95.9	85.4	91.4

City Cost Indexes

MAINE

DIVISION		HOULTON 047 MAT.	INST.	TOTAL	KITTERY 039 MAT.	INST.	TOTAL	LEWISTON 042 MAT.	INST.	TOTAL	MACHIAS 046 MAT.	INST.	TOTAL	PORTLAND 040-041 MAT.	INST.	TOTAL	ROCKLAND 048 MAT.	INST.	TOTAL
015433	CONTRACTOR EQUIPMENT		95.7	95.7		95.7	95.7		95.7	95.7		95.7	95.7		99.2	99.2		95.7	95.7
0241, 31 - 34	SITE & INFRASTRUCTURE, DEMOLITION	90.2	94.2	93.0	79.5	94.3	89.7	88.2	95.5	93.3	89.5	94.2	92.8	87.4	100.6	96.5	86.0	94.2	91.7
0310	Concrete Forming & Accessories	96.7	78.4	81.1	88.8	78.7	80.2	98.2	76.6	79.8	93.8	78.4	80.7	99.8	76.8	80.2	94.9	78.4	80.9
0320	Concrete Reinforcing	94.1	81.1	87.8	89.7	81.1	85.5	115.6	81.1	98.9	94.1	81.1	87.8	106.5	81.1	94.2	94.1	81.1	87.8
0330	Cast-in-Place Concrete	70.9	113.2	86.7	70.7	114.4	87.0	72.4	114.1	87.9	70.9	114.3	87.1	85.8	115.2	96.8	72.4	114.3	88.0
03	CONCRETE	89.5	91.5	90.4	82.1	92.1	86.5	88.1	91.0	89.4	89.0	91.9	90.3	91.5	91.4	91.5	86.0	91.9	88.6
04	MASONRY	97.9	94.1	95.6	106.5	94.1	98.9	98.5	93.7	95.5	97.9	94.1	95.6	104.7	93.7	98.0	92.5	94.1	93.5
05	METALS	92.3	92.8	92.4	86.8	93.1	88.6	97.1	93.0	95.9	92.3	92.8	92.5	103.0	92.5	99.9	92.2	92.9	92.4
06	WOOD, PLASTICS & COMPOSITES	94.5	75.4	84.0	88.2	75.4	81.2	96.3	73.0	83.5	91.5	75.4	82.7	95.8	73.1	83.3	94.7	75.4	83.3
07	THERMAL & MOISTURE PROTECTION	109.1	100.1	105.1	107.8	100.1	104.4	108.7	100.0	104.9	109.0	100.1	105.1	113.6	100.6	107.9	108.6	100.1	104.9
08	OPENINGS	96.5	78.4	92.3	96.1	82.0	92.8	99.3	77.1	94.1	96.5	78.4	92.3	96.8	77.2	92.2	96.4	78.4	92.2
0920	Plaster & Gypsum Board	119.0	74.5	89.0	106.1	74.5	84.8	121.8	71.9	88.2	117.4	74.5	88.5	108.0	71.9	83.7	117.4	74.5	88.5
0950, 0980	Ceilings & Acoustic Treatment	88.6	74.5	79.1	101.8	74.5	83.4	99.3	71.9	80.8	88.6	74.5	79.1	107.4	71.9	83.5	88.6	74.5	79.1
0960	Flooring	83.7	110.6	91.2	86.6	110.6	93.4	85.2	110.6	92.3	82.9	110.6	90.7	89.6	110.6	95.5	83.3	110.6	91.0
0970, 0990	Wall Finishes & Painting/Coating	90.6	88.9	89.6	81.8	102.1	93.8	90.6	88.9	89.6	90.6	88.9	89.6	94.6	88.9	91.2	90.6	88.9	89.6
09	FINISHES	90.2	84.6	87.2	91.6	86.0	88.6	92.6	83.0	87.4	89.7	84.6	86.9	96.2	83.2	89.1	89.4	84.6	86.8
COVERS	DIVS. 10 - 14, 25, 28, 41, 43, 44, 46	100.0	97.6	99.5	100.0	100.9	100.2	100.0	99.7	99.9	100.0	97.6	99.5	100.0	100.1	100.0	100.0	100.9	100.2
21, 22, 23	FIRE SUPPRESSION, PLUMBING & HVAC	96.7	74.3	87.9	96.7	80.4	90.3	100.3	74.1	90.0	96.7	74.3	87.9	100.0	74.1	89.9	96.7	74.3	87.9
26, 27, 3370	ELECTRICAL, COMMUNICATIONS & UTIL.	101.9	76.8	89.5	90.3	76.8	83.6	101.9	76.7	89.5	101.9	76.8	89.5	105.0	76.7	91.0	101.8	76.8	89.5
MF2018	WEIGHTED AVERAGE	95.6	85.2	91.2	92.6	87.0	90.2	97.5	84.9	92.2	95.4	85.3	91.1	99.6	85.4	93.6	94.6	85.4	90.7

MAINE / MARYLAND

DIVISION		MAINE WATERVILLE 049 MAT.	INST.	TOTAL	ANNAPOLIS 214 MAT.	INST.	TOTAL	BALTIMORE 210-212 MAT.	INST.	TOTAL	COLLEGE PARK 207-208 MAT.	INST.	TOTAL	CUMBERLAND 215 MAT.	INST.	TOTAL	EASTON 216 MAT.	INST.	TOTAL
015433	CONTRACTOR EQUIPMENT		95.7	95.7		104.5	104.5		104.1	104.1		107.8	107.8		102.4	102.4		102.4	102.4
0241, 31 - 34	SITE & INFRASTRUCTURE, DEMOLITION	90.0	94.2	92.9	100.0	93.5	95.5	100.1	97.7	98.4	100.1	93.6	95.6	92.0	90.3	90.8	98.8	87.7	91.1
0310	Concrete Forming & Accessories	88.6	78.4	79.9	101.0	78.1	81.5	98.4	76.2	79.5	85.1	76.2	77.5	92.3	81.9	83.5	90.4	73.6	76.1
0320	Concrete Reinforcing	94.1	81.1	87.8	108.6	94.9	102.0	101.2	87.3	94.5	103.5	94.0	98.9	93.7	87.4	90.6	92.9	87.1	90.1
0330	Cast-in-Place Concrete	70.9	114.3	87.1	114.1	79.4	101.2	113.8	77.1	100.1	109.4	76.4	97.2	80.8	82.9	82.4	108.8	64.8	92.4
03	CONCRETE	90.2	91.9	90.9	101.4	82.7	93.2	108.2	79.4	95.5	105.2	80.9	94.5	91.4	84.5	88.4	99.7	74.2	88.5
04	MASONRY	109.8	94.1	100.2	94.8	76.6	83.6	104.7	73.1	85.2	111.1	74.6	88.7	99.3	85.4	90.8	113.5	58.6	79.7
05	METALS	92.3	92.8	92.4	105.5	105.2	105.4	103.8	96.6	101.7	90.2	108.3	95.5	100.7	104.2	101.8	101.0	99.7	100.6
06	WOOD, PLASTICS & COMPOSITES	84.6	75.4	79.5	97.1	77.2	86.1	102.5	76.7	88.3	77.3	75.7	76.4	86.7	81.0	83.6	84.5	80.8	82.5
07	THERMAL & MOISTURE PROTECTION	109.1	100.1	105.2	103.5	83.4	94.8	102.7	80.7	93.2	105.3	81.7	95.0	100.9	81.9	92.6	101.0	73.3	89.0
08	OPENINGS	96.5	78.4	92.3	102.3	83.0	97.8	100.5	80.7	95.9	92.7	82.4	90.3	97.6	84.6	94.6	96.0	83.0	92.9
0920	Plaster & Gypsum Board	112.0	74.5	86.8	94.4	76.8	82.5	100.3	76.2	84.0	95.3	75.2	81.8	99.9	80.8	87.0	99.9	80.6	86.9
0950, 0980	Ceilings & Acoustic Treatment	88.6	74.5	79.1	89.1	76.8	80.8	100.5	76.2	84.1	115.1	75.2	88.2	100.9	80.8	86.7	99.0	80.6	86.6
0960	Flooring	80.6	110.6	89.0	90.7	79.5	87.6	92.8	75.7	88.0	87.9	78.4	85.2	85.8	93.1	87.9	85.0	75.7	82.4
0970, 0990	Wall Finishes & Painting/Coating	90.6	88.9	89.6	99.2	72.7	83.5	101.3	75.3	85.9	104.8	72.5	85.7	98.2	84.7	90.2	98.2	72.5	83.0
09	FINISHES	88.4	84.6	86.4	88.5	77.3	82.4	97.0	75.7	85.5	95.6	75.6	84.8	93.3	84.2	88.4	93.5	74.4	83.2
COVERS	DIVS. 10 - 14, 25, 28, 41, 43, 44, 46	100.0	97.5	99.4	100.0	90.3	97.8	100.0	86.8	97.1	100.0	86.7	97.0	100.0	91.3	98.1	100.0	82.9	96.2
21, 22, 23	FIRE SUPPRESSION, PLUMBING & HVAC	96.7	74.3	87.9	100.0	87.0	94.9	100.1	82.0	93.0	96.5	87.4	93.0	96.4	72.0	86.8	96.4	72.8	87.1
26, 27, 3370	ELECTRICAL, COMMUNICATIONS & UTIL.	101.9	76.8	89.5	101.6	88.9	95.3	99.6	86.9	93.3	97.3	101.9	99.5	98.2	80.8	89.6	97.7	62.3	80.3
MF2018	WEIGHTED AVERAGE	96.0	85.2	91.4	100.4	86.3	94.4	101.8	83.2	93.9	97.3	87.7	93.2	96.9	83.8	91.4	98.6	74.8	88.5

MARYLAND / MASSACHUSETTS

DIVISION		ELKTON 219 MAT.	INST.	TOTAL	HAGERSTOWN 217 MAT.	INST.	TOTAL	SALISBURY 218 MAT.	INST.	TOTAL	SILVER SPRING 209 MAT.	INST.	TOTAL	WALDORF 206 MAT.	INST.	TOTAL	BOSTON 020-022, 024 MAT.	INST.	TOTAL
015433	CONTRACTOR EQUIPMENT		102.4	102.4		102.4	102.4		102.4	102.4		99.7	99.7		99.7	99.7		103.0	103.0
0241, 31 - 34	SITE & INFRASTRUCTURE, DEMOLITION	86.3	89.0	88.1	90.6	90.4	90.4	98.8	87.6	91.1	88.9	85.7	86.7	95.1	85.7	88.6	92.7	102.1	99.2
0310	Concrete Forming & Accessories	96.2	92.3	92.9	91.5	78.8	80.7	104.5	50.2	58.2	93.0	75.5	78.1	99.7	75.4	79.0	105.1	139.7	134.6
0320	Concrete Reinforcing	92.9	117.3	104.7	93.7	87.4	90.6	92.9	65.2	79.5	102.3	93.9	98.3	103.0	93.9	98.6	117.7	156.9	136.7
0330	Cast-in-Place Concrete	88.1	72.2	82.2	93.3	82.9	89.5	108.8	62.9	91.7	112.1	76.8	98.9	125.6	76.7	107.4	99.2	141.8	115.1
03	CONCRETE	84.0	90.6	86.9	87.7	83.1	85.7	100.5	59.2	82.4	102.8	80.5	93.0	113.3	80.5	98.9	105.4	142.7	121.8
04	MASONRY	98.6	67.0	79.2	105.1	85.4	93.0	112.9	55.2	77.4	110.6	75.0	88.7	94.7	75.0	82.6	112.0	146.3	133.1
05	METALS	101.0	114.0	104.8	100.9	104.3	101.9	101.0	90.9	98.1	94.8	104.3	97.6	94.8	104.3	97.5	102.8	136.1	112.6
06	WOOD, PLASTICS & COMPOSITES	91.9	101.3	97.0	85.8	76.7	80.8	102.6	51.4	74.4	84.0	75.0	79.0	90.8	75.0	82.1	103.8	140.2	123.8
07	THERMAL & MOISTURE PROTECTION	100.5	79.4	91.3	101.2	83.0	93.3	101.3	68.5	87.0	108.4	86.5	98.9	109.0	86.5	99.2	107.2	137.1	120.2
08	OPENINGS	96.0	102.1	97.4	95.9	81.2	92.5	96.2	60.8	87.9	84.9	82.1	84.2	85.5	81.5	84.5	99.9	147.3	111.0
0920	Plaster & Gypsum Board	102.8	101.7	102.0	99.9	76.4	84.1	109.4	50.4	69.7	101.9	75.2	83.9	104.8	75.2	84.8	109.0	141.0	130.6
0950, 0980	Ceilings & Acoustic Treatment	99.0	101.7	100.8	101.7	76.4	84.6	99.0	50.4	66.2	123.2	75.2	90.8	123.2	75.2	90.8	100.5	141.0	127.8
0960	Flooring	87.2	75.7	84.0	85.4	93.1	87.6	90.7	75.7	86.5	93.8	78.4	89.5	97.2	78.4	92.0	90.5	163.5	111.0
0970, 0990	Wall Finishes & Painting/Coating	98.2	72.5	83.0	98.2	72.5	83.0	98.2	72.5	83.0	112.7	72.5	88.9	112.7	72.5	88.9	94.3	160.4	133.4
09	FINISHES	93.7	88.5	90.9	93.5	80.4	86.4	96.4	56.2	74.6	95.8	75.0	84.5	97.5	75.1	85.4	98.7	147.0	124.8
COVERS	DIVS. 10 - 14, 25, 28, 41, 43, 44, 46	100.0	56.7	90.4	100.0	90.8	98.0	100.0	76.5	94.8	100.0	85.2	96.7	100.0	83.1	96.2	100.0	119.1	104.3
21, 22, 23	FIRE SUPPRESSION, PLUMBING & HVAC	96.4	78.7	89.5	99.9	83.1	93.3	96.4	70.7	86.3	96.5	87.9	93.1	96.5	87.9	93.1	100.1	127.4	110.8
26, 27, 3370	ELECTRICAL, COMMUNICATIONS & UTIL.	99.4	86.9	93.2	98.0	80.8	89.5	96.6	60.1	78.6	94.6	101.9	98.2	92.2	101.9	97.0	102.6	129.5	115.9
MF2018	WEIGHTED AVERAGE	95.9	86.2	91.8	97.4	85.3	92.3	99.0	66.9	85.4	96.5	86.7	92.4	97.3	86.6	92.8	101.9	134.0	115.5

For customer support on your Electrical Change Order Costs with RSMeans data, call 800.448.8182.

549

MASSACHUSETTS

DIVISION		BROCKTON 023			BUZZARDS BAY 025			FALL RIVER 027			FITCHBURG 014			FRAMINGHAM 017			GREENFIELD 013		
		MAT.	INST.	TOTAL	MAT.	INST.	TOTAL	MAT.	INST.	TOTAL	MAT.	INST.	TOTAL	MAT.	INST.	TOTAL	MAT.	INST.	TOTAL
015433	CONTRACTOR EQUIPMENT		98.1	98.1		98.1	98.1		98.9	98.9		95.7	95.7		97.4	97.4		95.7	95.7
0241, 31 - 34	SITE & INFRASTRUCTURE, DEMOLITION	91.1	98.0	95.8	81.2	98.1	92.9	90.3	98.0	95.6	82.7	97.8	93.1	79.5	97.8	92.1	86.4	97.3	93.9
0310	Concrete Forming & Accessories	100.7	121.1	118.1	98.5	120.6	117.4	100.7	120.8	117.8	94.0	118.5	114.9	101.4	121.3	118.4	92.4	117.7	114.0
0320	Concrete Reinforcing	109.7	142.7	125.7	88.0	122.3	104.6	109.7	122.3	115.8	87.9	142.5	114.3	87.9	142.8	114.4	91.5	126.5	108.4
0330	Cast-in-Place Concrete	88.1	136.6	106.2	73.2	136.4	96.7	85.2	136.9	104.4	78.4	136.5	100.0	78.4	136.7	100.1	80.5	122.2	96.0
03	CONCRETE	95.1	129.8	110.3	80.7	126.0	100.6	93.8	126.2	108.0	79.5	128.3	101.0	82.3	130.0	103.2	83.0	120.3	99.4
04	MASONRY	105.4	135.1	123.6	97.9	135.1	120.8	105.9	135.1	123.8	98.2	131.9	118.9	104.7	136.1	124.0	102.7	119.1	112.8
05	METALS	98.8	125.6	106.7	93.6	117.0	100.5	98.8	117.3	104.3	95.3	122.1	103.2	95.3	125.8	104.3	97.8	113.8	102.5
06	WOOD, PLASTICS & COMPOSITES	98.1	119.7	109.9	94.7	119.7	108.4	98.1	119.9	110.1	94.5	116.5	106.6	101.0	119.4	111.2	92.5	119.7	107.4
07	THERMAL & MOISTURE PROTECTION	103.2	128.4	114.2	102.0	125.7	112.3	103.0	124.9	112.5	103.2	122.1	111.4	103.3	128.3	114.2	103.3	109.9	106.2
08	OPENINGS	97.4	128.4	104.6	93.4	118.9	99.3	97.4	118.3	102.2	99.2	126.7	105.6	89.8	128.3	98.7	99.4	120.3	104.3
0920	Plaster & Gypsum Board	92.3	119.9	110.9	87.3	119.9	109.3	92.3	119.9	110.9	111.3	116.7	114.9	114.4	119.9	118.1	112.6	119.9	117.5
0950, 0980	Ceilings & Acoustic Treatment	104.3	119.9	114.9	86.2	119.9	108.9	104.3	119.9	114.9	93.9	116.7	109.3	93.9	119.9	111.5	103.7	119.9	114.6
0960	Flooring	85.0	162.0	106.6	82.6	162.0	104.9	83.8	162.0	105.8	86.7	162.0	107.8	88.2	162.0	108.9	85.0	135.0	99.7
0970, 0990	Wall Finishes & Painting/Coating	85.0	139.2	117.1	85.0	139.2	117.1	85.0	139.2	117.1	86.2	139.2	117.6	87.1	139.2	117.9	86.2	109.3	99.9
09	FINISHES	90.8	130.8	112.5	85.2	130.8	109.9	90.5	131.0	112.4	89.5	128.9	110.9	90.2	130.6	112.1	91.7	120.9	107.5
COVERS	DIVS. 10 - 14, 25, 28, 41, 43, 44, 46	100.0	109.7	102.2	100.0	109.7	102.2	100.0	110.2	102.3	100.0	104.2	100.9	100.0	109.3	102.1	100.0	102.9	100.7
21, 22, 23	FIRE SUPPRESSION, PLUMBING & HVAC	100.4	103.9	101.7	96.8	103.5	99.4	100.4	103.5	101.6	97.0	104.5	99.9	97.0	121.1	106.4	97.0	98.5	97.6
26, 27, 3370	ELECTRICAL, COMMUNICATIONS & UTIL.	101.3	97.7	99.5	98.5	100.1	99.3	101.2	97.7	99.4	100.6	104.1	102.3	97.4	124.4	110.7	100.5	96.1	98.4
MF2018	WEIGHTED AVERAGE	98.5	116.7	106.2	93.2	115.2	102.5	98.3	114.9	105.3	94.6	116.2	103.7	94.0	124.2	106.8	95.9	109.0	101.4

MASSACHUSETTS

DIVISION		HYANNIS 026			LAWRENCE 019			LOWELL 018			NEW BEDFORD 027			PITTSFIELD 012			SPRINGFIELD 010 - 011		
		MAT.	INST.	TOTAL	MAT.	INST.	TOTAL	MAT.	INST.	TOTAL	MAT.	INST.	TOTAL	MAT.	INST.	TOTAL	MAT.	INST.	TOTAL
015433	CONTRACTOR EQUIPMENT		98.1	98.1		98.1	98.1		95.7	95.7		98.9	98.9		95.7	95.7		95.7	95.7
0241, 31 - 34	SITE & INFRASTRUCTURE, DEMOLITION	87.4	98.1	94.8	91.6	98.0	96.0	90.6	98.1	95.8	88.7	98.0	95.1	91.6	96.7	95.2	91.1	97.1	95.2
0310	Concrete Forming & Accessories	92.7	120.6	116.5	102.5	121.6	118.8	99.0	122.6	119.1	100.7	120.8	117.9	99.0	102.9	102.3	99.3	117.7	115.0
0320	Concrete Reinforcing	88.0	122.3	104.6	108.8	148.3	127.9	109.7	122.4	115.8	109.7	122.4	115.8	90.8	114.1	102.1	109.7	126.5	117.8
0330	Cast-in-Place Concrete	80.4	136.4	101.3	90.7	136.8	107.9	82.4	138.5	103.3	75.1	136.9	98.1	90.0	113.4	98.7	85.8	122.2	99.4
03	CONCRETE	86.7	126.0	104.0	96.2	131.1	111.5	88.0	131.9	107.3	89.1	126.3	105.4	89.0	108.3	97.5	89.5	120.2	103.0
04	MASONRY	104.3	135.1	123.2	110.1	135.9	125.9	97.3	134.9	120.4	104.2	135.7	123.6	98.0	111.5	106.3	97.6	119.1	110.8
05	METALS	95.1	117.0	101.5	98.2	128.4	107.1	98.2	125.0	106.1	98.8	117.4	104.3	98.0	108.2	101.0	100.9	113.7	104.6
06	WOOD, PLASTICS & COMPOSITES	88.0	119.7	105.4	101.5	119.7	111.5	100.8	119.7	111.2	98.1	119.9	110.1	100.8	102.5	101.8	100.8	119.7	111.2
07	THERMAL & MOISTURE PROTECTION	102.5	125.7	112.6	104.0	128.6	114.7	103.7	128.0	114.3	102.9	125.1	112.6	103.8	104.3	104.0	103.7	109.9	106.4
08	OPENINGS	93.9	118.9	99.8	93.4	129.9	101.9	100.4	129.9	107.3	97.4	123.0	103.3	100.4	107.6	102.1	100.4	120.3	105.1
0920	Plaster & Gypsum Board	83.2	119.9	107.9	117.3	119.9	119.1	117.3	119.9	119.1	92.3	119.9	110.9	117.3	102.3	107.2	117.3	119.9	119.1
0950, 0980	Ceilings & Acoustic Treatment	96.2	119.9	112.2	105.5	119.9	115.2	105.5	119.9	115.2	104.3	119.9	114.9	105.5	102.3	103.4	105.5	119.9	115.2
0960	Flooring	80.0	162.0	103.0	88.8	162.0	109.4	88.8	162.0	109.4	83.8	162.0	105.8	89.1	129.8	100.6	88.2	135.0	101.3
0970, 0990	Wall Finishes & Painting/Coating	85.0	139.2	117.1	86.3	139.2	117.6	86.2	139.2	117.6	85.0	139.2	117.1	86.2	109.3	99.9	87.0	109.3	100.2
09	FINISHES	86.3	130.8	110.4	93.8	130.8	113.8	93.7	131.6	114.2	90.4	131.0	112.3	93.8	108.6	101.8	93.6	120.9	108.4
COVERS	DIVS. 10 - 14, 25, 28, 41, 43, 44, 46	100.0	109.7	102.2	100.0	109.8	102.2	100.0	110.8	102.4	100.0	110.2	102.3	100.0	99.5	99.9	100.0	102.9	100.7
21, 22, 23	FIRE SUPPRESSION, PLUMBING & HVAC	100.4	104.2	101.9	100.1	119.7	107.8	100.1	121.2	108.4	100.4	103.4	101.5	100.1	94.2	97.8	100.1	98.5	99.5
26, 27, 3370	ELECTRICAL, COMMUNICATIONS & UTIL.	98.9	97.7	98.3	99.7	122.9	111.1	100.1	119.3	109.6	102.1	97.7	99.9	100.1	96.1	98.2	100.1	93.0	96.6
MF2018	WEIGHTED AVERAGE	95.7	115.0	103.8	98.4	124.2	109.3	97.5	123.8	108.6	97.7	115.2	105.1	97.7	102.6	99.7	98.2	108.5	102.5

DIVISION		MASSACHUSETTS WORCESTER 015 - 016			MICHIGAN ANN ARBOR 481			BATTLE CREEK 490			BAY CITY 487			DEARBORN 481			DETROIT 482		
		MAT.	INST.	TOTAL	MAT.	INST.	TOTAL	MAT.	INST.	TOTAL	MAT.	INST.	TOTAL	MAT.	INST.	TOTAL	MAT.	INST.	TOTAL
015433	CONTRACTOR EQUIPMENT		95.7	95.7		109.1	109.1		96.1	96.1		109.1	109.1		109.1	109.1		97.8	97.8
0241, 31 - 34	SITE & INFRASTRUCTURE, DEMOLITION	91.0	97.9	95.7	81.0	92.2	88.7	93.6	81.7	85.3	72.5	91.7	85.8	80.8	92.3	88.7	98.2	101.2	100.2
0310	Concrete Forming & Accessories	99.6	118.5	115.7	96.3	104.1	103.0	96.0	77.3	80.0	96.4	81.5	83.7	96.2	104.7	103.5	100.4	105.3	104.6
0320	Concrete Reinforcing	109.7	155.5	131.9	101.2	103.5	102.3	98.5	80.9	90.0	101.2	102.7	101.9	101.2	103.6	102.3	101.4	104.8	103.0
0330	Cast-in-Place Concrete	85.3	136.5	104.4	88.3	96.4	91.3	86.1	91.0	87.9	84.5	83.5	84.1	86.4	97.2	90.4	104.5	101.3	103.3
03	CONCRETE	89.3	130.6	107.5	89.8	102.2	95.2	85.7	82.6	84.3	88.0	87.3	87.7	88.9	102.8	95.0	103.1	103.2	103.2
04	MASONRY	97.2	131.9	118.5	99.6	96.7	97.8	99.1	78.4	86.4	99.2	77.7	86.0	99.5	98.2	98.7	102.6	99.8	100.9
05	METALS	100.9	127.4	108.7	102.7	115.7	106.5	103.7	83.7	97.9	103.3	114.0	106.4	102.8	115.8	106.6	104.3	94.3	101.4
06	WOOD, PLASTICS & COMPOSITES	101.2	116.5	109.6	89.2	106.6	98.8	88.4	75.6	81.3	89.2	81.9	85.2	89.2	106.6	98.8	97.1	106.7	102.4
07	THERMAL & MOISTURE PROTECTION	103.8	122.1	111.8	105.3	98.2	102.2	96.1	79.0	88.7	102.7	81.0	93.2	103.6	100.7	102.3	102.6	102.9	102.7
08	OPENINGS	100.4	130.2	107.4	94.2	100.8	95.7	86.5	74.1	83.6	94.2	83.5	91.7	94.2	101.1	95.8	97.2	101.9	98.3
0920	Plaster & Gypsum Board	117.3	116.7	116.9	107.4	106.5	106.8	91.7	71.7	78.2	107.4	81.1	89.7	107.4	106.5	106.8	104.1	106.5	105.7
0950, 0980	Ceilings & Acoustic Treatment	105.5	116.7	113.0	84.8	106.5	99.4	78.7	71.7	74.0	85.7	81.1	82.6	84.8	106.5	99.4	96.9	106.5	103.4
0960	Flooring	88.8	154.9	107.4	91.2	107.2	95.7	91.5	70.4	85.6	91.2	77.0	87.2	90.6	101.6	93.6	95.7	106.7	98.8
0970, 0990	Wall Finishes & Painting/Coating	86.2	139.2	117.6	82.7	95.5	90.3	82.3	76.3	78.7	82.7	79.1	80.6	82.7	93.8	89.3	93.1	100.1	97.3
09	FINISHES	93.7	127.5	112.0	90.8	104.3	98.1	84.7	75.2	79.6	90.6	79.9	84.8	90.6	103.5	97.6	97.7	105.4	101.9
COVERS	DIVS. 10 - 14, 25, 28, 41, 43, 44, 46	100.0	104.2	100.9	100.0	93.7	98.6	100.0	93.7	98.6	100.0	88.0	97.3	100.0	94.1	98.7	100.0	102.4	100.5
21, 22, 23	FIRE SUPPRESSION, PLUMBING & HVAC	100.1	104.7	101.9	100.0	91.3	96.6	100.1	82.6	93.2	100.0	78.6	91.6	100.0	100.0	100.0	99.9	102.7	101.0
26, 27, 3370	ELECTRICAL, COMMUNICATIONS & UTIL.	100.1	104.1	102.1	97.3	103.0	100.1	96.6	77.7	87.3	96.4	82.5	89.5	97.3	96.1	96.7	99.9	103.5	101.6
MF2018	WEIGHTED AVERAGE	98.1	117.0	106.1	97.1	99.8	98.2	95.5	80.4	89.1	96.6	85.3	91.8	96.9	100.9	98.6	100.7	102.0	101.3

For customer support on your Electrical Change Order Costs with RSMeans data, call 800.448.8182.

MICHIGAN

DIVISION		FLINT 484 - 485			GAYLORD 497			GRAND RAPIDS 493, 495			IRON MOUNTAIN 498 - 499			JACKSON 492			KALAMAZOO 491		
		MAT.	INST.	TOTAL	MAT.	INST.	TOTAL	MAT.	INST.	TOTAL	MAT.	INST.	TOTAL	MAT.	INST.	TOTAL	MAT.	INST.	TOTAL
015433	CONTRACTOR EQUIPMENT		109.1	109.1		103.6	103.6		98.3	98.3		91.1	91.1		103.6	103.6		96.1	96.1
0241, 31 - 34	SITE & INFRASTRUCTURE, DEMOLITION	70.2	91.8	85.1	87.4	79.6	82.0	92.9	85.9	88.1	96.4	88.5	90.9	111.0	81.2	90.4	93.9	81.6	85.4
0310	Concrete Forming & Accessories	99.1	81.8	84.4	95.0	73.9	77.0	95.6	75.7	78.6	86.6	78.6	79.8	91.6	82.3	83.7	96.0	77.1	79.9
0320	Concrete Reinforcing	101.2	103.1	102.1	91.9	93.0	92.4	103.4	80.7	92.4	91.6	85.0	88.4	89.3	103.1	96.0	98.5	76.2	87.8
0330	Cast-in-Place Concrete	88.9	85.8	87.8	85.8	78.4	83.1	90.4	89.0	89.9	101.7	67.6	89.0	85.7	89.8	87.2	87.8	91.0	89.0
03	CONCRETE	90.2	88.4	89.4	82.6	80.5	81.6	89.7	81.1	85.9	90.9	76.6	84.6	77.8	89.8	83.1	88.9	81.7	85.7
04	MASONRY	99.7	85.9	91.2	109.7	72.7	87.0	92.0	74.1	81.0	95.4	79.7	85.7	89.2	86.9	87.8	97.6	78.4	85.8
05	METALS	102.8	114.6	106.3	105.5	110.8	107.0	100.8	83.2	95.6	104.8	91.1	100.8	105.7	112.7	107.8	103.7	81.9	97.3
06	WOOD, PLASTICS & COMPOSITES	92.6	80.7	86.1	81.9	73.4	77.2	92.2	73.9	82.2	78.2	78.9	78.6	80.8	79.8	80.2	88.4	75.6	81.3
07	THERMAL & MOISTURE PROTECTION	103.0	84.5	95.0	94.7	74.7	86.0	98.0	71.8	86.6	98.4	74.8	88.1	94.0	88.3	91.5	96.1	79.0	88.7
08	OPENINGS	94.2	82.6	91.5	86.0	78.7	84.3	101.8	73.8	95.2	92.5	69.8	87.2	85.2	85.5	85.3	86.5	73.5	83.4
0920	Plaster & Gypsum Board	109.0	79.8	89.4	91.3	71.8	78.2	99.8	70.2	79.8	51.3	79.0	69.9	90.1	78.3	82.1	91.7	71.7	78.2
0950, 0980	Ceilings & Acoustic Treatment	84.8	79.8	81.5	77.1	71.8	73.5	91.1	70.2	77.0	76.0	79.0	78.0	77.1	78.3	77.9	78.7	71.7	74.0
0960	Flooring	91.2	86.0	89.7	84.9	85.6	85.1	95.7	76.1	90.2	102.0	91.0	98.9	83.4	78.7	82.1	91.5	70.4	85.6
0970, 0990	Wall Finishes & Painting/Coating	82.7	78.3	80.1	78.7	81.6	80.4	89.3	76.9	82.0	94.8	69.6	79.9	78.7	93.8	87.7	82.3	76.3	78.7
09	FINISHES	90.1	81.6	85.5	84.1	76.2	79.8	91.4	75.2	82.6	84.6	80.0	82.1	85.2	81.9	83.4	84.7	75.2	79.6
COVERS	DIVS. 10 - 14, 25, 28, 41, 43, 44, 46	100.0	88.8	97.5	100.0	77.9	95.1	100.0	93.0	98.4	100.0	85.9	96.8	100.0	91.6	98.1	100.0	93.7	98.6
21, 22, 23	FIRE SUPPRESSION, PLUMBING & HVAC	100.0	83.3	93.4	96.8	79.0	89.8	100.0	79.6	92.0	96.7	83.9	91.7	96.8	85.7	92.4	100.1	78.2	91.5
26, 27, 3370	ELECTRICAL, COMMUNICATIONS & UTIL.	97.3	89.1	93.2	94.5	76.3	85.5	102.2	82.9	92.7	100.8	80.7	90.9	98.3	103.0	100.6	96.4	74.1	85.4
MF2018	WEIGHTED AVERAGE	96.8	88.6	93.3	94.5	80.6	88.6	97.9	79.9	90.3	96.5	81.7	90.2	94.0	90.7	92.6	95.8	78.6	88.6

MICHIGAN / MINNESOTA

DIVISION		LANSING 488 - 489			MUSKEGON 494			ROYAL OAK 480, 483			SAGINAW 486			TRAVERSE CITY 496			BEMIDJI 566		
		MAT.	INST.	TOTAL	MAT.	INST.	TOTAL	MAT.	INST.	TOTAL	MAT.	INST.	TOTAL	MAT.	INST.	TOTAL	MAT.	INST.	TOTAL
015433	CONTRACTOR EQUIPMENT		111.0	111.0		96.1	96.1		89.8	89.8		109.1	109.1		91.1	91.1		97.7	97.7
0241, 31 - 34	SITE & INFRASTRUCTURE, DEMOLITION	90.7	96.5	94.7	91.5	81.7	84.8	84.8	92.6	90.2	73.5	91.7	86.1	81.8	87.5	85.7	95.1	95.1	95.1
0310	Concrete Forming & Accessories	93.8	74.9	77.7	96.3	79.4	81.9	92.3	105.2	103.2	96.3	79.7	82.1	86.6	72.4	74.5	88.0	82.8	83.6
0320	Concrete Reinforcing	104.8	102.9	103.8	99.3	80.7	90.3	91.7	104.8	98.0	101.2	102.7	101.9	93.1	80.2	86.9	97.4	101.8	99.5
0330	Cast-in-Place Concrete	97.9	85.1	93.1	85.8	90.5	87.5	77.3	97.5	84.8	87.3	83.4	85.8	79.3	76.7	78.3	101.8	104.5	102.8
03	CONCRETE	91.4	84.9	88.5	84.2	83.3	83.8	77.1	101.6	87.9	89.3	86.5	88.1	74.8	76.1	75.4	98.9	102.4	100.6
04	MASONRY	94.5	84.6	88.4	96.3	76.2	83.9	93.2	99.5	97.1	101.2	77.7	86.7	93.5	74.8	82.0	98.9	102.4	101.0
05	METALS	101.6	112.7	104.9	101.5	83.1	96.1	106.3	92.2	102.2	102.8	113.8	106.0	104.8	88.9	100.1	91.2	117.9	99.0
06	WOOD, PLASTICS & COMPOSITES	87.0	71.2	78.3	85.4	78.0	81.3	85.0	106.6	96.9	85.9	79.7	82.5	78.2	71.9	74.8	68.8	77.0	73.3
07	THERMAL & MOISTURE PROTECTION	102.0	83.8	94.1	95.0	73.5	85.7	101.2	100.9	101.1	103.9	80.9	93.9	97.3	74.2	87.3	106.3	90.7	99.5
08	OPENINGS	102.7	77.2	96.8	85.8	76.5	83.6	94.0	101.3	95.7	92.4	82.2	90.0	92.5	65.4	86.2	99.7	101.3	100.1
0920	Plaster & Gypsum Board	100.1	70.2	79.9	71.8	74.2	73.4	104.8	106.5	106.0	107.4	78.8	88.2	51.3	71.8	65.1	104.3	76.8	85.8
0950, 0980	Ceilings & Acoustic Treatment	83.9	70.2	74.6	78.7	74.2	75.6	84.1	106.5	99.2	84.8	78.8	80.8	76.0	71.8	73.2	128.8	76.8	93.7
0960	Flooring	97.2	78.7	92.0	90.2	72.3	85.1	88.2	106.7	93.4	91.2	77.0	87.2	102.0	85.6	97.4	90.7	91.4	90.9
0970, 0990	Wall Finishes & Painting/Coating	91.1	76.9	82.7	80.5	79.7	80.0	84.2	93.8	89.9	82.7	79.1	80.6	94.8	39.4	62.1	89.6	105.8	99.2
09	FINISHES	90.2	74.6	81.7	81.1	77.1	78.9	89.7	104.3	97.6	90.5	78.6	84.0	83.5	71.0	76.7	98.2	86.0	91.6
COVERS	DIVS. 10 - 14, 25, 28, 41, 43, 44, 46	100.0	88.8	97.5	100.0	94.5	98.8	100.0	100.0	100.0	100.0	87.7	97.3	100.0	84.4	96.5	100.0	95.6	99.0
21, 22, 23	FIRE SUPPRESSION, PLUMBING & HVAC	99.9	83.7	93.5	99.9	83.4	93.4	96.7	101.4	98.5	100.0	78.3	91.5	96.7	78.9	89.7	96.7	84.8	92.0
26, 27, 3370	ELECTRICAL, COMMUNICATIONS & UTIL.	99.1	87.6	93.4	96.8	73.0	85.1	99.4	103.1	101.3	95.4	84.5	90.0	96.2	76.2	86.4	105.0	100.7	102.8
MF2018	WEIGHTED AVERAGE	98.0	86.8	93.3	94.3	80.0	88.3	95.1	100.3	97.3	96.5	85.1	91.7	93.4	77.7	86.8	96.7	95.4	96.1

MINNESOTA

DIVISION		BRAINERD 564			DETROIT LAKES 565			DULUTH 556 - 558			MANKATO 560			MINNEAPOLIS 553 - 555			ROCHESTER 559		
		MAT.	INST.	TOTAL	MAT.	INST.	TOTAL	MAT.	INST.	TOTAL	MAT.	INST.	TOTAL	MAT.	INST.	TOTAL	MAT.	INST.	TOTAL
015433	CONTRACTOR EQUIPMENT		100.2	100.2		97.7	97.7		102.9	102.9		100.2	100.2		108.4	108.4		100.7	100.7
0241, 31 - 34	SITE & INFRASTRUCTURE, DEMOLITION	96.0	99.9	98.7	93.3	95.4	94.7	98.2	101.7	100.6	92.6	99.4	97.3	95.0	108.8	104.5	96.0	97.6	97.1
0310	Concrete Forming & Accessories	88.7	83.6	84.4	84.9	82.8	83.1	98.4	94.6	95.2	96.9	96.6	96.6	99.3	114.0	111.8	99.2	94.8	95.4
0320	Concrete Reinforcing	96.2	102.2	99.1	97.4	101.8	99.6	102.6	102.0	102.3	96.0	109.2	102.4	96.1	109.4	102.5	98.9	108.8	103.7
0330	Cast-in-Place Concrete	110.7	109.1	110.1	98.9	107.5	102.1	94.3	97.7	95.6	101.9	99.5	101.0	95.6	115.3	102.9	94.8	95.3	95.0
03	CONCRETE	94.8	97.1	95.8	88.6	96.1	91.9	95.3	98.1	96.5	90.5	100.8	95.0	97.6	114.3	104.9	91.4	98.6	94.6
04	MASONRY	122.5	116.3	118.7	122.6	109.8	114.8	95.3	105.7	101.7	111.3	110.8	111.0	114.5	119.5	117.6	102.9	103.5	103.3
05	METALS	92.2	118.3	99.9	91.1	117.5	98.9	100.2	119.1	105.8	92.1	121.6	100.7	98.7	124.5	106.3	99.3	123.1	106.3
06	WOOD, PLASTICS & COMPOSITES	84.9	74.2	79.1	65.8	74.4	70.6	93.9	91.9	92.8	94.5	95.4	95.0	96.1	110.8	104.2	95.2	92.9	93.9
07	THERMAL & MOISTURE PROTECTION	104.7	102.4	103.7	106.1	99.4	103.2	97.9	102.2	99.8	105.1	93.4	100.1	101.2	115.9	107.6	105.3	90.0	98.6
08	OPENINGS	86.4	99.8	89.5	99.6	99.9	99.7	103.3	104.6	103.6	90.9	112.2	95.9	99.3	121.5	104.5	98.6	112.5	101.8
0920	Plaster & Gypsum Board	91.5	74.1	79.8	103.7	74.1	83.8	94.5	92.1	92.9	95.6	95.9	95.8	101.1	111.2	107.9	103.2	93.2	96.5
0950, 0980	Ceilings & Acoustic Treatment	58.8	74.1	69.1	128.8	74.1	91.9	89.2	92.1	91.1	58.8	95.9	83.8	103.1	111.2	108.6	93.9	93.2	93.4
0960	Flooring	89.6	85.6	88.4	89.5	85.6	88.4	93.2	123.9	101.8	91.4	81.8	88.7	100.6	114.6	104.5	92.3	81.8	89.3
0970, 0990	Wall Finishes & Painting/Coating	84.0	105.8	96.9	89.6	83.4	85.9	86.0	104.4	96.9	95.0	103.7	100.2	101.3	123.8	114.6	85.3	98.2	92.9
09	FINISHES	82.1	85.2	83.8	97.6	82.5	89.4	89.4	100.2	95.3	83.7	94.9	89.7	98.9	115.0	107.6	90.0	93.1	91.7
COVERS	DIVS. 10 - 14, 25, 28, 41, 43, 44, 46	100.0	97.3	99.4	100.0	97.1	99.4	100.0	94.6	98.8	100.0	97.5	99.4	100.0	105.8	101.3	100.0	101.8	100.4
21, 22, 23	FIRE SUPPRESSION, PLUMBING & HVAC	96.0	88.6	93.1	96.7	87.5	93.1	99.8	94.5	97.7	96.0	84.6	91.5	100.0	110.6	104.2	100.0	91.6	96.7
26, 27, 3370	ELECTRICAL, COMMUNICATIONS & UTIL.	102.7	101.7	102.2	104.8	71.1	88.2	100.3	101.7	101.0	109.1	93.1	101.2	107.2	110.3	108.7	103.0	93.1	98.1
MF2018	WEIGHTED AVERAGE	95.6	98.6	96.8	97.4	92.4	95.3	98.6	101.3	99.7	95.7	98.5	96.9	100.7	114.1	106.3	98.3	98.7	98.5

For customer support on your Electrical Change Order Costs with RSMeans data, call 800.448.8182.

551

City Cost Indexes

MINNESOTA / MISSISSIPPI

DIVISION		SAINT PAUL 550 - 551 MAT.	INST.	TOTAL	ST. CLOUD 563 MAT.	INST.	TOTAL	THIEF RIVER FALLS 567 MAT.	INST.	TOTAL	WILLMAR 562 MAT.	INST.	TOTAL	WINDOM 561 MAT.	INST.	TOTAL	BILOXI 395 MAT.	INST.	TOTAL
015433	CONTRACTOR EQUIPMENT		102.9	102.9		100.2	100.2		97.7	97.7		100.2	100.2		100.2	100.2		100.1	100.1
0241, 31 - 34	SITE & INFRASTRUCTURE, DEMOLITION	96.5	102.5	100.7	91.4	100.8	97.9	94.0	94.9	94.7	90.6	99.7	96.9	84.5	98.9	94.4	105.2	85.9	91.9
0310	Concrete Forming & Accessories	98.6	117.0	114.3	86.0	113.5	109.5	88.8	81.9	82.9	85.8	86.8	86.7	90.1	82.2	83.4	92.8	64.1	68.4
0320	Concrete Reinforcing	105.4	109.9	107.6	96.2	109.4	102.6	97.8	101.6	99.6	95.8	109.2	102.3	95.8	108.1	101.8	93.4	48.8	71.8
0330	Cast-in-Place Concrete	99.9	116.3	106.0	97.7	115.0	104.1	100.9	81.3	93.6	99.2	79.5	91.9	85.8	83.3	84.8	114.2	66.3	96.4
03	CONCRETE	95.3	116.1	104.4	86.5	114.0	98.5	89.8	86.5	88.3	86.3	89.5	87.7	77.6	88.5	82.4	94.4	64.0	81.0
04	MASONRY	105.3	127.4	118.9	107.3	121.8	116.2	98.9	102.3	101.0	112.1	114.5	113.6	122.0	87.9	101.0	90.4	62.8	73.4
05	METALS	100.1	125.0	107.4	92.9	123.1	101.8	91.3	116.7	98.7	92.0	121.5	100.7	91.9	119.5	100.0	90.3	84.5	88.6
06	WOOD, PLASTICS & COMPOSITES	95.0	114.2	105.6	82.1	110.3	97.6	70.0	77.0	73.8	81.8	79.9	80.8	86.3	79.9	82.8	93.1	64.5	77.3
07	THERMAL & MOISTURE PROTECTION	102.0	120.5	110.1	104.9	111.7	107.9	107.1	88.1	98.8	104.7	101.0	103.1	104.6	83.0	95.2	100.0	63.1	83.9
08	OPENINGS	98.2	124.3	104.3	91.4	122.1	98.5	99.7	101.3	100.1	88.4	102.8	91.7	91.9	102.8	94.4	96.3	55.0	86.7
0920	Plaster & Gypsum Board	94.5	115.0	108.3	91.5	111.2	104.7	104.3	76.8	85.8	91.5	79.9	83.7	91.5	79.9	83.7	103.8	64.0	77.0
0950, 0980	Ceilings & Acoustic Treatment	91.6	115.0	107.4	58.8	111.2	94.1	128.8	76.8	93.7	58.8	79.9	73.0	58.8	79.9	73.0	86.4	64.0	71.3
0960	Flooring	92.4	122.8	101.0	86.4	120.9	96.1	90.4	85.6	89.0	87.8	85.6	87.2	90.3	85.6	89.0	94.0	65.0	85.8
0970, 0990	Wall Finishes & Painting/Coating	92.1	129.1	114.0	95.0	123.8	112.0	89.6	83.4	85.9	89.6	83.4	85.9	89.6	103.7	98.0	86.0	46.9	62.8
09	FINISHES	91.1	119.1	106.3	81.5	115.9	100.1	98.1	82.6	89.7	81.6	85.3	83.6	81.8	84.8	83.4	89.4	62.8	75.0
COVERS	DIVS. 10 - 14, 25, 28, 41, 43, 44, 46	100.0	108.7	101.9	100.0	103.2	100.7	100.0	95.5	99.0	100.0	97.3	99.4	100.0	93.7	98.6	100.0	71.9	93.7
21, 22, 23	FIRE SUPPRESSION, PLUMBING & HVAC	99.9	116.7	106.5	99.5	110.5	103.8	96.7	84.4	91.9	96.0	101.6	98.2	96.0	81.0	90.1	100.0	55.3	82.4
26, 27, 3370	ELECTRICAL, COMMUNICATIONS & UTIL.	102.9	115.1	108.9	102.7	115.1	108.8	102.1	71.0	86.8	102.7	83.7	93.4	109.1	93.1	101.2	101.0	53.7	77.7
MF2018	WEIGHTED AVERAGE	99.0	117.6	106.9	95.1	114.2	103.2	96.3	89.3	93.3	94.0	98.0	95.7	94.3	91.2	93.0	96.2	63.9	82.6

MISSISSIPPI

DIVISION		CLARKSDALE 386 MAT.	INST.	TOTAL	COLUMBUS 397 MAT.	INST.	TOTAL	GREENVILLE 387 MAT.	INST.	TOTAL	GREENWOOD 389 MAT.	INST.	TOTAL	JACKSON 390 - 392 MAT.	INST.	TOTAL	LAUREL 394 MAT.	INST.	TOTAL
015433	CONTRACTOR EQUIPMENT		100.1	100.1		100.1	100.1		100.1	100.1		100.1	100.1		102.6	102.6		100.1	100.1
0241, 31 - 34	SITE & INFRASTRUCTURE, DEMOLITION	101.4	84.4	89.7	103.9	85.2	91.0	107.5	86.1	92.7	104.5	84.1	90.4	101.1	90.6	93.9	109.6	84.5	92.2
0310	Concrete Forming & Accessories	83.8	43.9	49.8	81.9	46.3	51.6	80.7	62.7	65.4	92.5	44.3	51.4	92.0	64.6	68.6	82.0	59.5	62.8
0320	Concrete Reinforcing	104.2	66.5	85.9	99.9	66.6	83.8	104.7	66.7	86.3	104.2	66.5	86.0	104.0	51.7	78.7	100.6	32.4	67.6
0330	Cast-in-Place Concrete	104.1	58.8	87.2	116.5	61.1	95.9	107.2	65.7	91.8	112.0	58.9	92.2	99.6	66.5	87.3	113.9	60.1	93.9
03	CONCRETE	93.8	55.1	76.8	95.8	57.1	78.8	99.3	66.2	84.8	100.1	55.4	80.4	89.2	64.7	78.4	97.9	56.9	79.9
04	MASONRY	89.1	50.8	65.6	115.4	53.6	77.4	132.2	61.4	88.7	89.8	50.7	65.8	95.7	61.5	74.7	111.5	49.4	73.4
05	METALS	92.5	87.9	91.1	87.4	90.4	88.3	93.6	91.2	92.8	92.5	87.8	91.1	96.9	84.6	93.3	87.5	76.0	84.1
06	WOOD, PLASTICS & COMPOSITES	79.4	44.4	60.1	78.8	45.2	60.3	76.2	62.8	68.8	91.8	44.4	65.7	96.3	65.5	79.4	79.9	64.5	71.4
07	THERMAL & MOISTURE PROTECTION	97.7	52.0	77.8	100.0	55.4	80.6	98.2	62.1	82.5	98.1	54.2	79.0	98.3	62.9	82.9	100.2	56.5	81.2
08	OPENINGS	95.1	48.2	84.1	95.9	48.6	84.9	94.8	58.1	86.2	95.1	48.2	84.1	99.5	56.0	89.4	93.2	51.8	83.5
0920	Plaster & Gypsum Board	89.9	43.3	58.6	94.3	44.2	60.6	89.6	62.3	71.2	100.6	43.3	62.1	90.7	64.9	73.3	94.3	64.0	73.9
0950, 0980	Ceilings & Acoustic Treatment	82.3	43.3	56.0	81.5	44.2	56.3	85.0	62.3	69.7	82.3	43.3	56.0	89.7	64.9	73.0	81.5	64.0	69.7
0960	Flooring	97.8	45.0	83.0	87.9	65.0	81.4	96.2	65.0	87.4	103.5	65.0	92.7	92.1	65.0	84.5	86.6	65.0	80.5
0970, 0990	Wall Finishes & Painting/Coating	94.6	46.9	66.4	86.0	46.9	62.8	94.6	56.7	72.1	94.6	46.9	66.4	89.9	56.7	70.3	86.0	46.9	62.8
09	FINISHES	90.2	43.9	65.1	85.4	49.1	65.8	90.8	62.5	75.5	93.7	48.0	68.9	89.9	64.1	75.9	85.6	60.5	72.0
COVERS	DIVS. 10 - 14, 25, 28, 41, 43, 44, 46	100.0	48.0	88.4	100.0	49.0	88.6	100.0	71.3	93.6	100.0	48.0	88.4	100.0	71.9	93.7	100.0	34.2	85.3
21, 22, 23	FIRE SUPPRESSION, PLUMBING & HVAC	98.4	51.3	79.9	98.0	53.1	80.4	100.0	57.4	83.3	98.4	51.7	80.1	100.0	59.1	83.9	98.1	47.0	78.0
26, 27, 3370	ELECTRICAL, COMMUNICATIONS & UTIL.	96.8	41.6	69.6	98.6	55.1	77.2	96.8	55.9	76.6	96.8	39.0	68.3	102.7	55.9	79.6	100.1	57.9	79.3
MF2018	WEIGHTED AVERAGE	95.3	55.2	78.4	95.9	59.1	80.4	98.7	65.5	84.7	96.6	55.5	79.2	97.3	65.6	83.9	96.1	57.8	79.9

MISSISSIPPI / MISSOURI

DIVISION		MCCOMB 396 MAT.	INST.	TOTAL	MERIDIAN 393 MAT.	INST.	TOTAL	TUPELO 388 MAT.	INST.	TOTAL	BOWLING GREEN 633 MAT.	INST.	TOTAL	CAPE GIRARDEAU 637 MAT.	INST.	TOTAL	CHILLICOTHE 646 MAT.	INST.	TOTAL
015433	CONTRACTOR EQUIPMENT		100.1	100.1		100.1	100.1		100.1	100.1		107.1	107.1		107.1	107.1		101.5	101.5
0241, 31 - 34	SITE & INFRASTRUCTURE, DEMOLITION	96.3	84.3	88.0	100.7	86.1	90.7	98.9	84.3	88.8	87.6	90.0	89.3	89.3	89.9	89.7	101.2	88.4	92.3
0310	Concrete Forming & Accessories	81.9	45.5	50.9	79.4	62.8	65.2	81.2	45.9	51.1	96.0	94.0	94.3	88.9	82.0	83.0	86.2	95.7	94.3
0320	Concrete Reinforcing	101.2	34.3	68.8	99.9	51.6	76.6	101.9	66.5	84.8	93.4	96.5	94.9	94.6	79.6	87.3	100.4	102.6	101.5
0330	Cast-in-Place Concrete	101.1	58.4	85.2	108.2	66.0	92.5	104.1	67.9	90.6	90.7	95.5	92.5	89.8	86.6	88.6	91.7	86.0	89.6
03	CONCRETE	85.9	50.2	70.2	90.1	63.8	78.6	93.4	59.2	78.4	91.2	96.4	93.5	90.4	85.0	88.0	95.0	94.3	94.7
04	MASONRY	116.7	50.3	75.9	90.0	62.1	72.9	121.7	53.3	79.6	109.6	97.9	102.4	105.9	80.5	90.3	99.2	92.2	94.9
05	METALS	87.6	75.6	84.1	88.5	85.6	87.7	92.4	87.9	91.1	91.8	117.7	99.4	92.9	109.4	97.8	85.1	110.6	92.5
06	WOOD, PLASTICS & COMPOSITES	78.8	46.8	61.2	76.2	62.8	68.8	76.7	45.2	59.4	96.7	94.4	95.4	89.2	80.0	84.1	94.0	97.0	95.7
07	THERMAL & MOISTURE PROTECTION	99.4	54.2	79.8	99.6	62.4	83.4	97.6	53.9	78.6	96.4	99.2	97.6	95.9	85.3	91.3	92.7	92.7	92.7
08	OPENINGS	96.0	41.3	83.2	95.6	54.8	86.1	95.0	51.4	84.9	98.7	98.2	98.5	98.6	76.1	93.4	86.0	96.2	88.3
0920	Plaster & Gypsum Board	94.3	45.8	61.7	94.3	62.3	72.8	89.6	44.2	59.0	101.7	94.6	96.9	101.1	79.8	86.8	98.2	96.9	97.3
0950, 0980	Ceilings & Acoustic Treatment	81.5	45.8	57.4	82.4	62.3	68.8	82.3	44.2	56.6	88.9	94.6	92.8	88.9	79.8	82.8	87.4	96.9	93.8
0960	Flooring	87.9	65.0	81.4	86.5	65.0	80.5	96.5	65.0	87.6	95.7	96.9	96.0	92.4	84.5	90.2	94.9	99.7	96.2
0970, 0990	Wall Finishes & Painting/Coating	86.0	46.9	62.8	86.0	46.9	62.8	94.6	45.2	65.4	95.4	106.3	101.8	95.4	67.1	78.6	94.3	103.2	99.6
09	FINISHES	84.8	49.2	65.5	84.8	61.6	72.3	89.7	48.9	67.6	98.2	95.7	96.8	97.0	79.8	87.7	97.1	97.6	97.4
COVERS	DIVS. 10 - 14, 25, 28, 41, 43, 44, 46	100.0	50.9	89.1	100.0	71.5	93.6	100.0	49.0	88.6	100.0	81.4	95.8	100.0	94.3	98.7	100.0	81.3	95.8
21, 22, 23	FIRE SUPPRESSION, PLUMBING & HVAC	98.0	50.7	79.5	100.0	57.6	83.3	96.5	52.7	80.5	96.5	99.5	97.7	100.0	98.7	99.5	96.6	99.9	97.9
26, 27, 3370	ELECTRICAL, COMMUNICATIONS & UTIL.	97.3	56.0	76.9	100.1	55.6	78.2	96.5	55.0	76.1	99.2	76.8	88.2	99.2	98.5	98.9	94.9	75.8	85.5
MF2018	WEIGHTED AVERAGE	94.4	55.8	78.0	94.6	64.5	81.9	96.6	59.0	80.7	96.4	95.5	96.0	97.0	91.2	94.5	93.7	93.8	93.7

MISSOURI

DIVISION		COLUMBIA 652			FLAT RIVER 636			HANNIBAL 634			HARRISONVILLE 647			JEFFERSON CITY 650 - 651			JOPLIN 648		
		MAT.	INST.	TOTAL	MAT.	INST.	TOTAL	MAT.	INST.	TOTAL	MAT.	INST.	TOTAL	MAT.	INST.	TOTAL	MAT.	INST.	TOTAL
015433	CONTRACTOR EQUIPMENT		110.3	110.3		107.1	107.1		107.1	107.1		101.5	101.5		113.1	113.1		104.8	104.8
0241, 31 - 34	SITE & INFRASTRUCTURE, DEMOLITION	94.2	93.8	93.9	90.2	89.7	89.8	85.4	89.8	88.4	92.9	89.5	90.6	94.4	98.7	97.4	102.0	92.6	95.5
0310	Concrete Forming & Accessories	84.4	81.0	81.5	102.4	88.3	90.4	94.3	82.5	84.3	83.4	99.0	96.7	96.3	78.6	81.2	97.4	74.2	77.7
0320	Concrete Reinforcing	85.6	93.3	89.3	94.6	102.9	98.6	92.9	96.5	94.6	100.0	110.9	105.3	92.4	93.3	92.9	103.7	84.3	94.3
0330	Cast-in-Place Concrete	88.6	84.8	87.2	93.7	90.6	92.5	85.9	93.5	88.7	93.9	99.7	96.0	93.9	84.1	90.2	99.3	75.5	90.4
03	CONCRETE	80.8	86.3	83.2	94.1	93.3	93.8	87.6	90.5	88.9	91.1	102.0	95.9	87.5	84.8	86.3	93.9	77.7	86.7
04	MASONRY	136.9	86.4	105.9	106.5	76.4	88.0	101.6	94.6	97.3	94.0	99.4	97.3	96.0	86.5	90.1	92.4	80.2	84.9
05	METALS	96.3	115.4	101.9	91.7	119.2	99.8	91.8	117.2	99.3	85.4	115.1	94.1	95.8	114.0	101.1	87.9	98.8	91.1
06	WOOD, PLASTICS & COMPOSITES	80.3	78.6	79.3	105.2	89.6	96.6	94.9	81.1	87.3	90.4	98.7	95.0	97.1	75.6	85.2	106.1	73.5	88.2
07	THERMAL & MOISTURE PROTECTION	90.6	86.1	88.7	96.6	91.0	94.2	96.2	93.4	95.0	91.8	100.8	95.7	97.5	86.1	92.5	91.9	81.2	87.3
08	OPENINGS	94.4	81.7	91.4	98.6	97.4	98.4	98.7	83.7	95.2	85.7	103.2	89.8	94.5	80.0	91.1	86.9	76.3	84.5
0920	Plaster & Gypsum Board	84.7	78.2	80.3	107.7	89.7	95.6	101.4	81.0	87.7	94.1	98.6	97.1	96.2	74.9	81.9	105.4	72.7	83.4
0950, 0980	Ceilings & Acoustic Treatment	88.5	78.2	81.5	88.9	89.7	89.4	88.9	81.0	83.6	87.4	98.6	95.0	93.3	74.9	80.9	88.3	72.7	77.8
0960	Flooring	90.2	98.6	92.5	98.9	84.5	94.9	95.0	96.9	95.5	90.3	100.8	93.2	97.6	71.5	90.3	120.1	72.2	106.6
0970, 0990	Wall Finishes & Painting/Coating	93.1	81.9	86.4	95.4	71.7	81.4	95.4	94.1	94.6	98.6	107.8	104.0	91.2	81.9	85.7	93.9	75.9	83.2
09	FINISHES	84.0	83.5	83.7	100.0	85.1	92.0	97.7	85.3	91.0	94.8	100.2	97.7	91.5	77.1	83.7	103.8	74.4	87.9
COVERS	DIVS. 10 - 14, 25, 28, 41, 43, 44, 46	100.0	93.6	98.6	100.0	94.0	98.7	100.0	78.8	95.3	100.0	82.7	96.2	100.0	93.6	98.6	100.0	79.5	95.4
21, 22, 23	FIRE SUPPRESSION, PLUMBING & HVAC	100.0	96.9	98.8	96.5	96.8	96.6	96.5	97.7	97.0	96.6	100.8	98.2	100.0	96.9	98.8	100.1	71.1	88.7
26, 27, 3370	ELECTRICAL, COMMUNICATIONS & UTIL.	97.8	81.4	89.7	103.8	98.5	101.2	98.0	76.8	87.5	101.5	101.2	101.4	103.2	81.4	92.4	92.9	66.1	79.7
MF2018	WEIGHTED AVERAGE	96.1	90.7	93.8	97.3	94.2	96.0	95.3	91.6	93.7	93.2	100.8	96.4	96.4	89.9	93.7	95.1	77.7	87.7

MISSOURI

DIVISION		KANSAS CITY 640 - 641			KIRKSVILLE 635			POPLAR BLUFF 639			ROLLA 654 - 655			SEDALIA 653			SIKESTON 638		
		MAT.	INST.	TOTAL	MAT.	INST.	TOTAL	MAT.	INST.	TOTAL	MAT.	INST.	TOTAL	MAT.	INST.	TOTAL	MAT.	INST.	TOTAL
015433	CONTRACTOR EQUIPMENT		104.4	104.4		97.4	97.4		99.6	99.6		110.3	110.3		100.4	100.4		99.6	99.6
0241, 31 - 34	SITE & INFRASTRUCTURE, DEMOLITION	94.6	98.6	97.3	88.8	84.9	86.1	76.1	88.5	84.7	92.9	94.1	93.8	92.1	89.3	90.2	79.6	89.1	86.2
0310	Concrete Forming & Accessories	97.2	103.5	102.6	86.9	78.4	79.6	87.4	79.3	80.5	91.6	94.0	93.6	89.6	78.2	79.9	88.2	79.4	80.7
0320	Concrete Reinforcing	98.5	111.1	104.6	93.8	83.3	88.7	96.7	76.0	86.7	86.0	93.4	89.6	84.8	110.3	97.1	96.0	76.0	86.4
0330	Cast-in-Place Concrete	97.3	102.6	99.3	93.6	82.2	89.4	72.0	83.5	76.3	90.6	94.7	92.1	94.7	81.3	89.7	77.0	83.5	79.4
03	CONCRETE	93.8	104.9	98.7	106.9	81.8	95.9	81.5	81.4	81.5	82.5	95.6	88.2	95.7	86.1	91.5	85.2	81.5	83.6
04	MASONRY	99.6	103.3	101.9	113.2	84.8	95.8	104.5	74.1	85.8	110.8	85.3	95.2	117.1	81.6	95.3	104.2	74.1	85.7
05	METALS	94.9	112.8	100.2	91.5	101.0	94.3	92.1	97.8	93.7	95.7	115.9	101.6	94.5	112.8	99.9	92.4	97.9	94.0
06	WOOD, PLASTICS & COMPOSITES	100.1	103.8	102.2	82.2	76.5	79.1	81.3	80.0	80.6	88.0	95.6	92.2	81.6	76.7	78.9	82.9	80.0	81.3
07	THERMAL & MOISTURE PROTECTION	91.8	105.2	97.7	102.2	90.8	97.3	100.5	83.0	92.9	90.9	92.9	91.8	96.1	87.8	92.5	100.7	82.1	92.6
08	OPENINGS	96.5	106.0	98.7	104.1	78.8	98.2	105.0	75.2	98.1	94.4	91.1	93.6	99.4	87.0	96.5	105.0	75.2	98.1
0920	Plaster & Gypsum Board	100.9	104.0	103.0	96.3	76.2	82.8	96.6	79.8	85.3	86.9	95.5	92.8	80.5	76.2	77.6	98.5	79.8	85.9
0950, 0980	Ceilings & Acoustic Treatment	90.5	104.0	99.6	87.3	76.2	79.8	88.9	79.8	82.8	88.5	95.6	93.3	88.5	76.2	80.2	88.9	79.8	82.8
0960	Flooring	95.9	101.8	97.6	73.8	96.4	80.2	87.3	84.5	86.5	93.8	96.4	94.5	72.5	72.9	72.6	87.8	84.5	86.9
0970, 0990	Wall Finishes & Painting/Coating	97.2	107.8	103.5	91.3	78.3	83.6	90.6	67.1	76.7	93.1	90.2	91.4	93.1	103.2	99.1	90.6	67.1	76.7
09	FINISHES	98.1	103.6	101.1	97.4	81.1	88.6	96.9	78.4	86.9	85.4	94.0	90.1	82.9	79.1	80.8	97.5	78.7	87.3
COVERS	DIVS. 10 - 14, 25, 28, 41, 43, 44, 46	100.0	98.4	99.7	100.0	78.3	95.2	100.0	92.2	98.3	100.0	96.7	99.3	100.0	87.9	97.3	100.0	92.2	98.3
21, 22, 23	FIRE SUPPRESSION, PLUMBING & HVAC	100.0	103.4	101.4	96.6	97.1	96.8	96.6	94.9	95.9	96.5	98.6	97.3	96.4	94.9	95.8	96.6	95.0	95.9
26, 27, 3370	ELECTRICAL, COMMUNICATIONS & UTIL.	103.3	101.2	102.2	98.2	76.8	87.6	98.4	98.4	98.4	96.4	79.4	88.0	97.2	101.2	99.2	97.5	98.4	97.9
MF2018	WEIGHTED AVERAGE	97.9	103.8	100.4	98.9	86.5	93.7	95.1	87.7	92.0	94.1	94.3	94.2	96.4	91.5	94.3	95.7	87.8	92.3

		MISSOURI									MONTANA								
DIVISION		SPRINGFIELD 656 - 658			ST. JOSEPH 644 - 645			ST. LOUIS 630 - 631			BILLINGS 590 - 591			BUTTE 597			GREAT FALLS 594		
		MAT.	INST.	TOTAL	MAT.	INST.	TOTAL	MAT.	INST.	TOTAL	MAT.	INST.	TOTAL	MAT.	INST.	TOTAL	MAT.	INST.	TOTAL
015433	CONTRACTOR EQUIPMENT		102.8	102.8		101.5	101.5		108.6	108.6		97.9	97.9		97.7	97.7		97.7	97.7
0241, 31 - 34	SITE & INFRASTRUCTURE, DEMOLITION	94.4	91.4	92.4	96.7	87.4	90.3	94.4	97.8	96.7	91.8	93.3	92.9	97.3	93.3	94.5	101.0	93.3	95.7
0310	Concrete Forming & Accessories	97.7	75.4	78.7	96.3	89.1	90.2	99.5	102.5	102.1	99.3	68.3	72.9	86.5	68.2	70.9	99.2	68.2	72.8
0320	Concrete Reinforcing	82.5	92.9	87.5	97.3	110.4	103.6	86.5	105.2	95.5	98.0	81.8	90.2	106.3	81.7	94.4	98.0	81.8	90.2
0330	Cast-in-Place Concrete	96.3	73.7	87.9	92.2	95.5	93.4	101.3	102.5	101.7	118.7	72.7	101.6	130.6	72.5	109.0	138.1	72.5	113.7
03	CONCRETE	92.1	79.0	86.3	90.3	96.0	92.8	97.1	104.0	100.1	98.3	73.1	87.2	101.9	73.0	89.2	106.9	73.0	92.0
04	MASONRY	89.1	79.9	83.5	95.3	88.7	91.2	91.1	107.1	100.9	125.8	83.8	100.0	120.9	82.7	97.4	125.4	83.8	99.8
05	METALS	100.7	101.7	101.0	91.5	113.8	98.0	97.4	119.0	103.8	107.0	89.6	101.9	101.2	89.4	97.7	104.4	89.5	100.0
06	WOOD, PLASTICS & COMPOSITES	89.4	75.0	81.5	106.1	88.7	96.5	97.6	101.0	99.5	90.1	64.6	76.1	76.8	64.6	70.1	91.4	64.6	76.6
07	THERMAL & MOISTURE PROTECTION	94.6	75.5	86.3	92.2	88.7	90.7	94.3	105.5	99.2	108.3	73.7	93.2	107.8	73.3	92.8	108.5	73.7	93.4
08	OPENINGS	101.7	84.3	97.6	89.9	96.7	91.5	100.3	105.8	101.6	98.6	65.9	91.0	96.9	65.9	89.7	99.8	65.9	91.9
0920	Plaster & Gypsum Board	87.8	74.4	78.8	106.9	88.3	94.4	106.5	101.5	103.1	118.3	64.0	81.8	118.8	64.0	81.9	128.7	64.0	85.2
0950, 0980	Ceilings & Acoustic Treatment	88.5	74.4	79.0	94.8	88.3	90.4	90.5	101.5	97.9	96.6	64.0	74.6	104.1	64.0	77.0	105.7	64.0	77.6
0960	Flooring	92.8	71.5	86.8	99.4	99.0	99.3	98.0	99.1	98.3	88.1	80.8	86.0	85.4	90.9	87.0	91.9	80.8	88.8
0970, 0990	Wall Finishes & Painting/Coating	87.6	99.4	94.6	94.3	107.8	102.3	96.4	106.3	102.2	93.8	91.0	92.2	92.3	67.8	77.8	92.3	91.0	91.5
09	FINISHES	87.6	77.6	82.2	100.3	92.8	96.3	100.5	101.9	101.3	92.5	72.3	81.6	93.0	71.8	81.5	96.8	72.3	83.5
COVERS	DIVS. 10 - 14, 25, 28, 41, 43, 44, 46	100.0	90.5	97.9	100.0	94.0	98.7	100.0	102.8	100.6	100.0	91.6	98.1	100.0	91.6	98.1	100.0	91.6	98.1
21, 22, 23	FIRE SUPPRESSION, PLUMBING & HVAC	100.0	69.4	88.0	100.1	87.5	95.2	100.0	105.5	102.2	100.0	76.4	90.7	100.0	71.2	88.7	100.0	71.2	88.7
26, 27, 3370	ELECTRICAL, COMMUNICATIONS & UTIL.	101.3	68.8	85.3	101.5	75.8	88.9	101.8	98.5	100.2	102.3	75.2	88.9	109.2	71.3	90.5	101.7	70.8	86.5
MF2018	WEIGHTED AVERAGE	97.6	79.0	89.7	96.1	90.9	93.9	98.8	104.5	101.2	101.6	78.4	91.8	101.5	76.5	90.9	102.9	76.6	91.8

MONTANA

DIVISION		HAVRE 595 MAT.	INST.	TOTAL	HELENA 596 MAT.	INST.	TOTAL	KALISPELL 599 MAT.	INST.	TOTAL	MILES CITY 593 MAT.	INST.	TOTAL	MISSOULA 598 MAT.	INST.	TOTAL	WOLF POINT 592 MAT.	INST.	TOTAL
015433	CONTRACTOR EQUIPMENT		97.7	97.7		99.9	99.9		97.7	97.7		97.7	97.7		97.7	97.7		97.7	97.7
0241, 31 - 34	SITE & INFRASTRUCTURE, DEMOLITION	104.2	93.1	96.6	91.1	96.8	95.0	87.5	93.3	91.5	93.5	93.2	93.3	80.5	93.3	89.3	110.2	93.2	98.4
0310	Concrete Forming & Accessories	79.6	67.2	69.0	100.2	68.3	73.0	89.7	68.3	71.4	97.7	67.3	71.8	89.7	68.4	71.5	90.4	67.3	70.7
0320	Concrete Reinforcing	107.1	78.9	93.5	113.0	81.8	97.9	109.1	85.8	97.8	106.7	78.9	93.3	108.1	85.8	97.3	108.2	78.2	93.7
0330	Cast-in-Place Concrete	140.7	71.2	114.9	102.1	73.3	91.4	113.4	72.5	98.2	124.2	71.3	104.5	96.2	72.6	87.4	139.1	70.2	113.5
03	CONCRETE	109.8	71.6	93.0	94.7	73.3	85.3	91.7	73.7	83.8	98.8	71.7	86.9	80.5	73.8	77.6	113.2	71.2	94.8
04	MASONRY	121.9	80.6	96.5	114.2	82.7	94.8	119.6	82.7	96.9	127.6	80.6	98.7	144.5	82.7	106.5	128.9	80.6	99.2
05	METALS	97.3	88.2	94.6	103.2	88.5	98.9	97.2	90.8	95.3	96.5	88.4	94.1	97.7	91.0	95.7	96.6	88.1	94.1
06	WOOD, PLASTICS & COMPOSITES	68.1	64.6	66.2	93.8	64.7	77.8	80.2	64.6	71.6	88.4	64.6	75.3	80.2	64.6	71.6	79.5	64.6	71.3
07	THERMAL & MOISTURE PROTECTION	108.3	66.2	90.0	103.1	73.9	90.4	107.4	73.6	92.7	107.8	67.7	90.3	106.9	75.6	93.3	108.9	67.5	90.9
08	OPENINGS	97.4	65.2	89.9	96.3	65.9	89.2	97.4	66.8	90.3	96.9	65.2	89.5	96.9	66.8	89.9	96.9	65.1	89.5
0920	Plaster & Gypsum Board	114.1	64.0	80.4	113.5	64.0	80.2	118.8	64.0	81.9	127.8	64.0	84.9	118.8	64.0	81.9	121.8	64.0	82.9
0950, 0980	Ceilings & Acoustic Treatment	104.1	64.0	77.0	107.9	64.0	78.3	104.1	64.0	77.0	101.6	64.0	76.3	104.1	64.0	77.0	101.6	64.0	76.3
0960	Flooring	83.0	90.9	85.3	95.9	90.9	94.5	87.1	90.9	88.2	91.8	90.9	91.6	87.1	90.9	88.2	88.4	90.9	89.1
0970, 0990	Wall Finishes & Painting/Coating	92.3	67.8	77.8	97.6	67.8	79.9	92.3	91.0	91.5	92.3	67.8	77.8	92.3	91.0	91.5	92.3	67.8	77.8
09	FINISHES	92.3	71.3	80.9	100.1	71.8	84.8	92.9	74.3	82.8	95.4	71.3	82.3	92.3	74.3	82.6	95.0	71.3	82.1
COVERS	DIVS. 10 - 14, 25, 28, 41, 43, 44, 46	100.0	88.8	97.5	100.0	91.9	98.2	100.0	89.5	97.7	100.0	88.9	97.5	100.0	92.4	98.3	100.0	88.9	97.5
21, 22, 23	FIRE SUPPRESSION, PLUMBING & HVAC	96.5	68.9	85.6	100.0	71.2	88.7	96.5	69.4	85.8	96.5	75.2	88.1	100.0	71.2	88.7	96.5	75.2	88.1
26, 27, 3370	ELECTRICAL, COMMUNICATIONS & UTIL.	101.7	69.2	85.7	109.0	70.8	90.2	106.0	68.1	87.3	101.7	74.4	88.3	107.0	70.2	88.9	101.7	74.4	88.3
MF2018	WEIGHTED AVERAGE	100.4	74.8	89.6	100.9	76.7	90.7	98.1	76.2	88.9	99.2	77.0	89.8	98.6	77.1	89.5	101.4	76.9	91.1

NEBRASKA

DIVISION		ALLIANCE 693 MAT.	INST.	TOTAL	COLUMBUS 686 MAT.	INST.	TOTAL	GRAND ISLAND 688 MAT.	INST.	TOTAL	HASTINGS 689 MAT.	INST.	TOTAL	LINCOLN 683 - 685 MAT.	INST.	TOTAL	MCCOOK 690 MAT.	INST.	TOTAL
015433	CONTRACTOR EQUIPMENT		95.3	95.3		101.9	101.9		101.9	101.9		101.9	101.9		105.0	105.0		101.9	101.9
0241, 31 - 34	SITE & INFRASTRUCTURE, DEMOLITION	98.2	96.2	96.8	101.8	90.6	94.1	106.8	90.7	95.7	105.4	90.6	95.2	94.9	95.7	95.4	101.5	90.6	94.0
0310	Concrete Forming & Accessories	87.0	55.3	60.0	96.5	75.0	78.2	96.1	68.8	72.9	99.3	72.5	76.5	94.4	75.5	78.3	92.3	55.9	61.2
0320	Concrete Reinforcing	107.4	87.4	97.7	100.4	85.9	93.4	99.8	76.1	88.4	99.8	76.1	88.4	97.6	76.4	87.3	100.1	76.3	88.6
0330	Cast-in-Place Concrete	108.1	81.8	98.3	107.3	81.8	97.8	113.6	77.7	100.2	113.6	74.0	98.9	87.2	83.1	85.6	116.9	73.9	100.9
03	CONCRETE	114.3	70.9	95.2	100.2	80.4	91.5	104.9	74.5	91.5	105.1	74.9	91.8	87.5	79.4	83.9	103.7	67.4	87.7
04	MASONRY	106.0	77.4	88.4	111.1	77.4	90.4	104.3	73.8	85.6	112.8	74.0	88.9	93.6	78.2	84.1	101.9	77.4	86.8
05	METALS	99.1	85.2	95.0	91.4	98.0	93.3	93.2	93.7	93.3	94.0	93.7	93.9	95.0	93.3	94.5	93.9	93.8	93.9
06	WOOD, PLASTICS & COMPOSITES	82.6	48.7	64.0	94.3	74.9	83.6	93.4	66.3	78.5	97.2	72.1	83.4	99.0	75.0	85.8	90.7	49.7	68.1
07	THERMAL & MOISTURE PROTECTION	100.7	75.8	89.9	101.6	80.4	92.4	101.8	78.3	91.5	101.8	77.6	91.3	100.1	81.1	91.8	96.0	76.1	87.3
08	OPENINGS	90.8	58.1	83.2	91.3	73.7	87.2	91.3	66.8	85.6	91.3	69.8	86.3	105.4	68.8	96.9	91.2	55.8	83.0
0920	Plaster & Gypsum Board	79.8	47.5	58.1	93.4	74.3	80.6	92.5	65.5	74.3	94.4	71.4	78.9	102.7	74.3	83.6	91.0	48.4	62.4
0950, 0980	Ceilings & Acoustic Treatment	88.6	47.5	60.9	83.4	74.3	77.3	83.4	65.5	71.3	83.4	71.4	75.3	98.3	74.3	82.1	83.9	48.4	60.0
0960	Flooring	91.4	87.3	90.2	84.0	87.3	84.9	83.7	78.9	82.4	85.0	81.3	84.0	96.3	86.6	93.6	90.0	87.3	89.2
0970, 0990	Wall Finishes & Painting/Coating	155.8	51.7	94.2	74.9	60.1	66.1	74.9	60.1	66.1	74.9	60.1	66.1	94.4	79.3	85.5	87.6	45.0	62.4
09	FINISHES	90.8	59.2	73.7	84.9	75.5	79.8	84.9	69.3	76.5	85.6	72.7	78.6	95.6	77.9	86.0	88.4	59.1	72.5
COVERS	DIVS. 10 - 14, 25, 28, 41, 43, 44, 46	100.0	83.9	96.4	100.0	83.5	96.3	100.0	89.7	97.7	100.0	83.1	96.2	100.0	91.0	98.0	100.0	84.1	96.5
21, 22, 23	FIRE SUPPRESSION, PLUMBING & HVAC	96.7	75.0	88.2	96.6	75.3	88.2	100.1	79.5	92.0	96.6	74.6	87.9	100.0	79.6	92.0	96.5	75.2	88.1
26, 27, 3370	ELECTRICAL, COMMUNICATIONS & UTIL.	93.6	65.6	79.8	97.3	81.4	89.5	95.9	65.7	81.0	95.3	79.4	87.5	109.2	65.7	87.8	95.6	65.7	80.9
MF2018	WEIGHTED AVERAGE	98.7	73.3	87.9	96.0	80.7	89.6	97.4	76.8	88.7	97.1	78.2	89.1	98.3	79.7	90.5	96.3	73.1	86.5

NEBRASKA / NEVADA

DIVISION		NORFOLK 687 MAT.	INST.	TOTAL	NORTH PLATTE 691 MAT.	INST.	TOTAL	OMAHA 680 - 681 MAT.	INST.	TOTAL	VALENTINE 692 MAT.	INST.	TOTAL	CARSON CITY 897 MAT.	INST.	TOTAL	ELKO 898 MAT.	INST.	TOTAL
015433	CONTRACTOR EQUIPMENT		91.7	91.7		101.9	101.9		94.2	94.2		95.0	95.0		97.9	97.9		94.6	94.6
0241, 31 - 34	SITE & INFRASTRUCTURE, DEMOLITION	84.1	89.7	88.0	102.9	90.4	94.1	89.1	94.9	92.4	86.3	95.2	92.4	83.3	97.4	93.0	67.8	91.3	84.0
0310	Concrete Forming & Accessories	83.2	74.1	75.5	94.8	75.2	78.1	93.1	76.7	79.1	83.5	53.1	57.6	106.1	76.3	80.7	111.4	96.3	98.6
0320	Concrete Reinforcing	100.5	66.5	84.1	99.5	76.2	88.3	100.8	76.5	89.0	100.1	66.3	83.7	115.8	116.1	116.0	124.5	114.5	119.7
0330	Cast-in-Place Concrete	107.9	71.3	94.3	116.9	60.6	96.0	91.4	78.6	86.7	103.1	54.0	84.8	98.0	82.2	92.1	95.0	72.4	86.6
03	CONCRETE	98.8	72.4	87.2	103.8	71.4	89.6	89.8	77.8	84.5	101.2	56.9	81.7	101.0	85.3	94.1	98.5	90.8	95.1
04	MASONRY	117.1	77.3	92.7	90.4	76.9	82.1	93.0	80.0	85.0	101.4	76.8	86.3	118.2	68.8	87.8	125.0	67.7	89.8
05	METALS	94.8	80.4	90.6	93.1	93.2	93.2	95.0	83.9	91.7	105.1	79.4	97.5	111.4	94.1	106.3	115.0	93.6	108.7
06	WOOD, PLASTICS & COMPOSITES	78.4	74.4	76.2	92.7	76.9	84.0	93.4	76.6	84.2	77.5	47.4	61.0	93.6	74.4	83.0	104.7	102.3	103.4
07	THERMAL & MOISTURE PROTECTION	101.2	78.3	91.2	95.9	78.4	88.3	96.1	79.8	89.8	96.5	73.9	86.7	118.1	79.1	101.2	114.2	73.6	96.5
08	OPENINGS	92.7	68.7	87.1	90.6	72.1	86.3	99.8	76.8	94.4	92.7	53.4	83.6	101.1	77.6	95.6	102.5	92.6	100.2
0920	Plaster & Gypsum Board	93.2	74.3	80.5	91.0	76.4	81.2	102.9	76.4	85.1	92.5	46.6	61.6	100.8	73.7	82.6	106.9	102.5	103.9
0950, 0980	Ceilings & Acoustic Treatment	97.6	74.3	81.9	83.9	76.4	78.8	101.0	76.4	84.4	100.4	46.6	64.1	96.0	73.7	80.9	94.5	102.5	99.9
0960	Flooring	104.1	87.3	99.4	91.1	78.9	87.7	100.6	87.3	96.9	116.8	87.3	108.5	99.5	67.1	90.4	103.0	67.1	92.9
0970, 0990	Wall Finishes & Painting/Coating	132.5	60.1	89.7	87.6	58.9	70.6	104.2	61.0	78.6	155.3	61.0	99.5	95.8	79.3	86.1	94.2	79.3	85.4
09	FINISHES	101.8	75.2	87.4	88.7	74.8	81.2	100.4	76.9	87.7	109.1	59.3	82.1	96.5	74.0	84.3	94.0	90.4	92.0
COVERS	DIVS. 10 - 14, 25, 28, 41, 43, 44, 46	100.0	85.1	96.7	100.0	61.1	91.3	100.0	90.2	97.8	100.0	56.1	90.2	100.0	100.2	100.0	100.0	91.0	98.0
21, 22, 23	FIRE SUPPRESSION, PLUMBING & HVAC	96.3	74.3	88.0	96.3	75.3	88.0	100.0	86.3	92.3	96.1	73.5	87.2	100.0	77.0	91.0	98.3	76.9	89.9
26, 27, 3370	ELECTRICAL, COMMUNICATIONS & UTIL.	96.2	81.4	88.9	93.9	65.7	80.0	103.9	83.7	94.0	91.4	65.7	78.7	102.2	88.3	95.3	98.8	88.3	93.6
MF2018	WEIGHTED AVERAGE	97.4	77.5	89.0	96.3	75.6	87.6	97.6	81.5	90.8	98.6	69.2	86.2	102.9	82.5	94.3	102.3	85.2	95.0

DIVISION		NEVADA								NEW HAMPSHIRE								
		ELY			LAS VEGAS			RENO			CHARLESTON			CLAREMONT			CONCORD		
		893			889 - 891			894 - 895			036			037			032 - 033		
		MAT.	INST.	TOTAL	MAT.	INST.	TOTAL	MAT.	INST.	TOTAL	MAT.	INST.	TOTAL	MAT.	INST.	TOTAL	MAT.	INST.	TOTAL
015433	CONTRACTOR EQUIPMENT		94.6	94.6		94.6	94.6		94.6	94.6		95.7	95.7		95.7	95.7		98.4	98.4
0241, 31 - 34	SITE & INFRASTRUCTURE, DEMOLITION	73.4	92.5	86.6	76.6	95.5	89.7	73.4	92.6	86.7	80.6	96.1	91.3	74.6	96.1	89.5	88.7	101.2	97.3
0310	Concrete Forming & Accessories	104.3	101.6	102.0	105.2	105.6	105.5	100.6	76.3	79.9	86.6	83.1	83.6	92.1	83.2	84.5	97.2	94.8	95.1
0320	Concrete Reinforcing	123.1	114.7	119.0	113.5	125.9	119.5	116.4	124.2	120.1	89.7	88.6	89.1	89.7	88.6	89.1	100.9	88.7	95.0
0330	Cast-in-Place Concrete	102.0	95.2	99.4	98.7	108.2	102.2	107.9	81.1	97.9	84.7	116.1	96.4	77.8	116.1	92.1	100.9	117.8	107.2
03	CONCRETE	107.5	101.2	104.7	102.4	109.5	105.5	106.7	86.3	97.8	90.4	95.8	92.7	82.8	95.8	88.5	96.8	101.6	98.9
04	MASONRY	130.5	74.8	96.3	117.4	106.2	110.5	124.2	68.7	90.1	89.8	99.5	95.8	89.9	99.5	95.8	97.8	101.4	100.0
05	METALS	115.0	96.5	109.5	124.1	103.9	118.2	116.7	97.4	111.1	94.2	92.1	93.5	94.2	92.1	93.6	101.3	91.8	98.5
06	WOOD, PLASTICS & COMPOSITES	94.8	104.4	100.1	92.4	102.3	97.8	88.4	74.3	80.6	86.6	79.4	82.6	92.4	79.4	85.3	93.3	93.6	93.5
07	THERMAL & MOISTURE PROTECTION	114.7	92.2	104.9	129.4	101.9	117.4	114.2	78.3	98.6	107.3	106.5	107.0	107.1	106.5	106.8	111.3	109.5	110.5
08	OPENINGS	102.4	93.8	100.4	101.4	109.6	103.3	100.2	79.2	95.3	96.1	80.6	92.5	97.2	80.6	93.3	99.7	88.5	95.1
0920	Plaster & Gypsum Board	101.2	104.6	103.5	93.9	102.5	99.7	89.0	73.7	78.7	106.1	78.5	87.5	107.0	78.5	87.9	108.7	93.1	98.2
0950, 0980	Ceilings & Acoustic Treatment	94.5	104.6	101.3	101.0	102.5	102.0	97.8	73.7	81.5	101.8	78.5	86.1	101.8	78.5	86.1	106.1	93.1	97.4
0960	Flooring	100.5	67.1	91.1	92.0	104.1	95.4	97.3	67.1	88.8	85.1	113.1	92.9	87.6	113.1	94.7	96.4	113.1	101.1
0970, 0990	Wall Finishes & Painting/Coating	94.2	112.0	104.7	96.8	119.4	110.2	94.2	79.3	85.4	81.8	90.0	86.6	81.8	90.0	86.6	95.9	90.0	92.4
09	FINISHES	92.9	97.0	95.1	90.8	106.4	99.3	90.8	73.9	81.7	90.2	88.5	89.3	90.6	88.5	89.5	95.9	97.4	96.7
COVERS	DIVS. 10 - 14, 25, 28, 41, 43, 44, 46	100.0	70.7	93.5	100.0	105.1	101.1	100.0	99.8	100.0	100.0	88.5	97.4	100.0	88.5	97.4	100.0	109.5	102.1
21, 22, 23	FIRE SUPPRESSION, PLUMBING & HVAC	98.3	96.2	97.4	100.1	103.0	101.3	100.0	77.1	91.0	96.7	82.2	91.0	96.7	82.2	91.0	100.0	87.2	95.0
26, 27, 3370	ELECTRICAL, COMMUNICATIONS & UTIL.	99.0	93.0	96.0	103.5	107.8	105.6	99.4	88.3	93.9	91.4	75.2	83.4	91.4	75.2	83.4	90.2	75.2	82.8
MF2018	WEIGHTED AVERAGE	103.7	93.2	99.2	105.0	105.1	105.0	103.6	82.6	94.7	94.1	88.4	91.7	93.1	88.4	91.1	98.0	93.2	96.0

DIVISION		NEW HAMPSHIRE														NEW JERSEY		
		KEENE			LITTLETON			MANCHESTER			NASHUA			PORTSMOUTH			ATLANTIC CITY		
		034			035			031			030			038			082, 084		
		MAT.	INST.	TOTAL	MAT.	INST.	TOTAL	MAT.	INST.	TOTAL	MAT.	INST.	TOTAL	MAT.	INST.	TOTAL	MAT.	INST.	TOTAL
015433	CONTRACTOR EQUIPMENT		95.7	95.7		95.7	95.7		99.0	99.0		95.7	95.7		95.7	95.7		93.3	93.3
0241, 31 - 34	SITE & INFRASTRUCTURE, DEMOLITION	88.0	96.2	93.6	74.7	95.2	88.8	87.8	101.3	97.1	89.5	96.3	94.2	83.3	97.0	92.7	88.2	99.7	96.1
0310	Concrete Forming & Accessories	90.7	83.4	84.5	102.4	77.3	81.0	97.5	95.1	95.5	98.5	94.9	95.4	87.9	94.1	93.2	114.8	142.0	138.0
0320	Concrete Reinforcing	89.7	88.6	89.2	90.4	88.6	89.5	108.5	88.7	98.9	112.0	88.7	100.7	89.7	88.7	89.2	81.1	138.0	108.6
0330	Cast-in-Place Concrete	85.1	116.2	96.7	76.3	107.5	87.9	99.2	118.3	106.3	80.5	117.4	94.3	76.3	117.2	91.5	78.5	137.1	100.3
03	CONCRETE	89.9	95.9	92.6	82.2	90.1	85.7	97.0	101.9	99.2	90.0	101.6	95.1	82.4	101.2	90.7	87.7	138.1	109.8
04	MASONRY	92.5	99.5	96.8	100.9	84.1	90.6	94.4	101.4	98.7	94.4	101.4	98.7	90.2	99.8	96.1	104.1	140.0	126.1
05	METALS	94.8	92.5	94.2	94.9	92.2	94.1	102.0	92.2	99.1	100.5	92.7	98.2	96.4	94.3	95.8	98.8	115.0	103.6
06	WOOD, PLASTICS & COMPOSITES	90.9	79.4	84.6	101.9	79.4	89.5	93.6	93.7	93.7	100.2	93.6	96.6	87.8	93.6	91.0	122.8	143.7	134.3
07	THERMAL & MOISTURE PROTECTION	107.9	106.4	107.2	107.2	99.5	103.8	112.5	109.5	111.2	108.3	108.9	108.6	107.8	108.1	107.9	102.1	134.0	116.0
08	OPENINGS	94.8	84.1	92.3	98.1	80.6	94.0	95.7	93.3	95.2	99.0	91.3	97.2	99.6	81.9	95.5	96.4	139.2	106.4
0920	Plaster & Gypsum Board	106.4	78.5	87.6	120.6	78.5	92.3	99.7	93.1	95.3	115.6	93.1	100.5	106.1	93.1	97.4	114.9	144.6	134.9
0950, 0980	Ceilings & Acoustic Treatment	101.8	78.5	86.1	101.8	78.5	86.1	101.2	93.1	95.8	114.1	93.1	100.0	102.7	93.1	96.3	96.4	144.6	128.9
0960	Flooring	87.2	113.1	94.5	97.9	113.1	102.2	93.8	113.1	99.2	91.0	113.1	97.2	85.2	113.1	93.0	97.5	162.5	115.8
0970, 0990	Wall Finishes & Painting/Coating	81.8	103.4	94.6	81.8	90.0	86.6	95.3	119.0	109.3	81.8	102.2	93.9	81.8	102.2	93.9	81.9	141.8	117.4
09	FINISHES	92.1	89.9	90.9	95.5	84.6	89.6	94.8	100.6	97.9	96.8	98.7	97.8	91.1	98.2	95.0	93.5	147.2	122.6
COVERS	DIVS. 10 - 14, 25, 28, 41, 43, 44, 46	100.0	94.7	98.8	100.0	95.6	99.0	100.0	109.7	102.2	100.0	109.4	102.1	100.0	108.7	101.9	100.0	114.9	103.3
21, 22, 23	FIRE SUPPRESSION, PLUMBING & HVAC	96.7	82.3	91.0	96.7	74.1	87.8	100.1	87.3	95.1	100.2	87.3	95.1	100.2	86.3	94.7	99.7	133.1	112.8
26, 27, 3370	ELECTRICAL, COMMUNICATIONS & UTIL.	91.4	75.2	83.4	92.2	49.1	71.0	93.9	79.3	86.7	93.2	79.3	86.4	91.7	75.3	83.6	94.2	143.4	118.5
MF2018	WEIGHTED AVERAGE	94.5	89.0	92.2	94.4	80.2	88.4	98.2	94.5	96.6	97.5	93.7	95.9	94.9	92.4	93.8	96.8	133.2	112.2

DIVISION		NEW JERSEY																
		CAMDEN			DOVER			ELIZABETH			HACKENSACK			JERSEY CITY			LONG BRANCH		
		081			078			072			076			073			077		
		MAT.	INST.	TOTAL	MAT.	INST.	TOTAL	MAT.	INST.	TOTAL	MAT.	INST.	TOTAL	MAT.	INST.	TOTAL	MAT.	INST.	TOTAL
015433	CONTRACTOR EQUIPMENT		93.3	93.3		95.7	95.7		95.7	95.7		95.7	95.7		93.3	93.3		92.9	92.9
0241, 31 - 34	SITE & INFRASTRUCTURE, DEMOLITION	89.3	99.3	96.2	103.8	101.4	102.2	108.9	101.4	103.7	104.8	101.4	102.5	95.2	101.4	99.5	100.0	100.5	100.4
0310	Concrete Forming & Accessories	105.0	137.3	132.5	96.4	148.3	140.6	108.0	148.5	142.5	96.4	148.3	140.7	100.6	148.4	141.3	101.1	139.0	133.4
0320	Concrete Reinforcing	106.5	128.5	117.1	75.3	156.5	114.6	75.3	156.5	114.6	75.3	156.5	114.6	97.7	156.5	126.2	75.3	156.1	114.4
0330	Cast-in-Place Concrete	76.1	133.8	97.6	81.3	131.5	100.0	69.8	140.3	96.1	79.4	140.3	102.1	63.4	131.6	88.8	70.5	133.8	94.0
03	CONCRETE	88.2	133.2	108.0	86.1	142.3	110.8	83.3	145.5	110.6	84.5	145.4	111.2	81.0	142.2	107.9	85.1	138.6	108.6
04	MASONRY	94.2	135.0	119.3	91.2	145.6	124.7	107.2	145.6	130.8	95.3	145.6	126.2	85.3	145.6	122.4	99.6	135.5	121.6
05	METALS	104.7	111.0	106.5	96.5	125.1	104.9	98.0	125.3	106.0	96.5	125.1	104.9	102.4	122.2	108.2	96.6	121.0	103.7
06	WOOD, PLASTICS & COMPOSITES	110.1	137.9	125.4	94.9	148.3	124.3	109.7	148.3	131.0	94.9	148.3	124.3	95.8	148.3	124.7	96.9	139.3	120.2
07	THERMAL & MOISTURE PROTECTION	102.0	133.0	115.5	103.7	142.8	120.7	104.0	144.1	121.4	103.5	136.7	117.9	103.2	142.8	120.4	103.4	129.0	114.5
08	OPENINGS	98.4	132.0	106.3	102.0	145.1	112.1	100.4	145.1	110.8	99.8	145.1	110.4	98.5	145.1	109.4	94.6	138.5	104.8
0920	Plaster & Gypsum Board	110.9	138.6	129.6	111.5	149.4	137.0	119.0	149.4	139.5	111.5	149.4	137.0	114.6	149.4	138.0	113.7	140.2	131.5
0950, 0980	Ceilings & Acoustic Treatment	106.3	138.6	128.1	92.3	149.4	130.8	94.1	149.4	131.4	92.3	149.4	130.8	102.2	149.4	134.0	92.3	140.2	124.6
0960	Flooring	93.7	162.5	113.0	83.7	188.8	113.2	88.7	188.8	116.8	83.7	188.8	113.2	84.7	188.8	114.0	84.9	176.9	110.8
0970, 0990	Wall Finishes & Painting/Coating	81.9	141.8	117.4	83.8	148.6	122.1	83.8	148.6	122.1	83.8	148.6	122.1	83.9	148.6	122.2	83.9	141.8	118.2
09	FINISHES	93.7	143.7	120.8	90.0	156.2	125.8	93.4	156.2	127.4	89.9	156.2	125.8	92.4	156.6	127.1	91.0	147.2	121.4
COVERS	DIVS. 10 - 14, 25, 28, 41, 43, 44, 46	100.0	114.3	103.2	100.0	131.7	107.1	100.0	131.7	107.1	100.0	131.7	107.1	100.0	131.7	107.1	100.0	114.5	103.2
21, 22, 23	FIRE SUPPRESSION, PLUMBING & HVAC	100.0	126.3	110.3	99.7	135.7	113.8	100.0	137.1	114.6	99.7	135.7	113.8	100.0	135.7	114.0	99.7	131.1	112.1
26, 27, 3370	ELECTRICAL, COMMUNICATIONS & UTIL.	98.4	132.1	115.0	96.0	138.5	116.9	96.6	138.5	117.2	96.0	142.3	118.8	100.5	142.3	121.1	95.7	129.9	112.6
MF2018	WEIGHTED AVERAGE	98.0	127.8	110.6	96.4	137.5	113.7	97.5	138.3	114.7	96.1	138.3	114.0	96.5	137.8	114.0	95.8	130.9	110.7

NEW JERSEY

DIVISION		NEW BRUNSWICK 088-089			NEWARK 070-071			PATERSON 074-075			POINT PLEASANT 087			SUMMIT 079			TRENTON 085-086		
		MAT.	INST.	TOTAL	MAT.	INST.	TOTAL	MAT.	INST.	TOTAL	MAT.	INST.	TOTAL	MAT.	INST.	TOTAL	MAT.	INST.	TOTAL
015433	CONTRACTOR EQUIPMENT		92.9	92.9		98.3	98.3		95.7	95.7		92.9	92.9		95.7	95.7		97.0	97.0
0241, 31 - 34	SITE & INFRASTRUCTURE, DEMOLITION	101.1	100.9	100.9	110.0	106.2	107.4	106.6	101.4	103.0	102.6	100.5	101.2	106.3	101.4	102.9	88.6	105.4	100.2
0310	Concrete Forming & Accessories	108.5	148.0	142.2	100.7	148.7	141.6	98.3	148.3	140.9	102.4	139.1	133.6	99.0	148.5	141.2	103.5	138.8	133.6
0320	Concrete Reinforcing	82.1	156.1	117.9	96.2	156.6	125.4	97.7	156.5	126.2	82.1	156.1	117.9	75.3	156.5	114.6	107.7	116.2	111.8
0330	Cast-in-Place Concrete	97.0	136.9	111.8	89.8	140.9	108.8	80.8	140.2	102.9	97.0	135.8	111.5	67.5	140.3	94.6	95.6	134.4	110.0
03	CONCRETE	104.2	143.8	121.6	93.1	145.7	116.2	88.8	145.3	113.7	103.9	139.4	119.5	80.6	145.5	109.1	97.4	132.1	112.6
04	MASONRY	102.5	140.6	125.9	94.9	145.7	126.1	91.7	145.6	124.9	91.4	135.5	118.5	93.8	145.6	125.6	95.4	135.5	120.0
05	METALS	98.9	121.4	105.5	104.2	124.7	110.3	97.5	125.1	105.6	98.9	121.0	105.4	96.5	125.3	104.9	104.3	108.8	105.6
06	WOOD, PLASTICS & COMPOSITES	115.7	148.3	133.6	97.8	148.4	125.6	97.5	148.3	125.5	107.3	139.3	124.9	98.7	148.3	126.0	101.6	139.4	122.4
07	THERMAL & MOISTURE PROTECTION	102.5	138.2	118.0	105.2	144.7	122.4	103.8	136.7	118.1	102.5	131.3	115.0	104.1	144.1	121.5	102.9	135.2	117.0
08	OPENINGS	91.6	145.1	104.1	100.5	145.1	110.9	105.0	145.1	114.3	93.4	141.0	104.5	106.1	145.1	115.2	96.9	130.0	104.6
0920	Plaster & Gypsum Board	113.0	149.4	137.5	106.2	149.4	135.3	114.6	149.4	138.0	107.9	140.2	129.6	113.7	149.4	137.7	101.6	140.2	127.5
0950, 0980	Ceilings & Acoustic Treatment	96.4	149.4	132.2	103.0	149.4	134.3	102.2	149.4	134.0	96.4	140.2	125.9	92.3	149.4	130.8	101.2	140.2	127.5
0960	Flooring	95.1	188.8	121.4	96.7	188.8	122.6	84.7	188.8	114.0	92.5	162.5	112.2	84.9	188.8	114.1	100.1	176.9	121.7
0970, 0990	Wall Finishes & Painting/Coating	81.9	148.6	121.4	94.1	148.6	126.3	83.8	148.6	122.1	81.9	141.8	117.4	83.8	148.6	122.1	94.9	141.8	122.7
09	FINISHES	93.8	156.2	127.6	97.2	156.7	129.4	92.5	156.2	127.0	92.3	144.7	120.7	91.0	156.2	126.3	97.0	147.3	124.2
COVERS	DIVS. 10 - 14, 25, 28, 41, 43, 44, 46	100.0	131.5	107.0	100.0	131.8	107.1	100.0	131.7	107.1	100.0	111.6	102.6	100.0	131.7	107.1	100.0	114.7	103.3
21, 22, 23	FIRE SUPPRESSION, PLUMBING & HVAC	99.7	133.8	113.1	100.0	139.2	115.4	100.0	135.5	114.0	99.7	131.1	112.0	99.8	137.1	114.4	100.1	130.8	112.1
26, 27, 3370	ELECTRICAL, COMMUNICATIONS & UTIL.	94.8	133.7	114.0	105.0	142.3	123.4	100.5	138.5	119.2	94.2	129.7	111.8	96.6	138.5	117.2	102.6	128.7	115.4
MF2018	WEIGHTED AVERAGE	98.7	135.6	114.3	100.3	139.7	117.0	98.0	137.7	114.8	98.1	130.8	111.9	96.4	138.3	114.1	99.7	128.9	112.0

DIVISION		NEW JERSEY — VINELAND 080, 083			NEW MEXICO — ALBUQUERQUE 870-872			CARRIZOZO 883			CLOVIS 881			FARMINGTON 874			GALLUP 873		
		MAT.	INST.	TOTAL	MAT.	INST.	TOTAL	MAT.	INST.	TOTAL	MAT.	INST.	TOTAL	MAT.	INST.	TOTAL	MAT.	INST.	TOTAL
015433	CONTRACTOR EQUIPMENT		93.3	93.3		107.5	107.5		107.5	107.5		107.5	107.5		107.5	107.5		107.5	107.5
0241, 31 - 34	SITE & INFRASTRUCTURE, DEMOLITION	92.5	99.4	97.3	91.1	97.5	95.5	109.7	97.5	101.3	96.6	97.5	97.2	97.3	97.5	97.5	107.2	97.5	100.5
0310	Concrete Forming & Accessories	99.2	137.6	131.9	101.0	66.2	71.3	98.6	66.2	71.0	98.6	66.1	70.9	101.1	66.2	71.4	101.1	66.2	71.4
0320	Concrete Reinforcing	81.1	131.2	105.3	99.9	71.5	86.1	116.0	71.5	94.5	117.3	71.4	95.1	109.2	71.5	90.9	104.5	71.5	88.5
0330	Cast-in-Place Concrete	84.6	133.9	103.9	90.5	70.1	82.9	94.7	70.1	85.5	91.3	70.1	83.4	86.0	70.1	80.1	86.0	70.1	80.1
03	CONCRETE	92.2	133.8	110.5	92.9	69.8	82.7	116.7	69.8	96.1	104.3	69.7	89.1	96.4	69.8	84.7	103.3	69.8	88.6
04	MASONRY	92.6	135.5	118.9	106.0	60.5	78.1	108.5	60.5	79.0	108.5	60.5	79.0	113.8	60.5	81.1	101.6	60.5	76.4
05	METALS	98.8	112.6	102.8	108.1	90.6	102.9	106.6	90.6	101.9	106.3	90.5	101.6	105.7	90.6	101.3	104.8	90.6	100.7
06	WOOD, PLASTICS & COMPOSITES	103.7	137.9	122.5	102.1	67.2	82.9	93.5	67.2	79.0	93.5	67.2	79.0	102.2	67.2	82.9	102.2	67.2	82.9
07	THERMAL & MOISTURE PROTECTION	101.9	132.2	115.1	101.0	72.8	88.7	106.3	72.8	91.7	105.0	72.8	91.0	101.3	72.8	88.9	102.5	72.8	89.6
08	OPENINGS	92.9	134.9	102.7	98.5	67.0	91.1	96.5	67.0	89.6	96.6	67.0	89.7	100.8	67.0	92.9	100.8	67.0	93.0
0920	Plaster & Gypsum Board	106.4	138.6	128.1	113.0	66.1	81.5	80.2	66.1	70.7	80.2	66.1	70.7	99.4	66.1	77.0	99.4	66.1	77.0
0950, 0980	Ceilings & Acoustic Treatment	96.4	138.6	124.9	98.6	66.1	77.1	102.7	66.1	78.1	102.7	66.1	78.1	98.6	66.1	76.7	98.6	66.1	76.7
0960	Flooring	91.7	162.5	111.6	88.9	66.8	82.7	97.2	66.8	88.7	97.2	66.8	88.7	90.4	66.8	83.8	90.4	66.8	83.8
0970, 0990	Wall Finishes & Painting/Coating	81.9	141.8	117.4	97.8	52.6	71.1	92.6	52.6	68.9	92.6	52.6	68.9	92.1	52.6	68.7	92.1	52.6	68.7
09	FINISHES	91.1	143.9	119.6	92.3	64.7	77.4	94.2	64.7	78.3	92.9	64.7	77.7	90.7	64.7	76.6	92.1	64.7	77.3
COVERS	DIVS. 10 - 14, 25, 28, 41, 43, 44, 46	100.0	114.5	103.2	100.0	85.1	96.7	100.0	85.1	96.7	100.0	85.1	96.7	100.0	85.1	96.7	100.0	85.1	96.7
21, 22, 23	FIRE SUPPRESSION, PLUMBING & HVAC	99.7	126.6	110.3	100.3	69.0	88.0	97.9	69.0	86.5	97.9	68.6	86.4	100.2	69.0	88.0	98.1	69.0	86.6
26, 27, 3370	ELECTRICAL, COMMUNICATIONS & UTIL.	94.2	143.3	118.5	87.9	69.5	78.8	90.1	69.5	80.0	87.9	69.5	78.8	85.9	69.5	77.8	85.3	69.5	77.5
MF2018	WEIGHTED AVERAGE	96.3	129.9	110.5	98.6	72.5	87.6	101.6	72.5	89.3	99.3	72.4	87.9	99.0	72.5	87.8	99.0	72.5	87.8

NEW MEXICO

DIVISION		LAS CRUCES 880			LAS VEGAS 877			ROSWELL 882			SANTA FE 875			SOCORRO 878			TRUTH/CONSEQUENCES 879		
		MAT.	INST.	TOTAL	MAT.	INST.	TOTAL	MAT.	INST.	TOTAL	MAT.	INST.	TOTAL	MAT.	INST.	TOTAL	MAT.	INST.	TOTAL
015433	CONTRACTOR EQUIPMENT		83.5	83.5		107.5	107.5		107.5	107.5		110.4	110.4		107.5	107.5		83.5	83.5
0241, 31 - 34	SITE & INFRASTRUCTURE, DEMOLITION	97.8	77.5	83.8	96.2	97.5	97.1	99.0	97.5	98.0	99.1	102.8	101.6	92.8	97.5	96.1	112.5	77.5	88.3
0310	Concrete Forming & Accessories	95.4	65.2	69.7	101.1	66.2	71.4	98.6	66.2	71.0	100.0	66.3	71.2	101.1	66.2	71.4	98.8	65.1	70.1
0320	Concrete Reinforcing	113.1	71.3	92.9	106.3	71.5	89.5	117.3	71.5	95.1	99.3	71.5	85.9	108.4	71.5	90.5	101.7	71.3	87.0
0330	Cast-in-Place Concrete	89.5	63.1	79.7	88.8	70.1	81.9	94.7	70.1	85.6	94.3	70.9	85.6	87.0	70.1	80.7	95.4	63.1	83.4
03	CONCRETE	82.1	66.5	75.3	94.0	69.8	83.4	105.3	69.8	89.7	92.9	70.0	82.9	92.9	69.8	82.8	86.1	66.4	77.4
04	MASONRY	104.0	60.2	77.1	101.9	60.5	76.5	119.5	60.5	83.3	95.0	60.6	73.9	101.8	60.5	76.4	98.7	60.2	75.0
05	METALS	105.0	83.0	98.5	104.5	90.6	100.6	104.5	90.6	100.6	101.6	89.6	98.1	104.8	90.6	100.7	104.4	83.0	98.1
06	WOOD, PLASTICS & COMPOSITES	82.9	66.2	73.7	102.2	67.2	82.9	93.5	67.2	79.0	99.0	67.2	81.5	102.2	67.2	82.9	93.5	66.2	78.5
07	THERMAL & MOISTURE PROTECTION	92.2	68.2	81.7	100.8	72.8	88.6	105.2	72.8	91.1	103.5	73.7	90.5	100.8	72.8	88.6	90.1	68.2	80.5
08	OPENINGS	92.0	66.4	86.1	97.3	67.0	90.3	96.4	67.0	89.6	98.8	67.0	91.4	97.2	67.0	90.1	90.6	66.4	85.0
0920	Plaster & Gypsum Board	78.5	66.1	70.2	99.4	66.1	77.0	80.2	66.1	70.7	114.1	66.1	81.8	99.4	66.1	77.0	100.6	66.1	77.4
0950, 0980	Ceilings & Acoustic Treatment	87.3	66.1	73.0	98.6	66.1	76.7	102.7	66.1	78.1	95.2	66.1	75.6	98.6	66.1	76.7	86.2	66.1	72.7
0960	Flooring	128.1	66.8	110.9	90.4	66.8	83.8	97.2	66.8	88.7	100.9	66.8	91.3	90.4	66.8	83.8	119.2	66.8	104.5
0970, 0990	Wall Finishes & Painting/Coating	81.8	52.6	64.5	92.1	52.6	68.7	92.6	52.6	68.9	99.0	52.6	71.9	92.1	52.6	68.7	84.5	52.6	65.6
09	FINISHES	103.0	64.0	81.9	90.5	64.7	76.6	93.0	64.7	77.7	98.2	64.8	80.1	90.4	64.7	76.5	102.4	64.0	81.6
COVERS	DIVS. 10 - 14, 25, 28, 41, 43, 44, 46	100.0	82.7	96.1	100.0	85.1	96.7	100.0	85.1	96.7	100.0	85.2	96.7	100.0	85.1	96.7	100.0	82.6	96.1
21, 22, 23	FIRE SUPPRESSION, PLUMBING & HVAC	100.4	68.7	88.0	98.1	69.0	86.6	100.0	69.0	87.8	100.3	69.0	88.0	98.1	69.0	86.6	98.0	68.7	86.5
26, 27, 3370	ELECTRICAL, COMMUNICATIONS & UTIL.	90.1	83.8	87.0	87.4	69.5	78.6	89.2	69.5	79.5	100.3	69.5	85.2	85.7	69.5	77.7	89.5	69.5	79.6
MF2018	WEIGHTED AVERAGE	96.9	71.4	86.1	97.2	72.5	86.8	100.8	72.5	88.9	99.0	72.9	88.0	96.8	72.5	86.6	96.6	69.4	85.1

For customer support on your Electrical Change Order Costs with RSMeans data, call 800.448.8182.

DIVISION		NEW MEXICO TUCUMCARI 884			NEW YORK ALBANY 120-122			BINGHAMTON 137-139			BRONX 104			BROOKLYN 112			BUFFALO 140-142		
		MAT.	INST.	TOTAL	MAT.	INST.	TOTAL	MAT.	INST.	TOTAL	MAT.	INST.	TOTAL	MAT.	INST.	TOTAL	MAT.	INST.	TOTAL
015433	CONTRACTOR EQUIPMENT		107.5	107.5		115.6	115.6		117.6	117.6		104.4	104.4		109.4	109.4		100.5	100.5
0241, 31 - 34	SITE & INFRASTRUCTURE, DEMOLITION	96.2	97.5	97.1	80.0	105.1	97.4	94.3	89.3	90.8	97.3	109.7	105.9	118.1	120.6	119.8	97.3	102.3	100.7
0310	Concrete Forming & Accessories	98.6	66.1	70.9	97.3	107.3	105.8	99.4	93.4	94.3	95.7	189.1	175.3	105.1	189.1	176.7	101.3	117.1	114.7
0320	Concrete Reinforcing	115.0	71.4	93.9	101.3	115.2	108.0	95.8	106.3	100.9	94.8	182.0	136.9	97.2	242.5	167.5	98.9	116.4	107.3
0330	Cast-in-Place Concrete	94.7	70.1	85.5	81.0	116.4	94.2	106.3	106.9	106.5	85.6	172.0	117.8	108.3	170.5	131.5	109.1	123.5	114.4
03	CONCRETE	103.6	69.7	88.7	87.7	112.8	98.7	94.3	102.6	97.9	88.0	181.1	128.9	106.4	189.9	143.1	104.9	118.7	111.0
04	MASONRY	119.9	60.5	83.4	88.0	117.9	106.4	104.3	104.9	104.7	88.8	188.9	150.3	114.8	188.8	160.3	112.6	122.2	118.5
05	METALS	106.3	90.5	101.6	102.2	125.0	108.9	95.0	133.6	106.3	86.4	174.6	112.3	103.1	175.0	124.2	97.0	107.5	100.1
06	WOOD, PLASTICS & COMPOSITES	93.5	67.2	79.0	94.6	104.0	99.8	101.7	90.2	95.4	97.2	188.6	147.5	103.7	188.3	150.3	98.6	115.8	108.1
07	THERMAL & MOISTURE PROTECTION	105.0	72.8	91.0	105.4	110.4	107.6	109.2	94.8	102.9	102.5	168.8	131.3	109.3	168.2	134.9	102.5	111.7	106.5
08	OPENINGS	96.4	67.0	89.5	96.4	103.8	98.1	90.4	93.7	91.2	92.3	197.7	116.9	87.8	197.5	113.4	99.8	111.4	102.5
0920	Plaster & Gypsum Board	80.2	66.1	70.7	97.8	103.9	101.9	109.2	89.7	96.1	102.8	191.0	162.1	103.3	191.0	162.3	105.8	116.1	112.7
0950, 0980	Ceilings & Acoustic Treatment	102.7	66.1	78.1	97.1	103.9	101.7	96.6	89.7	92.0	89.1	191.0	157.8	91.8	191.0	158.7	104.2	116.1	112.2
0960	Flooring	97.2	66.8	88.7	90.9	110.3	96.4	103.1	102.4	102.9	99.9	182.5	123.1	110.2	182.5	130.5	96.3	116.6	102.0
0970, 0990	Wall Finishes & Painting/Coating	92.6	52.6	68.9	95.9	104.4	100.9	89.3	106.3	99.4	106.2	167.4	142.4	116.0	167.4	146.4	96.9	117.8	109.3
09	FINISHES	92.8	64.7	77.6	90.6	107.1	99.5	93.6	95.9	94.8	98.0	186.2	145.7	107.1	186.0	149.8	102.0	117.6	110.5
COVERS	DIVS. 10 - 14, 25, 28, 41, 43, 44, 46	100.0	85.1	96.7	100.0	103.8	100.9	100.0	96.6	99.2	100.0	141.5	109.3	100.0	140.8	109.1	100.0	106.5	101.4
21, 22, 23	FIRE SUPPRESSION, PLUMBING & HVAC	97.9	68.6	86.4	100.1	110.3	104.1	100.6	97.0	99.2	100.3	177.9	130.7	99.8	177.9	130.5	100.0	102.4	100.9
26, 27, 3370	ELECTRICAL, COMMUNICATIONS & UTIL.	90.1	69.5	80.0	100.0	108.4	104.1	98.7	101.0	99.8	90.6	184.5	136.9	98.5	184.5	140.9	99.5	105.3	102.4
MF2018	WEIGHTED AVERAGE	99.9	72.4	88.3	96.8	111.2	102.8	97.3	101.5	99.1	93.8	175.3	128.3	101.9	177.3	133.8	100.8	110.3	104.8

DIVISION		NEW YORK ELMIRA 148-149			FAR ROCKAWAY 116			FLUSHING 113			GLENS FALLS 128			HICKSVILLE 115, 117, 118			JAMAICA 114		
		MAT.	INST.	TOTAL	MAT.	INST.	TOTAL	MAT.	INST.	TOTAL	MAT.	INST.	TOTAL	MAT.	INST.	TOTAL	MAT.	INST.	TOTAL
015433	CONTRACTOR EQUIPMENT		120.0	120.0		109.4	109.4		109.4	109.4		112.6	112.6		109.4	109.4		109.4	109.4
0241, 31 - 34	SITE & INFRASTRUCTURE, DEMOLITION	97.4	89.6	92.0	121.2	120.6	120.8	121.3	120.6	120.8	71.4	99.6	90.9	111.1	119.1	116.6	115.5	120.6	119.0
0310	Concrete Forming & Accessories	85.0	96.3	94.6	92.2	189.1	174.8	95.9	189.1	175.3	81.5	99.9	97.2	88.8	157.2	147.1	95.9	189.1	175.3
0320	Concrete Reinforcing	98.6	105.3	101.8	97.2	242.5	167.5	98.9	242.5	168.4	96.9	112.7	104.5	97.2	174.4	134.6	97.2	242.5	167.5
0330	Cast-in-Place Concrete	98.8	106.4	101.6	117.2	170.5	137.1	117.2	170.5	137.1	78.1	111.6	90.6	99.5	161.5	122.6	108.3	170.5	131.5
03	CONCRETE	90.9	103.7	96.5	112.8	189.9	146.7	113.2	189.9	146.9	81.2	107.4	92.8	98.2	161.8	126.2	108.3	189.9	142.8
04	MASONRY	101.4	106.7	104.6	118.6	188.8	161.8	112.8	188.8	159.5	93.4	112.6	105.2	109.0	174.4	149.2	117.2	188.8	161.2
05	METALS	95.5	135.8	107.3	103.2	175.0	124.3	103.2	175.0	124.3	95.5	124.5	104.0	104.7	172.4	124.6	103.2	175.0	124.3
06	WOOD, PLASTICS & COMPOSITES	84.1	94.4	89.8	87.1	188.3	142.8	91.7	188.3	144.9	82.7	96.1	90.1	83.7	154.0	122.4	91.7	188.3	144.9
07	THERMAL & MOISTURE PROTECTION	108.9	95.2	102.9	109.2	168.3	134.9	109.3	168.3	134.9	98.6	107.1	102.3	108.9	157.8	130.2	109.1	168.3	134.8
08	OPENINGS	96.7	95.7	96.4	86.6	197.5	112.5	86.6	197.5	112.5	90.0	99.1	92.1	87.0	178.6	108.3	86.6	197.5	112.5
0920	Plaster & Gypsum Board	99.5	94.2	96.0	91.7	191.0	158.5	94.2	191.0	159.3	91.0	96.0	94.4	91.4	155.6	134.6	94.2	191.0	159.3
0950, 0980	Ceilings & Acoustic Treatment	106.0	94.2	98.1	81.2	191.0	155.2	81.2	191.0	155.2	86.4	96.0	92.9	80.3	155.6	131.1	81.2	191.0	155.2
0960	Flooring	87.6	102.4	91.8	105.0	182.5	126.8	106.5	182.5	127.9	82.3	107.7	89.5	104.0	179.7	125.2	106.5	182.5	127.9
0970, 0990	Wall Finishes & Painting/Coating	97.2	93.3	94.9	116.0	167.4	146.4	116.0	167.4	146.4	89.3	104.4	98.2	116.0	167.4	146.4	116.0	167.4	146.4
09	FINISHES	94.2	96.9	95.7	102.4	186.0	147.6	103.2	186.0	148.0	84.0	101.1	93.2	101.0	161.6	133.8	102.7	186.0	147.8
COVERS	DIVS. 10 - 14, 25, 28, 41, 43, 44, 46	100.0	97.6	99.5	100.0	140.8	109.1	100.0	140.8	109.1	100.0	96.9	99.3	100.0	132.1	107.1	100.0	140.8	109.1
21, 22, 23	FIRE SUPPRESSION, PLUMBING & HVAC	96.6	95.5	96.2	96.2	177.9	128.3	96.2	177.9	128.3	96.7	109.3	101.6	99.8	161.0	123.8	96.2	177.9	128.3
26, 27, 3370	ELECTRICAL, COMMUNICATIONS & UTIL.	96.4	104.1	100.2	105.3	184.5	144.4	105.3	184.5	144.4	94.3	105.0	99.6	98.0	141.0	119.2	97.0	184.5	140.1
MF2018	WEIGHTED AVERAGE	96.3	102.5	98.9	102.2	177.3	133.9	102.1	177.3	133.9	92.0	107.3	98.5	99.9	157.1	124.1	100.3	177.3	132.8

DIVISION		NEW YORK JAMESTOWN 147			KINGSTON 124			LONG ISLAND CITY 111			MONTICELLO 127			MOUNT VERNON 105			NEW ROCHELLE 108		
		MAT.	INST.	TOTAL	MAT.	INST.	TOTAL	MAT.	INST.	TOTAL	MAT.	INST.	TOTAL	MAT.	INST.	TOTAL	MAT.	INST.	TOTAL
015433	CONTRACTOR EQUIPMENT		90.6	90.6		109.4	109.4		109.4	109.4		109.4	109.4		104.4	104.4		104.4	104.4
0241, 31 - 34	SITE & INFRASTRUCTURE, DEMOLITION	98.8	89.9	92.6	139.6	115.9	123.2	119.2	120.6	120.2	134.9	115.8	121.7	102.7	105.5	104.5	102.2	105.3	104.3
0310	Concrete Forming & Accessories	85.1	90.1	89.3	82.7	130.7	123.7	100.0	189.1	176.0	89.3	130.8	124.7	86.7	138.8	131.1	100.5	135.3	130.1
0320	Concrete Reinforcing	98.9	111.2	104.8	97.3	160.4	127.8	97.2	242.5	167.5	96.5	160.4	127.4	93.5	180.8	135.7	93.6	180.6	135.7
0330	Cast-in-Place Concrete	102.4	105.2	103.4	105.5	146.3	120.7	111.8	170.5	133.7	98.7	146.3	116.5	95.6	150.0	115.8	95.6	149.7	115.7
03	CONCRETE	93.9	99.0	96.2	102.4	140.8	119.3	109.0	189.9	144.5	97.3	140.8	116.4	97.2	150.4	120.6	96.4	148.7	119.4
04	MASONRY	110.2	104.6	106.8	108.1	155.4	137.2	111.7	188.8	159.1	100.8	155.4	134.3	94.6	156.4	132.6	94.6	156.4	132.6
05	METALS	93.0	101.9	95.6	104.0	135.9	113.4	103.1	175.0	124.2	104.0	136.0	113.4	86.2	169.8	110.7	86.4	169.0	110.7
06	WOOD, PLASTICS & COMPOSITES	82.9	86.2	84.7	83.9	124.0	106.0	97.8	188.3	147.6	90.4	124.0	108.9	87.6	132.4	112.3	104.5	128.7	117.8
07	THERMAL & MOISTURE PROTECTION	108.5	95.4	102.8	121.5	145.0	131.7	109.2	168.3	134.9	121.1	145.0	131.5	103.4	147.3	122.5	103.5	144.7	121.4
08	OPENINGS	96.5	93.0	95.7	92.1	142.5	103.9	86.6	197.5	112.5	88.0	142.4	100.7	92.3	168.4	110.1	92.4	166.3	109.6
0920	Plaster & Gypsum Board	89.2	85.8	86.9	91.0	124.9	113.8	99.3	191.0	161.0	91.7	124.8	114.0	98.3	133.2	121.8	111.0	129.3	123.3
0950, 0980	Ceilings & Acoustic Treatment	102.7	85.8	91.3	77.5	124.9	109.4	81.2	191.0	155.2	77.5	124.8	109.4	87.5	133.2	118.3	87.5	129.3	115.7
0960	Flooring	90.3	102.4	93.7	98.6	160.5	116.0	108.2	182.5	129.1	101.0	160.5	117.8	91.2	179.7	116.1	99.2	164.0	117.4
0970, 0990	Wall Finishes & Painting/Coating	98.8	100.6	99.9	116.6	130.9	125.0	116.0	167.4	146.4	116.6	123.4	120.6	104.6	167.4	141.8	104.6	167.4	141.8
09	FINISHES	93.2	92.5	92.9	97.9	135.4	118.2	104.1	186.0	148.4	98.3	134.6	117.9	95.0	148.0	123.7	98.9	142.6	122.5
COVERS	DIVS. 10 - 14, 25, 28, 41, 43, 44, 46	100.0	96.8	99.3	100.0	121.2	104.7	100.0	140.8	109.1	100.0	121.3	104.7	100.0	129.2	106.5	100.0	135.0	102.9
21, 22, 23	FIRE SUPPRESSION, PLUMBING & HVAC	96.5	93.6	95.4	96.6	134.6	111.5	99.8	177.9	130.5	96.6	138.9	113.2	96.8	145.9	116.6	96.8	145.6	116.4
26, 27, 3370	ELECTRICAL, COMMUNICATIONS & UTIL.	95.4	94.9	95.1	95.7	116.3	105.9	97.5	184.5	140.4	95.7	116.3	105.9	88.8	170.6	129.1	88.8	141.7	114.9
MF2018	WEIGHTED AVERAGE	96.4	96.0	96.3	100.6	133.8	114.6	101.6	177.3	133.6	99.1	134.6	114.1	94.1	150.6	118.0	94.4	144.8	115.7

NEW YORK

DIVISION		NEW YORK 100 - 102			NIAGARA FALLS 143			PLATTSBURGH 129			POUGHKEEPSIE 125 - 126			QUEENS 110			RIVERHEAD 119		
		MAT.	INST.	TOTAL	MAT.	INST.	TOTAL	MAT.	INST.	TOTAL	MAT.	INST.	TOTAL	MAT.	INST.	TOTAL	MAT.	INST.	TOTAL
015433	CONTRACTOR EQUIPMENT		104.6	104.6		90.6	90.6		94.5	94.5		109.4	109.4		109.4	109.4		109.4	109.4
0241, 31 - 34	SITE & INFRASTRUCTURE, DEMOLITION	106.6	111.9	110.3	100.9	90.8	94.0	109.3	96.7	100.6	135.9	114.8	121.3	114.3	120.6	118.7	112.1	118.6	116.6
0310	Concrete Forming & Accessories	104.3	188.9	176.4	85.0	113.1	108.9	86.8	91.4	90.7	82.7	167.6	155.1	89.0	189.1	174.3	93.2	156.2	146.9
0320	Concrete Reinforcing	100.3	178.3	138.0	97.5	112.0	104.5	101.4	112.0	106.5	97.3	160.0	127.6	98.9	242.5	168.4	99.1	215.6	155.5
0330	Cast-in-Place Concrete	98.9	172.2	126.2	106.0	124.1	112.7	95.5	100.6	97.4	102.2	135.4	114.5	102.9	170.5	128.1	101.2	159.7	123.0
03	CONCRETE	101.6	180.3	136.2	96.2	116.1	105.0	94.3	98.0	95.9	99.7	153.6	123.4	101.5	189.9	140.3	98.7	166.5	128.5
04	MASONRY	102.1	188.8	155.4	117.5	125.1	122.2	88.3	97.4	93.9	100.6	139.5	124.5	106.0	188.8	156.9	114.3	173.3	150.6
05	METALS	97.5	171.0	119.1	95.6	102.4	97.6	99.8	99.5	99.7	104.0	135.6	113.3	103.1	175.0	124.2	105.1	160.6	121.4
06	WOOD, PLASTICS & COMPOSITES	102.0	188.6	149.7	82.8	108.2	96.8	88.8	88.2	88.5	83.9	179.3	136.4	83.8	188.3	141.3	88.7	154.0	124.6
07	THERMAL & MOISTURE PROTECTION	105.2	170.0	133.4	108.6	110.0	109.2	116.0	97.2	107.8	121.4	144.1	131.3	108.8	168.3	134.7	110.0	154.8	129.5
08	OPENINGS	96.0	197.5	119.7	96.5	104.7	98.4	97.4	94.7	96.8	92.1	170.5	110.4	86.6	197.5	112.5	87.0	172.1	106.8
0920	Plaster & Gypsum Board	108.6	191.0	164.0	89.2	108.4	102.1	108.2	87.4	94.2	91.0	181.7	152.0	91.4	191.0	158.4	92.5	155.6	135.0
0950, 0980	Ceilings & Acoustic Treatment	104.5	191.0	162.8	102.7	108.4	106.6	107.3	87.4	93.9	77.5	181.7	147.7	81.2	191.0	155.2	81.1	155.6	131.4
0960	Flooring	101.3	182.5	124.1	90.3	112.1	96.4	104.2	105.1	104.4	98.6	158.6	115.5	104.0	182.5	126.0	105.0	166.9	122.4
0970, 0990	Wall Finishes & Painting/Coating	102.7	167.4	141.0	98.8	111.5	106.3	111.2	95.1	101.7	116.6	123.4	120.6	116.0	167.4	146.4	116.0	167.4	146.4
09	FINISHES	102.9	186.2	148.0	93.3	112.9	103.9	95.0	93.5	94.2	97.7	164.2	133.7	101.3	186.0	147.1	101.5	158.6	132.4
COVERS	DIVS. 10 - 14, 25, 28, 41, 43, 44, 46	100.0	141.5	109.3	100.0	102.7	100.6	100.0	92.7	98.4	100.0	116.5	103.7	100.0	140.8	109.1	100.0	120.1	104.5
21, 22, 23	FIRE SUPPRESSION, PLUMBING & HVAC	100.2	178.3	130.9	96.5	105.1	99.9	96.6	100.3	98.1	96.6	120.3	105.9	99.8	177.9	130.5	100.0	155.9	121.9
26, 27, 3370	ELECTRICAL, COMMUNICATIONS & UTIL.	96.8	184.5	140.0	94.1	98.9	96.4	92.1	91.7	91.9	95.7	121.2	108.3	98.0	184.5	140.6	99.5	134.9	116.9
MF2018	WEIGHTED AVERAGE	99.7	175.2	131.6	97.4	107.5	101.6	97.0	96.6	96.8	99.8	136.8	115.4	100.0	177.3	132.6	100.6	153.5	122.9

DIVISION		ROCHESTER 144 - 146			SCHENECTADY 123			STATEN ISLAND 103			SUFFERN 109			SYRACUSE 130 - 132			UTICA 133 - 135		
		MAT.	INST.	TOTAL	MAT.	INST.	TOTAL	MAT.	INST.	TOTAL	MAT.	INST.	TOTAL	MAT.	INST.	TOTAL	MAT.	INST.	TOTAL
015433	CONTRACTOR EQUIPMENT		117.6	117.6		112.6	112.6		104.4	104.4		104.4	104.4		112.6	112.6		112.6	112.6
0241, 31 - 34	SITE & INFRASTRUCTURE, DEMOLITION	88.8	105.1	100.0	81.4	100.1	94.3	107.0	109.7	108.9	99.4	103.7	102.4	93.0	98.5	96.8	71.9	98.1	90.0
0310	Concrete Forming & Accessories	105.8	100.3	101.2	96.4	107.1	105.5	86.1	189.3	174.1	94.4	140.8	133.9	98.5	89.1	90.5	99.5	86.3	88.2
0320	Concrete Reinforcing	101.4	105.0	103.1	95.9	115.1	105.2	94.8	216.0	153.4	93.6	152.0	121.9	96.8	106.1	101.3	96.8	100.3	98.5
0330	Cast-in-Place Concrete	97.8	103.3	99.9	94.4	115.4	102.2	95.6	172.1	124.1	92.7	138.9	109.9	98.7	102.3	100.0	90.3	101.0	94.3
03	CONCRETE	99.5	103.3	101.2	94.9	112.4	102.6	99.1	186.4	137.5	93.8	141.3	114.7	97.6	98.1	97.8	95.8	95.4	95.6
04	MASONRY	93.6	103.4	99.7	90.7	117.9	107.4	100.9	188.9	154.9	94.5	142.2	123.8	97.0	101.0	99.5	88.4	99.4	95.2
05	METALS	104.6	119.3	108.9	99.7	126.1	107.4	84.6	174.8	111.1	84.6	131.9	98.5	98.6	119.8	104.9	96.7	117.5	102.9
06	WOOD, PLASTICS & COMPOSITES	108.4	100.2	103.9	100.1	103.8	102.1	86.2	188.6	142.6	96.9	142.5	122.0	98.2	86.0	91.5	98.2	82.7	89.7
07	THERMAL & MOISTURE PROTECTION	116.4	99.8	109.2	100.1	109.8	104.3	102.9	168.8	131.5	103.4	140.9	119.7	104.0	94.1	99.7	92.7	94.0	93.3
08	OPENINGS	100.7	99.3	100.3	95.3	103.7	97.3	92.3	197.7	116.9	92.4	150.7	106.0	92.4	89.5	91.8	95.2	86.3	93.1
0920	Plaster & Gypsum Board	103.8	100.1	101.3	100.3	103.9	102.7	98.4	191.0	160.7	102.1	143.5	130.0	96.8	85.6	89.3	96.8	82.2	87.0
0950, 0980	Ceilings & Acoustic Treatment	100.4	100.1	100.2	92.8	103.9	100.3	89.1	191.0	157.8	87.5	143.5	125.2	96.6	85.6	89.2	96.6	82.2	86.9
0960	Flooring	90.4	105.0	94.5	89.6	110.3	95.4	95.1	182.5	119.6	95.0	171.6	116.5	92.0	92.3	92.1	90.0	92.4	90.7
0970, 0990	Wall Finishes & Painting/Coating	97.7	98.4	98.1	89.3	104.4	98.2	106.2	167.4	142.4	104.6	126.9	117.8	91.9	98.7	95.9	85.6	98.7	93.4
09	FINISHES	95.7	101.4	98.8	89.2	107.0	98.9	96.7	186.2	145.2	96.3	145.8	123.1	92.4	90.0	91.1	90.8	87.9	89.2
COVERS	DIVS. 10 - 14, 25, 28, 41, 43, 44, 46	100.0	98.9	99.7	100.0	103.5	100.8	100.0	141.5	109.3	100.0	114.7	103.3	100.0	95.1	98.9	100.0	89.8	97.7
21, 22, 23	FIRE SUPPRESSION, PLUMBING & HVAC	99.9	88.1	95.3	100.2	106.2	102.6	100.3	177.9	130.8	96.8	123.7	107.4	100.3	94.4	98.0	100.3	92.5	97.2
26, 27, 3370	ELECTRICAL, COMMUNICATIONS & UTIL.	99.5	89.9	94.8	98.7	108.4	103.5	90.6	184.5	136.9	94.8	114.6	104.6	98.6	101.0	99.8	96.6	101.0	98.8
MF2018	WEIGHTED AVERAGE	100.3	99.2	99.8	97.0	109.9	102.4	95.6	176.1	129.6	94.1	130.2	109.4	97.8	98.3	98.0	96.0	96.6	96.2

DIVISION		NEW YORK									NORTH CAROLINA								
		WATERTOWN 136			WHITE PLAINS 106			YONKERS 107			ASHEVILLE 287 - 288			CHARLOTTE 281 - 282			DURHAM 277		
		MAT.	INST.	TOTAL	MAT.	INST.	TOTAL	MAT.	INST.	TOTAL	MAT.	INST.	TOTAL	MAT.	INST.	TOTAL	MAT.	INST.	TOTAL
015433	CONTRACTOR EQUIPMENT		112.6	112.6		104.4	104.4		104.4	104.4		99.0	99.0		100.4	100.4		104.1	104.1
0241, 31 - 34	SITE & INFRASTRUCTURE, DEMOLITION	79.5	98.6	92.7	97.1	105.4	102.8	104.5	105.4	105.1	96.4	77.8	83.5	98.6	81.9	87.1	98.5	86.3	90.1
0310	Concrete Forming & Accessories	85.3	93.0	91.9	99.2	149.3	141.9	99.4	149.1	141.8	90.5	61.4	65.7	96.6	61.5	66.6	96.2	61.4	66.5
0320	Concrete Reinforcing	97.5	106.1	101.6	93.6	180.8	135.8	97.3	180.8	137.7	95.4	66.0	81.2	99.6	66.8	83.7	104.4	66.4	86.2
0330	Cast-in-Place Concrete	105.1	104.7	105.0	84.9	150.0	109.1	95.0	150.0	115.4	105.9	71.5	93.1	108.0	71.6	94.4	106.3	71.2	93.2
03	CONCRETE	108.8	100.7	105.3	87.6	155.2	117.3	96.6	155.1	122.3	91.1	67.4	80.7	91.2	67.5	80.8	95.2	67.4	83.0
04	MASONRY	89.7	104.8	99.0	93.9	156.4	132.3	97.1	156.4	133.6	84.9	64.4	72.3	88.6	64.4	73.8	84.3	64.4	72.0
05	METALS	96.8	119.8	103.6	86.1	169.8	110.7	94.1	169.8	116.4	101.2	89.9	97.9	102.1	88.9	98.2	117.9	90.2	109.7
06	WOOD, PLASTICS & COMPOSITES	80.3	89.9	85.6	102.4	146.7	126.8	102.3	146.3	126.5	87.6	59.0	71.9	89.9	59.0	72.9	90.6	59.0	73.2
07	THERMAL & MOISTURE PROTECTION	93.0	97.2	94.8	103.2	148.7	123.0	103.5	149.3	123.4	99.1	63.8	83.8	93.7	64.5	81.0	102.8	63.8	85.8
08	OPENINGS	95.2	93.8	94.8	92.4	176.2	112.0	96.0	176.5	114.8	99.0	59.8	89.9	99.0	60.0	89.9	95.6	60.0	87.3
0920	Plaster & Gypsum Board	87.6	89.6	89.0	105.3	147.8	133.9	108.7	147.4	134.7	102.2	57.8	72.3	98.1	57.8	71.0	89.9	57.8	68.3
0950, 0980	Ceilings & Acoustic Treatment	96.6	89.6	91.9	87.5	147.8	128.2	103.7	147.4	133.2	79.4	57.8	64.9	83.5	57.8	66.2	82.8	57.8	66.0
0960	Flooring	83.1	92.4	85.7	97.5	179.7	120.6	96.9	182.5	121.0	94.9	66.8	87.0	95.3	66.8	87.3	102.0	66.8	92.1
0970, 0990	Wall Finishes & Painting/Coating	85.6	93.0	90.0	104.6	167.4	141.8	104.6	167.4	141.8	102.5	57.1	75.7	95.9	57.1	72.9	104.4	57.1	76.4
09	FINISHES	88.2	92.4	90.5	97.1	156.4	129.2	101.0	156.7	131.1	88.8	61.6	74.1	88.3	61.6	73.9	90.4	61.6	74.8
COVERS	DIVS. 10 - 14, 25, 28, 41, 43, 44, 46	100.0	96.4	99.2	100.0	127.5	106.1	100.0	130.7	106.8	100.0	82.5	96.1	100.0	82.5	96.1	100.0	82.5	96.1
21, 22, 23	FIRE SUPPRESSION, PLUMBING & HVAC	100.3	87.8	95.4	100.4	145.9	118.3	100.4	146.0	118.3	100.0	61.3	85.0	99.9	63.1	85.5	100.5	61.3	85.1
26, 27, 3370	ELECTRICAL, COMMUNICATIONS & UTIL.	98.6	90.2	94.4	88.8	170.6	129.1	94.9	170.6	132.2	100.8	57.8	79.6	100.0	60.3	80.4	96.5	57.4	77.2
MF2018	WEIGHTED AVERAGE	97.7	96.8	97.4	93.8	152.8	118.8	97.9	153.0	121.2	96.4	66.6	83.8	97.4	67.6	84.8	100.1	67.2	86.2

		NORTH CAROLINA																
	DIVISION	ELIZABETH CITY			FAYETTEVILLE			GASTONIA			GREENSBORO			HICKORY			KINSTON		
		279			283			280			270, 272 - 274			286			285		
		MAT.	INST.	TOTAL	MAT.	INST.	TOTAL	MAT.	INST.	TOTAL	MAT.	INST.	TOTAL	MAT.	INST.	TOTAL	MAT.	INST.	TOTAL
015433	CONTRACTOR EQUIPMENT		108.0	108.0		104.1	104.1		99.0	99.0		104.1	104.1		104.1	104.1		104.1	104.1
0241, 31 - 34	SITE & INFRASTRUCTURE, DEMOLITION	102.7	87.9	92.4	95.6	86.1	89.0	96.1	78.1	83.7	98.3	86.4	90.1	95.1	85.1	88.2	93.9	84.9	87.7
0310	Concrete Forming & Accessories	82.6	63.6	66.4	90.1	60.1	64.5	96.4	61.4	66.6	96.0	61.3	66.5	87.1	61.2	65.0	83.6	59.9	63.4
0320	Concrete Reinforcing	102.3	69.8	86.6	99.1	66.8	83.4	95.8	66.8	81.7	103.2	66.8	85.6	95.4	66.7	81.5	94.9	66.7	81.3
0330	Cast-in-Place Concrete	106.5	71.7	93.5	111.1	69.3	95.5	103.6	71.2	91.5	105.5	71.2	92.7	105.9	71.2	93.0	102.2	69.2	90.0
03	CONCRETE	95.4	69.1	83.8	93.1	66.2	81.3	89.8	67.5	80.0	94.7	67.4	82.7	90.9	67.3	80.5	88.1	66.1	78.4
04	MASONRY	95.4	61.0	74.2	88.3	61.0	71.5	89.4	64.4	74.0	81.8	64.4	71.1	74.6	64.4	68.3	81.2	61.0	68.8
05	METALS	103.5	92.3	100.2	122.2	90.2	112.8	101.8	90.2	98.4	110.0	90.2	104.1	101.3	90.0	98.0	100.0	90.0	97.1
06	WOOD, PLASTICS & COMPOSITES	75.1	63.4	68.7	86.5	59.0	71.4	95.4	59.0	75.4	90.2	59.0	73.1	82.4	59.0	69.6	79.3	59.0	68.2
07	THERMAL & MOISTURE PROTECTION	102.0	62.5	84.8	98.6	62.3	82.8	99.3	63.8	83.9	102.6	63.8	85.7	99.5	63.8	84.0	99.3	62.3	83.2
08	OPENINGS	92.9	63.1	86.0	89.1	60.0	82.3	92.3	60.0	84.8	95.6	60.0	87.3	89.0	60.0	82.3	89.1	60.0	82.3
0920	Plaster & Gypsum Board	84.4	61.7	69.1	105.8	57.8	73.5	107.9	57.8	74.2	91.1	57.8	68.7	102.2	57.8	72.3	102.2	57.8	72.3
0950, 0980	Ceilings & Acoustic Treatment	82.8	61.7	68.5	81.9	57.8	65.7	83.5	57.8	66.2	82.8	57.8	66.0	79.4	57.8	64.9	83.5	57.8	66.2
0960	Flooring	94.1	66.8	86.4	95.1	66.8	87.1	98.0	66.8	89.2	102.0	66.8	92.1	94.8	66.8	86.9	92.2	66.8	85.0
0970, 0990	Wall Finishes & Painting/Coating	104.4	57.1	76.4	102.5	57.1	75.7	102.5	57.1	75.7	104.4	57.1	76.4	102.5	57.1	75.7	102.5	57.1	75.7
09	FINISHES	87.7	63.4	74.6	89.7	60.8	74.0	91.1	61.6	75.2	90.6	61.6	74.9	88.9	61.6	74.1	88.7	60.8	73.6
COVERS	DIVS. 10 - 14, 25, 28, 41, 43, 44, 46	100.0	85.4	96.8	100.0	81.4	95.8	100.0	82.5	96.1	100.0	79.8	95.5	100.0	82.5	96.1	100.0	81.3	95.8
21, 22, 23	FIRE SUPPRESSION, PLUMBING & HVAC	96.9	58.6	81.9	100.2	59.5	84.2	100.4	60.1	84.6	100.4	61.3	85.0	96.9	60.2	82.5	96.9	58.3	81.7
26, 27, 3370	ELECTRICAL, COMMUNICATIONS & UTIL.	96.2	64.6	80.6	100.6	57.4	79.3	100.3	60.4	80.6	95.7	57.8	77.0	98.6	60.4	79.8	98.4	55.8	77.4
MF2018	WEIGHTED AVERAGE	96.9	68.3	84.8	100.2	66.1	85.8	97.0	66.8	84.2	98.5	67.2	85.3	94.7	67.3	83.1	94.4	65.5	82.2

		NORTH CAROLINA															NORTH DAKOTA		
	DIVISION	MURPHY			RALEIGH			ROCKY MOUNT			WILMINGTON			WINSTON-SALEM			BISMARCK		
		289			275 - 276			278			284			271			585		
		MAT.	INST.	TOTAL	MAT.	INST.	TOTAL	MAT.	INST.	TOTAL	MAT.	INST.	TOTAL	MAT.	INST.	TOTAL	MAT.	INST.	TOTAL
015433	CONTRACTOR EQUIPMENT		99.0	99.0		106.5	106.5		104.1	104.1		99.0	99.0		104.1	104.1		99.9	99.9
0241, 31 - 34	SITE & INFRASTRUCTURE, DEMOLITION	97.4	76.3	82.8	98.8	90.8	93.3	100.7	86.3	90.8	97.4	77.6	83.7	98.7	86.4	90.2	99.8	97.8	98.4
0310	Concrete Forming & Accessories	97.0	59.8	65.3	96.9	61.0	66.3	88.5	60.9	64.9	91.8	60.1	64.8	97.9	61.4	66.8	110.6	76.6	81.6
0320	Concrete Reinforcing	94.9	64.3	80.1	105.4	66.8	86.7	102.3	66.8	85.1	96.1	66.8	81.9	103.2	66.8	85.6	92.6	98.3	95.3
0330	Cast-in-Place Concrete	109.6	69.2	94.6	109.3	71.3	95.1	104.2	70.5	91.6	105.5	69.3	92.0	108.1	71.2	94.4	106.3	86.6	98.9
03	CONCRETE	93.9	65.6	81.4	93.9	67.2	82.2	95.9	66.9	83.2	91.1	66.2	80.1	96.0	67.4	83.5	93.3	84.4	89.4
04	MASONRY	77.3	61.0	67.3	80.5	63.1	69.8	74.4	63.1	67.4	75.1	61.0	66.4	82.0	64.4	71.2	105.3	83.4	91.8
05	METALS	99.0	89.1	96.1	102.2	89.1	98.3	102.6	90.2	99.0	100.7	90.2	97.6	107.0	90.2	102.1	95.6	94.2	95.2
06	WOOD, PLASTICS & COMPOSITES	96.1	58.9	75.6	90.9	59.2	73.4	82.0	59.0	69.4	89.2	59.0	72.6	90.2	59.0	73.1	106.6	72.7	88.0
07	THERMAL & MOISTURE PROTECTION	99.3	62.3	83.2	97.2	63.8	82.7	102.5	63.2	85.4	99.1	62.3	83.1	102.6	63.8	85.7	109.3	86.0	99.2
08	OPENINGS	89.0	59.4	82.1	97.6	60.1	88.8	92.2	60.0	84.7	89.1	60.0	82.3	95.6	60.0	87.3	103.7	81.9	98.6
0920	Plaster & Gypsum Board	106.9	57.7	73.8	86.5	57.8	67.2	85.8	57.8	67.0	103.7	57.8	72.8	91.1	57.8	68.7	102.4	72.2	82.1
0950, 0980	Ceilings & Acoustic Treatment	79.4	57.7	64.8	83.5	57.8	66.2	80.4	57.8	65.2	81.9	57.8	65.7	82.8	57.8	66.0	108.5	72.2	84.1
0960	Flooring	98.3	66.8	89.4	96.8	66.8	88.4	97.6	66.8	88.9	95.5	66.8	87.5	102.0	66.8	92.1	86.9	53.7	77.6
0970, 0990	Wall Finishes & Painting/Coating	102.5	57.1	75.7	97.8	57.1	73.7	104.4	57.1	76.4	102.5	57.1	75.7	104.4	57.1	76.4	90.9	59.8	72.5
09	FINISHES	90.6	60.7	74.4	90.0	61.4	74.5	89.6	61.3	73.8	89.6	60.8	74.0	90.6	61.6	74.9	96.1	71.1	82.5
COVERS	DIVS. 10 - 14, 25, 28, 41, 43, 44, 46	100.0	81.3	95.8	100.0	82.3	96.1	100.0	82.1	96.0	100.0	81.4	95.8	100.0	82.5	96.1	100.0	94.8	98.8
21, 22, 23	FIRE SUPPRESSION, PLUMBING & HVAC	96.9	58.3	81.7	100.0	60.6	84.5	96.9	59.5	82.2	100.4	59.5	84.3	100.4	61.3	85.0	99.9	77.0	90.9
26, 27, 3370	ELECTRICAL, COMMUNICATIONS & UTIL.	101.7	55.8	79.1	99.1	56.3	78.0	98.0	57.4	78.0	101.1	55.8	78.8	95.7	57.8	77.0	98.7	73.4	86.3
MF2018	WEIGHTED AVERAGE	95.5	64.7	82.5	97.4	67.0	84.5	96.0	66.6	83.6	96.0	65.2	83.0	98.2	67.3	85.2	98.9	81.6	91.6

		NORTH DAKOTA																	
	DIVISION	DEVILS LAKE			DICKINSON			FARGO			GRAND FORKS			JAMESTOWN			MINOT		
		583			586			580 - 581			582			584			587		
		MAT.	INST.	TOTAL	MAT.	INST.	TOTAL	MAT.	INST.	TOTAL	MAT.	INST.	TOTAL	MAT.	INST.	TOTAL	MAT.	INST.	TOTAL
015433	CONTRACTOR EQUIPMENT		97.7	97.7		97.7	97.7		99.9	99.9		97.7	97.7		97.7	97.7		97.7	97.7
0241, 31 - 34	SITE & INFRASTRUCTURE, DEMOLITION	107.5	93.7	98.0	115.7	93.7	100.5	101.3	97.0	98.8	111.5	93.7	99.2	106.5	93.7	97.7	108.9	94.1	98.7
0310	Concrete Forming & Accessories	107.0	71.6	76.8	95.5	71.5	75.1	99.4	72.1	76.1	99.7	71.6	75.7	97.1	71.6	75.4	95.1	71.8	75.2
0320	Concrete Reinforcing	94.8	98.0	96.4	95.7	98.2	96.9	95.7	98.7	97.2	93.4	98.0	95.6	95.4	98.7	97.0	96.6	98.1	97.4
0330	Cast-in-Place Concrete	126.2	81.6	109.6	114.1	81.5	102.0	102.5	85.6	96.2	114.1	81.5	102.0	124.6	81.6	108.6	114.1	81.8	102.1
03	CONCRETE	104.4	80.4	93.8	103.5	80.4	93.3	98.3	82.1	91.2	100.9	80.3	91.8	102.9	80.5	93.1	99.4	80.6	91.1
04	MASONRY	118.0	79.2	94.2	120.1	81.4	96.4	102.7	90.0	94.9	112.1	79.2	91.9	130.9	89.9	105.7	110.1	80.7	92.1
05	METALS	94.7	94.1	94.5	94.6	94.0	94.4	100.2	94.8	98.6	94.6	93.8	94.4	94.6	94.9	94.7	94.9	94.8	94.9
06	WOOD, PLASTICS & COMPOSITES	99.0	67.5	81.7	85.6	67.5	75.7	95.9	67.6	80.4	90.4	67.5	77.8	87.7	67.5	76.6	85.3	67.5	75.5
07	THERMAL & MOISTURE PROTECTION	107.1	82.3	96.3	107.7	83.7	97.2	104.1	87.4	96.8	107.3	83.0	96.7	106.9	85.7	97.7	107.0	83.9	97.0
08	OPENINGS	100.9	79.0	95.8	100.9	79.0	95.8	101.0	79.1	95.9	99.5	79.0	94.7	100.9	79.0	95.8	99.7	79.0	94.8
0920	Plaster & Gypsum Board	118.2	67.0	83.8	109.1	67.0	80.8	101.1	67.0	78.2	110.3	67.0	81.2	110.0	67.0	81.1	109.1	67.0	80.8
0950, 0980	Ceilings & Acoustic Treatment	105.6	67.0	79.6	105.6	67.0	79.6	95.4	67.0	76.3	105.6	67.0	79.6	105.6	67.0	79.6	105.6	67.0	79.6
0960	Flooring	93.8	53.7	82.5	87.4	53.7	77.9	101.2	53.7	87.9	89.2	53.7	79.2	88.0	53.7	78.4	87.1	53.7	77.7
0970, 0990	Wall Finishes & Painting/Coating	86.6	56.4	68.7	86.6	56.4	68.7	93.1	67.4	77.9	86.6	66.0	74.4	86.6	56.4	68.7	86.6	57.5	69.4
09	FINISHES	95.2	66.8	79.8	93.0	66.8	78.8	96.2	68.1	81.0	93.2	67.8	79.5	92.3	66.8	78.5	92.1	66.9	78.5
COVERS	DIVS. 10 - 14, 25, 28, 41, 43, 44, 46	100.0	87.2	97.2	100.0	87.2	97.2	100.0	93.6	98.6	100.0	87.2	97.2	100.0	87.2	97.2	100.0	93.3	98.5
21, 22, 23	FIRE SUPPRESSION, PLUMBING & HVAC	96.6	78.3	89.4	96.6	73.3	87.4	100.0	74.1	89.8	100.1	72.3	89.2	96.6	72.3	87.0	100.1	72.1	89.1
26, 27, 3370	ELECTRICAL, COMMUNICATIONS & UTIL.	94.3	68.3	81.5	101.9	67.9	85.1	98.5	69.4	84.2	97.3	68.3	83.0	94.3	68.3	81.5	100.1	72.6	86.6
MF2018	WEIGHTED AVERAGE	99.2	78.8	90.6	99.9	77.9	90.6	99.7	80.3	91.5	99.4	77.6	90.2	99.2	78.8	90.6	99.3	78.6	90.5

City Cost Indexes

		NORTH DAKOTA			OHIO														
		WILLISTON			AKRON			ATHENS			CANTON			CHILLICOTHE			CINCINNATI		
	DIVISION	588			442 - 443			457			446 - 447			456			451 - 452		
		MAT.	INST.	TOTAL	MAT.	INST.	TOTAL	MAT.	INST.	TOTAL	MAT.	INST.	TOTAL	MAT.	INST.	TOTAL	MAT.	INST.	TOTAL
015433	CONTRACTOR EQUIPMENT		97.7	97.7		88.7	88.7		84.9	84.9		88.7	88.7		95.5	95.5		95.9	95.9
0241, 31 - 34	SITE & INFRASTRUCTURE, DEMOLITION	109.2	91.5	97.0	96.2	94.1	94.8	107.3	85.3	92.1	96.3	93.9	94.7	93.8	94.9	94.6	90.0	98.2	95.7
0310	Concrete Forming & Accessories	101.7	71.3	75.8	102.8	82.8	85.8	95.3	78.9	81.4	102.8	74.3	78.5	98.0	81.5	84.0	101.5	80.2	83.3
0320	Concrete Reinforcing	97.5	98.1	97.8	94.7	91.0	92.9	86.9	89.2	88.0	94.7	75.0	85.2	84.0	88.8	86.3	89.1	77.8	83.6
0330	Cast-in-Place Concrete	114.1	81.4	101.9	102.6	89.0	97.6	111.2	95.8	105.5	103.6	87.2	97.5	100.9	92.4	97.8	96.4	77.5	89.3
03	CONCRETE	100.8	80.2	91.7	99.8	85.9	93.7	101.0	86.3	94.6	100.3	78.7	90.8	95.6	86.9	91.8	94.7	79.0	87.8
04	MASONRY	105.1	80.7	90.1	90.9	89.4	90.0	73.6	97.2	88.1	91.5	80.7	84.9	80.2	90.5	86.5	83.0	79.8	81.0
05	METALS	94.8	93.6	94.4	97.5	80.2	92.4	97.1	80.5	92.3	97.5	73.7	90.5	89.4	89.8	89.5	91.5	81.7	88.6
06	WOOD, PLASTICS & COMPOSITES	92.1	67.5	78.6	105.9	81.1	92.2	87.4	74.0	80.0	106.2	71.8	87.3	100.3	78.0	88.0	102.3	80.1	90.1
07	THERMAL & MOISTURE PROTECTION	107.3	83.5	96.9	104.9	91.1	98.9	102.9	91.5	98.0	106.0	87.2	97.8	104.7	88.4	97.6	102.7	80.9	93.2
08	OPENINGS	101.0	79.0	95.8	107.5	82.7	101.7	96.2	75.8	91.5	101.3	69.8	93.9	88.1	78.0	85.7	98.3	76.4	93.2
0920	Plaster & Gypsum Board	110.3	67.0	81.2	98.3	80.5	86.3	93.1	73.2	79.7	99.2	71.0	80.2	96.6	77.8	83.9	96.1	79.9	85.2
0950, 0980	Ceilings & Acoustic Treatment	105.6	67.0	79.6	90.6	80.5	83.8	105.9	73.2	83.8	90.6	71.0	77.4	98.7	77.8	84.6	93.6	79.9	84.3
0960	Flooring	90.1	53.7	79.9	91.7	83.3	89.3	120.7	75.3	108.0	91.8	74.1	86.8	97.5	75.3	91.3	98.7	78.5	93.0
0970, 0990	Wall Finishes & Painting/Coating	86.6	56.4	68.7	98.3	90.9	93.9	100.1	89.1	93.6	98.3	73.9	83.9	97.4	89.1	92.5	97.0	71.7	82.0
09	FINISHES	93.4	66.8	79.0	95.0	83.6	88.8	99.4	78.7	88.2	95.2	73.4	83.4	96.5	80.4	87.8	95.9	79.1	86.8
COVERS	DIVS. 10 - 14, 25, 28, 41, 43, 44, 46	100.0	87.2	97.1	100.0	90.3	97.8	100.0	87.9	97.3	100.0	88.4	97.4	100.0	86.8	97.1	100.0	88.4	97.4
21, 22, 23	FIRE SUPPRESSION, PLUMBING & HVAC	96.6	73.2	87.4	100.0	87.8	95.2	96.5	81.2	90.5	100.0	78.7	91.6	97.0	92.4	95.2	99.9	76.9	90.9
26, 27, 3370	ELECTRICAL, COMMUNICATIONS & UTIL.	97.7	68.7	83.4	98.6	81.8	90.4	97.9	89.7	93.9	97.9	85.0	91.5	97.2	89.7	93.5	96.3	74.2	85.4
MF2018	WEIGHTED AVERAGE	98.4	77.7	89.6	99.4	86.0	93.8	97.1	84.8	91.9	98.9	79.9	90.9	94.3	88.5	91.8	96.1	80.0	89.3

		OHIO																	
		CLEVELAND			COLUMBUS			DAYTON			HAMILTON			LIMA			LORAIN		
	DIVISION	441			430 - 432			453 - 454			450			458			440		
		MAT.	INST.	TOTAL	MAT.	INST.	TOTAL	MAT.	INST.	TOTAL	MAT.	INST.	TOTAL	MAT.	INST.	TOTAL	MAT.	INST.	TOTAL
015433	CONTRACTOR EQUIPMENT		92.0	92.0		92.3	92.3		88.9	88.9		95.5	95.5		88.3	88.3		88.7	88.7
0241, 31 - 34	SITE & INFRASTRUCTURE, DEMOLITION	94.9	96.2	95.8	101.8	92.3	95.3	90.0	94.2	92.9	90.0	94.5	93.1	101.1	84.9	89.9	95.5	94.3	94.7
0310	Concrete Forming & Accessories	102.3	90.2	92.0	98.7	78.1	81.1	99.9	77.2	80.6	99.9	77.4	80.7	95.3	76.3	79.1	102.8	74.6	78.7
0320	Concrete Reinforcing	95.2	91.4	93.4	99.9	78.7	89.7	89.1	79.1	84.2	89.1	77.2	83.3	86.9	79.3	83.2	94.7	91.3	93.1
0330	Cast-in-Place Concrete	99.5	97.2	98.6	102.1	80.9	94.2	86.5	80.6	84.3	92.6	80.9	88.3	102.1	89.5	97.4	97.7	91.4	95.4
03	CONCRETE	100.1	92.4	96.7	98.6	79.2	90.0	87.0	78.6	83.3	89.9	79.1	85.1	93.9	81.5	88.5	97.5	83.1	91.2
04	MASONRY	96.9	97.9	97.5	91.4	85.2	87.6	79.1	76.4	77.4	79.5	81.2	80.5	98.8	78.1	86.1	87.8	93.1	91.0
05	METALS	99.1	83.3	94.5	98.1	79.3	92.5	90.9	76.8	86.7	90.9	85.6	89.4	97.2	80.1	92.2	98.1	81.6	93.3
06	WOOD, PLASTICS & COMPOSITES	99.8	88.5	93.6	98.5	77.7	87.0	104.2	77.1	89.3	103.0	77.1	88.7	87.3	75.1	80.6	105.9	70.1	86.2
07	THERMAL & MOISTURE PROTECTION	102.0	96.9	99.8	93.6	83.8	89.3	108.8	79.5	96.1	104.8	79.9	94.0	102.4	83.7	94.3	105.9	91.0	99.4
08	OPENINGS	100.3	86.4	97.0	98.3	74.0	92.7	95.0	74.6	90.3	92.8	74.8	88.6	96.3	73.7	91.0	101.3	76.6	95.5
0920	Plaster & Gypsum Board	97.9	88.1	91.3	94.1	77.0	82.6	98.1	76.9	83.9	98.1	76.9	83.9	93.1	74.3	80.5	98.3	69.2	78.7
0950, 0980	Ceilings & Acoustic Treatment	86.3	88.1	87.5	93.5	77.0	82.4	99.6	76.9	84.3	98.7	76.9	84.0	105.9	74.3	84.6	90.6	69.2	76.2
0960	Flooring	93.1	89.8	92.2	95.1	75.3	89.5	101.2	72.4	93.1	98.5	78.5	92.9	119.8	77.2	107.8	91.8	89.8	91.2
0970, 0990	Wall Finishes & Painting/Coating	101.2	90.9	95.1	99.5	79.6	87.7	97.4	70.7	81.6	97.4	71.3	82.0	100.2	76.5	86.2	98.3	90.9	93.9
09	FINISHES	95.3	90.1	92.5	95.0	77.5	85.5	95.7	75.4	85.0	96.7	76.9	86.0	98.7	76.0	86.4	95.0	78.0	85.8
COVERS	DIVS. 10 - 14, 25, 28, 41, 43, 44, 46	100.0	96.4	99.2	100.0	88.2	97.4	100.0	85.1	96.7	100.0	85.3	96.7	100.0	84.4	96.5	100.0	90.9	98.0
21, 22, 23	FIRE SUPPRESSION, PLUMBING & HVAC	100.0	89.9	96.0	100.0	84.1	93.7	100.7	81.2	93.0	100.5	74.9	90.5	96.5	91.7	94.6	100.0	88.4	95.4
26, 27, 3370	ELECTRICAL, COMMUNICATIONS & UTIL.	98.3	92.7	95.5	99.9	80.2	90.2	94.9	75.7	85.4	95.2	75.5	85.5	98.2	75.7	87.1	98.0	78.2	88.2
MF2018	WEIGHTED AVERAGE	99.1	91.6	95.9	98.4	82.0	91.4	94.8	79.2	88.2	94.8	79.5	88.3	97.2	81.7	90.7	98.4	84.7	92.6

		OHIO																	
		MANSFIELD			MARION			SPRINGFIELD			STEUBENVILLE			TOLEDO			YOUNGSTOWN		
	DIVISION	448 - 449			433			455			439			434 - 436			444 - 445		
		MAT.	INST.	TOTAL	MAT.	INST.	TOTAL	MAT.	INST.	TOTAL	MAT.	INST.	TOTAL	MAT.	INST.	TOTAL	MAT.	INST.	TOTAL
015433	CONTRACTOR EQUIPMENT		88.7	88.7		88.7	88.7		88.9	88.9		92.4	92.4		92.4	92.4		88.7	88.7
0241, 31 - 34	SITE & INFRASTRUCTURE, DEMOLITION	91.7	94.0	93.3	95.7	90.6	92.2	90.3	94.2	93.0	141.5	98.2	111.6	99.8	90.9	93.6	96.1	93.9	94.6
0310	Concrete Forming & Accessories	92.1	73.5	76.3	96.6	79.4	81.9	99.9	77.1	80.5	97.6	79.0	81.7	100.3	84.9	87.1	102.8	76.6	80.5
0320	Concrete Reinforcing	86.2	79.0	82.7	92.1	79.0	85.8	89.1	79.1	84.2	89.7	94.9	92.2	99.9	82.9	91.7	94.7	85.2	90.1
0330	Cast-in-Place Concrete	95.1	87.8	92.3	88.5	88.1	88.3	88.8	80.3	85.6	96.1	89.3	93.6	96.9	90.2	94.4	101.7	87.2	96.3
03	CONCRETE	92.1	79.3	86.4	85.4	82.2	84.0	88.1	78.5	83.9	89.4	84.9	87.4	93.2	86.4	90.2	99.4	81.5	91.5
04	MASONRY	89.9	88.9	89.3	92.7	89.9	90.9	79.3	75.9	77.2	80.7	91.1	87.1	98.9	91.8	94.5	91.1	86.6	88.4
05	METALS	98.4	76.0	91.8	97.1	78.9	91.7	90.9	76.8	86.7	93.6	81.8	90.1	97.8	85.3	94.2	97.5	77.9	91.8
06	WOOD, PLASTICS & COMPOSITES	92.8	70.1	80.3	92.8	77.5	84.4	105.6	77.1	89.9	88.1	76.4	81.6	96.9	83.9	89.8	105.9	74.6	88.7
07	THERMAL & MOISTURE PROTECTION	104.1	88.4	97.3	90.5	88.8	89.8	108.7	79.2	95.9	102.0	87.4	95.7	91.0	91.2	91.1	106.1	88.2	98.3
08	OPENINGS	102.3	71.0	95.0	91.2	75.1	87.5	93.2	74.6	88.9	93.8	78.4	88.3	93.8	81.3	90.9	101.3	75.2	95.2
0920	Plaster & Gypsum Board	90.9	69.2	76.3	95.3	77.0	83.0	98.1	76.9	83.9	93.5	75.4	81.3	97.2	83.6	88.0	98.3	73.9	81.9
0950, 0980	Ceilings & Acoustic Treatment	91.4	69.2	76.4	99.5	77.0	84.3	99.6	76.9	84.3	96.7	75.4	82.3	99.5	83.6	88.8	90.6	73.9	79.3
0960	Flooring	86.6	92.7	88.3	94.4	92.7	93.9	101.2	72.4	93.1	122.5	92.7	114.1	94.8	94.4	94.7	91.8	90.0	91.3
0970, 0990	Wall Finishes & Painting/Coating	98.3	77.8	86.2	104.0	77.8	88.5	97.4	70.7	81.6	117.8	87.1	99.6	104.0	87.0	94.0	98.3	80.0	87.5
09	FINISHES	92.4	76.6	83.8	96.3	81.2	88.1	97.7	75.3	85.6	114.5	81.6	96.7	97.0	86.6	91.4	95.1	78.8	86.3
COVERS	DIVS. 10 - 14, 25, 28, 41, 43, 44, 46	100.0	88.3	97.4	100.0	85.9	96.8	100.0	84.9	96.6	100.0	86.2	96.9	100.0	91.3	98.1	100.0	88.5	97.4
21, 22, 23	FIRE SUPPRESSION, PLUMBING & HVAC	96.5	87.5	92.9	96.5	93.1	95.1	100.7	80.9	92.9	96.9	91.3	94.7	100.0	93.6	97.5	100.0	83.9	93.7
26, 27, 3370	ELECTRICAL, COMMUNICATIONS & UTIL.	95.7	90.8	93.3	94.0	90.8	92.4	94.9	80.2	87.7	88.7	106.4	97.4	100.0	103.4	101.7	98.0	74.3	86.3
MF2018	WEIGHTED AVERAGE	96.4	84.2	91.3	94.3	86.7	91.1	94.8	79.7	88.4	96.1	90.0	93.5	97.6	91.2	94.9	98.7	81.9	91.6

DIVISION		OHIO ZANESVILLE 437-438 MAT.	INST.	TOTAL	OKLAHOMA ARDMORE 734 MAT.	INST.	TOTAL	CLINTON 736 MAT.	INST.	TOTAL	DURANT 747 MAT.	INST.	TOTAL	ENID 737 MAT.	INST.	TOTAL	GUYMON 739 MAT.	INST.	TOTAL
015433	CONTRACTOR EQUIPMENT		88.7	88.7		80.7	80.7		79.9	79.9		79.9	79.9		79.9	79.9		79.9	79.9
0241, 31 - 34	SITE & INFRASTRUCTURE, DEMOLITION	98.9	90.5	93.1	96.9	92.6	94.0	98.1	91.2	93.3	93.4	88.6	90.1	100.0	91.2	93.9	102.4	90.5	94.2
0310	Concrete Forming & Accessories	93.7	78.3	80.6	88.0	55.4	60.2	86.8	55.4	60.1	82.9	55.0	59.1	89.9	55.6	60.7	92.6	55.1	60.7
0320	Concrete Reinforcing	91.6	92.6	92.1	79.5	67.7	73.8	80.0	67.7	74.0	88.6	63.6	76.5	79.5	67.7	73.8	80.0	63.3	74.7
0330	Cast-in-Place Concrete	93.2	86.5	90.7	96.6	70.5	86.9	93.4	70.5	84.9	88.6	70.3	81.8	93.4	70.6	84.9	93.4	70.0	84.7
03	CONCRETE	89.2	83.5	86.7	86.7	62.8	76.2	86.2	62.8	75.9	84.1	61.8	74.3	86.6	62.9	76.2	89.4	61.7	77.2
04	MASONRY	90.7	86.6	88.2	95.2	57.0	71.8	119.3	57.0	81.0	88.6	62.1	72.3	101.3	57.0	74.1	97.5	55.4	71.7
05	METALS	98.5	84.2	94.3	97.3	62.2	87.0	97.4	62.2	87.0	92.7	60.6	83.2	98.8	62.4	88.1	97.8	59.3	86.5
06	WOOD, PLASTICS & COMPOSITES	88.2	77.5	82.3	97.4	54.4	73.7	96.5	54.4	73.4	87.8	54.4	69.5	99.7	54.4	74.8	102.9	54.4	76.2
07	THERMAL & MOISTURE PROTECTION	90.6	84.5	88.0	100.7	64.8	85.1	100.8	64.8	85.1	96.4	64.6	82.6	100.9	64.8	85.2	101.3	62.2	84.3
08	OPENINGS	91.2	79.0	88.4	103.7	55.0	92.3	103.7	55.0	92.3	96.0	54.0	86.2	104.9	55.9	93.5	103.8	54.0	92.2
0920	Plaster & Gypsum Board	91.2	77.0	81.7	90.5	53.7	65.8	90.2	53.7	65.7	78.5	53.7	61.8	91.1	53.7	66.0	91.1	53.7	66.0
0950, 0980	Ceilings & Acoustic Treatment	99.5	77.0	84.3	89.1	53.7	65.3	89.1	53.7	65.3	83.0	53.7	63.3	89.1	53.7	65.3	89.1	53.7	65.3
0960	Flooring	92.6	75.3	87.7	82.7	57.3	75.6	81.7	57.3	74.9	89.7	51.7	79.0	83.3	72.9	80.4	84.8	57.3	77.1
0970, 0990	Wall Finishes & Painting/Coating	104.0	89.1	95.2	86.4	44.8	61.8	86.4	44.8	61.8	92.8	44.8	64.4	86.4	44.8	61.8	86.4	42.2	60.3
09	FINISHES	95.4	78.2	86.1	82.7	53.5	66.9	82.5	53.5	66.8	83.2	52.4	66.5	83.2	56.7	68.8	84.0	54.5	68.0
COVERS	DIVS. 10 - 14, 25, 28, 41, 43, 44, 46	100.0	83.3	96.3	100.0	79.4	95.4	100.0	79.4	95.4	100.0	79.3	95.4	100.0	79.4	95.4	100.0	79.4	95.4
21, 22, 23	FIRE SUPPRESSION, PLUMBING & HVAC	96.5	89.8	93.9	96.5	67.5	85.1	96.5	67.5	85.1	96.6	67.4	85.2	100.0	67.5	87.2	96.5	63.6	83.6
26, 27, 3370	ELECTRICAL, COMMUNICATIONS & UTIL.	94.1	89.7	92.0	95.0	71.0	83.2	95.9	71.0	83.7	97.3	69.2	83.4	95.9	71.0	83.7	97.6	65.6	81.8
MF2018	WEIGHTED AVERAGE	94.9	85.7	91.0	95.2	65.6	82.7	96.4	65.5	83.3	93.1	65.1	81.3	97.0	66.0	83.9	96.3	63.3	82.4

DIVISION		OKLAHOMA LAWTON 735 MAT.	INST.	TOTAL	MCALESTER 745 MAT.	INST.	TOTAL	MIAMI 743 MAT.	INST.	TOTAL	MUSKOGEE 744 MAT.	INST.	TOTAL	OKLAHOMA CITY 730-731 MAT.	INST.	TOTAL	PONCA CITY 746 MAT.	INST.	TOTAL
015433	CONTRACTOR EQUIPMENT		80.7	80.7		79.9	79.9		90.9	90.9		90.9	90.9		85.6	85.6		79.9	79.9
0241, 31 - 34	SITE & INFRASTRUCTURE, DEMOLITION	96.2	92.7	93.7	86.8	90.5	89.4	88.1	88.1	88.1	88.7	87.9	88.1	94.2	98.0	96.8	93.9	90.9	91.8
0310	Concrete Forming & Accessories	92.5	55.6	61.1	81.1	41.4	47.3	93.2	55.2	60.8	97.3	54.8	61.1	92.8	63.8	68.0	89.0	55.4	60.3
0320	Concrete Reinforcing	79.7	67.7	73.9	88.3	63.3	76.2	86.9	67.7	77.6	87.7	62.8	75.7	87.9	67.8	78.2	87.7	67.7	78.0
0330	Cast-in-Place Concrete	90.3	70.6	83.0	77.7	70.0	74.8	81.4	71.4	77.7	82.4	71.1	78.2	90.3	73.4	84.0	91.0	70.5	83.4
03	CONCRETE	83.0	62.9	74.2	75.3	55.4	66.6	79.7	64.0	72.8	81.2	62.8	73.1	86.1	67.6	78.0	86.1	62.7	75.8
04	MASONRY	97.4	57.0	72.6	106.1	56.9	75.9	91.3	57.1	70.3	108.1	49.2	71.9	98.9	57.4	73.8	84.1	57.0	67.5
05	METALS	102.5	62.3	90.7	92.6	59.4	82.9	92.6	77.0	88.0	94.1	73.9	88.2	95.7	63.7	86.3	92.6	62.1	83.6
06	WOOD, PLASTICS & COMPOSITES	102.0	54.4	75.8	85.5	36.2	58.4	100.0	54.6	75.0	104.3	54.6	76.9	92.5	65.1	77.4	95.6	54.4	72.9
07	THERMAL & MOISTURE PROTECTION	100.7	64.8	85.1	96.0	61.2	80.9	96.5	64.1	82.4	96.8	61.1	81.2	92.1	66.9	81.1	96.6	65.0	82.9
08	OPENINGS	106.6	55.9	94.8	96.0	43.9	83.9	96.0	55.0	86.5	97.2	54.0	87.1	100.0	61.8	91.1	96.0	55.0	86.4
0920	Plaster & Gypsum Board	92.8	53.7	66.5	77.6	35.0	48.9	84.2	53.7	63.7	86.4	53.7	64.4	94.7	64.6	74.4	82.9	53.7	63.3
0950, 0980	Ceilings & Acoustic Treatment	96.4	53.7	67.6	83.0	35.0	50.6	83.0	53.7	63.3	92.7	53.7	66.4	87.4	64.6	72.0	83.0	53.7	63.3
0960	Flooring	85.0	72.9	81.6	88.6	57.3	79.8	95.7	51.7	83.3	98.2	33.2	79.9	86.4	72.9	82.6	92.8	57.3	82.8
0970, 0990	Wall Finishes & Painting/Coating	86.4	44.8	61.8	92.8	42.2	62.9	92.8	42.2	62.9	92.8	42.2	62.9	89.5	44.8	64.4	98.6	44.8	64.4
09	FINISHES	84.9	56.7	69.6	82.2	42.4	60.7	85.1	52.2	67.3	88.1	48.8	66.9	86.3	63.1	73.7	84.9	54.4	68.4
COVERS	DIVS. 10 - 14, 25, 28, 41, 43, 44, 46	100.0	79.4	95.4	100.0	77.4	95.0	100.0	76.3	94.7	100.0	76.3	94.7	100.0	80.9	95.8	100.0	79.4	95.4
21, 22, 23	FIRE SUPPRESSION, PLUMBING & HVAC	100.0	67.5	87.2	96.6	63.5	83.6	96.6	63.6	83.7	100.1	61.3	84.9	99.9	67.8	87.3	96.6	63.6	83.7
26, 27, 3370	ELECTRICAL, COMMUNICATIONS & UTIL.	97.6	69.2	83.6	95.8	67.0	81.6	97.1	67.0	82.3	95.4	67.2	81.5	103.4	71.1	87.5	95.4	65.6	80.7
MF2018	WEIGHTED AVERAGE	97.3	65.8	84.0	92.4	60.5	78.9	92.7	65.1	81.1	95.0	62.9	81.4	96.4	68.7	84.7	93.1	64.0	80.8

DIVISION		OKLAHOMA POTEAU 749 MAT.	INST.	TOTAL	SHAWNEE 748 MAT.	INST.	TOTAL	TULSA 740-741 MAT.	INST.	TOTAL	WOODWARD 738 MAT.	INST.	TOTAL	OREGON BEND 977 MAT.	INST.	TOTAL	EUGENE 974 MAT.	INST.	TOTAL
015433	CONTRACTOR EQUIPMENT		89.7	89.7		79.9	79.9		90.9	90.9		79.9	79.9		97.3	97.3		97.3	97.3
0241, 31 - 34	SITE & INFRASTRUCTURE, DEMOLITION	74.8	83.8	81.0	96.9	90.9	92.7	94.8	87.2	89.5	98.5	91.2	93.5	105.9	98.5	100.8	96.6	98.5	97.7
0310	Concrete Forming & Accessories	87.0	54.9	59.7	82.8	55.3	59.4	97.4	56.0	62.1	86.9	43.7	50.1	107.5	96.6	98.2	104.2	96.6	97.7
0320	Concrete Reinforcing	88.7	67.7	78.5	87.7	65.4	76.9	88.0	67.7	78.2	79.5	67.7	73.8	91.0	112.8	101.5	94.9	112.8	103.5
0330	Cast-in-Place Concrete	81.4	71.3	77.7	93.9	70.5	85.2	89.7	73.2	83.6	93.4	70.5	84.9	117.6	100.6	111.3	113.9	100.6	108.9
03	CONCRETE	81.9	63.8	73.9	87.7	62.3	76.6	86.2	65.0	76.9	86.5	57.5	73.7	110.5	100.3	106.0	101.3	100.3	100.9
04	MASONRY	91.6	57.1	70.4	107.5	57.0	76.5	92.2	57.1	70.6	90.8	57.0	70.0	105.1	100.6	102.3	102.1	100.6	101.1
05	METALS	92.6	76.7	87.9	92.5	61.3	83.3	97.0	77.1	91.2	97.4	62.1	87.0	108.0	96.0	104.5	108.7	95.9	105.0
06	WOOD, PLASTICS & COMPOSITES	92.4	54.6	71.6	87.7	54.4	69.4	103.5	55.6	77.2	96.6	38.7	64.8	99.6	96.1	97.7	95.4	96.1	95.8
07	THERMAL & MOISTURE PROTECTION	96.6	64.1	82.5	96.6	63.9	82.4	96.8	64.3	82.7	100.9	63.3	84.5	118.3	100.2	110.4	117.4	102.7	111.0
08	OPENINGS	96.0	55.0	86.5	96.0	54.4	86.3	98.9	56.2	88.9	103.7	46.3	90.3	96.9	100.0	97.8	97.2	100.0	97.8
0920	Plaster & Gypsum Board	81.3	53.7	62.8	78.5	53.7	61.8	86.4	54.8	65.1	90.2	37.6	54.8	122.7	95.9	104.6	120.8	95.9	104.0
0950, 0980	Ceilings & Acoustic Treatment	83.0	53.7	63.3	83.0	53.7	63.3	92.7	54.8	67.1	89.1	37.6	54.4	85.7	95.9	92.6	86.6	95.9	92.9
0960	Flooring	92.2	51.7	80.8	89.7	57.3	80.6	97.0	61.6	87.1	81.7	54.6	74.1	105.8	105.7	105.7	104.2	105.7	104.6
0970, 0990	Wall Finishes & Painting/Coating	92.8	44.8	64.4	92.8	42.2	62.9	92.8	51.3	68.2	86.4	44.8	61.8	98.6	77.3	86.0	98.6	69.6	81.4
09	FINISHES	82.9	52.5	66.4	83.5	53.2	67.1	88.0	55.8	70.5	82.6	43.7	61.5	100.1	95.9	97.8	98.6	95.0	96.7
COVERS	DIVS. 10 - 14, 25, 28, 41, 43, 44, 46	100.0	79.6	95.5	100.0	79.4	95.4	100.0	76.4	94.7	100.0	77.7	95.0	100.0	102.0	100.4	100.0	102.0	100.4
21, 22, 23	FIRE SUPPRESSION, PLUMBING & HVAC	96.6	63.6	83.7	96.6	67.5	85.2	100.1	63.7	85.8	96.5	67.5	85.1	96.6	105.2	100.0	100.1	98.6	99.5
26, 27, 3370	ELECTRICAL, COMMUNICATIONS & UTIL.	95.5	67.0	81.5	97.4	71.0	84.4	97.3	67.0	82.4	97.4	71.0	84.4	102.0	99.8	100.9	100.7	99.8	100.2
MF2018	WEIGHTED AVERAGE	92.3	64.9	80.7	94.5	65.2	82.2	95.9	65.8	83.2	95.3	62.9	81.6	102.6	100.2	101.6	101.7	98.7	100.5

City Cost Indexes

OREGON

DIVISION		KLAMATH FALLS 976			MEDFORD 975			PENDLETON 978			PORTLAND 970 - 972			SALEM 973			VALE 979		
		MAT.	INST.	TOTAL	MAT.	INST.	TOTAL	MAT.	INST.	TOTAL	MAT.	INST.	TOTAL	MAT.	INST.	TOTAL	MAT.	INST.	TOTAL
015433	CONTRACTOR EQUIPMENT		97.3	97.3		97.3	97.3		94.9	94.9		97.3	97.3		99.9	99.9		94.9	94.9
0241, 31 - 34	SITE & INFRASTRUCTURE, DEMOLITION	109.8	98.4	102.0	104.0	98.4	100.2	103.0	91.9	95.4	99.1	98.5	98.7	92.6	102.1	99.2	90.6	91.8	91.5
0310	Concrete Forming & Accessories	100.7	96.3	97.0	99.7	96.3	96.8	101.2	96.6	97.3	105.3	96.8	98.1	105.8	96.8	98.1	107.5	95.3	97.1
0320	Concrete Reinforcing	91.0	112.7	101.5	92.6	112.7	102.3	90.3	112.8	101.2	95.6	112.8	103.9	102.7	112.8	107.6	88.1	112.5	99.9
0330	Cast-in-Place Concrete	117.6	97.0	110.0	117.6	100.4	111.2	118.4	97.5	110.7	117.0	100.6	110.9	107.6	101.7	105.4	93.2	97.8	94.9
03	CONCRETE	113.5	98.9	107.1	107.8	100.1	104.4	93.7	99.3	96.2	103.0	100.4	101.8	98.7	100.7	99.6	77.8	98.8	87.0
04	MASONRY	119.1	100.6	107.7	99.0	100.6	100.0	109.9	100.6	104.2	104.0	100.6	101.9	108.2	100.6	103.5	107.7	100.6	103.4
05	METALS	108.0	95.6	104.4	108.3	95.6	104.6	116.3	96.4	110.4	110.1	96.1	106.0	117.1	95.4	110.8	116.2	95.1	110.0
06	WOOD, PLASTICS & COMPOSITES	90.4	96.1	93.5	89.3	96.1	93.0	92.3	96.2	94.5	96.4	96.1	96.3	90.3	96.3	93.6	100.9	96.2	98.3
07	THERMAL & MOISTURE PROTECTION	118.5	96.8	109.1	118.1	96.1	108.5	110.7	95.4	104.1	117.3	100.2	109.9	114.1	100.9	108.3	110.1	91.4	102.0
08	OPENINGS	96.9	100.0	97.6	99.6	100.0	99.7	93.2	100.0	94.8	95.1	100.0	96.3	102.6	100.0	102.0	93.2	89.0	92.2
0920	Plaster & Gypsum Board	117.1	95.9	102.8	116.5	95.9	102.6	102.4	95.9	98.0	120.3	95.9	103.9	115.4	95.9	102.3	109.0	95.9	100.2
0950, 0980	Ceilings & Acoustic Treatment	93.0	95.9	95.0	98.5	95.9	96.7	61.9	95.9	84.8	88.5	95.9	93.5	96.0	95.9	95.9	61.9	95.9	84.8
0960	Flooring	102.8	105.7	103.6	102.2	105.7	103.2	70.8	105.7	80.6	101.9	105.7	102.9	107.4	105.7	106.9	72.9	105.7	82.1
0970, 0990	Wall Finishes & Painting/Coating	98.6	65.2	78.8	98.6	65.2	78.8	88.3	77.3	81.8	98.4	77.3	85.9	97.3	75.2	84.2	88.3	77.3	81.8
09	FINISHES	100.3	94.6	97.2	100.4	94.6	97.3	70.4	96.0	84.2	98.2	95.9	97.0	99.7	95.8	97.6	71.0	96.0	84.5
COVERS	DIVS. 10 - 14, 25, 28, 41, 43, 44, 46	100.0	101.9	100.4	100.0	101.9	100.4	100.0	96.5	99.2	100.0	102.1	100.5	100.0	102.4	100.5	100.0	102.2	100.5
21, 22, 23	FIRE SUPPRESSION, PLUMBING & HVAC	96.6	105.2	100.0	100.1	105.2	102.1	98.7	112.0	104.0	100.1	111.5	104.6	100.1	106.9	102.8	98.7	71.1	87.9
26, 27, 3370	ELECTRICAL, COMMUNICATIONS & UTIL.	100.7	82.6	91.8	104.1	82.6	93.5	92.8	96.8	94.8	100.9	108.7	104.7	109.3	99.8	104.6	92.8	68.1	80.6
MF2018	WEIGHTED AVERAGE	103.5	97.3	100.9	103.3	97.4	100.8	98.6	100.3	99.3	102.1	102.8	102.4	104.4	100.8	102.9	96.2	87.0	92.3

PENNSYLVANIA

DIVISION		ALLENTOWN 181			ALTOONA 166			BEDFORD 155			BRADFORD 167			BUTLER 160			CHAMBERSBURG 172		
		MAT.	INST.	TOTAL	MAT.	INST.	TOTAL	MAT.	INST.	TOTAL	MAT.	INST.	TOTAL	MAT.	INST.	TOTAL	MAT.	INST.	TOTAL
015433	CONTRACTOR EQUIPMENT		112.6	112.6		112.6	112.6		110.7	110.7		112.6	112.6		112.6	112.6		111.8	111.8
0241, 31 - 34	SITE & INFRASTRUCTURE, DEMOLITION	91.7	97.2	95.5	95.1	97.1	96.5	103.6	94.8	97.6	90.5	96.0	94.3	86.3	97.8	94.2	86.3	95.3	92.5
0310	Concrete Forming & Accessories	97.9	108.2	106.7	83.7	82.6	82.8	82.0	81.2	81.3	85.8	96.5	94.9	85.1	94.1	92.8	88.0	76.8	78.5
0320	Concrete Reinforcing	96.8	113.3	104.8	93.8	105.7	99.6	93.0	105.7	99.1	95.8	105.8	100.6	94.4	119.3	106.5	94.5	112.2	103.1
0330	Cast-in-Place Concrete	89.4	99.9	93.3	99.7	86.9	94.9	109.1	85.9	100.5	95.2	90.6	93.5	88.0	96.2	91.0	93.0	93.8	93.3
03	CONCRETE	92.3	107.0	98.8	87.5	89.5	88.4	98.5	86.2	96.2	94.2	97.0	95.5	79.5	100.3	88.7	97.1	90.5	94.2
04	MASONRY	92.1	93.1	92.7	95.7	83.4	88.1	107.1	80.2	90.6	93.1	82.0	86.3	97.6	93.0	94.8	93.8	79.1	84.8
05	METALS	98.9	120.5	105.3	92.9	114.8	99.3	102.2	113.4	105.5	96.7	113.2	101.6	92.6	121.7	101.2	98.3	118.8	104.3
06	WOOD, PLASTICS & COMPOSITES	97.6	111.5	105.3	76.7	81.8	79.5	79.4	81.7	80.7	82.7	101.2	92.9	78.1	94.1	86.9	84.3	75.6	79.5
07	THERMAL & MOISTURE PROTECTION	104.0	107.8	105.7	102.7	90.0	97.2	98.3	87.4	93.6	103.8	88.5	97.2	102.2	94.4	98.8	96.1	83.1	90.4
08	OPENINGS	92.4	107.3	95.9	86.3	85.6	86.1	93.4	85.6	91.6	92.4	95.3	93.1	86.2	99.3	89.3	89.2	81.4	87.4
0920	Plaster & Gypsum Board	94.9	111.8	106.3	86.3	81.2	82.9	97.5	81.2	86.6	87.1	101.2	96.6	86.3	93.9	91.4	108.9	74.9	86.0
0950, 0980	Ceilings & Acoustic Treatment	88.5	111.8	104.2	90.9	81.2	84.4	101.1	81.2	87.7	91.0	101.2	97.9	91.8	93.9	93.2	96.0	74.9	81.8
0960	Flooring	92.0	95.1	92.8	85.3	98.9	89.2	92.0	103.7	97.5	86.2	103.7	91.1	86.3	105.3	91.6	90.5	78.0	87.0
0970, 0990	Wall Finishes & Painting/Coating	91.9	101.8	97.8	87.2	106.2	98.4	97.6	105.7	102.4	91.9	105.7	100.0	87.2	106.2	98.4	92.0	100.7	97.1
09	FINISHES	90.6	105.8	98.8	88.2	87.5	87.8	100.6	87.4	93.5	88.4	99.4	94.3	88.1	97.2	93.0	90.8	78.8	84.3
COVERS	DIVS. 10 - 14, 25, 28, 41, 43, 44, 46	100.0	99.6	99.9	100.0	94.5	98.8	100.0	93.2	98.5	100.0	96.1	99.1	100.0	97.3	99.4	100.0	92.7	98.4
21, 22, 23	FIRE SUPPRESSION, PLUMBING & HVAC	100.3	114.5	105.9	99.8	83.6	93.4	96.6	84.4	91.8	96.8	90.6	94.4	96.3	95.4	96.0	96.7	89.7	94.0
26, 27, 3370	ELECTRICAL, COMMUNICATIONS & UTIL.	98.0	95.8	96.9	88.7	107.7	98.0	94.2	107.7	100.9	92.1	107.7	99.8	89.2	107.7	98.3	92.4	85.8	89.2
MF2018	WEIGHTED AVERAGE	96.7	105.8	100.6	93.3	92.8	93.1	98.8	92.1	96.0	94.9	97.1	95.8	91.3	100.6	95.2	95.1	89.4	92.7

PENNSYLVANIA

DIVISION		DOYLESTOWN 189			DUBOIS 158			ERIE 164 - 165			GREENSBURG 156			HARRISBURG 170 - 171			HAZLETON 182		
		MAT.	INST.	TOTAL	MAT.	INST.	TOTAL	MAT.	INST.	TOTAL	MAT.	INST.	TOTAL	MAT.	INST.	TOTAL	MAT.	INST.	TOTAL
015433	CONTRACTOR EQUIPMENT		91.5	91.5		110.7	110.7		112.6	112.6		110.7	110.7		114.7	114.7		112.6	112.6
0241, 31 - 34	SITE & INFRASTRUCTURE, DEMOLITION	104.8	85.9	91.7	108.4	95.2	99.3	92.1	97.4	95.7	99.6	96.7	97.6	86.6	100.3	96.1	85.0	97.1	93.3
0310	Concrete Forming & Accessories	82.9	126.8	120.3	81.6	83.7	83.4	97.2	85.8	87.5	87.8	88.6	88.5	100.5	85.7	87.9	80.6	88.8	87.6
0320	Concrete Reinforcing	93.6	150.9	121.3	92.3	119.4	105.4	95.8	107.3	101.4	92.3	119.0	105.2	103.4	111.0	107.1	94.0	113.1	103.2
0330	Cast-in-Place Concrete	84.5	129.7	101.3	105.2	94.1	101.1	98.0	88.7	94.5	101.3	95.7	99.2	91.7	96.9	93.6	84.5	95.6	88.6
03	CONCRETE	88.0	131.2	107.0	104.4	94.7	100.1	86.4	91.9	88.8	97.2	97.5	97.4	91.9	95.3	93.4	85.0	96.7	90.2
04	MASONRY	95.4	134.9	119.7	107.8	93.4	98.9	85.0	86.8	86.1	117.5	89.1	100.0	88.8	83.9	85.8	104.7	89.5	95.4
05	METALS	96.5	123.9	104.5	102.2	118.8	107.1	93.1	115.6	99.7	102.1	119.7	107.3	105.1	117.4	108.7	98.7	119.7	104.8
06	WOOD, PLASTICS & COMPOSITES	78.5	125.8	104.5	78.3	81.7	80.2	94.0	84.4	88.7	85.3	87.1	86.3	101.8	85.3	92.7	77.1	87.0	82.6
07	THERMAL & MOISTURE PROTECTION	101.1	131.7	114.4	98.7	92.9	96.1	103.2	88.6	96.9	98.2	92.2	95.6	98.9	100.4	99.6	103.3	100.7	102.2
08	OPENINGS	94.8	133.2	103.7	93.4	88.7	92.3	86.4	88.0	86.8	93.4	95.5	93.9	100.9	86.5	97.5	93.0	93.3	93.0
0920	Plaster & Gypsum Board	85.4	126.5	113.0	96.3	81.2	86.2	94.9	83.9	87.5	98.3	86.8	90.5	113.8	84.7	94.2	85.8	86.6	86.4
0950, 0980	Ceilings & Acoustic Treatment	87.7	126.5	113.8	101.1	81.2	87.7	88.5	83.9	85.4	100.3	86.8	91.2	104.0	84.7	91.0	89.4	86.6	87.5
0960	Flooring	76.2	139.3	93.9	94.8	103.7	97.3	92.2	98.9	94.1	98.4	78.0	92.7	94.7	90.4	93.5	83.5	89.3	85.1
0970, 0990	Wall Finishes & Painting/Coating	91.5	143.7	122.4	97.6	106.2	102.7	98.2	92.3	94.7	97.6	106.2	102.7	95.8	84.8	89.3	91.9	104.1	99.1
09	FINISHES	81.9	130.5	108.2	100.9	88.8	94.4	91.6	88.4	89.9	101.3	88.1	94.1	96.0	85.8	90.5	86.6	90.2	88.6
COVERS	DIVS. 10 - 14, 25, 28, 41, 43, 44, 46	100.0	113.0	102.9	100.0	95.2	98.9	100.0	96.6	99.2	100.0	96.4	99.2	100.0	95.4	99.0	100.0	95.5	99.0
21, 22, 23	FIRE SUPPRESSION, PLUMBING & HVAC	96.3	132.6	110.6	96.6	87.5	93.0	99.8	93.6	97.4	96.6	87.3	92.9	100.1	91.7	96.8	96.8	95.3	96.3
26, 27, 3370	ELECTRICAL, COMMUNICATIONS & UTIL.	91.6	132.4	111.7	94.8	107.7	101.1	90.3	93.1	91.7	94.8	107.7	101.2	99.7	85.8	92.9	92.9	90.0	91.5
MF2018	WEIGHTED AVERAGE	94.0	127.1	108.0	99.3	96.0	97.9	93.2	93.9	93.5	98.7	96.3	97.7	98.7	93.0	96.3	94.4	95.9	95.0

PENNSYLVANIA

DIVISION		INDIANA 157			JOHNSTOWN 159			KITTANNING 162			LANCASTER 175 - 176			LEHIGH VALLEY 180			MONTROSE 188		
		MAT.	INST.	TOTAL	MAT.	INST.	TOTAL	MAT.	INST.	TOTAL	MAT.	INST.	TOTAL	MAT.	INST.	TOTAL	MAT.	INST.	TOTAL
015433	CONTRACTOR EQUIPMENT		110.7	110.7		110.7	110.7		112.6	112.6		111.8	111.8		112.6	112.6		112.6	112.6
0241, 31 - 34	SITE & INFRASTRUCTURE, DEMOLITION	97.8	95.8	96.4	104.1	96.0	98.5	88.8	97.4	94.7	78.4	95.6	90.3	88.9	97.5	94.8	87.5	97.2	94.2
0310	Concrete Forming & Accessories	82.6	92.9	91.4	81.6	82.5	82.4	85.1	89.7	89.0	90.0	86.4	86.9	91.7	110.3	107.6	81.5	89.3	88.2
0320	Concrete Reinforcing	91.7	119.4	105.1	93.0	118.9	105.5	94.4	119.2	106.4	94.1	111.0	102.3	94.0	108.0	100.8	98.4	116.2	107.0
0330	Cast-in-Place Concrete	99.3	94.4	97.5	110.1	86.7	101.4	91.5	94.7	92.7	79.1	96.7	85.7	91.5	99.2	94.3	89.7	93.0	90.9
03	CONCRETE	94.7	99.1	96.6	103.2	91.6	98.1	82.1	97.7	88.9	85.2	95.6	89.8	91.4	106.9	98.2	90.1	96.6	93.0
04	MASONRY	103.8	93.5	97.5	104.7	83.7	91.8	100.1	88.8	93.1	99.6	86.7	91.7	92.1	96.5	94.8	92.0	91.5	91.7
05	METALS	102.3	119.9	107.4	102.2	117.9	106.8	92.7	120.6	100.9	98.3	118.7	104.3	98.6	119.0	104.6	96.8	120.4	103.7
06	WOOD, PLASTICS & COMPOSITES	80.1	94.0	87.7	78.3	81.7	80.2	78.1	89.3	84.3	87.4	85.1	86.2	89.1	112.6	102.0	77.8	87.0	82.9
07	THERMAL & MOISTURE PROTECTION	98.1	94.2	96.4	98.4	88.5	94.1	102.3	92.4	98.0	95.4	101.0	97.8	103.7	111.9	107.3	103.3	91.0	97.9
08	OPENINGS	93.4	95.4	93.9	93.4	88.7	92.3	86.3	94.7	88.7	89.2	86.4	88.5	92.9	106.6	96.1	89.8	91.9	90.3
0920	Plaster & Gypsum Board	97.8	93.9	95.2	96.1	81.2	86.1	86.3	89.0	88.1	111.8	84.7	93.6	88.0	112.9	104.7	86.2	86.6	86.5
0950, 0980	Ceilings & Acoustic Treatment	101.1	93.9	96.2	100.3	81.2	87.5	91.8	89.0	89.9	96.0	84.7	88.4	89.4	112.9	105.2	91.0	86.6	88.1
0960	Flooring	95.6	103.7	97.8	94.8	78.0	90.1	86.3	103.7	91.2	91.5	95.7	92.7	88.9	94.1	90.4	84.1	103.7	89.6
0970, 0990	Wall Finishes & Painting/Coating	97.6	106.2	102.7	97.6	106.2	102.7	87.2	106.2	98.4	92.0	86.6	88.8	91.9	101.0	97.3	91.9	104.1	99.1
09	FINISHES	100.4	96.1	98.1	100.3	83.9	91.4	88.3	93.4	91.0	90.9	87.3	88.9	88.8	106.5	98.4	87.4	92.7	90.3
COVERS	DIVS. 10 - 14, 25, 28, 41, 43, 44, 46	100.0	96.5	99.2	100.0	94.4	98.8	100.0	96.2	99.1	100.0	95.7	99.0	100.0	102.5	100.6	100.0	96.2	99.2
21, 22, 23	FIRE SUPPRESSION, PLUMBING & HVAC	96.6	85.4	92.2	96.6	78.9	89.6	96.3	90.7	94.1	96.7	92.7	95.1	96.8	117.6	104.9	96.8	94.8	96.0
26, 27, 3370	ELECTRICAL, COMMUNICATIONS & UTIL.	94.8	107.7	101.2	94.8	107.7	101.1	88.7	107.7	98.1	93.7	93.3	93.5	92.9	129.4	110.9	92.1	95.2	93.6
MF2018	WEIGHTED AVERAGE	97.6	97.7	97.7	98.9	91.9	95.9	91.8	97.9	94.4	93.8	94.5	94.1	94.9	111.6	102.0	93.8	96.8	95.1

PENNSYLVANIA

DIVISION		NEW CASTLE 161			NORRISTOWN 194			OIL CITY 163			PHILADELPHIA 190 - 191			PITTSBURGH 150 - 152			POTTSVILLE 179		
		MAT.	INST.	TOTAL	MAT.	INST.	TOTAL	MAT.	INST.	TOTAL	MAT.	INST.	TOTAL	MAT.	INST.	TOTAL	MAT.	INST.	TOTAL
015433	CONTRACTOR EQUIPMENT		112.6	112.6		97.5	97.5		112.6	112.6		99.8	99.8		99.7	99.7		111.8	111.8
0241, 31 - 34	SITE & INFRASTRUCTURE, DEMOLITION	86.7	97.7	94.3	97.3	95.1	95.8	85.2	96.3	92.9	99.8	101.1	100.7	103.7	96.1	98.5	81.3	95.8	91.3
0310	Concrete Forming & Accessories	85.1	93.6	92.3	83.7	125.3	119.1	85.1	92.7	91.6	99.8	142.4	136.2	97.5	97.5	97.5	82.0	87.7	86.8
0320	Concrete Reinforcing	93.3	100.2	96.7	95.9	150.8	122.5	94.4	94.9	94.7	100.4	143.1	121.0	93.5	122.9	107.7	93.4	114.1	103.4
0330	Cast-in-Place Concrete	88.8	94.3	90.8	86.1	127.4	101.5	86.3	94.4	89.3	89.4	135.6	106.8	108.6	101.2	105.8	84.3	96.4	88.8
03	CONCRETE	79.9	96.2	87.0	88.0	129.7	106.3	78.3	95.0	85.6	99.4	139.1	116.8	103.8	103.1	103.5	88.7	96.7	92.2
04	MASONRY	97.3	92.6	94.4	108.3	131.2	122.4	96.5	88.7	91.7	100.2	139.7	124.5	100.4	101.4	101.0	93.3	87.6	89.8
05	METALS	92.7	114.3	99.0	100.3	124.0	107.3	92.7	113.2	98.7	103.8	125.3	110.1	103.7	107.4	104.8	98.6	120.4	105.0
06	WOOD, PLASTICS & COMPOSITES	78.1	94.1	86.9	76.2	125.7	103.4	78.1	94.1	86.9	99.1	143.5	123.6	99.6	96.7	98.0	77.2	85.1	81.6
07	THERMAL & MOISTURE PROTECTION	102.2	92.0	97.8	108.7	130.1	118.0	102.1	90.7	97.1	103.4	139.7	119.2	98.6	99.3	98.9	95.6	99.6	97.3
08	OPENINGS	86.3	91.4	87.5	86.5	133.1	97.4	86.3	94.9	88.3	97.6	144.9	108.6	96.8	102.4	98.1	89.2	93.0	90.1
0920	Plaster & Gypsum Board	86.3	93.9	91.4	84.5	126.5	112.7	86.3	93.9	91.4	103.0	144.6	131.0	97.5	96.4	96.7	106.4	84.7	91.8
0950, 0980	Ceilings & Acoustic Treatment	91.8	93.9	93.2	91.1	126.5	114.9	91.8	93.9	93.2	104.2	144.6	131.5	95.2	96.4	96.0	96.0	84.7	88.4
0960	Flooring	86.3	105.3	91.6	86.7	139.3	101.5	86.3	103.7	91.2	97.8	156.2	114.2	103.1	107.3	104.3	87.5	103.7	92.1
0970, 0990	Wall Finishes & Painting/Coating	87.2	106.2	98.4	89.4	143.7	121.5	87.2	106.2	98.4	97.2	152.6	130.0	100.7	111.3	107.0	92.0	104.1	99.1
09	FINISHES	88.2	96.9	92.9	85.3	129.5	109.2	88.0	96.1	92.4	98.3	146.5	124.4	102.4	100.3	101.2	89.2	91.3	90.3
COVERS	DIVS. 10 - 14, 25, 28, 41, 43, 44, 46	100.0	97.3	99.4	100.0	109.1	102.0	100.0	96.7	99.3	100.0	119.2	104.3	100.0	103.0	100.7	100.0	97.5	99.4
21, 22, 23	FIRE SUPPRESSION, PLUMBING & HVAC	96.3	96.8	96.5	96.6	130.6	110.0	96.3	94.2	95.5	100.1	141.2	116.3	100.0	99.8	99.9	96.7	95.8	96.3
26, 27, 3370	ELECTRICAL, COMMUNICATIONS & UTIL.	89.2	98.2	93.7	91.9	142.0	116.6	90.9	107.7	99.2	98.8	159.0	128.4	97.1	111.3	104.1	92.0	91.3	91.6
MF2018	WEIGHTED AVERAGE	91.4	97.8	94.1	94.7	128.0	108.8	91.3	97.8	94.0	100.2	138.7	116.5	100.7	102.7	101.5	93.7	96.1	94.7

PENNSYLVANIA

DIVISION		READING 195 - 196			SCRANTON 184 - 185			STATE COLLEGE 168			STROUDSBURG 183			SUNBURY 178			UNIONTOWN 154		
		MAT.	INST.	TOTAL	MAT.	INST.	TOTAL	MAT.	INST.	TOTAL	MAT.	INST.	TOTAL	MAT.	INST.	TOTAL	MAT.	INST.	TOTAL
015433	CONTRACTOR EQUIPMENT		119.0	119.0		112.6	112.6		111.8	111.8		112.6	112.6		112.6	112.6		110.7	110.7
0241, 31 - 34	SITE & INFRASTRUCTURE, DEMOLITION	101.8	107.7	105.9	92.2	97.2	95.7	82.7	95.7	91.7	86.6	97.3	94.0	92.6	96.7	95.4	98.3	96.5	97.1
0310	Concrete Forming & Accessories	98.8	86.2	88.0	98.0	86.6	88.3	84.3	82.8	83.0	86.5	90.5	89.9	93.1	84.8	86.0	76.5	93.9	91.4
0320	Concrete Reinforcing	97.2	148.6	122.1	96.8	116.3	106.2	95.1	105.8	100.3	97.1	116.4	106.4	96.0	112.3	103.9	92.3	119.5	105.5
0330	Cast-in-Place Concrete	77.2	97.0	84.6	93.4	93.2	93.3	90.0	87.0	88.9	88.0	94.5	90.4	92.2	94.9	93.2	99.3	95.6	97.9
03	CONCRETE	87.1	102.1	93.7	94.1	95.4	94.7	94.6	89.7	92.5	88.8	97.7	92.7	92.0	94.5	93.1	94.4	100.0	96.9
04	MASONRY	97.5	91.5	93.8	92.5	93.8	93.3	97.8	83.9	89.3	90.3	96.3	94.0	93.6	81.0	85.8	119.3	95.4	104.6
05	METALS	100.6	133.0	110.1	100.9	120.5	106.6	96.6	115.1	102.0	98.7	120.9	105.2	98.3	119.4	104.5	102.0	120.2	107.3
06	WOOD, PLASTICS & COMPOSITES	93.7	82.4	87.5	97.6	83.3	89.8	84.5	81.8	83.0	83.5	87.0	85.5	85.4	85.1	85.2	72.8	94.0	84.4
07	THERMAL & MOISTURE PROTECTION	109.0	103.3	106.6	103.9	91.3	98.4	102.9	97.6	100.6	103.5	91.3	98.2	96.8	92.7	95.0	98.0	95.1	96.7
08	OPENINGS	90.9	99.4	92.9	92.4	89.8	91.8	89.6	85.6	88.7	93.0	94.5	93.3	89.3	89.9	89.4	93.3	99.3	94.7
0920	Plaster & Gypsum Board	94.1	81.9	85.9	96.8	82.8	87.4	87.8	81.2	83.4	86.7	86.6	86.6	105.9	84.7	91.6	94.2	93.9	94.0
0950, 0980	Ceilings & Acoustic Treatment	83.7	81.9	82.5	96.6	82.8	87.3	88.5	81.2	83.6	87.7	86.6	87.0	93.6	84.7	87.6	100.3	93.9	96.0
0960	Flooring	91.3	97.9	93.2	92.0	92.9	92.2	89.4	97.9	91.8	86.8	94.1	88.8	88.3	103.7	92.6	92.0	103.7	95.3
0970, 0990	Wall Finishes & Painting/Coating	88.2	101.8	96.3	91.9	108.3	101.6	91.9	106.2	100.4	91.9	104.1	99.1	92.0	104.1	99.1	97.6	110.0	104.9
09	FINISHES	86.8	88.4	87.7	92.4	89.3	90.7	88.0	87.3	87.6	87.5	91.8	89.8	90.2	89.6	89.9	98.7	97.0	97.8
COVERS	DIVS. 10 - 14, 25, 28, 41, 43, 44, 46	100.0	95.8	99.1	100.0	94.6	98.8	100.0	94.6	98.8	100.0	97.2	99.4	100.0	93.1	98.5	100.0	97.2	99.4
21, 22, 23	FIRE SUPPRESSION, PLUMBING & HVAC	100.3	106.8	102.8	100.3	96.0	98.8	96.3	91.4	94.7	96.8	87.4	93.0	96.7	87.4	93.0	96.6	91.7	94.6
26, 27, 3370	ELECTRICAL, COMMUNICATIONS & UTIL.	97.8	91.3	94.6	98.0	95.2	96.6	91.3	107.7	99.4	92.9	141.8	117.0	92.4	91.9	92.1	91.8	107.7	99.7
MF2018	WEIGHTED AVERAGE	96.5	101.5	98.6	97.4	96.5	97.1	94.5	94.6	94.5	94.3	104.7	98.7	94.6	92.7	93.8	97.8	99.8	98.6

City Cost Indexes

PENNSYLVANIA

DIVISION		WASHINGTON 153			WELLSBORO 169			WESTCHESTER 193			WILKES-BARRE 186 - 187			WILLIAMSPORT 177			YORK 173 - 174		
		MAT.	INST.	TOTAL	MAT.	INST.	TOTAL	MAT.	INST.	TOTAL	MAT.	INST.	TOTAL	MAT.	INST.	TOTAL	MAT.	INST.	TOTAL
015433	CONTRACTOR EQUIPMENT		110.7	110.7		112.6	112.6		97.5	97.5		112.6	112.6		112.6	112.6		111.8	111.8
0241, 31 - 34	SITE & INFRASTRUCTURE, DEMOLITION	98.3	96.8	97.3	94.0	96.6	95.8	103.4	96.1	98.4	84.5	97.2	93.3	83.7	96.8	92.8	82.4	95.6	91.5
0310	Concrete Forming & Accessories	82.8	94.0	92.4	85.2	84.2	84.3	90.0	126.6	121.2	89.1	85.3	85.8	90.1	85.6	86.3	85.1	86.3	86.1
0320	Concrete Reinforcing	92.3	119.5	105.5	95.1	116.1	105.5	95.0	150.8	122.0	95.8	116.3	105.7	95.3	110.9	102.8	96.0	111.0	103.3
0330	Cast-in-Place Concrete	99.3	95.9	98.0	94.4	88.2	92.1	95.4	129.5	108.1	84.5	93.0	87.6	78.0	90.0	82.5	84.7	96.7	89.2
03	CONCRETE	94.8	100.1	97.2	96.9	92.5	95.0	95.8	131.0	111.3	85.8	94.7	89.7	80.2	92.9	85.8	89.8	95.6	92.4
04	MASONRY	103.6	95.5	98.6	98.7	81.4	88.1	102.8	134.9	122.5	105.0	93.5	97.9	85.9	86.1	86.0	94.5	86.7	89.7
05	METALS	101.9	120.6	107.4	96.6	119.7	103.4	100.3	124.0	107.3	96.8	120.4	103.7	98.3	118.1	104.1	100.0	118.6	105.5
06	WOOD, PLASTICS & COMPOSITES	80.2	94.0	87.8	82.1	84.4	83.4	82.8	125.7	106.4	85.9	81.7	83.6	81.9	85.1	83.7	80.7	85.1	83.1
07	THERMAL & MOISTURE PROTECTION	98.1	95.1	96.8	104.1	86.0	96.2	109.1	127.0	116.9	103.3	90.9	97.9	96.1	88.3	92.7	95.6	101.0	98.0
08	OPENINGS	93.3	98.9	94.7	92.4	90.0	91.8	86.5	133.1	97.4	89.8	88.9	89.6	89.3	89.6	89.4	89.2	86.4	88.5
0920	Plaster & Gypsum Board	97.7	93.9	95.1	86.5	83.9	84.8	85.2	126.5	112.9	87.4	81.2	83.2	106.4	84.7	91.8	106.7	84.7	91.9
0950, 0980	Ceilings & Acoustic Treatment	100.3	93.9	96.0	88.5	83.9	85.4	91.1	126.5	114.9	91.0	81.2	84.4	96.0	84.7	88.4	95.1	84.7	88.1
0960	Flooring	95.7	103.7	97.9	85.8	103.7	90.9	89.6	139.3	103.6	87.7	92.9	89.1	87.2	86.8	87.1	88.9	95.7	90.8
0970, 0990	Wall Finishes & Painting/Coating	97.6	110.0	104.9	91.9	104.1	99.1	89.4	143.7	121.5	91.9	101.8	97.8	92.0	101.8	97.8	92.0	84.8	87.7
09	FINISHES	100.3	97.0	98.5	88.1	89.6	88.9	86.7	130.5	110.4	88.4	87.6	88.0	89.7	86.7	88.0	89.5	87.1	88.2
COVERS	DIVS. 10 - 14, 25, 28, 41, 43, 44, 46	100.0	97.2	99.4	100.0	96.2	99.2	100.0	115.3	103.4	100.0	94.3	98.7	100.0	95.5	99.0	100.0	95.6	99.0
21, 22, 23	FIRE SUPPRESSION, PLUMBING & HVAC	96.6	93.2	95.2	96.8	87.4	93.1	96.6	132.6	110.7	96.8	96.6	96.7	96.7	92.2	94.9	100.2	92.7	97.3
26, 27, 3370	ELECTRICAL, COMMUNICATIONS & UTIL.	94.2	107.7	100.9	92.1	95.2	93.6	91.7	132.4	111.8	92.9	90.0	91.5	92.8	83.4	88.2	93.7	82.9	88.4
MF2018	WEIGHTED AVERAGE	97.5	100.2	98.6	95.5	92.9	94.4	95.8	127.9	109.3	94.0	95.5	94.7	92.5	92.3	92.4	95.2	93.0	94.3

PUERTO RICO / RHODE ISLAND / SOUTH CAROLINA

DIVISION		PUERTO RICO SAN JUAN 009			RHODE ISLAND NEWPORT 028			RHODE ISLAND PROVIDENCE 029			SOUTH CAROLINA AIKEN 298			SOUTH CAROLINA BEAUFORT 299			SOUTH CAROLINA CHARLESTON 294		
		MAT.	INST.	TOTAL	MAT.	INST.	TOTAL	MAT.	INST.	TOTAL	MAT.	INST.	TOTAL	MAT.	INST.	TOTAL	MAT.	INST.	TOTAL
015433	CONTRACTOR EQUIPMENT		87.3	87.3		97.9	97.9		100.1	100.1		103.7	103.7		103.7	103.7		103.7	103.7
0241, 31 - 34	SITE & INFRASTRUCTURE, DEMOLITION	81.4	87.9	85.9	86.3	98.0	94.4	89.4	102.2	98.3	128.3	85.2	98.5	123.4	84.7	96.7	108.2	85.2	92.3
0310	Concrete Forming & Accessories	105.9	21.6	34.0	100.7	124.7	121.1	99.9	124.7	121.0	94.0	64.6	68.9	93.1	38.9	46.9	92.3	64.9	69.0
0320	Concrete Reinforcing	114.9	18.7	68.4	109.7	127.1	118.1	103.7	127.1	115.0	94.3	60.6	78.0	93.4	66.7	80.5	93.2	67.3	80.7
0330	Cast-in-Place Concrete	70.9	32.7	56.6	71.7	118.5	89.1	87.5	119.0	99.2	92.9	66.9	83.2	92.9	66.8	83.2	109.1	67.2	93.5
03	CONCRETE	89.8	26.1	61.8	87.5	122.4	102.9	91.8	122.5	105.3	101.0	66.4	85.8	98.4	55.7	79.7	94.9	67.8	83.0
04	MASONRY	110.0	24.2	57.3	99.2	123.1	113.8	103.4	123.1	115.5	79.9	64.8	70.6	94.9	64.8	76.4	96.4	67.5	78.6
05	METALS	116.8	40.6	94.4	98.8	116.7	104.1	103.2	115.9	106.9	102.6	89.1	98.7	102.7	90.7	99.2	104.6	92.1	100.9
06	WOOD, PLASTICS & COMPOSITES	145.4	20.5	76.6	98.0	124.0	112.3	98.4	124.0	112.5	89.7	66.8	77.1	88.4	32.5	57.6	87.2	66.8	76.0
07	THERMAL & MOISTURE PROTECTION	134.1	27.7	87.8	102.6	123.0	111.5	108.3	123.5	114.9	98.1	64.2	83.4	97.8	56.2	79.7	96.8	66.1	83.5
08	OPENINGS	96.8	19.1	78.7	97.4	124.4	103.7	99.0	124.4	104.9	94.0	62.5	86.7	94.1	44.9	82.6	97.6	65.1	90.0
0920	Plaster & Gypsum Board	76.4	25.2	48.0	91.4	124.3	113.6	104.9	124.3	118.0	83.4	65.8	71.6	87.0	30.5	49.0	88.1	65.8	73.1
0950, 0980	Ceilings & Acoustic Treatment	158.8	19.3	64.7	90.3	124.3	113.2	104.1	124.3	117.7	78.8	65.8	70.1	82.1	30.5	47.3	82.1	65.8	71.1
0960	Flooring	117.8	29.7	93.1	83.8	127.3	96.0	86.4	127.3	97.9	97.5	89.8	95.3	99.0	75.8	92.5	98.6	81.9	93.9
0970, 0990	Wall Finishes & Painting/Coating	90.2	24.6	51.4	85.0	118.9	105.0	90.8	118.9	107.4	95.2	66.0	77.9	95.2	56.9	72.5	95.2	70.0	80.3
09	FINISHES	106.0	23.2	61.2	87.7	125.1	107.9	93.7	125.1	110.7	89.0	69.7	78.5	90.1	46.0	66.2	88.2	68.9	77.8
COVERS	DIVS. 10 - 14, 25, 28, 41, 43, 44, 46	100.0	26.6	83.6	100.0	110.2	102.3	100.0	110.2	102.3	100.0	70.3	93.4	100.0	80.1	95.6	100.0	70.3	93.4
21, 22, 23	FIRE SUPPRESSION, PLUMBING & HVAC	88.5	18.8	61.1	100.4	114.8	106.0	100.2	114.8	105.9	96.9	55.2	80.5	96.9	55.6	80.7	100.4	58.4	83.9
26, 27, 3370	ELECTRICAL, COMMUNICATIONS & UTIL.	84.4	25.0	55.2	102.1	95.5	98.9	102.3	95.5	99.0	96.2	61.9	79.3	99.7	30.9	65.8	98.1	58.9	78.8
MF2018	WEIGHTED AVERAGE	98.5	29.8	69.4	97.0	114.7	104.5	99.3	115.0	105.9	97.6	67.1	84.7	98.3	57.5	81.1	98.7	68.2	85.8

SOUTH CAROLINA / SOUTH DAKOTA

DIVISION		SOUTH CAROLINA COLUMBIA 290 - 292			SOUTH CAROLINA FLORENCE 295			SOUTH CAROLINA GREENVILLE 296			SOUTH CAROLINA ROCK HILL 297			SOUTH CAROLINA SPARTANBURG 293			SOUTH DAKOTA ABERDEEN 574		
		MAT.	INST.	TOTAL	MAT.	INST.	TOTAL	MAT.	INST.	TOTAL	MAT.	INST.	TOTAL	MAT.	INST.	TOTAL	MAT.	INST.	TOTAL
015433	CONTRACTOR EQUIPMENT		105.6	105.6		103.7	103.7		103.7	103.7		103.7	103.7		103.7	103.7		97.7	97.7
0241, 31 - 34	SITE & INFRASTRUCTURE, DEMOLITION	105.9	89.5	94.6	117.6	84.9	95.1	112.6	85.4	93.8	110.0	84.3	92.2	112.4	85.4	93.8	99.2	93.4	95.2
0310	Concrete Forming & Accessories	93.2	64.9	69.1	81.1	64.8	67.2	91.8	64.9	68.8	90.1	64.0	67.9	94.4	64.9	69.2	98.9	75.7	79.1
0320	Concrete Reinforcing	98.3	67.3	83.3	92.8	67.2	80.4	92.8	64.4	79.1	93.5	65.8	80.1	92.8	67.3	80.4	101.0	71.4	86.7
0330	Cast-in-Place Concrete	110.3	67.5	94.4	92.9	67.0	83.2	92.9	67.2	83.3	92.8	66.4	83.0	92.9	67.2	83.3	109.5	78.8	98.1
03	CONCRETE	93.2	67.8	82.1	93.3	67.6	82.0	92.0	67.3	81.2	90.1	66.8	79.9	92.2	67.8	81.5	98.7	76.7	89.1
04	MASONRY	89.9	67.5	76.1	79.9	67.4	72.2	77.9	67.5	71.5	101.6	64.0	78.5	79.9	67.5	72.3	109.4	73.3	87.2
05	METALS	101.6	91.0	98.5	103.5	91.4	99.9	103.5	91.7	100.0	102.7	90.5	99.1	103.5	92.1	100.1	93.8	82.1	90.4
06	WOOD, PLASTICS & COMPOSITES	88.6	66.9	76.7	74.8	66.8	70.4	86.8	66.8	75.8	85.4	66.8	75.2	90.6	66.8	77.5	97.6	74.2	84.7
07	THERMAL & MOISTURE PROTECTION	93.3	66.3	81.5	97.2	66.1	83.7	97.1	66.1	83.6	96.9	59.2	80.5	97.1	66.1	83.6	103.1	77.2	91.8
08	OPENINGS	99.7	65.1	91.7	94.1	65.1	87.3	94.0	64.8	87.2	94.1	63.6	87.0	94.1	65.1	87.3	97.2	62.2	89.0
0920	Plaster & Gypsum Board	88.7	65.8	73.3	78.2	65.8	69.9	82.1	65.8	71.2	81.8	65.8	71.1	84.7	65.8	72.0	110.1	73.9	85.7
0950, 0980	Ceilings & Acoustic Treatment	86.2	65.8	72.5	79.6	65.8	70.3	78.8	65.8	70.1	78.8	65.8	70.1	78.8	65.8	70.1	98.8	73.9	82.0
0960	Flooring	90.7	80.4	87.8	89.9	80.4	87.2	96.3	80.4	91.8	95.2	75.8	89.7	97.6	80.4	92.8	91.1	43.8	77.8
0970, 0990	Wall Finishes & Painting/Coating	93.9	70.0	79.7	95.2	70.0	80.3	95.2	70.0	80.3	95.2	66.0	77.9	95.2	70.0	80.3	89.8	33.4	56.4
09	FINISHES	86.5	68.7	76.8	85.0	68.7	76.2	86.9	68.7	77.0	86.3	67.1	75.9	87.6	68.7	77.4	92.4	65.9	78.1
COVERS	DIVS. 10 - 14, 25, 28, 41, 43, 44, 46	100.0	70.5	93.4	100.0	70.3	93.4	100.0	70.3	93.4	100.0	70.0	93.3	100.0	70.3	93.4	100.0	84.8	96.6
21, 22, 23	FIRE SUPPRESSION, PLUMBING & HVAC	100.0	58.3	83.6	100.0	58.3	83.9	100.0	58.2	83.6	100.4	58.2	83.9	100.4	58.2	83.9	100.0	54.2	82.0
26, 27, 3370	ELECTRICAL, COMMUNICATIONS & UTIL.	98.5	62.8	80.9	96.1	62.8	79.7	98.2	59.8	79.3	98.2	59.9	79.3	98.2	59.8	79.3	101.1	65.3	83.4
MF2018	WEIGHTED AVERAGE	97.6	68.9	85.5	96.9	68.6	84.9	97.0	68.2	84.8	96.7	66.4	83.9	97.2	68.3	85.0	98.6	70.1	86.5

SOUTH DAKOTA

DIVISION		MITCHELL 573 MAT.	INST.	TOTAL	MOBRIDGE 576 MAT.	INST.	TOTAL	PIERRE 575 MAT.	INST.	TOTAL	RAPID CITY 577 MAT.	INST.	TOTAL	SIOUX FALLS 570-571 MAT.	INST.	TOTAL	WATERTOWN 572 MAT.	INST.	TOTAL
015433	CONTRACTOR EQUIPMENT		97.7	97.7		97.7	97.7		99.9	99.9		97.7	97.7		100.9	100.9		97.7	97.7
0241, 31 - 34	SITE & INFRASTRUCTURE, DEMOLITION	95.8	93.2	94.0	95.7	93.2	94.0	97.2	96.3	96.6	97.5	92.8	94.3	92.5	98.7	96.8	95.6	93.2	94.0
0310	Concrete Forming & Accessories	97.9	44.2	52.1	88.2	44.6	51.0	99.7	45.6	53.6	106.9	56.6	64.0	103.3	78.1	81.8	84.6	74.5	76.0
0320	Concrete Reinforcing	100.4	70.5	86.0	103.0	70.5	87.3	97.8	100.2	99.0	94.4	100.4	97.3	106.6	100.6	103.7	97.6	70.5	84.5
0330	Cast-in-Place Concrete	106.4	52.3	86.3	106.4	77.2	95.6	103.1	77.8	93.7	105.6	77.2	95.0	88.3	78.8	84.7	106.4	77.7	95.5
03	CONCRETE	96.5	53.1	77.4	96.2	62.0	81.2	91.8	67.4	81.1	95.9	72.3	85.5	88.6	82.7	86.0	95.3	75.5	86.6
04	MASONRY	97.8	75.5	84.1	106.5	72.8	85.8	106.5	72.6	85.7	106.0	74.0	86.3	91.1	75.6	81.5	132.0	73.7	96.2
05	METALS	92.9	81.6	89.6	92.9	82.0	89.7	96.3	89.5	94.3	95.5	91.2	94.2	95.3	92.6	94.5	92.9	81.9	89.6
06	WOOD, PLASTICS & COMPOSITES	96.6	33.7	62.0	84.4	33.4	56.4	104.2	34.0	65.6	102.2	48.0	72.3	101.6	76.5	87.8	80.5	74.2	77.0
07	THERMAL & MOISTURE PROTECTION	102.8	72.9	89.8	102.9	75.1	90.8	104.1	72.7	90.4	103.6	78.5	92.7	103.2	83.3	94.6	102.6	76.7	91.3
08	OPENINGS	96.3	39.6	83.1	98.9	39.0	84.9	99.8	58.4	90.2	101.0	66.1	92.9	104.2	81.8	99.0	96.3	61.5	88.2
0920	Plaster & Gypsum Board	109.0	32.2	57.3	102.2	32.0	54.9	108.1	32.5	57.2	109.8	46.9	67.5	103.6	76.1	85.1	100.5	73.9	82.6
0950, 0980	Ceilings & Acoustic Treatment	95.5	32.2	52.8	98.8	32.0	53.7	98.6	32.5	54.0	100.4	46.9	64.3	97.7	76.1	83.1	95.5	73.9	80.9
0960	Flooring	90.7	43.8	77.6	86.5	46.4	75.3	98.2	31.2	79.4	90.5	72.7	85.5	93.9	75.0	88.6	85.2	43.8	73.6
0970, 0990	Wall Finishes & Painting/Coating	89.8	36.6	58.3	89.8	37.5	58.8	97.1	95.6	96.2	89.8	95.6	93.2	100.3	95.6	97.5	89.8	33.4	56.4
09	FINISHES	91.4	41.1	64.2	89.9	41.6	63.7	98.4	45.2	69.6	92.5	61.8	75.9	96.4	79.1	87.1	88.6	64.7	75.7
COVERS	DIVS. 10 - 14, 25, 28, 41, 43, 44, 46	100.0	82.0	96.0	100.0	81.9	96.0	100.0	82.3	96.1	100.0	83.5	96.3	100.0	87.0	97.1	100.0	85.8	96.8
21, 22, 23	FIRE SUPPRESSION, PLUMBING & HVAC	96.5	49.0	77.8	96.5	69.0	85.7	100.0	76.8	90.9	100.0	76.8	90.9	99.9	70.2	88.2	96.5	52.1	79.0
26, 27, 3370	ELECTRICAL, COMMUNICATIONS & UTIL.	99.4	63.4	81.7	101.1	39.8	70.9	103.9	47.7	76.2	97.6	47.7	73.0	102.6	63.4	83.3	98.6	63.4	81.2
MF2018	WEIGHTED AVERAGE	96.3	60.9	81.3	96.9	62.9	82.5	99.0	68.5	86.1	98.3	72.0	87.2	97.7	78.4	89.5	97.3	69.0	85.4

TENNESSEE

DIVISION		CHATTANOOGA 373-374 MAT.	INST.	TOTAL	COLUMBIA 384 MAT.	INST.	TOTAL	COOKEVILLE 385 MAT.	INST.	TOTAL	JACKSON 383 MAT.	INST.	TOTAL	JOHNSON CITY 376 MAT.	INST.	TOTAL	KNOXVILLE 377-379 MAT.	INST.	TOTAL
015433	CONTRACTOR EQUIPMENT		104.7	104.7		99.5	99.5		99.5	99.5		105.5	105.5		98.8	98.8		98.8	98.8
0241, 31 - 34	SITE & INFRASTRUCTURE, DEMOLITION	106.6	93.8	97.8	91.5	84.6	86.7	97.3	81.6	86.5	100.9	94.0	96.2	113.6	81.1	91.2	92.6	83.8	86.5
0310	Concrete Forming & Accessories	95.8	57.3	63.0	80.9	62.4	65.1	81.1	31.9	39.2	87.5	41.0	47.8	82.7	57.5	61.2	94.3	61.7	66.5
0320	Concrete Reinforcing	101.6	65.1	83.9	91.4	65.0	78.6	91.4	64.6	78.4	91.3	67.2	79.7	102.2	61.9	82.7	101.6	61.8	82.3
0330	Cast-in-Place Concrete	97.5	64.1	85.1	92.8	63.8	82.0	105.2	56.7	87.2	102.7	66.0	89.0	78.4	58.0	70.8	91.6	64.0	81.3
03	CONCRETE	93.1	62.8	79.8	92.4	65.1	80.4	102.4	48.7	78.8	93.2	56.5	77.0	102.7	60.2	84.0	90.7	64.2	79.0
04	MASONRY	98.6	56.3	72.6	111.4	55.5	77.0	106.7	39.4	65.3	111.7	44.4	70.4	111.0	43.2	69.3	76.3	50.7	60.6
05	METALS	93.0	88.1	91.6	93.5	88.9	92.1	93.5	87.6	91.8	96.0	89.5	94.1	90.3	86.4	89.2	93.4	86.5	91.4
06	WOOD, PLASTICS & COMPOSITES	106.5	56.3	78.9	67.8	63.4	65.4	68.0	29.5	46.8	82.9	39.7	59.1	78.7	62.6	69.8	93.1	62.6	76.4
07	THERMAL & MOISTURE PROTECTION	99.2	61.7	82.9	95.6	62.9	81.4	96.1	52.6	77.2	98.0	55.5	79.5	95.3	55.6	78.0	92.8	60.8	78.9
08	OPENINGS	102.0	57.5	91.6	91.0	53.3	82.2	91.1	35.5	78.1	97.7	44.3	85.2	98.3	59.5	89.3	95.7	55.2	86.2
0920	Plaster & Gypsum Board	80.0	55.6	63.6	87.4	62.9	70.9	87.4	28.0	47.4	90.0	38.5	55.3	101.0	62.1	74.8	109.1	62.1	77.5
0950, 0980	Ceilings & Acoustic Treatment	97.9	55.6	69.4	75.8	62.9	67.1	75.8	28.0	43.6	83.9	38.5	53.3	93.5	62.1	72.3	94.3	62.1	72.6
0960	Flooring	98.3	55.2	86.2	81.7	53.9	73.9	81.7	50.0	72.8	81.0	54.2	73.5	93.3	43.9	79.4	98.4	49.7	84.7
0970, 0990	Wall Finishes & Painting/Coating	98.0	59.5	75.2	85.0	56.5	68.1	85.0	56.5	68.1	87.0	56.5	68.9	94.9	58.7	73.5	94.9	57.6	72.8
09	FINISHES	94.4	56.6	73.9	86.2	60.1	72.1	86.8	36.2	59.4	86.1	43.4	63.0	98.5	55.3	75.1	91.4	58.9	73.8
COVERS	DIVS. 10 - 14, 25, 28, 41, 43, 44, 46	100.0	67.8	92.8	100.0	68.0	92.9	100.0	35.9	85.7	100.0	62.0	91.5	100.0	75.8	94.6	100.0	79.3	95.4
21, 22, 23	FIRE SUPPRESSION, PLUMBING & HVAC	100.1	59.3	84.1	97.8	73.6	88.3	97.8	67.1	85.8	100.1	61.4	84.9	100.0	55.5	82.5	100.0	60.5	84.5
26, 27, 3370	ELECTRICAL, COMMUNICATIONS & UTIL.	100.8	82.9	92.0	94.0	50.6	72.6	95.5	61.3	78.6	100.5	65.7	83.3	91.5	40.8	66.5	97.0	54.9	76.2
MF2018	WEIGHTED AVERAGE	98.0	68.1	85.3	94.8	66.5	82.8	96.2	56.9	79.6	97.6	61.4	82.3	98.1	58.6	81.4	94.8	63.7	81.6

TENNESSEE / TEXAS

DIVISION		MCKENZIE 382 MAT.	INST.	TOTAL	MEMPHIS 375, 380-381 MAT.	INST.	TOTAL	NASHVILLE 370-372 MAT.	INST.	TOTAL	ABILENE 795-796 MAT.	INST.	TOTAL	AMARILLO 790-791 MAT.	INST.	TOTAL	AUSTIN 786-787 MAT.	INST.	TOTAL
015433	CONTRACTOR EQUIPMENT		99.5	99.5		102.0	102.0		105.5	105.5		90.9	90.9		93.9	93.9		92.5	92.5
0241, 31 - 34	SITE & INFRASTRUCTURE, DEMOLITION	96.9	81.8	86.5	91.1	95.4	94.0	102.8	98.1	99.6	91.9	87.4	88.8	91.3	91.3	91.3	95.5	90.1	91.7
0310	Concrete Forming & Accessories	88.4	34.1	42.1	94.6	64.7	69.1	95.9	64.8	72.2	94.0	59.2	64.3	95.6	51.7	58.2	86.3	53.0	59.3
0320	Concrete Reinforcing	91.5	65.5	78.9	110.9	66.0	89.2	107.1	66.8	87.6	88.5	51.7	70.7	95.3	49.6	73.2	86.8	47.2	67.6
0330	Cast-in-Place Concrete	102.9	57.9	86.2	96.5	76.7	89.1	88.2	69.0	81.1	84.6	64.9	77.3	83.2	65.7	76.7	94.0	65.9	83.5
03	CONCRETE	101.0	50.2	78.7	96.5	70.2	84.9	92.9	69.4	82.6	84.5	60.7	74.1	87.1	57.2	74.0	89.4	57.4	75.4
04	MASONRY	110.5	45.1	70.3	92.6	56.3	70.3	87.7	55.9	68.2	99.9	60.6	75.8	97.6	60.2	74.6	91.7	60.7	72.7
05	METALS	93.5	88.0	91.9	87.2	82.5	85.8	101.6	85.1	96.7	107.1	70.4	96.3	101.9	68.7	92.1	100.7	67.4	90.9
06	WOOD, PLASTICS & COMPOSITES	76.6	31.7	51.9	98.9	65.4	80.4	100.1	70.1	83.6	99.2	60.8	78.1	102.6	51.0	74.2	92.9	52.6	70.7
07	THERMAL & MOISTURE PROTECTION	96.1	53.6	77.6	93.0	66.4	81.4	96.9	65.0	83.0	98.3	63.4	83.1	100.1	62.3	83.6	95.1	63.8	81.5
08	OPENINGS	91.1	37.2	78.5	99.5	61.3	90.6	101.0	67.5	93.2	101.8	56.7	91.2	103.2	50.8	91.0	101.1	49.6	89.1
0920	Plaster & Gypsum Board	90.9	30.3	50.1	90.9	64.5	73.1	95.9	69.3	78.0	88.2	60.1	69.3	100.7	50.0	66.6	88.9	51.6	63.8
0950, 0980	Ceilings & Acoustic Treatment	75.8	30.3	45.1	90.8	64.5	73.0	94.1	69.3	77.4	99.2	60.1	72.9	100.7	50.0	66.5	93.3	51.6	65.2
0960	Flooring	84.5	53.9	75.9	97.8	55.2	85.9	97.1	57.5	86.0	93.2	67.6	86.0	93.7	63.8	85.3	94.1	63.8	85.6
0970, 0990	Wall Finishes & Painting/Coating	85.0	44.2	60.8	93.2	59.3	73.1	100.7	68.6	81.7	92.8	51.7	68.5	91.0	51.7	67.7	95.7	44.2	65.2
09	FINISHES	88.0	37.5	60.7	95.9	61.9	77.5	97.8	66.1	80.6	86.1	60.1	72.0	92.6	53.5	71.4	91.1	53.6	70.8
COVERS	DIVS. 10 - 14, 25, 28, 41, 43, 44, 46	100.0	24.0	83.1	100.0	80.5	95.7	100.0	81.1	95.8	100.0	78.9	95.3	100.0	72.7	93.9	100.0	76.5	94.8
21, 22, 23	FIRE SUPPRESSION, PLUMBING & HVAC	97.8	58.6	82.4	100.1	70.3	88.4	99.9	75.1	90.2	100.1	52.1	81.3	99.9	51.1	80.7	99.9	58.7	83.8
26, 27, 3370	ELECTRICAL, COMMUNICATIONS & UTIL.	95.3	55.4	75.6	102.9	63.6	83.5	93.5	63.2	78.6	97.9	54.0	76.3	100.9	58.9	80.2	97.5	57.2	77.6
MF2018	WEIGHTED AVERAGE	96.3	55.1	78.9	96.6	69.8	85.3	98.0	71.8	86.9	97.8	61.4	82.4	98.2	60.0	82.1	97.2	61.4	82.1

TEXAS

DIVISION		BEAUMONT 776 - 777			BROWNWOOD 768			BRYAN 778			CHILDRESS 792			CORPUS CHRISTI 783 - 784			DALLAS 752 - 753		
		MAT.	INST.	TOTAL	MAT.	INST.	TOTAL	MAT.	INST.	TOTAL	MAT.	INST.	TOTAL	MAT.	INST.	TOTAL	MAT.	INST.	TOTAL
015433	CONTRACTOR EQUIPMENT		95.3	95.3		90.9	90.9		95.3	95.3		90.9	90.9		100.6	100.6		106.9	106.9
0241, 31 - 34	SITE & INFRASTRUCTURE, DEMOLITION	90.6	91.8	91.4	101.3	87.3	91.7	81.5	91.9	88.7	101.8	85.8	90.8	144.2	82.8	101.8	106.0	97.6	100.2
0310	Concrete Forming & Accessories	102.3	54.6	61.7	97.6	58.7	64.4	82.1	58.2	61.7	92.6	58.6	63.6	98.4	52.2	59.0	95.0	62.3	67.1
0320	Concrete Reinforcing	85.7	62.0	74.2	81.9	51.1	67.0	87.9	51.2	70.2	88.7	51.0	70.5	79.8	50.9	65.8	91.4	52.3	72.5
0330	Cast-in-Place Concrete	90.7	65.2	81.2	96.8	58.6	82.6	72.6	59.6	67.8	86.7	58.6	76.3	114.1	67.1	96.6	92.6	71.4	84.7
03	CONCRETE	91.0	60.7	77.7	90.3	58.2	76.2	75.8	58.5	68.2	93.4	58.2	77.9	95.2	59.1	79.4	93.3	65.3	81.0
04	MASONRY	102.9	61.7	77.6	122.8	60.6	84.6	143.3	60.6	92.5	103.5	60.2	76.9	81.4	60.7	68.7	99.5	60.2	75.3
05	METALS	99.4	76.2	92.6	102.6	69.8	92.9	99.5	72.1	91.4	104.4	69.6	94.2	97.3	84.4	93.5	101.7	82.6	96.1
06	WOOD, PLASTICS & COMPOSITES	113.6	54.1	80.9	105.8	60.8	81.0	78.6	59.3	68.0	98.6	60.8	77.8	115.6	51.5	80.3	100.8	63.6	80.3
07	THERMAL & MOISTURE PROTECTION	92.8	64.3	80.4	92.0	61.9	78.9	85.2	63.8	75.9	99.0	60.4	82.2	97.7	62.2	82.2	87.3	67.2	78.6
08	OPENINGS	95.0	55.0	85.6	100.4	56.6	90.2	95.7	56.0	86.4	96.1	56.6	86.9	101.3	50.0	89.4	96.5	59.5	87.9
0920	Plaster & Gypsum Board	101.0	53.2	68.9	82.6	60.1	67.5	88.2	58.7	68.3	87.6	60.1	69.1	96.5	50.5	65.5	92.7	62.5	72.4
0950, 0980	Ceilings & Acoustic Treatment	105.3	53.2	70.2	86.6	60.1	68.7	96.7	58.7	71.1	96.8	60.1	72.1	98.8	50.5	66.2	102.4	62.5	75.5
0960	Flooring	116.9	74.6	105.0	75.7	54.6	69.8	86.7	67.0	81.1	91.6	54.6	81.2	107.4	67.7	96.3	94.2	69.4	87.2
0970, 0990	Wall Finishes & Painting/Coating	94.9	53.9	70.6	92.8	51.7	68.4	92.2	57.4	71.6	92.8	51.7	68.5	110.1	44.2	71.1	96.1	54.8	71.6
09	FINISHES	98.0	57.8	76.2	77.6	57.4	66.7	85.4	59.4	71.3	86.2	57.4	70.6	101.9	53.8	75.9	97.9	62.7	78.8
COVERS	DIVS. 10 - 14, 25, 28, 41, 43, 44, 46	100.0	77.8	95.0	100.0	73.4	94.1	100.0	77.0	94.9	100.0	73.4	94.1	100.0	77.8	95.0	100.0	82.5	96.1
21, 22, 23	FIRE SUPPRESSION, PLUMBING & HVAC	100.0	63.0	85.5	96.5	49.5	78.0	96.5	62.4	83.1	96.6	51.7	79.0	100.0	48.7	79.9	99.9	62.3	85.1
26, 27, 3370	ELECTRICAL, COMMUNICATIONS & UTIL.	100.1	63.4	82.0	98.6	46.1	72.7	97.8	63.4	80.8	97.9	55.8	77.1	94.0	63.4	78.9	95.2	63.2	79.4
MF2018	WEIGHTED AVERAGE	97.9	65.5	84.2	97.4	58.7	81.1	95.2	64.9	82.4	97.6	60.3	81.8	98.9	61.4	83.1	98.2	68.0	85.4

TEXAS

DIVISION		DEL RIO 788			DENTON 762			EASTLAND 764			EL PASO 798 - 799, 885			FORT WORTH 760 - 761			GALVESTON 775		
		MAT.	INST.	TOTAL	MAT.	INST.	TOTAL	MAT.	INST.	TOTAL	MAT.	INST.	TOTAL	MAT.	INST.	TOTAL	MAT.	INST.	TOTAL
015433	CONTRACTOR EQUIPMENT		89.9	89.9		100.2	100.2		90.9	90.9		93.9	93.9		93.9	93.9		107.9	107.9
0241, 31 - 34	SITE & INFRASTRUCTURE, DEMOLITION	122.2	85.5	96.9	101.4	79.7	86.4	104.0	85.5	91.2	89.7	92.7	91.8	97.4	92.7	94.2	108.4	90.6	96.1
0310	Concrete Forming & Accessories	95.1	51.0	57.5	105.0	58.8	65.7	98.4	58.6	64.4	95.6	56.7	62.5	98.2	59.5	65.2	90.9	55.8	61.0
0320	Concrete Reinforcing	80.4	47.6	64.5	83.2	51.1	67.7	82.1	51.1	67.1	94.2	51.1	73.4	88.1	50.9	70.1	87.5	59.7	74.0
0330	Cast-in-Place Concrete	122.9	58.4	98.9	74.9	59.7	69.2	102.3	58.5	86.0	73.9	68.0	71.7	89.3	65.9	80.6	96.4	60.6	83.1
03	CONCRETE	115.8	54.0	88.6	70.6	59.7	65.8	94.8	58.1	78.7	82.7	60.5	72.9	86.1	61.0	75.1	92.1	60.3	78.1
04	MASONRY	95.4	60.5	74.0	131.0	59.3	86.9	92.7	60.6	73.0	91.6	62.7	73.8	92.5	59.3	72.1	99.0	60.7	75.5
05	METALS	97.2	67.2	88.4	102.1	84.9	97.1	102.4	69.6	92.7	102.1	69.4	92.5	104.3	69.5	94.1	101.2	91.4	98.3
06	WOOD, PLASTICS & COMPOSITES	96.1	50.4	71.0	117.8	60.9	86.5	113.0	60.8	84.3	90.2	57.1	72.0	99.6	60.9	78.3	95.3	55.9	73.6
07	THERMAL & MOISTURE PROTECTION	94.7	62.0	80.5	90.0	62.4	78.0	92.4	61.9	79.1	91.9	64.4	80.0	89.7	64.4	78.7	84.6	64.4	75.8
08	OPENINGS	96.6	47.8	85.2	118.7	56.2	104.2	70.6	56.6	67.4	96.3	52.8	86.2	101.2	56.6	90.8	99.7	56.0	89.5
0920	Plaster & Gypsum Board	93.3	49.5	63.8	87.3	60.1	69.0	82.6	60.1	67.5	97.3	56.2	69.6	89.8	60.1	69.8	94.8	55.0	68.0
0950, 0980	Ceilings & Acoustic Treatment	95.3	49.5	64.4	90.7	60.1	70.1	86.6	60.1	68.7	94.2	56.2	68.6	92.7	60.1	70.7	100.8	55.0	69.9
0960	Flooring	91.9	54.6	81.4	71.9	54.6	67.1	95.9	54.6	84.3	97.3	69.7	89.5	94.5	64.5	86.0	102.4	67.0	92.5
0970, 0990	Wall Finishes & Painting/Coating	97.8	42.6	65.1	103.4	48.9	71.1	94.2	51.7	69.0	98.6	49.6	69.6	101.1	51.8	72.0	104.6	53.7	74.5
09	FINISHES	94.5	50.3	70.6	76.3	57.2	66.0	84.1	57.4	69.6	92.3	58.1	73.8	89.6	59.5	73.3	95.4	56.9	74.6
COVERS	DIVS. 10 - 14, 25, 28, 41, 43, 44, 46	100.0	73.6	94.1	100.0	79.2	95.4	100.0	71.7	93.7	100.0	76.6	94.8	100.0	79.2	95.4	100.0	78.3	95.2
21, 22, 23	FIRE SUPPRESSION, PLUMBING & HVAC	96.5	59.4	81.9	96.5	55.8	80.5	96.5	49.5	78.0	99.8	65.3	86.3	99.9	55.4	82.4	96.5	62.5	83.2
26, 27, 3370	ELECTRICAL, COMMUNICATIONS & UTIL.	95.9	59.8	78.1	101.0	58.7	80.1	98.5	58.6	78.8	97.5	50.5	74.3	100.2	58.6	79.7	99.6	63.5	81.8
MF2018	WEIGHTED AVERAGE	99.7	60.4	83.1	97.3	62.9	82.7	94.2	60.3	79.8	95.9	63.6	82.3	97.6	62.9	82.9	97.7	66.5	84.5

TEXAS

DIVISION		GIDDINGS 789			GREENVILLE 754			HOUSTON 770 - 772			HUNTSVILLE 773			LAREDO 780			LONGVIEW 756		
		MAT.	INST.	TOTAL	MAT.	INST.	TOTAL	MAT.	INST.	TOTAL	MAT.	INST.	TOTAL	MAT.	INST.	TOTAL	MAT.	INST.	TOTAL
015433	CONTRACTOR EQUIPMENT		89.9	89.9		100.4	100.4		103.3	103.3		95.3	95.3		89.9	89.9		91.8	91.8
0241, 31 - 34	SITE & INFRASTRUCTURE, DEMOLITION	106.8	85.8	92.3	96.4	82.4	86.7	107.7	95.5	99.3	96.8	91.6	93.2	100.2	86.2	90.5	93.7	88.6	90.2
0310	Concrete Forming & Accessories	92.6	51.1	57.3	86.9	57.1	61.5	96.0	54.9	64.4	89.2	54.1	59.2	95.1	52.1	58.5	82.9	58.8	62.3
0320	Concrete Reinforcing	80.9	48.4	65.2	91.9	51.1	72.2	87.3	52.2	70.3	88.2	51.1	70.3	80.4	50.9	66.1	90.9	50.7	71.5
0330	Cast-in-Place Concrete	104.1	58.7	87.2	92.1	59.6	80.0	89.6	68.9	81.9	99.9	59.5	84.9	87.8	65.0	79.3	107.1	58.5	89.0
03	CONCRETE	92.6	54.3	75.8	86.7	58.7	74.4	91.4	62.5	78.7	98.7	56.6	80.2	87.3	57.4	74.2	102.7	58.1	83.1
04	MASONRY	102.2	60.6	76.7	158.6	59.3	97.5	98.0	63.1	76.6	141.3	60.6	91.7	88.9	60.6	71.5	154.2	59.2	95.8
05	METALS	96.7	68.2	88.4	99.2	83.2	94.5	103.9	77.3	96.1	99.4	71.8	91.3	99.6	69.4	90.7	91.8	68.4	84.9
06	WOOD, PLASTICS & COMPOSITES	95.3	50.4	70.6	87.2	58.4	71.4	100.6	59.2	77.8	88.1	54.1	69.4	96.1	51.4	71.5	81.2	60.8	70.0
07	THERMAL & MOISTURE PROTECTION	95.0	62.7	81.0	87.3	61.5	76.1	86.3	68.5	78.5	86.5	63.2	76.4	93.8	63.0	80.4	89.1	60.9	76.8
08	OPENINGS	95.8	48.7	84.8	92.2	55.3	83.6	106.6	56.4	94.9	95.7	52.5	85.6	96.7	49.8	85.8	83.4	56.2	77.0
0920	Plaster & Gypsum Board	92.4	49.5	63.5	82.8	57.5	65.8	94.0	58.0	69.8	92.3	53.2	66.0	94.1	50.5	64.7	81.7	60.1	67.2
0950, 0980	Ceilings & Acoustic Treatment	95.3	49.5	64.4	98.1	57.5	70.7	97.9	58.0	71.0	96.7	53.2	67.4	98.6	50.5	66.1	93.2	60.1	70.9
0960	Flooring	91.7	54.6	81.2	87.7	54.6	78.4	103.5	71.2	94.4	91.3	54.6	81.0	91.8	63.8	83.9	93.7	54.6	82.7
0970, 0990	Wall Finishes & Painting/Coating	97.8	44.2	66.1	90.4	51.7	67.5	102.1	58.6	76.3	92.2	57.4	71.6	97.8	44.2	66.1	82.1	48.9	62.4
09	FINISHES	93.1	50.5	70.0	92.3	56.1	72.7	102.7	60.8	80.1	88.3	54.1	69.8	93.4	52.9	71.5	98.1	57.2	76.0
COVERS	DIVS. 10 - 14, 25, 28, 41, 43, 44, 46	100.0	70.3	93.4	100.0	76.5	94.8	100.0	83.1	96.2	100.0	69.8	93.3	100.0	75.7	94.6	100.0	79.0	95.3
21, 22, 23	FIRE SUPPRESSION, PLUMBING & HVAC	96.5	61.4	82.7	96.5	58.0	81.3	100.0	65.3	86.3	96.5	62.4	83.1	100.0	58.9	83.8	96.4	57.5	81.1
26, 27, 3370	ELECTRICAL, COMMUNICATIONS & UTIL.	93.1	55.6	74.6	92.1	58.7	75.6	101.8	67.8	85.1	97.8	59.1	78.7	96.0	57.1	76.8	92.6	50.9	72.0
MF2018	WEIGHTED AVERAGE	96.1	60.4	81.0	97.4	62.9	82.8	100.4	68.2	86.8	98.7	62.8	83.6	96.4	61.1	81.5	97.5	61.0	82.1

TEXAS

DIVISION		LUBBOCK 793-794 MAT.	INST.	TOTAL	LUFKIN 759 MAT.	INST.	TOTAL	MCALLEN 785 MAT.	INST.	TOTAL	MCKINNEY 750 MAT.	INST.	TOTAL	MIDLAND 797 MAT.	INST.	TOTAL	ODESSA 797 MAT.	INST.	TOTAL
015433	CONTRACTOR EQUIPMENT		102.9	102.9		91.8	91.8		100.7	100.7		100.4	100.4		102.9	102.9		90.9	90.9
0241, 31 - 34	SITE & INFRASTRUCTURE, DEMOLITION	115.0	86.7	95.5	88.6	90.0	89.5	149.0	82.9	103.3	92.8	82.4	85.6	117.7	85.8	95.6	92.1	88.1	89.3
0310	Concrete Forming & Accessories	92.9	53.0	58.9	85.9	54.1	58.8	99.1	51.2	58.3	86.0	57.1	61.4	96.4	59.1	64.6	93.9	59.2	64.3
0320	Concrete Reinforcing	89.7	51.8	71.4	92.6	66.7	80.1	80.0	50.9	65.9	91.9	51.1	72.2	90.7	51.7	71.8	88.5	51.6	70.7
0330	Cast-in-Place Concrete	84.7	68.2	78.6	95.8	58.3	81.8	124.0	59.7	100.1	86.3	59.6	76.4	90.0	65.9	81.1	84.6	64.9	77.3
03	CONCRETE	83.1	60.1	73.0	95.1	58.6	79.0	102.8	56.2	82.3	82.1	58.8	71.9	87.1	62.1	76.1	84.5	60.7	74.1
04	MASONRY	99.2	61.0	75.7	117.3	60.5	82.4	95.1	60.7	74.0	170.8	59.3	102.3	116.2	60.2	81.8	99.9	60.2	75.5
05	METALS	110.9	85.8	103.5	99.4	73.3	91.7	97.1	84.2	93.3	99.2	83.3	94.5	109.2	85.4	102.2	106.4	70.4	95.8
06	WOOD, PLASTICS & COMPOSITES	98.6	52.1	73.0	88.6	54.2	69.6	114.3	50.5	79.2	86.1	58.4	70.9	103.0	60.9	79.8	99.2	60.8	78.1
07	THERMAL & MOISTURE PROTECTION	88.2	63.5	77.5	88.8	60.0	76.3	97.6	62.9	82.5	87.1	61.5	76.0	88.5	63.3	77.5	98.3	62.7	82.8
08	OPENINGS	108.6	52.4	95.5	63.3	56.6	61.7	100.4	48.6	88.3	92.2	55.3	83.6	110.0	56.7	97.6	101.8	56.7	91.2
0920	Plaster & Gypsum Board	88.4	51.1	63.3	80.6	53.2	62.2	97.5	49.5	65.2	82.8	57.5	65.8	90.0	60.1	69.9	88.2	60.1	69.3
0950, 0980	Ceilings & Acoustic Treatment	100.0	51.1	67.0	88.3	53.2	64.7	98.6	49.5	65.5	98.1	57.5	70.7	97.6	60.1	72.3	93.2	60.1	72.9
0960	Flooring	87.5	69.0	82.3	126.2	54.6	106.1	106.9	72.8	97.3	87.2	54.6	78.1	88.6	63.8	81.7	93.2	63.8	84.9
0970, 0990	Wall Finishes & Painting/Coating	103.4	53.7	74.0	82.1	51.7	64.1	110.1	42.6	70.2	90.4	51.7	67.5	103.4	51.7	72.8	92.8	51.7	68.5
09	FINISHES	88.5	55.5	70.6	106.5	53.9	78.0	102.4	54.1	76.3	91.9	56.1	72.5	89.0	59.4	72.9	86.1	59.3	71.6
COVERS	DIVS. 10 - 14, 25, 28, 41, 43, 44, 46	100.0	78.3	95.2	100.0	70.1	93.3	100.0	75.8	94.6	100.0	79.1	95.3	100.0	73.8	94.1	100.0	73.5	94.1
21, 22, 23	FIRE SUPPRESSION, PLUMBING & HVAC	99.7	53.1	81.4	96.4	58.1	81.4	96.5	48.7	77.7	96.5	60.2	82.2	96.1	47.4	77.0	100.1	52.5	81.4
26, 27, 3370	ELECTRICAL, COMMUNICATIONS & UTIL.	96.7	60.1	78.6	93.9	61.4	77.8	93.8	31.6	63.1	92.2	58.7	75.6	96.7	58.9	78.1	98.0	58.9	78.7
MF2018	WEIGHTED AVERAGE	99.2	62.9	83.9	94.7	62.6	81.1	99.7	56.5	81.4	97.2	63.5	82.9	99.7	62.2	83.8	97.7	61.8	82.6

TEXAS

DIVISION		PALESTINE 758 MAT.	INST.	TOTAL	SAN ANGELO 769 MAT.	INST.	TOTAL	SAN ANTONIO 781-782 MAT.	INST.	TOTAL	TEMPLE 765 MAT.	INST.	TOTAL	TEXARKANA 755 MAT.	INST.	TOTAL	TYLER 757 MAT.	INST.	TOTAL
015433	CONTRACTOR EQUIPMENT		91.8	91.8		90.9	90.9		94.1	94.1		90.9	90.9		91.8	91.8		91.8	91.8
0241, 31 - 34	SITE & INFRASTRUCTURE, DEMOLITION	94.2	88.8	90.5	97.8	87.7	90.8	101.5	93.7	96.1	86.7	87.6	87.3	83.4	91.1	88.7	92.9	88.7	90.0
0310	Concrete Forming & Accessories	77.5	58.8	61.5	97.9	51.4	58.3	94.7	52.6	58.8	101.2	51.2	58.5	93.0	59.4	64.3	87.6	58.9	63.1
0320	Concrete Reinforcing	90.2	51.0	71.3	81.8	51.6	67.2	89.6	47.7	69.3	82.0	50.8	66.9	90.1	50.8	71.1	90.9	50.8	71.5
0330	Cast-in-Place Concrete	87.5	58.5	76.7	91.4	64.9	81.5	85.7	68.3	79.2	74.9	58.6	68.8	88.2	64.9	79.5	105.2	58.6	87.8
03	CONCRETE	97.1	58.1	80.0	86.1	57.2	73.4	88.0	58.0	74.8	73.4	54.7	65.2	88.0	60.6	75.9	102.3	58.2	82.9
04	MASONRY	111.9	59.2	79.5	119.4	60.6	83.3	89.7	60.8	71.9	130.1	60.6	87.4	175.3	59.2	103.9	164.4	59.2	99.7
05	METALS	99.1	68.5	90.1	102.7	70.3	93.2	102.0	65.4	91.3	102.4	69.2	92.7	91.7	69.0	85.0	98.6	68.5	90.1
06	WOOD, PLASTICS & COMPOSITES	79.3	60.8	69.1	106.1	50.4	75.5	98.1	51.7	72.5	115.7	50.4	79.8	93.7	60.8	75.6	90.3	60.8	74.1
07	THERMAL & MOISTURE PROTECTION	89.4	61.8	77.4	91.8	62.6	79.1	86.8	65.6	77.5	91.3	61.7	78.4	88.6	63.0	77.5	89.2	61.8	77.3
08	OPENINGS	63.3	56.6	61.7	100.5	51.0	88.9	102.1	49.5	89.8	67.3	50.8	63.4	83.3	56.5	77.1	63.2	56.5	61.7
0920	Plaster & Gypsum Board	79.0	60.1	66.3	82.6	49.5	60.3	93.3	50.5	64.5	82.6	49.5	60.3	86.8	60.1	68.8	80.6	60.1	66.8
0950, 0980	Ceilings & Acoustic Treatment	88.3	60.1	69.3	86.6	49.5	61.6	95.2	50.5	65.1	86.6	49.5	61.6	93.2	60.1	70.9	88.3	60.1	69.3
0960	Flooring	116.4	54.6	99.0	75.8	54.6	69.9	101.2	67.7	91.8	97.5	54.6	85.4	102.7	63.8	91.7	128.7	54.6	107.8
0970, 0990	Wall Finishes & Painting/Coating	82.1	51.7	64.1	92.8	51.7	68.4	99.4	44.2	66.7	94.2	44.2	64.6	82.1	51.7	64.1	107.5	57.5	80.4
09	FINISHES	104.0	57.5	78.8	77.4	51.3	63.3	104.2	54.0	77.0	83.3	50.5	65.5	100.9	59.3	78.4	107.5	57.5	80.4
COVERS	DIVS. 10 - 14, 25, 28, 41, 43, 44, 46	100.0	73.5	94.1	100.0	76.0	94.7	100.0	78.0	95.1	100.0	72.3	93.8	100.0	79.1	95.3	100.0	79.0	95.3
21, 22, 23	FIRE SUPPRESSION, PLUMBING & HVAC	96.4	56.7	80.8	96.5	52.2	79.1	100.0	59.8	84.2	96.5	53.2	79.5	96.4	55.5	80.3	96.4	58.2	81.4
26, 27, 3370	ELECTRICAL, COMMUNICATIONS & UTIL.	90.1	48.4	69.5	102.8	54.6	79.1	97.5	59.8	78.9	99.8	52.0	76.2	93.7	54.5	74.4	92.6	53.7	73.4
MF2018	WEIGHTED AVERAGE	94.2	60.4	79.9	97.1	59.4	81.2	98.3	62.3	83.1	92.5	58.5	78.2	96.7	62.0	82.1	97.8	61.6	82.5

DIVISION		TEXAS VICTORIA 779 MAT.	INST.	TOTAL	WACO 766-767 MAT.	INST.	TOTAL	WAXAHACHIE 751 MAT.	INST.	TOTAL	WHARTON 774 MAT.	INST.	TOTAL	WICHITA FALLS 763 MAT.	INST.	TOTAL	UTAH LOGAN 843 MAT.	INST.	TOTAL
015433	CONTRACTOR EQUIPMENT		106.6	106.6		90.9	90.9		100.4	100.4		107.9	107.9		90.9	90.9		93.9	93.9
0241, 31 - 34	SITE & INFRASTRUCTURE, DEMOLITION	113.4	88.0	95.8	95.1	88.0	90.2	94.7	82.5	86.3	118.7	90.0	98.8	95.8	88.1	90.5	99.8	89.6	92.7
0310	Concrete Forming & Accessories	90.6	51.3	57.1	99.8	59.2	65.2	86.0	59.0	63.0	85.7	53.1	57.9	99.8	59.2	65.2	103.5	67.2	72.6
0320	Concrete Reinforcing	84.0	50.9	68.0	81.6	47.2	65.0	91.9	51.1	72.2	87.4	50.9	69.7	81.6	51.7	67.1	108.0	84.4	96.6
0330	Cast-in-Place Concrete	108.3	60.5	90.5	81.0	64.9	75.0	91.1	59.8	79.5	111.3	59.5	92.0	86.5	65.0	78.5	86.4	74.3	81.9
03	CONCRETE	99.7	56.7	80.8	79.6	60.0	71.0	85.8	59.7	74.3	103.8	57.1	83.3	82.2	60.8	72.8	105.4	73.2	91.2
04	MASONRY	116.9	60.7	82.4	90.9	60.6	72.3	159.2	59.3	97.8	100.3	60.6	75.9	91.4	60.1	72.2	109.5	60.5	79.4
05	METALS	99.4	87.8	96.0	104.8	68.4	94.1	99.2	83.6	94.6	101.2	87.4	97.1	104.7	70.4	94.6	109.5	82.5	101.5
06	WOOD, PLASTICS & COMPOSITES	98.4	50.5	72.1	113.9	60.8	84.7	86.1	61.0	72.3	88.4	52.7	68.8	113.9	60.8	84.7	85.9	67.1	75.5
07	THERMAL & MOISTURE PROTECTION	88.0	61.1	76.3	92.2	63.7	79.8	87.2	61.6	76.1	85.0	63.7	75.7	92.2	62.9	79.5	103.0	68.8	88.1
08	OPENINGS	99.5	51.1	88.2	78.4	55.6	73.1	92.2	56.7	83.9	99.7	52.3	88.6	78.4	56.7	73.4	92.2	66.1	86.1
0920	Plaster & Gypsum Board	90.9	49.5	63.0	82.7	60.1	67.5	83.2	60.1	67.7	90.4	51.7	64.4	82.7	60.1	67.5	80.6	66.2	70.9
0950, 0980	Ceilings & Acoustic Treatment	101.6	49.5	66.5	87.4	60.1	69.0	99.7	60.1	73.0	100.8	51.7	67.7	87.4	60.1	69.0	104.3	66.2	78.7
0960	Flooring	100.7	54.6	87.7	96.8	63.8	87.5	87.2	54.6	78.1	99.5	66.6	90.2	97.4	71.9	90.2	97.2	60.5	86.9
0970, 0990	Wall Finishes & Painting/Coating	104.8	57.4	76.8	94.2	51.7	69.0	90.4	51.7	67.5	104.6	55.7	75.7	96.1	51.7	69.8	92.6	60.7	73.7
09	FINISHES	92.4	52.0	70.6	83.7	59.3	70.5	92.5	57.6	73.6	94.7	55.2	73.3	84.1	60.9	71.5	93.9	64.9	78.2
COVERS	DIVS. 10 - 14, 25, 28, 41, 43, 44, 46	100.0	69.7	93.2	100.0	78.9	95.3	100.0	79.4	95.4	100.0	70.0	93.3	100.0	73.5	94.1	100.0	84.4	96.5
21, 22, 23	FIRE SUPPRESSION, PLUMBING & HVAC	96.5	62.3	83.1	100.0	58.7	83.8	96.5	58.0	81.4	96.5	61.9	82.9	100.0	52.1	81.2	100.0	69.2	87.9
26, 27, 3370	ELECTRICAL, COMMUNICATIONS & UTIL.	104.9	54.0	79.8	103.0	54.1	78.9	92.1	58.7	75.6	103.9	59.2	81.9	105.0	56.0	80.8	95.2	71.3	83.4
MF2018	WEIGHTED AVERAGE	99.7	62.9	84.2	94.4	62.4	80.9	97.2	63.5	83.0	99.8	64.3	84.8	95.0	61.6	80.9	100.9	71.7	88.5

City Cost Indexes

UTAH / VERMONT

	DIVISION	OGDEN 842, 844 MAT.	INST.	TOTAL	PRICE 845 MAT.	INST.	TOTAL	PROVO 846 - 847 MAT.	INST.	TOTAL	SALT LAKE CITY 840 - 841 MAT.	INST.	TOTAL	BELLOWS FALLS 051 MAT.	INST.	TOTAL	BENNINGTON 052 MAT.	INST.	TOTAL
015433	CONTRACTOR EQUIPMENT		93.9	93.9		92.9	92.9		92.9	92.9		93.8	93.8		95.7	95.7		95.7	95.7
0241, 31 - 34	SITE & INFRASTRUCTURE, DEMOLITION	87.3	89.6	88.9	97.1	87.4	90.4	95.9	88.0	90.5	87.0	89.5	88.7	86.8	96.1	93.3	86.2	96.1	93.1
0310	Concrete Forming & Accessories	103.5	67.2	72.6	106.0	64.2	70.4	105.3	67.2	72.9	103.5	67.2	72.9	96.5	101.7	100.9	94.0	101.6	100.4
0320	Concrete Reinforcing	107.6	84.4	96.4	115.8	84.0	100.4	116.8	84.4	101.1	110.0	84.4	97.6	85.1	86.3	85.7	85.1	86.3	85.6
0330	Cast-in-Place Concrete	87.7	74.3	82.7	86.5	71.6	80.9	86.5	74.3	81.9	95.8	74.3	87.8	89.6	115.9	99.4	89.6	115.8	99.4
03	CONCRETE	94.4	73.2	85.1	106.6	70.8	90.8	104.9	73.2	91.0	114.2	73.2	96.2	88.7	103.7	95.3	88.5	103.6	95.2
04	MASONRY	104.0	60.5	77.3	115.1	65.1	84.4	115.3	60.5	81.6	118.1	60.5	82.7	96.3	86.2	90.1	105.1	86.2	93.5
05	METALS	110.0	82.5	101.9	106.5	82.5	99.5	107.5	82.5	100.1	113.7	82.5	104.5	95.2	91.1	94.0	95.1	90.9	93.9
06	WOOD, PLASTICS & COMPOSITES	85.9	67.1	75.5	89.2	64.0	75.3	87.5	67.1	76.3	87.8	67.1	76.4	102.8	106.8	105.0	100.1	106.8	103.8
07	THERMAL & MOISTURE PROTECTION	101.8	68.8	87.4	104.8	68.3	88.9	104.9	68.8	89.2	110.0	68.8	92.1	100.9	89.1	95.8	100.9	89.1	95.8
08	OPENINGS	92.2	66.1	86.1	96.1	76.3	91.5	96.1	66.1	89.1	94.1	66.1	87.5	100.1	97.5	99.5	100.1	97.5	99.5
0920	Plaster & Gypsum Board	80.6	66.2	70.9	83.4	63.1	69.8	81.2	66.2	71.1	90.9	66.2	74.3	108.9	106.7	107.4	107.9	106.7	107.1
0950, 0980	Ceilings & Acoustic Treatment	104.3	66.2	78.7	104.3	63.1	76.5	104.3	66.2	78.7	97.1	66.2	76.3	97.2	106.7	103.6	97.2	106.7	103.6
0960	Flooring	95.2	60.5	85.4	98.4	56.1	86.5	98.2	60.5	87.6	99.0	60.5	88.1	92.0	98.9	94.0	91.2	98.9	93.4
0970, 0990	Wall Finishes & Painting/Coating	92.6	60.7	73.7	92.6	60.7	73.7	92.6	60.7	73.7	95.7	60.7	75.0	86.4	101.5	95.3	86.4	101.5	95.3
09	FINISHES	92.0	64.9	77.3	95.0	61.7	77.0	94.5	64.9	78.5	93.7	64.9	78.1	90.6	101.8	96.7	90.3	101.8	96.5
COVERS	DIVS. 10 - 14, 25, 28, 41, 43, 44, 46	100.0	84.4	96.5	100.0	81.6	95.9	100.0	84.4	96.5	100.0	84.4	96.5	100.0	91.8	98.2	100.0	91.7	98.2
21, 22, 23	FIRE SUPPRESSION, PLUMBING & HVAC	100.0	69.2	87.9	98.1	63.2	84.4	100.0	69.2	87.9	100.1	69.2	88.0	96.7	87.6	93.1	96.7	87.6	93.1
26, 27, 3370	ELECTRICAL, COMMUNICATIONS & UTIL.	95.5	71.3	83.6	100.2	68.8	84.7	95.7	68.6	82.3	98.1	71.3	84.9	108.3	79.5	94.1	108.3	54.5	81.8
MF2018	WEIGHTED AVERAGE	98.8	71.7	87.4	101.4	70.0	88.1	101.2	71.2	88.5	103.5	71.7	90.1	96.7	92.1	94.8	97.0	88.6	93.4

VERMONT

	DIVISION	BRATTLEBORO 053 MAT.	INST.	TOTAL	BURLINGTON 054 MAT.	INST.	TOTAL	GUILDHALL 059 MAT.	INST.	TOTAL	MONTPELIER 056 MAT.	INST.	TOTAL	RUTLAND 057 MAT.	INST.	TOTAL	ST. JOHNSBURY 058 MAT.	INST.	TOTAL
015433	CONTRACTOR EQUIPMENT		95.7	95.7		99.0	99.0		95.7	95.7		99.0	99.0		95.7	95.7		95.7	95.7
0241, 31 - 34	SITE & INFRASTRUCTURE, DEMOLITION	87.8	96.1	93.5	92.6	101.0	98.4	85.8	91.9	90.0	89.7	100.8	97.3	90.4	96.1	94.3	85.9	95.2	92.3
0310	Concrete Forming & Accessories	96.7	101.7	101.0	99.4	80.2	83.0	94.6	95.1	95.1	97.3	101.3	100.7	97.0	80.0	82.5	93.5	95.6	95.2
0320	Concrete Reinforcing	84.1	86.3	85.2	107.3	86.2	97.1	85.8	86.1	86.0	101.6	86.2	94.2	106.4	86.1	96.6	84.1	86.2	85.1
0330	Cast-in-Place Concrete	92.5	115.9	101.2	109.6	116.7	112.2	86.9	107.2	94.4	104.5	116.7	109.0	87.8	115.6	98.2	86.9	107.3	94.5
03	CONCRETE	90.7	103.7	96.4	101.7	94.2	98.4	86.2	97.7	91.3	98.5	103.7	100.8	91.4	93.8	92.5	85.9	97.9	91.2
04	MASONRY	104.6	86.2	93.3	103.9	86.4	93.1	104.8	71.5	84.4	101.4	86.4	92.2	87.2	86.3	86.7	131.1	71.5	94.5
05	METALS	95.1	91.1	93.9	102.8	89.8	99.0	95.2	90.0	93.7	101.2	89.9	97.9	101.1	90.3	97.9	95.2	90.4	93.8
06	WOOD, PLASTICS & COMPOSITES	103.1	106.8	105.1	99.2	78.5	87.8	99.0	106.8	103.3	96.0	106.9	102.0	103.2	78.3	89.5	94.1	106.8	101.0
07	THERMAL & MOISTURE PROTECTION	101.0	89.1	95.8	107.4	86.6	98.3	100.7	82.4	92.8	109.4	89.7	100.8	101.1	86.1	94.6	100.6	82.4	92.7
08	OPENINGS	100.1	97.8	99.6	102.6	78.4	96.9	100.1	94.0	98.7	103.1	94.1	101.0	103.2	78.3	97.4	100.1	94.0	98.7
0920	Plaster & Gypsum Board	108.9	106.7	107.4	109.9	77.4	88.1	114.5	106.7	109.2	110.4	106.7	107.9	109.4	77.4	87.9	115.8	106.7	109.6
0950, 0980	Ceilings & Acoustic Treatment	97.2	106.7	103.6	103.9	77.4	86.0	97.2	106.7	103.6	104.9	106.7	106.1	103.2	77.4	85.8	97.2	106.7	103.6
0960	Flooring	92.2	98.9	94.1	101.8	98.9	101.0	95.0	98.9	96.1	98.5	98.9	98.6	92.0	98.9	94.0	98.3	98.9	98.5
0970, 0990	Wall Finishes & Painting/Coating	86.4	101.5	95.3	98.0	88.3	92.3	86.4	88.3	87.5	97.8	88.3	92.2	86.4	88.3	87.5	86.4	88.3	87.5
09	FINISHES	90.8	101.8	96.8	99.1	83.8	90.8	92.2	96.7	94.6	97.9	100.6	99.4	92.0	83.7	87.5	93.3	96.7	95.2
COVERS	DIVS. 10 - 14, 25, 28, 41, 43, 44, 46	100.0	91.8	98.2	100.0	89.1	97.6	100.0	86.7	97.0	100.0	92.2	98.3	100.0	88.6	97.5	100.0	86.8	97.0
21, 22, 23	FIRE SUPPRESSION, PLUMBING & HVAC	96.7	87.6	93.1	100.0	68.3	87.6	96.7	60.5	82.5	96.5	68.3	85.4	100.2	68.2	87.7	96.7	60.5	82.5
26, 27, 3370	ELECTRICAL, COMMUNICATIONS & UTIL.	108.3	79.5	94.1	107.6	53.6	81.0	108.3	54.5	81.8	108.1	53.6	81.3	108.4	53.6	81.4	108.3	54.5	81.8
MF2018	WEIGHTED AVERAGE	97.4	92.1	95.2	101.9	79.8	92.5	96.8	79.0	89.3	100.2	84.4	93.5	98.9	79.4	90.7	98.1	79.3	90.2

VERMONT / VIRGINIA

	DIVISION	WHITE RIVER JCT. 050 MAT.	INST.	TOTAL	ALEXANDRIA 223 MAT.	INST.	TOTAL	ARLINGTON 222 MAT.	INST.	TOTAL	BRISTOL 242 MAT.	INST.	TOTAL	CHARLOTTESVILLE 229 MAT.	INST.	TOTAL	CULPEPER 227 MAT.	INST.	TOTAL
015433	CONTRACTOR EQUIPMENT		95.7	95.7		105.3	105.3		104.1	104.1		104.1	104.1		108.1	108.1		104.1	104.1
0241, 31 - 34	SITE & INFRASTRUCTURE, DEMOLITION	90.3	95.2	93.7	113.7	89.0	96.7	123.9	87.0	98.5	108.2	86.0	92.8	112.7	88.3	95.9	111.4	86.5	94.2
0310	Concrete Forming & Accessories	91.4	96.1	95.4	91.5	71.9	74.8	90.5	71.6	74.4	86.6	64.1	67.5	85.1	61.7	65.2	82.4	71.0	72.7
0320	Concrete Reinforcing	85.1	86.2	85.6	86.6	85.0	85.9	97.8	85.1	91.6	97.7	83.7	91.0	97.1	73.0	85.4	97.8	84.9	91.6
0330	Cast-in-Place Concrete	92.5	108.1	98.3	108.1	77.7	96.8	105.2	77.2	94.8	104.8	44.5	82.3	109.0	75.5	96.5	107.7	74.4	95.3
03	CONCRETE	92.6	98.4	95.2	99.9	77.7	90.1	104.5	77.4	92.6	100.8	62.4	83.9	101.3	70.2	87.7	98.4	76.0	88.6
04	MASONRY	117.2	72.9	90.0	89.8	73.7	79.9	102.3	72.6	84.0	92.3	45.4	63.5	116.9	55.3	79.0	105.1	70.8	84.1
05	METALS	95.2	90.4	93.8	106.0	100.8	104.5	104.6	101.8	103.8	103.4	97.8	101.8	103.7	95.8	101.4	103.7	99.6	102.5
06	WOOD, PLASTICS & COMPOSITES	97.1	106.8	102.4	91.0	70.1	79.5	87.5	70.1	78.0	80.2	69.8	74.5	78.7	59.8	68.3	77.4	70.1	73.4
07	THERMAL & MOISTURE PROTECTION	101.1	83.0	93.2	102.5	81.1	93.2	104.6	80.5	94.1	103.9	58.8	84.3	103.5	69.7	88.8	103.7	78.9	92.9
08	OPENINGS	100.1	94.0	98.7	95.4	74.3	90.5	93.7	74.3	89.2	96.4	67.9	89.7	94.8	65.7	88.0	95.1	73.8	90.1
0920	Plaster & Gypsum Board	106.6	106.7	106.7	98.0	69.2	78.6	94.8	69.2	77.6	91.1	68.9	76.2	91.1	57.9	68.8	91.3	69.2	76.4
0950, 0980	Ceilings & Acoustic Treatment	97.2	106.7	103.6	92.4	69.2	76.8	90.8	69.2	76.2	90.0	68.9	75.8	90.0	57.9	68.4	90.8	69.2	76.2
0960	Flooring	90.2	98.9	92.6	97.1	77.4	91.5	95.6	75.7	90.0	92.3	55.7	82.0	90.8	55.7	80.9	90.8	74.4	86.2
0970, 0990	Wall Finishes & Painting/Coating	86.4	88.3	87.5	115.8	76.0	92.3	115.8	73.1	90.6	102.1	55.3	74.4	102.1	58.6	76.3	115.8	73.1	90.6
09	FINISHES	90.1	97.1	93.9	93.8	72.6	82.3	93.8	71.6	81.8	90.6	62.3	75.3	90.1	59.4	73.5	90.8	71.2	80.2
COVERS	DIVS. 10 - 14, 25, 28, 41, 43, 44, 46	100.0	87.2	97.1	100.0	86.6	97.0	100.0	84.4	96.5	100.0	75.1	94.5	100.0	83.2	96.2	100.0	83.6	96.3
21, 22, 23	FIRE SUPPRESSION, PLUMBING & HVAC	96.7	61.2	82.8	100.4	86.2	94.8	100.4	85.6	94.6	96.8	46.9	76.9	96.8	69.1	86.0	96.8	85.3	92.3
26, 27, 3370	ELECTRICAL, COMMUNICATIONS & UTIL.	108.3	54.3	81.7	96.6	100.2	98.4	94.4	100.2	97.2	96.2	36.3	66.7	96.2	70.4	83.5	98.5	96.9	97.7
MF2018	WEIGHTED AVERAGE	98.2	79.8	90.4	99.6	84.8	93.3	100.4	84.2	93.6	98.2	59.3	81.8	99.4	71.0	87.4	98.7	83.0	92.1

DIVISION		VIRGINIA																	
		FAIRFAX 220-221			FARMVILLE 239			FREDERICKSBURG 224-225			GRUNDY 246			HARRISONBURG 228			LYNCHBURG 245		
		MAT.	INST.	TOTAL	MAT.	INST.	TOTAL	MAT.	INST.	TOTAL	MAT.	INST.	TOTAL	MAT.	INST.	TOTAL	MAT.	INST.	TOTAL
015433	CONTRACTOR EQUIPMENT		104.1	104.1		108.1	108.1		104.1	104.1		104.1	104.1		104.1	104.1		104.1	104.1
0241, 31 - 34	SITE & INFRASTRUCTURE, DEMOLITION	122.4	86.7	97.8	108.5	87.3	93.9	110.9	86.4	93.9	106.0	85.3	91.7	119.3	85.0	95.6	106.8	86.7	92.9
0310	Concrete Forming & Accessories	85.1	63.8	67.0	98.9	60.3	66.0	85.1	65.4	68.3	89.6	38.2	45.8	81.4	56.6	60.2	86.6	70.5	72.9
0320	Concrete Reinforcing	97.8	90.3	94.2	92.9	67.9	80.8	98.5	80.1	89.6	96.4	44.1	71.1	97.8	57.3	78.2	97.1	84.5	91.0
0330	Cast-in-Place Concrete	105.2	78.5	95.2	106.8	83.9	98.3	106.8	74.2	94.7	104.8	50.9	84.7	105.2	52.6	85.6	104.8	74.3	93.4
03	CONCRETE	104.2	75.2	91.5	101.8	71.5	88.5	98.1	72.7	86.9	99.3	45.8	75.8	102.0	57.1	82.3	99.2	75.8	88.9
04	MASONRY	102.2	73.6	84.6	101.8	50.0	69.9	104.0	66.7	81.1	94.9	55.4	70.6	101.4	57.9	74.6	109.3	63.9	81.4
05	METALS	103.8	102.5	103.4	101.3	90.8	98.2	103.8	98.0	102.1	103.4	75.2	95.1	103.7	87.2	98.9	103.6	99.8	102.5
06	WOOD, PLASTICS & COMPOSITES	80.2	59.2	68.6	93.0	59.8	74.7	80.2	64.6	71.6	83.1	30.6	54.2	76.3	59.2	66.9	80.2	71.5	75.4
07	THERMAL & MOISTURE PROTECTION	104.4	79.9	93.7	105.2	69.4	89.6	103.7	76.3	91.8	103.9	57.3	83.6	104.1	69.6	89.1	103.7	70.4	89.2
08	OPENINGS	93.7	69.6	88.0	94.3	54.9	85.1	94.8	62.8	88.7	96.4	30.1	80.9	95.0	57.0	86.2	95.0	68.8	88.9
0920	Plaster & Gypsum Board	91.3	57.9	68.9	103.1	57.9	72.7	91.3	63.6	72.7	91.1	28.5	49.0	91.1	57.9	68.8	91.1	70.7	77.4
0950, 0980	Ceilings & Acoustic Treatment	90.8	57.9	68.6	86.5	57.9	67.2	90.8	63.6	72.4	90.0	28.5	48.6	90.0	57.9	68.4	90.0	70.7	77.0
0960	Flooring	92.5	76.7	88.1	95.6	50.5	82.9	92.5	72.4	86.9	93.5	28.2	75.2	90.6	74.4	86.1	92.3	65.4	84.7
0970, 0990	Wall Finishes & Painting/Coating	115.8	73.6	90.8	99.6	57.2	74.5	115.8	57.0	81.0	102.1	31.2	60.2	115.8	48.3	75.9	102.1	55.3	74.4
09	FINISHES	92.4	65.7	77.9	90.6	57.7	72.8	91.3	65.0	77.1	90.7	33.9	60.0	91.3	58.9	73.7	90.4	67.8	78.2
COVERS	DIVS. 10 - 14, 25, 28, 41, 43, 44, 46	100.0	84.2	96.5	100.0	77.1	94.9	100.0	78.3	95.2	100.0	73.6	94.1	100.0	62.8	91.7	100.0	78.8	95.3
21, 22, 23	FIRE SUPPRESSION, PLUMBING & HVAC	96.8	88.1	93.4	96.9	52.0	79.3	96.8	79.9	90.2	96.8	65.7	84.6	96.8	65.4	84.5	96.8	69.5	86.1
26, 27, 3370	ELECTRICAL, COMMUNICATIONS & UTIL.	97.3	101.1	99.1	91.2	68.3	79.9	94.5	91.8	93.2	96.2	68.3	82.5	96.4	86.4	91.5	97.2	67.8	82.7
MF2018	WEIGHTED AVERAGE	99.5	83.6	92.8	97.9	65.3	84.1	98.3	78.7	90.0	98.1	58.8	81.5	99.0	68.8	86.3	98.7	73.8	88.2

DIVISION		VIRGINIA																	
		NEWPORT NEWS 236			NORFOLK 233-235			PETERSBURG 238			PORTSMOUTH 237			PULASKI 243			RICHMOND 230-232		
		MAT.	INST.	TOTAL	MAT.	INST.	TOTAL	MAT.	INST.	TOTAL	MAT.	INST.	TOTAL	MAT.	INST.	TOTAL	MAT.	INST.	TOTAL
015433	CONTRACTOR EQUIPMENT		108.2	108.2		110.2	110.2		108.1	108.1		108.0	108.0		104.1	104.1		109.6	109.6
0241, 31 - 34	SITE & INFRASTRUCTURE, DEMOLITION	107.6	88.4	94.3	106.9	93.8	97.9	111.1	88.4	95.4	106.2	88.2	93.8	105.4	85.7	91.8	103.4	92.8	96.1
0310	Concrete Forming & Accessories	98.1	61.1	66.6	103.6	61.2	67.5	91.8	61.8	66.2	87.8	61.3	65.2	89.6	40.4	47.7	94.2	61.9	66.7
0320	Concrete Reinforcing	92.8	66.2	79.9	101.2	66.2	84.3	92.4	73.6	83.3	92.4	66.2	79.7	96.4	83.8	90.3	102.1	73.6	88.3
0330	Cast-in-Place Concrete	103.8	64.5	89.2	111.0	76.1	98.0	110.2	76.6	97.7	102.8	65.6	89.0	104.8	82.4	96.5	95.3	76.6	88.4
03	CONCRETE	98.9	65.0	84.0	102.1	69.0	87.6	104.5	70.8	89.7	97.7	65.5	83.6	99.3	64.7	84.1	94.5	70.7	84.0
04	MASONRY	97.3	55.0	71.3	97.1	55.0	71.3	110.2	64.6	82.2	103.0	55.6	73.9	87.9	54.3	67.3	93.3	64.7	75.7
05	METALS	103.8	93.1	100.6	105.3	91.8	101.3	101.4	96.7	100.0	102.7	93.1	99.9	103.5	96.0	101.3	105.9	95.3	102.8
06	WOOD, PLASTICS & COMPOSITES	91.8	59.8	74.2	99.3	59.7	77.5	83.2	59.5	70.2	79.6	59.8	68.7	83.1	32.0	55.0	93.7	59.7	75.0
07	THERMAL & MOISTURE PROTECTION	105.1	67.7	88.9	104.4	70.0	89.5	105.1	72.3	90.9	105.1	68.0	89.0	103.9	66.6	87.6	101.0	72.9	88.8
08	OPENINGS	94.6	60.1	86.6	95.2	60.1	87.0	94.0	65.6	87.4	94.7	56.0	85.7	96.4	42.9	83.9	101.7	65.7	93.3
0920	Plaster & Gypsum Board	104.3	57.9	73.1	98.5	57.9	71.2	97.8	57.7	70.8	98.6	57.9	71.2	91.1	30.0	50.0	99.5	57.9	71.5
0950, 0980	Ceilings & Acoustic Treatment	91.4	57.9	68.8	88.8	57.9	68.0	88.2	57.7	67.6	91.4	57.9	68.8	90.0	30.0	49.6	89.5	57.9	68.2
0960	Flooring	95.6	51.1	83.1	96.2	51.1	83.5	91.7	71.7	86.1	88.9	51.7	78.4	93.5	55.7	82.9	94.0	71.7	87.8
0970, 0990	Wall Finishes & Painting/Coating	99.6	59.7	76.0	101.2	59.7	76.6	99.6	58.6	75.3	99.6	59.7	76.0	102.1	48.3	70.3	95.4	58.6	73.6
09	FINISHES	91.5	58.4	73.6	90.5	58.3	73.1	89.3	62.5	74.8	88.8	58.6	72.5	90.7	40.9	63.8	91.9	62.5	76.0
COVERS	DIVS. 10 - 14, 25, 28, 41, 43, 44, 46	100.0	83.4	96.3	100.0	83.2	96.3	100.0	83.1	96.2	100.0	83.6	96.3	100.0	73.2	94.0	100.0	83.0	96.2
21, 22, 23	FIRE SUPPRESSION, PLUMBING & HVAC	100.4	66.8	87.2	100.1	66.8	87.0	96.9	69.1	86.0	100.4	67.1	87.4	96.8	66.0	84.7	100.0	69.1	87.9
26, 27, 3370	ELECTRICAL, COMMUNICATIONS & UTIL.	93.6	64.2	79.1	98.0	60.7	79.6	93.8	70.5	82.3	92.1	60.7	76.7	96.2	83.4	89.9	96.8	70.4	83.8
MF2018	WEIGHTED AVERAGE	98.8	68.3	85.9	99.9	68.7	86.7	98.7	72.6	87.7	98.3	67.8	85.4	97.8	67.1	84.8	99.2	72.8	88.1

DIVISION		VIRGINIA									WASHINGTON								
		ROANOKE 240-241			STAUNTON 244			WINCHESTER 226			CLARKSTON 994			EVERETT 982			OLYMPIA 985		
		MAT.	INST.	TOTAL	MAT.	INST.	TOTAL	MAT.	INST.	TOTAL	MAT.	INST.	TOTAL	MAT.	INST.	TOTAL	MAT.	INST.	TOTAL
015433	CONTRACTOR EQUIPMENT		104.1	104.1		108.1	108.1		104.1	104.1		90.2	90.2		99.3	99.3		101.6	101.6
0241, 31 - 34	SITE & INFRASTRUCTURE, DEMOLITION	106.0	86.8	92.7	109.3	87.3	94.1	118.1	86.4	96.2	98.1	86.1	89.8	92.1	105.6	101.4	92.8	108.2	103.5
0310	Concrete Forming & Accessories	95.7	71.2	74.8	89.3	61.4	65.5	83.6	66.4	68.9	107.0	64.2	70.5	111.6	104.6	105.7	101.6	104.3	103.9
0320	Concrete Reinforcing	97.5	84.6	91.2	97.1	83.8	90.7	97.1	80.2	88.9	108.4	95.9	102.3	109.2	114.5	111.8	119.5	114.4	117.0
0330	Cast-in-Place Concrete	119.0	86.2	106.8	109.0	85.5	100.2	105.2	63.9	89.8	82.3	82.9	82.5	102.6	109.4	105.1	95.3	109.8	100.7
03	CONCRETE	103.5	80.2	93.3	100.7	75.2	89.5	101.5	69.5	87.4	87.6	76.6	82.8	95.1	107.4	100.5	93.7	107.2	99.7
04	MASONRY	96.3	65.9	77.6	105.0	55.3	74.5	98.7	69.8	80.9	97.8	91.7	94.0	121.0	101.8	109.2	115.3	101.7	107.0
05	METALS	105.8	100.2	104.2	103.6	96.0	101.4	103.8	97.9	102.1	92.8	87.3	91.2	116.8	98.4	111.4	117.8	96.1	111.4
06	WOOD, PLASTICS & COMPOSITES	91.9	71.5	80.7	83.1	59.8	70.3	78.7	64.8	71.1	104.4	58.5	79.2	107.5	104.7	105.9	97.5	104.8	101.5
07	THERMAL & MOISTURE PROTECTION	103.6	76.0	91.6	103.4	67.5	87.8	104.2	75.9	91.9	159.7	82.9	126.3	112.4	105.9	109.6	108.3	103.0	106.0
08	OPENINGS	95.4	68.8	89.2	95.0	58.5	86.5	96.5	68.3	89.9	116.4	65.0	104.4	106.0	107.3	106.3	109.6	104.8	108.5
0920	Plaster & Gypsum Board	98.0	70.7	79.6	91.1	57.9	68.8	91.3	63.8	72.8	154.4	57.3	89.1	112.7	104.9	107.5	106.5	104.9	105.4
0950, 0980	Ceilings & Acoustic Treatment	92.4	70.7	77.8	90.0	57.9	68.4	90.8	63.8	72.6	103.8	57.3	72.4	101.2	104.9	103.7	98.7	104.9	102.9
0960	Flooring	97.1	67.3	88.7	93.1	33.1	76.2	92.0	74.4	87.1	85.4	77.5	83.2	109.9	96.0	106.0	97.6	96.0	97.1
0970, 0990	Wall Finishes & Painting/Coating	102.1	55.3	74.4	102.1	30.7	59.8	115.8	77.6	93.2	84.4	72.1	77.2	97.6	95.2	96.2	99.6	93.1	95.8
09	FINISHES	92.7	68.5	79.6	90.6	52.6	70.0	91.8	68.3	79.1	108.2	65.8	85.2	105.4	101.6	103.4	94.9	101.5	98.5
COVERS	DIVS. 10 - 14, 25, 28, 41, 43, 44, 46	100.0	79.2	95.4	100.0	78.2	95.1	100.0	82.7	96.1	100.0	89.3	97.6	100.0	103.7	100.8	100.0	99.9	100.0
21, 22, 23	FIRE SUPPRESSION, PLUMBING & HVAC	100.4	66.7	87.2	96.8	59.3	82.1	96.8	81.4	90.8	96.7	80.8	90.4	100.0	105.2	102.0	99.9	100.6	100.3
26, 27, 3370	ELECTRICAL, COMMUNICATIONS & UTIL.	96.2	60.8	78.8	95.2	68.4	82.0	94.9	91.9	93.4	89.4	95.6	92.5	107.0	104.1	105.6	105.4	97.7	101.6
MF2018	WEIGHTED AVERAGE	100.1	73.6	88.9	98.6	67.9	85.6	98.9	79.5	90.7	99.1	81.9	91.8	105.1	104.0	104.6	104.0	101.8	103.1

WASHINGTON

DIVISION		RICHLAND 993			SEATTLE 980 - 981, 987			SPOKANE 990 - 992			TACOMA 983 - 984			VANCOUVER 986			WENATCHEE 988		
		MAT.	INST.	TOTAL	MAT.	INST.	TOTAL	MAT.	INST.	TOTAL	MAT.	INST.	TOTAL	MAT.	INST.	TOTAL	MAT.	INST.	TOTAL
015433	CONTRACTOR EQUIPMENT		90.2	90.2		102.4	102.4		90.2	90.2		99.3	99.3		95.8	95.8		99.3	99.3
0241, 31 - 34	SITE & INFRASTRUCTURE, DEMOLITION	100.3	86.3	90.6	97.8	108.9	105.5	99.7	86.3	90.4	95.2	105.1	102.1	105.2	93.1	96.8	104.0	102.8	103.2
0310	Concrete Forming & Accessories	107.1	80.0	84.0	107.4	109.5	109.2	112.0	79.5	84.3	102.3	104.1	103.9	102.7	95.7	96.7	104.0	76.5	80.5
0320	Concrete Reinforcing	103.9	95.4	99.8	109.2	117.3	113.1	104.6	95.3	100.1	107.8	114.3	110.9	108.8	114.8	111.7	108.8	95.9	102.5
0330	Cast-in-Place Concrete	82.5	84.4	83.2	105.9	113.9	108.9	85.7	84.2	85.2	105.5	108.8	106.7	117.8	99.9	111.2	107.7	90.8	101.4
03	CONCRETE	87.2	84.2	85.9	102.8	111.8	106.7	89.2	83.9	86.8	97.0	106.9	101.4	107.2	100.2	104.1	105.3	84.9	96.3
04	MASONRY	98.8	83.7	89.5	123.5	107.2	113.5	99.4	83.7	89.7	117.6	100.2	106.9	118.8	105.5	110.6	120.9	94.1	104.4
05	METALS	93.2	87.1	91.4	116.1	101.9	111.9	95.4	86.6	92.8	118.7	96.5	112.2	116.0	99.3	111.1	116.0	87.3	107.6
06	WOOD, PLASTICS & COMPOSITES	104.6	77.9	89.9	103.4	108.7	106.3	113.5	77.9	93.9	97.1	104.7	101.2	89.5	94.9	92.5	98.7	73.3	84.8
07	THERMAL & MOISTURE PROTECTION	161.1	84.4	127.7	111.6	110.8	111.3	157.4	84.7	125.8	112.1	102.1	107.8	111.8	101.6	107.4	111.6	88.3	101.5
08	OPENINGS	114.4	74.7	105.2	106.1	110.7	107.2	115.0	74.7	105.6	106.7	104.7	106.2	102.8	101.5	102.5	106.2	73.0	98.5
0920	Plaster & Gypsum Board	154.4	77.2	102.5	111.2	109.1	109.8	144.2	77.2	99.1	110.6	104.9	106.8	109.0	95.2	99.7	113.9	72.7	86.2
0950, 0980	Ceilings & Acoustic Treatment	109.2	77.2	87.6	106.6	109.1	108.3	105.6	77.2	86.4	104.5	104.9	104.8	102.4	95.2	97.5	97.8	72.7	80.9
0960	Flooring	85.7	75.5	83.4	109.5	103.4	107.8	84.8	75.5	82.7	102.2	96.0	100.5	107.4	107.1	107.3	105.3	75.5	97.5
0970, 0990	Wall Finishes & Painting/Coating	84.4	75.1	78.9	107.1	95.8	100.4	84.6	76.4	79.8	97.6	93.1	95.0	100.3	79.9	88.2	97.6	70.9	81.8
09	FINISHES	109.5	78.1	92.5	108.9	106.4	107.6	107.3	78.2	91.6	103.6	101.4	102.4	101.3	95.6	98.3	104.7	74.9	88.5
COVERS	DIVS. 10 - 14, 25, 28, 41, 43, 44, 46	100.0	92.2	98.3	100.0	105.2	101.2	100.0	92.2	98.3	100.0	99.6	99.9	100.0	101.5	100.3	100.0	91.2	98.0
21, 22, 23	FIRE SUPPRESSION, PLUMBING & HVAC	100.3	108.6	103.6	99.9	123.0	109.0	100.3	83.7	93.8	100.0	100.6	100.2	100.0	110.7	104.2	96.5	92.0	94.7
26, 27, 3370	ELECTRICAL, COMMUNICATIONS & UTIL.	87.2	92.8	90.0	106.2	118.9	112.5	85.4	77.1	81.3	106.8	97.7	102.3	113.1	107.3	110.2	107.7	94.5	101.2
MF2018	WEIGHTED AVERAGE	99.8	90.0	95.6	106.3	112.5	108.9	100.1	82.4	92.6	105.4	101.3	103.7	106.6	102.8	105.0	105.6	88.7	98.5

DIVISION		WASHINGTON YAKIMA 989			WEST VIRGINIA BECKLEY 258 - 259			BLUEFIELD 247 - 248			BUCKHANNON 262			CHARLESTON 250 - 253			CLARKSBURG 263 - 264		
		MAT.	INST.	TOTAL	MAT.	INST.	TOTAL	MAT.	INST.	TOTAL	MAT.	INST.	TOTAL	MAT.	INST.	TOTAL	MAT.	INST.	TOTAL
015433	CONTRACTOR EQUIPMENT		99.3	99.3		104.1	104.1		104.1	104.1		104.1	104.1		106.8	106.8		104.1	104.1
0241, 31 - 34	SITE & INFRASTRUCTURE, DEMOLITION	97.6	103.8	101.9	100.8	87.4	91.6	99.9	87.4	91.3	106.0	87.4	93.1	100.4	93.0	95.3	106.8	87.4	93.4
0310	Concrete Forming & Accessories	102.9	99.2	99.8	85.0	87.1	86.8	86.6	87.1	87.0	86.1	83.7	84.0	96.0	88.6	89.7	83.6	83.8	83.7
0320	Concrete Reinforcing	108.3	95.4	102.1	98.0	87.7	93.1	96.0	82.5	89.5	96.7	82.4	89.8	106.1	87.8	97.2	96.7	95.3	96.0
0330	Cast-in-Place Concrete	112.7	83.0	101.7	99.2	92.5	96.7	102.5	92.5	98.8	102.1	92.0	98.4	95.9	96.6	96.2	111.9	87.6	102.9
03	CONCRETE	101.8	92.4	97.7	91.8	90.2	91.1	95.1	89.3	92.6	98.2	87.5	93.5	90.7	92.2	91.4	102.2	88.2	96.1
04	MASONRY	110.3	81.0	92.3	91.0	89.0	89.8	89.5	89.0	89.2	100.6	85.8	91.5	88.4	90.7	89.8	104.1	85.8	92.9
05	METALS	116.7	87.3	108.0	98.2	103.3	99.7	103.6	101.5	103.0	103.8	101.6	103.2	96.8	102.5	98.5	103.8	106.0	104.5
06	WOOD, PLASTICS & COMPOSITES	97.4	104.7	101.4	80.2	87.4	84.2	82.1	87.4	85.0	81.4	83.0	82.3	89.3	88.2	88.7	78.0	83.0	80.8
07	THERMAL & MOISTURE PROTECTION	112.3	86.1	100.9	105.0	86.5	96.9	103.6	86.5	96.1	104.0	86.6	96.4	101.2	88.1	95.5	103.8	86.3	96.2
08	OPENINGS	106.2	100.5	104.9	94.5	83.0	91.8	96.9	81.8	93.4	96.9	79.4	92.8	94.9	83.4	92.2	96.9	82.1	93.4
0920	Plaster & Gypsum Board	110.4	104.9	106.7	93.5	87.0	89.1	90.2	87.0	88.0	90.8	82.5	85.2	98.7	87.6	91.2	88.6	82.5	84.5
0950, 0980	Ceilings & Acoustic Treatment	99.6	104.9	103.2	78.1	87.0	84.1	87.5	87.0	87.2	90.0	82.5	84.9	89.6	87.6	88.2	90.0	82.5	84.9
0960	Flooring	103.2	77.5	96.0	91.0	98.9	93.2	90.1	98.9	92.6	89.8	95.2	91.3	97.1	98.9	97.6	88.8	95.2	90.6
0970, 0990	Wall Finishes & Painting/Coating	97.6	76.4	85.1	89.3	92.1	91.0	102.1	99.8	94.8	102.1	88.0	93.7	88.6	91.9	90.5	102.1	88.0	93.7
09	FINISHES	103.1	93.3	97.8	86.3	90.2	88.4	88.7	89.9	89.4	89.7	86.1	87.8	92.8	91.0	91.9	89.0	86.1	87.4
COVERS	DIVS. 10 - 14, 25, 28, 41, 43, 44, 46	100.0	96.1	99.1	100.0	90.3	97.8	100.0	90.3	97.8	100.0	90.6	97.9	100.0	92.5	98.3	100.0	90.6	97.9
21, 22, 23	FIRE SUPPRESSION, PLUMBING & HVAC	100.0	107.4	102.9	97.1	80.3	90.5	96.8	84.0	91.8	96.8	89.7	94.0	100.1	81.4	92.8	96.8	89.8	94.1
26, 27, 3370	ELECTRICAL, COMMUNICATIONS & UTIL.	109.9	92.9	101.5	92.9	84.1	88.6	95.2	84.1	89.8	96.6	89.3	93.0	98.0	87.1	92.6	96.6	89.3	93.0
MF2018	WEIGHTED AVERAGE	105.6	95.4	101.3	95.2	87.7	92.0	97.0	88.1	93.2	98.3	88.8	94.3	96.5	89.4	93.5	98.9	89.5	94.9

WEST VIRGINIA

DIVISION		GASSAWAY 266			HUNTINGTON 255 - 257			LEWISBURG 249			MARTINSBURG 254			MORGANTOWN 265			PARKERSBURG 261		
		MAT.	INST.	TOTAL	MAT.	INST.	TOTAL	MAT.	INST.	TOTAL	MAT.	INST.	TOTAL	MAT.	INST.	TOTAL	MAT.	INST.	TOTAL
015433	CONTRACTOR EQUIPMENT		104.1	104.1		104.1	104.1		104.1	104.1		104.1	104.1		104.1	104.1		104.1	104.1
0241, 31 - 34	SITE & INFRASTRUCTURE, DEMOLITION	103.5	87.4	92.4	105.4	88.5	93.7	115.5	87.4	96.1	104.7	88.0	93.2	100.9	88.2	92.1	109.3	88.4	94.9
0310	Concrete Forming & Accessories	85.5	86.9	86.7	96.0	89.3	90.3	83.9	86.6	86.2	85.0	79.5	80.3	83.9	83.8	83.9	88.0	86.0	86.3
0320	Concrete Reinforcing	96.7	91.6	94.2	99.5	93.1	96.4	96.7	82.4	89.8	98.0	95.1	96.6	96.7	95.3	96.0	96.0	82.3	89.4
0330	Cast-in-Place Concrete	106.9	94.0	102.1	108.2	94.0	102.9	102.5	92.3	98.7	104.0	88.3	98.1	102.1	92.0	98.4	104.4	89.0	98.7
03	CONCRETE	98.6	91.3	95.4	96.8	92.6	95.0	105.1	89.0	98.0	95.3	86.5	91.4	94.9	89.8	92.7	100.3	87.5	94.7
04	MASONRY	105.0	87.9	94.5	91.7	93.4	92.7	94.3	89.0	91.1	93.1	83.9	87.5	122.6	85.8	100.0	80.1	85.2	83.2
05	METALS	103.8	104.8	104.1	100.6	105.7	102.1	103.7	101.0	102.9	98.6	104.4	100.3	103.9	106.0	104.5	104.5	101.4	103.6
06	WOOD, PLASTICS & COMPOSITES	80.4	87.4	84.3	90.8	88.7	89.6	78.3	87.4	83.3	80.2	78.6	79.3	78.3	83.0	80.9	81.5	85.6	83.8
07	THERMAL & MOISTURE PROTECTION	103.6	87.9	96.8	105.2	88.4	97.9	104.6	86.9	96.7	105.3	80.8	94.6	103.7	86.6	96.2	103.9	86.7	96.4
08	OPENINGS	95.2	83.9	92.6	93.7	84.8	91.7	96.9	81.8	93.4	96.2	73.8	91.0	98.0	82.1	94.3	95.8	80.4	92.2
0920	Plaster & Gypsum Board	89.9	87.0	87.9	99.5	88.3	91.9	88.6	87.0	87.5	93.5	77.9	83.0	88.6	82.5	84.5	91.1	85.1	87.1
0950, 0980	Ceilings & Acoustic Treatment	90.0	87.0	88.0	78.1	88.3	85.0	90.0	87.0	88.0	78.1	77.9	78.0	90.0	82.5	84.9	90.0	85.1	86.7
0960	Flooring	89.6	98.9	92.2	98.4	103.3	99.7	88.9	98.9	91.7	91.0	98.9	93.2	88.9	95.2	90.7	92.6	91.9	92.4
0970, 0990	Wall Finishes & Painting/Coating	102.1	91.9	96.0	89.3	89.8	89.6	102.1	64.5	79.8	89.3	83.7	86.0	102.1	88.0	93.7	102.1	83.7	91.2
09	FINISHES	89.2	89.9	89.6	89.3	92.0	90.8	90.1	87.2	88.5	86.6	83.0	84.6	88.6	86.1	87.3	90.6	86.8	88.6
COVERS	DIVS. 10 - 14, 25, 28, 41, 43, 44, 46	100.0	91.1	98.0	100.0	92.3	98.3	100.0	65.4	92.3	100.0	81.4	95.8	100.0	83.5	96.3	100.0	91.3	98.1
21, 22, 23	FIRE SUPPRESSION, PLUMBING & HVAC	96.8	83.4	91.6	100.0	90.6	96.7	96.8	80.2	90.3	97.1	80.9	90.8	96.8	89.8	94.1	100.3	88.3	95.6
26, 27, 3370	ELECTRICAL, COMMUNICATIONS & UTIL.	96.6	87.1	91.9	96.6	89.6	93.1	92.8	84.1	88.5	98.6	76.9	87.9	96.7	89.3	93.1	96.7	88.3	92.5
MF2018	WEIGHTED AVERAGE	98.3	89.0	94.3	97.8	92.2	95.4	98.7	86.1	93.4	96.7	84.1	91.4	98.8	89.6	94.9	98.6	88.6	94.4

For customer support on your Electrical Change Order Costs with RSMeans data, call 800.448.8182.

DIVISION		WEST VIRGINIA									WISCONSIN								
		PETERSBURG			ROMNEY			WHEELING			BELOIT			EAU CLAIRE			GREEN BAY		
		268			267			260			535			547			541 - 543		
		MAT.	INST.	TOTAL	MAT.	INST.	TOTAL	MAT.	INST.	TOTAL	MAT.	INST.	TOTAL	MAT.	INST.	TOTAL	MAT.	INST.	TOTAL
015433	CONTRACTOR EQUIPMENT		104.1	104.1		104.1	104.1		104.1	104.1		99.0	99.0		99.9	99.9		97.7	97.7
0241, 31 - 34	SITE & INFRASTRUCTURE, DEMOLITION	100.2	88.3	92.0	103.0	88.3	92.9	109.9	88.2	94.9	97.1	101.7	100.3	96.0	99.0	98.1	99.5	95.3	96.6
0310	Concrete Forming & Accessories	87.0	86.4	86.5	83.3	86.8	86.3	89.7	83.8	84.7	99.8	96.2	96.7	98.5	94.7	95.3	108.2	97.5	99.1
0320	Concrete Reinforcing	96.0	91.4	93.8	96.7	95.4	96.1	95.5	95.3	95.4	99.1	137.1	115.4	94.2	114.2	103.9	92.3	109.4	100.5
0330	Cast-in-Place Concrete	102.1	91.8	98.3	106.9	83.5	98.2	104.4	92.0	99.8	105.8	102.3	103.7	99.9	94.8	98.0	103.3	99.9	102.1
03	CONCRETE	95.0	90.2	92.9	98.4	88.2	93.9	100.3	89.8	95.7	96.7	104.7	100.2	93.1	98.4	95.4	96.2	100.6	98.2
04	MASONRY	96.1	85.8	89.8	94.7	87.9	90.5	103.4	85.7	92.5	99.3	100.3	99.9	91.4	96.1	94.3	122.7	97.8	107.4
05	METALS	103.9	104.1	104.0	104.0	105.9	104.5	104.7	106.1	105.1	98.6	112.0	102.5	93.8	105.8	97.3	96.4	104.2	98.7
06	WOOD, PLASTICS & COMPOSITES	82.3	87.4	85.1	77.5	87.4	82.9	83.1	83.0	83.1	96.9	94.1	95.4	100.6	94.2	97.1	107.8	97.2	101.9
07	THERMAL & MOISTURE PROTECTION	103.7	82.5	94.5	103.8	83.8	95.1	104.1	86.7	96.6	105.4	93.4	100.2	104.0	94.6	99.9	106.1	97.4	102.3
08	OPENINGS	98.1	83.6	94.7	98.0	84.5	94.8	96.6	82.1	93.2	98.1	108.3	100.5	103.1	99.3	102.2	99.1	101.2	99.6
0920	Plaster & Gypsum Board	90.8	87.0	88.2	88.3	87.0	87.4	91.1	82.5	85.3	91.6	94.5	93.5	105.5	94.5	98.1	103.3	97.5	99.4
0950, 0980	Ceilings & Acoustic Treatment	90.0	87.0	88.0	90.0	87.0	88.0	90.0	82.5	84.9	84.4	94.5	91.2	90.9	94.5	93.3	83.6	97.5	93.0
0960	Flooring	90.6	95.2	91.9	88.7	98.9	91.6	93.5	95.2	94.0	94.8	118.5	101.4	80.5	106.5	87.8	96.0	114.8	101.3
0970, 0990	Wall Finishes & Painting/Coating	102.1	88.0	93.7	102.1	88.0	93.7	102.1	88.0	93.7	95.0	103.6	100.1	83.3	76.4	79.2	92.0	80.0	84.9
09	FINISHES	89.4	88.7	89.0	88.7	89.5	89.1	90.9	86.1	88.3	92.3	100.7	96.9	87.3	94.9	91.4	91.4	99.1	95.6
COVERS	DIVS. 10 - 14, 25, 28, 41, 43, 44, 46	100.0	55.0	90.0	100.0	91.0	98.0	100.0	84.4	96.5	100.0	98.6	99.7	100.0	94.1	98.7	100.0	95.7	99.0
21, 22, 23	FIRE SUPPRESSION, PLUMBING & HVAC	96.8	83.7	91.7	96.8	83.6	91.6	100.4	89.8	96.2	99.8	95.5	98.1	100.0	88.2	95.4	100.2	83.4	93.6
26, 27, 3370	ELECTRICAL, COMMUNICATIONS & UTIL.	99.7	76.9	88.4	99.0	76.9	88.1	94.1	89.3	91.7	100.6	84.3	92.5	104.3	84.4	94.5	99.0	81.2	90.2
MF2018	WEIGHTED AVERAGE	98.0	85.9	92.9	98.3	87.2	93.6	99.6	89.6	95.4	98.6	98.9	98.7	97.5	94.1	96.1	99.4	93.4	96.9

DIVISION		WISCONSIN																	
		KENOSHA			LA CROSSE			LANCASTER			MADISON			MILWAUKEE			NEW RICHMOND		
		531			546			538			537			530, 532			540		
		MAT.	INST.	TOTAL	MAT.	INST.	TOTAL	MAT.	INST.	TOTAL	MAT.	INST.	TOTAL	MAT.	INST.	TOTAL	MAT.	INST.	TOTAL
015433	CONTRACTOR EQUIPMENT		97.1	97.1		99.9	99.9		99.0	99.0		101.4	101.4		90.2	90.2		100.2	100.2
0241, 31 - 34	SITE & INFRASTRUCTURE, DEMOLITION	103.0	98.4	99.9	89.8	99.0	96.1	96.0	101.1	99.5	94.7	105.3	102.0	93.6	96.0	95.2	93.3	99.6	97.7
0310	Concrete Forming & Accessories	108.7	106.1	106.5	85.2	94.5	93.2	99.1	95.2	95.7	103.8	95.8	97.0	102.9	113.9	112.3	93.6	92.5	92.6
0320	Concrete Reinforcing	94.8	110.5	102.4	93.9	106.2	99.8	96.3	106.2	101.1	97.8	106.4	101.9	99.9	115.1	107.2	91.3	114.1	102.3
0330	Cast-in-Place Concrete	115.2	102.7	110.5	89.8	96.5	92.3	105.2	99.7	103.2	101.4	100.3	101.0	93.9	112.1	100.7	104.0	102.7	103.5
03	CONCRETE	101.5	105.5	103.3	84.6	97.5	90.3	96.4	98.8	97.4	96.3	99.2	97.6	94.5	112.5	102.4	90.8	100.1	94.9
04	MASONRY	97.1	108.0	103.8	90.5	96.1	93.9	99.4	100.1	99.9	97.4	99.1	98.5	102.3	117.2	111.4	117.9	98.4	105.9
05	METALS	99.6	103.3	100.7	93.7	102.8	96.4	96.0	100.3	97.3	102.8	100.4	102.1	97.5	97.5	97.5	94.1	105.4	97.4
06	WOOD, PLASTICS & COMPOSITES	103.9	105.7	104.9	85.2	94.2	90.2	96.1	94.1	95.0	99.4	94.3	96.6	101.2	112.3	107.3	89.5	90.2	89.9
07	THERMAL & MOISTURE PROTECTION	105.5	104.0	104.9	103.4	93.5	99.1	105.2	93.7	100.2	106.3	100.5	103.7	104.8	112.7	108.2	104.9	100.2	102.9
08	OPENINGS	92.5	105.4	95.5	103.1	92.0	100.5	94.0	91.9	93.5	101.6	98.9	100.9	103.0	111.5	105.0	88.6	90.3	89.0
0920	Plaster & Gypsum Board	80.2	106.4	97.8	100.9	94.5	96.6	90.9	94.5	93.3	101.7	94.5	96.8	96.4	112.6	107.3	91.7	90.5	90.9
0950, 0980	Ceilings & Acoustic Treatment	84.4	106.4	99.2	90.0	94.5	93.0	81.1	94.5	90.1	86.5	94.5	91.9	93.1	112.6	106.3	57.1	90.5	79.6
0960	Flooring	112.8	114.8	113.4	74.7	111.6	85.1	94.4	111.0	99.0	94.3	111.0	99.0	101.7	118.6	106.4	89.7	114.5	96.7
0970, 0990	Wall Finishes & Painting/Coating	106.4	118.7	113.7	83.3	81.7	82.4	95.0	97.7	96.6	94.9	103.6	100.0	104.5	124.3	116.3	95.0	81.7	87.1
09	FINISHES	97.1	109.4	103.7	84.4	96.5	90.9	91.4	98.4	95.2	93.7	99.3	96.7	99.5	116.1	108.5	82.5	95.2	89.4
COVERS	DIVS. 10 - 14, 25, 28, 41, 43, 44, 46	100.0	97.3	99.4	100.0	92.8	98.4	100.0	84.2	96.5	100.0	96.8	99.3	100.0	105.2	101.2	100.0	95.1	98.9
21, 22, 23	FIRE SUPPRESSION, PLUMBING & HVAC	100.0	95.9	98.4	100.0	88.1	95.3	96.3	89.2	93.5	99.8	95.3	98.0	99.8	107.0	102.6	96.0	89.0	93.2
26, 27, 3370	ELECTRICAL, COMMUNICATIONS & UTIL.	100.9	97.1	99.0	104.6	84.4	94.7	100.3	84.9	92.7	101.5	94.0	97.8	100.2	102.4	101.3	102.4	84.9	93.7
MF2018	WEIGHTED AVERAGE	99.4	102.0	100.5	95.9	93.5	94.9	96.7	94.3	95.7	99.7	98.1	99.0	99.2	107.9	102.9	95.4	94.6	95.0

DIVISION		WISCONSIN																	
		OSHKOSH			PORTAGE			RACINE			RHINELANDER			SUPERIOR			WAUSAU		
		549			539			534			545			548			544		
		MAT.	INST.	TOTAL	MAT.	INST.	TOTAL	MAT.	INST.	TOTAL	MAT.	INST.	TOTAL	MAT.	INST.	TOTAL	MAT.	INST.	TOTAL
015433	CONTRACTOR EQUIPMENT		97.7	97.7		99.0	99.0		99.0	99.0		97.7	97.7		100.2	100.2		97.7	97.7
0241, 31 - 34	SITE & INFRASTRUCTURE, DEMOLITION	91.0	95.3	94.0	86.9	100.9	96.6	96.9	102.1	100.5	103.2	95.3	97.7	90.1	99.3	96.5	86.9	95.8	93.0
0310	Concrete Forming & Accessories	90.7	94.7	94.1	91.3	95.7	95.1	100.1	106.1	105.2	88.3	94.6	93.7	91.6	88.6	89.0	90.1	97.4	96.3
0320	Concrete Reinforcing	92.4	109.3	100.6	96.4	106.4	101.2	95.0	110.5	102.5	92.6	106.4	99.3	91.3	110.2	100.4	92.6	106.2	99.2
0330	Cast-in-Place Concrete	95.8	98.0	96.6	90.3	98.3	93.3	103.8	102.5	103.3	108.6	97.9	104.6	98.0	98.3	98.1	89.3	92.4	90.5
03	CONCRETE	86.2	98.6	91.7	84.2	98.7	90.6	95.8	105.4	100.0	98.0	98.1	98.1	85.5	96.2	90.2	81.4	97.4	88.4
04	MASONRY	104.7	97.8	100.4	98.2	100.2	99.4	99.3	108.0	104.6	121.3	97.6	106.8	117.1	100.4	106.8	104.2	97.6	100.1
05	METALS	94.3	103.5	97.0	96.7	104.3	98.9	100.2	103.3	101.1	94.2	102.5	96.6	95.1	104.6	97.9	94.0	102.9	96.6
06	WOOD, PLASTICS & COMPOSITES	86.7	94.2	90.9	85.8	94.1	90.4	97.2	105.7	101.9	84.2	94.2	89.7	87.9	85.8	86.7	86.2	97.2	92.2
07	THERMAL & MOISTURE PROTECTION	105.0	83.3	95.6	104.5	100.1	102.6	105.5	102.7	104.3	106.0	83.1	96.0	104.5	95.9	100.8	104.8	95.1	100.6
08	OPENINGS	95.6	96.2	95.7	94.1	96.2	95.7	98.1	105.4	99.8	95.6	91.9	94.7	88.0	91.9	88.9	95.8	97.1	96.1
0920	Plaster & Gypsum Board	90.7	94.5	93.2	83.8	94.5	91.0	91.6	106.4	101.5	90.7	94.5	93.2	92.1	86.0	88.0	90.7	97.5	95.3
0950, 0980	Ceilings & Acoustic Treatment	83.6	94.5	90.9	83.6	94.5	90.9	84.4	106.4	99.2	83.6	94.5	90.9	58.8	86.0	77.1	83.6	97.5	93.0
0960	Flooring	87.8	114.8	95.4	90.5	114.5	97.3	94.8	114.8	100.4	87.2	114.5	94.9	90.8	123.1	99.9	87.7	114.5	95.2
0970, 0990	Wall Finishes & Painting/Coating	89.6	104.3	98.3	95.0	97.7	96.6	95.0	115.9	107.4	89.6	80.0	83.9	84.0	105.8	96.9	89.6	83.2	85.8
09	FINISHES	86.5	100.0	93.8	89.2	99.4	94.7	92.3	109.1	101.4	87.4	97.3	92.7	82.1	96.4	89.8	86.2	99.4	93.3
COVERS	DIVS. 10 - 14, 25, 28, 41, 43, 44, 46	100.0	85.6	96.8	100.0	85.6	96.8	100.0	97.3	99.4	100.0	85.9	96.8	100.0	93.9	98.6	100.0	95.7	99.0
21, 22, 23	FIRE SUPPRESSION, PLUMBING & HVAC	96.7	83.1	91.3	96.3	95.8	96.1	99.8	96.0	98.3	96.7	88.6	93.5	96.0	91.1	94.1	96.7	88.8	93.6
26, 27, 3370	ELECTRICAL, COMMUNICATIONS & UTIL.	103.0	79.1	91.2	104.0	94.0	99.1	100.4	98.0	99.2	102.3	78.1	90.4	107.2	98.5	102.9	104.1	78.1	91.3
MF2018	WEIGHTED AVERAGE	95.4	91.9	93.9	95.2	98.0	96.4	98.7	102.3	100.2	97.9	92.2	95.5	95.1	96.5	95.7	94.7	93.4	94.2

WYOMING

DIVISION		CASPER 826			CHEYENNE 820			NEWCASTLE 827			RAWLINS 823			RIVERTON 825			ROCK SPRINGS 829 - 831		
		MAT.	INST.	TOTAL	MAT.	INST.	TOTAL	MAT.	INST.	TOTAL	MAT.	INST.	TOTAL	MAT.	INST.	TOTAL	MAT.	INST.	TOTAL
015433	CONTRACTOR EQUIPMENT		97.3	97.3		94.6	94.6		94.6	94.6		94.6	94.6		94.6	94.6		94.6	94.6
0241, 31 - 34	SITE & INFRASTRUCTURE, DEMOLITION	97.8	94.7	95.7	91.6	90.1	90.6	83.6	89.3	87.6	97.5	89.3	91.9	91.1	89.3	89.8	87.4	89.5	88.9
0310	Concrete Forming & Accessories	101.3	53.6	60.6	102.5	64.6	70.2	93.7	70.5	74.0	97.6	70.8	74.7	92.7	60.3	65.0	99.7	64.7	69.9
0320	Concrete Reinforcing	111.1	83.4	97.7	106.4	83.4	95.3	114.5	83.2	99.3	114.2	83.2	99.2	115.2	83.2	99.7	115.2	81.6	99.0
0330	Cast-in-Place Concrete	103.8	77.8	94.1	97.9	77.4	90.3	99.0	72.8	89.2	99.0	72.9	89.3	99.0	72.8	89.2	99.0	75.0	90.0
03	CONCRETE	103.3	67.9	87.7	100.9	72.8	88.6	101.1	73.8	89.1	116.0	74.0	97.5	110.1	69.2	92.1	101.6	71.7	88.4
04	MASONRY	102.0	64.6	79.0	105.3	65.2	80.7	102.2	59.3	75.8	102.2	59.3	75.8	102.2	59.3	75.8	164.5	61.8	101.4
05	METALS	102.8	80.1	96.1	105.3	80.8	98.1	101.4	79.9	95.1	101.5	80.1	95.2	101.6	79.9	95.2	102.3	79.5	95.6
06	WOOD, PLASTICS & COMPOSITES	95.2	47.8	69.1	91.8	62.6	75.7	82.4	72.7	77.1	86.2	72.7	78.8	81.4	58.7	68.9	91.3	62.6	75.5
07	THERMAL & MOISTURE PROTECTION	112.7	65.9	92.3	107.2	68.0	90.2	108.8	63.9	89.3	110.3	74.9	94.9	109.7	67.7	91.4	108.9	69.5	91.8
08	OPENINGS	109.3	59.4	97.7	107.2	67.7	98.0	111.4	72.3	102.3	111.1	72.3	102.0	111.3	64.6	100.4	111.8	66.5	101.2
0920	Plaster & Gypsum Board	101.7	46.2	64.4	90.6	61.7	71.2	87.6	72.0	77.1	87.9	72.0	77.2	87.6	57.7	67.5	100.2	61.7	74.3
0950, 0980	Ceilings & Acoustic Treatment	105.3	46.2	65.5	98.5	61.7	73.7	100.3	72.0	81.2	100.3	72.0	81.2	100.3	57.7	71.6	100.3	61.7	74.3
0960	Flooring	102.2	73.8	94.2	100.3	73.8	92.8	94.2	44.5	80.2	97.0	44.5	82.3	93.6	60.3	84.2	99.3	59.9	88.3
0970, 0990	Wall Finishes & Painting/Coating	95.7	58.7	73.8	97.9	58.7	74.7	94.6	60.5	74.4	94.6	60.5	74.4	94.6	60.5	74.4	94.6	60.5	74.4
09	FINISHES	99.6	56.3	76.2	96.3	65.1	79.4	91.4	64.6	76.9	93.6	64.6	77.9	92.0	59.5	74.4	94.6	62.7	77.3
COVERS	DIVS. 10 - 14, 25, 28, 41, 43, 44, 46	100.0	92.5	98.3	100.0	87.4	97.2	100.0	91.8	98.2	100.0	91.8	98.2	100.0	77.3	94.9	100.0	86.9	97.1
21, 22, 23	FIRE SUPPRESSION, PLUMBING & HVAC	100.0	70.5	88.4	100.1	73.3	89.6	98.2	71.8	87.8	98.2	71.9	87.8	98.2	71.8	87.8	100.0	73.9	89.8
26, 27, 3370	ELECTRICAL, COMMUNICATIONS & UTIL.	94.1	60.3	77.5	95.2	69.0	82.3	94.0	62.3	78.4	94.0	62.3	78.4	94.0	59.1	76.8	92.5	66.0	79.4
MF2018	WEIGHTED AVERAGE	101.5	68.9	87.7	101.1	72.7	89.1	99.6	71.1	87.5	102.0	71.5	89.1	101.0	68.6	87.3	103.4	71.4	89.9

WYOMING / CANADA

DIVISION		SHERIDAN 828			WHEATLAND 822			WORLAND 824			YELLOWSTONE NAT'L PA 821			BARRIE, ONTARIO			BATHURST, NEW BRUNSWICK		
		MAT.	INST.	TOTAL	MAT.	INST.	TOTAL	MAT.	INST.	TOTAL	MAT.	INST.	TOTAL	MAT.	INST.	TOTAL	MAT.	INST.	TOTAL
015433	CONTRACTOR EQUIPMENT		94.6	94.6		94.6	94.6		94.6	94.6		94.6	94.6		100.1	100.1		99.7	99.7
0241, 31 - 34	SITE & INFRASTRUCTURE, DEMOLITION	91.5	90.1	90.5	88.1	89.3	88.9	85.5	89.3	88.1	85.6	89.3	88.1	117.0	94.7	101.6	105.1	90.6	95.1
0310	Concrete Forming & Accessories	100.5	61.6	67.4	95.4	48.4	55.3	95.5	61.1	66.1	95.5	61.1	66.2	125.6	82.3	88.7	106.1	57.8	64.9
0320	Concrete Reinforcing	115.2	83.2	99.7	114.5	82.6	99.1	115.2	83.2	99.7	117.2	82.3	100.3	175.2	86.9	132.5	140.5	57.5	100.4
0330	Cast-in-Place Concrete	102.3	77.4	93.0	103.3	72.7	91.9	101.3	72.7	89.2	99.0	72.8	89.2	157.7	82.1	129.6	114.0	56.2	92.5
03	CONCRETE	109.9	71.4	93.0	106.3	63.7	87.6	101.3	69.5	87.3	101.6	69.4	87.4	138.9	83.5	114.5	110.3	58.3	87.4
04	MASONRY	102.5	62.0	77.6	102.6	51.2	71.0	102.2	59.3	75.8	102.2	59.3	75.8	165.6	89.3	118.7	159.8	57.4	96.9
05	METALS	105.1	80.5	97.8	101.4	79.1	94.8	101.6	79.8	95.2	102.2	79.1	95.4	111.1	91.5	105.3	114.1	73.6	102.2
06	WOOD, PLASTICS & COMPOSITES	93.3	58.7	74.3	84.2	43.0	61.5	84.3	60.0	70.9	84.3	60.0	70.9	116.5	80.8	96.9	97.5	57.9	75.7
07	THERMAL & MOISTURE PROTECTION	110.1	66.7	91.2	109.1	58.1	86.9	108.9	64.8	89.7	108.3	64.8	89.4	115.3	86.1	102.6	111.8	58.4	88.6
08	OPENINGS	112.0	64.6	101.0	110.0	55.9	97.4	111.6	65.3	100.8	104.5	64.8	95.2	90.5	80.9	88.3	84.0	51.2	76.3
0920	Plaster & Gypsum Board	111.2	57.7	75.2	87.6	41.5	56.6	87.6	58.9	68.3	87.8	58.9	68.4	153.5	80.3	104.3	124.2	56.7	78.8
0950, 0980	Ceilings & Acoustic Treatment	103.9	57.7	72.7	100.3	41.5	60.6	100.3	58.9	72.4	101.1	58.9	72.7	94.3	80.3	84.9	116.8	56.7	76.3
0960	Flooring	97.9	60.3	87.4	95.6	43.8	81.0	95.6	44.5	81.3	95.6	44.5	81.3	119.3	87.3	110.3	99.6	41.5	83.3
0970, 0990	Wall Finishes & Painting/Coating	96.8	58.7	74.2	94.6	58.4	73.2	94.6	60.5	74.4	94.6	60.5	74.4	105.6	83.8	92.7	111.3	47.9	73.8
09	FINISHES	99.2	60.4	78.2	92.1	46.7	67.5	91.8	57.1	73.0	92.0	57.2	73.2	111.3	83.5	96.2	107.7	53.8	78.5
COVERS	DIVS. 10 - 14, 25, 28, 41, 43, 44, 46	100.0	93.4	98.5	100.0	81.4	95.9	100.0	76.1	94.7	100.0	76.2	94.7	139.2	65.0	122.7	131.1	57.9	114.8
21, 22, 23	FIRE SUPPRESSION, PLUMBING & HVAC	98.2	73.3	88.4	98.2	71.8	87.8	98.2	71.8	87.8	98.2	71.8	87.8	103.7	93.5	99.7	103.8	64.8	88.5
26, 27, 3370	ELECTRICAL, COMMUNICATIONS & UTIL.	96.1	59.1	77.9	94.0	76.6	85.4	94.0	76.6	85.4	93.1	86.1	89.7	117.0	82.9	100.2	113.5	55.9	85.1
MF2018	WEIGHTED AVERAGE	102.5	70.2	88.8	100.3	67.1	86.2	99.7	70.7	87.5	99.0	71.9	87.6	116.5	87.1	104.1	110.7	62.3	90.2

CANADA

DIVISION		BRANDON, MANITOBA			BRANTFORD, ONTARIO			BRIDGEWATER, NOVA SCOTIA			CALGARY, ALBERTA			CAP-DE-LA-MADELEINE, QUEBEC			CHARLESBOURG, QUEBEC		
		MAT.	INST.	TOTAL	MAT.	INST.	TOTAL	MAT.	INST.	TOTAL	MAT.	INST.	TOTAL	MAT.	INST.	TOTAL	MAT.	INST.	TOTAL
015433	CONTRACTOR EQUIPMENT		101.5	101.5		99.7	99.7		99.4	99.4		127.1	127.1		100.2	100.2		100.2	100.2
0241, 31 - 34	SITE & INFRASTRUCTURE, DEMOLITION	126.4	92.9	103.3	116.9	94.8	101.6	101.1	92.3	95.0	125.5	118.4	120.6	96.9	93.7	94.7	96.9	93.7	94.7
0310	Concrete Forming & Accessories	145.4	66.1	77.8	127.2	89.1	94.7	99.7	67.9	72.6	125.0	94.9	99.3	132.9	79.2	87.1	132.9	79.2	87.1
0320	Concrete Reinforcing	170.2	54.5	114.3	163.7	85.6	125.9	139.3	47.6	95.0	136.4	82.0	110.1	139.3	72.4	107.0	139.3	72.4	107.0
0330	Cast-in-Place Concrete	109.4	70.2	94.8	130.7	101.8	119.9	134.3	66.9	109.2	136.4	104.3	124.4	105.5	87.5	98.8	105.5	87.5	98.8
03	CONCRETE	118.1	66.6	95.5	121.8	93.2	109.3	120.1	65.1	95.9	126.3	96.8	113.3	107.5	81.5	96.1	107.5	81.5	96.1
04	MASONRY	212.8	60.7	119.3	166.1	93.4	121.4	161.8	65.6	102.6	212.4	89.8	137.1	162.4	76.3	109.5	162.4	76.3	109.5
05	METALS	127.2	78.2	112.8	112.3	92.0	106.3	111.6	76.0	101.1	131.0	102.4	122.6	110.7	85.0	103.1	110.7	85.0	103.1
06	WOOD, PLASTICS & COMPOSITES	150.5	66.8	104.4	120.3	87.7	102.4	89.6	67.4	77.4	102.6	94.4	98.1	131.6	79.0	102.7	131.6	79.0	102.7
07	THERMAL & MOISTURE PROTECTION	128.4	68.7	102.4	121.6	91.4	108.5	116.1	67.6	95.0	132.0	98.0	117.2	114.8	84.1	101.5	114.8	84.1	101.5
08	OPENINGS	100.3	59.9	90.8	87.7	86.6	87.5	82.4	61.3	77.5	82.3	83.7	82.6	89.1	72.3	85.2	89.1	72.3	85.2
0920	Plaster & Gypsum Board	115.4	65.6	81.9	115.8	87.4	96.7	124.3	66.5	85.4	126.4	93.3	104.1	147.6	78.4	101.0	147.6	78.4	101.0
0950, 0980	Ceilings & Acoustic Treatment	123.8	65.6	84.6	123.5	87.4	99.2	103.5	66.5	78.6	149.2	93.3	111.5	103.5	78.4	86.6	103.5	78.4	86.6
0960	Flooring	131.2	61.9	111.7	113.6	87.3	106.2	95.7	58.9	85.4	118.1	84.8	108.7	113.6	86.0	105.8	113.6	86.0	105.8
0970, 0990	Wall Finishes & Painting/Coating	117.9	54.1	80.1	110.3	92.0	99.5	110.3	59.3	80.1	113.7	106.4	109.4	110.3	83.6	94.5	110.3	83.6	94.5
09	FINISHES	122.9	64.3	91.2	108.7	89.3	98.2	103.8	65.5	83.1	123.6	94.8	108.0	112.2	81.2	95.4	112.2	81.2	95.4
COVERS	DIVS. 10 - 14, 25, 28, 41, 43, 44, 46	131.1	60.1	115.3	131.1	66.9	116.8	131.1	60.8	115.5	131.1	92.7	122.6	131.1	75.6	118.8	131.1	75.6	118.8
21, 22, 23	FIRE SUPPRESSION, PLUMBING & HVAC	104.0	78.9	94.1	103.8	96.3	100.9	103.8	79.5	94.3	105.1	90.2	99.2	104.2	84.2	96.4	104.2	84.2	96.4
26, 27, 3370	ELECTRICAL, COMMUNICATIONS & UTIL.	113.9	63.1	88.8	111.5	82.5	97.2	115.8	58.7	87.6	107.1	93.9	100.6	110.2	65.2	88.0	110.2	65.2	88.0
MF2018	WEIGHTED AVERAGE	120.6	70.5	99.4	113.1	90.7	103.6	111.3	70.3	94.0	119.7	95.6	109.5	110.6	80.0	97.7	110.6	80.0	97.7

CANADA

DIVISION		CHARLOTTETOWN, PRINCE EDWARD ISLAND			CHICOUTIMI, QUEBEC			CORNER BROOK, NEWFOUNDLAND			CORNWALL, ONTARIO			DALHOUSIE, NEW BRUNSWICK			DARTMOUTH, NOVA SCOTIA		
		MAT.	INST.	TOTAL	MAT.	INST.	TOTAL	MAT.	INST.	TOTAL	MAT.	INST.	TOTAL	MAT.	INST.	TOTAL	MAT.	INST.	TOTAL
015433	CONTRACTOR EQUIPMENT		116.8	116.8		100.3	100.3		100.0	100.0		99.7	99.7		99.7	99.7		98.9	98.9
0241, 31 - 34	SITE & INFRASTRUCTURE, DEMOLITION	134.6	101.5	111.7	102.0	93.6	96.2	130.4	90.5	102.8	115.1	94.3	100.7	101.0	90.6	93.8	117.9	92.0	100.0
0310	Concrete Forming & Accessories	123.7	53.1	63.5	134.9	89.8	96.4	121.9	75.6	82.4	125.0	82.4	88.7	105.9	58.0	65.1	111.8	67.8	74.3
0320	Concrete Reinforcing	156.7	46.6	103.5	106.0	94.5	100.4	156.0	48.8	104.2	163.7	85.3	125.8	142.5	57.6	101.4	163.2	47.6	107.3
0330	Cast-in-Place Concrete	152.1	57.6	116.9	107.4	94.7	102.7	127.3	63.6	103.6	117.5	92.8	108.3	110.5	56.3	90.3	122.9	66.8	102.0
03	CONCRETE	136.8	55.4	101.0	101.6	92.4	97.6	149.1	67.5	113.3	115.7	87.0	103.1	113.1	58.4	89.0	131.9	65.0	102.5
04	MASONRY	187.9	54.9	106.2	161.9	89.8	117.6	208.6	74.3	126.1	165.0	85.6	116.2	163.3	57.4	98.3	222.6	65.6	126.1
05	METALS	136.4	79.6	119.8	113.9	91.9	107.4	127.7	75.2	112.2	112.1	90.7	105.8	105.5	73.7	96.2	128.1	75.7	112.7
06	WOOD, PLASTICS & COMPOSITES	104.3	52.5	75.8	131.8	90.3	108.9	128.1	81.4	102.4	118.5	81.6	98.2	95.8	57.9	74.9	115.9	67.4	89.2
07	THERMAL & MOISTURE PROTECTION	137.8	57.4	102.8	112.3	96.4	105.4	133.0	65.9	103.8	121.4	86.2	106.1	121.0	58.4	93.8	130.1	67.6	102.9
08	OPENINGS	85.7	45.8	76.4	87.8	76.3	85.2	106.3	66.6	97.0	89.1	80.6	87.1	85.5	51.2	77.5	91.3	61.3	84.3
0920	Plaster & Gypsum Board	126.3	50.5	75.3	143.9	89.9	107.6	151.2	80.9	103.9	174.9	81.1	111.8	129.5	56.7	80.5	146.6	66.5	92.7
0950, 0980	Ceilings & Acoustic Treatment	128.9	50.5	76.1	116.0	89.9	98.4	124.6	80.9	95.1	107.6	81.1	89.8	106.6	56.7	72.9	131.1	66.5	87.5
0960	Flooring	115.1	55.3	98.3	115.7	86.0	107.3	112.8	49.5	95.0	113.6	86.0	105.8	101.9	63.4	91.1	107.6	58.9	93.9
0970, 0990	Wall Finishes & Painting/Coating	117.3	39.9	71.5	111.3	104.1	107.1	117.8	56.5	81.5	110.3	85.7	95.7	113.8	47.9	74.8	117.8	59.3	83.2
09	FINISHES	119.8	52.3	83.3	114.6	91.6	102.1	122.6	70.2	94.2	117.3	83.6	99.0	108.4	58.2	81.2	120.3	65.5	90.7
COVERS	DIVS. 10 - 14, 25, 28, 41, 43, 44, 46	131.1	58.4	114.9	131.1	81.5	120.1	131.1	60.6	115.4	131.1	64.6	116.3	131.1	57.8	114.8	131.1	60.8	115.5
21, 22, 23	FIRE SUPPRESSION, PLUMBING & HVAC	104.1	59.0	86.4	103.8	81.0	94.9	104.0	55.9	89.0	104.2	94.3	100.3	103.9	64.8	88.5	104.0	79.5	94.4
26, 27, 3370	ELECTRICAL, COMMUNICATIONS & UTIL.	110.0	47.5	79.2	109.9	84.1	97.2	110.9	51.5	81.6	111.6	83.4	97.7	113.9	52.8	83.8	115.0	58.7	87.2
MF2018	WEIGHTED AVERAGE	121.3	60.2	95.5	110.4	87.6	100.7	124.7	68.4	100.9	113.1	87.3	102.2	110.2	62.5	90.0	121.5	70.2	99.8

CANADA

DIVISION		EDMONTON, ALBERTA			FORT MCMURRAY, ALBERTA			FREDERICTON, NEW BRUNSWICK			GATINEAU, QUEBEC			GRANBY, QUEBEC			HALIFAX, NOVA SCOTIA		
		MAT.	INST.	TOTAL	MAT.	INST.	TOTAL	MAT.	INST.	TOTAL	MAT.	INST.	TOTAL	MAT.	INST.	TOTAL	MAT.	INST.	TOTAL
015433	CONTRACTOR EQUIPMENT		129.1	129.1		101.9	101.9		113.9	113.9		100.2	100.2		100.2	100.2		115.6	115.6
0241, 31 - 34	SITE & INFRASTRUCTURE, DEMOLITION	125.1	121.8	122.8	121.7	95.4	103.5	116.0	100.8	105.5	96.7	93.7	94.6	97.2	93.7	94.8	102.8	103.6	103.4
0310	Concrete Forming & Accessories	128.3	94.9	99.8	125.0	87.5	93.1	126.9	58.6	68.7	132.9	79.1	87.0	132.9	79.0	87.0	124.1	87.7	93.0
0320	Concrete Reinforcing	136.0	82.0	109.9	151.5	81.9	117.8	141.0	57.8	100.8	147.4	72.4	111.1	147.4	72.4	111.1	156.3	77.9	118.3
0330	Cast-in-Place Concrete	145.8	104.3	130.4	173.5	98.2	145.5	112.7	57.5	92.1	104.0	87.5	97.9	107.5	87.4	100.0	99.4	85.1	94.0
03	CONCRETE	130.8	96.8	115.9	139.8	90.6	118.2	116.9	59.7	91.8	107.9	81.4	96.2	109.5	81.4	97.1	111.9	86.1	100.6
04	MASONRY	212.7	89.8	137.2	206.1	86.3	132.5	189.1	58.8	109.0	162.3	76.3	109.5	162.6	76.3	109.6	185.4	86.4	124.6
05	METALS	134.6	102.4	125.1	139.8	90.9	125.5	136.6	83.3	121.0	110.7	84.8	103.1	110.9	84.8	103.2	136.3	99.2	125.4
06	WOOD, PLASTICS & COMPOSITES	101.6	94.4	97.6	114.5	86.7	99.2	106.5	58.2	79.9	131.6	79.0	102.7	131.6	79.0	102.7	103.7	87.8	95.0
07	THERMAL & MOISTURE PROTECTION	143.5	98.0	123.7	129.8	92.6	113.6	137.4	59.4	103.5	114.8	84.1	101.5	114.8	82.6	100.8	137.4	88.7	116.2
08	OPENINGS	80.9	83.7	81.5	89.1	79.5	86.9	88.0	50.3	79.2	89.1	67.9	84.2	89.1	67.9	84.2	88.9	79.2	86.7
0920	Plaster & Gypsum Board	129.8	93.3	105.2	116.6	86.1	96.1	130.3	56.7	80.8	114.4	78.4	90.2	114.4	78.4	90.2	117.8	87.1	97.1
0950, 0980	Ceilings & Acoustic Treatment	144.5	93.3	110.0	112.5	86.1	94.7	137.2	56.7	82.9	103.5	78.4	86.6	103.5	78.4	86.6	139.8	87.1	104.2
0960	Flooring	121.8	84.8	111.4	113.6	84.8	105.5	115.6	66.2	101.7	113.6	86.0	105.8	113.6	86.0	105.8	109.5	81.5	101.6
0970, 0990	Wall Finishes & Painting/Coating	115.6	107.4	110.8	110.4	88.3	97.3	118.8	61.1	84.6	110.3	83.6	94.5	110.3	83.6	94.5	118.4	90.9	102.1
09	FINISHES	123.6	94.9	108.0	111.8	87.4	98.6	125.1	60.4	90.1	107.8	81.2	93.4	107.8	81.2	93.4	117.7	87.5	101.3
COVERS	DIVS. 10 - 14, 25, 28, 41, 43, 44, 46	131.1	93.6	122.8	131.1	90.1	122.0	131.1	58.5	115.0	131.1	75.6	118.8	131.1	75.6	118.8	131.1	68.9	117.3
21, 22, 23	FIRE SUPPRESSION, PLUMBING & HVAC	104.9	90.4	99.2	104.2	93.2	99.9	104.2	73.7	92.2	104.2	84.2	96.3	103.8	84.2	96.1	104.9	82.3	96.0
26, 27, 3370	ELECTRICAL, COMMUNICATIONS & UTIL.	115.0	93.9	104.6	105.4	77.5	91.7	112.2	69.7	91.2	110.2	65.2	88.0	110.9	65.2	88.4	113.3	90.6	102.1
MF2018	WEIGHTED AVERAGE	121.8	95.9	110.9	121.9	88.4	107.7	119.3	69.0	98.0	110.3	79.8	97.4	110.5	79.7	97.5	117.9	88.0	105.2

CANADA

DIVISION		HAMILTON, ONTARIO			HULL, QUEBEC			JOLIETTE, QUEBEC			KAMLOOPS, BRITISH COLUMBIA			KINGSTON, ONTARIO			KITCHENER, ONTARIO		
		MAT.	INST.	TOTAL	MAT.	INST.	TOTAL	MAT.	INST.	TOTAL	MAT.	INST.	TOTAL	MAT.	INST.	TOTAL	MAT.	INST.	TOTAL
015433	CONTRACTOR EQUIPMENT		115.5	115.5		100.2	100.2		100.2	100.2		103.3	103.3		101.9	101.9		101.6	101.6
0241, 31 - 34	SITE & INFRASTRUCTURE, DEMOLITION	105.8	107.0	106.6	96.7	93.7	94.6	97.3	93.7	94.8	120.1	97.1	104.3	115.1	98.0	103.3	94.2	99.8	98.0
0310	Concrete Forming & Accessories	130.0	94.7	99.9	132.9	79.1	87.0	132.9	79.2	87.1	124.3	82.3	88.5	125.1	82.5	88.7	117.3	88.0	92.3
0320	Concrete Reinforcing	136.9	103.3	120.7	147.4	72.4	111.1	139.3	72.4	107.0	109.3	75.8	93.1	163.7	85.3	125.8	96.4	103.1	99.6
0330	Cast-in-Place Concrete	107.6	102.1	105.6	104.0	87.5	97.9	108.4	87.5	100.6	93.8	91.8	93.1	117.5	92.8	108.3	111.7	97.2	106.3
03	CONCRETE	110.2	99.3	105.4	107.9	81.4	96.2	108.8	81.5	96.8	117.2	85.0	103.0	117.5	87.0	104.1	97.1	94.1	95.8
04	MASONRY	173.2	100.5	128.5	162.3	76.3	109.5	162.7	76.3	109.6	169.4	84.0	116.9	171.9	85.7	118.9	142.2	98.6	115.4
05	METALS	134.7	106.4	126.4	110.9	84.8	103.2	110.9	85.0	103.3	112.8	87.1	105.3	113.5	90.7	106.8	124.1	98.0	116.5
06	WOOD, PLASTICS & COMPOSITES	111.4	94.0	101.8	131.6	79.0	102.7	131.6	79.0	102.7	102.2	80.7	90.3	118.5	81.7	98.3	110.3	85.9	96.9
07	THERMAL & MOISTURE PROTECTION	134.0	101.5	119.8	114.8	84.1	101.5	114.8	84.1	101.5	131.2	81.7	109.7	121.4	87.3	106.6	116.4	98.4	108.5
08	OPENINGS	85.3	93.2	87.1	89.1	67.9	84.2	89.1	72.3	85.2	86.1	78.5	84.3	89.1	80.3	87.1	80.0	87.1	81.7
0920	Plaster & Gypsum Board	131.0	93.1	105.5	114.4	78.4	90.2	147.6	78.4	101.0	101.1	79.8	86.8	177.6	81.2	112.8	107.2	85.6	92.6
0950, 0980	Ceilings & Acoustic Treatment	139.9	93.1	108.4	103.5	78.4	86.6	103.5	78.4	86.6	103.5	79.8	87.5	119.0	81.2	93.5	107.9	85.6	92.9
0960	Flooring	115.3	95.0	109.6	113.6	86.0	105.8	113.6	86.0	105.8	112.5	49.2	94.7	113.6	86.0	105.8	101.3	95.1	99.6
0970, 0990	Wall Finishes & Painting/Coating	113.0	104.1	107.8	110.3	83.6	94.5	110.3	83.6	94.5	110.3	76.4	90.3	110.3	79.2	91.9	104.4	93.9	98.2
09	FINISHES	121.0	95.6	107.3	107.8	81.2	93.4	112.2	81.2	95.4	108.3	75.8	90.7	119.9	82.9	99.9	102.1	89.4	95.2
COVERS	DIVS. 10 - 14, 25, 28, 41, 43, 44, 46	131.1	92.5	122.5	131.1	75.6	118.8	131.1	75.6	118.8	131.1	84.4	120.7	131.1	64.6	116.3	131.1	89.6	121.9
21, 22, 23	FIRE SUPPRESSION, PLUMBING & HVAC	105.2	92.1	100.0	103.8	84.2	96.1	103.8	84.2	96.1	103.8	87.5	97.4	104.2	94.4	100.4	104.2	91.3	99.1
26, 27, 3370	ELECTRICAL, COMMUNICATIONS & UTIL.	108.5	103.6	106.1	112.3	65.2	89.1	110.9	65.2	88.4	114.3	74.2	94.5	111.6	82.1	97.1	109.6	99.8	104.8
MF2018	WEIGHTED AVERAGE	116.3	98.8	108.9	110.5	79.8	97.5	110.8	80.0	97.8	113.0	83.5	100.5	114.1	87.4	102.8	108.6	94.6	102.7

CANADA

DIVISION		LAVAL, QUEBEC			LETHBRIDGE, ALBERTA			LLOYDMINSTER, ALBERTA			LONDON, ONTARIO			MEDICINE HAT, ALBERTA			MONCTON, NEW BRUNSWICK		
		MAT.	INST.	TOTAL	MAT.	INST.	TOTAL	MAT.	INST.	TOTAL	MAT.	INST.	TOTAL	MAT.	INST.	TOTAL	MAT.	INST.	TOTAL
015433	CONTRACTOR EQUIPMENT		100.2	100.2		101.9	101.9		101.9	101.9		116.7	116.7		101.9	101.9		99.7	99.7
0241, 31 - 34	SITE & INFRASTRUCTURE, DEMOLITION	97.2	93.9	94.9	114.5	96.0	101.7	114.5	95.4	101.3	103.5	106.8	105.8	113.2	95.5	101.0	104.5	92.5	96.2
0310	Concrete Forming & Accessories	133.2	81.1	88.8	126.4	87.6	93.3	124.6	78.3	85.1	129.1	89.1	95.0	126.4	78.2	85.3	106.1	66.1	72.0
0320	Concrete Reinforcing	147.4	74.4	112.1	151.5	81.9	117.8	151.5	81.8	117.8	126.7	102.1	114.8	151.5	81.8	117.8	140.5	69.8	106.3
0330	Cast-in-Place Concrete	107.5	89.4	100.7	130.1	98.2	118.2	120.7	94.6	111.0	120.2	100.8	113.0	120.7	94.6	111.0	109.8	68.0	94.3
03	CONCRETE	109.5	83.4	98.0	120.0	90.6	107.1	115.5	85.2	102.2	114.6	96.2	106.5	115.7	85.2	102.3	108.4	68.5	90.9
04	MASONRY	162.6	78.4	110.8	180.1	86.3	122.4	161.6	80.1	111.6	183.8	99.5	132.0	161.6	80.1	111.6	159.4	74.6	107.3
05	METALS	110.7	85.9	103.5	134.1	91.0	121.5	112.8	90.8	106.4	134.7	106.8	126.5	113.0	90.8	106.5	114.1	86.8	106.1
06	WOOD, PLASTICS & COMPOSITES	131.7	81.1	103.9	118.0	86.7	100.8	114.5	77.4	94.1	114.8	86.8	99.4	118.0	77.4	95.7	97.5	65.0	79.6
07	THERMAL & MOISTURE PROTECTION	115.4	86.0	102.6	127.1	92.6	112.1	123.5	88.2	108.1	129.9	99.0	116.5	130.5	88.2	112.1	116.6	72.4	97.4
08	OPENINGS	89.1	69.8	84.6	89.1	79.5	86.9	89.1	74.4	85.7	81.6	87.9	83.1	89.1	74.4	85.7	84.0	61.9	78.8
0920	Plaster & Gypsum Board	114.7	80.5	91.7	106.4	86.1	92.7	102.1	76.6	84.9	134.4	85.8	101.7	104.3	76.6	85.6	124.2	64.1	83.7
0950, 0980	Ceilings & Acoustic Treatment	103.5	80.5	88.0	112.5	86.1	94.7	103.5	76.6	85.3	138.5	85.8	102.9	103.5	76.6	85.3	116.8	64.1	81.2
0960	Flooring	113.6	88.4	106.5	113.6	84.8	105.5	113.6	84.8	105.5	109.9	95.1	105.7	113.6	84.8	105.5	99.6	65.6	90.1
0970, 0990	Wall Finishes & Painting/Coating	110.3	85.9	95.8	110.2	96.4	102.1	110.4	75.1	89.5	113.5	100.5	105.8	110.2	75.1	89.4	111.3	83.6	94.9
09	FINISHES	107.8	83.3	94.5	109.6	88.3	98.0	107.4	78.9	92.0	121.5	91.0	105.0	107.5	78.9	92.0	107.7	67.5	86.0
COVERS	DIVS. 10 - 14, 25, 28, 41, 43, 44, 46	131.1	77.5	119.2	131.1	89.2	121.8	131.1	87.1	121.3	131.1	91.6	122.3	131.1	86.1	121.1	131.1	61.0	115.5
21, 22, 23	FIRE SUPPRESSION, PLUMBING & HVAC	104.3	86.5	97.3	104.1	90.1	98.6	104.2	90.0	98.6	105.1	89.2	98.8	103.8	86.8	97.2	103.8	73.7	92.0
26, 27, 3370	ELECTRICAL, COMMUNICATIONS & UTIL.	110.9	67.0	89.3	107.1	77.5	92.5	104.6	77.5	91.2	103.5	102.0	102.8	104.6	77.5	91.2	117.3	88.5	103.1
MF2018	WEIGHTED AVERAGE	110.7	81.6	98.4	117.0	87.9	104.7	111.5	84.7	100.2	116.3	96.5	108.0	111.7	84.0	100.0	110.9	76.0	96.2

CANADA

DIVISION		MONTREAL, QUEBEC			MOOSE JAW, SASKATCHEWAN			NEW GLASGOW, NOVA SCOTIA			NEWCASTLE, NEW BRUNSWICK			NORTH BAY, ONTARIO			OSHAWA, ONTARIO		
		MAT.	INST.	TOTAL	MAT.	INST.	TOTAL	MAT.	INST.	TOTAL	MAT.	INST.	TOTAL	MAT.	INST.	TOTAL	MAT.	INST.	TOTAL
015433	CONTRACTOR EQUIPMENT		117.7	117.7		98.3	98.3		98.9	98.9		99.7	99.7		99.3	99.3		101.6	101.6
0241, 31 - 34	SITE & INFRASTRUCTURE, DEMOLITION	113.0	105.5	107.8	114.4	90.0	97.5	111.9	92.0	98.1	105.1	90.6	95.1	126.8	93.6	103.9	104.6	99.6	101.2
0310	Concrete Forming & Accessories	130.9	90.4	96.4	109.6	54.6	62.7	111.8	67.8	74.3	106.1	58.0	65.1	146.2	80.0	89.8	122.8	91.3	96.0
0320	Concrete Reinforcing	126.8	94.6	111.3	106.9	61.4	84.9	156.0	47.6	103.6	140.5	57.6	100.4	184.7	84.8	136.4	152.7	103.7	129.0
0330	Cast-in-Place Concrete	124.4	97.4	114.3	117.1	63.7	97.2	122.9	66.8	102.0	114.0	56.3	92.6	118.4	80.1	104.1	129.2	106.4	120.7
03	CONCRETE	116.7	94.3	106.8	101.4	60.0	83.2	131.0	65.0	102.0	110.3	58.4	87.5	132.7	81.4	110.2	118.3	98.8	109.8
04	MASONRY	178.8	89.8	124.1	160.3	56.4	96.4	208.2	65.6	120.5	159.8	57.4	96.9	214.4	81.9	133.0	145.1	101.9	118.6
05	METALS	144.3	102.2	131.9	109.6	75.0	99.4	125.6	75.7	111.0	114.1	73.8	102.3	126.5	90.2	115.8	114.2	98.6	109.6
06	WOOD, PLASTICS & COMPOSITES	111.3	90.8	100.0	99.4	53.2	74.0	115.9	67.4	89.2	97.5	57.9	75.7	153.5	80.2	113.2	117.6	89.4	102.1
07	THERMAL & MOISTURE PROTECTION	125.9	97.2	113.4	114.1	60.1	90.6	130.1	67.6	102.9	116.6	58.4	91.3	136.4	82.7	113.0	117.3	104.1	111.6
08	OPENINGS	87.4	78.7	85.4	85.3	51.4	77.4	91.3	61.3	84.3	84.0	51.2	76.3	98.6	78.2	93.8	85.0	90.2	86.2
0920	Plaster & Gypsum Board	126.2	89.9	101.8	98.2	51.9	67.1	145.0	66.5	92.2	124.2	56.7	78.8	140.6	79.8	99.7	110.7	89.2	96.2
0950, 0980	Ceilings & Acoustic Treatment	146.1	89.9	108.2	103.5	51.9	68.7	123.8	66.5	85.2	116.8	56.7	76.3	123.8	79.8	94.1	103.9	89.2	94.0
0960	Flooring	112.4	90.5	106.2	103.6	53.7	89.6	107.6	58.9	93.9	99.6	63.4	89.5	131.2	86.0	118.5	104.2	97.6	102.4
0970, 0990	Wall Finishes & Painting/Coating	115.0	104.1	108.5	110.3	61.0	81.1	117.8	59.3	83.2	111.3	47.9	73.8	117.8	85.0	98.4	104.4	108.1	106.6
09	FINISHES	121.6	92.8	106.0	103.7	54.8	77.2	118.7	65.5	89.9	107.7	58.2	80.9	125.9	81.9	102.1	103.2	93.7	98.1
COVERS	DIVS. 10 - 14, 25, 28, 41, 43, 44, 46	131.1	82.8	120.4	131.1	57.5	114.7	131.1	60.8	115.5	131.1	57.9	114.8	131.1	63.4	116.0	131.1	90.4	122.0
21, 22, 23	FIRE SUPPRESSION, PLUMBING & HVAC	105.1	81.2	95.7	104.2	70.3	90.9	104.0	79.5	94.4	103.8	64.8	88.5	104.0	92.4	99.4	104.2	91.9	99.3
26, 27, 3370	ELECTRICAL, COMMUNICATIONS & UTIL.	107.9	84.1	96.2	113.4	56.3	85.3	111.3	58.7	85.4	112.9	55.9	84.8	112.6	83.3	98.1	110.6	103.8	107.3
MF2018	WEIGHTED AVERAGE	119.1	90.1	106.9	109.0	63.9	89.9	119.6	70.2	98.7	110.7	62.9	90.5	122.5	85.2	106.8	110.8	97.2	105.0

CANADA

DIVISION		OTTAWA, ONTARIO			OWEN SOUND, ONTARIO			PETERBOROUGH, ONTARIO			PORTAGE LA PRAIRIE, MANITOBA			PRINCE ALBERT, SASKATCHEWAN			PRINCE GEORGE, BRITISH COLUMBIA		
		MAT.	INST.	TOTAL	MAT.	INST.	TOTAL	MAT.	INST.	TOTAL	MAT.	INST.	TOTAL	MAT.	INST.	TOTAL	MAT.	INST.	TOTAL
015433	CONTRACTOR EQUIPMENT		117.8	117.8		100.1	100.1		99.7	99.7		101.9	101.9		98.3	98.3		103.3	103.3
0241, 31 - 34	SITE & INFRASTRUCTURE, DEMOLITION	107.0	107.2	107.1	117.0	94.6	101.5	116.9	94.2	101.3	115.4	93.2	100.1	110.2	90.1	96.3	123.4	97.1	105.3
0310	Concrete Forming & Accessories	124.8	88.4	93.7	125.6	78.7	85.6	127.2	81.0	87.8	126.6	65.7	74.7	109.6	54.4	62.5	115.1	77.5	83.0
0320	Concrete Reinforcing	137.0	102.0	120.1	175.2	86.9	132.5	163.7	85.3	125.8	151.5	54.5	104.6	111.7	61.4	87.3	109.3	75.8	93.1
0330	Cast-in-Place Concrete	122.3	103.0	115.1	157.7	76.6	127.5	130.7	81.8	112.5	120.7	69.8	101.8	106.2	63.6	90.3	117.6	91.8	108.0
03	CONCRETE	116.6	96.7	107.9	138.9	79.9	113.0	121.8	82.5	104.6	109.5	66.2	90.5	96.9	59.9	80.7	127.5	82.8	107.9
04	MASONRY	173.2	99.4	127.8	165.6	87.1	117.3	166.1	88.0	118.1	164.8	59.7	100.2	159.5	56.4	96.1	171.4	84.0	117.7
05	METALS	135.5	108.0	127.7	111.1	91.3	105.3	112.3	90.8	106.0	113.0	78.4	102.8	109.7	74.8	99.4	112.8	87.2	105.3
06	WOOD, PLASTICS & COMPOSITES	108.4	85.7	95.9	116.5	77.3	94.9	120.3	79.0	97.6	118.0	66.8	89.8	99.4	53.2	74.0	102.2	74.1	86.7
07	THERMAL & MOISTURE PROTECTION	137.3	99.7	121.0	115.3	83.5	101.5	121.6	88.0	107.0	114.6	68.2	94.4	114.0	59.0	90.1	124.5	81.0	105.6
08	OPENINGS	90.3	87.5	89.7	90.5	77.6	87.5	87.7	79.9	85.9	89.1	59.9	82.3	84.3	51.4	76.6	86.1	74.9	83.5
0920	Plaster & Gypsum Board	129.1	84.6	99.1	153.5	76.7	101.8	115.8	78.5	90.7	104.0	65.6	78.2	98.2	51.9	67.1	101.1	73.0	82.2
0950, 0980	Ceilings & Acoustic Treatment	142.3	84.6	103.4	94.3	76.7	82.4	103.5	78.5	86.7	103.5	65.6	78.0	103.5	51.9	68.7	103.5	73.0	83.0
0960	Flooring	108.3	90.6	103.4	119.3	87.3	110.3	113.6	86.0	105.8	113.6	61.9	99.1	103.6	53.7	89.6	109.0	67.4	97.3
0970, 0990	Wall Finishes & Painting/Coating	113.5	95.9	103.1	105.6	83.8	92.7	110.3	87.1	96.6	110.4	54.1	77.1	110.3	52.0	75.8	110.3	76.4	90.3
09	FINISHES	122.3	89.0	104.3	111.3	80.8	94.8	108.7	82.6	94.5	107.4	64.1	84.0	103.7	53.8	76.7	107.2	75.0	89.8
COVERS	DIVS. 10 - 14, 25, 28, 41, 43, 44, 46	131.1	89.5	121.9	139.2	63.9	122.4	131.1	64.8	116.3	131.1	59.8	115.2	131.1	57.5	114.7	131.1	83.7	120.6
21, 22, 23	FIRE SUPPRESSION, PLUMBING & HVAC	105.1	90.7	99.4	103.7	92.4	99.3	103.8	95.8	100.7	103.8	78.4	93.8	104.2	63.4	88.2	103.8	87.5	97.4
26, 27, 3370	ELECTRICAL, COMMUNICATIONS & UTIL.	104.4	101.9	103.2	118.5	82.1	100.6	111.5	82.9	97.4	113.0	54.4	84.2	113.4	56.3	85.3	111.5	74.2	93.1
MF2018	WEIGHTED AVERAGE	117.5	96.8	108.7	116.7	85.4	103.4	113.1	87.1	102.1	111.6	69.1	93.6	108.2	62.2	88.8	114.0	82.8	100.8

CANADA

	DIVISION	QUEBEC CITY, QUEBEC			RED DEER, ALBERTA			REGINA, SASKATCHEWAN			RIMOUSKI, QUEBEC			ROUYN-NORANDA, QUEBEC			SAINT HYACINTHE, QUEBEC		
		MAT.	INST.	TOTAL	MAT.	INST.	TOTAL	MAT.	INST.	TOTAL	MAT.	INST.	TOTAL	MAT.	INST.	TOTAL	MAT.	INST.	TOTAL
015433	CONTRACTOR EQUIPMENT		117.4	117.4		101.9	101.9		128.3	128.3		100.2	100.2		100.2	100.2		100.2	100.2
0241, 31 - 34	SITE & INFRASTRUCTURE, DEMOLITION	114.1	104.7	107.6	113.2	95.5	101.0	126.4	119.6	121.7	97.0	93.6	94.6	96.7	93.7	94.6	97.2	93.7	94.8
0310	Concrete Forming & Accessories	127.4	90.5	95.9	139.7	78.2	87.3	131.0	91.5	97.4	132.9	89.8	96.2	132.9	79.1	87.0	132.9	79.1	87.0
0320	Concrete Reinforcing	124.2	94.6	109.9	151.5	81.8	117.8	123.8	88.8	106.9	105.1	94.5	100.0	147.4	72.4	111.1	147.4	72.4	111.1
0330	Cast-in-Place Concrete	131.6	97.0	118.7	120.7	94.6	111.0	153.8	97.7	132.9	109.4	94.7	103.9	104.0	87.5	97.9	107.5	87.5	100.0
03	CONCRETE	119.4	94.3	108.4	116.5	85.2	102.7	131.3	94.0	114.9	105.0	92.4	99.5	107.9	81.4	96.2	109.5	81.4	97.1
04	MASONRY	173.6	89.8	122.1	161.6	80.1	111.6	205.2	86.7	132.4	162.1	88.9	117.6	162.3	76.3	109.5	162.6	76.3	109.6
05	METALS	138.0	103.3	127.8	113.0	90.8	106.5	139.1	101.4	128.0	110.4	91.8	104.9	110.9	84.8	103.2	110.9	84.8	103.2
06	WOOD, PLASTICS & COMPOSITES	114.2	90.7	101.3	118.0	77.4	95.7	108.0	92.2	99.3	131.6	90.3	108.9	131.6	79.0	102.7	131.6	79.0	102.7
07	THERMAL & MOISTURE PROTECTION	125.1	97.2	112.9	141.4	88.2	118.3	143.8	86.6	118.9	114.8	96.4	106.8	114.8	84.1	101.5	115.2	84.1	101.7
08	OPENINGS	88.4	86.0	87.8	89.1	74.4	85.7	87.9	79.8	86.1	88.7	76.3	85.8	89.1	67.9	84.2	89.1	67.9	84.2
0920	Plaster & Gypsum Board	128.5	89.9	102.6	104.3	76.6	85.6	140.8	91.2	107.4	147.4	89.9	108.7	114.2	78.4	90.1	114.2	78.4	90.1
0950, 0980	Ceilings & Acoustic Treatment	131.7	89.9	103.5	103.5	76.6	85.3	154.3	91.2	111.7	102.7	89.9	94.1	102.7	78.4	86.3	102.7	78.4	86.3
0960	Flooring	108.9	90.5	103.8	116.0	84.8	107.2	123.6	95.1	115.6	114.8	86.0	106.7	113.6	86.0	105.8	113.6	86.0	105.8
0970, 0990	Wall Finishes & Painting/Coating	120.8	104.1	111.0	110.2	75.1	89.4	118.0	89.0	100.9	113.4	104.1	107.9	110.3	83.6	94.5	110.3	83.6	94.5
09	FINISHES	117.4	92.7	104.0	108.2	78.9	92.4	130.9	92.9	110.3	112.6	91.6	101.2	107.6	81.2	93.3	107.6	81.2	93.3
COVERS	DIVS. 10 - 14, 25, 28, 41, 43, 44, 46	131.1	82.5	120.3	131.1	86.1	121.1	131.1	70.9	117.7	131.1	81.5	120.1	131.1	75.6	118.8	131.1	75.6	118.8
21, 22, 23	FIRE SUPPRESSION, PLUMBING & HVAC	105.1	81.2	95.7	103.8	86.8	97.2	104.8	86.9	97.8	103.8	81.0	94.9	103.8	84.2	96.1	100.5	84.2	94.1
26, 27, 3370	ELECTRICAL, COMMUNICATIONS & UTIL.	114.7	84.1	99.6	104.6	77.5	91.2	110.8	91.5	101.3	110.9	84.1	97.7	110.9	65.2	88.4	111.5	65.2	88.7
MF2018	WEIGHTED AVERAGE	118.7	90.4	106.7	112.1	84.0	100.2	123.1	92.4	110.2	110.2	87.6	100.6	110.3	79.8	97.4	109.8	79.8	97.1

CANADA

	DIVISION	SAINT JOHN, NEW BRUNSWICK			SARNIA, ONTARIO			SASKATOON, SASKATCHEWAN			SAULT STE MARIE, ONTARIO			SHERBROOKE, QUEBEC			SOREL, QUEBEC		
		MAT.	INST.	TOTAL	MAT.	INST.	TOTAL	MAT.	INST.	TOTAL	MAT.	INST.	TOTAL	MAT.	INST.	TOTAL	MAT.	INST.	TOTAL
015433	CONTRACTOR EQUIPMENT		99.7	99.7		99.7	99.7		98.4	98.4		99.7	99.7		100.2	100.2		100.2	100.2
0241, 31 - 34	SITE & INFRASTRUCTURE, DEMOLITION	105.2	92.5	96.4	115.5	94.3	100.9	112.7	92.8	98.9	105.6	93.9	97.5	97.2	93.7	94.8	97.3	93.7	94.8
0310	Concrete Forming & Accessories	126.2	63.4	72.7	125.9	87.6	93.3	109.7	90.7	93.5	115.1	86.2	90.5	132.9	79.1	87.0	132.9	79.2	87.1
0320	Concrete Reinforcing	140.5	69.8	106.3	116.5	86.7	102.1	114.2	88.7	101.9	105.1	85.4	95.6	147.4	72.4	111.1	139.3	72.4	107.0
0330	Cast-in-Place Concrete	112.4	68.0	95.9	120.6	94.2	110.8	113.5	93.1	105.9	108.3	80.5	98.0	107.5	87.5	100.0	108.4	87.5	100.6
03	CONCRETE	110.9	67.3	91.7	111.1	90.0	101.8	101.3	91.2	96.9	97.4	84.7	91.8	109.5	81.4	97.1	108.8	81.5	96.8
04	MASONRY	180.8	74.4	115.4	177.6	90.5	124.1	172.1	86.6	119.5	162.9	90.1	118.2	162.6	76.3	109.6	162.7	76.3	109.6
05	METALS	114.0	86.8	106.0	112.3	91.2	106.1	106.5	88.6	101.2	111.5	95.2	106.7	110.7	84.8	103.1	110.9	85.0	103.3
06	WOOD, PLASTICS & COMPOSITES	120.7	61.4	88.0	119.4	86.7	101.4	96.0	91.3	93.4	106.5	87.9	96.3	131.6	79.0	102.7	131.6	79.0	102.7
07	THERMAL & MOISTURE PROTECTION	116.9	72.2	97.5	121.7	91.5	108.6	117.0	84.7	103.0	120.4	87.3	106.0	114.8	84.1	101.5	114.8	84.1	101.5
08	OPENINGS	83.9	58.9	78.0	90.5	83.5	88.8	85.3	79.4	83.9	82.4	86.1	83.2	89.1	67.9	84.2	89.1	72.3	85.2
0920	Plaster & Gypsum Board	137.9	60.3	85.7	143.2	86.4	105.0	115.6	91.2	99.1	107.1	87.6	94.0	114.2	78.4	90.1	147.4	78.4	101.0
0950, 0980	Ceilings & Acoustic Treatment	121.7	60.3	80.3	109.2	86.4	93.9	124.5	91.2	102.0	103.5	87.6	92.8	102.7	78.4	86.3	102.7	78.4	86.3
0960	Flooring	111.4	65.6	98.5	113.6	94.1	108.1	106.5	95.1	103.3	106.5	92.1	102.4	113.6	86.0	105.8	113.6	86.0	105.8
0970, 0990	Wall Finishes & Painting/Coating	111.3	83.6	94.9	110.3	98.5	103.3	113.8	89.0	99.2	110.3	91.3	99.1	110.3	83.6	94.5	110.3	83.6	94.5
09	FINISHES	114.0	65.4	87.7	113.5	90.2	100.9	112.1	92.3	101.4	104.7	88.2	95.8	107.6	81.2	93.3	112.0	81.2	95.3
COVERS	DIVS. 10 - 14, 25, 28, 41, 43, 44, 46	131.1	61.0	115.5	131.1	66.1	116.6	131.1	69.3	117.4	131.1	87.9	121.5	131.1	75.6	118.8	131.1	75.6	118.8
21, 22, 23	FIRE SUPPRESSION, PLUMBING & HVAC	103.8	75.1	92.6	103.8	101.5	102.9	104.4	86.7	97.5	103.8	90.5	98.6	104.2	84.2	96.3	103.8	84.2	96.1
26, 27, 3370	ELECTRICAL, COMMUNICATIONS & UTIL.	119.9	88.5	104.4	114.7	85.0	100.0	114.6	91.4	103.2	113.2	83.3	98.5	110.9	65.2	88.4	110.9	65.2	88.4
MF2018	WEIGHTED AVERAGE	113.2	75.7	97.3	113.2	91.3	103.9	109.9	88.5	100.9	108.6	88.7	100.2	110.6	79.8	97.6	110.8	80.0	97.8

CANADA

	DIVISION	ST. CATHARINES, ONTARIO			ST JEROME, QUEBEC			ST. JOHN'S, NEWFOUNDLAND			SUDBURY, ONTARIO			SUMMERSIDE, PRINCE EDWARD ISLAND			SYDNEY, NOVA SCOTIA		
		MAT.	INST.	TOTAL	MAT.	INST.	TOTAL	MAT.	INST.	TOTAL	MAT.	INST.	TOTAL	MAT.	INST.	TOTAL	MAT.	INST.	TOTAL
015433	CONTRACTOR EQUIPMENT		99.5	99.5		100.2	100.2		123.4	123.4		99.5	99.5		98.8	98.8		98.9	98.9
0241, 31 - 34	SITE & INFRASTRUCTURE, DEMOLITION	94.2	96.1	95.6	96.7	93.7	94.6	115.4	111.8	112.9	94.4	95.7	95.3	121.8	89.4	99.5	108.0	92.0	96.9
0310	Concrete Forming & Accessories	115.3	93.8	97.0	132.9	79.1	87.0	125.3	85.2	91.1	111.0	88.9	92.1	112.0	52.5	61.3	111.8	67.8	74.3
0320	Concrete Reinforcing	97.2	103.2	100.1	147.4	72.4	111.1	172.1	82.4	128.7	98.0	101.4	99.7	154.0	46.5	102.0	156.0	47.6	103.6
0330	Cast-in-Place Concrete	106.7	98.7	103.7	104.0	87.5	97.9	137.1	98.5	122.8	107.6	95.3	103.0	116.7	54.7	93.6	94.8	66.8	84.4
03	CONCRETE	94.8	97.2	95.8	107.9	81.4	96.2	132.7	90.3	114.1	95.0	93.5	94.3	139.1	53.4	101.5	118.1	65.0	94.8
04	MASONRY	141.8	100.8	116.6	162.3	76.3	109.5	200.1	89.9	132.4	141.9	94.9	113.0	207.4	54.8	113.7	205.8	65.6	119.6
05	METALS	114.2	97.9	109.4	110.9	84.8	103.2	137.9	99.1	126.5	113.6	97.1	108.8	125.6	69.2	109.1	125.6	75.7	111.0
06	WOOD, PLASTICS & COMPOSITES	107.8	93.2	99.8	131.6	79.0	102.7	106.9	83.4	94.0	103.5	87.9	94.9	116.4	51.9	80.9	115.9	67.4	89.2
07	THERMAL & MOISTURE PROTECTION	116.4	102.0	110.1	114.8	84.1	101.5	140.2	94.7	120.4	115.7	96.3	107.3	129.4	57.3	98.0	130.1	67.6	102.9
08	OPENINGS	79.5	91.4	82.3	89.1	67.9	84.2	85.4	74.3	82.8	80.2	86.7	81.7	102.5	45.4	89.2	91.3	61.3	84.3
0920	Plaster & Gypsum Board	100.8	93.1	95.6	114.2	78.4	90.1	150.0	82.1	104.3	100.0	87.6	91.7	145.9	50.5	81.8	145.0	66.5	92.2
0950, 0980	Ceilings & Acoustic Treatment	103.9	93.1	96.6	102.7	78.4	86.3	143.7	82.1	102.2	98.2	87.6	91.0	123.8	50.5	74.4	123.8	66.5	85.2
0960	Flooring	100.2	91.9	97.8	113.6	86.0	105.8	116.8	52.2	98.7	98.3	92.1	96.6	107.6	55.3	92.9	107.6	58.9	93.9
0970, 0990	Wall Finishes & Painting/Coating	104.4	104.1	104.2	110.3	83.6	94.5	120.8	98.6	107.7	104.4	95.3	99.0	117.8	39.9	71.7	117.8	59.3	83.2
09	FINISHES	100.1	94.6	97.1	107.6	81.2	93.3	129.2	80.4	102.8	98.3	89.9	93.7	119.8	51.9	83.1	118.7	65.5	89.9
COVERS	DIVS. 10 - 14, 25, 28, 41, 43, 44, 46	131.1	69.7	117.4	131.1	75.6	118.8	131.1	68.6	117.2	131.1	89.6	121.9	131.1	57.0	114.6	131.1	60.8	115.5
21, 22, 23	FIRE SUPPRESSION, PLUMBING & HVAC	104.2	91.1	99.0	103.8	84.2	96.1	104.8	83.0	96.2	103.6	89.4	98.1	104.0	58.9	86.2	104.0	79.5	94.4
26, 27, 3370	ELECTRICAL, COMMUNICATIONS & UTIL.	111.3	101.8	106.6	111.6	65.2	88.7	115.6	81.4	98.7	109.5	102.1	105.9	110.2	47.4	79.2	111.3	58.7	85.4
MF2018	WEIGHTED AVERAGE	106.6	95.6	101.9	110.4	79.8	97.4	122.6	87.4	107.7	106.1	93.6	100.8	121.9	57.9	94.9	117.8	70.2	97.7

City Cost Indexes

		CANADA																
	DIVISION	THUNDER BAY, ONTARIO			TIMMINS, ONTARIO			TORONTO, ONTARIO			TROIS RIVIERES, QUEBEC			TRURO, NOVA SCOTIA			VANCOUVER, BRITISH COLUMBIA		
		MAT.	INST.	TOTAL	MAT.	INST.	TOTAL	MAT.	INST.	TOTAL	MAT.	INST.	TOTAL	MAT.	INST.	TOTAL	MAT.	INST.	TOTAL
015433	CONTRACTOR EQUIPMENT		99.5	99.5		99.7	99.7		117.5	117.5		99.7	99.7		99.4	99.4		135.3	135.3
0241, 31 - 34	SITE & INFRASTRUCTURE, DEMOLITION	98.8	96.0	96.9	116.9	93.9	101.0	108.4	107.7	107.9	108.2	93.4	97.9	101.3	92.3	95.1	117.0	122.8	121.0
0310	Concrete Forming & Accessories	122.8	92.5	97.0	127.2	80.0	87.0	128.7	102.5	106.3	153.7	79.1	90.1	99.7	67.9	72.6	131.0	91.6	97.4
0320	Concrete Reinforcing	86.9	102.6	94.5	163.7	84.8	125.6	136.9	103.9	120.9	156.0	72.4	115.6	139.3	47.6	95.0	132.5	95.2	114.5
0330	Cast-in-Place Concrete	117.5	98.0	110.2	130.7	80.2	111.9	108.4	113.7	110.4	98.2	87.4	94.2	135.7	66.9	110.1	123.5	90.6	111.3
03	CONCRETE	101.9	96.2	99.4	121.8	81.5	104.1	110.5	107.1	109.0	120.3	81.4	103.2	120.8	65.1	96.3	119.7	93.0	108.0
04	MASONRY	142.4	100.5	116.7	166.1	81.9	114.4	173.3	108.7	133.6	209.8	76.3	127.8	161.9	65.6	102.7	165.3	85.6	116.3
05	METALS	114.2	97.0	109.1	112.3	90.4	105.9	135.3	109.4	127.7	124.9	84.7	113.1	111.6	76.0	101.1	133.5	109.8	126.6
06	WOOD, PLASTICS & COMPOSITES	117.6	91.3	103.1	120.3	80.3	98.3	114.8	100.5	107.0	168.2	79.0	119.1	89.6	67.4	77.4	110.3	91.0	99.7
07	THERMAL & MOISTURE PROTECTION	116.6	99.3	109.1	121.6	82.7	104.7	135.6	109.4	124.2	128.6	84.1	109.2	116.1	67.6	95.0	141.7	88.6	118.6
08	OPENINGS	78.9	89.8	81.4	87.7	78.2	85.5	83.9	98.9	87.4	100.3	72.3	93.7	82.4	61.3	77.5	84.5	88.7	85.5
0920	Plaster & Gypsum Board	128.0	91.2	103.2	115.8	79.8	91.6	129.1	99.9	109.5	174.3	78.4	109.8	124.3	66.5	85.4	127.2	89.7	102.0
0950, 0980	Ceilings & Acoustic Treatment	98.2	91.2	93.4	103.5	79.8	87.5	139.9	99.9	112.9	122.2	78.4	92.6	103.5	66.5	78.6	151.0	89.7	109.7
0960	Flooring	104.2	98.5	102.6	113.6	86.0	105.8	109.1	100.8	106.8	131.2	86.0	118.5	95.7	58.9	85.4	124.7	88.5	114.6
0970, 0990	Wall Finishes & Painting/Coating	104.4	96.2	99.5	110.3	85.0	95.3	109.8	108.1	108.8	117.8	83.6	97.6	110.3	59.3	80.1	113.7	100.3	105.8
09	FINISHES	104.1	93.9	98.6	108.7	81.9	94.2	117.8	102.8	109.7	129.3	81.1	103.2	103.8	65.5	83.1	128.0	92.1	108.6
COVERS	DIVS. 10 - 14, 25, 28, 41, 43, 44, 46	131.1	69.7	117.4	131.1	63.4	116.0	131.1	95.4	123.2	131.1	75.6	118.8	131.1	60.8	115.5	131.1	90.8	122.1
21, 22, 23	FIRE SUPPRESSION, PLUMBING & HVAC	104.2	91.2	99.1	103.8	92.4	99.3	105.1	99.6	103.0	104.0	84.2	96.2	103.8	79.5	94.3	105.2	87.4	98.2
26, 27, 3370	ELECTRICAL, COMMUNICATIONS & UTIL.	109.6	101.2	105.4	113.2	83.3	98.5	106.8	103.8	105.3	111.6	65.2	88.8	110.5	58.7	85.0	108.7	78.6	93.9
MF2018	WEIGHTED AVERAGE	107.8	95.0	102.4	113.3	85.3	101.4	116.0	104.2	111.0	120.2	79.9	103.2	110.8	70.3	93.7	117.9	92.4	107.2

		CANADA																	
	DIVISION	VICTORIA, BRITISH COLUMBIA			WHITEHORSE, YUKON			WINDSOR, ONTARIO			WINNIPEG, MANITOBA			YARMOUTH, NOVA SCOTIA			YELLOWKNIFE, NWT		
		MAT.	INST.	TOTAL	MAT.	INST.	TOTAL	MAT.	INST.	TOTAL	MAT.	INST.	TOTAL	MAT.	INST.	TOTAL	MAT.	INST.	TOTAL
015433	CONTRACTOR EQUIPMENT		106.3	106.3		132.8	132.8		99.5	99.5		125.5	125.5		98.9	98.9		130.7	130.7
0241, 31 - 34	SITE & INFRASTRUCTURE, DEMOLITION	124.3	101.3	108.4	135.7	119.6	124.6	90.8	96.0	94.4	113.7	115.3	114.8	111.7	92.0	98.1	146.3	123.6	130.6
0310	Concrete Forming & Accessories	114.9	88.9	92.7	135.0	56.7	68.2	122.8	90.6	95.3	130.4	65.4	75.0	111.8	67.8	74.3	137.3	75.8	84.8
0320	Concrete Reinforcing	111.7	95.0	103.6	167.7	63.3	117.2	95.2	101.9	98.5	127.3	61.0	95.3	156.0	47.6	103.6	138.6	65.8	103.4
0330	Cast-in-Place Concrete	117.6	87.0	106.2	152.4	71.2	122.2	109.3	99.7	105.7	143.6	72.0	116.9	121.6	66.8	101.2	174.9	87.6	142.4
03	CONCRETE	129.4	89.5	111.9	144.0	64.8	109.1	96.2	95.9	96.0	127.0	68.4	101.3	130.4	65.0	101.7	150.7	79.5	119.4
04	MASONRY	175.0	85.5	120.0	246.5	58.1	130.7	141.9	100.1	116.3	194.6	65.3	115.2	208.1	65.6	120.5	234.9	68.9	132.9
05	METALS	107.7	92.2	103.2	145.1	90.5	129.1	114.2	97.3	109.2	141.6	88.0	125.8	125.6	75.7	111.0	140.5	92.7	126.4
06	WOOD, PLASTICS & COMPOSITES	100.0	88.3	93.6	121.3	55.3	85.0	117.6	88.0	101.9	107.7	66.0	84.8	115.9	67.4	89.2	130.1	77.1	100.9
07	THERMAL & MOISTURE PROTECTION	126.9	86.3	109.2	145.8	63.3	109.9	116.4	99.1	108.9	136.1	70.3	107.5	130.1	67.6	102.9	143.7	78.8	115.5
08	OPENINGS	86.8	82.9	85.9	99.5	53.0	88.6	78.7	88.9	81.1	87.5	59.7	81.0	91.3	61.3	84.3	93.9	65.5	87.3
0920	Plaster & Gypsum Board	109.1	87.7	94.7	178.3	53.0	94.0	111.4	88.8	96.2	133.0	64.2	86.7	145.0	66.5	92.2	192.2	75.5	113.7
0950, 0980	Ceilings & Acoustic Treatment	105.8	87.7	93.6	169.3	53.0	90.9	98.2	88.8	91.8	145.3	64.2	90.6	123.8	66.5	85.2	169.0	75.5	105.9
0960	Flooring	112.0	67.4	99.5	122.4	55.2	103.5	104.2	95.9	101.9	115.9	68.5	102.6	107.6	58.9	93.9	121.9	82.9	111.0
0970, 0990	Wall Finishes & Painting/Coating	113.8	100.3	105.8	120.7	53.1	80.7	104.4	97.1	100.1	113.9	53.5	78.1	117.8	59.3	83.2	123.3	75.9	95.2
09	FINISHES	110.7	86.8	97.8	143.7	55.8	96.1	101.6	92.1	96.5	126.9	65.0	93.4	118.7	65.5	89.9	147.5	76.6	109.1
COVERS	DIVS. 10 - 14, 25, 28, 41, 43, 44, 46	131.1	65.8	116.6	131.1	60.6	115.4	131.1	69.2	117.3	131.1	63.8	116.1	131.1	60.8	115.5	131.1	63.9	116.1
21, 22, 23	FIRE SUPPRESSION, PLUMBING & HVAC	103.9	88.7	97.9	104.3	71.6	91.5	104.2	90.7	98.9	105.2	64.6	89.2	104.0	79.5	94.4	104.8	88.5	98.4
26, 27, 3370	ELECTRICAL, COMMUNICATIONS & UTIL.	112.8	82.1	97.7	130.5	57.1	94.3	114.5	100.4	107.6	113.1	63.4	88.6	111.3	58.7	85.4	125.8	77.4	102.0
MF2018	WEIGHTED AVERAGE	114.1	87.6	102.9	132.2	69.2	105.6	107.1	94.4	101.8	122.0	71.2	100.5	119.5	70.2	98.7	131.4	83.2	111.0

Location Factors - Commercial

Costs shown in RSMeans cost data publications are based on national averages for materials and installation. To adjust these costs to a specific location, simply multiply the base cost by the factor and divide by 100 for that city. The data is arranged alphabetically by state and postal zip code numbers. For a city not listed, use the factor for a nearby city with similar economic characteristics.

STATE/ZIP	CITY	MAT.	INST.	TOTAL
ALABAMA				
350-352	Birmingham	97.4	71.8	86.6
354	Tuscaloosa	97.1	71.4	86.3
355	Jasper	97.8	69.8	86.0
356	Decatur	97.1	69.2	85.3
357-358	Huntsville	97.1	71.7	86.4
359	Gadsden	97.3	70.2	85.9
360-361	Montgomery	96.3	73.1	86.5
362	Anniston	95.9	67.7	84.0
363	Dothan	96.2	72.7	86.3
364	Evergreen	95.8	68.3	84.2
365-366	Mobile	96.7	68.7	84.9
367	Selma	95.9	72.5	86.0
368	Phenix City	96.8	70.8	85.8
369	Butler	96.1	70.6	85.3
ALASKA				
995-996	Anchorage	118.5	113.1	116.2
997	Fairbanks	118.8	113.0	116.3
998	Juneau	116.8	113.1	115.2
999	Ketchikan	128.4	113.4	122.1
ARIZONA				
850,853	Phoenix	100.6	72.3	88.6
851,852	Mesa/Tempe	98.7	71.5	87.2
855	Globe	99.5	71.4	87.6
856-857	Tucson	97.7	70.2	86.1
859	Show Low	99.7	71.5	87.8
860	Flagstaff	102.4	70.5	88.9
863	Prescott	99.9	71.3	87.8
864	Kingman	98.3	70.4	86.5
865	Chambers	98.3	71.8	87.1
ARKANSAS				
716	Pine Bluff	96.0	63.6	82.3
717	Camden	94.0	64.7	81.6
718	Texarkana	94.8	63.3	81.5
719	Hot Springs	93.5	64.4	81.2
720-722	Little Rock	94.5	65.4	82.2
723	West Memphis	93.3	69.8	83.3
724	Jonesboro	93.8	66.6	82.3
725	Batesville	91.6	63.6	79.8
726	Harrison	92.9	62.8	80.2
727	Fayetteville	90.5	64.1	79.3
728	Russellville	91.7	63.1	79.6
729	Fort Smith	94.1	62.6	80.8
CALIFORNIA				
900-902	Los Angeles	98.5	129.8	111.7
903-905	Inglewood	93.8	128.8	108.6
906-908	Long Beach	95.4	128.8	109.5
910-912	Pasadena	94.8	128.6	109.1
913-916	Van Nuys	97.8	128.6	110.8
917-918	Alhambra	96.7	128.6	110.2
919-921	San Diego	99.7	121.7	109.0
922	Palm Springs	97.4	125.3	109.2
923-924	San Bernardino	95.0	126.3	108.2
925	Riverside	99.3	126.6	110.8
926-927	Santa Ana	97.0	126.3	109.4
928	Anaheim	99.3	126.6	110.9
930	Oxnard	98.0	126.7	110.1
931	Santa Barbara	97.2	126.1	109.4
932-933	Bakersfield	98.9	124.6	109.8
934	San Luis Obispo	98.2	125.8	109.9
935	Mojave	95.4	124.4	107.7
936-938	Fresno	98.5	129.4	111.5
939	Salinas	99.1	137.2	115.2
940-941	San Francisco	107.4	158.4	129.0
942,956-958	Sacramento	100.8	133.0	114.4
943	Palo Alto	99.2	153.8	122.3
944	San Mateo	101.5	152.0	122.9
945	Vallejo	100.2	141.6	117.7
946	Oakland	103.4	152.0	123.9
947	Berkeley	103.0	151.9	123.7
948	Richmond	102.4	146.2	120.9
949	San Rafael	104.5	149.1	123.4
950	Santa Cruz	104.6	137.5	118.5

STATE/ZIP	CITY	MAT.	INST.	TOTAL
CALIFORNIA (CONT'D)				
951	San Jose	102.6	153.5	124.2
952	Stockton	100.7	131.3	113.7
953	Modesto	100.6	130.6	113.3
954	Santa Rosa	101.2	148.4	121.2
955	Eureka	102.5	134.1	115.9
959	Marysville	101.7	131.3	114.2
960	Redding	107.6	131.5	117.7
961	Susanville	107.3	131.6	117.5
COLORADO				
800-802	Denver	103.0	75.4	91.3
803	Boulder	98.7	77.0	89.6
804	Golden	100.9	74.7	89.8
805	Fort Collins	102.2	73.5	90.1
806	Greeley	99.3	73.4	88.4
807	Fort Morgan	99.3	73.5	88.4
808-809	Colorado Springs	101.2	71.5	88.6
810	Pueblo	101.8	70.0	88.3
811	Alamosa	103.6	70.3	89.5
812	Salida	103.1	70.9	89.5
813	Durango	103.9	65.6	87.8
814	Montrose	102.6	65.8	87.1
815	Grand Junction	105.7	70.5	90.8
816	Glenwood Springs	103.9	66.8	88.2
CONNECTICUT				
060	New Britain	98.1	116.7	106.0
061	Hartford	99.6	118.0	107.3
062	Willimantic	98.7	117.1	106.5
063	New London	95.2	117.2	104.5
064	Meriden	97.2	117.0	105.6
065	New Haven	99.9	117.0	107.1
066	Bridgeport	99.3	117.0	106.8
067	Waterbury	98.9	117.2	106.6
068	Norwalk	98.8	116.7	106.4
069	Stamford	99.0	123.2	109.2
D.C.				
200-205	Washington	101.1	87.5	95.4
DELAWARE				
197	Newark	98.3	109.7	103.1
198	Wilmington	97.9	109.7	102.9
199	Dover	99.0	109.6	103.5
FLORIDA				
320,322	Jacksonville	96.9	66.6	84.1
321	Daytona Beach	97.2	69.5	85.5
323	Tallahassee	98.2	67.7	85.3
324	Panama City	98.6	68.3	85.8
325	Pensacola	101.2	65.8	86.2
326,344	Gainesville	98.8	65.8	84.8
327-328,347	Orlando	99.0	66.6	85.3
329	Melbourne	100.1	70.3	87.5
330-332,340	Miami	96.7	68.0	84.6
333	Fort Lauderdale	96.1	66.4	83.6
334,349	West Palm Beach	95.1	64.1	82.0
335-336,346	Tampa	97.6	68.4	85.3
337	St. Petersburg	99.7	66.8	85.8
338	Lakeland	97.1	67.6	84.6
339,341	Fort Myers	96.5	67.7	84.3
342	Sarasota	98.8	67.0	85.4
GEORGIA				
300-303,399	Atlanta	98.2	75.0	88.4
304	Statesboro	97.8	69.7	85.9
305	Gainesville	96.2	65.6	83.2
306	Athens	95.6	66.9	83.5
307	Dalton	97.5	71.4	86.5
308-309	Augusta	96.0	73.9	86.6
310-312	Macon	95.2	73.1	85.8
313-314	Savannah	97.3	72.7	86.9
315	Waycross	96.8	70.0	85.5
316	Valdosta	96.7	66.2	83.8
317,398	Albany	96.5	71.9	86.1
318-319	Columbus	96.5	72.2	86.2

For customer support on your Electrical Change Order Costs with RSMeans data, call 800.448.8182.

577

Location Factors - Commercial

STATE/ZIP	CITY	MAT.	INST.	TOTAL	STATE/ZIP	CITY	MAT.	INST.	TOTAL
HAWAII					**KANSAS (CONT'D)**				
967	Hilo	114.8	118.4	116.3	678	Dodge City	98.6	73.3	87.9
968	Honolulu	119.1	118.5	118.9	679	Liberal	96.5	71.7	86.0
STATES & POSS.					**KENTUCKY**				
969	Guam	137.1	51.4	100.9	400-402	Louisville	93.9	79.6	87.8
					403-405	Lexington	93.1	78.1	86.7
IDAHO					406	Frankfort	95.4	80.3	89.0
832	Pocatello	102.0	78.4	92.0	407-409	Corbin	90.8	78.0	85.4
833	Twin Falls	103.2	76.8	92.1	410	Covington	94.3	77.8	87.3
834	Idaho Falls	100.5	79.9	91.8	411-412	Ashland	93.2	89.7	91.7
835	Lewiston	108.2	83.5	97.8	413-414	Campton	94.4	78.0	87.5
836-837	Boise	101.0	79.9	92.1	415-416	Pikeville	95.8	85.4	91.4
838	Coeur d'Alene	108.1	84.4	98.1	417-418	Hazard	93.8	78.7	87.4
					420	Paducah	92.4	83.2	88.6
ILLINOIS					421-422	Bowling Green	94.6	77.2	87.3
600-603	North Suburban	98.3	142.9	117.1	423	Owensboro	94.7	81.0	88.9
604	Joliet	98.3	138.8	115.4	424	Henderson	92.2	81.9	87.8
605	South Suburban	98.3	142.8	117.1	425-426	Somerset	91.6	79.5	86.5
606-608	Chicago	100.1	145.6	119.3	427	Elizabethtown	91.3	75.6	84.7
609	Kankakee	95.0	134.9	111.9					
610-611	Rockford	96.6	125.5	108.8	**LOUISIANA**				
612	Rock Island	94.6	99.5	96.7	700-701	New Orleans	99.0	68.5	86.1
613	La Salle	95.9	127.0	109.0	703	Thibodaux	96.0	67.0	83.8
614	Galesburg	95.7	108.5	101.1	704	Hammond	93.7	64.4	81.3
615-616	Peoria	97.6	108.8	102.3	705	Lafayette	95.4	67.9	83.8
617	Bloomington	95.0	109.5	101.1	706	Lake Charles	95.6	68.2	84.1
618-619	Champaign	98.5	110.5	103.6	707-708	Baton Rouge	96.6	69.3	85.1
620-622	East St. Louis	93.8	110.4	100.9	710-711	Shreveport	97.1	66.3	84.1
623	Quincy	95.7	104.2	99.3	712	Monroe	95.8	64.4	82.5
624	Effingham	95.1	109.1	101.0	713-714	Alexandria	95.9	65.6	83.1
625	Decatur	96.6	110.1	102.3					
626-627	Springfield	97.5	108.3	102.1	**MAINE**				
628	Centralia	92.8	111.2	100.5	039	Kittery	92.6	87.0	90.2
629	Carbondale	92.5	109.6	99.7	040-041	Portland	99.6	85.4	93.6
					042	Lewiston	97.5	84.9	92.2
INDIANA					043	Augusta	100.4	85.6	94.2
460	Anderson	96.0	79.8	89.1	044	Bangor	97.1	83.9	91.5
461-462	Indianapolis	98.7	83.2	92.1	045	Bath	95.9	85.4	91.4
463-464	Gary	97.5	108.1	102.0	046	Machias	95.4	85.3	91.1
465-466	South Bend	98.4	81.8	91.4	047	Houlton	95.6	85.2	91.2
467-468	Fort Wayne	96.6	76.3	88.0	048	Rockland	94.6	85.4	90.7
469	Kokomo	94.2	81.1	88.6	049	Waterville	96.0	85.2	91.4
470	Lawrenceburg	92.5	79.6	87.0					
471	New Albany	93.9	77.3	86.9	**MARYLAND**				
472	Columbus	96.1	80.5	89.5	206	Waldorf	97.3	86.6	92.8
473	Muncie	96.2	78.2	88.6	207-208	College Park	97.3	87.7	93.2
474	Bloomington	97.8	81.1	90.7	209	Silver Spring	96.5	86.7	92.4
475	Washington	94.6	84.8	90.4	210-212	Baltimore	101.8	83.2	93.9
476-477	Evansville	95.5	82.7	90.1	214	Annapolis	100.4	86.3	94.4
478	Terre Haute	96.2	81.7	90.1	215	Cumberland	96.9	83.8	91.4
479	Lafayette	95.8	78.1	88.3	216	Easton	98.6	74.8	88.5
					217	Hagerstown	97.4	85.3	92.3
IOWA					218	Salisbury	99.0	66.9	85.4
500-503,509	Des Moines	97.2	90.2	94.3	219	Elkton	95.9	86.2	91.8
504	Mason City	95.3	74.0	86.3					
505	Fort Dodge	95.5	73.7	86.3	**MASSACHUSETTS**				
506-507	Waterloo	96.8	77.4	88.6	010-011	Springfield	98.2	108.5	102.5
508	Creston	95.7	84.2	90.8	012	Pittsfield	97.7	102.6	99.7
510-511	Sioux City	97.7	81.5	90.9	013	Greenfield	95.9	109.0	101.4
512	Sibley	96.8	62.4	82.3	014	Fitchburg	94.6	116.2	103.7
513	Spencer	98.2	62.3	83.0	015-016	Worcester	98.1	117.0	106.1
514	Carroll	95.5	82.4	89.9	017	Framingham	94.0	124.2	106.8
515	Council Bluffs	98.9	79.2	90.6	018	Lowell	97.5	123.8	108.6
516	Shenandoah	96.1	81.9	90.1	019	Lawrence	98.4	124.2	109.3
520	Dubuque	97.2	79.9	89.9	020-022, 024	Boston	101.9	134.0	115.5
521	Decorah	96.4	73.7	86.8	023	Brockton	98.5	116.7	106.2
522-524	Cedar Rapids	98.0	83.7	92.0	025	Buzzards Bay	93.2	115.2	102.5
525	Ottumwa	96.3	78.7	88.9	026	Hyannis	95.7	115.0	103.8
526	Burlington	95.7	84.3	90.9	027	New Bedford	97.7	115.2	105.1
527-528	Davenport	97.1	94.5	96.0					
					MICHIGAN				
KANSAS					480,483	Royal Oak	95.1	100.3	97.3
660-662	Kansas City	97.3	98.6	97.8	481	Ann Arbor	97.1	99.8	98.2
664-666	Topeka	98.6	76.5	89.3	482	Detroit	100.7	102.0	101.3
667	Fort Scott	96.0	77.1	88.0	484-485	Flint	96.8	88.6	93.3
668	Emporia	96.1	76.5	87.8	486	Saginaw	96.5	85.1	91.7
669	Belleville	97.9	70.7	86.4	487	Bay City	96.6	85.3	91.8
670-672	Wichita	97.0	72.6	86.7	488-489	Lansing	98.0	86.8	93.3
673	Independence	97.2	76.0	88.2	490	Battle Creek	95.5	80.4	89.1
674	Salina	97.2	73.5	87.2	491	Kalamazoo	95.8	78.6	88.6
675	Hutchinson	92.5	71.9	83.8	492	Jackson	94.0	90.7	92.6
676	Hays	96.5	72.0	86.1	493,495	Grand Rapids	97.9	79.9	90.3
677	Colby	97.3	74.0	87.4	494	Muskegon	94.3	80.0	88.3

Location Factors - Commercial

STATE/ZIP	CITY	MAT.	INST.	TOTAL
MICHIGAN (CONT'D)				
496	Traverse City	93.4	77.7	86.8
497	Gaylord	94.5	80.6	88.6
498-499	Iron Mountain	96.5	81.7	90.2
MINNESOTA				
550-551	Saint Paul	99.0	117.6	106.9
553-555	Minneapolis	100.7	114.1	106.3
556-558	Duluth	98.6	101.3	99.7
559	Rochester	98.3	98.7	98.5
560	Mankato	95.7	98.5	96.9
561	Windom	94.3	91.2	93.0
562	Willmar	94.0	98.0	95.7
563	St. Cloud	95.1	114.2	103.2
564	Brainerd	95.6	98.6	96.8
565	Detroit Lakes	97.4	92.4	95.3
566	Bemidji	96.7	95.4	96.1
567	Thief River Falls	96.3	89.3	93.3
MISSISSIPPI				
386	Clarksdale	95.3	55.2	78.4
387	Greenville	98.7	65.5	84.7
388	Tupelo	96.6	59.0	80.7
389	Greenwood	96.6	55.5	79.2
390-392	Jackson	97.3	65.6	83.9
393	Meridian	94.6	64.5	81.9
394	Laurel	96.1	57.8	79.9
395	Biloxi	96.2	63.9	82.6
396	McComb	94.4	55.8	78.0
397	Columbus	95.9	59.1	80.4
MISSOURI				
630-631	St. Louis	98.8	104.5	101.2
633	Bowling Green	96.4	95.5	96.0
634	Hannibal	95.3	91.6	93.7
635	Kirksville	98.9	86.5	93.7
636	Flat River	97.3	94.2	96.0
637	Cape Girardeau	97.0	91.2	94.5
638	Sikeston	95.7	87.8	92.3
639	Poplar Bluff	95.1	87.7	92.0
640-641	Kansas City	97.9	103.8	100.4
644-645	St. Joseph	96.1	90.9	93.9
646	Chillicothe	93.7	93.8	93.7
647	Harrisonville	93.2	100.8	96.4
648	Joplin	95.1	77.7	87.7
650-651	Jefferson City	96.4	89.9	93.7
652	Columbia	96.1	90.7	93.8
653	Sedalia	96.4	91.5	94.3
654-655	Rolla	94.1	94.3	94.2
656-658	Springfield	97.6	79.0	89.7
MONTANA				
590-591	Billings	101.6	78.4	91.8
592	Wolf Point	101.4	76.9	91.1
593	Miles City	99.2	77.0	89.8
594	Great Falls	102.9	76.6	91.8
595	Havre	100.4	74.8	89.6
596	Helena	100.9	76.7	90.7
597	Butte	101.5	76.5	90.9
598	Missoula	98.6	77.1	89.5
599	Kalispell	98.1	76.2	88.9
NEBRASKA				
680-681	Omaha	97.6	81.5	90.8
683-685	Lincoln	98.3	79.7	90.5
686	Columbus	96.0	80.7	89.6
687	Norfolk	97.4	77.5	89.0
688	Grand Island	97.4	76.8	88.7
689	Hastings	97.1	78.2	89.1
690	McCook	96.3	73.1	86.5
691	North Platte	96.3	75.6	87.6
692	Valentine	98.6	69.2	86.2
693	Alliance	98.7	73.3	87.9
NEVADA				
889-891	Las Vegas	105.0	105.1	105.0
893	Ely	103.7	93.2	99.2
894-895	Reno	103.6	82.6	94.7
897	Carson City	102.9	82.5	94.3
898	Elko	102.3	85.2	95.0
NEW HAMPSHIRE				
030	Nashua	97.5	93.7	95.9
031	Manchester	98.2	94.5	96.6

STATE/ZIP	CITY	MAT.	INST.	TOTAL
NEW HAMPSHIRE (CONT'D)				
032-033	Concord	98.0	93.2	96.0
034	Keene	94.5	89.0	92.2
035	Littleton	94.4	80.2	88.4
036	Charleston	94.1	88.4	91.7
037	Claremont	93.1	88.4	91.1
038	Portsmouth	94.9	92.4	93.8
NEW JERSEY				
070-071	Newark	100.3	139.7	117.0
072	Elizabeth	97.5	138.3	114.7
073	Jersey City	96.5	137.8	114.0
074-075	Paterson	98.0	137.7	114.8
076	Hackensack	96.1	138.3	114.0
077	Long Branch	95.8	130.9	110.7
078	Dover	96.4	137.5	113.7
079	Summit	96.4	138.3	114.1
080,083	Vineland	96.3	129.8	110.5
081	Camden	98.0	127.8	110.6
082,084	Atlantic City	96.8	133.2	112.2
085-086	Trenton	99.7	128.9	112.0
087	Point Pleasant	98.1	130.8	111.9
088-089	New Brunswick	98.7	135.6	114.3
NEW MEXICO				
870-872	Albuquerque	98.6	72.5	87.6
873	Gallup	99.0	72.5	87.8
874	Farmington	99.0	72.5	87.8
875	Santa Fe	99.0	72.9	88.0
877	Las Vegas	97.2	72.5	86.8
878	Socorro	96.8	72.5	86.6
879	Truth/Consequences	96.6	69.4	85.1
880	Las Cruces	96.9	71.4	86.1
881	Clovis	99.3	72.4	87.9
882	Roswell	100.8	72.5	88.9
883	Carrizozo	101.6	72.5	89.3
884	Tucumcari	99.9	72.4	88.3
NEW YORK				
100-102	New York	99.7	175.2	131.6
103	Staten Island	95.6	176.1	129.6
104	Bronx	93.8	175.3	128.3
105	Mount Vernon	94.1	150.6	118.0
106	White Plains	93.8	152.8	118.8
107	Yonkers	97.9	153.0	121.2
108	New Rochelle	94.4	144.8	115.7
109	Suffern	94.1	130.2	109.4
110	Queens	100.0	177.3	132.6
111	Long Island City	101.6	177.3	133.6
112	Brooklyn	101.9	177.3	133.8
113	Flushing	102.1	177.3	133.9
114	Jamaica	100.3	177.3	132.8
115,117,118	Hicksville	99.9	157.1	124.1
116	Far Rockaway	102.2	177.3	133.9
119	Riverhead	100.6	153.5	122.9
120-122	Albany	96.8	111.2	102.8
123	Schenectady	97.0	109.9	102.4
124	Kingston	100.6	133.8	114.6
125-126	Poughkeepsie	99.8	136.8	115.4
127	Monticello	99.1	134.6	114.1
128	Glens Falls	92.0	107.3	98.5
129	Plattsburgh	97.0	96.6	96.8
130-132	Syracuse	97.8	98.3	98.0
133-135	Utica	96.0	96.6	96.2
136	Watertown	97.7	96.8	97.4
137-139	Binghamton	97.3	101.5	99.1
140-142	Buffalo	100.8	110.3	104.8
143	Niagara Falls	97.4	107.5	101.6
144-146	Rochester	100.3	99.2	99.8
147	Jamestown	96.4	96.0	96.3
148-149	Elmira	96.3	102.5	98.9
NORTH CAROLINA				
270,272-274	Greensboro	98.5	67.2	85.3
271	Winston-Salem	98.2	67.3	85.2
275-276	Raleigh	97.4	67.0	84.5
277	Durham	100.1	67.2	86.2
278	Rocky Mount	96.0	66.6	83.6
279	Elizabeth City	96.9	68.3	84.8
280	Gastonia	97.0	66.8	84.2
281-282	Charlotte	97.4	67.6	84.8
283	Fayetteville	100.2	66.1	85.8
284	Wilmington	96.0	65.2	83.0
285	Kinston	94.4	65.5	82.2

Location Factors - Commercial

STATE/ZIP	CITY	MAT.	INST.	TOTAL
NORTH CAROLINA (CONT'D)				
286	Hickory	94.7	67.3	83.1
287-288	Asheville	96.4	66.6	83.8
289	Murphy	95.5	64.7	82.5
NORTH DAKOTA				
580-581	Fargo	99.7	80.3	91.5
582	Grand Forks	99.4	77.6	90.2
583	Devils Lake	99.2	78.8	90.6
584	Jamestown	99.2	78.8	90.6
585	Bismarck	98.9	81.6	91.6
586	Dickinson	99.9	77.9	90.6
587	Minot	99.3	78.6	90.5
588	Williston	98.4	77.7	89.6
OHIO				
430-432	Columbus	98.4	82.0	91.4
433	Marion	94.3	86.7	91.1
434-436	Toledo	97.6	91.2	94.9
437-438	Zanesville	94.9	85.7	91.0
439	Steubenville	96.1	90.0	93.5
440	Lorain	98.4	84.7	92.6
441	Cleveland	99.1	91.6	95.9
442-443	Akron	99.4	86.0	93.8
444-445	Youngstown	98.7	81.9	91.6
446-447	Canton	98.9	79.9	90.9
448-449	Mansfield	96.4	84.2	91.3
450	Hamilton	94.8	79.5	88.3
451-452	Cincinnati	96.1	80.0	89.3
453-454	Dayton	94.8	79.2	88.2
455	Springfield	94.8	79.7	88.4
456	Chillicothe	94.3	88.5	91.8
457	Athens	97.1	84.8	91.9
458	Lima	97.2	81.7	90.7
OKLAHOMA				
730-731	Oklahoma City	96.4	68.7	84.7
734	Ardmore	95.2	65.6	82.7
735	Lawton	97.3	65.8	84.0
736	Clinton	96.4	65.5	83.3
737	Enid	97.0	66.0	83.9
738	Woodward	95.3	62.9	81.6
739	Guymon	96.3	63.3	82.4
740-741	Tulsa	95.9	65.8	83.2
743	Miami	92.7	65.1	81.1
744	Muskogee	95.0	62.9	81.4
745	McAlester	92.4	60.5	78.9
746	Ponca City	93.1	64.0	80.8
747	Durant	93.1	65.1	81.3
748	Shawnee	94.5	65.2	82.2
749	Poteau	92.3	64.9	80.7
OREGON				
970-972	Portland	102.1	102.8	102.4
973	Salem	104.4	100.8	102.9
974	Eugene	101.7	98.7	100.5
975	Medford	103.3	97.4	100.8
976	Klamath Falls	103.5	97.3	100.9
977	Bend	102.6	100.2	101.6
978	Pendleton	98.6	100.3	99.3
979	Vale	96.2	87.0	92.3
PENNSYLVANIA				
150-152	Pittsburgh	100.7	102.7	101.5
153	Washington	97.5	100.2	98.6
154	Uniontown	97.8	99.8	98.6
155	Bedford	98.8	92.1	96.0
156	Greensburg	98.7	96.3	97.7
157	Indiana	97.6	97.7	97.7
158	Dubois	99.3	96.0	97.9
159	Johnstown	98.9	91.9	95.9
160	Butler	91.3	100.6	95.2
161	New Castle	91.4	97.8	94.1
162	Kittanning	91.8	97.9	94.4
163	Oil City	91.3	97.8	94.0
164-165	Erie	93.2	93.9	93.5
166	Altoona	93.3	92.8	93.1
167	Bradford	94.9	97.1	95.8
168	State College	94.5	94.6	94.5
169	Wellsboro	95.5	92.9	94.4
170-171	Harrisburg	98.7	93.0	96.3
172	Chambersburg	95.1	89.4	92.7
173-174	York	95.2	93.0	94.3
175-176	Lancaster	93.8	94.5	94.1

STATE/ZIP	CITY	MAT.	INST.	TOTAL
PENNSYLVANIA (CONT'D)				
177	Williamsport	92.5	92.3	92.4
178	Sunbury	94.6	92.7	93.8
179	Pottsville	93.7	96.1	94.7
180	Lehigh Valley	94.9	111.6	102.0
181	Allentown	96.7	105.8	100.6
182	Hazleton	94.4	95.9	95.0
183	Stroudsburg	94.3	104.7	98.7
184-185	Scranton	97.4	96.5	97.1
186-187	Wilkes-Barre	94.0	95.5	94.7
188	Montrose	93.8	96.8	95.1
189	Doylestown	94.0	127.1	108.0
190-191	Philadelphia	100.2	138.7	116.5
193	Westchester	95.8	127.9	109.3
194	Norristown	94.7	128.0	108.8
195-196	Reading	96.5	101.5	98.6
PUERTO RICO				
009	San Juan	98.5	29.8	69.4
RHODE ISLAND				
028	Newport	97.0	114.7	104.5
029	Providence	99.3	115.0	105.9
SOUTH CAROLINA				
290-292	Columbia	97.6	68.9	85.5
293	Spartanburg	97.2	68.3	85.0
294	Charleston	98.7	68.2	85.8
295	Florence	96.9	68.6	84.9
296	Greenville	97.0	68.2	84.8
297	Rock Hill	96.7	66.4	83.9
298	Aiken	97.6	67.1	84.7
299	Beaufort	98.3	57.5	81.1
SOUTH DAKOTA				
570-571	Sioux Falls	97.7	78.4	89.5
572	Watertown	97.3	69.0	85.4
573	Mitchell	96.3	60.9	81.3
574	Aberdeen	98.6	70.1	86.5
575	Pierre	99.0	68.5	86.1
576	Mobridge	96.9	62.9	82.5
577	Rapid City	98.3	72.0	87.2
TENNESSEE				
370-372	Nashville	98.0	71.8	86.9
373-374	Chattanooga	98.0	68.1	85.3
375,380-381	Memphis	96.6	69.8	85.3
376	Johnson City	98.1	58.6	81.4
377-379	Knoxville	94.8	63.7	81.6
382	McKenzie	96.3	55.1	78.9
383	Jackson	97.6	61.4	82.3
384	Columbia	94.8	66.5	82.8
385	Cookeville	96.2	56.9	79.6
TEXAS				
750	McKinney	97.2	63.5	82.9
751	Waxahachie	97.2	63.5	83.0
752-753	Dallas	98.2	68.0	85.4
754	Greenville	97.4	62.9	82.8
755	Texarkana	96.7	62.0	82.1
756	Longview	97.5	61.0	82.1
757	Tyler	97.8	61.6	82.5
758	Palestine	94.2	60.4	79.9
759	Lufkin	94.7	62.6	81.1
760-761	Fort Worth	97.6	62.9	82.9
762	Denton	97.3	62.9	82.7
763	Wichita Falls	95.0	61.6	80.9
764	Eastland	94.2	60.3	79.8
765	Temple	92.5	58.5	78.2
766-767	Waco	94.4	62.4	80.9
768	Brownwood	97.4	58.7	81.1
769	San Angelo	97.1	59.4	81.2
770-772	Houston	100.4	68.2	86.8
773	Huntsville	98.7	62.8	83.6
774	Wharton	99.8	64.3	84.8
775	Galveston	97.7	66.5	84.5
776-777	Beaumont	97.9	65.5	84.2
778	Bryan	95.2	64.9	82.4
779	Victoria	99.7	62.9	84.2
780	Laredo	96.4	61.1	81.5
781-782	San Antonio	98.3	62.3	83.1
783-784	Corpus Christi	98.9	61.4	83.1
785	McAllen	99.7	56.5	81.4
786-787	Austin	97.2	61.4	82.1

STATE/ZIP	CITY	MAT.	INST.	TOTAL
TEXAS (CONT'D)				
788	Del Rio	99.7	60.4	83.1
789	Giddings	96.1	60.4	81.0
790-791	Amarillo	98.2	60.0	82.1
792	Childress	97.6	60.3	81.8
793-794	Lubbock	99.2	62.9	83.9
795-796	Abilene	97.8	61.4	82.4
797	Midland	99.7	62.2	83.8
798-799,885	El Paso	95.9	63.6	82.3
UTAH				
840-841	Salt Lake City	103.5	71.7	90.1
842,844	Ogden	98.8	71.7	87.4
843	Logan	100.9	71.7	88.5
845	Price	101.4	70.0	88.1
846-847	Provo	101.2	71.2	88.5
VERMONT				
050	White River Jct.	98.2	79.8	90.4
051	Bellows Falls	96.7	92.1	94.8
052	Bennington	97.0	88.6	93.4
053	Brattleboro	97.4	92.1	95.2
054	Burlington	101.9	79.8	92.5
056	Montpelier	100.2	84.4	93.5
057	Rutland	98.9	79.4	90.7
058	St. Johnsbury	98.1	79.3	90.2
059	Guildhall	96.8	79.0	89.3
VIRGINIA				
220-221	Fairfax	99.5	83.6	92.8
222	Arlington	100.4	84.2	93.6
223	Alexandria	99.6	84.8	93.3
224-225	Fredericksburg	98.3	78.7	90.0
226	Winchester	98.9	79.5	90.7
227	Culpeper	98.7	83.0	92.1
228	Harrisonburg	99.0	68.8	86.3
229	Charlottesville	99.4	71.0	87.4
230-232	Richmond	99.2	72.8	88.1
233-235	Norfolk	99.9	68.7	86.7
236	Newport News	98.8	68.3	85.9
237	Portsmouth	98.3	67.8	85.4
238	Petersburg	98.7	72.6	87.7
239	Farmville	97.9	65.3	84.1
240-241	Roanoke	100.1	73.6	88.9
242	Bristol	98.2	59.3	81.8
243	Pulaski	97.8	67.1	84.8
244	Staunton	98.6	67.9	85.6
245	Lynchburg	98.7	73.8	88.2
246	Grundy	98.1	58.8	81.5
WASHINGTON				
980-981,987	Seattle	106.3	112.5	108.9
982	Everett	105.1	104.0	104.6
983-984	Tacoma	105.4	101.3	103.7
985	Olympia	104.0	101.8	103.1
986	Vancouver	106.6	102.8	105.0
988	Wenatchee	105.6	88.7	98.5
989	Yakima	105.6	95.4	101.3
990-992	Spokane	100.1	82.4	92.6
993	Richland	99.8	90.0	95.6
994	Clarkston	99.1	81.9	91.8
WEST VIRGINIA				
247-248	Bluefield	97.0	88.1	93.2
249	Lewisburg	98.7	86.1	93.4
250-253	Charleston	96.5	89.4	93.5
254	Martinsburg	96.7	84.1	91.4
255-257	Huntington	97.8	92.2	95.4
258-259	Beckley	95.2	87.7	92.0
260	Wheeling	99.6	89.6	95.4
261	Parkersburg	98.6	88.6	94.4
262	Buckhannon	98.3	88.8	94.3
263-264	Clarksburg	98.9	89.5	94.9
265	Morgantown	98.8	89.6	94.9
266	Gassaway	98.3	89.0	94.3
267	Romney	98.3	87.2	93.6
268	Petersburg	98.0	85.9	92.9
WISCONSIN				
530,532	Milwaukee	99.2	107.9	102.9
531	Kenosha	99.4	102.0	100.5
534	Racine	98.7	102.3	100.2
535	Beloit	98.6	98.9	98.7
537	Madison	99.7	98.1	99.0

STATE/ZIP	CITY	MAT.	INST.	TOTAL
WISCONSIN (CONT'D)				
538	Lancaster	96.7	94.3	95.7
539	Portage	95.2	98.0	96.4
540	New Richmond	95.4	94.6	95.0
541-543	Green Bay	99.4	93.4	96.9
544	Wausau	94.7	93.4	94.2
545	Rhinelander	97.9	92.2	95.5
546	La Crosse	95.9	93.5	94.9
547	Eau Claire	97.5	94.1	96.1
548	Superior	95.1	96.5	95.7
549	Oshkosh	95.4	91.9	93.9
WYOMING				
820	Cheyenne	101.1	72.7	89.1
821	Yellowstone Nat'l Park	99.0	71.9	87.6
822	Wheatland	100.3	67.1	86.2
823	Rawlins	102.0	71.5	89.1
824	Worland	99.7	70.7	87.5
825	Riverton	101.0	68.6	87.3
826	Casper	101.5	68.9	87.7
827	Newcastle	99.6	71.1	87.5
828	Sheridan	102.5	70.2	88.8
829-831	Rock Springs	103.4	71.4	89.9
CANADIAN FACTORS (reflect Canadian currency)				
ALBERTA				
	Calgary	119.7	95.6	109.5
	Edmonton	121.8	95.9	110.9
	Fort McMurray	121.9	88.4	107.7
	Lethbridge	117.0	87.9	104.7
	Lloydminster	111.5	84.7	100.2
	Medicine Hat	111.7	84.0	100.0
	Red Deer	112.1	84.0	100.2
BRITISH COLUMBIA				
	Kamloops	113.0	83.5	100.5
	Prince George	114.0	82.8	100.8
	Vancouver	117.9	92.4	107.2
	Victoria	114.1	87.6	102.9
MANITOBA				
	Brandon	120.6	70.5	99.4
	Portage la Prairie	111.6	69.1	93.6
	Winnipeg	122.0	71.2	100.5
NEW BRUNSWICK				
	Bathurst	110.7	62.3	90.2
	Dalhousie	110.2	62.5	90.0
	Fredericton	119.3	69.0	98.0
	Moncton	110.9	76.0	96.2
	Newcastle	110.7	62.9	90.5
	Saint John	113.2	75.7	97.3
NEWFOUNDLAND				
	Corner Brook	124.7	68.4	100.9
	St. John's	122.6	87.4	107.7
NORTHWEST TERRITORIES				
	Yellowknife	131.4	83.2	111.0
NOVA SCOTIA				
	Bridgewater	111.3	70.3	94.0
	Dartmouth	121.5	70.2	99.8
	Halifax	117.9	88.0	105.2
	New Glasgow	119.6	70.2	98.7
	Sydney	117.8	70.2	97.7
	Truro	110.8	70.3	93.7
	Yarmouth	119.5	70.2	98.7
ONTARIO				
	Barrie	116.5	87.1	104.1
	Brantford	113.1	90.7	103.6
	Cornwall	113.1	87.3	102.2
	Hamilton	116.3	98.8	108.9
	Kingston	114.1	87.4	102.8
	Kitchener	108.6	94.6	102.7
	London	116.3	96.5	108.0
	North Bay	122.5	85.2	106.8
	Oshawa	110.8	97.2	105.0
	Ottawa	117.5	96.8	108.7
	Owen Sound	116.7	85.4	103.4
	Peterborough	113.1	87.1	102.1
	Sarnia	113.2	91.3	103.9

For customer support on your Electrical Change Order Costs with RSMeans data, call 800.448.8182.

581

STATE/ZIP	CITY	MAT.	INST.	TOTAL
ONTARIO (CONT'D)				
	Sault Ste. Marie	108.6	88.7	100.2
	St. Catharines	106.6	95.6	101.9
	Sudbury	106.1	93.6	100.8
	Thunder Bay	107.8	95.0	102.4
	Timmins	113.3	85.3	101.4
	Toronto	116.0	104.2	111.0
	Windsor	107.1	94.4	101.8
PRINCE EDWARD ISLAND				
	Charlottetown	121.3	60.2	95.5
	Summerside	121.9	57.9	94.9
QUEBEC				
	Cap-de-la-Madeleine	110.6	80.0	97.7
	Charlesbourg	110.6	80.0	97.7
	Chicoutimi	110.4	87.6	100.7
	Gatineau	110.3	79.8	97.4
	Granby	110.5	79.7	97.5
	Hull	110.5	79.8	97.5
	Joliette	110.8	80.0	97.8
	Laval	110.7	81.6	98.4
	Montreal	119.1	90.1	106.9
	Quebec City	118.7	90.4	106.7
	Rimouski	110.2	87.6	100.6
	Rouyn-Noranda	110.3	79.8	97.4
	Saint-Hyacinthe	109.8	79.8	97.1
	Sherbrooke	110.6	79.8	97.6
	Sorel	110.8	80.0	97.8
	Saint-Jerome	110.4	79.8	97.4
	Trois-Rivieres	120.2	79.9	103.2
SASKATCHEWAN				
	Moose Jaw	109.0	63.9	89.9
	Prince Albert	108.2	62.2	88.8
	Regina	123.1	92.4	110.2
	Saskatoon	109.9	88.5	100.9
YUKON				
	Whitehorse	132.2	69.2	105.6

For customer support on your Electrical Change Order Costs with RSMeans data, call 800.448.8182.

Abbreviations

A	Area Square Feet; Ampere	Brk., brk	Brick	Csc	Cosecant
AAFES	Army and Air Force Exchange Service	brkt	Bracket	C.S.F.	Hundred Square Feet
		Brs.	Brass	CSI	Construction Specifications Institute
ABS	Acrylonitrile Butadiene Stryrene; Asbestos Bonded Steel	Brz.	Bronze		
		Bsn.	Basin	CT	Current Transformer
A.C., AC	Alternating Current;	Btr.	Better	CTS	Copper Tube Size
	Air-Conditioning;	BTU	British Thermal Unit	Cu	Copper, Cubic
	Asbestos Cement;	BTUH	BTU per Hour	Cu. Ft.	Cubic Foot
	Plywood Grade A & C	Bu.	Bushels	cw	Continuous Wave
ACI	American Concrete Institute	BUR	Built-up Roofing	C.W.	Cool White; Cold Water
ACR	Air Conditioning Refrigeration	BX	Interlocked Armored Cable	Cwt.	100 Pounds
ADA	Americans with Disabilities Act	°C	Degree Centigrade	C.W.X.	Cool White Deluxe
AD	Plywood, Grade A & D	c	Conductivity, Copper Sweat	C.Y.	Cubic Yard (27 cubic feet)
Addit.	Additional	C	Hundred; Centigrade	C.Y./Hr.	Cubic Yard per Hour
Adh.	Adhesive	C/C	Center to Center, Cedar on Cedar	Cyl.	Cylinder
Adj.	Adjustable	C-C	Center to Center	d	Penny (nail size)
af	Audio-frequency	Cab	Cabinet	D	Deep; Depth; Discharge
AFFF	Aqueous Film Forming Foam	Cair.	Air Tool Laborer	Dis., Disch.	Discharge
AFUE	Annual Fuel Utilization Efficiency	Cal.	Caliper	Db	Decibel
AGA	American Gas Association	Calc	Calculated	Dbl.	Double
Agg.	Aggregate	Cap.	Capacity	DC	Direct Current
A.H., Ah	Ampere Hours	Carp.	Carpenter	DDC	Direct Digital Control
A hr	Ampere-hour	C.B.	Circuit Breaker	Demob.	Demobilization
A.H.U., AHU	Air Handling Unit	C.C.A.	Chromate Copper Arsenate	d.f.t.	Dry Film Thickness
A.I.A.	American Institute of Architects	C.C.F.	Hundred Cubic Feet	d.f.u.	Drainage Fixture Units
AIC	Ampere Interrupting Capacity	cd	Candela	D.H.	Double Hung
Allow.	Allowance	cd/sf	Candela per Square Foot	DHW	Domestic Hot Water
alt., alt	Alternate	CD	Grade of Plywood Face & Back	DI	Ductile Iron
Alum.	Aluminum	CDX	Plywood, Grade C & D, exterior glue	Diag.	Diagonal
a.m.	Ante Meridiem			Diam., Dia	Diameter
Amp.	Ampere	Cefi.	Cement Finisher	Distrib.	Distribution
Anod.	Anodized	Cem.	Cement	Div.	Division
ANSI	American National Standards Institute	CF	Hundred Feet	Dk.	Deck
		C.F.	Cubic Feet	D.L.	Dead Load; Diesel
APA	American Plywood Association	CFM	Cubic Feet per Minute	DLH	Deep Long Span Bar Joist
Approx.	Approximate	CFRP	Carbon Fiber Reinforced Plastic	dlx	Deluxe
Apt.	Apartment	c.g.	Center of Gravity	Do.	Ditto
Asb.	Asbestos	CHW	Chilled Water;	DOP	Dioctyl Phthalate Penetration Test (Air Filters)
A.S.B.C.	American Standard Building Code		Commercial Hot Water		
Asbe.	Asbestos Worker	C.I., CI	Cast Iron	Dp., dp	Depth
ASCE	American Society of Civil Engineers	C.I.P., CIP	Cast in Place	D.P.S.T.	Double Pole, Single Throw
A.S.H.R.A.E.	American Society of Heating, Refrig. & AC Engineers	Circ.	Circuit	Dr.	Drive
		C.L.	Carload Lot	DR	Dimension Ratio
ASME	American Society of Mechanical Engineers	CL	Chain Link	Drink.	Drinking
		Clab.	Common Laborer	D.S.	Double Strength
ASTM	American Society for Testing and Materials	Clam	Common Maintenance Laborer	D.S.A.	Double Strength A Grade
		C.L.F.	Hundred Linear Feet	D.S.B.	Double Strength B Grade
Attchmt.	Attachment	CLF	Current Limiting Fuse	Dty.	Duty
Avg., Ave.	Average	CLP	Cross Linked Polyethylene	DWV	Drain Waste Vent
AWG	American Wire Gauge	cm	Centimeter	DX	Deluxe White, Direct Expansion
AWWA	American Water Works Assoc.	CMP	Corr. Metal Pipe	dyn	Dyne
Bbl.	Barrel	CMU	Concrete Masonry Unit	e	Eccentricity
B&B, BB	Grade B and Better; Balled & Burlapped	CN	Change Notice	E	Equipment Only; East; Emissivity
		Col.	Column	Ea.	Each
B&S	Bell and Spigot	CO₂	Carbon Dioxide	EB	Encased Burial
B.&W.	Black and White	Comb.	Combination	Econ.	Economy
b.c.c.	Body-centered Cubic	comm.	Commercial, Communication	E.C.Y	Embankment Cubic Yards
B.C.Y.	Bank Cubic Yards	Compr.	Compressor	EDP	Electronic Data Processing
BE	Bevel End	Conc.	Concrete	EIFS	Exterior Insulation Finish System
B.F.	Board Feet	Cont., cont	Continuous; Continued, Container	E.D.R.	Equiv. Direct Radiation
Bg. cem.	Bag of Cement	Corkbd.	Cork Board	Eq.	Equation
BHP	Boiler Horsepower; Brake Horsepower	Corr.	Corrugated	EL	Elevation
		Cos	Cosine	Elec.	Electrician; Electrical
B.I.	Black Iron	Cot	Cotangent	Elev.	Elevator; Elevating
bidir.	bidirectional	Cov.	Cover	EMT	Electrical Metallic Conduit; Thin Wall Conduit
Bit., Bitum.	Bituminous	C/P	Cedar on Paneling		
Bit., Conc.	Bituminous Concrete	CPA	Control Point Adjustment	Eng.	Engine, Engineered
Bk.	Backed	Cplg.	Coupling	EPDM	Ethylene Propylene Diene Monomer
Bkrs.	Breakers	CPM	Critical Path Method		
Bldg., bldg	Building	CPVC	Chlorinated Polyvinyl Chloride	EPS	Expanded Polystyrene
Blk.	Block	C.Pr.	Hundred Pair	Eqhv.	Equip. Oper., Heavy
Bm.	Beam	CRC	Cold Rolled Channel	Eqlt.	Equip. Oper., Light
Boil.	Boilermaker	Creos.	Creosote	Eqmd.	Equip. Oper., Medium
bpm	Blows per Minute	Crpt.	Carpet & Linoleum Layer	Eqmm.	Equip. Oper., Master Mechanic
BR	Bedroom	CRT	Cathode-ray Tube	Eqol.	Equip. Oper., Oilers
Brg., brng.	Bearing	CS	Carbon Steel, Constant Shear Bar Joist	Equip.	Equipment
Brhe.	Bricklayer Helper			ERW	Electric Resistance Welded
Bric.	Bricklayer				

Abbreviation	Meaning
E.S.	Energy Saver
Est.	Estimated
esu	Electrostatic Units
E.W.	Each Way
EWT	Entering Water Temperature
Excav.	Excavation
excl	Excluding
Exp., exp	Expansion, Exposure
Ext., ext	Exterior; Extension
Extru.	Extrusion
f.	Fiber Stress
F	Fahrenheit; Female; Fill
Fab., fab	Fabricated; Fabric
FBGS	Fiberglass
F.C.	Footcandles
f.c.c.	Face-centered Cubic
f'c.	Compressive Stress in Concrete; Extreme Compressive Stress
F.E.	Front End
FEP	Fluorinated Ethylene Propylene (Teflon)
F.G.	Flat Grain
F.H.A.	Federal Housing Administration
Fig.	Figure
Fin.	Finished
FIPS	Female Iron Pipe Size
Fixt.	Fixture
FJP	Finger jointed and primed
Fl. Oz.	Fluid Ounces
Flr.	Floor
Flrs.	Floors
FM	Frequency Modulation; Factory Mutual
Fmg.	Framing
FM/UL	Factory Mutual/Underwriters Labs
Fdn.	Foundation
FNPT	Female National Pipe Thread
Fori.	Foreman, Inside
Foro.	Foreman, Outside
Fount.	Fountain
fpm	Feet per Minute
FPT	Female Pipe Thread
Fr	Frame
F.R.	Fire Rating
FRK	Foil Reinforced Kraft
FSK	Foil/Scrim/Kraft
FRP	Fiberglass Reinforced Plastic
FS	Forged Steel
FSC	Cast Body; Cast Switch Box
Ft., ft	Foot; Feet
Ftng.	Fitting
Ftg.	Footing
Ft lb.	Foot Pound
Furn.	Furniture
FVNR	Full Voltage Non-Reversing
FVR	Full Voltage Reversing
FXM	Female by Male
Fy.	Minimum Yield Stress of Steel
g	Gram
G	Gauss
Ga.	Gauge
Gal., gal.	Gallon
Galv., galv	Galvanized
GC/MS	Gas Chromatograph/Mass Spectrometer
Gen.	General
GFI	Ground Fault Interrupter
GFRC	Glass Fiber Reinforced Concrete
Glaz.	Glazier
GPD	Gallons per Day
gpf	Gallon per Flush
GPH	Gallons per Hour
gpm, GPM	Gallons per Minute
GR	Grade
Gran.	Granular
Grnd.	Ground
GVW	Gross Vehicle Weight
GWB	Gypsum Wall Board
H	High Henry
HC	High Capacity
H.D., HD	Heavy Duty; High Density
H.D.O.	High Density Overlaid
HDPE	High Density Polyethylene Plastic
Hdr.	Header
Hdwe.	Hardware
H.I.D., HID	High Intensity Discharge
Help.	Helper Average
HEPA	High Efficiency Particulate Air Filter
Hg	Mercury
HIC	High Interrupting Capacity
HM	Hollow Metal
HMWPE	High Molecular Weight Polyethylene
HO	High Output
Horiz.	Horizontal
H.P., HP	Horsepower; High Pressure
H.P.F.	High Power Factor
Hr.	Hour
Hrs./Day	Hours per Day
HSC	High Short Circuit
Ht.	Height
Htg.	Heating
Htrs.	Heaters
HVAC	Heating, Ventilation & Air-Conditioning
Hvy.	Heavy
HW	Hot Water
Hyd.; Hydr.	Hydraulic
Hz	Hertz (cycles)
I.	Moment of Inertia
IBC	International Building Code
I.C.	Interrupting Capacity
ID	Inside Diameter
I.D.	Inside Dimension; Identification
I.F.	Inside Frosted
I.M.C.	Intermediate Metal Conduit
In.	Inch
Incan.	Incandescent
Incl.	Included; Including
Int.	Interior
Inst.	Installation
Insul., insul	Insulation/Insulated
I.P.	Iron Pipe
I.P.S., IPS	Iron Pipe Size
IPT	Iron Pipe Threaded
I.W.	Indirect Waste
J	Joule
J.I.C.	Joint Industrial Council
K	Thousand; Thousand Pounds; Heavy Wall Copper Tubing, Kelvin
K.A.H.	Thousand Amp. Hours
kcmil	Thousand Circular Mils
KD	Knock Down
K.D.A.T.	Kiln Dried After Treatment
kg	Kilogram
kG	Kilogauss
kgf	Kilogram Force
kHz	Kilohertz
Kip	1000 Pounds
KJ	Kilojoule
K.L.	Effective Length Factor
K.L.F.	Kips per Linear Foot
Km	Kilometer
KO	Knock Out
K.S.F.	Kips per Square Foot
K.S.I.	Kips per Square Inch
kV	Kilovolt
kVA	Kilovolt Ampere
kVAR	Kilovar (Reactance)
KW	Kilowatt
KWh	Kilowatt-hour
L	Labor Only; Length; Long; Medium Wall Copper Tubing
Lab.	Labor
lat	Latitude
Lath.	Lather
Lav.	Lavatory
lb.; #	Pound
L.B., LB	Load Bearing; L Conduit Body
L. & E.	Labor & Equipment
lb./hr.	Pounds per Hour
lb./L.F.	Pounds per Linear Foot
lbf/sq.in.	Pound-force per Square Inch
L.C.L.	Less than Carload Lot
L.C.Y.	Loose Cubic Yard
Ld.	Load
LE	Lead Equivalent
LED	Light Emitting Diode
L.F.	Linear Foot
L.F. Hdr	Linear Feet of Header
L.F. Nose	Linear Foot of Stair Nosing
L.F. Rsr	Linear Foot of Stair Riser
Lg.	Long; Length; Large
L & H	Light and Heat
LH	Long Span Bar Joist
L.H.	Labor Hours
L.L., LL	Live Load
L.L.D.	Lamp Lumen Depreciation
lm	Lumen
lm/sf	Lumen per Square Foot
lm/W	Lumen per Watt
LOA	Length Over All
log	Logarithm
L-O-L	Lateralolet
long.	Longitude
L.P., LP	Liquefied Petroleum; Low Pressure
L.P.F.	Low Power Factor
LR	Long Radius
L.S.	Lump Sum
Lt.	Light
Lt. Ga.	Light Gauge
L.T.L.	Less than Truckload Lot
Lt. Wt.	Lightweight
L.V.	Low Voltage
M	Thousand; Material; Male; Light Wall Copper Tubing
M²CA	Meters Squared Contact Area
m/hr.; M.H.	Man-hour
mA	Milliampere
Mach.	Machine
Mag. Str.	Magnetic Starter
Maint.	Maintenance
Marb.	Marble Setter
Mat; Mat'l.	Material
Max.	Maximum
MBF	Thousand Board Feet
MBH	Thousand BTU's per hr.
MC	Metal Clad Cable
MCC	Motor Control Center
M.C.F.	Thousand Cubic Feet
MCFM	Thousand Cubic Feet per Minute
M.C.M.	Thousand Circular Mils
MCP	Motor Circuit Protector
MD	Medium Duty
MDF	Medium-density fibreboard
M.D.O.	Medium Density Overlaid
Med.	Medium
MF	Thousand Feet
M.F.B.M.	Thousand Feet Board Measure
Mfg.	Manufacturing
Mfrs.	Manufacturers
mg	Milligram
MGD	Million Gallons per Day
MGPH	Million Gallons per Hour
MH, M.H.	Manhole; Metal Halide; Man-Hour
MHz	Megahertz
Mi.	Mile
MI	Malleable Iron; Mineral Insulated
MIPS	Male Iron Pipe Size
mj	Mechanical Joint
m	Meter
mm	Millimeter
Mill.	Millwright
Min., min.	Minimum, Minute

Misc.	Miscellaneous	PCM	Phase Contrast Microscopy	SBS	Styrene Butadiere Styrene
ml	Milliliter, Mainline	PDCA	Painting and Decorating	SC	Screw Cover
M.L.F.	Thousand Linear Feet		Contractors of America	SCFM	Standard Cubic Feet per Minute
Mo.	Month	P.E., PE	Professional Engineer;	Scaf.	Scaffold
Mobil.	Mobilization		Porcelain Enamel;	Sch., Sched.	Schedule
Mog.	Mogul Base		Polyethylene; Plain End	S.C.R.	Modular Brick
MPH	Miles per Hour	P.E.C.I.	Porcelain Enamel on Cast Iron	S.D.	Sound Deadening
MPT	Male Pipe Thread	Perf.	Perforated	SDR	Standard Dimension Ratio
MRGWB	Moisture Resistant Gypsum	PEX	Cross Linked Polyethylene	S.E.	Surfaced Edge
	Wallboard	Ph.	Phase	Sel.	Select
MRT	Mile Round Trip	P.I.	Pressure Injected	SER, SEU	Service Entrance Cable
ms	Millisecond	Pile.	Pile Driver	S.F.	Square Foot
M.S.F.	Thousand Square Feet	Pkg.	Package	S.F.C.A.	Square Foot Contact Area
Mstz.	Mosaic & Terrazzo Worker	Pl.	Plate	S.F. Flr.	Square Foot of Floor
M.S.Y.	Thousand Square Yards	Plah.	Plasterer Helper	S.F.G.	Square Foot of Ground
Mtd., mtd., mtd	Mounted	Plas.	Plasterer	S.F. Hor.	Square Foot Horizontal
Mthe.	Mosaic & Terrazzo Helper	plf	Pounds Per Linear Foot	SFR	Square Feet of Radiation
Mtng.	Mounting	Pluh.	Plumber Helper	S.F. Shlf.	Square Foot of Shelf
Mult.	Multi; Multiply	Plum.	Plumber	S4S	Surface 4 Sides
MUTCD	Manual on Uniform Traffic Control	Ply.	Plywood	Shee.	Sheet Metal Worker
	Devices	p.m.	Post Meridiem	Sin.	Sine
M.V.A.	Million Volt Amperes	Pntd.	Painted	Skwk.	Skilled Worker
M.V.A.R.	Million Volt Amperes Reactance	Pord.	Painter, Ordinary	SL	Saran Lined
MV	Megavolt	pp	Pages	S.L.	Slimline
MW	Megawatt	PP, PPL	Polypropylene	Sldr.	Solder
MXM	Male by Male	P.P.M.	Parts per Million	SLH	Super Long Span Bar Joist
MYD	Thousand Yards	Pr.	Pair	S.N.	Solid Neutral
N	Natural; North	P.E.S.B.	Pre-engineered Steel Building	SO	Stranded with oil resistant inside
nA	Nanoampere	Prefab.	Prefabricated		insulation
NA	Not Available; Not Applicable	Prefin.	Prefinished	S-O-L	Socketolet
N.B.C.	National Building Code	Prop.	Propelled	sp	Standpipe
NC	Normally Closed	PSF, psf	Pounds per Square Foot	S.P.	Static Pressure; Single Pole; Self-
NEMA	National Electrical Manufacturers	PSI, psi	Pounds per Square Inch		Propelled
	Assoc.	PSIG	Pounds per Square Inch Gauge	Spri.	Sprinkler Installer
NEHB	Bolted Circuit Breaker to 600V.	PSP	Plastic Sewer Pipe	spwg	Static Pressure Water Gauge
NFPA	National Fire Protection Association	Pspr.	Painter, Spray	S.P.D.T.	Single Pole, Double Throw
NLB	Non-Load-Bearing	Psst.	Painter, Structural Steel	SPF	Spruce Pine Fir; Sprayed
NM	Non-Metallic Cable	P.T.	Potential Transformer		Polyurethane Foam
nm	Nanometer	P. & T.	Pressure & Temperature	S.P.S.T.	Single Pole, Single Throw
No.	Number	Ptd.	Painted	SPT	Standard Pipe Thread
NO	Normally Open	Ptns.	Partitions	Sq.	Square; 100 Square Feet
N.O.C.	Not Otherwise Classified	Pu	Ultimate Load	Sq. Hd.	Square Head
Nose.	Nosing	PVC	Polyvinyl Chloride	Sq. In.	Square Inch
NPT	National Pipe Thread	Pvmt.	Pavement	S.S.	Single Strength; Stainless Steel
NQOD	Combination Plug-on/Bolt on	PRV	Pressure Relief Valve	S.S.B.	Single Strength B Grade
	Circuit Breaker to 240V.	Pwr.	Power	sst, ss	Stainless Steel
N.R.C., NRC	Noise Reduction Coefficient/	Q	Quantity Heat Flow	Sswk.	Structural Steel Worker
	Nuclear Regulator Commission	Qt.	Quart	Sswl.	Structural Steel Welder
N.R.S.	Non Rising Stem	Quan., Qty.	Quantity	St.; Stl.	Steel
ns	Nanosecond	Q.C.	Quick Coupling	STC	Sound Transmission Coefficient
NTP	Notice to Proceed	r	Radius of Gyration	Std.	Standard
nW	Nanowatt	R	Resistance	Stg.	Staging
OB	Opposing Blade	R.C.P.	Reinforced Concrete Pipe	STK	Select Tight Knot
OC	On Center	Rect.	Rectangle	STP	Standard Temperature & Pressure
OD	Outside Diameter	recpt.	Receptacle	Stpi.	Steamfitter, Pipefitter
O.D.	Outside Dimension	Reg.	Regular	Str.	Strength; Starter; Straight
ODS	Overhead Distribution System	Reinf.	Reinforced	Strd.	Stranded
O.G.	Ogee	Req'd.	Required	Struct.	Structural
O.H.	Overhead	Res.	Resistant	Sty.	Story
O&P	Overhead and Profit	Resi.	Residential	Subj.	Subject
Oper.	Operator	RF	Radio Frequency	Subs.	Subcontractors
Opng.	Opening	RFID	Radio-frequency Identification	Surf.	Surface
Orna.	Ornamental	Rgh.	Rough	Sw.	Switch
OSB	Oriented Strand Board	RGS	Rigid Galvanized Steel	Swbd.	Switchboard
OS&Y	Outside Screw and Yoke	RHW	Rubber, Heat & Water Resistant;	S.Y.	Square Yard
OSHA	Occupational Safety and Health		Residential Hot Water	Syn.	Synthetic
	Act	rms	Root Mean Square	S.Y.P.	Southern Yellow Pine
Ovhd.	Overhead	Rnd.	Round	Sys.	System
OWG	Oil, Water or Gas	Rodm.	Rodman	t.	Thickness
Oz.	Ounce	Rofc.	Roofer, Composition	T	Temperature; Ton
P.	Pole; Applied Load; Projection	Rofp.	Roofer, Precast	Tan	Tangent
p.	Page	Rohe.	Roofer Helpers (Composition)	T.C.	Terra Cotta
Pape.	Paperhanger	Rots.	Roofer, Tile & Slate	T & C	Threaded and Coupled
P.A.P.R.	Powered Air Purifying Respirator	R.O.W.	Right of Way	T.D.	Temperature Difference
PAR	Parabolic Reflector	RPM	Revolutions per Minute	TDD	Telecommunications Device for
P.B., PB	Push Button	R.S.	Rapid Start		the Deaf
Pc., Pcs.	Piece, Pieces	Rsr	Riser	T.E.M.	Transmission Electron Microscopy
P.C.	Portland Cement; Power Connector	RT	Round Trip	temp	Temperature, Tempered, Temporary
P.C.F.	Pounds per Cubic Foot	S.	Suction; Single Entrance; South	TFFN	Nylon Jacketed Wire

Abbreviations

TFE	Tetrafluoroethylene (Teflon)	U.L., UL	Underwriters Laboratory	w/	With		
T. & G.	Tongue & Groove;	Uld.	Unloading	W.C., WC	Water Column; Water Closet		
	Tar & Gravel	Unfin.	Unfinished	W.F.	Wide Flange		
Th., Thk.	Thick	UPS	Uninterruptible Power Supply	W.G.	Water Gauge		
Thn.	Thin	URD	Underground Residential	Wldg.	Welding		
Thrded	Threaded		Distribution	W. Mile	Wire Mile		
Tilf.	Tile Layer, Floor	US	United States	W-O-L	Weldolet		
Tilh.	Tile Layer, Helper	USGBC	U.S. Green Building Council	W.R.	Water Resistant		
THHN	Nylon Jacketed Wire	USP	United States Primed	Wrck.	Wrecker		
THW.	Insulated Strand Wire	UTMCD	Uniform Traffic Manual For Control	WSFU	Water Supply Fixture Unit		
THWN	Nylon Jacketed Wire		Devices	W.S.P.	Water, Steam, Petroleum		
T.L., TL	Truckload	UTP	Unshielded Twisted Pair	WT., Wt.	Weight		
T.M.	Track Mounted	V	Volt	WWF	Welded Wire Fabric		
Tot.	Total	VA	Volt Amperes	XFER	Transfer		
T-O-L	Threadolet	VAT	Vinyl Asbestos Tile	XFMR	Transformer		
tmpd	Tempered	V.C.T.	Vinyl Composition Tile	XHD	Extra Heavy Duty		
TPO	Thermoplastic Polyolefin	VAV	Variable Air Volume	XHHW	Cross-Linked Polyethylene Wire		
T.S.	Trigger Start	VC	Veneer Core	XLPE	Insulation		
Tr.	Trade	VDC	Volts Direct Current	XLP	Cross-linked Polyethylene		
Transf.	Transformer	Vent.	Ventilation	Xport	Transport		
Trhv.	Truck Driver, Heavy	Vert.	Vertical	Y	Wye		
Trlr	Trailer	V.F.	Vinyl Faced	yd	Yard		
Trlt.	Truck Driver, Light	V.G.	Vertical Grain	yr	Year		
TTY	Teletypewriter	VHF	Very High Frequency	Δ	Delta		
TV	Television	VHO	Very High Output	%	Percent		
T.W.	Thermoplastic Water Resistant	Vib.	Vibrating	~	Approximately		
	Wire	VLF	Vertical Linear Foot	Ø	Phase; diameter		
UCI	Uniform Construction Index	VOC	Volatile Organic Compound	@	At		
UF	Underground Feeder	Vol.	Volume	#	Pound; Number		
UGND	Underground Feeder	VRP	Vinyl Reinforced Polyester	<	Less Than		
UHF	Ultra High Frequency	W	Wire; Watt; Wide; West	>	Greater Than		
U.I.	United Inch			Z	Zone		

Index

For customer support on your Electrical Change Order Costs with RSMeans data, call 800.448.8182.

For customer support on your Electrical Change Order Costs with RSMeans data, call 800.448.8182.

Notes

Division Notes

	CREW	DAILY OUTPUT	LABOR-HOURS	UNIT	BARE COSTS				TOTAL INCL O&P
					MAT.	LABOR	EQUIP.	TOTAL	

Division Notes

	CREW	DAILY OUTPUT	LABOR-HOURS	UNIT	BARE COSTS				TOTAL INCL O&P
					MAT.	LABOR	EQUIP.	TOTAL	

Division Notes

	CREW	DAILY OUTPUT	LABOR-HOURS	UNIT	BARE COSTS				TOTAL INCL O&P
					MAT.	LABOR	EQUIP.	TOTAL	

Division Notes

	CREW	DAILY OUTPUT	LABOR-HOURS	UNIT	BARE COSTS				TOTAL INCL O&P
					MAT.	LABOR	EQUIP.	TOTAL	

Division Notes

	CREW	DAILY OUTPUT	LABOR-HOURS	UNIT	BARE COSTS				TOTAL INCL O&P
					MAT.	LABOR	EQUIP.	TOTAL	

Division Notes

	CREW	DAILY OUTPUT	LABOR-HOURS	UNIT	BARE COSTS				TOTAL INCL O&P
					MAT.	LABOR	EQUIP.	TOTAL	

Division Notes

	CREW	DAILY OUTPUT	LABOR-HOURS	UNIT	BARE COSTS				TOTAL INCL O&P
					MAT.	LABOR	EQUIP.	TOTAL	

Other Data & Services

A tradition of excellence in construction cost information and services since 1942

For more information visit our website at RSMeans.com

Unit prices according to the latest MasterFormat®

Cost Data Selection Guide

The following table provides definitive information on the content of each cost data publication. The number of lines of data provided in each unit price or assemblies division, as well as the number of crews, is listed for each data set. The presence of other elements such as reference tables, square foot models, equipment rental costs, historical cost indexes, and city cost indexes, is also indicated. You can use the table to help select the RSMeans data set that has the quantity and type of information you most need in your work.

Unit Cost Divisions	Building Construction	Mechanical	Electrical	Commercial Renovation	Square Foot	Site Work Landsc.	Green Building	Interior	Concrete Masonry	Open Shop	Heavy Construction	Light Commercial	Facilities Construction	Plumbing	Residential
1	609	444	465	564	0	533	198	365	495	608	550	310	1078	450	217
2	754	278	87	710	0	970	181	397	219	753	737	479	1197	285	274
3	1745	341	232	1265	0	1537	1043	355	2274	1745	1930	538	2028	317	445
4	960	22	0	920	0	724	180	613	1158	928	614	532	1175	0	446
5	1890	158	155	1094	0	853	1788	1107	729	1890	1026	980	1907	204	746
6	2462	18	18	2121	0	110	589	1544	281	2458	123	2151	2135	22	2671
7	1593	215	128	1633	0	580	761	532	523	1590	26	1326	1693	227	1046
8	2140	80	3	2733	0	255	1138	1813	105	2142	0	2328	2966	0	1552
9	2125	86	45	1943	0	313	464	2216	424	2062	15	1779	2379	54	1544
10	1088	17	10	684	0	232	32	898	136	1088	34	588	1179	237	224
11	1096	199	166	540	0	135	56	924	29	1063	0	230	1116	162	108
12	539	0	2	297	0	219	147	1546	14	506	0	272	1565	23	216
13	740	149	157	252	0	365	124	250	77	716	266	109	756	115	103
14	273	36	0	223	0	0	0	257	0	273	0	12	293	16	6
21	127	0	41	37	0	0	0	293	0	127	0	121	665	685	259
22	1165	7543	160	1226	0	2010	1061	849	20	1154	2109	875	7505	9400	719
23	1170	6906	546	940	0	157	865	775	38	1153	98	887	5143	1919	486
25	0	0	14	14	0	0	0	0	0	0	0	0	0	0	0
26	1513	491	10465	1293	0	860	646	1159	55	1439	649	1361	10246	399	636
27	95	0	448	102	0	0	0	71	0	95	39	67	389	0	56
28	143	79	223	124	0	0	28	97	0	127	0	70	209	57	41
31	1511	733	610	807	0	3263	286	7	1216	1456	3280	607	1568	660	616
32	896	49	8	937	0	4523	408	417	361	867	1941	486	1800	140	533
33	1255	1088	565	260	0	3078	33	0	241	532	3213	135	1726	2101	161
34	107	0	47	4	0	190	0	0	31	62	221	0	136	0	0
35	18	0	0	0	0	327	0	0	0	18	442	0	84	0	0
41	63	0	0	34	0	8	0	22	0	62	31	0	69	14	0
44	75	79	0	0	0	0	0	0	0	0	0	0	75	75	0
46	23	16	0	0	0	274	261	0	0	23	264	0	33	33	0
48	8	0	36	2	0	0	21	0	0	8	15	8	21	0	8
Totals	26183	19027	14631	20759	0	21516	10310	16507	8426	24945	17623	16251	51136	17595	13113

Assem Div	Building Construction	Mechanical	Electrical	Commercial Renovation	Square Foot	Site Work Landscape	Assemblies	Green Building	Interior	Concrete Masonry	Heavy Construction	Light Commercial	Facilities Construction	Plumbing	Asm Div	Residential
A		15	0	188	164	577	598	0	0	536	571	154	24	0	1	378
B		0	0	848	2554	0	5661	56	329	1976	368	2094	174	0	2	211
C		0	0	647	954	0	1334	0	1641	146	0	844	251	0	3	591
D		1057	941	712	1858	72	2538	330	824	0	0	1345	1104	1088	4	851
E		0	0	85	261	0	301	0	5	0	0	258	5	0	5	391
F		0	0	0	114	0	143	0	0	0	0	114	0	0	6	357
G		527	447	318	312	3378	792	0	0	535	1349	205	293	677	7	307
															8	760
															9	80
															10	0
															11	0
															12	0
Totals		1599	1388	2798	6217	4027	11367	386	2799	3193	2288	5014	1851	1765		3926

Reference Section	Building Construction Costs	Mechanical	Electrical	Commercial Renovation	Square Foot	Site Work Landscape	Assem.	Green Building	Interior	Concrete Masonry	Open Shop	Heavy Construction	Light Commercial	Facilities Construction	Plumbing	Resi.
Reference Tables	yes	yes	yes	yes	no	yes	yes	yes	yes	yes	yes	yes	yes	yes	yes	yes
Models					111			25					50			28
Crews	582	582	582	561		582		582	582	582	560	582	560	561	582	560
Equipment Rental Costs	yes	yes	yes	yes		yes		yes	yes	yes	yes	yes	yes	yes	yes	yes
Historical Cost Indexes	yes	yes	yes	yes	yes	yes	yes	yes	yes	yes	yes	yes	yes	yes	yes	no
City Cost Indexes	yes	yes	yes	yes	yes	yes	yes	yes	yes	yes	yes	yes	yes	yes	yes	yes

2020 Seminar Schedule ☏ 877.620.6245

Note: call for exact dates, locations, and details as some cities are subject to change.

Location	Dates	Location	Dates
Seattle, WA	January and August	San Francisco, CA	June
Dallas/Ft. Worth, TX	January	Bethesda, MD	June
Austin, TX	February	Dallas, TX	September
Jacksonville, FL	February	Raleigh, NC	October
Anchorage, AK	March and September	Baltimore, MD	November
Las Vegas, NV	March	Orlando, FL	November
Washington, D.C.	April and September	San Diego, CA	December
Charleston, SC	April	San Antonio, TX	December
Toronto	May		
Denver, CO	May		

Gordian also offers a suite of online RSMeans data self-paced offerings.
Check our website at RSMeans.com/products/training.aspx for more information.

Facilities Construction Estimating

In this two-day course, professionals working in facilities management can get help with their daily challenges to establish budgets for all phases of a project.

Some of what you'll learn:
- Determining the full scope of a project
- Understanding of Means data and what is included in prices
- Identifying appropriate factors to be included in your estimate
- Creative solutions to estimating issues
- Organizing estimates for presentation and discussion
- Special estimating techniques for repair/remodel and maintenance projects
- Appropriate use of contingency, city cost indexes, and reference notes
- Techniques to get to the correct estimate quickly

Who should attend: facility managers, engineers, contractors, facility tradespeople, planners, and project managers.

Mechanical & Electrical Estimating

This two-day course teaches attendees how to prepare more accurate and complete mechanical/electrical estimates, avoid the pitfalls of omission and double-counting, and understand the composition and rationale within the RSMeans mechanical/electrical database.

Some of what you'll learn:
- The unique way mechanical and electrical systems are interrelated
- M&E estimates—conceptual, planning, budgeting, and bidding stages
- Order of magnitude, square foot, assemblies, and unit price estimating
- Comparative cost analysis of equipment and design alternatives

Who should attend: architects, engineers, facilities managers, mechanical and electrical contractors, and others who need a highly reliable method for developing, understanding, and evaluating mechanical and electrical contracts.

Construction Cost Estimating: Concepts and Practice

This one or two day introductory course to improve estimating skills and effectiveness starts with the details of interpreting bid documents and ends with the summary of the estimate and bid submission.

Some of what you'll learn:
- Using the plans and specifications to create estimates
- The takeoff process—deriving all tasks with correct quantities
- Developing pricing using various sources; how subcontractor pricing fits in
- Summarizing the estimate to arrive at the final number
- Formulas for area and cubic measure, adding waste and adjusting productivity to specific projects
- Evaluating subcontractors' proposals and prices
- Adding insurance and bonds
- Understanding how labor costs are calculated
- Submitting bids and proposals

Who should attend: project managers, architects, engineers, owners' representatives, contractors, and anyone who's responsible for budgeting or estimating construction projects.

RSMeans data Training

Training for our Online Estimating Solution

Construction estimating is vital to the decision-making process at each state of every project. Our online solution works the way you do. It's systematic, flexible and intuitive. In this one-day class you will see how you can estimate any phase of any project faster and better.

Some of what you'll learn:
- Customizing our online estimating solution
- Making the most of RSMeans "Circle Reference" numbers
- How to integrate your cost data
- Generating reports, exporting estimates to MS Excel, sharing, collaborating and more

Also offered as a self-paced or on-site training program!

Training for our CD Estimating Solution

This one-day course helps users become more familiar with the functionality of the CD. Each menu, icon, screen, and function found in the program is explained in depth. Time is devoted to hands-on estimating exercises.

Some of what you'll learn:
- Searching the database using all navigation methods
- Exporting RSMeans data to your preferred spreadsheet format
- Viewing crews, assembly components, and much more
- Automatically regionalizing the database

This training session requires you to bring a laptop computer to class.

When you register for this course you will receive an outline for your laptop requirements.

Also offered as a self-paced or on-site training program!

Site Work Estimating with RSMeans data

This one-day program focuses directly on site work costs. Accurately scoping, quantifying, and pricing site preparation, underground utility work, and improvements to exterior site elements are often the most difficult estimating tasks on any project. Some of what you'll learn:
- Evaluation of site work and understanding site scope including: site clearing, grading, excavation, disposal and trucking of materials, backfill and compaction, underground utilities, paving, sidewalks, and seeding & planting.
- Unit price site work estimates—Correct use of RSMeans site work cost data to develop a cost estimate.
- Using and modifying assemblies—Save valuable time when estimating site work activities using custom assemblies.

Who should attend: Engineers, contractors, estimators, project managers, owner's representatives, and others who are concerned with the proper preparation and/or evaluation of site work estimates.

Please bring a laptop with ability to access the internet.

Facilities Estimating Using the CD

This two-day class combines hands-on skill-building with best estimating practices and real-life problems. You will learn key concepts, tips, pointers, and guidelines to save time and avoid cost oversights and errors.

Some of what you'll learn:
- Estimating process concepts
- Customizing and adapting RSMeans cost data
- Establishing scope of work to account for all known variables
- Budget estimating: when, why, and how
- Site visits: what to look for and what you can't afford to overlook
- How to estimate repair and remodeling variables

This training session requires you to bring a laptop computer to class.

Who should attend: facility managers, architects, engineers, contractors, facility tradespeople, planners, project managers, and anyone involved with JOC, SABRE or IDIQ.

Registration Information

Register early to save up to $100!!!
Register 45+ days before the date of a class and save $50 off each class. This savings cannot be combined with any other promotion or discounting of the regular price of classes!

How to register
By Phone
Register by phone at 877.620.6245

Online
Register online at RSMeans.com/products/seminars.aspx

Note: Purchase Orders or Credits Cards are required to register.

Two-day seminar registration fee - $1,200*.

One-Day Construction Cost Estimating or Building Systems and the Construction Process - $765*.

Government pricing
All federal government employees save off the regular seminar price. Other promotional discounts cannot be combined with the government discount. Call 781.422.5115 for government pricing.

CANCELLATION POLICY:
If you are unable to attend a seminar, substitutions may be made at any time before the session starts by notifying the seminar registrar at 781.422.5115 or your sales representative.
If you cancel twenty-one (21) days or more prior to the seminar, there will be no penalty and your registration fees will be refunded. These cancellations must be received by the seminar registrar or your sales representative and will be confirmed to be eligible for cancellation.
If you cancel fewer than twenty-one (21) days prior to the seminar, you will forfeit the registration fee.
In the unfortunate event of an RSMeans cancellation, RSMeans will work with you to reschedule your attendance in the same seminar at a later date or will fully refund your registration fee. RSMeans cannot be responsible for any non-refundable travel expenses incurred by you or another as a result of your registration, attendance at, or cancellation of an RSMeans seminar.
Any on-demand training modules are not eligible for cancellation, substitution, transfer, return or refund.

AACE approved courses
Many seminars described and offered here have been approved for 14 hours (1.4 recertification credits) of credit by the AACE International Certification Board toward meeting the continuing education requirements for recertification as a Certified Cost Engineer/Certified Cost Consultant.

AIA Continuing Education
We are registered with the AIA Continuing Education System (AIA/CES) and are committed to developing quality learning activities in accordance with the CES criteria. Many seminars meet the AIA/CES criteria for Quality Level 2. AIA members may receive 14 learning units (LUs) for each two-day RSMeans course.

Daily course schedule
The first day of each seminar session begins at 8:30 a.m. and ends at 4:30 p.m. The second day begins at 8:00 a.m. and ends at 4:00 p.m. Participants are urged to bring a hand-held calculator since many actual problems will be worked out in each session.

Continental breakfast
Your registration includes the cost of a continental breakfast and a morning and afternoon refreshment break. These informal segments allow you to discuss topics of mutual interest with other seminar attendees. (You are free to make your own lunch and dinner arrangements.)

Hotel/transportation arrangements
We arrange to hold a block of rooms at most host hotels. To take advantage of special group rates when making your reservation, be sure to mention that you are attending the RSMeans Institute data seminar. You are, of course, free to stay at the lodging place of your choice. (Hotel reservations and transportation arrangements should be made directly by seminar attendees.)

Important
Class sizes are limited, so please register as soon as possible.

Note: Pricing subject to change.

[As]sessing Scope of Work for Facilities [Con]struction Estimating

[Thi]s two-day practical training program addresses the vital [im]portance of understanding the scope of projects in order [to] produce accurate cost estimates for facilities repair and [r]emodeling.

[S]ome of what you'll learn:
- Discussions of site visits, plans/specs, record drawings of facilities, and site-specific lists
- Review of CSI divisions, including means, methods, materials, and the challenges of scoping each topic
- Exercises in scope identification and scope writing for accurate estimating of projects
- Hands-on exercises that require scope, take-off, and pricing

Who should attend: corporate and government estimators, planners, facility managers, and others who need to produce accurate project estimates.

Maintenance & Repair Estimating for Facilities

This two-day course teaches attendees how to plan, budget, and estimate the cost of ongoing and preventive maintenance and repair for existing buildings and grounds.

Some of what you'll learn:
- The most financially favorable maintenance, repair, and replacement scheduling and estimating
- Auditing and value engineering facilities
- Preventive planning and facilities upgrading
- Determining both in-house and contract-out service costs
- Annual, asset-protecting M&R plan

Who should attend: facility managers, maintenance supervisors, buildings and grounds superintendents, plant managers, planners, estimators, and others involved in facilities planning and budgeting.

Practical Project Management for Construction Professionals

In this two-day course, acquire the essential knowledge and develop the skills to effectively and efficiently execute the day-to-day responsibilities of the construction project manager.

Some of what you'll learn:
- General conditions of the construction contract
- Contract modifications: change orders and construction change directives
- Negotiations with subcontractors and vendors
- Effective writing: notification and communications
- Dispute resolution: claims and liens

Who should attend: architects, engineers, owners' representatives, and project managers.

Life Cycle Cost Estimating for Facilities Asset Managers

Life Cycle Cost Estimating will take the attendee through choosing the correct RSMeans database to use and then correctly applying RSMeans data to their specific life cycle application. Conceptual estimating through RSMeans' new building models, conceptual estimating of major existing building projects through RSMeans' renovation models, pricing specific renovation elements, estimating repair, replacement and preventive maintenance costs today and forward up to 30 years will be covered.

Some of what you'll learn:
- Cost implications of managing assets
- Planning projects and initial & life cycle costs
- How to use RSMeans data online

Who should attend: facilities owners and managers and anyone involved in the financial side of the decision making process in the planning, design, procurement, and operation of facilities real assets.

Please bring a laptop with ability to access the internet.

Building Systems and the Construction Process

This one-day course was written to assist novices and those outside the industry in obtaining a solid understanding of the construction process - from both a building systems and construction administration approach.

Some of what you'll learn:
- Various systems used and how components come together to create a building
- Start with foundation and end with the physical systems of the structure such as HVAC and Electrical
- Focus on the process from start of design through project closeout

This training session requires you to bring a laptop computer to class.

Who should attend: building professionals or novices to help make the crossover to the construction industry; suited for anyone responsible for providing high level oversight on construction projects.

RSMeans data Training